T0191701

Lecture Notes in Computer Science 12353

More information about this series at http://www.springer.com/series/7412

Andrea Vedaldi · Horst Bischof ·
Thomas Brox · Jan-Michael Frahm (Eds.)

Computer Vision – ECCV 2020

16th European Conference
Glasgow, UK, August 23–28, 2020
Proceedings, Part VIII

Editors
Andrea Vedaldi
University of Oxford
Oxford, UK

Thomas Brox
University of Freiburg
Freiburg im Breisgau, Germany

Horst Bischof
Graz University of Technology
Graz, Austria

Jan-Michael Frahm
University of North Carolina at Chapel Hill
Chapel Hill, NC, USA

ISSN 0302-9743 ISSN 1611-3349 (electronic)
Lecture Notes in Computer Science
ISBN 978-3-030-58597-6 ISBN 978-3-030-58598-3 (eBook)
https://doi.org/10.1007/978-3-030-58598-3

LNCS Sublibrary: SL6 – Image Processing, Computer Vision, Pattern Recognition, and Graphics

This Springer imprint is published by the registered company Springer Nature Switzerland AG
The registered company address is: Gewerbestrasse 11, 6330 Cham, Switzerland

Foreword

Hosting the European Conference on Computer Vision (ECCV 2020) was certainly an exciting journey. From the 2016 plan to hold it at the Edinburgh International Conference Centre (hosting 1,800 delegates) to the 2018 plan to hold it at Glasgow's Scottish Exhibition Centre (up to 6,000 delegates), we finally ended with moving online because of the COVID-19 outbreak. While possibly having fewer delegates than expected because of the online format, ECCV 2020 still had over 3,100 registered participants.

Although online, the conference delivered most of the activities expected at a face-to-face conference: peer-reviewed papers, industrial exhibitors, demonstrations, and messaging between delegates. In addition to the main technical sessions, the conference included a strong program of satellite events with 16 tutorials and 44 workshops.

Furthermore, the online conference format enabled new conference features. Every paper had an associated teaser video and a longer full presentation video. Along with the papers and slides from the videos, all these materials were available the week before the conference. This allowed delegates to become familiar with the paper content and be ready for the live interaction with the authors during the conference week. The live event consisted of brief presentations by the oral and spotlight authors and industrial sponsors. Question and answer sessions for all papers were timed to occur twice so delegates from around the world had convenient access to the authors.

As with ECCV 2018, authors' draft versions of the papers appeared online with open access, now on both the Computer Vision Foundation (CVF) and the European Computer Vision Association (ECVA) websites. An archival publication arrangement was put in place with the cooperation of Springer. SpringerLink hosts the final version of the papers with further improvements, such as activating reference links and supplementary materials. These two approaches benefit all potential readers: a version available freely for all researchers, and an authoritative and citable version with additional benefits for SpringerLink subscribers. We thank Alfred Hofmann and Aliaksandr Birukou from Springer for helping to negotiate this agreement, which we expect will continue for future versions of ECCV.

August 2020

Vittorio Ferrari
Bob Fisher
Cordelia Schmid
Emanuele Trucco

Preface

Welcome to the proceedings of the European Conference on Computer Vision (ECCV 2020). This is a unique edition of ECCV in many ways. Due to the COVID-19 pandemic, this is the first time the conference was held online, in a virtual format. This was also the first time the conference relied exclusively on the Open Review platform to manage the review process. Despite these challenges ECCV is thriving. The conference received 5,150 valid paper submissions, of which 1,360 were accepted for publication (27%) and, of those, 160 were presented as spotlights (3%) and 104 as orals (2%). This amounts to more than twice the number of submissions to ECCV 2018 (2,439). Furthermore, CVPR, the largest conference on computer vision, received 5,850 submissions this year, meaning that ECCV is now 87% the size of CVPR in terms of submissions. By comparison, in 2018 the size of ECCV was only 73% of CVPR.

The review model was similar to previous editions of ECCV; in particular, it was double blind in the sense that the authors did not know the name of the reviewers and vice versa. Furthermore, each conference submission was held confidentially, and was only publicly revealed if and once accepted for publication. Each paper received at least three reviews, totalling more than 15,000 reviews. Handling the review process at this scale was a significant challenge. In order to ensure that each submission received as fair and high-quality reviews as possible, we recruited 2,830 reviewers (a 130% increase with reference to 2018) and 207 area chairs (a 60% increase). The area chairs were selected based on their technical expertise and reputation, largely among people that served as area chair in previous top computer vision and machine learning conferences (ECCV, ICCV, CVPR, NeurIPS, etc.). Reviewers were similarly invited from previous conferences. We also encouraged experienced area chairs to suggest additional chairs and reviewers in the initial phase of recruiting.

Despite doubling the number of submissions, the reviewer load was slightly reduced from 2018, from a maximum of 8 papers down to 7 (with some reviewers offering to handle 6 papers plus an emergency review). The area chair load increased slightly, from 18 papers on average to 22 papers on average.

Conflicts of interest between authors, area chairs, and reviewers were handled largely automatically by the Open Review platform via their curated list of user profiles. Many authors submitting to ECCV already had a profile in Open Review. We set a paper registration deadline one week before the paper submission deadline in order to encourage all missing authors to register and create their Open Review profiles well on time (in practice, we allowed authors to create/change papers arbitrarily until the submission deadline). Except for minor issues with users creating duplicate profiles, this allowed us to easily and quickly identify institutional conflicts, and avoid them, while matching papers to area chairs and reviewers.

Papers were matched to area chairs based on: an affinity score computed by the Open Review platform, which is based on paper titles and abstracts, and an affinity score computed by the Toronto Paper Matching System (TPMS), which is based on the paper's full text, the area chair bids for individual papers, load balancing, and conflict avoidance. Open Review provides the program chairs a convenient web interface to experiment with different configurations of the matching algorithm. The chosen configuration resulted in about 50% of the assigned papers to be highly ranked by the area chair bids, and 50% to be ranked in the middle, with very few low bids assigned.

Assignments to reviewers were similar, with two differences. First, there was a maximum of 7 papers assigned to each reviewer. Second, area chairs recommended up to seven reviewers per paper, providing another highly-weighed term to the affinity scores used for matching.

The assignment of papers to area chairs was smooth. However, it was more difficult to find suitable reviewers for all papers. Having a ratio of 5.6 papers per reviewer with a maximum load of 7 (due to emergency reviewer commitment), which did not allow for much wiggle room in order to also satisfy conflict and expertise constraints. We received some complaints from reviewers who did not feel qualified to review specific papers and we reassigned them wherever possible. However, the large scale of the conference, the many constraints, and the fact that a large fraction of such complaints arrived very late in the review process made this process very difficult and not all complaints could be addressed.

Reviewers had six weeks to complete their assignments. Possibly due to COVID-19 or the fact that the NeurIPS deadline was moved closer to the review deadline, a record 30% of the reviews were still missing after the deadline. By comparison, ECCV 2018 experienced only 10% missing reviews at this stage of the process. In the subsequent week, area chairs chased the missing reviews intensely, found replacement reviewers in their own team, and managed to reach 10% missing reviews. Eventually, we could provide almost all reviews (more than 99.9%) with a delay of only a couple of days on the initial schedule by a significant use of emergency reviews. If this trend is confirmed, it might be a major challenge to run a smooth review process in future editions of ECCV. The community must reconsider prioritization of the time spent on paper writing (the number of submissions increased a lot despite COVID-19) and time spent on paper reviewing (the number of reviews delivered in time decreased a lot presumably due to COVID-19 or NeurIPS deadline). With this imbalance the peer-review system that ensures the quality of our top conferences may break soon.

Reviewers submitted their reviews independently. In the reviews, they had the opportunity to ask questions to the authors to be addressed in the rebuttal. However, reviewers were told not to request any significant new experiment. Using the Open Review interface, authors could provide an answer to each individual review, but were also allowed to cross-reference reviews and responses in their answers. Rather than PDF files, we allowed the use of formatted text for the rebuttal. The rebuttal and initial reviews were then made visible to all reviewers and the primary area chair for a given paper. The area chair encouraged and moderated the reviewer discussion. During the discussions, reviewers were invited to reach a consensus and possibly adjust their ratings as a result of the discussion and of the evidence in the rebuttal.

After the discussion period ended, most reviewers entered a final rating and recommendation, although in many cases this did not differ from their initial recommendation. Based on the updated reviews and discussion, the primary area chair then made a preliminary decision to accept or reject the paper and wrote a justification for it (meta-review). Except for cases where the outcome of this process was absolutely clear (as indicated by the three reviewers and primary area chairs all recommending clear rejection), the decision was then examined and potentially challenged by a secondary area chair. This led to further discussion and overturning a small number of preliminary decisions. Needless to say, there was no in-person area chair meeting, which would have been impossible due to COVID-19.

Area chairs were invited to observe the consensus of the reviewers whenever possible and use extreme caution in overturning a clear consensus to accept or reject a paper. If an area chair still decided to do so, she/he was asked to clearly justify it in the meta-review and to explicitly obtain the agreement of the secondary area chair. In practice, very few papers were rejected after being confidently accepted by the reviewers.

This was the first time Open Review was used as the main platform to run ECCV. In 2018, the program chairs used CMT3 for the user-facing interface and Open Review internally, for matching and conflict resolution. Since it is clearly preferable to only use a single platform, this year we switched to using Open Review in full. The experience was largely positive. The platform is highly-configurable, scalable, and open source. Being written in Python, it is easy to write scripts to extract data programmatically. The paper matching and conflict resolution algorithms and interfaces are top-notch, also due to the excellent author profiles in the platform. Naturally, there were a few kinks along the way due to the fact that the ECCV Open Review configuration was created from scratch for this event and it differs in substantial ways from many other Open Review conferences. However, the Open Review development and support team did a fantastic job in helping us to get the configuration right and to address issues in a timely manner as they unavoidably occurred. We cannot thank them enough for the tremendous effort they put into this project.

Finally, we would like to thank everyone involved in making ECCV 2020 possible in these very strange and difficult times. This starts with our authors, followed by the area chairs and reviewers, who ran the review process at an unprecedented scale. The whole Open Review team (and in particular Melisa Bok, Mohit Unyal, Carlos Mondragon Chapa, and Celeste Martinez Gomez) worked incredibly hard for the entire duration of the process. We would also like to thank René Vidal for contributing to the adoption of Open Review. Our thanks also go to Laurent Charling for TPMS and to the program chairs of ICML, ICLR, and NeurIPS for cross checking double submissions. We thank the website chair, Giovanni Farinella, and the CPI team (in particular Ashley Cook, Miriam Verdon, Nicola McGrane, and Sharon Kerr) for promptly adding material to the website as needed in the various phases of the process. Finally, we thank the publication chairs, Albert Ali Salah, Hamdi Dibeklioglu, Metehan Doyran, Henry Howard-Jenkins, Victor Prisacariu, Siyu Tang, and Gul Varol, who managed to compile these substantial proceedings in an exceedingly compressed schedule. We express our thanks to the ECVA team, in particular Kristina Scherbaum for allowing open access of the proceedings. We thank Alfred Hofmann from Springer who again serve as the publisher. Finally, we thank the other chairs of ECCV 2020, including in particular the general chairs for very useful feedback with the handling of the program.

August 2020

Andrea Vedaldi
Horst Bischof
Thomas Brox
Jan-Michael Frahm

Organization

General Chairs

Vittorio Ferrari	Google Research, Switzerland
Bob Fisher	University of Edinburgh, UK
Cordelia Schmid	Google and Inria, France
Emanuele Trucco	University of Dundee, UK

Program Chairs

Andrea Vedaldi	University of Oxford, UK
Horst Bischof	Graz University of Technology, Austria
Thomas Brox	University of Freiburg, Germany
Jan-Michael Frahm	University of North Carolina, USA

Industrial Liaison Chairs

Jim Ashe	University of Edinburgh, UK
Helmut Grabner	Zurich University of Applied Sciences, Switzerland
Diane Larlus	NAVER LABS Europe, France
Cristian Novotny	University of Edinburgh, UK

Local Arrangement Chairs

Yvan Petillot	Heriot-Watt University, UK
Paul Siebert	University of Glasgow, UK

Academic Demonstration Chair

Thomas Mensink	Google Research and University of Amsterdam, The Netherlands

Poster Chair

Stephen Mckenna	University of Dundee, UK

Technology Chair

Gerardo Aragon Camarasa	University of Glasgow, UK

Tutorial Chairs

Carlo Colombo	University of Florence, Italy
Sotirios Tsaftaris	University of Edinburgh, UK

Publication Chairs

Albert Ali Salah	Utrecht University, The Netherlands
Hamdi Dibeklioglu	Bilkent University, Turkey
Metehan Doyran	Utrecht University, The Netherlands
Henry Howard-Jenkins	University of Oxford, UK
Victor Adrian Prisacariu	University of Oxford, UK
Siyu Tang	ETH Zurich, Switzerland
Gul Varol	University of Oxford, UK

Website Chair

Giovanni Maria Farinella	University of Catania, Italy

Workshops Chairs

Adrien Bartoli	University of Clermont Auvergne, France
Andrea Fusiello	University of Udine, Italy

Area Chairs

Lourdes Agapito	University College London, UK
Zeynep Akata	University of Tübingen, Germany
Karteek Alahari	Inria, France
Antonis Argyros	University of Crete, Greece
Hossein Azizpour	KTH Royal Institute of Technology, Sweden
Joao P. Barreto	Universidade de Coimbra, Portugal
Alexander C. Berg	University of North Carolina at Chapel Hill, USA
Matthew B. Blaschko	KU Leuven, Belgium
Lubomir D. Bourdev	WaveOne, Inc., USA
Edmond Boyer	Inria, France
Yuri Boykov	University of Waterloo, Canada
Gabriel Brostow	University College London, UK
Michael S. Brown	National University of Singapore, Singapore
Jianfei Cai	Monash University, Australia
Barbara Caputo	Politecnico di Torino, Italy
Ayan Chakrabarti	Washington University, St. Louis, USA
Tat-Jen Cham	Nanyang Technological University, Singapore
Manmohan Chandraker	University of California, San Diego, USA
Rama Chellappa	Johns Hopkins University, USA
Liang-Chieh Chen	Google, USA

Yung-Yu Chuang	National Taiwan University, Taiwan
Ondrej Chum	Czech Technical University in Prague, Czech Republic
Brian Clipp	Kitware, USA
John Collomosse	University of Surrey and Adobe Research, UK
Jason J. Corso	University of Michigan, USA
David J. Crandall	Indiana University, USA
Daniel Cremers	University of California, Los Angeles, USA
Fabio Cuzzolin	Oxford Brookes University, UK
Jifeng Dai	SenseTime, SAR China
Kostas Daniilidis	University of Pennsylvania, USA
Andrew Davison	Imperial College London, UK
Alessio Del Bue	Fondazione Istituto Italiano di Tecnologia, Italy
Jia Deng	Princeton University, USA
Alexey Dosovitskiy	Google, Germany
Matthijs Douze	Facebook, France
Enrique Dunn	Stevens Institute of Technology, USA
Irfan Essa	Georgia Institute of Technology and Google, USA
Giovanni Maria Farinella	University of Catania, Italy
Ryan Farrell	Brigham Young University, USA
Paolo Favaro	University of Bern, Switzerland
Rogerio Feris	International Business Machines, USA
Cornelia Fermuller	University of Maryland, College Park, USA
David J. Fleet	Vector Institute, Canada
Friedrich Fraundorfer	DLR, Austria
Mario Fritz	CISPA Helmholtz Center for Information Security, Germany
Pascal Fua	EPFL (Swiss Federal Institute of Technology Lausanne), Switzerland
Yasutaka Furukawa	Simon Fraser University, Canada
Li Fuxin	Oregon State University, USA
Efstratios Gavves	University of Amsterdam, The Netherlands
Peter Vincent Gehler	Amazon, USA
Theo Gevers	University of Amsterdam, The Netherlands
Ross Girshick	Facebook AI Research, USA
Boqing Gong	Google, USA
Stephen Gould	Australian National University, Australia
Jinwei Gu	SenseTime Research, USA
Abhinav Gupta	Facebook, USA
Bohyung Han	Seoul National University, South Korea
Bharath Hariharan	Cornell University, USA
Tal Hassner	Facebook AI Research, USA
Xuming He	Australian National University, Australia
Joao F. Henriques	University of Oxford, UK
Adrian Hilton	University of Surrey, UK
Minh Hoai	Stony Brooks, State University of New York, USA
Derek Hoiem	University of Illinois Urbana-Champaign, USA

Timothy Hospedales	University of Edinburgh and Samsung, UK
Gang Hua	Wormpex AI Research, USA
Slobodan Ilic	Siemens AG, Germany
Hiroshi Ishikawa	Waseda University, Japan
Jiaya Jia	The Chinese University of Hong Kong, SAR China
Hailin Jin	Adobe Research, USA
Justin Johnson	University of Michigan, USA
Frederic Jurie	University of Caen Normandie, France
Fredrik Kahl	Chalmers University, Sweden
Sing Bing Kang	Zillow, USA
Gunhee Kim	Seoul National University, South Korea
Junmo Kim	Korea Advanced Institute of Science and Technology, South Korea
Tae-Kyun Kim	Imperial College London, UK
Ron Kimmel	Technion-Israel Institute of Technology, Israel
Alexander Kirillov	Facebook AI Research, USA
Kris Kitani	Carnegie Mellon University, USA
Iasonas Kokkinos	Ariel AI, UK
Vladlen Koltun	Intel Labs, USA
Nikos Komodakis	Ecole des Ponts ParisTech, France
Piotr Koniusz	Australian National University, Australia
M. Pawan Kumar	University of Oxford, UK
Kyros Kutulakos	University of Toronto, Canada
Christoph Lampert	IST Austria, Austria
Ivan Laptev	Inria, France
Diane Larlus	NAVER LABS Europe, France
Laura Leal-Taixe	Technical University Munich, Germany
Honglak Lee	Google and University of Michigan, USA
Joon-Young Lee	Adobe Research, USA
Kyoung Mu Lee	Seoul National University, South Korea
Seungyong Lee	POSTECH, South Korea
Yong Jae Lee	University of California, Davis, USA
Bastian Leibe	RWTH Aachen University, Germany
Victor Lempitsky	Samsung, Russia
Ales Leonardis	University of Birmingham, UK
Marius Leordeanu	Institute of Mathematics of the Romanian Academy, Romania
Vincent Lepetit	ENPC ParisTech, France
Hongdong Li	The Australian National University, Australia
Xi Li	Zhejiang University, China
Yin Li	University of Wisconsin-Madison, USA
Zicheng Liao	Zhejiang University, China
Jongwoo Lim	Hanyang University, South Korea
Stephen Lin	Microsoft Research Asia, China
Yen-Yu Lin	National Chiao Tung University, Taiwan, China
Zhe Lin	Adobe Research, USA

Haibin Ling	Stony Brooks, State University of New York, USA
Jiaying Liu	Peking University, China
Ming-Yu Liu	NVIDIA, USA
Si Liu	Beihang University, China
Xiaoming Liu	Michigan State University, USA
Huchuan Lu	Dalian University of Technology, China
Simon Lucey	Carnegie Mellon University, USA
Jiebo Luo	University of Rochester, USA
Julien Mairal	Inria, France
Michael Maire	University of Chicago, USA
Subhransu Maji	University of Massachusetts, Amherst, USA
Yasushi Makihara	Osaka University, Japan
Jiri Matas	Czech Technical University in Prague, Czech Republic
Yasuyuki Matsushita	Osaka University, Japan
Philippos Mordohai	Stevens Institute of Technology, USA
Vittorio Murino	University of Verona, Italy
Naila Murray	NAVER LABS Europe, France
Hajime Nagahara	Osaka University, Japan
P. J. Narayanan	International Institute of Information Technology (IIIT), Hyderabad, India
Nassir Navab	Technical University of Munich, Germany
Natalia Neverova	Facebook AI Research, France
Matthias Niessner	Technical University of Munich, Germany
Jean-Marc Odobez	Idiap Research Institute and Swiss Federal Institute of Technology Lausanne, Switzerland
Francesca Odone	Università di Genova, Italy
Takeshi Oishi	The University of Tokyo, Tokyo Institute of Technology, Japan
Vicente Ordonez	University of Virginia, USA
Manohar Paluri	Facebook AI Research, USA
Maja Pantic	Imperial College London, UK
In Kyu Park	Inha University, South Korea
Ioannis Patras	Queen Mary University of London, UK
Patrick Perez	Valeo, France
Bryan A. Plummer	Boston University, USA
Thomas Pock	Graz University of Technology, Austria
Marc Pollefeys	ETH Zurich and Microsoft MR & AI Zurich Lab, Switzerland
Jean Ponce	Inria, France
Gerard Pons-Moll	MPII, Saarland Informatics Campus, Germany
Jordi Pont-Tuset	Google, Switzerland
James Matthew Rehg	Georgia Institute of Technology, USA
Ian Reid	University of Adelaide, Australia
Olaf Ronneberger	DeepMind London, UK
Stefan Roth	TU Darmstadt, Germany
Bryan Russell	Adobe Research, USA

Mathieu Salzmann	EPFL, Switzerland
Dimitris Samaras	Stony Brook University, USA
Imari Sato	National Institute of Informatics (NII), Japan
Yoichi Sato	The University of Tokyo, Japan
Torsten Sattler	Czech Technical University in Prague, Czech Republic
Daniel Scharstein	Middlebury College, USA
Bernt Schiele	MPII, Saarland Informatics Campus, Germany
Julia A. Schnabel	King's College London, UK
Nicu Sebe	University of Trento, Italy
Greg Shakhnarovich	Toyota Technological Institute at Chicago, USA
Humphrey Shi	University of Oregon, USA
Jianbo Shi	University of Pennsylvania, USA
Jianping Shi	SenseTime, China
Leonid Sigal	University of British Columbia, Canada
Cees Snoek	University of Amsterdam, The Netherlands
Richard Souvenir	Temple University, USA
Hao Su	University of California, San Diego, USA
Akihiro Sugimoto	National Institute of Informatics (NII), Japan
Jian Sun	Megvii Technology, China
Jian Sun	Xi'an Jiaotong University, China
Chris Sweeney	Facebook Reality Labs, USA
Yu-wing Tai	Kuaishou Technology, China
Chi-Keung Tang	The Hong Kong University of Science and Technology, SAR China
Radu Timofte	ETH Zurich, Switzerland
Sinisa Todorovic	Oregon State University, USA
Giorgos Tolias	Czech Technical University in Prague, Czech Republic
Carlo Tomasi	Duke University, USA
Tatiana Tommasi	Politecnico di Torino, Italy
Lorenzo Torresani	Facebook AI Research and Dartmouth College, USA
Alexander Toshev	Google, USA
Zhuowen Tu	University of California, San Diego, USA
Tinne Tuytelaars	KU Leuven, Belgium
Jasper Uijlings	Google, Switzerland
Nuno Vasconcelos	University of California, San Diego, USA
Olga Veksler	University of Waterloo, Canada
Rene Vidal	Johns Hopkins University, USA
Gang Wang	Alibaba Group, China
Jingdong Wang	Microsoft Research Asia, China
Yizhou Wang	Peking University, China
Lior Wolf	Facebook AI Research and Tel Aviv University, Israel
Jianxin Wu	Nanjing University, China
Tao Xiang	University of Surrey, UK
Saining Xie	Facebook AI Research, USA
Ming-Hsuan Yang	University of California at Merced and Google, USA
Ruigang Yang	University of Kentucky, USA

Kwang Moo Yi	University of Victoria, Canada
Zhaozheng Yin	Stony Brook, State University of New York, USA
Chang D. Yoo	Korea Advanced Institute of Science and Technology, South Korea
Shaodi You	University of Amsterdam, The Netherlands
Jingyi Yu	ShanghaiTech University, China
Stella Yu	University of California, Berkeley, and ICSI, USA
Stefanos Zafeiriou	Imperial College London, UK
Hongbin Zha	Peking University, China
Tianzhu Zhang	University of Science and Technology of China, China
Liang Zheng	Australian National University, Australia
Todd E. Zickler	Harvard University, USA
Andrew Zisserman	University of Oxford, UK

Technical Program Committee

Sathyanarayanan N. Aakur
Wael Abd Almgaeed
Abdelrahman Abdelhamed
Abdullah Abuolaim
Supreeth Achar
Hanno Ackermann
Ehsan Adeli
Triantafyllos Afouras
Sameer Agarwal
Aishwarya Agrawal
Harsh Agrawal
Pulkit Agrawal
Antonio Agudo
Eirikur Agustsson
Karim Ahmed
Byeongjoo Ahn
Unaiza Ahsan
Thalaiyasingam Ajanthan
Kenan E. Ak
Emre Akbas
Naveed Akhtar
Derya Akkaynak
Yagiz Aksoy
Ziad Al-Halah
Xavier Alameda-Pineda
Jean-Baptiste Alayrac
Samuel Albanie
Shadi Albarqouni
Cenek Albl
Hassan Abu Alhaija
Daniel Aliaga
Mohammad S. Aliakbarian
Rahaf Aljundi
Thiemo Alldieck
Jon Almazan
Jose M. Alvarez
Senjian An
Saket Anand
Codruta Ancuti
Cosmin Ancuti
Peter Anderson
Juan Andrade-Cetto
Alexander Andreopoulos
Misha Andriluka
Dragomir Anguelov
Rushil Anirudh
Michel Antunes
Oisin Mac Aodha
Srikar Appalaraju
Relja Arandjelovic
Nikita Araslanov
Andre Araujo
Helder Araujo
Pablo Arbelaez
Shervin Ardeshir
Sercan O. Arik
Anil Armagan
Anurag Arnab
Chetan Arora
Federica Arrigoni
Mathieu Aubry
Shai Avidan
Angelica I. Aviles-Rivero
Yannis Avrithis
Ismail Ben Ayed
Shekoofeh Azizi
Ioan Andrei Bârsan
Artem Babenko
Deepak Babu Sam
Seung-Hwan Baek
Seungryul Baek
Andrew D. Bagdanov
Shai Bagon
Yuval Bahat
Junjie Bai
Song Bai
Xiang Bai
Yalong Bai
Yancheng Bai
Peter Bajcsy
Slawomir Bak

Mahsa Baktashmotlagh
Kavita Bala
Yogesh Balaji
Guha Balakrishnan
V. N. Balasubramanian
Federico Baldassarre
Vassileios Balntas
Shurjo Banerjee
Aayush Bansal
Ankan Bansal
Jianmin Bao
Linchao Bao
Wenbo Bao
Yingze Bao
Akash Bapat
Md Jawadul Hasan Bappy
Fabien Baradel
Lorenzo Baraldi
Daniel Barath
Adrian Barbu
Kobus Barnard
Nick Barnes
Francisco Barranco
Jonathan T. Barron
Arslan Basharat
Chaim Baskin
Anil S. Baslamisli
Jorge Batista
Kayhan Batmanghelich
Konstantinos Batsos
David Bau
Luis Baumela
Christoph Baur
Eduardo Bayro-Corrochano
Paul Beardsley
Jan Bednavr'ik
Oscar Beijbom
Philippe Bekaert
Esube Bekele
Vasileios Belagiannis
Ohad Ben-Shahar
Abhijit Bendale
Róger Bermúdez-Chacón
Maxim Berman
Jesus Bermudez-cameo
Florian Bernard
Stefano Berretti
Marcelo Bertalmio
Gedas Bertasius
Cigdem Beyan
Lucas Beyer
Vijayakumar Bhagavatula
Arjun Nitin Bhagoji
Apratim Bhattacharyya
Binod Bhattarai
Sai Bi
Jia-Wang Bian
Simone Bianco
Adel Bibi
Tolga Birdal
Tom Bishop
Soma Biswas
Mårten Björkman
Volker Blanz
Vishnu Boddeti
Navaneeth Bodla
Simion-Vlad Bogolin
Xavier Boix
Piotr Bojanowski
Timo Bolkart
Guido Borghi
Larbi Boubchir
Guillaume Bourmaud
Adrien Bousseau
Thierry Bouwmans
Richard Bowden
Hakan Boyraz
Mathieu Brédif
Samarth Brahmbhatt
Steve Branson
Nikolas Brasch
Biagio Brattoli
Ernesto Brau
Toby P. Breckon
Francois Bremond
Jesus Briales
Sofia Broomé
Marcus A. Brubaker
Luc Brun
Silvia Bucci
Shyamal Buch
Pradeep Buddharaju
Uta Buechler
Mai Bui
Tu Bui
Adrian Bulat
Giedrius T. Burachas
Elena Burceanu
Xavier P. Burgos-Artizzu
Kaylee Burns
Andrei Bursuc
Benjamin Busam
Wonmin Byeon
Zoya Bylinskii
Sergi Caelles
Jianrui Cai
Minjie Cai
Yujun Cai
Zhaowei Cai
Zhipeng Cai
Juan C. Caicedo
Simone Calderara
Necati Cihan Camgoz
Dylan Campbell
Octavia Camps
Jiale Cao
Kaidi Cao
Liangliang Cao
Xiangyong Cao
Xiaochun Cao
Yang Cao
Yu Cao
Yue Cao
Zhangjie Cao
Luca Carlone
Mathilde Caron
Dan Casas
Thomas J. Cashman
Umberto Castellani
Lluis Castrejon
Jacopo Cavazza
Fabio Cermelli
Hakan Cevikalp
Menglei Chai
Ishani Chakraborty
Rudrasis Chakraborty
Antoni B. Chan

Kwok-Ping Chan
Siddhartha Chandra
Sharat Chandran
Arjun Chandrasekaran
Angel X. Chang
Che-Han Chang
Hong Chang
Hyun Sung Chang
Hyung Jin Chang
Jianlong Chang
Ju Yong Chang
Ming-Ching Chang
Simyung Chang
Xiaojun Chang
Yu-Wei Chao
Devendra S. Chaplot
Arslan Chaudhry
Rizwan A. Chaudhry
Can Chen
Chang Chen
Chao Chen
Chen Chen
Chu-Song Chen
Dapeng Chen
Dong Chen
Dongdong Chen
Guanying Chen
Hongge Chen
Hsin-yi Chen
Huaijin Chen
Hwann-Tzong Chen
Jianbo Chen
Jianhui Chen
Jiansheng Chen
Jiaxin Chen
Jie Chen
Jun-Cheng Chen
Kan Chen
Kevin Chen
Lin Chen
Long Chen
Min-Hung Chen
Qifeng Chen
Shi Chen
Shixing Chen
Tianshui Chen
Weifeng Chen
Weikai Chen
Xi Chen
Xiaohan Chen
Xiaozhi Chen
Xilin Chen
Xingyu Chen
Xinlei Chen
Xinyun Chen
Yi-Ting Chen
Yilun Chen
Ying-Cong Chen
Yinpeng Chen
Yiran Chen
Yu Chen
Yu-Sheng Chen
Yuhua Chen
Yun-Chun Chen
Yunpeng Chen
Yuntao Chen
Zhuoyuan Chen
Zitian Chen
Anchieh Cheng
Bowen Cheng
Erkang Cheng
Gong Cheng
Guangliang Cheng
Jingchun Cheng
Jun Cheng
Li Cheng
Ming-Ming Cheng
Yu Cheng
Ziang Cheng
Anoop Cherian
Dmitry Chetverikov
Ngai-man Cheung
William Cheung
Ajad Chhatkuli
Naoki Chiba
Benjamin Chidester
Han-pang Chiu
Mang Tik Chiu
Wei-Chen Chiu
Donghyeon Cho
Hojin Cho
Minsu Cho
Nam Ik Cho
Tim Cho
Tae Eun Choe
Chiho Choi
Edward Choi
Inchang Choi
Jinsoo Choi
Jonghyun Choi
Jongwon Choi
Yukyung Choi
Hisham Cholakkal
Eunji Chong
Jaegul Choo
Christopher Choy
Hang Chu
Peng Chu
Wen-Sheng Chu
Albert Chung
Joon Son Chung
Hai Ci
Safa Cicek
Ramazan G. Cinbis
Arridhana Ciptadi
Javier Civera
James J. Clark
Ronald Clark
Felipe Codevilla
Michael Cogswell
Andrea Cohen
Maxwell D. Collins
Carlo Colombo
Yang Cong
Adria R. Continente
Marcella Cornia
John Richard Corring
Darren Cosker
Dragos Costea
Garrison W. Cottrell
Florent Couzinie-Devy
Marco Cristani
Ioana Croitoru
James L. Crowley
Jiequan Cui
Zhaopeng Cui
Ross Cutler
Antonio D'Innocente

Rozenn Dahyot
Bo Dai
Dengxin Dai
Hang Dai
Longquan Dai
Shuyang Dai
Xiyang Dai
Yuchao Dai
Adrian V. Dalca
Dima Damen
Bharath B. Damodaran
Kristin Dana
Martin Danelljan
Zheng Dang
Zachary Alan Daniels
Donald G. Dansereau
Abhishek Das
Samyak Datta
Achal Dave
Titas De
Rodrigo de Bem
Teo de Campos
Raoul de Charette
Shalini De Mello
Joseph DeGol
Herve Delingette
Haowen Deng
Jiankang Deng
Weijian Deng
Zhiwei Deng
Joachim Denzler
Konstantinos G. Derpanis
Aditya Deshpande
Frederic Devernay
Somdip Dey
Arturo Deza
Abhinav Dhall
Helisa Dhamo
Vikas Dhiman
Fillipe Dias Moreira de Souza
Ali Diba
Ferran Diego
Guiguang Ding
Henghui Ding
Jian Ding
Mingyu Ding
Xinghao Ding
Zhengming Ding
Robert DiPietro
Cosimo Distante
Ajay Divakaran
Mandar Dixit
Abdelaziz Djelouah
Thanh-Toan Do
Jose Dolz
Bo Dong
Chao Dong
Jiangxin Dong
Weiming Dong
Weisheng Dong
Xingping Dong
Xuanyi Dong
Yinpeng Dong
Gianfranco Doretto
Hazel Doughty
Hassen Drira
Bertram Drost
Dawei Du
Ye Duan
Yueqi Duan
Abhimanyu Dubey
Anastasia Dubrovina
Stefan Duffner
Chi Nhan Duong
Thibaut Durand
Zoran Duric
Iulia Duta
Debidatta Dwibedi
Benjamin Eckart
Marc Eder
Marzieh Edraki
Alexei A. Efros
Kiana Ehsani
Hazm Kemal Ekenel
James H. Elder
Mohamed Elgharib
Shireen Elhabian
Ehsan Elhamifar
Mohamed Elhoseiny
Ian Endres
N. Benjamin Erichson
Jan Ernst
Sergio Escalera
Francisco Escolano
Victor Escorcia
Carlos Esteves
Francisco J. Estrada
Bin Fan
Chenyou Fan
Deng-Ping Fan
Haoqi Fan
Hehe Fan
Heng Fan
Kai Fan
Lijie Fan
Linxi Fan
Quanfu Fan
Shaojing Fan
Xiaochuan Fan
Xin Fan
Yuchen Fan
Sean Fanello
Hao-Shu Fang
Haoyang Fang
Kuan Fang
Yi Fang
Yuming Fang
Azade Farshad
Alireza Fathi
Raanan Fattal
Joao Fayad
Xiaohan Fei
Christoph Feichtenhofer
Michael Felsberg
Chen Feng
Jiashi Feng
Junyi Feng
Mengyang Feng
Qianli Feng
Zhenhua Feng
Michele Fenzi
Andras Ferencz
Martin Fergie
Basura Fernando
Ethan Fetaya
Michael Firman
John W. Fisher

Matthew Fisher
Boris Flach
Corneliu Florea
Wolfgang Foerstner
David Fofi
Gian Luca Foresti
Per-Erik Forssen
David Fouhey
Katerina Fragkiadaki
Victor Fragoso
Jean-Sébastien Franco
Ohad Fried
Iuri Frosio
Cheng-Yang Fu
Huazhu Fu
Jianlong Fu
Jingjing Fu
Xueyang Fu
Yanwei Fu
Ying Fu
Yun Fu
Olac Fuentes
Kent Fujiwara
Takuya Funatomi
Christopher Funk
Thomas Funkhouser
Antonino Furnari
Ryo Furukawa
Erik Gärtner
Raghudeep Gadde
Matheus Gadelha
Vandit Gajjar
Trevor Gale
Juergen Gall
Mathias Gallardo
Guillermo Gallego
Orazio Gallo
Chuang Gan
Zhe Gan
Madan Ravi Ganesh
Aditya Ganeshan
Siddha Ganju
Bin-Bin Gao
Changxin Gao
Feng Gao
Hongchang Gao
Jin Gao
Jiyang Gao
Junbin Gao
Katelyn Gao
Lin Gao
Mingfei Gao
Ruiqi Gao
Ruohan Gao
Shenghua Gao
Yuan Gao
Yue Gao
Noa Garcia
Alberto Garcia-Garcia
Guillermo Garcia-Hernando
Jacob R. Gardner
Animesh Garg
Kshitiz Garg
Rahul Garg
Ravi Garg
Philip N. Garner
Kirill Gavrilyuk
Paul Gay
Shiming Ge
Weifeng Ge
Baris Gecer
Xin Geng
Kyle Genova
Stamatios Georgoulis
Bernard Ghanem
Michael Gharbi
Kamran Ghasedi
Golnaz Ghiasi
Arnab Ghosh
Partha Ghosh
Silvio Giancola
Andrew Gilbert
Rohit Girdhar
Xavier Giro-i-Nieto
Thomas Gittings
Ioannis Gkioulekas
Clement Godard
Vaibhava Goel
Bastian Goldluecke
Lluis Gomez
Nuno Gonçalves
Dong Gong
Ke Gong
Mingming Gong
Abel Gonzalez-Garcia
Ariel Gordon
Daniel Gordon
Paulo Gotardo
Venu Madhav Govindu
Ankit Goyal
Priya Goyal
Raghav Goyal
Benjamin Graham
Douglas Gray
Brent A. Griffin
Etienne Grossmann
David Gu
Jiayuan Gu
Jiuxiang Gu
Lin Gu
Qiao Gu
Shuhang Gu
Jose J. Guerrero
Paul Guerrero
Jie Gui
Jean-Yves Guillemaut
Riza Alp Guler
Erhan Gundogdu
Fatma Guney
Guodong Guo
Kaiwen Guo
Qi Guo
Sheng Guo
Shi Guo
Tiantong Guo
Xiaojie Guo
Yijie Guo
Yiluan Guo
Yuanfang Guo
Yulan Guo
Agrim Gupta
Ankush Gupta
Mohit Gupta
Saurabh Gupta
Tanmay Gupta
Danna Gurari
Abner Guzman-Rivera

JunYoung Gwak
Michael Gygli
Jung-Woo Ha
Simon Hadfield
Isma Hadji
Bjoern Haefner
Taeyoung Hahn
Levente Hajder
Peter Hall
Emanuela Haller
Stefan Haller
Bumsub Ham
Abdullah Hamdi
Dongyoon Han
Hu Han
Jungong Han
Junwei Han
Kai Han
Tian Han
Xiaoguang Han
Xintong Han
Yahong Han
Ankur Handa
Zekun Hao
Albert Haque
Tatsuya Harada
Mehrtash Harandi
Adam W. Harley
Mahmudul Hasan
Atsushi Hashimoto
Ali Hatamizadeh
Munawar Hayat
Dongliang He
Jingrui He
Junfeng He
Kaiming He
Kun He
Lei He
Pan He
Ran He
Shengfeng He
Tong He
Weipeng He
Xuming He
Yang He
Yihui He
Zhihai He
Chinmay Hegde
Janne Heikkila
Mattias P. Heinrich
Stéphane Herbin
Alexander Hermans
Luis Herranz
John R. Hershey
Aaron Hertzmann
Roei Herzig
Anders Heyden
Steven Hickson
Otmar Hilliges
Tomas Hodan
Judy Hoffman
Michael Hofmann
Yannick Hold-Geoffroy
Namdar Homayounfar
Sina Honari
Richang Hong
Seunghoon Hong
Xiaopeng Hong
Yi Hong
Hidekata Hontani
Anthony Hoogs
Yedid Hoshen
Mir Rayat Imtiaz Hossain
Junhui Hou
Le Hou
Lu Hou
Tingbo Hou
Wei-Lin Hsiao
Cheng-Chun Hsu
Gee-Sern Jison Hsu
Kuang-jui Hsu
Changbo Hu
Di Hu
Guosheng Hu
Han Hu
Hao Hu
Hexiang Hu
Hou-Ning Hu
Jie Hu
Junlin Hu
Nan Hu
Ping Hu
Ronghang Hu
Xiaowei Hu
Yinlin Hu
Yuan-Ting Hu
Zhe Hu
Binh-Son Hua
Yang Hua
Bingyao Huang
Di Huang
Dong Huang
Fay Huang
Haibin Huang
Haozhi Huang
Heng Huang
Huaibo Huang
Jia-Bin Huang
Jing Huang
Jingwei Huang
Kaizhu Huang
Lei Huang
Qiangui Huang
Qiaoying Huang
Qingqiu Huang
Qixing Huang
Shaoli Huang
Sheng Huang
Siyuan Huang
Weilin Huang
Wenbing Huang
Xiangru Huang
Xun Huang
Yan Huang
Yifei Huang
Yue Huang
Zhiwu Huang
Zilong Huang
Minyoung Huh
Zhuo Hui
Matthias B. Hullin
Martin Humenberger
Wei-Chih Hung
Zhouyuan Huo
Junhwa Hur
Noureldien Hussein
Jyh-Jing Hwang
Seong Jae Hwang

Sung Ju Hwang
Ichiro Ide
Ivo Ihrke
Daiki Ikami
Satoshi Ikehata
Nazli Ikizler-Cinbis
Sunghoon Im
Yani Ioannou
Radu Tudor Ionescu
Umar Iqbal
Go Irie
Ahmet Iscen
Md Amirul Islam
Vamsi Ithapu
Nathan Jacobs
Arpit Jain
Himalaya Jain
Suyog Jain
Stuart James
Won-Dong Jang
Yunseok Jang
Ronnachai Jaroensri
Dinesh Jayaraman
Sadeep Jayasumana
Suren Jayasuriya
Herve Jegou
Simon Jenni
Hae-Gon Jeon
Yunho Jeon
Koteswar R. Jerripothula
Hueihan Jhuang
I-hong Jhuo
Dinghuang Ji
Hui Ji
Jingwei Ji
Pan Ji
Yanli Ji
Baoxiong Jia
Kui Jia
Xu Jia
Chiyu Max Jiang
Haiyong Jiang
Hao Jiang
Huaizu Jiang
Huajie Jiang
Ke Jiang
Lai Jiang
Li Jiang
Lu Jiang
Ming Jiang
Peng Jiang
Shuqiang Jiang
Wei Jiang
Xudong Jiang
Zhuolin Jiang
Jianbo Jiao
Zequn Jie
Dakai Jin
Kyong Hwan Jin
Lianwen Jin
SouYoung Jin
Xiaojie Jin
Xin Jin
Nebojsa Jojic
Alexis Joly
Michael Jeffrey Jones
Hanbyul Joo
Jungseock Joo
Kyungdon Joo
Ajjen Joshi
Shantanu H. Joshi
Da-Cheng Juan
Marco Körner
Kevin Köser
Asim Kadav
Christine Kaeser-Chen
Kushal Kafle
Dagmar Kainmueller
Ioannis A. Kakadiaris
Zdenek Kalal
Nima Kalantari
Yannis Kalantidis
Mahdi M. Kalayeh
Anmol Kalia
Sinan Kalkan
Vicky Kalogeiton
Ashwin Kalyan
Joni-kristian Kamarainen
Gerda Kamberova
Chandra Kambhamettu
Martin Kampel
Meina Kan
Christopher Kanan
Kenichi Kanatani
Angjoo Kanazawa
Atsushi Kanehira
Takuhiro Kaneko
Asako Kanezaki
Bingyi Kang
Di Kang
Sunghun Kang
Zhao Kang
Vadim Kantorov
Abhishek Kar
Amlan Kar
Theofanis Karaletsos
Leonid Karlinsky
Kevin Karsch
Angelos Katharopoulos
Isinsu Katircioglu
Hiroharu Kato
Zoltan Kato
Dotan Kaufman
Jan Kautz
Rei Kawakami
Qiuhong Ke
Wadim Kehl
Petr Kellnhofer
Aniruddha Kembhavi
Cem Keskin
Margret Keuper
Daniel Keysers
Ashkan Khakzar
Fahad Khan
Naeemullah Khan
Salman Khan
Siddhesh Khandelwal
Rawal Khirodkar
Anna Khoreva
Tejas Khot
Parmeshwar Khurd
Hadi Kiapour
Joe Kileel
Chanho Kim
Dahun Kim
Edward Kim
Eunwoo Kim
Han-ul Kim

Hansung Kim
Heewon Kim
Hyo Jin Kim
Hyunwoo J. Kim
Jinkyu Kim
Jiwon Kim
Jongmin Kim
Junsik Kim
Junyeong Kim
Min H. Kim
Namil Kim
Pyojin Kim
Seon Joo Kim
Seong Tae Kim
Seungryong Kim
Sungwoong Kim
Tae Hyun Kim
Vladimir Kim
Won Hwa Kim
Yonghyun Kim
Benjamin Kimia
Akisato Kimura
Pieter-Jan Kindermans
Zsolt Kira
Itaru Kitahara
Hedvig Kjellstrom
Jan Knopp
Takumi Kobayashi
Erich Kobler
Parker Koch
Reinhard Koch
Elyor Kodirov
Amir Kolaman
Nicholas Kolkin
Dimitrios Kollias
Stefanos Kollias
Soheil Kolouri
Adams Wai-Kin Kong
Naejin Kong
Shu Kong
Tao Kong
Yu Kong
Yoshinori Konishi
Daniil Kononenko
Theodora Kontogianni
Simon Korman
Adam Kortylewski
Jana Kosecka
Jean Kossaifi
Satwik Kottur
Rigas Kouskouridas
Adriana Kovashka
Rama Kovvuri
Adarsh Kowdle
Jedrzej Kozerawski
Mateusz Kozinski
Philipp Kraehenbuehl
Gregory Kramida
Josip Krapac
Dmitry Kravchenko
Ranjay Krishna
Pavel Krsek
Alexander Krull
Jakob Kruse
Hiroyuki Kubo
Hilde Kuehne
Jason Kuen
Andreas Kuhn
Arjan Kuijper
Zuzana Kukelova
Ajay Kumar
Amit Kumar
Avinash Kumar
Suryansh Kumar
Vijay Kumar
Kaustav Kundu
Weicheng Kuo
Nojun Kwak
Suha Kwak
Junseok Kwon
Nikolaos Kyriazis
Zorah Lähner
Ankit Laddha
Florent Lafarge
Jean Lahoud
Kevin Lai
Shang-Hong Lai
Wei-Sheng Lai
Yu-Kun Lai
Iro Laina
Antony Lam
John Wheatley Lambert
Xiangyuan lan
Xu Lan
Charis Lanaras
Georg Langs
Oswald Lanz
Dong Lao
Yizhen Lao
Agata Lapedriza
Gustav Larsson
Viktor Larsson
Katrin Lasinger
Christoph Lassner
Longin Jan Latecki
Stéphane Lathuilière
Rynson Lau
Hei Law
Justin Lazarow
Svetlana Lazebnik
Hieu Le
Huu Le
Ngan Hoang Le
Trung-Nghia Le
Vuong Le
Colin Lea
Erik Learned-Miller
Chen-Yu Lee
Gim Hee Lee
Hsin-Ying Lee
Hyungtae Lee
Jae-Han Lee
Jimmy Addison Lee
Joonseok Lee
Kibok Lee
Kuang-Huei Lee
Kwonjoon Lee
Minsik Lee
Sang-chul Lee
Seungkyu Lee
Soochan Lee
Stefan Lee
Taehee Lee
Andreas Lehrmann
Jie Lei
Peng Lei
Matthew Joseph Leotta
Wee Kheng Leow

Gil Levi
Evgeny Levinkov
Aviad Levis
Jose Lezama
Ang Li
Bin Li
Bing Li
Boyi Li
Changsheng Li
Chao Li
Chen Li
Cheng Li
Chenglong Li
Chi Li
Chun-Guang Li
Chun-Liang Li
Chunyuan Li
Dong Li
Guanbin Li
Hao Li
Haoxiang Li
Hongsheng Li
Hongyang Li
Houqiang Li
Huibin Li
Jia Li
Jianan Li
Jianguo Li
Junnan Li
Junxuan Li
Kai Li
Ke Li
Kejie Li
Kunpeng Li
Lerenhan Li
Li Erran Li
Mengtian Li
Mu Li
Peihua Li
Peiyi Li
Ping Li
Qi Li
Qing Li
Ruiyu Li
Ruoteng Li
Shaozi Li
Sheng Li
Shiwei Li
Shuang Li
Siyang Li
Stan Z. Li
Tianye Li
Wei Li
Weixin Li
Wen Li
Wenbo Li
Xiaomeng Li
Xin Li
Xiu Li
Xuelong Li
Xueting Li
Yan Li
Yandong Li
Yanghao Li
Yehao Li
Yi Li
Yijun Li
Yikang LI
Yining Li
Yongjie Li
Yu Li
Yu-Jhe Li
Yunpeng Li
Yunsheng Li
Yunzhu Li
Zhe Li
Zhen Li
Zhengqi Li
Zhenyang Li
Zhuwen Li
Dongze Lian
Xiaochen Lian
Zhouhui Lian
Chen Liang
Jie Liang
Ming Liang
Paul Pu Liang
Pengpeng Liang
Shu Liang
Wei Liang
Jing Liao
Minghui Liao
Renjie Liao
Shengcai Liao
Shuai Liao
Yiyi Liao
Ser-Nam Lim
Chen-Hsuan Lin
Chung-Ching Lin
Dahua Lin
Ji Lin
Kevin Lin
Tianwei Lin
Tsung-Yi Lin
Tsung-Yu Lin
Wei-An Lin
Weiyao Lin
Yen-Chen Lin
Yuewei Lin
David B. Lindell
Drew Linsley
Krzysztof Lis
Roee Litman
Jim Little
An-An Liu
Bo Liu
Buyu Liu
Chao Liu
Chen Liu
Cheng-lin Liu
Chenxi Liu
Dong Liu
Feng Liu
Guilin Liu
Haomiao Liu
Heshan Liu
Hong Liu
Ji Liu
Jingen Liu
Jun Liu
Lanlan Liu
Li Liu
Liu Liu
Mengyuan Liu
Miaomiao Liu
Nian Liu
Ping Liu
Risheng Liu

Sheng Liu
Shu Liu
Shuaicheng Liu
Sifei Liu
Siqi Liu
Siying Liu
Songtao Liu
Ting Liu
Tongliang Liu
Tyng-Luh Liu
Wanquan Liu
Wei Liu
Weiyang Liu
Weizhe Liu
Wenyu Liu
Wu Liu
Xialei Liu
Xianglong Liu
Xiaodong Liu
Xiaofeng Liu
Xihui Liu
Xingyu Liu
Xinwang Liu
Xuanqing Liu
Xuebo Liu
Yang Liu
Yaojie Liu
Yebin Liu
Yen-Cheng Liu
Yiming Liu
Yu Liu
Yu-Shen Liu
Yufan Liu
Yun Liu
Zheng Liu
Zhijian Liu
Zhuang Liu
Zichuan Liu
Ziwei Liu
Zongyi Liu
Stephan Liwicki
Liliana Lo Presti
Chengjiang Long
Fuchen Long
Mingsheng Long
Xiang Long
Yang Long
Charles T. Loop
Antonio Lopez
Roberto J. Lopez-Sastre
Javier Lorenzo-Navarro
Manolis Lourakis
Boyu Lu
Canyi Lu
Feng Lu
Guoyu Lu
Hongtao Lu
Jiajun Lu
Jiasen Lu
Jiwen Lu
Kaiyue Lu
Le Lu
Shao-Ping Lu
Shijian Lu
Xiankai Lu
Xin Lu
Yao Lu
Yiping Lu
Yongxi Lu
Yongyi Lu
Zhiwu Lu
Fujun Luan
Benjamin E. Lundell
Hao Luo
Jian-Hao Luo
Ruotian Luo
Weixin Luo
Wenhan Luo
Wenjie Luo
Yan Luo
Zelun Luo
Zixin Luo
Khoa Luu
Zhaoyang Lv
Pengyuan Lyu
Thomas Möllenhoff
Matthias Müller
Bingpeng Ma
Chih-Yao Ma
Chongyang Ma
Huimin Ma
Jiayi Ma
K. T. Ma
Ke Ma
Lin Ma
Liqian Ma
Shugao Ma
Wei-Chiu Ma
Xiaojian Ma
Xingjun Ma
Zhanyu Ma
Zheng Ma
Radek Jakob Mackowiak
Ludovic Magerand
Shweta Mahajan
Siddharth Mahendran
Long Mai
Ameesh Makadia
Oscar Mendez Maldonado
Mateusz Malinowski
Yury Malkov
Arun Mallya
Dipu Manandhar
Massimiliano Mancini
Fabian Manhardt
Kevis-kokitsi Maninis
Varun Manjunatha
Junhua Mao
Xudong Mao
Alina Marcu
Edgar Margffoy-Tuay
Dmitrii Marin
Manuel J. Marin-Jimenez
Kenneth Marino
Niki Martinel
Julieta Martinez
Jonathan Masci
Tomohiro Mashita
Iacopo Masi
David Masip
Daniela Massiceti
Stefan Mathe
Yusuke Matsui
Tetsu Matsukawa
Iain A. Matthews
Kevin James Matzen
Bruce Allen Maxwell
Stephen Maybank

Helmut Mayer
Amir Mazaheri
David McAllester
Steven McDonagh
Stephen J. Mckenna
Roey Mechrez
Prakhar Mehrotra
Christopher Mei
Xue Mei
Paulo R. S. Mendonca
Lili Meng
Zibo Meng
Thomas Mensink
Bjoern Menze
Michele Merler
Kourosh Meshgi
Pascal Mettes
Christopher Metzler
Liang Mi
Qiguang Miao
Xin Miao
Tomer Michaeli
Frank Michel
Antoine Miech
Krystian Mikolajczyk
Peyman Milanfar
Ben Mildenhall
Gregor Miller
Fausto Milletari
Dongbo Min
Kyle Min
Pedro Miraldo
Dmytro Mishkin
Anand Mishra
Ashish Mishra
Ishan Misra
Niluthpol C. Mithun
Kaushik Mitra
Niloy Mitra
Anton Mitrokhin
Ikuhisa Mitsugami
Anurag Mittal
Kaichun Mo
Zhipeng Mo
Davide Modolo
Michael Moeller
Pritish Mohapatra
Pavlo Molchanov
Davide Moltisanti
Pascal Monasse
Mathew Monfort
Aron Monszpart
Sean Moran
Vlad I. Morariu
Francesc Moreno-Noguer
Pietro Morerio
Stylianos Moschoglou
Yael Moses
Roozbeh Mottaghi
Pierre Moulon
Arsalan Mousavian
Yadong Mu
Yasuhiro Mukaigawa
Lopamudra Mukherjee
Yusuke Mukuta
Ravi Teja Mullapudi
Mario Enrique Munich
Zachary Murez
Ana C. Murillo
J. Krishna Murthy
Damien Muselet
Armin Mustafa
Siva Karthik Mustikovela
Carlo Dal Mutto
Moin Nabi
Varun K. Nagaraja
Tushar Nagarajan
Arsha Nagrani
Seungjun Nah
Nikhil Naik
Yoshikatsu Nakajima
Yuta Nakashima
Atsushi Nakazawa
Seonghyeon Nam
Vinay P. Namboodiri
Medhini Narasimhan
Srinivasa Narasimhan
Sanath Narayan
Erickson Rangel Nascimento
Jacinto Nascimento
Tayyab Naseer
Lakshmanan Nataraj
Neda Nategh
Nelson Isao Nauata
Fernando Navarro
Shah Nawaz
Lukas Neumann
Ram Nevatia
Alejandro Newell
Shawn Newsam
Joe Yue-Hei Ng
Trung Thanh Ngo
Duc Thanh Nguyen
Lam M. Nguyen
Phuc Xuan Nguyen
Thuong Nguyen Canh
Mihalis Nicolaou
Andrei Liviu Nicolicioiu
Xuecheng Nie
Michael Niemeyer
Simon Niklaus
Christophoros Nikou
David Nilsson
Jifeng Ning
Yuval Nirkin
Li Niu
Yuzhen Niu
Zhenxing Niu
Shohei Nobuhara
Nicoletta Noceti
Hyeonwoo Noh
Junhyug Noh
Mehdi Noroozi
Sotiris Nousias
Valsamis Ntouskos
Matthew O'Toole
Peter Ochs
Ferda Ofli
Seong Joon Oh
Seoung Wug Oh
Iason Oikonomidis
Utkarsh Ojha
Takahiro Okabe
Takayuki Okatani
Fumio Okura
Aude Oliva
Kyle Olszewski

Björn Ommer
Mohamed Omran
Elisabeta Oneata
Michael Opitz
Jose Oramas
Tribhuvanesh Orekondy
Shaul Oron
Sergio Orts-Escolano
Ivan Oseledets
Aljosa Osep
Magnus Oskarsson
Anton Osokin
Martin R. Oswald
Wanli Ouyang
Andrew Owens
Mete Ozay
Mustafa Ozuysal
Eduardo Pérez-Pellitero
Gautam Pai
Dipan Kumar Pal
P. H. Pamplona Savarese
Jinshan Pan
Junting Pan
Xingang Pan
Yingwei Pan
Yannis Panagakis
Rameswar Panda
Guan Pang
Jiahao Pang
Jiangmiao Pang
Tianyu Pang
Sharath Pankanti
Nicolas Papadakis
Dim Papadopoulos
George Papandreou
Toufiq Parag
Shaifali Parashar
Sarah Parisot
Eunhyeok Park
Hyun Soo Park
Jaesik Park
Min-Gyu Park
Taesung Park
Alvaro Parra
C. Alejandro Parraga
Despoina Paschalidou
Nikolaos Passalis
Vishal Patel
Viorica Patraucean
Badri Narayana Patro
Danda Pani Paudel
Sujoy Paul
Georgios Pavlakos
Ioannis Pavlidis
Vladimir Pavlovic
Nick Pears
Kim Steenstrup Pedersen
Selen Pehlivan
Shmuel Peleg
Chao Peng
Houwen Peng
Wen-Hsiao Peng
Xi Peng
Xiaojiang Peng
Xingchao Peng
Yuxin Peng
Federico Perazzi
Juan Camilo Perez
Vishwanath Peri
Federico Pernici
Luca Del Pero
Florent Perronnin
Stavros Petridis
Henning Petzka
Patrick Peursum
Michael Pfeiffer
Hanspeter Pfister
Roman Pflugfelder
Minh Tri Pham
Yongri Piao
David Picard
Tomasz Pieciak
A. J. Piergiovanni
Andrea Pilzer
Pedro O. Pinheiro
Silvia Laura Pintea
Lerrel Pinto
Axel Pinz
Robinson Piramuthu
Fiora Pirri
Leonid Pishchulin
Francesco Pittaluga
Daniel Pizarro
Tobias Plötz
Mirco Planamente
Matteo Poggi
Moacir A. Ponti
Parita Pooj
Fatih Porikli
Horst Possegger
Omid Poursaeed
Ameya Prabhu
Viraj Uday Prabhu
Dilip Prasad
Brian L. Price
True Price
Maria Priisalu
Veronique Prinet
Victor Adrian Prisacariu
Jan Prokaj
Sergey Prokudin
Nicolas Pugeault
Xavier Puig
Albert Pumarola
Pulak Purkait
Senthil Purushwalkam
Charles R. Qi
Hang Qi
Haozhi Qi
Lu Qi
Mengshi Qi
Siyuan Qi
Xiaojuan Qi
Yuankai Qi
Shengju Qian
Xuelin Qian
Siyuan Qiao
Yu Qiao
Jie Qin
Qiang Qiu
Weichao Qiu
Zhaofan Qiu
Kha Gia Quach
Yuhui Quan
Yvain Queau
Julian Quiroga
Faisal Qureshi
Mahdi Rad

Filip Radenovic
Petia Radeva
Venkatesh B. Radhakrishnan
Ilija Radosavovic
Noha Radwan
Rahul Raguram
Tanzila Rahman
Amit Raj
Ajit Rajwade
Kandan Ramakrishnan
Santhosh K. Ramakrishnan
Srikumar Ramalingam
Ravi Ramamoorthi
Vasili Ramanishka
Ramprasaath R. Selvaraju
Francois Rameau
Visvanathan Ramesh
Santu Rana
Rene Ranftl
Anand Rangarajan
Anurag Ranjan
Viresh Ranjan
Yongming Rao
Carolina Raposo
Vivek Rathod
Sathya N. Ravi
Avinash Ravichandran
Tammy Riklin Raviv
Daniel Rebain
Sylvestre-Alvise Rebuffi
N. Dinesh Reddy
Timo Rehfeld
Paolo Remagnino
Konstantinos Rematas
Edoardo Remelli
Dongwei Ren
Haibing Ren
Jian Ren
Jimmy Ren
Mengye Ren
Weihong Ren
Wenqi Ren
Zhile Ren
Zhongzheng Ren
Zhou Ren
Vijay Rengarajan
Md A. Reza
Farzaneh Rezaeianaran
Hamed R. Tavakoli
Nicholas Rhinehart
Helge Rhodin
Elisa Ricci
Alexander Richard
Eitan Richardson
Elad Richardson
Christian Richardt
Stephan Richter
Gernot Riegler
Daniel Ritchie
Tobias Ritschel
Samuel Rivera
Yong Man Ro
Richard Roberts
Joseph Robinson
Ignacio Rocco
Mrigank Rochan
Emanuele Rodolà
Mikel D. Rodriguez
Giorgio Roffo
Grégory Rogez
Gemma Roig
Javier Romero
Xuejian Rong
Yu Rong
Amir Rosenfeld
Bodo Rosenhahn
Guy Rosman
Arun Ross
Paolo Rota
Peter M. Roth
Anastasios Roussos
Anirban Roy
Sebastien Roy
Aruni RoyChowdhury
Artem Rozantsev
Ognjen Rudovic
Daniel Rueckert
Adria Ruiz
Javier Ruiz-del-solar
Christian Rupprecht
Chris Russell
Dan Ruta
Jongbin Ryu
Ömer Sümer
Alexandre Sablayrolles
Faraz Saeedan
Ryusuke Sagawa
Christos Sagonas
Tonmoy Saikia
Hideo Saito
Kuniaki Saito
Shunsuke Saito
Shunta Saito
Ken Sakurada
Joaquin Salas
Fatemeh Sadat Saleh
Mahdi Saleh
Pouya Samangouei
Leo Sampaio Ferraz Ribeiro
Artsiom Olegovich Sanakoyeu
Enrique Sanchez
Patsorn Sangkloy
Anush Sankaran
Aswin Sankaranarayanan
Swami Sankaranarayanan
Rodrigo Santa Cruz
Amartya Sanyal
Archana Sapkota
Nikolaos Sarafianos
Jun Sato
Shin'ichi Satoh
Hosnieh Sattar
Arman Savran
Manolis Savva
Alexander Sax
Hanno Scharr
Simone Schaub-Meyer
Konrad Schindler
Dmitrij Schlesinger
Uwe Schmidt
Dirk Schnieders
Björn Schuller
Samuel Schulter
Idan Schwartz

William Robson Schwartz
Alex Schwing
Sinisa Segvic
Lorenzo Seidenari
Pradeep Sen
Ozan Sener
Soumyadip Sengupta
Arda Senocak
Mojtaba Seyedhosseini
Shishir Shah
Shital Shah
Sohil Atul Shah
Tamar Rott Shaham
Huasong Shan
Qi Shan
Shiguang Shan
Jing Shao
Roman Shapovalov
Gaurav Sharma
Vivek Sharma
Viktoriia Sharmanska
Dongyu She
Sumit Shekhar
Evan Shelhamer
Chengyao Shen
Chunhua Shen
Falong Shen
Jie Shen
Li Shen
Liyue Shen
Shuhan Shen
Tianwei Shen
Wei Shen
William B. Shen
Yantao Shen
Ying Shen
Yiru Shen
Yujun Shen
Yuming Shen
Zhiqiang Shen
Ziyi Shen
Lu Sheng
Yu Sheng
Rakshith Shetty
Baoguang Shi
Guangming Shi
Hailin Shi
Miaojing Shi
Yemin Shi
Zhenmei Shi
Zhiyuan Shi
Kevin Jonathan Shih
Shiliang Shiliang
Hyunjung Shim
Atsushi Shimada
Nobutaka Shimada
Daeyun Shin
Young Min Shin
Koichi Shinoda
Konstantin Shmelkov
Michael Zheng Shou
Abhinav Shrivastava
Tianmin Shu
Zhixin Shu
Hong-Han Shuai
Pushkar Shukla
Christian Siagian
Mennatullah M. Siam
Kaleem Siddiqi
Karan Sikka
Jae-Young Sim
Christian Simon
Martin Simonovsky
Dheeraj Singaraju
Bharat Singh
Gurkirt Singh
Krishna Kumar Singh
Maneesh Kumar Singh
Richa Singh
Saurabh Singh
Suriya Singh
Vikas Singh
Sudipta N. Sinha
Vincent Sitzmann
Josef Sivic
Gregory Slabaugh
Miroslava Slavcheva
Ron Slossberg
Brandon Smith
Kevin Smith
Vladimir Smutny
Noah Snavely
Roger D. Soberanis-Mukul
Kihyuk Sohn
Francesco Solera
Eric Sommerlade
Sanghyun Son
Byung Cheol Song
Chunfeng Song
Dongjin Song
Jiaming Song
Jie Song
Jifei Song
Jingkuan Song
Mingli Song
Shiyu Song
Shuran Song
Xiao Song
Yafei Song
Yale Song
Yang Song
Yi-Zhe Song
Yibing Song
Humberto Sossa
Cesar de Souza
Adrian Spurr
Srinath Sridhar
Suraj Srinivas
Pratul P. Srinivasan
Anuj Srivastava
Tania Stathaki
Christopher Stauffer
Simon Stent
Rainer Stiefelhagen
Pierre Stock
Julian Straub
Jonathan C. Stroud
Joerg Stueckler
Jan Stuehmer
David Stutz
Chi Su
Hang Su
Jong-Chyi Su
Shuochen Su
Yu-Chuan Su
Ramanathan Subramanian
Yusuke Sugano

Masanori Suganuma
Yumin Suh
Mohammed Suhail
Yao Sui
Heung-Il Suk
Josephine Sullivan
Baochen Sun
Chen Sun
Chong Sun
Deqing Sun
Jin Sun
Liang Sun
Lin Sun
Qianru Sun
Shao-Hua Sun
Shuyang Sun
Weiwei Sun
Wenxiu Sun
Xiaoshuai Sun
Xiaoxiao Sun
Xingyuan Sun
Yifan Sun
Zhun Sun
Sabine Susstrunk
David Suter
Supasorn Suwajanakorn
Tomas Svoboda
Eran Swears
Paul Swoboda
Attila Szabo
Richard Szeliski
Duy-Nguyen Ta
Andrea Tagliasacchi
Yuichi Taguchi
Ying Tai
Keita Takahashi
Kouske Takahashi
Jun Takamatsu
Hugues Talbot
Toru Tamaki
Chaowei Tan
Fuwen Tan
Mingkui Tan
Mingxing Tan
Qingyang Tan
Robby T. Tan
Xiaoyang Tan
Kenichiro Tanaka
Masayuki Tanaka
Chang Tang
Chengzhou Tang
Danhang Tang
Ming Tang
Peng Tang
Qingming Tang
Wei Tang
Xu Tang
Yansong Tang
Youbao Tang
Yuxing Tang
Zhiqiang Tang
Tatsunori Taniai
Junli Tao
Xin Tao
Makarand Tapaswi
Jean-Philippe Tarel
Lyne Tchapmi
Zachary Teed
Bugra Tekin
Damien Teney
Ayush Tewari
Christian Theobalt
Christopher Thomas
Diego Thomas
Jim Thomas
Rajat Mani Thomas
Xinmei Tian
Yapeng Tian
Yingli Tian
Yonglong Tian
Zhi Tian
Zhuotao Tian
Kinh Tieu
Joseph Tighe
Massimo Tistarelli
Matthew Toews
Carl Toft
Pavel Tokmakov
Federico Tombari
Chetan Tonde
Yan Tong
Alessio Tonioni
Andrea Torsello
Fabio Tosi
Du Tran
Luan Tran
Ngoc-Trung Tran
Quan Hung Tran
Truyen Tran
Rudolph Triebel
Martin Trimmel
Shashank Tripathi
Subarna Tripathi
Leonardo Trujillo
Eduard Trulls
Tomasz Trzcinski
Sam Tsai
Yi-Hsuan Tsai
Hung-Yu Tseng
Stavros Tsogkas
Aggeliki Tsoli
Devis Tuia
Shubham Tulsiani
Sergey Tulyakov
Frederick Tung
Tony Tung
Daniyar Turmukhambetov
Ambrish Tyagi
Radim Tylecek
Christos Tzelepis
Georgios Tzimiropoulos
Dimitrios Tzionas
Seiichi Uchida
Norimichi Ukita
Dmitry Ulyanov
Martin Urschler
Yoshitaka Ushiku
Ben Usman
Alexander Vakhitov
Julien P. C. Valentin
Jack Valmadre
Ernest Valveny
Joost van de Weijer
Jan van Gemert
Koen Van Leemput
Gul Varol
Sebastiano Vascon
M. Alex O. Vasilescu

Subeesh Vasu
Mayank Vatsa
David Vazquez
Javier Vazquez-Corral
Ashok Veeraraghavan
Erik Velasco-Salido
Raviteja Vemulapalli
Jonathan Ventura
Manisha Verma
Roberto Vezzani
Ruben Villegas
Minh Vo
MinhDuc Vo
Nam Vo
Michele Volpi
Riccardo Volpi
Carl Vondrick
Konstantinos Vougioukas
Tuan-Hung Vu
Sven Wachsmuth
Neal Wadhwa
Catherine Wah
Jacob C. Walker
Thomas S. A. Wallis
Chengde Wan
Jun Wan
Liang Wan
Renjie Wan
Baoyuan Wang
Boyu Wang
Cheng Wang
Chu Wang
Chuan Wang
Chunyu Wang
Dequan Wang
Di Wang
Dilin Wang
Dong Wang
Fang Wang
Guanzhi Wang
Guoyin Wang
Hanzi Wang
Hao Wang
He Wang
Heng Wang
Hongcheng Wang
Hongxing Wang
Hua Wang
Jian Wang
Jingbo Wang
Jinglu Wang
Jingya Wang
Jinjun Wang
Jinqiao Wang
Jue Wang
Ke Wang
Keze Wang
Le Wang
Lei Wang
Lezi Wang
Li Wang
Liang Wang
Lijun Wang
Limin Wang
Linwei Wang
Lizhi Wang
Mengjiao Wang
Mingzhe Wang
Minsi Wang
Naiyan Wang
Nannan Wang
Ning Wang
Oliver Wang
Pei Wang
Peng Wang
Pichao Wang
Qi Wang
Qian Wang
Qiaosong Wang
Qifei Wang
Qilong Wang
Qing Wang
Qingzhong Wang
Quan Wang
Rui Wang
Ruiping Wang
Ruixing Wang
Shangfei Wang
Shenlong Wang
Shiyao Wang
Shuhui Wang
Song Wang
Tao Wang
Tianlu Wang
Tiantian Wang
Ting-chun Wang
Tingwu Wang
Wei Wang
Weiyue Wang
Wenguan Wang
Wenlin Wang
Wenqi Wang
Xiang Wang
Xiaobo Wang
Xiaofang Wang
Xiaoling Wang
Xiaolong Wang
Xiaosong Wang
Xiaoyu Wang
Xin Eric Wang
Xinchao Wang
Xinggang Wang
Xintao Wang
Yali Wang
Yan Wang
Yang Wang
Yangang Wang
Yaxing Wang
Yi Wang
Yida Wang
Yilin Wang
Yiming Wang
Yisen Wang
Yongtao Wang
Yu-Xiong Wang
Yue Wang
Yujiang Wang
Yunbo Wang
Yunhe Wang
Zengmao Wang
Zhangyang Wang
Zhaowen Wang
Zhe Wang
Zhecan Wang
Zheng Wang
Zhixiang Wang
Zilei Wang
Jianqiao Wangni

Anne S. Wannenwetsch
Jan Dirk Wegner
Scott Wehrwein
Donglai Wei
Kaixuan Wei
Longhui Wei
Pengxu Wei
Ping Wei
Qi Wei
Shih-En Wei
Xing Wei
Yunchao Wei
Zijun Wei
Jerod Weinman
Michael Weinmann
Philippe Weinzaepfel
Yair Weiss
Bihan Wen
Longyin Wen
Wei Wen
Junwu Weng
Tsui-Wei Weng
Xinshuo Weng
Eric Wengrowski
Tomas Werner
Gordon Wetzstein
Tobias Weyand
Patrick Wieschollek
Maggie Wigness
Erik Wijmans
Richard Wildes
Olivia Wiles
Chris Williams
Williem Williem
Kyle Wilson
Calden Wloka
Nicolai Wojke
Christian Wolf
Yongkang Wong
Sanghyun Woo
Scott Workman
Baoyuan Wu
Bichen Wu
Chao-Yuan Wu
Huikai Wu
Jiajun Wu
Jialin Wu
Jiaxiang Wu
Jiqing Wu
Jonathan Wu
Lifang Wu
Qi Wu
Qiang Wu
Ruizheng Wu
Shangzhe Wu
Shun-Cheng Wu
Tianfu Wu
Wayne Wu
Wenxuan Wu
Xiao Wu
Xiaohe Wu
Xinxiao Wu
Yang Wu
Yi Wu
Yiming Wu
Ying Nian Wu
Yue Wu
Zheng Wu
Zhenyu Wu
Zhirong Wu
Zuxuan Wu
Stefanie Wuhrer
Jonas Wulff
Changqun Xia
Fangting Xia
Fei Xia
Gui-Song Xia
Lu Xia
Xide Xia
Yin Xia
Yingce Xia
Yongqin Xian
Lei Xiang
Shiming Xiang
Bin Xiao
Fanyi Xiao
Guobao Xiao
Huaxin Xiao
Taihong Xiao
Tete Xiao
Tong Xiao
Wang Xiao
Yang Xiao
Cihang Xie
Guosen Xie
Jianwen Xie
Lingxi Xie
Sirui Xie
Weidi Xie
Wenxuan Xie
Xiaohua Xie
Fuyong Xing
Jun Xing
Junliang Xing
Bo Xiong
Peixi Xiong
Yu Xiong
Yuanjun Xiong
Zhiwei Xiong
Chang Xu
Chenliang Xu
Dan Xu
Danfei Xu
Hang Xu
Hongteng Xu
Huijuan Xu
Jingwei Xu
Jun Xu
Kai Xu
Mengmeng Xu
Mingze Xu
Qianqian Xu
Ran Xu
Weijian Xu
Xiangyu Xu
Xiaogang Xu
Xing Xu
Xun Xu
Yanyu Xu
Yichao Xu
Yong Xu
Yongchao Xu
Yuanlu Xu
Zenglin Xu
Zheng Xu
Chuhui Xue
Jia Xue
Nan Xue

Tianfan Xue
Xiangyang Xue
Abhay Yadav
Yasushi Yagi
I. Zeki Yalniz
Kota Yamaguchi
Toshihiko Yamasaki
Takayoshi Yamashita
Junchi Yan
Ke Yan
Qingan Yan
Sijie Yan
Xinchen Yan
Yan Yan
Yichao Yan
Zhicheng Yan
Keiji Yanai
Bin Yang
Ceyuan Yang
Dawei Yang
Dong Yang
Fan Yang
Guandao Yang
Guorun Yang
Haichuan Yang
Hao Yang
Jianwei Yang
Jiaolong Yang
Jie Yang
Jing Yang
Kaiyu Yang
Linjie Yang
Meng Yang
Michael Ying Yang
Nan Yang
Shuai Yang
Shuo Yang
Tianyu Yang
Tien-Ju Yang
Tsun-Yi Yang
Wei Yang
Wenhan Yang
Xiao Yang
Xiaodong Yang
Xin Yang
Yan Yang
Yanchao Yang
Yee Hong Yang
Yezhou Yang
Zhenheng Yang
Anbang Yao
Angela Yao
Cong Yao
Jian Yao
Li Yao
Ting Yao
Yao Yao
Zhewei Yao
Chengxi Ye
Jianbo Ye
Keren Ye
Linwei Ye
Mang Ye
Mao Ye
Qi Ye
Qixiang Ye
Mei-Chen Yeh
Raymond Yeh
Yu-Ying Yeh
Sai-Kit Yeung
Serena Yeung
Kwang Moo Yi
Li Yi
Renjiao Yi
Alper Yilmaz
Junho Yim
Lijun Yin
Weidong Yin
Xi Yin
Zhichao Yin
Tatsuya Yokota
Ryo Yonetani
Donggeun Yoo
Jae Shin Yoon
Ju Hong Yoon
Sung-eui Yoon
Laurent Younes
Changqian Yu
Fisher Yu
Gang Yu
Jiahui Yu
Kaicheng Yu
Ke Yu
Lequan Yu
Ning Yu
Qian Yu
Ronald Yu
Ruichi Yu
Shoou-I Yu
Tao Yu
Tianshu Yu
Xiang Yu
Xin Yu
Xiyu Yu
Youngjae Yu
Yu Yu
Zhiding Yu
Chunfeng Yuan
Ganzhao Yuan
Jinwei Yuan
Lu Yuan
Quan Yuan
Shanxin Yuan
Tongtong Yuan
Wenjia Yuan
Ye Yuan
Yuan Yuan
Yuhui Yuan
Huanjing Yue
Xiangyu Yue
Ersin Yumer
Sergey Zagoruyko
Egor Zakharov
Amir Zamir
Andrei Zanfir
Mihai Zanfir
Pablo Zegers
Bernhard Zeisl
John S. Zelek
Niclas Zeller
Huayi Zeng
Jiabei Zeng
Wenjun Zeng
Yu Zeng
Xiaohua Zhai
Fangneng Zhan
Huangying Zhan
Kun Zhan

Xiaohang Zhan
Baochang Zhang
Bowen Zhang
Cecilia Zhang
Changqing Zhang
Chao Zhang
Chengquan Zhang
Chi Zhang
Chongyang Zhang
Dingwen Zhang
Dong Zhang
Feihu Zhang
Hang Zhang
Hanwang Zhang
Hao Zhang
He Zhang
Hongguang Zhang
Hua Zhang
Ji Zhang
Jianguo Zhang
Jianming Zhang
Jiawei Zhang
Jie Zhang
Jing Zhang
Juyong Zhang
Kai Zhang
Kaipeng Zhang
Ke Zhang
Le Zhang
Lei Zhang
Li Zhang
Lihe Zhang
Linguang Zhang
Lu Zhang
Mi Zhang
Mingda Zhang
Peng Zhang
Pingping Zhang
Qian Zhang
Qilin Zhang
Quanshi Zhang
Richard Zhang
Rui Zhang
Runze Zhang
Shengping Zhang
Shifeng Zhang
Shuai Zhang
Songyang Zhang
Tao Zhang
Ting Zhang
Tong Zhang
Wayne Zhang
Wei Zhang
Weizhong Zhang
Wenwei Zhang
Xiangyu Zhang
Xiaolin Zhang
Xiaopeng Zhang
Xiaoqin Zhang
Xiuming Zhang
Ya Zhang
Yang Zhang
Yimin Zhang
Yinda Zhang
Ying Zhang
Yongfei Zhang
Yu Zhang
Yulun Zhang
Yunhua Zhang
Yuting Zhang
Zhanpeng Zhang
Zhao Zhang
Zhaoxiang Zhang
Zhen Zhang
Zheng Zhang
Zhifei Zhang
Zhijin Zhang
Zhishuai Zhang
Ziming Zhang
Bo Zhao
Chen Zhao
Fang Zhao
Haiyu Zhao
Han Zhao
Hang Zhao
Hengshuang Zhao
Jian Zhao
Kai Zhao
Liang Zhao
Long Zhao
Qian Zhao
Qibin Zhao
Qijun Zhao
Rui Zhao
Shenglin Zhao
Sicheng Zhao
Tianyi Zhao
Wenda Zhao
Xiangyun Zhao
Xin Zhao
Yang Zhao
Yue Zhao
Zhichen Zhao
Zijing Zhao
Xiantong Zhen
Chuanxia Zheng
Feng Zheng
Haiyong Zheng
Jia Zheng
Kang Zheng
Shuai Kyle Zheng
Wei-Shi Zheng
Yinqiang Zheng
Zerong Zheng
Zhedong Zheng
Zilong Zheng
Bineng Zhong
Fangwei Zhong
Guangyu Zhong
Yiran Zhong
Yujie Zhong
Zhun Zhong
Chunluan Zhou
Huiyu Zhou
Jiahuan Zhou
Jun Zhou
Lei Zhou
Luowei Zhou
Luping Zhou
Mo Zhou
Ning Zhou
Pan Zhou
Peng Zhou
Qianyi Zhou
S. Kevin Zhou
Sanping Zhou
Wengang Zhou
Xingyi Zhou

Yanzhao Zhou
Yi Zhou
Yin Zhou
Yipin Zhou
Yuyin Zhou
Zihan Zhou
Alex Zihao Zhu
Chenchen Zhu
Feng Zhu
Guangming Zhu
Ji Zhu
Jun-Yan Zhu
Lei Zhu
Linchao Zhu
Rui Zhu
Shizhan Zhu
Tyler Lixuan Zhu
Wei Zhu
Xiangyu Zhu
Xinge Zhu
Xizhou Zhu
Yanjun Zhu
Yi Zhu
Yixin Zhu
Yizhe Zhu
Yousong Zhu
Zhe Zhu
Zhen Zhu
Zheng Zhu
Zhenyao Zhu
Zhihui Zhu
Zhuotun Zhu
Bingbing Zhuang
Wei Zhuo
Christian Zimmermann
Karel Zimmermann
Larry Zitnick
Mohammadreza Zolfaghari
Maria Zontak
Daniel Zoran
Changqing Zou
Chuhang Zou
Danping Zou
Qi Zou
Yang Zou
Yuliang Zou
Georgios Zoumpourlis
Wangmeng Zuo
Xinxin Zuo

Additional Reviewers

Victoria Fernandez Abrevaya
Maya Aghaei
Allam Allam
Christine Allen-Blanchette
Nicolas Aziere
Assia Benbihi
Neha Bhargava
Bharat Lal Bhatnagar
Joanna Bitton
Judy Borowski
Amine Bourki
Romain Brégier
Tali Brayer
Sebastian Bujwid
Andrea Burns
Yun-Hao Cao
Yuning Chai
Xiaojun Chang
Bo Chen
Shuo Chen
Zhixiang Chen
Junsuk Choe
Hung-Kuo Chu
Jonathan P. Crall
Kenan Dai
Lucas Deecke
Karan Desai
Prithviraj Dhar
Jing Dong
Wei Dong
Turan Kaan Elgin
Francis Engelmann
Erik Englesson
Fartash Faghri
Zicong Fan
Yang Fu
Risheek Garrepalli
Yifan Ge
Marco Godi
Helmut Grabner
Shuxuan Guo
Jianfeng He
Zhezhi He
Samitha Herath
Chih-Hui Ho
Yicong Hong
Vincent Tao Hu
Julio Hurtado
Jaedong Hwang
Andrey Ignatov
Muhammad Abdullah Jamal
Saumya Jetley
Meiguang Jin
Jeff Johnson
Minsoo Kang
Saeed Khorram
Mohammad Rami Koujan
Nilesh Kulkarni
Sudhakar Kumawat
Abdelhak Lemkhenter
Alexander Levine
Jiachen Li
Jing Li
Jun Li
Yi Li
Liang Liao
Ruochen Liao
Tzu-Heng Lin
Phillip Lippe
Bao-di Liu
Bo Liu
Fangchen Liu

Hanxiao Liu
Hongyu Liu
Huidong Liu
Miao Liu
Xinxin Liu
Yongfei Liu
Yu-Lun Liu
Amir Livne
Tiange Luo
Wei Ma
Xiaoxuan Ma
Ioannis Marras
Georg Martius
Effrosyni Mavroudi
Tim Meinhardt
Givi Meishvili
Meng Meng
Zihang Meng
Zhongqi Miao
Gyeongsik Moon
Khoi Nguyen
Yung-Kyun Noh
Antonio Norelli
Jaeyoo Park
Alexander Pashevich
Mandela Patrick
Mary Phuong
Bingqiao Qian
Yu Qiao
Zhen Qiao
Sai Saketh Rambhatla
Aniket Roy
Amelie Royer
Parikshit Vishwas Sakurikar
Mark Sandler
Mert Bülent Sarıyıldız
Tanner Schmidt
Anshul B. Shah
Ketul Shah
Rajvi Shah
Hengcan Shi
Xiangxi Shi
Yujiao Shi
William A. P. Smith
Guoxian Song
Robin Strudel
Abby Stylianou
Xinwei Sun
Reuben Tan
Qingyi Tao
Kedar S. Tatwawadi
Anh Tuan Tran
Son Dinh Tran
Eleni Triantafillou
Aristeidis Tsitiridis
Md Zasim Uddin
Andrea Vedaldi
Evangelos Ververas
Vidit Vidit
Paul Voigtlaender
Bo Wan
Huanyu Wang
Huiyu Wang
Junqiu Wang
Pengxiao Wang
Tai Wang
Xinyao Wang
Tomoki Watanabe
Mark Weber
Xi Wei
Botong Wu
James Wu
Jiamin Wu
Rujie Wu
Yu Wu
Rongchang Xie
Wei Xiong
Yunyang Xiong
An Xu
Chi Xu
Yinghao Xu
Fei Xue
Tingyun Yan
Zike Yan
Chao Yang
Heran Yang
Ren Yang
Wenfei Yang
Xu Yang
Rajeev Yasarla
Shaokai Ye
Yufei Ye
Kun Yi
Haichao Yu
Hanchao Yu
Ruixuan Yu
Liangzhe Yuan
Chen-Lin Zhang
Fandong Zhang
Tianyi Zhang
Yang Zhang
Yiyi Zhang
Yongshun Zhang
Yu Zhang
Zhiwei Zhang
Jiaojiao Zhao
Yipu Zhao
Xingjian Zhen
Haizhong Zheng
Tiancheng Zhi
Chengju Zhou
Hao Zhou
Hao Zhu
Alexander Zimin

Contents – Part VIII

Weakly-Supervised Crowd Counting Learns from Sorting Rather Than Locations

Yifan Yang[1], Guorong Li[1,2](✉), Zhe Wu[1], Li Su[1,2], Qingming Huang[1,2,3], and Nicu Sebe[4]

[1] School of Computer Science and Technology, UCAS, Beijing, China
liguorong@ucas.ac.cn
[2] Key Lab of Big Data Mining and Knowledge Management, UCAS, Beijing, China
[3] Key Lab of Intelligent Information Processing, ICT, CAS, Beijing, China
[4] University of Trento, Trento, Italy

Abstract. In crowd counting datasets, the location labels are costly, yet, they are not taken into the evaluation metrics. Besides, existing multi-task approaches employ high-level tasks to improve counting accuracy. This research tendency increases the demand for more annotations. In this paper, we propose a weakly-supervised counting network, which directly regresses the crowd numbers without the location supervision. Moreover, we train the network to count by exploiting the relationship among the images. We propose a soft-label sorting network along with the counting network, which sorts the given images by their crowd numbers. The sorting network drives the shared backbone CNN model to obtain density-sensitive ability explicitly. Therefore, the proposed method improves the counting accuracy by utilizing the information hidden in crowd numbers, rather than learning from extra labels, such as locations and perspectives. We evaluate our proposed method on three crowd counting datasets, and the performance of our method plays favorably against the fully supervised state-of-the-art approaches.

Keywords: Weakly-supervised · Sorting · Multi-frames · Crowd counting

1 Introduction

Counting objects is a hot topic in computer vision because of its wide applications in many areas. Significant effort has been devoted to this task [2,11,15,18,21,30,34,38]. These approaches either employ a detection framework [18,22] or a regression framework [15,18,34]. However, in congested scenes, there are many occlusions, and it is difficult for the detection approaches to recognize the person. Therefore, the density estimation based methods [15,18,34], in particular, have received increasing research focus.

A. Vedaldi et al. (Eds.): ECCV 2020, LNCS 12353, pp. 1–17, 2020.
https://doi.org/10.1007/978-3-030-58598-3_1

However, there are still several drawbacks. Firstly, the annotations of the crowd counting are generally expensive. The existing counting datasets [3,9,45, 46] provide the location of each instance to train the counting networks, while in the evaluation stage, these location labels are not taken into account, and the performance metrics only evaluate the estimation accuracy of the crowd number. In fact, without the demand for locations, the crowd numbers can be obtained in other economical ways. For instance, with an already collected dataset, the crowd numbers can be obtained by gathering the environmental information, e.g., detection of disturbances in spaces, or estimation of the number of moving crowds. Chan et al. [3] segment the scene by crowd motions and estimate the crowd number by calculating the area of the segmented regions. To collect a novel counting dataset, we can employ sensor technology to obtain the crowd number in constrained scenes, such as mobile crowd sensing technology [10]. Moreover, Sheng et al. [35] propose a GPS-less energy-efficient sensing scheduling to acquire the crowd number more economically. On the other hand, several approaches [4,14,17,23] prove that, in the estimated results, there is no tight bond between the crowd number and the location. Finally, in the existing multi-task approaches, high level tasks are employed to improve the counting accuracy, for instance, tracking [23], detection [18,22], segmentation [37,47], localization [17,25], depth prediction [47] and scene analysis [19,20,36,44,45]. This research tendency increases the demand for more annotations.

In this work, we propose a weakly-supervised framework to directly regress the crowd number without the supervision of location labels. To our best knowledge, we are the first to train a counting network without location supervision. Moreover, we train the network to count by exploiting the relationship among the images. We propose an end-to-end trainable soft-label sorting network along with the counting network, which sorts the given images by their crowd numbers. The sorting network drives the shared backbone CNN model to obtain density-sensitive ability explicitly. Therefore, the proposed method improves the counting accuracy by utilizing the relationship among crowd numbers, rather than learning from extra labels, such as locations and perspectives. More concretely, the proposed sorting network processes several images and employs a soft-sort layer to generate dense order matrixes. The previous sorting works [6,8,16] employ hard-labels to train the sorting network, for instance, one-hot vectors [8,16] or indexes which are real integers [6]. However, we find that hard-labels are incapable of capturing the complexity of sorting task. As the candidates may have limited variations or even have the same values, the hard-labels introduce ambiguous supervision to the training stage. Therefore, we propose an informative soft-label, which introduces the Rayleigh distribution to characterize the sorting complexity. The proposed soft-labels have high entropy. They provide much more information per training case than hard-labels and much less variance in the gradient between training cases.

The main contributions of our method are summarized as follows:

- We propose a weakly-supervised counting network, which directly regresses the crowd number without the supervision of location labels.

- We propose a soft-label sorting network to facilitate the counting task, which sorts the images by their crowd numbers. The proposed framework improves the counting task without extra labels, especially costly semantic labels.
- The proposed weakly-supervised approach plays favorably against fully supervised state-of-the-art approaches on three datasets.

2 Related Work

2.1 Density Estimation Based Methods

The counting datasets provide a location label for each person. The fully supervised density estimation based methods have to generate density maps with various strategies. Several approaches [15,46] coarsely estimate the instance scales by the interval distances and employ Gaussian kernels with various scales to represent the objects. Wan et al. [41] propose a network to adaptively generate density maps, and train the generative network with the regression network. The obtained density maps are better recognized by the counting network. However, as proved by several approaches [4,14,17,23], in the estimated density maps, there is no tight bond between the crowd number and the location. In the scenes with large perspective distortions, regardless of the low regression errors, dense-crowd regions are usually underestimated, while sparse-crowd regions are over-estimated. Du et al. [23] prove a similar phenomenon in the scenes with limited scale variations.

The existing approaches employ various strategies to improve counting accuracy. Several approaches employ multiple receptive fields to evaluate the instances with various scales. To obtain the multiple receptive fields, Zhang et al. [46], Deb et al. [7] and Sam et al. [29] employ multi-column networks; several approaches [2,13,19] utilize inception blocks. Besides changing the convolution kernels, a deep network can also obtain various receptive fields from its different layers. Several counting approaches [2,13,20,34] utilize similar architectures with U-net [28]. However, Li et al. [15] prove that the multiple receptive fields deliver similar results. Moreover, several methods employ extra supervision to improve evaluation accuracy. For instance, Liu et al. [20], Shi et al. [36], Yan et al. [44] and Zhang et al. [45] employ perspective maps to smooth the final density maps. However, the perspective maps are delivered from extra annotations. Besides, the existing multi-task approaches utilize high-level tasks to improve the estimation accuracy, for instance, tracking [23], detection [18,22], segmentation [37,47], localization [17,25], and depth prediction [47]. These multi-task approaches boost the counting task with extra semantic labels.

In this work, we propose a weakly-supervised counting approach, which is trained without location labels. Moreover, we propose a soft-label sorting network to improve the counting accuracy, which sorts the images with various crowd numbers. The proposed framework improves the counting task without extra labels, especially costly semantic labels.

2.2 Methods Dealing with the Lack of Labelled Data

Several approaches are proposed to relieve the expensive labeling work in crowd counting. One of the most relevant works for our method is L2R [21], which facilitates the counting task by ranking the image patches. However, L2R is fully supervised by using the location labels. Moreover, the ranking network only operates on the image patch and one of its sub-patches. Our proposed network is trained without location labels. Besides, the sorting network processes the whole image, and the number of the candidate images are not fixed.

Wang et al. [42] generate synthetic crowd scenes and simultaneously annotate them. The proposed network is pre-trained on the synthetic dataset and then fine-tuned with real data. Although, this approach improves the counting performance with less expensive labels. Labeling the real data is still expensive and challenging. Loy et al. [24] employ active learning to label more representative frames of the videos. This strategy efficiently releases the laborious work. However, it only works on the video counting task, where the video data are assumed to lie along a low-dimensional manifold. Sam et al. [31] pre-train the feature extractor with several restricted Boltzmann machines progressively in an unsupervised way, but the training of the top regression layers is fully-supervised.

Out of these mentioned approaches, only our method and Loy et al. [24] employ fewer labels to train the networks. Still, only our approach trains the network without the supervision of locations.

2.3 Learning from Sorting

Sorting is used pervasively in machine learning. However, it is also a poor match for the differentiable pipelines of deep learning. Currently, several approaches combine the sorting layers with deep networks. Several works [26,33,43] encode the permutations into indexes and train the network to regress the index. While others algorithms [6,8,16] propose differentiable operators to directly regress the order.

Several self-supervised approaches employ a sorting task to pre-train the feature extractor. Noroozi et al. [26] first propose a self-supervised network to learn a feature domain by solving Jigsaw puzzles. Inspired by this work, Sermanet et al. [33] propose a similar framework to sort the shuffled video frames, and the learned features are used as video representation. Xu et al. [43] also learn video representations by sorting shuffled frames. However, they employ video clips rather than single frames to train the network. These approaches have several restrictions. Firstly, they employ indexes to represent all the possible permutations and train the network to regress the index. This strategy reduces the information embedded in the supervision labels. For example, in [26], there are 9 elements to sort and the number of possible permutations is $9! = 362{,}880$. The massive numbers inhibit the methods to sort more elements; for instance, Xu et al. [43] restrict the number of clips between 2 to 5. Moreover, with the casual encoding strategy, a slight variation between two permutations may cause a dramatic difference in their indexes. Therefore, sorting networks cannot learn a representation efficiently. Otherwise, the feature extractors are pre-trained. Our proposed method employs a dense order matrix to capture the possible permutations and end-to-end trains the sorting network with the counting network.

On the other hand, several approaches propose differentiable soft-sort methods to tackle these issues. Sinkhorn distance [5] has been initially proposed to tackle optimal transportation, while Linderman et al. [16] employ it to address sorting issues. Grover et al. [8] propose an attractive task, which sorts n numbers between 0 and 9999 given as four concatenated MNIST images. They also propose a differentiable neural sort method to tackle this task. Based on the Sinkhorn method, Cuturi et al. [6] further propose a differentiable soft-sort algorithm, which directly generates the sort and rank indexes of a vector. However, these approaches employ hard-labels to train the network, and this leads to the loss of valuable information.

In this paper, instead of pre-training a feature extractor, we train the sorting network with counting network in an end-to-end manner. In the final layer, we also employ a differentiable soft-sort operator to generate dense order matrixes. Moreover, we propose an informative soft-label to train the sorting network.

3 Method

In this paper, we propose a weakly-supervised counting approach, which does not rely on location supervision for training. Besides, to improve the counting task, we exploit the relationship among images. We propose a novel soft-label sorting network along with the regression network, which sorts the images by their various crowd numbers. Both the regression network and the sorting network share a same backbone, and both networks are end-to-end trained. As both networks estimate the crowd numbers of the given images, they promote each other without extra labels, such as location and perspective.

More concretely, the regression network employs several adaptive pooling layers to formulate a pyramidal feature vector. Besides, the sorting network employs a network to formulate the comparison and uses a differentiable soft-sort layer to generate dense order predictions. Moreover, to train the network effectively, we propose soft-labels, which have high entropy and provide much more information. The soft-labels employ the Rayleigh distribution to characterize the complexity of sorting tasks. We will elaborate on the details of the regression network, the sorting network, and the training method in the following subsections.

3.1 Regression Network

As shown in Fig. 1, the regression network directly regresses the crowd number from the whole frame. Moreover, the front-end of the network, which delivers the pyramidal feature vector, is shared with the sorting network.

In the front-end network, we first employ the first 13 layers of VGG-16 [39] as the backbone of our network, similar to previous methods [1,15,32,40]. The front-end is marked as $G_b(\gamma)$, where γ stands for the parameters, and the output

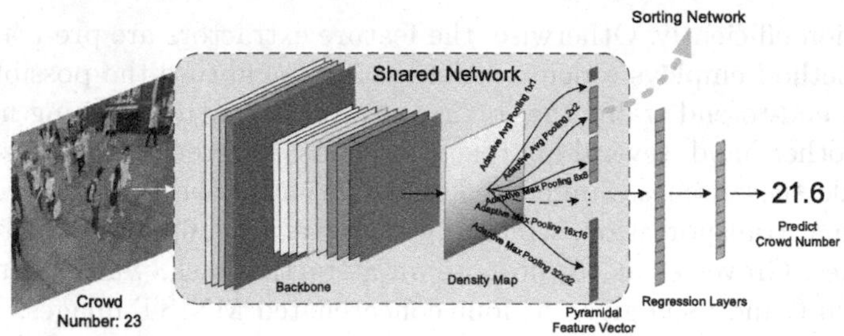

Fig. 1. Framework of the regression network, which contains a shared front-end and a back-end to regress the crowd number.

size is 1/8 of the original input size. Then the front-end network regresses a single channel density map based on the extracted features, which is formulated as:

$$f_d = F_d(G_b(x, \gamma), \zeta), \tag{1}$$

where $f_d \in \mathbb{R}^{1WH}$ represents the estimated density map, W, H are the width and height of the feature map respectively. Moreover, $F_d(\cdot)$ denotes the convolution operation, and ζ stands for the parameters of the convolution layers.

The network needs the front-end to be sensitive to both the densities of the local and global crowds. Therefore, we propose to use adaptive pooling layers, denoted as $\mathcal{P}$, to extract a pyramidal feature vector from f_d. The adaptive pooling layers consist of global sub-cluster layers and local sub-cluster layers. Each global sub-cluster layer is denoted as $\mathbb{P}_{\mathrm{G}}^{i,S(i)}$, which employs an adaptive average pooling with a high sampling rate $S(i)$ to integrate the global information. Here, $i \in \{1, \cdots, N\}$, N is the number of the global sub-cluster layers. Besides, each local sub-cluster layer is denoted as $\mathbb{P}_{\mathrm{L}}^{j,S(j+N)}$, which employs an adaptive max pooling with a lower sampling rate $S(j+N)$ to extract the most discriminative features. Here, $j \in \{1, \cdots, T-N\}$, T is the total number of the pooling layers. The sampling rates of these pooling layers belong to the sampling rate set, which is denoted as S. We concatenate the outputs of adaptive pooling layers to formulate the pyramidal feature vector:

$$f_{pfv} = \mathcal{P}(f_d, S). \tag{2}$$

The back-end of the regression network employs several fully-connected layers to predict the crowd number:

$$c = F_r(f_{pfv}, \psi), \tag{3}$$

where c is the predicted crowd number, and $F_r(\cdot)$ stands for the regression layers with parameters ψ.

3.2 Sorting Network

To process multi-frames, the sorting network employs a multi-branch network to extract the pyramidal feature vectors. Each branch is shared with the regression

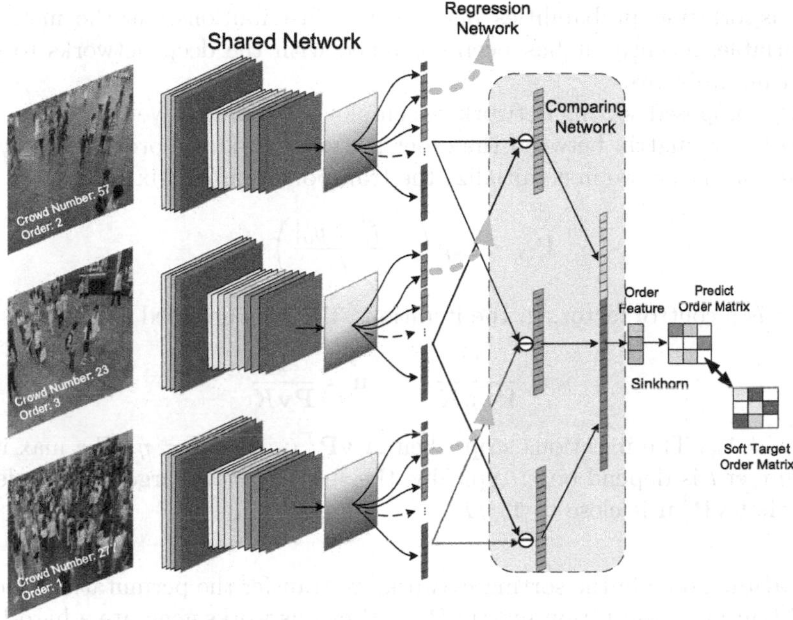

Fig. 2. Framework of the sorting network, which contains several branches to extract the pyramidal feature vectors and a comparing network to regress the order features. The Sinkhorn layer transfers the order feature to an order matrix. We employ the proposed soft-labels to train the sorting network.

network, while all the branches share the same parameters. The multi-branch network is formulated as $G_s(\omega)$, and the pyramidal feature vectors are denoted as $\{f_{pfv}^1, f_{pfv}^2, \cdots, f_{pfv}^K\}$. Here, ω represents the shared parameters, and K is the number of the given images.

The sorting network then utilizes a comparing network to regress the order feature. The comparing network first organizes $K(K-1)/2$ non-repeating tuples, each of which has two elements. Then the network calculates the difference between each pair: $f_{\mathrm{m}}^{ij} = f_{pfv}^i - f_{pfv}^j$, and concatenates the difference features: $f_{diff} = f_{\mathrm{m}}^{12}||f_{\mathrm{m}}^{13}|| \cdots ||f_{\mathrm{m}}^{1K}||f_{\mathrm{m}}^{23}||f_{\mathrm{m}}^{24}|| \cdots ||f_{\mathrm{m}}^{2K} \cdots ||f_{\mathrm{m}}^{K-1,K}$, where $||$ denotes the concatenation operation. Finally, the sorting network regresses the order feature of the given images. The order feature is formulated as f_o, where $f_o \in \mathbb{R}^K$.

We employ the Sinkhorn layer [5] to transfer f_o into an order matrix $\mathbf{P}$, where $\mathbf{P}_{ij}$ is the probability that the i-th element is ordered in j, and $\mathbf{P} \in \mathbb{R}^{KK}$. More importantly, we propose a soft-label to characterize the complexity of the sorting task, which is more informative than the hard-label. We elaborate on this in the following subsections.

Sinkhorn Operator. The Sinkhorn method is proposed to solve the optimal transportation issue. After several iterations, it generates a matrix to capture the transportation probabilities between two distributions. As the method is differentiable, recently, it has been combined with the deep networks to solve the sorting problems.

In the proposed sorting network, we employ a Sinkhorn layer to generate the transportation matrix between the order feature f_o and the order vector y_o. In the Sinkhorn layer, we first initialize the transportation matrix $\mathbf{P}$ by:

$$\mathbf{P}_{ij} = exp\left(-\frac{|f_o^i - y_o^j|}{\epsilon}\right), \tag{4}$$

where, ϵ is a control factor. In the iterations, the $\mathbf{P}$ is updated as:

$$\mathbf{v} = \frac{1}{\mathbf{P}^\top \mathbf{u} K}, \qquad \mathbf{u} = \frac{1}{\mathbf{P}\mathbf{v} K}, \tag{5}$$

where $\mathbf{u} = \mathbf{1}_K$. The iterations stop when $\Delta(\mathbf{v}\mathbf{P}^\top\mathbf{u}, \mathbf{1}_K/K) < \eta$. The max iteration number l is depend on ϵ: typically, the smaller ϵ, the larger l is needed to ensure that $\mathbf{v}\mathbf{P}^\top\mathbf{u}$ is close to $\mathbf{1}_K/K$.

Soft-Label. To train the sorting network, we transfer the permutation σ to the ground truth transportation matrix $\mathbf{P}^{GT}$. Previous works generate a hard label, where $\mathbf{P}^{GT}(i, \sigma(i)) = 1$. However, the sorting task is complex, and the hard-labels are unable to cover all the situations. For instance, there may be several candidates with similar or even identical values. Therefore, we propose soft-labels with high entropy to capture transportation probabilities. They provide not only much more information per training case than hard-labels but also much less variance in the gradient between training cases.

The soft-labels introduce the Rayleigh distribution to capture the relations between one element and its neighbors in the permutation. We denote the differences between one element and its neighbours in permutation as $\Delta_{i+1}, \Delta_{i-1}$, where $\Delta_{i+1} = |c_{\sigma(i)} - c_{\sigma(i)+1}|$. We set a threshold, which is denoted as Δ_{thr}, as the sensitivity of the network. If the difference between the two elements is less than the threshold, the network considers them as being similar instances. The elements of the transportation matrix are calculated as:

$$\widehat{\mathbf{P}}^{GT}(i, \sigma(i) + j) = \frac{(h(j) + \mu)}{\sigma^2} e^{\frac{-(h(j)+\mu)^2}{2\sigma^2}}, j \in \{-1, 0, 1\}. \tag{6}$$

To ensure the correct calculation in the edges, before calculating each element, we pad the matrix, and then crop it after calculations. The rate of both operations is 1.

The $\mu, \sigma, h(x)$ are determined by the differences between neighbours in permutation:

$$\begin{cases} \mu = 1, \sigma = 0.5, h(x) = x; & \Delta_{i+1} > \Delta_{thr}, \Delta_{i-1} > \Delta_{thr}, \\ \mu = 1, \sigma = 1.0, h(x) = x; & \Delta_{i+1} \leqslant \Delta_{thr}, \Delta_{i-1} > \Delta_{thr}, \\ \mu = 1, \sigma = 1.0, h(x) = -x; & \Delta_{i+1} > \Delta_{thr}, \Delta_{i-1} \leqslant \Delta_{thr}, \\ \mu = 2, \sigma = 2.0, h(x) = x; & \Delta_{i+1} \leqslant \Delta_{thr}, \Delta_{i-1} \leqslant \Delta_{thr}. \end{cases} \tag{7}$$

Finally, we obtain the soft-label transportation matrix $\mathbf{P}^{GT}$ by normalizing $\widehat{\mathbf{P}}^{GT}$:

$$\mathbf{P}^{GT}(i,\sigma(j)) = \frac{\widehat{\mathbf{P}}^{GT}(i,\sigma(j))}{\sum_{j=1}^{K} \widehat{\mathbf{P}}^{GT}(i,\sigma(j))} \tag{8}$$

3.3 Training Method

We use a straightforward way to train both the regression network and the sorting network as an end-to-end structure. The first 10 convolutional layers are fine-tuned from a pre-trained VGG-16. For the other layers, the initial values come from a Gaussian initialization with 0.01 standard deviation. Stochastic gradient descent (SGD) is applied with a fixed learning rate.

While training on the image dataset and the video dataset, we employ various sampling strategies. This is because the video surveillance scene pays more attention to the variation of pedestrian flow in a constrained scene. With the image dataset, we randomly select images from the dataset and train the network with their crowd numbers. With the video dataset, we first randomly select the video fragments of the same scene. We then randomly choose images within these clips.

We utilize MSE loss to train the regression network:

$$\mathcal{L}_r = \frac{1}{n} \sum_{i=1}^{n} (c_i^{GT} - c_i)^2, \tag{9}$$

where c_i^{GT} is the ground truth crowd number of i-th image.

We employ the cross-entropy loss to supervise the sorting network:

$$\mathcal{L}_s = -\frac{1}{K^2} \sum_{i=1}^{K} \sum_{j=1}^{K} (\mathbf{P}_{ij}^{GT}\ log(\mathbf{P}_{ij}) + (1 - \mathbf{P}_{ij}^{GT})\ log(1 - \mathbf{P}_{ij})). \tag{10}$$

The total loss is formulated as: $\mathcal{L}_{total} = \mathcal{L}_r + \xi\ \mathcal{L}_s$, where ξ is a scalar to balance the two losses.

4 Experiments

We evaluate our approach on three datasets: WorldExpo10 [12], UCSD [3], and ShanghaiTech [46]. In this section, we first provide the implementation details and evaluation metrics. We then evaluate and compare our method with the previous fully-supervised state-of-the-art approaches [2,13,15,19,27,46] on all these datasets. In the last subsection, we present ablation study results on the WorldExpo10 dataset.

Table 1. The evaluation results on WorldExpo10, UCSD, and ShanghaiTech datasets. SHA represents ShanghaiTech Part A, while SHB represents ShanghaiTech Part B. The results reported on WorldExpo10 are only evaluated with the MAE metric.

Method	Label		WorldExpo10						UCSD		SHA		SHB	
	Location	Crowd Number	Sce.1	Sce.2	Sce.3	Sce.4	Sce.5	Avg.	MAE	MSE	MAE	MSE	MAE	MSE
MCMM [46]	✓	✓	3.4	20.6	12.9	13.0	8.1	11.6	1.07	1.35	110.2	173.2	26.4	41.3
SANet [2]	✓	✓	2.6	13.2	9.0	13.3	3.0	8.2	1.02	1.29	67.0	104.5	8.4	13.6
CSRNet [15]	✓	✓	2.9	11.5	8.6	16.6	3.4	8.6	1.16	1.47	68.2	115.0	10.6	16.0
IG-CNN [27]	✓	✓	2.6	16.1	10.15	20.2	7.6	11.3	-	-	72.5	118.2	13.6	21.1
TEDnet [13]	✓	✓	2.3	10.1	11.3	13.8	2.6	**8.0**	-	-	64.2	109.1	8.2	**12.8**
ADCrowdNet [19]	✓	✓	1.6	15.8	11.0	10.9	3.2	8.5	**0.98**	**1.25**	**63.2**	**98.9**	**8.2**	15.7
Ours	-	✓	3.5	13.2	12.4	13.5	5.4	9.6	1.8	2.8	104.6	145.2	12.3	21.2

4.1 Implementation Details

To avoid over-fitting, we employ dropout layers in the fully-connected layers of both the regression network and sorting network. The ratio of dropout is 0.5. In the regression network, we set the sampling rate set as $S = \{\{1,2\}_{Avg}, \{8,16,32\}_{Max}\}$. As the order feature $f_o \in \mathbb{R}^K$, the sorting network employs different regression networks while sorting the various number of candidates. However, we organize each sorting network with the same structure. In the evaluate and compare subsection, we report the regression results of the proposed method, and the candidate number K is 3. In the ablation study subsection, we report the results of sorting the various number of candidates. In the Sinkhorn layer, we set ϵ as 1e−1, and η as 1e−3. When generating the soft-label, we set Δ_{thr} as 5.0. In the training stage, the learning rate is set to 1e−7, the ξ is set to 1e2, and the batch size is set to 10.

4.2 Evaluation Metrics

Similar to Sam et al. [30], we use the MAE and the MSE for evaluation:

$$\text{MAE} = \frac{1}{N}\sum_{i=1}^{N}\left|c_i - c_i^{GT}\right|, \tag{11}$$

$$\text{MSE} = \sqrt{\frac{1}{N}\sum_{i=1}^{N}\left|c_i - c_i^{GT}\right|^2}, \tag{12}$$

where N is the number of images in one test set, and c_i^{GT} is the ground-truth crowd number.

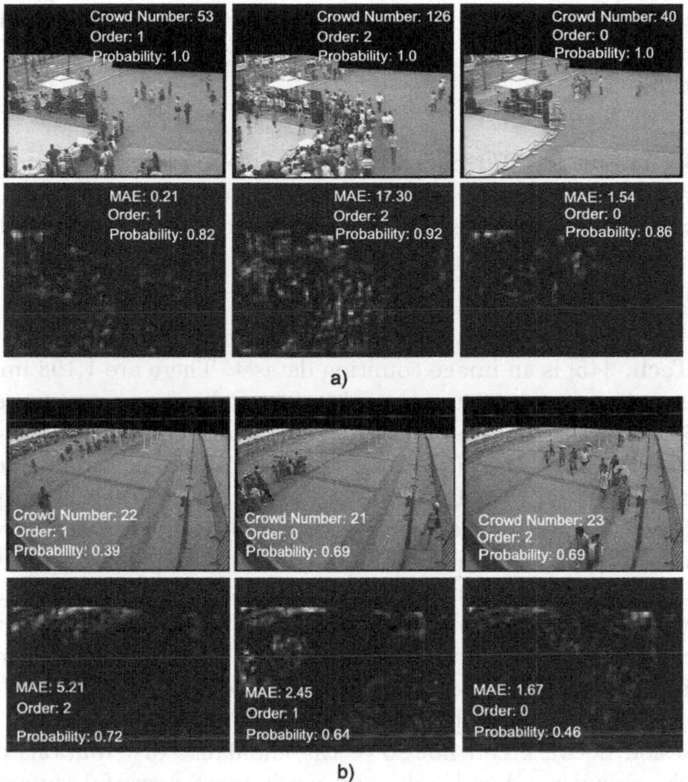

Fig. 3. In the upper row of each example, each image is labeled with its crowd number. Moreover, it is labeled with the order in the tuple and the corresponding probability. In the lower row of each example, we report the MAE of each estimation, and the predicted order with corresponding probability.

4.3 Evaluation and Comparison

WorldExpo10. [46] is a video counting dataset. The training set has 3,380 videos in 106 scenes, in which 3,380 frames are labeled with point labels. Besides, the testing set has 5 videos in 5 scenes, and 600 frames are labeled. In each training clip, we randomly select 3 videos with the same scene to ensure that the crowd numbers of chosen images have enough diversity. We list the result comparisons of MAE in Table 1, where our method achieves 9.6 average MAE. Without location supervision, our method plays favorably against the fully-supervised approaches. We visualize the density maps delivered from the internal layer and show a successful example in Fig. 3 (a), where the regression results have low errors. Moreover, the predicted orders are correct, and each prediction has high confidence. When processing the image clip in Fig. 3 (b), although the regression network delivers accurate estimations, the sorting network encounters a failure. This is because the images have similar crowd numbers. The experiments prove that the proposed counting network can accurately estimate the crowd numbers without location supervision. Moreover, the sorting network is capable of sorting the image clips.

UCSD. [3] contains 2,000 frames which are captured by surveillance cameras, and the frames have the same perspective. The comparison between existing approaches and our method is summarized in Table 1. Proved by the results on UCSD and WorldExpo10 datasets, our method overall performs comparably with the fully-supervised approaches in the video surveillance scene.

ShanghaiTech. [45] is an image counting dataset. There are 1,198 images with different perspectives and resolutions. This dataset has two parts named Part A and Part B. We report the comparison between our method and state-of-arts in Table 1. Compared with the supervised approaches, our method achieves comparable performance on Part B. While in Part A, our method has a particular gap with other methods. As there is a significant gap between the crowd number distributions of testing set and training set of Part A. More concretely, in the testing set, the mean and standard variance are 354.7 and 433.9, while in the training set, the mean and standard variance are 505.3 and 542.4. On the contrary, the crowd numbers in both sub-sets of Part B have a similar distribution. In the testing set, the mean and standard variance are 95.3 and 124.1, while in the training set, the mean and standard variance are 94.0 and 123.2. The non-linear regression network can not solve the unbalance distributions of dataset only with the crowd number labels, and needs more powerful supervision, for instance, the location and perspective labels.

Table 2. We conduct experiments to verify the efficiency of proposed framework and soft-label on WorldExpo10.

Sorting network	Regression network	Soft-label	MAE	Sorting accuracy (%)
✓	-	✓	-	50.6
-	✓	✓	20.1	-
✓	✓	-	13.2	**89.1**
✓	✓	✓	**9.6**	78.2

4.4 Ablation Study

In this section, we conduct several experiments to study the effect of different aspects of our method on WorldExpo10 and show the results in Table 2 and Table 3. In the first part, we conduct experiments to verify the dependency between the sorting and regression tasks. In the second part, we experiment to verify the efficiency of the proposed soft-label. While in the last part, we test several modifications to the proposed method.

Table 3. We evaluate several modifications to the proposed method, and report the results on WorldExpo10.

Backbone	Pooling cluster	Frame number	MAE	Sorting accuracy(%)
R3D	MAx & Avg	3	10.1	72.2
VGG	Avg	3	11.4	58.5
VGG	MAx & Avg	4	12.5	30.1
VGG	MAx & Avg	5	13.1	16.3
VGG	MAx & Avg	3	**9.6**	78.2

Only Sorting and Only Regression. To verify the dependency between the sorting and regression networks, we train the two networks separately and report the results in Table 2. When we train the sorting network alone, the sorting accuracy decreases by 35.3%. While we train the regression network alone, the regression accuracy decreases by 109.4%. The experiments demonstrate that the two tasks can promote each other. This is because the two tasks both estimate the crowd numbers and are closely related.

Soft-Label and Hard-Label. We employ the hard-label to train the proposed framework and report the results in Table 2. Each line of the hard-label is a one-hot vector. The sorting accuracy obtains 13.4% improvement, while the performance of counting task drops by 28.1%. This is because soft-labels have high entropy, and it is hard for the sorting network to predict accurate transportation probabilities. However, the soft-labels also contain much more information. Thus, they facilitate the counting task to improve performance.

Different Backbones. When evaluating on the video datasets, we employ images and clips to train the network, respectively. Each video clip contains 5 frames, which are sampled every 10 frames. To process the video clips, we employ the R3D network [43] as the backbone, which uses 3D convolutional layers and residual connections. The R3D network obtains improvements on several tasks, for instance, action recognition. However, as shown in Table 3, the performance of the modified method drops by 5.2%. The experiments show that the time dimension has a limited influence on the counting results. In the counting dataset, there is no regular pattern for the crowd movement. Therefore, the 3D convolutional layers extract less discriminative features.

Different Frames Numbers. We employ various numbers of frames to train the sorting network and report the results in Table 3. As mentioned above, the regression networks of the candidate sorting networks have various structures. However, regression networks have the same structure. The sorting task is more difficult while sorting more candidates. For instance, when sorting 3 images, a randomly guess has a probability of 1/6 to be right, while sorting 5 images,

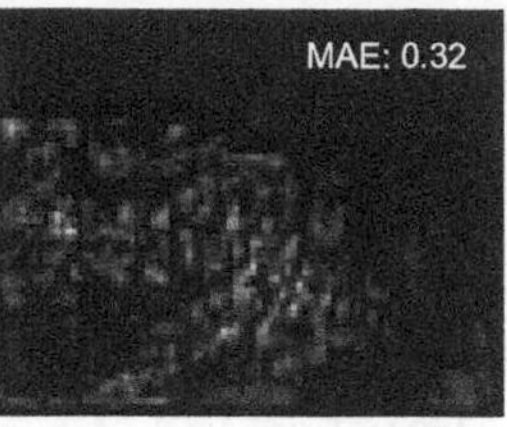

Fig. 4. In the left image, we show an example frame and label its crowd number. While in the right image, we show the corresponding density map, which is an internal feature map of the network. Moreover, we label the density map with the estimation error.

the right probability drops to 1/120. Therefore, the sorting accuracies of 4-candidates network and 5-candidates decrease by 61.5% and 79.1%, respectively. The regression accuracies of the two networks drop by 30.2% and 36.4%. This phenomenon affirms that more accurate sorting facilitates the counting task more. However, the regression accuracies of both candidates are still higher than that of the one without the assistant of sorting network.

Different Sampling Methods. In the counting network, we employ various pooling cluster operations to extract the features. The candidate method employs the same sampling rates, yet all the layers use adaptive average pooling layers. The performance of the modified network drops by 18.8%. This result suggests that max-pooling layers are more efficient while extracting the local features. In Fig. 4, we show an example, which is an internal density map delivered from the proposed method. As the density map is noisy, the max-pooling layers extract most discriminative features. Meanwhile, the internal feature map is not supervised by the artificial density maps. Therefore, the responses of this density map are not ideal Gaussian signals. The obtained density map maintains the original semantic information. This phenomenon affirms that without the demand for regressing the instance locations with Gaussian kernels, the counting network concentrates on regressing the crowd number. Therefore, the weakly-supervised crowd counting is a promising research tendency.

5 Conclusions

In this paper, we propose a weakly-supervised counting method, which is trained without location supervision. Moreover, we exploit the relationship among the images to improve the counting accuracy. We propose a novel soft-label sorting network along with the counting network, which sorts the given images by their crowd numbers. We train end-to-end both the sorting network and the regression network. During training, the sorting network drives the shared backbone CNN model to obtain density-sensitive ability explicitly. Therefore, the proposed method improves the counting accuracy by using the information among crowd numbers, rather than learning from extra labels. In the proposed sorting network, we propose a more informative soft-label to capture the complexity of the sorting task. We conduct experiments on three datasets and compare the proposed weakly-supervised approach with the fully-supervised methods. Extensive experimental results demonstrate the state-of-the-art performance of our method. In future work, we will propose a corresponding weakly-supervised benchmark to facilitate this task.

Acknowledgements. This work was supported in part by the Italy-China collaboration project TALENT:2018YFE0118400, in part by National Natural Science Foundation of China: 61620106009, 61772494, 61931008, U1636214, 61836002 and 61976069, in part by Key Research Program of Frontier Sciences, CAS: QYZDJ-SSW-SYS013, in part by Youth Innovation Promotion Association CAS.

References

1. Boominathan, L., Kruthiventi, S.S.S., Babu, R.V.: Crowdnet: a deep convolutional network for dense crowd counting. In: ACM Multimedia, pp. 640–644 (2016)
2. Cao, X., Wang, Z., Zhao, Y., Su, F.: Scale aggregation network for accurate and efficient crowd counting. In: The IEEE Conference on Computer Vision and Pattern Recognition (CVPR), pp. 757–773 (2018)
3. Chan, A.B., Liang, Z.S.J., Vasconcelos, N.: Privacy preserving crowd monitoring: counting people without people models or tracking. In: The IEEE Conference on Computer Vision and Pattern Recognition (CVPR), pp. 1–7 (2008)
4. Cheng, Z., Li, J., Dai, Q., Wu, X., Hauptmann, A.G.: Learning spatial awareness to improve crowd counting. In: The IEEE Conference on Computer Vision and Pattern Recognition (CVPR), pp. 6152–6161 (2019)
5. Cuturi, M.: Sinkhorn distances: lightspeed computation of optimal transportation distances. In: Neural Information Processing Systems, pp. 2292–2300 (2013)
6. Cuturi, M., Teboul, O., Vert, J.: Differentiable ranks and sorting using optimal transport. In: Conference on Neural Information Processing Systems (2019)
7. Deb, D., Ventura, J.: An aggregated multicolumn dilated convolution network for perspective-free counting. In: The IEEE Conference on Computer Vision and Pattern Recognition (CVPR), pp. 195–204 (2018)
8. Grover, A., Wang, E.H., Zweig, A., Ermon, S.: Stochastic optimization of sorting networks via continuous relaxations. In: International Conference on Learning Representations (2019)
9. Guerrerogomezolmedo, R., Torrejimenez, B., Lopezsastre, R.J., Maldonadobascon, S., Onororubio, D.: Extremely overlapping vehicle counting. In: Iberian Conference on Pattern Recognition and Image Analysis, pp. 423–431 (2015)
10. Guo, B., et al.: Mobile crowd sensing and computing: the review of an emerging human-powered sensing paradigm. ACM Comput. Surv. **48**(1), 7:1–7:31 (2015)
11. Huang, S., et al.: Body structure aware deep crowd counting. IEEE Trans. Image Process. **27**(3), 1049–1059 (2018)
12. Idrees, H., Saleemi, I., Seibert, C., Shah, M.: Multi-source multi-scale counting in extremely dense crowd images. In: The IEEE Conference on Computer Vision and Pattern Recognition (CVPR), pp. 2547–2554 (2013)
13. Jiang, X., et al.: Crowd counting and density estimation by trellis encoder-decoder network. In: The IEEE Conference on Computer Vision and Pattern Recognition (CVPR) (2019)

14. Lempitsky, V.S., Zisserman, A.: Learning to count objects in images. In: NIPS (2010)
15. Li, Y., Zhang, X., Chen, D.: Csrnet: Dilated convolutional neural networks for understanding the highly congested scenes. In: The IEEE Conference on Computer Vision and Pattern Recognition (CVPR), pp. 1091–1100 (2018)
16. Linderman, S.W., Mena, G., Cooper, H., Paninski, L., Cunningham, J.P.: Reparameterizing the birkhoff polytope for variational permutation inference. In: International Conference on Artificial Intelligence and Statistics (2017)
17. Liu, C., Wen, X., Mu, Y.: Recurrent attentive zooming for joint crowd counting and precise localization. In: The IEEE Conference on Computer Vision and Pattern Recognition (CVPR) (2019)
18. Liu, J., Gao, C., Meng, D., Hauptmann, A.G.: Decidenet: counting varying density crowds through attention guided detection and density estimation. In: The IEEE Conference on Computer Vision and Pattern Recognition (CVPR), pp. 5197–5206 (2018)
19. Liu, N., Long, Y., Zou, C., Niu, Q., Pan, L., Wu, H.: Adcrowdnet: an attention-injective deformable convolutional network for crowd understanding. In: The IEEE Conference on Computer Vision and Pattern Recognition (CVPR) (2018)
20. Liu, W., Salzmann, M., Fua, P.: Context-aware crowd counting. In: The IEEE Conference on Computer Vision and Pattern Recognition (CVPR) (2018)
21. Liu, X., De Weijer, J.V., Bagdanov, A.D.: Leveraging unlabeled data for crowd counting by learning to rank. In: The IEEE Conference on Computer Vision and Pattern Recognition (CVPR), pp. 7661–7669 (2018)
22. Liu, Y., Shi, M., Zhao, Q., Wang, X.: Point in, box out: beyond counting persons in crowds. In: The IEEE Conference on Computer Vision and Pattern Recognition (CVPR) (2019)
23. Longyin, W., et al.: Drone-based joint density map estimation, localization and tracking with space-time multi-scale attention network. arxiv (2020)
24. Loy, C.C., Gong, S., Xiang, T.: From semi-supervised to transfer counting of crowds. In: International Conference on Computer Vision, pp. 2256–2263 (2013)
25. Ma, Z., Wei, X., Hong, X., Gong, Y.: Bayesian loss for crowd count estimation with point supervision. In: The IEEE Conference on Computer Vision and Pattern Recognition (CVPR) (2019)
26. Noroozi, M., Favaro, P.: Unsupervised learning of visual representations by solving Jigsaw puzzles. In: Leibe, B., Matas, J., Sebe, N., Welling, M. (eds.) ECCV 2016. LNCS, vol. 9910, pp. 69–84. Springer, Cham (2016). https://doi.org/10.1007/978-3-319-46466-4_5
27. Ranjan, V., Le, H., Hoai, M.: Iterative crowd counting. In: Ferrari, V., Hebert, M., Sminchisescu, C., Weiss, Y. (eds.) ECCV 2018. LNCS, vol. 11211, pp. 278–293. Springer, Cham (2018). https://doi.org/10.1007/978-3-030-01234-2_17
28. Ronneberger, O., Fischer, P., Brox, T.: U-net: convolutional networks for biomedical image segmentation. In: Medical Image Computing and Computer Assisted Intervention, pp. 234–241 (2015)
29. Sam, D.B., Babu, R.V.: Top-down feedback for crowd counting convolutional neural network. In: National Conference on Artificial Intelligence, pp. 7323–7330 (2018)
30. Sam, D.B., Sajjan, N., Babu, R.V., Srinivasan, M.: Divide and grow: capturing huge diversity in crowd images with incrementally growing CNN. In: The IEEE Conference on Computer Vision and Pattern Recognition (CVPR), pp. 3618–3626 (2018)
31. Sam, D.B., Sajjan, N.N., Maurya, H., Radhakrishnan, V.B.: Almost unsupervised learning for dense crowd counting. Assoc. Adv. Artif. Intell. **33**, 8868–8875 (2019)

32. Sam, D.B., Surya, S., Babu, R.V.: Switching convolutional neural network for crowd counting. In: The IEEE Conference on Computer Vision and Pattern Recognition (CVPR), pp. 4031–4039 (2017)
33. Sermanet, P., et al.: Time-contrastive networks: self-supervised learning from video. In: The IEEE Conference on Computer Vision and Pattern Recognition (CVPR) (2017)
34. Shen, Z., Xu, Y., Ni, B., Wang, M., Hu, J., Yang, X.: Crowd counting via adversarial cross-scale consistency pursuit. In: The IEEE Conference on Computer Vision and Pattern Recognition (CVPR), June 2018
35. Sheng, X., Tang, J., Xiao, X., Xue, G.: Leveraging GPS-less sensing scheduling for green mobile crowd sensing. IEEE Internet Things J. **1**(4), 328–336 (2014)
36. Shi, M., Yang, Z., Xu, C., Chen, Q.: Revisiting perspective information for efficient crowd counting. In: The IEEE Conference on Computer Vision and Pattern Recognition (CVPR) (2018)
37. Shi, Z., Mettes, P., Snoek, C.G.M.: Counting with focus for free. In: International Conference on Computer Vision, pp. 4200–4209 (2019)
38. Shi, Z., et al.: Crowd counting with deep negative correlation learning. In: The IEEE Conference on Computer Vision and Pattern Recognition (CVPR), pp. 5382–5390 (2018)
39. Simonyan, K., Zisserman, A.: Very deep convolutional networks for large-scale image recognition. In: International Conference on Learning Representations (2015)
40. Sindagi, V.A., Patel, V.M.: Generating high-quality crowd density maps using contextual pyramid CNNs. In: International Conference on Computer Vision, pp. 1879–1888 (2017)
41. Wan, J., Chan, A.B.: Adaptive density map generation for crowd counting. In: International Conference on Computer Vision, pp. 1130–1139 (2019)
42. Wang, Q., Gao, J., Lin, W., Yuan, Y.: Learning from synthetic data for crowd counting in the wild. In: The IEEE Conference on Computer Vision and Pattern Recognition (CVPR) (2019)
43. Xu, D., Xiao, J., Zhao, Z., Shao, J., Xie, D., Zhuang, Y.: Self-supervised spatiotemporal learning via video clip order prediction. In: The IEEE Conference on Computer Vision and Pattern Recognition (CVPR), pp. 10334–10343 (2019)
44. Yan, Z., et al.: Perspective-guided convolution networks for crowd counting. In: The IEEE Conference on Computer Vision and Pattern Recognition (CVPR) (2019)
45. Zhang, C., Li, H., Wang, X., Yang, X.: Cross-scene crowd counting via deep convolutional neural networks. In: The IEEE Conference on Computer Vision and Pattern Recognition (CVPR), pp. 833–841 (2015)
46. Zhang, Y., Zhou, D., Chen, S., Gao, S., Ma, Y.: Single-image crowd counting via multi-column convolutional neural network. In: The IEEE Conference on Computer Vision and Pattern Recognition (CVPR), pp. 589–597 (2016)
47. Zhao, M., Zhang, J., Zhang, C., Zhang, W.: Leveraging heterogeneous auxiliary tasks to assist crowd counting. In: The IEEE Conference on Computer Vision and Pattern Recognition (CVPR), pp. 12736–12745 (2019)

Unsupervised Domain Attention Adaptation Network for Caricature Attribute Recognition

Wen Ji[1], Kelei He[2,3(✉)], Jing Huo[1(✉)], Zheng Gu[1], and Yang Gao[1,3]

[1] State Key Laboratory for Novel Software Technology, Nanjing, China
{jiwen,guzheng}@smail.nju.edu.cn, {huojing,gaoy}@nju.edu.cn
[2] Medical School of Nanjing University, Nanjing, China
hkl@nju.edu.cn
[3] National Institute of Healthcare Data Science at Nanjing University, Nanjing, China

Abstract. Caricature attributes provide distinctive facial features to help research in Psychology and Neuroscience. However, unlike the facial photo attribute datasets that have a quantity of annotated images, the annotations of caricature attributes are rare. To facility the research in attribute learning of caricatures, we propose a caricature attribute dataset, namely WebCariA. Moreover, to utilize models that trained by face attributes, we propose a novel unsupervised domain adaptation framework for cross-modality (i.e., photos to caricatures) attribute recognition, with an integrated inter- and intra-domain consistency learning scheme. Specifically, the inter-domain consistency learning scheme consisting an image-to-image translator to first fill the domain gap between photos and caricatures by generating intermediate image samples, and a label consistency learning module to align their semantic information. The intra-domain consistency learning scheme integrates the common feature consistency learning module with a novel attribute-aware attention-consistency learning module for a more efficient alignment. We did an extensive ablation study to show the effectiveness of the proposed method. And the proposed method also outperforms the state-of-the-art methods by a margin. The implementation of the proposed method is available at https://github.com/KeleiHe/DAAN.

Keywords: Unsupervised domain adaptation · Caricature · Attribute recognition · Attention

W. Ji and K. He—These authors contributed equally as co-first authors.

Electronic supplementary material The online version of this chapter (https://doi.org/10.1007/978-3-030-58598-3_2) contains supplementary material, which is available to authorized users.

A. Vedaldi et al. (Eds.): ECCV 2020, LNCS 12353, pp. 18–34, 2020.
https://doi.org/10.1007/978-3-030-58598-3_2

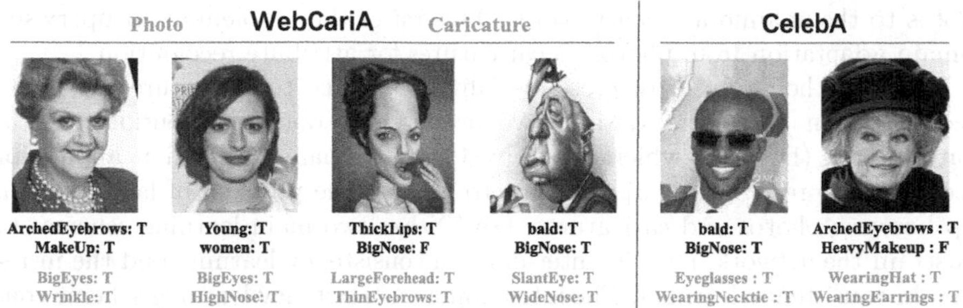

Fig. 1. The comparison of typical cases in WebCariA and CelebA datasets. First two rows denote the same attributes annotated in the WebCariA dataset and the CelebA dataset. The last two rows indicate the distinctive attributes annotated in the two datasets. 'T' indicates 'True, 'F' indicates 'False'.

1 Introduction

Caricatures are facial drawings of human faces with exaggerating facial features. Studying the latent information conveyed by caricatures has been long to the neurologists, psychologists, and also the computer scientists. The recognition of caricatures indicates the mechanism of human thoughts, and the knowledge learned by human during this task. Compared with photos, recognize the identities of caricatures may easier for humans [26,27]. This indicates the most representative face features are not destroyed even the shape and appearance of faces have been largely changed. By contrast, they are usually harder for the machine learning methods to recognize and comprehend.

Recently, the research area of analyzing the caricatures in machine learning society has been raised, with several datasets are publicly released [1,14,18]. Most of the existing researches focus on face recognition [14,18,26,33] and image generation [3,4,17] using these datasets. However, as lacked by the annotations of the caricature attributes, a more valuable task of attribute recognition, has rarely been touched. To solve this problem, in this paper, we introduce the WebCariA dataset, by extending a large caricature dataset 'WebCaricture' [14] with the annotation of fifty intrinsic face attributes. We hope it can boost this research area. Specifically, to help understand the intrinsic facial characteristics that are felt by human, the face attributes on WebCariA are purely facial characteristics without the non-face attributes, compared with the existing face attribute in photo datasets (e.g., the attribute of 'Eyeglasses' and 'WearingHat' in CelebA dataset). (See the comparisons of typical examples in Fig. 1) The details of the dataset will be further introduced in Sect. 3.

The face attribute recognition task has been solved well by deep convolutional networks [7,16,19,21,31]. It can be concluded as large-scale annotated face attribute datasets are already established, as the facial photos are easy to acquire. By contrast, the number of attribute annotated caricatures is small. Therefore, a natural idea is adapting a method that is trained on the annotated photos to the unannotated caricatures. This raises the problem of unsupervised domain adaptation from photos to caricatures for attribute recognition.

To solve the problem of cross-modality (i.e., face to caricature) attribute recognition on WebCariA dataset, we propose the domain attention adaptation network (DAAN), which has robust cross-domain adaptation ability for face attribute recognition. Specifically, to address the problem of large domain gap between photos and caricatures, DAAN has two main learning schemes to constrain the network, i.e., the inter-domain consistency learning and the intra-domain consistency learning. The inter-domain consistency learning scheme consisting of a pixel-level cross-domain image-to-image translator and the corresponding inter-domain consistency losses. The inter-domain consistency losses force the network to have consistent face attribute predictions on a certain image and its translated one. On the other hand, to align a certain image and a transferred image generated by the image of a different identity and domain property, we further propose the intra-domain consistency learning. It leverages both the feature and the attribute-aware attention map to build up the consistency between the two domains. Herein, we propose the attribute-aware attention-consistency learning by two observations: (1) the conventional feature-based consistency learning that align the distribution of the two domains in a global perspective is not efficient. (2) As a fine-grained classification task, face attributes are often revealed in a small region of the face, using attribute-aware attention will eliminate the noise conveyed by other parts. We did extensive ablation studies to show our proposed components, which can improve the discriminate ability of the network. The experiments also show our proposed method can outperform the state-of-the-art methods by a margin.

In this paper, our contributions are three-fold:

- We introduce the WebCariA dataset with the annotation of fifty intrinsic face attributes on caricatures.
- We propose a novel unsupervised attention adaptation framework for the recognition of attributes on unlabeled caricatures, which outperforms the state-of-the-art methods.
- We propose the attention-consistency learning which transfers the most task-discriminate features to achieve a more efficient adaptation.

2 Related Work

This work proposes an unsupervised domain adaptation method for attribute recognition in unpaired facial photos and caricatures. The related work can be concluded into three aspects: (1) The methods for face attribute recognition; (2) The methods for unsupervised domain adaptation in classification; (3) The generative adversarial learning-based image-to-image translation.

2.1 Face Attribute Recognition Methods

Face attribute recognition is a fine-grained classification task aiming to estimate facial characteristics. Previous works [7,21,24,25,41] based on convolutional neural networks have achieved satisfactory results. In the literature, early works mostly focus on single attribute learning [10,36]. After several large face attribute datasets being released, e.g., CelebA and LFWA [21], more works attempt to study on multi-attribute estimation, formulating the inter-attribute relationships in one method. For example, Liu *et al.* [21] and Ding *et al.* [7] proposed to localize face regions for multiple attribute recognition. Typically, multi-task learning (MTL) with multiple classifiers is a common technique to boost the generalization ability of the method by exploiting the inter-attribute relationships. Han *et al.* proposed a deep MTL approach for multi-attribute estimation in heterogeneous faces. Lu *et al.* [23] learn a deep MTL framework that dynamically groups similar tasks together. Zhang *et al.* [38] divide the attributes into different groups according to the location of the attribute. The multi-task learning have been proved to be very effective for facilitating the prediction of face attributes [2,8,12,23,28,37].

2.2 Unsupervised Domain Adaptation Methods in Classification

The scenario of unsupervised domain adaptation (UDA) arises when we aiming at constructing a model that learns from an annotated data distribution (i.e., source domain), and need it to generalize well on a different (but related) unannotated data distribution (i.e., target domain). Plenty of UDA methods for classification have been proposed in recent years. [6] We can roughly divide them into two categories: (1) Non-adversarial learning methods: A metric is often used as the objective to directly measure and minimize the discrepancy of high-level features between the source domain and the target domain. For instance, the work in [22] uses the maximum mean discrepancy (MMD) as the metric. Vazquez *et al.* [35] proposed to use a transductive SVM algorithm to solve the UDA problem. The non-adversarial methods have the ability to strongly align two different domains that is efficient for simple data distributions. (2) Adversarial learning methods: The discrepancy between the source and the target domain is usually large. Therefore, the adversarial learning-based methods try to minimize such discrepancy with the learning of the domain distributions. For example, Ganin *et al.* [9] proposes a domain adversarial neural network (DANN) for image classification which contains a loss for label prediction and a loss for domain classification. The maximum classifier discrepancy (MCD) [30] method builds up two classifiers to classify the source samples. Specifically, the two classifiers are forced to have different task-specific decision boundaries for the target sample. To make the method aware of a specific task, the Drop to Adapt (DTA) [20] method is proposed to learn robust and discriminate features by leveraging the adversarial dropout strategy, which supports the cluster assumption. However, these methods still cannot get satisfying performance as they only directly align the two domain distributions. Therefore, several works [13,29] have adopted image-to-image translation into the UDA framework, with the assumption of the generated intermediate data helps domain alignment. Besides, previous methods align the features learned in two domains, that are redundant and not efficient.

2.3 Image-to-Image Translation

The goal of image-to-image translation is to learn the mapping between input images and output images. GAN has been widely utilized to solve the image-to-image translation problem, as it can learn the latent distribution of the images. For example, Isola *et al.* [15] have proposed the conditional adversarial networks (conditional GAN) for the image-to-image translation problem. Zhu *et al.* [40] used the cycle-consistent adversarial networks (CycleGAN) to tackle the problem without paired training examples. Kim *et al.* proposes a method namely U-GAT-IT [17] which tried to combine the attention mechanism with the generative adversarial networks. It prompts the generator to focus on areas that specifically distinguish between the two domains, and the discriminator to focus on the difference between an original image and transferred image in the target domain.

3 The WebCariA Dataset

To date, several caricature datasets [1,14,18] are publicly available for caricature recognition. However, due to lack of caricature attribute annotations, there are little attempts to solve the task of caricature attribute recognition. In order to promote the research of caricature attribute recognition and generation, we construct a face attribute dataset, namely WebCariA, by labeling all images in the WebCaricature [14] dataset. The dataset contains 6024 caricatures and 5974 photos of 252 people. Each image was labeled more than three times with fifty attributes of intrinsic face characters, and the final labels are determined by a voting strategy. The dataset is released at[1].

Different from the photo dataset CelebA [21] and LFWA [21], we did not mark the attributes that are changeable, such as 'WearingJewelry', 'WearingHats', etc. Instead, we labeled the attributes that can reflect the intrinsic characteristics of the faces such as 'BigNose', 'Bald', etc. (See typical cases compared to CelebA in Fig. 1). Besides, we hope the WebCariA dataset can also promote the research area of caricature generation.

4 Method

In this section, we first provide the pipeline of our unsupervised domain adaptation framework. Secondly, we introduce the integrated-domain generalization learning paradigm which consists of the inter-domain consistency learning and the intra-domain consistency learning. Finally, we describe the multi-task learning setting for more precise face attribute recognition.

[1] https://cs.nju.edu.cn/huojing/WebCariA.htm.

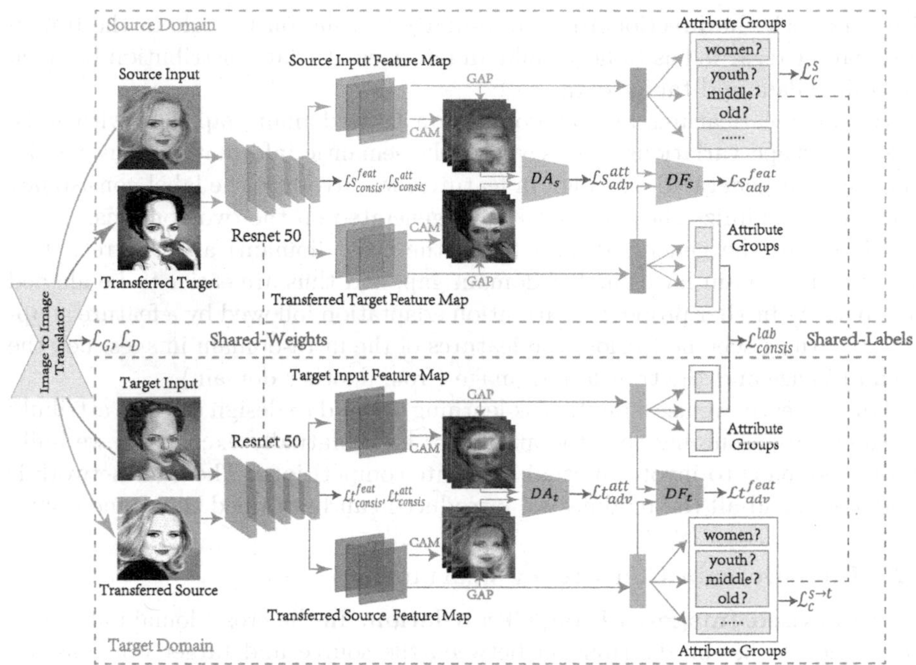

Fig. 2. The overall framework of our proposed unsupervised domain attention adaptation network (DAAN) for unsupervised face attribute recognition on caricatures.

4.1 The Unsupervised Domain Adaptation Framework

The overall framework is shown in Fig. 2. The goal of our method is to estimate face attributes on caricatures (i.e., the target domain) with the annotations of the attributes that are only given on photos (i.e., the source domain). Moreover, the images in source and target domain are unpaired, for which forms a typical unsupervised domain adaptation setting. Formally, let us denote the labeled source domain as $S = \{X_s, Y_s^i\}$, $i \in [1, N]$, where $\mathbf{x}_s \in X_s$ and $\mathbf{y}_s \in Y_s$ denote the source image and the corresponding label pair, N denotes the number of face attributes. The unlabeled target domain can be therefore denoted as $T = \{X_t\}$, where $\mathbf{x}_t \in X_t$ denotes the target image.

The domain gap and domain shift are the major problems for the cross-domain learning method. To solve the problems, our method adopted an integrated learning scene for learning both the inter- and intra-domain consistencies. The consistencies are learned by given constraints on the features. The inter-domain consistency learning firstly uses an unsupervised image-to-image translator to generate images from one domain to another by learning the influence of the style discrepancy of the two domains. The semantic information is preserved and the style information is transformed on the transferred images. Specifically, in this work, the image translation is bi-directional, where each image of the source domain will generate one transferred image in the target domain, and vise versa. The bi-directional image-to-image translation module in the framework can be regarded as to help build an intermediate data distribution between the source data and target data.

However, caricatures and photos have a large domain gap, i.e., with large style and shape variations. To constrain the semantic information conveyed by image and its generated one align in features, we introduce the label-consistency learning to minimize the prediction error generated on the two domains.

Obviously, the generated data (from the other domain) and the real data in a certain domain have smaller domain gap, and thus are easier to be aligned together. We further propose an attention adaptation followed by a feature adaptation learning scheme to close the features of the intra-domain images, i.e., the original image and the transferred image (from another domain).

For inference, we use a multi-task learning method to design the face attribute classifiers. By assuming that the attributes have mutually exclusions, we build attribute-groups to involve mutual attribute competitions. The final network is end-to-end, and all the modules we introduced can be trained simultaneously.

4.2 Inter-Domain Consistency Learning

Unsupervised Image-to-Image Translation. In the cross-domain attribute recognition task, the distribution between the source and target domains are large. Photos are shots of the real-world that often contain rich illumination, texture and noise. And caricatures are facial drawings are often concise, with clear outlines and exaggerated facial features. In order to eliminate the domain gap between the photo and caricatures, we first utilize an unsupervised image-to-image translator to alleviate the style discrepancy between these two domains. The translator takes a source image $\mathbf{x}_s$ and a target image $\mathbf{x}_t$ as inputs, and translate them from one domain to the other. The obtained corresponding translated images are denoted as $\mathbf{x}_{s\to t}$ and $\mathbf{x}_{t\to s}$. Here we use the generator and discriminator proposed in [17] as the image-to-image translator $\mathbf{G}_{s\to t}$ and $\mathbf{G}_{t\to s}$. Therefore, the loss are denoted as $\mathcal{L}_G$, $\mathcal{L}_D$, respectively. In this way, the style difference between the source domain and the target domain is eliminated, making their domain distribution more consistent with each other. Then, the features of the four images are extracted by a feature extractor. The feature extractor *Resnet* takes $\mathbf{x}_s$ and $\mathbf{x}_{t\to s}$ as inputs, then the feature vectors $\mathbf{F}_s$ and $\mathbf{F}_{t\to s}$ is obtained after the global average pooling (GAP). The features $\mathbf{F}_t$ and $\mathbf{F}_{s\to t}$ in the target domain are acquired accordingly.

Label-Consistency Learning. By assuming the semantic information is not destroyed during image translation, we give $\mathbf{x}_s$ and $\mathbf{x}_{s\to t}$ the same labels to calculate the classification loss. Here, we use the cross-entropy loss to calculate the attribute estimating errors with the input of $\mathbf{x}_s$ and $\mathbf{x}_{s\to t}$,

$$\mathcal{L}_c^s = -\mathbb{E}_{\mathbf{x}_s,\mathbf{y}_s\sim S}\left[\mathbf{y}_s^\top \log C\left(Resnet\left(\mathbf{x}_s\right)\right)\right] \tag{1}$$

$$\mathcal{L}_c^{s\to t} = -\mathbb{E}_{\mathbf{x}_{s\to t},\mathbf{y}_s\sim T'}\left[\mathbf{y}_s^\top \log C\left(Resnet\left(\mathbf{x}_{s\to t}\right)\right)\right] \tag{2}$$

where C is the classifier, S' and T' are the distributions of the transferred image features. This also helps to make full use of annotations given in the source domain for semantic attribute consistency.

On the other hand, the labels of $\mathbf{x}_t$ and $\mathbf{x}_{t\to s}$ should be predicted as consistent as possible, therefore, we design the label-consistency loss for $\mathbf{x}_t$ and $\mathbf{x}_{t\to s}$ to control semantic attribute consistency,

$$\mathcal{L}_{consis}^{lab} = -\mathbb{E}_{\mathbf{x}_t \sim T, \mathbf{x}_{t\to s} \sim S'} \left[\| C\left(Resnet\left(\mathbf{x}_t\right)\right) - C\left(Resnet\left(\mathbf{x}_{t\to s}\right)\right) \|_2 \right] \quad (3)$$

$$\mathcal{L}_{inter} = \lambda_G \mathcal{L}_G + \lambda_D \mathcal{L}_D + \lambda_l \mathcal{L}_{consis}^{lab} \quad (4)$$

The proposed method achieves style consistency and semantic consistency between the source domain and the target domain by adopting the above-mentioned techniques. However, attribute recognition is a fine-grained classification task, these consistencies are built on a global perspective. They are not sensitive to the tiny differences between the two domains in the cross-domain face attribute recognition task. And the network still lack of the constraints between the transferred image and original image in one domain, e.g., $\mathbf{x}_{\mathbf{s}\to\mathbf{t}}$ and $\mathbf{x}_\mathbf{t}$. We propose the intra-domain consistency learning to solve the problem.

4.3 Intra-domain Consistency Learning

To make the source (/target) domain image and the transferred target (/source) image have stronger consistency constraints, we use the generative adversarial strategy to make the distribution of features align between the source domain and target domain. We introduce two discriminators D_{Fs} and D_{Ft} for the output features of the two domains, to construct the feature-level domain adaptation.

Under the adversarial learning setting, the loss $\mathcal{L}_{consis}^{feat}$ of the features can be written as,

$$\mathcal{L}_{consis}^{feat} = \mathbb{E}\left[\sum \mathbf{Y}_s \log\left(D_{Fs}\left(\mathbf{F}_{t\to s}\right)\right)\right] + \mathbb{E}\left[\sum \mathbf{Y}_{s\to t} \log\left(D_{Ft}\left(\mathbf{F}_t\right)\right)\right] \quad (5)$$

where $\mathbf{Y}$ is the originated domain label of the samples, and have the same shape to $\mathbf{F}$. If the samples come from source domain the elements of Y are 1 else 0. As the two pair of cross-domain features are jointly optimized, the final consistency loss in Eq. 5 aggregate the errors raised by the two features. And in the discriminate process, the intra-domain feature consistency for source domain try to achieve feature-level alignment between $\mathbf{F}_s$ and $\mathbf{F}_{t\to s}$, which can be defined as follows,

$$\mathcal{L}s_{adv}^{feat} = \mathbb{E}\left[\log\left(D_{Fs}\left(\mathbf{F}_s\right)\right)\right] + \mathbb{E}\left[\log\left(1 - D_{Fs}\left(\mathbf{F}_{t\to s}\right)\right)\right] \quad (6)$$

For the features in the target domain, the alignment can be written as,

$$\mathcal{L}t_{adv}^{feat} = \mathbb{E}\left[\log\left(D_{Ft}\left(\mathbf{F}_{s\to t}\right)\right)\right] + \mathbb{E}\left[\log\left(1 - D_{Ft}\left(\mathbf{F}_{t}\right)\right)\right] \quad (7)$$

The features convey the whole information of an image, that is rough and not efficient. It is proved that discriminative features often make more contribution to the classification task [34]. And a slight difference between the source and target domain can seriously affect the classification results in face attributes classification, because it is a fine-grained classification problem. Moreover, the distortion and exaggerations of caricature are often beyond realism which makes the gap of photos and caricatures very large. Therefore, we propose the attribute-attention alignment between the original images and translated images (from the other domain). The attention maps $\mathbf{A}_s$, $\mathbf{A}_{t\to s}$, $\mathbf{A}_t$ and $\mathbf{A}_{s\to t}$ of all inputs are calculated by class activation map (CAM) [39]. Take the source original image as example, $\mathbf{A}_s^i = \sum_{j=1}^{n} w_j^i \mathbf{F}'_{s,j}$, where $\mathbf{F}'_{s,j}$ is the jth channel of output feature maps $\mathbf{F}'_s$ which is a metric before GAP and w_j^i is the jth weight of the fully connected layer for predicting the ith category. Similar to the feature domain adaptation, we propose two attention-based discriminators, i.e., D_{As} which is performed on $\mathbf{A}_s$ and $\mathbf{A}_{t\to s}$, and D_{At} which is performed on $\mathbf{A}_t$ and $\mathbf{A}_{s\to t}$ to construct the attribute-attention domain adaptation.

The attribute attention-based adaptation can locate the decisive image region for classification. And with the help of the discriminators, the network will pay more attention to the inconsistent regions of the two domains to eliminate their tiny differences. Therefore, attention is suitable for domain alignment in face attributes classification.

The attention-consistency loss can be defined as follow:

$$\mathcal{L}_{consis}^{att} = \mathbb{E}\left[\sum \mathbf{Y'}_s \log\left(D_{As}\left(\mathbf{A}_{t\to s}\right)\right)\right] + \mathbb{E}\left[\sum \mathbf{Y'}_{s\to t} \log\left(D_{At}\left(\mathbf{A}_{t}\right)\right)\right], \quad (8)$$

where $\mathbf{Y}'$ is the label of the domain where the sample is coming from, its shape is same to $\mathbf{A}$. If the samples are drawn from the source domain, the elements of $\mathbf{Y}'$ is 1 else 0. And the attention-based intra-domain consistencies $\mathbf{A}_s$ and $\mathbf{A}_{t\to s}$ for the source domain are defined as,

$$\mathcal{L}s_{adv}^{att} = \mathbb{E}\left[\log\left(D_{As}\left(\mathbf{A}_{s}\right)\right)\right] + \mathbb{E}\left[\log\left(1 - D_{As}\left(\mathbf{A}_{t\to s}\right)\right)\right] \quad (9)$$

Similarly, D_{At} aligns the distributions between $\mathbf{A}_t$ and $\mathbf{A}_{s\to t}$,

$$\mathcal{L}t_{adv}^{att} = \mathbb{E}\left[\log\left(D_{At}\left(\mathbf{A}_{s\to t}\right)\right)\right] + \mathbb{E}\left[\log\left(1 - D_{At}\left(\mathbf{A}_{t}\right)\right)\right] \quad (10)$$

Overall, the final intra-domain consistency loss can be written as,

$$\mathcal{L}_{intra} = \lambda_f(\mathcal{L}s_{consis}^{feat} + \mathcal{L}t_{consis}^{feat}) + \mathcal{L}s_{adv}^{feat} + \mathcal{L}t_{adv}^{feat} \quad (11)$$

$$+\lambda_a(\mathcal{L}s_{consis}^{att} + \mathcal{L}t_{consis}^{att}) + \mathcal{L}s_{adv}^{att} + \mathcal{L}t_{adv}^{att} \quad (12)$$

4.4 Multi-task Attribute Recognition

For the task of caricature attribute recognition, the attributes are not independent. It is not suitable to treat them as a pure multi-class classification problem, as done by the conventional attribute recognition methods. Thus, the work in [38] partitioned all the attributes into different groups based on the global and local spatial regions.

The idea is natural but not very reasonable to the learning theory. By contrast, we observe that in most face attribute datasets, there are mutually exclusive relationships among the face attributes. For example, the attributes of young, middle-aged and old are related to the group of age. Obviously, the mutually exclusive relationships among the attributes provides stronger and more stable constraints to build up the classifiers. Herein, for the classifiers, we use one fully-connected layer to map the features to a certain number of classes for each attribute group. The cross-entropy loss is used to calculate the error. The results of each classifier is concatenated to obtain the final prediction result.

Finally, the objective function involving all the previous mentioned losses can be written as,

$$\mathcal{L} = \mathcal{L}_c^s + \mathcal{L}_c^{s\rightarrow t} + \mathcal{L}_{inter} + \mathcal{L}_{intra} \tag{13}$$

4.5 Implementation Details

We implement the proposed method using the open-source framework *PyTorch*. We use ResNet-50 [11] without the layer after the global average pooling layer as the feature extractor, and share the extractor through all inputs in this work. The discriminators are randomly initialized with a structure similar to [32], consisting five convolutional layers with size of $\{3 \times 3{,}1{,}1\}$ for {kernel, stride, padding}. We use the Stochastic Gradient Descent (SGD) algorithm to optimize the network, with a batch size of 40. The learning rate for the feature extractor and the classifiers is 0.05, with momentum of 0.9 and weight decay of 5×10^{-4}. The learning rate is decayed by the polynomial strategy [5] with the power of 0.75. For the discriminators, we use Adam optimizer with the learning rate of 1×10^{-4}. For the adversarial losses, we use the learning rates of 0.02, 0.1 and 0.1 for λ_l, λ_f and λ_a, respectively.

5 Experimental Results

Because annotated photos are often easy to acquire, in this paper, the method is built to estimate caricature attributes by the annotated photos, and using the photos in WebCariA dataset as the source domain. We use the metrics of accuracy (Acc.) and F1 score to measure the performance through the experiments. Please note that the attribute recognition task is very task-imbalanced for each attribute, the F1 score is more representative to show the real performance of the methods.

5.1 Ablation Study

We did extensive experiments for DAAN, including (1) the evaluation of the effectiveness of the MTL method, (2) the performance comparison of different network configurations of DAAN. The parameter analysis of DAAN is reported in *Supplementary Material*.

Firstly, we construct the 'Source Only' method, which is only trained on the source domain and evaluate on the target domain; and the 'Target Only' method, which is trained on the target domain and also evaluates on the target domain, forms a fully-supervised fashion. These two methods are usually used as the baselines to indicate the lower and upper bound of the performance of domain adaptation. For the proposed DAAN, we construct different configurations of DAAN, with or without the consistency learning of label, feature and attribute-aware attention.

Table 1. The effectiveness of the proposed multi-task learning strategy. 'MultiTask' denotes the network is learned with the attribute groups.

Method	Avg. Acc	Avg. F1
Source only	0.8054	0.6770
Source only (MultiTask)	0.8050	0.6922
DTA [20]	0.8100	0.6941
DTA(MultiTask) [20]	0.8076	0.7000
Target only	0.8474	0.7358
Target only (MultiTask)	0.8526	0.7601

The Effectiveness of Multi-task Learning. Table 1 shows the performance in average accuracy and average F1 score through all attributions on the WebCariA dataset. The table suggests that using the multi-task learning strategy consistently improves the performance of the network. For the two baseline networks 'Source Only' and 'Target Only' which are purely unsupervised and supervised methods, group attributes with multi-task learning improve the average F1 score by 1.52% and 2.43%, respectively. To evaluate the generalization of the proposed MTL, we implement the state-of-the-art unsupervised domain adaptation method in [20] with a single classifier, denote as DTA, and with the proposed MTL, denote as DTA(MultiTask). The experiments show that, for the well designed unsupervised domain adaptation method, our proposed MTL still improves the F1 score by 0.59%.

Evaluation of Different Adaptation Constraints. To evaluate the effectiveness of different adaptation constraints proposed by the method, we first

Table 2. Performance comparison with different configurations of DAAN. (Bests are in Bold)

Method		Label	Feat.	Att.	Avg. Acc	Avg. F1
DAAN	**-L**	✓			0.8122	0.7038
	-F		✓		0.8169	0.7094
	-A			✓	0.8147	0.7107
	-LF	✓	✓		0.8219	0.7181
	-LA	✓		✓	0.8212	0.7192
	-LFA	✓	✓	✓	**0.8239**	**0.7215**

construct DAAN with only one consistency constraint of label, feature and attribute-aware attention as 'DAAN-L', 'DAAN-F' and 'DAAN-A', respectively. Then, we compose two and three constraints into DAAN. The evaluations in average accuracy and F1 score are reported in 2. We make three conclusions according to the table: (1) All the constraints help to regularize the features under the UDA setting. (2) Among one factor constrained DAANs, the attention-consistency constraint performs better than the other two constraints, indicates the attention is more efficient. (3) The combination of three constraints can further improve the model, and achieve the overall best performance. This reveals multiple constraints are useful to the UDA setting, which has more freedom in parameter space compared with fully-supervised learning.

5.2 Comparison with the State-of-the-Art Methods

We compare DAAN with several recent state-of-the-art methods for image classification. The compared methods include DAN [22], DANN [9], MCD [30] and DTA [20]. The compared methods are reimplemented on the WebCariA dataset without our proposed MTL by mutually exclusive attribute groups. The performance in average accuracy and F1 score is reported in Table 3, and the attribute-wise performance in the F1 score is illustrated in Fig. 3. The table suggests that the four state-of-the-art methods can both improve the performance of the network when adapting to another domain, and DTA performs best among the four methods in the second set of rows. So we further implement it with our MTL setting (denoted as 'DTA(MultiTask)'). As shown by the table, compared DAAN-LF with DTA, our proposed method gets a 1.19% improvement in average accuracy and a 2.4% improvement in average F1 score. Compared with DAAN-LFA with DTA and DTA(MultiTask), the improvements are 1.39%, 1.63% in average accuracy and 2.74%, 2.15% in average F1 score, respectively. This shows the effectiveness of DAAN compared with the current state-of-the-art methods. Furthermore, as shown in Fig. 3, DAAN-LFA performs consistently best in most attributes, when compared with Source Only, DAN and DTA method. It is worth noting that DAAN performs especially well in attributes that are hard to estimate (i.e., with F1 score under 0.6), e.g., SquareFace, HookNose, etc. This demonstrates the robustness of the proposed DAAN. We also analyze the generalization ability for DAAN on small benchmark datasets, which is reported in *Supplementary Material.*

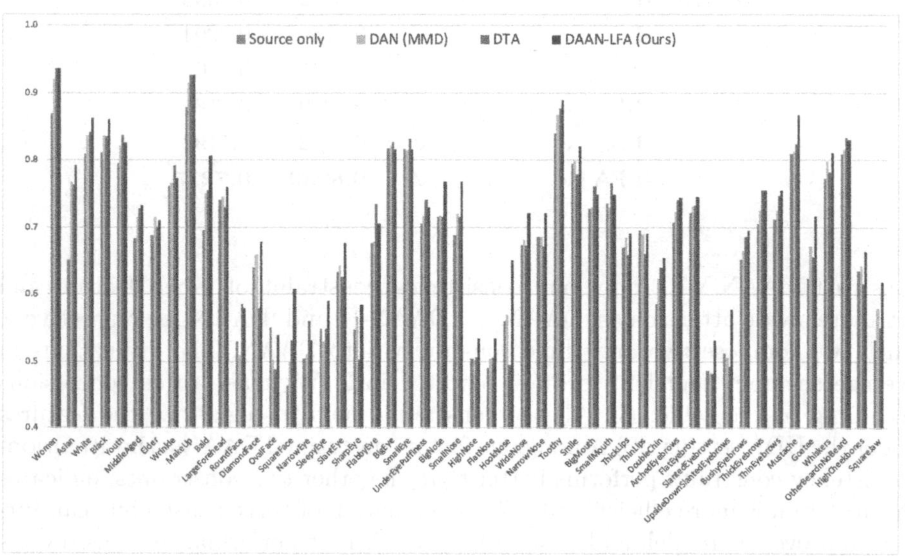

Fig. 3. The performance comparison in F1 score with Source Only, DAN (MMD), DTA and the proposed DAAN-LFA.

Table 3. Comparison with the state-of-the-art methods for performance in average Acc. and F1 score. As MCD method proposed two classifiers in the network, we report both of them separately.

Method	Avg. Acc	Avg. F1
Source only	0.8054	0.6770
Source only (MultiTask)	0.8050	0.6922
DAN [22]	0.8098	0.6921
DANN [9]	0.7783	0.6604
MCD [30]	0.7967 (0.7993)	0.6920 (0.6914)
DTA [20]	0.8100	0.6941
DTA (MultiTask) [20]	0.8076	0.7000
DAAN-LF (Ours)	0.8219	0.7181
DAAN-LA (Ours)	0.8212	0.7192
DAAN-LFA (Ours)	**0.8239**	**0.7215**
Target only	0.8474	0.7358
Target only (MultiTask)	0.8526	0.7601

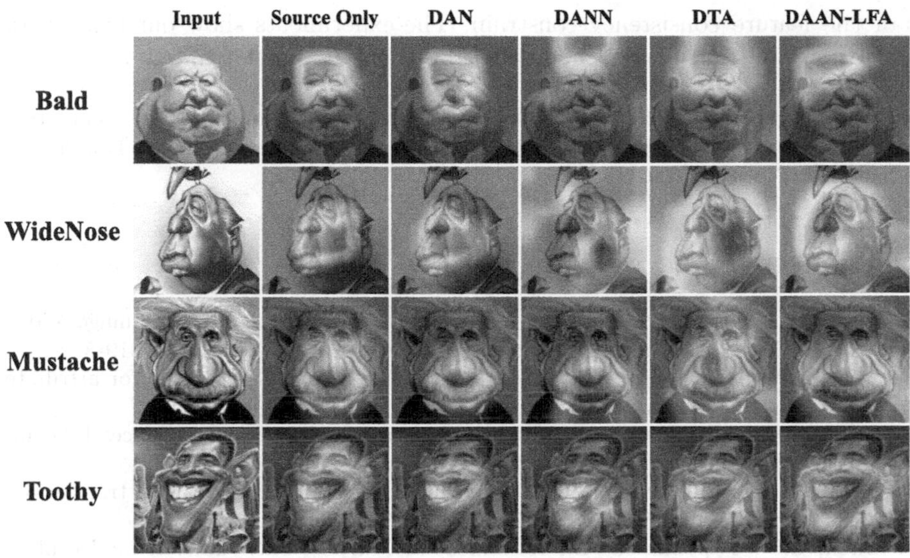

Fig. 4. The visualization of CAM for the proposed compared with the state-of-the-art methods on typical cases. Obviously, the proposed DAAN-LFA consistently generates high quality CAMs of accurate positions and active region sizes.

5.3 Visualization of the Attribute-Wise Attention Maps

The CAM reveals the attention of the network when it makes decisions on a certain class. We visualize the CAMs for the proposed method compared with the state-of-the-art methods on the test set in Fig. 4. As the face attributes are often spatial-related, more precise attention regions for the specific attribute indicates higher predict performance. It is suggested by the figure that DAAN-LFA generates consistently best region activation through the cases. (More visualization results for DAAN is shown in *Supplementary Material.*)

6 Conclusions

In this paper, we extend the 'WebCaricature' with fifty intrinsic face attributes to prepare the WebCariA dataset, for boosting the research in caricature attribute recognition. Then, to solve the problem of lacking annotated caricature attributes, we propose an unsupervised domain adaptation framework, i.e., DAAN, to estimate the caricature attributes by training the network with only given the face attribute labels. DAAN integrated with both inter- and intra-domain consistency learning paradigms. Specifically, in DAAN, an image-to-image translator and a label-consistency learning constrain are contained to fill the gap between photos and caricatures. To align the domain between the transferred images and original images in one certain domain, we propose the attribute-aware attention-consistency learning constrain, which is more efficient than the feature consistency constrain. The experiments show our framework can outperform the state-of-the-art methods by a reasonable margin.

Acknowledgement. This work is supported in part by National Science Foundation of China under Grant No. 61806092, and in part by Jiangsu Natural Science Foundation under Grant No. BK20180326.

References

1. Abaci, B., Akgul, T.: Matching caricatures to photographs. Signal Image Video Process. **9**(1), 295–303 (2015). https://doi.org/10.1007/s11760-015-0819-8
2. Abdulnabi, A.H., Wang, G., Lu, J., Jia, K.: Multi-task CNN model for attribute prediction. IEEE Trans. Multimed. **17**(11), 1949–1959 (2015)
3. Brennan, S.E.: Caricature generator: the dynamic exaggeration of faces by computer. Leonardo **40**(4), 392–400 (2007)
4. Cao, K., Liao, J., Yuan, L.: Carigans: unpaired photo-to-caricature translation. ACM Trans. Graph. **37**(6), 244 (2018)
5. Chen, L.C., Papandreou, G., Kokkinos, I., Murphy, K., Yuille, A.L.: Deeplab: semantic image segmentation with deep convolutional nets, atrous convolution, and fully connected CRFs. IEEE Trans. Pattern Anal. Mach. Intell. **40**(4), 834–848 (2017)
6. Csurka, G.: Domain adaptation for visual applications: a comprehensive survey. arXiv preprint arXiv:1702.05374 (2017)
7. Ding, H., Zhou, H., Zhou, S.K., Chellappa, R.: A deep cascade network for unaligned face attribute classification. In: Thirty-Second AAAI Conference on Artificial Intelligence (2018)
8. Ehrlich, M., Shields, T.J., Almaev, T., Amer, M.R.: Facial attributes classification using multi-task representation learning. In: Proceedings of the IEEE Conference on Computer Vision and Pattern Recognition Workshops, pp. 47–55 (2016)
9. Ganin, Y., Lempitsky, V.: Unsupervised domain adaptation by backpropagation. In: International Conference on Machine Learning, pp. 1180–1189 (2015)
10. Geng, X., Yin, C., Zhou, Z.H.: Facial age estimation by learning from label distributions. IEEE Trans. Pattern Anal. Mach. Intell. **35**(10), 2401–2412 (2013)
11. He, K., Zhang, X., Ren, S., Sun, J.: Deep residual learning for image recognition. In: Proceedings of the IEEE Conference on Computer Vision and Pattern Recognition, pp. 770–778 (2016)
12. He, K., Wang, Z., Fu, Y., Feng, R., Jiang, Y.G., Xue, X.: Adaptively weighted multi-task deep network for person attribute classification. In: Proceedings of the 25th ACM International Conference on Multimedia, pp. 1636–1644. ACM (2017)
13. Hoffman, J., et al.: Cycada: cycle-consistent adversarial domain adaptation. In: Proceedings of the 35th International Conference on Machine Learning (2018)
14. Huo, J., Li, W., Shi, Y., Gao, Y., Yin, H.: Webcaricature: a benchmark for caricature recognition. arXiv preprint arXiv:1703.03230 (2017)
15. Isola, P., Zhu, J.Y., Zhou, T., Efros, A.A.: Image-to-image translation with conditional adversarial networks. In: Proceedings of the IEEE Conference on Computer Vision and Pattern Recognition, pp. 1125–1134 (2017)
16. Jacob, L., Philippe Vert, J., Bach, F.R.: Clustered multi-task learning: a convex formulation. In: Koller, D., Schuurmans, D., Bengio, Y., Bottou, L. (eds.) Advances in Neural Information Processing Systems, vol. 21, pp. 745–752. Curran Associates, Inc. (2009). http://papers.nips.cc/paper/3499-clustered-multi-task-learning-a-convex-formulation.pdf

17. Kim, J., Kim, M., Kang, H., Lee, K.H.: U-gat-it: unsupervised generative attentional networks with adaptive layer-instance normalization for image-to-image translation. In: International Conference on Learning Representations (2019)
18. Klare, B.F., Bucak, S.S., Jain, A.K., Akgul, T.: Towards automated caricature recognition. In: 2012 5th IAPR International Conference on Biometrics (ICB), pp. 139–146. IEEE (2012)
19. Kumar, A., Daume III, H.: Learning task grouping and overlap in multi-task learning. In: ICML (2012)
20. Lee, S., Kim, D., Kim, N., Jeong, S.G.: Drop to adapt: learning discriminative features for unsupervised domain adaptation. In: Proceedings of the IEEE International Conference on Computer Vision, pp. 91–100 (2019)
21. Liu, Z., Luo, P., Wang, X., Tang, X.: Deep learning face attributes in the wild. In: Proceedings of the IEEE International Conference on Computer Vision, pp. 3730–3738 (2015)
22. Long, M., Cao, Y., Wang, J., Jordan, M.: Learning transferable features with deep adaptation networks. In: International Conference on Machine Learning, pp. 97–105 (2015)
23. Lu, Y., Kumar, A., Zhai, S., Cheng, Y., Javidi, T., Feris, R.: Fully-adaptive feature sharing in multi-task networks with applications in person attribute classification. In: Proceedings of the IEEE Conference on Computer Vision and Pattern Recognition, pp. 5334–5343 (2017)
24. Luo, P., Wang, X., Tang, X.: Hierarchical face parsing via deep learning. In: 2012 IEEE Conference on Computer Vision and Pattern Recognition, pp. 2480–2487. IEEE (2012)
25. Luo, P., Wang, X., Tang, X.: A deep sum-product architecture for robust facial attributes analysis. In: Proceedings of the IEEE International Conference on Computer Vision, pp. 2864–2871 (2013)
26. Mauro, R., Kubovy, M.: Caricature and face recognition. Mem. Cogn. **20**(4), 433–440 (1992)
27. Perkins, D.: A definition of caricature and caricature and recognition. Stud. Vis. Commun. **2**(1), 1–24 (1975)
28. Rudd, E.M., Günther, M., Boult, T.E.: MOON: a mixed objective optimization network for the recognition of facial attributes. In: Leibe, B., Matas, J., Sebe, N., Welling, M. (eds.) ECCV 2016. LNCS, vol. 9909, pp. 19–35. Springer, Cham (2016). https://doi.org/10.1007/978-3-319-46454-1_2
29. Russo, P., Carlucci, F.M., Tommasi, T., Caputo, B.: From source to target and back: symmetric bi-directional adaptive GAN. In: Proceedings of the IEEE Conference on Computer Vision and Pattern Recognition (CVPR), June 2018
30. Saito, K., Watanabe, K., Ushiku, Y., Harada, T.: Maximum classifier discrepancy for unsupervised domain adaptation. In: Proceedings of the IEEE Conference on Computer Vision and Pattern Recognition, pp. 3723–3732 (2018)
31. Smith, V., Chiang, C.K., Sanjabi, M., Talwalkar, A.S.: Federated multi-task learning. In: Advances in Neural Information Processing Systems, pp. 4424–4434 (2017)
32. Tsai, Y.H., Hung, W.C., Schulter, S., Sohn, K., Yang, M.H., Chandraker, M.: Learning to adapt structured output space for semantic segmentation. In: Proceedings of the IEEE Conference on Computer Vision and Pattern Recognition, pp. 7472–7481 (2018)
33. Valentine, T., Lewis, M.B., Hills, P.J.: Face-space: a unifying concept in face recognition research. Quart. J. Exp. Psychol. **69**(10), 1996–2019 (2016)
34. Vaswani, A., et al.: Attention is all you need. In: Advances in Neural Information Processing Systems, pp. 5998–6008 (2017)

35. Vázquez, D., López, A.M., Ponsa, D.: Unsupervised domain adaptation of virtual and real worlds for pedestrian detection. In: Proceedings of the 21st International Conference on Pattern Recognition (ICPR 2012), pp. 3492–3495. IEEE (2012)
36. Wang, X., Guo, R., Kambhamettu, C.: Deeply-learned feature for age estimation. In: 2015 IEEE Winter Conference on Applications of Computer Vision, pp. 534–541. IEEE (2015)
37. Wang, Z., He, K., Fu, Y., Feng, R., Jiang, Y.G., Xue, X.: Multi-task deep neural network for joint face recognition and facial attribute prediction. In: Proceedings of the 2017 ACM on International Conference on Multimedia Retrieval, pp. 365–374. ACM (2017)
38. Zhang, Y., Shen, W., Sun, L., Li, Q.: Position-squeeze and excitation module for facial attribute analysis. In: BMVC (2018)
39. Zhou, B., Khosla, A., Lapedriza, A., Oliva, A., Torralba, A.: Learning deep features for discriminative localization. In: Proceedings of the IEEE Conference on Computer Vision and Pattern Recognition, pp. 2921–2929 (2016)
40. Zhu, J.Y., Park, T., Isola, P., Efros, A.A.: Unpaired image-to-image translation using cycle-consistent adversarial networks. In: Proceedings of the IEEE International Conference on Computer Vision, pp. 2223–2232 (2017)
41. Zhu, Z., Luo, P., Wang, X., Tang, X.: Multi-view perceptron: a deep model for learning face identity and view representations. In: Advances in Neural Information Processing Systems, pp. 217–225 (2014)

Many-Shot from Low-Shot: Learning to Annotate Using Mixed Supervision for Object Detection

Carlo Biffi[1(✉)], Steven McDonagh[1], Philip Torr[3], Aleš Leonardis[1], and Sarah Parisot[1,2]

[1] Huawei Noah's Ark Lab, London, UK
carlo.biffi@gmail.com
[2] Mila Montréal, Montreal, Canada
[3] University of Oxford, Oxford, UK

Abstract. Object detection has witnessed significant progress by relying on large, manually annotated datasets. Annotating such datasets is highly time consuming and expensive, which motivates the development of weakly supervised and few-shot object detection methods. However, these methods largely underperform with respect to their strongly supervised counterpart, as weak training signals *often* result in partial or oversized detections. Towards solving this problem we introduce, for the first time, an online annotation module (OAM) that learns to generate a many-shot set of *reliable* annotations from a larger volume of weakly labelled images. Our OAM can be jointly trained with any fully supervised two-stage object detection method, providing additional training annotations on the fly. This results in a fully end-to-end strategy that only requires a low-shot set of fully annotated images. The integration of the OAM with Fast(er) R-CNN improves their performance by 17% mAP, 9% AP50 on PASCAL VOC 2007 and MS-COCO benchmarks, and significantly outperforms competing methods using mixed supervision.

1 Introduction

Object detection is an essential building block of many computer vision systems [26]. State-of-the-art (SOTA) methods mainly rely on large scale datasets with manually annotated bounding boxes to train fully supervised CNN-based models [3,8,12,16,17]. However, the prohibitive cost and time requirements associated with data annotation reduce the applicability of SOTA detection models in real life scenarios. This has motivated research on object detection strategies with reduced data annotation requirements. Amongst the most popular low data regimes, we distinguish Weakly Supervised Object Detection (WSOD), which aims to train object detectors using only image-level annotations [1,2,18,21,25],

Electronic supplementary material The online version of this chapter (https://doi.org/10.1007/978-3-030-58598-3_3) contains supplementary material, which is available to authorized users.

A. Vedaldi et al. (Eds.): ECCV 2020, LNCS 12353, pp. 35–50, 2020.
https://doi.org/10.1007/978-3-030-58598-3_3

and Few-Shot or Low-Shot Object Detection (FSOD/LSOD), training supervised models with only a handful of training examples on all (LSOD) or only a subset of novel test classes (FSOD) [5,10,24]. FSOD and in particular WSOD have been the focus of a large body of work with innovative strategies obtaining promising performance. Nonetheless, these models typically fall far short of their strongly supervised counterparts. Numerical performance gaps are attributed to the low quality of bounding-box annotations produced, *e.g.* by WSOD methods, that often manifest as partial or oversized boxes. Such results are not reliable enough for use in real-world scenarios and can be observed to cause deterioration of detection performance when used in fully supervised models training. This can be attributed to weak training signals requiring very large and curated datasets (WSOD) or very representative and carefully selected annotated examples (FSOD).

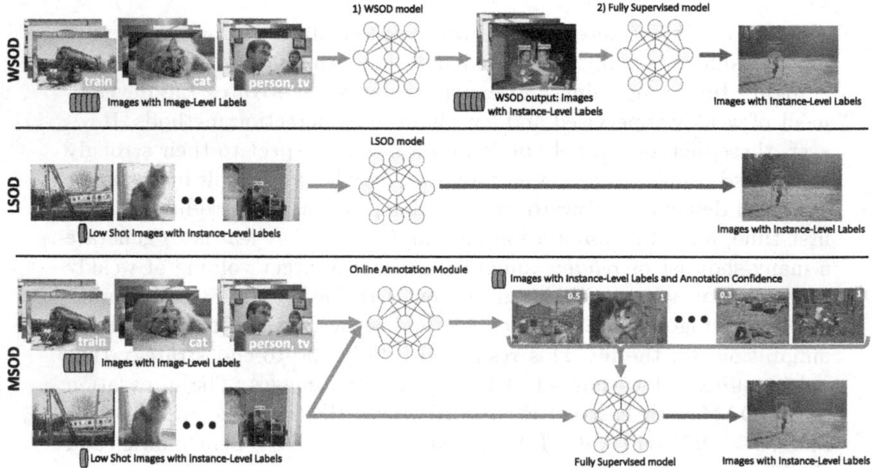

Fig. 1. Weak and low data detection strategies and our proposed mixed supervision-based setting. First row: Weakly Supervised Object Detection (WSOD) models learn to annotate images with image-level annotations which are then used to train fully supervised models. WSOD annotations are often partial or oversized, resulting in poor detector performance. Second row: Few-shot or low-shot object detection (LSOD) trains models on only a handful of training examples. Research mainly focuses on situations where only a specific subset of novel classes have limited training data. Bottom row: Mixed Supervision for Object Detection (MSOD) combines a low shot set of images containing object annotations with a large volume of images comprising only image-level annotations. We train an online annotation module to generate a many-shots set which, at the same time, is used to train a fully supervised model.

To address the aforementioned challenges, we focus on a recent training paradigm relying on Mixed Supervision for Object Detection (MSOD) [7,14]. The distinction between this protocol and the previously introduced weak and low data settings is illustrated in Fig. 1. The objective of MSOD is to exploit and combine the complementary advantages provided by WSOD and LSOD; weak (image-level) supervision affords the construction of large databases with minimal effort, while low-shot supervision provides information rich, fully annotated ground truth examples. The MSOD paradigm has, only very recently, been initially investigated in two related works. Fang *et al.* [7] propose a cascaded architecture yielding performance competitive with fully supervised counterparts yet using a significant fraction of the full training data to achieve comparable performance. Pan *et al.* [14] use low-shot examples to refine bounding box annotations obtained from a pre-trained WSOD model [19], resulting in a method intrinsically linked to the performance and drawbacks of WSOD techniques.

In this work, we approach the MSOD scenario from a different angle. Due to the sparsity of rich training information provided, we expect a MSOD model to output annotations of variable quality, especially for images containing crowded scenes or objects with appearance substantially dissimilar to the training data. In contrast to existing MSOD models we introduce an Online Annotation Module (OAM), trained with mixed supervision, that can be used in conjunction with any two-stage fully supervised object detection method to improve its performance (*e.g.*Fast(er) R-CNN family [8,17]). Our OAM generates, on the fly, additional reliable automated annotations obtained from a larger set of weakly annotated images (containing only image-level class labels). Furthermore, we exploit prediction stability to reason about annotation reliability resulting in associated confidence scores. Generated annotations are used to train, concurrently to the OAM, a fully supervised detector that shares the same encoding features. This produces an intrinsic training curriculum for the standard detector model; only simple images, labelled with high confidence will be presented to the model at the outset. Compared to previous MSOD work, our OAM strategy provides increased robustness against mislabeled crowded and ambiguous training images as only confident MSOD annotations are exploited for fully supervised training. Furthermore, our joint MSOD and fully supervised training provides intrinsic regularisation for both tasks, allowing the learning of higher quality and more discriminative feature extractors.

Experiments show that our strategy allows effective training of standard detection algorithms with only minimal annotation requirements and significantly outperforms WSOD and competitive MSOD approaches on PASCAL VOC 2007 and MS-COCO benchmarks. Additionally, we report competitive performance in comparison to fully supervised alternatives, illustrating the ability of our OAM to annotate a many-shot set of (weakly labelled) images that can be leveraged to improve the fully supervised model performance.

In summary, we propose a new direction using Mixed Supervision for Object Detection (MSOD). Our main contributions are the following:

- We introduce a novel Online Annotation Module (OAM), trained using mixed supervision. This module allows expansion of the low-shot training set of fully annotated images by generating *reliable* annotations from a larger volume of weakly labelled images.

- Training our OAM concurrently with any two-stage object detection model introduces a strategy for object detection performance improvement due to the generated annotation. We report on the benefits of intrinsic regularisation afforded to both tasks when common encoding features are shared.
- The integration of the OAM with Fast(er) R-CNN improves their performance by 17% mAP, 9% AP50 on PASCAL VOC 2007 and MS-COCO benchmarks, and significantly outperforms MSOD approaches.

2 Related Work

Weakly Supervised Object Detection. A large body of recent work, considering WSOD, couples CNN feature extractors with Multiple Instance Learning (MIL) frameworks, thus casting weakly supervised object detection as a multi-label classification problem. Each image is typically represented as a bag of pre-computed proposals (*e.g.*Selective Search [20], Edge Boxes [27], etc.) and the objective is to identify proposals that are most relevant for bag classification [2,18,21,25]. Being framed as a classification task, MIL WSOD models typically focus on proposals that comprise of either the most discriminative object parts or image regions that define the presence of an object category. They therefore struggle to detect full object extent (*e.g.*human faces in contrast to an entire human body) or group multiple object instances of the same object within a single bounding box [14,25]. In order to address this issue, recent work has focused on bounding-box refinement strategies using cascaded refinements of MIL classifications [18,19], using saliency maps [23,25], adopting continuation strategies [21,22] and modelling uncertainty [1]. However, the ill-posed nature of the WSOD problem and insufficient statistics provided by the PASCAL VOC dataset (on which these approaches are usually evaluated) has lead to the development of ad-hoc training strategies and parameter sensitive methods to cope with the weak training signal, which substantially reduce generalisability across datasets. In this paper, we argue that including a handful of labelled samples yields accuracy and stability model improvements at only minimal annotation cost. Usually, all the images annotated by MIL WSOD methods are used, in a second step, to train fully supervised models [18,21,25]. Further previous work has also focused on alternating between the pseudo-labelling of images and, in conjunction, training a fully supervised model [5,9]. In this work, we generate bounding box annotations on the fly from mixed supervision and we concurrently train a fully supervised detector *only* on the images annotated with high confidence.

Few-Shot and Mixed Supervision Object Detection. Few-Shot Object Detection (FSOD) considers a fully supervised training set, and aims to achieve strong performance on a set of novel classes comprising of only K annotated training images per class. To date only a handful of works have focused on FSOD [4,10,11,24]. Such approaches typically adapt few-shot classification techniques to the object detection setting, exploring metric learning [11] or meta-learning [24] strategies. Mixed Supervision for Object Detection (MSOD) enhances a WSOD training set containing only image-level labels with a small set of fully annotated (strong) images (*e.g.*K images per class, analogous to an FSOD scenario) and aims to achieve strong performance on *all* the training classes. Pan *et al.* recently propose BCNet [14], which learns to refine the output of a pre-trained WSOD model using a small set of strong images. The definition of small set explored in their work ranges from 10 shots to 20% of the entire dataset (~1000 images in PASCAL VOC 2007 training set). This approach provides a strong performance increase with respect to WSOD methods, however remains highly dependent on the original WSOD model detections as input. If detections are originally missed by the pre-trained model, the approach cannot recover. Moreover, BCNet requires the training of two independent models which makes the adaption of WSOD parameters, *i.e.*training for new datasets, challenging. In this work, we instead propose a one-stage approach relying on an adaptive pool of annotations, updated dynamically as training progresses. EHSOD [7] and BAOD [15] focus on larger data regimes (*e.g.* 10% to 90%) and aim to reduce the data required to reach fully supervised performance using a cascaded MIL model and a student-teacher setup trained on weak and strong annotations, respectively. In contrast to all outlined methods, we propose instead to learn, and annotate on the fly, only a subset of weak images that can be labelled with high confidence. These additional samples are then used together with strong images to train an object detector and thus improve performance.

3 Method

Let $\mathcal{I}$ be a set of training images annotated with image-level supervision. Under our mixed supervision paradigm, a subset of these images, $\mathcal{S} \subset \mathcal{I}$ with $|\mathcal{S}| \ll |\mathcal{I}|$, is further annotated with bounding box annotations. We refer to the images contained in $\mathcal{S}$ as *strong* training images, while the images in $\mathcal{W} = \mathcal{I} \setminus \mathcal{S}$, that have only image-level annotations, are referred to as *weak* training images. An overview of our proposed method is reported in Fig. 2. Our model comprises two branches with shared encoder backbone, and employs an ROI pooling layer to compute a fixed-length feature representation for each image bounding box proposal. The first branch of our model employs both weak and strong training images to learn an Online Annotation Module (OAM) for weak training images. The OAM generates bounding box annotations, with associated confidence scores, on the fly, for every weak training image. Annotated weak images are added to a third set of images, $\mathcal{P} \subset \mathcal{W}$, if they have been annotated with high confidence, and can be subsequently removed if their annotation confidence drops during training. Images contained in $\mathcal{P}$ are referred to as *semi-strong* training images throughout the paper. The second branch of our model is designed as a standard fully supervised component and trained, in parallel, in an end-to-end manner using strong and semi-strong images. At testing, only the fully supervised model is used for object detection.

Given an input image, we first compute a set of B candidate proposals $\{b_r\}_{r=1}^{B}$, using either an unsupervised method (*e.g.* Selective Search [20] or Edge

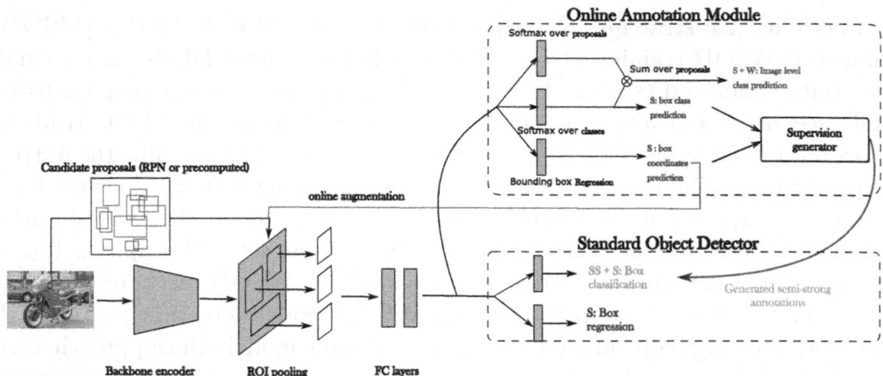

Fig. 2. Architecture of the proposed approach. Our model comprises two branches with a shared encoder. Our Online Annotation Module (OAM) is trained on weakly ($\mathcal{W}$) and strongly ($\mathcal{S}$) annotated images to generate, on the fly, confident annotations for $\mathcal{W}$ images which are added to a pool of semi-strong ($\mathcal{SS}$) images. The model's second branch uses $\mathcal{SS}$ and $\mathcal{S}$ images to train a standard fully supervised detection model.

Boxes [27]) or a Region Proposal Network (RPN) [17], and their associated feature vectors $\{\boldsymbol{\xi}_r\}_{r=1}^{B}$. These feature vectors $\{\boldsymbol{\xi}_r\}_{r=1}^{B}$ are obtained using a standard CNN backbone and ROI Pooling layer and provide a common input to both of our model branches: the OAM and the fully supervised branch.

3.1 Online Annotation Module

Our OAM is designed to jointly exploit weak and strong supervision in an efficient manner. It comprises three main components: 1) a joint detection module exploiting weak and strong labels in a single, common architecture to predict bounding boxes and their classes, 2) an online bounding box augmentation step that generates refined bounding box proposals, 3) a supervision generator, identifying confident annotations to be used as supervision. We next describe all three components in detail.

Joint Detection Module. Similarly to the strategy proposed in [7], we combine a multiple instance learning (MIL) type image-level classification task with a fully supervised joint classification and regression task. Our joint detection module hence comprises three parallel, fully connected layers focusing on three different subtasks: proposal scoring, classification and regression (Fig. 2, online annotation module block). Proposal scoring $\gamma_{C}(c,l)$ and classification $\gamma_{R}(c,l) \in \mathbb{R}^{C \times B}$ are obtained by applying the softmax function to the output of their layers along both dimensions, independently, (classes for γ_{C}, proposals for γ_{R}). After this operation, $\gamma_{C}(c,l)$ represents the probability that the l-th proposal belongs to class c, while $\gamma_{R}(c,l)$ represents the proportional contribution that proposal l provides to the image being classified as class c. Following [2], these layers are trained by exploiting the image-level supervision of both strong and weak images. In particular, a proposal score $\phi_P = \gamma_C \odot \gamma_R$, per class, is obtained by combining them where $\odot$ is a Hadamard product. Then, summing these scores over proposals, $\alpha_c = \sum_{r=1}^{R} \phi_P$, enables the use of a binary cross-entropy loss as image-level loss function:

$$L_{gc}(\alpha_c, y_c) = -\sum_{c=1}^{C}[(1-y_c)log(1-\alpha_c) + y_c log(\alpha_c)] \tag{1}$$

where y_c is the label indicating the presence or absence of class c in an image.

Similar to traditional object detectors, we use strong images to compute bounding box regression and classification via the corresponding fully connected layers. We therefore combine weak and strong supervision by providing direct supervision to proposal-level class prediction γ_C. For regression, each bounding box b is parametrised as a four-tuple (x, y, h, w) that specifies its center coordinate (x, y) and its height and width (h, w). For each proposal classified as foreground in a strong image, this regression branch predicts the offset of these coordinates $t^k = (t_x, t_y, t_h, t_w)$. Hence, for every strong image, the following additional loss is computed on a batch of M proposals:

$$L_p(\gamma, u, t, v) = L_{cls}(\gamma, u) + 1[u \geq 1]L_{reg}(t, v) \tag{2}$$

where:

$$L_{cls} = -\frac{1}{M}\sum_{r=1}^{M}\sum_{c=1}^{C+1} u_{cr}\log(\gamma_{cr}), \quad L_{reg}(t, v) = \sum_{i\in(x,y,h,w)} \text{smooth}_{L1}(t_i - v_i) \tag{3}$$

Parameters γ and u constitute the predicted and target proposal classes respectively, t and v the predicted and target bounding box offsets respectively and smooth_{L1} is a smooth $L1$ loss function [8].

The loss function of the joint detection module is hence $L_{I_s} = L_p + L_{gc}$ on strong images, while the loss function on weak images is $L_{I_w} = L_{gc}$. Enforcing synergy between the two types of supervision regularises the low-shot task thanks to the statistical information provided by weak images. Moreover, due to the instance-level annotations provided by strong images, this also constrains the MIL task and encoder to learn stronger discriminative features between full and partial-extent object proposals.

Online Bounding Box Augmentation Strategy. Learning to update and improve bounding box spatial regions via low-shot regression is highly challenging. When initial inference and ground-truth box overlap is small, large corrections (spatial offsets) are required. Previous work (BCNet [14]) actively elects to exclude such challenging samples, further reducing already highly limited data. We alternatively fully exploit available annotations and push our regression branch output through a second forward pass of our OAM (red arrow in Fig. 2).

More specifically; after the first forward pass, we select the M top scoring proposals, per class, corresponding to image-level ground-truth. M is defined

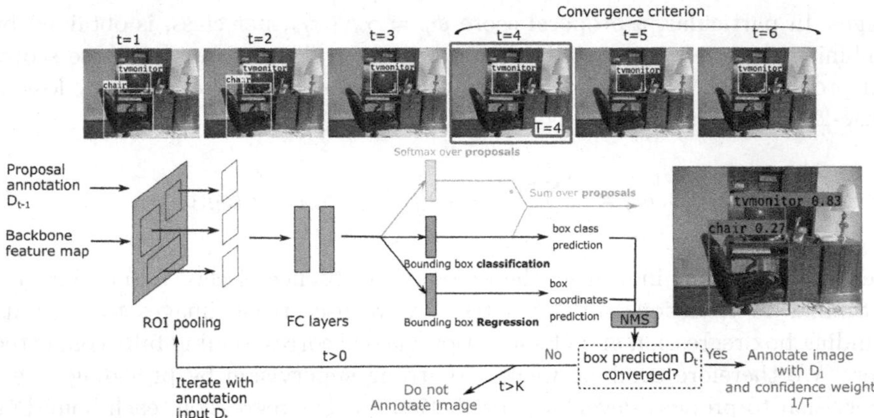

Fig. 3. Proposed online pseudo-supervision generation strategy. At each iteration, a new set of bounding boxes D_t is computed via classification, regression and NMS of the features from previous set D_{t-1}. If bounding box predictions converge at iteration $T < K$, and all proposal classes agree with the image-level label, the weak image is annotated.

as half the size of the proposal batch used to train the strongly supervised component. This accounts for the presence of irrelevant background proposals and allows us to fix this hyperparameter. Once regression branch offsets have been applied, our ROI pooling layer ingests the proposals and yields a new set of bounding box features. Loss functions are evaluated using the updated boxes features and combined with the first pass loss. The overall loss function of our OAM branch is then: $L_{1B} = L^{I}_{I_s} + L^{I}_{I_w} + L^{II}_{I_s} + L^{II}_{I_w}$, where superscripts I and II indicate the first and second pass, respectively. At every iteration, a batch with the same number of weak and strong images is used.

Motivation for our second pass is two-fold. Firstly augmentation is intrinsically provided as new sets of proposal candidates are generated for regression and classification task training. In contrast pre-computed proposals (predominant in WSOD), that lack additional external augmentation strategies, provide only static input, reducing sample variability during training. Secondly, our regression task is regularised as any weak proposals receiving modifications that hinder correct image-level classification are penalised.

Online Pseudo-Supervision Generation. The key objective of our OAM is to generate reliable annotations on a large set of weakly labelled images in order to guide the training of a fully supervised second branch. As the OAM is trained concurrently with the second branch, it is critical to identify and add only reliable annotations to the pool of training images. Our rationale is that only these images should be used to train the final supervised detection network, while images that the joint detection module struggles to annotate with high confidence should not be used for model training, as they may hurt the training process and deteriorate detector performance.

Fig. 4. Examples of semi-strong images. First row: annotated semi-strong images at epoch E, with T iterations required for convergence (see text for details). Second row: examples of semi-strong annotation at pairs of early and late epochs. Magenta color: OAM annotation (class, bounding box score). Yellow: OICR [19] annotation. The results are obtained from a model trained on PASCAL VOC 2007 with 10 shot strong supervision. (Color figure online)

During early stages of the training process, uncertainty regarding both the class of bounding box proposals and the related regression refinement of box coordinates will be high. As training progresses and model predictive quality improves, confidence, accuracy and stability will increase. This results in an increasingly difficult set of images being accurately annotated. We propose to exploit this behaviour by introducing a supervision generator that is able to reliably identify annotated images, creating a set we refer to as *semi-strong* images $\mathcal{P} \subset \mathcal{W}$, that are used to train the fully supervised branch. Intuitively, $\mathcal{P}$ will comprise "easy" images in early stages of training (*e.g.*single instances, uniform colour backgrounds) and sample diversity will progressively increase as the model becomes more accurate (examples of images annotated by our OAM at different training epochs are reported in Fig. 4).

In order to build a set of semi-strong images $\mathcal{P}$, with bounding boxes and associated annotation confidence scores, we propose the following mechanism. Given a weak image I, we obtain a set of N_1 bounding boxes $D_1 = \{c_r, p_r\}_{r=1}^{N_1}$ after Non-Maximum Supression (NMS) is performed on the output of the joint detection module, where c_r and p_r correspond to the class label and coordinates of box r respectively. $D_t = \{c_r, p_r\}_{r=1}^{N_t}$ at every iteration $t > 1$, using D_{t-1} as input candidate proposals. More specifically, the bounding boxes D_{t-1} obtained at the previous iteration are fed again to the RoI Pooling Layer, providing a new set of image features allowing to compute new proposal coordinates. The process iterates until bounding box prediction stabilises and is stopped when $D_t = D_{t-1}$ for three consecutive iterations, i.e. for each bounding box $b_t \in D_t$, there exists a corresponding box $b_{t-1} \in D_{t-1}$ such that b_t and b_{t-1} have intersection-over-union (IoU) ≥ 0.5 and possess matching class predictions (*i.e.*a standard criterion for characterising object equivalence in detection methods).

We assign a global confidence weight $1/T$, per image, where T is defined as the first of three iterations in which $D_t = D_{t-1}$. Pseudo-code for the OAM algorithm is found in Supplementary Materials A.

The set of proposals D_1 obtained at iteration 1 constitute the final bounding box annotations. Each box is weighted (box level confidence) by its average IoU with the best matching box at all subsequent iterations. Boxes absent at a given iteration (IoU < 0.5) are, by definition, down weighted due to being assigned an overlap of 0 at that iteration (Fig. 3 shows an example). Images that do not reach convergence by K iterations, or that fail to find any foreground proposals at iteration t, are considered to be annotated with low confidence and are not added to the semi-strong pool. We set the maximum number of updates $K = 30$, to prevent large sets of iterations and observe that large T (*e.g.* $T > K$) would only occur during early stage training in practice. Finally, the image is only added to the semi-strong pool if the set of obtained annotations contains *all* classes pertaining to the image-level label. We highlight that images requiring large iteration count T for convergence are assigned low confidence scores by design and therefore have limited influence on the training procedure of the second branch. As weak images get annotated by the proposed OAM during training; the semi-strong set expands, while at the same time refining annotations and confidence as the model improves. At a given training step, a weak image that is not successfully annotated, and yet was present in the pool of semi-strong images, will be removed. In this way, the set of semi-strong images has the ability to both expand and contract during training.

3.2 Fully Supervised Branch

Concurrently to OAM training, the obtained strong and semi-strong sets of images are used to train a fully supervised second branch, that comprises both bounding box classification and regression modules on the proposal features ξ_{rf} in a similar fashion to Fast(er) R-CNN [8] style methods. In particular, at every training iteration a batch with the same number of strong and semi-strong images is used. The loss function for this branch is:

$$L_{2B}(p, u, t, v) = L_{cls}(p, u) + L_{reg}(t, v), \tag{4}$$

where p is the ROI class predictions, t is the predicted offset between ROIs and targets, u is the class label and v is the target offset. Only ROIs with foreground labels contribute to the regression loss, L_{reg}. The L_{cls} loss constitutes a weighted cross-entropy for each image:

$$L_{cls}(p, u) = -\frac{1}{T} \sum_i \omega_i p_i log(u_i) \tag{5}$$

where the proposals in each batch, contributing to the loss, are indexed by i, the confidence for GT proposal u_i is denoted ω_i and the image-level annotation confidence score is denoted $\frac{1}{T}$. Strong images are assigned image and

Table 1. Detailed detection performance (%) on VOC07 dataset. In all the setting, the same BCNet data splits were employed [14].

method	backbone	aero	bike	bird	boat	bottle	bus	car	cat	chair	cow	table	dog	horse	moto	person	plant	sheep	sofa	train	tv	mAP(%)
10% images																						
Fast R-CNN	VGG	47.9	62.9	45.5	34.2	23.0	54.6	70.8	65.5	27.2	61.1	39.8	60.6	70.0	63.3	64.2	14.7	52.9	43.0	55.7	49.5	50.3
BAOD	VGG	51.6	50.7	52.6	41.7	36.0	52.9	63.7	69.7	34.4	65.4	22.1	66.1	63.9	53.5	59.8	24.5	60.2	43.3	59.7	46.0	50.9
BCNet	VGG	64.7	**73.1**	55.2	37.0	39.1	**73.3**	74.0	75.4	**35.9**	69.8	56.3	**74.7**	77.6	71.6	66.9	25.4	61.0	**61.4**	**73.8**	**69.3**	61.8
Ours	VGG	**65.6**	**73.1**	**59.0**	**49.4**	**42.5**	72.5	**78.3**	**76.4**	35.4	**72.3**	**57.6**	73.6	**80.0**	**72.5**	**71.1**	**28.3**	**64.6**	55.3	71.4	66.2	**63.3**
EHSOD	ResNet	60.6	65.2	55.0	35.4	32.8	66.1	71.3	75.3	38.4	54.1	26.5	71.7	65.0	67.8	63.0	27.7	52.6	48.6	70.9	57.3	55.3
BCNet	ResNet	**68.3**	72.0	61.2	48.1	40.8	73.3	73.4	77.8	37.0	69.7	58.3	78.2	80.0	67.5	70.5	**27.4**	62.9	**63.6**	**73.4**	63.6	63.4
Ours	ResNet	62.3	**73.2**	**61.8**	**56.2**	**44.3**	**75.4**	**76.7**	**80.5**	**39.5**	**73.7**	**61.7**	**78.8**	**82.8**	**71.5**	**74.3**	27.0	**67.4**	62.7	71.2	**64.4**	**65.3**
10 shots																						
BCNet	VGG	59.7	69.1	44.6	29.4	40.1	69.2	73.2	**72.9**	32.9	58.1	53.3	66.7	71.3	66.0	61.7	**24.6**	53.0	**62.0**	**67.2**	**67.4**	57.1
Ours	VGG	**60.2**	**71.6**	**51.5**	**45.6**	**43.5**	**71.1**	**75.8**	72.2	**33.8**	**62.9**	**54.0**	**70.0**	**72.9**	**67.5**	**67.4**	23.6	**61.5**	59.1	63.6	66.7	**59.7**
BCNet	ResNet	**63.4**	69.4	54.7	39.5	35.9	70.6	71.8	71.8	33.5	64.6	50.0	65.3	72.7	62.5	61.6	**29.2**	54.5	**63.3**	66.7	**69.4**	58.5
Ours	ResNet	61.7	**72.3**	**56.5**	**52.0**	**37.2**	**71.3**	**74.6**	**77.8**	**36.0**	**67.1**	**58.3**	**78.1**	**77.6**	**68.0**	**71.8**	25.5	**63.6**	62.4	**72.7**	61.2	**62.3**

proposal-level weights of 1. In the early stages of the training process, the semi-strong annotations present some localisation inaccuracies, but are nonetheless highly informative to learn foreground vs background proposals. As training progresses, our OAM improves annotation quality with tighter object coverage and these additional high accuracy annotations will more often contain proposals of exactly full object extent. Such annotations reinforce and strengthen a base signal, provided by strong images alone, towards better bounding-box classification. We also explored utilising semi-strong images to improve bounding-box regression, analogously. In practice, however, this produced slightly worse results. We hypothesise that the discrete problem, associated with the bounded classification loss, affords more robustness to (early-stage) imperfect semi-strong annotations and therefore compute bounding box regression on only strong images in our final model. To conclude, collecting the introduced components results in the complete loss function for our model: $L_{tot} = L_{1B} + L_{2B}$. At testing, only this fully supervised model is deployed.

4 Results

4.1 Datasets and Implementation Details

We evaluate the performance of our proposed method on two common detection benchmarks: the PASCAL VOC 2007 [6] and the MS-COCO 14 dataset [13], referred to as VOC07 and COCO14. VOC07 has 5011 training and 4952 testing images across 20 categories. COCO14 has $82k$ training and $5k$ testing images across 80 categories. Following evaluation strategies used in the literature, we evaluate detection accuracy on VOC07 using mean Average Precision (mAP), while we employ the COCO metrics, AP_{50} and $AP_{50:95}$, on the COCO dataset. In the reported experiments, reference to 10% of labelled images dictates that 10% of all images have bounding box annotations while the remaining 90% have image-level labels. This corresponds to 500 images in VOC07, $8.2k$ images in COCO14. With reference to our "N-shot" experimental setup, we define each class to have access to N images possessing bounding box annotations. All the experiments on VOC07 use the same data splits provided by BCNet [14], experiments on COCO14 use random selection.

Table 2. Comparison to SOTA on VOC07 dataset. A VGG backbone is used unless specified. Gray rows correspond to methods learning an RPN (vs methods using pre-computed proposals).

Method type	Method	10-shots/WSOD		10% images	
		AP (%) person class	mAP (%)	AP (%) person class	mAP (%)
fully supervised	Fast R-CNN	58.0	42.1	64.2	50.3
fully supervised	Faster R-CNN	54.3	37.7	55.7	46.7
WSOD	PCL	17.8	43.5	-	-
WSOD	PCL + Fast R-CNN	15.8	44.2	-	-
WSOD	WSOD²	21.9	53.6	-	-
MSOD	BAOD	-	-	59.8	50.9
MSOD	EHSOD (ResNet + FPN)	-	-	63.0	55.3
MSOD	BCNet	61.7	57.1	66.9	61.8
MSOD	Ours	**67.4**	**59.7**	**71.1**	**63.3**
MSOD	Ours + RPN	64.3	54.6	68.9	60.5
fully supervised	Fast-RCNN 100 % images (Ours upper bound)	76.8 (person), 71.6			
fully supervised	Faster-RCNN 100 % images (Ours + RPN upper bound)	75.6 (person), 67.0			

We employ popular network backbones VGG16 and ResNet101 in our experiments to retain consistency with recent approaches. We combine our OAM with Fast R-CNN [8] (using Edge Boxes [27]) and Faster R-CNN using a trainable RPN [17]. Optimisation of all models is performed using SGD with weight decay 0.0001 and momentum 0.9. For experiments concerning the VOC07 dataset, models are trained for 60 epochs. The initial learning rate is 0.001 (first 40 epochs) and reduced to 0.0001 for the final 20 epochs. Analogously for MS COCO experiments; models are trained for 12 epochs, with learning rate 0.001 in the first 9 epochs and then reduced to 0.0001 for the final 3 epochs. Remaining model hyper-parameters follow the values reported in [14]. For data augmentation, we apply the same augmentation strategy as BCNet [14] for fair comparison, *i.e.*we bilinearly resize images to induce a minimum side length $\in \{400, 600, 750\}$ and, for fully supervised training, uniformly crop image regions with a fixed 600×600 window. All experiments are implemented in PyTorch using a single GeForce GTX 1080 GPU.

4.2 Comparisons with State-of-the-Art

Baselines: We evaluate our model with respect to two SOTA WSOD methods, PCL [18] and WSOD² [25], that were evaluated on both VOC07 and COCO14. We further compare to three MSOD approaches: the two level approach of BCNet [14], end-to-end methods BAOD [15] and EHSOD [7]. To the best of our knowledge, these are the only three methods adopting mixed supervision. All three methods were evaluated on VOC07. Results for BCNet, the best performing baseline on VOC07, were not available for the COCO dataset. The approach requires training two models (OICR and BCNet) with two separate sets of parameters that need to be adapted to the new dataset, making it highly challenging and time consuming to provide a fair comparison, hence we were not able to provide it. Similarly, EHSOD was evaluated only on the COCO 2017 database with a much larger set of annotated training images (approx. $12k$), making results not directly comparable to our experiments and different from

Table 3. Comparison with the SOTA on MS-COCO14 with 10-shot training examples (VGG backbone). Gray rows correspond to methods learning an RPN (vs methods using precomputed proposals).

Method type	Method	AP@.50	AP@[.50,.95]
fully supervised	Fast R-CNN - 10 shots	22.1	10.0
fully supervised	Faster R-CNN - 10 shots	16.1	6.7
WSOD	PCL	19.4	8.5
WSOD	PCL+ Fast R-CNN	19.6	9.2
WSOD	WSOD^2	22.7	10.8
MSOD	Ours - 10 shots	**31.2**	**14.9**
MSOD	Ours + RPN - 10 shots	**24.9**	**10.2**
fully supervised	Fast R-CNN - 100% data	49.9	29.0
fully supervised	Faster R-CNN - 100% data	42.1	20.5

the low-shot setting studied in this work. Finally, we compare our results with respect to Fast R-CNN and Faster R-CNN trained with full supervision (our upper bounds) and low-shot supervision (*i.e.* 10% and 10-shot training data), using the same augmentation strategy as all previous models.

PASCAL VOC 2007: We report detailed per-class results, compared to competing MSOD approaches in Table 1 using 10% annotated training images, and 10 shots. We consistently outperform all competing methods in terms of mAP, with an improvement of up to 4% with respect to BCNet in the 10 shot scenario (ResNet), and 10% with respect to EHSOD in the 10% images scenario. We further highlight that BCNet constitutes a two-level WSOD dependent method. The influence of the chosen WSOD component is clearly visible; object classes where their method excels, and surpasses our per-class performance, are the same classes for which their adopted WSOD component (OICR) provides best initial bounding box estimations [19]. In Table 2, we provide more comparisons in the 10 shots and 10% images scenarios using precomputed proposals (white rows) and an RPN [17] (grey rows). We highlight that we use an off-the-shelf RPN without parameter optimisation, and expect performance to be worse, and not directly comparable to strategies relying on pre-computed proposals. We further compare with top performing WSOD methods and Fast(er)-RCNN approaches and highlight our performance on the "person" class, often reported as one of the most challenging classes for WSOD methods due to the large intra-class variability in terms of appearance [14,25]. We significantly outperform all SOTA methods, and substantially improve with respect to WSOD methods, in particular for the person class, with only minor additional labelling cost. Comparing to Fast(er)-RCNN methods, we highlight that our OAM improves upon models trained on 10% data and 10 shots by a large margin (13% and 17% respectively), reaching performance close to the fully supervised upper bound.

MS-COCO14: We provide further comparison to additional benchmark datasets in order to highlight model generalisability. We note that contemporary WSOD methods mainly focus on detection datasets of modest size such

Table 4. Ablative analysis of our method on VOC07 for the 10 shot scenario. SE: shared encoder, OAM: second branch training also on OAM generated semi-strong images, BBA: bounding box augmentation strategy. 1B: first branch output, 2B: second branch output.

10 shots					AP (%)																				
SE	BBA	OAM	1B	2B	aero	bike	bird	boat	bottle	bus	car	cat	chair	cow	table	dog	horse	moto	person	plant	sheep	sofa	train	tv	mAP(%)
	✓		✓		42.0	57.1	40.2	34.2	30.3	62.6	69.0	62.5	23.2	63.8	33.0	58.5	72.2	63.3	62.9	20.8	54.9	44.2	54.3	55.2	50.2
				✓	30.9	53.2	35.8	27.8	19.9	51.6	65.8	54.7	19.3	48.3	27.8	46.3	57.7	54.3	58.0	14.9	49.1	37.5	43.8	44.7	42.1
✓	✓		✓		44.3	60.2	40.4	37.8	28.1	67.0	72.8	64.1	24.2	64.6	40.9	60.5	70.5	61.6	63.5	16.1	55.0	46.2	57.5	58.0	51.7
✓	✓			✓	47.3	62.1	42.4	35.2	28.2	67.0	72.8	65.1	21.7	65.3	43.4	61.4	70.6	63.5	63.0	16.5	57.6	45.8	58.7	54.7	52.1
✓		✓	✓		50.3	67.3	49.8	44.1	35.9	64.3	72.7	70.3	32.6	57.7	44.5	66.3	65.6	68.3	62.8	25.2	60.0	48.8	62.6	64.5	55.7
✓		✓		✓	61.4	71.0	48.5	42.9	37.8	69.8	75.6	72.8	34.0	63.2	47.6	71.9	71.1	71.1	64.6	25.7	63.4	55.6	61.9	65.8	58.8
✓	✓	✓	✓		57.9	71.4	48.2	42.7	38.0	71.4	75.5	75.5	34.0	67.1	54.0	71.4	74.3	69.4	65.7	23.7	61.6	56.1	61.0	65.0	59.2
✓	✓	✓		✓	60.2	71.6	51.5	45.6	43.5	71.1	75.8	72.2	33.8	62.9	54.0	70.0	72.9	67.5	67.4	23.6	61.5	59.1	63.6	66.7	59.7

as VOC07. COCO14 is significantly larger, and constitutes a more challenging dataset due to both the increased size and variability expressed in image content. Table 3 reports comparisons between our method (precomputed and RPN proposals) and WSOD approaches PCL and WSOD^2 on COCO14 using 10 shots labelled images. As we compare solely to WSOD methods, we limit our experiments to the 10 shots setting, as 10% annotated examples provide a very significant advantage compared to WSOD methods. We additionally provide comparison to Fast(er) R-CNN methods trained on 10 shots as well as their fully supervised equivalent on 100% images. We highlight that our method maintains robust performance and significantly outperforms the WSOD methods and 10 shots Fast(er)-RCNN models (9%). This provides evidence in support of our claim that the strategy of providing mixed supervision significantly improves generalisation ability in settings that entail more difficult tasks with higher variability.

4.3 Ablation Studies

We conduct experiments to understand the different contributions and assignment of credit for our OAM components using the VOC07 dataset and a VGG backbone. Table 4 shows ablative results for the 10 shots scenario while additional results for the 10% images scenario are reported in supplementary materials. Studied components are: *SE*: shared encoder (*i.e.*no SE entails independent branch training); *OAM:* fully supervised branch is also trained on semi-strong images generated by the OAM; *BBA:* online bounding box augmentation strategy. For each configuration, we report mAP with respect to the output of the OAM (first branch; 1B) as well as the output of the fully supervised branch (second branch; 2B). We experimentally verify the importance of each component; performance consistently improves as new components are integrated. We note that the shared encoder strongly improves the fully supervised branch, while the OAM, and communication between branches, affords mutual branch improvement. Both performance gains can be attributed to the more discriminative full *vs.*partial object proposal features learned by the shared encoder.

5 Conclusion

We have introduced a novel online annotation module (OAM), trained using mixed supervision, that learns to generate annotations on the fly and thus affords concurrent training for fully supervised object detection. The OAM can be combined with any two-stage object detector and provides an intrinsic curriculum to improve the training procedure. Extensive experiments on two popular benchmarks show SOTA performance in the mixed supervision scenario, and significant improvement of two-stage detection methods in low-shot settings. Moreover, our method has the potential to increase performance on rare, long tail classes that typically only possess a handful of annotated examples.

References

1. Arun, A., Jawahar, C., Kumar, M.P.: Dissimilarity coefficient based weakly supervised object detection. In: Proceedings of the IEEE Conference on Computer Vision and Pattern Recognition, pp. 9432–9441 (2019)
2. Bilen, H., Vedaldi, A.: Weakly supervised deep detection networks. In: Proceedings of the IEEE Conference on Computer Vision and Pattern Recognition, pp. 2846–2854 (2016)
3. Cai, Z., Vasconcelos, N.: Cascade R-CNN: high quality object detection and instance segmentation. IEEE Trans. Pattern Anal. Mach. Intell. (2019). https://doi.org/10.1109/TPAMI.2019.2956516
4. Chen, H., Wang, Y., Wang, G., Qiao, Y.: LSTD: a low-shot transfer detector for object detection. In: Thirty-Second AAAI Conference on Artificial Intelligence (2018)
5. Dong, X., Zheng, L., Ma, F., Yang, Y., Meng, D.: Few-example object detection with model communication. IEEE Trans. Pattern Anal. Mach. Intell. **41**(7), 1641–1654 (2018)
6. Everingham, M., Eslami, S.A., Van Gool, L., Williams, C.K., Winn, J., Zisserman, A.: The pascal visual object classes challenge: a retrospective. Int. J. Comput. Vis. **111**(1), 98–136 (2015)
7. Fang, L., Xu, H., Liu, Z., Parisot, S., Li, Z.: EHSOD: CAM-Guided End-to-End Hybrid-Supervised Object Detection with cascade refinement. In: Proceedings of the 29th International Joint Conference on Artificial Intelligence. AAAI Press (2020)
8. Girshick, R.: Fast R-CNN. In: Proceedings of the IEEE International Conference on Computer Vision, pp. 1440–1448 (2015)
9. Jie, Z., Wei, Y., Jin, X., Feng, J., Liu, W.: Deep self-taught learning for weakly supervised object localization. In: Proceedings of the IEEE Conference on Computer Vision and Pattern Recognition, pp. 1377–1385 (2017)
10. Kang, B., Liu, Z., Wang, X., Yu, F., Feng, J., Darrell, T.: Few-shot object detection via feature reweighting. In: Proceedings of the IEEE International Conference on Computer Vision, pp. 8420–8429 (2019)
11. Karlinsky, L., et al.: Repmet: representative-based metric learning for classification and few-shot object detection. In: Proceedings of the IEEE Conference on Computer Vision and Pattern Recognition, pp. 5197–5206 (2019)

12. Lin, T.Y., Goyal, P., Girshick, R., He, K., Dollár, P.: Focal loss for dense object detection. In: Proceedings of the IEEE International Conference on Computer Vision, pp. 2980–2988 (2017)
13. Lin, T.Y., et al.: Microsoft COCO: common objects in context. In: Fleet, D., Pajdla, T., Schiele, B., Tuytelaars, T. (eds.) ECCV 2014. LNCS, vol. 8693, pp. 740–755. Springer, Cham (2014). https://doi.org/10.1007/978-3-319-10602-1_48
14. Pan, T., Wang, B., Ding, G., Han, J., Yong, J.: Low shot box correction for weakly supervised object detection. In: Proceedings of the 28th International Joint Conference on Artificial Intelligence, pp. 890–896. AAAI Press (2019)
15. Pardo, A., Xu, M., Thabet, A., Arbelaez, P., Ghanem, B.: Baod: budget-aware object detection. arXiv preprint arXiv:1904.05443 (2019)
16. Redmon, J., Divvala, S., Girshick, R., Farhadi, A.: You only look once: unified, real-time object detection. In: Proceedings of the IEEE Conference on Computer Vision and Pattern Recognition, pp. 779–788 (2016)
17. Ren, S., He, K., Girshick, R., Sun, J.: Faster R-CNN: towards real-time object detection with region proposal networks. In: Advances in Neural Information Processing Systems, pp. 91–99 (2015)
18. Tang, P., et al.: Pcl: proposal cluster learning for weakly supervised object detection. IEEE Trans. Pattern Anal. Mach. Intell. **42**(1), 176–191 (2018)
19. Tang, P., Wang, X., Bai, X., Liu, W.: Multiple instance detection network with online instance classifier refinement. In: Proceedings of the IEEE Conference on Computer Vision and Pattern Recognition, pp. 2843–2851 (2017)
20. Uijlings, J.R., Van De Sande, K.E., Gevers, T., Smeulders, A.W.: Selective search for object recognition. Int. J. Comput. Vis. **104**(2), 154–171 (2013)
21. Wan, F., Liu, C., Ke, W., Ji, X., Jiao, J., Ye, Q.: C-mil: continuation multiple instance learning for weakly supervised object detection. In: Proceedings of the IEEE Conference on Computer Vision and Pattern Recognition, pp. 2199–2208 (2019)
22. Wan, F., Wei, P., Jiao, J., Han, Z., Ye, Q.: Min-entropy latent model for weakly supervised object detection. IEEE Trans. Pattern Anal. Mach. Intell. (2019)
23. Wei, Y., et al.: Ts2c: tight box mining with surrounding segmentation context for weakly supervised object detection. In: Proceedings of the European Conference on Computer Vision (ECCV), pp. 434–450 (2018)
24. Yan, X., Chen, Z., Xu, A., Wang, X., Liang, X., Lin, L.: Meta R-CNN: towards general solver for instance-level low-shot learning. In: Proceedings of the IEEE International Conference on Computer Vision, pp. 9577–9586 (2019)
25. Zeng, Z., Liu, B., Fu, J., Chao, H., Zhang, L.: Wsod2: learning bottom-up and top-down objectness distillation for weakly-supervised object detection. In: Proceedings of the IEEE International Conference on Computer Vision, pp. 8292–8300 (2019)
26. Zhao, Z.Q., Zheng, P., Xu, S.T., Wu, X.: Object detection with deep learning: a review. IEEE Trans. Neural Netw. Learn. Syst. **30**(11), 3212–3232 (2019)
27. Zitnick, C.L., Dollár, P.: Edge boxes: locating object proposals from edges. In: Fleet, D., Pajdla, T., Schiele, B., Tuytelaars, T. (eds.) ECCV 2014. LNCS, vol. 8693, pp. 391–405. Springer, Cham (2014). https://doi.org/10.1007/978-3-319-10602-1_26

Curriculum DeepSDF

Yueqi Duan[1(✉)], Haidong Zhu[2], He Wang[1], Li Yi[3], Ram Nevatia[2], and Leonidas J. Guibas[1]

[1] Stanford University, Stanford, USA
duanyq19@stanford.edu
[2] University of Southern California, Los Angeles, USA
[3] Google Research, Menlo Park, USA

Abstract. When learning to sketch, beginners start with simple and flexible shapes, and then gradually strive for more complex and accurate ones in the subsequent training sessions. In this paper, we design a "shape curriculum" for learning continuous Signed Distance Function (SDF) on shapes, namely *Curriculum DeepSDF*. Inspired by how humans learn, Curriculum DeepSDF organizes the learning task in ascending order of difficulty according to the following two criteria: *surface accuracy* and *sample difficulty*. The former considers stringency in supervising with ground truth, while the latter regards the weights of hard training samples near complex geometry and fine structure. More specifically, Curriculum DeepSDF learns to reconstruct coarse shapes at first, and then gradually increases the accuracy and focuses more on complex local details. Experimental results show that a carefully-designed curriculum leads to significantly better shape reconstructions with the same training data, training epochs and network architecture as DeepSDF. We believe that the application of shape curricula can benefit the training process of a wide variety of 3D shape representation learning methods.

1 Introduction

In recent years, 3D shape representation learning has aroused much attention [16,26,30,31,33]. Compared with images indexed by regular 2D grids, there has not been a single standard representation for 3D shapes in the literature. Existing 3D shape representations can be cast into several categories including: point-based [1,10,31,33,34,49], voxel-based [7,26,32,51], mesh-based [16,17,42,48], and multi-view [32,43,45].

More recently, implicit function representations have gained an increasing amount of interest due to their high fidelity and efficiency. An implicit function depicts a shape through assigning a gauge value to each point in the object space [6,27,28,30]. Typically, a negative, a positive or a zero gauge value represents that the corresponding point lies inside, outside or on the surface of the 3D shape. Hence, the shape is implicitly encoded by the iso-surface (e.g., zero-level-set) of the function, which can then be rendered by Marching Cubes [24] or

Y. Duan and H. Zhu–Equal contribution.

A. Vedaldi et al. (Eds.): ECCV 2020, LNCS 12353, pp. 51–67, 2020.
https://doi.org/10.1007/978-3-030-58598-3_4

Fig. 1. 3D reconstruction results of shapes with complex local details. From top to bottom: ground truth, DeepSDF [30], and Curriculum DeepSDF. We observe that the network benefits from the designed shape curriculum so as to better reconstruct local details. It is worth noting that the training data, training epochs and network architecture are the same for both methods.

similar methods. Implicit functions can also be considered as a shape-conditioned binary classifier whose decision boundary is the surface of the 3D shape. As each shape is represented by a continuous field, it can be evaluated at arbitrary resolution, irrespective of the resolution of the training data and limitations in the memory footprint.

One of the main challenges in implicit function learning lies in accurate reconstruction of shape surfaces, especially around complex or fine structure. Figure 1 shows some 3D shape reconstruction results where we can observe that DeepSDF [30] fails to precisely reconstruct complex local details. Note that the implicit function is less smooth in these areas and hence difficult for the network to parameterize precisely. Furthermore, as the magnitudes of SDF values inside small parts are usually close to zero, a tiny mistake may lead to a wrong sign, resulting in inaccurate surface reconstruction.

Inspired by the works on curriculum learning [3,11], we aim to address this problem in learning SDF by *starting small*: starting from easier geometry and gradually increasing the difficulty of learning. In this paper, we propose a Curriculum DeepSDF method for shape representation learning. We design a shape curriculum where we first teach the network using coarse shapes, and gradually move on to more complex geometry and fine structure once the network becomes more experienced. In particular, our shape curriculum is designed according to two criteria: *surface accuracy* and *sample difficulty*. We consider these two criteria both important and complementary to each other for shape representation learning: *surface accuracy* cares about the stringency in supervising with training loss, while *sample difficulty* focuses on the weights of hard training samples containing complex geometry.

Surface Accuracy. We design a tolerance parameter ε that allows small errors in estimating the surfaces. Starting with a relatively large ε, the network aims for a smooth approximation, focusing on the global structure of the target shape and ignoring hard local details. Then, we gradually decrease ε to expose more shape details until $\varepsilon = 0$. We also use a shallow network to reconstruct coarse shapes at the beginning and then progressively add more layers to learn more accurate details.

Sample Difficulty. Signs greatly matter in implicit function learning. The points with incorrect sign estimations lead to significant errors in shape reconstruction, suggesting that we treat these as hard samples during training. We gradually increase the weights of hard and semi-hard[1] training samples to make the network more and more focused on difficult local details.

One advantage of curriculum shape representation learning is that, it provides a training path for the network to start from coarse shapes and finally reach fine-grained geometries. At the beginning, it is substantially more stable for the network to reconstruct coarse surfaces with the complex details omitted. Then, we continuously ask for more accurate shapes which are relatively simple tasks, benefiting from the previous reconstruction results. Lastly, we focus on hard samples to obtain complete reconstruction with precise shape details. This training process can help avoid poor local minima as compared with learning to reconstruct the precise complex shapes directly. Figure 1 shows that Curriculum DeepSDF obtains better reconstruction accuracy than DeepSDF. Experimental results illustrate the effectiveness of the designed shape curriculum. Code will be available at https://github.com/haidongz-usc/Curriculum-DeepSDF.

In summary, the key contributions of this work are:

1) We design a shape curriculum for shape representation learning, starting from coarse shapes to complex details. The curriculum includes two aspects of *surface accuracy* and *sample difficulty*.
2) For surface accuracy, we introduce a tolerance parameter ε in the training objective to control the smoothness of the learned surfaces. We also progressively grow the network according to different training stages.
3) For sample difficulty, we define hard, semi-hard and easy training samples for SDF learning based on sign estimations. We re-weight the samples to make the network gradually focus more on hard local details.

2 Related Work

Implicit Function. Different from point-based, voxel-based, mesh-based and multi-view methods which explicitly represent shape surfaces, implicit functions aim to learn a continuous field and represent the shape with the iso-surface. Conventional implicit function based methods include [4,29,41,46,47]. For example, Carr *et al.* [4] used polyharmonic Radial Basis Functions (RBFs) to implicitly model the surfaces from point clouds. Shen *et al.* [41] created

[1] Here, semi-hard samples are with the correct sign estimations but close to the boundary. In practice, we also decrease the weights of easy samples to avoid overshooting.

implicit surfaces by moving least squares. In recent years, several deep learning based methods have been proposed to capture more complex topologies [6,12,13,15,22,23,27,28,30,36,52]. For example, Park *et al.* [30] proposed DeepSDF by learning an implicit field where the magnitude represents the distance to the surface and the sign shows whether the point lies inside or outside of the shape. Mescheder *et al.* [27] presented Occupancy Networks by approximating the 3D continuous occupancy function of the shape, which indicates the occupancy probability of each point. Chen and Zhang [6] proposed IM-NET by only encoding the signs of SDF, which can be used for representation learning (IM-AE) and shape generation (IM-GAN). Saito *et al.* [36] and Liu *et al.* [23] learned implicit surfaces of 3D shapes from 2D images. These methods show promising results in 3D shape representation. However, the challenges still remains to reconstruct the local details accurately. Instead of proposing new implicit functions, our approach studies how to design a curriculum of shapes for more effective model training.

Curriculum Learning. The idea of curriculum learning can be at least traced back to [11]. Inspired by the learning system of humans, Elman [11] demonstrated the importance of starting small in neural network training. Sanger [37] extended the idea to robotics by gradually increasing the difficulty of the task. Bengio *et al.* [3] further formalized this training strategy and explored curriculum learning in various cases including vision and language tasks. They introduced one formulation of curriculum learning by using a family of functions $L_\mu(\theta)$, where L_0 is the highly smoothed version and L_1 is the real objective. One could start with L_0 and gradually increase μ to 1, keeping θ at a local minimum of $L_\mu(\theta)$. They also explained the advantage of curriculum learning as a continuation method [2], which could benefit the optimization of a non-convex training criterion to find better local minima. Graves *et al.* [14] designed an automatic curriculum learning method by automatically selecting the training path to address the sensitivity of progression mode. Recently, curriculum learning has been successfully applied to varying tasks [8,18–21,38,40,50]. For example, deep metric learning methods learn hierarchical mappings by gradually selecting hard training samples [9,25,44]. FaceNet [38] proposed an online negative sample mining strategy for face recognition, which was improved by DE-DSP [8] to learn a discriminative sampling policy. Progressive growing of GANs [21,40] learned to sequentially generate images from low-resolution to high-resolution, and also grew both generator and discriminator symmetrically. Although curriculum learning has improved the performance of many tasks, the problem of how to design a curriculum for 3D shape representation learning still remains. Unlike 2D images where the pixels are regularly arranged, 3D shapes usually have irregular structures, which makes the effective curriculum design more challenging.

3 Proposed Approach

Our shape curriculum is designed based on DeepSDF [30], which is a popular implicit function based 3D shape representation learning method. In this section, we first review DeepSDF and then describe the proposed Curriculum DeepSDF approach. Finally, we introduce the implementation details.

3.1 Review of DeepSDF [30]

DeepSDF is trained on a set of N shapes $\{X_i\}$, where K points $\{x_j\}$ are sampled around each shape X_i with the corresponding SDF values $\{s_j\}$ precomputed. This results in K (point, SDF value) pairs:

$$X_i := \{(x_j, s_j) : s_j = SDF^i(x_j)\}, \tag{1}$$

A deep neural network $f_\theta(z_i, x)$ is trained to approximate SDF values of points x, with an input latent code z_i representing the target shape.

The loss function given z_i, x_j and s_j is defined by the L_1-norm between the estimated and ground truth SDF values:

$$L(f_\theta(z_i, x_j), s_j) = |\text{clamp}_\delta(f_\theta(z_i, x_j)) - \text{clamp}_\delta(s_j)|, \tag{2}$$

where $\text{clamp}_\delta(s) := \min(\delta, \max(-\delta, s))$ uses a parameter δ to clamp an input value s. For simplicity, we use $\bar{s}$ to represent a clamping function with $\delta = 0.1$ in the rest of the paper.

DeepSDF also designs an auto-decoder structure to directly pair a latent code z_i with a target shape X_i without an encoder. Please refer to [30] for more details. At training time, z_i is randomly initialized from $\mathcal{N}(0, 0.01^2)$ and optimized along with the parameters θ of the network through back-propagation:

$$\arg\min_{\theta, z_i} \sum_{i=1}^{N} \left(\sum_{j=1}^{K} L(f_\theta(z_i, x_j), s_j) + \frac{1}{\sigma^2} ||z_i||_2^2 \right), \tag{3}$$

where $\sigma = 10^{-2}$ is the regularization parameter.

At inference time, an optimal z can be estimated with the network fixed:

$$\hat{z} = \arg\min_{z} \sum_{j=1}^{K} L(f_\theta(z, x_j), s_j) + \frac{1}{\sigma^2} ||z||_2^2. \tag{4}$$

3.2 Curriculum DeepSDF

Different from DeepSDF which trains the network with a fixed objective all the time, Curriculum SDF starts from learning smooth shape approximations and then gradually strives for more local details. We carefully design the curriculum from the following two aspects: *surface accuracy* and *sample difficulty*.

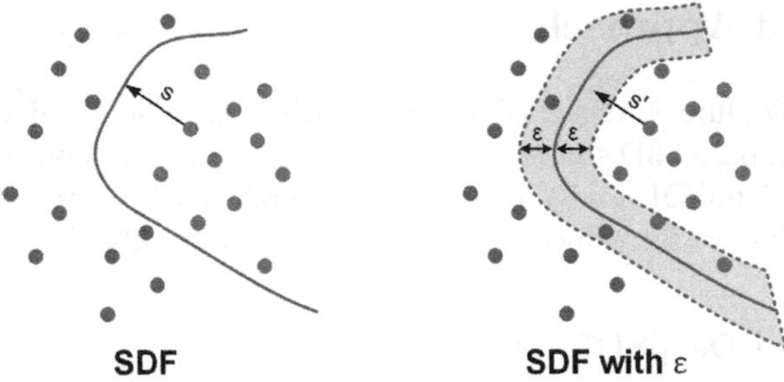

Fig. 2. The comparison between original SDF and SDF with the tolerance parameter ε. With the tolerance parameter ε, all the surfaces inside the tolerance zone are considered correct. The training of Curriculum DeepSDF starts with a relative large ε and then gradually reduces it until $\varepsilon = 0$.

Surface Accuracy. A smoothed approximation for a target shape could capture the global shape structure without focusing too much on local details, and thus is a good starting point for the network to learn. With a changing smoothness level at different training stages, more and more local details can be exposed to improve the network. Such smoothed approximations could be generated by traditional geometry processing algorithms. However, the generation process is time-consuming, and it is also not clear whether such fixed algorithmic routines could meet the needs of network training. In this paper, we address the problem from another view by introducing surface error tolerance ε which represents the upper bound of the allowed errors in the predicted SDF values. We observe that starting with relatively high surface error tolerance, the network tends to omit complex details and aims for a smooth shape approximation. Then, we gradually reduce the tolerance to expose more details.

More specifically, we allow small mistakes for the SDF estimation within the range of $[-\varepsilon, \varepsilon]$ for Curriculum DeepSDF. In other words, all the estimated SDF values whose errors are smaller than ε are considered correct without any punishment, and we can control the difficulty of the task by changing ε. Figure 2 illustrates the physical meaning of the tolerance parameter ε. Compared with DeepSDF which aims to reconstruct the exact surface of the shape, Curriculum DeepSDF provides a tolerance zone with the thickness of 2ε, and the objective becomes to reconstruct any surface in the zone. At the beginning of network training, we set a relatively large ε which allows the network to learn general and smooth surfaces in a wide tolerance zone. Then, we gradually decrease ε to expose more details and finally set $\varepsilon = 0$ to predict the exact surface.

We can formulate the objective function with ε as follows:

$$L_\varepsilon(f_\theta(z_i, x_j), s_j) = \max\{|\bar{f}_\theta(z_i, x_j) - \bar{s}_j| - \varepsilon, 0\}, \tag{5}$$

where (5) will degenerate to (2) if $\varepsilon = 0$.

Unlike most recent curriculum learning methods that rank training samples by difficulty [18,50], our designed curriculum on shape accuracy directly modifies the training loss. It follows the formulation in [3] and also has a clear physical

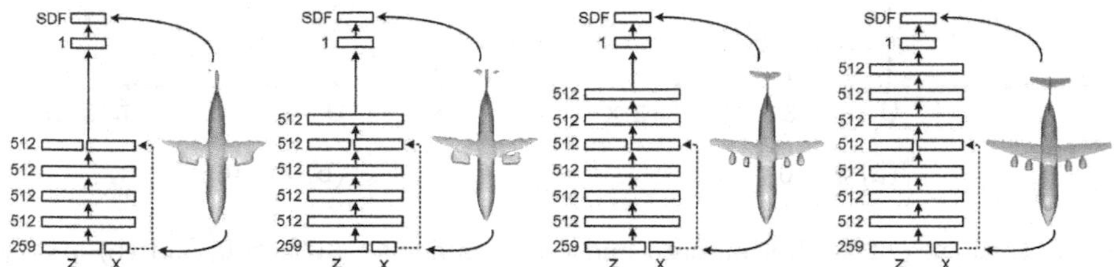

Fig. 3. The network architecture of Curriculum DeepSDF. We apply the same final network architecture with DeepSDF for fair comparisons, which contains 8 fully connected layers followed by hyperbolic tangent non-linear activation to obtain SDF value. The input is the concatenation of latent vector z and 3D point x, which is also concatenated to the output of the fourth layer. When ε decreases during training, we add one more layer to learn more precise shape surface.

meaning for the task of SDF estimation. It is also relevant to label smoothing methods, where our curriculum has clear geometric meanings by gradually learning more precise shapes. We summarize the two advantages of the tolerance parameter based shape curriculum as follows:

1) We only need to change the hyperparameter ε to control the surface accuracy, instead of manually creating series of smooth shapes. The network automatically finds the surface that is easy to learn in the tolerance zone.
2) For any ε, the ground truth surface of the original shape is always an optimal solution of the objective, which has good optimization consistency.

In addition to controlling the surface accuracy by the tolerance parameter, we also use a shallow network to learn coarse shapes with a large ε, and gradually add more layers to improve the surface accuracy when ε decreases. This idea is mainly inspired by [21]. Figure 3 shows the network architecture of the proposed Curriculum DeepSDF, where we employ the same network as DeepSDF for fair comparisons. After adding a new layer with random initialization to the network, the well-trained lower layers may suffer from sudden shocks if we directly train the new network in an end-to-end manner. Inspired by [21], we treat the new layer as a residual block with a weight of α, where the original link has a weight of $1 - \alpha$. We linearly increase α from 0 to 1, so that the new layer can be faded in the original network smoothly.

Sample Difficulty. In DeepSDF, the sampled points $\{x_j\}$ in X_i all share the same weights in training, which presumes that every point is equally important. However, this assumption may result in the following two problems for reconstructing complex local details:

1) Points depicting local details are usually undersampled, and they could be ignored by the network during training due to their small population. We take the second lamp in Fig. 1 as an example. The number of sampled points around the lamp rope is nearly 1/100 of all the sampled points, which is too small to affect the network training.

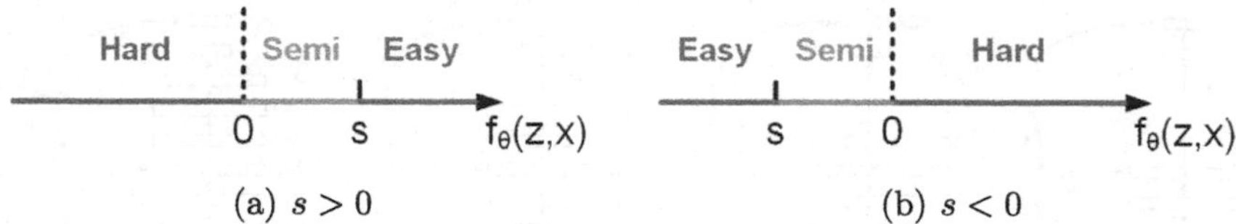

(a) $s > 0$ (b) $s < 0$

Fig. 4. Examples of hard, semi-hard and easy samples for (a) $s > 0$, and (b) $s < 0$. In the figure, s is the ground truth SDF, and we define the difficulty of each sample according to its estimation $f_\theta(z, x)$.

2) In these areas, the magnitudes of SDF values are small as the points are close to surfaces (e.g. points inside the lamp rope). Without careful emphasis, the network could easily predict the wrong signs. Followed by a surface reconstruction method like Marching Cubes, the wrong sign estimations will further lead to inaccurate surface reconstructions.

To address these issues, we weight the sampled points differently during training. An intuitive idea is to locate all the complex local parts at first, and then weight or sort the training samples according to some difficulty measurement [3,18,50]. However, it is difficult to detect complex regions and rank the difficulty of points exactly. In this paper, we propose an adaptive difficulty measurement based upon the SDF estimation of each sample and re-weight the samples to gradually emphasize more on hard and semi-hard samples on the fly.

Most deep embedding learning methods judge the difficulty of samples according to the loss function [8,38]. However, the L_1-norm loss can be very small for the points with wrong sign estimations. As signs play an important role in implicit representations, we directly define the hard and semi-hard samples based on their sign estimations. More specifically, we consider the points with wrong sign estimations as hard samples, with the estimated SDF values between zero and ground truth values as semi-hard samples, and the others as easy samples. Figure 4 shows the examples. For the semi-hard samples, although currently they obtain correct sign estimations, they are still at high risk of becoming wrong as their predictions are closer to the boundary than the ground truth positions.

To increase the weights of both hard and semi-hard samples, and also decrease the weights of easy samples, we formulate the objective function as below:

$$L_{\varepsilon,\lambda}(f_\theta(z_i, x_j), s_j) = \left(1 + \lambda sgn(\bar{s}_j) sgn(\bar{s}_j - \bar{f}_\theta(z_i, x_j))\right) L_\varepsilon(f_\theta(z_i, x_j), s_j), \quad (6)$$

where $0 \leq \lambda < 1$ is a hyperparameter controlling the importance of the hard and semi-hard samples, $sgn(v) = 1$ if $v \geq 0$ and -1 otherwise.

The physical meaning of (6) is that we increase the weights of hard and semi-hard samples to $1 + \lambda$, and also decrease the weights of easy samples to $1 - \lambda$. Although we treat hard and semi-hard samples similarly, their properties are different due to the varying physical meanings as we will demonstrate in the experiments. Our hard sample mining strategy always targets at the weakness of the current network rather than using the predefined weights. Still, (6) will degenerate to (5) if we set $\lambda = 0$. Another understanding of (6) is that $sgn(\bar{s}_j)$

Table 1. The training details of our method. *Layer* shows the number of fully connected layers. *Residual* represents whether we use a residual block to add layers smoothly.

Epoch	0–200	200–400	400–600	600–800	800–1000	1000–1200	1200–2000
Layer	5	6	6	7	7	8	8
Residual	×	✓	×	✓	×	✓	×
ε	0.025	0.01	0.01	0.0025	0.0025	0	0
λ	0	0.1	0.1	0.2	0.2	0.5	0.5

shows the ground truth sign while $sgn(\bar{s}_j - \bar{f}_\theta(z_i, x_j))$ indicates the direction of optimization. We increase the weights if this direction matches the ground truth sign and decrease the weights otherwise.

We also design a curriculum for sample difficulty by controlling λ at different training stages. At the beginning of training, we aim to teach the network global structures and allow small errors in shape geometry. To this end, we set a relatively small λ to make the network equally focused on all training samples. Then, we gradually increase λ to emphasize more on hard and semi-hard samples, which helps the network to address its weaknesses and reconstruct better local details. Strictly speaking, the curriculum of sample difficulty is slightly different from the formulation in [3], as it starts from the original task and gradually increases the difficulty to a harder objective. However, they share similar thoughts and the ablation study also shows the effectiveness of the designed curriculum.

3.3 Implementation Details

In order to make fair comparisons, we applied the same training data, training epochs and network architecture as DeepSDF [30]. More specifically, we prepared the input samples X_s from each shape mesh which was normalized to a unit sphere. We sampled 500,000 points from each shape. The points were sampled more aggressively near the surface to capture more shape details. The learning rate for training the network was set as $N_b \times 10^{-5}$ where N_b is the batch size and 10^{-3} for the latent vectors. We trained the models for 2,000 epochs. Table 1 presents the training details, which will degenerate to DeepSDF if we train all the 8 fully connected layers by setting $\varepsilon = \lambda = 0$ from beginning to the end.

4 Experiments

In this section, we perform a thorough comparison of our proposed Curriculum DeepSDF to DeepSDF along with comprehensive ablation studies for the shape reconstruction task on the ShapeNet dataset [5]. We use the missing part recovery task as an application to demonstrate the usage of our method.

Following [30], we report the standard distance metrics of mesh reconstruction including the mean and the median of Chamfer distance (CD), mean Earth Mover's distance (EMD) [35], and mean mesh accuracy [39]. For evaluating CD, we sample 30,000 points from mesh surfaces. For evaluating EMD, we follow [30] by sampling 500 points from mesh surfaces due to a high computation cost. For evaluating mesh accuracy, following [30,39], we sample 1,000 points from mesh surfaces and compute the minimum distance d such that 90% of the points lie within d of the ground truth surface.

4.1 Shape Reconstruction

We conducted experiments on the ShapeNet dataset [5] for the shape reconstruction task. In the following, we will introduce quantitative results, ablation studies and visualization results.

Quantitative Results. We compare our method to the state-of-the-art methods, including AtlasNet [16] and DeepSDF [30] in Table 2. We also include several variants of our own method for ablation studies. *Ours*, representing the proposed Curriculum DeepSDF method, performs a complete curriculum learning considering both surface accuracy and sample difficulty. As variants of our method, *ours-sur* and *ours-sur w/o* only employ the surface accuracy based curriculum learning with/without progressively growth of the network layers, where *ours-sur w/o* uses the fixed architecture with the deepest size; *ours-sam* only employs sample difficulty based curriculum learning. For a fair comparison, we evaluated all SDF-based methods following the same training and testing protocols as DeepSDF, including training/test split, the number of training epochs, and network architecture, etc. For AtlasNet-based methods, we directly report the numbers from [30]. Here are the three key observations from Table 2:

1) Compared to vanilla DeepSDF, curriculum learning on either surface accuracy or sample difficulty can lead to a significant performance gain. The best performance is achieved by simultaneously performing both curricula.
2) In general, the curriculum of sample difficulty helps more on lamp and plane as these categories suffer more from reconstructing slender or thin structures. The curriculum of surface accuracy is more effective for the categories of chair, sofa and table where shapes are more regular.
3) As we only sample 500 points for computing EMD, even the ground truth mesh has non-zero EMD to itself rising from the randomness in point sampling. Our performance is approaching the upper bound on plane and sofa.

Hard Sample Mining Strategies. We conducted ablation studies for a more detailed analysis of different hard sample mining strategies on the lamp category due to its large variations and complex shape details. In the curriculum of sample difficulty, we gradually increase λ to make the network more and more focused on the hard samples. We compared it with the simple strategy by fixing a single λ. Table 3 shows that the performance improves as λ increases until reaching a sweet spot, after which further increasing λ could hurt the performance. The best result is achieved by our method which gradually increases λ as it encourages the network to focus more and more on hard details.

For hard sample mining, we increase the weights of hard and semi-hard samples to $1 + \lambda$ and also decrease the weights of easy samples to $1 - \lambda$. As various similar strategies can be used, we demonstrate the effectiveness of our design in Table 4. We observe that both increasing the weights of semi-hard samples and

Table 2. Reconstructing shapes from the ShapeNet test set. Here we report shape reconstruction errors in term of several distance metrics on five ShapeNet classes. Note that we multiply CD by 10^3 and mesh accuracy by 10^1. The *average* column shows the average distance and the *relative* column shows the relative distance reduction compared to DeepSDF. For all metrics except for *relative*, the lower, the better.

CD, mean	Lamp	Plane	Chair	Sofa	Table	Average	Relative
AtlasNet-Sph	2.381	0.188	0.752	0.445	0.725	0.730	-
AtlasNet-25	1.182	0.216	0.368	0.411	0.328	0.391	-
DeepSDF	0.776	0.143	0.243	0.117	0.424	0.319	-
Ours-Sur w/o	0.743	0.109	0.162	0.110	0.343	0.257	19.4%
Ours-Sur	0.639	0.086	0.157	0.108	0.327	0.239	25.1%
Ours-Sam	0.592	0.078	0.175	0.113	0.342	0.246	22.9%
Ours	**0.473**	**0.070**	**0.156**	**0.105**	**0.304**	**0.216**	**32.3%**
CD, median							
AtlasNet-Sph	2.180	0.079	0.511	0.330	0.389	0.490	-
AtlasNet-25	0.993	0.065	0.276	0.311	0.195	0.267	-
DeepSDF	0.178	0.061	0.098	0.081	0.052	0.078	-
Ours-Sur w/o	0.172	0.048	0.071	0.077	0.047	0.066	15.4%
Ours-Sur	0.147	0.045	**0.064**	0.077	0.051	0.063	19.2%
Ours-Sam	0.139	0.040	0.066	0.080	0.050	0.063	19.2%
Ours	**0.105**	**0.033**	**0.064**	**0.069**	**0.048**	**0.056**	**28.2%**
EMD, mean							
GT	0.034	0.026	0.041	0.044	0.041	0.039	-
AtlasNet-Sph	0.085	0.038	0.071	0.050	0.060	0.060	-
AtlasNet-25	0.062	0.041	0.064	0.063	0.073	0.064	-
DeepSDF	0.066	0.035	0.055	0.051	0.057	0.053	-
Ours-Sur w/o	0.057	0.032	**0.048**	0.046	0.049	0.046	13.2%
Ours-Sur	0.055	0.027	**0.048**	0.046	**0.048**	0.045	15.1%
Ours-Sam	0.055	0.027	0.053	0.050	0.051	0.048	9.4%
Ours	**0.052**	**0.026**	**0.048**	**0.044**	**0.048**	**0.044**	**17.0%**
Mesh acc, mean							
AtlasNet-Sph	0.540	0.130	0.330	0.170	0.320	0.290	-
AtlasNet-25	0.420	0.130	0.180	0.170	0.140	0.172	-
DeepSDF	0.155	0.044	0.104	0.041	0.120	0.097	-
Ours-Sur w/o	0.133	0.035	0.089	0.040	0.104	0.083	14.4%
Ours-Sur	0.121	0.034	0.082	0.039	0.098	0.078	19.6%
Ours-Sam	0.135	**0.031**	0.083	**0.036**	0.087	0.074	23.7%
Ours	**0.103**	**0.031**	**0.080**	**0.036**	**0.087**	**0.071**	**26.8%**

Table 3. Experimental comparisons with using fixed λ for hard sample mining. The method degenerates to *ours-sur* when $\lambda = 0$. CD is multiplied by 10^3.

λ	0	0.05	0.10	0.25	0.50	0.75	Ours
CD, mean	0.639	0.606	0.549	0.538	0.508	0.567	**0.473**

Table 4. Experimental comparisons of different hard sample mining strategies. In the table, *H*, *S* and *E* are the hard, semi-hard and easy samples, respectively. For the symbols, $\uparrow$ is to increase the weights to $1 + \lambda$, $\downarrow$ is to decrease the weights to $1 - \lambda$ and - is to maintain the weights. $H(\uparrow)S(\uparrow)E(\downarrow)$ is the sampling strategy used in our method, while *H(-)S(-)E(-)* degenerates to *ours-sur*. CD is multiplied by 10^3.

Strategy	H(-)S(-)E(-)	H(-)S(-)E(↓)	H(-)S(↑)E(-)	H(-)S(↑)E(↓)
CD, mean	0.639	0.563	0.587	0.508
Strategy	H(↑)S(-)E(-)	H(↑)S(-)E(↓)	H(↑)S(↑)E(-)	H(↑)S(↑)E(↓)
CD, mean	0.676	0.661	0.512	**0.473**

decreasing the weights of easy samples can boost the performance. However, it is risky to only increase weights for hard samples excluding semi-hard ones in which case the performance drops. One possible reason is that focusing too much on hard samples may lead to more wrong sign estimations for the semi-hard ones as they are close to the boundary. Hence, it is necessary to increase the weights of semi-hard samples as well to maintain their correct sign estimations. The best performance is achieved by simultaneously increasing the weights of hard and semi-hard samples and decreasing the weights of easy ones.

Number of Points for EMD. In Table 2, we followed [30] by sampling 500 points to compute accurate EMD, which would lead to relatively large distance even for ground truth meshes. To this end, we increase the number of sampled points during EMD computation and tested the performance on lamps. Results in Table 5 show that the number of sampled points can affect EMD due to the

Table 5. Comparison of mean of EMD on the lamp category of the ShapeNet dataset with varying numbers of sampled points.

Number of points	500	2000	5000	10000
GT	0.034	0.008	0.008	0.004
DeepSDF	0.066	0.056	0.052	0.051
Ours-Sur w/o	0.057	0.053	0.050	0.048
Ours-Sur	0.055	0.052	0.049	0.048
Ours-Sam	0.055	0.053	0.050	0.048
Ours	**0.052**	**0.051**	**0.047**	**0.046**

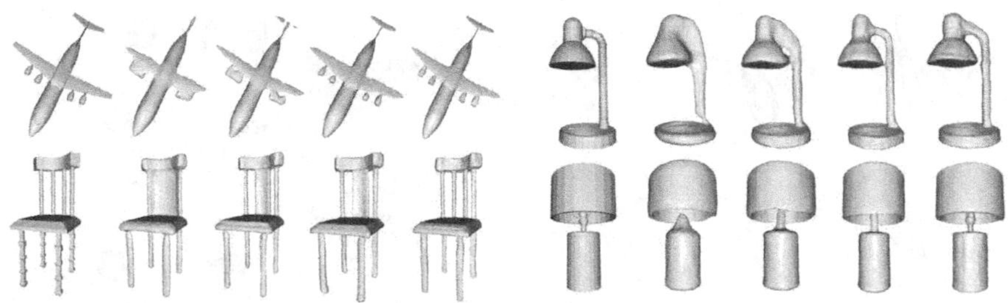

Fig. 5. The visualization of shape reconstruction at the end of each training stage. From left to right: ground truth, 200 epochs, 600 epochs, 1000 epochs, and 2000 epochs.

Table 6. Experimental comparisons under different ratios of removed points. CD and mesh accuracy are multiplied by 10^3 and 10^1, respectively.

Method	5%		10%		15%		20%		25%	
\Metric	CD	Mesh	CD	Mesh	CD	Mesh	CD	Mesh	CD	Mesh
Plane										
DeepSDF	0.163	0.056	0.229	0.066	0.217	0.067	0.224	0.069	0.233	0.080
Ours	**0.095**	**0.032**	**0.124**	**0.044**	**0.149**	**0.052**	**0.163**	**0.062**	**0.192**	**0.072**
Sofa										
DeepSDF	0.133	0.045	0.137	0.047	0.149	0.050	0.169	0.058	**0.196**	0.066
Ours	**0.110**	**0.037**	**0.120**	**0.041**	**0.143**	**0.046**	**0.165**	**0.053**	**0.196**	**0.061**
Lamp										
DeepSDF	2.08	0.230	3.10	0.241	3.50	0.286	4.18	0.307	4.79	0.331
Ours	**1.96**	**0.167**	**2.87**	**0.195**	**3.27**	**0.231**	**3.52**	**0.277**	**4.07**	**0.320**

randomness in sampling, and the EMD of resampled ground truth decreases when using more points. Our method continuously obtains better results.

Visualization Results. We visualize the shape reconstruction results in Fig. 1 to qualitatively compare DeepSDF and Curriculum DeepSDF. We observe that Curriculum DeepSDF reconstructs more accurate shape surfaces. The curriculum of surface accuracy helps to better capture the general structure, and sample difficulty encourages the recovery of complex local details. We also provide the reconstructed shapes at key epochs in Fig. 5. Curriculum DeepSDF learns coarse shapes at early stages which omits complex details. Then, it gradually refines local parts based on the learned coarse shapes. This training procedure improves the performance of the learned shape representation.

4.2 Missing Part Recovery

One of the main advantages of the DeepSDF framework is that we can optimize a shape code based upon a partial shape observation, and then render the complete shape through the learned network. In this subsection, we compare DeepSDF with Curriculum DeepSDF on the task of missing part recovery.

Fig. 6. The visualization results of missing part recovery. The green points are the remaining points that we use to recover the whole mesh. From top to bottom: ground truth, DeepSDF, and Curriculum DeepSDF. (Color figure online)

To create partial shapes with missing parts, we remove a subset of points from each shape X_i As random point removal may still preserve the holistic structures, we remove all the points in a local area to create missing parts. More specifically, we randomly select a point from the shape and then remove a certain quantity of its nearest neighbor points including itself, so that all the points within a local range can be removed. We conducted the experiments on three ShapeNet categories: plane, sofa and lamp. In these categories, plane and sofa have more regular and symmetric structures, while lamp is more complex and contains large variations. Table 6 shows that part removal largely affects the performance on the lamp category compared with plane and sofa, and Curriculum DeepSDF continuously obtains better results than DeepSDF under different ratios of removed points. A visual comparison is provided in Fig. 6.

5 Conclusion

In this paper, we have proposed Curriculum DeepSDF by designing a shape curriculum for shape representation learning. Inspired by the learning principle of humans, we organize the learning task into a series of difficulty levels from surface accuracy and sample difficulty. For surface accuracy, we design a tolerance parameter to control the global smoothness, which gradually increases the accuracy of the learned shape with more layers. For sample difficulty, we define hard, semi-hard and easy training samples in SDF learning, and gradually re-weight the samples to focus more and more on difficult local details. Experimental results show that our method largely improves the performance of DeepSDF with the same training data, training epochs and network architecture.

Acknowledgements. This research was supported by a Vannevar Bush Faculty Fellowship, a grant from the SAIL-Toyota Center for AI Research, and gifts from Amazon Web Services and the Dassault Foundation.

References

1. Achlioptas, P., Diamanti, O., Mitliagkas, I., Guibas, L.: Learning representations and generative models for 3D point clouds. In: ICML, pp. 40–49 (2018)
2. Allgower, E.L., Georg, K.: Introduction to numerical continuation methods. In: SIAM, vol. 45 (2003)
3. Bengio, Y., Louradour, J., Collobert, R., Weston, J.: Curriculum learning. In: ICML, pp. 41–48 (2009)
4. Carr, J.C., et al.: Reconstruction and representation of 3D objects with radial basis functions. In: SIGGRAPH, pp. 67–76 (2001)
5. Chang, A.X., et al.: ShapeNet: an information-rich 3D model repository. arXiv preprint arXiv:1512.03012 (2015)
6. Chen, Z., Zhang, H.: Learning implicit fields for generative shape modeling. In: CVPR, pp. 5939–5948 (2019)
7. Choy, C.B., Xu, D., Gwak, J.Y., Chen, K., Savarese, S.: 3D-R2N2: a unified approach for single and multi-view 3D object reconstruction. In: Leibe, B., Matas, J., Sebe, N., Welling, M. (eds.) ECCV 2016. LNCS, vol. 9912, pp. 628–644. Springer, Cham (2016). https://doi.org/10.1007/978-3-319-46484-8_38
8. Duan, Y., Chen, L., Lu, J., Zhou, J.: Deep embedding learning with discriminative sampling policy. In: CVPR, pp. 4964–4973 (2019)
9. Duan, Y., Lu, J., Zheng, W., Zhou, J.: Deep adversarial metric learning. TIP **29**(1), 2037–2051 (2020)
10. Duan, Y., Zheng, Y., Lu, J., Zhou, J., Tian, Q.: Structural relational reasoning of point clouds. In: CVPR, pp. 949–958 (2019)
11. Elman, J.L.: Learning and development in neural networks: the importance of starting small. Cognition **48**(1), 71–99 (1993)
12. Genova, K., Cole, F., Sud, A., Sarna, A., Funkhouser, T.: Deep structured implicit functions. arXiv preprint arXiv:1912.06126 (2019)
13. Genova, K., Cole, F., Vlasic, D., Sarna, A., Freeman, W.T., Funkhouser, T.: Learning shape templates with structured implicit functions. In: ICCV, pp. 7154–7164 (2019)
14. Graves, A., Bellemare, M.G., Menick, J., Munos, R., Kavukcuoglu, K.: Automated curriculum learning for neural networks. In: ICML, pp. 1311–1320 (2017)
15. Gropp, A., Yariv, L., Haim, N., Atzmon, M., Lipman, Y.: Implicit geometric regularization for learning shapes. arXiv preprint arXiv:2002.10099 (2020)
16. Groueix, T., Fisher, M., Kim, V.G., Russell, B.C., Aubry, M.: A papier-mâché approach to learning 3D surface generation. In: CVPR, pp. 216–224 (2018)
17. Guo, K., Zou, D., Chen, X.: 3D mesh labeling via deep convolutional neural networks. TOG **35**(1), 1–12 (2015)
18. Hacohen, G., Weinshall, D.: On the power of curriculum learning in training deep networks. In: ICML, pp. 2535–2544 (2019)
19. Ilg, E., Mayer, N., Saikia, T., Keuper, M., Dosovitskiy, A., Brox, T.: Flownet 2.0: evolution of optical flow estimation with deep networks. In: CVPR, pp. 2462–2470 (2017)
20. Jiang, L., Zhou, Z., Leung, T., Li, L.J., Fei-Fei, L.: MentorNet: learning data-driven curriculum for very deep neural networks on corrupted labels. In: ICML, pp. 2304–2313 (2018)
21. Karras, T., Aila, T., Laine, S., Lehtinen, J.: Progressive growing of gans for improved quality, stability, and variation. arXiv preprint arXiv:1710.10196 (2017)

22. Liao, Y., Donne, S., Geiger, A.: Deep marching cubes: learning explicit surface representations. In: CVPR, pp. 2916–2925 (2018)
23. Liu, S., Saito, S., Chen, W., Li, H.: Learning to infer implicit surfaces without 3D supervision. In: NeurIPS, pp. 8293–8304 (2019)
24. Lorensen, W.E., Cline, H.E.: Marching cubes: a high resolution 3D surface construction algorithm. In: SIGGRAPH, pp. 163–169 (1987)
25. Lu, J., Hu, J., Tan, Y.P.: Discriminative deep metric learning for face and kinship verification. TIP **26**(9), 4269–4282 (2017)
26. Maturana, D., Scherer, S.: Voxnet: a 3D convolutional neural network for real-time object recognition. In: IROS, pp. 922–928 (2015)
27. Mescheder, L., Oechsle, M., Niemeyer, M., Nowozin, S., Geiger, A.: Occupancy networks: learning 3D reconstruction in function space. In: CVPR, pp. 4460–4470 (2019)
28. Michalkiewicz, M., Pontes, J.K., Jack, D., Baktashmotlagh, M., Eriksson, A.: Deep level sets: implicit surface representations for 3D shape inference. arXiv preprint arXiv:1901.06802 (2019)
29. Ohtake, Y., Belyaev, A., Alexa, M., Turk, G., Seidel, H.P.: Multi-level partition of unity implicits. In: SIGGRAPH, pp. 173–180 (2005)
30. Park, J.J., Florence, P., Straub, J., Newcombe, R., Lovegrove, S.: DeepSDF: learning continuous signed distance functions for shape representation. In: CVPR, pp. 165–174 (2019)
31. Qi, C.R., Su, H., Mo, K., Guibas, L.J.: Pointnet: deep learning on point sets for 3D classification and segmentation. In: CVPR, pp. 652–660 (2017)
32. Qi, C.R., Su, H., Nießner, M., Dai, A., Yan, M., Guibas, L.J.: Volumetric and multi-view CNNs for object classification on 3D data. In: CVPR, pp. 5648–5656 (2016)
33. Qi, C.R., Yi, L., Su, H., Guibas, L.J.: Pointnet++: deep hierarchical feature learning on point sets in a metric space. In: NeurIPS, pp. 5099–5108 (2017)
34. Rao, Y., Lu, J., Zhou, J.: Global-local bidirectional reasoning for unsupervised representation learning of 3D point clouds. In: CVPR, pp. 5376–5385 (2020)
35. Rubner, Y., Tomasi, C., Guibas, L.J.: The earth mover's distance as a metric for image retrieval. IJCV **40**(2), 99–121 (2000)
36. Saito, S., Huang, Z., Natsume, R., Morishima, S., Kanazawa, A., Li, H.: PIFu: pixel-aligned implicit function for high-resolution clothed human digitization. In: ICCV, pp. 2304–2314 (2019)
37. Sanger, T.D.: Neural network learning control of robot manipulators using gradually increasing task difficulty. IEEE Trans. Rob. Autom. **10**(3), 323–333 (1994)
38. Schroff, F., Kalenichenko, D., Philbin, J.: FaceNet: a unified embedding for face recognition and clustering. In: CVPR, pp. 815–823 (2015)
39. Seitz, S.M., Curless, B., Diebel, J., Scharstein, D., Szeliski, R.: A comparison and evaluation of multi-view stereo reconstruction algorithms. In: CVPR, pp. 519–528 (2006)
40. Sharma, R., Barratt, S., Ermon, S., Pande, V.: Improved training with curriculum GANs. arXiv preprint arXiv:1807.09295 (2018)
41. Shen, C., O'Brien, J.F., Shewchuk, J.R.: Interpolating and approximating implicit surfaces from polygon soup. In: SIGGRAPH, pp. 896–904 (2004)
42. Sinha, A., Bai, J., Ramani, K.: Deep learning 3D shape surfaces using geometry images. In: Leibe, B., Matas, J., Sebe, N., Welling, M. (eds.) ECCV 2016. LNCS, vol. 9910, pp. 223–240. Springer, Cham (2016). https://doi.org/10.1007/978-3-319-46466-4_14

43. Su, H., Maji, S., Kalogerakis, E., Learned-Miller, E.: Multi-view convolutional neural networks for 3D shape recognition. In: ICCV, pp. 945–953 (2015)
44. Sun, Y., et al.: Circle loss: a unified perspective of pair similarity optimization. In: CVPR, pp. 6398–6407 (2020)
45. Tulsiani, S., Zhou, T., Efros, A.A., Malik, J.: Multi-view supervision for single-view reconstruction via differentiable ray consistency. In: CVPR, pp. 2626–2634 (2017)
46. Turk, G., O'brien, J.F.: Shape transformation using variational implicit functions. In: SIGGRAPH, pp. 14–20 (1999)
47. Turk, G., O'brien, J.F.: Modelling with implicit surfaces that interpolate. TOG **21**(4), 855–873 (2002)
48. Wang, N., Zhang, Y., Li, Z., Fu, Y., Liu, W., Jiang, Y.G.: Pixel2Mesh: generating 3D mesh models from single RGB images. In: Ferrari, V., Hebert, M., Sminchisescu, C., Weiss, Y. (eds.) Computer Vision - ECCV 2018. LNCS, vol .11215, pp. 55–75. Springer, Cham (2018). https://doi.org/10.1007/978-3-030-01252-6_4
49. Wang, Y., Sun, Y., Liu, Z., Sarma, S.E., Bronstein, M.M., Solomon, J.M.: Dynamic graph CNN for learning on point clouds. TOG **38**(5), 1–12 (2019)
50. Weinshall, D., Cohen, G., Amir, D.: Curriculum learning by transfer learning: theory and experiments with deep networks. In: ICML, pp. 5238–5246 (2018)
51. Wu, Z., et al.: 3D ShapeNets: a deep representation for volumetric shapes. In: CVPR, pp. 1912–1920 (2015)
52. Xu, Q., Wang, W., Ceylan, D., Mech, R., Neumann, U.: DISN: deep implicit surface network for high-quality single-view 3D reconstruction. In: NeurIPS, pp. 490–500 (2019)

Meshing Point Clouds with Predicted Intrinsic-Extrinsic Ratio Guidance

Minghua Liu(✉), Xiaoshuai Zhang, and Hao Su

University of California, San Diego, USA
{minghua,zxs,haosu}@ucsd.edu

Abstract. We are interested in reconstructing the mesh representation of object surfaces from point clouds. Surface reconstruction is a prerequisite for downstream applications such as rendering, collision avoidance for planning, animation, etc. However, the task is challenging if the input point cloud has a low resolution, which is common in real-world scenarios (e.g., from LiDAR or Kinect sensors). Existing learning-based mesh generative methods mostly predict the surface by first building a shape embedding that is at the whole object level, a design that causes issues in generating fine-grained details and generalizing to unseen categories. Instead, we propose to leverage the input point cloud as much as possible, by only adding connectivity information to existing points. Particularly, we predict which triplets of points should form faces. Our key innovation is a surrogate of local connectivity, calculated by comparing the intrinsic/extrinsic metrics. We learn to predict this surrogate using a deep point cloud network and then feed it to an efficient post-processing module for high-quality mesh generation. We demonstrate that our method can not only preserve details, handle ambiguous structures, but also possess strong generalizability to unseen categories by experiments on synthetic and real data.

Keywords: Mesh reconstruction · Point cloud

1 Introduction

Among various 3D representations (e.g., polygonal meshes, voxels, point clouds, multi-view 2D images, part-based primitives, and implicit field functions), polygonal meshes capture the geometric details of the shape in an efficient way, which prevents high memory footprints and artifacts caused by discretization. Reconstructing high-quality 3D meshes from point clouds thus has been studied for quite a long time and serves as a prerequisite for numerous real-world applications, including autonomous driving, augmented reality, and robotics.

Despite its long history, the mesh reconstruction problem remains unresolved. Traditional methods [2,24,30] typically reconstruct the mesh either by explicitly

Electronic supplementary material The online version of this chapter (https://doi.org/10.1007/978-3-030-58598-3_5) contains supplementary material, which is available to authorized users.

A. Vedaldi et al. (Eds.): ECCV 2020, LNCS 12353, pp. 68–84, 2020.
https://doi.org/10.1007/978-3-030-58598-3_5

connecting the points or implicitly approximating the surface, both of which resort to local geometric hints. Without reasoning about the shape, traditional methods may be hard to handle the ambiguous structures when the resolution of the input point cloud is limited. For example, the ambiguous structures may include thin structures consisting of two very close surfaces, independent but spatially adjacent parts, and corners. Traditional methods tend to produce distortion or connect independent parts incorrectly when facing these structures. However, the reconstruction of these fine-grained structures may be essential for many downstream applications such as robotics grasping which needs an accurate understanding of part-level mobility. Moreover, traditional methods are typically sensitive to hyper-parameters. For most of these methods, a dedicated parameter-tuning is required for each input, making batch processing of point clouds impractical.

With the rapid development of 3D deep learning and the availability of large-scale 3D datasets, people tend to learn geometric or semantic priors from data. Unlike 2D images and 3D voxels, polygon meshes is an irregular geometric representation, which prevents it from being generated by the neural network directly. However, there are still lots of attempts to explore the neural-network-compatible representations for mesh generation, including template meshes with deformation [14,16,18,23,29,35,43,44], 2D squares with folding [10,19,46], primitives with assembly [6,39,41], implicit field function [7,15,32,36], and meshlets with optimization [1,45]. Existing learning-based methods typically follow the "encoder-decoder" paradigm. The limited capability of the network prevents existing methods from generating fine-grained structures and details. Also, since most existing methods learn the priors at the object level, they tend to memorize the overall shapes and typically cannot generalize to unseen categories.

To this end, we propose a novel method that reconstructs meshes from point clouds by leveraging the intermediate representation of triangle faces. Unlike existing methods, our method fully utilizes the input point clouds, which are on the ground truth surface in most cases, and then estimate the local connectivity with the help of learned guidance. More specifically, we first propose a set of candidate triangle faces, which could be the elements of the reconstructed mesh, by constructing a k-nearest neighbor (k-NN) graph on the input point cloud. We then utilize the neural network to filter out the incorrect candidates and provide cues for sorting the remaining candidates. We find that the ratio of geodesic distance (intrinsic metric) and Euclidean distance (extrinsic metric) between two vertices may provide strong cues for inferring the connectivity and can naturally serve as the supervision for the candidate classification task. Since there are multiple ways to triangulate a surface, we only filter out those candidates that should never appear in the reconstructed mesh, such as the candidates linking two independent parts. A greedy post-processing algorithm is then used to sort all the remaining candidates and merge them into the final mesh.

We demonstrate that our algorithm can preserve fine-grained details and handle ambiguous structures with the help of learned intrinsic-extrinsic guidance. Since our method reconstructs meshes by estimating local connectivity, which relies mainly on the local geometric information, it can well generalize to unseen categories. In experiments on the ShapeNet dataset, our method outperforms both the existing traditional methods and learning-based mesh generative methods with regard to all commonly used metrics, including the F-score, Chamfer distance, and normal consistency. We also provide extensive ablation studies on different sampling densities, sampling strategies, noisy levels, and real scans to demonstrate our generalizability and robustness.

2 Related Work

3D mesh reconstruction is a core problem for many applications. Yet despite its long history, the problem is still far from being solved. In this section, we review the existing methods and the remaining difficulties of the problem.

2.1 Traditional Mesh Reconstruction

Traditional mesh reconstruction methods mainly include two paradigms: explicit reconstruction and implicit reconstruction.

Explicit reconstruction methods, such as ball-pivoting algorithm (BPA) [2], Delaunay triangulation [3], alpha shapes [12], and zippering [42], resort to the local surface connectivity estimation and connect the sampled points directly by triangles. For example, the principle of BPA is simple: three points form a triangle if a ball of a user-specified radius touches them without containing any other point. However, the radius of BPA matters a lot: a small radius can lead to holes while a large radius may cause incorrect connections. Although there are some following works trying to utilize multiple radii [11], they still fail to handle ambiguous structures well.

Implicit reconstruction methods [4,20,22,24,25,34] try to find a field function (e.g., signed distance function) approximating the point cloud and then employ the marching cube algorithm [30] to extract the iso-surface of the field function. For example, Poisson surface reconstruction (PSR) [24,25] reconstructs the surface by solving a Poisson problem for the oriented points. However, solving large-scale equations is time-consuming. Also, it is difficult for traditional algorithms to determine the consistent direction of the normals based only on the coordinates of the point cloud. Without correct vertex normal directions, PSR tends to generate poor results. Moreover, implicit reconstruction methods utilize marching cube [30] to generate the mesh, which may lead to expensive voxelization and the artifacts caused by the discretization.

Under the limited resolution, ambiguities of the input point cloud require the integration of strong geometric or semantic priors about our 3D world. With such priors and reasoning, our learning-based methods are expected to handle those ambiguous structures and avoid results with distortion and artifacts. In addition, traditional algorithms heavily rely on selecting a set of proper hyper-parameters, which may require a case-by-case parameter tuning, while our learning-based algorithm should be applied to all cases adaptively and thus enable automatic batch processing.

2.2 Learning-Based Mesh Generation

The recent success of 3D deep learning [37,38] and the availability of large 3D datasets [5,33] nourish the tasks of 3D analysis and 3D synthesis. However, unlike 2D images, 3D polygon meshes are irregular geometric formats and are difficult to be directly generated from the neural networks. Existing learning-based mesh generative methods mainly follow five paradigms: deformation-based methods [14,16,18,23,29,35,43,44], folding-based methods [10,19,46], primitive-based methods [6,39,41], optimization-based methods [1,45], and implicit-field-function-based methods [7,15,26,32,36].

Deformation-based methods resort to deform a template mesh (e.g., a sphere mesh) into the desired shape. However, since they only deform the position of the vertices without changing the connectivity, the topology of the template mesh may restrict the methods from generating shapes of a specific topology. Folding-based methods learn a set of mappings from 2D squares to 3D patches, which are then used to form the mesh. Primitive-based methods utilize a set of primitives (e.g., planes and convex patches) to form the final mesh, and learn the parameters of the primitives. The simplicity of the primitives may prevent the methods from generating fine-grained details. As for the implicit-field-function-based methods, they employ neural networks to learn an implicit field function and then utilize marching cube algorithms [30] to extract the iso-surface, and thus face similar problems as the traditional implicit reconstruction methods. There are also some recent optimization-based methods, which either utilize a deep neural network as local geometric prior [45] or learn some local shape priors from the data [1]. They formulate mesh reconstruction as an optimization problem and are thus computational expensive.

Most existing learning-based methods do not make full use of the input point cloud that processing should be grounded upon. Although they may be able to generate the coarse-grained shapes, they may fail to capture some of the structures and details. Also, most existing methods learn the priors at the object level, which makes them category-specific, and even sensitive to the pose of the object. In contrast, our method will be fully based on the grounded point clouds to preserve all the structures and generate fine-grained details. Moreover, our method relies on local priors and can thus generalize to unseen categories.

3 Method

Given a 3D point cloud $P = \{(x_i, y_i, z_i)\}$, we aim to reconstruct a polygon mesh, which consists of a vertex set and a face set, approximating the underlying surface. Unlike existing learning-based mesh generative methods, we fully utilize the grounded input point cloud and let P serve as the vertex set of the reconstructed mesh, since they are usually on or near the surface of the shape and provide lots of cues for the structures and details. We then reconstruct the mesh by predicting the local connectivity between the vertices. Before presenting our reconstruction method, we would like to introduce a motivating remeshing algorithm, where the ground truth mesh is known. We then extend the remeshing algorithm to mesh reconstruction by introducing a neural network module.

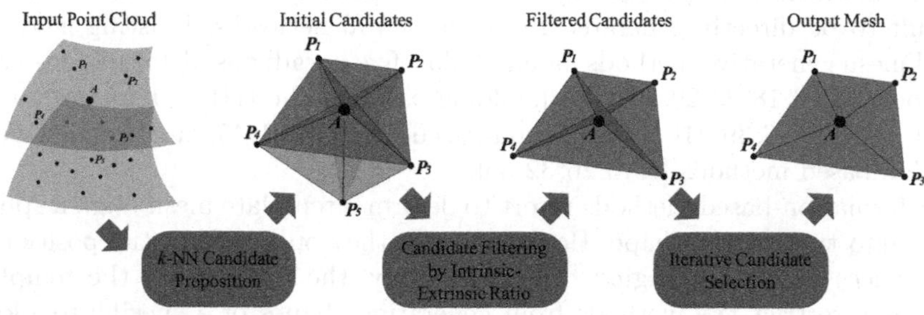

Fig. 1. Our remeshing pipeline from a local view. The candidate triangles near the underlying surfaces are colored in pink, while others are colored in green.

3.1 A Motivating Remeshing Algorithm

Given a reference triangle mesh M_R and a point cloud P sampled on M_R as input, the remeshing algorithm aims to generate a new mesh M_N whose vertices come from the point cloud P. As shown in Fig. 1, the algorithm first proposes a set of candidate triangle faces and then uses a subset of candidate triangles to form the mesh M_N.

Candidate Proposition. Since each vertex should only connect to its neighbors on the surface, the algorithm first constructs a k-nearest neighbor (k-NN) graph for each of the points in P based on the Euclidean distance, and then each vertex form candidate triangle faces with each two of its k-NN neighbors. We expect the union of candidate triangles to cover the whole surface of M_R, which means for an even surface with uniformly distributed vertices, a small k would be enough, while for complex surfaces and nonuniformly distributed vertices, we may need a larger k. However, we could select a k that is large enough to cover most practical cases. Among all the candidates, some of them are on or near the surface of mesh M_R, and we denote them as *correct candidates*, while the others are away from the surface and are denoted as *incorrect candidates*. The incorrect candidates may appear in areas such as (a) thin structures consisting of two very close surfaces, (b) independent but spatially adjacent parts, and (c) surfaces with large curvature. We would like to filter out the incorrect candidates, and use some of the remaining candidate faces to form the mesh M_N.

Candidate Filtering. Since the reference mesh M_R is given, we can calculate the geodesic distance between two vertices, which is defined to be the length

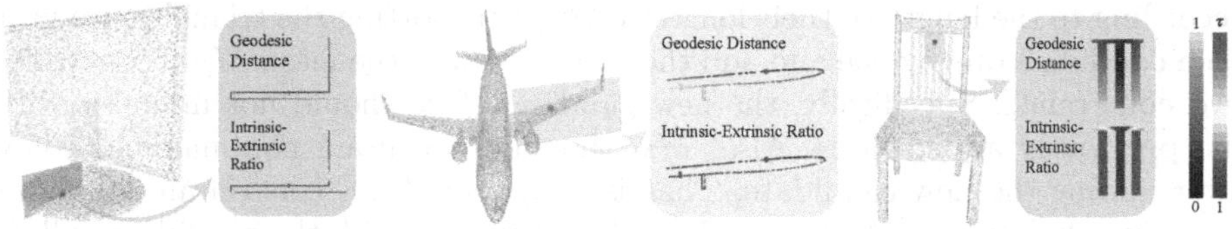

Fig. 2. In each example, we sample a slice from the input point cloud, and demonstrate the geodesic distance and the intrinsic-extrinsic ratio (IER) to the key point (marked in red) (Color figure online).

of the shortest path over the surface of M_R. As the intrinsic metric of the surface manifold, geodesic distance provides strong cues for inferring the connectivity between two vertices. The geodesic distance could be inconsistent with the Euclidean distance (extrinsic metric). However, in a small neighborhood, if the geodesic distance between two vertices is equal or close to their Euclidean distance, the connecting line between them is likely to be on or close to the surface of mesh M_R. We thus define the intrinsic-extrinsic ratio (IER) for a pair of vertices u, v as:

$$\mathrm{IER}(u,v) = \frac{d_G(u,v)}{d_E(u,v)}, \tag{1}$$

where d_G and d_E indicate the geodesic distance and the Euclidean distance respectively. Figure 2 shows some examples of how geodesic distance and the IER to a key point change within a slice. The key point should not be connected with the points in the red region where IER is much higher than 1. The cases in Fig. 2 demonstrate that IER can effectively handle corners (see display), thin structures (see jet), and adjacent parts (see chair). We thus propose to employ the IER to filter out the incorrect triangle candidates. For a triangle face with vertices u, v, and w, the IER is extended to be:

$$\mathrm{IER}(u,v,w) = \frac{d_G(u,v) + d_G(u,w) + d_G(v,w)}{d_E(u,v) + d_E(u,w) + d_E(v,w)}. \tag{2}$$

With the definition above, we filter out candidates whose IER is greater than τ, and $\tau > 1$ is a preset threshold. After filtering out the incorrect candidates, the remaining candidates should be on or near the surface of M_R.

Sort and Merge. Since there is no canonical way to triangulate a surface, we sort the remaining candidate triangles and merge them into mesh M_N in a greedy way. Specifically, we prefer triangles that are closer to the surface of M_R. Also, we prefer candidate triangles with three short edges, which typically correspond to the equilateral triangles when the input point cloud is uniformly distributed. We thus sort the remaining candidates with respect to their distance to M_R and the length of their longest edge. Specifically, we first divide the remaining candidate triangles into l bins based on the distance to M_R, and then sort them in each bin according to the length of their longest edge. After sorting the triangles, we visit each candidate one by one and add the candidate into the mesh M_N if it satisfies two constraints. Specifically, the new candidate face should not intersect with the previously added faces. Also, M_N should not contain non-manifold edges after adding the new candidate. That is, M_N should not contain an edge that has more than two incident faces. If both constraints hold, the candidate will be added into M_N; otherwise, it will be discarded. After visiting all the candidates, we get the final mesh M_N.

Please refer to our supplementary materials for the pseudo code of the remeshing algorithm. Some remeshing results of the algorithm are shown in Fig. 3.

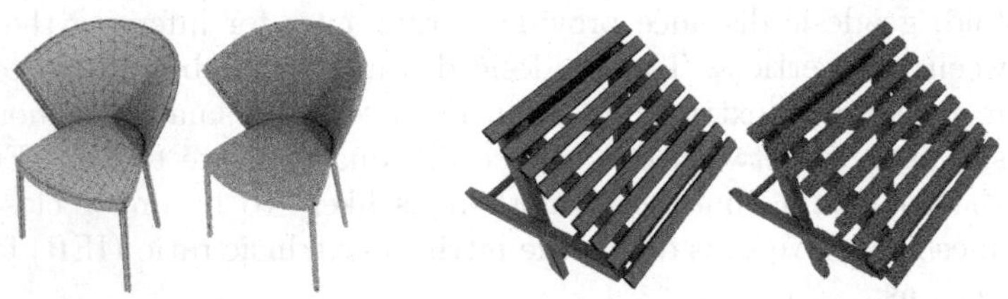

Fig. 3. In each pair, the left one is the original reference mesh (M_R), and the right one is the result of our remeshing algorithm (M_N).

3.2 From Remeshing to Reconstruction

In the mesh reconstruction setting, we are only given a point cloud P as input and aim to reconstruct the mesh M_N. As shown in Fig. 4, we follow the remeshing algorithm to construct a k-NN graph on P and propose m candidate triangle faces. Unlike in the remeshing algorithm, the reference mesh is not available, we thus cannot directly calculate the intrinsic-extrinsic ratio to serve as the guidance. At this point, the neural network may be helpful for estimating the local connectivity and geometry of the input point cloud. We thus resort to the neural network to filter out the incorrect candidate triangles and provide cues for sorting the remaining candidates.

In training, we have ground truth mesh and we follow the above remeshing algorithm to generate a label for each candidate, which serves as dense supervision for the candidate classification network. Specifically, the candidates are divided into $l + 1$ categories. The incorrect candidates with IER $\geq \tau$ are in category 0. The remaining candidates are near the ground truth surface and are divided into l categories according to their distances to the surface. Empirically, we set $l = 2$ in all our experiments, which means candidates that are very close to the ground truth surface are in category 1, and other correct ones are in category 2. The network is thus trained to predict a 3-class label for each candidate.

As shown in Fig. 4, our network consists of several parts. We first utilize SparseConvNet [17], which are designed to process spatially-sparse data with

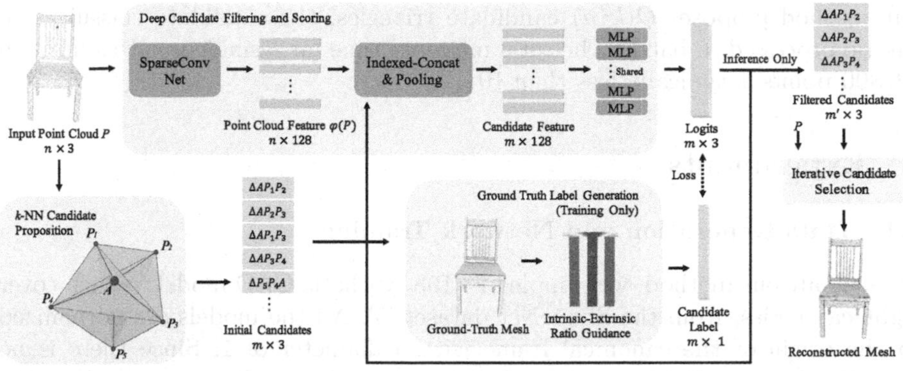

Fig. 4. Full pipeline of our reconstruction algorithm: given a point cloud as input, we first propose a set of candidate triangles. During training, the network is trained to classify the candidate triangles with the supervision of intrinsic-extrinsic ratio. During inference, the predicted label is used to filter out and score the remaining candidate triangles, which are then merged into the output mesh by our iterative selection algorithm.

convolutional operations, for point cloud feature extraction. Specifically, it maps the input point cloud into a feature set $\{\varphi(p_i) \,|\, p_i \in P\}$, where $\varphi(p_i) \in \mathbb{R}^{3+C}$ is a concatenation of the xyz coordinates and the C dimensional embedding of a point p_i. The three latent features $\varphi(u)$, $\varphi(v)$, and $\varphi(w)$ are then concatenated together for every candidate triangle face with vertices u, v and w. A symmetry function (e.g., a max-pooling layer) then takes the concatenated feature as input to aggregates the information from the three vertices. The resulting tensors are fed into a shared weight multiple layer perceptron (MLP) to predict the final label for each triangle. The convolutional operation and the shared weight MLP are designed to inspire the learning of generalizable local priors across different parts and shapes.

During inference, we utilize the predicted labels to filter out the incorrect candidates and then sort the remaining candidates according to their labels and the length of their longest edges. We finally merge them into the output mesh through a greedy post-processing, as in the remeshing algorithm.

Although we utilize the ratio between the geodesic distance and the Euclidean distance (IER) to determine the label of the candidate triangles, the neural network does not need to regress the geodesic distances directly, since it could learn to utilize local geometric and semantic cues to recognize those incorrect triangles. In our experiments, the network achieves high accuracy classifying the candidate triangles, and thus enables high-quality mesh reconstruction. Please refer to the supplementary materials for the experiment of directly inferring the geodesic distance, which produces less effective results.

We find our model can transfer well to unseen categories. Since the estimation of local connectivity relies more on local inductive biases, which encourages the generalizability. Similar phenomenons are also observed in [1,27,31]. Although our method proposes $O(k^2n)$ candidate triangles, the candidate classification can be processed in batch. The total inference time for a single point cloud with 12, 800 points is typically less than 10 s.

4 Experiments

4.1 Data Generation and Network Training

To evaluate our method, we sampled 23,108 synthetic CAD models, which cover eight categories, from the ShapeNet dataset [5]. All the models are normalized to the origin of the canonical frame with a diameter of 1. Since there is no watertight or manifold guarantee for the ShapeNet meshes, we pre-clean the meshes to facilitate the calculation of geodesic distance. Specifically, we merge the vertices that are within a small distance of 0.001, remove all the duplicate faces, and split the edges that go through the non-endpoint vertices. For each model, we then utilize the Poisson-disk sampling [9] to uniformly sample 10,000 ~12,800 points on the mesh surface. The points are then unified into the size of 12,800 by randomly replicating, and serve as the input point cloud. For each vertex, we construct a k-NN graph ($k = 50$). We follow the idea of the MMP algorithm [40] to calculate the exact geodesic distance between each vertex and its small neighborhood over the mesh surface. For calculating the Euclidean distance between a candidate triangle and the ground truth mesh, we randomly sample 10 points on the candidate face and average the distances of the sampled points to the ground truth mesh surface. To filter out the incorrect candidates, we empirically set τ to be 1.3 in our experiments. The correct candidates are then divided into two categories according to their distances to the mesh surface with a threshold of 0.005.

We reserve 3,146 models for testing, and the rest is used for training. All the 8 categories are trained together in a single model. The resolution of the SparseConvNet is set to be 150. In each training iteration, we randomly sample 25,000 candidate triangles per shape. For the following evaluations, our models were trained on a single Nvidia 2080Ti GPU for 50 epochs with a batch size of 24. The Adam optimizer was used. The initial learning rate was set to 1e−3 and decayed by 0.7 per 5 epochs.

4.2 Comparison with Existing Methods

We compare our method to both traditional surface reconstruction methods and learning-based mesh generative methods. Traditional methods include screened Poisson surface reconstruction (PSR) [24,25], marching cube (APSS variant) [20,30], and ball-pivoting algorithm (BPA) [2]. Learning-based methods include AtlasNet [19], Deep Geometric Prior (DGP) [45], Deep Marching Cubes (DMC) [26], and DeepSDF [36], which are representatives of different paradigms. Since meshes in ShapeNet are not manifolds, it's not trivial to calculate point normals with consistent directions, but many algorithms rely on correct normals. We thus employ PCPNet [21] to predict normals for the input

Table 1. Quantitative results on the ShapeNet test set: F-score with two different thresholds, Chamfer distance, and normal consistency score.

	F-score (μ) ↑								F-score (2 μ) ↑							
Category	PSR	MC	BPA	ATLAS	DMC	DSDF	DGP	OURS	PSR	MC	BPA	ATLAS	DMC	DSDF	DGP	OURS
Display	0.468	0.495	0.834	0.071	0.108	0.632	0.417	**0.903**	0.666	0.669	0.929	0.179	0.246	0.787	0.607	**0.975**
Lamp	0.455	0.518	0.826	0.029	0.047	0.268	0.405	**0.855**	0.648	0.681	0.934	0.077	0.113	0.478	0.662	**0.951**
Airplane	0.415	0.442	0.788	0.070	0.050	0.350	0.249	**0.844**	0.619	0.639	0.914	0.179	0.289	0.566	0.515	**0.946**
Cabinet	0.392	0.392	0.553	0.077	0.154	0.573	0.513	**0.860**	0.598	0.591	0.706	0.195	0.128	0.694	0.738	**0.946**
Vessel	0.415	0.466	0.789	0.058	0.055	0.323	0.387	**0.862**	0.633	0.647	0.906	0.153	0.120	0.509	0.648	**0.956**
Table	0.233	0.287	0.772	0.080	0.095	0.577	0.307	**0.880**	0.442	0.462	0.886	0.195	0.221	0.743	0.494	**0.963**
Chair	0.382	0.433	0.802	0.050	0.088	0.447	0.481	**0.875**	0.617	0.615	0.913	0.134	0.345	0.665	0.693	**0.964**
Sofa	0.499	0.535	0.786	0.058	0.129	0.577	0.638	**0.895**	0.725	0.708	0.895	0.153	0.208	0.734	0.834	**0.972**
Average	0.407	0.446	0.769	0.062	0.091	0.468	0.425	**0.872**	0.618	0.626	0.885	0.158	0.209	0.647	0.649	**0.959**
	Chamfer Distance (×100) ↓								Normal Consistency ↑							
Category	PSR	MC	BPA	ATLAS	DMC	DSDF	DGP	OURS	PSR	MC	BPA	ATLAS	DMC	DSDF	OURS	
Display	0.273	0.269	0.093	1.094	0.662	0.317	0.293	**0.069**	0.889	0.842	0.952	0.828	0.882	0.932	**0.974**	
Lamp	0.227	0.244	0.060	1.988	3.377	0.955	0.167	**0.053**	0.876	0.872	0.951	0.593	0.725	0.864	**0.963**	
Airplane	0.217	0.171	0.059	1.011	2.205	1.043	0.200	**0.049**	0.848	0.835	0.926	0.737	0.716	0.872	**0.955**	
Cainet	0.363	0.373	0.292	1.661	0.766	0.921	0.237	**0.112**	0.880	0.827	0.836	0.682	0.845	0.872	**0.957**	
Vessel	0.254	0.228	0.078	0.997	2.487	1.254	0.199	**0.061**	0.861	0.831	0.917	0.671	0.706	0.841	**0.953**	
Table	0.383	0.375	0.120	1.311	1.128	0.660	0.333	**0.076**	0.833	0.809	0.919	0.783	0.831	0.901	**0.962**	
Chair	0.293	0.283	0.099	1.575	1.047	0.483	0.219	**0.071**	0.850	0.818	0.938	0.638	0.794	0.886	**0.962**	
Sofa	0.276	0.266	0.124	1.307	0.703	0.496	0.174	**0.080**	0.892	0.851	0.940	0.633	0.850	0.906	**0.971**	
Average	0.286	0.276	0.116	1.368	1.554	0.766	0.228	**0.071**	0.866	0.836	0.923	0.695	0.794	0.884	**0.962**	

point clouds. We then utilize MeshLab [8] to reconstruct the meshes for the three traditional methods. Specifically, for Poisson surface reconstruction, an outlier removal as post-processing is applied. Since the ball-pivoting algorithm is sensitive to the radius, we tried the auto-guess mode of the MeshLab and also selected 3 radii manually to choose the best radius. Please refer to the supplementary materials for more details about the baseline algorithms.

We use F-score [43], Chamfer distance [13], and normal consistency score [32] to evaluate the methods. Specifically, we uniformly sample 10^6 points on the reconstructed mesh and the ground truth mesh respectively, and calculate a normal for each point. For F-score, the precision and recall are calculated by checking the percentage of points in one point set that can find a neighbor from the other point set within a threshold μ. The F-score is then calculated as the harmonic mean of precision and recall. To align with different sampling densities, μ is set to $\sqrt{S/10^6}$ for each shape and S is the surface area of the ground truth mesh. For Chamfer Distance (CD), it measures the mean distance between each point in one point set to its nearest neighbor in the other point set. The normal consistency score is defined as the average of the absolute dot product between the normals in one mesh and the normals of the corresponding nearest points in the other mesh.

The quantitative results are shown in Table 1. Our method outperforms all the baseline algorithms with regard to the three metrics across all categories with a large margin. Figure 5 demonstrates some of the representative cases. As shown in the figure, existing learning-based mesh generative methods (e.g., AtlasNet, DMC, and DeepSDF) are difficult to generate fine-grained details and

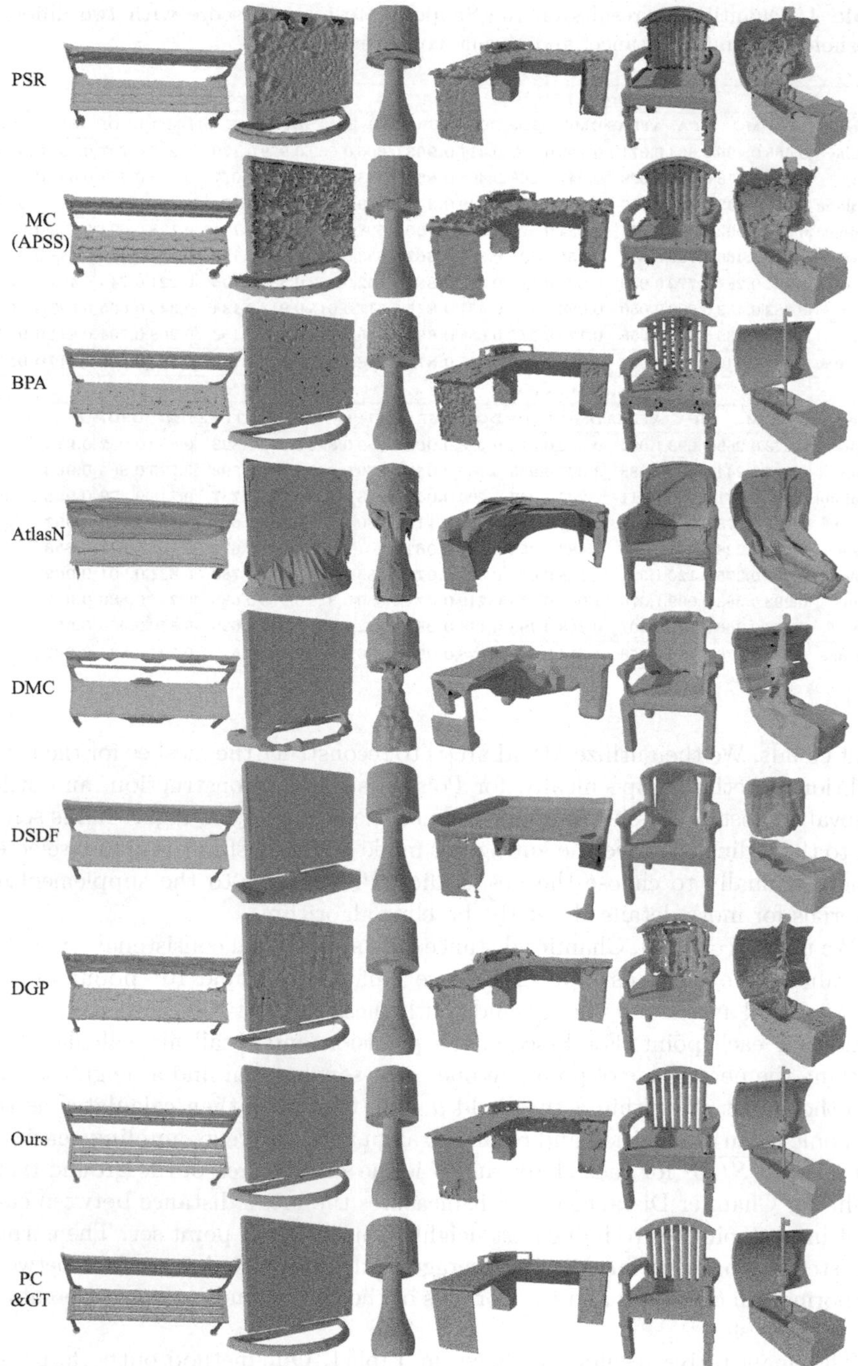

Fig. 5. Poisson surface reconstruction, marching cube (APSS), ball-pivoting algorithm, AtlasNet, Deep Marching Cubes, DeepSDF, Deep Geometric Prior, our method, and ground-truth meshes with input point clouds are shown from top to bottom.

Fig. 6. Our method can transfer to unseen categories. The first row shows the results of our method where the shape categories are unseen during training. The second row shows the results of ball-pivoting algorithm for comparison.

Table 2. First row: the results of our leave-one-out cross-validation. Second row: the results of feeding all candidates directly to the post-processing algorithm without filtering. Last row: our original results for comparison.

	F-score (μ) ↑	F-score (2μ) ↑	CD($\times$100) ↓	normal similarity ↑
Leave-one-out	0.870	0.959	0.072	0.961
W/o filtering	0.728	0.882	0.110	0.862
Ours	0.872	0.959	0.071	0.962

generalize to unseen shapes (see the vessel of AtlasNet and the desk of DeepSDF). They even fail to preserve all the structures which are revealed in the input (see the bench and the desk of DMC for missing parts). This is due to the fact that they generate meshes based only on an object-level shape embedding. Also, since DMC utilizes 3D convolutional networks to predict the surface, the generated meshes are limited to a low resolution (i.e., $32 \times 32 \times 32$). As for the three traditional methods, they generally preserve the overall structures from the input point clouds. However, it may be difficult for them to handle ambiguous structures when the resolution of the input point cloud is limited. For example, they failed to distinguish the thin structures consisting of spatially close surfaces, such as the display and the desk, and produced much distortion. Without priors and reasoning about the shape, nor can they distinguish those independent but spatially adjacent parts, such as the long strips of the bench and the armchair. In contrast, our method fully utilizes the input point cloud, which enables the generation of fine-grained structures. The learned local priors also help us to better estimate the local connectivity and generalize to unseen categories.

4.3 Ablation Studies

We would like to evaluate the transferability of our method, the importance of the candidate filtering, and the robustness with regard to various situations.

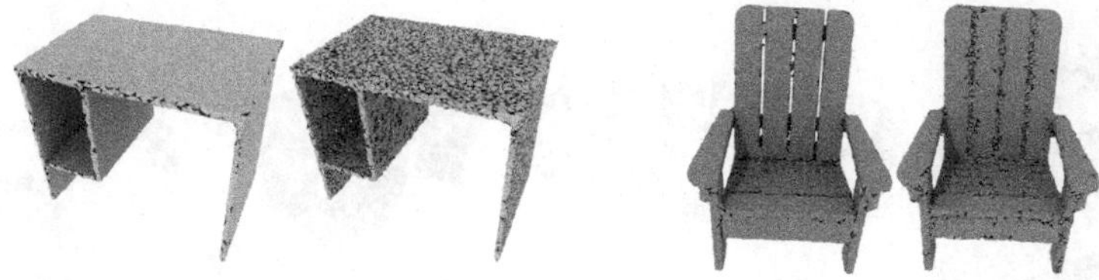

Fig. 7. In each pair, the left one is the result of method, and the right one is the result without candidate filtering.

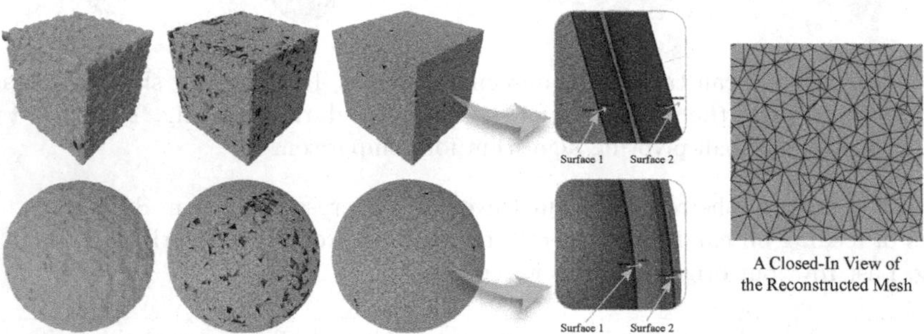

Fig. 8. Qualitative results on uniformly randomly sampled point clouds. The ground truth meshes are concentric dual spheres and concentric dual cubes. From left to right: results from Poisson surface reconstruction (PSR) [24,25], results from the ball-pivoting algorithm [2], and results from our method. The second to the last column is a zoomed-in sliced view showing the interior of our results. The last column is a zoomed-in view of the reconstructed surface.

Category Transferability. We utilize leave-one-out cross-validation to evaluate the category transferability of our method. Specifically, we trained a separate model for each of the eight categories with the training data of all the rest seven categories, and a test is made for that category. Table 2 reports the quantitative results where the numbers are averaged across all the eight categories. Compared to the model that trained on all categories and tests on all categories (the third row), the performances are quite similar, from which we can infer that our method does not heavily rely on category-specific priors and thus enable strong generalizability. Examples in Fig. 6 have further confirmed our belief that local priors can be transferred across different categories.

Effect of Filtering. To verify the importance of the candidate filtering, we also test a variant where all the proposed candidates are directly fed into the post-processing algorithm without filtering. The quantitative results are shown in Table 2, from which we find that the performance drops dramatically. Figure 7 also shows some of the qualitative comparisons. Without passing the candidates through the network and filtering out all the incorrect candidates, the method cannot handle ambiguous structures anymore.

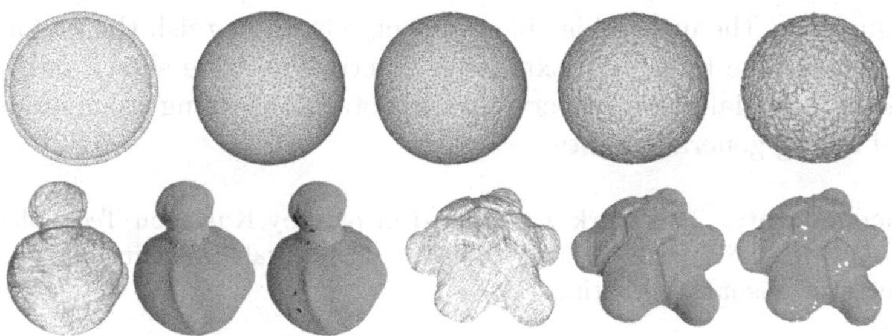

Fig. 9. First row: a point cloud of a concentric dual sphere and reconstructed meshes with different levels of input noise. Second row: two real-world LiDAR scans from Aim@Shape and the reconstructed meshes by our method.

Distribution of Point Cloud. In general, our method favors evenly distributed point clouds, and applying a Poisson-disk sampling as pre-processing could improve the performance. To examine the robustness to other point cloud distributions, we test our method on uniformly randomly sampled point clouds, virtual scanned point clouds, as well as Poisson-disk sampling with different density. Due to the space limits, we only include the results of uniformly randomly sampled point clouds in the text. Please refer to the supplementary materials for the results of other experiments. Note that though misleading, the uniformly randomly sampled points are typically not evenly distributed over the surface, as shown in the last column of Fig. 8. Two simple shapes are tested, namely the concentric dual spheres and concentric dual cubes. It can be seen from Fig. 8 that although we only trained on evenly distributed Poisson-disk sampled point clouds (on ShapeNet), our method can process the uniformly randomly sampled point clouds effectively, and outperforms both PSR and BPA by a large margin. This further proves the strong generalizability of our method.

Noisy Data and Real Scans. Since our method directly interpolate triangles upon input point clouds, the algorithm may be sensitive to the noise. However, as shown in Fig. 9, without explicit denoising and data augmentation mechanisms, our method is still resistant to the noise to a certain extent. As for real scans, there may be more issues, such as part missing and uneven distribution of the points. With the point set consolidation network [47] as pre-processing, our method can generate satisfying meshes from real-world LiDAR scans (see Fig. 9). In the future, we would like to explore explicit ways to propose the position of the vertices and compensate for the structural loss of input [28].

5 Conclusions

In this paper, we proposed a novel learning-based framework for mesh reconstruction that is based on the grounded point clouds and explicitly estimates the local connectivity of the points. By leveraging the intrinsic-extrinsic ratio as training guidance, the method is able to effectively distinguish the surface triangles and non-surface triangles. Extensive experiments have shown our superior performance, especially for preserving the details, handling ambiguous structures, and strong generalizability.

Acknowledgements. This work was funded in part by Kuaishou Technology, NSF grant IIS-1764078, NSF grant 1703957, the Ronald L. Graham chair and the UC San Diego Center for Visual Computing.

References

1. Badki, A., Gallo, O., Kautz, J., Sen, P.: Meshlet priors for 3D mesh reconstruction. arXiv preprint arXiv:2001.01744 (2020)
2. Bernardini, F., Mittleman, J., Rushmeier, H., Silva, C., Taubin, G.: The ball-pivoting algorithm for surface reconstruction. IEEE Trans. Vis. Comput. Graph. **5**(4), 349–359 (1999)
3. Boissonnat, J.D., Geiger, B.: Three-dimensional reconstruction of complex shapes based on the delaunay triangulation. In: Biomedical Image Processing and Biomedical Visualization, vol. 1905, pp. 964–975. International Society for Optics and Photonics (1993)
4. Carr, J.C., et al.: Reconstruction and representation of 3D objects with radial basis functions. In: Proceedings of the 28th Annual Conference on Computer Graphics and Interactive Techniques, pp. 67–76 (2001)
5. Chang, A.X., et al.: Shapenet: an information-rich 3d model repository. arXiv preprint arXiv:1512.03012 (2015)
6. Chen, Z., Tagliasacchi, A., Zhang, H.: BSP-Net: generating compact meshes via binary space partitioning. arXiv preprint arXiv:1911.06971 (2019)
7. Chen, Z., Zhang, H.: Learning implicit fields for generative shape modeling. In: Proceedings of the IEEE Conference on Computer Vision and Pattern Recognition, pp. 5939–5948 (2019)
8. Cignoni, P., Callieri, M., Corsini, M., Dellepiane, M., Ganovelli, F., Ranzuglia, G.: MeshLab: an open-source mesh processing tool. In: Eurographics Italian Chapter Conference 2008, pp. 129–136 (2008)
9. Corsini, M., Cignoni, P., Scopigno, R.: Efficient and flexible sampling with blue noise properties of triangular meshes. IEEE Trans. Vis. Comput. Graph. **18**(6), 914–924 (2012)
10. Deprelle, T., Groueix, T., Fisher, M., Kim, V., Russell, B., Aubry, M.: Learning elementary structures for 3D shape generation and matching. In: Advances in Neural Information Processing Systems, pp. 7433–7443 (2019)
11. Digne, J.: An analysis and implementation of a parallel ball pivoting algorithm. Image Process. Line **4**, 149–168 (2014)
12. Edelsbrunner, H., Mücke, E.P.: Three-dimensional alpha shapes. ACM Trans. Graph. (TOG) **13**(1), 43–72 (1994)
13. Fan, H., Su, H., Guibas, L.J.: A point set generation network for 3d object reconstruction from a single image. In: Proceedings of the IEEE Conference on Computer Vision and Pattern Recognition, pp. 605–613 (2017)
14. Gao, L., et al.: SDM-NET: deep generative network for structured deformable mesh. ACM Trans. Graph. (TOG) **38**(6), 1–15 (2019)

15. Genova, K., Cole, F., Vlasic, D., Sarna, A., Freeman, W.T., Funkhouser, T.: Learning shape templates with structured implicit functions. In: Proceedings of the IEEE International Conference on Computer Vision, pp. 7154–7164 (2019)
16. Gkioxari, G., Malik, J., Johnson, J.: Mesh R-CNN. In: Proceedings of the IEEE International Conference on Computer Vision, pp. 9785–9795 (2019)
17. Graham, B., Engelcke, M., van der Maaten, L.: 3D semantic segmentation with submanifold sparse convolutional networks. In: Proceedings of the IEEE Conference on Computer Vision and Pattern Recognition, pp. 9224–9232 (2018)
18. Groueix, T., Fisher, M., Kim, V.G., Russell, B.C., Aubry, M.: 3D-CODED: 3D correspondences by deep deformation. In: Ferrari, V., Hebert, M., Sminchisescu, C., Weiss, Y. (eds.) ECCV 2018. LNCS, vol. 11206, pp. 235–251. Springer, Cham (2018). https://doi.org/10.1007/978-3-030-01216-8_15
19. Groueix, T., Fisher, M., Kim, V.G., Russell, B.C., Aubry, M.: A papier-mâché approach to learning 3D surface generation. In: Proceedings of the IEEE Conference on Computer Vision and Pattern Recognition, pp. 216–224 (2018)
20. Guennebaud, G., Germann, M., Gross, M.: Dynamic sampling and rendering of algebraic point set surfaces. In: Computer Graphics Forum, vol. 27, pp. 653–662. Wiley Online Library (2008)
21. Guerrero, P., Kleiman, Y., Ovsjanikov, M., Mitra, N.J.: PCPNET learning local shape properties from raw point clouds. In: Computer Graphics Forum, vol. 37, pp. 75–85. Wiley Online Library (2018)
22. Hoppe, H., DeRose, T., Duchamp, T., McDonald, J., Stuetzle, W.: Surface reconstruction from unorganized points. In: Proceedings of the 19th Annual Conference on Computer Graphics and Interactive Techniques, pp. 71–78 (1992)
23. Kanazawa, A., Tulsiani, S., Efros, A.A., Malik, J.: Learning category-specific mesh reconstruction from image collections. In: Ferrari, V., Hebert, M., Sminchisescu, C., Weiss, Y. (eds.) ECCV 2018. LNCS, vol. 11219, pp. 386–402. Springer, Cham (2018). https://doi.org/10.1007/978-3-030-01267-0_23
24. Kazhdan, M., Bolitho, M., Hoppe, H.: Poisson surface reconstruction. In: Proceedings of the Fourth Eurographics Symposium on Geometry Processing, vol. 7 (2006)
25. Kazhdan, M., Hoppe, H.: Screened Poisson surface reconstruction. ACM Trans. Graph. (ToG) **32**(3), 1–13 (2013)
26. Liao, Y., Donne, S., Geiger, A.: Deep marching cubes: learning explicit surface representations. In: Proceedings of the IEEE Conference on Computer Vision and Pattern Recognition, pp. 2916–2925 (2018)
27. Liu, H.T.D., Kim, V.G., Chaudhuri, S., Aigerman, N., Jacobson, A.: Neural subdivision. arXiv preprint arXiv:2005.01819 (2020)
28. Liu, M., Sheng, L., Yang, S., Shao, J., Hu, S.M.: Morphing and sampling network for dense point cloud completion. arXiv preprint arXiv:1912.00280 (2019)
29. Liu, S., Chen, W., Li, T., Li, H.: Soft rasterizer: differentiable rendering for unsupervised single-view mesh reconstruction. arXiv preprint arXiv:1901.05567 (2019)
30. Lorensen, W.E., Cline, H.E.: Marching cubes: a high resolution 3D surface construction algorithm. ACM SIGGRAPH Comput. Graph. **21**(4), 163–169 (1987)
31. Luo, T., et al.: Learning to group: a bottom-up framework for 3d part discovery in unseen categories. In: International Conference on Learning Representations (2020). https://openreview.net/forum?id=rkl8dlHYvB
32. Mescheder, L., Oechsle, M., Niemeyer, M., Nowozin, S., Geiger, A.: Occupancy networks: Learning 3D reconstruction in function space. In: Proceedings of the IEEE Conference on Computer Vision and Pattern Recognition, pp. 4460–4470 (2019)

33. Mo, K., et al.: Partnet: a large-scale benchmark for fine-grained and hierarchical part-level 3d object understanding. In: Proceedings of the IEEE Conference on Computer Vision and Pattern Recognition, pp. 909–918 (2019)
34. Öztireli, A.C., Guennebaud, G., Gross, M.: Feature preserving point set surfaces based on non-linear kernel regression. In: Computer Graphics Forum, vol. 28, pp. 493–501. Wiley Online Library (2009)
35. Pan, J., Han, X., Chen, W., Tang, J., Jia, K.: Deep mesh reconstruction from single RGB images via topology modification networks. In: Proceedings of the IEEE International Conference on Computer Vision, pp. 9964–9973 (2019)
36. Park, J.J., Florence, P., Straub, J., Newcombe, R., Lovegrove, S.: DeepSDF: learning continuous signed distance functions for shape representation. In: Proceedings of the IEEE Conference on Computer Vision and Pattern Recognition, pp. 165–174 (2019)
37. Qi, C.R., Su, H., Mo, K., Guibas, L.J.: PointNet: deep learning on point sets for 3D classification and segmentation. In: Proceedings of the IEEE Conference on Computer Vision and Pattern Recognition, pp. 652–660 (2017)
38. Qi, C.R., Yi, L., Su, H., Guibas, L.J.: Pointnet++: deep hierarchical feature learning on point sets in a metric space. In: Advances in Neural Information Processing Systems, pp. 5099–5108 (2017)
39. Sharma, G., Goyal, R., Liu, D., Kalogerakis, E., Maji, S.: CSGNet: neural shape parser for constructive solid geometry. In: Proceedings of the IEEE Conference on Computer Vision and Pattern Recognition, pp. 5515–5523 (2018)
40. Surazhsky, V., Surazhsky, T., Kirsanov, D., Gortler, S.J., Hoppe, H.: Fast exact and approximate geodesics on meshes. ACM Trans. Graph. (TOG) **24**(3), 553–560 (2005)
41. Tulsiani, S., Su, H., Guibas, L.J., Efros, A.A., Malik, J.: Learning shape abstractions by assembling volumetric primitives. In: Proceedings of the IEEE Conference on Computer Vision and Pattern Recognition, pp. 2635–2643 (2017)
42. Turk, G., Levoy, M.: Zippered polygon meshes from range images. In: Proceedings of the 21st Annual Conference on Computer Graphics and Interactive Techniques, pp. 311–318 (1994)
43. Wang, N., Zhang, Y., Li, Z., Fu, Y., Liu, W., Jiang, Y.-G.: Pixel2Mesh: generating 3D mesh models from single RGB images. In: Ferrari, V., Hebert, M., Sminchisescu, C., Weiss, Y. (eds.) ECCV 2018. LNCS, vol. 11215, pp. 55–71. Springer, Cham (2018). https://doi.org/10.1007/978-3-030-01252-6_4
44. Wen, C., Zhang, Y., Li, Z., Fu, Y.: Pixel2mesh++: multi-view 3D mesh generation via deformation. In: Proceedings of the IEEE International Conference on Computer Vision, pp. 1042–1051 (2019)
45. Williams, F., Schneider, T., Silva, C., Zorin, D., Bruna, J., Panozzo, D.: Deep geometric prior for surface reconstruction. In: Proceedings of the IEEE Conference on Computer Vision and Pattern Recognition, pp. 10130–10139 (2019)
46. Yang, Y., Feng, C., Shen, Y., Tian, D.: Foldingnet: interpretable unsupervised learning on 3D point clouds. arXiv preprint arXiv:1712.07262 (2017)
47. Yu, L., Li, X., Fu, C.-W., Cohen-Or, D., Heng, P.-A.: EC-Net: an edge-aware point set consolidation network. In: Ferrari, V., Hebert, M., Sminchisescu, C., Weiss, Y. (eds.) ECCV 2018. LNCS, vol. 11211, pp. 398–414. Springer, Cham (2018). https://doi.org/10.1007/978-3-030-01234-2_24

Improved Adversarial Training via Learned Optimizer

Yuanhao Xiong(✉) and Cho-Jui Hsieh

University of California, Los Angeles, CA 90024, USA
{yhxiong,chohsieh}@cs.ucla.edu

Abstract. Adversarial attack has recently become a tremendous threat to deep learning models. To improve the robustness of machine learning models, adversarial training, formulated as a minimax optimization problem, has been recognized as one of the most effective defense mechanisms. However, the non-convex and non-concave property poses a great challenge to the minimax training. In this paper, we empirically demonstrate that the commonly used PGD attack may not be optimal for inner maximization, and improved inner optimizer can lead to a more robust model. Then we leverage a learning-to-learn (L2L) framework to train an optimizer with recurrent neural networks, providing update directions and steps adaptively for the inner problem. By co-training optimizer's parameters and model's weights, the proposed framework consistently improves over PGD-based adversarial training and TRADES.

Keywords: Optimization · Adversarial training · Learning to learn

1 Introduction

It has been widely acknowledged that deep neural networks (DNN) have made tremendous breakthroughs benefiting both academia and industry. Despite being effective, many DNN models trained with benign inputs are vulnerable to small and undetectable perturbation added to original data and tend to make wrong predictions under such threats. Those perturbed examples, also known as adversarial examples, can be easily constructed by algorithms such as DeepFool [23], Fast Gradient Sign Method (FGSM) [11], and Carlini-Wagner (C&W) attack [4]. Moreover, such adversarial attacks can also be conducted in the black-box setting [3,5,6] and can appear naturally in the physical world [12,16]. This phenomenon can bring about serious consequences in domains such as face recognition and autonomous-driving. Therefore, how to train a model resistant to adversarial inputs has become an important topic.

A variety of defense methods have been proposed to improve the performance of DNNs against adversarial attacks [17,27,30,31,35,38]. Among them,

Electronic supplementary material The online version of this chapter (https://doi.org/10.1007/978-3-030-58598-3_6) contains supplementary material, which is available to authorized users.

A. Vedaldi et al. (Eds.): ECCV 2020, LNCS 12353, pp. 85–100, 2020.
https://doi.org/10.1007/978-3-030-58598-3_6

adversarial training [17] stands out for its effectiveness. Moreover, [21] shows that adversarial training can be formulated as a minimax optimization problem, resembling a game between the attacker and the defender. The formulation is so intuitive that the inner problem aims at generating adversarial examples by maximizing the training loss while the outer one guides the network in the direction that minimizes the loss to resist attacks. However, directly obtaining the optimal value of the inner maximization is infeasible, so one has to run an iterative optimization algorithm for a fixed number (often 10) iterations to get an approximate inner maximizer.

Existing adversarial training often uses hand-designed general purpose optimizers, such as PGD attack, to (approximately) solve the inner maximization. However, there is an essential property of adversarial training that is rarely explored: the maximization problems associated with each sample share very similar structure, and a good inner maximizer for adversarial training only needs to work well for this set of data-dependent problems. To be specific, there are a finite of n maximization problems need to be solved (where n is number of training samples), and those maximization problems share the same objective function along with identical network structure and weights, and the only difference is their input $\boldsymbol{x}$. Based on this observation, can we have a better optimizer that in particular works well for these very similar and data-dependent problems?

Motivated by this idea, we propose a learned optimizer for improved adversarial training. Instead of using an existing optimizer with a fixed update rule (such as PGD), we aim at learning the inner maximizer that could be faster and more effective for this particular set of maximization problems. We have noticed that two works have already put forward algorithms to combine learning to learn with adversarial training [13,14]. Both of them adopt a convolutional neural network (CNN) generator to produce malicious perturbations whereas CNN structure might complicate the training process and cannot grasp the essence of the update rule in the long term. In contrast, we propose an L2L-based adversarial training method with recurrent neural networks (RNN). RNN is capable of capturing long-term dependencies and has shown great potentials in predicting update directions and steps adaptively [20]. Thus, following the framework in [1], we leverage RNN as the optimizer to generate perturbations in a coordinate-wise manner. Based on the properties of the inner problem, we tailor our RNN optimizer with removed bias and weighted loss for further elaborations to ameliorate issues like short-horizon in L2L [33].

Specifically, our main contributions in this paper are summarized as follows:

- We first investigate and confirm the improvement in the model robustness from stronger attacks by searching a suitable step size for PGD.
- In replacement of hand-designed algorithms like PGD, an RNN-based optimizer based on the properties of the inner problem is designed to learn a better update rule. In addition to standard adversarial training, the proposed algorithm can also be applied to any other minimax defense objectives such as TRADES [38].

- Comprehensive experimental results show that the proposed method can noticeably improve the robust accuracy of both adversarial training [21] and TRADES [38]. Furthermore, our RNN-based adversarial training significantly outperforms previous CNN-based L2L adversarial training and requires much less number of trainable parameters.

2 Related Work

2.1 Adversarial Attack and Defense

Model robustness has recently become a great concern for deploying deep learning models in real-world applications. Goodfellow *et al.*[11] succeeded in fooling the model to make wrong predictions by Fast Gradient Sign Method (FGSM). Subsequently, to produce adversarial examples, IFGSM and Projected Gradient Descent (PGD) [11,21] accumulate attack strength through running FGSM iteratively, and Carlini-Wagner (C&W) attack [4] designs a specific objective function to increase classification errors. Besides these conventional optimization-based methods, there are several algorithms [25,34] focusing on generating malicious perturbations via neural networks. For instance, Xiao *et al.*[34] exploit GAN, which is originally designed for crafting deceptive images, to output corresponding noises added to benign input data. The appearance of various attacks has pushed forward the development of effective defense algorithms to train neural networks that are resistant to adversarial examples. The seminal work of adversarial training has significantly improved adversarial robustness [21]. It has inspired the emergence of various advanced defense algorithms: TRADES [38] is designed to minimize a theoretically-driven upper bound and GAT [19] takes generator-based outputs to train the robust classifier. All these methods can be formulated as a minimax problem [21], where the defender makes efforts to mitigate negative effects (outer minimization) brought by adversarial examples from the attacker (inner maximization). Whereas, performance of such an adversarial game is usually constrained by the quality of solutions to the inner problem [13,14]. Intuitively, searching a better maxima for the inner problem can improve the solution of minimax training, leading to improved defensive models.

2.2 Learning to Learn

Recently, learning to learn emerges as a novel technique to efficiently address a variety of problems such as automatic optimization [1], few-shot learning [10], and neural architecture search [8]. In this paper, we emphasize on the subarea of L2L: how to learn an optimizer for better performance. Rather than using human-defined update rules, learning to learn makes use of neural networks for designing optimization algorithms automatically. It is developed originally from [7] and [36], in which early attempts are made to model adaptive algorithms on simple convex problems. More recently, [1] proposes an LSTM optimizer for some complex optimization problems, such as training a convolutional neural network classifier. Based on this work, elaborations in [20] and [32] further improve the generalization and scalability for learned optimizers. Moreover, [26] demonstrates that a zeroth order optimizer can also be learned using L2L. Potentials of learning-to-learn motivates a line of L2L-based defense which replaces hand-designed methods for solving the inner problem with neural network optimizers. [14] uses a CNN generator mapping clean images into corresponding perturbations. Since it only makes one-step and deterministic attack like FGSM, [13] modifies the algorithm and produces stronger and more diverse attacks iteratively. Unfortunately, due to the large number of parameters and the lack of ability to capture the long-term dependencies, the CNN generator adds too much difficulty in the optimization, especially for the minimax problem in adversarial training. Therefore, we adopt an RNN optimizer in our method for a more stable training process as well as a better grasp of the update rule.

3 Preliminaries

3.1 Notations

We use bold lower-case letters $\boldsymbol{x}$ and $\boldsymbol{y}$ to represent clean images and their corresponding labels. An image classification task is considered in this paper with the classifier f parameterized by θ. $\mathsf{sign}(\cdot)$ is an elementwise operation to output the sign of a given input with $\mathsf{sign}(0) = 1$. $\mathbb{B}(\boldsymbol{x}, \epsilon)$ denotes the neighborhood of $\boldsymbol{x}$ as well as the set of admissible perturbed images: $\{\boldsymbol{x}' : \|\boldsymbol{x}' - \boldsymbol{x}\|_\infty \leq \epsilon\}$, where the infinity norm is adopted as the distance metric. We denote by Π the projection operator that maps perturbed data to the feasible set. Specifically, $\Pi_{\mathbb{B}(\boldsymbol{x},\epsilon)}(\boldsymbol{x}') = \max(\boldsymbol{x} - \epsilon, \min(\boldsymbol{x}', \boldsymbol{x} + \epsilon))$, which is an elementwise operator. $\mathcal{L}(\cdot, \cdot)$ is a multi-class loss like cross-entropy.

3.2 Adversarial Training

In this part, we present the formulation of adversarial training, together with some hand-designed optimizers to solve this problem. To obtain a robust classifier against adversarial attacks, an intuitive idea is to minimize the robust loss, defined as the worst-case loss within a small neighborhood $\mathbb{B}(\boldsymbol{x}, \epsilon)$. Adversarial training, which aims to find the weights that minimize the robust loss, can be formulated as a minimax optimization problem in the following way [21]:

$$\min_\theta \mathbb{E}_{(\boldsymbol{x},\boldsymbol{y})\sim D} \left\{ \max_{\boldsymbol{x}' \in \mathbb{B}(\boldsymbol{x},\epsilon)} \mathcal{L}(f(\boldsymbol{x}'), \boldsymbol{y}) \right\} \tag{1}$$

where D is the empirical distribution of input data. However, (1) only focuses on accuracy over adversarial examples and might cause severe over-fitting issues on the training set. To address this problem, TRADES [38] investigates the trade-off between natural and robust errors and theoretically puts forward a different objective function for adversarial training:

$$\min_{\theta} \mathbb{E}_{(\boldsymbol{x},\boldsymbol{y})\sim D} \left\{ \mathcal{L}(f(\boldsymbol{x}), \boldsymbol{y}) + \max_{\boldsymbol{x}'\in\mathbb{B}(\boldsymbol{x},\epsilon)} \mathcal{L}(f(\boldsymbol{x}), f(\boldsymbol{x}'))/\lambda \right\}. \tag{2}$$

Note that (1) and (2) are both defined as minimax optimization problems, and to solve such saddle point problems, a commonly used approach is to first get an approximate solution $\boldsymbol{x}'$ of inner maximization based on the current θ, and then use $\boldsymbol{x}'$ to conduct updates on model weights θ. The adversarial training procedure then iteratively runs this on each batch of samples until convergence. Clearly, the quality and efficiency of inner maximization is crucial to the performance of adversarial training. The most commonly used inner maximizer is the projected gradient descent algorithm, which conducts a fixed number of updates:

$$\boldsymbol{x}'_{t+1} = \Pi_{\mathbb{B}(\boldsymbol{x},\epsilon)}(\alpha \mathsf{sign}(\nabla_{\boldsymbol{x}'}\mathcal{L}(\boldsymbol{x}'_t)) + \boldsymbol{x}'_t). \tag{3}$$

Here $\mathcal{L}(\boldsymbol{x}'_t)$ represents the maximization term in (1) or (2) with abuse of notation.

3.3 Effects of Adaptive Step Sizes

We found that the performance of adversarial training crucially depends on the optimization algorithm used for inner maximization, and the current widely used PGD algorithm may not be the optimal choice. Here we demonstrate that even a small modification of PGD and without any change to the adversarial training objective can boost the performance of model robustness. We use the CNN structure in [38] to train a classifier on MNIST dataset. When 10-step PGD (denoted by PGD for simplicity) is used for the inner maximization, a constant step size is always adopted, which may not be suitable for the subsequent update. Therefore, we make use of backtracking line search (BLS) to select a step size adaptively for adversarial training (AdvTrain as abbreviation). Starting with a maximum candidate step size value α_0, we iteratively decrease it by $\alpha_t = \rho\alpha_{t-1}$ until the following condition is satisfied:

$$\mathcal{L}(\boldsymbol{x}' + \alpha_t \boldsymbol{p}) \geq \mathcal{L}(\boldsymbol{x}') + c\alpha_t \boldsymbol{p}^{\mathrm{T}}\boldsymbol{p} \tag{4}$$

where $\boldsymbol{p} = \nabla_{\boldsymbol{x}'}\mathcal{L}(\boldsymbol{x}')$ is a search direction. Based on a selected control parameter $c \in (0, 1)$, the condition tests whether the update with step size α_t leads to sufficient increase in the objective function, and it is guaranteed that a sufficiently small α will satisfy the condition so line search will always stop in finite steps. This is standard in gradient ascent (descent) optimization, and see more discussions in [24]. Following the convention, we set $\rho = 0.5$ and $c = 10^{-4}$. As shown in Table 1, defense with AdvTrain+BLS leads to a more robust model than solving the inner problem only by PGD (88.71% vs 87.33%). At the same time the attacker combined with BLS generates stronger adversarial examples: the robust accuracy of the model trained from vanilla adversarial training drops over 1.2% with PGD+BLS, compared to merely PGD attack. This experiment motivates our efforts to find a better inner maximizer for adversarial training.

Table 1. Effects of the inner solution quality on robust accuracy (%)

Defense	Attack		
	Natural	PGD	PGD+BLS
AdvTrain	96.43	87.33	86.09
AdvTrain+BLS	**96.70**	**88.71**	**88.00**

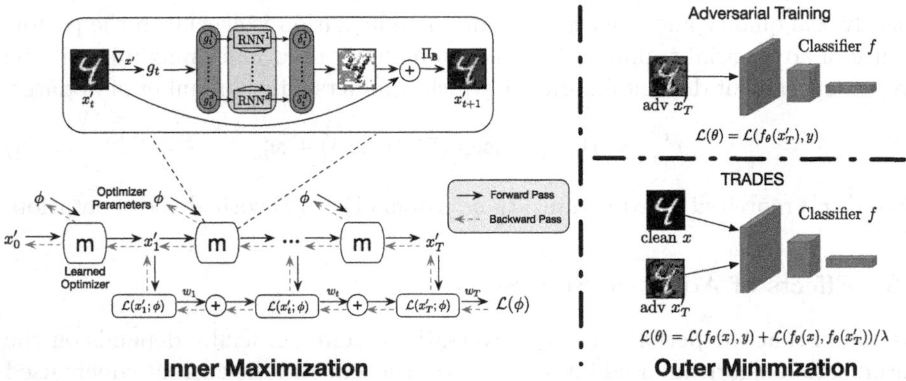

Fig. 1. Model architecture of our defense method with an RNN optimizer

4 Proposed Algorithm

4.1 Learning to Learn for Adversarial Training

As mentioned in the previous section, it can be clearly seen that the inner maximizer plays an important role in the performance of adversarial training. However, despite the effectiveness of BLS introduced in Subsect. 3.3, it is impractical to combine it with adversarial training as multiple line searches together with loss calculation in this algorithms increase its computational burden significantly. Then a question arises naturally: is there any automatic way for determining a good step size for inner maximization without too much computation overhead? Moreover, apart from the step size, the question can be extended to whether such a maximizer can be learned for a particular dataset and model in replacement of a general optimizer like PGD. Recently, as a subarea of learning-to-learn, researchers have been investigating whether it is possible to use machine learning, especially neural networks, to learn improved optimizer to replace the hand-designed optimizer [1,20,32]. However, it is commonly believed that those ML-learned general-purpose optimizers are still not practically useful due to several unsolved issues. For instance, the exploded gradient [22] in unrolled optimization impedes generalization of these learned optimizers to longer steps and truncated optimization on the other hand induces short-horizon bias [33].

In this paper, we show it is possible and practical to learn an optimizer for inner maximization in adversarial training. Note that in adversarial training, the maximization problems share very similar form: $\max_{\boldsymbol{x}' \in \mathbb{B}(\boldsymbol{x},\epsilon)} \mathcal{L}(f(\boldsymbol{x}'), \boldsymbol{y})$, where they all have the same loss function $\mathcal{L}$ and the same network (structure and weights) f, and the only difference is their input $\boldsymbol{x}$ and label $\boldsymbol{y}$. Furthermore, we only need the maximizer to perform well on a fixed set of n optimization problems for adversarial training. These properties thus enable us to learn a better optimizer that outperforms PGD.

To allow a learned inner maximizer, we parameterize the learned optimizer by an RNN network. This is following the literature of learning-to-learn [1], but we propose several designs as shown below that works better for our inner maximization problem which is a constrained optimization problem instead of a standard unconstrained training task in [1]. We then jointly optimize the classifier parameters (θ) as well as the parameters of the inner maximizer (ϕ). The overall framework can be found in Fig. 1.

Specifically, the inner problem is to maximize vanilla adversarial training loss in (1) or TRADES loss in (2), with a constraint that $\boldsymbol{x}' \in \mathbb{B}(\boldsymbol{x}, \epsilon)$. We expand on adversarial training here and more details about TRADES can be found in Appendix A. With an RNN optimizer m parameterized by ϕ, we propose the following parameterized update rule to mimic the PGD update rule in (3):

$$\boldsymbol{\delta}_t, \boldsymbol{h}_{t+1} = m_\phi(\boldsymbol{g}_t, \boldsymbol{h}_t), \quad \boldsymbol{x}'_{t+1} = \Pi_{\mathbb{B}(\boldsymbol{x},\epsilon)} \left(\boldsymbol{x}'_t + \boldsymbol{\delta}_t\right). \tag{5}$$

Here, $\boldsymbol{g}_t$ is the gradient $\nabla_{\boldsymbol{x}'} \mathcal{L}(f(\boldsymbol{x}'), \boldsymbol{y})$ and $\boldsymbol{h}_t$ is the hidden state representation. It has to be emphasized that our RNN optimizer generates perturbations coordinate-wisely, in contrast to other L2L based methods which take as input the entire image. This property reduces trainable parameters significantly, making it much easier and faster for training. In addition, note that the hidden state of our RNN optimizer plays an important role in the whole optimization. A separate hidden state for each coordinate guarantees the different update behavior. And it contains richer information like the trajectory of loss gradients mentioned in [13] but can produce a recursive update with a simpler structure.

For the RNN design, we mainly follow the structure in [1] but with some modifications to make it more suitable to adversarial training. We can expand the computation of perturbation for each step as:

$$\boldsymbol{\delta}_t = \tanh(\boldsymbol{V}\boldsymbol{h}_t + \boldsymbol{b}_1), \tag{6}$$

$$\boldsymbol{h}_{t+1} = \tanh(\boldsymbol{U}\boldsymbol{g}_t + \boldsymbol{W}\boldsymbol{h}_t + \boldsymbol{b}_2) \tag{7}$$

where $\boldsymbol{h}_t \in \mathbb{R}^d$, $\boldsymbol{V} \in \mathbb{R}^{1\times d}$, $\boldsymbol{U} \in \mathbb{R}^{d\times 1}$, $\boldsymbol{W} \in \mathbb{R}^{d\times d}$, $\boldsymbol{b}_1 \in \mathbb{R}$ and $\boldsymbol{b}_2 \in \mathbb{R}^d$ in the coordinate-wise update manner. As the optimization proceeds, the gradient will become much smaller when approaching the local maxima. At that time, a stable value of the perturbation is expected without much change between two consecutive iterations. However, from (6) and (7), we can clearly see that despite small $\boldsymbol{g}_t$, the update rule will still produce an update with magnitude proportional to $\tanh(\boldsymbol{b}_1)$. Imagine the case where the exact optimal value is found with an all-zero hidden state ($\boldsymbol{b}_2$ needs to be zero as well), $\boldsymbol{\delta}_t = \tanh(\boldsymbol{b}_1)$ with a non-zero bias will push the adversarial example away from the optimal one. Thus, two bias terms $\boldsymbol{b}_1$ and $\boldsymbol{b}_2$ are problematic for optimization close to the optimal solution. Due to the short horizon of the inner maximization in adversarial training, it is unlikely for the network to learn zero bias terms. Therefore, to ensure stable training, we remove the bias terms in the vanilla RNN in all implementations.

With an L2L framework, we simultaneously train the RNN optimizer parameters ϕ and the classifier weights θ together. The joint optimization problem can be formulated as follows:

$$\min_{\theta} \ \mathbb{E}_{(\boldsymbol{x},\boldsymbol{y})\sim D}\{\mathcal{L}(f_\theta(\boldsymbol{x}'_T(\phi^*)), \boldsymbol{y})\} \tag{8}$$

$$\text{s.t.} \ \ \phi^* = \arg\max \mathcal{L}(\phi) \tag{9}$$

where $\boldsymbol{x}'_T(\phi^*)$ is computed by running Eq. (5) T times iteratively. Since the learned optimizer aims at finding a better solution to the inner maximization term, the objective function for training it in the horizon T is defined as:

$$\mathcal{L}(\phi) = \sum_{t=1}^{T} w_t \mathcal{L}(f_\theta(\boldsymbol{x}'_t(\phi)), \boldsymbol{y}). \tag{10}$$

Note that if we set $w_t = 0$ for all $t < T$ and $w_T = 1$, then (10) implies that our learned maximizer m_ϕ will maximize the loss after T iterations. However, in practice we found that considering intermediate iterations can further improve the performance since it will make the maximizer converges faster even after conducting one or few iterations. Therefore in the experiments we set an increased weights $w_t = t$ for $t = 1, \ldots, T$. Note that [22] showed that this kind of unrolled optimization may lead to some issues such as exploded gradients which is still an unsolved problem in L2L. However, in adversarial training we only need to set a relative small T (e.g., $T = 10$) so we do not encounter that issue.

While updating the learned optimizer, corresponding adversarial examples are produced together. We can then train the classifier by minimizing the loss accordingly. The whole algorithm is presented in Algorithm 1.

4.2 Advantages over Other L2L-Based Methods

Previous methods have proposed to use a CNN generator [13,14] to produce perturbations in adversarial training. However, CNN-based generator has a larger number of trainable parameters, which makes it hard to train. In Table 2, the detailed properties including the number of parameter and training time per epoch are provided for different learning-to-learn based methods. We can observe that our proposed RNN approach stands out with the smallest parameters as well as efficiency in training. Specifically, our RNN optimizer only has 120 parameters, almost 5000 times fewer than L2LDA while the training time per epoch is 268.50 s (RNN-TRADES only consumes 443.52 s per training epoch) v.s. 1972.41 s. Furthermore, our method also leads to better empirical performance, as shown in our main comparison in Tables 3, 4 and 5. Comparison of our variants and original adversarial training methods can be found in Appendix B.

Algorithm 1. RNN-based adversarial training

Input: clean data $\{(\boldsymbol{x}, \boldsymbol{y})\}$, batch size B, step sizes α_1 and α_2, number of inner iterations T, classifier parameterized by θ, RNN optimizer parameterized by ϕ
Output: Robust classifier f_θ, learned optimizer m_ϕ
Randomly initialize f_θ and m_ϕ, or initialize them with pre-trained configurations
repeat
  Sample a mini-batch M from clean data.
  for $(\boldsymbol{x}, \boldsymbol{y})$ **in** B **do**
    Initialization: $\boldsymbol{h}_0 \leftarrow 0$, $\mathcal{L}_\theta \leftarrow 0$, $\mathcal{L}_\phi \leftarrow 0$
    Gaussian augmentation: $\boldsymbol{x}'_0 \leftarrow \boldsymbol{x} + 0.001 \cdot \mathcal{N}(\boldsymbol{0}, \boldsymbol{I})$
    for $t = 0, \ldots, T-1$ **do**
      $\boldsymbol{g}_t \leftarrow \nabla_{\boldsymbol{x}'} \mathcal{L}(f_\theta(\boldsymbol{x}'_t), \boldsymbol{y})$
      $\boldsymbol{\delta}_t, \boldsymbol{h}_{t+1} \leftarrow m_\phi(\boldsymbol{g}_t, \boldsymbol{h}_t)$, where coordinate-wise update is applied
      $\boldsymbol{x}'_{t+1} \leftarrow \Pi_{\mathbb{B}(\boldsymbol{x},\epsilon)}(\boldsymbol{x}'_t + \boldsymbol{\delta}_t)$
      $\mathcal{L}_\phi \leftarrow \mathcal{L}_\phi + w_{t+1}\mathcal{L}(f_\theta(\boldsymbol{x}'_{t+1}), \boldsymbol{y})$, where $w_{t+1} = t+1$
    end for
    $\mathcal{L}_\theta \leftarrow \mathcal{L}_\theta + \mathcal{L}(f_\theta(\boldsymbol{x}'_T), \boldsymbol{y})$
  end for
  Update ϕ by $\phi \leftarrow \phi + \alpha_1 \nabla_\phi \mathcal{L}_\phi / B$
  Update θ by $\theta \leftarrow \theta - \alpha_2 \nabla_\theta \mathcal{L}_\theta / B$
until training converged

Table 2. Comparion among different L2L-based methods

	Number of parameters	Training time per epoch (s)
RNN-Adv	**120**	**268.50**
RNN-TRADES	**120**	443.52
L2LDA	500944	1972.41

5 Experimental Results

In this section, we present experimental results of our proposed RNN-based adversarial training. We compare our method with various baselines against both white-box and black-box attack. In addition, different datasets and network architectures are also evaluated.

5.1 Experimental Settings

- **Datasets and classifier networks.** We mainly use MNIST [18] and CIFAR-10 [15] datasets for performance evaluation in our experiments. For MNIST, the CNN architecture with four convolutional layers in [4] is adopted as the classifier. For CIFAR-10, we use both the standard VGG-16 [28] and Wide ResNet [37], which has been used in most of the previous defense papers including adversarial training [21] and TRADES [38]. We also conduct an additional experiment on Restricted ImageNet [29] with ResNet-18 and results are presented in Appendix C.

- **Baselines for Comparison.** Note that our method is an optimization framework which is irrelevant to what minimax objective function is used. Therefore we choose two most popular minimax formulations, AdvTrain[1] [21] and TRADES[2] [38], and substitute the proposed L2L-based optimization for their original PGD-based algorithm. Moreover, we also compare with a previous L2L defense mechanism L2LDA[3] [13] which outperforms other L2L-based methods for thorough comparison. We use the source code provided by the authors on github with their recommended hyper-parameters for all these baseline methods.
- **Evaluation and implementation details.** Defense algorithms are usually evaluated by classification accuracy under different attacks. Effective attack algorithms including PGD, C&W and the attacker of L2LDA are used for evaluating the model robustness, with the maximum ℓ_∞ perturbation strength $\epsilon = 0.3$ for MNIST and $\epsilon = 8/255$ for CIFAR-10. For PGD, we run 10 and 100 iterations (PGD-10 and -100) with the step size $\eta = \epsilon/4$, as suggested in [13]. C&W is implemented with 100 iterations in the infinity norm. For L2LDA attacker, it is learned from L2LDA [13] under different settings with 10 attack steps. In addition, we also uses the learned optimizer of RNN-Adv to conduct 10-step attacks.

For our proposed RNN-based defense, we use a one-layer vanilla RNN with the hidden size of 10 as the optimizer for the inner maximization. Since we test our method under two different minimax losses, we name them as RNN-Adv and RNN-TRADES respectively. The classifier and the optimizer are updated alternately according to the Algorithm 1. All algorithms are implemented in PyTorch-1.1.0 with four NVIDIA 1080Ti GPUs. Note that all adversarial training methods adopt 10-step inner optimization for fair comparison. We run each defense method five times with different random seeds and report the lowest classification accuracy.

5.2 Performance on White-Box Attacks

We demonstrate the robustness of models trained from different defense methods under the white-box setting in this part. Experimental results are shown in Tables 3, 4 and 5. From these three tables, we can observe that our proposed L2L-based adversarial training with RNN always outperforms its counterparts.

To be specific, our method achieves 95.80% robust accuracy among various attacks on MNIST dataset. On CIFAR-10, RNN-TRADES reaches 47.23% and 54.11 for VGG-16 and Wide ResNet with 1.28% and 1.43% gain over other baselines. It should be stressed that our method surpasses L2LDA (the previous CNN-based L2L method) noticeably. For conventional defense algorithms, our L2L-based variant improves the original method by 1%–2% percents under different attacks from comparison of robust accuracy in AdvTrain and RNN-Adv. A

[1] https://github.com/xuanqing94/BayesianDefense.
[2] https://github.com/yaodongyu/TRADES.
[3] https://github.com/YunseokJANG/l2l-da.

Table 3. Robust accuracy under white-box attacks (MNIST, 4-layer CNN)

Defense	Attack						
	Natural	PGD-10	PGD-100	CW100	L2LDA	RNN-Adv	Min
Plain	99.46	1.04	0.42	83.63	5.94	0.79	0.42
AdvTrain	99.17	94.89	94.28	98.38	95.83	94.39	94.28
TRADES	**99.52**	95.77	95.50	98.72	96.03	95.50	95.50
L2LDA	98.76	94.73	93.22	97.69	95.28	93.16	93.16
RNN-Adv	99.20	95.80	95.62	98.75	96.05	95.51	95.51
RNN-TRADES	99.46	**96.09**	**95.83**	**98.85**	**96.56**	**95.80**	**95.80**

Table 4. Robust accuracy under white-box attacks (CIFAR-10, VGG-16)

Defense	Attack						
	Natural	PGD-10	PGD-100	CW100	L2LDA	RNN-Adv	Min
Plain	**93.66**	0.74	0.09	0.08	0.89	0.43	0.08
AdvTrain	81.11	42.32	40.75	42.26	43.55	41.07	40.75
TRADES	78.08	48.83	48.30	45.94	49.94	48.38	45.95
L2LDA	77.47	35.49	34.27	35.31	36.27	34.54	34.27
RNN-Adv	81.22	44.98	42.89	43.67	46.20	43.21	42.89
RNN-TRADES	80.76	**50.23**	**49.42**	**47.23**	**51.29**	**49.49**	**47.23**

Table 5. Robust accuracy under white-box attacks (CIFAR-10, WideResNet)

Defense	Attack						
	Natural	PGD-10	PGD-100	CW100	L2LDA	RNN-Adv	Min
Plain	**95.14**	0.01	0.00	0.00	0.02	0.00	0.00
AdvTrain	86.28	46.64	45.13	46.64	48.46	45.41	45.13
TRADES	85.89	54.28	52.68	53.68	56.49	53.00	52.68
L2LDA	85.30	45.47	44.35	44.19	47.16	44.54	44.19
RNN-Adv	85.92	47.62	45.98	47.26	49.40	46.23	45.98
RNN-TRADES	84.21	**56.35**	**55.68**	**54.11**	**58.86**	**55.80**	**54.11**

similar phenomenon can also be observed in TRADES and RNN-TRADES. Since previous works of L2L-based defense only concentrate on PGD-based adversarial training, the substantial performance gain indicates that the learned optimizer can contribute to the minimax problem in TRADES as well. Furthermore, apart from traditional attack algorithms, we leverage our RNN optimizer learned from adversarial training as the attacker (the column RNN-Adv). Results in three experiments show that compared with other general attackers when conducting 10 iterations such as PGD-10 and L2LDA, ours is capable of producing much stronger perturbations which lead to low robust accuracy.

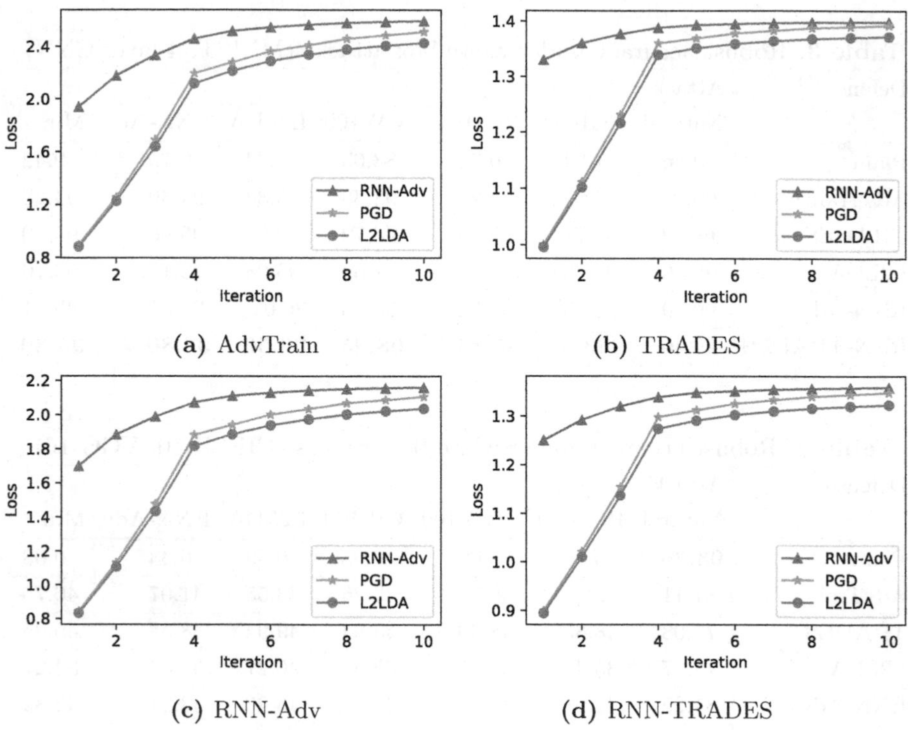

(a) AdvTrain (b) TRADES

(c) RNN-Adv (d) RNN-TRADES

Fig. 2. Comparison of optimization trajectories among various attack algorithms. We evaluate four defense mechanisms, AdvTrain TRADES, RNN-Adv and RNN-TRADES, under three attackers including PGD, L2LDA and our proposed RNN-Adv. All attackers conduct 10-step perturbing process

5.3 Analysis

Learned Optimizer. As mentioned in Subsect. 5.2, the optimizer learned from PGD-based adversarial training can be regarded as an special attacker. Thus, we primarily investigate the update trajectories of different attackers to obtain an in-depth understanding of our RNN optimizer. For VGG-16 models trained from four defense methods, three attacker are used to generate perturbations in 10 steps respectively and losses are recorded as shown in Fig. 2.

We can see clearly from these four figures that the losses obtained from RNN-Adv are always larger than others within 10 iterations, reflecting stronger attacks produced by our proposed optimizer. Moreover, it should be noted that the loss gap between RNN-Adv and other attackers is much more prominent at some very beginning iterations. This in fact demonstrates an advantage of the learning-to-learn framework that the optimizer can converge faster than hand-designed algorithms.

Table 6. Generalization to more steps of learned optimizer

Defense	Step	
	10	40
Plain	0.43	**0.03**
AdvTrain	41.07	**40.70**
TRADES	48.38	**48.27**
L2LDA	34.54	**34.19**
RNN-Adv	43.21	**42.89**
RNN-TRADES	49.49	**49.28**

Table 7. Robust accuracy under black-box attack settings

Defense	Surrogate	
	Plain-Net	PGD-Net
AdvTrain	79.94	62.57
TRADES	77.01	65.41
L2LDA	76.37	60.32
RNN-Adv	**80.58**	63.17
RNN-TRADES	79.54	**67.09**

Generalization to More Attack Steps. Although our learned RNN optimizer is only trained under 10 steps, we show that it can generalize to more steps as an attacker. From Table 6, we can observe that the attacker is capable of producing much stronger adversarial examples by extending its attack steps to 40. Performance of our attacker is even comparable with that of PGD-100, which further demonstrates the superiority of our proposed method.

5.4 Performance on Black-Box Transfer Attacks

We further test the robustness of the proposed defense method under transfer attack. As suggested by [2], this can be served as a sanity check to see whether our defense leads to obfuscated gradients and gives a false sense of model robustness. Following procedures in [2], we first train a surrogate model with the same architecture of the target model using a different random seed, and then generate adversarial examples from the surrogate model to attack the target model.

Specifically, we choose VGG-16 models obtained from various defense algorithms as our target models. In the meanwhile, we train two surrogate models: one is Plain-Net with natural training and the other is PGD-Net with 10-step PGD-based adversarial training. Results are presented in Table 7. We can observe that our method outperforms all other baselines, with RNN-PGD and RNN-TRADES standing out in defending attacks from Plain-Net and PGD-Net respectively. It suggests great resistance of our L2L defense to transfer attacks.

5.5 Loss Landscape Exploration

To further verify the superior performance of the proposed algorithm, we visualize the loss landscapes of VGG-16 models trained under different defense strategies, as shown in Fig. 3. According to the implementation in [9], we modify the input along a linear space defined by the sign of the gradient and a random Rademacher vector, where the x and y axes represent the magnitude of the perturbation added in each direction and the z axis represents the loss. It can be

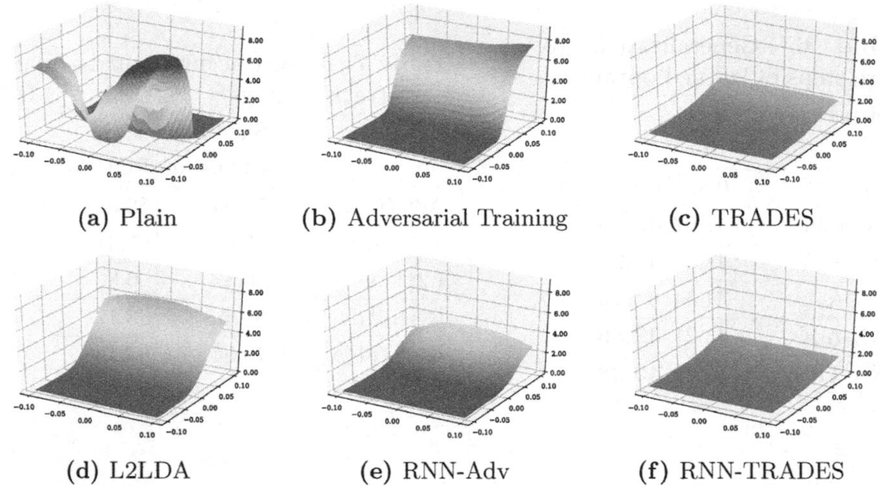

(a) Plain (b) Adversarial Training (c) TRADES

(d) L2LDA (e) RNN-Adv (f) RNN-TRADES

Fig. 3. Comparison of loss landscapes among different training methods. The color gradually changes from blue (low loss) to red (high loss) (Color figure online)

observed that loss surfaces of models trained from RNN-Adv and RNN-TRADES in Fig. 3e and f are much smoother than those of their counterparts in Fig. 3b and c. Besides, our method significantly reduces the loss value of perturbed data close to the original input. In particular, the maximum loss decreases roughly from 7.61 in adversarial training to 3.66 in RNN-Adv. Compared with L2LDA in Fig. 3d, the proposed RNN optimizer can contribute to less bumpier loss landscapes with smaller variance, which further demonstrates the stability and superiority of our L2L-based adversarial training.

6 Conclusion

For defense mechanisms that can be formulated as a minimax optimization problem, we propose to replace the inner PGD-based maximizer with a automatically learned RNN maximizer, and show that jointly training the RNN maximizer and classifier can significantly improve the defense performance. Empirical results demonstrate that the proposed approach can be combined with several minimax defense objectives, including adversarial training and TRADES.

For future work, it can be a worthwhile direction to address the inadequacy of L2L in dealing with a long-horizon problem. Then we can substitute the learned optimizer for hand-designed algorithms in both inner and outer problems, which enables an entirely automatic process for adversarial training.

References

1. Andrychowicz, M., et al.: Learning to learn by gradient descent by gradient descent. In: Advances in Neural Information Processing Systems, pp. 3981–3989 (2016)
2. Athalye, A., Carlini, N., Wagner, D.: Obfuscated gradients give a false sense of security: circumventing defenses to adversarial examples. In: International Conference on Machine Learning, pp. 274–283 (2018)
3. Brendel, W., Rauber, J., Bethge, M.: Decision-based adversarial attacks: reliable attacks against black-box machine learning models. arXiv preprint arXiv:1712.04248 (2017)
4. Carlini, N., Wagner, D.: Towards evaluating the robustness of neural networks. In: 2017 IEEE Symposium on Security and Privacy (SP), pp. 39–57. IEEE (2017)
5. Cheng, M., Le, T., Chen, P.Y., Yi, J., Zhang, H., Hsieh, C.J.: Query-efficient hard-label black-box attack: an optimization-based approach. arXiv preprint arXiv:1807.04457 (2018)
6. Cheng, M., Singh, S., Chen, P.H., Chen, P.Y., Liu, S., Hsieh, C.J.: Sign-OPT: a query-efficient hard-label adversarial attack. In: ICLR (2020)
7. Cotter, N.E., Conwell, P.R.: Fixed-weight networks can learn. In: 1990 IJCNN International Joint Conference on Neural Networks, pp. 553–559. IEEE (1990)
8. Elsken, T., Metzen, J.H., Hutter, F.: Neural architecture search: a survey. arXiv preprint arXiv:1808.05377 (2018)
9. Engstrom, L., Ilyas, A., Athalye, A.: Evaluating and understanding the robustness of adversarial logit pairing. arXiv preprint arXiv:1807.10272 (2018)
10. Finn, C., Abbeel, P., Levine, S.: Model-agnostic meta-learning for fast adaptation of deep networks. In: Proceedings of the 34th International Conference on Machine Learning, vol. 70, pp. 1126–1135. JMLR.org (2017)
11. Goodfellow, I.J., Shlens, J., Szegedy, C.: Explaining and harnessing adversarial examples. arXiv preprint arXiv:1412.6572 (2014)
12. Hendrycks, D., Zhao, K., Basart, S., Steinhardt, J., Song, D.: Natural adversarial examples. arXiv preprint arXiv:1907.07174 (2019)
13. Jang, Y., Zhao, T., Hong, S., Lee, H.: Adversarial defense via learning to generate diverse attacks. In: Proceedings of the IEEE International Conference on Computer Vision, pp. 2740–2749 (2019)
14. Jiang, H., Chen, Z., Shi, Y., Dai, B., Zhao, T.: Learning to defense by learning to attack. arXiv preprint arXiv:1811.01213 (2018)
15. Krizhevsky, A., Nair, V., Hinton, G.: CIFAR-10 (Canadian Institute for Advanced Research), 8 (2010). http://www.cs.toronto.edu/~kriz/cifar.html
16. Kurakin, A., Goodfellow, I., Bengio, S.: Adversarial examples in the physical world. arXiv preprint arXiv:1607.02533 (2016)
17. Kurakin, A., Goodfellow, I., Bengio, S.: Adversarial machine learning at scale. arXiv preprint arXiv:1611.01236 (2016)
18. LeCun, Y., Bottou, L., Bengio, Y., Haffner, P.: Gradient-based learning applied to document recognition. Proc. IEEE **86**(11), 2278–2324 (1998)
19. Lee, H., Han, S., Lee, J.: Generative adversarial trainer: defense to adversarial perturbations with GAN. arXiv preprint arXiv:1705.03387 (2017)
20. Lv, K., Jiang, S., Li, J.: Learning gradient descent: better generalization and longer horizons. In: Proceedings of the 34th International Conference on Machine Learning, vol. 70, pp. 2247–2255. JMLR.org (2017)
21. Madry, A., Makelov, A., Schmidt, L., Tsipras, D., Vladu, A.: Towards deep learning models resistant to adversarial attacks. arXiv preprint arXiv:1706.06083 (2017)

22. Metz, L., Maheswaranathan, N., Nixon, J., Freeman, D., Sohl-Dickstein, J.: Understanding and correcting pathologies in the training of learned optimizers. In: International Conference on Machine Learning, pp. 4556–4565 (2019)
23. Moosavi-Dezfooli, S.M., Fawzi, A., Frossard, P.: DeepFool: a simple and accurate method to fool deep neural networks. In: Proceedings of the IEEE Conference on Computer Vision and Pattern Recognition, pp. 2574–2582 (2016)
24. Nocedal, J., Wright, S.: Numerical Optimization. Springer Science & Business Media, New York (2006). https://doi.org/10.1007/978-0-387-40065-5
25. Reddy Mopuri, K., Ojha, U., Garg, U., Venkatesh Babu, R.: NAG: network for adversary generation. In: Proceedings of the IEEE Conference on Computer Vision and Pattern Recognition, pp. 742–751 (2018)
26. Ruan, Y., Xiong, Y., Reddi, S., Kumar, S., Hsieh, C.J.: Learning to learn by zeroth-order oracle. arXiv preprint arXiv:1910.09464 (2019)
27. Samangouei, P., Kabkab, M., Chellappa, R.: Defense-GAN: protecting classifiers against adversarial attacks using generative models. arXiv preprint arXiv:1805.06605 (2018)
28. Simonyan, K., Zisserman, A.: Very deep convolutional networks for large-scale image recognition. arXiv preprint arXiv:1409.1556 (2014)
29. Tsipras, D., Santurkar, S., Engstrom, L., Turner, A., Madry, A.: Robustness may be at odds with accuracy. arXiv preprint arXiv:1805.12152 (2018)
30. Wang, H., Yu, C.N.: A direct approach to robust deep learning using adversarial networks. arXiv preprint arXiv:1905.09591 (2019)
31. Wang, J., Zhang, H.: Bilateral adversarial training: towards fast training of more robust models against adversarial attacks. In: Proceedings of the IEEE International Conference on Computer Vision, pp. 6629–6638 (2019)
32. Wichrowska, O., et al.: Learned optimizers that scale and generalize. In: Proceedings of the 34th International Conference on Machine Learning, vol. 70, pp. 3751–3760. JMLR.org (2017)
33. Wu, Y., Ren, M., Liao, R., Grosse, R.: Understanding short-horizon bias in stochastic meta-optimization. arXiv preprint arXiv:1803.02021 (2018)
34. Xiao, C., Li, B., Zhu, J.Y., He, W., Liu, M., Song, D.: Generating adversarial examples with adversarial networks. In: Proceedings of the 27th International Joint Conference on Artificial Intelligence, pp. 3905–3911 (2018)
35. Xie, C., Wu, Y., van der Maaten, L., Yuille, A.L., He, K.: Feature denoising for improving adversarial robustness. In: Proceedings of the IEEE Conference on Computer Vision and Pattern Recognition, pp. 501–509 (2019)
36. Younger, A.S., Hochreiter, S., Conwell, P.R.: Meta-learning with backpropagation. In: Proceedings of the International Joint Conference on Neural Networks, IJCNN 2001 (Cat. No. 01CH37222), vol. 3. IEEE (2001)
37. Zagoruyko, S., Komodakis, N.: Wide residual networks. arXiv preprint arXiv:1605.07146 (2016)
38. Zhang, H., Yu, Y., Jiao, J., Xing, E.P., Ghaoui, L.E., Jordan, M.I.: Theoretically principled trade-off between robustness and accuracy. arXiv preprint arXiv:1901.08573 (2019)

Component Divide-and-Conquer for Real-World Image Super-Resolution

Pengxu Wei[1], Ziwei Xie[1], Hannan Lu[2], Zongyuan Zhan[1], Qixiang Ye[3], Wangmeng Zuo[2], and Liang Lin[1,4](✉)

[1] Sun Yat-sen University, Guangzhou, China
weipx3@mail.sysu.edu.cn, xiezw5@mail2.sysu.edu.cn, zhanzy178@gmail.com, linliang@ieee.org

[2] Harbin Institute of Technology, Harbin, China
hannanlu@hit.edu.cn, wmzuo@hit.edu.cn

[3] University of Chinese Academy of Sciences, Beijing, China
qxye@ucas.ac.cn

[4] DarkMatter AI, Abu Dhabi, United Arab Emirates

Abstract. In this paper, we present a large-scale Diverse Real-world image Super-Resolution dataset, *i.e.*, DRealSR, as well as a divide-and-conquer Super-Resolution (SR) network, exploring the utility of guiding SR model with low-level image components. DRealSR establishes a new SR benchmark with diverse real-world degradation processes, mitigating the limitations of conventional simulated image degradation. In general, the targets of SR vary with image regions with different low-level image components, *e.g.*, smoothness preserving for flat regions, sharpening for edges, and detail enhancing for textures. Learning an SR model with conventional pixel-wise loss usually is easily dominated by flat regions and edges, and fails to infer realistic details of complex textures. We propose a Component Divide-and-Conquer (CDC) model and a Gradient-Weighted (GW) loss for SR. Our CDC parses an image with three components, employs three Component-Attentive Blocks (CABs) to learn attentive masks and intermediate SR predictions with an intermediate supervision learning strategy, and trains an SR model following a divide-and-conquer learning principle. Our GW loss also provides a feasible way to balance the difficulties of image components for SR. Extensive experiments validate the superior performance of our CDC and the challenging aspects of our DRealSR dataset related to diverse real-world scenarios. Our dataset and codes are publicly available at https://github.com/xiezw5/Component-Divide-and-Conquer-for-Real-World-Image-Super-Resolution.

Keywords: Real-world image super-resolution · Image degradation · Corner point · Component divide-and-conquer · Gradient-weighted loss

Electronic supplementary material The online version of this chapter (https://doi.org/10.1007/978-3-030-58598-3_7) contains supplementary material, which is available to authorized users.

A. Vedaldi et al. (Eds.): ECCV 2020, LNCS 12353, pp. 101–117, 2020.
https://doi.org/10.1007/978-3-030-58598-3_7

1 Introduction

Single image Super-Resolution (SR) is an inherently ill-posed inverse problem that reconstructs High-Resolution (HR) images from Low-Resolution (LR) counterparts with image quality degradations. As a fundamental research topic, it has attracted a long-standing and considerable attention in the computer vision community [7,28]. SR methods based on Convolutional Neural Network (CNN) (*e.g.*, SRCNN [5], SRGAN [13], EDSR [14], ESRGAN [26] and RCAN [33]) have achieved a remarkable improvement over conventional SR methods [7].

However, such improvements remain limited for real-world SR applications. The first reason is that SR models have to be trained on datasets with simulated image degradation, as LR images are obtained by simplified downsampling methods (*e.g.*, bicubic) due to the difficulty of HR-LR pair collection. Such simulated degradation usually deviates from real ones, making the learned model not applicable to many real-world applications [2,32]. The second reason is that homogenous pixel-wise losses (*e.g.*, MSE) would lead to model overfitting or attend to regions for easy reconstruction. Intuitively, the targets of SR vary with LR regions with different low-level image elements, *e.g.*, smoothness preserving for flat regions, sharpening for edges, and detail enhancing for textures. Considering that flat regions and edges are the most frequent in an image, the models learned by homogeneous pixel-wise loss prefer addressing flat regions and edges, but usually fail to infer realistic details of complex textures. In Fig. 1, an SR image from EDSR [14] trained with L_1 loss presents different reconstruction difficulties in different regions, specifically in flat, edge and corner point regions. In Fig. 2, we analyze proportions of three components (flat, edges and corners) for L_1 loss in EDSR [14] and evaluate their respective effects for SR recon-

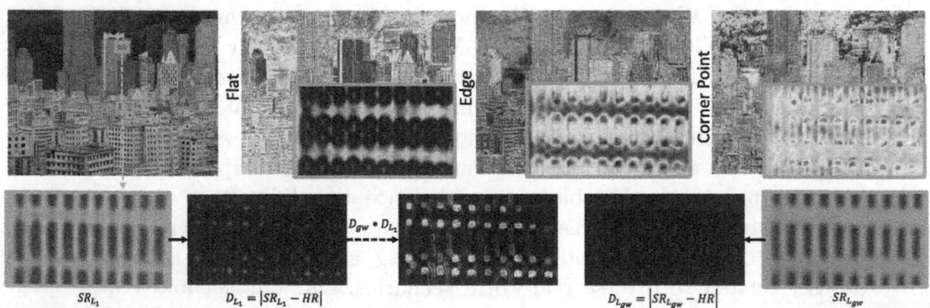

Fig. 1. Image degradation reflected by loss values. SR_{L_1} and $SR_{L_{gw}}$ are SR results trained with L_1 loss and our GW loss, respectively. The difference map D_{L_1} which presents different reconstruction difficulties demonstrates the complex image degradation. It is observed that regions from small to large values in D_{L_1} are relatively consistent with flat, edges and corner regions, which motivates us to explore these components and introduce a weighting strategy for D_{L_1} which drives models to attend to hard regions. The first row shows three attentive masks learnt by our CDC which well predict the confidence of flat, edges and corner regions, respectively.

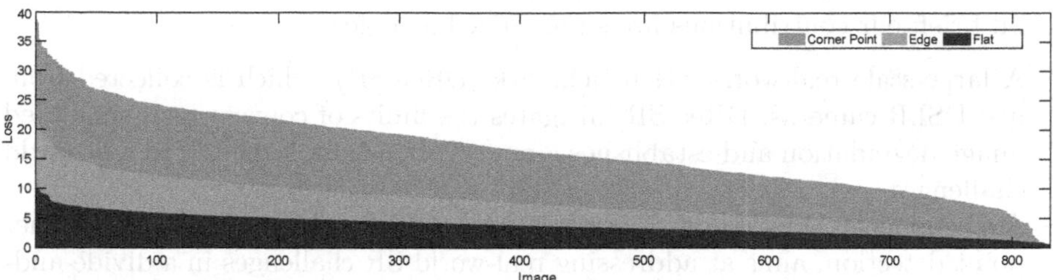

Fig. 2. Component analysis for real SR. To investigate the challenging aspects, we analyze proportions of three components (flat, edges and corners) for L_1 loss in EDSR [14] and evaluate their respective effects for SR reconstruction with averaged pixel-wise loss. Three components are observed to have different recovery difficulties: smooth regions and edges have a lower loss while corner points have a higher loss.

struction with averaged pixel-wise loss. Three components are observed to have different recovery difficulties: smooth regions and edges have lower loss while corner points have higher loss. Thus, these observations inspire us to investigate the utility of these three components in the SR task.

In this paper, we establish a large-scale Diverse Real-world SR benchmark, DRealSR, and propose a Component Divide-and-Conquer model (CDC) to address real-world SR challenges, *i.e.*, (i) the gap between simulated and real degradation processes and (ii) the diverse image degradation processes caused by different devices. CDC is inspired by the mechanism of Harris corner point detection algorithm [19]. An image can be disentangled into three low-level components (*i.e.*, flat, edge and corner) with respect to the importance of the information they convey. Flat regions have almost constant pixel values, edges can be regarded as the boundary of different flat regions, and multiple edges interweave into corners. In CDC, three low-level elements which facilitate an implicit composition optimization are treated as guidance to regularize the SR task.

Specifically, we first develop a base model, named HGSR, based on a stacked hourglass network. HGSR learns multi-scale features with repeated bottom-up and top-down inference across all scales. With HGSR, CDC builds three Component-Attentive Blocks (CABs) which are associated with flat, edges and corners, respectively. Each CAB focuses on learning one of the three low-level components with the Intermediate Supervision (IS) strategy. CDC takes the flat regions, edges and corners extracted from HR images only in the training stage and then incorporates them separately into three different branches with CABs. These three CABs form a progressive paradigm and are aggregated to yield the final SR reconstruction. Considering that different image regions convey different gradients in all directions, we propose a Gradient-Weighted (GW) loss function for SR reconstruction. More complex a region is, larger impacts on the loss function it has. Our GW loss, in a way like Focal loss [15] for training object detectors, adapts the model training based on different image reconstruction difficulties.

In brief, our contributions are summarized as follows:

- A large-scale real-world SR benchmark (DRealSR), which is collected from five DSLR cameras. DRealSR mitigates the limits of conventional simulated image degradation and establishes a new SR benchmark related to real-world challenges.
- A Component Divide-and-Conquer model (CDC), which, inspired by corner point detection, aims at addressing real-world SR challenges in a divide-and-conquer manner. CDC employs three Component-Attentive Blocks to learn attentive masks for different components and predicts intermediate SRs with an intermediate supervision learning strategy.
- A Gradient-Weighted loss, which fully utilizes image structural information for fine-detailed image SR. GW loss explores the imbalance learning issue for different image regions in the pixel-wise SR task and provides a promising solution, which can be extended to other low-level vision tasks.

2 Related Work

Datasets. In the area of SR, widely-used SR datasets include Set5 [1], Set14 [29], BSD300 [17], Urban100 [9] and DIV2K [24]. Due to the difficulty of collecting HR-LR pairs, non-blind SISR approaches usually adopt a simulated image degradation for training and testing, *e.g.*, bicubic downsampling. Consequently, images in those SR datasets are usually regarded as HR images whose LR counterparts are obtained by HR downsampling. However, the real-world degradation process can be much more complex and even nonlinear. This simulated degradation limits related SR researches to a rather ideal SR simulation with approximately linear kernels and causes a great gap for practical SR applications [2,3,30,32].

To fill this gap, City100 dataset [3] with 100 aligned image pairs is built for SR modeling in the realistic imaging system. However, City100 is captured for the printed postcards under an indoor environment. To capture real-world natural scenes, SR-RAW dataset is introduced for super-resolution from raw data via optical zoom [32]. RealSR dataset [2] provides a well-prepared benchmark for real-world single image super-resolution, which is captured with two DSLR cameras. In this work, we build a larger and more challenging real SR dataset with five DSLR cameras, with the target to further explore SR degradation in real-world scenarios.

Methods. Recent years have witnessed an evolution of image super-resolution research with widely-explored deep learning, which has significantly improved SR performance against traditional methods [5]. Sequentially, deep SR networks derived from various CNN models, *e.g.*, VDSR [10], EDSR [14], SRRestNet [13], LapSR [12] and RCAN [33], are presented to further improve the SR performance. To regularize the model for ill-posed SR problem, several works are suggested to incorporate image priors, *e.g.*, edge detection [4][6], texture synthesis [20] or semantic segmentation [25]. Despite their progress, most existing approaches are still tested on synthesized SR datasets with bicubic downsampling or downsampling after Gaussian blurring, while few researches are devoted to real-world SR problems.

Recently, a contextual bilateral loss (CoBi) is introduced to mitigate the misalignment issue in a real-world SR-RAW dataset [32]. Besides, LP-KPN [2] proposes a Laplacian pyramid based method to deal with the non-uniform blur kernels for SR. However, it remains limited in considering the complexity and diversity of real degradation processes among different devices, hindering the applications of real-world SR. In this work, we neither train an SR model by treating uniformly all the pixels/regions/components in an image nor bias towards only edges or textures. We parse an image into three low-level components (flat, edge and corner), explore their different importance, develop a CDC model in a divide-and-conquer learning framework and propose a GW loss to adaptively balance the pixel-wise reconstruction difficulties.

3 DRealSR: A Large-Scale Real-World SR Dataset

To further explore complex real-world SR degradation, we build a large-scale diverse SR benchmark, named DRealSR, by zooming DSLR cameras to collect real LR and HR images. DRealSR covers 4 scaling factors (*i.e.*, ×1–×4).

Dataset Collection. These images are captured from five DSLR cameras (*i.e.*, Canon, Sony, Nikon, Olympus and Panasonic) in natural scenes and cover indoor and outdoor scenes avoiding moving objects, *e.g.*, advertising posters, plants, offices, buildings, *etc.* For each scaling factor, we adopt SIFT method [16] for image registration to crop an LR image to match the content of its HR counterpart. To refine the registration results, an image is cropped into patches and an iterative registration algorithm and brightness match are employed. To better facilitate the model training, considering that their image sizes are 4,000 × 6,000 or 3,888 × 5,184, these training images are cropped into 380 × 380, 272 × 272, 192 × 192 patches for ×2–×4, respectively. Since the misalignment between HR and LR possibly induces to severely blurry SR results, after each step of registration, we conduct a careful manual selection for patches. More details on the dataset construction are provided in the supplementary file.

Challenges in Real-World SR. Due to the difficulty to capture high-resolution and low-resolution image pairs in real world, extensive SR methods are demonstrated on datasets with simulated image degradation (*e.g.*, bicubic downsampling). Compared with simulated image degradation, real-world SR exhibits the following new challenges.

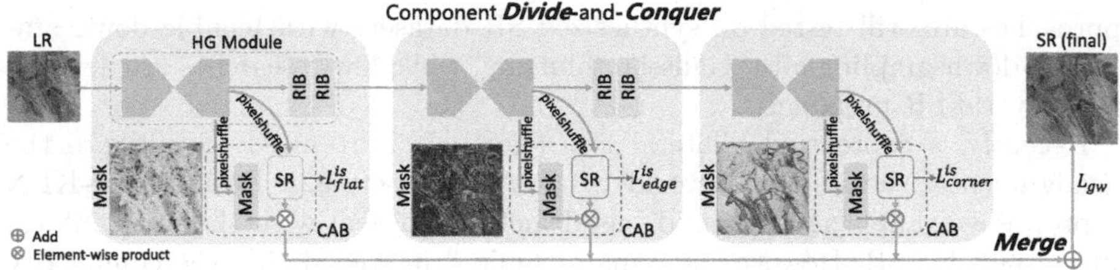

Fig. 3. CDC framework. The stacked architecture enables CDC to incorporate flat regions, edges and corners in three CABs separately. Each CAB branch produces an attentive mask and an intermediate SR. This mask regularizes the produced intermediate SR to collaborate with other branches and seamlessly blend them to yield natural SR images.

- **More complex degradation against bicubic downsampling**. Bicubic downsampling simply applies the bicubic downsampler to an HR image to obtain the LR image. In real scenarios, however, downsampling usually is performed after anisotropic blurring, and signal-dependent noise may also be added. Thus, the acquisition of LR images suffers from both blurring, downsampling and noise. Also it is affected by in-camera signal processing (ISP) pipeline. This non-uniform property of realistic image degradation can be verified based on the reconstruction difficulty analysis of different image regions/components, Fig. 2. Usually, SR models trained on bicubic degradation exhibit poor performance when handling real-world degradation.
- **Diverse degradation processes among devices.** In practical scenarios, differences among lens and sensors of cameras determine the different imaging patterns, which is the primary reason for explaining the diverse degradation processes in real-world SR. Consequently, SR models learned on a real LR dataset may generalize poorly to other datasets and real-world LR images, raising another challenging issue for applications SR.

Nonetheless, different kinds of regions exhibit robustness characteristics to degradation. For example, a flat region is less affected by the diversity of the degradation process while changes in degradation settings produce quite different results for regions with edges and corners, Fig. 2. Accordingly, this motivates us to parse an image into flat, edges and corners for easing model training.

4 Real-Word Image Super-Resolution

To handle diverse image degradation, one intuitive solution is to learn anisotropic blur kernels. Due to complex contents in a natural image, however, it is hard to propose a universal solution to estimate anisotropic kernels. Inspired by Harris corner points [8], image contents are divided into three primary visual components: flat, edges and corner points according to their gradient changes. These components are considered in our work. Because they represent the complexity of image contents, which indicates their reconstruction difficulty, as demonstrated in Fig. 1 and Fig. 2. For example, corners possess crucial orientation cues that control the shape or appearance of edges or textures [8] and are potentially beneficial for image reconstruction. Thus, these three components, *i.e.*, flat, edges and corners, are explored to facilitate SR model training being free from the limits to diverse degradation processes.

Specifically, in this work, considering the reconstruction difficulty of different components, we build an HGSR network with a stacked architecture and propose a Component Divide-and-Conquer model with a Gradient-Weighted loss to address the real SR problem. *Divide* arranges the introduction order as flat, edges and corners to facilitate the feature learning in the network from easy to hard; *conquer* separately produces intermediate SR results for each component which are *merged* into the final SR prediction (Fig. 3).

- *Divide*: Consider the complexity of image contents in flat, edges and corner point regions, we guide three HG modules to emphatically learn component-attentive masks from LR images respectively, with the component parsing guidance from HR images. It is noted that, we do not directly detect three components from LR images with off-the-shelf methods but predict their maps coherent with the HR image. The main reason is that the low quality of LR images hinders the more accurate corner point detection and yields undesirable detection results. Another reason is that this strategy avoids corner point detection for each image in the test stage.
- *Conquer*: Three Component-Attentive Blocks produce different component-attentive masks and intermediate SR predictions. The generated attentive maps present remarkable characteristics of three components. Meanwhile, intermediate SR results are consistent with the characteristics of three regions.
- *Merge*: To yield the final SR result, we collaboratively aggregate three intermediate SR outputs weighted by the corresponding component-attentive maps. In particular, a GW loss is proposed to drive the model to adapt learning objectives to their reconstruction difficulties.

4.1 Formulation

In the real SR, given N LR-HR pairs, we estimate an SR image $\hat{\mathbf{x}}_i$ by minimizing the loss function $\mathcal{L} = \frac{1}{N}\sum_{i=1}^{N} \mathcal{L}_{rec}(\hat{\mathbf{x}}_i, \mathbf{x}_i)$, where $\mathcal{L}_{rec}(\cdot)$ is a reconstruction loss function. The network learns a mapping function $\mathcal{F}$ from the LR image $\mathbf{y}_i$ to the HR image $\mathbf{x}_i$; namely, $\hat{\mathbf{x}}_i = \mathcal{F}(\mathbf{y}_i; \Theta)$, where Θ is the network model parameter. In general, the realistic image degradation is complex and diverse as claimed above. To make it relatively tractable, our CDC employs three CABs and learn models with the intermediate supervision in a divide-and-conquer manner, rather than directly learning LR-HR mapping function or estimating blur kernels. Thus, our loss function is defined as follows,

$$\mathcal{L} = \frac{1}{N}\sum\nolimits_{i=1}^{N} [\mathcal{L}_{rec}(\hat{\mathbf{x}}_i, \mathbf{x}_i) + \sum\nolimits_{e=1}^{3} \mathcal{L}_{is}(\tilde{\mathbf{x}}_i^e, \mathbf{x}_i)], \tag{1}$$

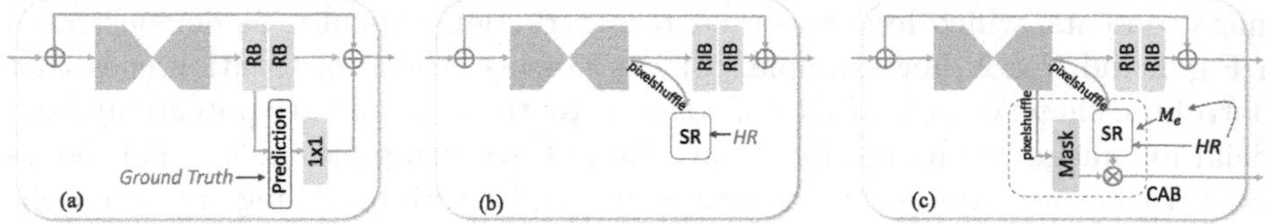

Fig. 4. CAB and intermediate supervision in the HG module. We compare (a) the IS in the basic HG [18] with those in HGSR (b) and our CDC (c). In [18], an intermediate prediction recursively joins into the next HG module after 1×1 convolution for human pose estimation. For SR, our IS strategy in (b) and (c) avoids the recursive operation which tends to invite large disturbance for feature learning in the backbone.

where $\mathcal{L}_{is}$ is the intermediate loss function, the index e represents an CAB module that is specific to either *flat*, *edge* or *corner*, and $\tilde{\mathbf{x}}_i^e$ is the intermediate SR prediction ($\tilde{\mathbf{x}}_e$ for simplicity in the following sections).

4.2 Hourglass Super-Resolution Network

We propose a basemodel, Hourglass Super-Resolution network (HGSR), which has a stacked hourglass architecture [18] followed by a pixelshuffle layer [21]. The hourglass (HG) architecture is motivated to capture information at every scale and has a superior performance for keypoint detection [18]. Its hourglass module can be regarded as an encoder-decoder with skip connections to preserve spatial information at each resolution and bring them together to predict pixel-wise outputs. In the HG module, an input passes through a convolutional layer firstly and then is downsampled to a lower resolution by a maximum pooling layer. During Top-Down inference, it repeats this procedure until reaching the lowest resolution. Next, a Bottom-Up inference performs constantly upsampling by nearest neighbor interpolation and combines features across scales by skip-connection layers until the original resolution is restored.

The conventional connection between two HG modules are two Residual Blocks (RBs) [18], as shown in Fig. 4(a). HGSR replaces RBs with Residual Inception Blocks (RIBs) [23]. RIBs have a parallel cascade structure and concatenate feature maps produced by filters of different sizes. Besides, HGSR utilizes the Intermediate Supervision (IS) strategy for model learning. The main difference of the IS module in [18] is that HGSR does not recursively feed the IS prediction to the next HG module, as shown in Fig. 4(b). The intermediate loss function $\mathcal{L}_{is}$ in HGSR is the $\mathcal{L}_1$ loss.

4.3 Component Divide-and-Conquer Model

Our Component Divide-and-Conquer model takes HGSR as the backbone and follows the divide-and-conquer principle to learn the model. Specifically, CDC focuses on three image components, *i.e.*, flat, edges and corners, rather than edges or/and complex textures. This makes it relatively tractable to solve the ill-posed real SR problem. These components are explicitly extracted from HR images with Harris corner detection algorithm, separately in CABs and are implicitly blended seamlessly to yield natural SR results by minimizing a GW loss. Although the guidance of three components is from HR images, CDC infers the component probability maps in the test stage without any detection.

Component-Attentive Block. CDC has three CABs which respectively correspond to either flat, edges or corners. Since it inherits the advantages of HGSR with a cascaded hourglass network, CDC is suitable to incorporate the intermediate supervision. As shown in Fig. 4(c), CAB consists of two pixel-shuffle layers. One is used to generate a coarse intermediate SR result. The other one is used to generate a mask which indicates the component probability map. It weights this coarse SR for the final SR reconstruction together with outputs of other CABs. In the training stage, CDC leverages the HR image as intermediate supervision to generate an IS loss weighted by the guidance of the component mask from HR. Accordingly, the intermediate loss function in an CAB is defined as

$$\mathcal{L}_{is} = l(\boldsymbol{M}_e * \mathbf{x}, \boldsymbol{M}_e * \tilde{\mathbf{x}}_e), \tag{2}$$

where e is similar to that defined in Eq. 1 and $*$ denotes the entry-wise product; $\boldsymbol{M}_e$ is the component guidance mask extracted from HR images. In general, $l(\cdot)$ can be any loss functions; we adopt widely-used L_1 loss function in the CDC.

As shown in Fig. 5, the learned component-attentive masks in three CABs exhibit their own characteristics in indicating flat regions, edges and textures, respectively. Accordingly, their intermediate SR results are also consistent with these characteristics. To further aggregate these three types of information, we will describe how to collaboratively aggregate them to yield the final SR result with a gradient-weighted loss.

Gradient-Weighted Loss. For conventional pixel-wise loss, regions in an image are treated identically. However, flat regions and edges dominate the loss function due to their large quantity in images. Thus, the learned SR models incline to address flat regions and edges, but fail to infer realistic details of complex textures. Inspired by Focal loss [15], we propose to suppress a large number of simple regions while emphasizing the hard ones. Notably, this strategy is also crucial for low-level vision tasks. In our work, the solution of flat, edge and corner point detection provides a plausible disentanglement of images according to their importance, which can thus be used to determine the easy and hard regions and obtain the final SR prediction $\hat{\mathbf{x}}$ as the sum of outputs from three CABs, namely, $\hat{\mathbf{x}} = \sum_e \boldsymbol{A}_e * \tilde{\mathbf{x}}_e$, where $\boldsymbol{A}_e$ is a component-attentive mask.

We propose a Gradient-Weighted loss to dynamically adjust their roles for minimizing the SR reconstruction loss. Following this philosophy, the flat and single edge regions are naturally classified as simple regions. Corners are categorized as difficult regions since they possess the fine-details in images. Considering

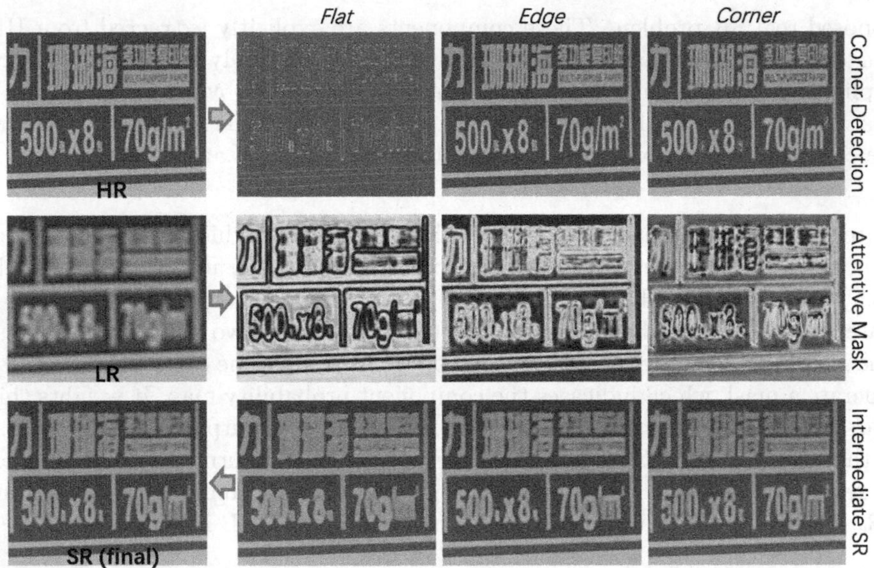

Fig. 5. Harris corner point detection, learned component-attentive masks and intermediate SR images from three CABs. Component-attentive masks from each CAB present a high similarity to flat, edges and corners, respectively.

the diversity in the first-order gradient of different regions, the new reconstruction loss function for SR, named GW loss, is defined as

$$\mathcal{L}_{gw} = l(D_{gw} * \mathbf{x}, D_{gw} * \hat{\mathbf{x}}), \tag{3}$$

where $D_{gw} = (1+\alpha D_x)(1+\alpha D_y)$; $D_x = |G_x^{sr} - G_x^{hr}|$ and $D_y = |G_y^{sr} - G_y^{hr}|$ represent gradient difference maps between SR and HR in the horizontal and vertical directions; α is a scalar factor to determine the quantity for this weighting in the loss function. Generally, $l(\cdot)$ can be also any loss function and we adopt L_1 loss in this paper. If $\alpha = 0$, GW loss becomes the original loss $l(\mathbf{x}, \hat{\mathbf{x}})$. α is 4 in our experiments. This GW loss is regarded as the reconstruction loss $\mathcal{L}_{rec}$.

5 Experiments

5.1 Experimental Settings

Dataset. We conduct experiments on an existing real-world SR dataset, RealSR, and our DRealSR. **RealSR** [2] has 595 HR-LR image pairs captured from two DSLR cameras. 15 image pairs at each scaling factor of each camera are selected randomly for building the testing set and the rest pairs are training set. Their image sizes are in the range of [700, 3100] and [600, 3500] and each training image is cropped in 192×192 patches. For $\times 2$–$\times 4$, our **DRealSR** has 884, 783 and 840 image pairs respectively, where 83, 84 and 93 image pairs are randomly selected for testing respectively and the rest are for training at each scaling factor.

Network Architecture. CDC cascades six HG modules. In each HG module, a residual block followed by a max-pooling layer for the top-down process and nearest neighbor method for the bottom-up process. For shortcut connection across two adjacent resolutions, we also use a residual block consisting of three convolution layers: 1×1, 3×3 and 1×1 filters. Between two hourglass modules, there are two connection layers using Residual Inception Block [23] for multi-scale processing. To introducing the intermediate supervision, those six HG modules are divided into three groups and the last HG in a group generates a coarse SR image by an upsampling layer of pixelshuffle.

Implementation Details. Harris Corner detection method [8] with OpenCV is used. In our experiments, we use Adam optimizer [11] and set 0.9 and 0.999 for its exponential decay rates. The initial learning rate is set to 2e−4 and then reduced to half every 100 epochs. For each training batch, we randomly extract 16 LR patches with the size of 48×48. All of our experiments are conducted in PyTorch. Three common image quality metrics are used to evaluate SR models, *i.e.*, peak signal-to-noise ratio (PSNR) and structural similarity index (SSIM) [27] and Learned Perceptual Image Patch Similarity (LPIPS) [31]. PSNR is evaluated on the Y channel; SSIM and LPIPS are on RGB images.

5.2 Model Ablation Study

Evaluation on HG Blocks. Our base model, HGSR, adopts a stacked hourglass network [18] as the backbone. We provide experimental evaluations on the number of HG blocks in Table 1. It is observed that the SR performance has a sustaining boost when the number of HG blocks in HGSR increases from 2 to 4 while it has a stable performance when the number of HG blocks is larger than 6. Thus, the number of HG blocks is set 6 in our experiments.

Table 1. Evaluation of the number of HG blocks

Method	HG blocks	PSNR	SSIM	LPIPS
HGSR	2	31.55	0.847	0.336
	4	31.80	0.854	0.312
	6	**31.95**	**0.854**	0.304
	8	31.94	0.854	**0.303**

Table 2. Ablation study of the proposed CDC

Method	PSNR	SSIM	LPIPS
HGSR(baseline, w/o IS)	31.95	0.854	0.304
HGSR(baseline)	32.13	0.855	0.310
HGSR+RIB	32.15	0.857	0.310
HGSR+RIB+CAB	32.27	0.858	0.302
HGSR+RIB+CAB+GW	**32.42**	**0.861**	**0.300**

Table 3. Evaluation of CDC in flat, edge, and corner regions

Method	Regions			PSNR	SSIM	LPIPS
	Flat	*Edge*	*Corner*			
CDC	✓			32.03	0.856	0.310
		✓		32.25	0.858	0.307
			✓	32.37	0.861	0.302
	✓	✓		32.23	0.860	0.301
	✓		✓	32.39	0.861	0.300
		✓	✓	32.40	0.861	**0.298**
	✓	✓	✓	**32.42**	**0.861**	0.300

Table 4. Comparison results of our proposed GW loss with L_1 loss

Method	Loss	PSNR	SSIM	LPIPS
SRResNet [13]	L_1	31.63	0.847	0.341
	L_{gw}	**31.93**	**0.853**	**0.321**
EDSR [14]	L_1	32.03	**0.855**	0.307
	L_{gw}	**32.27**	0.857	**0.304**
HGSR(Our baseline)	L_1	32.15	0.857	**0.310**
	L_{gw}	**32.25**	**0.857**	0.313
CDC(Ours)	L_1	32.27	0.858	0.302
	L_{gw}	**32.42**	**0.861**	**0.300**

Evaluation on IS and RIB. We leverage the intermediate supervision strategy to hierarchically supervise the model learning. As shown in Table 2, HGSR with IS achieves 0.18 dB PSNR gains. In the following parts, if no special claim, HGSR denotes the base model with IS. Besides, two convolution layers are added between the two HG modules in HGSR to build their connections. To aggravate multi-scale information, these two layers are substituted by two RIBs. This modification slightly improves the base model with 0.02 dB PSNR gains. Besides, our CAB and the GW loss have 0.12 and 0.15 dB PSNR improvements, respectively. Particularly, our final version, *i.e.*, CDC, exhibits an impressive improvement (*i.e.*, 0.29 dB) compared with the base model HGSR.

Evaluation on Component-Attentive Block. As demonstrated in Table 2, our CAB brings an improvement of 0.12 dB in PSNR. In order to analyze CAB, we conduct experiments on different guidance from flat, edge and corner regions, as shown in Table 3. Without corner branches, our model has a significant drop of 0.19 dB by PSNR. Among three components (*i.e.*, flat, edges and corners), corners that represent important information play a crucial role in the SR task, although they have a small quantity in an image. This observation is encouraging to pay more attention to exploring corner points in the SR task, as well as directly on edges or/and textures.

Evaluation on Gradient-Weighted Loss. In Table 4, in comparison with L_1 loss, the GW loss respectively introduces 0.30 dB, 0.25 dB, 0.10 dB and 0.15 dB improvement in PSNR for EDSR, SRResNet, HGSR, and our CDC. This indicates that our proposed GW loss can be applied to other SR models to further improve their performance. Notably, the GW loss rooted in L_1 loss achieves a greater improvement than L_1. Therefore, our GW loss provides a new way to understand the SR model learning and can be explored in other loss functions and other low level vision tasks.

Table 5. Performance comparison on RealSR [2] and DRealSR datasets

Method	Scale	Training Set: DRealSR						Training Set: RealSR					
		Test on RealSR [2]			Test on DRealSR			Test on RealSR [2]			Test on DRealSR		
		PSNR	SSIM	LPIPS	PSNR	SSIM	LPIPS	PSNR	SSIM	LPIPS	PSNR	SSIM	LPIPS
Bicubic	×2	31.67	0.887	0.223	32.67	0.877	0.201	31.67	0.887	0.223	32.67	0.877	0.201
SRResNet [13]		32.65	0.907	0.169	33.56	0.900	0.163	33.17	0.918	0.158	32.85	0.890	0.172
EDSR [14]		32.71	0.906	0.172	34.24	0.908	0.155	33.88	0.920	0.145	32.86	**0.891**	0.170
ESRGAN [26]		32.25	0.900	0.185	33.89	0.906	0.155	33.80	0.922	0.146	32.70	0.889	0.172
RCAN [33]		**32.88**	0.908	0.173	34.34	0.908	0.158	33.83	0.923	0.147	**32.93**	0.889	0.169
LP-KPN [2]		32.14	-	-	33.88	-	-	-	-	-	-	-	-
DDet [22]		32.58	-	-	33.92	-	-	33.22	-	-	32.77	-	-
CDC (Ours)		32.81	**0.910**	**0.167**	**34.45**	**0.910**	**0.146**	**33.96**	**0.925**	**0.142**	32.80	0.888	**0.167**
Bicubic	×3	28.63	0.809	0.388	31.50	0.835	0.362	28.61	0.810	0.389	31.50	0.835	0.362
SRResNet [13]		28.85	0.832	0.290	31.16	0.859	0.272	30.65	0.862	0.228	31.25	0.841	0.267
EDSR [14]		29.50	0.841	0.266	32.93	0.876	0.241	30.86	0.867	0.219	31.20	0.843	**0.264**
ESRGAN [26]		29.57	0.841	0.266	32.39	0.873	0.243	30.72	0.866	0.219	31.25	0.842	0.268
RCAN [33]		**29.68**	0.841	0.267	33.03	0.876	**0.241**	30.90	0.864	0.225	31.76	0.847	0.268
LP-KPN [2]		29.20	-	-	32.64	-	-	30.60	-	-	**31.79**	-	-
DDet [22]		29.48	-	-	32.13	-	-	30.62	-	-	31.77	-	-
CDC (Ours)		29.57	**0.841**	**0.261**	**33.06**	**0.876**	0.244	**30.99**	**0.869**	**0.215**	31.65	**0.847**	0.276
Bicubic	×4	27.24	0.764	0.476	30.56	0.820	0.438	27.24	0.764	0.476	30.56	0.820	0.438
SRResNet [13]		27.63	0.785	0.368	31.63	0.847	0.341	28.99	0.825	0.281	29.98	0.822	0.347
EDSR [14]		27.77	0.792	0.339	32.03	0.855	0.307	29.09	0.827	0.278	30.21	0.817	**0.344**
ESRGAN [26]		27.82	0.794	0.340	31.92	0.857	0.308	29.15	0.826	0.279	30.18	0.821	0.353
RCAN [33]		27.93	0.795	0.341	31.85	0.857	0.305	29.21	0.824	0.287	30.37	0.825	0.349
LP-KPN [2]		27.79	-	-	31.58	-	-	28.65	-	-	**30.75**	-	-
DDet [22]		27.83	-	-	31.57	-	-	28.94	-	-	30.12	-	-
CDC (Ours)		**28.11**	**0.800**	**0.330**	**32.42**	**0.861**	**0.300**	**29.24**	**0.827**	**0.278**	30.41	**0.827**	0.357

5.3 Comparison with State-of-the-arts on Real SR Datasets

We compare our method with several state-of-the-art SR methods, including SRResNet [13], EDSR [14], ESRGAN [26], RCAN [33], LP-KPN [2] and DDet [22]. Among these SR methods, LP-KPN [2] and DDet [22] are the only two designed to solve the real-world SR problem. Quantitative comparison results are given in Table 5. LP-KPN [2] and DDet [22] are trained on the Y channel and other methods are trained on RGB images. Considering this difference, LPIPS and SSIM of LP-KPN and DDet are not provided since these two metrics are evaluated on RGB images. Our CDC outperforms the state-of-the-art algorithms on two real-world SR datasets. On DRealSR, CDC achieves the best results in all scales and notably improves the performance by about 0.4 dB for ×4. Similar to the performance on DRealSR, PSNR and SSIM of CDC on RealSR are also superior to the others, validating the effectiveness of our method.

Figure 6 visualizes SR results of the competing methods and ours. It is observed that existing SR methods (*e.g.*, EDSR, RCAN, LP-KPN) are prone to generate realistic detailed textures with visual aliasing and artifacts. For instance, in the second example in Fig. 6, SRRestNet, EDSR and LP-KPN produce blurry details of the building and the result of RCAN has obvious aliasing effects. In comparison, our proposed CDC reconstructs sharp and natural details.

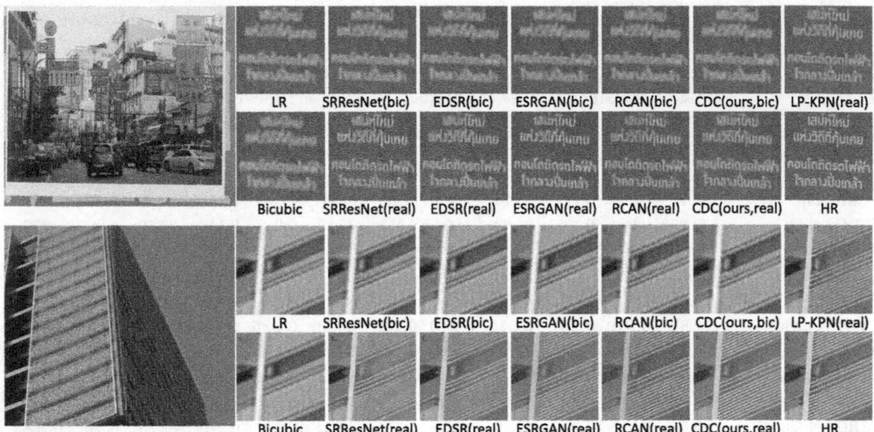

Fig. 6. SR results for ×4 on DRealSR in comparison with state-of-the-art approaches. 'real' indicates models trained on DRealSR while 'bic' indicates those trained on a dataset version by bicubic downsampling DRealSR HR images.

To further validate challenges of RealSR and our DRealSR, we also conduct the cross-testing on the two datasets, *i.e.*, training models on one of them and then testing on the other one. In Table 5, CDC trained on DRealSR maintains a superior performance in all scales when tested on RealSR. However, for models trained on RealSR, the testing performance drops greatly on DRealSR, especially for ×4, which indicates that DRealSR is more challenging than RealSR.

Table 6. SR performance comparison upon the image degradation

Method	Training set		Testing set (DRealSR)					
			Bicubic			Real		
	Dataset	Degradation	PSNR	SSIM	LPIPS	PSNR	SSIM	LPIPS
SRResNet [13]	DRealSR	Bicubic	41.28	0.954	0.103	30.61	0.822	0.422
EDSR [14]	DRealSR	Bicubic	41.49	0.956	0.099	30.60	0.822	**0.421**
RRDB [26]	DRealSR	Bicubic	41.66	0.957	0.097	30.60	0.822	0.425
CDC (Ours)	DRealSR	Bicubic	**41.78**	**0.957**	**0.096**	**30.63**	**0.822**	0.425
SRResNet [13]	DRealSR	Real	31.43	0.864	0.249	31.63	0.847	0.341
EDSR [14]	DRealSR	Real	32.53	0.880	0.231	32.03	0.855	0.307
RRDB [26]	DRealSR	Real	32.37	0.877	0.234	31.92	0.857	0.308
CDC (Ours)	DRealSR	Real	**32.54**	**0.883**	**0.215**	**32.42**	**0.861**	**0.300**

5.4 Analysis on Real and Simulated SR Results

In this section, we analyze the bicubic and real image degradation on DRealSR. In Table 6, the performance of real image degradation on our dataset is very close to that of bicubic image degradation even if the training set is different. Actually, the performance on PSNR, SSIM and LPIPS is close to that of bicubic upsampling method, as shown in Table 5. Thus, no matter which model is used, it is not useful to restore the real image with the model trained on bicubic images. This demonstrates the limited generalization of simulated bicubic degradation. On the other hand, our proposed CDC still achieves an improvement on bicubic images and outperforms most of state-of-the-arts methods. This is also the evidence to prove the superiority and generalization of our method. Figure 6 visualizes SR results from models trained on simulated SR datasets. One can see that models trained on bicubic degradation produce blurry and poor SR results. This clearly demonstrates that the image degradation of the simulated SR dataset greatly hinders the performance of SR methods in real-world scenarios.

6 Conclusion

In this paper, we establish a large-scale real-world image super-resolution dataset, named DRealSR, to facilitate the further researches on realistic image degradation. To mitigate the complex and diverse image degradation, considering reconstruction difficulty of different components, we build a HGSR network with a stacked architecture and propose a Component Divide-and-Conquer model (CDC) to address the real SR problems. CDC employs three Component-Attentive Blocks (CABs) to learn attentive masks and intermediate SR predictions with an intermediate supervision learning strategy. Meanwhile, a Gradient-Weighted loss is proposed to drive the model to adapt learning objectives to their reconstruction difficulties. Extensive experiments validate the challenging aspects of our DRealSR dataset related to real-world scenarios, while our divide-and-conquer solution and GW loss provide a novel impetus for the challenging real-world SR task or other low-level vision tasks.

References

1. Bevilacqua, M., Roumy, A., Guillemot, C., Alberi-Morel, M.: Low-complexity single-image super-resolution based on nonnegative neighbor embedding. In: Proceedings of the British Machine Vision Conference, pp. 1–10 (2012)
2. Cai, J., Zeng, H., Yong, H., Cao, Z., Zhang, L.: Toward real-world single image super-resolution: a new benchmark and a new model. In: Proceedings of the International Conference on Computer Vision (2019)
3. Chen, C., Xiong, Z., Tian, X., Zha, Z., Wu, F.: Camera lens super-resolution. In: Proceedings of the IEEE Conference on Computer Vision and Pattern Recognition, pp. 1652–1660 (2019)
4. Dai, S., Han, M., Wu, Y., Gong, Y.: Bilateral back-projection for single image super resolution. In: Proceedings of the IEEE International Conference on Multimedia and Expo, pp. 1039–1042 (2007)
5. Dong, C., Loy, C.C., He, K., Tang, X.: Learning a deep convolutional network for image super-resolution. In: Fleet, D., Pajdla, T., Schiele, B., Tuytelaars, T. (eds.) ECCV 2014. LNCS, vol. 8692, pp. 184–199. Springer, Cham (2014). https://doi.org/10.1007/978-3-319-10593-2_13

6. Fan, Y., Gan, Z., Qiu, Y., Zhu, X.: Single image super resolution method based on edge preservation. In: Proceedings of the International Conference on Image and Graphics, pp. 394–399 (2011)
7. Glasner, D., Bagon, S., Irani, M.: Super-resolution from a single image. In: Proceedings of the 2009 IEEE 12th International Conference on Computer Vision, pp. 349–356 (2009)
8. Harris, C.G., Stephens, M., et al.: A combined corner and edge detector. In: Alvey Vision Conference (1988)
9. Huang, J.B., Singh, A., Ahuja, N.: Single image super-resolution from transformed self-exemplars. In: Proceedings of the IEEE Conference on Computer Vision and Pattern Recognition, pp. 5197–5206 (2015)
10. Kim, J., Kwon Lee, J., Mu Lee, K.: Accurate image super-resolution using very deep convolutional networks. In: Proceedings of the IEEE Conference on Computer Vision and Pattern Recognition, pp. 1646–1654 (2016)
11. Kingma, D.P., Ba, J.: Adam: a method for stochastic optimization. In: ICLR (2017)
12. Lai, W.S., Huang, J.B., Ahuja, N., Yang, M.H.: Deep Laplacian pyramid networks for fast and accurate super-resolution. In: Proceedings of the IEEE Conference on Computer Vision and Pattern Recognition, pp. 624–632 (2017)
13. Ledig, C., et al.: Photo-realistic single image super-resolution using a generative adversarial network. In: Proceedings of the IEEE Conference on Computer Vision and Pattern Recognition, pp. 105–114 (2017)
14. Lim, B., Son, S., Kim, H., Nah, S., Lee, K.M.: Enhanced deep residual networks for single image super-resolution. In: IEEE Conference on Computer Vision and Pattern Recognition Workshops, pp. 1132–1140 (2017)
15. Lin, T.Y., Goyal, P., Girshick, R., He, K., Dollár, P.: Focal loss for dense object detection. In: Proceedings of the IEEE International Conference on Computer Vision, pp. 2980–2988 (2017)
16. Lowe, D.G.: Distinctive image features from scale-invariant keypoints. Int. J. Comput. Vision **60**(2), 91–110 (2004)
17. Martin, D.R., Fowlkes, C.C., Tal, D., Malik, J.: A database of human segmented natural images and its application to evaluating segmentation algorithms and measuring ecological statistics. In: Proceedings of the Eighth International Conference On Computer Vision, pp. 416–425 (2001)
18. Newell, A., Yang, K., Deng, J.: Stacked hourglass networks for human pose estimation. In: Leibe, B., Matas, J., Sebe, N., Welling, M. (eds.) ECCV 2016. LNCS, vol. 9912, pp. 483–499. Springer, Cham (2016). https://doi.org/10.1007/978-3-319-46484-8_29
19. Rosten, E., Porter, R., Drummond, T.: Faster and better: a machine learning approach to corner detection. IEEE Trans. Pattern Anal. Mach. Intell. **32**(1), 105–119 (2008)
20. Sajjadi, M.S., Scholkopf, B., Hirsch, M.: EnhanceNet: single image super-resolution through automated texture synthesis. In: Proceedings of the IEEE International Conference on Computer Vision, pp. 4491–4500 (2017)
21. Shi, W., et al.: Real-time single image and video super-resolution using an efficient sub-pixel convolutional neural network. In: Proceedings of the IEEE Conference on Computer Vision and Pattern Recognition, pp. 1874–1883 (2016)
22. Shi, Y., Zhong, H., Yang, Z., Yang, X., Lin, L.: DDet: dual-path dynamic enhancement network for real-world image super-resolution. IEEE Signal Process. Lett. **27**, 481–485 (2020)

23. Szegedy, C., Ioffe, S., Vanhoucke, V., Alemi, A.A.: Inception-v4, inception-resnet and the impact of residual connections on learning. In: Proceedings of the AAAI Conference on Artificial Intelligence (2017)
24. Timofte, R., Agustsson, E., Van Gool, L., Yang, M.H., Zhang, L.: NTIRE 2017 challenge on single image super-resolution: methods and results. In: Proceedings of the IEEE Conference on Computer Vision and Pattern Recognition Workshops, pp. 114–125 (2017)
25. Wang, X., Yu, K., Dong, C., Change Loy, C.: Recovering realistic texture in image super-resolution by deep spatial feature transform. In: Proceedings of the IEEE Conference on Computer Vision and Pattern Recognition, pp. 606–615 (2018)
26. Wang, X., et al.: ESRGAN: enhanced super-resolution generative adversarial networks. In: Leal-Taixé, L., Roth, S. (eds.) ECCV 2018. LNCS, vol. 11133, pp. 63–79. Springer, Cham (2019). https://doi.org/10.1007/978-3-030-11021-5_5
27. Wang, Z., Bovik, A.C., Sheikh, H.R., Simoncelli, E.P.: Image quality assessment: from error visibility to structural similarity. IEEE Trans. Image Process. **13**(4), 600–612 (2004)
28. Yang, J., Wright, J., Huang, T.S., Ma, Y.: Image super-resolution via sparse representation. IEEE Trans. Image Process. **19**(11), 2861–2873 (2010)
29. Zeyde, R., Elad, M., Protter, M.: On single image scale-up using sparse-representations. In: Boissonnat, J.-D., et al. (eds.) Curves and Surfaces 2010. LNCS, vol. 6920, pp. 711–730. Springer, Heidelberg (2012). https://doi.org/10.1007/978-3-642-27413-8_47
30. Zhang, K., Zuo, W., Zhang, L.: Learning a single convolutional super-resolution network for multiple degradations. In: Proceedings of the IEEE Conference on Computer Vision and Pattern Recognition, pp. 3262–3271 (2018)
31. Zhang, R., Isola, P., Efros, A.A., Shechtman, E., Wang, O.: The unreasonable effectiveness of deep features as a perceptual metric. In: Proceedings of the IEEE Conference on Computer Vision and Pattern Recognition, pp. 586–595 (2018)
32. Zhang, X., Chen, Q., Ng, R., Koltun, V.: Zoom to learn, learn to zoom. In: Proceedings of the IEEE Conference on Computer Vision and Pattern Recognition, pp. 3762–3770 (2019)
33. Zhang, Y., Li, K., Li, K., Wang, L., Zhong, B., Fu, Y.: Image super-resolution using very deep residual channel attention networks. In: Ferrari, V., Hebert, M., Sminchisescu, C., Weiss, Y. (eds.) ECCV 2018. LNCS, vol. 11211, pp. 294–310. Springer, Cham (2018). https://doi.org/10.1007/978-3-030-01234-2_18

Enabling Deep Residual Networks for Weakly Supervised Object Detection

Yunhang Shen[1], Rongrong Ji[1(✉)], Yan Wang[2], Zhiwei Chen[1], Feng Zheng[3], Feiyue Huang[4], and Yunsheng Wu[4]

[1] Media Analytics and Computing Lab, Department of Artificial Intelligence, School of Informatics, Xiamen University, Xiamen 361005, China
shenyunhang01@gmail.com, rrji@xmu.edu.cn, zhiweichen@stu.xmu.edu.cn
[2] Pinterest, San Francisco, USA
yanw@pinterest.com
[3] CSE, Southern University of Science and Technology, Shenzhen, China
zhengf@sustech.edu.cn
[4] Tencent Youtu Lab, Tencent Technology (Shanghai) Co., Ltd., Shanghai, China
garyhuang@tencent.com, wuyunsheng@gmail.com

Abstract. Weakly supervised object detection (WSOD) has attracted extensive research attention due to its great flexibility of exploiting large-scale image-level annotation for detector training. Whilst deep residual networks such as ResNet and DenseNet have become the standard backbones for many computer vision tasks, the cutting-edge WSOD methods still rely on plain networks, *e.g.*, VGG, as backbones. It is indeed not trivial to employ deep residual networks for WSOD, which even shows significant deterioration of detection accuracy and non-convergence. In this paper, we discover the intrinsic root with sophisticated analysis and propose a sequence of design principles to take full advantages of deep residual learning for WSOD from the perspectives of adding redundancy, improving robustness and aligning features. First, a redundant adaptation neck is key for effective object instance localization and discriminative feature learning. Second, small-kernel convolutions and MaxPool down-samplings help improve the robustness of information flow, which gives finer object boundaries and make the detector more sensitivity to small objects. Third, dilated convolution is essential to align the proposal features and exploit diverse local information by extracting high-resolution feature maps. Extensive experiments show that the proposed principles enable deep residual networks to establishes new state-of-the-arts on PASCAL VOC and MS COCO.

1 Introduction

Different from fully supervised object detection (FSOD) [19,39,41,42] that requires bounding-box-level annotations, weakly supervised object detection

Electronic supplementary material The online version of this chapter (https://doi.org/10.1007/978-3-030-58598-3_8) contains supplementary material, which is available to authorized users.

A. Vedaldi et al. (Eds.): ECCV 2020, LNCS 12353, pp. 118–136, 2020.
https://doi.org/10.1007/978-3-030-58598-3_8

Table 1. Comparisons of different backbones for WSDDN [7] on VOC 2007 [15].

Arch.	Backbone	Combination	Depth	Stride	CorLoc (%)	mAP (%)
Plain	AlexNet[1]	C5 [19]	8	16	53.8	32.6
	VGG F[28]		8	16	54.2	34.5
	VGG M[28]		8	16	56.1	34.9
	VGG S[28]		8	12	56.0	34.2
	VGG 16[52]		16	16	53.5	34.8
Residual	ResNet[26]	C4 [26]	18	16	56.8	31.5
			50	16	55.6	30.3
		FPN[36]	18	4/8/16/32	52.3	30.3
			50	4/8/16/32	50.1	30.1
			101	4/8/16/32	46.9	27.7
		C5 [19]	18	32	49.7	28.4
			50	32	50.5	26.5
			101	32	50.9	25.7
	DenseNet[23]	C5 [19]	121	32	55.3	29.7
			161	32	53.0	28.5
	ResNet-WS	C5 [19]	22	8	63.1	43.4
			54	8	63.6	44.0
			105	8	64.0	44.1
	DenseNet-WS	C5 [19]	125	8	**66.3**	**44.8**
			173	8	66.1	44.3

(WSOD) only needs image-level labels. Such relaxation significantly saves the labelling cost and brings large flexibility to many real-world applications.

In a standard pipeline, state-of-the-art WSOD methods first crop region proposals using methods such as RoIPool [19] from backbone networks. Then task-specific heads, *i.e.*, WSOD heads, are built on top of the backbones to localize object instances and learn proposal features jointly. Despite the promising progress made in recent years, there is still a large performance gap from WSOD to FSOD. Prevailing methods generally focus on designing WSOD heads and seldom touch the design of backbone networks, and most state-of-the-art WSOD methods are still built on plain network architectures, *e.g.*, VGG16 [52], VGG-F (M, S) [28] and AlexNet [1], leaving deep residual networks under-explored.

In contrast, it is well known that backbones are important for FSOD in both detection accuracy and inference speed. For accuracy, by simple replacing VGG16 with ResNet [26], Faster R-CNN [42] can increase the mAP@0.5 from 41.5%/75.9% (VGG16) to 48.4%/83.8% (ResNet-101) on COCO and PASCAL VOC 2012, respectively. For speed, light-weight backbones [44,62] significantly reduce the model size and computational complexity. And backbones proposed in [49,77] also enable training detectors from scratch.

However, the direct replacement of residual network to plain networks in WSOD has led to significant performance drop. As an investigation, we first

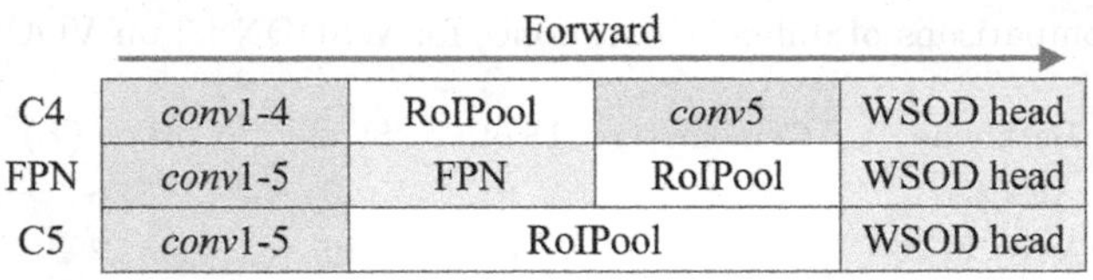

Forward →

C4	*conv*1-4	RoIPool	*conv*5	WSOD head
FPN	*conv*1-5	FPN	RoIPool	WSOD head
C5	*conv*1-5	RoIPool		WSOD head

Fig. 1. Various schemes to adapt deep residual networks for WSOD.

quantize the performance of deep residual networks to WSOD under various combinational schemes, as shown in Fig. 1. We build WSDDN [7] head on various plain and residual backbones and evaluate them on PASCAL VOC 2007 [15]. As shown in Table 1, ResNet [26] and DenseNet [23] deteriorate detection performance, which is even inferior to AlexNet [1] in terms of mAP. Moreover, some state-of-the-art methods [27,55,59] are unable to converge as shown in Table 4.

In this paper, we investigate the intrinsic nature towards enabling residual networks to be workable in WSOD. The underlying problem is that WSOD heads are sensitive to model initialization [5,9,11,32] and suffer from instability [38], which may back-propagate uncertain and erroneous gradient to backbones and deteriorate the visual representation learning. Specifically, we propose a sequence of design principles to take full advantage of deep residual networks in three perspectives, *i.e.*, adding redundancy, improving robustness and aligning features.

1. Redundant adaptation neck. Directly employing ResNet backbones to train WSOD deteriorates the discriminability of proposal features, which also fails to localize object instances accurately. The shortcut connections in residual blocks also enlarge the uncertain and erroneous gradient, which overwhelms the direction of optimization steps. Therefore, our first principle is proposing a redundancy adaptation neck with high-dimension proposal representation between deep residual backbones and WSOD heads, which serves as the key to localize object instances and learn discriminative features jointly.
2. Robust information flow. We have also found that ResNet suffers from uncertainty around object boundaries and imperceptibility of small instances under weak supervision. This is mainly caused by the large-kernel (7×7) convolution and non-maximum down-sampling, *i.e.*, 2×2 strided convolution and AveragePool, which lose highly informative features from the raw images. We show that small-kernel convolutions and MaxPool down-samplings provide finer object boundaries and preserve the information of small instances, which enhances the robustness of information flow through the networks.
3. Proposal feature alignment. Modern residual networks commonly achieve large receptive fields by applying an overall stride with $32\times$ sub-sampling. However, such coarse feature maps lead to feature misalignment due to the quantizations in RoIPool [19] layer, which introduces confusing context and lacking diversity. By exploiting dilated convolution to extracts high-resolution feature maps for WSOD, we are able to support the efficient alignment of proposal features and exploit diverse local information, as well as to detect small objects.

We implement two instantiations of the proposed principles: ResNet-WS and DenseNet-WS. Extensive experiments are conducted on PASCAL VOC [15] and MS COCO [37]. We show that the proposed principles enable deep residual networks to achieve significant improvement compared with plain networks for various WSOD methods, which also establishes new state-of-the-arts.

2 Related Work

2.1 Weakly Supervised Object Detection

Prevailing WSOD work generally focuses on two successive stages, object discovery and instance refinement.

Object discovery stage combines multiple instance learning (MIL) and CNNs to implicitly model latent object locations with image-level labels. Several different strategies to train the MIL model had been proposed in the literature [6,8,17,51,61,63]. Bilen *et al.* [7] selected proposals by parallel detection and classification branches. Contextual information [27], attention mechanism [58], saliency map [31,46,48] and semantic segmentation [64] are leveraged to learn outstanding proposals. High-precision object proposals for WSOD are generated in [30,57]. Some methods focused on proposal-free paradigms with deep feature maps [3,4,78], class activation maps [12,21,69,70,75] and generative adversarial learning [13,45]. Some work also used additional information to improve the performance, *e.g.*, object-size estimation [51], instance-count annotations [16], video-motion cue [30,53] and human verification [40]. Knowledge transfer has also been exploited for cross-domain adaptation w.r.t. data [50] and task [24].

Instance refinement stage aims at explicitly learning the object location by making use of the predictions from the object discovery stage. The top-scoring proposals generated from the object discovery stage are used as supervision to train the instance refinement classifier [16,25,32,56,65]. Other different strategies [29,43,55,71] are also proposed to generate pseudo-ground-truth boxes and label proposals. Some methods exploit to improve the optimization of the overall framework that jointly learn the two-stage modules with min-entropy prior [34,60], multi-view learning [72] and continuation MIL [59]. Collaboration mechanism between segmentation and detection is proposed to take advantages of the complementary interpretations of weakly supervised tasks [33,47].

With the output of the above two stages, a fully-supervised detector can also be trained. Many efforts [18,74] have been made to mine high-quality bounding boxes. Zhang *et al.* [73] proposed a self-directed optimization to propagate object priors of the reliable instances to unreliable ones.

2.2 Network Architectures for Object Detection

Significant efforts have been devoted to the design of network architectures for the task of FSOD. DSOD [49] and Root-ResNet [77] exploit to train single-shot detectors, *i.e.*, SSD [39], from scratch, whilst PeleeNet [62] is proposed to train

Table 2. Result of freezing different number of stages in ResNet for WSDDN [7] on VOC 2007 [15]. "NAN" indicates that the training is non-convergent.

Backbone	ResNet18										ResNet50									
Learning Rate	0.001					0.01					0.001					0.01				
#Frozen stages	0	2	3	4	5	0	2	3	4	5	0	2	3	4	5	0	2	3	4	5
mAP (%)	25.2	25.6	26.2	**27.5**	14.4	NAN	NAN	26.7	**28.4**	NAN	23.0	24.9	**25.5**	24.9	NAN	21.0	26.3	**26.5**	26.0	NAN
CorLoc (%)	**46.7**	45.0	42.3	44.6	24.8	NAN	NAN	**50.4**	49.7	NAN	**43.3**	41.8	40.2	37.1	NAN	43.8	**51.3**	50.5	45.9	NAN

SSD for mobile devices. Li *et al.* [35] proposed DetNet backbone for FSOD. Fine feature maps are also useful for detecting small objects as observed in FPN [36].

In conclusion, most traditional backbone networks are usually designed for image classification or FSOD. We have not found one that explores the backbone networks for WSOD. Moreover, the cutting-edge WSOD methods follow the pipeline of ImageNet pre-trained plain networks, *i.e.*, VGG-style networks. Undoubtedly, the advanced modules in recent deep residual architectures have not been explored in WSOD.

3 Baseline WSOD

Without loss of generality, we consider building WSOD models on the pre-trained backbones and fine-tuning its parameters on the target data. We use the popular WSDDN [7] method as baseline WSOD head, which is also a basic module in many state-of-the-art approaches [27,55,56,59,65].

We first investigate several common combination schemes in FSOD to build WSOD heads on ResNet and DenseNet, which are widely used for Faster R-CNN [42], as illustrated in Fig. 1. The C4 [42] combination performs RoIPool [19] on the full-image feature maps from previous 4 stages. All layers in *conv*5 stage and WSOD heads are stacked sequentially on the RoIPooled features. The FPN [36] combination learns full-image feature pyramids from backbones. Then RoIPool is performed to extract 7×7 proposal features followed by two hidden 1,024-d fully-connected (FC) layers before the WSOD heads. Besides, we also consider a solution, termed C5 [19] combination. C5 combination computes full-image feature maps using all convolutional layers (all 5 stages), followed by a RoIPool layer and later layers.

As shown in Table 1, directly employing ResNet and DenseNet for WSOD task reduces the performance dramatically in various combinations. The best performance of 31.5 mAP is obtained from C4 combination, which is still inferior to the shallow AlexNet backbone in terms of mAP. Moreover, some state-of-the-art methods [27,55,59] are unable to converge according to further experiments in Table 4. We focus on the C5 combination in the rest of the paper, as C4 and FPN combinations have their drawbacks in WSOD setting. C4 combination computes entire *conv*5 stage for each proposal. Thus, it will is cost additional $10\times$ training time and 100% memory usage compared with the C5 combination when each image has about 2,000 proposals. FPN combination imposes

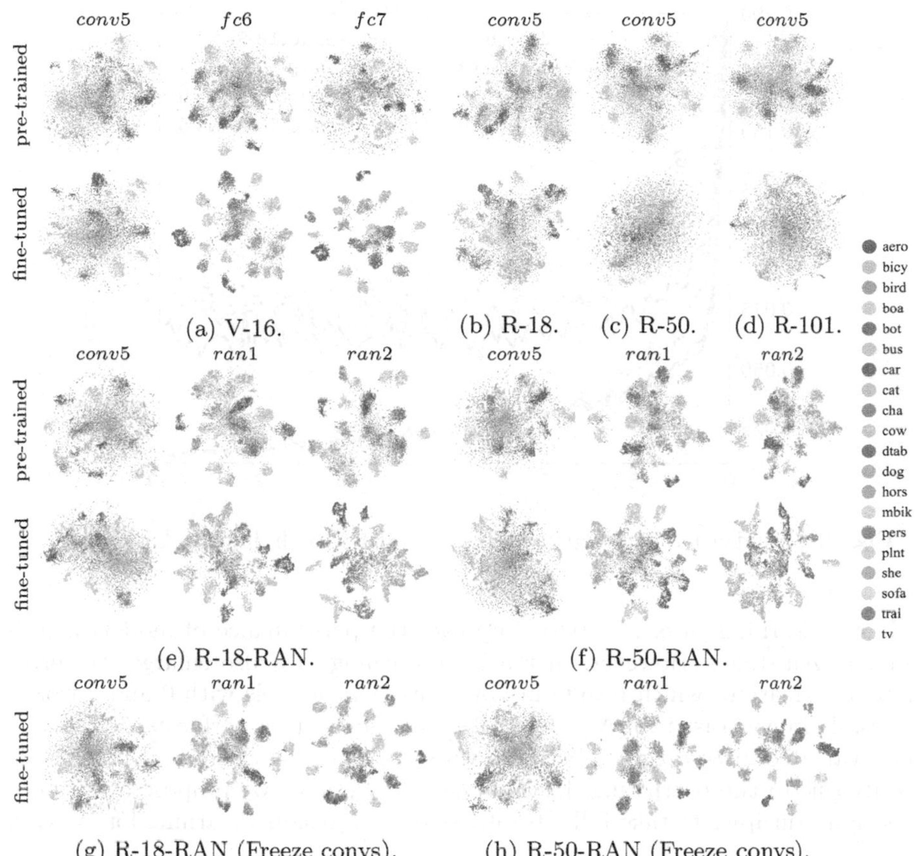

Fig. 2. Visualization of proposal features on PASCAL VOC using t-SNE [20].

an extra burden of learning top-down full-image feature pyramids with lateral connections.

Different from FSOD, WSOD has insufficient supervision and is often formulated via multiple instance learning (MIL) [14], which is sensitive to model initialization [5,9,11,32] and suffers from instability [38]. In this sense, WSOD heads may back-propagate uncertain and erroneous gradient to backbones, whilst deep residual networks enlarge the erroneous information and deteriorate the visual representation learning, which results in dramatically reduced detection performance. To further verify the above analysis, we freeze different number of stages in ResNet, and show the results in Table 2. We summarize: 1) The detection performance mAP is improved progressively by freezing pre-trained layers up to 4 stages, because it prevents convolutional layers in backbones from receiving the erroneous information from WSOD heads. 2) When freezing entire backbones, *i.e.*, all 5 frozen stages, the models has not enough capacity for representation learning (mAP drops dramatically) and even fails to converge.

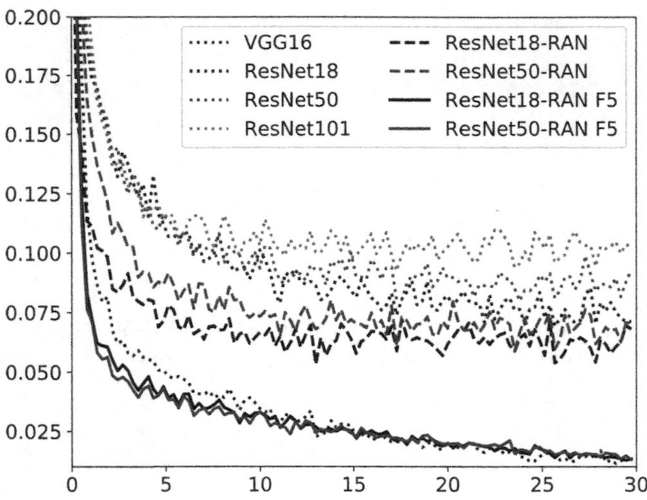

Fig. 3. Optimization landscape analysis of WSDDN with different backbones.

3) Larger learning rate, *i.e.*, 0.01, improves the performance of models with 3 and 4 frozen stages. However, such a large learning rate also enlarges the erroneous information, which results in non-convergent models with 0 and 2 frozen stages. 4) In contrast to mAP of *test* set, the localization performance *CorLoc* that evaluated in *trainval* set becomes worse as more stages are frozen, which is mainly due to the overfitting. In the following sections, we propose a sequence of design principles to take full advantages of deep residual learning for WSOD.

4 Redundant Adaptation Neck

We visualize the distribution of proposal features uniformly sampled from the PASCAL VOC 2007 trainval set [15] using t-SNE [20] in Fig. 2. We compared VGG16 (V-16) with ResNet of 18 (R-18), 50 (R-50) and 101 (R-101) layers. Proposal features from RoIPool and subsequent layers, *i.e.*, *conv*5, $fc6$ and $fc7$ for VGG16 and *conv*5 for ResNet, are shown. In Fig. 2a, we observe that the proposal features of FC layers from fine-tuned VGG16 are more discriminative than that of the pre-trained ones, whilst the distribution of the features from *conv*5 only changes slightly. However, Fig. 2b, 2c and 2d show that the proposal features of ResNet are not discriminative enough to distinguish different categories. Even more, the proposal features of ResNet50 and ResNet101 are deteriorated compared with the pre-trained counterparts. To further explore the training procedure, we also draw the optimization landscape analysis curves for different backbones in Fig. 3. Generally, optimization loss indicates how well the models reason the relationship of proposals to satisfied the imposed constraints in WSOD. VGG16 demonstrates faster convergence and has lower loss than ResNet backbones, which converge to undesirable local minimums.

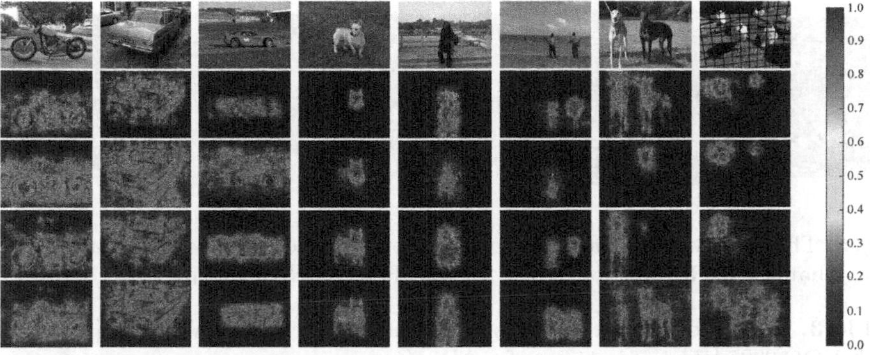

Fig. 4. The first row shows input images. The rest rows show gradient maps of VGG16, R-18-RAN, R-18-RAN-SK and R-18-RAN-SK-MP, respectively.

In conclusion, we observe indiscriminative proposal representation and poor convergence when directly employing ResNet backbones in WSOD task, which cause deteriorated detection performance. As WSOD is required to localize object instances and learn proposal feature jointly with only image-level labels. Therefore, directly stacking WSOD heads on top of residual networks has a large negative impact on the convolutional feature learning. And shortcut connections in residual blocks also enlarge the uncertain and erroneous gradient from WSOD heads throughout the backbones during back-propagation, which overwhelms the direction of optimization steps and fails to infer the proposal-level classifier.

From the perspective of adding redundancy, we propose the first principle that a Redundant Adaptation Neck (RAN), which learns high-dimension visual representation of proposals between deep residual network backbones and the WSOD heads, is the key to localize object instances and learn discriminative features jointly. Our intuition is that the redundant feature representation ensures various WSOD constraints under weak supervision and decreases the negative impact of uncertain and erroneous gradient from WSOD heads, whilst the convolutional layers focus on full-image feature learning. We implement and visualize this principle for ResNet (ResNet-RAN). Instead of instantiating the RAN by stacking convolutional layers, we use multiple perception layers, which are memory-feasible to extract high-dimension features for about $2,000$ proposals. Specifically, the last global pooling layer in ResNet is replaced by two FC layers with high dimension 2,048–4096 before the WSDDN heads. We show the proposal features from $conv5$ and two FC layers from the RAN, *i.e.*, $ran1$ and $ran2$, in Fig. 2e and 2f. ResNet18-RAN and ResNet50-RAN obtain discriminative proposal features in $ran1$ and $ran2$ layers. Figure 3 shows that ResNet-RAN also converges to better minimum. It demonstrates that the entangled tasks of localizing object instances and learning proposal features are optimized jointly.

To further explore the limit of the RAN, we freeze all convolutional layers in the backbones, which completely removes the effect of WSOD heads to the convolutional layers. Figure 2g and 2h show that the proposal features are even

Fig. 5. The two rows show sampling locations of 7^2 discrete bins with maximum values along channels in RoIPool for R-18-RAN and R-18-RAN-DC, respectively.

Table 3. Comparison of various proposal feature extractors for WSDDN [7] on PASCAL VOC 2007 [15] *test* in terms of mAP (%).

Extractor	VGG16	ResNet18	ResNet50	ResNet101
RoIPool [19]	**34.8**	**28.4**	**26.5**	**25.7**
RoIAlign [22]	29.4	24.2	23.7	24.8

more discriminative. Meanwhile, the optimization landscape in Fig. 3 is also improved (ResNet-RAN F5). This interesting observation shows that RAN has elastic capacity to accommodate the entangled tasks.

5 Robust Information Flow

Residual learning greatly alleviates the problem of vanishing gradient in deep networks by enhancing information flow with the skip connections. However, there still exist two main drawbacks in deep residual networks, *i.e.*, ResNet and DenseNet, that hinder the robustness of information flow to uncertain and erroneous gradient under weak supervision. First, the large-kernel (7×7) convolutions in the stem block weaken the information of object boundaries, resulting in uncertainty around the object boundaries. Second, non-maximum down-sampling, *i.e.*, 2×2 strided convolutions and AveragePool, may also hurt the flow of information, which makes small instances imperceptible, as the non-maximum down-sampling may not preserve the informative activations and gradient flowing through the network under weak supervision.

From the perspective of improving robustness, we propose a principle that using small-kernel (SK) convolution and MaxPool (MP) down-sampling in the backbones to improve the robustness of information flow, which give finer object boundaries and more sensitivity on small objects. Specifically, we replace the original stem block with three conservative 3×3 convolutions, with the first and third convolutions followed by 2×2 MaxPool layers. For down-sampling, we change the strided convolution or AveragePool operation with MaxPool, which is set to 2×2 with 2×2 stride to avoid the overlapping between input activations.

We utilize the gradient maps of input images to observe how information flows through the networks. In the second and third rows of Fig. 4, we observe that

Table 4. Ablation study on PASCAL VOC 2007 *test*.

	Backbone	Method	RAN	SM	MP	DC	CorLoc (%)	*m*AP (%)
a	ResNet18	WSDDN					49.7	28.4
b			√				57.7	37.5
c				√			53.7	32.9
d					√		54.4	33.7
e						√	51.9	30.9
f			√	√			60.2	40.2
g			√		√		59.2	39.1
h			√	√	√		62.3	42.5
i			√	√		√	61.9	42.1
j			√	√	√	√	**63.1**	**43.4**
k		ContextLocNet					NAN	NAN
l			√	√	√	√	**64.7**	**45.4**
m		OICR					55.3	34.7
n			√	√	√	√	**68.7**	**51.0**
o		PCL					NAN	NAN
p			√	√	√	√	**67.1**	**50.2**
q		C-MIL					NAN	NAN
r			√	√	√	√	**68.5**	**52.6**
s	ResNet50	WSDDN					50.5	26.5
t			√	√	√	√	**63.6**	**44.0**
u	ResNet101	WSDDN					50.9	25.7
v			√	√	√	√	**64.0**	**44.1**
w	DenseNet121	WSDDN					55.5	29.7
x			√	√	√	√	**66.3**	**44.8**
y	DenseNet169	WSDDN					53.0	28.5
z			√	√	√	√	**66.1**	**44.3**

the gradients of object boundaries in R-18-RAN are more blurry than that of VGG16. And the gradients of some object parts and small instances are missed in R-18-RAN. However, gradient maps of R-18-RAN-SK provide finer object boundaries, and R-18-RAN-SK-MP responses to multiple small objects.

6 Proposal Feature Alignment

Modern deep residual networks commonly use 5 stages to extract full-image feature maps with 32× sub-sampling. This brings large effective receptive fields, which are critical for high classification accuracy. However, the large stride may cause misalignment between region proposals and pooled features from the RoIPool [19] layer. The feature misalignment is caused by two quantization operations: coordinate rounding after being divided by the stride, and projected proposals segmentation into discrete bins. Although the misalignment has little negative impact on FSOD, it introduces serious features ambiguity in WSOD, which further raises the instability problem.

Table 5. Comparison with SotAs on PASCAL VOC 2007 *test* in terms of AP.

Method	Backbone	aero	bicy	bird	boa	bot	bus	car	cat	cha	cow	dtab	dog	hors	mbik	pers	plnt	she	sofa	trai	tv	Av.
Object Discovery																						
WCCN[12]	VGG16	49.5	60.6	38.6	29.2	16.2	70.8	56.9	42.5	10.9	44.1	29.9	42.2	47.9	64.1	13.8	23.5	45.9	**54.1**	60.8	54.5	42.8
Jie *et al.* [25]	VGG16	52.2	47.1	35.0	26.7	15.4	61.3	**66.0**	54.3	3.0	**53.6**	24.7	43.6	48.4	**65.8**	6.6	18.8	**51.9**	43.6	53.6	**62.4**	41.7
SGWSOD[31]	VGG16	48.4	61.5	33.3	**30.0**	15.3	**72.4**	62.4	59.1	10.9	42.3	34.3	53.1	48.4	65.0	20.5	16.6	40.6	46.5	54.6	55.1	43.5
TS^2C[64]	VGG16	**59.3**	57.5	43.7	27.3	13.5	63.9	61.7	**59.9**	**24.1**	46.9	36.7	45.6	39.9	62.6	10.3	23.6	41.7	52.4	58.7	56.6	44.3
CSC C5[48]	VGG16	51.4	62.0	35.2	18.7	**27.9**	66.7	53.5	51.4	16.2	43.6	43.0	46.7	20.0	58.4	**31.1**	23.8	43.6	48.8	65.4	53.5	43.0
WS-JDS[47]	VGG16	52.0	**64.5**	**45.5**	26.7	**27.9**	60.5	47.8	59.7	13.0	50.4	**46.4**	**56.3**	**49.6**	60.7	25.4	**28.2**	50.0	51.4	**66.5**	29.7	**45.6**
WSDDN[7]	VGG16	39.4	50.1	31.5	16.3	12.6	64.5	42.8	42.6	10.1	35.7	24.9	38.2	34.4	55.6	9.4	14.7	30.2	40.7	54.7	46.9	34.8
	ResNet18-WS	47.9	56.8	40.2	17.6	**29.9**	67.2	54.6	**49.6**	8.7	46.6	**47.0**	34.8	**52.0**	**61.4**	**17.0**	24.3	42.2	49.3	60.5	**59.9**	43.4
	ResNet50-WS	**50.4**	56.7	**41.8**	**24.9**	**29.9**	64.0	55.8	47.8	**21.5**	**50.3**	35.0	49.5	49.5	58.1	13.9	**24.5**	44.7	40.7	**65.3**	55.8	44.0
	ResNet101-WS	47.0	**58.6**	40.4	21.1	28.4	**68.4**	**57.1**	46.5	20.1	49.5	35.5	**51.8**	48.1	55.8	12.2	19.6	**45.4**	**53.8**	63.2	58.1	**44.1**
ContextLocNet[27]	VGG-F	57.1	52.0	31.5	7.6	11.5	55.0	53.1	34.1	1.7	33.1	**49.2**	42.0	47.3	56.6	15.3	12.8	24.8	48.9	44.4	47.8	36.3
	ResNet18-WS	58.1	56.3	41.6	**31.4**	22.9	66.1	**57.3**	64.1	11.0	34.2	45.0	**59.7**	**58.9**	60.2	12.9	22.0	30.2	**56.6**	68.5	50.6	45.4
	ResNet50-WS	54.9	**62.8**	41.5	19.1	**28.5**	**67.3**	55.1	52.4	17.9	**48.3**	39.4	45.7	55.3	61.2	**31.1**	22.2	**44.4**	46.7	64.6	45.9	45.3
	ResNet101-WS	**60.6**	53.5	**50.3**	26.1	26.4	66.9	55.8	**73.1**	**18.0**	35.7	19.2	54.7	56.0	**65.6**	25.5	**24.3**	30.3	51.9	**69.4**	**54.4**	**45.9**
Object Discovery + Instance Refinement																						
MELM[60]	VGG16	55.6	66.9	34.2	29.1	16.4	68.8	68.1	43.0	25.0	**65.6**	45.3	53.2	49.6	68.6	2.0	25.4	52.5	56.8	62.1	57.1	47.3
ZLDN[71]	VGG16	55.4	68.5	50.1	16.8	20.8	62.7	66.8	56.5	2.1	57.8	47.5	40.1	69.7	68.2	21.6	27.2	53.4	56.1	52.5	58.2	47.6
GAL-fWSD512[45]	VGG16	58.4	63.8	45.8	24.0	22.7	67.7	65.7	58.9	15.0	58.1	47.0	53.7	23.8	64.3	**36.2**	22.3	46.7	50.3	**70.8**	55.1	47.5
ML-LocNet[72]	VGG16	59.3	68.9	45.7	29.0	24.5	64.8	68.4	59.3	18.6	49.1	50.2	43.1	65.8	70.2	19.9	24.3	48.1	54.2	62.8	41.8	48.4
WSRPN[57]	VGG16	57.9	**70.5**	37.8	5.7	21.0	66.1	69.2	59.4	3.4	57.1	**57.3**	35.2	64.2	68.6	32.8	**28.6**	50.8	49.5	41.1	30.0	45.3
Kosugi *et al.* [29]	VGG16	61.5	64.8	43.7	26.4	17.1	67.4	62.4	67.8	25.4	51.0	33.7	47.6	51.2	65.2	19.3	24.4	44.6	54.1	65.6	59.5	47.6
Pred Net[2]	VGG16	**66.7**	69.5	52.8	31.4	**24.7**	**74.5**	**74.1**	67.3	14.6	53.0	46.1	52.9	69.9	**70.8**	18.5	28.4	54.6	**60.7**	67.1	60.4	52.9
OICR W-RPN[30]	VGG16	-	-	-	-	-	-	-	-	-	-	-	-	-	-	-	-	-	-	-	-	46.9
SDCN[33]	VGG16	59.8	67.1	32.0	34.7	22.8	67.1	63.8	**67.9**	22.5	48.9	47.8	**60.5**	51.7	65.2	11.8	20.6	42.1	54.7	60.8	64.3	48.3
WSOD[2][68]	VGG16	65.1	64.8	**57.2**	**39.2**	24.3	69.8	66.2	61.0	**29.8**	64.6	42.5	60.1	**71.2**	70.7	21.9	28.1	**58.6**	59.7	52.2	**64.8**	**53.6**
OICR[56]	VGG16	58.0	**62.4**	31.1	19.4	13.0	65.1	62.2	28.4	**24.8**	44.7	**30.6**	25.3	37.8	**65.5**	15.7	**24.1**	41.7	46.9	64.3	62.6	41.2
	ResNet18-WS	61.3	54.5	52.4	30.1	34.9	**68.9**	65.0	75.0	22.5	57.4	19.7	66.6	64.8	64.9	**16.8**	22.3	**53.2**	54.9	69.9	64.8	51.0
	ResNet50-WS	61.2	50.9	**55.0**	33.2	**36.2**	68.6	**65.7**	**79.2**	17.3	58.1	19.3	69.1	65.7	64.8	15.1	18.9	50.1	55.1	69.8	64.4	50.9
	ResNet101-WS	**63.2**	51.1	51.9	**33.7**	32.4	67.9	65.0	78.9	19.0	**59.4**	21.9	**70.6**	**68.3**	64.4	15.2	20.8	49.3	**55.3**	**72.5**	**66.6**	**51.4**
PCL[55]	VGG16	54.4	69.0	39.3	19.2	15.7	62.9	64.4	30.0	**25.1**	52.5	44.4	19.6	39.3	67.7	17.8	22.9	46.6	**57.5**	58.6	**63.0**	43.5
	ResNet18-WS	54.4	**69.5**	48.7	29.7	**33.2**	70.7	69.7	57.2	11.5	62.4	37.2	39.3	66.3	67.5	23.7	**30.9**	**60.1**	52.0	65.3	55.3	50.2
	ResNet18-WS F2	54.5	67.6	48.1	**31.6**	32.6	**71.5**	**72.3**	67.7	3.3	**64.2**	**58.7**	45.4	67.3	**68.4**	27.7	30.8	56.7	50.6	67.4	51.1	51.9
	ResNet50-WS	55.4	60.7	50.8	30.1	31.0	69.8	69.0	66.6	9.6	62.0	25.0	56.4	68.2	65.5	35.7	28.1	57.2	52.9	67.0	54.2	50.8
	ResNet101-WS	**56.5**	65.4	**54.2**	27.8	30.2	70.8	67.5	**74.8**	3.2	60.4	56.0	**68.0**	**70.6**	65.4	**35.8**	23.1	53.1	53.0	**70.7**	60.4	**53.3**
C-MIL[59]	VGG16	62.5	**58.4**	49.5	32.1	19.8	70.5	66.1	63.4	20.0	**60.5**	**52.9**	53.5	57.4	68.9	8.4	24.6	51.8	58.7	66.7	63.5	50.5
	ResNet18-WS	57.0	54.9	43.6	**39.9**	32.2	70.9	**69.8**	75.2	14.2	59.9	28.5	66.3	67.5	65.3	**37.6**	21.8	**56.7**	49.8	71.1	**68.9**	52.6
	ResNet50-WS	**67.5**	45.2	62.9	33.4	41.6	**73.9**	66.7	76.2	**26.4**	54.8	11.6	**71.4**	71.9	72.9	20.6	31.9	42.5	58.8	**77.1**	61.3	53.4
	ResNet101-WS	66.7	41.4	**64.7**	35.5	**42.2**	73.7	67.3	**76.3**	23.4	56.0	12.1	68.7	**74.5**	**75.1**	22.6	**34.1**	43.6	**60.5**	76.2	64.2	**53.9**
OICR+REG[65]	VGG16	55.2	66.5	40.1	31.1	16.9	69.8	64.3	67.8	**27.8**	52.9	**47.0**	33.0	60.8	64.4	13.8	**26.0**	44.0	**55.7**	68.9	**65.5**	48.6
	ResNet101-WS	**67.3**	**72.1**	**55.8**	**31.8**	**31.3**	**71.6**	**70.0**	**76.7**	19.4	**58.7**	21.1	**68.5**	**74.6**	**69.9**	**19.1**	18.8	**48.4**	55.1	**71.9**	53.2	**52.8**

To address the misalignment of proposal features, we exploit dilated convolution (DC) [66,76] to extract high-resolution full-image feature maps for WSOD. Specifically, we fix the spatial size after stage 3 and use dilated convolution with a rate of 2 in the subsequent stages, which results in only 8× sub-sampling. We visualize the sampling locations of RoIPool in Fig. 5. In the first three columns, the sampling locations of RoIPool in R-18-RAN may exceed the border of proposals, due to the rounded coordinates, whilst R-18-RAN-DC constraints the regions of the sampling inside the proposals. Quantizing proposals into discrete bins in low-resolution feature maps also causes less diversity of sampling locations, as shown in the last five columns of Fig. 5, while high-resolution feature maps from dilated convolution provide more diverse information.

It is worth noting that RoIAlign [22] uses bilinear interpolation to compute the exact values at sampled locations in discrete bins, which aims to address the quantization errors. However, RoIAlign samples activation in a fixed position, which results in inferior performance as shown in Table 3.

7 Quantitative Results

Datasets. We evaluate the proposed design principles on PASCAL VOC 2007, 2012 [15] and MS COCO [37], which are widely-used benchmark datasets.

Evaluation Protocols. The CorLoc indicates the percentage of images in which a method correctly localizes an object of the target category according to the

Table 6. Comparison with SotAs on VOC 2007 *trainval* in terms of CorLoc.

Method	Backbone	aero	bicy	bird	boa	bot	bus	car	cat	cha	cow	dtab	dog	hors	mbik	pers	plnt	she	sofa	trai	tv	Av.
Object Discovery																						
WCCN[12]	VGG16	83.9	72.8	64.5	44.1	40.1	65.7	82.5	58.9	33.7	72.5	25.6	53.7	67.4	77.4	26.8	49.1	68.1	27.9	64.5	55.7	56.7
Jie *et al.*'17[25]	VGG16	72.7	55.3	53.0	27.8	35.2	68.6	81.9	60.7	11.6	71.6	29.7	54.3	64.3	88.2	22.2	53.7	72.2	52.6	68.9	75.5	56.1
SP-VGGNet[78]	VGG16	85.3	64.2	67.0	42.0	16.4	71.0	64.7	88.7	20.7	63.8	58.0	84.1	84.7	80.0	60.0	29.4	56.3	68.1	77.4	30.5	60.6
TST[50]	AlexNet	–	–	–	–	–	–	–	–	–	–	–	–	–	–	–	–	–	–	–	–	59.5
SGWSOD[31]	VGG16	71.0	76.5	54.9	49.7	54.1	78.0	87.4	68.8	32.4	75.2	29.5	58.0	67.3	84.5	41.5	49.0	78.1	60.3	62.8	78.9	62.9
TS^2C[64]	VGG16	84.2	74.1	61.3	52.1	32.1	76.7	82.9	66.6	42.3	70.6	39.5	57.0	61.2	88.4	9.3	54.6	72.2	60.0	65.0	70.3	61.0
CSC C5[48]	VGG16	76.1	75.3	61.8	42.0	54.1	74.7	78.8	67.4	32.8	73.1	46.5	59.9	37.6	78.0	56.0	42.5	71.9	67.3	82.4	65.6	62.2
WS-JDS[47]	VGG16	82.9	74.0	73.4	47.1	60.9	80.4	77.5	78.8	18.6	70.0	56.7	67.0	64.5	84.0	47.0	50.1	71.9	57.6	83.3	43.5	64.5
WSDDN[7]	VGG16	65.1	58.8	58.5	33.1	39.8	68.3	60.2	59.6	34.8	64.5	30.5	43.0	56.8	82.4	25.5	41.6	61.5	55.9	65.9	63.7	53.5
	ResNet18-WS	75.0	63.7	65.0	37.0	60.4	80.4	81.1	70.5	21.8	60.8	65.9	44.3	74.8	85.0	37.5	56.3	68.7	63.9	74.0	75.0	63.1
	ResNet50-WS	74.1	68.9	69.4	39.5	64.0	79.3	84.3	66.2	42.4	73.9	38.1	52.7	69.7	83.3	27.2	54.8	68.7	57.6	81.8	75.7	63.6
	ResNet101-WS	72.3	63.7	67.7	49.3	61.8	77.3	85.1	63.8	36.1	68.1	45.3	52.2	71.2	87.5	27.9	58.6	66.6	66.6	76.3	81.2	64.0
ContextLocNet[27]	VGG-F	83.3	68.6	54.7	23.4	18.3	73.6	74.1	54.1	8.6	65.1	47.1	59.5	67.0	83.5	35.3	39.9	67.0	49.7	63.5	65.2	55.1
	ResNet18-WS	82.1	62.9	66.6	44.4	53.2	80.4	84.5	82.2	22.7	60.8	60.8	68.4	76.2	84.1	29.5	55.6	64.5	68.4	76.3	70.3	64.7
	ResNet50-WS	74.0	77.8	63.5	43.6	58.0	81.6	79.6	68.0	25.2	79.5	59.5	60.0	61.3	81.1	54.5	47.3	82.5	62.3	75.7	65.9	65.1
	ResNet101-WS	81.7	78.7	65.3	56.0	56.2	77.5	82.2	73.0	32.1	81.9	36.9	62.8	67.6	84.0	55.2	57.1	70.8	64.4	731.	57.0	65.7
Object Discovery + Instance Refinement																						
ZLDN[71]	VGG16	74.0	77.8	65.2	37.0	46.7	75.8	83.7	58.8	17.5	73.1	49.0	51.3	76.7	87.4	30.6	47.8	75.0	62.5	64.8	68.8	61.2
GAL-fWSD512[45]	VGG16	78.6	81.9	63.6	40.3	48.8	80.7	85.3	76.3	30.3	78.0	54.5	65.3	48.4	86.5	56.3	46.9	76.0	68.1	83.9	73.1	66.1
ML-LocNet[72]	VGG16	78.6	82.3	68.2	42.0	53.3	78.5	88.5	70.3	36.4	70.2	60.5	58.0	80.5	88.2	38.8	59.2	75.0	69.0	78.2	64.5	67.0
WSRPN[57]	VGG16	77.5	81.2	55.3	19.7	44.3	80.2	86.6	69.5	10.1	87.7	68.4	52.1	84.4	91.6	57.4	63.4	77.3	58.1	57.0	53.8	63.8
Kosugi *et al.* [29]	VGG16	85.5	79.6	68.1	55.1	33.6	83.5	83.1	78.5	42.7	79.8	37.8	61.5	74.4	88.6	32.6	55.7	77.9	63.7	78.4	74.1	66.7
Pred Net VGG16[2]	VGG16	88.6	86.3	71.8	53.4	51.2	87.6	89.0	65.3	33.2	86.6	58.8	65.9	87.7	93.3	30.9	58.9	83.4	67.8	78.7	80.2	70.9
OICR W-RPN[30]	VGG16	-	-	-	-	-	-	-	-	-	-	-	-	-	-	-	-	-	-	-	-	66.5
SDCN[33]	VGG16	85.8	83.1	56.2	58.5	44.7	80.2	85.0	77.9	29.6	78.8	53.6	74.2	73.1	88.4	18.2	57.5	74.2	60.8	76.1	79.2	66.8
$WSOD^2$[68]	VGG16	87.1	80.0	74.8	60.1	36.6	79.2	83.8	70.6	43.5	88.4	46.0	74.7	87.4	90.8	44.2	52.4	81.4	61.8	67.7	79.9	69.5
OICR[56]	VGG16	81.7	80.4	48.7	49.5	32.8	81.7	85.4	40.1	40.6	79.5	35.7	33.7	60.5	88.8	21.8	57.9	76.3	59.9	75.3	81.4	60.6
	ResNet18-WS	82.1	60.3	81.1	49.3	67.6	81.4	87.2	84.0	33.4	76.8	21.6	78.8	87.0	87.5	30.8	52.6	81.2	66.6	81.8	82.8	68.7
	ResNet50-WS	75.9	65.6	70.9	56.0	50.0	81.6	80.8	83.8	33.0	79.5	27.3	79.9	81.7	81.0	30.4	45.0	85.5	72.2	79.1	81.2	67.4
	ResNet101-WS	83.3	68.6	71.3	53.5	54.7	83.3	88.6	87.3	33.8	80.3	31.6	82.5	85.8	83.8	26.7	42.0	82.5	73.5	80.3	84.7	68.9
PCL[55]	VGG16	79.6	85.5	62.2	47.9	37.0	83.8	83.4	43.0	38.3	80.1	50.6	30.9	57.8	90.8	27.0	58.2	75.3	68.5	75.7	78.9	62.7
	ResNet18-WS	76.7	81.9	74.4	48.1	53.9	84.5	87.7	86.5	25.4	68.1	36.0	67.4	84.8	86.6	52.5	51.1	81.2	54.9	78.7	62.5	67.1
	ResNet18-WS F2	79.4	86.2	75.0	54.3	53.2	87.6	88.8	80.9	10.2	81.1	68.0	59.6	89.2	87.5	41.7	59.4	83.3	62.1	80.3	74.2	70.1
	ResNet50-WS	75.8	82.7	73.3	48.1	60.4	88.6	88.5	74.2	28.1	71.0	46.3	55.6	88.4	88.3	29.3	56.3	81.2	69.3	79.5	71.8	67.8
	ResNet101-WS	84.9	77.2	71.3	60.0	44.8	76.4	86.4	87.9	16.7	86.1	67.0	84.4	86.5	88.8	53.1	50.0	81.3	72.9	85.8	78.9	72.0
C-MIL[59]	VGG16	–	–	–	–	–	–	–	–	–	–	–	–	–	–	–	–	–	–	–	–	65.0
	ResNet18-WS	80.3	64.6	68.3	53.0	56.8	84.5	89.1	86.5	28.1	72.4	28.8	77.3	84.1	79.1	56.8	51.8	85.4	62.1	81.1	80.4	68.5
	ResNet50-WS	80.8	70.7	74.4	53.4	56.6	85.6	88.0	85.4	35.2	84.2	27.8	78.4	82.4	79.7	31.0	50.0	89.6	73.0	79.1	80.3	69.3
	ResNet101-WS	78.0	75.3	69.6	63.4	52.2	85.3	83.3	81.1	37.8	79.4	44.5	79.9	78.3	85.1	51.8	55.7	85.5	68.9	74.9	79.8	70.4
OICR+REG[65]	VGG16	81.7	81.2	58.9	54.3	37.8	83.2	86.2	77.0	42.1	83.6	51.3	44.9	78.2	90.8	20.5	56.8	74.2	66.1	81.0	86.0	66.8
	ResNet101-WS	88.8	86.6	66.6	57.0	48.5	78.6	91.1	91.3	34.3	88.8	29.1	78.9	90.5	89.6	34.1	41.0	77.0	74.5	87.3	66.4	70.1

PASCAL criterion. The mAP follows standard PASCAL VOC protocol to report the mAP at 50% Intersection-over-Union (IoU) of the detected boxes with the ground-truth ones. For MS COCO data, we report the standard COCO metrics, including AP at different IoU thresholds and instance scales.

Implementation Details. All backbone networks are initialized with the weights pre-trained on ImageNet ILSVRC [10]. We use synchronized SGD training on 4 GPUs. A mini-batch involves 1 images per GPU. In the multi-scale setting, we use scales of $\{480, 576, 688, 864, 1200\}$. We set the maximum number of proposals in an image to be $2,000$. We freeze all pre-trained convolutional layers in backbones unless specified otherwise. The test scores are the average of all scales and flips. Detection results are post-processed by non-maximum suppression using a threshold of 0.3.

7.1 Ablation Study

We validate the contribution of each design principle on PASCAL VOC 2007 in Table 4. For rows (b–e), we report the results of applying each principle to ResNet18, which show consistent improvements over the original backbone (a). Especially, RAN (b) provides the largest performance gain among all principles. It demonstrates that RAN is key to localize object instances and learn proposal features jointly. Rows (f–j) show integrating different principles further improve

Table 7. Comparison with SotAs on VOC 2012 in terms of mAP and CorLoc.

Method	Backbone	mAP (%)	CorLoc (%)
	Object Discovery		
WCCN[12]	VGG16	37.9	–
Jie *et al.* [25]	VGG16	38.3	58.8
SGWSOD[31]	VGG16	39.6	62.9
TS^{2}C[64]	VGG16	**40.0**	**64.4**
CSC[48]	VGG16	37.1	61.4
WS-JDS[47]	VGG16	39.1	63.5
ContextLocNet[27]	VGG-F	35.3	54.8
	ResNet18-WS	**42.0**	**66.7**
	Object Discovery + Instance Refinement		
MELM[60]	VGG16	42.4	–
ZLDN[71]	VGG16	42.9	61.5
WSRPN[57]	VGG16	40.8	64.9
GAL-fWSD300[45]	VGG16	43.1	67.2
Kosugi *et al.* [29]	VGG16	43.4	66.7
ML-LocNet[72]	VGG16	42.2	66.3
Pred Net VGG16[2]	VGG16	**48.4**	69.5
OICR + W-RPN[30]	VGG16	43.2	67.5
SDCN[33]	VGG16	43.5	67.9
WSOD^2[68]	VGG16	47.2	**71.9**
OICR[56]	VGG16	37.9	62.1
	ResNet101-WS	**50.4**	**72.5**
	DenseNet121-WS	48.6	70.3
PCL[55]	VGG16	40.6	63.2
	ResNet101-WS	**51.2**	**73.5**
C-MIL[59]	VGG16	46.7	67.4
	ResNet18-WS	**50.6**	**73.0**
OICR+REG[65]	VGG16	46.8	69.5
	ResNet101-WS	**51.1**	**73.2**

detection performance. Compared with the baseline (a), the best performances are improved by 15.0% mAP significantly. It demonstrates that the proposed principles are orthogonal to each other. Rows (k–r) show that more state-of-the-art WSOD methods [27,55,56,59] also have significant performance boost. Thus, the proposed principles for backbones are orthogonal to WSOD methods. Finally, rows (s–z) show that with different deep residual backbones, our models also outperform corresponding baselines, with ResNet50 and ResNet101 having more gains compared with ResNet18 (15.0% *vs.* 17.5% *vs.* 18.4% mAP).

7.2 Comparison with State of the Arts

To fully compare with other backbones, we separately report the detection results for two successive stages, *i.e.*, object discovery and instance refinement. Table 5 and Table 6 show the results on VOC 2007 in terms of mAP and CorLoc, respectively. For object discovery methods, our models with ResNet-WS obtains 43.4–44.1% mAP and 63.1–64.0% CorLoc for WSDDN [7], which significantly outperform the previous result with VGG16 by 8.6–9.3% mAP and 9.6–11.5% CorLoc. The improvements of ResNet101-WS for ContextLocNet [27] are 9.6% mAP and 10.6% CorLoc. For the instance refinement methods, replacing the backbones of

Table 8. Comparison with the state-of-the-art methods on COCO minival set.

Method	Bakcbone	Avg. Precision, IoU:			Avg. Precision, Area:		
		0.5:0.95	0.5	0.75	S	M	L
WSDDN[7]	VGG-M	8.1	16.0	7.3	1.0	7.8	14.3
	VGG16	9.5	19.2	8.2	2.1	10.4	17.2
	ResNet18-WS	10.7	21.9	**9.1**	2.6	**10.9**	**19.7**
	ResNet101-WS	**10.8**	**22.0**	9.0	**2.7**	10.8	19.6

OICR [56], PCL [55] and C-MIL [59] with ResNet-WS sets the new state-of-the-art results with improvements of 10.8–5.4 mAP.

For CorLoc, our ResNet101-WS backbone surpasses all single-model detectors with improvements of 8.3%, 7.3% and 7.4%, respectively. It is noted that we freeze all convolutional layers in our backbones when fine-tuning on target data. When only freezing the first two stages (ResNet18-WS F2) during training, the performances of PCL achieve further gains with 1.7% mAP and 3.0% CorLoc. Table 7 shows the results on VOC 2012. It can be observed that ResNet-WS models outperform all counterparts with different WSOD methods and achieve new state-of-the-art results. The superiority of ResNet-WS mainly benefits from successfully optimizing the entangled tasks of jointly localizing object instances and learning discriminative features. Table 8 shows the result on MS COCO. We find that ResNet18-WS backbone surpasses existing models on all metrics. For $AP_{0.5:0.95}$, our models outperforms compared works by at least 1.8%. The performance are significantly improved for small instances (44.8% relative improvement for ContextLocNet [27]). This also indicates the efficiency of improving robustness and aligning features.

8 Conclusion

In this paper, we propose a sequence of design principles to take full advantages of deep residual learning for WSOD task. Extensive experiments show that the proposed principles enable deep residual networks to achieve significant performance improvements compared with plain networks for various WSOD methods, which also establishes new state-of-the-arts. Note that our contributions are not specific to ResNet or DenseNet – other backbones (*e.g.*, GoogLeNet [54], WideResNet [67]) can also benefit from the proposed principles for WSOD task.

Acknowledgment. This work is supported by the Nature Science Foundation of China (No. U1705262, No. 61772443, No. 61572410, No. 61802324 and No. 61702136), National Key R&D Program (No. 2017YFC0113000, and No. 2016YFB1001503), Key R&D Program of Jiangxi Province (No. 20171ACH80022) and Natural Science Foundation of Guangdong Province in China (No. 2019B1515120049).

References

1. Alex, K., Sutskever, I., Hinton, G.E.: ImageNet classification with deep convolutional neural networks. In: Conference on Neural Information Processing Systems (NeurIPS) (2012)
2. Arun, A., Jawahar, C.V., Kumar, M.P.: Dissimilarity coefficient based weakly supervised object detection. In: IEEE/CVF Conference on Computer Vision and Pattern Recognition (CVPR) (2019)
3. Bazzani, L., Bergamo, A., Anguelov, D., Torresani, L.: Self-taught object localization with deep networks. In: WACV (2016)
4. Bency, A.J., Kwon, H., Lee, H., Karthikeyan, S., Manjunath, B.S.: Weakly supervised localization using deep feature maps. In: Leibe, B., Matas, J., Sebe, N., Welling, M. (eds.) ECCV 2016. LNCS, vol. 9905, pp. 714–731. Springer, Cham (2016). https://doi.org/10.1007/978-3-319-46448-0_43
5. Bilen, H., Pedersoli, M., Tuytelaars, T.: Weakly supervised object detection with posterior regularization. In: The British Machine Vision Conference (BMVC) (2014)
6. Bilen, H., Pedersoli, M., Tuytelaars, T.: Weakly supervised object detection with convex clustering. In: IEEE/CVF Conference on Computer Vision and Pattern Recognition (CVPR) (2015)
7. Bilen, H., Vedaldi, A.: Weakly supervised deep detection networks. In: IEEE/CVF Conference on Computer Vision and Pattern Recognition (CVPR) (2016)
8. Cinbis, R.G., Verbeek, J., Schmid, C.: Multi-fold MIL training for weakly supervised object localization. In: IEEE/CVF Conference on Computer Vision and Pattern Recognition (CVPR) (2014)
9. Cinbis, R.G., Verbeek, J., Schmid, C.: Weakly supervised object localization with multi-fold multiple instance learning. IEEE Trans. Pattern Anal. Mach. Intell. (TPAMI) **39**, 189–203 (2015)
10. Deng, J., Dong, W., Socher, R., Li, L.J., Li, K., Fei-Fei, L.: ImageNet: a large-scale hierarchical image database. In: IEEE/CVF Conference on Computer Vision and Pattern Recognition (CVPR) (2009)
11. Deselaers, T., Alexe, B., Ferrari, V.: Localizing objects while learning their appearance. In: Daniilidis, K., Maragos, P., Paragios, N. (eds.) ECCV 2010. LNCS, vol. 6314, pp. 452–466. Springer, Heidelberg (2010). https://doi.org/10.1007/978-3-642-15561-1_33
12. Diba, A., Sharma, V., Pazandeh, A., Pirsiavash, H., Van Gool, L.: Weakly supervised cascaded convolutional networks. In: IEEE/CVF Conference on Computer Vision and Pattern Recognition (CVPR) (2017)
13. Diba, A., Sharma, V., Stiefelhagen, R., Van Gool, L.: Weakly supervised object discovery by generative adversarial and ranking networks. In: CVPR Workshop (2019)
14. Dietterich, T.G., Lathrop, R.H., Lozano-Pérez, T.: Solving the multiple instance problem with axis-parallel rectangles. Artif. Intell. (AI) **89**, 31–71 (1997)
15. Everingham, M., Van Gool, L., Williams, C.K.I., Winn, J., Zisserman, A.: The pascal visual object classes (VOC) challenge. Int. J. Comput. Vis. (IJCV) **88**, 303–338 (2010). https://doi.org/10.1007/s11263-009-0275-4
16. Gao, M., Li, A., Yu, R., Morariu, V.I., Davis, L.S.: C-WSL: count-guided weakly supervised localization. In: European Conference on Computer Vision (ECCV) (2018)

17. Ge, C., Wang, J.: Fewer is more : image segmentation based weakly supervised object detection with partial aggregation. In: The British Machine Vision Conference (BMVC) (2018)
18. Ge, W., Yang, S., Yu, Y.: Multi-evidence filtering and fusion for multi-label classification, object detection and semantic segmentation based on weakly supervised learning. In: IEEE/CVF Conference on Computer Vision and Pattern Recognition (CVPR) (2018)
19. Girshick, R.: Fast R-CNN. In: IEEE International Conference on Computer Vision (ICCV) (2015)
20. Graham-Rowe, D.: Visualizing data using t-SNE. JMLR **9**, 2579–2605 (2008)
21. Gudi, A., van Rosmalen, N., Loog, M., van Gemert, J.: Object-extent pooling for weakly supervised single-shot localization. In: The British Machine Vision Conference (BMVC) (2017)
22. He, K., Gkioxari, G., Dollár, P., Girshick, R.: Mask R-CNN. In: IEEE International Conference on Computer Vision (ICCV) (2017)
23. Huang, G., Liu, Z., Weinberger, K.Q.: Densely connected convolutional networks. In: IEEE/CVF Conference on Computer Vision and Pattern Recognition (CVPR) (2017)
24. Inoue, N., Furuta, R., Yamasaki, T., Aizawa, K.: Cross-domain weakly-supervised object detection through progressive domain adaptation. In: IEEE/CVF Conference on Computer Vision and Pattern Recognition (CVPR) (2018)
25. Jie, Z., Wei, Y., Jin, X., Feng, J., Liu, W.: Deep self-taught learning for weakly supervised object localization. In: IEEE/CVF Conference on Computer Vision and Pattern Recognition (CVPR) (2017)
26. Kaiming He, Zhang, X., Ren, S., Sun, J.: Deep residual learning for image recognition. In: IEEE/CVF Conference on Computer Vision and Pattern Recognition (CVPR) (2016)
27. Kantorov, V., Oquab, M., Cho, M., Laptev, I.: ContextLocNet: context-aware deep network models for weakly supervised localization. In: Leibe, B., Matas, J., Sebe, N., Welling, M. (eds.) ECCV 2016. LNCS, vol. 9909, pp. 350–365. Springer, Cham (2016). https://doi.org/10.1007/978-3-319-46454-1_22
28. Ken, C., Karen, S., Andrea, V., Andrew, Z.: Return of the devil in the details delving deep into convolutional nets. In: The British Machine Vision Conference (BMVC) (2014)
29. Kosugi, S., Yamasaki, T., Aizawa, K.: Object-aware instance labeling for weakly supervised object detection. In: IEEE International Conference on Computer Vision (ICCV) (2019)
30. Kumar Singh, K., Jae Lee, Y., Singh, K.K., Lee, Y.J.: You reap what you sow: using videos to generate high precision object proposals for weakly-supervised object detection. In: IEEE/CVF Conference on Computer Vision and Pattern Recognition (CVPR) (2019)
31. Lai, B., Gong, X.: Saliency guided end-to-end learning for weakly supervised object detection. In: International Joint Conferences on Artificial Intelligence (IJCAI) (2017)
32. Li, D., Huang, J.B., Li, Y., Wang, S., Yang, M.H.: Weakly supervised object localization with progressive domain adaptation. In: IEEE/CVF Conference on Computer Vision and Pattern Recognition (CVPR) (2016)
33. Li, X., Kan, M., Shan, S., Chen, X.: Weakly supervised object detection with segmentation collaboration. In: IEEE International Conference on Computer Vision (ICCV) (2019)

34. Li, Y., Liu, L., Shen, C., van den Hengel, A.: Image co-localization by mimicking a good detector's confidence score distribution. In: Leibe, B., Matas, J., Sebe, N., Welling, M. (eds.) ECCV 2016. LNCS, vol. 9906, pp. 19–34. Springer, Cham (2016). https://doi.org/10.1007/978-3-319-46475-6_2
35. Li, Z., Peng, C., Yu, G., Zhang, X., Deng, Y., Sun, J.: DetNet: a backbone network for object detection. In: European Conference on Computer Vision (ECCV) (2018)
36. Lin, T.Y., Dollár, P., Girshick, R., He, K., Hariharan, B., Belongie, S.: Feature pyramid networks for object detection. In: IEEE/CVF Conference on Computer Vision and Pattern Recognition (CVPR) (2017)
37. Lin, T.-Y., et al.: Microsoft COCO: common objects in context. In: Fleet, D., Pajdla, T., Schiele, B., Tuytelaars, T. (eds.) ECCV 2014. LNCS, vol. 8693, pp. 740–755. Springer, Cham (2014). https://doi.org/10.1007/978-3-319-10602-1_48
38. Liu, B., Gao, Y., Guo, N., Ye, X., You, H., Fan, D.: Utilizing the instability in weakly supervised object detection. In: CVPR Workshop (2019)
39. Liu, W., et al.: SSD: single shot multibox detector. In: Leibe, B., Matas, J., Sebe, N., Welling, M. (eds.) ECCV 2016. LNCS, vol. 9905, pp. 21–37. Springer, Cham (2016). https://doi.org/10.1007/978-3-319-46448-0_2
40. Papadopoulos, D.P., Uijlings, J.R.R., Keller, F., Ferrari, V.: We don't need no bounding-boxes: training object class detectors using only human verification. In: IEEE/CVF Conference on Computer Vision and Pattern Recognition (CVPR) (2016)
41. Redmon, J., Divvala, S., Girshick, R., Farhadi, A.: You only look once: unified, real-time object detection. In: IEEE/CVF Conference on Computer Vision and Pattern Recognition (CVPR) (2016)
42. Ren, S., He, K., Girshick, R., Sun, J.: Faster R-CNN: towards real-time object detection with region proposal networks. In: Conference on Neural Information Processing Systems (NeurIPS) (2015)
43. Ren, Z., et al.: instance-aware, context-focused, and memory-efficient weakly supervised object detection. In: IEEE/CVF Conference on Computer Vision and Pattern Recognition (CVPR) (2020)
44. Sandler, M., Howard, A., Zhu, M., Zhmoginov, A., Chen, L.C.: MobileNetV2: inverted residuals and linear bottlenecks. In: IEEE/CVF Conference on Computer Vision and Pattern Recognition (CVPR) (2018)
45. Shen, Y., Ji, R., Zhang, S., Zuo, W., Wang, Y.: Generative adversarial learning towards fast weakly supervised detection. In: IEEE/CVF Conference on Computer Vision and Pattern Recognition (CVPR) (2018)
46. Shen, Y., Ji, R., Wang, C., Li, X., Li, X.: Weakly supervised object detection via object-specific pixel gradient. IEEE Trans. Neural Netw. Learn. Syst. (TNNLS) **29**, 5960–5970 (2018)
47. Shen, Y., Ji, R., Wang, Y., Wu, Y., Cao, L.: Cyclic guidance for weakly supervised joint detection and segmentation. In: IEEE/CVF Conference on Computer Vision and Pattern Recognition (CVPR) (2019)
48. Shen, Y., Ji, R., Yang, K., Deng, C., Wang, C.: Category-aware spatial constraint for weakly supervised detection. IEEE Trans. Image Process. (TIP) **29**, 843–858 (2019)
49. Shen, Z., Liu, Z., Li, J., Jiang, Y.G., Chen, Y., Xue, X.: DSOD: learning deeply supervised object detectors from scratch. In: IEEE International Conference on Computer Vision (ICCV) (2017)
50. Shi, M., Caesar, H., Ferrari, V.: Weakly supervised object localization using things and stuff transfer. In: IEEE International Conference on Computer Vision (ICCV) (2017)

51. Shi, M., Ferrari, V.: Weakly supervised object localization using size estimates. In: Leibe, B., Matas, J., Sebe, N., Welling, M. (eds.) ECCV 2016. LNCS, vol. 9909, pp. 105–121. Springer, Cham (2016). https://doi.org/10.1007/978-3-319-46454-1_7
52. Simonyan, K., Zisserman, A.: Very deep convolutional networks for large-scale image recognition. In: The International Conference on Learning Representations (ICLR) (2015)
53. Singh, K.K., Xiao, F., Lee, Y.J.: Track and transfer: watching videos to simulate strong human supervision for weakly-supervised object detection. In: IEEE/CVF Conference on Computer Vision and Pattern Recognition (CVPR) (2016)
54. Szegedy, C., et al.: Going deeper with convolutions. In: IEEE/CVF Conference on Computer Vision and Pattern Recognition (CVPR) (2015)
55. Tang, P., et al.: PCL: proposal cluster learning for weakly supervised object detection. IEEE Trans. Pattern Anal. Mach. Intell. (TPAMI) **42**, 176–91 (2018)
56. Tang, P., Wang, X., Bai, X., Liu, W.: Multiple instance detection network with online instance classifier refinement. In: IEEE/CVF Conference on Computer Vision and Pattern Recognition (CVPR) (2017)
57. Tang, P., et al.: Weakly supervised region proposal network and object detection. In: European Conference on Computer Vision (ECCV) (2018)
58. Teh, E.W., Wang, Y.: Attention networks for weakly supervised object localization. In: The British Machine Vision Conference (BMVC) (2016)
59. Wan, F., Liu, C., Ke, W., Ji, X., Jiao, J., Ye, Q.: C-MIL: continuation multiple instance learning for weakly supervised object detection. In: IEEE/CVF Conference on Computer Vision and Pattern Recognition (CVPR) (2019)
60. Wan, F., Wei, P., Jiao, J., Han, Z., Ye, Q.: Min-entropy latent model for weakly supervised object detection. In: IEEE/CVF Conference on Computer Vision and Pattern Recognition (CVPR) (2018)
61. Wang, C., Ren, W., Huang, K., Tan, T.: Weakly supervised object localization with latent category learning. In: Fleet, D., Pajdla, T., Schiele, B., Tuytelaars, T. (eds.) ECCV 2014. LNCS, vol. 8694, pp. 431–445. Springer, Cham (2014). https://doi.org/10.1007/978-3-319-10599-4_28
62. Wang, R.J., Li, X., Ao, S., Ling, C.X.: Pelee: a real-time object detection system on mobile devices. In: Conference on Neural Information Processing Systems (NeurIPS) (2018)
63. Wang, X., Zhu, Z., Yao, C., Bai, X.: Relaxed multiple-instance SVM with application to object discovery. In: IEEE International Conference on Computer Vision (ICCV) (2015)
64. Wei, Y., et al.: TS2C: tight box mining with surrounding segmentation context for weakly supervised object detection. In: European Conference on Computer Vision (ECCV) (2018)
65. Yang, K., Li, D., Dou, Y.: Towards precise end-to-end weakly supervised object detection network. In: IEEE International Conference on Computer Vision (ICCV) (2019)
66. Yu, F., Koltun, V., Funkhouser, T.: Dilated residual networks. In: IEEE/CVF Conference on Computer Vision and Pattern Recognition (CVPR) (2017)
67. Zagoruyko, S., Komodakis, N.: Wide residual networks. In: The British Machine Vision Conference (BMVC) (2016)
68. Zeng, Z., Liu, B., Fu, J., Chao, H., Zhang, L.: WSOD$\wedge$2: learning bottom-up and top-down objectness distillation for weakly-supervised object detection. In: IEEE International Conference on Computer Vision (ICCV) (2019)

69. Zhang, X., Wei, Y., Feng, J., Yang, Y., Huang, T.: Adversarial complementary learning for weakly supervised object localization. In: IEEE/CVF Conference on Computer Vision and Pattern Recognition (CVPR) (2018)
70. Zhang, X., Wei, Y., Kang, G., Yang, Y., Huang, T.: Self-produced guidance for weakly-supervised object localization. In: European Conference on Computer Vision (ECCV) (2018)
71. Zhang, X., Feng, J., Xiong, H., Tian, Q.: Zigzag learning for weakly supervised object detection. In: IEEE/CVF Conference on Computer Vision and Pattern Recognition (CVPR) (2018)
72. Zhang, X., Yang, Y., Feng, J.: ML-LocNet: improving object localization with multi-view learning network. In: European Conference on Computer Vision (ECCV) (2018)
73. Zhang, X., Yang, Y., Feng, J.: Learning to localize objects with noisy labeled instances. In: AAAI Conference on Artificial Intelligence (AAAI) (2019)
74. Zhang, Y., Li, Y., Ghanem, B.: W2F : a weakly-supervised to fully-supervised framework for object detection. In: IEEE/CVF Conference on Computer Vision and Pattern Recognition (CVPR) (2018)
75. Zhou, B., Khosla, A., Lapedriza, A., Oliva, A., Torralba, A.: Learning deep features for discriminative localization. In: IEEE/CVF Conference on Computer Vision and Pattern Recognition (CVPR) (2016)
76. Zhou, B., et al.: Semantic understanding of scenes through the ADE20K dataset. Int. J. Comput. Vis. (IJCV) **127**, 302–321 (2019). https://doi.org/10.1007/s11263-018-1140-0
77. Zhu, R., et al.: ScratchDet: exploring to train single-shot object detectors from scratch. In: IEEE/CVF Conference on Computer Vision and Pattern Recognition (CVPR) (2019)
78. Zhu, Y., Zhou, Y., Ye, Q., Qiu, Q., Jiao, J.: Soft proposal networks for weakly supervised object localization. In: IEEE International Conference on Computer Vision (ICCV) (2017)

Deep Near-Light Photometric Stereo for Spatially Varying Reflectances

Hiroaki Santo(✉), Michael Waechter, and Yasuyuki Matsushita

Graduate School of Information Science and Technology, Osaka University, Osaka, Japan
{santo.hiroaki,waechter.michael,yasumat}@ist.osaka-u.ac.jp

Abstract. This paper presents a near-light photometric stereo method for spatially varying reflectances. Recent studies in photometric stereo proposed learning-based approaches to handle diverse real-world reflectances and achieve high accuracy compared to conventional methods. However, they assume distant (*i.e.*, parallel) lights, which can in practical settings only be approximately realized, and they fail in near-light conditions. Near-light photometric stereo methods address near-light conditions but previous works are limited to over-simplified reflectances, such as Lambertian reflectance. The proposed method takes a hybrid approach of distant- and near-light models, where the surface normal of a small area (corresponding to a pixel) is computed locally with a distant light assumption, and the reconstruction error is assessed based on a near-light image formation model. This paper is the first work to solve unknown, spatially varying, diverse reflectances in near-light photometric stereo.

Keyword: Near-light photometric stereo

1 Introduction

Photometric stereo estimates surface normals of a scene from multiple images captured by a fixed camera under varying light conditions. The basic idea of photometric stereo was introduced in 1980 by Woodham [34] assuming Lambertian reflectance under distant light. In practice, these assumptions typically do not hold; therefore, a photometric stereo method that can deal with *diverse and spatially varying reflectances* in a *nearby light* setting is wanted.

Recent studies have shown that a deep learning-based approach [6,11,26] can effectively deal with diverse and spatially varying reflectances by establishing a mapping from observed images to a surface normal map. These methods assume a distant light setting for ease of learning. In a different thread, nearby light

Electronic supplementary material The online version of this chapter (https://doi.org/10.1007/978-3-030-58598-3_9) contains supplementary material, which is available to authorized users.

A. Vedaldi et al. (Eds.): ECCV 2020, LNCS 12353, pp. 137–152, 2020.
https://doi.org/10.1007/978-3-030-58598-3_9

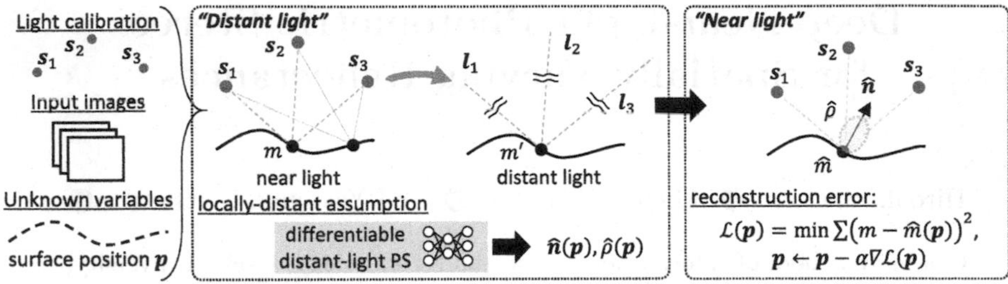

Fig. 1. Overview of the proposed method. The inputs are the given light calibration and observations m, and the unknown is the surface position $\mathbf{p}$. Based on assuming light to be distant (*i.e.*, parallel) locally, we employ a near-light effect cancellation (Eqs. (1) and (4)) to create a pseudo (distant-light) observation m', and compute the surface normal $\hat{\mathbf{n}}$ and reflectance $\hat{\rho}$ using distant-light photometric stereo. We then assess the reconstruction of the observation $\hat{m}$ based on a near-light image formation model.

photometric stereo has been studied [12,17,22] to explicitly eliminate the distant light assumption. These works have shown to be effective for Lambertian or simple parametric reflectances, but still suffer from diverse and spatially varying reflectances. Since these studies for relaxing the Lambertian reflectance and distant light assumptions have been developed rather independently, it is still unclear how these two distinct studies can benefit from each other.

In this work, we present a hybrid approach of distant- and near-light models for simultaneously removing the assumptions of both Lambertian reflectance and distant lighting. Specifically, we assume that a single pixel covers a small surface area within which incoming light emitted from a nearby light source can be modeled as distant (*i.e.*, parallel), although different pixels may be illuminated by different light directions and strengths. Based on this locally-distant assumption, our method predicts a surface normal per pixel using a deep learning-based distant-light photometric stereo method that can deal with spatially varying reflectances. Based on the surface normal estimates, we assess the reconstruction error by re-rendering based on a near-light image formation model that explicitly considers the light fall-off effect. The whole procedure is designed in a differentiable manner with respect to the surface positions so that our method can benefit from a gradient-based method to efficiently predict the surface positions.

To sum up, our paper offers the following contributions:

- We propose a near-light photometric stereo method that can deal with spatially varying reflectances.
- Compared to previous near-light photometric stereo methods, the proposed method does not depend on a simplified parametric reflectance model.
- Compared to existing deep learning based photometric stereo methods, the proposed method explicitly takes nearby light conditions into account.

As a result, the proposed method can handle scenes with diverse materials in contrast to existing near-light photometric stereo methods. At the same time, in contrast to most deep learning-based methods, the proposed method can handle near-light conditions, which should always be considered in a practical setting.

2 Related Work

In this section, we describe previous works of photometric stereo on both distant- and near-light assumptions. For distant-light photometric stereo, we mainly discuss the recent deep learning-based methods.

Deep Learning-Based Photometric Stereo. Early works of photometric stereo [30, 34] assume Lambertian reflectance and many extended works study the use of more flexible parametric models, such as the Torrance-Sparrow model [9], microfacet-based models [7,32], and bi-polynomial models [29]. Although these methods have greater flexibility in representing reflectances, they still cannot represent real-world reflectances well enough, introducing large estimation errors.

Unlike conventional photometric stereo based on parametric reflectance models, deep learning-based photometric stereo does not explicitly assume a specific reflectance model, but learns it from a synthesized training dataset. Santo *et al.* [26] proposed a fully-connected photometric stereo network, called DPSN, which directly learns the mapping from observations to the corresponding surface normal direction. While DPSN assumes pre-defined light conditions for testing, the newer methods CNN-PS [11] and PS-FCN [6] relax this limitation by handling an arbitrary number of lights and their directions in an order-agnostic way. CNN-PS proposed a new representation for a photometric observation, called an observation map, which represents single-pixel observations under an arbitrary number of light sources by a fixed-shape map representation. In PS-FCN, to handle an arbitrary number of input images, Chen *et al.*used a feature fusion technique with max-pooling to extract a fixed-shape feature map. These methods use a synthesized dataset rendered with realistic bidirectional reflectance distribution functions (BRDFs), such as the MERL BRDF database [16] and the Disney principled BRDF [5] for training.

Unlike these methods, Taniai and Maehara [31] proposed an unsupervised approach. Specifically, they use two networks, a photometric stereo network that outputs a prediction of surface normals and an image reconstruction network that estimates reflectances and outputs re-rendered images, and train the networks by re-rendering loss, which is defined as the difference between input and re-rendered images. We use a similar approach of [31] and minimize the re-rendering loss, but our setting explicitly assumes a near-light setting that cannot be directly addressed by Taniai's work.

Near-Light Photometric Stereo. While early works of photometric stereo [30,34] assume ideal distant-light sources, explicit treatment of a near-light setting in photometric stereo began with the work of Iwahori *et al.* [12]. They consider the effects of spatially varying light directions and light fall-off that occur in near-light settings. These effects pose a challenge in photometric stereo because the image formation model becomes non-linear even with a Lambertian assumption.

The non-linear image formation w.r.t. surface normal results is a non-convex optimization problem. One line of approaches to this difficulty is based on iterative optimization [2,4,8,10,20]. These methods alternatingly estimate the scene's shape and albedo based on the image formation model using the prediction of the previous step. Although each step of the optimization can be made convex, the whole objective is non-convex; therefore, a good initial guess is needed for these methods to work well.

Another class of approaches is based on a variational method, yielding non-linear partial differential equations (PDEs) [17–19]. For example, Mecca *et al.* [19] consider the intensity ratios of two images and formulate the problem as a quasi-linear PDE. They extend their work in [17,18] to relax the reflectance model from the Lambertian to the Blinn–Phong model [3]. More recently, Quéau *et al.* [22] reviewed iterative and PDE-based methods. To solve the problems that (1) the convergence of iterative methods is not established and (2) PDE-based methods are sensitive to the initialization, they proposed a provably convergent alternating reweighted least-squares scheme for solving the near-light photometric stereo problem. Although they assume Lambertianness, they show that their method can deal with non-Lambertian observations, such as shadows and specularities, by a robust variational approach [24].

To sum up, a major limitation of existing near-light photometric stereo methods is their dependency on simplified reflectance models. Specifically, most of them rely on the Lambertian model, which limits their applicability in practice. Our proposed method eliminates this restriction and works with diverse, spatially varying BRDFs.

3 Image Formation Model

We first explain our forward model of how images are formed given a scene's parameters with arbitrary BRDFs and nearby light. The actual (and in fact often ill-posed) task of photometric stereo is then the reverse problem, *i.e.*, inferring scene parameters from given images. In Sect. 4 we explain how our algorithm does this. Throughout this paper, function $\boldsymbol{u}_1(\cdot) : \mathbb{R}^3 \rightarrow \mathcal{S}^2\ (\subset \mathbb{R}^3)$ represents vector normalization, *i.e.*, $\boldsymbol{u}_1(\mathbf{x}) = \mathbf{x}/\|\mathbf{x}\|_2$.

We denote the 3D position of the jth light source by $\mathbf{s}_j \in \mathbb{R}^3$ and the surface position and surface normal corresponding to the ith pixel by $\mathbf{p}_i \in \mathbb{R}^3$ and $\mathbf{n}_i \in \mathcal{S}^2$. Let us use $\mathbf{l}_{ij}$ to represent the light direction from the ith scene point to the jth light source, *i.e.*,

$$\mathbf{l}_{ij} = \boldsymbol{u}_1(\mathbf{s}_j - \mathbf{p}_i)\,. \tag{1}$$

The observed intensity $m_{ij} \in \mathbb{R}$ at the ith pixel under the jth light without global illumination effects (cast shadows, inter-reflection, *etc.*) can be written as [22]

$$m_{ij} = \Phi_{ij} \frac{1}{\|\mathbf{s}_j - \mathbf{p}_i\|_2^2} \max\left(0, \mathbf{l}_{ij}^\top \mathbf{n}_i\right) \rho_{ij}, \tag{2}$$

where Φ_{ij} is the radiant intensity of the jth light at the surface point corresponding to the ith pixel. ρ_{ij} is the reflectance at the ith point under the jth light, expressed as a function $\rho_{ij} : \mathcal{S}^2 \times \mathcal{S}^2 \to \mathbb{R}$ taking surface normal $\mathbf{n}_i$ and incoming light direction $\mathbf{l}_{ij}$ as input. The term $\frac{1}{\|\mathbf{s}_j - \mathbf{p}_i\|_2^2}$ accounts for light fall-off and the max(·) operator accounts for attached shadows.

Let $[u_i, v_i]^\top \in \mathbb{R}^2$ be a pixel position in image coordinates. Its corresponding 3D surface point $\mathbf{p}_i = [x_i, y_i, z_i]^\top$ in world coordinates is

$$\mathbf{p}_i = [u_i, v_i, z_i]^\top$$

under orthographic camera projection, and

$$\mathbf{p}_i = \left[\frac{z_i}{f} u_i, \frac{z_i}{f} v_i, z_i\right]^\top$$

under perspective camera projection, where z_i is the depth, and f is the camera's focal length, which can be obtained through camera calibration.

The surface normal $\mathbf{n}_i$ at surface point $\mathbf{p}_i$ is $\mathbf{n}_i = \boldsymbol{u}_1(\partial_x \mathbf{p}_i \times \partial_y \mathbf{p}_i)$, in which $\times$ is the cross product, and ∂_* represents partial gradient with respect to $*$. Therefore, in orthographic and perspective projection models, the surface normal $\mathbf{n}_i$ can be respectively written as

$$\mathbf{n}_i = \boldsymbol{u}_1([\partial_u z_i, \partial_v z_i, -1])^\top,$$

and

$$\mathbf{n}_i = \boldsymbol{u}_1([f\partial_u z_i, f\partial_v z_i, -z - u_i\partial_u z_i - v_i\partial_v z_i])^\top.$$

In this paper, we use $\mathbf{n}_i = \boldsymbol{\nu}(\mathbf{p}_i)$ to represent conversion from a surface point $\mathbf{p}_i$ to its surface normal $\mathbf{n}_i$ for representing either projection model.

We model the light source's radiant intensity Φ_{ij} as anisotropic point light, which is a common assumption in existing near-light photometric stereo methods [19,22]. It can be written as

$$\Phi_{ij} = \psi_j \left[\mathbf{l}_{ij}^\top \boldsymbol{\omega}_j\right]^{\mu_j}, \tag{3}$$

where $\psi_j \in \mathbb{R}$ is the light source intensity, $\boldsymbol{\omega}_j \in \mathcal{S}^2$ is the principal direction of a light source, and $\mu_j \in \mathbb{R}$ is an anisotropy parameter. In our setting, we assume these parameters as well as the light source positions $\mathbf{s}_j$ are known from a light calibration method [1,15,21,27].

4 Proposed Method

Our goal is to determine surface positions $\mathbf{p}_i$, corresponding surface normal $\mathbf{n}_i$ and reflectances ρ_{ij} from a set of observations m_{ij}, given light source positions $\mathbf{s}_j$ and their radiant intensity parameters $(\psi_j, \mu_j, \boldsymbol{\omega}_j)$ in Φ_{ij}. To alleviate the difficulty of the nearby light setting, we cast the problem into a *per-point distant light* setting, where individual surface points receive different strengths of light from different directions. It allows us to use pre-trained learning-based photometric stereo networks, that are trained under a distant light assumption. Once the prediction of surface normal $\mathbf{n}_i$ and reflectances ρ_{ij} are obtained via the photometric stereo networks, we re-render the scene observations based on the image formation model Eq. (2) and estimate the scene shape $\mathbf{p}_i$ by minimizing the re-rendering loss. Figure 1 illustrates an overview of the proposed method.

4.1 Formulation

We first define a pseudo observation m'_{ij} as

$$m'_{ij} = m_{ij} \frac{\|\mathbf{s}_j - \mathbf{p}_i\|_2^2}{\Phi_{ij}}, \tag{4}$$

in which the light fall-off $\frac{1}{\|\mathbf{s}_j - \mathbf{p}_i\|_2^2}$ and anisotropic radiant intensity Φ_{ij} are discounted from the actual observation m_{ij} in Eq. (2). With this expression, Eq. (2) can be rewritten as

$$m'_{ij} = \rho_{ij} \max(0, \mathbf{l}_{ij}^\top \mathbf{n}_i), \tag{5}$$

which is equivalent to the distant light image formation model except that we do not know the surface point $\mathbf{p}_i$ included in both m'_{ij} and $\mathbf{l}_{ij}$. Under f point light sources, measurements at the ith surface point form a pseudo-observation vector $\mathbf{m}'_i = [m'_{i1}, \cdots, m'_{if}]^\top$, and the corresponding light matrix $\mathbf{L}_i$ can be defined as $\mathbf{L} = [\mathbf{l}_{i1}, \cdots, \mathbf{l}_{if}]^\top$.

Now suppose that we have a guess about the surface position $\mathbf{p}_i$. Then we can compute both the light matrix $\mathbf{L}_i$ and the measurement vector $\mathbf{m}'_i$ from Eqs. (1) and (4), respectively. With the light matrix $\mathbf{L}_i$ and measurement vector $\mathbf{m}'_i$, our method solves for surface normal $\mathbf{n}_i$ and reflectance ρ_{ij} at the ith surface point using two differentiable networks; namely the surface normal estimation network PS and the reflectance estimation network R:

$$\begin{cases} \mathbf{n}_i^* &= \mathtt{PS}(\mathbf{m}'_i, \mathbf{L}_i), \\ \rho_{ij}^* &= \mathtt{R}_j(\mathbf{m}'_i, \mathbf{L}_i), \end{cases} \tag{6}$$

where $\mathbf{n}_i^* \in \mathcal{S}^2$ and $\rho_{ij}^* \in \mathbb{R}$ are the prediction of the surface normal and the reflectance under the jth light, respectively. Unlike previous works which depend on a parametric reflectance model such as the Lambertian model, the capability of handling a variety of BRDFs in the proposed method stems from Eq. (6), whose detail is explained in the next section. Here, we assumed a given guess of the surface position $\mathbf{p}_i$ as input for the networks. However, since the surface normal estimation network PS and reflectance estimation network R are pre-trained and treated as deterministic functions, by substituting Eqs. (1) and (4) into Eq. (6), the prediction of the surface normal $\mathbf{n}_i^*$ and reflectance ρ_{ij}^* become (differentiable) functions of the surface position $\mathbf{p}_i$.

The partial derivative of the surface position $\mathbf{p}_i$, written as $\boldsymbol{\nu}(\mathbf{p}_i)$, also represents surface normal prediction. While the partial derivative $\boldsymbol{\nu}(\mathbf{p}_i)$ is directly calculated from the prediction of the surface position $\mathbf{p}_i$, the estimated normal $\mathbf{n}_i^*$ is constrained by learned prior knowledge in the estimation network PS. For robust estimation, we define the estimated surface normal as the weighted mean of the partial derivative of the surface point $\mathbf{p}_i$ and the surface normal $\mathbf{n}_i^*$ obtained by the estimation network:

$$\hat{\mathbf{n}}_i(\mathbf{p}_i) = \boldsymbol{u}_1((1-\kappa)\boldsymbol{u}_1(\boldsymbol{\nu}(\mathbf{p}_i)) + \kappa\mathbf{n}_i^*(\mathbf{p}_i)), \tag{7}$$

where κ (set to 0.5 in our implementation) balances the two surface normals.

In addition, to ensure that the estimates of position $\mathbf{p}_i$ and surface normal $\mathbf{n}_i^*$ are consistent with each other, we use an objective function $\mathcal{L}_n$ defined over the partial derivative of the surface point $\mathbf{p}_i$ and the predicted surface normal $\mathbf{n}_i^*$,

$$\mathcal{L}_n(\{\mathbf{p}_i|\forall i\}) = \sum_i \left\{1 - \boldsymbol{u}_1(\boldsymbol{\nu}(\mathbf{p}_i))^\top \mathbf{n}_i^*(\mathbf{p}_i)\right\}.$$

Once we obtain the predicted surface normal $\hat{\mathbf{n}}_i$ and reflectances ρ_{ij}^*, we reconstruct the re-rendered observations $\hat{m}_{ij} \in \mathbb{R}$ using Eq. (2) as

$$\hat{m}_{ij} = \frac{\Phi_{ij}}{\|\mathbf{s}_j - \mathbf{p}_i\|_2^2} \max\left(0, \mathbf{l}_{ij}^\top \hat{\mathbf{n}}_i\right) \rho_{ij}^*. \tag{8}$$

The re-rendering $\hat{m}_{ij}$ is differentiable[1] with respect to the surface position $\mathbf{p}_i$ because the networks in Eq. (6) as well as Eqs. (3), (1), (4), (7) are all differentiable. Therefore, we minimize the following objective function for estimating surface point $\mathbf{p}_i$ for all i and j starting with an initial guess for $\mathbf{p}_i$:

$$\mathcal{L}_m(\{\mathbf{p}_i|\forall i\}) = \frac{1}{f}\sum_i\sum_j \{u_2(m_{ij}) - u_2(\hat{m}_{ij}(\mathbf{p}_i))\}^2,$$

in which $u_2(\cdot)$ represents the normalization operation for observations and is defined as $u_2(x_{ij}) = x_{ij}/\|\mathbf{X}\|_F$, $\mathbf{X} = [x_{ij}]$, taking care of the global scaling in observations, as used in [31].

As a result, the final form of the objective function becomes

$$\mathcal{L} = (1-\lambda)\mathcal{L}_m + \lambda\mathcal{L}_n, \tag{9}$$

where the scalar weight $\lambda \in (0,1)$ balances the two objective functions. Finally, we obtain estimates of the position $\hat{\mathbf{p}}_i = \operatorname{argmin}_{\mathbf{p}}\mathcal{L}$, the surface normal $\hat{\mathbf{n}}_i(\hat{\mathbf{p}}_i)$, and the reflectance $\rho_{ij}^*(\hat{\mathbf{p}}_i)$. Since the objective function $\mathcal{L}$ is non-convex, the proposed method requires an initial guess as with most existing near-light photometric stereo methods. For initialization, we assume that the distance from the camera to the scene is given by a rough measurement and we use a plane as initial scene shape. Equation (8) is defined for grayscale observations. For multi-channel observations, we calculate the re-rendered observations for each color channel and take the sum of the re-rendering loss $\mathcal{L}_m$ from each channel.

[1] $\max(0, x)$ is differentiable everywhere except at $x = 0$, which is in practice not a problem with numerical differentiation as in many other works.

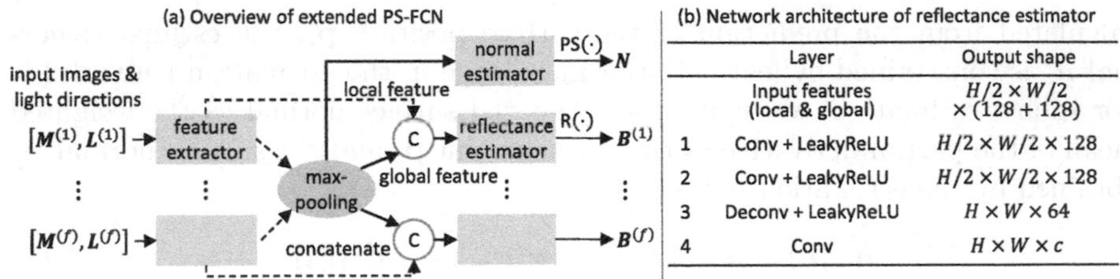

	Layer	Output shape
	Input features (local & global)	$H/2 \times W/2 \times (128+128)$
1	Conv + LeakyReLU	$H/2 \times W/2 \times 128$
2	Conv + LeakyReLU	$H/2 \times W/2 \times 128$
3	Deconv + LeakyReLU	$H \times W \times 64$
4	Conv	$H \times W \times c$

Fig. 2. *(a)* Architecture of the surface normal and reflectance estimation networks. The feature extractor and normal estimator are the same as in PS-FCN [6] and we add the reflectance estimator shown in the yellow box. *(b)* Details of the reflectance estimator. "Conv", "Deconv", and "LeakyReLU" mean a convolution layer with a 3×3 kernel, a deconvolution layer with a 3×3 kernel and stride 2, and a Leaky ReLU with a scale factor of $\alpha = 0.1$, respectively. $H \times W$ is the input image size. The input is the concatenated features of the local and global features, where both features have a size of $H/2 \times W/2 \times 128$ and the output is c channel reflectance maps $\mathbf{B}_m^{(j)}$ in the form of the image shape $H \times W \times c$. The weights are shared for all lightings.

4.2 Normal and Reflectance Estimation Networks

The proposed method uses a surface normal network PS and a reflectance estimation network R. For the surface normal estimation network PS we adopt PS-FCN [6]. PS-FCN is an end-to-end differentiable network that takes input images concatenated with vectors of light directions and outputs the corresponding surface normal map. Its authors showed that PS-FCN works well for scenes with spatially varying, diverse real-world BRDFs through a benchmark comparison on a real-world dataset. We extend the original PS-FCN for simultaneous estimation of surface normals and reflectances.

Figure 2 shows an overview of the extended PS-FCN. We add the reflectance estimator R to the original PS-FCN, which estimates the reflectances $\mathbf{B} \in \mathbb{R}^{p \times f}$, in which an element $B_i^{(j)} \in \mathbb{R}$ represents the reflectance at the ith point (corresponding to the pixel) under the jth lighting. p and f are the numbers of pixels and light sources, respectively. We denote the reflectance map for all pixels under the jth lighting as $\mathbf{B}^{(j)} \in \mathbb{R}^p$ and the reflectances at the ith pixel under all lightings as $\mathbf{B}_i \in \mathbb{R}^f$. The reflectance estimator takes as input the local features that are concatenated with the global feature, and outputs the prediction of reflectance map $\mathbf{B}^{(\cdot)}$ for all lightings. The global feature provides global information such as the object's shape, while the local feature accounts for the reflectances under individual lightings. For the network architecture of the reflectance estimator R, we use an architecture identical to the one for surface normal estimation except for the normalization and output shape.

For training, in addition to the original cosine similarity loss for the surface normals, we use the following re-rendering loss $\mathcal{L}_\mathbf{B}$ for the reflectance estimator that is defined with the estimated reflectance $B_i^{(j)}$:

$$\mathcal{L}_{\mathbf{B}} = \sum_i \sum_j \left\{ \check{m}_{ij} - B_i^{(j)} \max\left(0, \mathbf{l}_{ij}^\top \check{\mathbf{n}}_i\right) \right\}^2. \tag{10}$$

In the above equation, $\check{m}_{ij} \in \mathbb{R}$ and $\check{\mathbf{n}}_i \in \mathcal{S}^2$ represent the ground truth measurements and the surface normal at the ith surface point under the jth lighting. Since this network assumes distant lighting, $\mathbf{l}_{ij}$ represents the lighting direction.

We use the same training dataset as the original PS-FCN, but normalize the input images to remove a global scaling ambiguity. Specifically, a ground truth measurement $\check{\mathbf{m}}_i = [\check{m}_{i1}, \cdots, \check{m}_{if}]$ is normalized for each pixel by two factors: its original norm $\|\check{\mathbf{m}}_i\|_2$ and the number of lights/images f. We normalize the measurements with a scaling factor $s = f^{-\frac{1}{2}} \|\check{\mathbf{m}}_i\|_2^{-1}$. Since scaling the observations also scales the reflectances, we need to undo that by inversely scaling them with s^{-1} to keep them consistent with the original reflectances in the images. We show the evaluation of our extended PS-FCN on a distant-light photometric stereo dataset in our supplementary material. Although Eq. (10) is for grayscale observations, for multi-channel observations, we change the output shape of the reflectance estimator to $\mathbf{B}_m^{(j)} \in \mathbb{R}^{p \times c}$ where c is the number of color channels.

4.3 Implementation

The proposed method obtains predictions of the scene's shape by minimizing the objective function of Eq. (9). In our implementation, to minimize the objective $\mathcal{L}$ we use the Adam optimizer [14] with default settings ($\beta_1 = 0.9$ and $\beta_2 = 0.99$) and set the balancing weight λ to 0.05. We stop iterating when one of the following is met: (1) $\mathcal{L}^{(t+1)} - \mathcal{L}^{(t)} < \tau$ or (2) $t > T_{\text{iter}}$, where $\mathcal{L}^{(t)}$ is the value of the loss $\mathcal{L}$ after the tth iteration. We use $\tau = 10^{-6}$ and $T_{\text{iter}} = 10^4$.

During the iterations, we randomly sample light sources to construct mini-batch data. In our implementation, each iteration uses 32 randomly selected images to reduce the usage of computational resources. The problem would otherwise not fit into the GPU memory if there is a large number of light sources. Too small mini-batches, on the other hand, result in unstable predictions.

Following Quéau [22], for better and faster convergence we use hierarchical scaling optimization, *i.e.*, we reduce the input image resolution and use the resulting solution to initialize the optimization at a higher resolution. In our experiments, we use a coarse-to-fine approach starting from $1/8\times$, $1/4\times$, $1/2\times$, to $1\times$ of the input image resolution. The surface position $\mathbf{p}_i$ at the lowest resolution is initialized with a planar depth map.

The learning rate depends on the scaling of the shape $\mathbf{p}_i$. Using the initial depth d and focal length f, we set the initial learning rate to $0.5 \times \frac{d}{f}$ where the second term corresponds to the physical size of one pixel. For each finer resolution in the coarse-to-fine approach we then set the learning rate to half of the previous coarser resolution's.

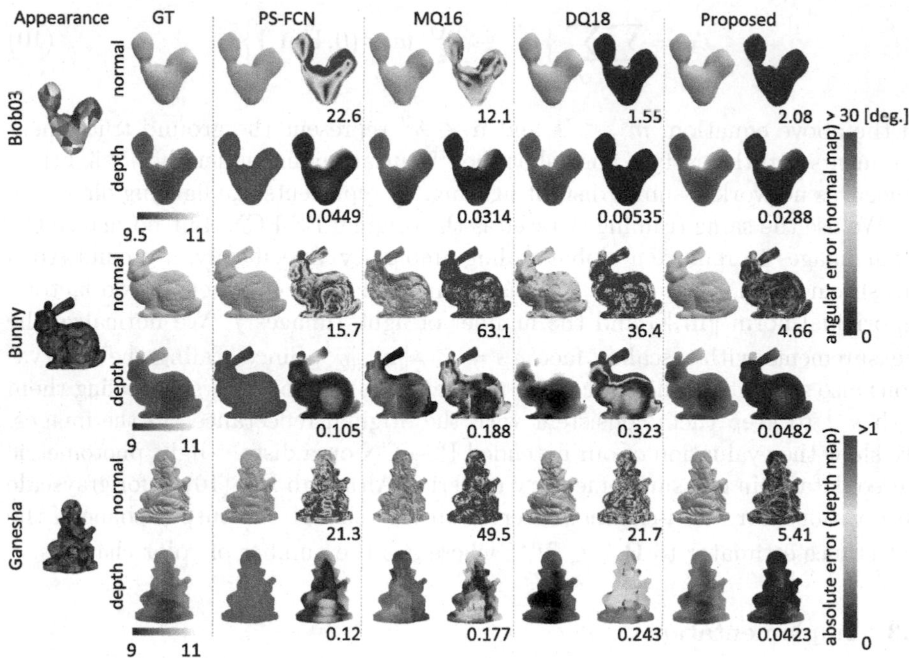

Fig. 3. Estimated results for our synthetic dataset. For each scene and each method, we show the estimated surface normal map, the depth map, the normal error map, and the depth error map. "GT" shows the ground truth of normals and depth. The numbers underneath the error maps show mean angular error in degrees for the surface normal maps, and mean absolute error for the depth maps.

5 Experiments

To evaluate the proposed method, we conducted experiments using both synthetic and real-world scenes. For comparison, we used MQ16 [17] and DQ18 [22] as existing methods for near-light photometric stereo, and PS-FCN [6] for distant-light photometric stereo. For the input of PS-FCN, we calculated the directions of each light source using the distance between the camera and the target object, which is used for the initial guess in our near-light photometric stereo method. Since PS-FCN only estimates the surface normal, we used a quadratic integration-based method [23] to obtain depth maps from estimated normal maps. As for MQ16, the surface normal maps are calculated from the estimated depth maps. Our method is implemented in PyTorch[2]. On an NVIDIA Quadro RTX 8000 GPU it took about 0.6 s per iteration and 30–60 min until full convergence for a scene with 256×256 px.

[2] PyTorch v1.1.0: http://pytorch.org.

Fig. 4. Real-world experiment setup. *(a):* Our capture device has 256 LEDs on a printed circuit board (500×500 mm) at known positions. The CCD camera (FLIR Blackfly S; 3072×2048 resolution) is fixed to the board at a known position by a 3D-printed frame. We put the target objects about 300 mm away from the device. *(b):* The three target objects used in our experiments: *CAN*, *CUP*, and *TURTLE*.

5.1 Evaluation with Synthetic Data

We first evaluated our method using a synthetic dataset. For the evaluation, we used three scenes: *Blob03* [13], the Stanford *Bunny* [33], and *Ganesha* [25]. *Blob03* is rendered using spatially varying Lambertian reflectances, and *Bunny* and *Ganesha* consist of five different BRDFs, respectively, that are sampled from measured BRDFs (MERL BRDF database [16]).

We rendered the scenes using the Mitsuba renderer[3]. The camera's focal length was set to 120 mm 35 mm-equivalent), and the image resolution was set to 256×256. We defined the scene size so that the distance from camera to target is 10, which means $(X, Y, Z) = (0, 0, 10)$ in camera coordinates, and put 100 point light sources randomly in the range $(X, Y, Z) = (\pm 5, \pm 5, 6 \pm 1)$. The light sources had identical intensity and ideal uniform radiant patterns ($\psi_j = 1$ and $\mu_j = 0$ for all j in Eq. (3)). For the initialization of the shape $\mathbf{p}_i$, we used a planar depth map whose distance from the camera to the target was obtained by the mean distance of the ground truth.

Figure 3 shows the scene and estimation results of our method and the comparison methods PS-FCN [6], MQ16 [17], and DQ18 [22]. While MQ16 and DQ18 work better for the Lambertian *Blob03*, for the other two scenes with more general BRDFs our method shows superior accuracy in both surface normals and depth estimates. MQ16's mean absolute errors for the depth maps are slightly better than DQ18's; however, MQ16 exhibits unstable results in scenes with diverse BRDFs, which can also be seen in the accuracy of the surface normal maps. One of the reasons is that, as discussed by Quéau [22], PDE-based methods are sensitive to the initialization and MQ16 depends on the initialization of both the depth map and the reflectance parameter, *i.e.*, the shininess of the Blinn–Phong model. Although PS-FCN can handle non-Lambertian BRDFs such as the MERL BRDFs, the estimated results are flatter than the ground truth because, in contrast to the proposed method, it ignores the near-light effects.

[3] Mitsuba v0.5.0: http://mitsuba-renderer.org.

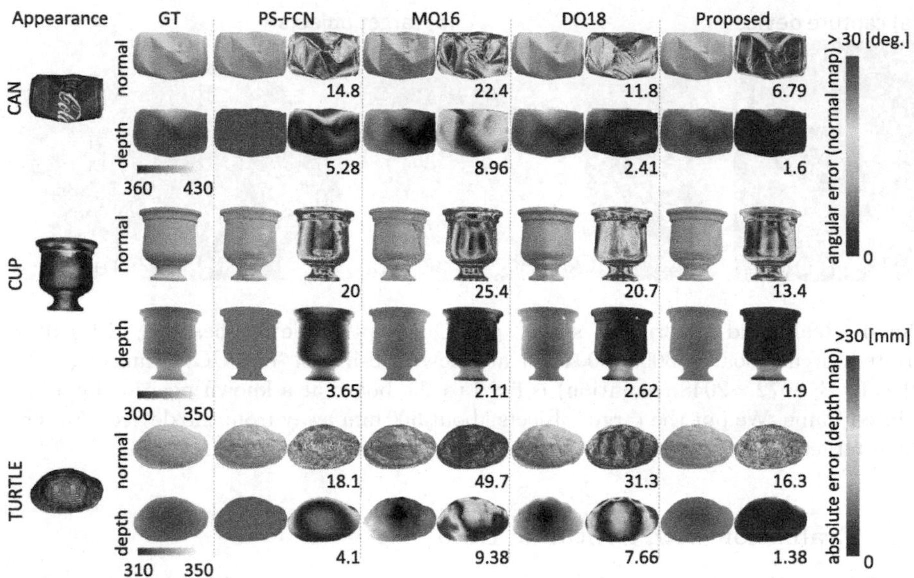

Fig. 5. Estimation results for our real-world dataset. For each scene and each method, we show the estimated surface normal map, the depth map, the normal error map, and the depth error map. "GT" shows the ground truth of normals and depth. The numbers underneath the error maps show mean angular error in degrees for the surface normal maps, and mean absolute error for the depth maps.

5.2 Evaluation with Real-World Data

To evaluate the performance of our method in real-world scenes, we performed a real-world experiment in the setup shown in Fig. 4. We carefully designed our capture device so that the LEDs and the camera are fixed at known positions. We assumed that all LEDs have identical intensity $\psi_j = 1$ and used the radiant intensity distribution obtained from the LED datasheet to calibrate the light emission ($\mu_j = 1$ in Eq. (3)). For the target objects, we used (1) a crushed aluminum can (*CAN*), (2) a plastic cup (*CUP*), and (3) a brass turtle shell (*TURTLE*). All of them are made of different materials and include specular reflectances. To obtain ground truth, we used a structured-light scanner (EinScan Pro) and followed the alignment procedures of the DiLiGenT dataset [28]. Note that, in the evaluation of depth maps we align the estimated depth map to the ground truth because the estimate may have a shift even when we use the initialization calculated from the ground truth. Unlike in the synthetic experiments, we obtained the ground truth by shape-to-image alignment and the absolute depth value of the ground-truth depth map is sensitive to this alignment. The input images are first cropped based on the object mask to avoid redundant computation and are then resized so that the image resolution does not exceed 600×600 pixels due to GPU memory limitations. A more detailed discussion about this limitation can be found in Sect. 6.

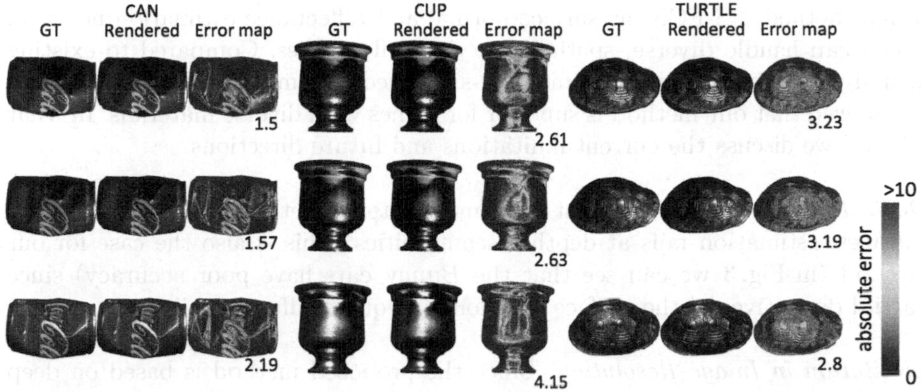

Fig. 6. Rerendering results of our method for our real-world dataset. For each scene, we selected 3 lighting conditions out of the 256 available ones. Each row corresponds to one lighting condition. "GT" and "Rendered" are the ground truth observations (*i.e.*, input images) and rendered images using the estimated reflectances, respectively. The error maps visualize the absolute error and the numbers underneath the error maps are the mean absolute errors in a scaled intensity of 0 to 255. For better visualization, we applied the same brightness correction to both the ground truth and rendered images.

Figure 5 shows the evaluation results on our dataset. As can be seen, MQ16 and DQ18 are heavily affected by the specular reflections whereas our method handles them significantly better. For example on the *CUP*, MQ16 is slightly better than DQ18, especially around the center part of the object, because it can handle the non-Lambertian reflectances with the Blinn Phong model, but it still exhibits large errors due to the instability of the optimization. In contrast, the proposed method works consistently better on all scenes. Although the proposed method utilizes PS-FCN for normal estimation, it achieves more accurate estimations than PS-FCN by taking near-light effects into account.

To demonstrate the performance of reflectance estimation, Fig. 6 shows renderings of the scenes using the estimated reflectances. The proposed method estimates per-pixel and per-light reflectances, and can therefore handle the spatially varying real-world BRDFs. We can see that the obtained reflectance estimations are quite good in all scenes. The estimated reflectances can potentially be used for applications such as a material recognition and parametric BRDF estimation, as well as rendering.

6 Discussion

In this paper, we presented a near-light photometric stereo method for spatially varying reflectances using deep neural networks. Based on the assumption that lighting can be regarded as distant (*i.e.*, parallel) in a small surface area, our formulation allows us to use distant-light photometric stereo in near-light settings. The proposed method uses a state-of-the-art deep learning-based photometric stereo method, PS-FCN, as surface normal and reflectance estimation network, which can handle diverse, spatially varying reflectances. Compared to existing near-light methods which assume over-simplified parametric reflectance model, we showed that our method is superior for scenes with diverse materials. In what follows, we discuss the current limitations and future directions.

Depth Discontinuity. Since most photometric stereo methods assume continuous surfaces, estimation fails at depth discontinuities. This is also the case for our method (in Fig. 3 we can see that the *Bunny* ears have poor accuracy) since partial derivatives of the surface position $\mathbf{p}_i$ require differentiability.

Limitation in Image Resolution. Since the proposed method is based on deep neural networks, it has a limitation due to the available GPU memory. In our implementation, the optimization for a scene with 460×630 px consumes about 40 GB GPU memory which fits on the *NVIDIA Quadro RTX 8000*'s 48 GB. Since we use mini-batch training with respect to the light sources as described in Sect. 4.3, the number of light sources does not matter. The most memory intensive block in our network is the reflectance estimation in a per-light manner. To reduce GPU memory consumption, one possible approach would be to lower the resolution of the reflectance maps, assuming that scenes do not have high spatial frequency in the reflectances.

Effect of Perspective Projection. While the light source conditions, distant or nearby, and the camera projection model, orthographic or perspective, are independent configurations, near-light photometric stereo typically assumes perspective projection. However, our network, the extended PS-FCN from Fig. 2, assumes orthographic projection in a small patch area. In Sect. 5.1, we only showed results with a fixed focal length. In our supplemental material we show the effects of perspective projection and demonstrate that the result deterioration is not very significant. To handle perspective projection better, one possible extension would be to use a surface normal and reflectance estimation network that works in a per-pixel manner as discussed below.

Alternative Networks for Surface Normal and Reflectance Estimation. In this paper, we presented our framework based on PS-FCN. However, our method can use any differentiable photometric stereo method for the surface normal and reflectance estimation network (Eq. (6)).

One possible alternative would be CNN-PS [11], which estimates surface normals from an observation map which represents per-pixel observations in a fixed shape and achieves the best accuracy in the DiLiGenT benchmark for distant-light photometric stereo [28]. Since CNN-PS works in a per-pixel manner, it is more suitable for the assumptions in Eq. (6). However, to use CNN-PS in our method, in future work we would have to develop (1) a differentiable representation of the observation map with respect to both observations and lighting directions and (2) a simultaneous estimation of reflectances.

Acknowledgment. This work was supported by JSPS KAKENHI Grant Number JP19H01123. Hiroaki Santo and Michael Waechter are grateful for support through a JSPS Research Fellowship for Young Scientists (JP19J10326) and JSPS Postdoctoral Fellowship (JP17F17350), respectively.

References

1. Ackermann, J., Fuhrmann, S., Goesele, M.: Geometric point light source calibration. In: Vision, Modeling, and Visualization, pp. 161–168 (2013)
2. Ahmad, J., Sun, J., Smith, L., Smith, M.: An improved photometric stereo through distance estimation and light vector optimization from diffused maxima region. Pattern Recogn. Lett. **50**, 15–22 (2014)
3. Blinn, J.F.: Models of light reflection for computer synthesized pictures. In: SIGGRAPH (1977)
4. Bony, A., Bringier, B., Khoudeir, M.: Tridimensional reconstruction by photometric stereo with near spot light sources. In: European Signal Processing Conference (2013)
5. Burley, B.: Physically-based shading at Disney. In: SIGGRAPH 2012 Course Notes (2012)
6. Chen, G., Han, K., Wong, K.Y.K.: PS-FCN: a flexible learning framework for photometric stereo. In: European Conference on Computer Vision (ECCV) (2018)
7. Chen, L., Zheng, Y., Shi, B., Subpa-Asa, A., Sato, I.: A microfacet-based reflectance model for photometric stereo with highly specular surfaces. In: International Conference on Computer Vision (ICCV) (2017)
8. Collins, T., Bartoli, A.: 3D reconstruction in laparoscopy with close-range photometric stereo. In: Ayache, N., Delingette, H., Golland, P., Mori, K. (eds.) MICCAI 2012. LNCS, vol. 7511, pp. 634–642. Springer, Heidelberg (2012). https://doi.org/10.1007/978-3-642-33418-4_78
9. Georghiades, A.S.: Incorporating the Torrance and Sparrow model of reflectance in uncalibrated photometric stereo. In: International Conference on Computer Vision (ICCV) (2003)
10. Huang, X., Walton, M., Bearman, G., Cossairt, O.: Near light correction for image relighting and 3D shape recovery. In: 2015 Digital Heritage (2015)
11. Ikehata, S.: CNN-PS: CNN-based photometric stereo for general non-convex surfaces. In: European Conference on Computer Vision (ECCV) (2018)
12. Iwahori, Y., Sugie, H., Ishii, N.: Reconstructing shape from shading images under point light source illumination. In: International Conference on Pattern Recognition (ICPR) (1990)
13. Johnson, M.K., Adelson, E.H.: Shape estimation in natural illumination. In: Computer Vision and Pattern Recognition (CVPR) (2011)
14. Kingma, D., Ba, J.: Adam: a method for stochastic optimization (2014)
15. Ma, L., Liu, J., Pei, X., Hu, Y., Sun, F.: Calibration of position and orientation for point light source synchronously with single image in photometric stereo. Opt. Express **27**(4), 4024–4033 (2019)
16. Matusik, W., Pfister, H., Brand, M., McMillan, L.: A data-driven reflectance model. Trans. Graph. (TOG) **22**(3), 759–769 (2003)
17. Mecca, R., Quéau, Y.: Unifying diffuse and specular reflections for the photometric stereo problem. In: Winter Conference on Applications of Computer Vision (WACV) (2016)

18. Mecca, R., Rodolà, E., Cremers, D.: Realistic photometric stereo using partial differential irradiance equation ratios. Comput. Graph. **51**, 8–16 (2015)
19. Mecca, R., Wetzler, A., Bruckstein, A.M., Kimmel, R.: Near field photometric stereo with point light sources. SIAM J. Imaging Sci. **7**(4), 2732–2770 (2014)
20. Nie, Y., Song, Z.: A novel photometric stereo method with nonisotropic point light sources. In: International Conference on Pattern Recognition (ICPR), pp. 1737–1742. IEEE (2016)
21. Park, J., Sinha, S.N., Matsushita, Y., Tai, Y., Kweon, I.: Calibrating a non-isotropic near point light source using a plane. In: Computer Vision and Pattern Recognition (CVPR), pp. 2267–2274 (2014)
22. Quéau, Y., Durix, B., Wu, T., Cremers, D., Lauze, F., Durou, J.D.: LED-based photometric stereo: modeling, calibration and numerical solution. J. Math. Imaging Vis. **60**(3), 313–340 (2018). https://doi.org/10.1007/s10851-017-0761-1
23. Quéau, Y., Durou, J.D., Aujol, J.F.: Variational methods for normal integration. J. Math. Imaging Vis. **60**(4), 609–632 (2018). https://doi.org/10.1007/s10851-017-0777-6
24. Quéau, Y., Wu, T., Lauze, F., Durou, J.D., Cremers, D.: A non-convex variational approach to photometric stereo under inaccurate lighting. In: Computer Vision and Pattern Recognition (CVPR) (2017)
25. Rodolà, E., Albarelli, A., Bergamasco, F., Torsello, A.: A scale independent selection process for 3D object recognition in cluttered scenes. Int. J. Comput. Vis. (IJCV) **102**(1–3), 129–145 (2013). https://doi.org/10.1007/s11263-012-0568-x
26. Santo, H., Samejima, M., Sugano, Y., Shi, B., Matsushita, Y.: Deep photometric stereo network. In: ICCV Workshop on Physics Based Vision meets Deep Learning (PBDL) (2017)
27. Santo, H., Waechter, M., Lin, W.Y., Sugano, Y., Matsushita, Y.: Light structure from pin motion: geometric point light source calibration. Int. J. Comput. Vis. (IJCV) **128**(7), 1889–1912 (2020). https://doi.org/10.1007/s11263-020-01312-3
28. Shi, B., Mo, Z., Wu, Z., Duan, D., Yeung, S.K., Tan, P.: A benchmark dataset and evaluation for non-Lambertian and uncalibrated photometric stereo. Trans. Pattern Anal. Machi. Intell. (PAMI) **41**(2), 271–284 (2019)
29. Shi, B., Tan, P., Matsushita, Y., Ikeuchi, K.: Bi-polynomial modeling of low-frequency reflectances. Trans. Pattern Anal. Machi. Intell. (PAMI) **36**(6), 1078–1091 (2014)
30. Silver, W.M.: Determining shape and reflectance using multiple images. Master's Thesis, Massachusetts Institute of Technology (1980)
31. Taniai, T., Maehara, T.: Neural inverse rendering for general reflectance photometric stereo. In: International Conference on Machine Learning (ICML) (2018)
32. Torrance, K.E., Sparrow, E.M.: Theory for off-specular reflection from roughened surfaces. JOSA **57**(9), 1105–1114 (1967)
33. Turk, G., Levoy, M.: Zippered polygon meshes from range images. In: SIGGRAPH. ACM (1994)
34. Woodham, R.J.: Photometric method for determining surface orientation from multiple images. Opt. Eng. **19**(1), 139–144 (1980)

Learning Visual Representations with Caption Annotations

Mert Bulent Sariyildiz(✉), Julien Perez, and Diane Larlus

NAVER LABS Europe, Meylan, France
mertbulent.sariyildiz@naverlabs.com

Abstract. Pretraining general-purpose visual features has become a crucial part of tackling many computer vision tasks. While one can learn such features on the extensively-annotated ImageNet dataset, recent approaches have looked at ways to allow for noisy, fewer, or even no annotations to perform such pretraining. Starting from the observation that captioned images are easily crawlable, we argue that this overlooked source of information can be exploited to supervise the training of visual representations. To do so, motivated by the recent progresses in language models, we introduce *image-conditioned masked language modeling* (ICMLM) – a proxy task to learn visual representations over image-caption pairs. ICMLM consists in predicting masked words in captions by relying on visual cues. To tackle this task, we propose hybrid models, with dedicated visual and textual encoders, and we show that the visual representations learned as a by-product of solving this task transfer well to a variety of target tasks. Our experiments confirm that image captions can be leveraged to inject global and localized semantic information into visual representations. Project website: https://europe.naverlabs.com/ICMLM.

1 Introduction

Large-scale manually annotated datasets [11,62] have been fueling the rapid development of deep learning-based methods in computer vision. Training supervised models over such datasets not only leads to state-of-the-art results, but also enables networks to learn useful image representations that can be exploited on downstream tasks. However, this approach has major limitations. First, the cost and complexity of annotating datasets is considerable, especially when the class taxonomy is fine-grained requiring expert knowledge [11,41,55]. Second, retraining from scratch dedicated models for every new task is inefficient.

Electronic supplementary material The online version of this chapter (https://doi.org/10.1007/978-3-030-58598-3_10) contains supplementary material, which is available to authorized users.

A. Vedaldi et al. (Eds.): ECCV 2020, LNCS 12353, pp. 153–170, 2020.
https://doi.org/10.1007/978-3-030-58598-3_10

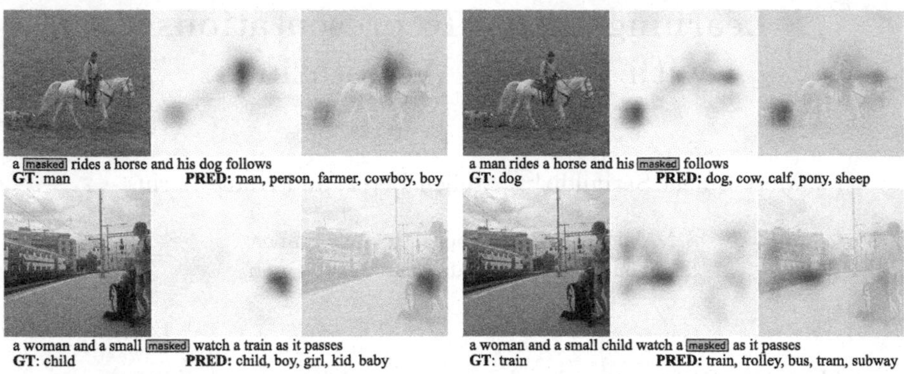

a masked rides a horse and his dog follows
GT: man
PRED: man, person, farmer, cowboy, boy
a man rides a horse and his masked follows
GT: dog
PRED: dog, cow, calf, pony, sheep
a woman and a small masked watch a train as it passes
GT: child
PRED: child, boy, girl, kid, baby
a woman and a small child watch a masked as it passes
GT: train
PRED: train, trolley, bus, tram, subway

Fig. 1. We introduce *image-conditioned masked language modeling* (ICMLM), a proxy task to learn visual representations from scratch given image-caption pairs. This task masks tokens in captions and predicts them by fusing visual and textual cues. The figure shows how the *visual attention* changes as we mask different tokens in a caption (produced by our $\text{ICMLM}_{\texttt{tfm}}$ trained on COCO).

Some alternative approaches address these issues and require less curated or fewer annotations [38,59]. At the other extreme of visual representation learning, self-supervised learning methods [6,14,15,17,60,61] do not require annotations and fabricate proxy labels from the data itself. They induce regularities of the data itself, decorrelated from any specific downstream task annotations. Unfortunately, recent findings show that these approaches are not data efficient, *i.e.* they require either extremely large training sets (up to a hundred million images) [6,23] or need to be trained much longer with larger networks to express their full potential [8,25]. Hence they demand huge computational resources.

Interestingly, data often comes with informative metadata for free. For instance, user tags associated with images can be used as image labels [31,38]. Even richer, companion text for images, is sometimes available for free. Using recent sanitation procedures [44], high-quality large-scale captioned datasets can automatically be constructed.

In this paper, we argue that learning visual representations with captions should significantly reduce the scale of the training sets required for pretraining visual representations. If no text is available, in some context it is still easier to acquire short captions than expert-quality-level fine-grained class labels over thousands of categories like in ImageNet [11]. Yet, caption annotations have rarely been used to train visual representations from scratch. Notable exceptions are [19,31,45] which learn image features by training to predict words in a caption or topic probabilities estimated from an associated text. However, none of these approaches use the structure of the entire sentences, *i.e.* they treat words individually. Recent studies [13,43] have shown the superiority of word representations which are conditioned by their surrounding, where the same word has different representations depending on the sentence. We believe such caption representations should also be beneficial for learning image representations.

This paper focuses on the following research questions. *Can we train transferable visual representations from limited sets of image-caption pairs?* If so, *how should we formulate the interaction between images and captions?* To address these questions, we propose several proxy tasks involving images and their captions which we use to train visual representations from scratch. The first one (Sect. 3.1) is intuitive and requires only extracting *image tags* from captions. We propose several ways to do so, and we show that predicting image tags is already competitive compared to other pretraining strategies. Then, to utilize the captions more effectively, and inspired by the recent advances in natural language processing [13], we propose a second proxy task (Sect. 3.2) which employs masked language modeling to learn visual representations. Similar to the first proxy task, it also leverages both images and captions, but it additionally allows visual representations to learn to *localize semantic concepts* in captions. Qualitative results show that the architecture proposed to tackle this second proxy task effectively leverages the text and attends to relevant image regions (see Fig. 1).

Our contributions are threefold. First, we empirically validate that simple tag prediction tasks, where tags are obtained from captions, already learn transferable visual representations. Second, in an attempt to benefit from captions more, we introduce a new task called **i**mage-**c**onditioned **m**asked **l**anguage **m**odeling (ICMLM) and propose two multi-modal architectures to solve this task. Third, we show that solving ICMLM leads to useful visual representations as a by-product. These visual representations, which we obtain using only a hundred thousand captioned images, are competitive with recent self-supervised approaches leveraging a hundred million images, and, in some cases, even fully-supervised approaches showing how powerful a cue text is.

2 Related Work

Pretraining CNNs on an external dataset has become standard practice in computer vision [7,21,46,48], especially for domains or tasks for which data is scarce. The most common strategy is to train a CNN for the ImageNet-1K classification task [47] and then to use it as a feature extractor or to fine-tune it on a target task or domain. Although this scheme has proven to be quite useful, designing fully-annotated datasets represents a significant effort requiring prior knowledge and domain expertise [11]. Thus, alternative research directions have gained interest. We review the ones closest to our work.

Weakly/Webly-Supervised Learning. Two main research lines have prospered recently. The first line focuses on using *metadata* associated to web data, such as tags or captions for images or videos [53]. Although the signal-to-noise ratio of samples crawled from the web may arguably be lower than carefully-constructed datasets, significant progress has been made leveraging this type of data to pretrain models [9,27]. Among those, to learn visual representations, [31] extracts the most common hashtags and words from the captions and titles of 99 million images in the YFCC100M dataset [53] and train to predict these words using CNNs. Similarly, [38] uses hashtags associated with images from Instagram to construct datasets containing up to 3.5 billion images.

The second line upscales ImageNet. Leveraging ImageNet labels, those approaches produce *pseudo-labels* for additional unlabeled images [58,59]. We note that these methods require initial annotations and extremely large-scale sets of images. In contrary, our models need far less images, 118 thousand images at most, but companion captions to learn visual representations.

Unsupervised Representation Learning. Self-supervised approaches build a *pretext task* to learn image representations which are decorrelated from any downstream task and they do not require any manual annotations. Often, *proxy tasks* consist in predicting missing pieces in manipulated images, for instance context prediction [14], colorization [12,34,60], inpainting of missing portions [42], prediction of image rotations [18], spotting artifacts [30], or cross-channel prediction [61]. Besides, recently, contrastive learning-based unsupervised methods [2,25,40,57] have showed significant improvements. However, computational and data efficiency of these methods are still inferior to supervised models.

It is important to note that most unsupervised approaches are trained on *curated datasets* such as ImageNet for which images were carefully selected to form a well-balanced collection for a diverse set of fine-grained categories. Although these approaches do not directly use ImageNet labels, they implicitly benefit from this careful selection and the resulting underlying structure of the dataset. Indeed, [5,14] show that the feature quality drops when raw data are used instead of ImageNet. Yet, assuming that a curated dataset such as ImageNet is readily available is a strong assumption. Consequently, some works [6,23,39] have evaluated unsupervised methods trained on *uncurated data* [53]. They have concluded that large amounts of raw data (*e.g.* 96 millions images) is required to express the full potential of these approaches. In this work, we focus on learning from a much smaller set of images by leveraging textual information.

Vision and Language. Vision and language (VL) have been jointly leveraged to learn cross-modal representations for various VL tasks, such as cross-modal retrieval [20,56], visual question answering [24], captioning [51] or visual grounding [10,29]. Building on the recent advances in natural language processing [13,54], several works have fine-tuned BERT [13] to fuse visual and textual information [37,50–52,64] for VL tasks. However, while learning cross-modal representations, such approaches rely on pretrained feature extractors, *i.e.* they use visual features pooled from regions of interest produced by a state-of-the-art detector such as Faster-RCNN [46]. Therefore, their objectives are formulated under the assumption that discriminative visual features are readily available for a list of relevant objects. We note that such feature extractors are already state-of-the-art for most vision tasks, requiring *expensive bounding box annotations* to train. Our approach follows a different path. We focus on learning visual representations *from scratch* for purely visual tasks by leveraging captions.

Learning Visual Features Using Text. Only few works have taken advantage of the companion text to learn image representations. [45] creates and solves auxiliary prediction tasks from images with associated captions. [35] constructs label sets out of caption n-grams, and trains CNNs by predicting these labels. [19] extracts topic models for Wikipedia pages using latent Dirichlet allocation and trains a CNN to embed their associated images in this topic space. [22] uses captions to learn image representations for the specific task of semantic retrieval.

We argue that language has a complex structure which cannot be reduced to computing n-grams statistics in a text. Motivated by this, we differ and propose to use a pretrained language model - which can be trained in an unsupervised manner for large text corpora - to represent captions and individual words in them. In our experiments, we show that by doing so it is possible to learn visual representations that are useful for a broad range of tasks.

3 Method

We argue that captions associated with images can provide semantic information about some *observable concepts* that can be captured by image representations. Such concepts can be objects, attributes, or actions that visually appear in images. With this motivation, given a dataset composed of image-caption pairs, we want to formulate non-trivial proxy tasks conditioned on both images and captions such that solving these tasks produce generic visual representations as a by-product. In particular, we want such tasks to properly use the structure of caption sentences, and not only treat them as orderless sets of words.

To this end, we propose two proxy tasks focusing on two distinct objectives to train CNNs to recognize a predefined set of concepts in images. The first proxy task captures *global* semantics in images by predicting image-level tags and is presented in Sect. 3.1. The second proxy task, the image-conditioned masked language modeling task, focuses on *local* semantics in images and is detailed in Sect. 3.2. Experiments show that both proxy tasks are complementary.

Notations. We assume that our dataset $\mathcal{D} = \{(I^i, c^i)\}_1^N$ is composed of N image-caption pairs. We denote by $O = \{o_i\}_1^K$ the set of concepts to be recognized in images. As there can be multiple concepts in an image, we use binary label vectors $\mathbf{y} \in \{0,1\}^K$ to denote the presence of concepts in images, *i.e.* $\mathbf{y}_k = 1$ if concept o_k appears in image I and 0 otherwise. We define two parametric functions ϕ and ψ which respectively embed images and text. More precisely, $\phi : I \rightarrow \mathbf{X} \in \mathbb{R}^{H \times W \times d_x}$ takes an image I as input and produces $\mathbf{X}$ which is composed of d_{x}-dimensional visual features over a spatial grid of size $H \cdot W$. Similarly, $\psi : c \rightarrow \mathbf{W} \in \mathbb{R}^{T \times d_{\mathrm{w}}}$ transforms a caption (a sequence of T tokens) into a set of d_{w}-dimensional vectors, one for each token. In our models, we train only ϕ, which is a CNN producing visual representations, and we use a pretrained language model as ψ that we freeze during training.

3.1 Capturing Image-Level Semantics

A straightforward way to build a proxy task given image-caption pairs is to formulate a multi-label image classification problem, where, according to its

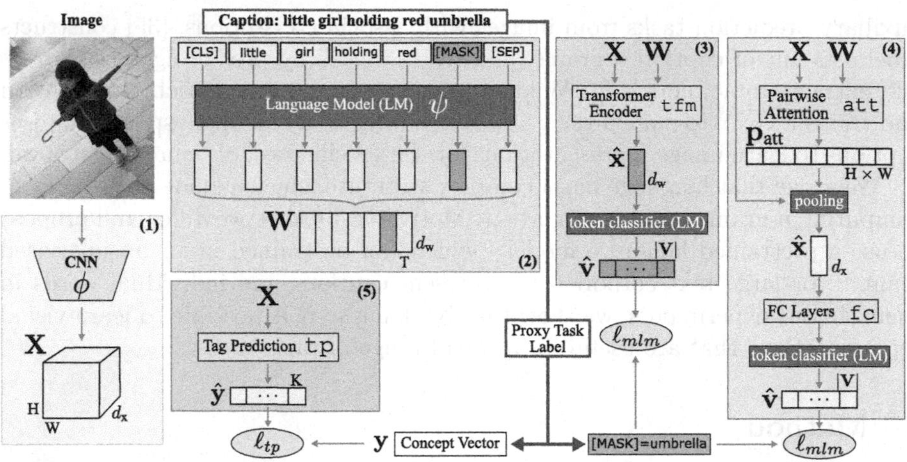

Fig. 2. Modules used in our models. **(1)** a CNN to extract visual features; **(2)** a language model to extract token features; **(3)**, **(4)** and **(5)** respectively correspond to our `tfm`, `att` + `fc` and `tp` modules. Our $\text{TP}_\star$, $\text{ICMLM}_{\texttt{tfm}}$ and $\text{ICMLM}_{\texttt{att-fc}}$ models combine these modules: **(1)** + **(5)**, **(1)** + **(2)** + **(3)** and **(1)** + **(2)** + **(4)**, respectively. Trainable (and frozen) components are colored in blue (and black). Only the CNN is used during target task evaluations. (Color figure online)

caption, multiple concepts may appear in an image [31,45]. For this setup, we create a label vector $\mathbf{y} \in \{0,1\}^K$ for each image I such that $\mathbf{y}_j = 1$ if concept o_j appears in the image, and 0 otherwise. We denote these labels as *tags*, and name this task as *tag prediction* (TP), illustrated in Fig. 2 (modules **(1)** + **(5)**).

One of the contributions of this work is to consider different ways to define concept sets O from captions. Ground-truth concept vectors can be easily obtained by considering the most frequent bi-grams [31] or even n-grams [35] in captions. More sophisticated ways to obtain artificial labels include using LDA [4] to discover latent topics in captions [19]. In addition to these existing methods, we look for ways to exploit semantics of tokens in captions.

$\text{TP}_{\text{Postag}}$. As a first approach, we simply propose to construct label sets by taking into account the *part-of-speech* (POS) tags of tokens in captions. Concretely, we use the off-the-shelf language parser [28] to determine POS tags of tokens in captions and gather three label sets of size K, including (i) only nouns, (ii) nouns and adjectives, (iii) nouns, adjectives and verbs. These three label sets are used to train three separate $\text{TP}_{\text{Postag}}$ models.

$\text{TP}_{\text{Cluster}}$. As mentioned above, we believe it would be beneficial to use the structure of the full caption and not just treat it as an orderless set of tokens as the previously proposed $\text{TP}_{\text{Postag}}$. To this end, we use the pretrained $\text{BERT}_{\text{base}}$ [13] model to extract sentence-level caption representations. We do this by feeding the caption into $\text{BERT}_{\text{base}}$ and taking the representation for the [CLS] token, which is used as a special token to encode sentence-level text representations in $\text{BERT}_{\text{base}}$. Then, we cluster the sentence-level representations of all captions using the k-means algorithm and apply hard cluster assignment. This way, the labels are the cluster indices and we train ϕ by learning to predict the cluster assignments of captions from their associated image. K-means learns K cluster centroids $\xi^\star \in \mathbb{R}^{d_w \times K}$ in the caption representation space by minimizing:

$$\xi^\star, \{\mathbf{y}^{i\star}\}_{i=1}^N = \underset{\substack{\xi \in \mathbb{R}^{d_w \times K}, \\ \{\mathbf{y}^i \in \{0,1\}^K, \mathbf{1}_K^\top \mathbf{y}^i = 1\}_{i=1}^N}}{\arg\min} \sum_{i=1}^N \|\psi(c^i)_{[\text{CLS}]} - \xi \mathbf{y}^i\|_2^2, \tag{1}$$

where $\psi(c)_{[\text{CLS}]}$ and $\mathbf{y}^\star$ denote the [CLS] representation of the caption c and of the one-hot cluster assignment vector obtained for c. Note that $\mathbf{y}^\star$ is used as the label for image I. In case there are multiple captions for an image, we simply aggregate the cluster labels of all captions associated to that image.

Training $\text{TP}_\star$ Models. Once we have crafted image labels over a chosen set of concepts (either using POS tags or cluster assignments), following [38], we normalize the binary label vectors to sum up to one, *i.e.* $\mathbf{y}^\top \mathbf{1}_K = 1$, for all samples. Then we train models by minimizing the categorical cross-entropy:

$$\ell_{\text{tp}} = -\mathbb{E}_{(I,c)\in\mathcal{D}} \left[\sum_{k=1}^K \mathbf{y}_k \log(p(\hat{\mathbf{y}}_k \mid I)) \right], \tag{2}$$

where $p(\hat{\mathbf{y}}_k \mid I) = \frac{\exp(\hat{\mathbf{y}}_k)}{\sum_j \exp(\hat{\mathbf{y}}_j)}$, $\hat{\mathbf{y}}_k = \texttt{tp}(\phi(I))_k$, and $\texttt{tp} : \mathbb{R}^{H\times W\times d_x} \to \mathbb{R}^K$ is a parametric function performing tag predictions.

3.2 Capturing Localized Semantics

The previous section presents a cluster prediction task where the structure of the sentence is leveraged through the use of the [CLS] output of the pretrained $\text{BERT}_{\text{base}}$. Yet, this has a major limitation: token-level details may largely be ignored especially when captions are long [3]. Our experiments also support this argument, *i.e.* $\text{TP}_{\text{Cluster}}$ performs on par with or worse than $\text{TP}_{\text{Postag}}$. To address this issue, we propose a second learning protocol that learns to explicitly *relate* individual concepts appearing in both an image and its caption.

To this end, we extend the natural language processing task known as Masked Language Model (MLM) [13] into an *image-conditioned* version. The MLM task trains a language model by masking a subset of the tokens in an input sentence, and then by predicting these masked tokens. Inspired by this idea, we introduce the Image-Conditioned Masked Language Model (ICMLM) task. Compared to MLM, we propose to predict masked tokens in a caption by using the visual information computed by ϕ. This way, we learn visual representations that should be informative enough to reconstruct the missing information in captions.

For this task, for each image-caption pair (I, c), we assume that there is at least one concept appearing in the caption c. Since c describes the visual scene in I, we assume that concepts appearing in c are observable in I as well. This allows us to define ICMLM as a concept set recognition problem in images. More precisely, we use the pretrained $\text{BERT}_{\text{base}}$ model [13] as the textual embedding function ψ and define the learning protocol as follows. First, we segment the caption c into a sequence of tokens $(t_1, \ldots, t_T)$, and mask one of the tokens t_m, which belongs to the concept set. Masking is simply done by replacing the token t_m with a special token reserved for this operation, for instance $\text{BERT}_{\text{base}}$ [13] uses "[MASK]". Then, *contextualized* representations of the tokens are computed as $\mathbf{W} = \psi((t_1, \ldots, t_T))$. Meanwhile, the visual representation of the image I is computed by $\phi(I) = \mathbf{X}$. Since our goal is to predict the masked token by using both visual and textual representations, we need to merge them. A naive way to accomplish that is to (i) pool the representations of each modality into a global vector, (ii) aggregate (*i.e.* concatenate) these vectors, (iii) use the resulting vector to predict the label of the masked token. However, the representations obtained in this way could only focus on the global semantics, and the local information for both modalities might be lost during the pooling stage. To address this concern, we describe two possible designs for ICMLM relying on individual visual (in the spatial grid) and textual (in the sequence) features.

$\text{ICMLM}_{\texttt{tfm}}$. Here, we contextualize token representations among visual ones by fusing them in a data-driven manner (similar to [37]). Concretely, we spatially flatten and project $\mathbf{X}$ to the token embedding space, concatenate it with $\mathbf{W}$ and apply a transformer encoder module [54], `tfm`, on top of the stacked representations. Finally, as done in $\text{BERT}_{\text{base}}$ [13], the label of the masked token t_m can be predicted by feeding the representation of the *transformed* masked token into the pretrained token classification layer of $\text{BERT}_{\text{base}}$. We call this ICMLM flavor $\text{ICMLM}_{\texttt{tfm}}$(modules **(1)** + **(2)** + **(3)** in Fig. 2).

$\text{ICMLM}_{\texttt{att-fc}}$. Transformer networks employ a self-attention mechanism with respect to their inputs. Therefore they can learn the pairwise relationships of both the visual and the textual representations. This allows them, for instance, to fuse different domains quite effectively [37,51]. We also verify this powerful aspect of the transformers in our experiments, *e.g.* even a single-layered transformer network is enough to perform significantly well at predicting masked tokens on the MS-COCO dataset [36]. However, the rest of the caption is already a powerful cue to predict the masked token and this transformer-based architecture might rely too much on the text, potentially leading to weaker visual representations. As an alternative, we propose to predict the label of the masked token by using the visual features alone. Since the masked token is a concept that we want to recognize in the image, we divide the prediction problem into two sub-problems: localizing the concept in the image and predicting its label. To do that we define two additional trainable modules: `att` and `fc` modules that we describe in detail below. This ICMLM flavor is referred to as $\text{ICMLM}_{\texttt{att-fc}}$ (modules **(1)** + **(2)** + **(4)** in Fig. 2).

The goal of the `att` module is to create a 2D attention map on the spatial grid of the visual feature tensor $\mathbf{X}$ such that high energy values correspond to the location of the concept masked in the caption c. It takes as input the spatially-flattened visual features $\mathbf{X} \in \mathbb{R}^{H \cdot W \times d_x}$ and the textual features $\mathbf{W}$. First, $\mathbf{X}$ and $\mathbf{W}$ are mapped to a common d_z-dimensional space and then pairwise attention scores between visual and textual vectors are computed:

$$\tilde{\mathbf{X}} = \lfloor \text{norm}(\mathbf{X}\Sigma_x) \rfloor_+, \tilde{\mathbf{W}} = \lfloor \text{norm}(\mathbf{W}\Sigma_w) \rfloor_+, \mathbf{S} = \frac{\tilde{\mathbf{X}}\tilde{\mathbf{W}}^\top}{\sqrt{d_z}}, \quad (3)$$

where $\Sigma_x \in \mathbb{R}^{d_x \times d_z}$ and $\Sigma_w \in \mathbb{R}^{d_w \times d_z}$ are parameters to learn, norm is LayerNorm [1] and $\lfloor . \rfloor_+$ is ReLU operator. Note that $\mathbf{S}_{i,j}$ denotes the attention of visual vector i (a particular location in the flattened spatial-grid of the image) to textual vector j (a particular token in the caption). To be able to suppress attention scores of vague tokens such as "about" or "through", we compute *soft maximum* of the textual attentions for each visual feature:

$$\mathbf{s}_i = \log \sum_{j=1}^{T} \exp\left(\mathbf{S}_{i,j}\right). \quad (4)$$

We note that operations in Eqs. (3) and (4) are performed for a single attention head and the multi-headed attention mechanism [54] can easily be adopted by learning a weighted averaging layer: $\mathbf{s}_i = \left[\mathbf{s}_i^1 | \cdots | \mathbf{s}_i^H\right] \Sigma_h + b_h$, where $\Sigma_h \in \mathbb{R}^H$ and $b_h \in \mathbb{R}$ are the parameters of the averaging layer, $\mathbf{s}_i^h$ is the aggregated textual attention score for the ith visual feature coming from the hth attention head, and $[.|.]$ denotes concatenation. Finally, attention probabilities are obtained by applying softmax, and used to pool $\mathbf{X}$ into a single visual feature $\hat{\mathbf{x}}$:

$$\mathbf{p}_{\mathtt{att}_i} = \frac{\exp(\mathbf{s}_i)}{\sum_{j=1}^{H \cdot W} \exp(\mathbf{s}_j)}, \hat{\mathbf{x}} = \mathbf{X}^\top \mathbf{p}_{\mathtt{att}}, \quad (5)$$

where $\mathbf{p}_{\mathtt{att}} \in [0, 1]^{H \cdot W}$ such that $\mathbf{p}_{\mathtt{att}}{}^\top \mathbf{1}_{H \cdot W} = 1$.

After localizing the concept of interest in image I by means of pooling $\mathbf{X}$ into $\hat{\mathbf{x}}$, we feed $\hat{\mathbf{x}}$ into the `fc` module, which consists in a sequence of fully-connected layers, each composed of linear transformation, LayerNorm and ReLU operator. Finally, we map the output of the `fc` module to the $\text{BERT}_{\text{base}}$'s token vocabulary $\mathbf{V}$ and compute prediction probabilities as follows:

$$p_{\mathbf{V}}\left(k | I, c, t_m\right) = \frac{\exp(\hat{\mathbf{v}}_k)}{\sum_j \exp(\hat{\mathbf{v}}_j)}, \quad (6)$$

where $\hat{\mathbf{v}}_k = \mathtt{fc}(\hat{\mathbf{x}})^\top \mathbf{V}_k$ and $\mathbf{V}_k \in d_w$ are the prediction score and the pretrained distributed representation of the kth token in the pretrained candidate lexicon of $\text{BERT}_{\text{base}}$. As we compute dot-products between post-processed $\hat{\mathbf{x}}$ and the pretrained representations of the tokens in $\text{BERT}_{\text{base}}$'s vocabulary, it is possible to leverage the structure in $\text{BERT}_{\text{base}}$'s hidden representation space. Indeed, we observe that such probability estimation of a candidate token is more effective than learning a fully connected layer which projects $\mathtt{fc}(\hat{\mathbf{x}})$ onto the vocabulary.

Training $\text{ICMLM}_\star$ Models. To train both model flavors, for each masked token t_m in all (I, c) pairs in $\mathcal{D}$, we minimize the cross-entropy loss between the

Table 1. Proxy *vs.* **target task performances.** We report top-1 and top-5 masked token prediction scores (as proxy, on VG and COCO) and mAP scores obtained using features from various layers (as target, on VOC-07), on validation sets. T-1/5: top-1/5 scores, C-$\star$: conv. layer from which features are extracted.

Method	Proxy			Target			Proxy			Target		
	Dataset	**T-1**	**T-5**	C-11	C-12	C-13	**Dataset**	**T-1**	**T-5**	C-11	C-12	C-13
BERT$_{\text{base}}$	VG	17.4	36.9	–	–	–	COCO	25.7	40.3	–	–	–
ICMLM$_{\texttt{tfm}}$	VG	**49.7**	**79.2**	71.3	75.8	80.5	COCO	**70.3**	**91.5**	70.2	74.2	77.5
ICMLM$_{\texttt{att-fc}}$	VG	41.1	71.3	**73.7**	**78.7**	**83.1**	COCO	59.4	83.4	**72.3**	**77.5**	**82.2**

probability distribution over the BERT$_{\text{base}}$'s vocabulary as computed in Eq. (6) and the label of the masked token t_m (index of t_m in $\mathbf{V}$):

$$\ell_{\text{mlm}} = - \mathop{\mathbb{E}}_{(I,c)\in\mathcal{D}} \left[\mathop{\mathbb{E}}_{t_m \in c} \left[\log(p_{\mathbf{V}}(k|I,c,t_m)) \right] \right], \tag{7}$$

where k is the index of t_m in BERT$_{\text{base}}$'s vocabulary. The expectation over captions implies that there can be multiple concepts in a caption and we can mask and predict each of them separately. For ICMLM$_{\texttt{tfm}}$, $\hat{\mathbf{x}}$ is computed by the `tfm` module, and it corresponds to the representation of the masked token. For ICMLM$_{\texttt{att-fc}}$, $\hat{\mathbf{x}}$ corresponds to the output from the `fc` module.

We also note that ℓ_{tp} and ℓ_{mlm} are complementary, enforcing ϕ to focus on global and local semantics in images, respectively. Therefore, in both ICMLM$_{\texttt{att-fc}}$ and ICMLM$_{\texttt{tfm}}$ we minimize the weighted combination of ℓ_{tp} and ℓ_{mlm}:

$$\ell_{\text{icmlm}} = \ell_{\text{mlm}} + \lambda \ell_{\text{tp}}. \tag{8}$$

4 Experiments

This section evaluates **(i)** how the performance on the masked language modeling (MLM) proxy task translates to target tasks (Sect. 4.1), **(ii)** how several types of supervision associated to a set of images (*i.e.* full, weak and self-supervision) compare to each other (Sect. 4.2), **(iii)** if the gains of ICMLM$_\star$ models are consistent across backbone architectures (Sect. 4.3), **(iv)** if ICMLM$_\star$ models attend to relevant regions in images (Figs. 1 and 3). First, we introduce our experimental setup (remaining details are in the supplementary material).

Datasets. We train our models on the image-caption pairs of either the 2017 split of MS-COCO [36] (COCO) or the Visual Genome [33] (VG) datasets. COCO has 123K images (118K and 5K for train and val) and 5 captions for each image while VG has 108K images (we randomly split 103K and 5K for train and val) and 5.4M captions. We remove duplicate captions and those with more than 25 or less than 3 tokens. We construct several concept sets using the captions of COCO or VG, to be used as tags for TP$_{\text{Postag}}$ and as maskable tokens for ICMLM$_\star$ models (an ablative study is provided in the supplementary material). Note that depending on the concept set, the number of tags and the (image, caption, maskable token) triplets vary, therefore, we specify which concept set is used in all TP_{Postag} and $ICMLM_{\star}$ experiments.

Networks. To be comparable with the state-of-the-art self-supervised learning method DeeperCluster [6], we mainly use VGG16 [49] backbones. We also evaluate $ICMLM_{\star}$ models using ResNet50 [26] in Sect. 4.3. Note that $ICMLM_{\star}$ models operate on a set of visual tensors, therefore, for $TP_{\star}$ and $ICMLM_{\star}$ models we remove the FC layers from VGG16. To compensate, we use 4-layered CNNs combined with global average pooling and linear layer for tag predictions as `tp` modules. For `tfm`, `att` and `fc` modules, we cross-validated the number of hidden layers and attention heads on the validation set of Pascal VOC-07 dataset, and found that 1 hidden layer (in `tfm` and `fc`) and 12 attention heads (in `tfm` and `att`) works well. While training $ICMLM_{\star}$ models we set $\lambda = 1$ in Eq. (8).

Target Task. Once a model is trained, we discard its additional modules used during training (*i.e.* all but ϕ) and evaluate ϕ on image classification tasks, to test how well pretrained representations generalize to new tasks. To do that, following [6], we train linear logistic regression classifiers attached to the last three convolutional layers of the frozen backbones ϕ with SGD updates and data augmentation. We perform these analyses on the Pascal-VOC07 dataset [16] (VOC) for multi-label classification, and ImageNet-1K (IN-1K) [11] and Places-205 [63] datasets for large-scale categorization, using the publicly available code of [6] with slight modifications: We apply heavier data augmentations [8] and train the classifiers for more iterations, which we found useful in our evaluations.

Additional $TP_{\star}$ Models. We note that the TP model defined in Sect. 3.1 can be used for predicting any type of image tags, with slight modifications. We use it to predict topics as proposed in [19] and denote this approach as TP_{LDA}. To do so, we only modify Eq. (2) to minimize binary cross-entropy loss instead, where K denotes the number of hidden topics. Similarly, we denote TP_{Label} as the supervised approach which uses the annotated image labels as tags.

4.1 Ablative Study on the Proxy Task

We first study the interplay between ICMLM and target tasks. To do so, we train several $ICMLM_{\star}$ models, and monitor their performance on both the proxy and target tasks, *i.e.* we report masked token prediction (MTP) scores on VG and COCO, and mAP scores on VOC, respectively. For reference, we also report MTP scores obtained by a single $BERT_{base}$ model, where masked tokens are predicted using only the remainder of the captions. In this study, we used the 1K most frequent nouns and adjectives in the captions as maskable tokens.

Results are shown in Table 1. We observe that $ICMLM_{\star}$ models significantly improve MTP scores compared to $BERT_{base}$ model, showing that visual cues are useful for MLM tasks. Moreover, $ICMLM_{\texttt{tfm}}$ is better than $ICMLM_{\texttt{att-fc}}$ on the proxy task, indicating that blending visual and textual cues, which is effectively done by the `tfm` module, is beneficial for MLM. However, $ICMLM_{\texttt{att-fc}}$ generalizes better than $ICMLM_{\texttt{tfm}}$ to VOC. We believe that, as $ICMLM_{\texttt{att-fc}}$ predicts

Table 2. Fully-, weakly- and self-supervised methods trained with VGG16 backbones. We report mAP on VOC and top-1 on IN-1K and Places. For VOC, we report the mean of 5 runs (std. ≤0.2). We use pretrained models for ImageNet and DeeperCluster, and train other models from scratch. #I: number of images in training sets. C-⋆: Conv. layer from which features are extracted. Red and orange numbers denote the first and second best numbers in columns. Blue numbers are not transfer tasks (*i.e.* they use the same dataset for proxy/target).

Method	*Proxy tasks*				*Target tasks*								
					VOC			IN-1K			Places		
	Dataset	Supervision		# I	C-11	C-12	C-13	C-11	C-12	C-13	C-11	C-12	C-13
ImageNet	$IN-1K_{full}$	Labels	1K classes	1.3M	77.5	81.0	84.7	59.8	65.7	71.8	43.0	43.5	47.3
$\mathcal{S}$-ImageNet	$IN-1K_{sub}$	Labels	1K classes	100K	69.3	72.4	74.1	50.5	52.5	53.8	40.9	41.6	41.1
$\mathcal{S}$-ImageNet	$IN-1K_{sub}$	Labels	100 classes	100K	67.4	69.6	70.5	47.4	48.4	46.3	39.3	39.3	35.8
TP_{Label}	COCO	Labels	80 classes	118K	72.4	76.3	79.9	50.4	50.6	49.9	44.5	45.0	44.5
DeeperCluster [6]	YFCC	Self	-	96M	71.4	73.3	73.1	48.0	48.8	45.1	43.1	44.1	41.0
RotNet [18]	COCO	Self	-	118K	60.3	61.1	58.6	41.8	40.1	33.3	39.5	38.4	34.7
RotNet [18]	VG	Self	-	103K	59.9	60.9	59.2	39.5	38.4	34.7	39.7	38.9	34.9
TP_{LDA} [19]	COCO	Text	40 topics	118K	70.6	73.9	76.3	48.7	48.4	46.7	43.7	44.1	43.0
$TP_{Cluster}$ *(Ours)*	COCO	Text	1K clusters	118K	71.5	74.5	77.0	49.5	49.8	48.1	44.1	44.6	43.7
$TP_{Cluster}$ *(Ours)*	COCO	Text	10K clusters	118K	72.1	75.0	77.2	50.2	50.3	48.7	45.1	45.3	44.2
TP_{Postag} *(Ours)*	COCO	Text	1K tokens	118K	73.3	76.4	79.3	50.6	51.1	50.0	45.9	46.5	45.8
TP_{Postag} *(Ours)*	COCO	Text	10K tokens	118K	73.6	77.0	79.4	51.2	51.7	50.5	46.1	47.0	46.1
$ICMLM_{tfm}$ *(Ours)*	COCO	Text	sentences	118K	74.8	77.8	80.5	52.0	52.0	50.8	46.8	47.3	46.2
$ICMLM_{att\text{-}fc}$ *(Ours)*	COCO	Text	sentences	118K	75.4	79.1	82.5	52.2	52.2	49.4	46.4	47.0	44.6
TP_{LDA} [19]	VG	Text	40 topics	103K	71.5	74.6	77.7	49.3	49.2	47.8	44.4	44.9	44.0
$TP_{Cluster}$ *(Ours)*	VG	Text	1K clusters	103K	73.0	76.2	79.4	50.0	49.8	47.3	45.4	45.8	44.5
$TP_{Cluster}$ *(Ours)*	VG	Text	10K clusters	103K	73.9	77.8	81.3	50.8	50.7	48.5	46.2	46.9	45.6
TP_{Postag} *(Ours)*	VG	Text	1K tokens	103K	72.9	76.4	79.6	49.9	49.8	49.1	46.0	46.5	46.4
TP_{Postag} *(Ours)*	VG	Text	10K tokens	103K	73.5	76.9	80.1	50.9	51.3	50.0	46.1	46.7	46.7
$ICMLM_{tfm}$ *(Ours)*	VG	Text	sentences	103K	75.5	79.3	82.6	52.4	52.2	51.1	47.3	47.8	47.5
$ICMLM_{att\text{-}fc}$ *(Ours)*	VG	Text	sentences	103K	76.9	81.2	85.0	52.2	52.2	47.8	47.4	47.9	47.7

masked tokens using visual cues only, it learns semantic concepts from the given training set better than $ICMLM_{tfm}$. A similar study which uses ResNet50 backbones [26] leads to similar observations (see the supplementary material).

4.2 Comparison of Fully-, Weakly- and Self-supervised Methods

Next, we compare the visual representations learned by different state-of-the-art fully-, weakly- and self-supervised learning (SSL) models. We do this by training the models explained below on COCO or VG, then using their backbones ϕ to perform the target tasks, *i.e.* image classification on VOC, IN-1K and Places-205.

Supervised. For reference, we report the results obtained by three supervised classifiers trained on different subsets of IN-1K: **(i)** "ImageNet" on the full IN-1K, **(ii)** "$\mathcal{S}$-ImageNet with 1K classes" on randomly-sampled 100 images per class, **(iii)** "$\mathcal{S}$-ImageNet with 100 classes" on 1K images for each of 100 randomly sampled classes. The latter two contain 100K images each *i.e.* the same order of magnitude as COCO or VG. For the models trained on these three subsets, we

Table 3. Fully- and weakly-supervised methods trained with ResNet50 backbones. We use the pretrained ImageNet model and train other models from scratch. We report mAP and top-1 obtained by linear SVMs (on VOC) and logistic regression classifiers (on IN-1K) using pre-extracted features (avg. of 5 runs, std. ≤0.2). Blue numbers are not transfer tasks.

Model	Dataset	Sup.	VOC	IN-1K
ImageNet	IN-1K	Labels	**87.9**	74.7
TP_{Label}	COCO	Labels	80.2	34.0
TP_{Postag}	COCO	Text	82.6	43.9
$ICMLM_{tfm}$	COCO	Text	87.3	**51.9**
$ICMLM_{att-fc}$	COCO	Text	87.5	47.9

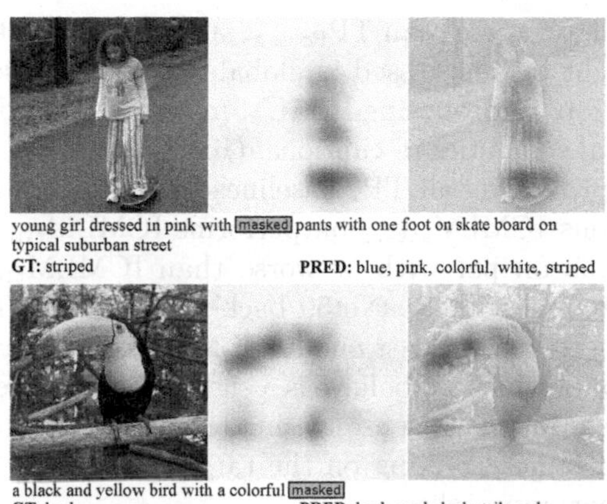

young girl dressed in pink with masked pants with one foot on skate board on typical suburban street
GT: striped
PRED: blue, pink, colorful, white, striped
a black and yellow bird with a colorful masked
GT: beak
PRED: beak, neck, body, tail, neck

Fig. 3. Attention maps for masked tokens produced by $ICMLM_{tfm}$ model with ResNet50 backbone trained on COCO (darker red means stronger attention). (Color figure online)

repeat the sampling 4 times and report their mean target task results. We also report TP_{Label} which is trained to predict ground-truth labels.

Weakly-supervised. We compare TP_{LDA}, $TP_{Cluster}$, TP_{Postag} and $ICMLM_{\star}$ methods, for which image-level tags are extracted from the captions of COCO or VG. For TP_{LDA} we use the publicly-available code of [19] to find 40 latent topics among all captions (the number of topics was validated on the validation set of VOC). Then, probabilities over caption topics define the tag labels for each image. For $TP_{Cluster}$, we cluster the captions (finding 1K or 10K clusters) and assign the cluster IDs of the captions associated to images as their tag labels. For TP_{Postag}, the tag labels are the most frequent 1K or 10K nouns, adjectives and verbs in the captions. For $ICMLM_{\star}$ models the maskable tokens are the most frequent 1K nouns, adjectives and verbs in the captions.

Self-supervised. For reference, we also provide results for two self-supervised approaches: RotNet [18] and DeeperCluster [6]. We train RotNet models from scratch on COCO or VG. For DeeperCluster, we use a model pretrained on the large-scale YFCC-100M dataset [53] (96M images).

Results are reported in Table 2. We observe the following. **(i)** We see that the good results of "ImageNet" are mostly due to its scale. Reducing it to 100K images, either by reducing the number of classes or the number of images per class significantly hurt the performance. Similarly, the supervised TP_{Label}, which uses an order of magnitude fewer categories and images performs far worse than ImageNet. **(ii)** The proposed $TP_{Cluster}$ outperforms the current state of the art for training with captions, TP_{LDA} [19], for all three datasets. Exploiting both the structure and the semantics of captions with the $BERT_{base}$ language model, it improves over a topic model. However, we see that $TP_{Cluster}$ performs on par with or worse than TP_{Postag}, suggesting that the importance of individual tokens might be suppressed in global caption representations. This validates our motivation for proposing ICMLM in Sect. 3.2: models should leverage both global and local semantics in captions. **(iii)** We see that both $ICMLM_{tfm}$ and $ICMLM_{att-fc}$ improve over all $TP_\star$ baselines by significant margins. Moreover, on VOC evaluations $ICMLM_{att-fc}$ outperforms $ICMLM_{tfm}$ while on IN-1K and Places it performs on par with or worse than $ICMLM_{tfm}$. Note that we observe a similar outcome with ResNet50 backbones (Sect. 4.3). **(iv)** Surprisingly, for VOC and Places-205, at least one ICMLM flavor outperforms the full ImageNet pretrained model which we believe is a significant achievement. For IN-1K, such comparison does not make sense as, in this setting, the proxy and the target datasets are the same. Training on the target set clearly confers an unfair advantage w.r.t. other approaches.

4.3 Additional Results with ResNet50

Some self-supervised proxy tasks might favor certain network architectures (*e.g.* see [32]). This section provides additional results where $ICMLM_\star$ models use ResNet50 [26] backbone architectures. To this end, we train TP_{Label}, TP_{Postag} and $ICMLM_\star$ models on COCO and perform image classification on VOC and IN-1K. To reduce computational costs, following [23], we train linear SVMs (on VOC) and logistic regression classifiers (on IN-1K) using image features pre-extracted from frozen backbones. Note that ResNet50 is a fully-convolutional network being more expressive compared to VGG16 (thanks to its residual connections and higher number of parameters). Consequently, in this analysis, we use a 2-layered MLPs as `tp` module, a single attention head, and $\lambda = 0.1$ in Eq. (8). We also move to a bigger concept set for TP_{Postag} and $ICMLM_\star$ models, *i.e.* the 5K most frequent nouns, adjectives and verbs.

Results are shown in Table 3. We observe larger improvements of TP_{Postag} over TP_{Label} and of $ICMLM_\star$ over TP_{Postag}. $ICMLM_\star$ outperforms TP_{Postag} by at least **4.7%**, **4.0%** and TP_{Label} by at least **7.1%**, **13.9%** on VOC and IN-1K. These results indicate that more complex CNNs are better at suppressing noise in weak labels and at learning cross-modal representations. Besides, similar to our previous analyses, we see that $ICMLM_{att-fc}$ learns semantic concepts from the training set slightly better (see the VOC results). However, $ICMLM_{tfm}$ performs better on IN-1K, suggesting that the ResNet50 backbone learns more discriminative features when guided by the same language model.

Qualitative Results. Our goal in ICMLM is to perform MLM task by *looking at* images. To see if they can attend to relevant parts in images, we visualize *attention maps* corresponding to the attention weights of visual features to masked tokens. Figures 1 and 3 present such visualizations produced by our $ICMLM_{tfm}$ model with ResNet50 backbone trained on COCO. We see that not only the model is able to detect possible concepts of interest, it can also understand which concept is asked in the captions (see the supplementary for more visualizations).

5 Conclusion

Until recently, carefully collected and manually annotated image sets have provided the most efficient way of learning general purpose visual representations. To address the annotation cost, weakly-, webly-, and self-supervised learning approaches have traded quality – a clean supervisory signal – with quantity, requiring up to hundreds of million images. Although, in some cases, large quantities of unlabeled data are readily available, processing such large volumes is far from trivial. In this paper, we seek for a cheaper alternative to ground-truth labels to train visual representations. First, starting from the observation that captions for images are often easier to collect compared to *e.g.* fine-grained category annotations, we have defined a new proxy task on image-caption pairs, namely image-conditioned masked language modeling (ICMLM), where image labels are automatically produced thanks to an efficient and effective way of leveraging their captions. Second, we have proposed a novel approach to tackle this proxy task which produces general purpose visual representations that perform on par with state-of-the-art self-supervised learning approaches on a variety of tasks, using a fraction of the data. This approach even rivals, on some settings, with a fully supervised pretraining on ImageNet. Such results are particularly relevant for domains where images are scarce but companion text is abundant.

References

1. Ba, J.L., Kiros, J.R., Hinton, G.E.: Layer normalization. arXiv:1607.06450 (2016)
2. Bachman, P., Hjelm, R.D., Buchwalter, W.: Learning representations by maximizing mutual information across views. In: Proceedings of the NeurIPS (2019)
3. Bahdanau, D., Cho, K., Bengio, Y.: Neural machine translation by jointly learning to align and translate. In: Proceedings of the ICLR (2015)
4. Blei, D.M., Ng, A.Y., Jordan, M.I.: Latent Dirichlet allocation. JMLR **3**(Jan), 993–1022 (2003)
5. Caron, M., Bojanowski, P., Joulin, A., Douze, M.: Deep clustering for unsupervised learning of visual features. In: Ferrari, V., Hebert, M., Sminchisescu, C., Weiss, Y. (eds.) Computer Vision – ECCV 2018. LNCS, vol. 11218, pp. 139–156. Springer, Cham (2018). https://doi.org/10.1007/978-3-030-01264-9_9
6. Caron, M., Bojanowski, P., Mairal, J., Joulin, A.: Unsupervised Pre-training of Image Features on Non-curated Data. In: Proceedings of the ICCV (2019)
7. Chen, L., Papandreou, G., Kokkinos, I., Murphy, K., Yuille, A.L.: DeepLab: semantic image segmentation with deep convolutional nets atrous convolution and fully connected CRFs. PAMI **40**(4), 834–848 (2018)
8. Chen, T., Kornblith, S., Norouzi, M., Hinton, G.: A simple framework for contrastive learning of visual representations. In: Proceedings of the ICML (2020)
9. Chen, X., Gupta, A.: Webly supervised learning of convolutional networks. In: Proceedings of the ICCV (2015)
10. Deng, C., Wu, Q., Wu, Q., Hu, F., Lyu, F., Tan, M.: Visual grounding via accumulated attention. In: Proceedings of the CVPR (2018)
11. Deng, J., Dong, W., Socher, R., Li, L.J., Li, K., Fei-Fei, L.: ImageNet: a large-scale hierarchical image database. In: Proceedings of the CVPR (2009)

12. Deshpande, A., Rock, J., Forsyth, D.: Learning large-scale automatic image colorization. In: Proceedings of the ICCV (2015)
13. Devlin, J., Chang, M.W., Lee, K., Toutanova, K.: BERT: Pre-training of deep bidirectional transformers for language understanding. In: ACL (2019)
14. Doersch, C., Gupta, A., Efros, A.A.: Unsupervised visual representation learning by context prediction. In: Proceedings of the ICCV (2015)
15. Doersch, C., Zisserman, A.: Multi-task self-supervised visual learning. In: Proceedings of the ICCV (2017)
16. Everingham, M., Van Gool, L., Williams, C.K.I., Winn, J., Zisserman, A.: The PASCAL Visual Object Classes Challenge 2007 (VOC2007) Results (2007)
17. Fernando, B., Bilen, H., Gavves, E., Gould, S.: Self-supervised video representation learning with odd-one-out networks. In: Proceedings of the CVPR (2017)
18. Gidaris, S., Singh, P., Komodakis, N.: Unsupervised representation learning by predicting image rotations. In: Proceedings of the ICLR (2018)
19. Gomez, L., Patel, Y., Rusiñol, M.R., Karatzas, D., Jawahar, C.: Self-supervised learning of visual features through embedding images into text topic spaces. In: Proceedings of the CVPR (2017)
20. Gomez, R., Gomez, L., Gibert, J., Karatzas, D.: Self-supervised learning from web data for multimodal retrieval. In: Multimodal Scene Understanding, chap. 9 (2019)
21. Gordo, A., Almazan, J., Revaud, J., Larlus, D.: End-to-end learning of deep visual representations for image retrieval. IJCV **124**, 237–254 (2017)
22. Gordo, A., Larlus, D.: Beyond instance-level image retrieval: Leveraging captions to learn a global visual representation for semantic retrieval. In: Proceedings of the CVPR (2017)
23. Goyal, P., Mahajan, D., Gupta, A., Misra, I.: Scaling and benchmarking self-supervised visual representation learning. In: Proceedings of the ICCV (2019)
24. Goyal, Y., Khot, T., Summers-Stay, D., Batra, D., Parikh, D.: Making the V in VQA matter: elevating the role of image understanding in Visual Question Answering. In: Proceedings of the CVPR (2017)
25. He, K., Fan, H., Wu, Y., Xie, S., Girshick, R.: Momentum contrast for unsupervised visual representation learning. In: Proceedings of the CVPR (2020)
26. He, K., Zhang, X., Ren, S., Sun, J.: Deep residual learning for image recognition. In: Proceedings of the CVPR (2016)
27. Hong, S., Yeo, D., Kwak, S., Lee, H., Han, B.: Weakly supervised semantic segmentation using web-crawled videos. In: Proceedings of the CVPR (2017)
28. Honnibal, M., Montani, I.: spaCy 2: natural language understanding with bloom embeddings, convolutional neural networks and incremental parsing (2017, to appear). https://spacy.io
29. Hu, R., Xu, H., Rohrbach, M., Feng, J., Saenko, K., Darrell, T.: Natural language object retrieval. In: Proceedings of the CVPR (2016)
30. Jenni, S., Favaro, P.: Self-supervised feature learning by learning to spot artifacts. In: Proceedings of the CVPR (2018)
31. Joulin, A., van der Maaten, L., Jabri, A., Vasilache, N.: Learning visual features from large weakly supervised data. In: Leibe, B., Matas, J., Sebe, N., Welling, M. (eds.) ECCV 2016. LNCS, vol. 9911, pp. 67–84. Springer, Cham (2016). https://doi.org/10.1007/978-3-319-46478-7_5
32. Kolesnikov, A., Zhai, X., Beyer, L.: Revisiting self-supervised visual representation learning. In: Proceedings of the CVPR (2019)
33. Krishna, R., et al.: Visual genome: Connecting language and vision using crowdsourced dense image annotations. IJCV **123**, 32–73 (2017)

34. Larsson, G., Maire, M., Shakhnarovich, G.: Colorization as a proxy task for visual understanding. In: Proceedings of the CVPR (2017)
35. Li, A., Jabri, A., Joulin, A., van der Maaten, L.: Learning visual n-grams from web data. In: Proceedings of the ICCV (2017)
36. Lin, T.-Y., et al.: Microsoft COCO: common objects in context. In: Fleet, D., Pajdla, T., Schiele, B., Tuytelaars, T. (eds.) ECCV 2014. LNCS, vol. 8693, pp. 740–755. Springer, Cham (2014). https://doi.org/10.1007/978-3-319-10602-1_48
37. Lu, J., Batra, D., Parikh, D., Lee, S.: ViLBERT: pretraining task-agnostic visiolinguistic representations for vision-and-language tasks. In: Proceedings of the NeurIPS (2019)
38. Mahajan, D., et al.: Exploring the limits of weakly supervised pretraining. In: Ferrari, V., Hebert, M., Sminchisescu, C., Weiss, Y. (eds.) ECCV 2018. LNCS, vol. 11206, pp. 185–201. Springer, Cham (2018). https://doi.org/10.1007/978-3-030-01216-8_12
39. Mahendran, A., Thewlis, J., Vedaldi, A.: Cross pixel optical-flow similarity for self-supervised learning. In: Jawahar, C.V., Li, H., Mori, G., Schindler, K. (eds.) ACCV 2018. LNCS, vol. 11365, pp. 99–116. Springer, Cham (2019). https://doi.org/10.1007/978-3-030-20873-8_7
40. Oord, A.v.d., Li, Y., Vinyals, O.: Representation learning with contrastive predictive coding. arXiv:1807.03748 (2018)
41. Parkhi, O.M., Vedaldi, A., Zisserman, A., Jawahar, C.: Cats and dogs. In: Proceedings of the CVPR (2012)
42. Pathak, D., Krahenbuhl, P., Donahue, J., Darrell, T., Efros, A.A.: Context encoders: feature learning by inpainting. In: Proceedings of the CVPR (2016)
43. Peters, M.E., et al.: Deep contextualized word representations. In: Proceedings of the NAACL-HLT (2018)
44. Qi, D., Su, L., Song, J., Cui, E., Bharti, T., Sacheti, A.: ImageBERT: cross-modal pre-training with large-scale weak-supervised image-text data. arXiv:2001.07966 (2020)
45. Quattoni, A., Collins, M., Darrell, T.: Learning visual representations using images with captions. In: Proceedings of the CVPR (2007)
46. Ren, S., He, K., Girshick, R., Sun, J.: Faster R-CNN: towards real-time object detection with region proposal networks. In: Proceedings of the NeurIPS (2015)
47. Russakovsky, O., et al.: ImageNet large scale visual recognition challenge. IJCV **115**(3), 211–252 (2015)
48. Sariyildiz, M.B., Cinbis, R.G.: Gradient matching generative networks for zero-shot learning. In: Proceedings of the CVPR (2019)
49. Simonyan, K., Zisserman, A.: Very deep convolutional networks for large-scale image recognition. In: Proceedings of the ICLR (2015)
50. Su, W., et al.: VL-BERT: pre-training of generic visual-linguistic representations. In: Proceedings of the ICLR (2020)
51. Sun, C., Myers, A., Vondrick, C., Murphy, K., Schmid, C.: VideoBERT: a joint model for video and language representation learning. In: Proceedings of the ICCV (2019)
52. Tan, H., Bansal, M.: LXMERT: learning cross-modality encoder representations from transformers. In: Proceedings of the EMNLP (2019)
53. Thomee, B., et al.: YFCC100M: the new data in multimedia research. arXiv:1503.01817 (2015)
54. Vaswani, A., et al.: Attention is all you need. In: Proceedings of the NeurIPS (2017)

55. Wah, C., Branson, S., Welinder, P., Perona, P., Belongie, S.: The Caltech-UCSD Birds-200-2011 dataset. Technical report CNS-TR-2011-001, California Institute of Technology (2011)
56. Wang, L., Li, Y., Lazebnik, S.: Learning deep structure-preserving image-text embeddings. In: Proceedings of the CVPR (2016)
57. Wu, Z., Xiong, Y., Yu, S.X., Lin, D.: Unsupervised feature learning via non-parametric instance discrimination. In: Proceedings of the CVPR (2018)
58. Xie, Q., Luong, M.T., Hovy, E., Le, Q.V.: Self-training with noisy student improves imagenet classification. In: Proceedings of the CVPR (2020)
59. Yalniz, I.Z., Jégou, H., Chen, K., Paluri, M., Mahajan, D.: Billion-scale semi-supervised learning for image classification. arXiv:1905.00546 (2019)
60. Zhang, R., Isola, P., Efros, A.A.: Colorful image colorization. In: Leibe, B., Matas, J., Sebe, N., Welling, M. (eds.) ECCV 2016. LNCS, vol. 9907, pp. 649–666. Springer, Cham (2016). https://doi.org/10.1007/978-3-319-46487-9_40
61. Zhang, R., Isola, P., Efros, A.A.: Split-brain autoencoders: unsupervised learning by cross-channel prediction. In: Proceedings of the CVPR (2017)
62. Zhou, B., Lapedriza, A., Khosla, A., Oliva, A., Torralba, A.: Places: a 10 million image database for scene recognition. PAMI **40**, 1452–1464 (2017)
63. Zhou, B., Lapedriza, A., Xiao, J., Torralba, A., Oliva, A.: Learning deep features for scene recognition using places database. In: Proceedings of the NeurIPS (2014)
64. Zhou, L., Palangi, H., Zhang, L., Hu, H., Corso, J.J., Gao, J.: Unified vision-language pre-training for image captioning and VQA. In: Proceedings of the AAAI (2020)

Solving Long-Tailed Recognition with Deep Realistic Taxonomic Classifier

Tz-Ying Wu(✉), Pedro Morgado, Pei Wang, Chih-Hui Ho, and Nuno Vasconcelos

University of California, San Diego, USA
{tzw001,pmaravil,pew062,chh279,nvasconcelos}@ucsd.edu

Abstract. Long-tail recognition tackles the natural non-uniformly distributed data in real-world scenarios. While modern classifiers perform well on populated classes, its performance degrades significantly on tail classes. Humans, however, are less affected by this since, when confronted with uncertain examples, they simply opt to provide coarser predictions. Motivated by this, a *deep realistic taxonomic classifier* (Deep-RTC) is proposed as a new solution to the long-tail problem, combining realism with hierarchical predictions. The model has the option to reject classifying samples at different levels of the taxonomy, once it cannot guarantee the desired performance. Deep-RTC is implemented with a stochastic tree sampling during training to simulate all possible classification conditions at finer or coarser levels and a rejection mechanism at inference time. Experiments on the long-tailed version of four datasets, CIFAR100, AWA2, Imagenet, and iNaturalist, demonstrate that the proposed approach preserves more information on all classes with different popularity levels. Deep-RTC also outperforms the state-of-the-art methods in long-tailed recognition, hierarchical classification, and learning with rejection literature using the proposed *correctly predicted bits* (CPB) metric.

Keywords: Realistic predictor · Taxonomic classifier · Long-tail recognition

1 Introduction

Recent advances in computer vision can be attributed to large datasets [16] and deep convolutional neural networks (CNN) [32,42,53]. While these models have achieved great success on balanced datasets, with approximately the same number of images per class, real world data tends to be highly imbalanced, with a very long-tailed class distribution. In this case, classes are frequently split into many-shot, medium-shot and few-shot, based on the number of examples [46]. Since deep CNNs tend to overfit in the small data regime, they frequently underperform for medium and few-shot classes. Popular attempts to overcome this

Electronic supplementary material The online version of this chapter (https://doi.org/10.1007/978-3-030-58598-3_11) contains supplementary material, which is available to authorized users.

A. Vedaldi et al. (Eds.): ECCV 2020, LNCS 12353, pp. 171–189, 2020.
https://doi.org/10.1007/978-3-030-58598-3_11

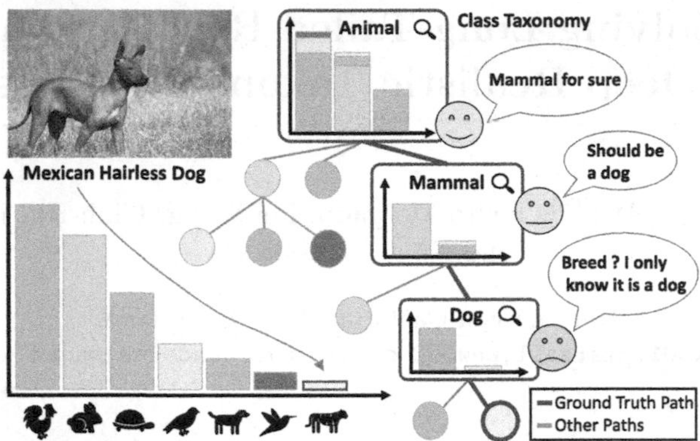

Fig. 1. Real-world datasets have class imbalance and long tails (left). Humans deal with these problems by combining class taxonomies and self-awareness (right). When faced with rare objects, like a "Mexican Hairless Dog", they push the decision to a coarser taxonomic level, e.g., simply recognizing a "Dog", of which they feel confident. This is denoted as *realistic taxonomic classification* to guarantee that all samples are processed with a high level of confidence.

limitation include data resampling [6,8,20,30], cost-sensitive losses [14], knowledge transfer from high to low population classes [46,60], normalization [38], or margin-based methods [7]. All these approaches seek to improve the classification performance of the standard softmax CNN architecture.

There is, however, little evidence that this architecture is optimally suited to deal with long-tailed recognition. For example, humans do not use this model. Rather than striving for discrimination between all objects in the world, they adopt class *taxonomies* [4,5,36,37,56], where classes are organized hierarchically at different levels of granularity, e.g. ranging from coarse domains to fine-grained 'species' or 'breeds,' as shown in Fig. 1. Classification with taxonomies is broadly denoted as *hierarchical*. The standard softmax, also known as the *flat,* classifier is a hierarchical classifier of a single taxonomic level. The use of deeper taxonomies has been shown advantageous for classification by allowing feature sharing [2, 27,39,45,61,64] and information transfer across classes [17,48,51,52,63]. While most previous works on either flat or hierarchical classification attempt to classify all images at the leaves of the taxonomic tree, independently of how difficult this is, the introduction of a taxonomy enables alternate strategies.

In this work, we explore a strategy inspired by human cognition and suited for long-tailed recognition. When humans feel insufficiently trained to answer a question at a certain level of granularity, they simply provide an answer to a coarser level, for which they feel *confident.* For example, most people do not recognize the animal of Fig. 1 as a "Mexican Hairless Dog". Instead, they change the problem from classifying dog breeds into classifying mammals and simply say it is a "Dog". Hence, a long-tailed recognition strategy more consistent with human cognition is to adopt hierarchical classification and allow decisions at *intermediate* tree levels, to achieve two goals: 1) classify all examples with high confidence, and 2) classify each example as deep in the tree as possible without violating the first goal. Since examples from low-shot classes are harder to classify confidently than those of popular classes, they tend to be classified at earlier tree levels. This can be seen as a soft version of *realistic classification* [13,57] where a classifier refuses to process examples of low-classification confidence and is denoted *realistic taxonomic classification* (RTC). The taxonomic extension enables multiple "exit levels" for the classification, at different taxonomic levels.

RTC recognizes that, while classification at the leaves uncovers full label information, *partial* label information can still be recovered when this is not feasible, by performing the classification at intermediate taxonomic stages. The goal is then to maximize the *average* information recovered per sample, favoring correct decisions of intermediate level over incorrect decisions at the leaves. We introduce a new measure of classifier performance, denoted *correctly predicted bits* (CPB), to capture this average information and propose it as a new performance measure for long-tailed recognition. Rather than simply optimizing classification accuracy at the leaves, high CPB scores require learning algorithms that produce calibrated estimates of class probabilities at *all* tree levels. This is critical to enable accurate determination of when examples should leave the tree. For long-tailed recognition, where different images can be classified at different taxonomic levels, this calibration is particularly challenging.

We address this problem with two new contributions to the training of deep CNNs for RTC. The first is a new regularization procedure based on *stochastic tree sampling,* (STS) which allows the consideration of all possible cuts of the taxonomic tree during training. RTC is then trained with a procedure similar to dropout [55], which considers the CNNs consistent with all these cuts. The second contribution addresses the challenge that RTC requires a *dynamic* CNN, capable of generating predictions at different taxonomic levels for each input example. This is addressed with a novel *dynamic predictor synthesis* procedure inspired by parameter inheritance, a regularization strategy commonly used in hierarchical classification [51,52]. To the best of our knowledge, these contributions enable the first implementation of RTC with deep CNNs and dynamic predictors. This is denoted as *Deep-RTC*, which achieves leaf classification accuracy comparable to state of the art long-tail recognition methods, but recovers much more average label information per sample.

Overall, the paper makes three contributions. 1) we propose RTC as a new solution to the long-tailed problem. 2) the *Deep-RTC* architecture, which implements a combination of stochastic taxonomic regularization and dynamic taxonomic prediction, for implementation of RTC with deep CNNs. 3) an alternative setup for the evaluation of long-tailed recognition, based on CPB scores, that accounts for the amount of information in class predictions.

2 Related Work

This work is related to several previously explored topics.

Long-Tailed Recognition: Several strategies have been proposed to address class unbalance in recognition. One possibility is to perform data resampling [31], by undersampling head and oversampling tail classes [6,8,20,30]. Sample synthesis [29,65] has also been proposed to increase the population of tail classes. Unlike Deep-RTC, these methods do not seek improved classification architectures for long-tailed recognition. An alternative is to transfer knowledge from head to tail classes. Inspired by meta-learning [58,59], these methods learn how to leverage knowledge from head classes to improve the generalization of tail classes [60]. [46] introduces memory features that encapsulate knowledge from head classes and uses an attention mechanism to discriminate between head and tail classes. This has some similarity with Deep-RTC, which also transfers knowledge from head to tail classes, but does so by leveraging hierarchical relations between them. Long-tailed recognition has also been addressed with cost-sensitive losses, which assign different weights to different samples. A typical approach is to weight classes by their frequency [34,47] or treat tail classes as hard examples [19]. [14] proposed a class balanced loss that can be directly applied to a softmax layer and focal loss [44]. These approaches can underperform for very low-frequency classes. [7] addressed this problem by enforcing large margins for few-shot classes, where the margin is inversely proportional to the number of class samples. While effective losses for long-tailed recognition are a goal of this work, we seek losses for calibration of taxonomic classifiers, which cost-sensitive losses do not address. Finally, inspired by the correlation between the weight norm of a class and its number of samples, [38] proposed to adjust the former after classifier training. All these approaches use the flat softmax classifier architecture and do not address the design of RTC.

Hierarchical Classification: Hierarchical classification has received substantial attention in computer vision. For example, sharing information across classes has been used for object recognition on large and unbalanced datasets [17,48,63], and defining a common hierarchical semantic space across classes has been explored for zero-shot learning [3,50]. Some of the ideas used in this work, e.g. parameter inheritance, are from this literature [18,51,52]. However, most of them precede deep learning and cannot be directly applied to modern CNNs. More recently, the ideas of sharing parameters or features hierarchically have inspired the design of CNN architectures [2,27,39,45,61,64]. Some of these do not support class taxonomies, e.g. learning hierarchical feature structures for flat classification [39,50]. Others are only applicable to a somewhat rigid two-level hierarchy [2,61]. Closer to this work are architectures that complement a flat classifier with convolutional branches that regularize its features to enforce hierarchical structure [27,45,64]. These branches can be based on hierarchies of feature pooling operators [27], or classification layers [45,64] supervised with labels from intermediate taxonomic levels. However, the use of additional layers makes the comparison to flat classifier unfair, which would undermine an important goal of the paper: to investigate the benefit of hierarchical (over flat) classification for long-tailed recognition. Hence, we avoid hierarchical architectures that add parameters to the backbone network. These methods also fail to address a central challenge of RTC, namely the need for simultaneous optimization with respect to many label sets, associated with the different levels of the class taxonomy. This requires a dynamic network, whose architecture can change on-the-fly to enable 1) the use of different label sets to classify different samples, and 2) optimization with respect to many label sets.

Learning with Rejection. The idea of learning with rejection dates back to at least [9]. Subsequent works derive theoretical results on the error-rejection trade-off [10,21], and explore alternative rejection criteria that avoid computation of class posterior probabilities [12,13,22]. Since the introduction of deep learning has made the estimation of the posterior distribution central to classification, most recent rejection functions consist of thresholding posteriors or derived quantities, such as the posterior entropy [24,25,57]. Alternative rejection methods have also been proposed, including the use of relative distances between samples [35], Monte-Carlo dropout [23], or classification model with a routing or rejection network [11,25,57]. We adopt the simple threshold based rejection rule of [24,25,57] in our implementation of RTC. However, rejection is applied to each level of a hierarchical classifier, instead of once for a flat classifier. This resembles the hedge your bets strategy of [15,18], in that it aims to maximize the average label information recovered per sample. However, while [15,18] accumulate the class probabilities of a flat classifier, our Deep-RTC addresses the calibration of probabilities *throughout* the tree. Our experiments show that this significantly outperforms the accumulation of flat classifier probabilities. [15] further calibrates class probabilities before rejection, but calibration is only conducted a posteriori (at test time). Instead, we propose STS for training hierarchical classifiers whose predictions are inherently calibrated at all taxonomic levels.

3 Long-Tailed Recognition and RTC

This section motivates the need for RTC as a solution to long-tailed recognition.

Long-Tailed Recognition. Existing approaches formulate long-tailed recognition as flat classification, solved by some variant of the softmax classifier. This combines a feature extractor $h(\mathbf{x};\mathbf{\Phi}) \in \mathbb{R}^k$, implemented by a CNN of parameters $\mathbf{\Phi}$, and a softmax regression layer composed by a linear transformation $\mathbf{W}$ and a softmax function $\sigma(\cdot)$

$$f(\mathbf{x};\mathbf{W},\mathbf{\Phi}) = \sigma(z(\mathbf{x};\mathbf{W},\mathbf{\Phi})) \qquad z(\mathbf{x};\mathbf{W},\mathbf{\Phi}) = \mathbf{W}^T h(\mathbf{x};\mathbf{\Phi}). \tag{1}$$

These networks are trained to minimize classification errors. Since samples are limited for mid and low-shot classes, performance can be weak. Long-tailed recognition approaches address the problem with example resampling, cost-sensitive losses, parameter sharing across classes, or post-processing. These strategies are not free of drawbacks. For example, cost-sensitive or resampling methods face a "whack-a-mole" dilemma, where performance improvements in low-shot classes (e.g. by giving them more weight) imply decreased performance in more populated ones (less weight). They are also very different from the recognition strategies of human cognition, which relies extensively on class taxonomies.

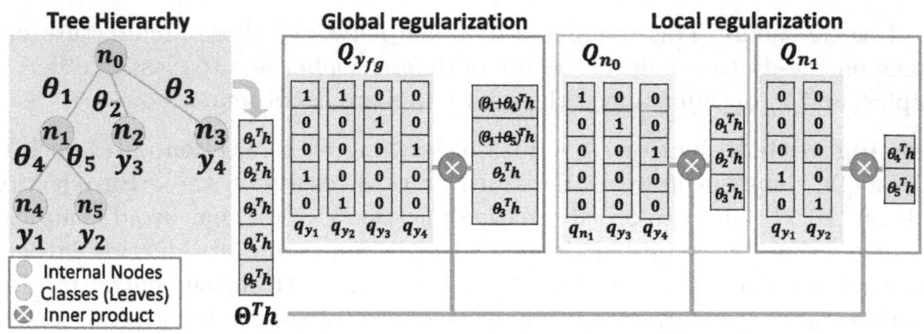

Fig. 2. Parameter sharing based on the tree hierarchy are implemented through the codeword matrices Q. The training is regularized globally from the stochastically selected label set and locally from the node-conditional consistency loss.

Many cognitive science studies have attempted to determine taxonomic levels at which humans categorize objects [4,5,36,37,56]. This has shown that most object classes have a default level, which is used by most humans to label the object (e.g. "dog" or "cat"). However, this so-called basic level is known to vary from person to person, depending on the person's training, also known as *expertise*, on the object class [4,37,56]. For example, a dog owner naturally refers to his/her pet as a "labrador" instead of as "dog." This suggests that even humans are not great long-tail recognizers. Unless they are experts (i.e. have been extensively trained in a class), they instead perform the classification at a higher taxonomic level. From a machine learning point of view, this is sensible in two ways. First, by moving up the taxonomic tree, it is always possible to find a node with sufficient training examples for accurate classification. Second, while not providing full label information for all examples, this is likely to produce a higher average label information per sample than the all-or-nothing strategy of the flat classifier [15,18]. In summary, when faced with low-shot classes, humans *trade-off classification granularity for class popularity,* choosing a classification level where their training has enough examples to guarantee accurate recognition. This does not mean that they cannot do fine-grained recognition, only that this is reserved for classes where they are experts. For example, because all humans are extensively trained on face recognition, they excel in this very fine-grained task. These observations motivate the RTC approach to long-tailed recognition.

Realistic Taxonomic Classification. A taxonomic classifier maps images $\mathbf{x} \in \mathcal{X}$ into a set of C classes $y \in \mathcal{Y} \in \{1, \ldots, C\}$, organized into a taxonomic structure where classes are recursively grouped into parent nodes according to a tree-type hierarchy $\mathcal{T}$. It is defined by a set of classification nodes $\mathcal{N} = \{n_1, \cdots, n_N\}$ and a set of taxonomic relations $\mathcal{A} = \{\mathcal{A}(n_1), \cdots, \mathcal{A}(n_N)\}$, where $\mathcal{A}(n)$ is the set of ancestor nodes of n. The finest-grained classification decisions admitted by the taxonomy occur at the leaves. We denote this set of fine-grained classes $\mathcal{Y}_{fg} = \text{Leaves}(\mathcal{T})$. Figure 2 gives an example for a classification problem with $|\mathcal{Y}_{fg}| = 4, |\mathcal{N}| = 5$, $\mathcal{A}(n_4) = \mathcal{A}(n_5) = \{n_1\}$ and

$\mathcal{A}(n_i) = \emptyset, i \in \{1, 2, 3\}$. Classes y_1, y_2 belong to parent class n_1 and the root n_0 is a dummy node containing all classes. Note that we use n to represent nodes and y to represent leaf labels. In RTC, *different samples can be classified at different hierarchy levels.* For example, a sample of class y_2 can be rejected at the root, classified at node n_1, or classified into one of the leaf classes. These options assign successively finer-grained labels to the sample. Samples rejected at the root can belong to any of the four classes, while those classified at node n_1 belong to classes y_1 or y_2. Classification at the leaves assigns the sample to a single class. Hence, RTC can predict any sub-class in the taxonomy $\mathcal{T}$. Given a training set $\mathcal{D} = \{(\mathbf{x}_i, y_i)\}_{i=1}^M$ of images and class labels, and a class taxonomy $\mathcal{T}$, the *goal* is to learn a pair of classifier $f(\mathbf{x})$ and rejection function $g(\mathbf{x})$ that work together to assign each input image $\mathbf{x}$ to the finest grained class $\hat{y}$ possible, while guaranteeing certain confidence in this assignment.

The depth at which the class prediction $\hat{y}$ is made depends on the sample difficulty and the *competence-level* γ of the classification. This is a lower bound for the confidence with which $\mathbf{x}$ can be classified. A confidence score $s(f(\mathbf{x}))$ is defined for $f(\mathbf{x})$, which is declared competent (at the γ level) for classification of $\mathbf{x}$, if $s(f(\mathbf{x})) \geq \gamma$. RTC has competence level γ if all its intermediate node decisions have this competence level. While this may be impossible to guarantee for classification with the leaf label set $\mathcal{Y}_{fg}$, it can always be guaranteed by rejecting samples at intermediate nodes of the hierarchy, i.e. defining

$$g_v(\mathbf{x}; \gamma) = 1_{[s(f_v(\mathbf{x})) \geq \gamma]} \tag{2}$$

per classification node v, where $1_{[.]}$ is the Kroneker delta. This prunes the hierarchy $\mathcal{T}$ *dynamically* per sample $\mathbf{x}$, producing a customized cut $\mathcal{T}_p$ for which the hierarchical classifier is competent at a competence level γ. This pruning is illustrated on the right of Fig. 3. Samples that are hard to classify, e.g. from few-shot classes, induce low confidence scores and are rejected earlier in the hierarchy. Samples that match the classifier expertise, e.g. from highly populated classes, progress until the leaves. This is a generalization of flat realistic classifiers [57], which simply accept or reject samples. RTC mimics human behavior in that, while $\mathbf{x}$ may not be classified at the finest-grained level, confident predictions can usually be made at intermediate or coarse levels. The competence level γ offers a guarantee for the quality of these decisions. Since larger values of γ require decisions of higher confidence, they encourage sample classification early in the hierarchy, avoiding the harder decisions that are more error-prone. The trade-off between accuracy and fine-grained labeling is controlled by adjusting γ. The confidence score $s(\cdot)$ can be implemented in various ways [11,25,57]. While RTC is compatible with any of these, we adopt the popular maximum posterior probability criterion, i.e. $s(f(\mathbf{x})) = \max_i f^i(\mathbf{x})$, where $f^i(\cdot)$ is the ith entry of $f(.)$. In our experience, the calibration of the node predictors $f_v(\mathbf{x})$ is more important than the particular implementation of the confidence score function.

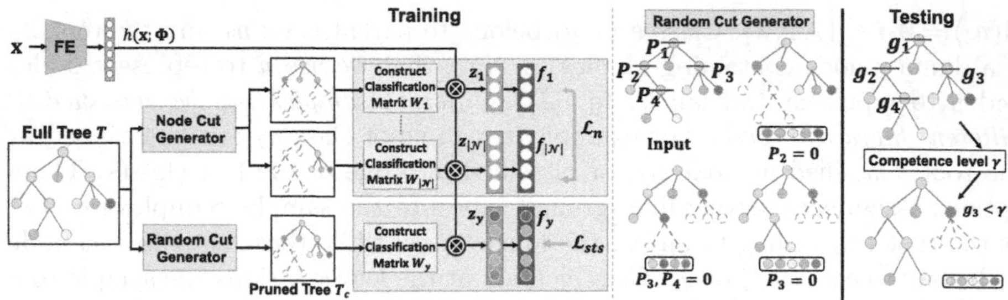

Fig. 3. Left: Deep-RTC is composed of a feature extractor, a node cut generator producing $\mathcal{Y}_n = \mathcal{C}(n)$ for all internal nodes and a random cut generator producing a potential label sets $\mathcal{Y}_c$ from $\mathcal{T}_c$. Classification matrix $\mathbf{W}_{\mathcal{Y}_c}$ is constructed for each label set and loss of (12) is imposed. Right: Rejecting samples at certain level during inference time.

4 Taxonomic Probability Calibration

In this section, we introduce the architecture of Deep-RTC.

Taxonomic Calibration. Since RTC requires decisions at all levels of the taxonomic tree, samples can be classified into any potential label set $\mathcal{Y}$ containing leaf nodes of any cut of $\mathcal{T}$. For example, the taxonomy of Fig. 2 admits two label sets, namely, $\mathcal{Y}_{fg} = \{y_1, y_2, y_3, y_4\}$ containing all classes and $\mathcal{Y} = \{n_1, y_3, y_4\}$ obtained by pruning the children of node n_1. For long-tailed recognition, where different images can be classified at very different taxonomic levels, it is important to calibrate the posterior probability distributions of *all* these label sets. We address this problem by optimizing the ensemble of *all* classifiers implementable with the hierarchy, i.e., minimize the loss

$$\mathcal{L}_{ens} = \frac{1}{|\Omega|} \sum\nolimits_{\mathcal{Y} \in \Omega} L_{\mathcal{Y}}, \tag{3}$$

where Ω is the set of all target label sets $\mathcal{Y}$ that can be derived from $\mathcal{T}$ by pruning the tree and $L_{\mathcal{Y}}$ is a loss function associated with label set $\mathcal{Y}$. While feasible for small taxonomies, this approach does not scale with taxonomy size, since the set Ω increases exponentially with $|\mathcal{T}|$. Instead, we introduce a mechanism, inspired by dropout [55], for *stochastic tree sampling (STS)* during training. At each training iteration, a random cut $\mathcal{T}_c$ of the taxonomy $\mathcal{T}$ is sampled, and the predictor $f_{\mathcal{Y}_c}(\mathbf{x}; \mathbf{W}_{\mathcal{Y}_c}, \mathbf{\Phi})$ associated with the corresponding label set $\mathcal{Y}_c$ is optimized. For this, random cuts are generated by sampling a Bernoulli random variable $P_v \sim Bernoulli(p)$ for each internal node v with a given dropout rate p. The subtree rooted at v is pruned if $P_v = 0$. Examples of these taxonomy cuts are shown in Fig. 3. The predictor $f_{\mathcal{Y}_c}$ of (1) consistent with the target label set $\mathcal{Y}_c$ associated with the cut $\mathcal{T}_c$ is then synthesized, and the loss computed as

$$\mathcal{L}_{sts} = \frac{1}{M} \sum\nolimits_{i=1}^{M} L_{\mathcal{Y}_c}(\mathbf{x}_i, y_i). \tag{4}$$

By considering different cuts at different iterations, the learning algorithm forces the hierarchical classifier to produce well calibrated decisions for all label sets.

Parameter Sharing. The procedure above requires *on-the-fly* synthesis of predictors $f_{\mathcal{Y}_c}$ for all possible label sets $\mathcal{Y}_c$ that can be derived from taxonomy $\mathcal{T}$. This implies a *dynamic* CNN architecture, where (1) changes with the sample $\mathbf{x}$. Deep-RTC is one such architecture, inspired by the fact that, for long-tailed recognition, the predictors $f_{\mathcal{Y}_c}$ should share parameters, so as to enable information transfer from head to tail classes [46,60]. This is implemented with a combination of two parameter sharing mechanisms. First, the backbone feature extractor $h(\mathbf{x}; \mathbf{\Phi})$ is shared across all label sets. Since this enables the implementation of Deep-RTC with a single network and no additional parameters, it is also critical for fair comparisons with the flat classifier. More complex hierarchical network architectures [27,45,64] would compromise these comparisons and are not investigated. Second, the predictor of (1) should reflect the hierarchical structure of each label set $\mathcal{Y}_c$. A popular implementation of this constraint, denoted *parameter inheritance (PI)*, reuses parameters of ancestors nodes $\mathcal{A}(n)$ in the predictor of node n. The column vector $\mathbf{w}_n$ of $\mathbf{W}_\mathcal{Y}$ is then defined as

$$\mathbf{w}_n = \theta_n + \sum\nolimits_{p \in \mathcal{A}(n)} \theta_p, \quad \forall n \in \mathcal{Y} \tag{5}$$

where θ_n are non-hierarchical node parameters. This compositional structure has two advantages. First, it leverages the parameters of parent nodes (more training data) to regularize the parameters of their low-level descendants (less training data). Second, the parameter vector θ_n of node n only needs to model the residuals between n and its parent, in order to be discriminative of its siblings. In summary, low-level decisions are simultaneously simplified and robustified.

Dynamic Predictor Synthesis. Deep-RTC is a novel architecture to enable the *dynamic* synthesis of predictors $f_{\mathcal{Y}_c}$ that comply with (5). This is achieved by introducing a codeword vector $\mathbf{q}_n \in \{0,1\}^{|\mathcal{N}|}$ per node n, containing binary flags that identify the ancestors $\mathcal{A}(n)$ of n

$$\mathbf{q}_n(v) = 1_{[v \in \mathcal{A}(n) \cup \{n\}]}. \tag{6}$$

For example, in the taxonomy of Fig. 2, $\mathbf{q}_{n_1} = (1,0,0,0,0)$ since $\mathcal{A}(n_1) = \varnothing$, and $\mathbf{q}_{n_4} = (1,0,0,1,0)$ since $\mathcal{A}(n_4) = \{n_1\}$. Codeword $\mathbf{q}_n$ encodes which nodes of $\mathcal{T}$ contribute to the prediction of node n under the PI strategy, thus providing a recipe for composing predictors for any label set $\mathcal{Y}$. A matrix of node-specific parameters $\mathbf{\Theta} = [\theta_1, \ldots, \theta_{|\mathcal{N}|}]$ where $\theta_n \in \mathbb{R}^k$ for all $n \in \mathcal{N}$ is then introduced, and $\mathbf{w}_n$ can be reformulated as

$$\mathbf{w}_n = \mathbf{\Theta}\mathbf{q}_n. \tag{7}$$

The codeword vectors of all nodes $n \in \mathcal{Y}$ are then written into the columns of a codeword matrix $\mathbf{Q}_\mathcal{Y} \in \{0,1\}^{|\mathcal{N}| \times |\mathcal{Y}|}$, to define a predictor as in (1),

$$f_\mathcal{Y}(\mathbf{x}; \mathbf{\Theta}, \mathbf{\Phi}) = \sigma(z_\mathcal{Y}(\mathbf{x}; \mathbf{\Theta}, \mathbf{\Phi})) \qquad z_\mathcal{Y}(\mathbf{x}; \mathbf{\Theta}, \mathbf{\Phi}) = \mathbf{W}_\mathcal{Y}^T h(\mathbf{x}; \mathbf{\Phi}), \tag{8}$$

where $\mathbf{W}_{\mathcal{Y}} = \mathbf{\Theta}\mathbf{Q}_{\mathcal{Y}}$. This enables the classification of sample $\mathbf{x}$ with respect to *any* label set $\mathcal{Y}_c$ by simply making $\mathbf{Q}_{\mathcal{Y}}$ a dynamic matrix $\mathbf{Q}_{\mathcal{Y}}(\mathbf{x}) = \mathbf{Q}_{\mathcal{Y}_c}$, as illustrated in Fig. 3.

Loss Function. Deep-RTC is trained with a cross-entropy loss

$$L_{\mathcal{Y}}(\mathbf{x}_i, y_i) = -\mathbf{y}_i^T \log f_{\mathcal{Y}}(\mathbf{x}; \mathbf{\Theta}, \mathbf{\Phi}), \tag{9}$$

where $\mathbf{y}_i$ is the one-hot encoding of $y_i \in \mathcal{Y}$. When this is used in (4), the CNN is globally optimized with respect to the label set $\mathcal{Y}_c$ associated with taxonomic cut $\mathcal{T}_c$. The regularization of the many classifiers associated with different cuts of $\mathcal{T}$ is a *global* regularization, guaranteeing that all classifiers are well calibrated. Beyond this, it is also possible to calibrate the internal node-conditional decisions. Given that a sample $\mathbf{x}$ has been assigned to node n, the node-conditional decisions are *local* and determine which of the children $\mathcal{C}(n)$ the sample should be assigned to. They consider only the target label set $\mathcal{Y}_n = \mathcal{C}(n)$ defined by the children of n. For these label sets, all nodes $v \in \mathcal{C}(n)$ share the same ancestor set $\mathcal{A}_v$ and thus the second term of (5). Hence, after softmax normalization, (5) is equivalent to $\mathbf{w}_v = \theta_v$ and the node-conditional classifier $f_n(\cdot)$ reduces to

$$f_n(\mathbf{x}; \mathbf{\Theta}, \mathbf{\Phi}) = \sigma(\mathbf{Q}_n^T\, \mathbf{\Theta}^T\, h(\mathbf{x}; \mathbf{\Phi})), \tag{10}$$

where, as illustrated in Fig. 2, the codeword matrix $\mathbf{Q}_n$ contains zeros for all ancestor nodes. Internal node decisions can thus be calibrated by noting that sample $\mathbf{x}_i$ provides supervision for all node-conditional classifiers in its ground-truth ancestor path $\mathcal{A}(y_i)$. This allows the definition of a node-conditional consistency loss per node n of the form

$$\mathcal{L}_n = \frac{1}{M} \sum\nolimits_{i=1}^{M} \frac{1}{|\mathcal{A}(y_i)|} \sum\nolimits_{n \in \mathcal{A}(y_i)} L_{\mathcal{Y}_n}(\mathbf{x}_i, y_{n,i}) \tag{11}$$

where $L_{\mathcal{Y}_n}$ is the loss of (9) for the label set $\mathcal{Y}_n$ and $y_{n,i}$ the label of $\mathbf{x}_i$ for the decision at node n. Deep-RTC is trained by minimizing a combination of these local node-conditional consistency losses and the global ensemble loss of (4)

$$\mathcal{L}_{cls} = \mathcal{L}_n + \lambda \mathcal{L}_{sts}, \tag{12}$$

where λ weights the contribution of the two terms.

Performance Evaluation. Due to the universal adoption of the flat classifier, previous long-tailed recognition works equate performance to recognition accuracy. Under the taxonomic setting, this is identical to measuring leaf node accuracy $\mathbb{E}\{1_{[\hat{y}_i = y_i]}\}$ and fails to reward trade-offs between classification granularity and accuracy. In the example of Fig. 1, it only rewards the "Mexican Hairless Dog" label, making no distinction between the labels "Dog" or "Tarantula," which are both considered errors. A taxonomic alternative is to rely on hierarchical accuracy $\mathbb{E}\{1_{[\hat{y}_i \in \mathcal{A}(y_i)]}\}$ [18]. This has the limitation of rewarding "Dog" and "Mexican Hairless Dog" equally, i.e. does not encourage finer-grained decisions. In this work, we propose that a better performance measure should capture the amount of class label information captured by the classification. While a correct classification at the leaves captures all the information, a rejection at an intermediate node can only capture partial information. To measure this, we propose to use the number of *correctly predicted bits* (CPB) by the classifier, under the assumption that each class at the leaves of the taxonomy contributes one bit of information. This is defined as

$$\text{CPB} = \frac{1}{M}\sum\nolimits_{i=1}^{M} 1_{[\hat{y}_i \in \mathcal{A}(y_i)]}\left(1 - \frac{|\text{Leaves}(\mathcal{T}_{\hat{y}_i})|}{|\text{Leaves}(\mathcal{T})|}\right) \tag{13}$$

where $\mathcal{T}_{\hat{y}_i}$ is the sub-tree rooted at $\hat{y}_i$. This assigns a score of 1 to correct classification at the leaves, and smaller scores to correct classification at higher tree levels. Note that any correct prediction of intermediate level is preferred to an incorrect prediction at the leaves, but scores less than a correct prediction of finer-grain. Finally, for flat classifiers, CPB is equal to classification accuracy.

5 Experiments

This section presents the long-tailed recognition performance of Deep-RTC.

5.1 Experimental Setup

Datasets. We consider 4 datasets. **CIFAR100-LT** [14] is a long-tailed version of [40] with "imbalance factor" 0.01 (i.e. most populated class 100× larger than rarest class). **AWA2-LT** is a long-tailed version, curated by ourselves, of [41]. It contains 30 475 images from 50 animal classes and hierarchical relations extracted from WordNet [49], leading to a 7-level imbalanced tree. The training set has an imbalance factor of 0.01, the testing set is balanced. **ImageNet-LT** [46] is a long-tailed version of [16], with 1000 classes of more than 5 and less than 1280 images per class, and a balanced test set. **iNaturalist (2018)** [1,33] is a large-scale dataset of 8 142 classes with the class imbalance factor of 0.001, and a balanced test set. While the full iNaturalist dataset is used for comparisons to previous work, a more manageable subset, iNaturalist-sub, containing 55 929 images for training and 8 142 for testing, is used for ablation studies. Please refer to supplementary material for more details.

Data Partitions for Long-Tail Evaluation. The evaluation protocol of [46] is adopted by splitting the classes into many-shot, medium-shot, and few-shot. The splitting rule of [46] is used on iNaturalist. On CIFAR100-LT and AWA2-LT, the top and bottom 1/3 populated classes belong to many-shot and few-shot respectively, and the remaining to medium-shot.

Backbone Architectures. CIFAR100-LT and iNaturalist use the setup of [14], where ResNet32 [32] and ImageNet pre-trained ResNet50 are used respectively. For ImageNet-LT, ResNet10 is chosen as in [46]. For AWA2-LT, we use ResNet18.

Competence Level. Unless otherwise noted, the value of γ is cross-validated, i.e. the value of best performance on the validation set is applied to the test set.

Table 1. Ablations on iNaturalist-sub.

Method	leaf acc.	depth	hier. acc.	CPB	inference
Flat classifier	.163	1	.163	.163	-
RHC	.163	.58	.754	.537	BU
PI+STS	.174	.46	.913	.601	TD
PI+NCL	.185	.48	.904	.563	TD
PI+STS+NCL (Deep-RTC)	.181	.50	.899	.619	TD

Table 2. Comparisons to hierarchical classifiers.

Method	CIFAR100-LT	AWA2-LT	ImageNet-LT	inference
CNN-RNN [28]	.379	.882	.514	TD
B-CNN [64]	.366	.805	.511	TD/BU
HND [43]	.374	-	-	TD
NofE [2]	.373	.770	.463	BU
Deep-RTC	**.397**	**.894**	**.529**	TD

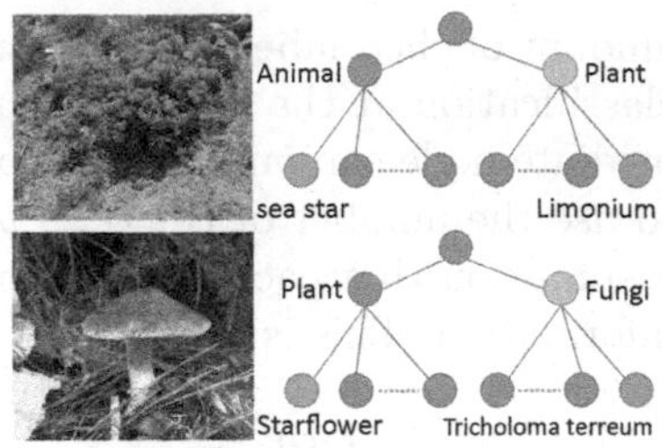

Fig. 4. Prediction of Deep-RTC (yellow) and flat classifier (gray) on two iNaturalist-sub images (orange: ground truth) (Color figure online).

5.2 Ablations

We started by evaluating how the different components of Deep-RTC - parameter inheritance (PI) regularization of (5), node-consistency loss (NCL) of (11), and stochastic tree sampling (STS) of (9) - affect the performance. Two baselines were used in this experiment. The first is a flat classifier, implemented with the standard softmax architecture, and trained to optimize classification accuracy. This is a representative baseline for the architectures used in the long-tailed recognition literature. The second is a hierarchical classifier derived from this flat classifier, by recursively adding class probabilities as dictated by the class taxonomy. This is denoted as the *recursive hierarchical classifier* (RHC). We refer to this computation of probabilities as *bottom-up* (BU) inference. This is opposite to the *top-down* (TD) inference used by most hierarchical approaches, where probabilities are sequentially computed from the root (top) to leaves (bottom) of the tree. The performance of the different classifiers was measured in multiple ways. CPB is the metric of (13). Leaf acc. is the classification accuracy at the leaves of the taxonomy. For a flat classifier, this is the standard performance measure. For a hierarchical classifier, it is the accuracy when intermediate rejections are not allowed. Hier acc. is the accuracy of a classifier that supports rejections, measured at the point where the decision is taken. In the example of Fig. 1 a decision of "dog" is considered accurate under this metric. Finally, depth is the average depth at which images are rejected, normalized by tree depth (e.g. 1 when no intermediate rejections are allowed).

CPB Performance: Table 1 shows that the flat classifier has very poor CPB performance because prediction at leaves requires the classifier to make decisions on tail classes where it is poorly trained. The result is a very large number of errors, i.e. images for which no label information is preserved. RHC, its bottom-up hierarchical extension, is a much better solution to long-tailed recognition. While most images are not classified at the leaves, both hierarchical accuracy and CPB increases dramatically. Nevertheless, RHC has weaker CPB performance than the combination of the PI architecture of Fig. 2 with either STS or NCL.

Table 3. Results on iNaturalist. Classes are discussed with popularity classes (many, medium and few- shot).

Method	metric	Many	Medium	Few	All
Softmax	CPB	0.76	0.67	0.62	0.66
CBLoss [14]		0.61	0.62	0.61	0.61
LDAM-SGD [7]		-	-	-	0.65
LDAM-DRW [7]		-	-	-	0.68
NCM [38]		0.61	0.64	0.63	0.63
cRT [38]		0.73	0.69	0.66	0.68
τ-norm [38]		0.71	0.69	0.69	0.69
Deep-RTC	CPB	**0.84**	**0.79**	**0.75**	**0.78**
	hier. acc.	0.92	0.91	0.89	0.90
	leaf freq.	0.71	0.56	0.48	0.54
	leaf acc.	0.76	0.67	0.60	0.64

Table 4. Results on ImageNet-LT.

Method	CPB
FSLwF [26]	0.28
Focal Loss [44]	0.31
Range Loss [62]	0.31
Lifted Loss [54]	0.31
OLTR [46]	0.36
Softmax	0.35
NCM [38]	0.36
cRT [38]	0.42
τ-norm [38]	0.41
Deep-RTC	**0.53**

Table 5. Comparisons to learning with rejection under different rejection rates (CPB).

		CIFAR100-LT				AWA2-LT			
Rej. Rate	Method	Many	Medium	Few	All	Many	Medium	Few	All
5%	RP [57]	**.779**	**.722**	.306	.404	**.977**	.963	.887	.914
	Deep-RTC	.773	.719	**.335**	**.416**	.975	**.978**	**.907**	**.931**
10%	RP [57]	**.793**	**.734**	.315	.416	**.980**	.966	.900	.924
	Deep-RTC	.789	.7314	**.344**	**.439**	.975	**.984**	**.929**	**.947**
20%	RP [57]	.816	.751	.328	.433	**.985**	.970	.916	.939
	Deep-RTC	**.833**	**.770**	**.393**	**.491**	.969	**.975**	**.943**	**.954**

Among these, the global regularization of STS is more effective than the local regularization of L_n. However, by combining two regularizations, they lead to the classifier (Deep-RTC) that preserves most information about the class label.

Performance Measures: The long-tailed recognition literature has focused on maximizing the accuracy of flat classifiers. Table 1 shows some limitations of this approach. First, all classifiers have very poor performance under this metric, with leaf acc. between 16% and 18%. Furthermore, as shown in Fig. 4, the labels can be totally uninformative of the object class. In the example, the flat classifier assigns the label of "sea star" ("star flower") to the image of the plant (mushroom) shown on the top (at the bottom). We are aware of *no* application that would find such labels useful. Second, all classifiers perform dramatically better in terms of hier. acc. For the practitioner, this means that they are accurate classifiers. Not expert enough to always carry the decision to the bottom of the tree, but reliable in their decisions. In the same example, Deep-RTC instead correctly assigns the images to the broader classes of "Plant" (top) and "Fungi" (bottom). Furthermore, Deep-RTC classifies 90% of the images correctly at this level! This could make it useful for many applications. For example, it could be used to automatically route the images to experts in these particular classes, for further labeling. Third, among TD classifiers, Deep-RTC pushes decisions furthest down the tree (e.g. 4% deeper than PI+STS). This makes it a better *expert* on iNaturalist-sub than its two variants, a fact captured by the proposed

CPB measure. Given all this, we believe that CPB optimality is much more meaningful than leaf acc. as a performance measure for long-tailed recognition.

5.3 Comparisons to Hierarchical Classifiers

We next performed a comparison to prior works in hierarchical classification with CPB in Table 2. These experiments show that prior methods have similar performance, without discernible advantage for TD or BU inference; however, they all underperform Deep-RTC. This is particularly interesting because these methods use networks more complex than Deep-RTC, adding branches (and parameters) to the backbone in order to regularize features according to the taxonomy. Deep-RTC simply implements a dynamic softmax classifier with the label encoding of Fig. 2. Instead, it leverages its dynamic ability and stochastic sampling to simultaneously optimize decisions for many tree cuts. The results suggest that this optimization over label sets is more important than shaping the network architecture according to the taxonomy. This is sensible since, under the Deep-RTC strategy, feature regularization is learned end-to-end, instead of hard-coded. Details of the compared methods are in the supplementary material.

5.4 Comparisons to Long-Tail Recognizers

A comparison to the state of the art methods from the long-tailed recognition is presented in Tables 3–4 for iNaturalist and ImageNet-LT respectively. More comparisons for other datasets are provided in the supplementary material. In all cases, Deep-RTC predicts *more bits* correctly (i.e. higher CPB), which beats the state of the art flat classifier by 9% on iNaturalist and 11% on ImageNet-LT. For iNaturalist, we also discuss other metrics by class popularity, where leaf freq. represents the frequency that samples are classified to leaves. A comparison to the standard softmax classifier shows that prior long-tailed methods improve performance CPB on few-shot classes but *degrade* for popular classes. Deep-RTC is the only method to consistently improve CPB performance for all levels of class popularity. It is also noted that, unlike the state of the art flat classifier, Deep-RTC does not have to sacrifice leaf acc. for the many-shot classes in order to accommodate few-shot classes where its performance will not be great anyway. Instead, it exits early for about half of the images of the few-shot classes and guarantees highly accurate answers for all classes (around 90% hier. acc.). This is similar to how humans treat the long-tail recognition problem.

5.5 Comparisons to Learning with Rejection

While the classifiers of the previous sections were allowed to reject examples at intermediate nodes, whenever feasible, they were not explicitly optimized for such rejection. Table 5 shows a comparison to a state-of-the-art flat realistic predictor (RP) [57], on CIFAR100-LT and AWA2-LT. In these comparisons, the percentage of rejected examples (rejection rate) is kept the same. The rejection rate of Deep-RTC is the percent of examples rejected at the root node. Deep-RTC achieves the best performance for all rejection rates on both datasets, because it has the option of soft-rejecting, i.e. letting examples propagate until some intermediate tree node. This is not possible for the flat RP, which always faces an all or nothing decision. In terms of class popularity, Deep-RTC always has higher CPB for few-shot classes, and frequently considerable gains. For many and medium-shot classes, the two methods have the comparable performance on CIFAR100-LT. On AWA2-LT, RP has an advantage for many and Deep-RTC for medium-shot classes. This shows that the gains of Deep-RTC are mostly due to its ability to push images of low-shot classes as far down the tree as possible without forcing decisions for which the classifier is poorly trained.

6 Conclusion

In this work, a *realistic taxonomic classifier* (RTC) is proposed to address the long-tail recognition problem. Instead of seeking the finest-grained classification for each sample, we propose to classify each sample up to the level that the classifier is competent. Deep-RTC architecture is then introduced for implementing RTC with deep CNN and is able to 1) share knowledge between head and tail classes 2) align data hierarchy with model design in order to predict at all levels in the taxonomy, and 3) guarantee high prediction performance by opting to provide coarser predictions when samples are too hard. Extensive experiments validate the effectiveness of the proposed method on 4 long-tailed datasets using the proposed tree metric. This indicates that RTC is well suited for solving long-tail problem. We believe this opens up a new direction for long-tailed literature.

Acknowledgments. This work was partially funded by NSF awards IIS-1637941, IIS-1924937, and NVIDIA GPU donations.

References

1. iNaturalist 2018 Competition. https://github.com/visipedia/inat_comp
2. Ahmed, K., Baig, M.H., Torresani, L.: Network of experts for large-scale image categorization. In: Leibe, B., Matas, J., Sebe, N., Welling, M. (eds.) ECCV 2016. LNCS, vol. 9911, pp. 516–532. Springer, Cham (2016). https://doi.org/10.1007/978-3-319-46478-7_32
3. Akata, Z., Reed, S., Walter, D., Lee, H., Schiele, B.: Evaluation of output embeddings for fine-grained image classification. In: IEEE Conference on Computer Vision and Pattern Recognition (CVPR) (2015)
4. Anaki, D., Bentin, S.: Familiarity effects on categorization levels of faces and objects. Cognition **111**, 144–149 (2009)
5. Anderson, J.: The adaptive nature of human categorization. Psychol. Rev. **98**, 409–429 (1991)
6. Buda, M., Maki, A., Mazurowski, M.: A systematic study of the class imbalance problem in convolutional neural networks. Neural Netw. **106** (2017). https://doi.org/10.1016/j.neunet.2018.07.011

7. Cao, K., Wei, C., Gaidon, A., Arechiga, N., Ma, T.: Learning imbalanced datasets with label-distribution-aware margin loss. In: Advances in Neural Information Processing Systems (NIPS) (2019)
8. Chawla, N.V., Bowyer, K.W., Hall, L.O., Kegelmeyer, W.P.: Smote: Synthetic minority over-sampling technique. J. Artif. Int. Res. **16**(1), 321–357 (2002). http://dl.acm.org/citation.cfm?id=1622407.1622416
9. Chow, C.K.: An optimum character recognition system using decision functions. IRE Trans. Electron. Comput. **EC-6**, 247–254 (1957)
10. Chow, C.K.: On optimum recognition error and reject tradeoff. IEEE Trans. Inf. Theory **16**, 41–46 (1970)
11. Corbiére, C., Thome, N., Bar-Hen, A., Cord, M., Pérez, P.: Addressing failure detection by learning model confidence. In: Advances in Neural Information Processing Systems (NIPS) (2019)
12. Cortes, C., DeSalvo, G., Mohri, M.: Boosting with abstention. In: Advances in Neural Information Processing Systems (NIPS) (2016)
13. Cortes, C., DeSalvo, G., Mohri, M.: Learning with rejection. In: Ortner, R., Simon, H.U., Zilles, S. (eds.) ALT 2016. LNCS (LNAI), vol. 9925, pp. 67–82. Springer, Cham (2016). https://doi.org/10.1007/978-3-319-46379-7_5
14. Cui, Y., Jia, M., Lin, T.Y., Song, Y., Belongie, S.: Class-balanced loss based on effective number of samples. In: IEEE Conference on Computer Vision and Pattern Recognition (CVPR) (2019)
15. Davis, J., Liang, T., Enouen, J., Ilin, R.: Hierarchical semantic labeling with adaptive confidence. In: International Symposium on Visual Computing (2019)
16. Deng, J., Dong, W., Socher, R., Li, L.J., Li, K., Fei-Fei, L.: ImageNet: a large-scale hierarchical image database. In: IEEE Conference on Computer Vision and Pattern Recognition (CVPR) (2009)
17. Deng, J., et al.: Large-scale object classification using label relation graphs. In: Fleet, D., Pajdla, T., Schiele, B., Tuytelaars, T. (eds.) ECCV 2014. LNCS, vol. 8689, pp. 48–64. Springer, Cham (2014). https://doi.org/10.1007/978-3-319-10590-1_4
18. Deng, J., Krause, J., Berg, A.C., Fei-Fei, L.: Hedging your bets: optimizing accuracy-specificity trade-offs in large scale visual recognition. In: IEEE Conference on Computer Vision and Pattern Recognition (CVPR) (2012)
19. Dong, Q., Gong, S., Zhu, X.: Class rectification hard mining for imbalanced deep learning. In: International Conference on Computer Vision (ICCV) (10 2017)
20. Drummond, C., Holte, R.: C4.5, class imbalance, and cost sensitivity: why under-sampling beats oversampling. Proceedings of the ICML 2003 Workshop on Learning from Imbalanced Datasets (2003)
21. El-Yaniv, R., Wiener, Y.: On the foundations of noise-free selective classification. J. Mach. Learn. Res. **11**, 1605–1641 (2010)
22. Fumera, G., Roli, F.: Support vector machines with embedded reject option. In: Lee, S.-W., Verri, A. (eds.) SVM 2002. LNCS, vol. 2388, pp. 68–82. Springer, Heidelberg (2002). https://doi.org/10.1007/3-540-45665-1_6
23. Gal, Y., Ghahramani, Z.: Dropout as a Bayesian approximation: representing model uncertainty in deep learning. In: International Conference on Machine Learning (ICML) (2016)
24. Geifman, Y., El-Yaniv, R.: Selective classification for deep neural networks. In: Advances in Neural Information Processing Systems (NIPS) (2017)
25. Geifman, Y., El-Yaniv, R.: SelectiveNet: a deep neural network with an integrated reject option. In: International Conference on Machine Learning (ICML) (2019)

26. Gidaris, S., Komodakis, N.: Dynamic few-shot visual learning without forgetting. In: IEEE Conference on Computer Vision and Pattern Recognition (CVPR) (2018)
27. Goo, W., Kim, J., Kim, G., Hwang, S.J.: Taxonomy-regularized semantic deep convolutional neural networks. In: Leibe, B., Matas, J., Sebe, N., Welling, M. (eds.) ECCV 2016. LNCS, vol. 9906, pp. 86–101. Springer, Cham (2016). https://doi.org/10.1007/978-3-319-46475-6_6
28. Guo, Y., Liu, Y., Bakker, E.M., Guo, Y., Lew, M.S.: CNN-RNN: a large-scale hierarchical image classification framework. Multimedia Tools Appl. **77**, 10251–10271 (2018)
29. He, H., Bai, Y., Garcia, E.A., Li, S.: ADASYN: adaptive synthetic sampling approach for imbalanced learning. In: 2008 IEEE International Joint Conference on Neural Networks (IEEE World Congress on Computational Intelligence), pp. 1322–1328 (2008)
30. Han, H., Wang, W.-Y., Mao, B.-H.: Borderline-SMOTE: a new over-sampling method in imbalanced data sets learning. In: Huang, D.-S., Zhang, X.-P., Huang, G.-B. (eds.) ICIC 2005. LNCS, vol. 3644, pp. 878–887. Springer, Heidelberg (2005). https://doi.org/10.1007/11538059_91
31. He, H., Garcia, E.A.: Learning from imbalanced data. IEEE Trans. Knowl. Data Eng. **21**(9), 1263–1284 (2009). https://doi.org/10.1109/TKDE.2008.239
32. He, K., Zhang, X., Ren, S., Sun, J.: Deep residual learning for image recognition. In: IEEE Conference on Computer Vision and Pattern Recognition (CVPR) (2016)
33. Horn, G.V., et al.: The iNaturalist species classification and detection dataset. In: IEEE Conference on Computer Vision and Pattern Recognition (CVPR) (2018)
34. Huang, C., Li, Y., Loy, C.C., Tang, X.: Learning deep representation for imbalanced classification. In: IEEE Conference on Computer Vision and Pattern Recognition (CVPR) (2016)
35. Jiang, H., Kim, B., Guan, M., Gupta, M.: To trust or not to trust a classifier. In: Advances in Neural Information Processing Systems (NIPS), pp. 5541–5552 (2018)
36. Johnson, K.: Impact of varying levels of expertise on decisions of category typicality. Memory Cogn. **29**, 1036–1050 (2001)
37. Johnson, K., Mervis, C.: Effects of varying levels of expertise on the basic level of categorization. J. Exp. Psychol. Gen. **126**(3), 248–77 (1997)
38. Kang, B., et al.: Decoupling representation and classifier for long-tailed recognition. In: International Conference on Learning Representations (ICLR) (2020)
39. Kim, H.J., Frahm, J.-M.: Hierarchy of alternating specialists for scene recognition. In: Ferrari, V., Hebert, M., Sminchisescu, C., Weiss, Y. (eds.) ECCV 2018. LNCS, vol. 11215, pp. 471–488. Springer, Cham (2018). https://doi.org/10.1007/978-3-030-01252-6_28
40. Krizhevsky, A., Hinton, G.: Learning multiple layers of features from tiny images. Technical report, Citeseer (2009)
41. Krizhevsky, A., Hinton, G.: Zero-shot learning-a comprehensive evaluation of the good, the bad and the ugly. IEEE Trans. Pattern Anal. Mach. Intell. **41**, 2251–2265 (2019)
42. Krizhevsky, A., Sutskever, I., Hinton, G.E.: Imagenet classification with deep convolutional neural networks. In: Advances in Neural Information Processing Systems (NIPS) (2012)
43. Lee, K., Lee, K., Min, K., Zhang, Y., Shin, J., Lee, H.: Hierarchical novelty detection for visual object recognition. In: IEEE Conference on Computer Vision and Pattern Recognition (CVPR) (2018)
44. Lin, T.Y., Goyal, P., Girshick, R., He, K., Dollar, P.: Focal loss for dense object detection. IEEE Trans. Pattern Anal. Mach. Intell. **1**, 1 (2018)

45. Liu, Y., Dou, Y., Jin, R., Qiao, P.: Visual tree convolutional neural network in image classification. In: International Conference on Pattern Recognition (ICPR) (2018)
46. Liu, Z., Miao, Z., Zhan, X., Wang, J., Gong, B., Yu, S.X.: Large-scale long-tailed recognition in an open world. In: IEEE Conference on Computer Vision and Pattern Recognition (CVPR) (2019)
47. Mahajan, D., et al.: Exploring the limits of weakly supervised pretraining. In: Ferrari, V., Hebert, M., Sminchisescu, C., Weiss, Y. (eds.) ECCV 2018. LNCS, vol. 11206, pp. 185–201. Springer, Cham (2018). https://doi.org/10.1007/978-3-030-01216-8_12
48. Marszałek, M., Schmid, C.: Semantic hierarchies for visual object recognition. In: IEEE Conference on Computer Vision and Pattern Recognition (CVPR) (2007)
49. Miller, G.A.: WordNet: a lexical database for English. Commun. ACM **38**, 39–41 (1995)
50. Morgado, P., Vasconcelos, N.: Semantically consistent regularization for zero-shot recognition. In: IEEE Conference on Computer Vision and Pattern Recognition (CVPR) (2017)
51. Salakhutdinov, R., Torralba, A., Tenenbaum, J.: Learning to share visual appearance for multiclass object detection. In: IEEE Conference on Computer Vision and Pattern Recognition (CVPR) (2011)
52. Shahbaba, B., Neal, R.M.: Improving classification when a class hierarchy is available using a hierarchy-based prior. Bayesian Anal. **2**(1), 221–238 (2007)
53. Simonyan, K., Zisserman, A.: Very deep convolutional networks for large-scale image recognition. CoRR abs/1409.1556 (2014)
54. Song, H.O., Xiang, Y., Jegelka, S., Savarese, S.: Deep metric learning via lifted structured feature embedding. In: IEEE Computer Vision and Pattern Recognition (CVPR) (2016)
55. Srivastava, N., Hinton, G., Krizhevsky, A., Sutskever, I., Salakhutdinov, R.: Dropout: a simple way to prevent neural networks from overfitting. J. Mach. Learn. Res. **15**(56), 1929–1958 (2014). http://jmlr.org/papers/v15/srivastava14a.html
56. Tanaka, J., Taylor, M.: Object categories and expertise: is the basic level in the eye of the beholder. Cogn. Psychol. (1991). https://doi.org/10.1016/0010-0285(91)90016-H
57. Wang, P., Vasconcelos, N.: Towards realistic predictors. In: Ferrari, V., Hebert, M., Sminchisescu, C., Weiss, Y. (eds.) ECCV 2018. LNCS, vol. 11217, pp. 37–53. Springer, Cham (2018). https://doi.org/10.1007/978-3-030-01261-8_3
58. Wang, Y.X., Hebert, M.: Learning from small sample sets by combining unsupervised meta-training with CNNs. In: Advances in Neural Information Processing Systems (NIPS) (2016)
59. Wang, Y.-X., Hebert, M.: Learning to learn: model regression networks for easy small sample learning. In: Leibe, B., Matas, J., Sebe, N., Welling, M. (eds.) ECCV 2016. LNCS, vol. 9910, pp. 616–634. Springer, Cham (2016). https://doi.org/10.1007/978-3-319-46466-4_37
60. Wang, Y.X., Ramanan, D., Hebert, M.: Learning to model the tail. In: Advances in Neural Information Processing Systems (NIPS) (2017)
61. Yan, Z., et al.: HD-CNN: hierarchical deep convolutional neural networks for large scale visual recognition. In: International Conference on Computer Vision (ICCV) (2015)
62. Zhang, X., Fang, Z., Wen, Y., Li, Z., Qiao, Y.: Range loss for deep face recognition with long-tailed training data. In: International Conference on Computer Vision (ICCV) (2017)

63. Zhao, B., Fei-Fei, L., Xing, E.P.: Large-scale category structure aware image categorization. In: Advances in Neural Information Processing Systems (NIPS) (2011)
64. Zhu, X., Bain, M.: B-CNN: branch convolutional neural network for hierarchical classification. CoRR abs/1709.09890 (2017)
65. Zou, Y., Yu, Z., Vijaya Kumar, B.V.K., Wang, J.: Unsupervised domain adaptation for semantic segmentation via class-balanced self-training. In: Ferrari, V., Hebert, M., Sminchisescu, C., Weiss, Y. (eds.) ECCV 2018. LNCS, vol. 11207, pp. 297–313. Springer, Cham (2018). https://doi.org/10.1007/978-3-030-01219-9_18

Regression of Instance Boundary by Aggregated CNN and GCN

Yanda Meng[1], Wei Meng[1], Dongxu Gao[1], Yitian Zhao[2], Xiaoyun Yang[3], Xiaowei Huang[4], and Yalin Zheng[1](✉)

[1] Department of Eye and Vision Science, Institute of Life Course and Medical Sciences, University of Liverpool, Liverpool, UK
yalin.zheng@liverpool.ac.uk

[2] Cixi Institute of Biomedical Engineering, Ningbo Institute of Industrial Technology, Chinese Academy of Sciences, Ningbo, China

[3] China Science IntelliCloud Technology Co., Ltd., Shanghai, China

[4] Department of Computer Science, University of Liverpool, Liverpool, UK

Abstract. This paper proposes a straightforward, intuitive deep learning approach for (biomedical) image segmentation tasks. Different from the existing dense pixel classification methods, we develop a novel multi-level aggregation network to directly regress the coordinates of the boundary of instances in an end-to-end manner. The network seamlessly combines standard convolution neural network (CNN) with Attention Refinement Module (ARM) and Graph Convolution Network (GCN). By iteratively and hierarchically fusing the features across different layers of the CNN, our approach gains sufficient semantic information from the input image and pays special attention to the local boundaries with the help of ARM and GCN. In particular, thanks to the proposed aggregation GCN, our network benefits from direct feature learning of the instances' boundary locations and the spatial information propagation across the image. Experiments on several challenging datasets demonstrate that our method achieves comparable results with state-of-the-art approaches but requires less inference time on the segmentation of fetal head in ultrasound images and of optic disc and optic cup in color fundus images.

Keywords: Regression · Semantic segmentation · CNN · GCN · Attention · Aggregation

1 Introduction

The accurate assessment of anatomic structures in biomedical images plays an important role in the management of many medical conditions or diseases. For

Electronic supplementary material The online version of this chapter (https://doi.org/10.1007/978-3-030-58598-3_12) contains supplementary material, which is available to authorized users.

A. Vedaldi et al. (Eds.): ECCV 2020, LNCS 12353, pp. 190–207, 2020.
https://doi.org/10.1007/978-3-030-58598-3_12

instance, fetal head (FH) circumference in ultrasound images is a critical indicator for prenatal diagnosis and can be used to estimate the gestational age and to monitor the growth of the fetus [25]. Similarly, the size of the optic disc (OD) and optic cup (OC) in color fundus images is of great importance for the diagnosis of glaucoma, an irreversible eye disease [35]. Manual annotation of this kind of structures by delineating their boundaries in clinics is unrealistic as it is costly, time consuming, labor intensive, and subject to human experience and errors. Automatic segmentation of biomedical images is believed to be able to help improve the efficiency of workflow in clinical scenarios. Inspired by the way clinicians annotate images, we propose an aggregated network to solve the segmentation tasks through directly regressing the locations of objects' boundaries, and demonstrate the effectiveness of the network in the segmentation of FH in ultrasound and OD & OC in color fundus images, respectively.

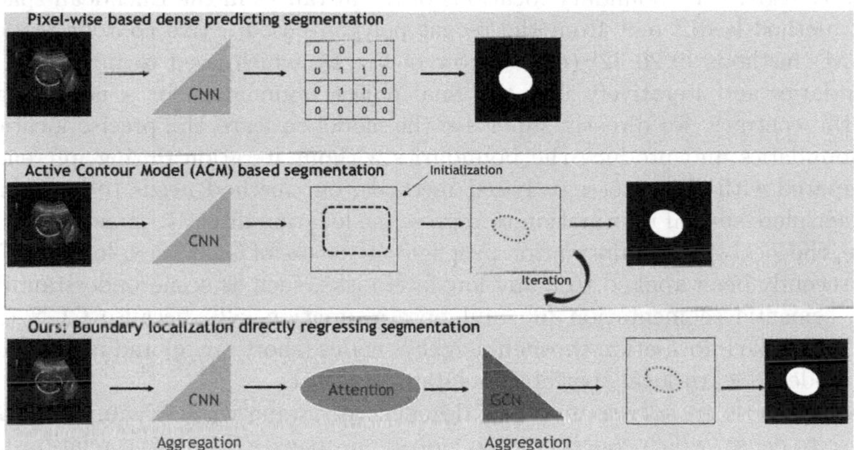

Fig. 1. Three different segmentation paradigms by deep learning. Top row: pixel-wise based methods [6,14,22] that classify each pixel into objects or background. Middle row: active contour based methods [9,32] that need iterative optimization in action to find the final contours. Bottom row: our proposed method that directly regresses the locations of object boundaries by information aggregation through CNN and GCN, enhanced by an attention module.

The biomedical image semantic segmentation task remains a challenging problem in the field of computer vision. The commonly-used deep learning-based semantic segmentation methods [6,22,41] (top row of Fig. 1) classify each pixel of an image into a category or class. These methods benefit from Convolution Neural Networks (CNN)'s excellent ability to extract high-level semantic features. Being a part of the understanding of scenes or global contexts, these methods need to learn the object location, object boundary, and object category from the high-level semantic information and local location information [31].

However, they suffer from the loss of local location information at the pixel-level [8], because a large receptive field corresponds to a small feature map, and this dilemma has increased the difficulties of dense prediction tasks. In order to solve this problem, approaches in [4,50] either maintain the resolution of the input image with dilated convolution, or capture sufficient receptive fields with pyramid pooling modules. The insights behind these methods indicate that the spatial information and the receptive field are both important to achieving high accuracy. However, it is hard to meet these two requirements simultaneously with CNN [45]. In particular, it is often challenging to maintain enough spatial information of the input image.

To address the aforementioned challenges, we follow a straightforward and intuitive methodology that human operators take to segment objects and regard segmentation as a regression task. Compared with the preserving abstraction of spatial details [4,50], we use a combination of CNN, ARM, and GCN to directly regress the boundary locations of the instances in the Euclidean space. Our method is different from the recent polygon-based active contour models (ACM) methods [9,20,32] (middle row of Fig. 1), which need to initialize the boundaries and iteratively find the final object boundaries for a new image. On the contrary, we directly supervise the model to learn the precise location of boundaries and produce the boundaries without iteration during inference. Compared with the pixel-wise based methods, our method needs to learn and extract more spatial information to regress the location directly. To address this issue, the local spatial information propagation nature of GCN is exploited. GCN has recently been applied to many low-level tasks, such as scene understanding [29], semantic segmentation [6], and pose estimation [52], because GCN can propagate the information through neighbor nodes (short range) and hence allow the model to learn local spatial correlation structure.

We propose an aggregated GCN decoder with graph vertices sampling from sparse to dense, which contributes to globally propagate the spatial relationship information across the whole image. This will provide greater representational power and more sufficient information propagation than previous segmentation methods based on Conditional Random Fields or Markov Random Fields [2,30]. Thus, we can directly regress explicit boundary location with the Euclidean space coordinate representation. This strategy addresses the concerns of most recent works [43,44], which share the similar idea but convert the Euclidean space representation into polar representation, and regressing the low-level distance between the center point and boundary points. They found that CNN cannot regress the Euclidean space coordinate representation of the boundary well as some more noise may be added, and the CNN may not maintain enough spatial information [43,44]. Our proposed aggregation GCN can handle this issue well, and our experiment results prove that. Besides, those methods' performance may suffer from the low-quality of center point, so, Xie et al. [43] utilized center sample methods to classify and selected high-quality center points to improve the segmentation result. In contrast, our methods can directly regress the boundary location without any further center selection process. As for the proposed CNN aggregation mechanism, some low-level features are unnecessarily over-extracted while object boundaries are simultaneously under-sampled. In order to extract more useful and representative features, we apply the ARM working as a filter between CNN encoder and GCN decoder, which cooperates with the GCN to gain more effective semantic and spatial features, especially the boundary location information from CNN.

In summary, this work makes the following contributions:

- We take a straightforward and intuitive approach to (biomedical) image semantic segmentation and regard it as a direct boundary regression problem in an end-to-end fashion.
- We propose aggregating mechanisms on both CNN and GCN modules, to enable them to reuse and fuse the contextual and spatial information. The additional attention mechanism helps the GCN decoder to gain more useful semantic and spatial information from the CNN encoder.
- We apply a new loss function tailored for object boundary localization that will help to make update step size adaptive to the error values during the training stage.

It is envisaged that the proposed framework may serve as a fundamental and strong baseline in future studies of biomedical semantic segmentation tasks.

2 Related Work

2.1 Pixel-Based Methods

Fully Convolution Neural Networks (FCNs) [31] and U-Net architectures [38] are widely used in semantic segmentation tasks [6,22]. These methods are aimed at extracting more spatial information or extending the receptive field that is of pivotal importance in semantic segmentation tasks. However, it is still difficult to capture longer-range correspondence between pixels in an image [48].

Aggregation Module. In order to gain global contextual dependencies of an image, methods like [41,47,50,53] proposed to fuse multi-scale or multi-level features through aggregating across semantic and spatial feature domains. Zhao et al. [50] proposed a pyramid network that utilizes multiple dilated convolution blocks [46] to aggregating global feature maps on different scales. Other approaches such as Deeplab methods [4–6] exploited parallel dilated convolution with different rates to extract features at an arbitrary resolution and preserve the spatial information. However, it is still hard to efficiently learn the discriminative feature representation as many low-level features are unnecessarily over-extracted. Therefore, these aggregation methods may result in an excessive use of information flow.

Attention Mechanism. Alternatively, some other algorithms exploited the benefits of attention mechanism to integrate local discriminative representation and global contextual features. For example, DANet and CSNet [15,34] used the attentions in spatial and channel dimensions respectively to adaptively integrate local features with their global dependencies. Furthermore, Zhao et al. proposed the point-wise spatial attention network [51], which connected each position in the feature map with all the others through self-adaptive attention maps to harvest local and long-range contextual information flexibly and dynamically. In this work, an ARM module is also used to supervise our model to learn discriminate features from input images.

2.2 Polygon-Based Methods

Instead of assigning each pixel with a class, some recent methods [9,20,32,43,44] started to predict the position of all vertices of the polygon around the boundary of the target objects. The recent work [43,44] used polar coordinates to represent object contours. Both methods achieved comparable results with pixel-based segmentation methods in instance segmentation tasks. Also, the combination of FCNs and Active Contour Models (ACMs) [27] has been exploited. Some methods formulated new loss functions that were inspired by the ACMs principles [7,21] to tackle the task of ventricle segmentation in cardiac MRI. Other approaches used the ACMs as a post-processor of the output of an FCN, for example, Marcos et al. [32] proposed a Deep Structured Active Contours model that combined ACMs and pre-trained FCNs to learn the energy surface of the reference map. These ACM-based methods achieved state-of-the-art performance in many segmentation tasks. However, there are still two main limitations. First, the contour curve must be initialized, while the initialized curve is far away from the ground truth, it may be insufficient to optimize or make an inference. Second, due to the iterative inference mechanism of ACMs, they require a relatively longer running time during training and testing.

2.3 GCNs in Segmentation

GCNs have been applied to image segmentation tasks recently, as they can propagate and exchange the local short-range information through the whole image to learn the semantic relations between objects [39,48]. In 2D image semantic segmentation tasks, Li et al. proposed a Dual Graph Convolutional Network (DGCNet) [48], which applied two orthogonal graphs frameworks to compute the global relational reasoning of the whole image and the reasoning process can help the whole network to gain rich global contextual information. Another work [39] proposed by Shin et al. shared the similar idea, and utilized GCN to learn the global structure of the shape of the object, which reflected the connectivity of neighbouring vertices. Apart from using GCN to learn global contextual information from 2D input, our approach also exploits spatial and local location information. Compared with a recent similar work [33], our method further exploit the relations between low-level and much more high-level vertex information in GCN decoder and perform a 'skip up sampling' in terms of Graph convolutions between two layers. This operation helps our model further extract feature correlations among different layers.

3 Method

3.1 Graph Representation

The manually annotated object boundaries are extracted from the binary image and equally sampled into N vertices with the same angle interval $\Delta\theta$ (e.g. N = 360, $\Delta\theta = 1°$). The geometric center of the boundary represents the center vertex. We describe the object contour with vertices and edges as $B = (V, E)$, where V has $N+1$ vertices in the Euclidean space, $V \in \mathbb{R}^{N\times 2}$, and $E \in \{0,1\}^{(N+1)\times(N+1)}$ is a sparse adjacency matrix, representing the edge connections between vertices, where $E_{i,j} = 1$ means vertices V_i and V_j are connected by an edge, and $E_{i,j} = 0$ otherwise. Every two continuous vertices on the contour are connected with an edge and are both connected to the center vertices with another two edges to form a triangle. For the OD and OC segmentation, their contours are sampled separately while the geometric centre of the OC is shared as the centre vertex. Thus, there are 360 triangles and 361 vertices for instances in FH images and 720 triangles and 721 vertices for OD and OC images. For more details, please refer to the supplementary material.

We directly use the coordinates in the Euclidean space to represent all the vertices and exploit the semantic and spatial correspondence between the inputs' instance and boundaries. Besides, our boundary representation method is not sensitive to the center point as the boundary does not have too many correlations with the center point.

3.2 Graph Fourier Transform and Convolution

According to [10], the normalized Laplacian matrix is $L = I - D^{-\frac{1}{2}}ED^{-\frac{1}{2}}$, where I is the identity matrix, and D is a diagonal matrix that represents the degree of each vertex in V, such that $D_{i,i} = \sum_{j=1}^{N} E_{i,j}$. The Laplacian of the graph is a symmetric and positive semi-definite matrix, so L can be diagonalized by the Fourier basis $U \in \mathbb{R}^{N\times N}$, such that $L = U\Lambda U^T$. The columns of U are the orthogonal eigenvectors $U = [u_1, \ldots, u_n]$, and $\Lambda = diag([\lambda_1, \ldots, \lambda_n]) \in \mathbb{R}^{N\times N}$ is a diagonal matrix with non-negative eigenvalues. The graph Fourier transform of the vertices representation $x \in \mathbb{R}^{N\times 3}$ is defined as $\hat{x} = U^T x$, and the inverse Fourier transform as $x = U\hat{x}$. The spectral graph convolution of i and j is defined as $i * j = U((U^T i) \odot (U^T j))$ in the Fourier space. Since U is not a sparse matrix, this operation is computationally expensive. To reduce the computation, Defferrard et al. [12] proposed that the convolution operation on a graph can be defined in Fourier space by formulating spectral filtering with a kernel g_θ using a recursive Chebyshev polynomial [12]. The filter g_θ is parametrized as a Chebyshev polynomial expansion of order K, such that

$$g_\theta(L) = \sum_{k=1}^{K} \theta_k T_k(\hat{L}) \tag{1}$$

where $\theta \in \mathbb{R}^K$ is a vector of Chebyshev coefficients, and $\hat{L} = 2L/\lambda_{max} - I_N$ represents the rescaled Laplacian. $T_k \in \mathbb{R}^{N\times N}$ is the Chebyshev polynomial of order K, that can be recursively computed as $T_k(x) = 2xT_{k-1}(x) - T_{k-2}(x)$ with $T_0 = 1$ and $T_1 = x$. Therefore, the spectral convolution can be defined as

$$y_j = \sum_{i=1}^{F_{in}} g_{\theta_{i,j}}(L)x_i \tag{2}$$

where x_i is the i-th feature of input $x \in \mathbb{R}^{N \times F_{in}}$, which has F_{in} features, with $F_{in} = 2$ in this work and $y \in \mathbb{R}^{N \times F_{out}}$ is the output. The entire filter operation is computationally faster and the complexity drops from $\mathcal{O}(n^2)$ to $\mathcal{O}(n)$ [3].

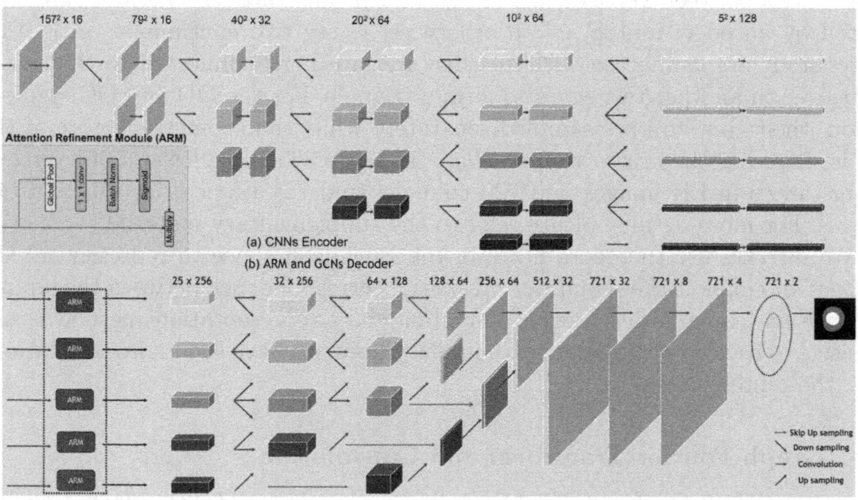

Fig. 2. Overview of our proposed network structure. The size of feature maps of the CNN encoder and vertex maps of the GCN decoder for each stage (columns) are shown. In the CNN encoder, the horizontal black arrow represents CNN convolutional operations that are achieved by a standard CNN Residual Block [24] with kernel size 3×3, stride 1, followed by a Batch Normalization (BN) layer [26] and Leaky ReLU as the activation function. The down-sampling is conducted by setting stride size as 2, the lower level feature is bi-linearly up-sampled by a factor 2. In the GCN decoder, down-sampling and up-sampling are conducted by graph vertices sampling, which is described in Sect. 3.3, and the horizontal black arrow represents residual graph convolution (Res-GCN) blocks [28] with polynomial order 4. The horizontal blue arrow achieves 'skip up sampling' with vertices number four times up sampled in terms of graph vertices sampling method via retained vertices. In this figure, the example is for OD and OC segmentation, and for FH segmentation, the convolution operation will be the same. Still, the feature map and vertex map size will be different because of different input size and number of contours of instances.

3.3 Graph Vertices Sampling

To achieve multi-level aggregated graph convolutions on different vertex resolutions, we follow [36] to form a new topology and neighbour relationships of vertices. More specifically, we use the permutation matrix $Q_d \in \{0,1\}^{m\times n}$ to down-sample m vertices, m = 360 or 720 in our work. Q_d is gained by iteratively decreasing vertices, which uses a quadratic matrix to keep the approximations of the surface error [17]. The down-sampling is a pre-processing, and the discarded vertices are saved with barycentric coordinates. We conduct up-sampling with another transformation matrix $Q_u \in \mathbb{R}^{m\times n}$. The up-sampled vertices V_u can be obtained by a sparse matrix multiplication, i.e., $V_u = Q_u V_d$, where V_d are down-sampled vertices.

3.4 Proposed Aggregation Network

Our novel aggregation graph regression network is motivated by fusing features hierarchically and iteratively [41,47,53], which consists of an image context encoder, an attention refinement module and a vertex location decoder. Both the encoder and decoder contain aggregation mechanisms through up-samplings and down-samplings, which provide improvements in extracting the full spectrum of semantic and spatial information across stages and resolutions. Besides, the attention module plays an essential role to guide the feature learning and refine the output from the CNN encoder, then passes to the GCN decoder through multi-paths. In Sect. 5.3, our ablation study demonstrates that the proposed aggregation module helps to extract more useful information, and the attention module helps to refine the extracted features from the encoder to guide feature learning better.

Semantic Encoder: Figure 2 (a) shows the detailed structure of our semantic encoder, which maintains high-resolution representations by connecting low-to-high resolution convolutions in parallel, where multi-scale fusions are repeated across different levels (rows). Our encoder is designed to lessen the spatial information loss and extract a wider spectrum of semantic features through different receptive fields. The encoder takes input images of shape $314 \times 314 \times 3$ (Fundus OD & OC images) or $140 \times 140 \times 1$ (Ultrasound FH images), with operations of up-sampling and down-sampling. The aggregation block can extract and reuse more features across various resolutions and scales, which helps to reduce spatial information loss during the encoding process.

Attention Module: We propose an Attention Refinement Module (ARM) to refine the features from the outputs of the encoder. As Fig. 2 (a) and (b) shows, ARM contains five attention blocks, and each block employs global average pooling to capture global context through the different channels, and conducts an attention tensor to lead the emphasis of feature learning through a convolution layer followed by a BN layer and sigmoid as the activation function. For the filter, the kernel size is 1×1, and the stride is 1. This design can refine the output features of each stage in the Semantic Encoder, which easily integrates the global context information.

Spatial Decoder: The decoder takes refined multi-paths outputs from the attention module, then employ ResGCN blocks [28] through different stages and levels, which has been shown that as layers go deeper, ResGCN blocks can prevent vanishing gradient problems. As Fig. 2 (b) shows, our decoder fuses and reuses the features extracted by ResGCN blocks through different stages. Benefits from the graph Vertices sampling, our decoder can regress the location of the vertices from sparse to dense, which allows the ResGCN blocks to hierarchically extract spatial location information from refined outputs of the attention module. For each ResGCN Block, it consists of 4 graph convolution layers, and each graph convolution layer is followed by a Batch Normalization layer [26] and Leaky ReLU as the activation function. After ResGCN blocks and graph vertices up-samplings, the number of vertices is up-sampled from 25 to 721, and each vertex is represented by a vector of length 32. Different from [33], Our decoder further explored the relations between low and high level resolution of vertices features, which improves the performance and is shown in Sect. 4. At last, three graph convolution layers are added to generate 2D object contour vertices, which reduces the output channels to 2, as each contour vertex has two dimensions: x and y. With the output from the decoder, we connect every two consecutive vertices on the boundary to form a polygon contour as the final segmentation result.

3.5 Loss Function

L2 and L1 loss have been widely used in regression tasks, such as object detection [19,23] and human pose estimation [42]. However, it is difficult for the L1 loss to find the global minimization in the late training stage without fine-tuning of the learning rate. L2 loss is sensitive to outliers which may result in unstable training in the early training stage.

In this work we solve segmentation as a contour vertices location regression problem. Following Wing-loss [13] and Smooth-L1 loss [18], we adopt a new loss function, Fan-loss (Fig. 3) that can take small update steps when reaching small range errors in the late training stage and can remain stable training during the early training stage. This loss function is defined as:

$$L(x) = \begin{cases} W[e^{(|x|/\epsilon)} - 1] & \text{if } |x| < W \\ |x| - C & \text{otherwise} \end{cases} \tag{3}$$

Where W is non-negative and decide the range of the non-linear part, ϵ limits the curvature between $(-W, W)$ and $C = W - W[e^{(|w|/\epsilon)} - 1]$ connects the linear and non-linear parts. After several evaluation experiments, the parameter W is set to 8 and ϵ to 5 for FH segmentation and W $= 6$, $\epsilon = 5$ for OD & OC segmentation. For the OD & OC segmentation tasks, we integrate a weight mask and assign more weights to the vertices that belong to the OC, as OC is usually difficult to segment because of poor image quality or low color contrast.

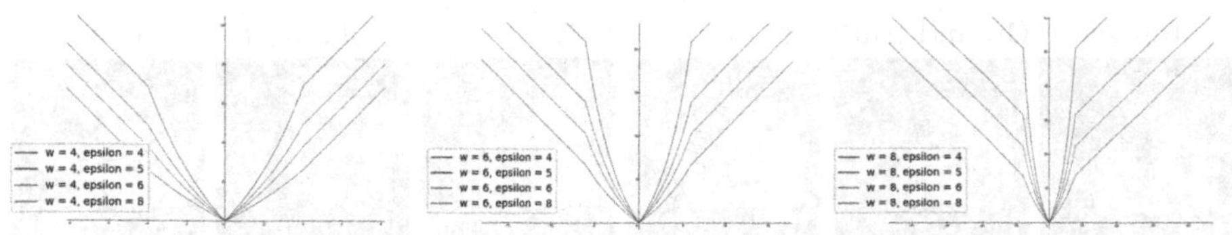

Fig. 3. The loss function plotted with different parameter settings, where w controls the non-linear part and epsilon (ϵ) limits the curvature.

4 Experiments

4.1 Datasets

We evaluate our approach with two major types of biomedical images on two segmentation tasks respectively: fundus images of retinal for OD & OC segmentation, and ultrasound images of the fetus for FH segmentation.

Fudus OD & OC Images: 2068 images from five datasets are merged together. 190 fundus images are randomly selected as the retina test dataset, the rest 1878 fundus images are used for the training. Considering the negative influence of non-target areas in fundus retina images, we first localize the disc centers by detector [37] and crop to 314×314 pixels and then transmit into our network. **Refuge** [35] consists of 400 training images and 400 validation images. The pixel-wise OD & OC gray-scale annotations are provided. **Drishti-GS** [40] contains 50 training images and 51 validation images. All images are taken centered on OD & OC with a field-of-view of 30°. The annotations are provided in the form of average boundaries. **ORIGA** [49] contains 650 fundus images. The OD & OC boundaries were manually marked by experienced graders from the Singapore Eye Research Institute. **RIGA** [1] contains 750 fundus images from **MESSIDOR** [11] database. The OD and OC are labeled manually by six ophthalmologists and the mean OD and OC are used ad the ground truth. **RIM-ONE** [16] contains 169 fundus images, annotated by five different experts. **Ultrasound FH Images:** The HC18-Challenge dataset are used which contains 999 two-dimensional (2D) ultrasound images with size of 800×540 pixels collected from the database of Radboud University Medical Center [25]. We apply zero-padding to each image to 840×840 pixels, and then resize into 140×140 as the input image, then we randomly select 94 images as the test dataset, and the model is trained on the rest 905 images.

4.2 Implementation Details

To augment the dataset, we randomly rotating the input image of training dataset for both segmentation tasks. To be specific, the rotation ranges from −15 to 15°. We randomly select 10% of training dataset as the validation dataset. We use stochastic gradient descent with a momentum of 0.9 to optimize the Fan-loss. The number of graph vertices for FH is sampled to 361, 256, 128, 64,

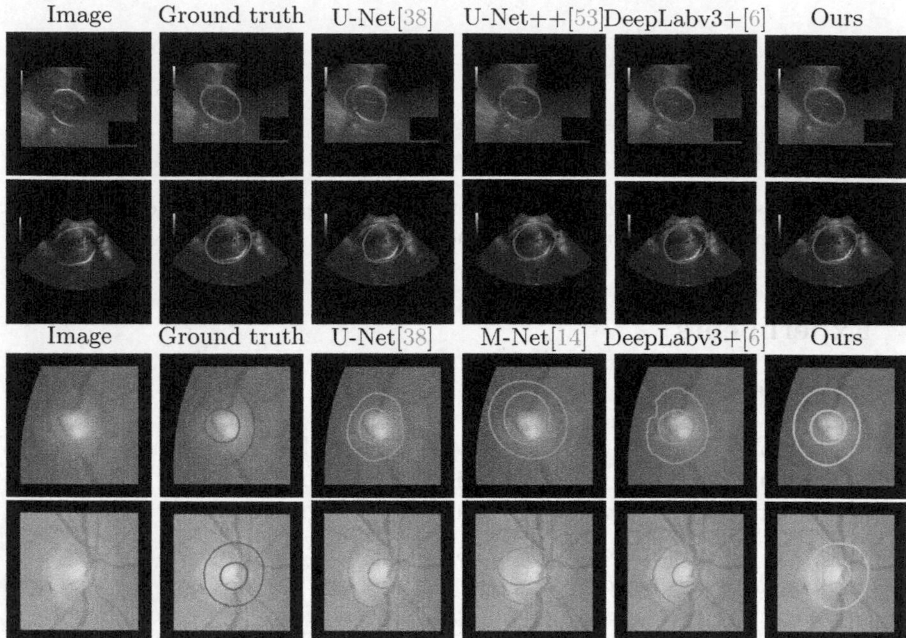

Fig. 4. Qualitative results of segmentation on the testing images of the fundus dataset and HC18-Challenge [25]. Top two rows are the ultrasound FH segmentation results, and the bottom two rows are the fundus OD & OC segmentation results.

32, 25 crosses five stages with Graph Vertices Sampling introduced in Sect. 3.3. We trained our model for 300 epochs for all the experiments, with a learning rate of 1e−2 and decay rate of 0.997 every epoch. The batch size is set as 48. All the training processes are performed on a server with 8 TESLA V100 and 4 TESLA P100, and all the test experiments are conducted on a local workstation with Geforce RTX 2080Ti.

5 Results

In this section, we present our experimental results on the OD & OC and FH segmentation task in comparison to other state-of-the-art methods. We compare our model with other state-of-the-art methods, including U-Net [38], PolarMask [43], M-Net [14], U-Net++ [53], DANet [15], DARNet [9], DeepLabv3+ [6], CGRNet [33] through running their open public source code. Dice score and Area Under the Curve (AUC) are used as the segmentation accuracy metrics. The results of an ablation study are shown in order to demonstrate the effectiveness of the proposed aggregation mechanism, attention mechanism and loss function, respectively.

Table 1. Segmentation results on retina test dataset for OD & OC and on HC18-Challenge [25] for FH. The performance is reported as Dice score (%), AUC (%), mean absolute error of Hausdorff distance (HD) for FH and mean absolute error of the vertical cup-to-disc ratio (vCDR) for OD & OC. The top three results in each category are highlighted in bold.

Methods	Tasks							
	OC		OD			FH		
	Dice score	AUC	Dice score	AUC	vCDR	Dice score	AUC	HD (mm)
U-Net [38]	0.9016	0.9186	0.9522	0.9648	0.0674	0.9625	0.9688	1.79
M-Net[14]	**0.9335**	**0.9417**	0.9230	0.9332	0.0488	-	-	-
U-Net++ [53]	0.9198	0.9285	0.9626	0.9777	0.0469	0.9701	0.9789	1.73
DANet [15]	0.9232	0.9327	0.9654	0.9726	0.0450	0.9719	0.9786	1.69
DARNet [9]	0.9235	0.9339	0.9617	0.9684	0.0455	0.9719	0.9790	**1.52**
PolarMask [43]	0.9238	0.9366	**0.9670**	**0.9782**	**0.0419**	0.9723	0.9780	1.66
DeepLabv3+ [6]	**0.9308**	**0.9406**	0.9669	0.9779	0.0467	**0.9779**	**0.9819**	**1.58**
CGRNet [33]	0.9246	0.9376	**0.9688**	**0.9784**	**0.0438**	**0.9738**	**0.9796**	**1.58**
Our method	**0.9255**	**0.9385**	**0.9697**	**0.9791**	**0.0421**	**0.9746**	**0.9801**	**1.47**

5.1 Optic Disc and Cup Segmentation

The retinal dataset we used is merged from five different fundus OD & OC images datasets. In terms of different dataset sources, they may contain different annotation standards for ground truths by different doctors. However, our model still achieve good performance, which shows the robustness and generalizability of our model. Figure 4 shows some qualitative results. We achieve 0.9697 and 0.9255 Dice similarity score on OD & OC segmentation respectively, which are comparable with other pixel-wise based state-of-the-art methods even without any bells and whistles (e.g. multi-scale training, ellipse fitting, longer training epochs, etc.). Table 1 provides the results of ours and the other methods. As for the inference speed, our model uses 64.1 ms per image that is faster than PolarMask [43] (72.1 ms) and DeepLabv3 [6] (323.9 ms). In the supplementary material, we also show some 'failed' cases compared with the ground truth. According to the comments from an anonymous expert at the Liverpool Reading Center, our model produces more accurate results than the ground truth. This highlights the potential issue of imperfect ground truth in many deep learning applications.

5.2 Fetal Head Segmentation

Table 1 and Figure 4 shows the quantitative and qualitative results, our model achieves 0.9746 Dice similarity score and 0.9801 % AUC, which outperforms DARNet [9] and DANet [15] by 0.3%. Our model (59.1 ms) is faster than PolarMask [43] (65.5 ms) and Deeplabv3+ [6] (290.3 ms) for per image inference.

Table 2. Performance comparisons between different loss function and weight mask parameter settings on the OD & OC segmentation and the FH segmentation respectively. For weight mask = 5, our model achieves best performance on the OD & OC segmentation.

Loss function	Tasks					
	OC		OD		FH	
	Dice score	AUC	Dice score	AUC	Dice score	AUC
L1	0.9111	0.9259	0.9546	0.9639	0.9505	0.9688
L2	0.9105	0.9210	0.9551	0.9666	0.9440	0.9568
Smooth-L1 [18]	0.9088	0.9114	0.9523	0.9655	0.9394	0.9454
Fan-loss						
Weight mask = 0	0.9184	0.9220	0.9618	0.9739		
Weight mask = 3	0.9221	0.9337	0.9649	0.9769		
Weight mask = 5	**0.9255**	**0.9385**	**0.9697**	**0.9791**	**0.9746**	**0.9801**
Weight mask = 7	0.9175	0.9240	0.9624	0.9720		
Weight mask = 9	0.9107	0.9213	0.9600	0.9705		

5.3 Ablation Study

We investigate the effect of each component in our proposed model. All the ablation experiments are performed with the same setting as Sect. 4.2 described. The performance in the form of Dice score and AUC are reported in Fig. 5, Table 2 and 3. The best performance in each experiment is highlighted in bold. For more qualitative results, please refer to the supplementary material.

Ablation on Parameters of Loss Function. We perform Experiments to evaluate the effect of parameter settings of Fan-loss function. When w = 6, ϵ = 5, our model achieve the best performance on OD & OC segmentation test dataset, and w = 6, ϵ = 7, for FH segmentation test dataset. For more details, please refer to Fig. 5.

Ablation on Loss Function. We conduct experiments to evaluate the effectiveness of the loss function. We compare with L1, L2, Smooth-L1 [18] loss functions, which are commonly used in the regression problem. Table 2 shows the quantitative results on OD & OC and FH segmentation tasks respectively. As illustrated, Fan-loss function attains a superior performance over the other three loss functions. In particular, it achieves a mean Dice score that is 1.6% relatively better than that of L1 loss function on OD & OC and 2.7% relatively better than L1 loss function on FH segmentation. Table 2 shows comparing with no-weight mask loss function, our proposed weight mask helps to improve OD & OC segmentation results by 0.79% when weight mask = 5 is used.

Ablation on Angle Interval. Experiments are conducted to evaluate the effect of different angle intervals $\Delta\theta$ for vertices sampling. The larger angle interval indicates the smaller number of vertices sampled on the contour. With $\Delta\theta = 1^\circ$,

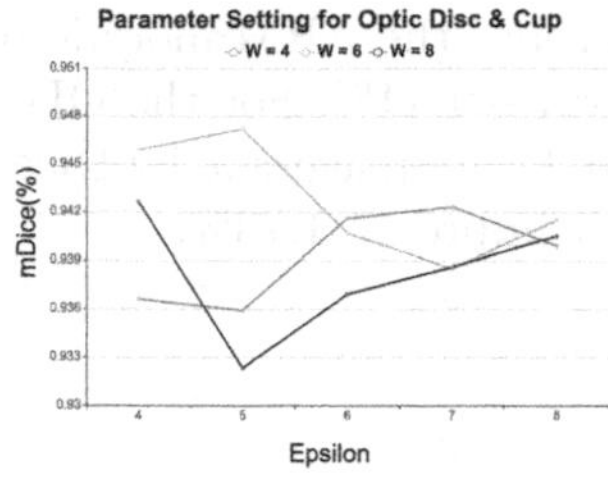

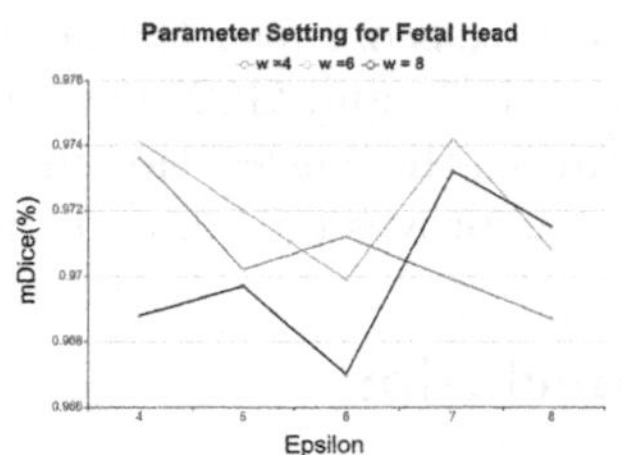

Fig. 5. A comparison of different parameter settings (w and ϵ) for Fan-loss function, measured in terms of the mean Dice score on the fundus dataset for OD & OC. With w = 6, $\epsilon = 5$, our model achieves the best performance (0.9255 & 0.9697). On the HC18-Challenge test dataset [25] for FH segmentation, with w = 6, $\epsilon = 7$, our model gains the best results 0.9746). It shows that our network is not sensitive to these parameters as no significantly different results are found.

Table 3. Ablation study on different structure components of the loss function (w = 6, $\epsilon = 5$ for FH segmentation and w = 6, $\epsilon = 7$ for OD & OC).

Methods	Tasks					
	OC		OD		FH	
	Dice score	AUC	Dice score	AUC	Dice score	AUC
No aggregation (Encoder + Decoder)	0.9025	0.9065	0.9589	0.9665	0.9567	0.9690
Aggregation	0.9207	0.9303	0.9624	0.9660	0.9700	0.9776
Aggregation + ARM (with CNN decoder)	0.9099	0.9178	0.9529	0.9635	0.9639	0.9758
Aggregation + ARM (our method)	**0.9255**	**0.9385**	**0.9697**	**0.9791**	**0.9746**	**0.9801**

our model achieves best performance on both the FH segmentation and the OD & OC segmentation. The results are shown in supplementary material.

Ablation on Structure Components. In this section, we evaluate the effectiveness of our aggregation module, attention module and GCN decoder. First, we compare with no-aggregation structure network, in which we remove all the aggregation parts and attention modules to form a standard encoder-decoder network structure. Then we add aggregated CNN and GCN module to form an aggregation network. To further improve the performance, we design an attention module, and the effect of the attention module is presented in Table 3. Furthermore, we evaluate the effectiveness of proposed GCN decoder and replace the GCN with CNN, which are the same as we used in the encoder. As illustrated, for the FH segmentation, the proposed aggregation module helps to improve

1.83% on Dice score over the no-aggregation method, the ARM module further improves 0.47%, and GCN decoder further improves 1.11%. For the OD & OC segmentation, the aggregation module improves 1.17 % on average by Dice score, the ARM improves 0.64%, and the GCN decoder improves 1.73%.

6 Conclusion

We propose a straightforward regression method for segmentation tasks by directly regressing the boundary of the instances instead of pixel-wise dense predictions. We have demonstrated its potentials on the segmentation problems of the fetal head and optic disc & cup. In the future work, we will study to extend the proposed model to tackle 3D biomedical image segmentation tasks.

Acknowledgement. Y. Meng thanks the China Science IntelliCloud Technology Co., Ltd. for the studentship. D. Gao is supported by EPSRC Grant (EP/R014094/1). We thank NVIDIA for the donation of GPU cards. This work was undertaken on Barkla, part of the High Performance Computing facilities at the University of Liverpool, UK.

References

1. Almazroa, A., et al.: Retinal fundus images for glaucoma analysis: the RIGA dataset. In: Imaging Informatics for Healthcare, Research, and Applications, Medical Imaging 2018, vol. 10579, p. 105790B. International Society for Optics and Photonics (2018)
2. Arbab, A., et al.: Conditional random fields meet deep neural networks for semantic segmentation: combining probabilistic graphical models with deep learning for structured prediction. IEEE Sig. Process. Mag. **35**(1), 37–52 (2018)
3. Bruna, J., Zaremba, W., Szlam, A., LeCun, Y.: Spectral networks and locally connected networks on graphs. arXiv preprint arXiv:1312.6203 (2013)
4. Chen, L.C., Papandreou, G., Kokkinos, I., Murphy, K., Yuille, A.L.: DeepLab: semantic image segmentation with deep convolutional nets, atrous convolution, and fully connected CRFs. IEEE Trans. Pattern Anal. Mach. Intell. **40**(4), 834–848 (2017)
5. Chen, L.C., Papandreou, G., Schroff, F., Adam, H.: Rethinking atrous convolution for semantic image segmentation. arXiv preprint arXiv:1706.05587 (2017)
6. Chen, L.-C., Zhu, Y., Papandreou, G., Schroff, F., Adam, H.: Encoder-decoder with atrous separable convolution for semantic image segmentation. In: Ferrari, V., Hebert, M., Sminchisescu, C., Weiss, Y. (eds.) ECCV 2018. LNCS, vol. 11211, pp. 833–851. Springer, Cham (2018). https://doi.org/10.1007/978-3-030-01234-2_49
7. Chen, X., Williams, B.M., Vallabhaneni, S.R., Czanner, G., Williams, R., Zheng, Y.: Learning active contour models for medical image segmentation. In: Proceedings of the IEEE Conference on Computer Vision and Pattern Recognition, pp. 11632–11640 (2019)
8. Chen, Y., Zhao, D., Lv, L., Zhang, Q.: Multi-task learning for dangerous object detection in autonomous driving. Inf. Sci. **432**, 559–571 (2018)
9. Cheng, D., Liao, R., Fidler, S., Urtasun, R.: DARNet: deep active ray network for building segmentation. In: Proceedings of the IEEE Conference on Computer Vision and Pattern Recognition, pp. 7431–7439 (2019)

10. Chung, F.R., Graham, F.C.: Spectral Graph Theory, vol. 92. American Mathematical Society, Rhode Island (1997)
11. Decencière, E., et al.: Feedback on a publicly distributed image database: the Messidor database. Image Anal. Stereol. **33**(3), 231–234 (2014)
12. Defferrard, M., Bresson, X., Vandergheynst, P.: Convolutional neural networks on graphs with fast localized spectral filtering. In: Advances in Neural Information Processing Systems, pp. 3844–3852 (2016)
13. Feng, Z.H., Kittler, J., Awais, M., Huber, P., Wu, X.J.: Wing loss for robust facial landmark localisation with convolutional neural networks. In: Proceedings of the IEEE Conference on Computer Vision and Pattern Recognition, pp. 2235–2245 (2018)
14. Fu, H., Cheng, J., Xu, Y., Wong, D.W.K., Liu, J., Cao, X.: Joint optic disc and cup segmentation based on multi-label deep network and polar transformation. IEEE Trans. Med. Imaging **37**(7), 1597–1605 (2018)
15. Fu, J., et al.: Dual attention network for scene segmentation. In: Proceedings of the IEEE Conference on Computer Vision and Pattern Recognition, pp. 3146–3154 (2019)
16. Fumero, F., Alayón, S., Sanchez, J.L., Sigut, J., Gonzalez-Hernandez, M.: RIM-ONE: an open retinal image database for optic nerve evaluation. In: 2011 24th International Symposium on Computer-Based Medical Systems (CBMS), pp. 1–6. IEEE (2011)
17. Garland, M., Heckbert, P.S.: Surface simplification using quadric error metrics. In: Proceedings of the 24th Annual Conference on Computer Graphics and Interactive Techniques, pp. 209–216. ACM Press/Addison-Wesley Publishing Co. (1997)
18. Girshick, R.: Fast R-CNN. In: Proceedings of the IEEE International Conference on Computer Vision, pp. 1440–1448 (2015)
19. Girshick, R., Donahue, J., Darrell, T., Malik, J.: Rich feature hierarchies for accurate object detection and semantic segmentation. In: Proceedings of the IEEE Conference on Computer Vision and Pattern Recognition, pp. 580–587 (2014)
20. Gur, S., Shaharabany, T., Wolf, L.: End to end trainable active contours via differentiable rendering. arXiv preprint arXiv:1912.00367 (2019)
21. Gur, S., Wolf, L., Golgher, L., Blinder, P.: Unsupervised microvascular image segmentation using an active contours mimicking neural network. In: Proceedings of the IEEE International Conference on Computer Vision, pp. 10722–10731 (2019)
22. He, K., Gkioxari, G., Dollár, P., Girshick, R.: Mask R-CNN. In: Proceedings of the IEEE international Conference on Computer Vision, pp. 2961–2969 (2017)
23. He, K., Zhang, X., Ren, S., Sun, J.: Spatial pyramid pooling in deep convolutional networks for visual recognition. IEEE Trans. Pattern Anal. Mach. Intell. **37**(9), 1904–1916 (2015)
24. He, K., Zhang, X., Ren, S., Sun, J.: Deep residual learning for image recognition. In: Proceedings of the IEEE Conference on Computer Vision and Pattern Recognition, pp. 770–778 (2016)
25. van den Heuvel, T.L., de Bruijn, D., de Korte, C.L., van Ginneken, B.: Automated measurement of fetal head circumference using 2D ultrasound images. PLoS ONE **13**(8), e0200412 (2018)
26. Ioffe, S., Szegedy, C.: Batch normalization: Accelerating deep network training by reducing internal covariate shift. arXiv preprint arXiv:1502.03167 (2015)
27. Kass, M., Witkin, A., Terzopoulos, D.: Snakes: active contour models. Int. J. Comput. Vis. **1**(4), 321–331 (1988). https://doi.org/10.1007/BF00133570
28. Li, G., Müller, M., Thabet, A., Ghanem, B.: Can GCNs go as deep as CNNs? arXiv preprint arXiv:1904.03751 (2019)

29. Li, Y., Gupta, A.: Beyond grids: learning graph representations for visual recognition. In: Advances in Neural Information Processing Systems, pp. 9225–9235 (2018)
30. Liu, Z., Li, X., Luo, P., Loy, C.C., Tang, X.: Semantic image segmentation via deep parsing network. In: Proceedings of the IEEE International Conference on Computer Vision, pp. 1377–1385 (2015)
31. Long, J., Shelhamer, E., Darrell, T.: Fully convolutional networks for semantic segmentation. In: Proceedings of the IEEE Conference on Computer Vision and Pattern Recognition, pp. 3431–3440 (2015)
32. Marcos, D., et al.: Learning deep structured active contours end-to-end. In: Proceedings of the IEEE Conference on Computer Vision and Pattern Recognition, pp. 8877–8885 (2018)
33. Meng, Y., et al.: CNN-GCN aggregation enabled boundary regression for biomedical image segmentation. In: International Conference on Medical Image Computing and Computer-Assisted Intervention (2020, in press)
34. Mou, L., et al.: CS-Net: channel and spatial attention network for curvilinear structure segmentation. In: Shen, D., et al. (eds.) MICCAI 2019. LNCS, vol. 11764, pp. 721–730. Springer, Cham (2019). https://doi.org/10.1007/978-3-030-32239-7_80
35. Orlando, J.I., et al.: REFUGE challenge: a unified framework for evaluating automated methods for glaucoma assessment from fundus photographs. Med. Image Anal. **59**, 101570 (2020)
36. Ranjan, A., Bolkart, T., Sanyal, S., Black, M.J.: Generating 3D faces using convolutional mesh autoencoders. In: Ferrari, V., Hebert, M., Sminchisescu, C., Weiss, Y. (eds.) ECCV 2018. LNCS, vol. 11207, pp. 725–741. Springer, Cham (2018). https://doi.org/10.1007/978-3-030-01219-9_43
37. Ren, S., He, K., Girshick, R., Sun, J.: Faster R-CNN: towards real-time object detection with region proposal networks. In: Advances in Neural Information Processing Systems, pp. 91–99 (2015)
38. Ronneberger, O., Fischer, P., Brox, T.: U-Net: convolutional networks for biomedical image segmentation. In: Navab, N., Hornegger, J., Wells, W.M., Frangi, A.F. (eds.) MICCAI 2015. LNCS, vol. 9351, pp. 234–241. Springer, Cham (2015). https://doi.org/10.1007/978-3-319-24574-4_28
39. Shin, S.Y., Lee, S., Yun, I.D., Lee, K.M.: Deep vessel segmentation by learning graphical connectivity. Med. Image Anal. **58**, 101556 (2019)
40. Sivaswamy, J., Krishnadas, S., Joshi, G.D., Jain, M., Tabish, A.U.S.: Drishti-GS: retinal image dataset for optic nerve head (ONH) segmentation. In: 2014 IEEE 11th International Symposium on Biomedical Imaging (ISBI), pp. 53–56. IEEE (2014)
41. Sun, K., et al.: High-resolution representations for labeling pixels and regions. arXiv preprint arXiv:1904.04514 (2019)
42. Toshev, A., Szegedy, C.: DeepPose: human pose estimation via deep neural networks. In: Proceedings of the IEEE Conference on Computer Vision and Pattern Recognition, pp. 1653–1660 (2014)
43. Xie, E., et al.: PolarMask: Single shot instance segmentation with polar representation. arXiv preprint arXiv:1909.13226 (2019)
44. Xu, W., Wang, H., Qi, F., Lu, C.: Explicit shape encoding for real-time instance segmentation. In: Proceedings of the IEEE International Conference on Computer Vision, pp. 5168–5177 (2019)

45. Yu, C., Wang, J., Peng, C., Gao, C., Yu, G., Sang, N.: BiSeNet: bilateral segmentation network for real-time semantic segmentation. In: Ferrari, V., Hebert, M., Sminchisescu, C., Weiss, Y. (eds.) ECCV 2018. LNCS, vol. 11217, pp. 334–349. Springer, Cham (2018). https://doi.org/10.1007/978-3-030-01261-8_20
46. Yu, F., Koltun, V.: Multi-scale context aggregation by dilated convolutions. arXiv preprint arXiv:1511.07122 (2015)
47. Yu, F., Wang, D., Shelhamer, E., Darrell, T.: Deep layer aggregation. In: Proceedings of the IEEE Conference on Computer Vision and Pattern Recognition, pp. 2403–2412 (2018)
48. Zhang, L., Li, X., Arnab, A., Yang, K., Tong, Y., Torr, P.H.: Dual graph convolutional network for semantic segmentation. arXiv preprint arXiv:1909.06121 (2019)
49. Zhang, Z., et al.: ORIGA-light: an online retinal fundus image database for glaucoma analysis and research. In: 2010 Annual International Conference of the IEEE Engineering in Medicine and Biology, pp. 3065–3068. IEEE (2010)
50. Zhao, H., Shi, J., Qi, X., Wang, X., Jia, J.: Pyramid scene parsing network. In: Proceedings of the IEEE Conference on Computer Vision and Pattern Recognition, pp. 2881–2890 (2017)
51. Zhao, H., et al.: PSANet: point-wise spatial attention network for scene parsing. In: Ferrari, V., Hebert, M., Sminchisescu, C., Weiss, Y. (eds.) ECCV 2018. LNCS, vol. 11213, pp. 270–286. Springer, Cham (2018). https://doi.org/10.1007/978-3-030-01240-3_17
52. Zhao, L., Peng, X., Tian, Y., Kapadia, M., Metaxas, D.N.: Semantic graph convolutional networks for 3D human pose regression. In: Proceedings of the IEEE Conference on Computer Vision and Pattern Recognition, pp. 3425–3435 (2019)
53. Zhou, Z., Rahman Siddiquee, M.M., Tajbakhsh, N., Liang, J.: UNet++: a nested U-Net architecture for medical image segmentation. In: Stoyanov, D., et al. (eds.) DLMIA/ML-CDS -2018. LNCS, vol. 11045, pp. 3–11. Springer, Cham (2018). https://doi.org/10.1007/978-3-030-00889-5_1

Social Adaptive Module for Weakly-Supervised Group Activity Recognition

Rui Yan[1], Lingxi Xie[2], Jinhui Tang[1(✉)], Xiangbo Shu[1], and Qi Tian[2]

[1] School of Computer Science and Engineering, Nanjing University of Science and Technology, Nanjing, China
{ruiyan,inhuitang,shuxb}@njust.edu.cn
[2] Huawei Inc., Shenzhen, China
198808xc@gmail.com, tian.qi1@huawei.com

Abstract. This paper presents a new task named weakly-supervised group activity recognition (GAR) which differs from conventional GAR tasks in that only video-level labels are available, yet the important persons within each frame are not provided even in the training data. This eases us to collect and annotate a large-scale NBA dataset and thus raise new challenges to GAR. To mine useful information from weak supervision, we present a key insight that key instances are likely to be related to each other, and thus design a social adaptive module (SAM) to reason about key persons and frames from noisy data. Experiments show significant improvement on the NBA dataset as well as the popular volleyball dataset. In particular, our model trained on video-level annotation achieves comparable accuracy to prior algorithms which required strong labels.

Keywords: Group activity recognition · Video analysis · Scene understanding

1 Introduction

Group activity recognition (GAR) has a variety of applications in video understanding, such as sports analysis, video surveillance, and public security. Compared with traditional individual actions [14,23,27,30,38], group activities (*a.k.a*, collective activities) [10,18,42,45] are performed by multiple persons cooperating with each other. Thus, the models for GAR require to understand not only the individual behaviors but also the relationship between each person.

Previous fully-supervised methods that require person-level annotation (*i.e.* ground-truth bounding boxes and individual action label for each person, even interaction label for person-person pairs) have achieved promising performance on group activity recognition. Typically, these methods [3,4,18,35,39,40,42,44,45] extract feature for each people according to the corresponding bounding boxes supervised by individual action label, and then fuse person-level feature

A. Vedaldi et al. (Eds.): ECCV 2020, LNCS 12353, pp. 208–224, 2020.
https://doi.org/10.1007/978-3-030-58598-3_13

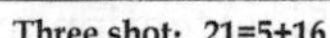

Fig. 1. *Best viewed in color.* Illustration of the uncertain input issue under weakly-supervised setting. For different activities, the off-the-shelf detector will generate varying numbers of proposals, most of which (in red boxes) are useless for recognizing group activities. For instance, "**Three shot: 21** = 5 + 16" means that the detector generates a total of **21** proposals, but only 5 of them are players and other 16 proposals are outliers in an activity of three-shot (Color figure online)

into a single representation for each frame. However, previous methods are sensitive to the varying number of people in each frame and require the explicit locations of them, which is limited in practical applications.

To this end, we investigate GAR in a weakly-supervised setting that only provides video-level labels for each video clip. This setting not only is practical to real-world scenarios but also provides a simpler and lower-cost way for the annotation of new benchmarks. Benefiting from it, we collect a larger and more challenging benchmark, NBA, consisting of 181 basketball games which involve more long-term temporal and fast-moving activities. Meanwhile, the weakly-supervised setting also brings uncertain input issue in each frame, as illustrated in Fig. 1. Under this setting, lots of useless proposals will be fed into the approach. Besides, numerous irrelevant frames will also appear in the video clip, if the temporal structure of activities (*e.g.*, in NBA) is long.

To tackle these issues, we further propose a simple yet effective module, namely Social Adaptive Module (SAM), which can adaptively select discriminative proposals and frames from the video for weakly-supervised GAR. SAM aims at assisting the weakly-supervised training by leveraging a social assumption that **key instances (people/frames) are highly related to each other**. Specifically, we firstly construct a dense relation graph on all possible input feature to measure the relatedness between each other, then pick the top ones according to their relatedness. Based on the selected feature, a sparse relation graph is built to perform relational embedding for them. Benefiting from SAM, our approach trained without fully-supervision still obtains the comparable performance to previous methods on the popular volleyball dataset [18].

Our contributions include: (a) The weakly-supervised setting that only provides video-level labels is introduced for GAR. (b) Thanks to this setting, a larger and more challenging benchmark, NBA, is collected from the web at a low cost. (c) To ease the weakly-supervised training, a SAM is proposed to adaptively find the effective person-level and frame-level representation based on the social assumption that key instances are usually closely related to each other.

2 Related Work

Group Activity Recognition. Initial approaches [18,40,42,45] for recognizing group activities adopted the two-stage pipeline. They pre-extracted feature for each person from a set of patch images and then fuse them into a single vector for each frame by various methods (*e.g.*, pooling strategies [18,35], attention mechanism [31,40,45], recurrent models [13,39,42,45], graphical models [2,24,25], and AND-OR grammar models [1,36]). Nevertheless, these two-stage methods separate feature aggregation from representation learning, which is not conducive to a deep understanding of group activities. To this end, Bagautdinov et al. [4] introduced an end-to-end framework to jointly detect multiple individuals, infer their individual actions, and estimate the group activity. Wu et al. [44] extended [4] by stacking multiple graph convolutional layer to infer the latent relation between each person. Azar et al. [3] constructed an activity map based on bounding boxes and explore the spatial relationships among people by iteratively refining the map. However, all of the above methods still require the action-level supervision (action labels and bounding boxes for each person), which is time-consuming to tag. Ramanathan et al. [32] detected events and key actors in multi-person videos without individual action labels, but they still needed to annotate the bounding boxes of all the players in a subset of 9,000 frames for training a detector. This work introduces a more practical weakly-supervised setting that only provides video-level labels for group activity recognition.

Existing Datasets Related to GAR. Limited by the time-consuming tagging, there are currently only four datasets for understanding group activities, as shown in Table 1. Choi et al. [10] proposed the first dataset, Collective Activity Dataset (CAD), consisting of real-world pedestrian sequences. Then, Choi et al. [11] extended CAD to CAED by adding two new actions (*i.e.*, "Dancing" and "Jogging") and removing the ill-defined action (*i.e.*, "Walking"). There is no specific group activity defined in CAD and CAED, in which the scenarios are assigned group activities based on majority voting. Moreover, Choi and Savarese [9] collected a Choi's New Dataset (CND) composed of many artificial pedestrian sequences. Recently, Ibrahim et al. [18] introduced a sports video dataset, Volleyball Dataset (VD), which contains numerous volleyball games. However, as the largest and most popular dataset, VD contains quite a few wrong labels that directly affect the evaluation of proposed approaches. In addition, Ramanathan et al. [32] released NCAA but few researchers have used it for GAR since only YouTube video links are provided and many of them are dead now. Some activities (*e.g.*, "steal", "slam dunk *" and "free-throw *") in NCAA can be recognized using one key frame, which actually evades from some key challenges of GAR. Limited by the size and quality of the above datasets, the recent studies of group activity recognition have encountered the bottleneck. In this work, we collect a larger and more challenging dataset from the basketball games and do not provide any person-level information (*i.e.*, the bounding boxes and action labels for each person), thanks to the weakly-supervised setting. Moreover, compared with previous benchmarks, our NBA contains more activities that involve long-term temporal structure and are fast-moving.

Relational Reasoning. Recently, relationships among entities (*i.e.*, pixels, objects or persons) have been widely leveraged in various computer vision tasks, such as Visual Question Answering [5,20,34], Scene Graph Generation [21,26,46], Object Detection [8,17], and Video Understanding [29,43,47]. Santoro et al. [34] presented a relational network module to infer the potential relationships among objects for improving the performance of visual question answering. Hu et al. [17] embedded a relation module into existing object detection systems for simultaneously detecting a set of objects and interactions between their appearance and geometry. Besides the spatial relationship among objects in the image, some recent works also explored the temporal relational structure of the video. Liu et al. [29] proposed a novel neural network to learn video representations by capturing potential correspondences for each feature point. Moreover, some recent methods [13,31,44] explored the spatial relationships between each people in group activities. In this work, we apply relational reasoning to choose the most relevant people from a number of proposals for weakly-supervised GAR.

3 Weakly-Supervised Group Activity Recognition

3.1 Weakly-Supervised Setting

For a more practical group activity recognition, i) the number of people in the scene varies over different activities even time, and ii) the person-level annotations cannot be provided in real-world applications. Therefore, we introduce a weakly-supervised setting that *only video-level labels are available, yet the location and action label of each person are not provided.*

In this work, the task of recognizing group activity under this setting is called weakly-supervised GAR that aims to directly recognize the activity performed by multiple collectively from the video with only a video-level label during training. Apparently, weakly-supervised GAR can be applied to more complex and real-world applications (*e.g.*, real-time sports analysis and video surveillance) which cannot provide fine-grained supervision. Besides, the weakly-supervised setting eases the annotation of benchmarks for the task. Without annotating the person-level supervision, we only require $\frac{1}{2K+1}$ tagging labor[1] as before where K is the number of people in the scene.

3.2 The NBA Dataset for Weakly-Supervised GAR

Under the weakly-supervised setting, we introduce a new video-based dataset, the NBA dataset. It describes the group activities that are common in basketball games. There is no annotation for each person and only a group activity label assigned to each clip. To the best of our knowledge, it is currently the largest

[1] The fully-supervised setting requires K boxes, K actions, and 1 group activity, but the weakly-supervised setting only needs 1 group activity label. We roughly assumed the same labor for each annotation.

Table 1. Comparison of the existing datasets for group activity recognition

Dataset	# Videos	# Clips	# Individual actions	# Group activities	Activity speed	Camera moving
CAD [10]	44	$\approx$2,500	5	5	Slow	N
CAED [11]	30	$\approx$3,300	6	6	Slow	N
CND [9]	32	$\approx$2,000	3	6	Slow	N
VD [18]	55	4,830	9	8	Medium	Y
NBA (ours)	181	9,172	-	9	Fast	Y

and most challenging benchmark for group activity analysis, as shown in Table 1. We will introduce the NBA dataset from the following aspects: the source of the video data, the effective annotation strategy, and the statistics of this dataset.

Data Source. It is a natural choice to collect videos of team sports for studying group activity recognition. In this work, we collect a subset of the 181 NBA games of 2019 periods from the web. Compared with the activities in volleyball games [18], the ones in basketball games have more long-term temporal structure and fast moving-speed, which brings up new challenges to group activity analysis. For one thing, the number of players may vary over different frames. On the other hand, the activity is so fast that the single-frame based person-level annotation is useless to track these players. Therefore, it is difficult to label all people in these videos which differs from volleyball games, thus we annotate this benchmark under the weakly-supervised setting. Due to the copyright restriction, this dataset is available upon request.

Annotation. Given a video, the goal of annotation is to assign the group activities to the corresponding segments. It is time-consuming to manually label such a huge dataset with conventional annotation tools. To improve the annotation efficiency, we take full advantage of the logs provided by the NBA's official website and design a simple and automatic pipeline to label our dataset. There are three steps: i) Filter out some unwanted records in the log file corresponding with a video. ii) Identify the timer in each frame by Tesseract-OCR [37] and match it with the valid records generated from step i. iii) Save the segments with a fixed length according to the time points obtained from step ii.

Statistics. We collect a total of 181 videos with a high resolution of 1920×1080. Then we divide each video into 6-second clips by the above-mentioned annotation method and sub-sample them to 12fps. Besides, we remove some abnormal clips that contain close-up shots of players or instant replays. Ultimately, there are a total of 9, 172 video clips, each of which belongs to one of the 9 activities. Here, we drop some activities such as "dunk" and "turnover" due to the limited sample size, and do not use "free-throw" that is easy to be distinguished. We randomly select 7, 624 clips for training and 1, 548 clips for testing. Table 2 shows the sample distributions across different categories of group activities and the corresponding average number of people in the scene.

Table 2. Statistics of the group activity labels in NBA. "2p", "3p", "succ", "fail", "def" and "off" are abbreviations of "two points", "three points", "success", "failure", "defensive rebound" and "offensive rebound", respectively

Group activity		2p -succ.	2p -fail. -off.	2p- fail. -def.	2p -layup -succ.	2p -layup -fail.-off.	2p -layup fail.-def.	3p -succ.	3p -fail -off.	3p -fail. -def.
# clips	Train	798	434	1316	822	455	702	728	519	1850
	Test	163	107	234	172	89	157	183	83	360

4 Approach

4.1 Mining Key Instances via Social Relationship

In general, the key and difficult point in obtaining category information from visual input is to construct and learn their intermediate representation. For the task of group activity recognition, such intermediate representation made up of individual features and underlying relationships among them, refers to *social-representation* in this paper. The previous fully-supervised setting [18,42,45] provides a variety of extra fine-grained supervision information (*e.g.*, ground-truth bounding box and action label for each person, and even the interaction label for each person-person pair) to ensure that social-representation can be constructed and learned stably during training. However, under the weakly-supervised setting that only provides video-level labels, it is difficult for models to define and learn discriminative social-representation stably.

To this end, we propose a simple yet effective framework, as illustrated in Fig. 2, to stabilize the weakly-supervised training for GAR. The core idea of our approach is to firstly construct all possible social-representation and then find the effective ones based on the social assumption that **key instances (people/frames) are closely related to each other.** Formally, given a sequence of frames $(V_1, V_2, \cdots, V_T)$, our approach models them as follow:

$$O = \mathcal{O}(\mathcal{F}(V_1; \mathcal{D}(V_1); \mathbf{W}), \mathcal{F}(V_2; \mathcal{D}(V_2); \mathbf{W}), \cdots, \mathcal{F}(V_T; \mathcal{D}(V_T); \mathbf{W})). \quad (1)$$

Here, $\mathcal{D}(V_t)$ represents detecting N^{p} proposals from each frame. There are two choices to determine the value of N^{p} as follows, i) **Quantity-aware:** empirically select top-N^{p} boxes from numerous proposals; ii) **Probability-aware:** choose the boxes whose probability is larger than a threshold θ.

The spatial modeling function $\mathcal{F}(V_t; \mathcal{D}(V_t); \mathbf{W}))$ represents that i) adopt CNN with parameters $\mathbf{W}$ to extract the convolutional feature map for frame V_t, ii) apply RoIAlign [15] to extract person-level features according to the corresponding proposals from $\mathcal{D}(V_t)$, and iii) fuse person-level features into a single frame-level vector. However, without person-level annotation, it is unavoidable for $\mathcal{D}(\cdot)$ to get many useless proposals from each frame. Moreover, the number of proposals (N^{p}) varies over samples in practical applications. Thus, $\mathcal{F}(\cdot)$ needs to

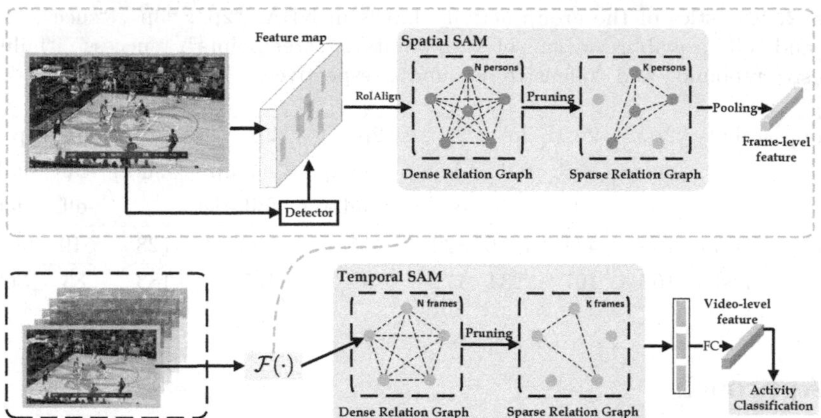

Fig. 2. Overview of our approach for weakly-supervised GAR. The inputs are a set of frames and the associating pre-detected bounding boxes for people. We apply SAM to concurrently select discriminative person-level features in the spatial domain and effective frame-level representations in the temporal domain (Best viewed in color) (Color figure online)

be able to choose K^{p} discriminative person-level features in the spatial domain. $\mathcal{O}(\cdot)$ is a temporal modeling function that samples a set of N^{f} frames from the entire video sequence (T frames) as the input of our approach according to the sampling strategy used in [41]. However, the long temporal structure of the activities in our NBA dataset will bring numerous irrelevant frames that may affect the construction of social-representation. Therefore, we also hope $\mathcal{O}(\cdot)$ can select K^{f} effective frame-level representations in the temporal domain.

It is clear that $\mathcal{F}(\cdot)$ and $\mathcal{O}(\cdot)$ need to have similar properties that attending to effective person/frame-level features in the spatial and temporal domains, respectively. Therefore, this work aims at endowing the function $\mathcal{F}(\cdot)$ and $\mathcal{O}(\cdot)$ with the ability of feature selection according to the social assumption that key instances are highly related to each other.

4.2 Social Adaptive Module (SAM)

Inspired by relational reasoning [34,43,47], we build a generic module, namely Social Adaptive Module, to implement the idea of assisting weakly-supervised training with the social assumption. Specifically, we abstract $\mathcal{F}(\cdot)$ and $\mathcal{O}(\cdot)$ into a unified form as

$$\begin{aligned}\mathbf{Z} = \mathcal{M}(\mathbf{X}) = \{\mathbf{a} \mid \mathbf{a} \in \{\lambda_1\mathcal{E}(\mathbf{x}_1), \lambda_2\mathcal{E}(\mathbf{x}_2), \cdots, \lambda_N\mathcal{E}(\mathbf{x}_N)\}, \\ \mathbf{a} \neq \mathbf{0}, \mathbf{x}_i \in \mathbf{X}, \lambda_i \in \{0,1\}, \|\boldsymbol{\lambda}\|_1 = K\},\end{aligned} \tag{2}$$

where $\mathbf{X} \in \mathbb{R}^{N\times D}$ and $\mathbf{Z} \in \mathbb{R}^{K\times D}$ are the input and output of $\mathcal{M}(\cdot)$, respectively, and $K \leq N$. Put simply, $\mathcal{M}(\cdot)$ aims to learn the parameter $\boldsymbol{\lambda} \in \mathbb{R}^N$, a zero-one vector used to select K discriminative ones from N input feature nodes. $\mathcal{E}(\cdot)$ is the embedding function for input and is optional. We hold that $\boldsymbol{\lambda}$ will be effective for feature selection only if driven by $\mathbf{X}$. Moreover, not only $N \neq K$ but also the value of N varies over samples. Therefore, directly replacing the function $\mathcal{F}(\cdot)$ and $\mathcal{O}(\cdot)$ in Eq.(1) with $\mathcal{M}(\cdot)$ is difficult for our approach to be optimized.

In this work, we approximate the solution of $\boldsymbol{\lambda}$ via pruning a Dense Relation Graph with N nodes to a Sparse Relation Graph with K nodes. Specifically, we build a dense relation graph on N input features to measure relationships between each other. During the process of pruning, we aim at maintaining the top-K feature nodes of the graph according to their relatedness. Based on the K selected features, a sparse relation graph is built to perform relational embedding for them. The details are described as follows.

Dense Relation Graph. We first build dense relationships between each input node, based on their visual features. More specifically, given a set of feature vectors as $\{\mathbf{x}_1, \mathbf{x}_2, \cdots, \mathbf{x}_N\}$, we compute the directional relation between them as $r_{ij} = g(\mathbf{x}_i, \mathbf{x}_j)$ where i, j are indices and $g(\cdot,\cdot)$ is the relation function. There are several common implementations [29,34] of $g(\cdot,\cdot)$. For instance, we can measure the L_2 distance between each feature, but which is not a data-driven and learnable method. Besides that, we can treat the concatenation $[x_i, x_j]$ as the input of a multi-layer perceptron to get the relation score. However, as the number of pairs increases, this approach will consume a lot of memory and computation. In this work, we adopt a learnable and low-cost function to measure the relation between i-th and j-th feature node as $g(\mathbf{x}_i, \mathbf{x}_j) = \Phi(\mathbf{x}_i)^\top \Psi(\mathbf{x}_j)$, where $\Phi(\cdot)$ and $\Psi(\cdot)$ are two embeddings of i-th and j-th feature node, respectively. Based on this formulation, the calculation of relation matrices, $\mathbf{R} = \{r_{ij}\}^{N \times N}$, can be implemented by only two embedding processes and a matrix multiplication. We also apply a *softmax* computation along the dimension j of the matrix $\mathbf{R}$.

Pruning Operation. To approximate the solution of $\boldsymbol{\lambda}$ in Eq.(2), this paper select the K most relevant nodes from the above dense graph based on the social assumption that key instances are likely to be related to each other. Concretely, after obtaining the $N \times N$ relation matrix for all feature pairs, we construct the *relatedness* for each feature node as $\alpha_i = \sum_{j=1}^{N}(r_{ij} + r_{ji})$, where r_{i*} and r_{*i} denote the out-edges and in-edges of i-th feature node in the dense relation graph. Intuitively, the nodes with strong connections can be easily retained in the graph. Thus, we hold that the sum of a specific node's corresponding connections can depict the importance (*relatedness*) of itself. Based on the social assumption, we sort the values of $\boldsymbol{\alpha} \in \mathbb{R}^N$ in descending order and select the top-K values denoted as $\texttt{topk}(\boldsymbol{\alpha}) \in \mathbb{R}^K$. Thus, the satisfactory $\boldsymbol{\lambda}$ can be expressed as

$$\lambda_i = \begin{cases} 1, & \alpha_i \in \texttt{topk}(\boldsymbol{\alpha}), \\ 0, & \text{otherwise.} \end{cases} \tag{3}$$

Sparse Relation Graph. According to $\boldsymbol{\lambda}$, we can get the corresponding K selected feature, $\hat{\mathbf{X}} \in \mathbb{R}^{K \times D}$, namely sparse feature. However, $\boldsymbol{\lambda}$ is driven by $\mathbf{R}$, but $\hat{\mathbf{X}}$ is unrelated to it. Therefore, $\boldsymbol{\lambda}$ will be unlearnable if we directly regard $\hat{\mathbf{X}}$

as the output of this module. To tackle this problem, we construct a relational embedding $\mathcal{E}(\cdot)$ for the sparse feature $\hat{\mathbf{X}}$ by combing with relation matrix $\mathbf{R}$. Similarly, we obtain a sparse relation matrix $\hat{\mathbf{R}} = \{\hat{r}_{ij}\}^{K\times K}$ associating to the K selected feature, and then perform relational embedding as

$$\mathbf{z}_i = \mathcal{E}(\hat{\mathbf{x}}_i) = \mathbf{W}_z\Big(\sum_{j=1}^{K} \hat{r}_{ij}\Omega(\hat{\mathbf{x}}_j)\Big) + \hat{\mathbf{x}}_i. \tag{4}$$

Here "+" denotes a residual connection, $\Omega(\cdot)$ is the embedding of sparse feature $\hat{\mathbf{x}}_j$, and $\mathbf{W}_z$ is a weight vector that projects the relational feature to the new representation with the same dimension as the sparse feature $\hat{\mathbf{x}}_i$.

SAM is the first to introduce the "social assumption" that helps a lot in the GAR scenario where many uncertain inputs are involved. More importantly, this module makes our method more appropriate to work in the weakly-supervised setting and can also be used upon previous methods [13,31,39,40,45].

4.3 Implementation Details

Person Detection and Feature Extraction. For each frame, we first adopt Faster-RCNN [33] pre-trained on the MS-COCO [28] to detect possible persons in the scene, based on the mmdetection toolbox [7]. Then, we track them over all frames by correlation tracker [12] implemented by Dlib [22]. After that, we adopt ResNet-18 [16] as the backbone to extract the convolutional feature map for each frame. Finally, we get the aligned feature for each proposal from the map by RoIAlign [15] with the crop size of 5×5 and embed it to 1024 dimensional feature vector by a fully connected layer.

Social Adaptive Module. This module is designed to select out K effective feature from N input ones. However, the values of N and K depend on the situation and will be explained in experiments. If N varies over samples (*e.g.*, different numbers of proposals are generated by the Probability-aware strategy mentioned in Sect. 4.1), we feed data into this module with a batch-size of 1 but do not change the batch-size of the entire framework. The $\Phi(\cdot)$, $\Psi(\cdot)$, and $\Omega(\cdot)$ used to embed input feature are implemented by 1×1 convolutional layers.

Optimization. We adopt the ADAM to optimize our approach with fixed hyper-parameters ($\beta_1 = \beta_2 = 0.9$, $\varepsilon = 10^{-4}$) and train it in 30 epochs with an initial learning rate of 0.0001 that is reduced to 1/10 of the previous value for every 5 epochs. Compared with SSU [4] and ARG [44] that require pre-training the CNN backbone and fine-tuning the top model separately, our approach excluding detection can be optimized in an end-to-end fashion.

5 Experiments

5.1 Quantitative Analysis on the NBA Dataset

We first evaluate our approach on the new benchmark by compared with several variants and baseline methods. For this dataset, we sample $N^{\mathrm{f}} = 20$ frames from

Table 3. Ablation studies on NBA. Quan-$N^{\rm p}$ and Prob-$N^{\rm p}$ are two different strategies of deciding the number of input proposals, as described in Sect. 4.1. θ is the probability threshold used in Prob-$N^{\rm p}$, $N^{\rm f}$ is the number of input frames, and K^* denote the number of feature selected by our SAM

Type	Options of Our Approach	Acc (%)	Mean Acc (%)
Quan-$N^{\rm p}$	B1: w/o SAM ($N^{\rm p}=8$)	44.6	39.5
	B2: w/Spatial-SAM ($N^{\rm p}=14, K^{\rm p}=14$)	46.8	41.3
	B3: w/Spatial-SAM ($N^{\rm p}=14, K^{\rm p}=8$)	**50.3**	43.6
	B4: w/Spatial-SAM ($N^{\rm p}=8, K^{\rm p}=8$)	47.4	41.4
	B5: w/ Spatial-SAM ($N^{\rm p}=14, K^{\rm p}=8$) + w/ Temporal-SAM ($N^{\rm f}=20, K^{\rm f}=6$)	49.1	**47.5**
Prob-$N^{\rm p}$	B6: w/Spatial-SAM ($\theta=0.9, K^{\rm p}=8$)	47.5	42.6

the entire video clip as the input for all methods and train them with a batch-size of 16. Because of the fast speed of activities in this benchmark, we do not track pre-detected proposals over frames. Moreover, we do not apply any strategy to handle the class-imbalance issue in this benchmark.

Ablation Study. To evaluate the effectiveness of SAM, different variants of our approach are performed on NBA and the results are reported in Table 3. B1 that does not use the proposed SAM achieves the base accuracy of 44.6% and 39.5% on Acc and Mean Acc, respectively. Compared with B1, B2 that employs the SAM to build relational embeddings among $N^{\rm p}=14$ proposals in the spatial domain but does not prune useless ones, only obtains 2.2% and 1.8% improvement on Acc and Mean Acc. Similarly, B4 that directly adapts SAM to generate relational representations from $N^{\rm p}=8$ proposals has small improvement. However, by selecting $K^{\rm p}=8$ persons from $N^{\rm p}=14$ proposals and modeling relationships among them, B3 improves Acc and Mean Acc by 5.7% and 4.1%, respectively, compared with B1. Moreover, our Quan-$N^{\rm p}$ based approach (B6) suffering an uncertain number of proposals also gets a satisfactory Mean Acc of 42.6%. Based on B3, B5 obtains the best Mean Acc by applying SAM on the temporal domain, suggesting that the ability of feature selection of SAM can also be used to capture the long temporal structure. The further analysis on the parameters of N^* and K^* are present in Sect. 5.2.

Comparison with the Baselines. We also compare our approach with recent works in the video classification domain, including TSN [41], TRN [47], I3D [6], I3D+NLN [43]. To be fair, all these baseline methods are built on ResNet-18 and the input modality is RGB. The results are reported in Table 4. We see that "Ours w/o SAM" is hardly improved or worse due to noisy input (irrelevant pre-detected proposals), compared with methods ("TSN" and "TRN") only using frame-level information. By introducing SAM to select discriminative proposals in the spatial domain, "Ours w/SAM (S)" achieves significant improvement on Mean Acc but still overfits on some classes. As expected, "Ours w/SAM (S+T)" outperforms all baselines by a good margin and obtained the best Mean Acc by

Table 4. Comparison on NBA. "Ours w/o SAM", "Ours w/SAM (S)", and "Ours w/SAM (S+T)" are the B1, B3, and B5 reported in Table 3, respectively

Group activity	Frame classification				Our approach		
	TSN [41]	TRN [47]	I3D [6]	I3D+NLN [43]	w/o SAM	w/SAM (S)	w/SAM (S+T)
2p-succ	38.7	44.8	33.1	22.1	46.6	39.3	**47.2**
2p-fail.-off	30.8	23.4	14.0	20.6	28.0	25.2	**42.1**
2p-fail.-def	49.1	50.0	39.3	45.3	49.6	**53.4**	48.3
2p-layup-succ	52.9	54.7	50.6	48.8	44.2	**57.6**	53.5
2p-layup-fail.-off	10.1	22.5	22.5	22.5	20.2	19.1	**32.6**
2p-layup-fail.-def	44.6	46.5	43.3	31.2	44.6	51.6	**59.9**
3p-succ	39.3	37.7	31.1	26.8	39.9	**41.0**	30.1
3p-fail.-off	10.8	20.5	4.8	12.0	24.1	38.6	**55.4**
3p-fail.-def	63.9	62.8	55.3	61.7	58.6	**66.9**	58.1
Mean acc (%)	37.8	40.3	32.7	32.3	39.5	43.6	**47.5**

applying SAM to both the spatial and temporal domains. Nevertheless, "Ours w/SAM (S+T)" performs poorly on the activity of "3p-succ." that does not have long-term temporal structure. Moreover, "I3D" and "I3D+NLN" that depend on dense frames perform poorly on this benchmark.

5.2 Qualitative Analysis on the NBA Dataset

Analysis of parameters. We first diagnose N, the number of nodes of the dense relation graph. Limited by the computation resource, we only analyze the N^{p} of Spatial-SAM and it indicates how many pre-detected proposals should be fed into our approach. It can be decided by two strategies as mentioned in Sect. 4.1. Thus, we first run our Quan-N^{p} based approach on the NBA dataset by fixing $K^{\mathrm{p}} = 8$ and changing N^{p} from 8 to 64 with a step of 4. As shown in Fig. 3(a), although N^{p} is increasing, the performance of our approach has been persistently higher than the baseline. Moreover, we also conduct our Prob-N^{p} based approach on NBA by using fixed $K^{\mathrm{p}} = 8$ and adjust θ from 0.05 to 0.95 with a mini-step of 0.05. As shown in Fig. 3(b), our approach can achieve promising results when $\theta \geq 0.3$ and is more likely to get high performance when θ around 0.4. Overall, our Spatial-SAM is not sensitive to N^{p} whether decided by Quan-N^{p} or Prob-N^{p}. We also diagnose K, the number of nodes of the sparse relation graph, and it decides how many feature nodes need to be selected for modeling. As shown in Fig. 3(c), the performance of Spatial-SAM maintains over the baseline and it obtains the best result at $K^{\mathrm{p}} = 1$. Therefore, we hold that Spatial-SAM is not sensitive to K^{p}. By contrast, the performance of Temporal-SAM cannot get satisfactory performance when the K^{f} is too small or large, due to the different temporal length of activities in NBA. However, our approach with Temporal-SAM significantly improves Mean Acc when $4 < K^{\mathrm{f}} < 10$.

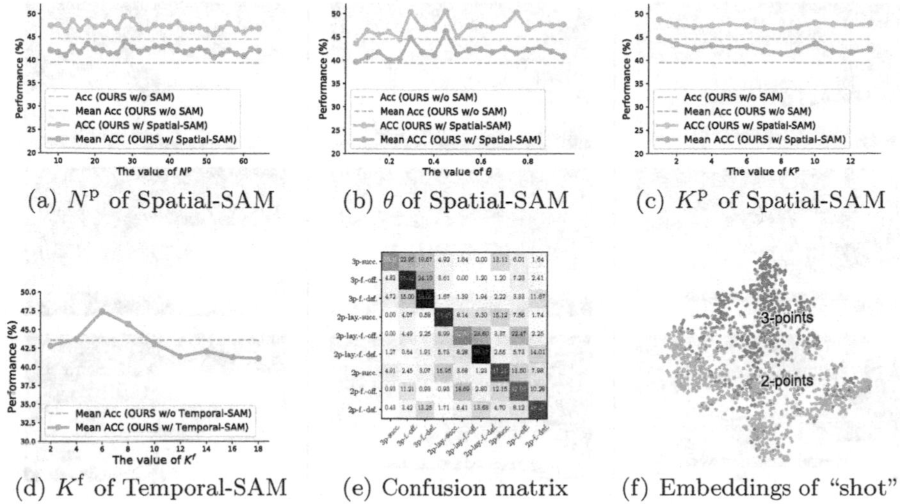

(a) N^{p} of Spatial-SAM (b) θ of Spatial-SAM (c) K^{p} of Spatial-SAM

(d) K^{f} of Temporal-SAM (e) Confusion matrix (f) Embeddings of "shot"

Fig. 3. (a)–(d) Experimental analysis on parameters. (e) The confusion matrix of OURS w/Spatial-SAM and Temporal-SAM. (f) t-SNE visualization of embeddings of 2/3-points based activities. These experiments are carried out on the NBA dataset

Confusion Matrix. To figure out the confusion between each activity in the NBA dataset, we report the confusion matrix of our approach in Fig. 3(d). We can see that the activities involving "defense" and "offense" are easily confused, due to the class-imbalance issue between these two kinds of activities. However, it is relatively easy to distinguish 2-points and 3-points, as embeddings shown in Fig. 3(f). Because 3-point players usually jump to shot behind the 3-point line without blocking. By contrast, 2-point players are often blocked by others.

Visualization. To further understand the discriminative learning process of SAM, we show some typical cases of NBA in Fig. 4. The group activities in NBA have long-term temporal information, thus top-K proposals vary over time. Take the rightmost one as an example, a "3p-failure-defense" has 3 parts: (1) preparation, (2) shooting, (3) defensive rebound. For (1) and (2), the players controlling the ball are the key instances, but for (3), the players that quickly turn back are the key instances. It is not hard to find that SAM aims at focusing on the players who are controlling the basketball or close to it and these people can form a group semantically.

5.3 Quantitative Analysis on Volleyball Dataset

We also evaluate our approach on the existing largest and most widely-used benchmark, Volleyball Dataset (VD) [18] consisting of 4830 volleyball game sequences. The middle frame of each sequence is labeled with 9 action labels (not used in our approach) and 8 group activity labels. However, we find that there are many wrong annotations between "pass" and "set", which seriously affects the evaluation for models, thus we merged them into "pass-set". To be fair, we follow the train/test split provided in [18] and sample $N^{\mathrm{f}} = 3$ from the

Fig. 4. A visualization of the top-K proposals focused by SAM over time on the NBA dataset, where $K = 3$. Each column shows three different frames of an activity. We highlight the top-K players (in cyan boxes) at three time steps of different activities. The people in red boxes are treat as noisy data by our model (Color figure online)

Table 5. Results on VD. (a) Ablation studies. (b) Comparison with SOTA. "Ours" represents "Ours w/Spatial-SAM" with $N^{\mathrm{p}} = 16$ and $K^{\mathrm{p}} = 12$ based on Quan-N^{p}

(a)

Type	Our Approach	Acc (%)
Quan-N^{p}	B1: w/o SAM	91.5
	B2: w/Spatial-SAM (N^{p}=16, K^{p}=16)	92.4
	B3: w/Spatial-SAM (N^{p}=16, K^{p}=12)	**93.1**
	B4: w/Spatial-SAM (N^{p}=12, K^{p}=12)	91.9
Prob-N^{p}	B5: w/Spatial-SAM ($\theta = 0.9, K^{\mathrm{p}} = 12$)	**93.1**

(b)

Method	Supervision	Acc (%)
HTDM	Fully	89.7
PCTDM	Fully	90.2
CCGL	Fully	91.0
StagNet	Fully	90.0
‡ARG	Fully	**94.0**
‡ARG	Weakly	90.7
†Ours	Weakly	93.1
‡Ours	Weakly	**94.0**

† ResNet-18
‡ Inception-v3

video clip similar to [44]. Because the activities in VD always occur in the middle frame, we do not apply our SAM to the temporal domain for this benchmark.

Ablation Study. We also perform ablation study on VD and the experimental results are reported in Table 5(a). All these variants do not use person-level supervision information (bounding boxes and action labels) provided by [18] and are built on ResNet-18. Compared with the baseline method B1, our B2 and B3 that only apply SAM to generate relational embedding for proposals but do not prune the irrelevant ones, only improve the accuracy by 0.9% and 0.4%, respectively. Besides, by using SAM to build relationships among $N = 16$ proposals and choosing $K = 12$ effective proposals from them, B3 and B5 improve the accuracy of 1.6% based on whether Quan-N or Prob-N. This observation indicates again that useless proposals will affect the weakly-supervised training and SAM is effective for pruning them.

Comparison with the State-of-the-art. Referring to [42], we report the results of HTDM [18,19], PCTDM [45], CCGL [39], and StagNet [31] by computing their corresponding confusion matrices. We reproduce the state-of-the-art method, ARG [44], with fully-supervised and weakly-supervised settings, respectively. As shown in Table 5(b), our weakly-supervised approach with the backbone of ResNet-18 is superior to almost all previous fully-supervised methods, except ARG that is built on Inception-v3. But our approach goes far beyond ARG under the weakly-supervised setting, suggesting that useless pre-detected proposals seriously affect the construction of relation graphs in ARG. Furthermore, our approach with Inception-v3 can achieve the best performance.

6 Conclusions

In this work, we introduce a weakly-supervised setting for GAR, which is more practical and friendly for real-world scenarios. To investigate this problem, we collect a larger and more challenging dataset from high-resolution basketball videos of NBA. Furthermore, we propose a social adaptive module (SAM) for assisting the weakly-supervised training by leveraging the social assumption that discriminative features are highly related to each other. As demonstrated on two datasets, our approach achieves state-of-the-art results while it can attend to key proposals/frames automatically.

This work reveals that social relationship among visual entities is helpful for high-level semantic understanding. We look forward to applying this method to more challenging scenarios, in particular, for mining semantic knowledge from weakly-annotated or un-annotated visual data.

Acknowledgements. This work was supported by the National Key Research and Development Program of China under Grant 2018AAA0102002, the National Natural Science Foundation of China under Grants 61732007, 61702265, and 61932020.

References

1. Amer, M.R., Xie, D., Zhao, M., Todorovic, S., Zhu, S.-C.: Cost-sensitive top-down/bottom-up inference for multiscale activity recognition. In: Fitzgibbon, A., Lazebnik, S., Perona, P., Sato, Y., Schmid, C. (eds.) ECCV 2012. LNCS, vol. 7575, pp. 187–200. Springer, Heidelberg (2012). https://doi.org/10.1007/978-3-642-33765-9_14

2. Amer, M.R., Lei, P., Todorovic, S.: HiRF: hierarchical random field for collective activity recognition in videos. In: Fleet, D., Pajdla, T., Schiele, B., Tuytelaars, T. (eds.) ECCV 2014. LNCS, vol. 8694, pp. 572–585. Springer, Cham (2014). https://doi.org/10.1007/978-3-319-10599-4_37
3. Azar, S.M., Atigh, M.G., Nickabadi, A., Alahi, A.: Convolutional relational machine for group activity recognition. In: CVPR (2019)
4. Bagautdinov, T., Alahi, A., Fleuret, F., Fua, P., Savarese, S.: Social scene understanding: end-to-end multi-person action localization and collective activity recognition. In: CVPR (2017)
5. Cadene, R., Ben-Younes, H., Cord, M., Thome, N.: MUREL: multimodal relational reasoning for visual question answering. In: CVPR (2019)
6. Carreira, J., Zisserman, A.: Quo vadis, action recognition? A new model and the kinetics dataset. In: CVPR (2017)
7. Chen, K., et al.: MMDetection: Open MMLab detection toolbox and benchmark. arXiv preprint arXiv:1906.07155 (2019)
8. Chen, X., Gupta, A.: Spatial memory for context reasoning in object detection. In: ICCV (2017)
9. Choi, W., Savarese, S.: A unified framework for multi-target tracking and collective activity recognition. In: Fitzgibbon, A., Lazebnik, S., Perona, P., Sato, Y., Schmid, C. (eds.) ECCV 2012. LNCS, vol. 7575, pp. 215–230. Springer, Heidelberg (2012). https://doi.org/10.1007/978-3-642-33765-9_16
10. Choi, W., Shahid, K., Savarese, S.: What are they doing?: collective activity classification using spatio-temporal relationship among people. In: ICCV Workshops (2009)
11. Choi, W., Shahid, K., Savarese, S.: Learning context for collective activity recognition. In: CVPR (2011)
12. Danelljan, M., Häger, G., Khan, F., Felsberg, M.: Accurate scale estimation for robust visual tracking. In: BMVC (2014)
13. Deng, Z., Vahdat, A., Hu, H., Mori, G.: Structure inference machines: recurrent neural networks for analyzing relations in group activity recognition. In: CVPR (2016)
14. Gan, C., Wang, N., Yang, Y., Yeung, D.Y., Hauptmann, A.G.: DevNet: a deep event network for multimedia event detection and evidence recounting. In: CVPR (2015)
15. He, K., Gkioxari, G., Dollár, P., Girshick, R.: Mask R-CNN. In: ICCV (2017)
16. He, K., Zhang, X., Ren, S., Sun, J.: Deep residual learning for image recognition. In: CVPR (2016)
17. Hu, H., Gu, J., Zhang, Z., Dai, J., Wei, Y.: Relation networks for object detection. In: CVPR (2018)
18. Ibrahim, M.S., Muralidharan, S., Deng, Z., Vahdat, A., Mori, G.: A hierarchical deep temporal model for group activity recognition. In: CVPR (2016)
19. Ibrahim, M.S., Muralidharan, S., Deng, Z., Vahdat, A., Mori, G.: Hierarchical deep temporal models for group activity recognition. arXiv preprint arXiv:1607.02643 (2016)
20. Jang, Y., Song, Y., Yu, Y., Kim, Y., Kim, G.: TGIF-QA: toward spatio-temporal reasoning in visual question answering. In: CVPR (2017)
21. Johnson, J., Krishna, R., Stark, M., Li, L.J., Shamma, D., Bernstein, M., Fei-Fei, L.: Image retrieval using scene graphs. In: CVPR (2015)
22. King, D.E.: Dlib-ml: a machine learning toolkit. JMLR **10**, 1755–1758 (2009)
23. Kuehne, H., Jhuang, H., Garrote, E., Poggio, T., Serre, T.: HMDB: a large video database for human motion recognition. In: ICCV (2011)

24. Lan, T., Sigal, L., Mori, G.: Social roles in hierarchical models for human activity recognition. In: CVPR (2012)
25. Lan, T., Wang, Y., Yang, W., Robinovitch, S.N., Mori, G.: Discriminative latent models for recognizing contextual group activities. TPAMI **34**, 1549–1562 (2012)
26. Li, Y., Ouyang, W., Zhou, B., Wang, K., Wang, X.: Scene graph generation from objects, phrases and caption regions. In: ICCV (2017)
27. Lin, J., Gan, C., Han, S.: TSM: temporal shift module for efficient video understanding. In: ICCV (2019)
28. Lin, T.-Y., et al.: Microsoft COCO: common objects in context. In: Fleet, D., Pajdla, T., Schiele, B., Tuytelaars, T. (eds.) ECCV 2014. LNCS, vol. 8693, pp. 740–755. Springer, Cham (2014). https://doi.org/10.1007/978-3-319-10602-1_48
29. Liu, X., Lee, J.Y., Jin, H.: Learning video representations from correspondence proposals. In: CVPR (2019)
30. Long, X., Gan, C., De Melo, G., Wu, J., Liu, X., Wen, S.: Attention clusters: purely attention based local feature integration for video classification. In: CVPR (2018)
31. Qi, M., Qin, J., Li, A., Wang, Y., Luo, J., Van Gool, L.: stagNet: an attentive semantic RNN for group activity recognition. In: Ferrari, V., Hebert, M., Sminchisescu, C., Weiss, Y. (eds.) ECCV 2018. LNCS, vol. 11214, pp. 104–120. Springer, Cham (2018). https://doi.org/10.1007/978-3-030-01249-6_7
32. Ramanathan, V., Huang, J., Abu-El-Haija, S., Gorban, A., Murphy, K., Fei-Fei, L.: Detecting events and key actors in multi-person videos. In: CVPR, pp. 3043–3053 (2016)
33. Ren, S., He, K., Girshick, R., Sun, J.: Faster R-CNN: towards real-time object detection with region proposal networks. In: NeurIPS (2015)
34. Santoro, A., et al.: A simple neural network module for relational reasoning. In: NeurIPS (2017)
35. Shu, T., Todorovic, S., Zhu, S.C.: CERN: confidence-energy recurrent network for group activity recognition. In: CVPR (2017)
36. Shu, T., Xie, D., Rothrock, B., Todorovic, S., Chun Zhu, S.: Joint inference of groups, events and human roles in aerial videos. In: CVPR (2015)
37. Smith, R.: An overview of the Tesseract OCR engine. In: ICDAR (2007)
38. Soomro, K., Zamir, A.R., Shah, M.: Ucf101: a dataset of 101 human actions classes from videos in the wild. arXiv preprint arXiv:1212.0402 (2012)
39. Tang, J., Shu, X., Yan, R., Zhang, L.: Coherence constrained graph LSTM for group activity recognition. TPAMI (2019)
40. Tang, Y., Wang, Z., Li, P., Lu, J., Yang, M., Zhou, J.: Mining semantics-preserving attention for group activity recognition. In: ACM MM (2018)
41. Wang, L., et al.: Temporal segment networks: towards good practices for deep action recognition. In: Leibe, B., Matas, J., Sebe, N., Welling, M. (eds.) ECCV 2016. LNCS, vol. 9912, pp. 20–36. Springer, Cham (2016). https://doi.org/10.1007/978-3-319-46484-8_2
42. Wang, M., Ni, B., Yang, X.: Recurrent modeling of interaction context for collective activity recognition. In: CVPR (2017)
43. Wang, X., Girshick, R., Gupta, A., He, K.: Non-local neural networks. In: CVPR (2018)
44. Wu, J., Wang, L., Wang, L., Guo, J., Wu, G.: Learning actor relation graphs for group activity recognition. In: CVPR (2019)
45. Yan, R., Tang, J., Shu, X., Li, Z., Tian, Q.: Participation-contributed temporal dynamic model for group activity recognition. In: ACM MM (2018)

46. Yang, J., Lu, J., Lee, S., Batra, D., Parikh, D.: Graph R-CNN for scene graph generation. In: Ferrari, V., Hebert, M., Sminchisescu, C., Weiss, Y. (eds.) ECCV 2018. LNCS, vol. 11205, pp. 690–706. Springer, Cham (2018). https://doi.org/10.1007/978-3-030-01246-5_41
47. Zhou, B., Andonian, A., Oliva, A., Torralba, A.: Temporal relational reasoning in videos. In: Ferrari, V., Hebert, M., Sminchisescu, C., Weiss, Y. (eds.) ECCV 2018. LNCS, vol. 11205, pp. 831–846. Springer, Cham (2018). https://doi.org/10.1007/978-3-030-01246-5_49

RGB-D Salient Object Detection with Cross-Modality Modulation and Selection

Chongyi Li[1], Runmin Cong[2(✉)], Yongri Piao[3], Qianqian Xu[4], and Chen Change Loy[1]

[1] Nanyang Technological University, Singapore, Singapore
lichongyi25@gmail.com, ccloy@ntu.edu.sg
[2] Beijing Jiaotong University, Beijing, China
rmcong@bjtu.edu.cn
[3] Dalian University of Technology, Dalian, China
yrpiao@dlut.edu.cn
[4] Institute of Computing Technology, Chinese Academy of Sciences, Beijing, China
xuqianqian@ict.ac.cn

https://li-chongyi.github.io/Proj_ECCV20

Abstract. We present an effective method to progressively integrate and refine the cross-modality complementarities for RGB-D salient object detection (SOD). The proposed network mainly solves two challenging issues: 1) how to effectively integrate the complementary information from RGB image and its corresponding depth map, and 2) how to adaptively select more saliency-related features. *First*, we propose a cross-modality feature modulation (cmFM) module to enhance feature representations by taking the depth features as prior, which models the complementary relations of RGB-D data. *Second*, we propose an adaptive feature selection (AFS) module to select saliency-related features and suppress the inferior ones. The AFS module exploits multi-modality spatial feature fusion with the self-modality and cross-modality interdependencies of channel features are considered. *Third*, we employ a saliency-guided position-edge attention (sg-PEA) module to encourage our network to focus more on saliency-related regions. The above modules as a whole, called cmMS block, facilitates the refinement of saliency features in a coarse-to-fine fashion. Coupled with a bottom-up inference, the refined saliency features enable accurate and edge-preserving SOD. Extensive experiments demonstrate that our network outperforms state-of-the-art saliency detectors on six popular RGB-D SOD benchmarks.

C. Li and R. Cong—equal contribution.

Electronic supplementary material The online version of this chapter (https://doi.org/10.1007/978-3-030-58598-3_14) contains supplementary material, which is available to authorized users.

A. Vedaldi et al. (Eds.): ECCV 2020, LNCS 12353, pp. 225–241, 2020.
https://doi.org/10.1007/978-3-030-58598-3_14

1 Introduction

Depth maps provide useful cues such as depth of field, shape, and boundary to complement RGB images for SOD [3,4,6,32,33,39,46]. However, depth maps are inherently noisy and the cues provided can be inconsistent or misaligned with the RGB modality. The issues make designing an RGB-D algorithm challenging. Contemporary RGB-D SOD detectors, CPFP [46] (Fig. 1(d)) and A2dele [33] (Fig. 1(e)), could still miss salient objects due to cluttered backgrounds or yield incomplete or serrated boundaries of saliency maps.

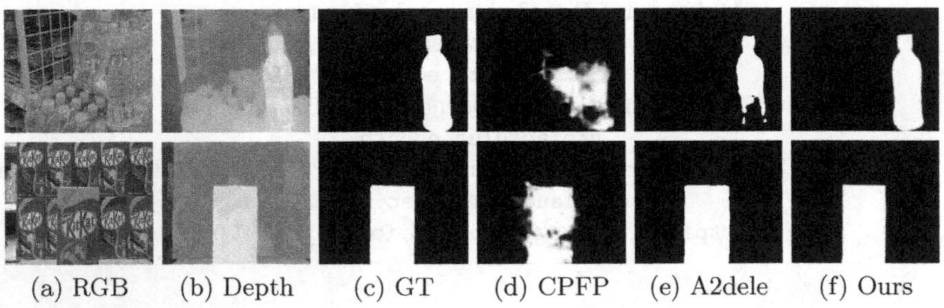

(a) RGB (b) Depth (c) GT (d) CPFP (e) A2dele (f) Ours

Fig. 1. Two motivating examples of SOD. (a)–(c) represent the input images, the corresponding depth maps, and the ground truth (GT), respectively. (d) and (e) are the results of state-of-the-art RGB-D SOD detectors CPFP (**CVPR'19**) [46] and A2dele (**CVPR'20**) [33], respectively. (f) are our results. *Compared with the latest CPFP and A2dele, our method can yield more complete, sharp, and edge-preserving saliency detection results by effectively integrating cross-modality complementaries and adaptively selecting saliency-related features.*

In this work, we consider addressing the aforementioned problem through more careful investigation on the integration of cross-modality complementaries from RGB image and depth map as well as the selection of saliency-related features. To this end, we present an effective network that achieves complete, sharp, and edge-preserving saliency detection, as shown in Fig. 1(f).

First, we propose a cross-modality feature modulation (cmFM) module that enhances RGB feature representations by taking the corresponding depth features as prior. This is in contrast to popular strategies that perform either input fusion [30], early fusion [19], or late fusion [18], that crudely concatenate or add the multi-modality information. The proposed modulation design enables effective integration of multi-modality information through feature transformation, distinctly models the inseparable cross-modality relations, and reduces the interference caused by the inherent inconsistency of multi-modality data.

Second, we devise an adaptive feature selection (AFS) module that highlights the importance of different channel features in self- and cross-modalities, while fusing multi-modality spatial features in a gated manner. This is different from previous RGB-D SOD algorithms [3–6,22,46] that treat channel features from different modalities equally and independently. Relaxing such assumptions allows our method to adaptively select more saliency-related features and suppress the inferior ones from both spatial features and channel features. It also mitigates the negative influence of poorly captured depth maps. Hence, our network equips additional flexibility in dealing with different information. We also emphasize the saliency-related positions and edges by introducing a saliency-guided position-edge attention (sg-PEA) module, which collects its attention weights from the predicted saliency maps and saliency edge maps.

Our method is unique in that the feature modulation and attention mechanism are closely coupled in a coarse-to-fine manner. Specifically, fusion is first performed by the cmFM module to provide rich features representations. Coordinated with our AFS module, saliency-related features are emphasized while redundant features are suppressed. The saliency-related features are further refined by the sg-PEA module. A careful design to place the cmFM, AFS, and sg-PEA modules allows the cross-modality complementarities to go through modulation, selection, and refinement in a coarse-to-fine fashion, providing our network with precise saliency features. Coupled with a bottom-up inference, the precise saliency features enable us to perform more accurate and robust SOD.

Contributions. We present an effective approach for RGB-D SOD. Cross-modality complementarities are effectively integrated and saliency-related features are adaptively selected. This is made possible by designing a coarse-to-fine fusion that consists of 1) a cross-modality feature modulation module that enhances RGB feature representations by taking the corresponding depth features as prior, and 2) an adaptive feature selection module that progressively emphasizes the importance of channel features in self- and cross-modalities while fusing the significant multi-modality spatial features. Our method consistently outperforms state-of-the-art SOD methods on six popular RGB-D SOD benchmarks.

2 Related Work

Salient Object Detection. SOD methods range from bottom-up [25,29,42] to top-down models [14,17,19,26,34,47]. In addition to the color appearance, depth maps can provide useful cues such as depth of field, shape, and boundary. The depth map is implicitly used in the unsupervised methods [9,10,21,27,30,37,48]. Whereas for the supervised methods, the discriminative and complementary features are learned from RGB-D images [3–6,12,16,18,22,32,35,43,44,46]. Our work differs from recent works [12,16,32,33,43,44,46], mainly in two aspects: 1) we use depth features as prior to learn optimal affine transformation parameters, which can flexibly modulate multi-level RGB features, and 2) we consider both self-modality and cross-modality channel features as well as multi-modality spatial features, thus effectively capturing relations among different modalities.

Feature Modulation. Inspired by FiLM [31] that first applies linear feature modulation for visual reasoning, feature modulation has been used in few-shot learning [28] and image super-resolution [40]. In our studies, we modulate the

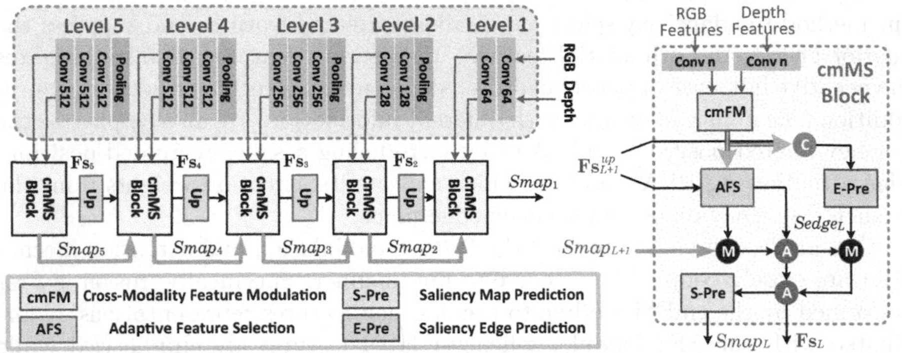

Fig. 2. Overview of our network architecture. The inputs are the RGB image and its depth map. The cmMS block consists of a cmFM module, an AFS module, and an sg-PEA module. Here, the sg-PEA module further contains an S-Pre unit and an E-Pre unit. 'Conv n' represents the convolutional layer that outputs n feature maps, where n is the half number of input feature maps. 'A', 'M', and 'C' represent element-wise addition, element-wise multiplication, and concatenation along with the channel dimension, respectively. 'Up' represents the up-sampling block. Pink line indicates $2\times$ linear interpolation. $\mathbf{Fs}$ represent the refined features after the cmMS block while $\mathbf{Fs}^{up}$ are the up-sampled $\mathbf{Fs}$ by the 'Up' block. In this figure, each convolutional layer is followed by the ReLU activation. Our network finally produces five saliency maps ($Smap_L$) and five saliency edge maps ($Sedge_L$) with the resolutions, ranging from 14×14 to 224×224 by a scale of 2. L indicates the level. We treat $Smap_1$ as the final result.

multi-level feature representations conditioned on the corresponding depth features. Besides, we design the cross-modality feature modulation in a pixel-wise manner, which provides elaborate and fine-grained control to the features.

Attention Mechanism. Attention mechanism is increasingly applied in diverse forms such as spatial attention [7], dual-attention [15], self-attention [38], multi-level attention [41], and channel attention [45]. In contrast, we employ the attention mechanism in our adaptive feature selection module, which explores the interdependencies of channel features in the self- and cross-modalities while fusing the significant multi-modality spatial features in a gated manner.

3 Our Method

We first present an overview of our network architecture. Then, we describe the key components including the cross-modality feature modulation module, adaptive feature selection module, and saliency-guided position-edge attention module. At last, we introduce the loss functions.

3.1 Overview of Network Architecture

The overview of our network architecture is illustrated in Fig. 2. After the top-down features extraction from VGG-16 backbone [36], the multi-level RGB features and depth features are fed to a convolutional layer for halving the number of feature maps, respectively. Then, the dimension reduced RGB-D features are forwarded to the corresponding cmMS block. In each cmMS block, the RGB-D features go through cmFM module, AFS module, and sg-PEA module for feature modulation, selection, and refinement, respectively. Specifically, we introduce modulated features by using our proposed cross-modality feature modulation (cmFM) module. The purpose of cmFM module is to effectively integrate the cross-modality complementarities in a flexible and trainable fashion. After that, RGB features, depth features, modulated features, and up-sampled features from the higher level (if any) are independently forwarded to our proposed adaptive feature selection (AFS) module for selectively emphasizing the informative channel features and fusing the significant spatial features. The AFS module models the relations between different levels and accelerates task-oriented feature integration. Meanwhile, the concatenation of RGB features, depth features, modulated features, and up-sampled features (if any) is applied to predict the saliency edge map via a saliency edge prediction (E-Pre) unit. Then, with the saliency map up-sampled from the higher level (if any) and saliency edge map, we highlight the saliency position and edge regions of the features after the AFS module. After that, we predict the saliency map in the current level via a saliency map prediction (S-Pre) unit by using the refined features. At last, in the bottom-up inference, we progressively integrate and highlight multi-level features to predict the fine-scaled saliency map (*i.e.*, the $Smap_1$ in Fig. 2). We adopt 3×3 kernels for all convolutional layers in our network, except the cmFM module that employs the multi-scale convolutions to enlarge receptive field.

3.2 Cross-Modality Feature Modulation (cmFM)

Inspired by the unsupervised RGB-D SOD algorithms [10,13] which take the depth map as prior information to enrich the saliency cues, we propose a cmFM module conditioned on the depth features. The cmFM module learns pixel-wise affine transformation parameters from the conditioning depth features then modulates the corresponding RGB feature representations in each level of our network. The detailed cmFM module is illustrated in Fig. 3.

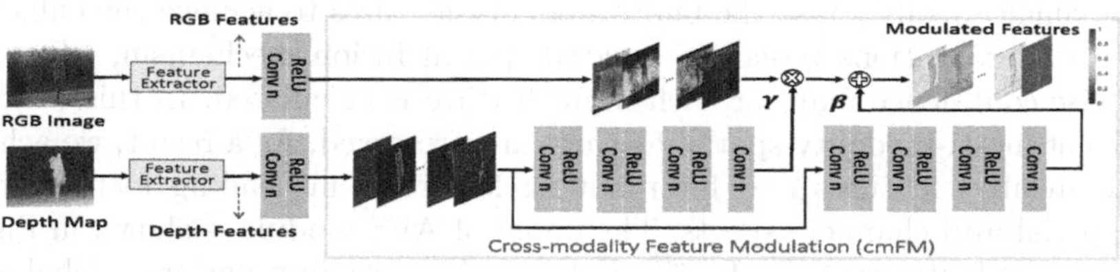

Fig. 3. The proposed cmFM module. For the estimation of both γ and β, the kernels of convolutional layers are 7×7, 5×5, 3×3, and 3×3. The feature extractor represents VGG-16 backbone. The feature maps are illustrated as heatmaps.

Given the dimension halved RGB features $\mathbf{F}_L^{rgb} \in \mathbb{R}^{N\times H\times W}$ and depth features $\mathbf{F}_L^{depth} \in \mathbb{R}^{N\times H\times W}$, the cmFM module learns a mapping function $\mathcal{M}$ conditioned on the depth features to yield a set of affine transformation parameters $(\gamma_L,\beta_L) \in \mathbb{R}^{N\times H\times W}$. Here, N is the number of feature maps; H and W are the height and width of the feature maps, respectively. It can be expressed as:

$$(\gamma_L, \beta_L) = \mathcal{M}(\mathbf{F}_L^{depth}), \tag{1}$$

where the superscript indicates the modality while the subscript represents the level. The mapping function $\mathcal{M}$ is built on two parallel stacked convolutional layers as shown in Fig. 3. With the estimated affine transformation parameters (γ_L, β_L), we conduct pixel-wise scaling and shifting on the RGB feature representations, which can be expressed as:

$$\mathbf{F}_L^{mod} = \mathbf{F}_L^{rgb} \otimes \gamma_L \oplus \beta_L, \tag{2}$$

where $\mathbf{F}_L^{mod}$ represent the modulated features; $\otimes$ and $\oplus$ indicate the element-wise multiplication and element-wise addition, respectively. As shown in Fig. 3, the cluttered backgrounds of RGB features become clear and the salient object is highlighted with the modulation of depth features.

3.3 Adaptive Feature Selection (AFS)

To make our network focus more on informative features, we propose an AFS module to progressively re-scale channel-wise features. Simultaneously, the AFS module fuses significant spatial features of multi-modalities. To be specific, we first explore the interdependencies of channel features in the self-modality, then further determine the relevance in the cross-modality. After squeezing by a convolutional layer that reduces the redundant features, we achieve the channel attention-on-channel attention features. Such a self-modality and cross-modality channel attention mechanism can model relations of the channel features among different modalities well and adaptively select the informative channel features. The advantages of our channel attention-on-channel attention than the conventional channel attention are verified in the ablation studies.

We simultaneously fuse the multi-modality features to achieve the enhanced feature representations based on a gated spatial fusion mechanism, where the pixel-wise confidence map for each input feature is calculated. In this way, the significant multi-modality spatial features are preserved. As a result, we achieve saliency-related features and filter out irrelevant or misleading features from both spatial and channel aspects. The detail of AFS module is shown in Fig. 4.

Given the features $(\mathbf{F}_L^{rgb}, \mathbf{F}_L^{depth}, \mathbf{F}_L^{mod}, \mathbf{Fs}_{L+1}^{up})$, we first perform global average pooling on each set of features separately, leading to a channel descriptor $\mathbf{z} \in \mathbb{R}^{N\times 1}$ for each one, which is an embedded global distribution of channel-wise feature responses. $\mathbf{Fs}_{L+1}^{up}$ indicate $2\times$ up-sampled features from the $L+1$ level by using the 'Up' block that consists of one $2\times$ linear interpolation followed by

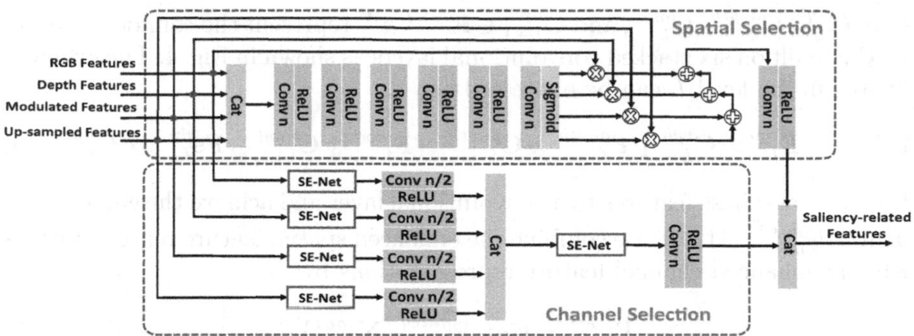

Fig. 4. The detail of AFS module. 'Cat' represents the concatenation operation. 'SE-Net' is the squeeze-and-excitation network.

two convolutional layers, where each convolutional layer is followed by the ReLU activation and outputs n feature maps. The k-th entry of $\mathbf{z}$ is expressed as:

$$z_k = \frac{1}{H \times W} \sum_{i}^{H} \sum_{j}^{W} \mathbf{F}_k(i, j), \tag{3}$$

where $k \in [1, N]$. Then, a self-gating mechanism is used to fully capture channel-wise dependencies $\mathbf{s} \in \mathbb{R}^{N \times 1}$:

$$\mathbf{s} = \sigma(\mathbf{W}_2 * (\delta(\mathbf{W}_1 * \mathbf{z}))), \tag{4}$$

where $\sigma(\cdot)$ represents the Sigmoid activation, $\delta(\cdot)$ represents the ReLU activation, $*$ denotes the convolution operation, and $\mathbf{W}_1$ and $\mathbf{W}_2$ are the weights of two fully-connected layers with their numbers of output channels being $\frac{N}{16}$ and N, respectively. At last, these weights are applied to each set of input features $\mathbf{F}$ to generate re-scaled features $\mathbf{U} \in \mathbb{R}^{N \times H \times W}$: $\mathbf{U} = \mathbf{F} \otimes \mathbf{s}$. This processing is mathematically expressed as an SE mapping function in this paper and can also be implemented by the squeeze-and-excitation network [20]. However, the highlighted channel features may become relatively useless among all channel attention results from multi-modalities.

To emphasize the informative channel features, we first halve the number of feature maps in each channel attention result by a convolutional layer, then concatenate them: $\mathbf{V}_L = Cat\{\mathbf{U}_L^{rgb}, \mathbf{U}_L^{depth}, \mathbf{U}_L^{mod}, \mathbf{Us}_{L+1}^{up}\}$. After that, we further explore the interdependencies of channel features by $\mathbf{Y}_L = SE(\mathbf{V}_L)$. We finally squeeze the number of channel features by a convolutional layer and achieve the results of channel attention-on-channel attention $\mathbf{Y}_L^{caca}$.

Meanwhile, we fuse the multi-modality input features to achieve enhanced spatial feature representations. First, the input features are concatenated $\mathbf{F}_L^{cat} = Cat\{\mathbf{F}_L^{rgb}, \mathbf{F}_L^{depth}, \mathbf{F}_L^{mod}, \mathbf{Fs}_{L+1}^{up}\}$, and fed to a plain CNN network (indicated as $\mathcal{G}$) to estimate their pixel-wise confidence maps:

$$(\mathbf{C}_L^{rgb}, \mathbf{C}_L^{depth}, \mathbf{C}_L^{mod}, \mathbf{C}_{L+1}^{up}) = \mathcal{G}(\mathbf{F}_L^{cat}), \tag{5}$$

where $\mathbf{C}_L^{rgb}$, $\mathbf{C}_L^{depth}$, $\mathbf{C}_L^{mod}$, and $\mathbf{C}_{L+1}^{up} \in \mathbb{R}^{N\times H\times W}$ represent the confidence maps. The $\mathcal{G}$ is built on six stacked convolutional layers as shown in Fig. 4. The achieved features in the level L can be expressed as:

$$\mathbf{F}_L^{gated} = \mathbf{F}_L^{rgb} \otimes \mathbf{C}_L^{rgb} \oplus \mathbf{F}_L^{depth} \otimes \mathbf{C}_L^{depth} \oplus \mathbf{F}_L^{mod} \otimes \mathbf{C}_L^{mod} \oplus \mathbf{Fs}_{L+1}^{up} \otimes \mathbf{C}_{L+1}^{up} \quad (6)$$

Then, we pass these features to a convolutional layer and achieve the gated fusion features $\mathbf{F}_L^{gated'}$. At last, we combine the enhanced spatial feature representations with the enhanced channel feature representations by:

$$\mathbf{F}_L^{AFS} = Cat\{\mathbf{F}_L^{gated'}, \mathbf{Y}_L^{caca}\}, \quad (7)$$

where the final results $\mathbf{F}_L^{AFS}$ enjoy the most informative features towards saliency detection, called saliency-related features in this paper. The visual examples are presented in Fig. 5. As shown, the saliency-related spatial features and channel features are preserved and highlighted.

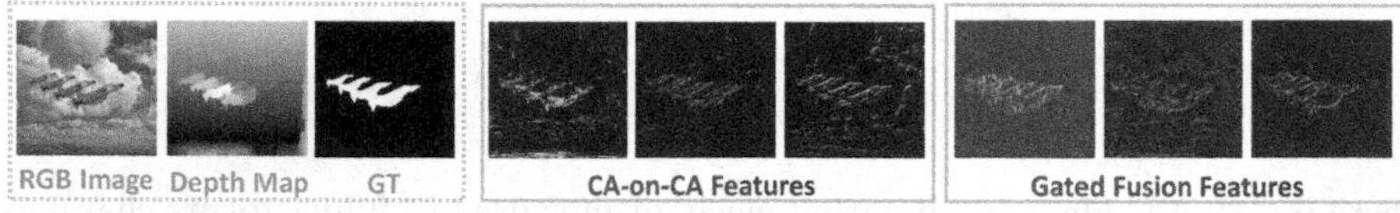

Fig. 5. Visual results of the intermediate features in our AFS module. 'CA-on-CA Features' indicates the features after our channel selection while 'Gated Fusion Features' represents the features after our spatial selection.

3.4 Saliency-Guided Position-Edge Attention (sg-PEA)

After selecting the saliency-related features, we also encourage the network to focus on those positions and edges most essential to the nature of salient objects. The benefits are illustrated as follows: 1) the saliency position attention can better locate the salient objects and accelerate the network convergence, and 2) the saliency edge attention can alleviate the problem of edge blur caused by the repeated pooling operations, which is vital for the pixel-wise saliency prediction.

To the end, we propose a saliency-guided position-edge attention (sg-PEA) module to locate and sharpen salient objects. The sg-PEA module further includes a saliency map prediction (S-Pre) unit and a saliency edge prediction (E-Pre) unit as shown in Fig. 2. The details are provided in Fig. 6, where S-Pre unit and E-Pre unit share the same structure, but different weights.

Position Attention. We employ the up-sampled saliency map from the higher level as the attention weights. Here, the up-sampling is implemented by the simple 2× linear interpolation. In our method, the saliency map is predicted by the S-Pre unit in each level in a supervised learning manner. The benefits of such a side supervision manner lie in four aspects: 1) the convolutional layers in each level have explicit objective towards saliency detection, 2) the side supervision can accelerate gradient back-propagation, 3) the predicted saliency map works as a guidance and can steer the convolutional layers of lower level to focus more on saliency positions in a low-computational manner, and 4) the multiple side outputs can provide diverse choices based on accuracy and inference speed. We provide more analysis on the side outputs in the supplementary material.

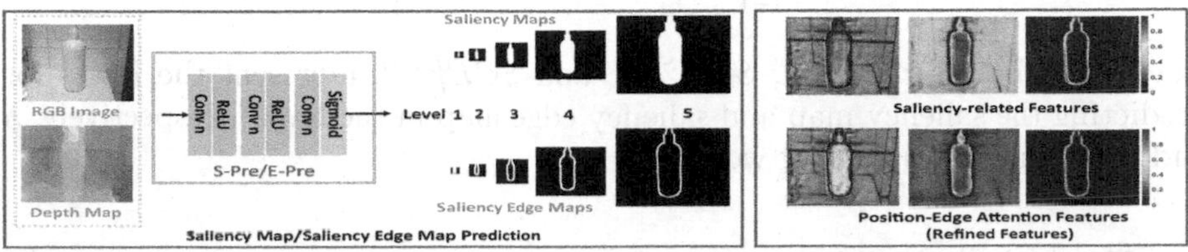

Fig. 6. Visual results of sg-PEA module. Left panel shows the structure of S-Pre/E-Pre unit, and the predicted saliency maps and saliency edge maps in different levels. Right panel shows the intermediate features before and after the sg-PEA module. After the sg-PEA module, the background of features are suppressed, and the edge and position details are assigned more focuses.

To be specific, with the saliency-related features $\mathbf{F}_L^{AFS}$ and the up-sampled saliency map $Smap_{L+1}^{up}$, the position attention results $\mathbf{F}_L^{poa}$ can be expressed as:

$$\mathbf{F}_L^{poa} = \mathbf{F}_L^{AFS} \oplus \mathbf{F}_L^{AFS} \otimes Smap_{L+1}^{up} \tag{8}$$

In contrast to treating all positions of saliency features equally, the position attention can quickly and efficiently employ the saliency property of higher level and enhance the saliency representations of the current level. To avoid gradient diffusion induced by successive attention (the values of feature maps are close to zero), we adopt an identical mapping manner as shown in Eq. (8).

Edge Attention. To obtain the edge attention weights, we first concatenate the RGB-D features, the modulated features, and up-sampled features, then forward them to the E-Pre unit to predict the saliency edge map in each level. The saliency edge maps, also estimated by supervised learning, can be used to emphasize the salient edges of the features by simple element-wise multiplication. For level L, the output features of edge attention can be expressed as:

$$\mathbf{Fs}_L = \mathbf{F}_L^{poa} \oplus \mathbf{F}_L^{poa} \otimes Sedge_L, \tag{9}$$

where $Sedge_L$ is the predicted saliency edge map in the level L. We call $\mathbf{Fs}_L$ as the refined features. At last, with the refined features, the final result (*i.e.*, $Smap_1$) with the same size as the input RGB image can be achieved in a bottom-up manner. In Fig. 6, we present the changes of features before and after sg-PEA module. As shown, the features increasingly focus on the saliency position and edge details, while the cluttered backgrounds are concurrently reduced.

3.5 Loss Function

We employ the standard cross-entropy (SCE) loss [1] to jointly optimize our network for the saliency prediction and saliency edge prediction:

$$Loss = \sum_{i=1}^{L}(\lambda_i SCE_i^{SPre} + \eta_i SCE_i^{EPre}), \tag{10}$$

where L indicates the level, SCE_i^{SPre} and SCE_i^{EPre} represent the losses for predicting the saliency map and saliency edge map in the level i, respectively. λ and η are the corresponding weights.

4 Experiments

4.1 Benchmark Datasets and Evaluation Metrics

We conduct experiments on six popular RGB-D SOD datasets, including **NJUD** [21] (1985 RGB-D images), **NLPR** [30] (1000 RGB-D images), **STEREO** [27] (797 RGB-D images), **LFSD** [24] (100 RGB-D images), **SSD** [23] (80 RGB-D images), and **DUT** [32] (1200 RGB-D images). For quantitative evaluations, Precision-Recall (P-R) curve, F-measure [2], MAE score [8], and S-measure [11] are employed. P-R curve depicts the different combinations of precision and recall scores; the closer the P-R curve is to (1, 1), the better the performance of the method. F-measure is the weighted harmonic mean of precision and recall; it is a comprehensive measurement, with a larger value indicating a better performance. MAE score measures the difference between the continuous saliency map and ground truth; a smaller value indicates a smaller gap hence better. S-measure calculates the structural similarity between the saliency map and ground truth; a larger value indicates a better performance. Additionally, we compare the model sizes of different methods in the supplementary material.

4.2 Implementation Details

We adopt the same training, validation, and testing sets as described in [32,33]. The ground truth of saliency edge map prediction is obtained by using the Canny edge detector on the saliency mask. We implement our network with TensorFlow on a PC with an Nvidia Tesla V100 GPU. During training, the batch size is set to 4, the filter weights of each layer are initialized by Gaussian distribution, and the bias is initialized as a constant. We use ADAM and fix the learning rate to $1e^{-4}$. The weight λ_1 for predicting the final saliency map is set to 1.2 while other weights are set to 1 in Eq. (10). For a pair of RGB-D images of size 224×224, the average runtime of our method is 0.037 s on the aforementioned PC.

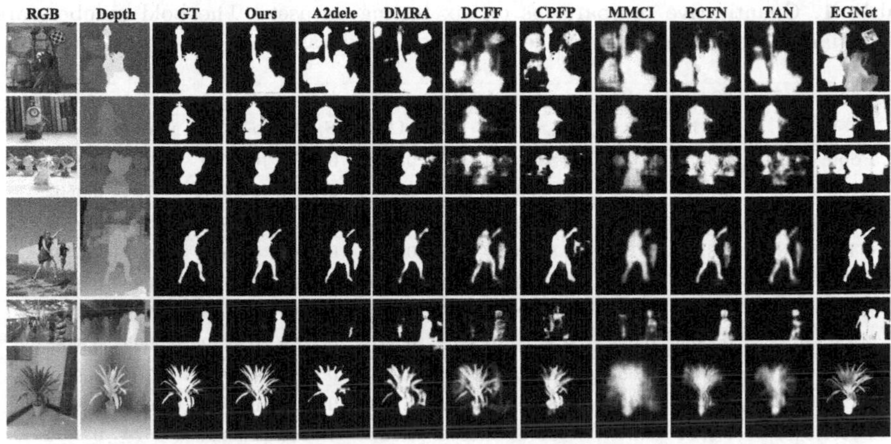

Fig. 7. Visual examples of different methods.

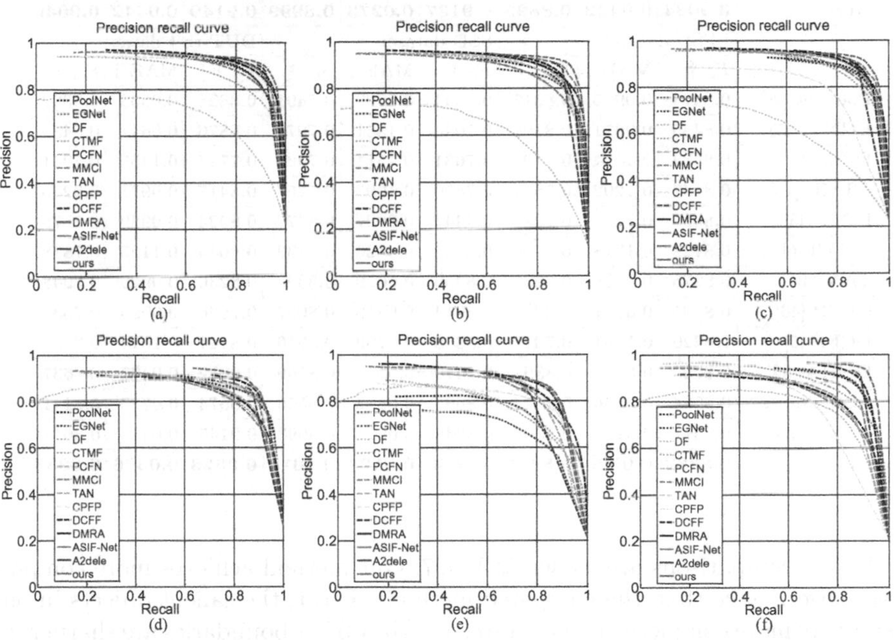

Fig. 8. P-R curves of different methods on the testing datasets. (a)-(f) correspond to STEREO, NLPR-Test, NJUD-Test, LFSD, SSD, and DUT-Test datasets.

4.3 Comparisons with State-of-the-art Methods

We compare our method with 12 state-of-the-art learning-based SOD methods, including two latest RGB-induced SOD methods (*i.e.*, PoolNet [26] and EGNet [47]), and ten RGB-D SOD methods (*i.e.*, DF [35], CTMF [18], MMCI [6], PCFN [3], TAN [4], CPFP [46], DCFF [5], DMRA [32], ASIF-Net [22], and A2dele [33]).

Table 1. Quantitative comparisons on six testing datasets. The bold numbers are performance of our method, also the best across all datasets

	STEREO dataset			NLPR-test dataset			NJUD-test dataset		
	$F_\beta \uparrow$	MAE $\downarrow$	$S_m \uparrow$	$F_\beta \uparrow$	MAE $\downarrow$	$S_m \uparrow$	$F_\beta \uparrow$	MAE $\downarrow$	$S_m \uparrow$
PoolNet [26]	0.8757	0.0655	0.8359	0.8627	0.0448	0.8573	0.8740	0.0676	0.8600
EGNet [47]	0.8717	0.0671	0.8363	0.8452	0.0504	0.8497	0.8667	0.0704	0.8562
DF [35]	0.6961	0.1738	0.6279	0.6480	0.1079	0.6710	0.6355	0.1987	0.5930
CTMF [18]	0.8265	0.1023	0.8230	0.8407	0.0561	0.8549	0.8572	0.0847	0.8493
PCFN [3]	0.8838	0.0606	0.8722	0.8635	0.0437	0.8592	0.8875	0.0592	0.8768
MMCI [6]	0.8610	0.0796	0.8504	0.8412	0.0591	0.8524	0.8684	0.0789	0.8588
TAN [4]	0.8865	0.0591	0.8701	0.8765	0.0410	0.8736	0.8882	0.0605	0.8785
CPFP [46]	0.8856	0.0537	0.8702	0.8878	0.0359	0.8760	0.7994	0.0794	0.7984
DCFF [5]	0.8867	0.0638	0.8706	0.8779	0.0439	0.8695	0.8910	0.0646	0.8774
DMRA [32]	0.8953	0.0474	0.8778	0.8870	0.0339	0.8646	0.9003	0.0529	0.8804
ASIF-Net [22]	0.8939	0.0493	0.8686	0.9002	0.0298	0.8844	0.9007	0.0471	0.8887
A2dele [33]	0.8997	0.0431	0.8713	0.8976	0.0285	0.8770	0.8939	0.0510	0.8704
Ours	**0.9084**	**0.0422**	**0.8895**	**0.9137**	**0.0273**	**0.8999**	**0.9149**	**0.0442**	**0.9040**
	LFSD dataset			SSD dataset			DUT-test dataset		
	$F_\beta \uparrow$	MAE $\downarrow$	$S_m \uparrow$	$F_\beta \uparrow$	MAE $\downarrow$	$S_m \uparrow$	$F_\beta \uparrow$	MAE $\downarrow$	$S_m \uparrow$
PoolNet [26]	0.8474	0.0945	0.8217	0.7644	0.1099	0.7491	0.8828	0.0669	0.8392
EGNet [47]	0.8445	0.0871	0.8300	0.7040	0.1351	0.7072	0.8876	0.0641	0.8439
DF [35]	0.8534	0.1424	0.7791	0.7631	0.1511	0.7422	0.7747	0.1455	0.7051
CTMF [18]	0.8147	0.1202	0.7883	0.7550	0.1003	0.7757	0.8417	0.0971	0.8226
PCFN [3]	0.8290	0.1118	0.7919	0.8447	0.0627	0.8427	0.8094	0.0999	0.7878
MMCI [6]	0.8128	0.1318	0.7793	0.8230	0.0820	0.8133	0.8044	0.1125	0.7818
TAN [4]	0.8275	0.1108	0.7935	0.8350	0.0629	0.8393	0.8236	0.0926	0.7948
CPFP [46]	0.8495	0.0881	0.8200	0.8014	0.0818	0.8067	0.7866	0.0995	0.7335
DCFF [5]	0.8220	0.1191	0.7917	0.8388	0.0769	0.8316	0.8141	0.1014	0.7835
DMRA [32]	0.8723	0.0754	0.8391	0.8579	0.0583	0.8569	0.9082	0.0477	0.8637
ASIF-Net [22]	0.8584	0.0896	0.8144	0.8633	0.0562	0.8566	0.8574	0.0725	0.8141
A2dele [33]	0.8577	0.0740	0.8306	0.8248	0.0691	0.8093	0.9145	0.0426	0.8611
Ours	**0.8882**	**0.0720**	**0.8465**	**0.8650**	**0.0524**	**0.8615**	**0.9328**	**0.0366**	**0.8853**

Visual comparisons are shown in Fig. 7. Our method achieves more competitive performance than the compared methods. **First**, the salient objects in our results are more complete and accurate, and the object boundaries are sharper. In the first image, only our method can accurately and completely detect the salient toy in front, while the competing methods incorrectly reserve the background regions (*e.g.*, Android doll and checkerboard). In the fourth image that comes with an unsatisfactory depth map, our method can still accurately locate salient target with a complete structure and clear boundaries. **Second**, our method preserves more details in the saliency result. In the sixth image, more details of plant leaves are better conserved. **Third**, our method can address some challenging cases, such as a complex background and small object. In the third image, the cat dolls in the back row are successfully suppressed by our method, the

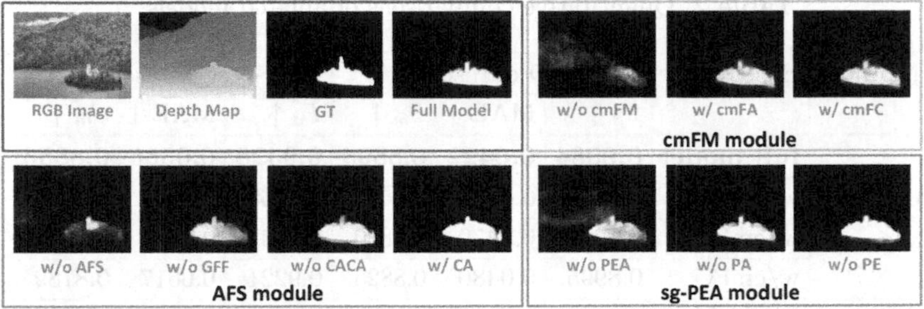

Fig. 9. Visual comparison with different baselines. **(1)** The baseline *w/o cmFM* represents our full model without the cmFM module (*i.e.*, no modulated features); and the baselines *w/cmFA* and *w/cmFC* refer to that the cmFM module is replaced by the cmFA or cmFC module (*i.e.*, the depth and RGB features are integrated by the element-wise addition or concatenation). **(2)** The baseline *w/o AFS* represents our full model without the AFS module (*i.e.*, the features after cmFM module are directly concatenated with the up-sampled saliency-related features); the baselines *w/o GFF* and *w/o CACA* correspond to removing the fused spatial features and the channel attention-on-channel attention features, respectively; and the baseline *w/CA* refers to that the AFS module is replaced by the conventional channel attention module [20]. **(3)** The baselines *w/o PEA*, *w/o PA*, and *w/o PE* correspond to our full model without the sg-PEA module, the position attention unit, and the edge attention unit, respectively.

detected salient boundaries are sharper, and the structure is more complete. In the fifth image illustrating a case of complex background, our method can still completely detect a small salient object (*i.e.*, the human).

The P-R curves of different methods are shown in Fig. 8. Our method (*i.e.*, the red solid line) achieves the highest precision compared to other methods on all datasets. The numerical results are reported in Table 1. Our method achieves the best quantitative results across all metrics, outperforming all competing methods. Compared with the **second best method** on the NJUD-Test dataset, the percentage gain reaches 1.6% for F-measure, 6.2% for MAE score, and 1.7% for S-measure. On the DUT-test dataset, the **minimum percentage gain** reaches 2.0% for F-measure, 14.1% for MAE score, and 2.5% for S-measure. All these measures demonstrate the superiority and effectiveness of our method.

4.4 Ablation Studies

To verify the impact of our key modules, we conduct experiments on the STEREO dataset and DUT-Test dataset. The quantitative results are shown in Table 2. An example of visual comparison is illustrated in Fig. 9.

Cross-Modality Feature Modulation (cmFM). We compare three variants: *w/o cmFM*, *w/cmFA*, and *w/cmFC*. In Fig. 9, the baseline *w/o cmFM* cannot effectively detect the salient object while the baselines *w/cmFA* and *w/cmFC* achieve the similar detection result. The same quantitative trend also reflects in Table 2. Compared with the full model, the results indicate that the proposed

Table 2. Quantitative comparisons of ablated models

Modules	Baselines	STEREO dataset			DUT-test dataset		
		$F_\beta \uparrow$	MAE $\downarrow$	$S_m \uparrow$	$F_\beta \uparrow$	MAE $\downarrow$	$S_m \uparrow$
	full model	**0.9084**	**0.0422**	**0.8895**	**0.9328**	**0.0366**	**0.8853**
cmFM	w/o cmFM	0.8727	0.0722	0.8573	0.8968	0.0616	0.8599
	w/cmFA	0.9020	0.0479	0.8820	0.9237	0.0429	0.8771
	w/cmFC	0.8995	0.0480	0.8825	0.9221	0.0617	0.8789
AFS	w/o AFS	0.8990	0.0546	0.8762	0.9165	0.0503	0.8666
	w/o GFF	0.9012	0.0690	0.8826	0.9212	0.0458	0.8777
	w/o CACA	0.9017	0.0517	0.8797	0.9276	0.0470	0.8742
	w/CA	0.9027	0.0503	0.8780	0.9216	0.0468	0.8747
sg-PEA	w/o PEA	0.9057	0.0450	0.8854	0.9205	0.0427	0.8796
	w/o PA	0.9064	0.0442	0.8857	0.9234	0.0409	0.8827
	w/o PE	0.9065	0.0481	0.8862	0.9296	0.0385	0.8806

cmFM module is important for improving the SOD performance. Besides, the simple addition and concatenation can only boost a little performance.

Adaptive Feature Selection (AFS). We compare with four baselines: *w/o AFS*, *w/o GFF*, *w/o CACA*, and *w/CA*. Observing Fig. 9 and Table 2, we found that the performance of the baseline *w/o AFS* is obviously worse than the baselines *w/o GFF*, *w/o CACA*, and *w/CA*. The visual results reflect that the baseline *w/o GFF* produces incomplete salient object while the baseline *w/o CACA* yields the result with an unclear boundary. Collectively, these results underscore the importance of progressive self-modality and cross-modality channel attention while fusing important spatial features of multi-modalities.

Saliency-guided Position-Edge Attention (sg-PEA). We compare with three baselines: *w/o PEA*, *w/o PA*, and *w/o EA*. In Fig. 9, the baseline *w/o PEA* fails to highlight the position and edge of salient object. The baseline *w/o PA* has a sharper boundary of partial complete object while the baseline *w/o PE* shows a more complete object but unclear boundary. In contrast, our full model achieves better performance than these three baselines as presented in Table 2.

In summary, the ablation studies demonstrate the effectiveness and advantages of the proposed three modules qualitatively and quantitatively. In addition, the ablation studies also demonstrate that careful feature modulation, selection, and refinement can effectively improve the performance of RGB-D SOD.

5 Conclusion

We propose an RGB-D SOD network equipped with cross-modality feature modulation and adaptive feature selection. The former effectively integrates the multi-modality complementarities while the latter adaptively highlights saliency-related features. We demonstrate that both elaborate integration of cross-modality features and adaptive selection of multi-modality spatial and channel features can boost the performance of SOD. Experiment results also demonstrate that our method achieves new state-of-the-art performance on six benchmarks.

Acknowledgments. This research was supported by SenseTime-NTU Collaboration Project, Singapore MOE AcRF Tier 1 (2018-T1-002-056), NTU NAP, in part by the Fundamental Research Funds for the Central Universities under Grant 2019RC039, and in part by China Postdoctoral Science Foundation Grant 2019M660438.

References

1. Boer, P.T.D., Kroese, D.P., Mannor, S., Rubinstein, R.Y.: A tutorial on the cross-entropy method. Ann. Oper. Res. **134**(1), 19–67 (2005)
2. Borji, A., Cheng, M.M., Jiang, H., Li, J.: Salient object detection: a benchmark. IEEE Trans. Image Process. **24**(12), 5706–5722 (2015)
3. Chen, H., Li, Y.: Progressively complementarity-aware fusion network for RGB-D salient object detection. In: CVPR, pp. 3051–3060 (2018)
4. Chen, H., Li, Y.: Three-stream attention-aware network for RGB-D salient object detection. IEEE Trans. Image Process. **28**(6), 2825–2835 (2019)
5. Chen, H., Li, Y., Su, D.: Discriminative cross-modal transfer learning and densely cross-level feedback fusion for RGB-D salient object detection. IEEE Trans. Cybern., 1–13 (2019)
6. Chen, H., Li, Y., Su, D.: Multi-modal fusion network with multiscale multi-path and cross-modal interactions for RGB-D salient object detection. Pattern Recognit. **86**, 376–385 (2019)
7. Chen, L., et al.: SCA-CNN: spatial and channel-wise attention in convolutional networks for image captioning. In: CVPR, pp. 5659–5667 (2017)
8. Cong, R., Lei, J., Fu, H., Cheng, M.M., Lin, W., Huang, Q.: Review of visual saliency detection with comprehensive information. IEEE Trans. Circuits Syst. Video Technol. **29**(10), 2941–2959 (2019)
9. Cong, R., Lei, J., Fu, H., Hou, J., Huang, Q., Kwong, S.: Going from RGB to RGBD saliency: a depth-guided transformation model. IEEE Trans. Cybern. **50**(8), 3627–3639 (2020)
10. Cong, R., Lei, J., Zhang, C., Huang, Q., Cao, X., Hou, C.: Saliency detection for stereoscopic images based on depth confidence analysis and multiple cues fusion. IEEE Sig. Process. Lett. **23**(6), 819–823 (2016)
11. Fan, D.P., Cheng, M.M., Liu, Y., Li, T., Borji, A.: Structure-measure: a new way to evaluate foreground maps. In: ICCV, pp. 4548–4557 (2017)
12. Fan, D.P., Zhai, Y., Borji, A., Yang, J., Shao, L.: BBS-Net: RGB-D salient object detection with a bifurcated backbone strategy network. In: Vedaldi, A., Bischof, H., Brox, T., Frahm, J.M. (eds.) ECCV 2020. LNCS, vol. 12357. Springer, Cham (2020). https://doi.org/10.1007/978-3-030-58610-2_17
13. Feng, D., Barnes, N., You, S., McCarthy, C.: Local background enclosure for RGB-D salient object detection. In: CVPR, pp. 2343–2350 (2016)
14. Feng, M., Lu, H., Ding, E.: Attentive feedback network for boundary-aware salient object detection. In: CVPR, pp. 1623–1632 (2019)

15. Fu, J., Liu, J., Tian, H., Li, Y.: Dual attention network for scene segmentation. In: CVPR, pp. 3146–3154 (2019)
16. Fu, K.F., Fan, D.P., Ji, G.P., Zhao, Q.: JL-DCF: joint learning and densely-cooperative fusion framework for RGB-D salient object detection. In: CVPR, pp. 3052–3062 (2020)
17. Guan, W., Wang, T., Qi, J., Zhang, L., Lu, H.: Edge-aware convolutional neural network based salient object detection. IEEE Sig. Process. Lett. **26**, 114–118 (2018)
18. Han, J., Chen, H., Liu, N., Yan, C., Li, X.: CNNs-based RGB-D saliency detection via cross-view transfer and multiview fusion. IEEE Trans. Cybern. **48**(11), 3171–3183 (2018)
19. Hou, Q., Cheng, M.M., Hu, X., Borji, A., Tu, Z., Torr, P.H.: Deeply supervised salient object detection with short connections. IEEE Trans. Pattern Anal. Mach. Intell. **41**(4), 815–828 (2019)
20. Hu, J., Shen, L., Sun, G.: Squeeze-and-excitation networks. In: CVPR, pp. 7132–7141 (2018)
21. Ju, R., Liu, Y., Ren, T., Ge, L., Wu, G.: Depth-aware salient object detection using anisotropic center-surround difference. Sig. Process. Image Commun. **38**, 115–126 (2015)
22. Li, C., et al.: ASIF-Net: attention steered interweave fusion network for RGBD salient object detection. IEEE Trans. Cybern., 1–13 (2020)
23. Li, G., Zhu, C.: A three-pathway psychobiological framework of salient object detection using stereoscopic technology. In: ICCVW, pp. 3008–3014 (2017)
24. Li, N., Ye, J., Ji, Y., Ling, H., Yu, J.: Saliency detection on light field. In: CVPR, pp. 2806–2813 (2014)
25. Li, X., Lu, H., Zhang, L., Ruan, X., Yang, M.H.: Saliency detection via dense and sparse reconstruction. In: ICCV, pp. 2976–2983 (2013)
26. Liu, J., Hou, Q., Cheng, M.M., Feng, J., Jiang, J.: A simple pooling-based design for real-time salient object detection. In: CVPR, pp. 3917–3926 (2019)
27. Niu, Y., Geng, Y., Li, X., Liu, F.: Leveraging stereopsis for saliency analysis. In: CVPR, pp. 454–461 (2012)
28. Oreshkin, B.N., Rodriguez, P., Lacoste, A.: TADAM: task dependent adaptive metric for improved few-shot learning. In: NeurIPS, pp. 721–731 (2018)
29. Peng, H., Li, B., Ling, H., Hu, W., Xiong, W., Maybank, S.J.: Salient object detection via structured matrix decomposition. IEEE Trans. Pattern Anal. Mach. Intell. **39**(4), 818–832 (2017)
30. Peng, H., Li, B., Xiong, W., Hu, W., Ji, R.: RGBD salient object detection: a benchmark and algorithms. In: Fleet, D., Pajdla, T., Schiele, B., Tuytelaars, T. (eds.) ECCV 2014. LNCS, vol. 8691. Springer, Cham (2014). https://doi.org/10.1007/978-3-319-10578-9_7
31. Perez, E., Strub, F., de Vries, H., Dumoulin, V., Courville, A.: FiLM: Visual reasoning with a general conditioning layer. In: AAAI, pp. 3942–3951 (2018)
32. Piao, Y., Ji, W., Li, J., Zhang, M., Lu, H.: Depth-induced multi-scale recurrent attention network for saliency detection. In: ICCV, pp. 7254–7263 (2019)
33. Piao, Y., Rong, Z., Zhang, M., Ren, W., Lu, H.: A2dele: adaptive and attentive depth distiller for efficient RGB-D salient object detection. In: CVPR, pp. 9060–9069 (2020)
34. Qin, X., Zhang, Z., Huang, C., Gao, C., Dehghan, M., Jagersand, M.: BASNet: boundary-aware salient object detection. In: CVPR, pp. 7479–7489 (2019)
35. Qu, L., He, S., Zhang, J., Tian, J., Tang, Y., Yang, Q.: RGBD salient object detection via deep fusion. IEEE Trans. Image Process. **26**(5), 2274–2285 (2017)

36. Simonyan, K., Zisserman, A.: Very deep convolutional networks for large-scale image recognition. arXiv preprint arXiv:1409.1556 (2014)
37. Song, H., Liu, Z., Du, H., Sun, G., Le Meur, O., Ren, T.: Depth-aware salient object detection and segmentation via multiscale discriminative saliency fusion and bootstrap learning. IEEE Trans. Image Process. **26**(9), 4204–4216 (2017)
38. Vaswani, A., et al.: Attention is all you need. In: NeurIPS, pp. 5998–6008 (2017)
39. Wang, W., Lai, Q., Fu, H., Shen, J., Ling, H.: Salient object detection in the deep learning era: An in-depth survey. arXiv preprint arXiv:1904.09146 (2019)
40. Wang, X., Yu, K., Dong, C., Loy, C.C.: Recovering realistic texture in image super-resolution by deep spatial feature transform. In: CVPR, pp. 606–615 (2018)
41. Yu, D., Fu, J., Mei, T., Rui, Y.: Multi-level attention networks for visual question answering. In: CVPR, pp. 4709–4717 (2017)
42. Yuan, Y., Li, C., Kim, J., Cai, W., Feng, D.D.: Reversion correction and regularized random walk ranking for saliency detection. IEEE Trans. Image Process. **27**(3), 1311–1322 (2018)
43. Zhang, J., et al.: UC-Net: uncertainty inspired RGB-D saliency detection via conditional variational autoencoders. In: CVPR, pp. 8582–8591 (2020)
44. Zhang, M., Ren, W., Piao, Y., Rong, Z., Lu, H.: Select, supplement and focus for RGB-D saliency detection. In: CVPR, pp. 3472–3481 (2020)
45. Zhang, Y., Li, K., Li, K., Wang, L., Zhong, B., Fu, Y.: Image super-resolution using very deep residual channel attention networks. In: Ferrari, V., Hebert, M., Sminchisescu, C., Weiss, Y. (eds.) ECCV 2018. LNCS, vol. 11211. Springer, Cham (2018). https://doi.org/10.1007/978-3-030-01234-2_18
46. Zhao, J., Cao, Y., Fan, D.P., Cheng, M.M., Li, X.Y., Zhang, L.: Contrast prior and fluid pyramid integration for RGBD salient object detection. In: CVPR, pp. 3927–3936 (2019)
47. Zhao, J., Liu, J., Fan, D.P., Cao, Y., Yang, J., Cheng, M.M.: EGNet: edge guidance network for salient object detection. In: ICCV, pp. 8779–8788 (2019)
48. Zhu, C., Li, G.: A multilayer backpropagation saliency detection algorithm and its applications. Multimed. Tools Appl. **77**(19), 25181–25197 (2018). https://doi.org/10.1007/s11042-018-5780-4

RetrieveGAN: Image Synthesis via Differentiable Patch Retrieval

Hung-Yu Tseng[2](✉), Hsin-Ying Lee[2], Lu Jiang[1], Ming-Hsuan Yang[1,2,3], and Weilong Yang[1]

[1] Google Research, Mountain View, USA
[2] University of California, Merced, USA
htseng6@ucmerced.edu
[3] Yonsei University, Seoul, South Korea

Abstract. Image generation from scene description is a cornerstone technique for the controlled generation, which is beneficial to applications such as content creation and image editing. In this work, we aim to synthesize images from scene description with retrieved patches as reference. We propose a differentiable retrieval module. With the differentiable retrieval module, we can (1) make the entire pipeline end-to-end trainable, enabling the learning of better feature embedding for retrieval; (2) encourage the selection of mutually compatible patches with additional objective functions. We conduct extensive quantitative and qualitative experiments to demonstrate that the proposed method can generate realistic and diverse images, where the retrieved patches are reasonable and mutually compatible.

1 Introduction

Image generation from scene descriptions has received considerable attention. Since the description often requests multiple objects in a scene with complicated relationships between objects, it remains challenging to synthesize images from scene descriptions. The task requires not only the ability to generate realistic images but also the understanding of the mutual relationships among different objects in the same scene. The usage of the scene description provides flexible user-control over the generation process and enables a wide range of applications in content creation [18] and image editing [24] (Fig. 1).

Taking advantage of generative adversarial networks (GANs) [5], recent research employs conditional GAN for the image generation task. Various conditional signals have been studied, such as scene graph [13], bounding box [40],

H.-Y. Tseng, H.-Y. Lee—Equal contribution. Work done during their internships at Google Research.

Electronic supplementary material The online version of this chapter (https://doi.org/10.1007/978-3-030-58598-3_15) contains supplementary material, which is available to authorized users.

A. Vedaldi et al. (Eds.): ECCV 2020, LNCS 12353, pp. 242–257, 2020.
https://doi.org/10.1007/978-3-030-58598-3_15

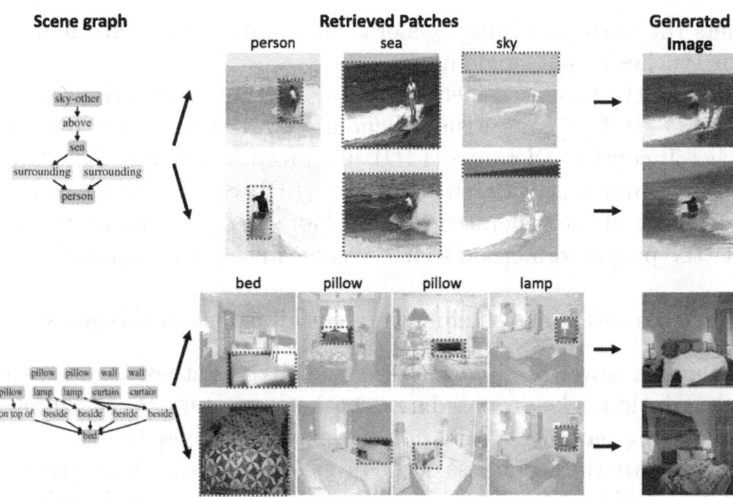

Fig. 1. Image synthesize from retrieved examples. We propose the RetrieveGAN model that takes as input the scene graph description and learns to 1) select mutually compatible image patches via a differentiable retrieval process and 2) synthesize the output image from the retrieved patches.

semantic segmentation map [24], audio [19], and text [36]. A stream of work has been driven by parametric models that rely on the deep neural network to capture and model the appearance of objects [13,36]. Another stream of work has recently emerged to explore the semi-parametric model that leverages a memory bank to retrieve the objects for synthesizing the image [25,37].

In this work, we focus on the semi-parametric model in which a memory bank is provided for the retrieval purpose. Despite the promising results, existing retrieval-based image synthesis methods face two issues. First, the current models require pre-defined embeddings since the retrieval process is non-differentiable. The pre-defined embeddings are independent of the generation process and thus cannot guarantee the retrieved objects are suitable for the surrogate generation task. Second, oftentimes there are multiple objects to be retrieved given a scene description. However, the conventional retrieval process selects each patch independently and thus neglect the subtle mutual relationship between objects.

We propose *RetrieveGAN*, a conditional image generation framework with a differentiable retrieval process to address the issues. First, we adopt the Gumbel-softmax [11] trick to make the retrieval process differentiable, thus enable optimizing the embedding through the end-to-end training. Second, we design an iterative retrieval process to select a set of compatible patches (i.e., objects) for synthesizing a single image. Specifically, the retrieval process operates iteratively to retrieve the image patch that is most compatible with the already selected patches. We propose a co-occurrence loss function to boost the mutual compatibility between the selected patches. With the proposed differentiable retrieval design, the proposed RetrieveGAN is capable of retrieving image patches that

1) considers the surrogate image generation quality, and 2) are mutually compatible for synthesizing a single image.

We evaluate the proposed method through extensive experiments conducted on the COCO-stuff [2] and Visual Genome [16] datasets. We use three metrics, Fréchet Inception Distance (FID) [9], Inception Score (IS) [26], and the Learned Perceptual Image Patch Similarity (LPIPS) [39], to measure the realism and diversity of the generated images. Moreover, we conduct the user study to validate the proposed method's effectiveness in selecting mutually compatible patches.

To summarize, we make the following contributions in this work:

- We propose a novel semi-parametric model to synthesize images from the scene description. The proposed model takes advantage of the complementary strength of the parametric and non-parametric techniques.
- We demonstrate the usefulness of the proposed differentiable retrieval module. The differentiable retrieval process can be jointly trained with the image synthesis module to capture the relationships among the objects in an image.
- Extensive qualitative and quantitative experiments demonstrate the efficacy of the proposed method to generate realistic and diverse images where retrieved objects are mutually compatible.

2 Related Work

Conditional Image Synthesis. The goal of the generative models is to model a data distribution given a set of samples from that distribution. The data distribution is either modeled explicitly (e.g., variational autoencoder [15]) or implicitly (e.g., generative adversarial networks [5]). On the basis of unconditional generative models, conditional generative models target synthesizing images according to additional context such as image [4,17,23,30,41], segmentation mask [10,24,33,43], and text. The text conditions are often expressed in two formats: natural language sentences [36,38] or scene graphs [13]. Particularly, the scene graph description is in a well-structured format (i.e., a graph with a node representing objects and edges describing their relationship), which mitigates the ambiguity in natural language sentences. In this work, we focus on using the scene graph description as our input for the conditional image synthesis.

Image Synthesis from Scene Descriptions. Most existing methods employ parametric generative models to tackle this task. The appearance of objects and relationships among objects are captured via a graph convolution network [13,21] or a text embedding network [20,29,36,38,42], then images are synthesized with the conditional generative approach. However, current parametric models synthesize objects at pixel-level, thus failing to generate realistic images for complicated scene descriptions. More recent frameworks [25,37] adopt semi-parametric models to perform generation at patch-level based on reference object patches. These schemes retrieve reference patches from an external bank and use them to synthesize the final images. Although the retrieval module is a crucial component, existing works all use predefined retrieval modules that cannot be optimized during the training stage. In contrast, we propose a novel semi-parametric model with a differentiable retrieval process that is end-to-end trainable with the conditional generative model.

Image Retrieval. Image retrieval has been a classical vision problem with numerous applications such as product search [1,6,22], multimodal image retrieval [3,32], image geolocalization [7], event detection [12], among others. Solutions based on deep metric learning use the triplet loss [8,31] or softmax cross-entropy objective [32] to learn a joint embedding space between the query (e.g., text, image, or audio) and the target images. However, there is no prior work studying learning retrieval models for the image synthesis task. Different from the existing semi-parametric generative models [29,37] that use the pre-defined (or fixed) embedding to retrieve image patches, we propose a differentiable retrieval process that can be jointly optimized with the conditional generative model.

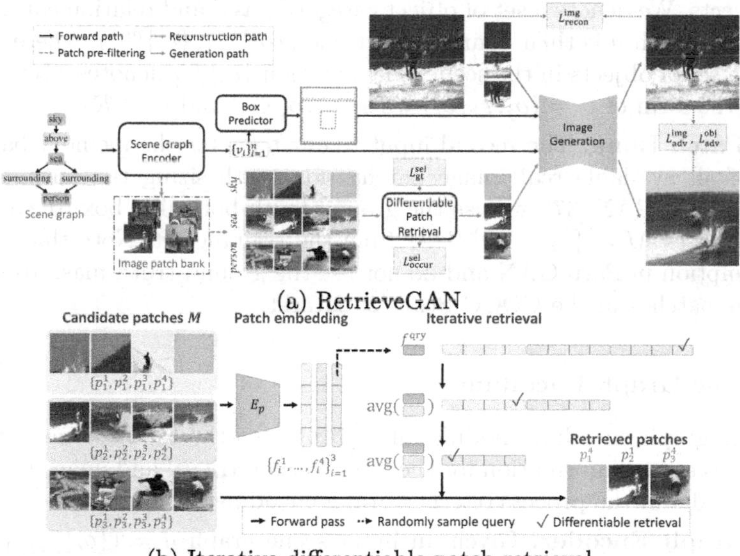

(a) RetrieveGAN

(b) Iterative differentiable patch retrieval

Fig. 2. Method overview. (a) Our model takes as input the scene graph description and sequentially performs scene graph encoding, patch retrieval, and image generation to synthesize the desired image. (b) Given a set of candidate patches, we first extract the features using the patch embedding function. We then randomly select a patch feature as the query feature for the iterative retrieval process. At each step of the iterative procedure, we select the patch that is most compatible with the already selected patches. The iteration ends as all the objects are assigned with a selected patch.

3 Methodology

3.1 Preliminaries

Our goal is to synthesize an image $x \in \mathbb{R}^{H\times W\times 3}$ from the input scene graph g by compositing appropriate image patches retrieved from the image patch bank. As the overview shown in Fig. 2, the proposed RetrieveGAN framework consists of three stages: scene graph encoding, patch retrieval, and image generation. The scene graph encoding module processes the input scene graph g, extracts features, and predicts bounding box coordinates for each object o_i defined in the scene graph. The patch retrieval module then retrieves an image patch for each object o_i from the image patch bank. The goal of the retrieval module is to maximize the compatibility of all retrieved patches, thus improving the quality of the image synthesized by the subsequent image generation module. Finally, the image generation module takes as input the selected patches along with the predicted bounding boxes to synthesize the final image.

Scene Graph. Serving as the input data to our framework, the scene graph representation [14] describes the objects in a scene and the relationships between these objects. We denote a set of object categories as $\mathcal{C}$ and relation categories as $\mathcal{R}$. A scene graph g is then defined as a tuple $(\{o_i\}_{i=1}^n, \{e_i\}_{i=1}^m)$, where $\{o_i|o_i \in \mathcal{C}\}_{i=1}^n$ is a set of objects in the scene. The notation $\{e_i\}_{i=1}^m$ denotes a set of direct edges in the form of $e_i = (o_j, r_k, o_t)$ where $o_j, o_t \in \mathcal{C}$ and $r_k \in \mathcal{R}$.

Image Patch Bank. The second input to our model is the memory bank consisting of all available real image patches for synthesizing the output image. Following PasteGAN [37], we use the ground-truth bounding box to extract the images patches $M = \{p_i \in \mathbb{R}^{h\times w\times 3}\}$ from the training set. Note that we relax the assumption in PasteGAN and do not use the ground-truth mask to segment the image patches in the COCO-Stuff [2] dataset.

3.2 Scene Graph Encoding

The scene graph encoding module aims to process the input scene graph and provides necessary information for the later patch retrieval and image generation stages. We detail the process of scene graph encoding as follows:

Scene Graph Encoder. Given an input scene graph $g = (\{o_i\}_{i=1}^n, \{e_i\}_{i=1}^m)$, the scene graph encoder E_g extracts the object features, namely $\{v_i\}_{i=1}^n = E((\{o_i\}_{i=1}^n, \{e_i\}_{i=1}^m))$. Adopting the strategy in sg2im [13], we construct the scene graph encoder with a series of graph convolutional networks (GCNs). We further discuss the detail of the scene graph encoder in the supplementary document.

Bounding Box Predictor. For each object o_i, the bounding box predictor learns to predict the bounding box coordinates $\hat{b}_i = (x_0, y_0, x_1, y_1)$ from the object features v_i. We use a series of fully-connected layers to build the predictor.

Patch Pre-filtering. Since there are a large number of image patches in the image patch bank, performing the retrieval on the entire bank online is intractable in practice due to the memory limitation. We address this problem by pre-filtering a set of k candidate patches $M(o_i) = \{p_i^1, p_i^2, \cdots, p_i^k\}$ for each object o_i. And the later patch retrieval process is conducted on the pre-filtered candidate patches as opposed to the entire patch bank. To be more specific, we use the pre-trained GCN in sg2im [13] to obtain the candidate patches for each object. We use the corresponding scene graph to compute the GCN feature. The computed GCN feature is used to select similar candidate patches $M(o_i)$ with respect to the negative ℓ_2 distance.

3.3 Patch Retrieval

The patch retrieval aims to select a number of mutually compatible patches for synthesizing the output image. We illustrate the overall process on the bottom side of Fig. 2. Given the pre-filtered candidate patches $\{M(o_i)\}_{i=1}^n$, we first use a patch embedding function E_p to extract the patch features. Starting with a randomly sampled patch feature as a query, we propose an iterative retrieval process to select compatible patches for all objects. In the following, we 1) describe how a single retrieval is operated, 2) introduce the proposed iterative retrieval process, and 3) discuss the objective functions used to facilitate the training of the patch retrieval module.

Differentiable Retrieval for a Single Object. Given the query feature f^{qry}, we aim to sample a single patch from the candidate set $M(o) = \{p^1, p^2, \cdots, p^k\}$ for object o. Let $\pi \in \mathbb{R}^k_{>0}$ be the categorical variable with probabilities $P(x = i) \propto \pi_i$ which indicates the probability of selecting the i-th patch from the bank. To compute π_i, we calculate the ℓ_2 distance between the query feature and the corresponding patch feature, namely $\pi_i = e^{-\|f_{\text{qry}} - E_p(p^i;\theta_{E_p})\|_2}$, where E_p is the embedding function and θ_{E_p} is the learnable mode parameter. The intuition is that the candidate patch with smaller feature distance to the query feature should be sampled with higher probability. By optimizing θ_{E_p} with our loss functions, we hope our model is capable of retrieving compatible patches. As we are sampling from a categorical distribution, we use the Gumbel-Max trick [11] to sample a single patch:

$$\arg\max_i [P(x = i)] = \arg\max_i [g_i + \log \pi_i] = \arg\max_i [\hat{\pi}_i], \tag{1}$$

where $g_i = -\log(-\log(u_i))$ is the re-parameterization term and $u_i \sim \text{Uniform}(0, 1)$. To make the above process differentiable, the argmax operation is approximated with the continuous softmax operation:

$$s = \text{softmax}(\hat{\pi}) = \frac{\exp(\hat{\pi}_i/\tau)}{\sum_{q=1}^k \exp(\hat{\pi}_q/\tau)}, \tag{2}$$

where τ is the temperature controlling the degree of the approximation.[1]

[1] When τ is small, we found it is useful to make the selection variable s uni-modal. This can also be achieved by post-processing (e.g., thresholding) the softmax outputs.

Iterative Differentiable Retrieval for Multiple Objects. Rather than retrieving only a single image patch, the proposed framework needs to select a subset of n patches for the n objects defined in the input scene graph. Therefore, we adopt the weighted reservoir sampling strategy [35] to perform the subset sampling from the candidate patch sets. Let $M = \{p_i | i = 1, \ldots, n \times k\}$ denote a multiset (with possible duplicated elements) consisting of all candidates patches in which n is the number of objects and k is the size of each candidate patch set. We leave the preliminaries on weighted reservoir sampling in the supplementary materials. In our problem, we first compute the vector $\hat{\pi}_i$ defined in (1) for all patches. We then iteratively apply n softmax operations over $\hat{\pi}$ to approximate the top-k selection. Let $\hat{\pi}_i^{(j)}$ denote the probability of sampling patch p_i at iteration j and $\hat{\pi}_i^{(1)} \leftarrow \hat{\pi}_i$. The probability is iteratively updated by:

$$\hat{\pi}_i^{(j+1)} \leftarrow \hat{\pi}_i^{(j)} + \log(1 - s_i^{(j)}), \tag{3}$$

where $s_i^{(j)} = \text{softmax}(\hat{\pi}^{(j)})_i$ computed by (2). Essentially, (3) sets the probability of the selected patch to negative infinity, thus ensures this patch will not be chosen again. After n iterations, we compute the relaxed n-hot vector $s = \sum_{j=1}^{n} s^{(j)}$, where $s_i \in [0, 1]$ indicates the score of selecting the i-th patch and $\sum_{i=1}^{|M|} s_i = n$. The entire process is differentiable with respect to the model parameters.

We make two modifications to the above iterative process based on practical consideration. First, our candidate multiset $M = \{p_i\}_{i=1}^{n \times k}$ is formed by n groups of pre-filtered patches where every object has a group k patches. Since we are only allowed to retrieve a single patch from a group, we modify (3) by:

$$\hat{\pi}_i^{(j+1)} \leftarrow \hat{\pi}_i^{(j)} + \log(1 - \max_t [s_t^{(j)}]) \quad \forall t \text{ such that } m^{-1}(p_i) = t, \tag{4}$$

where we denote $m^{-1}(p_j) = i$ if patch p_j in M is pre-fetched by the object o_i. (4) uses max pooling to disable selecting multiple patches from the same group. Second, to incorporate the prior knowledge that compatible images patches tend to lie closer in the embedding space, we use a greedy strategy to encourage selecting image patches that are compatible with the already selected ones. We detail this process in Fig. 2(b). To be more specific, at each iteration, the features of the selected patches are aggregated by average pooling to update the query f^{qry}. π and $\hat{\pi}$ is also recomputed accordingly after the query update. This leads to a greedy strategy encouraging the selected patches to be visually or semantically similar in the feature space. We summarize the overall retrieval process in Algorithm 1.

As the retrieval process is differentiable, we can optimize the retrieval module (i.e., patch embedding function E_p) with the loss functions (e.g., adversarial loss) applied to the following image generation module. Moreover, we incorporate two additional objectives to facilitate the training of iterative retrieval process: ground-truth selection loss $L_{\text{gt}}^{\text{sel}}$ and co-occurrence loss $L_{\text{occur}}^{\text{sel}}$.

Ground-Truth Selection Loss. As the ground-truth patches are available at the training stage, we add them to the candidate set M. Given one of the ground-truth patch features as the query feature f^{qry}, the ground-truth selection loss

Algorithm 1. Iterative Differential Retrieval

Input : Candidate patches $M = \{p_i\}_{i=1}^{n \times k}$ for n objects and each object has k pre-filtered patches.

Output: relaxed n-hot vector s where $\sum_{i=1}^{|M|} s_i = n$ and $0 \leq s_i \leq 1$.

for $i = 1, \ldots, |M|$ **do** $f_i = E_p(p_i)$ `// Get patch features`

Randomly select a patch feature to initialize the query f^{qry}

`// Iterative patch retrieval`

for $t = 1, \ldots, n$ **do**

  for $i = 1, \ldots, |M|$ **do**

    $\pi_i = e^{-\|f_i - f^{\text{qry}}\|_2}$ `// Calculate` π `according to the query`

    `// Gumbel-Max trick`

    $u_i \leftarrow \text{Uniform}(0, 1)$

    $\hat{\pi}_i \leftarrow -\log(-\log(u_i)) + \log(\pi_i)$

    `// Disable other patches in the selected group`

    $\hat{\pi}_i \leftarrow \hat{\pi}_i + \log(1 - \max_j[s_j^{(t-1)}]) \quad \forall j$ such that $m^{-1}(p_i) = j$

  end

  for $i = 1, \ldots, |M|$ **do** $s_i^{(t)} = \frac{\exp(\hat{\pi}_i/\tau)}{\sum_{q=1}^{|M|} \exp(\hat{\pi}_q/\tau)}$ `// Softmax operation`

  $f^{\text{qry}} = \text{avg}(f^{\text{qry}}, \sum_{i=1}^{|M|} s_i^{(t)} f_i)$ `// Update the query`

end

return the relaxed n-hot vector $s^{(n)}$

$L_{\text{gt}}^{\text{sel}}$ encourages the retrieval process to select the ground-truth patches for the other objects in the input scene graph.

Co-occurrence Penalty. We design a co-occurrence loss to ensure the mutual compatibility between the retrieved patches. The core idea is to minimize the distances between the retrieved patches in a co-occurrence embedding space. Specifically, we first train a co-occurrence embedding function F_{occur} using the patches cropped from the training images with the triplet loss [34]. The distance on the co-occurrence embedding space between the patches sampled from the same image is minimized, while the distance between the patches cropped from the different images is maximized. Then the proposed co-occurrence loss is the pairwise distance between the retrieved patches on the co-occurrence embedding space:

$$L_{\text{occur}}^{\text{sel}} = \sum_{i,j} d(F_{\text{occur}}(p_i), F_{\text{occur}}(p_j)), \tag{5}$$

where p_i and p_j are the patches retrieved by the iterative retrieval process.

Limitations vs. Advantages. The size of the candidate patches considered by the proposed retrieval process is currently limited by the GPU memory. Therefore, we cannot perform the differentiable retrieval over the entire memory bank.

Nonetheless, the differentiable mechanism and iterative design enable us to train the retrieval process using the abovementioned loss functions that maximize the mutual compatibility of the selected patches.

3.4 Image Generation

Given selected patches after the differentiable patch retrieval process, the image generation module synthesizes the realistic image with the selected patches as reference. We adopt a similar architecture to PasteGAN [37] as our image generation module. Please refer to the supplementary materials for details regarding the image generation module. We use two discriminators D_{img} and D_{obj} to encourage the realism of the generated images on the image-level and object-level, respectively. Specifically, the adversarial loss can be expressed as:

$$\begin{aligned} L_{\text{adv}}^{\text{img}} &= \mathbb{E}_x[\log D_{\text{img}}(x)] + \mathbb{E}_{\hat{x}}[\log(1 - D_{\text{img}}(\hat{x}))], \\ L_{\text{adv}}^{\text{obj}} &= \mathbb{E}_p[\log D_{\text{obj}}(p)] + \mathbb{E}_{\hat{p}}[\log(1 - D_{\text{obj}}(\hat{p}))], \end{aligned} \tag{6}$$

where x and p denote the real image and patch, whereas $\hat{x}$ and $\hat{p}$ represent the generated image and the patch crop from the generated image, respectively.

3.5 Training Objective Functions

In addition to the abovementioned loss functions, we use the following loss functions during the training phase:

Bounding Box Regression Loss. We penalize the prediction of the bounding box coordinates with $\ell 1$ distance $L_{\text{bbx}} = \sum_{i=1}^{n} \|b_i - \hat{b}_i\|_1$.

Image Reconstruction Loss. Given the ground-truth patches and the ground-truth bounding box coordinates, the image generation module should recover the ground-truth image. The loss $L_{\text{recon}}^{\text{img}}$ is an $\ell 1$ distance measuring the difference between the recovered and ground-truth images.

Auxiliary Classification Loss. We adopt the auxiliary classification loss $L_{\text{ac}}^{\text{obj}}$ to encourage the generated patches to be correctly classified by the object discriminator D_{obj}.

Perceptual Loss. The perceptual loss is computed as the distance in the pre-trained VGG [27] feature space. We apply the perceptual losses $L_{\text{p}}^{\text{img}}, L_{\text{p}}^{\text{obj}}$ on both image and object levels to stabilize the training procedure.

The full loss functions for training our model is:

$$\begin{aligned} L = &\lambda_{\text{gt}}^{\text{sel}} L_{\text{gt}}^{\text{sel}} + \lambda_{\text{occur}}^{\text{sel}} L_{\text{occur}}^{\text{sel}} + \lambda_{\text{adv}}^{\text{img}} L_{\text{adv}}^{\text{img}} + \lambda_{\text{recon}}^{\text{img}} L_{\text{recon}}^{\text{img}} + \lambda_{\text{p}}^{\text{img}} L_{\text{p}}^{\text{img}} + \\ &\lambda_{\text{adv}}^{\text{obj}} L_{\text{adv}}^{\text{obj}} + \lambda_{\text{ac}}^{\text{obj}} L_{\text{ac}}^{\text{obj}} + \lambda_{\text{p}}^{\text{obj}} L_{\text{p}}^{\text{obj}} + \lambda_{\text{bbx}} L_{\text{bbx}}, \end{aligned} \tag{7}$$

where λ controls the importance of each loss term. We describe the implementation detail of the proposed approach in the supplementary document.

4 Experimental Results

Datasets. The COCO-Stuff [2] and Visual Genome [16] datasets are standard benchmark datasets for evaluating scene generation models [13,37,40]. We use the image resolution of 128×128 for all the experiments. Except for the image resolution, we follow the protocol in sg2im [13] to pre-process and split the dataset. Different from the PasteGAN [37] approach, we do not access the ground-truth mask for segmenting the image patches.

Evaluated Methods. We compare the proposed approach to three parametric generation models and one semi-parametric model in the experiments:

- **sg2im** [13]: The sg2im framework takes as input a scene graph and learns to synthesize the corresponding image.
- **AttnGAN** [36]: As the AttnGAN method synthesizes the images from text, we convert the scene graph to the corresponding text description. Specifically, we convert each relationship in the graph into a sentence, and link every sentence via the conjunction word "and". We train the AttnGAN model on these converted sentences.
- **layout2im** [40]: The layout2im scheme takes as input the ground-truth bounding boxes to perform the generation. For a fair comparison, we use the ground-truth bounding box coordinate as the input data for other methods, which we denote GT in the experimental results.
- **PasteGAN** [37]: The PasteGAN approach is most related to our work as it uses the pre-trained embedding function to retrieve candidate patches.

Evaluation Metrics. We use the following metrics to measure the realism and diversity of the generated images:

Table 1. Quantitative Comparisons. We evaluate all methods on the COCO-Stuff and Visual Genome datasets using the FID, IS, and DS metrics. The first row shows the results of models that predict bounding boxes during the inference time. The second row shows the results of models that take ground-truth bounding as inputs during the inference time.

Datasets	COCO-stuff			Visual genome		
	FID ↓	IS ↑	DS ↑	FID ↓	IS ↑	DS ↑
sg2im [13]	136.8	4.1±0.1	0.02±0.0	126.9	5.1±0.1	0.11±0.1
AttnGAN [36]	72.8	8.4±0.2	0.14±0.1	114.6	**10.4**±0.2	<u>0.27</u>±0.2
PasteGAN [37]	<u>59.8</u>	<u>8.8</u>±0.3	**0.43**±0.1	<u>81.8</u>	6.7±0.2	**0.30**±0.1
RetrieveGAN (Ours)	**43.2**	**10.6**±0.6	<u>0.34</u>±0.1	**70.3**	<u>7.7</u>±0.1	0.24±0.1
sg2im (GT)	79.9	8.5±0.1	0.02±0.0	111.9	5.8±0.1	0.13±0.1
layout2im [40]	<u>45.3</u>	<u>10.2</u>±0.6	<u>0.29</u>±0.1	**44.0**	**9.3**±0.4	**0.29**±0.1
PasteGAN (GT)	54.9	9.6±0.2	**0.38**±0.1	68.1	6.7±0.1	<u>0.28</u>±0.1
RetrieveGAN (GT)	**42.7**	**10.7**±0.1	0.21±0.1	<u>46.3</u>	<u>9.1</u>±0.1	0.23±0.1
Real data	6.8	24.3±0.3	-	6.9	24.1±0.4	-

Table 2. Ablation studies. We conduct ablation studies on two loss functions added upon the proposed retrieval module.

Datasets	$\mathcal{L}_{gt}^{sel}$	$\mathcal{L}_{occur}^{sel}$	COCO-stuff		
			FID ↓	IS ↑	DS ↑
RetrieveGAN	-	-	56.8	8.8±0.3	0.30±0.1
RetrieveGAN	✓	-	47.8	9.7±0.2	**0.36**±0.1
RetrieveGAN	-	✓	52.8	9.8±0.2	0.29±0.1
RetrieveGAN	✓	✓	**43.2**	**10.6**±0.6	0.34±0.1
Real data			6.8	24.3±0.3	-

- **Inception Score (IS).** Inception Score [26] uses the Inception V3 [28] model to measure the visual quality of the generated images.
- **Fréchet Inception Distance (FID).** Fréchet Inception Distance [9] measures the visual quality and diversity of the synthesized images. We use the Inception V3 model as the feature extractor.
- **Diversity (DS).** We use the AlexNet model to explicitly evaluate the diversity by measuring the distances between the features of the images using the Learned Perceptual Image Patch Similarity (LPIPS) [39] metric.

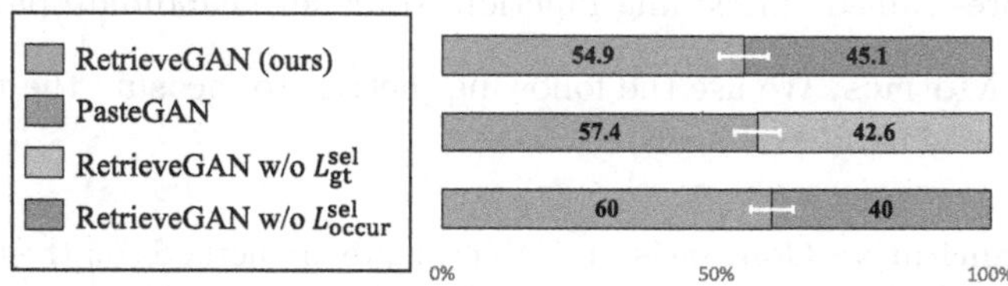

Fig. 3. User study. We conduct the user study to evaluate the mutual compatibility of the selected patches.

4.1 Quantitative Evaluation

Realism and Diversity. We evaluate the realism and diversity of all methods using the IS, FID, and DS metrics. To have a fair comparison with different methods, we conduct the evaluation using two different settings. First, bounding boxes of objects are predicted by models. Second, ground-truth bounding boxes are given as inputs in addition to the scene graph. The results of these two settings are shown in the first and second row of Table 1, respectively. Since the patch retrieval process is optimized to consider the generation quality during the training stage, our approach performs favorably against the other algorithms in terms of realism. On the other hand, as we can sample different query features

Fig. 4. Sample generation results. We show example results on the COCO-Stuff (*left*) and Visual Genome (*right*) datasets. The object locations in each image are predicted by models.

for the proposed retrieval process, our model synthesizes comparably diverse images compared to the other schemes.

Moreover, there are two noteworthy observations. First, the proposed RetrieveGAN has similar performance in both settings on the COCO-Stuff dataset, but has significant improvement using ground-truth bounding boxes on the Visual Genome dataset. The reason for the inferior performance on the Visual Genome dataset without using ground-truth bounding boxes is due to the existence of lots of isolated objects (i.e., objects that have no relationships to other objects) in the scene graph annotation (e.g., the last scene graph in Fig. 6), which greatly increase the difficult of predicting reasonable bounding boxes. Second, on the Visual Genome dataset, AttnGAN outperforms the proposed method on the IS and DS metrics, while performs significantly worse than the proposed method on the FID metric. Compared to the FID metric, the IS score has the limitation that it is less sensitive to the mode collapse problem. The DS metric only measures the feature distance without considering visual quality. The results from AttnGAN shown in Fig. 4 also support our observation.

Patch Compatibility. The proposed differentiable retrieval process aims to improve the mutual compatibility among the selected patches. We conduct a user study to evaluate the patch compatibility. For each scene graph, we present two sets of patches selected by different methods, and ask users "which set of patches are more mutually compatible and more likely to coexist in the same image?". Figure 3 presents the results of the user study. The proposed method outperforms PasteGAN, which uses a pre-defined patch embedding function for retrieval. The results also validate the benefits of the proposed ground-truth selection loss and co-occurrence loss.

Ablation Study. We conduct an ablation study on the COCO-Stuff dataset to understand the impact of each component in the proposed design. The results are shown in Table 2. As the ground-truth selection loss and the co-occurrence penalty maximize the mutual compatibility of the selected patches, they both improve the visual quality of the generated images.

4.2 Qualitative Evaluation

Image Generation. We qualitatively compare the visual results generated by different methods. We show the results on the COCO-Stuff (left column) and the Visual Genome (right column) datasets under two settings of using predicted (Fig. 4) and ground-truth (Fig. 5) bounding boxes. The sg2im and layout2im methods can roughly capture the appearance of objects and mutual relationships among objects. However, the quality of generated images in complicated scenes is limited. Similarly, the AttnGAN model cannot handle scenes with complex relationships well. The overall image quality generated by the PasteGAN scheme is similar to that by the proposed approach, yet the quality is affected by the compatibility of the selected patches (e.g., the third result on COCO-Stuff in Fig. 5).

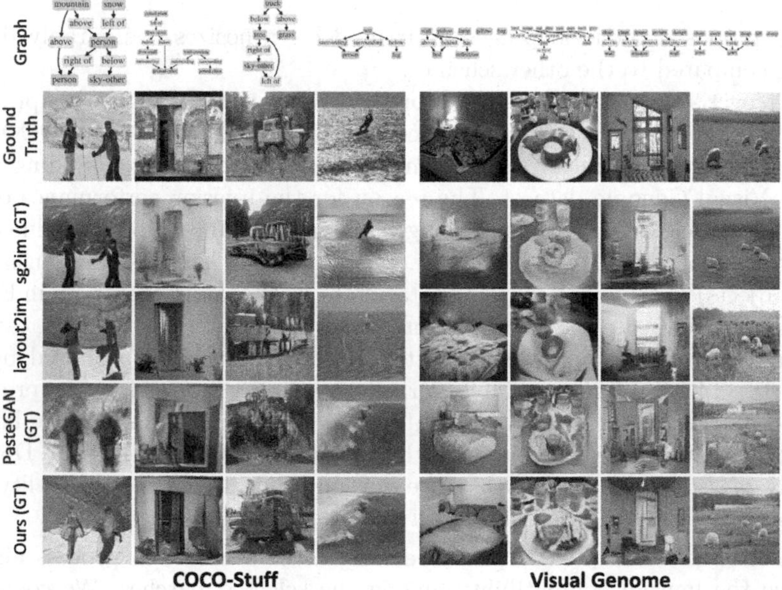

Fig. 5. Sample generation results. We show example results on the COCO-Stuff (*left*) and Visual Genome (*right*) datasets. The object locations in each image are given as additional inputs.

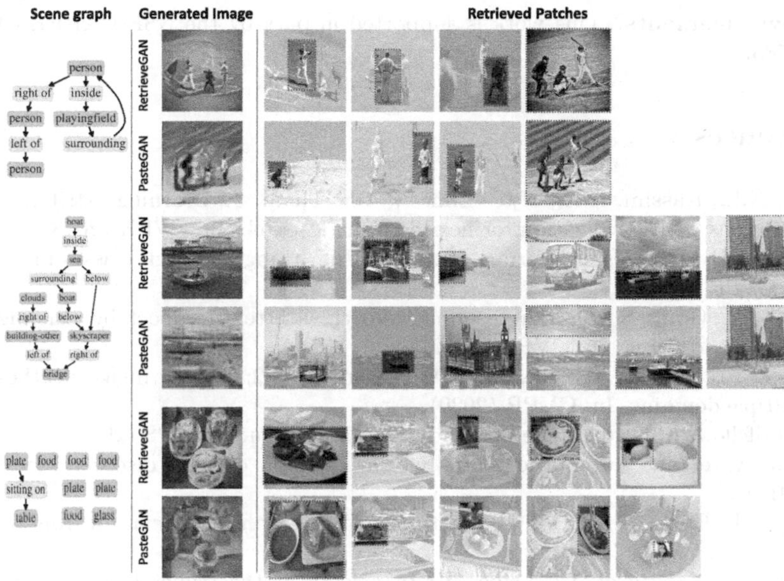

Fig. 6. Retrieved patches. For each sample, we show the retrieved patches which are used to guide the following image generation process. We also show the original image of each selected patch for more clear visualization.

Patch Retrieval. To better visualize the source of retrieved patches, we present the generated images as well as the original images of selected patches in Fig. 6. The proposed method can tackle complex scenes where multiple objects are present. With the help of selected patches, each object in the generated images has a clear and reasonable appearance (e.g., the boat in the second row and the food in the third row). Most importantly, the retrieved patches are mutually compatible, thanks to the proposed iterative and differentiable retrieval process. As shown in the first example in Fig. 6, the selected patches are all related to baseball, while the PasteGAN method, which uses random selection, has chances to select irrelevant patches (i.e., the boy on the soccer court).

5 Conclusions and Future Work

In this work, we present a differentiable retrieval module to aid the image synthesis from the scene description. Qualitative and quantitative evaluations validate that the synthesized images are realistic and diverse, while the retrieved patches are reasonable and compatible. The proposed approach points out a new direction in the content creation research field. It can be trained with the image generation or manipulation models to learn to select real reference patches that improves the generation or manipulation quality.

Acknowledgements. This work is supported in part by the NSF CAREER Grant #1149783.

References

1. Ak, K.E., Kassim, A.A., Hwee Lim, J., Yew Tham, J.: Learning attribute representations with localization for flexible fashion search. In: CVPR (2018)
2. Caesar, H., Uijlings, J., Ferrari, V.: Coco-stuff: thing and stuff classes in context. In: CVPR (2018)
3. Chen, Y., Gong, S., Bazzani, L.: Image search with text feedback by visiolinguistic attention learning. In: CVPR (2020)
4. Choi, Y., Uh, Y., Yoo, J., Ha, J.W.: StarGAN v2: diverse image synthesis for multiple domains. In: CVPR (2020)
5. Goodfellow, I., et al.: Generative adversarial nets. In: NeurIPS (2014)
6. Han, X., et al.: Automatic spatially-aware fashion concept discovery. In: ICCV (2017)
7. Hays, J., Efros, A.A.: IM2GPS: estimating geographic information from a single image. In: CVPR (2008)
8. Hermans, A., Beyer, L., Leibe, B.: In defense of the triplet loss for person re-identification. arXiv preprint arXiv:1703.07737 (2017)
9. Heusel, M., Ramsauer, H., Unterthiner, T., Nessler, B., Hochreiter, S.: GANs trained by a two time-scale update rule converge to a local nash equilibrium. In: NeurIPS (2017)
10. Huang, H.-P., Tseng, H.-Y., Lee, H.-Y., Huang, J.-B.: Semantic view synthesis. In: Vedaldi, A., Bischof, H., Brox, T., Frahm, J.-M. (eds.) ECCV 2020. LNCS, vol. 12357, pp. 592–608. Springer, Cham (2020). https://doi.org/10.1007/978-3-030-58610-2_35
11. Jang, E., Gu, S., Poole, B.: Categorical reparameterization with Gumbel-Softmax. In: ICLR (2017)
12. Jiang, L., Meng, D., Mitamura, T., Hauptmann, A.G.: Easy samples first: self-paced reranking for zero-example multimedia search. In: ACM MM (2014)
13. Johnson, J., Gupta, A., Fei-Fei, L.: Image generation from scene graphs. In: CVPR (2018)
14. Johnson, J., Krishna, R., Stark, M., Li, L.J., Shamma, D., Bernstein, M., Fei-Fei, L.: Image retrieval using scene graphs. In: CVPR (2015)
15. Kingma, D.P., Welling, M.: Auto-encoding variational Bayes. In: ICLR (2014)
16. Krishna, R., et al.: Visual genome: connecting language and vision using crowdsourced dense image annotations. IJCV **123**(1), 32–73 (2017)
17. Lee, H.Y., et al.: Drit++: diverse image-to-image translation via disentangled representations. IJCV **128**, 2402–2417 (2020). https://doi.org/10.1007/s11263-019-01284-z
18. Lee, H.Y., Yang, W., Jiang, L., Le, M., Essa, I., Gong, H., Yang, M.H.: Neural design network: Graphic layout generation with constraints. In: ECCV. Springer, Heidelberg (2020)
19. Lee, H.Y., et al.: Dancing to music. In: NeurIPS (2019)
20. Li, W., et al.: Object-driven text-to-image synthesis via adversarial training. In: CVPR (2019)
21. Li, Y., Jiang, L., Yang, M.H.: Controllable and progressive image extrapolation. arXiv preprint arXiv:1912.11711 (2019)

22. Liu, Z., Luo, P., Qiu, S., Wang, X., Tang, X.: DeepFashion: powering robust clothes recognition and retrieval with rich annotations. In: CVPR (2016)
23. Mao, Q., Lee, H.Y., Tseng, H.Y., Ma, S., Yang, M.H.: Mode seeking generative adversarial networks for diverse image synthesis. In: CVPR (2019)
24. Park, T., Liu, M.Y., Wang, T.C., Zhu, J.Y.: Semantic image synthesis with spatially-adaptive normalization. In: CVPR (2019)
25. Qi, X., Chen, Q., Jia, J., Koltun, V.: Semi-parametric image synthesis. In: CVPR (2018)
26. Salimans, T., Goodfellow, I., Zaremba, W., Cheung, V., Radford, A., Chen, X.: Improved techniques for training GANs. In: NeurIPS (2016)
27. Simonyan, K., Zisserman, A.: Very deep convolutional networks for large-scale image recognition. In: ICLR (2015)
28. Szegedy, C., Vanhoucke, V., Ioffe, S., Shlens, J., Wojna, Z.: Rethinking the inception architecture for computer vision. In: CVPR (2016)
29. Tan, F., Feng, S., Ordonez, V.: Text2Scene: generating compositional scenes from textual descriptions. In: CVPR (2019)
30. Tseng, H.Y., Fisher, M., Lu, J., Li, Y., Kim, V., Yang, M.H.: Modeling artistic workflows for image generation and editing. In: ECCV. Springer, Heidelberg (2020)
31. Vo, N.N., Hays, J.: Localizing and orienting street views using overhead imagery. In: Leibe, B., Matas, J., Sebe, N., Welling, M. (eds.) ECCV 2016. LNCS, vol. 9905, pp. 494–509. Springer, Cham (2016). https://doi.org/10.1007/978-3-319-46448-0_30
32. Vo, N., et al.: Composing text and image for image retrieval-an empirical odyssey. In: CVPR (2019)
33. Wang, T.C., Liu, M.Y., Zhu, J.Y., Tao, A., Kautz, J., Catanzaro, B.: High-resolution image synthesis and semantic manipulation with conditional GANs. In: CVPR (2018)
34. Wang, X., He, K., Gupta, A.: Transitive invariance for self-supervised visual representation learning. In: ICCV (2017)
35. Xie, S.M., Ermon, S.: Reparameterizable subset sampling via continuous relaxations. In: IJCAI (2019)
36. Xu, T., et al.: AttnGAN: fine-grained text to image generation with attentional generative adversarial networks. In: CVPR (2018)
37. Yikang, L., Ma, T., Bai, Y., Duan, N., Wei, S., Wang, X.: PasteGAN: a semi-parametric method to generate image from scene graph. In: NeurIPS (2019)
38. Zhang, H., et al.: StackGAN: text to photo-realistic image synthesis with stacked generative adversarial networks. In: ICCV (2017)
39. Zhang, R., Isola, P., Efros, A.A., Shechtman, E., Wang, O.: The unreasonable effectiveness of deep features as a perceptual metric. In: CVPR (2018)
40. Zhao, B., Meng, L., Yin, W., Sigal, L.: Image generation from layout. In: CVPR (2019)
41. Zhu, J.Y., Park, T., Isola, P., Efros, A.A.: Unpaired image-to-image translation using cycle-consistent adversarial networkss. In: ICCV (2017)
42. Zhu, M., Pan, P., Chen, W., Yang, Y.: DM-GAN: dynamic memory generative adversarial networks for text-to-image synthesis. In: ICCV (2019)
43. Zhu, P., Abdal, R., Qin, Y., Wonka, P.: SEAN: image synthesis with semantic region-adaptive normalization. In: CVPR (2020)

Cheaper Pre-training Lunch: An Efficient Paradigm for Object Detection

Dongzhan Zhou[1], Xinchi Zhou[1], Hongwen Zhang[2], Shuai Yi[3], and Wanli Ouyang[1(✉)]

[1] The University of Sydney, SenseTime Computer Vision Research Group, Sydney, Australia
{d.zhou,xinchi.zhou1,wanli.ouyang}@sydney.edu.au

[2] Institute of Automation, Chinese Academy of Sciences and University of Chinese Academy of Sciences, Beijing, China

[3] Sensetime Research, Hong Kong, China
yishuai@sensetime.com

Abstract. In this paper, we propose a general and efficient pre-training paradigm, Montage pre-training, for object detection. Montage pre-training needs only the target detection dataset while taking only 1/4 computational resources compared to the widely adopted ImageNet pre-training. To build such an efficient paradigm, we reduce the potential redundancy by carefully extracting useful samples from the original images, assembling samples in a Montage manner as input, and using an ERF-adaptive dense classification strategy for model pre-training. These designs include not only a new input pattern to improve the spatial utilization but also a novel learning objective to expand the effective receptive field of the pre-trained model. The efficiency and effectiveness of Montage pre-training are validated by extensive experiments on the MS-COCO dataset, where the results indicate that the models using Montage pre-training are able to achieve on-par or even better detection performances compared with the ImageNet pre-training.

Keywords: Pre-training · Object detection · Acceleration · Deep neural networks · Deep learning

1 Introduction

Pre-training on the classification dataset (*e.g.* ImageNet [11]) is a common practice to achieve better network initialization for object detection. Under this paradigm, deep networks benefit from useful feature representations learned from

Electronic supplementary material The online version of this chapter (https://doi.org/10.1007/978-3-030-58598-3_16) contains supplementary material, which is available to authorized users.

A. Vedaldi et al. (Eds.): ECCV 2020, LNCS 12353, pp. 258–274, 2020.
https://doi.org/10.1007/978-3-030-58598-3_16

large-scale data, which promotes the convergence of models during fine-tuning stage. Despite the benefits, the burdens caused by extra data should not be neglected.

Previous works [7,33,46] have proposed alternative solutions to directly train detection models from scratch with random initialization. However, there is always no free lunch. Training from scratch suffers from slower convergence, namely, additional training iterations are needed to obtain competitive models. *Can we incorporate the merit of fast convergence via pre-training without paying for the extra data or expensive training cost?*

The answer is **Yes**. We find the cheaper lunch for pre-training. In this work, we propose a new pre-training paradigm, Montage pre-training, which is based only on the detection dataset. Compared with ImageNet pre-training, Montage pre-training takes only 1/4 computational resources without extra data while achieving on-par or even better performance on the target object detection task.

Montage pre-training is built upon the observation that a large number of pixels seen by the model during naive training are invalid or less informative, *i.e.* most pixels/neurons in background regions would not fire during the learning process. Those excessive background pixels inevitably lead to redundant computational costs. To tackle this issue, we carefully extract positive and negative samples from original images in the detection dataset for pre-training. Before being fed into the backbone network, these samples will be assembled in a Montage manner in consideration of their aspect ratios to improve the spatial utilization. To further improve the pixel level utilization, we design an ERF-adaptive dense classification strategy to leverage the Effective Receptive Field (ERF) via assigning soft labels in the learning objective. Our Montage pre-training largely takes every pixel seen by the model into account, which greatly reduces the redundancy and provides an efficient and general pre-training solution for object detection.

Our major contributions can be summarized as follows.

(1) We propose an efficient and general pre-training paradigm based only on detection dataset, which eliminates the burdens of additional data.
(2) We design rules of sample extraction, the Montage assembly strategy, and the ERF-adaptive dense classification for efficient pre-training, which largely considers the network utilization and improves the learning efficiency and final performance.
(3) We validate the effectiveness of our Montage pre-training on various detection frameworks and backbones and demonstrate the versatility of the proposed pre-training strategy. We hope this work would inspire more discussions about the pre-training of object detectors.

2 Related Work

Classification-Based Pre-training for Object Detector. Recent years have witnessed the significant breakthroughs of deep learning-based object detectors on various scenarios [1,5,8,17,18,20,22,23,25,27,31,38,45]. Most of these frameworks follow the standard 'pre-training followed by fine-tuning' training procedure, where networks are first pre-trained on the large-scale dataset (*e.g.* ImageNet [11]) and then fine-tuned on the target detection dataset. This pre-training paradigm is mainly classification-based and aims to learn strong or universal representations, which speed up the convergence of detection models. Many efforts have been devoted to push the boundary of transferability further through different learning modes such as supervised [10], weakly supervised [24,40], unsupervised [6] learning, or exploiting larger scale training data such as Instagram-17k [24] and JFT-300M [36]. Despite the improvements for transferability, the corresponding expensive training cost of large scale data should not be neglected. Our Montage pre-training is entirely based on detection dataset which eliminates the burden of using external data. Meanwhile, the pre-training process is 4× faster than ImageNet-1k classification training.

Redundancy in Object Detector. Sample imbalance is a common source of redundancy for object detection, where many background pixels belonging to easy negative samples contribute no useful information for training. To alleviate this issue, several attempts have been made to improve the efficiency of detection training. OHEM [34] tries to solve the imbalance sampling by discarding easy negative samples. Focal loss [17] adopts a weighting factor to reduce loss weight for easy samples. Chen *et al.* design a more reasonable method for sample evaluation in [2]. Libra R-CNN [28] proposes the IoU-balanced sampling strategy to augment the hard cases. SNIPER [35] reduces the calculation burden of multi-scale training by only training on selected chips rather than the entire images. All these works mainly focus on the efficiency and performance within detection frameworks, but they still provide inspirations on sample selection in our work. By carefully selecting positive and negative samples for pre-training, the redundancy is significantly reduced, which eventually speeds up the classification pre-training process.

Object Detector Trained from Scratch. Many works [7,12,14,26,33,37,46] have proposed another possible training paradigm which is to train the detector from scratch. For instance, DSOD [33] is motivated by designing a pre-training free detector, but limited to the structure they designed. CornerNet [12] and Det-Net [14] present the results of their models trained from scratch. These efforts indicate that pre-training might be unnecessary when adequate data is available. Furthermore, doubts on ImageNet pre-training are also raised recently. He *et al.* [7] and Zhu *et al.* [46] suggest that ImageNet pre-training might be a historical workaround. However, although these solutions get rid of the burdens for large-scale external data, the random initialized detection models suffer from the problem of low convergence speed, which comes at the cost of extending training iterations by 4–5 times to obtain competitive models. Inspired by these works, we move steps forward to exploit an efficient pre-training paradigm for pre-training on detection data, which takes the advantages of both fast convergence and no extra data at the same time.

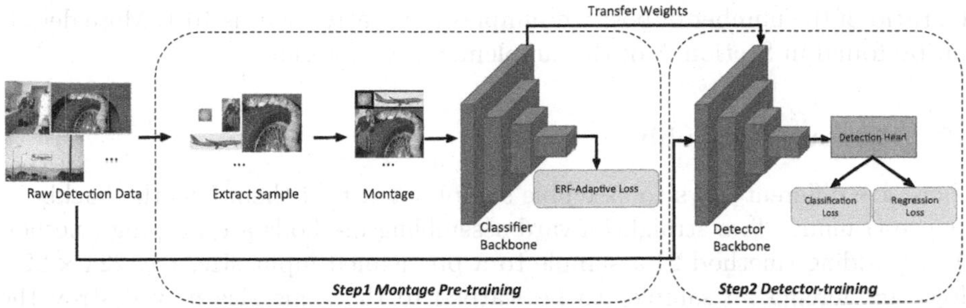

Fig. 1. Pipeline of the proposed Montage pre-training scheme. Firstly, we will extract positive and negative samples from detection data to build classification dataset. The pre-training process is conducted by Montage, which assembles four objects into a single image, and optimized via ERF-adaptive loss. Finally, the backbones will be fine-tuned on target detection task

3 Methodology

The pipeline of using the proposed Montage pre-training scheme is shown in Fig. 1. Given a detection dataset $\mathcal{D}$, positive and negative samples will be extracted from the images of $\mathcal{D}$ and saved as classification dataset beforehand (Sect. 3.1). These samples will be assembled in a Montage manner (Sect. 3.2) and fed into the detector backbone for pre-training, where an ERF-adaptive loss is used as the loss function (Sect. 3.3). After pre-training, the object detector will be fine-tuned on $\mathcal{D}$ under the detection task. Note that our pre-training scheme is flexible and can be applied to object detectors with diverse detection head and backbone architectures.

3.1 Sample Selection

As demonstrated in previous works [28,34], balanced sample selection is critical during the training of object detectors. For efficient pre-training, we carefully select regions extracted from original images as positive and negative samples, which will be further assembled and fed into the detector backbone.

The positive samples are regions that should be classified as one of the C foreground categories in the detection dataset, while the negative samples are background regions. To effectively select diverse and important samples, we set up following rules for the sample extraction. (1) For **positive samples**, we extract regions from the original images according to the ground-truth bounding boxes. The bounding boxes will be randomly enlarged to involve more context information, which is under the consideration that contextual information is beneficial to learn better feature representations [4,43]. (2) **Negative samples** are proposals randomly generated from the background regions. To avoid ambiguity, we require that all negative samples meet the requirement $IoU\,(pos, neg) = 0$, where IoU indicates Intersection-over-Union. In our pre-training experiments, the ratio of the number of positive samples to negative ones is 10:1. More details can be found in Section A of the supplementary material.

3.2 Montage Assembly

There are different ways to assemble samples and feed them into the backbone for pre-training. Two straightforward assembling methods are warping (method 1) or padding (method 2) a sample to a pre-defined input size, *e.g.* 224×224. However, forcing all samples to be warped to the same size may destroy the texture information and distort the original shapes, while padding would introduce many uninformative padded pixels and hence bring additional costs in both training time and computational resources. These two straightforward methods are either harmful or wasteful for the pre-training process. For more efficient pre-training, we propose to assemble samples in a Montage manner in consideration of the scale and aspect ratio of objects. Specifically, four samples will be stitched into a new image and then taken as input for pre-training.

As depicted in Fig. 2, compared to warping and padding, our Montage assembly can not only preserve original texture information but also eliminate the uninformative padded pixels.

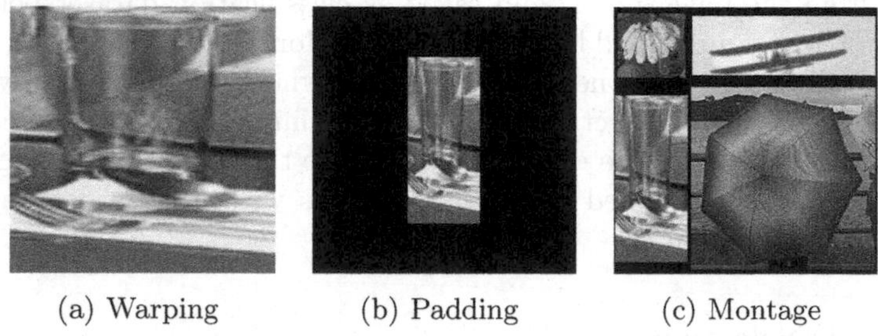

(a) Warping (b) Padding (c) Montage

Fig. 2. Different methods to adjust sample to pre-defined input size. (a) Warping distorts the original shape or texture. (b) Padding introduces many uninformative pixels. (c) Montage preserves original information while improving space utilization

Objects vary in aspect ratio. Montage assembly takes this property into consideration so that samples could be stitched together more naturally according to their aspect ratios. To this end, samples will be first divided into three Groups according to their aspect ratios, *i.e.* Group S (square), T (tall), and W (wide). Samples in Group S should have the aspect ratios between 0.5 and 1.5, while samples in Group T and W should respectively have aspect ratio smaller than 0.5 and larger than 1.5. For simplicity, samples from Group S, T, and W are referred to as S-samples, T-samples, and W-samples, respectively.

As shown in Fig. 3, for every Montage assembled image, 2 S-samples, 1 T-sample, and 1 W-sample will be selected randomly from above three groups and stitched into four regions accordingly. Specifically, the S-sample with smaller bounding box area is at the top-left region, while the larger S-sample is at the bottom-right region. The T-sample and W-sample will be respectively assigned to bottom-left and top-right regions. Details about sample size adjustment (to fit the template) can be found in Section B of the supplementary material.

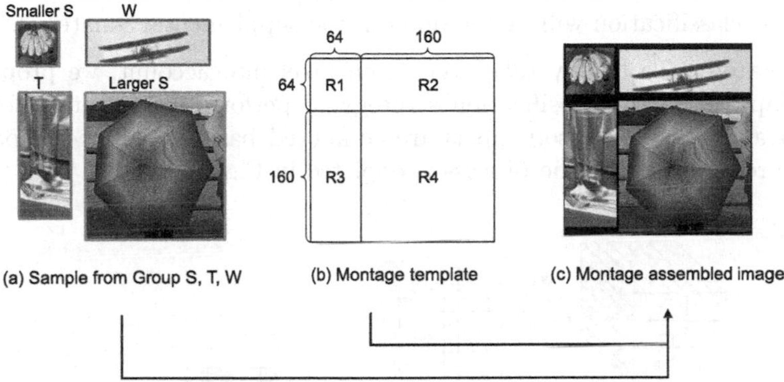

Fig. 3. Pipeline for Montage assembled image generation. We first randomly select 2 S-samples, 1 T-sample and 1 W-sample, respectively as shown in (a), then assemble the samples according to the template (b) and get an assembled image (c). The numbers on Montage template denote the height/width of each region

3.3 ERF-Adaptive Dense Classification

During pre-training, the Montage assembled images will be fed into the backbone network to obtain the feature maps $\mathbf{X} \in \mathbb{R}^{C \times \alpha H \times \alpha W}$ before the final average pooling. Here we omit the number of samples in $\mathbf{X}$ for simplicity. Compared to the conventional classification pre-training, Montage pre-training should have different learning strategy since there are four samples stitched in one assembled image. In the following, we discuss two alternative strategies and then introduce our proposed ERF-adaptive Dense Classification.

Global Classification. As shown in Fig. 3, an image contains four objects in our Montage assembled image. As an intuitive strategy, we can assign the whole image a single global label, which is the weighted sum of the labels of the four objects according to their region areas. This strategy could be reminiscent of the CutMix [42], where certain region of the original image will be replaced by a patch from another image and the corresponding label will also be mixed proportionally with the label of the new patch. The visualization of global classification will be provided in the supplementary material.

Block-Wise Classification. Another intuitive strategy would perform individually for each block/region, that is, the average pooling is independently applied to the four blocks of feature maps $\mathbf{X}$ corresponding to four samples, followed by individual classification according to the label of each sample. However, these two intuitive strategies confine the learning of each block to the corresponding sample. As can be seen in Fig. 5(a) and 5(b), the Effective Receptive Field (ERF) [21] of the top-left region in $\mathbf{X}$ mainly concentrates on the area of the corresponding smaller S-sample. The confined receptive field may empircally degrade the performance of deep models, as illustrated in [13,19,30]. The visualization of block-wise classification will be provided in the supplementary material.

Our Strategy. To largely take every seen pixel into account, we propose an ERF-adaptive Dense Classification strategy to perform classification for each position at $\mathbf{X}$, where its soft labels are computed based on the corresponding effective receptive field. The process is depicted in Fig. 4.

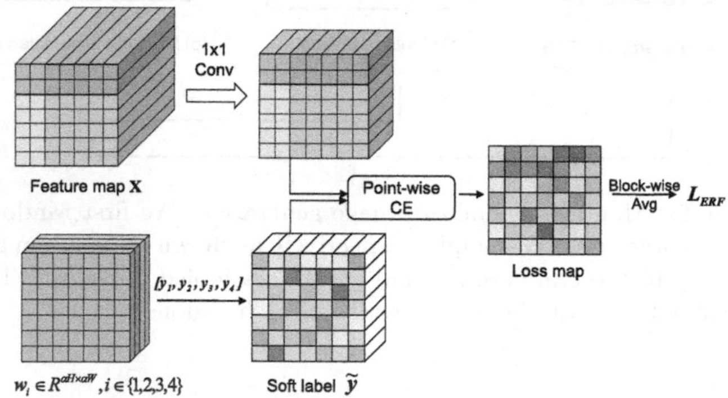

Fig. 4. Process of our Dense Classification Strategy. We use different colors to distinguish different regions, *e.g.* green for R_1, and the brightness difference in soft label and loss map represents different values. The feature map $\mathbf{X}$ is convolved by a 1×1 kernel to reduce the number of channels to C (category number). Given the weight w_i of label $\mathbf{y}_i$, we obtain the soft label tensor for each point. Then the cross-entropy loss is densely imposed on the feature map and we will get the loss value at each point (denoted as loss map). In the mean time, the loss weight for a position is calculated according to the ERF of the position. After that, block-wise average is exerted on the loss map to generate average losses for each region. The final ERF-adaptive loss is the mean of four region losses. Best viewed in color (Color figure online)

Specifically, for the four regions in the Montage template as shown in Fig. 3(b), we denote $\mathbf{y}_i$ as the original label for the region R_i, $i = 1, 2, 3, 4$.

At the position (j, k) of feature map $\mathbf{X}$ ($j = 1, \ldots, \alpha H, k = 1, \ldots, \alpha W$), the soft label $\tilde{\mathbf{y}}^{j,k}$ is the weighted sum of four labels:

$$\tilde{\mathbf{y}}^{j,k} = \sum_{i=1}^{4} w_i^{j,k} \mathbf{y}_i, \tag{1}$$

where the weight $w_i^{j,k}$ is dependent on its ERF. At the position (j, k) of feature map $\mathbf{X}$ ($j = 1, \ldots, \alpha H, k = 1, \ldots, \alpha W$), we obtain the corresponding ERF

$\mathbf{G}^{j,k} \in \mathbb{R}^{H \times W}$ on the input space. Then, the weight $w_i^{j,k}$ for the label $\mathbf{y}_i$ at position (j,k) should be proportional to the ratio of the summed activation within the region R_i to the whole summed activation. Moreover, if the position (j,k) is in region R_i, we empirically set a threshold τ for $w_i^{j,k}$ to make sure that $\mathbf{y}_i$ is dominant at region R_i. Hence, for the position (j,k) at region R_r, we have the weight $w_i^{j,k}$ of the label $\mathbf{y}_i$ as follows:

$$w_i^{j,k} = \begin{cases} \max(\tau, \frac{\sum_{h=1,w=1}^{H,W} g_{h,w}^{j,k} \cdot m_{h,w}^i}{\sum_{h=1,w=1}^{H,W} g_{h,w}^{j,k}}), \text{if } i = r, \\ (1 - w_r^{j,k}) \frac{\sum_{h=1,w=1}^{H,W} g_{h,w}^{j,k} \cdot m_{h,w}^i}{\sum_{h=1,w=1}^{H,W} g_{h,w}^{j,k} \cdot (1 - m_{h,w}^r)}, \text{if } i \neq r, \end{cases} \tag{2}$$

where $g_{h,w}^{j,k}$ is the element at position (h,w) on the corresponding ERF matrix $\mathbf{G^{j,k}}$, $m_{h,w}^i$ refers to the value at position (h,w) on binary mask $\mathbf{M}^i \in \{0,1\}^{H \times W}$. The binary mask $\mathbf{M}^i$ is used to select the region R_i in ERF.

Denote $\mathbf{x}^{j,k} \in \mathbb{R}^C$ as the features at the position (j,k) of $\mathbf{X}$ $(j = 1,\ldots,\alpha H, k = 1,\ldots,\alpha W)$. After obtaining the weights $\{w_i^{j,k}\}_{i=1}^4$, we perform dense classification upon the feature $\mathbf{x}^{j,k}$, where its soft label $\tilde{\mathbf{y}}^{j,k}$ is defined in Eq. (1). In our implementation, the final fully connected layer is replaced by a 1×1 convolution layer and the cross-entropy loss is imposed on the category prediction at every position. To make a balance among different regions, the final ERF-adaptive loss is the block-wise average of the loss map, as the last step in Fig. 4. We also need to clarify that the weights of soft label Eq. (1) are updated at every 5k iterations instead of at each iteration. Thus, even if dense classification is adopted, its effect on training time is negligible. Correspondingly, since ERF will be updated regularly during the whole training process, different initialization choices of ERF will not affect the final results. We choose the method to calculate ERF based on the randomly initialized network parameters.

The effective receptive field of the top-left region for the different pre-training strategies is visualized in Fig. 5. Our strategy in Fig. 5(c) has the largest ERF among the above three strategies.

Relationships among Different Strategies. The above three strategies perform classification at different scale levels, where the proposed ERF-adaptive classification is the most fine-grained one while the global classification is the coarsest one. Compared with the other two alternative strategies, the proposed one has different soft labels for each position at $\mathbf{X}$. The ERF-adaptive dense classification would be equivalent to the block-wise classification with threshold τ set to 1. The block-wise classification would be also equivalent to the global classification if the region losses are re-weighted in a CutMix manner. Under different label assignment strategies, the pre-trained model has different pixel level utilization and hence behaves differently. Comparison of performance for the strategies can be found in the supplementary material.

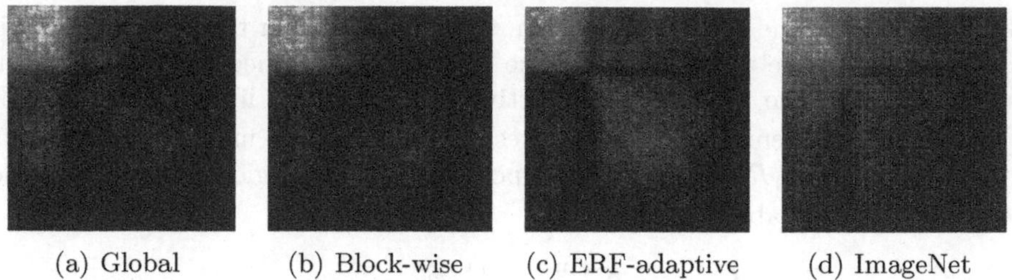

(a) Global (b) Block-wise (c) ERF-adaptive (d) ImageNet

Fig. 5. Visualization of Effective Receptive Field of the top-left region for different pre-trained models. (a) Global represents conducting global average pooling and set global label as weighted sum of labels from four regions. (b) Block-wise refers to the intuitive strategy which performs classification individually on each region. (c) ERF-adaptive refers to adopting ERF-adaptive dense classification. (d) ImageNet stands for the ImageNet pre-trained model officially provided by PyTorch [29]

4 Experiments

4.1 Implementation Details

This section introduces the implementation details of the classification pre-training. Details of data augmentation in pre-training and detector training settings will be provided in the supplementary.

Unless otherwise specified, the models are pre-trained for 64k iterations on 8 Tesla V100 GPUs with the total batch size of 512. Note that the batch size 512 is for Montage assembled images, so the total number of individual samples in each batch is 2048 (an assembled image consists of 4 samples). Warm-up is used during the first 1250 iterations, where the learning rate starts from 0.2 and then linearly increases to 0.8. Afterwards, the learning rate decreases to 0.0 following a cosine scheduler. The weight decay is 1e−4. We update weights $w_i^{j,k}$ of soft labels in Eq. 2 for every 5k iterations. The threshold τ in Eq. 2 for ERF-adaptive classification is set to 0.7. The data augmentation implementation can be found in the supplementary.

4.2 Main Results

We conduct the Montage pre-training process based on samples extracted from MS-COCO train2017 split, and fine-tune the detection models on the same dataset. The backbone is ResNet-50. Note that the Montage pre-training process only consumes **1/4** computation resources compared with ImageNet pre-training. As reported in Table 1, the results show that the models using our Montage pre-training strategy are able to achieve on-par or even better performances compared with the ImageNet pre-training counterparts for various detection frameworks. For original Faster R-CNN [32], the AP increases from 34.8% to 36.3% (+1.5%), for Faster R-CNN with FPN [16], AP increases from

36.2% to 36.5% (+0.3%), for Mask R-CNN with FPN [8], AP increases from 37.2% to 37.4%(+0.2%).

We notice that the improvement is most significant in the original Faster R-CNN structure (denoted as C4 in Table 1). We suspect the possible reason is that, compared with FPN structure, the backbone accounts for a larger proportion in C4. In other words, for detection models with FPN structure, the lateral connections and entire structures at the second stage will be randomly initialized without being transferred from pre-trained model. But for the original Faster R-CNN, the main part of the second stage is still transferred from the pre-trained network. We speculate that the improvement of detection models with FPN may be consistent with that of C4 if the FPN structure is incorporated into pre-training, and we will leave this exploration for future work.

Table 1. Results on different detection frameworks with backbone ResNet-50. The cost refers to pre-training cost and the unit is GPU days. The AP results are evaluated on COCO val2017. 'C4' denotes original Faster R-CNN without FPN [32], 'FPN' denotes Faster R-CNN with FPN [16], 'Mask' denotes Mask R-CNN with FPN [8]. '+ ImageNet' means the backbone is pre-trained on ImageNet dataset. '+ Montage' denotes that the backbone is pre-trained with our Montage strategy. Δ measures the difference in absolute AP or cost between adopting Montage and ImageNet pre-trained backbones, respectively

Method	Cost	AP	AP_{50}	AP_{75}	AP_s	AP_m	AP_l
C4 [32] + ImageNet	6.80	34.8	55.5	36.8	18.3	38.7	48.4
C4 [32] + Montage	1.73	36.3	56.5	38.9	18.9	40.8	49.7
Δ	−5.07	+1.5	+1.0	+2.1	+0.6	+2.1	+1.3
FPN [16] + ImageNet	6.80	36.2	58.0	39.2	21.2	39.9	45.6
FPN [16] + Montage	1.73	36.5	58.3	39.2	22.2	40.4	45.8
Δ	−5.07	+0.3	+0.3	0.0	+1.0	+0.5	+0.2
Mask [8] + ImageNet	6.80	37.3	59.0	40.3	21.9	40.6	46.2
Mask [8] + Montage	1.73	37.5	58.9	40.6	22.8	41.2	46.9
Δ	−5.07	+0.2	−0.1	+0.3	+0.9	+0.6	+0.7

4.3 Ablation Study

Threshold for ERF-Adaptive Dense Classification. When ERF-adaptive Dense Classification is used, there is a threshold τ in Eq. (1) to make sure that the original label y_i is dominant at its corresponding region i. We explore the effects of this threshold and the results are depicted in Fig. 6(a). Although using mixed labels is beneficial, relatively low proportion of original label (*e.g.* 0.5) may still hinder the pre-training. As the threshold becomes higher, the loss is gradually approaching the use of single hard label for each point, which may suffer from relatively confined receptive field, as analyzed in Sect. 3.3. Therefore, it is important to choose proper threshold and we find 0.7 is an ideal choice. Figure 6(a) also shows that setting the threshold in [0.6 0.8] will not cause much variation in mAP. Therefore, the experimental results are not so sensitive to this hyper-parameter.

Iterations for Pre-training. We also investigate the influences of changing the pre-training iterations and visualize the results in Fig. 6(b). Naturally, increasing training iterations will provide better pre-trained models, which leads to better detection performance. But we also observe that the gains from longer iterations are not so significant after 64k iterations (4× in Fig. 6(b)). Considering the trade-off between performance and computation, we choose to train 64k iterations during pre-training, which consumes only 1/4 computation resources but achieve 1.5% higher mAP compared with ImageNet pre-training.

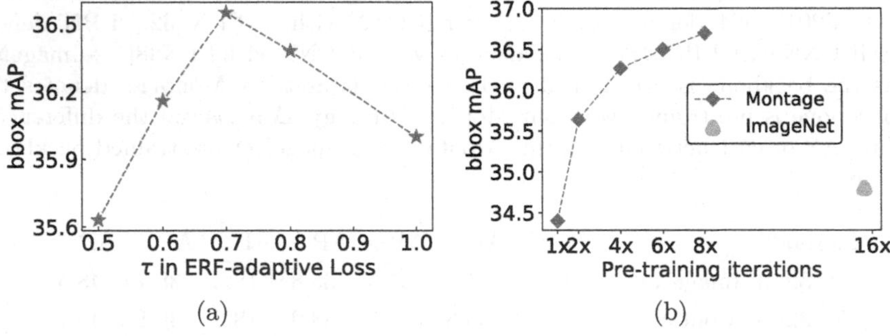

Fig. 6. (a) Trend of bbox mAP with different τ settings in ERF-adaptive loss. (b) bbox mAP for different pre-training iterations, the values on x-axis stand for multiple of 16k iterations. Our pre-training approach uses 4× as the default setting, which requires 1/4 the number of iterations but achieves 1.5% higher mAP when compared with ImageNet counterpart. The results are evaluated on COCO val2017

Different Backbone Structures. We also implement our pre-training strategy on different backbone structures to evaluate the versatility. The results in Table 2 show that Montage pre-training strategy does not rely on specific network structures but will consistently keep on-par performance or obtain improvements.

4.4 Compatibility to Other Designs

We also examine the compatibility of Montage pre-training strategy with commonly used designs in object detection, including longer training iterations (2× schedule), deformable convolution [3], multi-scale augmentation, etc. The results in Fig. 7 indicate that Montage pre-training can still achieve comparable or even higher performance even on various enhanced baselines.

Table 2. Results for different backbone structures evaluated on COCO val2017. The detection framework is original Faster R-CNN. ImageNet means the backbone is trained on ImageNet dataset. Montage denotes that the backbone is pre-trained with Montage strategy. X101-32x4d refers to ResNeXt101-32x4d [41]

Method	AP	AP_{50}	AP_{75}	AP_s	AP_m	AP_l
ResNet-101 + ImageNet	38.3	58.9	41.1	20.0	42.8	53.0
ResNet-101 + Montage	39.2	59.6	42.0	20.6	43.3	54.8
Δ	+0.9	+0.7	+0.9	+0.6	+0.5	+1.8
X101-32x4d + ImageNet	40.2	61.2	43.2	21.2	44.6	55.7
X101-32x4d + Montage	40.2	61.0	43.0	21.3	44.5	55.7
Δ	0.0	−0.2	−0.2	+0.1	−0.1	0.0

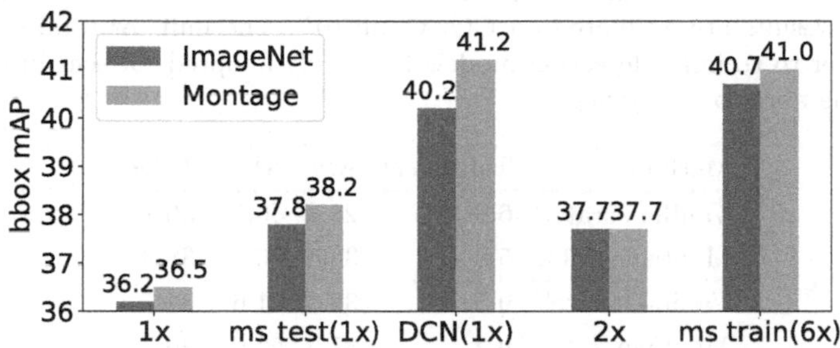

Fig. 7. Comparison between **ImageNet** pre-training and **Montage** pre-training of various strategies on Faster R-CNN FPN framework with ResNet-50 backbone, the results are evaluated on COCO val2017. Strategies include: (1) 1×: serving as original strategy that train for 1× schedule (13 epochs), (2) ms test: adding multi-scale augmentation during test stage, (3) DCN: replace the 3 × 3 convolution layers of stage 2–4 in backbone with 3 × 3 deformable convolution layer [3], (4) 2×: extending the training time to 2× schedule, (5) ms train: implementing multi-scale augmentation during train and test stage and extending training epochs to 6× schedule

It is worth noting that the most obvious improvement has been achieved when replacing some convolution layers to deformable convolution layers. We suspect that this improvement may come from the relief of domain shift between pre-training dataset and detection dataset. Therefore, our approach has the potential of further boosting the performance gains from new designs on backbones.

4.5 Comparison with Vanilla Training Detection from Scratch

We also compare our Montage pre-training strategy with the vanilla training detection from scratch method (denoted as **vanilla scratch** for simplicity). Vanilla scratch and our strategy share an advantage that the entire training process is only based on detection dataset without introducing any external data. However, adopting pre-training process will speed up the convergence of detection models, which helps the models achieve better performance under common training iterations, such as 1× or 2× schedules. The results are presented in Table 3. To make a fair comparison, we keep total training costs similar for the two methods, that is, the total costs in our method include both pre-training and detection training consumptions. We follow the experimental settings for vanilla scratch in [7] where all batch norm layers in the network are replaced by group norm [39]. The batch norm layers are frozen at detection stage when transferring from our pre-trained backbones. The results indicate that even with group normalization, which is proven to improve the performance of detection models, vanilla scratch still shows suboptimal performance compared with our pre-training strategy.

Table 3. Comparison of vanilla scratch and Montage pre-training under similar computation costs. The detection framework is Faster R-CNN with backbone ResNet-50, and the AP results are evaluated on COCO val2017. The unit for total cost is GPU days. 1× refer to training detection models for widely adopted 1× schedule, while 2× to extend the iterations to twice

Method	Total cost	AP	AP_{50}	AP_{75}
Vanilla scratch	6.0	28.6	46.5	30.1
Montage + 1×	5.8	36.3	56.5	38.9
Vanilla scratch	9.5	32.6	51.6	34.7
Montage + 2×	9.2	37.5	57.6	40.7

5 Discussion

The possible reasons for the Montage pre-training being effective are as follows:

First, there is domain gap between the ImageNet dataset and objective detection dataset, such as the data distribution and category. Directly pre-training on the target detection dataset will alleviate the domain gap and help obtain better initialization. However, simply changing the pre-training dataset is not enough and the specially designed pre-training strategy is necessary. Table 4 shows the comparison between simply replacing dataset from ImageNet to MS-COCO and Montage pre-training. We can see that pre-training on MS-COCO classification for the same training time as ours (1.73 GPU days) performs worse than ImageNet classification and our approach. Thus, directly training on MS-COCO by saving the training computational costs leads to drop in detection accuracy. If the training time on MS-COCO is extended to the same as that on ImageNet (6.80 GPU days), the final AP will be similar to ImageNet, but still worse than our approach. Therefore, preserving detection accuracy and saving computational cost at the same time cannot be simply brought by adopting MS-COCO classification dataset, but our Montage pre-training strategy is able to preserve detection performance under lower computation costs.

Table 4. Experimental results on ImageNet pre-training, Montage Pre-training using the same and higher computational costs. The detection framework is Faster R-CNN with backbone ResNet50, and the AP results are evaluated on COCO val2017. The unit of cost is GPU days. ImageNet refers to pre-training on ImageNet dataset and MS-COCO represents training on samples extracted from MS-COCO dataset, as illustrated in Sect. 3.1

Method	Total cost	AP
ImageNet	6.80	34.8
MS-COCO *w/o* Montage (higher cost)	6.80	34.4
MS-COCO *w/o* Montage	1.73	33.5
MS-COCO *w.* Montage (ours)	1.73	36.3

Second, we speculate that the improved training efficiency of Montage pre-training comes from the reduction of redundancy in training dataset. In the training dataset, the amount of foreground and background pixels are imbalanced, especially for the detection dataset. Thus, we design a reasonable sampling strategy to compose training data, which makes the pre-trained networks focus more on positive samples. By discarding the useless pixels and effectively assembling training samples, our Montage pre-training strategy contributes to the reduction of redundancy, which explains the improvements of training efficiency. By assigning soft labels at the regions that overlap with multiple objects leads, more supervised signals are provided for learning better features.

Finally, the ERF-adaptive loss has positive effects on expanding the effective receptive field of the pre-trained models, which provides stronger supervision signals and obtains pre-trained model more appropriate for the detection task. Larger receptive field helps to promote the performance of detection models, as demonstrated in previous works [3,13,19].

6 Conclusions

In this work, we present a choice to obtain cheaper lunch on pre-training for object detection, which is able to reduce the consumption of pre-training to 1/4 compared with the original ImageNet pre-training, while achieving on-par or even higher performance. We define a novel pre-training paradigm based only on detection dataset, which eliminates the burdens of extra training data while retaining the advantage of fast convergence. Our efficient Montage Pre-training facilitates training from scratch, which can reduce the computational cost when directly using network compression and neural architecture search [9,44] for target tasks like object detection. We expect this work would help researchers reduce the trial-and-error cost, inspire more future research on pre-training process, and facilitate new backbone CNN architecture design/search [15] tailored for object detection.

Acknowledgement. This work was supported by SenseTime, the Australian Research Council Grant DP200103223, and Australian Medical Research Future Fund MRFAI000085.

References

1. Brazil, G., Liu, X.: M3D-RPN: monocular 3D region proposal network for object detection. In: The IEEE International Conference on Computer Vision (ICCV), October 2019
2. Chen, K., et al.: Towards accurate one-stage object detection with AP-loss. In: Proceedings of the IEEE Conference on Computer Vision and Pattern Recognition, pp. 5119–5127 (2019)
3. Dai, J., et al.: Deformable convolutional networks. In: Proceedings of the IEEE International Conference on Computer Vision, pp. 764–773 (2017)
4. Divvala, S.K., Hoiem, D., Hays, J.H., Efros, A.A., Hebert, M.: An empirical study of context in object detection. In: 2009 IEEE Conference on Computer Vision and Pattern Recognition, pp. 1271–1278. IEEE (2009)
5. Girshick, R., Donahue, J., Darrell, T., Malik, J.: Rich feature hierarchies for accurate object detection and semantic segmentation. In: Proceedings of the IEEE Conference on Computer Vision and Pattern Recognition, pp. 580–587 (2014)
6. He, K., Fan, H., Wu, Y., Xie, S., Girshick, R.: Momentum contrast for unsupervised visual representation learning. arXiv preprint arXiv:1911.05722 (2019)
7. He, K., Girshick, R., Dollár, P.: Rethinking ImageNet pre-training. In: Proceedings of the IEEE International Conference on Computer Vision, pp. 4918–4927 (2019)
8. He, K., Gkioxari, G., Dollár, P., Girshick, R.: Mask R-CNN. In: Proceedings of the IEEE International Conference on Computer Vision, pp. 2961–2969 (2017)
9. Jiang, C., Xu, H., Zhang, W., Liang, X., Li, Z.: SP-NAS: serial-to-parallel backbone search for object detection. In: Proceedings of the IEEE/CVF Conference on Computer Vision and Pattern Recognition, pp. 11863–11872 (2020)
10. Kornblith, S., Shlens, J., Le, Q.V.: Do better imagenet models transfer better? In: Proceedings of the IEEE Conference on Computer Vision and Pattern Recognition, pp. 2661–2671 (2019)
11. Krizhevsky, A., Sutskever, I., Hinton, G.E.: ImageNet classification with deep convolutional neural networks. In: Advances in Neural Information Processing Systems, pp. 1097–1105 (2012)
12. Law, H., Deng, J.: CornerNet: detecting objects as paired keypoints. In: Proceedings of the European Conference on Computer Vision (ECCV), pp. 734–750 (2018)
13. Li, Y., Chen, Y., Wang, N., Zhang, Z.: Scale-aware trident networks for object detection. In: Proceedings of the IEEE International Conference on Computer Vision, pp. 6054–6063 (2019)
14. Li, Z., Peng, C., Yu, G., Zhang, X., Deng, Y., Sun, J.: DetNet: a backbone network for object detection. arXiv preprint arXiv:1804.06215 (2018)
15. Liang, F., et al.: Computation reallocation for object detection. arXiv preprint arXiv:1912.11234 (2019)
16. Lin, T.Y., Dollár, P., Girshick, R., He, K., Hariharan, B., Belongie, S.: Feature pyramid networks for object detection. In: Proceedings of the IEEE Conference on Computer Vision and Pattern Recognition, pp. 2117–2125 (2017)
17. Lin, T.Y., Goyal, P., Girshick, R., He, K., Dollár, P.: Focal loss for dense object detection. In: Proceedings of the IEEE International Conference on Computer Vision, pp. 2980–2988 (2017)

18. Liu, L., Ouyang, W., Wang, X., Fieguth, P., Chen, J., Liu, X., Pietikäinen, M.: Deep learning for generic object detection: a survey. Int. J. Comput. Vis. **128**(2), 261–318 (2020)
19. Liu, S., Huang, D., et al.: Receptive field block net for accurate and fast object detection. In: Proceedings of the European Conference on Computer Vision (ECCV), pp. 385–400 (2018)
20. Lu, X., Li, B., Yue, Y., Li, Q., Yan, J.: Grid R-CNN. In: Proceedings of the IEEE Conference on Computer Vision and Pattern Recognition, pp. 7363–7372 (2019)
21. Luo, W., Li, Y., Urtasun, R., Zemel, R.: Understanding the effective receptive field in deep convolutional neural networks. In: Advances in Neural Information Processing Systems, pp. 4898–4906 (2016)
22. Ma, X., Liu, S., Xia, Z., Zhang, H., Zeng, X., Ouyang, W.: Rethinking pseudo-lidar representation. In: Proceedings of the European Conference on Computer Vision (ECCV) (2020)
23. Ma, X., Wang, Z., Li, H., Zhang, P., Ouyang, W., Fan, X.: Accurate monocular 3D object detection via color-embedded 3D reconstruction for autonomous driving. In: Proceedings of the IEEE/CVF International Conference on Computer Vision (ICCV), October 2019
24. Mahajan, D., et al.: Exploring the limits of weakly supervised pretraining. In: Proceedings of the European Conference on Computer Vision (ECCV), pp. 181–196 (2018)
25. Manhardt, F., Kehl, W., Gaidon, A.: ROI-10D: monocular lifting of 2D detection to 6D pose and metric shape. In: The IEEE Conference on Computer Vision and Pattern Recognition (CVPR), June 2019
26. Matan, O., Burges, C.J., LeCun, Y., Denker, J.S.: Multi-digit recognition using a space displacement neural network. In: Advances in Neural Information Processing Systems, pp. 488–495 (1992)
27. Ouyang, W., Wang, K., Zhu, X., Wang, X.: Chained cascade network for object detection. In: Proceedings of the IEEE International Conference on Computer Vision, pp. 1938–1946 (2017)
28. Pang, J., Chen, K., Shi, J., Feng, H., Ouyang, W., Lin, D.: Libra R-CNN: towards balanced learning for object detection. In: Proceedings of the IEEE Conference on Computer Vision and Pattern Recognition, pp. 821–830 (2019)
29. Paszke, A., et al.: Pytorch: an imperative style, high-performance deep learning library. In: Advances in Neural Information Processing Systems, pp. 8024–8035 (2019)
30. Peng, J., Sun, M., Zhang, Z.X., Tan, T., Yan, J.: Efficient neural architecture transformation search in channel-level for object detection. In: Advances in Neural Information Processing Systems, pp. 14290–14299 (2019)
31. Redmon, J., Divvala, S., Girshick, R., Farhadi, A.: You only look once: unified, real-time object detection. In: Proceedings of the IEEE Conference on Computer Vision And Pattern Recognition, pp. 779–788 (2016)
32. Ren, S., He, K., Girshick, R., Sun, J.: Faster R-CNN: towards real-time object detection with region proposal networks. In: Advances in Neural Information Processing Systems, pp. 91–99 (2015)
33. Shen, Z., Liu, Z., Li, J., Jiang, Y.G., Chen, Y., Xue, X.: DSOD: learning deeply supervised object detectors from scratch. In: Proceedings of the IEEE International Conference on Computer Vision, pp. 1919–1927 (2017)
34. Shrivastava, A., Gupta, A., Girshick, R.: Training region-based object detectors with online hard example mining. In: Proceedings of the IEEE Conference on Computer Vision and Pattern Recognition, pp. 761–769 (2016)

35. Singh, B., Najibi, M., Davis, L.S.: SNIPER: efficient multi-scale training. In: Advances in Neural Information Processing Systems, pp. 9310–9320 (2018)
36. Sun, C., Shrivastava, A., Singh, S., Gupta, A.: Revisiting unreasonable effectiveness of data in deep learning era. In: Proceedings of the IEEE International Conference on Computer Vision, pp. 843–852 (2017)
37. Szegedy, C., Toshev, A., Erhan, D.: Deep neural networks for object detection. In: Advances in Neural Information Processing Systems, pp. 2553–2561 (2013)
38. Tan, M., Pang, R., Le, Q.V.: EfficientDet: scalable and efficient object detection. arXiv preprint arXiv:1911.09070 (2019)
39. Wu, Y., He, K.: Group normalization. In: Proceedings of the European Conference on Computer Vision (ECCV), pp. 3–19 (2018)
40. Xie, Q., Hovy, E., Luong, M.T., Le, Q.V.: Self-training with noisy student improves imagenet classification. arXiv preprint arXiv:1911.04252 (2019)
41. Xie, S., Girshick, R., Dollár, P., Tu, Z., He, K.: Aggregated residual transformations for deep neural networks. In: Proceedings of the IEEE Conference on Computer Vision and Pattern Recognition, pp. 1492–1500 (2017)
42. Yun, S., Han, D., Oh, S.J., Chun, S., Choe, J., Yoo, Y.: CutMix: regularization strategy to train strong classifiers with localizable features. In: The IEEE International Conference on Computer Vision (ICCV), October 2019
43. Zheng, W.S., Gong, S., Xiang, T.: Quantifying and transferring contextual information in object detection. IEEE Trans. Pattern Anal. Mach. Intell. **34**(4), 762–777 (2011)
44. Zhou, D., et al.: EcoNAS: finding proxies for economical neural architecture search. In: Proceedings of the IEEE/CVF Conference on Computer Vision and Pattern Recognition, pp. 11396–11404 (2020)
45. Zhu, C., He, Y., Savvides, M.: Feature selective anchor-free module for single-shot object detection. In: Proceedings of the IEEE Conference on Computer Vision and Pattern Recognition, pp. 840–849 (2019)
46. Zhu, R., Zhang, S., Wang, X., Wen, L., Shi, H., Bo, L., Mei, T.: ScratchDet: training single-shot object detectors from scratch. In: Proceedings of the IEEE Conference on Computer Vision and Pattern Recognition, pp. 2268–2277 (2019)

Faster Person Re-identification

Guan'an Wang[1,3], Shaogang Gong[2], Jian Cheng[3,4,5], and Zengguang Hou[1,4,5](✉)

[1] The State Key Laboratory of Management and Control of Complex System, Institute of Automation, Chinese Academy of Sciences (CAS), Beijing, China
{wangguanan2015,zengguang.hou}@ia.ac.cn

[2] Queen Mary University of London, London, UK
s.gong@qmul.ac.uk

[3] School of Artificial Intelligence, University of Chinese Academy of Sciences, Beijing, China
jcheng@nlpr.ia.ac.cn

[4] National Laboratory of Pattern Recognition, Institute of Automation, CAS, Beijing, China

[5] CAS Center for Excellence in Brain Science and Intelligence Technology, Beijing, China

Abstract. Fast person re-identification (ReID) aims to search person images quickly and accurately. The main idea of recent fast ReID methods is the hashing algorithm, which learns compact binary codes and performs fast Hamming distance and counting sort. However, a very long code is needed for high accuracy (*e.g.* 2048), which compromises search speed. In this work, we introduce a new solution for fast ReID by formulating a novel Coarse-to-Fine (CtF) hashing code search strategy, which complementarily uses short and long codes, achieving both faster speed and better accuracy. It uses shorter codes to coarsely rank broad matching similarities and longer codes to refine only a few top candidates for more accurate instance ReID. Specifically, we design an All-in-One (AiO) framework together with a Distance Threshold Optimization (DTO) algorithm. In AiO, we simultaneously learn and enhance multiple codes of different lengths in a single model. It learns multiple codes in a pyramid structure, and encourage shorter codes to mimic longer codes by self-distillation. DTO solves a complex threshold search problem by a simple optimization process, and the balance between accuracy and speed is easily controlled by a single parameter. It formulates the optimization target as a F_β score that can be optimised by Gaussian cumulative distribution functions. Experimental results on 2 datasets show that our proposed method (CtF) is not only 8% more accurate but also 5× faster than contemporary hashing ReID methods. Compared with non-hashing ReID methods, CtF is 50× faster with comparable accuracy. Code is available at https://github.com/wangguanan/light-reid.

G. Wang—This work was done when Guan'an Wang was at QMUL supervised by *Prof.* Shaogang Gong.

A. Vedaldi et al. (Eds.): ECCV 2020, LNCS 12353, pp. 275–292, 2020.
https://doi.org/10.1007/978-3-030-58598-3_17

1 Introduction

Person re-identification (ReID) [8,50] aims to match images of a person across disjoint cameras, which is widely used in video surveillance, security and smart city. Many methods [11,16,21,22,26,32,43,50,51] have been proposed for person ReID. However, for higher accuracy, most of them utilize a large deep network to learn high-dimensional real-value features for computing similarities by Euclidean distance and returning a rank list by quick-sort [13]. Quick-sort of high-dimensional deep features can be slow, especially when the gallery set is large. Table 1 shows that the query time per ReID probe image increases massively with the increase of the ReID gallery size; and counting-sort [1] is much more efficient than quick-sort, in which the former has a linear complexity w.r.t the gallery size ($O(n)$) whilst the latter has a logarithm complexity ($O(nlogn)$).

Several fast ReID methods [4,5,7,24,41,47,55,56] have been proposed to increase ReID speed whist retaining ReID accuracy. The common main idea is hashing, which learns a binary code instead of real-value features. To sort binary codes, the inefficient Euclidean distance and quick-sort are replaced by the Hamming-distance and counting-sort [1]. Table 2 shows that computing a Hamming distance between 2048-dimensional binary-codes is 229× faster than that of a Euclidean distance between real-value features.

Table 1. ReID search time per probe image by quick-sort (real-value) and counting-sort (binary). The latter is much faster.

Gallery size	Query time (s)	
	Quick-sort	Counting-sort
1×10^3	3.4×10^{-3}	4.7×10^{-4}
1×10^4	1.0×10^{-1}	2.7×10^{-3}
1×10^5	4.3×10^{-1}	2.7×10^{-2}
1×10^6	6.4×10^0	2.6×10^{-1}
1×10^7	1.1×10^2	2.7×10^0
Per Sample	-	2.6×10^{-7}
Complexity	$O(nlogn)$	$O(n)$

Table 2. Comparing Euclidean and Hamming distances, Euclidean and longer lengths are slow to compute.

Code length	Computation time (s)	
	Euclidean	Hamming
32	6.8×10^{-5}	2.4×10^{-6}
64	1.3×10^{-4}	2.7×10^{-6}
128	2.6×10^{-4}	2.8×10^{-6}
256	5.0×10^{-4}	3.3×10^{-6}
512	1.0×10^{-3}	4.4×10^{-6}
1,024	2.0×10^{-3}	7.1×10^{-6}
2,048	3.9×10^{-3}	1.7×10^{-5}

Different from common image retrieval tasks, which are category-level matching in a close-set, ReID is instance-level matching in an open-set (zero-shot setting). For image retrieval in the ImageNet [28], the classes of training and test sets are the same and imagery appearances of different classes diverse a lot, such as dog, car, and airplane. In contrast, the training and test ReID images have completely different ID classes without any overlap (ZSL) whilst the appearances of different persons can be very similar to subtle changes (fine-grained) on clothing, body characteristics, gender, and carried-objects. The ZSL and fine-grained

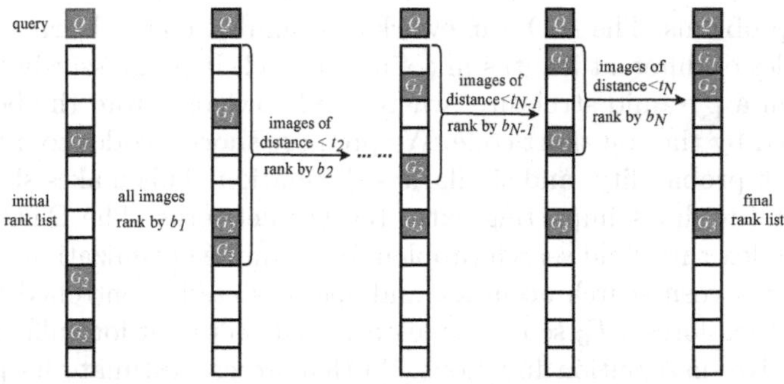

Fig. 1. A Coarse-to-Fine (CtF) hashing code search strategy to speed up ReID, where Q is a query image, $\{G_i\}_{i=1}^{3}$ are the positive images in the gallery set, $B = \{b_k\}_{k=1}^{N}$ are binary codes of lengths $L = \{l_k\}_{k=1}^{N}$, $T = \{k_i\}_{k=2}^{N}$ are Hamming distance thresholds where gallery images are selected by each t_k for further comparison by increasingly longer codes b_k.

characteristics of ReID require state-of-the-art hashing-based fast ReID models [24] to employ very long binary codes, *e.g.* 2048, in order to retain competitive ReID accuracy. However, the binary code length affects significantly the cost of computing Hamming distance. Table 2 shows that computing a Hamming distance between two 2048-dimensional binary codes takes 1.7×10^{-5} s, which is 7× slower than computing that of 32-dimensional binary codes at 2.4×10^{-6} s. This motivates us to solve the following problem: How to yield higher accuracy from hashing-based ReID using shorter binary codes.

To that end, we propose a novel Coarse-to-Fine (CtF) search strategy for faster ReID whilst also retaining competitive accuracy. At test time, our model (CtF) first uses shorter codes to coarsely rank a gallery, then iteratively utilises longer codes to further rank selected top candidates where the top-ranked candidates are defined iteratively by a set of Hamming distance thresholds. Thus, the long codes are only used for a decreasingly fewer matches in ranking in order to reduce the overall search time whilst retaining ReID accuracy. This is an intuitively straightforward idea but not easily computable for ReID due to three difficulties: (1) Coarse-to-fine search requires multiple codes of different lengths. Asymmetrically, computing them with multiple models is both time-consuming and sub-optimal. (2) The coarse ranking must be accurate enough to minimise missing true-match candidates in fine-grained ranking whilst keeping their numbers small, thus reduce the total search time. Paradoxically, shorter codes perform much worse than longer codes in ReID task therefore hard to be sufficiently accurate. (3) The set of distance thresholds for guiding the coarse search affect both final accuracy and overall speed. How to determine *automatically* these thresholds to balance optimally accuracy and speed is both important and nontrivial.

In this work, we propose a novel All-in-One (AiO) framework together with a Distance Threshold Optimization (DTO) algorithm to simultaneously solve these three problems. The AiO framework can simultaneously learn and enhance multiple codes of different lengths in a single model. It progressively learns multiple codes in a pyramid structure, where the knowledge from the bottom long code is shared by the top short code. We promote shorter codes to mimic longer codes by both probability- and similarity- distillation. This makes shorter codes more powerful without importing extra teacher networks. The DTO algorithm solves a complex threshold search problem by a simple optimization process and the balance between search accuracy and speed is easily controlled by a single parameter. It explores a F_β score as the optimization target formulated as Gaussian cumulative distribution functions. So that we can estimate its parameters by the statistics of Gaussian probability distributions modeling the distances of positive and negative pairs. Finally, by maximizing the F_β score, we compute iteratively optimal distance thresholds.

Our contributions are: (1) We propose a novel Coarse-to-Fine (CtF) search strategy for Faster ReID, not only speeding up hashing ReID, but also improving their accuracy. To the best of our knowledge, this is the first work to introduce such search strategy into ReID. (2) A novel All-in-One (AiO) framework is proposed to learn and enhance multiple codes of different lengths in a single framework by viewing it as a multi-channel self-distillation problem. In the framework, the multiple codes are learned in a pyramid structure and shorter codes mimic longer codes via probability- and similarity- distillation loss. (3) A novel Distance Threshold Optimization (DTO) algorithm is proposed to find the optimal thresholds for coarse-to-fine search by concluding the threshold search task to a F_β distance optimization problem. The F_β score is represented with Gaussian cumulative distribution functions, whose mean and variance can be estimated by fitting a small validation set. (4) Extensive experimental results on two datasets show that, our proposed method is 50× faster than non-hashing ReID methods, 5× faster and 8% more accurate than hashing ReID methods.

2 Related Works

In this work, we wish to solve the fast ReID problem under the framework of hashing by proposing an All-in-One (AiO) hashing learning module and a Distance Threshold Optimization (DTO) algorithm. Thus, we mainly discuss the related works including non-fast person re-identification (ReID) task, fast ReID task and hashing algorithm.

Person Re-identification. Person re-identification addresses the problem of matching pedestrian images across disjoint cameras [8]. The key challenges lie in the large intra-class and small inter-class variation caused by different views, poses, illuminations, and occlusions. Existing methods can be grouped into hand-crafted descriptors [21,26,43], metric learning methods [16,22,51] and deep learning algorithms [11,32,35–38,50]. The goal of hand-crafted descriptors is to design robust features. Metric learning aims to make a pair of true matches have a relatively smaller distance than that of a wrong match pair in a discriminant manner. Deep learning algorithms adopt deep neural networks to straightly learn robust and discriminative features in an end-to-end manner and achieve the best performance. However, all the ReID methods above learn real-value features for high accuracy, which is slow.

Hashing Algorithm. Hashing algorithm mainly divided into unsupervised and (semi-)supervised ones. Unsupervised hashing methods (LSH [6], SH [40], ITQ [19]) employ unlabeled data even no data. (Semi-)Supervised (SSH [39], BRE [17], KSH [23], SDH [30], SSGAH [34]) utilize labeled information to improve binary codes. Recently, inspired by powerful deep networks, some deep hashing methods (CNNH [42], NINH [18], DPSH [20]) have been proposed and achieve much better performance. They usually utilize a CNN to extract meaningful features, formulate the hashing function as a fully-connected layer with *tanh*/*sigmoid* activation function, and quantize features by *signature* function. The framework can be optimized with a related layer or some iteration strategies. However, all the hashing methods are designed for close-set category-level retrieval tasks, which cannot be directly used for person ReID, an open-set fine-grained search problem.

Fast Person Re-identification. Fast ReID methods aims to search in a fast speed meanwhile obtaining accuracy as high as possible. The main idea of those methods is hashing algorithm, which learns binary code instead of real-value features. Based on the binary codes, the inefficient Euclidean distance and quick-sorting can be replaced by efficient Hamming distance and counting sort. Zheng *et al.* [47] learn cross-view binary codes using two hash functions for two different views. Wu *et al.* [41] simultaneously learn both CNN feature and hash functions to get robust yet discriminative features and similarity-preserving binary codes. CSBT [4] solves the cross-camera variations problem by employing a subspace projection to maximize intra-person similarity and inter-person discrepancies. In [55] integrate spatial information for discriminative features by representing horizontal parts to binary codes. ABC [24] improves binary codes by implicitly fits the feature distribution to a pre-defined binary one with Wasserstein distance. However, all the fast ReID methods take very long binary codes (*e.g.* 2048) for high accuracy. Different from them, we propose a coarse-to-fine search strategy which complementarily uses codes of different lengths, obtaining not only faster speed but also higher accuracy.

3 Proposed Method

In this work, we propose a coarse-to-fine (CtF) search strategy for fast and accurate ReID. For effectively implementing the strategy, we design an All-in-One (AiO) framework together with a Distance Threshold Optimization (DTO) algorithm. The former learns and enhances multiple codes of different lengths in a single framework. The latter finds the optimal distance thresholds to balance time and accuracy.

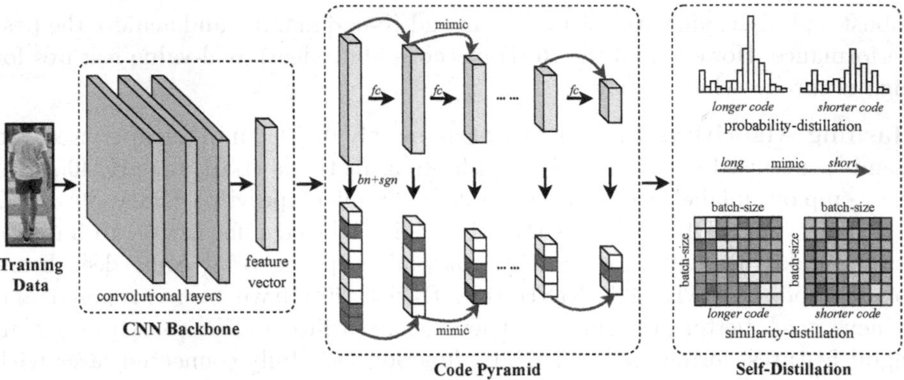

Fig. 2. All-in-One framework. It learns and enhances multiple codes of different lengths in a single framework with a code pyramid structure and self-distillation learning.

3.1 Coarse-to-Fine Search

As we illustrated in the introduction section, although the long binary codes can get high accuracy, it takes much longer time than short codes. This motivates us to think about that can we reduce the usage of long codes to further speed hashing ReID methods up. Thus, a simple but efficient solution is complementarily using both short and long codes. Here, shorter codes fast return a rough rank list of gallery, and longer codes carefully refine a small number of top candidates. Figure 1 shows its procedures.

Although the idea is straightforward, there are still three difficulties preventing it being applied to ReID. (1) Coarse-to-fine search requires multiple codes of different lengths. Asymmetrically, computing them with multiple models is both time-consuming and sub-optimal. (2) The coarse ranking must be accurate enough to minimise missing true-match candidates in fine-grained ranking whilst keeping their numbers small, thus reduce the total search time. Paradoxically, shorter codes perform much worse than longer codes in ReID task. (3) The set of distance thresholds for guiding the coarse search affect both final accuracy and overall speed. How to determine *automatically* these thresholds to balance optimally accuracy and speed is both important and nontrivial. To solve the problems, we propose an All-in-One (AiO) framework and a Distance Threshold Optimization (DTO) algorithm. Please see the next two parts for more details.

3.2 All-in-One Framework

The All-in-One (AiO) framework aims to simultaneously learn and enhance multiple codes of different lengths in a single model, whose architecture can be seen in Fig. 2. Specifically, it first utilizes a convolutional network to extract the real-value feature vectors, then learns multiple codes of different lengths in a pyramid structure, finally enhances the codes by encouraging shorter codes mimic longer codes via self-distillation.

Learn Multiple Codes in a Pyramid Structure. The code pyramid learns multiple codes of different lengths, where the shorter codes are based on the longer codes. With such a structure, we can not only learn many codes in one shot, but also share the knowledge of longer codes with shorter codes. The equations are as below:

$$v_0 = F(x), \quad v_k = FC_k(v_{k-1}), \quad k \in 1, 2, ..., N, \tag{1}$$

where x is input image, F is the CNN backbone, N is the code number, $V = \{v_k\}_{k=1}^N$ are the real-value feature vectors with different lengths $L = \{l_k\}_{k=1}^N$, FC_k is the fully-connected layers with l_{k-1} input- and l_k output-sizes. After getting real-value features of different lengths, we can obtain their binary codes $B = \{b_k\}_{k=1}^N$ in the following equation.

$$b_k = sgn(bn(v_k)), \tag{2}$$

where bn is the batch normalization layer, sgn is the symbolic function. We use the batch normalization layer because it normalizes the real-value features to be symmetric to 0 and reduces the quantization loss.

Enhance Codes with Self-distillation Learning. As we discussed in the introduction section, the coarse ranking must be accurate enough to minimise missing true-match candidates in fine-grained ranking. Inspired by [12,33], we introduce self-distillation learning to enhance the multiple codes in a single framework without importing extra teacher network. Different from conventional distillation models, which imports an extra large teacher network to supervise a small student network, we perform distillation learning in a single network and achieve better performance, which is important for fast ReID.

Specifically, our self-distillation learning is composed of a probability- and a similarity- distillation. The probability-distillation transfers the instance-level knowledge in a from of softened class scores. Its formulation is given by

$$\mathcal{L}_{pro} = \frac{1}{N-1} \sum_{k=1}^{N-1} \mathcal{L}_{ce}(\sigma(\frac{z_{k+1}}{T}), \sigma(\frac{\hat{z}_k}{T})), \tag{3}$$

where $\mathcal{L}_{ce}(\cdot, \cdot)$ denotes the cross-entropy loss, σ is the softmax function, $\hat{z}_k/z_{k+1}$ means the output logits of the binary code b_k/b_{k+1}, $\hat{z}_k$ means it act as a teacher and fixed during training, T is a temperature hyperparameter, which is set 1.0 empirically. The similarity-distillation transfers the knowledge of relationship from longer codes to shorter one, whose formulation is in Eq. (4). This is motivated by that as an image search task, ReID features should also focus on the relationship among samples, *i.e.* to what extent the sample A is similar/dissimilar to sample B.

$$\mathcal{L}_{sim} = \frac{1}{N-1} \sum_{k=1}^{N-1} \sum_{i,j} || \frac{1}{l_{k+1}} G_{k+1}^{i,j} - \frac{1}{l_k} \hat{G}_k^{i,j} ||^2, \tag{4}$$

Algorithm 1. Distance Threshold Optimization

Input: Trained Model in Eq.(2), Validation Data (X_v, Y_v)

Output: Thresholds $\{T_i\}_{i=2}^{N}$

```
for k = {1, 2, ..., n − 1} do
    B_k: Extract binary codes of validation set with length l_k via Eq.(2)
    D^r: Hamming distances of relevant pairs (b_k^i, b_k^j), where y^i = y^j
    D^n: Hamming distances of non-relevant pairs (b_{k−1}^i, b_{k−1}^j), where y^i ≠ y^j
    PDF^r, PDF^n: Probability distribution function of D^r and D^n of in Eq.(7)
    CDF^r, CDF^n: Cumulative Distribution Function of D^r and D^n in Eq.(7)
    t_{n+1}: Maximize F_β score in Eq.(8) and return t_{n+1}
return T = {t_i}_{i=2}^N
```

where $G_k^{i,j}/G_{k+1}^{i,j}$ is the Hamming distance between b_k^i/b_{k+1}^i and b_k^j/b_{k+1}^j, $b_{k/k+1}^{i/j}$ is the binary code of image x_i/x_j with length l_k/l_{k+1}, the $\hat{G}$ means that G acts as a label and is fixed during the optimization process, thus contributes nothing to the gradients.

Overall Objective Function and Training. Recent progresses on ReID have shown the effectiveness of the classification [50] and triplet [11] losses. Thus, our final objective function includes our proposed probability- and similarity-distillation losses together with the classification and triplet losses as the final objective function. The formulation can be found in Eq. (5),

$$\mathcal{L} = \mathcal{L}_{ce} + \mathcal{L}_{tri} + \lambda_1 \mathcal{L}_{prob} + \lambda_2 \mathcal{L}_{sim} \tag{5}$$

Considering that the mapping function sgn in Eq. (2) is discrete and Hamming distance in Eq. (2) is not differentiable, a natural relaxation [20] is utilised in Eq. (5) by replacing sgn with $tanh$ and changing the Hamming distance to the inner-product distance. Finally, our All-in-One framework can be optimized in an end-to-end way by minimizing the loss in Eq. (5).

3.3 Distance Threshold Optimization

After getting the multiple codes of different lengths $B = \{b_i\}_{i=1}^{N}$, we can perform the Coarse-to-Fine (CtF) search. There are two tips in CtF search, *i.e.* high accuracy and fast speed. For fast speed, the candidate number returned by coarse search should be small. For high accuracy, the candidates returned by coarse search should include relevant images as more as possible. But the two requirements are naturally conflicting. Thus, it is important to find the proper thresholds to optimally balance the two targets, *i.e.* both high accuracy and fast speed. One simple solution is brute search via cross-validation. However, the search space is too large. For example, if we have multiple binary codes of lengths $L = \{32, 128, 512, 2048\}$, the complexity of the brute search will be $\prod_L > 4 \times 10^9$ times.

In this part, we propose a novel Distance Threshold Optimization (DTO) algorithm which solves the time-consuming brute parameter search task with a simple optimization process. Specifically, inspired by [9], we first explicitly formulate the two sub-targets as two scores in Eq. (6), *i.e.* precision (P) and recall (R) scores. Then we balance the two sub-targets by mixing the two scores with a single parameter β and get F_β score in Eq. (6).

$$P = \frac{TP}{TP+FP}, \quad R = \frac{TP}{TP+FN}, \quad F_\beta = (\beta^2+1)\frac{PR}{\beta^2 P + R} \tag{6}$$

Here, TP is the number of relevant images in the candidates, FP is the number of non-relevant images in the candidates and FN is not retrieved relevant samples. As we can see, the precision score P means the rate of relevant images in the candidates. Usually a high P means a small candidate number, which is good for fast speed. The recall score R represents the rate of returned relevant samples in the total relevant samples. A high R score means more returned relevant samples, which is important for high accuracy. The F_β mixed the precision and recall scores with a parameter β, which considers both speed and accuracy.

$$PDF(t) = \frac{1}{\sigma\sqrt{2\pi}} exp(-\frac{(t-u)^2}{\sigma\sqrt{2}}), \quad CDF(t) = \frac{1}{2}(1 + erf\frac{t-u}{\sigma\sqrt{2}}) \tag{7}$$

$$F_\beta = \frac{CDF^r(\beta^2+1)}{CDF^n + CDF^r + \beta^2(1 - CDF^n + CDF^r)} \tag{8}$$

Considering that TP/FP/FN are statistics which cannot be optimized, we replace them with two Gaussian cumulative distribution functions in form of Eq. (7) (right), whose parameters u and σ are estimated by fitting a validation set using the Gaussian probability distribution function in Eq. (7) (left). Finally, by maximizing the F_β in Eq. (8), we can get the optimal distance thresholds $T = \{t_k\}_{k=2}^N$ balanced by β.

4 Experiments

4.1 Dataset and Evaluation Protocols

Datasets. We extensively evaluate our proposed method on two common datasets (Market-1501 [49] and DukeMTMC-reID [52]) and one large-scale dataset (Market-1501+500k [49]). The Market-1501 dataset contains 1,501 identities observed under 6 cameras, which are splited into 12,936 training, 3,368 query and 15,913 gallery images. The Market-1501+500k enlarges the gallery of Market-1501 with extra 500,000 distractors, making it more challenging for both accuracy and speed. DukeMTMC-reID contains 1,404 identities with 16,5522 training, 2,228 query and 17,661 gallery images.

Evaluation Protocols. For accuracy, we use standard metrics including Cumulative Matching Characteristic (CMC) curves and mean average precision (mAP). All the results are from a single query setting. To evaluate speed, we use average query time per image, including distance computation and sorting time. For fair evaluation, we do not use any parallel algorithm for distance computation and sorting.

Table 3. Comparisons with non-hashing ReID methods using real-value features of different lengths on Market-1501 and DukeTMTC-reID. **B**: binary code, **R**: real-value feature. Longer real-value features have higher accuracy but slower query speed. Our model CtF (including AiO) has very fast query speed (two orders of magnitude faster) and comparable accuracy with non-hashing ReID methods.

Methods	Code		Market-1501			DukeMTMC-reID		
	Type	Length	R1(%)	mAP(%)	***Q.Time***(s)	R1(%)	mAP(%)	***Q.Time***(s)
PSE [29]	**R**	1,536	78.7	56.0	–	–	–	–
PN-GAN [27]	**R**	1,024	89.4	72.6	–	73.6	53.2	–
IDE [50]	**R**	2,048	88.1	72.8	–	69.4	55.4	–
Camstyle [54]	**R**	2,048	88.1	68.7	–	75.3	53.5	–
PIE [48]	**R**	2,062	87.7	69.0	–	79.8	62.0	–
BoT [25]	**R**	2,048	94.1	85.7	2.2×10^0	86.4	76.4	2.0×10^0
SPReID [14]	**R**	10,240	92.5	81.3	–	84.4	71.0	–
PCB [32]	**R**	12,288	93.8	81.6	6.9×10^0	83.3	69.2	6.3×10^0
VPM [31]	**R**	14,336	93.0	80.8	–	83.6	72.6	–
CtF (ours)	**B**	2,048	**93.7**	**84.9**	$\mathbf{4.6 \times 10^{-2}}$	**87.6**	**74.8**	$\mathbf{3.7 \times 10^{-2}}$

4.2 Implementation Details

We implemented our method with Pytorch on a PC with 2.6Ghz Intel Core i5 CPUs, 10 GB memory, and a NVIDIA RTX 2080Ti GPU. For a fair comparison and following [24,25], we use ResNet50 [10] as the CNN backbone. In training stage, each image is resized to 256×128 and augmented by horizontal flip and random erasing [53]. A batch data includes 64 images from 16 different persons, where every person includes 4 images. The lengths $L = \{l_k\}_{k=1}^{N}$ of multiple codes are empirically set $\{32, 128, 512, 2048\}$. The margin in the triplet loss in Eq. (5) is 0.3. The framework is optimized by Adam [15] with total epochs 120. Its initial learning rate is 0.00035, which is warmed up for 10 epochs and decayed to its $0.1\times$ and $0.01\times$ at 40 and 70 epochs. We randomly split the training data into a training and a validation set according to 6 : 4, then decide the parameters via cross-validation, After that, we train our method with all training data. λ_1 and λ_2 in Eq. (5) are set as 1.0 and 1,000, and β in Eq. (8) is set 2.0. The three paramters are decided via cross validation. Code is available at github[1].

4.3 Comparisons with Non-hashing ReID Methods

Non-hashing ReID use longer real-value features, such as 2048-dimensional $float64$ features, for a better accuracy. This significantly affects their speed, *i.e.* query time. Table 3 shows that our proposed CtF (including AiO) method is significantly faster than non-hashing ReID methods (two orders of magnitude). CtF also achieves very competitive accuracy with close Rank-1 (93.7% vs. 94.1%) and mAP (87.6% vs. 86.4%) scores of the best non-hashing ReID mehtod BoT

[1] https://github.com/wangguanan/light-reid.

Table 4. Comparisons with state-of-the-art hashing ReID methods on Market-1501 and DukeTMTC-reID. AiO+k means learning multiple codes with all-in-one framework, but querying with only the code of length l_k. Aio+CtF not only learns multiple codes with all-in-one framework, but also query with coarse-to-fine search strategy. Our AiO+CtF achieve a good balance between accuracy and speed.

Methods	Code length	Market-1501			DukeMTMC-reID		
		R1(%)	mAP(%)	**Q.Time**(s)	R1(%)	mAP(%)	**Q.Time**(s)
DRSCH [44]	512	17.1	11.5	-	19.3	13.6	-
DSRH [45]	512	27.1	17.7	-	25.6	18.6	-
HashNet [3]	512	29.2	19.1	-	40.8	28.6	-
DCH [2]	512	40.7	20.2	-	57.4	37.3	-
CSBT [4]	512	42.9	20.3	-	47.2	33.1	-
PDH [55]	512	44.6	24.3	-	-	-	-
DeepSSH [46]	512	46.5	24.1	-	-	-	-
ABC [24]	512	69.4	48.5	9.8×10^{-2}	69.9	52.6	7.5×10^{-2}
ABC [24]	2,048	81.4	64.7	2.8×10^{-1}	82.5	61.2	2.0×10^{-1}
CtF (ours)	AiO+32	60.0	37.7	3.4×10^{-2}	49.5	28.7	2.3×10^{-2}
	AiO+128	88.9	71.0	4.2×10^{-2}	78.6	59.4	3.2×10^{-2}
	AiO+512	92.8	82.2	9.8×10^{-2}	85.4	71.6	7.5×10^{-2}
	AiO+2,048	93.7	85.4	2.8×10^{-1}	87.7	75.7	2.0×10^{-1}
	AiO+CtF	**93.7**	**84.0**	$\mathbf{4.6 \times 10^{-2}}$	**87.6**	**74.8**	$\mathbf{3.7 \times 10^{-2}}$

[25] on Market-1501 and DukeMTMC-reID, and better than all the other non-hashing methods using different feature length, of which 5 methods have features shorter than 2,062 (PSE [29], IDE [50], PN-GAN [27], CamStyle [54], PIE [48]) and 3 methods have features longer than 10,240 (SPReID [14], PCB [32], VPM [31]). Overall, longer feature usually contributes to higher accuracy but with slower speed. For example, SPReID, PCB and VPM take features longer than 10,240 and achieves 92%–93% and 83%-84% Rank-1 scores on Market-1501 and DukeMTMC-reID datasets, respectively. The others utilize features no longer than 2,048 achieving Rank-1 score less than 92% and 80%. On the other hand, the query speed of those methods with long features is much slower. For example, PCB takes 6.9s and 6.3s for query each image on the two datasets respectively. This is 3-4× slower than IDE with 2s on either dataset. Specifically, CtF performs much faster than non-hashing methods and significantly, it achieves much better accuracy than comparable length real-value feature model. For example, CtF achieves 93.7%/87.6% Rank-1 scores on Market-1501/DukeMTMC-reID, as compared to BoT having 94.1%/86.4% respectively. This is because CtF (including AiO) utilizes all-in-one framework together with coarse-to-fine search strategy, which not only learns powerful binary code, but also complementarily uses short and long codes for both high accuracy and fast speed.

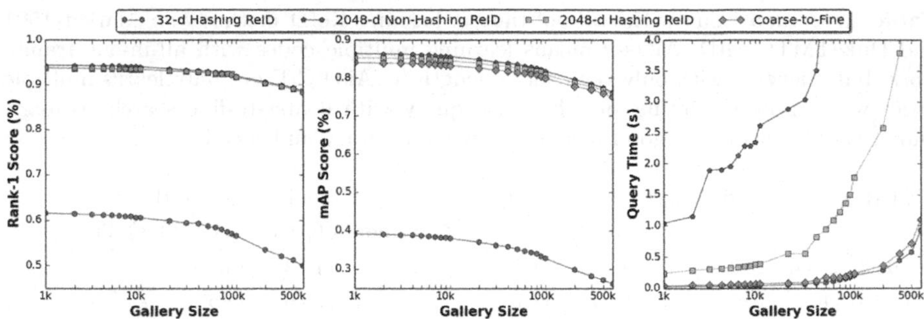

Fig. 3. Experimental results on large-scale ReID dataset Market-1501+500k. Our Coarse-to-Fine (CtF) get a high accuracy comparable with non-hashing ReID method of long code and fast speed comparable with hashing ReID method of short code.

4.4 Comparisons with Hashing ReID Methods

Hashing ReID methods learn binary codes using a hashing algorithm. Binary codes are good for speed but sacrifice model accuracy. To mitigate this problem, the state-of-the-art hashing ReID methods usually employ long codes such as 2048. In binary coding, 2048 is relatively very long as compared to the more commonly used 512 length, unlike in real-value feature length compared above. Table 4 shows that CtF (with AiO) not only achieves the best accuracy (even compared to much shorter code length used by other hashing methods), but also is significantly faster than existing hashing ReID methods (even compared to the same code length used by other hashing methods). Overall, hashing ReID methods usually perform much worse than non-hashing methods. For example, best non-hashing ReID methods achieves 94.1% and 86.4% Rank-1 scores on Market-1501 and DukeMTMC-reID respectively. But the best hashing ReID method only obtains 81.4% and 82.5% Rank-1 scores. Moreover, existing hashing ReID models can increase accuracy by using longer code length and compromising speed. For example, ABC with 512-dimensional binary codes achieves 69.4%/69.9% Rank-1 scores and $9.8/7.5 \times 10^{-2}s$ query time per probe image. When using 2048 binary codes, its Rank-1 scores increase to 81.4%/82.5% with query time slow down to $2.8/2.0 \times 10^{-1}s$. This observation is also verified with our method CtF (with AiO) using different code lengths. Importantly, our method CtF (with AiO) significantly outperforms all existing hashing ReID methods in terms of both accuracy (R1 12.3% or 5.1% better) and speed (5× faster). Specifically, CtF with AiO achieves high accuracy very close to AiO without CtF using 2048 code length, but yields significant speed advantage that is comparable to much shorter 128 binary code length. CtF obtains 93.7% and 87.6% Rank-1 scores, similar to AiO without CtF of a fixed 2048 length at 93.7% and 87.7%.

Table 5. Analysis of the All-in-One (AiO) framework. **CP**: learn multiple codes in a pyramid structure, otherwise separate models. **SD**: enhance binary codes via self-distillation. **B** and **R** mean binary codes and real-value features, respectively.

AiO	CP	SD	Feature type	Rank-1(%)					mAP(%)				
				32	128	512	2048	CtF	32	128	512	2048	CtF
×	×	×	**B**	-	-	-	-	-	-	-	-	-	
✓	×	×	**B**	25.5	84.8	92.3	93.8	92.5	33.9	67.5	81.4	85.3	75.1
✓	✓	×	**B**	54.4	87.8	92.7	93.8	93.0	35.0	72.2	81.7	85.3	80.2
✓	✓	✓	**B**	60.0	88.9	92.9	93.8	93.7	37.7	71.0	82.0	85.3	84.0
upper bound			**R**	82.7	90.9	93.4	94.2	-	66.7	78.9	84.3	85.4	-

4.5 Evaluation on Large-Scale ReID

Gallery size affects significantly ReID search accuracy and speed. To show the effectiveness of our proposed Coarse-to-Fine (CtF) search strategy, we evaluated it on a large-scale ReID dataset Market1501+500k. The dataset is based on the Market-1501 and enlarged with 500,000 distractors. The experimental results are shown in Fig. 3. We can observe the following phenomenons.

Firstly, with the increase of gallery size, for all methods, the Rank-1 and mAP scores decrease, and the ReID speed per probe image slows down gradually. The reason is that more gallery images is more likely to contain more difficult samples. They make ReID search more challenging. Also, the extra gallery images significantly increase the time for computing all the distance comparisons and sorting required for ReID each probe image. Secondly, the non-hashing method with 2048-D real-value feature achieves the best accuracy but the worst time. This is because the real-value feature is more discriminative but slow to compute and sort. Thirdly, for hashing ReID methods, the 2048-D binary code obtains comparable ReID accuracy to that of the non-hashing model, but 10× faster. This is because Hamming distances and counting sort are faster to compute. ReID speed of 32-D binary code is 5× faster than that of 2048-D binary codes, but its accuracy drops dramatically. Finally, the proposed CtF model achieves a comparable accuracy to that of the non-hashing method but the advantage of similar speed to that of a hashing ReID method of 32-D binary code. Critically, the advantage is independent of the gallery size. Overall, these experiments demonstrate the effectiveness of CtF for a large-scale ReID task.

4.6 Model Analysis

Analysis of AiO. The All-in-One (AiO) framework aims to learn and enhance multiple codes of different lengths in a single model. It uses code pyramid (CP) structure and self-distillation (SD) learning. Results are in Table 5. Firstly, longer codes contribute to better accuracy. This can be seen in all settings no matter whether CP or SD is used and what code type is. Secondly, when using short codes, real-value features is much better than binary ones. But for long codes,

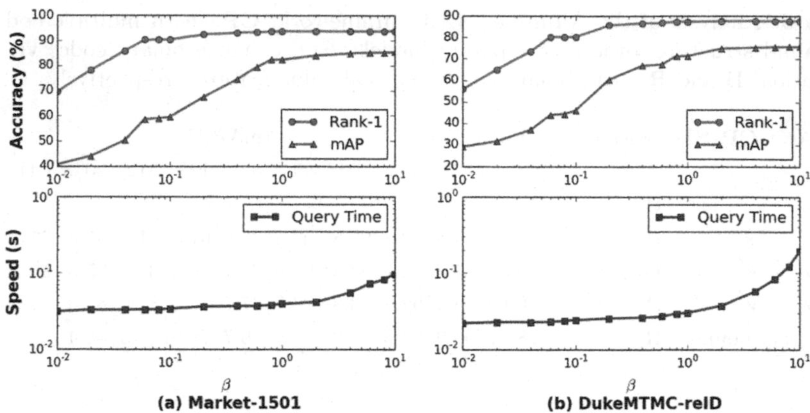

Fig. 4. Accuracy and speed controlled by β. With the increase of β, the accuracy increases and speed becomes slow gradually.

they obtain similar accuracy. For example, the 32-dimensional real-value feature obtains 82.7% Rank-1 score, outperforming the 32-dimensional binary code by 60%, where the latter achieved only 25.5%. But when using 2048 code length, binary codes and real-valure features both achieve approx. Rank-1 94% and mAP 84%. This suggests that the quantization loss of short codes is significantly worse than that of longer codes. Thirdly, learning with code pyramid (CP) structure or self-distillation (SD) improves short codes significantly. For example, CP+SD boosts the 32-dimensional binary codes from 25.5% to 60.0% in Rank-1 score, upto 35% gain. It is evident that both code pyramid (CP) structure and self-distillation (SD) learning contribute to the effectiveness of the coarse-to-fine (CtF) search strategy, and significantly improve model performance.

Analysis of DTO. We further analyzed parameter β of the Distance Threshold Optimization (DTO) algorithm, which controls the balance between ReID accuracy and speed. Figure 4 show the model accuracy and speed using different β value on Market-1501 and DukeMTMC-reID. Firstly, it is evident that the value of β has a good control of accuracy and speed, increasing β slows down the speed but improves accuracy. For example, when $\beta = 10^{-2}$, ReID is fastest at approx. 0.03 and 0.02 s to ReID each probe image on Market-1501 and DukeMTMC-reID, but with mAP scores only at 40% and 30%. In contrast, $\beta = 10^1$ gives high mAP 85% and 75%, but the query speed is 5× slower at approx. 0.1 and 0.2 s. Secondly, when β is close to 10^0, Rank-1 and mAP are almost peaked with a good balance on speed.

5 Conclusion

In this work, we proposed a novel Coarse-to-Fine (CtF) search strategy for faster person re-identification whilst also improve accuracy on conventional hashing

ReID. Extensive experiments show that our method is 5× faster than existing hashing ReID methods but achieves comparable accuracy with non-hashing ReID models that are 50× slower.

Acknowledgement. This work was supported in part by the National Key R&D Program of China (Grant 2018YFC2001700), by the National Natural Science Foundation of China (Grants 61720106012, and U1913601), by the Beijing Natural Science Foundation (Grants L172050), by the Strategic Priority Research Program of Chinese Academy of Sciences (Grant XDB32040000), by the Youth Innovation Promotion Association of CAS (2020140), the Alan Turing Institute Turing Fellowship, and Vision Semantics Ltd.

References

1. Bajpai, K., Kots, A.: Implementing and analyzing an efficient version of counting sort (e-counting sort). Int. J. Comput. Appl.**98**(9) (2014)
2. Cao, Y., Long, M., Liu, B., Wang, J.: Deep cauchy hashing for hamming space retrieval. In: 2018 IEEE/CVF Conference on Computer Vision and Pattern Recognition, pp. 1229–1237 (2018)
3. Cao, Z., Long, M., Wang, J., Yu, P.S.: HashNet: deep learning to hash by continuation. In: 2017 IEEE International Conference on Computer Vision (ICCV), pp. 5609–5618 (2017)
4. Chen, J., Wang, Y., Qin, J., Liu, L., Shao, L.: Fast person re-identification via cross-camera semantic binary transformation. In: 2017 IEEE Conference on Computer Vision and Pattern Recognition (CVPR), pp. 5330–5339 (2017)
5. Chen, J., Wang, Y., Wu, R.: Person re-identification by distance metric learning to discrete hashing. In: 2016 IEEE International Conference on Image Processing (ICIP), pp. 789–793 (2016)
6. Datar, M., Immorlica, N., Indyk, P., Mirrokni, V.S.: Locality-sensitive hashing scheme based on p-stable distributions. In: Scg 2004: Proceedings of the Twentieth Symposium on Computational Geometry, pp. 253–262 (2004)
7. Fang, W., Hu, H.M., Hu, Z., Liao, S., Li, B.: Perceptual hash-based feature description for person re-identification. Neurocomputing **272**(1), 520–531 (2018)
8. Gong, S., Cristani, M., Yan, S., Loy, C.C.: Person re-identification (2014)
9. Goutte, C., Gaussier, E.: A probabilistic interpretation of precision, recall and F-score, with implication for evaluation. In: Losada, D.E., Fernández-Luna, J.M. (eds.) ECIR 2005. LNCS, vol. 3408, pp. 345–359. Springer, Heidelberg (2005). https://doi.org/10.1007/978-3-540-31865-1_25
10. He, K., Zhang, X., Ren, S., Sun, J.: Deep residual learning for image recognition. In: 2016 IEEE Conference on Computer Vision and Pattern Recognition (CVPR), pp. 770–778 (2016)
11. Hermans, A., Beyer, L., Leibe, B.: In defense of the triplet loss for person re-identification. arXiv preprint arXiv:1703.07737 (2017)
12. Hinton, G.E., Vinyals, O., Dean, J.: Distilling the knowledge in a neural network. arXiv preprint arXiv:1503.02531 (2015)
13. Hoare, C.A.: Quicksort. Comput. J. **5**(1), 10–16 (1962)
14. Kalayeh, M.M., Basaran, E., Gökmen, M., Kamasak, M.E., Shah, M.: Human semantic parsing for person re-identification. In: Proceedings of the IEEE Conference on Computer Vision and Pattern Recognition, pp. 1062–1071 (2018)

15. Kingma, D.P., Ba, J.: Adam: a method for stochastic optimization. arXiv preprint arXiv:1412.6980 (2014)
16. Koestinger, M., Hirzer, M., Wohlhart, P., Roth, P.M., Bischof, H.: Large scale metric learning from equivalence constraints. In: 2012 IEEE Conference on Computer Vision and Pattern Recognition, pp. 2288–2295. IEEE (2012)
17. Kulis, B., Darrell, T.: Learning to hash with binary reconstructive embeddings. In: International Conference on Neural Information Processing Systems, pp. 1042–1050 (2009)
18. Lai, H., Pan, Y., Liu, Y., Yan, S.: Simultaneous feature learning and hash coding with deep neural networks. In: The IEEE Conference on Computer Vision and Pattern Recognition (CVPR), June 2015
19. Lazebnik, S.: Iterative quantization: a procrustean approach to learning binary codes. In: IEEE Conference on Computer Vision and Pattern Recognition, pp. 817–824 (2011)
20. Li, W.J., Wang, S., Kang, W.C.: Feature learning based deep supervised hashing with pairwise labels. In: International Joint Conference on Artificial Intelligence, pp. 1711–1717 (2016)
21. Liao, S., Hu, Y., Zhu, X., Li, S.Z.: Person re-identification by local maximal occurrence representation and metric learning. In: Proceedings of the IEEE Conference on Computer Vision and Pattern Recognition, pp. 2197–2206 (2015)
22. Liao, S., Li, S.Z.: Efficient PSD constrained asymmetric metric learning for person re-identification. In: Proceedings of the IEEE International Conference on Computer Vision, pp. 3685–3693 (2015)
23. Liu, W., Wang, J., Ji, R., Jiang, Y.G.: Supervised hashing with kernels. In: Computer Vision and Pattern Recognition, pp. 2074–2081 (2012)
24. Liu, Z., Qin, J., Li, A., Wang, Y., Gool, L.V.: Adversarial binary coding for efficient person re-identification. In: 2019 IEEE International Conference on Multimedia and Expo (ICME), pp. 700–705 (2019)
25. Luo, H., Gu, Y., Liao, X., Lai, S., Jiang, W.: Bag of tricks and a strong baseline for deep person re-identification. In: Proceedings of the IEEE Conference on Computer Vision and Pattern Recognition Workshops (2019)
26. Ma, B., Su, Y., Jurie, F.: Covariance descriptor based on bio-inspired features for person re-identification and face verification. Image Vis. Comput. **32**(6–7), 379–390 (2014)
27. Qian, X., et al.: Pose-normalized image generation for person re-identification. In: Proceedings of the European Conference on Computer Vision (ECCV), pp. 661–678 (2018)
28. Russakovsky, O., et al.: ImageNet large scale visual recognition challenge. Int. J. Comput. Vis. **115**(3), 211–252 (2015)
29. Sarfraz, M.S., Schumann, A., Eberle, A., Stiefelhagen, R.: A pose-sensitive embedding for person re-identification with expanded cross neighborhood re-ranking. In: 2018 IEEE/CVF Conference on Computer Vision and Pattern Recognition, pp. 420–429 (2018)
30. Shen, F., Shen, C., Liu, W., Shen, H.T.: Supervised discrete hashing. In: The IEEE Conference on Computer Vision and Pattern Recognition (CVPR) pp. 37–45 (2015)
31. Sun, Y., et al.: Perceive where to focus: learning visibility-aware part-level features for partial person re-identification, pp. 393–402 (2019)
32. Sun, Y., Zheng, L., Yang, Y., Tian, Q., Wang, S.: Beyond part models: person retrieval with refined part pooling (and a strong convolutional baseline). In: Proceedings of the European Conference on Computer Vision (ECCV), pp. 480–496 (2018)

33. Tung, F., Mori, G.: Similarity-preserving knowledge distillation. In: Proceedings of the IEEE International Conference on Computer Vision, pp. 1365–1374 (2019)
34. Wang, G., Hu, Q., Cheng, J., Hou, Z.: Semi-supervised generative adversarial hashing for image retrieval. In: Proceedings of the European Conference on Computer Vision (ECCV), pp. 469–485 (2018)
35. Wang, G., et al.: High-order information matters: learning relation and topology for occluded person re-identification. arXiv preprint arXiv:2003.08177 (2020)
36. Wang, G., Yang, Y., Cheng, J., Wang, J., Hou, Z.: Color-sensitive person re-identification. In: IJCAI 2019 Proceedings of the 28th International Joint Conference on Artificial Intelligence, pp. 933–939 (2019)
37. Wang, G., Zhang, T., Cheng, J., Liu, S., Yang, Y., Hou, Z.: RGB-infrared cross-modality person re-identification via joint pixel and feature alignment. In: 2019 IEEE/CVF International Conference on Computer Vision (ICCV), pp. 3622–3631 (2019)
38. Wang, G., Zhang, T., Yang, Y., Cheng, J., Chang, J., Hou, Z.: Cross-modality paired-images generation for RGB-infrared person re-identification. In: AAAI 2020 : The Thirty-Fourth AAAI Conference on Artificial Intelligence (2020)
39. Wang, J., Kumar, S., Chang, S.F.: Semi-supervised hashing for large-scale search. IEEE Trans. Pattern Anal. Mach. Intell. **34**(12), 2393–2406 (2012)
40. Weiss, Y., Torralba, A., Fergus, R.: Spectral hashing. In: International Conference on Neural Information Processing Systems, pp. 1753–1760 (2008)
41. Wu, L., Wang, Y., Ge, Z., Hu, Q., Li, X.: Structured deep hashing with convolutional neural networks for fast person re-identification. Comput. Vis. Image Underst. **167**, 63–73 (2017)
42. Xia, R., Pan, Y., Lai, H., Liu, C., Yan, S.: Supervised hashing for image retrieval via image representation learning (2014)
43. Yang, Y., Yang, J., Yan, J., Liao, S., Yi, D., Li, S.Z.: Salient color names for person re-identification. In: Fleet, D., Pajdla, T., Schiele, B., Tuytelaars, T. (eds.) ECCV 2014. LNCS, vol. 8689, pp. 536–551. Springer, Cham (2014). https://doi.org/10.1007/978-3-319-10590-1_35
44. Zhang, R., Lin, L., Zhang, R., Zuo, W., Zhang, L.: Bit-scalable deep hashing with regularized similarity learning for image retrieval and person re-identification. IEEE Trans. Image Process. Publ. IEEE Signal Process. Soc. **24**(12), 4766 (2015)
45. Zhao, F., Huang, Y., Wang, L., Tan, T.: Deep semantic ranking based hashing for multi-label image retrieval. In: 2015 IEEE Conference on Computer Vision and Pattern Recognition (CVPR), pp. 1556–1564 (2015)
46. Zhao, Y., Luo, S., Yang, Y., Song, M.: DeepSSH: deep semantic structured hashing for explainable person re-identification. In: 2018 25th IEEE International Conference on Image Processing (ICIP), pp. 1653–1657 (2018)
47. Zheng, F., Shao, L.: Learning cross-view binary identities for fast person re-identification. In: IJCAI 2016 Proceedings of the Twenty-Fifth International Joint Conference on Artificial Intelligence, pp. 2399–2406 (2016)
48. Zheng, L., Huang, Y., Lu, H., Yang, Y.: Pose-invariant embedding for deep person re-identification. IEEE Trans. Image Process. **28**(9), 4500–4509 (2019)
49. Zheng, L., Shen, L., Tian, L., Wang, S., Wang, J., Tian, Q.: Scalable person re-identification: a benchmark. In: Proceedings of the IEEE International Conference on Computer Vision, pp. 1116–1124 (2015)
50. Zheng, L., Yang, Y., Hauptmann, A.G.: Person re-identification: past, present and future. arXiv preprint arXiv:1610.02984 (2016)
51. Zheng, W.S., Gong, S., Xiang, T.: Reidentification by relative distance comparison. IEEE Trans. Pattern Anal. Mach. Intell. **35**(3), 653–668 (2013)

52. Zheng, Z., Zheng, L., Yang, Y.: Unlabeled samples generated by GAN improve the person re-identification baseline in vitro. arXiv preprint arXiv:1701.07717 (2017), https://academic.microsoft.com/paper/2949257576
53. Zhong, Z., Zheng, L., Kang, G., Li, S., Yang, Y.: Random erasing data augmentation. arXiv preprint arXiv:1708.04896 (2017)
54. Zhong, Z., Zheng, L., Zheng, Z., Li, S., Yang, Y.: Camera style adaptation for person re-identification. In: 2018 IEEE/CVF Conference on Computer Vision and Pattern Recognition, pp. 5157–5166 (2018). https://academic.microsoft.com/paper/2963289251
55. Zhu, F., Kong, X., Zheng, L., Fu, H., Tian, Q.: Part-based deep hashing for large-scale person re-identification. IEEE Trans. Image Process. **26**(10), 4806–4817 (2017)
56. Zhu, X., Wu, B., Huang, D., Zheng, W.S.: Fast open-world person re-identification. IEEE Trans. Image Process. **27**(5), 2286–2300 (2018)

Quantization Guided JPEG Artifact Correction

Max Ehrlich[1,2(✉)], Larry Davis[1], Ser-Nam Lim[2], and Abhinav Shrivastava[1]

[1] University of Maryland, College Park, MD 20742, USA
{maxehr,lsd}@umiacs.umd.edu, abhinav@cs.umd.edu
[2] Facebook AI, New York, NY 10003, USA
sernamlim@fb.com

Abstract. The JPEG image compression algorithm is the most popular method of image compression because of it's ability for large compression ratios. However, to achieve such high compression, information is lost. For aggressive quantization settings, this leads to a noticeable reduction in image quality. Artifact correction has been studied in the context of deep neural networks for some time, but the current methods delivering state-of-the-art results require a different model to be trained for each quality setting, greatly limiting their practical application. We solve this problem by creating a novel architecture which is parameterized by the JPEG file's quantization matrix. This allows our single model to achieve state-of-the-art performance over models trained for specific quality settings. ...

Keywords: JPEG · Discrete Cosine Transform · Artifact correction · Quantization

1 Introduction

The JPEG image compression algorithm [43] is ubiquitous in modern computing. Thanks to its high compression ratios, it is extremely popular in bandwidth constrained applications. The JPEG algorithm is a lossy compression algorithm, so by using it, some information is lost for a corresponding gain in saved space. This is most noticable for low quality settings.

For highly space-constrained scenarios, it may be desirable to use aggressive compression. Therefore, algorithmic restoration of the lost information, referred to as artifact correction, has been well studied both in classical literature and in the context of deep neural networks.

While these methods have enjoyed academic success, their practical application is limited by a single architectural defect: they train a single model per JPEG quality level. The JPEG quality level is an integer between 0 and 100,

Electronic supplementary material The online version of this chapter (https://doi.org/10.1007/978-3-030-58598-3_18) contains supplementary material, which is available to authorized users.

A. Vedaldi et al. (Eds.): ECCV 2020, LNCS 12353, pp. 293–309, 2020.
https://doi.org/10.1007/978-3-030-58598-3_18

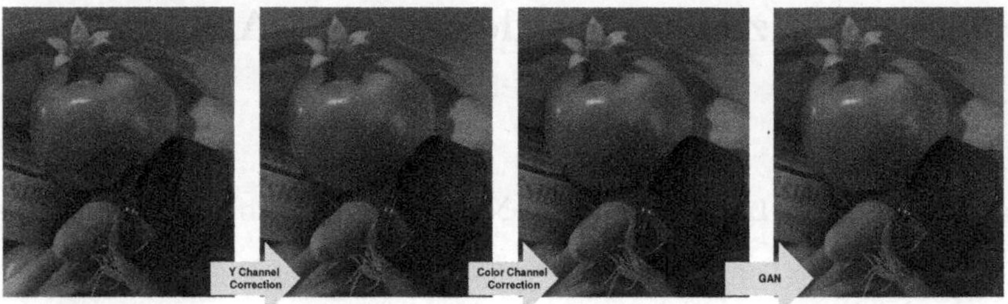

Fig. 1. Correction process. Excerpt from ICB RGB 8bit dataset "hdr.ppm". Input was compressed at quality 10.

where 100 indicates very little loss of information and 0 indicates the maximum loss of information. Not only is this expensive to train and deploy, but the quality setting is not known at inference time (it is not stored with the JPEG image [43]) making it impossible to use these models in practical applications. Only recently have methods begun considering the "blind" restoration scenario [23,24] with a single network, with mixed results compared to non-blind methods.

We solve this problem by creating a single model that uses quantization data, which is stored in the JPEG file. Our CNN model processes the image entirely in the DCT [2] domain. While previous works have recognized that the DCT domain is less likely to spread quantization errors [45,49], DCT domain-based models alone have historically not been successful unless combined with pixel domain models (so-called "dual domain" models). Inspired by recent methods [7–9,16], we formulate fully DCT domain regression. This allows our model to be parameterized by the quantization matrix, an 8×8 matrix that directly determines the quantization applied to each DCT coefficient. We develop a novel method for parameterizing our network called Convolution Filter Manifolds, an extension of the Filter Manifold technique [22]. By adapting our network weights to the input quantization matrix, our single network is able to handle a wide range of quality settings. Finally, since JPEG images are stored in the YCbCr color space, with the Y channel containing more information than the subsampled color channels, we use the reconstructed Y channel to guide the color channel reconstructions. As in [53], we observe that using the Y channel in this way achieves good color correction results. Finally, since regression results for artifact correction are often blurry, as a result of lost texture information, we fine-tune our model using a GAN loss specifically designed to restore texture. This allows us to generate highly realistic reconstructions. See Fig. 1 for an overview of the correction flow.

To summarize, our contributions are:

1. A single model for artifact correction of JPEG images at any quality, parameterized by the quantization matrix, which is state-of-the-art in color JPEG restoration.
2. A formulation for fully DCT domain image-to-image regression.
3. Convolutional Filter Manifolds for parameterizing CNNs with spatial side-channel information.

2 Prior Work

Pointwise Shape-Adaptive DCT [10] is a standard classical technique which uses thresholded DCT coefficients reconstruct local estimates of the input signal. Yang *et al.* [47] use a lapped transform to approximate the inverse DCT on the quantized coefficients.

More recent techniques use convolutional neural networks [26,39]. ARCNN [8] is a regression model inspired by superresolution techniques; L4/L8 [40] continues this work. CAS-CNN [5] adds hierarchical skip connections and a multi-scale loss function. Liu *et al.* [27] use a wavelet-based network for general denoising and artifact correction, which is extended by Chen *et al.* [6]. Galteri *et al.* [12] use a GAN formulation to achieve more visually appealing results. S-Net [52] introduces a scalable architecture that can produce different quality outputs based on the desired computation complexity. Zhang *et al.* [50] use a dense residual formulation for image enhancement. Tai *et al.* [42] use persistent memory in their restoration network.

Liu *et al.* [28] introduce the dual domain idea in the sparse coding setting. Guo and Chao [17] use convolutional autoencoders for both domains. DMCNN [49] extends this with DCT rectifier to constrain errors. Zheng *et al.* [51] target color images and use an implicit DCT layer to compute DCT domain loss using pixel information. D3 [45] extends Liu *et al.* [28] by using a feed-forward formulation for parameters which were assumed in [28]. Jin *et al.* [20] extend the dual domain concept to separate streams processing low and high frequencies, allowing them to achieve competitive results with a fraction of the parameters.

The latest works examine the "blind" scenario that we consider here. Zhang *et al.* [48] formulate general image denoising and apply it to JPEG artifact correction with a single network. DCSC uses convolution features in their sparse coding scheme [11] with a single network. Galteri *et al.* [13] extend their GAN work with an ensemble of GANs where each GAN in the ensemble is trained to correct artifacts of a specific quality level. They train an auxiliary network to classify the image into the quality level that it was compressed with. The resulting quality level is used to pick a GAN from the ensemble to use for the final artifact correction. Kim *et al.* [24] also use an ensemble method based on quality factor estimation. AGARNET [23] uses a single network by learning a per-pixel quality factor extending the concept [13] from a single quality factor to a per-pixl map. This allows them to avoid the ensemble method and using a single network with two inputs.

3 Our Approach

Our goal is to design a single model capable of JPEG artifact correction at any quality. Towards this, we formulate an architecture that is parameterized by the quantization matrix.

Recall that a JPEG quantization matrix captures the amount of rounding applied to DCT coefficients and is indicative of information lost during compression. A key contribution of our approach is utilizing this quantization matrix

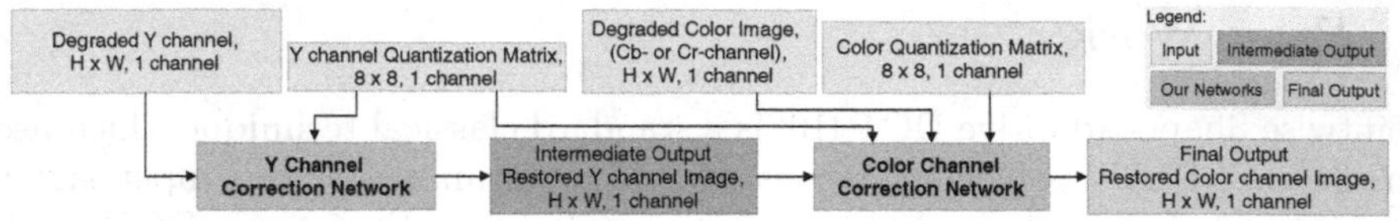

Fig. 2. Overview. We first restore the Y channel of the input image, then use the restored Y channel to correct the color channels which have much worse input quality.

directly to guide the restoration process using a fully DCT domain image-to-image regression network. JPEG stores color data in the YCbCr colorspace. The compressed Y channel is much higher quality compared to CbCr channels since human perception is less sensitive to fine color details than to brightness details. Therefore, we follow a staged approach: first restoring artifacts in the Y channel and then using the restored Y channel as guidance to restore the CbCr channels.

An illustrative overview of our approach is presented in Fig. 2. Next, we present building blocks utilized in our architecture in Sect. 3.1, that allow us to parameterize our model using the quantization matrix and operate entirely in the DCT domain. Our Y channel and color artifact correction networks are described in Sect. 3.2 and Sect. 3.3 respectively, and finally the training details in Sect. 3.4.

3.1 Building Blocks

By creating a single model capable of JPEG artifact correction at any quality, our model solves a significantly harder problem than previous works. To solve it, we parameterize our network using the 8×8 quantization matrix available with every JPEG file. We first describe Convolutional Filter Manifolds (CFM), our solution for adaptable convolutional kernels parameterized by the quantization matrix. Since the quantization matrix encodes the amount of rounding per each DCT coefficient, this parameterization is most effective in the DCT domain, a domain where CNNs have previously struggled. Therefore, we also formulate artifact correction as fully DCT domain image-to-image regression and describe critical frequency-relationships-preserving operations.

Convolutional Filter Manifold (CFM). Filter Manifolds [22] were introduced as a way to parameterize a deep CNN using side-channel scalar data. The method learns a manifold of convolutional kernels, which is a function of a scalar input. The manifold is modeled as a three-layer multilayer perceptron. The input to this network is the scalar side-channel data, and the output vector is reshaped to the shape of the desired convolutional kernel and then convolved with the input feature map for that layer.

Recall that in the JPEG compression algorithm, a quantization matrix is derived from a scalar quality setting to determine the amount of rounding to apply, and therefore the amount of information removed from the original image. This quantization matrix is then stored in the JPEG file to allow for correct scaling of the DCT coefficients at decompression time. This quantization matrix is

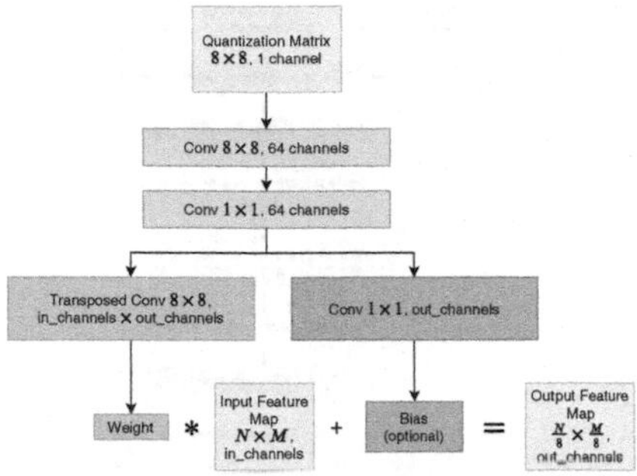

Fig. 3. Convolutional filter manifold, as used in our network. Note that the convolution with the input feature map is done with stride-8.

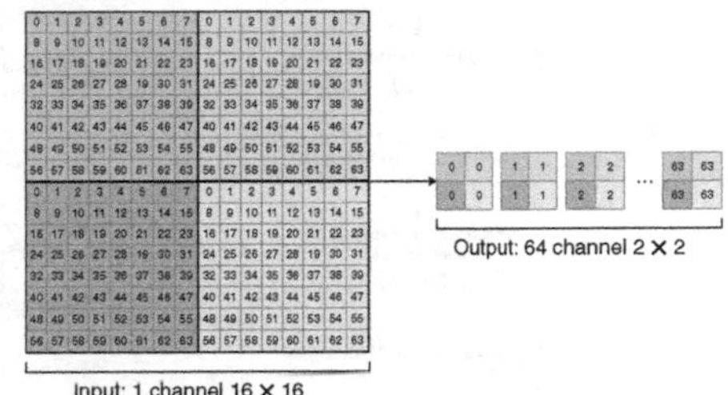

Fig. 4. Coefficient rearrangement. Frequencies are arranged channelwise giving an image with 64 times the number of channels at $\frac{1}{8}$th the size. This can then be convolved with 64 groups per convolution to learn per-frequency filters.

then a strong signal for the amount of information lost. However, the quantization matrix is an 8×8 matrix with spatial structure, applying the Filter Manifold technique to it has the same drawbacks as processing images with multilayer perceptrons, *e.g.*, a large number of parameters and a lack of spatial relationships.

To solve this, we propose an extension to create Convolutional Filter Manifolds (CFM), replacing the multilayer perceptron by a lightweight three-layer CNN. The input to the CNN is our quantization matrix, and the output is reshaped to the desired convolutional kernel shape and convolved with the input feature map as in the Filter Manifold method. For our problem, we follow the network structure in Fig. 3 for each CFM layer. However, this is a general technique and can be used with a different architecture when spatially arranged side-channel data is available.

Coherent Frequency Operations. In prior works, DCT information has been used in dual-domain models [45,49]. These models used standard 3×3 convolutional kernels with U-Net [35] structures to process the coefficients. Although the DCT is a linear map on image pixels [9,38], ablation studies in prior work show that the DCT network alone is not able to surpass even classical artifact correction techniques.

Although the DCT coefficients are arranged in a grid structure of the same shape as the input image, that spatial structure does not have the same meaning as pixels. Image pixels are samples of a continuous signal in two dimensions. DCT coefficients, however, are samples from different, orthogonal functions and the two-dimensional arrangement indexes them. This means that a 3×3 convolutional kernel is trying to learn a relationship not between spatially related samples of the same function as it was designed to do, but rather between samples from completely unrelated functions. Moreover, it must maintain this structure throughout the network to produce a valid DCT as output. This is the root cause

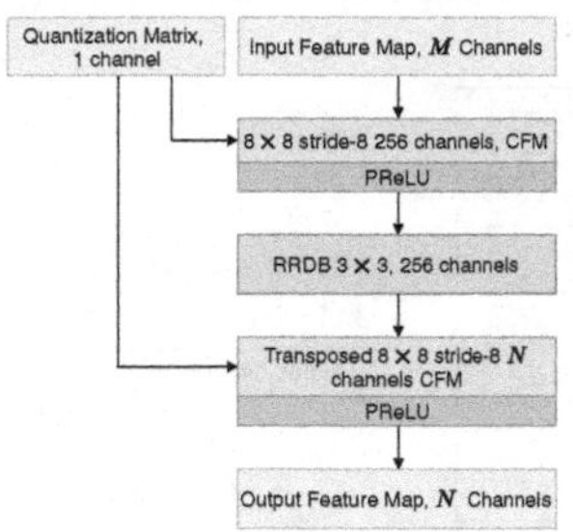

Fig. 5. BlockNet. Both the block generator and decoder are parameterized by the quanitzation matrix.

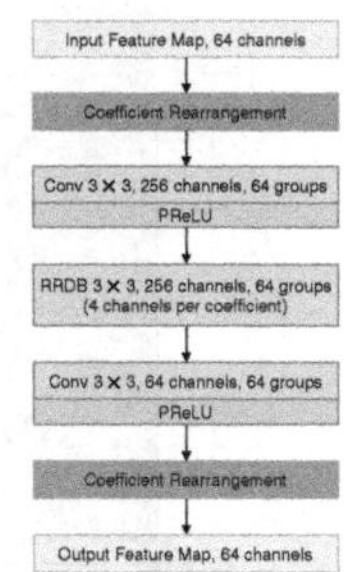

Fig. 6. FrequencyNet. Note that the 256 channels in the RRDB layer actually compute 4 channels per frequency.

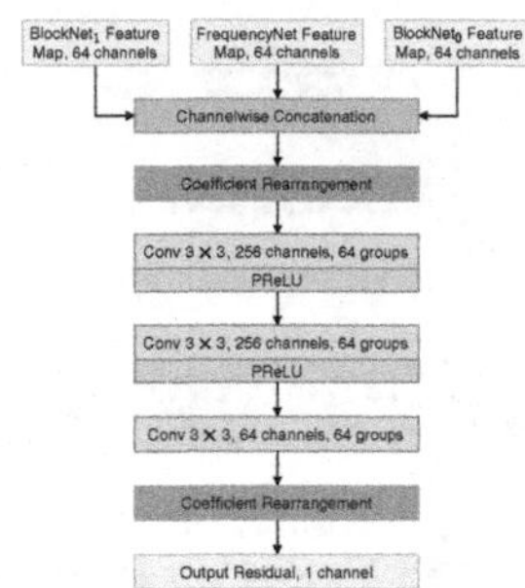

Fig. 7. Fusion subnetwork. Outputs from all three subnetworks are fused to produce the final residual.

of CNN's poor performance on DCT coefficients for image-to-image regression, semantic segmentation, and object detection (Note that this should not affect whole image classification performance as in [14,16]).

A class of recent techniques [7,29], which we call Coherent Frequency Operations for their preservation of frequency relationships, are used as the building block for our regression network. The first layer is an 8×8 stride-8 layer [7], which computes a representation for each block (recall that JPEG blocks are non-overlapping 8×8 DCT coefficients). This block representation, which is one eighth the size of the input, can then be processed with a standard CNN.

The next layer is designed to process each frequency in isolation. Since each of the 64 coefficients in an 8×8 JPEG block corresponds to a different frequency, the input DCT coefficients are first rearranged so that the coefficients corresponding to different frequencies are stored channelwise (see Fig. 4). This gives an input, which is again one eighth the size of the original image, but this time with 64 channels (one for each frequency). This was referred to as Frequency Component Rearrangement in [29]. We then use convolutions with 64 groups to learn per-frequency convolutional weights.

Combining these two operations (block representation using 8×8 8-stride and frequency component rearrangement) allows us to match state-of-the-art pixel and dual-domain results using only DCT coefficients as input and output.

3.2 Y Channel Correction Network

Our primary goal is artifact correction of full color images, and we again leverage the JPEG algorithm to do this. JPEG stores color data in the YCbCr colorspace. The color channels, which contribute less to the human visual response, are both subsampled and more heavily quantized. Therefore, we employ a larger network to correct only the Y channel, and a smaller network which uses the restored Y channel to more effectively correct the Cb and Cr color channels.

Subnetworks. Utilizing the building blocks developed earlier, our network design proceeds in two phases: block enhancement, which learns a quantization invariant representations for each JPEG block, and frequency enhancement, which tries to match each frequency reconstruction to the regression target. These phases are fused to produce the final residual for restoring the Y channel. We employ two purpose-built subnetworks: the block network (BlockNet) and the frequency network (FrequencyNet). Both of these networks can be thought of as separate image-to-image regression models with a structure inspired by ESRGAN [44], which allows sufficient low-level information to be preserved as well as allowing sufficient gradient flow to train these very deep networks. Following recent techniques [44], we remove batch normalization layers. While recent works have largely replaced PReLU [18] with LeakyReLU [12,13,31,44], we find that PReLU activations give much higher accuracy.

BlockNet. This network processes JPEG blocks to restore the Y channel (refer to Fig. 5). We use the 8×8 stride-8 coherent frequency operations to create a block representation. Since this layer is computing a block representation from all the input DCT coefficients, we use a Convolutional Filter Manifold (CFM) for this layer so that it has access to quantization information. This allows the layer to learn the quantization table entry to DCT coefficient correspondence with the goal to output a quantization-invariant block representation. Since there is a one to one correspondence between the quantization table entry and rounding applied to a DCT coefficient, this motivates our choice to operate entirely in the DCT domain. We then process these quantization-invariant block representations with Residual-in-Residual Dense Blocks (RRDB) from [44]. RRDB layers are an extension of the commonly used residual block [19] and define several recursive and highly residual layers. Each RRDB has 15 convolution layers, and we use a single RRDB for the block network with 256 channels. The network terminates with another 8×8 stride-8 CFM, this time transposed, to reverse the block representation back to its original form so that it can be used for later tasks.

FrequencyNet. This network, shown in Fig. 6, processes the individual frequency coefficients using the Frequency Component Rearrangement technique (Fig. 4). The architecture of this network is similar to BlockNet. We use a single 3×3 convolution to change the number of channels from the 64 input channels to the 256 channels used by the RRDB layer. The single RRDB layers processes feature maps with 256 channels and 64 groups yielding 4 channels per frequency. An output 3×3 convolution transforms the 4 channel output to the 64 output channels, and the coefficients are rearranged back into blocks for later tasks.

Final Network. The final Y channel artifact correction network is shown in Fig. 8. We observe that since the FrequencyNet processes frequency coefficients in isolation, if those coefficients were zeroed out by the compression process, then it can make no attempt at restoring them (since they are zero valued they would be set to the layer bias). This is common with high frequencies by design, since they have larger quanitzation table entries and they contribute less to the human visual response. We, therefore, lead with the BlockNet to restore

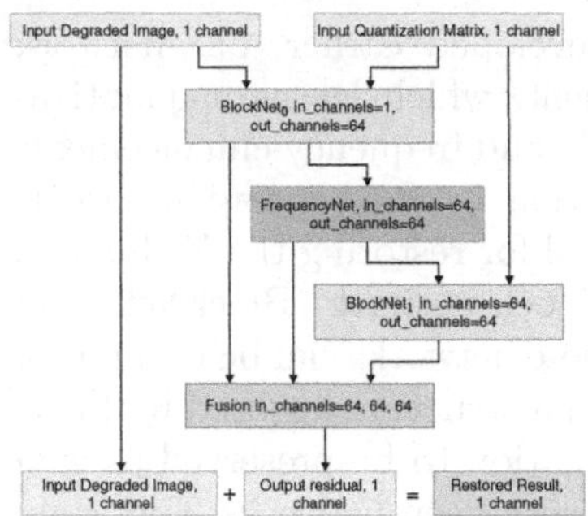

Fig. 8. Y channel network. We include two copies of the BlockNet, one to perform early restoration of high frequency coefficients, and one to work on the restored frequencies. All three subnetworks contribute to the final result using the fusion subnetwork.

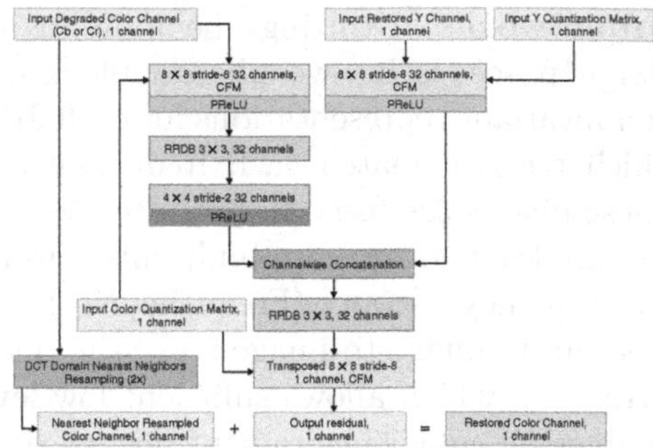

Fig. 9. Color channel network. Color channels are downsampled, so the block representation is upsampled using a learned upsampling. The Y and color channel block representations are then concatenated to guide the color channel restoration. Cb and Cr channels are processed independently with the same network.

high frequencies. We then pass the result to the FrequencyNet, and its result is then processed by a second block network to restore more information. Finally, a three-layer fusion network (see Fig. 7 and 8) fuses the output of all three subnetworks into a final result. Having all three subnetworks contribute to the final result in this way allows for better gradient flow. The effect of fusion, as well as the three subnetworks, is tested in our ablation study. The fusion output is treated as a residual and added to the input to produce the final corrected coefficients for the Y channel.

3.3 Color Correction Network

The color channel network (Fig. 9) processes the Cb and Cr DCT coefficients. Since the color channels are subsampled with respect to the Y channel by half, they incur a much higher loss of information and lose the structural information which is preserved in the Y channel. We first compute the block representation of the downsampled color channel coefficients using a CFM layer, then process them with a single RRDB layer. The block representation is then upsampled using a 4×4 stride-2 convolutional layer. We compute the block representation of the restored Y channel, again using a CFM layer. The block representations are concatenated channel-wise and processed using a single RRDB layer before being transformed back into coefficient space using a transposed 8×8 stride-8 CFM. By concatenating the Y channel restoration, we give the network structural information that may be completely missing in the color channels. The result of this network is the color channel residual. This process is repeated individually for each color channel with a single network learned on Cb and Cr. The output residual is added to nearest-neighbor upsampled input coefficients to give the final restoration.

3.4 Training

Objective. We use two separate objective functions to train, an error loss and a GAN loss. Our error loss is based on prior works which minimize the l_1 error of the result and the target image. We additionally maximize the Structural Similarity (SSIM) [46] of the result since SSIM is generally regarded as a closer metric to human perception than PSNR. This gives our final objective function as

$$\mathcal{L}_{\text{JPEG}}(x, y) = \|y - x\|_1 - \lambda \text{SSIM}(x, y) \tag{1}$$

where x is the network output, y is the target image, and λ is a balancing hyperparameter.

A common phenomenon in JPEG artifact correction and superresolution is the production of a blurry or textureless result. To correct for this, we fine tune our fully trained regression network with a GAN loss. For this objective, we use the relativistic average GAN loss $\mathcal{L}_G^{Ra}$ [21], we use l_1 error to prevent the image from moving too far away from the regression result, and we use preactivation network-based loss [44]. Instead of a perceptual loss that tries to keep the outputs close in ImageNet-trained VGG feature space used in prior works, we use a network trained on the MINC dataset [4], for material classification. This texture loss provided only marginal benefit in ESRGAN [44] for super-resolution. We find it to be critical in our task for restoring texture to blurred regions, since JPEG compression destroys these fine details. The texture loss is defined as

$$\mathcal{L}_{\text{texture}}(x, y) = \|\text{MINC}_{5,3}(y) - \text{MINC}_{5,3}(x)\|_1 \tag{2}$$

where $\text{MINC}_{5,3}$ indicates that the output is from layer 5 convolution 3. The final GAN loss is

$$\mathcal{L}_{\text{GAN}}(x, y) = \mathcal{L}_{\text{texture}}(x, y) + \gamma \mathcal{L}_G^{Ra}(x, y) + \nu \|x - y\|_1 \tag{3}$$

with γ and ν balancing hyperparameters. We note that the texture restored using the GAN model is, in general, not reflective of the regression target at inference time and actually produces worse numerical results than the regression model despite the images looking more realistic.

Staged Training. Analogous to our staged restoration, Y channel followed by color channels, we follow a staged training approach. We first train the Y channel correction network using $\mathcal{L}_{\text{JPEG}}$. We then train the color correction network using $\mathcal{L}_{\text{JPEG}}$ keeping the Y channel network weights frozen. Finally, we train the entire network (Y and color correction) with $\mathcal{L}_{\text{GAN}}$.

4 Experiments

We validate the theoretical discussion in the previous sections with experimental results. We first describe the datasets we used along with the training procedure we followed. We then show artifact correction results and compare them with previous state-of-the-art methods. Finally, we perform an ablation study. Please see our supplementary material for further results and details.

Table 1. Color artifact correction results. PSNR/PSNR-B/SSIM format. Best result in bold, second best underlined. JPEG column gives input error. For ICB, we used the RGB 8bit dataset.

Dataset	Quality	JPEG	ARCNN[8]	MWCNN [27]	IDCN [51]	DMCNN [49]	Ours
Live-1	10	25.60 / 23.53 / 0.755	26.66 / 26.54 / 0.792	27.21 / 27.02 / 0.805	27.62 / 27.32 / 0.816	27.18 / 27.03 / 0.810	**27.65 / 27.40 / 0.819**
	20	27.96 / 25.77 / 0.837	28.97 / 28.65 / 0.860	29.54 / 29.23 / 0.873	**30.01** / 29.49 / 0.881	29.45 / 29.08 / 0.874	29.92 / **29.51 / 0.882**
	30	29.25 / 27.10 / 0.872	30.29 / 29.97 / 0.891	30.82 / 30.45 / 0.901	-	-	**31.21 / 30.71 / 0.908**
BSDS500	10	25.72 / 23.44 / 0.748	26.83 / 26.65 / 0.783	27.18 / 26.93 / 0.794	27.61 / 27.22 / 0.805	27.16 / 26.95 / 0.799	**27.69 / 27.36 / 0.810**
	20	28.01 / 25.57 / 0.833	29.00 / 28.53 / 0.853	29.45 / 28.96 / 0.866	**29.90** / 29.20 / 0.873	29.35 / 28.84 / 0.866	29.89 / **29.29 / 0.876**
	30	29.31 / 26.85 / 0.869	30.31 / 29.85 / 0.887	30.71 / 30.09 / 0.895	-	-	**31.15 / 30.37 / 0.903**
ICB	10	29.31 / 28.07 / 0.749	30.06 / 30.38 / 0.744	30.76 / 31.21 / 0.779	31.71 / 32.02 / 0.809	30.85 / 31.31 / 0.796	**32.11 / 32.47 / 0.815**
	20	31.84 / 30.63 / 0.804	32.24 / 32.53 / 0.778	32.79 / 33.32 / 0.812	33.99 / 34.37 / 0.838	32.77 / 33.26 / 0.830	**34.23 / 34.67 / 0.845**
	30	33.02 / 31.87 / 0.830	33.31 / 33.72 / 0.807	34.11 / 34.69 / 0.845	-	-	**35.20 / 35.67 / 0.860**

4.1 Experimental Setup

Datasets and Metrics. For training, we use the DIV2k and Flickr2k [1] datasets. DIV2k consists of 900 images, and the Flickr2k dataset contains 2650 images. We preextract 256 × 256 patches from these images taking 30 random patches from each image and compress them using quality in [10,100] in steps of 10. This gives a total training set of 1,065,000 patches. For evaluation, we use the Live1 [36,37], Classic-5 [10], BSDS500 [3], and ICB datasets [34]. ICB is a new dataset which provides 15 high-quality lossless images designed specifically to measure compression quality. It is our hope that the community will gradually begin including ICB dataset results. Where previous works have provided code and models, we reevaluate their methods and provide results here for comparison. As with all prior works, we report PSNR, PSNR-B [41], and SSIM [46].

Implementation Details. All training uses the Adam [25] optimizer with a batch size of 32 patches. Our network is implemented using the PyTorch [32] library. We normalize the DCT coefficients using per-frequency and per-channel mean and standard deviations. Since the DCT coefficients are measurements of different signals, by computing the statistics per-frequency we normalize the distributions so that they are all roughly the same magnitude. We find that this greatly speeds up the convergence of the network. Quantization table entries are normalized to [0, 1], with 1 being the most quantization and 0 the least. We use libjpeg [15] for compression with the baseline quantization setting.

Training Procedure. As described in Sect. 3.4, we follow a staged training approach by first training the Y channel or grayscale artifact correction network, then training the color (CbCr) channel network, and finally training both networks using the GAN loss.

For the first stage, the Y channel artifact correction network, the learning rate starts at 1×10^{-3} and decays by a factor of 2 every 100,000 batches. We stop training after 400,000 batches. We set λ in Eq. 1 to 0.05.

For the next stage, all color channels are restored. The weights for the Y channel network are initialized from the previous stage and frozen during training.

Table 2. Y channel correction results. PSNR/PSNR-B/SSIM format, the best result is highlighted in bold, second best is underlined. The JPEG column gives with input error of the images. For ICB, we used the Grayscale 8bit dataset. We add Classic-5, a grayscale only dataset.

Dataset	Quality	JPEG	ARCNN[8]	MWCNN [27]	IDCN [51]	DMCNN [49]	Ours
Live-1	10	27.76 / 25.32 / 0.790	28.96 / 28.68 / 0.821	29.68 / 29.30 / 0.839	29.68 / 29.32 / 0.838	**29.73** / **29.43** / 0.839	29.53 / 29.15 / **0.840**
	20	30.05 / 27.55 / 0.868	31.26 / 30.73 / 0.887	32.00 / 31.47 / **0.901**	32.05 / 31.46 / 0.900	**32.07** / **31.49** / **0.901**	31.86 / 31.27 / **0.901**
	30	31.37 / 28.90 / 0.900	32.64 / 32.11 / 0.916	**33.40** / **32.76** / **0.926**	-	-	33.23 / 32.50 / 0.925
Classic-5	10	27.82 / 25.21 / 0.780	29.03 / 28.76 / 0.811	**30.01** / 29.59 / **0.837**	29.83 / 29.48 / 0.833	29.98 / **29.65** / 0.836	29.84 / 29.43 / **0.837**
	20	30.12 / 27.50 / 0.854	31.15 / 30.59 / 0.869	**32.16** / **31.52** / **0.886**	31.99 / 31.46 / 0.884	32.11 / 31.48 / 0.885	31.98 / 31.37 / 0.885
	30	31.48 / 28.94 / 0.884	32.51 / 31.98 / 0.896	**33.43** / **32.62** / **0.907**	-	-	33.22 / 32.42 / **0.907**
BSDS500	10	27.86 / 25.18 / 0.785	29.14 / 28.76 / 0.816	29.63 / 29.16 / 0.831	29.60 / 29.13 / 0.829	**29.66** / **29.27** / 0.831	29.54 / 29.04 / **0.833**
	20	30.08 / 27.28 / 0.864	31.27 / 30.52 / 0.881	31.88 / 31.12 / **0.894**	31.88 / 31.05 / 0.893	**31.91** / **31.13** / **0.894**	31.79 / 30.96 / **0.894**
	30	31.37 / 28.56 / 0.896	32.64 / 31.90 / 0.912	**33.23** / **32.29** / **0.920**	-	-	33.12 / 32.07 / **0.920**
ICB	10	32.08 / 29.92 / 0.856	31.13 / 30.97 / 0.794	34.12 / 34.06 / 0.884	32.50 / 32.42 / 0.826	34.18 / 34.15 / 0.874	**34.73** / **34.58** / **0.806**
	20	35.04 / 32.72 / 0.905	32.62 / 32.31 / 0.821	36.50 / 36.44 / 0.902	34.30 / 34.18 / 0.851	35.93 / 35.79 / 0.918	**37.12** / **36.88** / **0.924**
	30	36.66 / 34.22 / 0.927	33.79 / 33.52 / 0.841	38.20 / 37.96 / 0.927	-	-	**38.43** / **38.05** / **0.938**

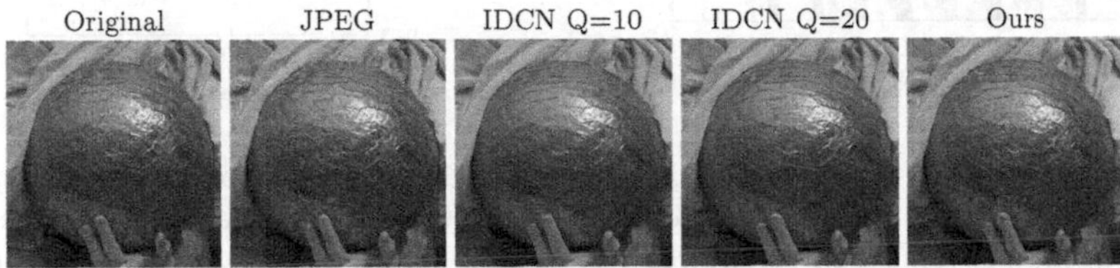

Fig. 10. Generalization Example. Input was compressed at quality 50. Please zoom in to view details.

The color channel network weights are trained using a cosine annealing learning rate schedule [30] decaying from 1×10^{-3} to 1×10^{-6} over 100,000 batches.

Finally, we train both Y and color channel artifact correction networks (jointly referred to as the generator model) using a GAN loss to improve qualitative textures. The generator model weights are initialized to the pre-trained models from the previous stages. We use the DCGAN [33] discriminator. The model is trained for 100,000 iterations using cosine annealing [30] with the learning rate starting from 1×10^{-4} and ending at 1×10^{-6}. We set γ and ν in Eq. 3 to 5×10^{-3} and 1×10^{-2} respectively.

4.2 Results: Artifact Correction

Color Artifact Correction. We report the main results of our approach, color artifact correction, on Live1, BSDS500, and ICB in Table 1. Our model consistently outperforms recent baselines on all datasets. Note that of all the approaches, only ours and IDCN [51] include native processing of color channels. For the other models, we convert input images to YCbCr and process the channels independently.

For quantitative comparisons to more methods on Live-1 dataset, at compression quality 10, refer to Fig. 12. We present qualitative results from a mix of all three datasets in Fig. 13 ("Ours"). Since our model is not restricted by

Table 3. Generalization Capabilities. Live-1 dataset (PSNR/PSNR-B/SSIM).

Model quality	Image quality	JPEG	IDCN [51]	Ours
10	50	30.91/28.94/0.905	30.19/30.14/0.889	**32.78/32.19/0.932**
20		30.91/28.94/0.905	31.91/31.65/0.916	
10	20	27.96/25.77/0.837	29.25/29.08/0.863	**29.92/29.51/0.882**
20	10	25.60/23.53/0.755	26.95/26.24/0.804	**27.65/27.40/0.819**

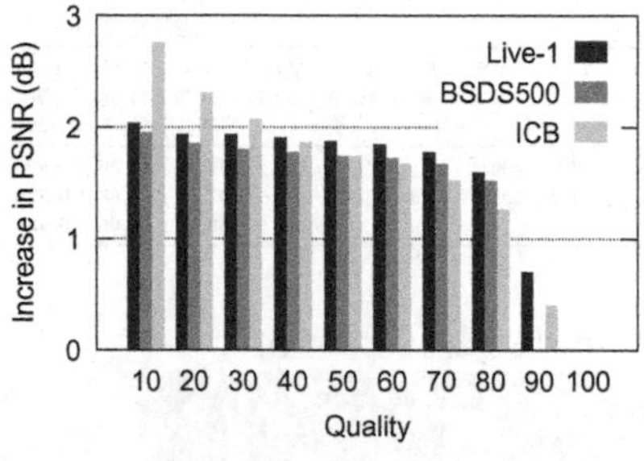

Fig. 11. Increase in PSNR on color datasets. For all three datasets we show the average improvement in PSNR values on qualities 10–100. Improvement drops off steeply at quality 90.

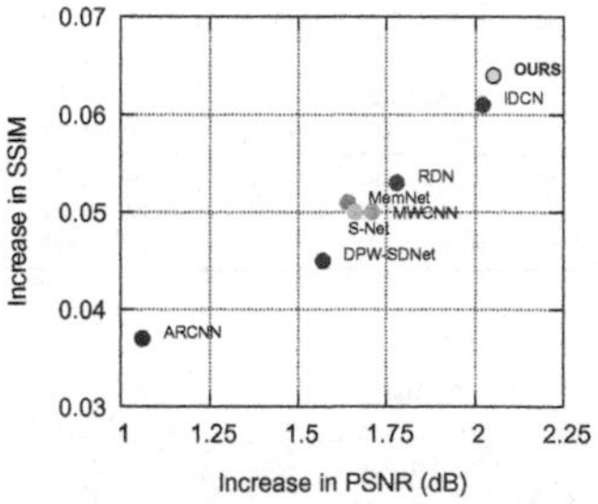

Fig. 12. Comparison for Live-1 quality 10. Where code was available we reevaluated, otherwise we used published numbers.

which quality settings it can be run on, we also show the increase in PSNR for qualities 10–100 in Fig. 11.

Intermediate Results on Y Channel Artifact Correction. Since the first stage of our approach trains for grayscale or Y channel artifact correction, we can also compare the intermediate results from this stage with other approaches. We report results in Table 2 for Live1, Classic-5, BSDS500, and ICB. As the table shows, intermediate results from our model can match or outperform previous state-of-the-art models in many cases, consistently providing high SSIM results using a single model for all quality factors.

GAN Correction Finally, we show results from our model trained using GAN correction. We use model interpolation [44] and show qualitative results for the interpolation parameter (α) set to 0.7 in Fig. 13. ("Ours-GAN") Notice that the GAN loss is able to restore texture to blurred, flat regions and sharpen edges, yielding a more visually pleasing result. We provide additional qualitative results in the supplementary material. Note that we do not show error metrics using the GAN model as it produces higher quality images, at the expense of quantitative metrics, by adding texture details that are not present in the original images. We instead show FID scores for the GAN model compared to our regression model in Table 4, indicating that the GAN model generates significantly more realistic images.

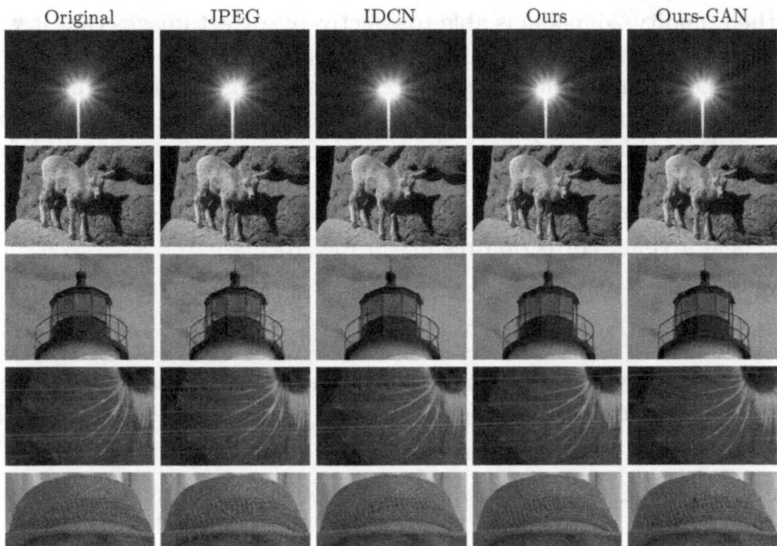

Fig. 13. Qualitative Results. All images were compressed at Quality 10. Please zoom in to view details.

Table 4. GAN FID Scores.

Dataset	Quality	Ours	Ours-GAN
Live-1	10	69.57	**35.86**
	20	36.32	**16.99**
	30	24.72	**12.20**
BSDS500	10	75.15	**34.80**
	20	42.46	**18.74**
	30	29.04	**13.03**
ICB	10	33.37	**26.08**
	20	17.23	**13.53**
	30	11.66	**10.13**

Table 5. Ablation Results. (refer to Sect. 4.4 for details).

Experiment	Model	PSNR	PSNR-B	SSIM
CFM	None	29.38	28.9	0.825
	Concat	29.32	28.94	0.823
	CFM	**29.46**	**29.05**	**0.827**
Subnetworks	FrequencyNet	28.03	25.58	0.787
	BlockNet	**29.45**	**29.04**	**0.827**
Fusion	No Fusion	27.82	25.21	0.78
	Fusion	**29.22**	**28.76**	**0.822**

4.3 Results: Generalization Capabilities

The major advantage of our method is that it uses a single model to correct JPEG images at any quality, while prior works train a model for each quality factor. Therefore, we explore if other methods are capable of generalizing or if they really require this ensemble of quality-specific models. To evaluate this, we use our closest competitor and prior state-of-the-art, IDCN [51]. IDCN does not provide a model for quality higher than 20, we explore if their model generalizes by using their quality 10 and quality 20 models to correct quality 50 Live-1 images. We also use the quality 20 model to correct quality 10 images and use the quality 10 model to correct quality 20 images. These results are shown in Table 3 along with our result.

As the table shows, the choice of model is critical for IDCN, and there is a significant quality drop when choosing the wrong model. Neither their quality

10 nor their quality 20 model is able to effectively correct images that it was not trained on, scoring significantly lower than if the correct model were used. At quality 50, the quality 10 model produces a result worse than the input JPEG, and the quality 20 model makes only a slight improvement. In comparison, our single model provides consistently better results across image quality factors. We stress that the quality setting is not stored in the JPEG file, so a deployed system has no way to pick the correct model. We show an example of a quality 50 image and artifact correction results in Fig. 10.

4.4 Design and Ablation Analysis

Here we ablate many of our design decisions and observe their effect on network accuracy. The results are reported in Table 5, we report metrics on quality 10 classic-5.

Implementation Details: For all ablation experiments, we keep the number of parameters approximately the same between tested models to alleviate the concern that a network performs better simply because it has a higher capacity. All models are trained for 100,000 batches on the grayscale training patch set using cosine annealing [30] from a learning rate of 1×10^{-3} to 1×10^{-6}.

Importance of CFM Layers. We emphasized the importance of adaptable weights in the CFM layers, which can be adapted using the quantization matrix. However, there are other simpler methods of using side-channel information. We could simply concatenate the quantization matrix channelwise with the input, or we could ignore the quantization matrix altogether. As shown in the "CFM" experiment in Table 5, the CFM unit performs better than both of these alternatives by a considerable margin. We further visualize the filters learned by the CFM layers and the underlying embeddings in the supplementary material which validate that the learned filters follow a manifold structure.

BlockNet *vs.* FrequencyNet. We noted that the FrequencyNet should not be able to perform without a preceding BlockNet because high-frequency information will be zeroed out from the compression process. To test this claim, we train individual BlockNet and FrequencyNet in isolation and report the results in Table 5 ("Subnetworks"). We can see that BlockNet alone attains significantly higher performance than FrequencyNet alone.

Importance of the Fusion Layer. Finally, we study the necessity of the fusion layer presented. We posited that the fusion layer was necessary for gradient flow to the early layers of our network. As demonstrated in Table 5 ("Fusion"), the network without fusion fails to learn, matching the input PSNR of classic-5 after full training, whereas the network with fusion makes considerable progress.

5 Conclusion

We showed a design for a quantization guided JPEG artifact correction network. Our single network is able to achieve state-of-the-art results, beating methods which train a different network for each quality level. Our network relies only on information that is available at inference time, and solves a major practical problem for the deployment of such methods in real-world scenarios.

Acknowledgement. This project was partially supported by Facebook AI and Defense Advanced Research Projects Agency (DARPA) MediFor program (FA87501620191). There is no collaboration between Facebook and DARPA.

References

1. Agustsson, E., Timofte, R.: Ntire 2017 challenge on single image super-resolution: Dataset and study. In: Proceedings of the IEEE Conference on Computer Vision and Pattern Recognition Workshops, 2017, pp. 126–135 (2017)
2. Ahmed, N., Natarajan, T., Rao, K.R.: Discrete cosine transform. IEEE Trans. Comput. **100**(1), 90–93 (1974)
3. Arbelaez, P., et al.: Contour detection and hierarchical image segmentation. IEEE Trans. Pattern Anal. Mach. Intell. **33**(5), 898–916 (2010)
4. Bell, S., et al.: Material recognition in the wild with the materials in context database. In: Proceedings of the IEEE Conference on Computer Vision and Pattern Recognition, pp. 3479–3487 (2015)
5. Cavigelli, L., Hager, P., Benini, L.: CAS-CNN: a deep convolutional neural network for image compression artifact suppression. In: 2017 International Joint Conference on Neural Networks (IJCNN), pp. 752–759. IEEE (2017)
6. Chen, H., et al.: DPW-SDNet: dual pixel-wavelet domain deep CNNs for soft decoding of JPEG-compressed images. In: Proceedings of the IEEE Conference on Computer Vision and Pattern Recognition Workshops, pp. 711–720 (2018)
7. Deguerre, B., Chatelain, C., Gasso, G.: Fast object detection in compressed JPEG Images. In: arXiv preprint arXiv:1904.08408 (2019)
8. Dong, C., et al.: Compression artifacts reduction by a deep convolutional network. In: Proceedings of the IEEE International Conference on Computer Vision, pp. 576–584 (2015)
9. Ehrlich, M., Davis, L.S.: Deep residual learning in the JPEG transform domain. In: Proceedings of the IEEE International Conference on Computer Vision, pp. 3484–3493 (2019)
10. Foi, A., Katkovnik, V., Egiazarian, K: Pointwise shape-adaptive DCT for high-quality deblocking of compressed color images. In: 14th European Signal Processing Conference, pp. 1–5. IEEE (2006)
11. Fu, X., et al.: JPEG artifacts reduction via deep convolutional sparse coding. In: Proceedings of the IEEE International Conference on Computer Vision, pp. 2501–2510 (2019)
12. Galteri, L., et al.: Deep generative adversarial compression artifact removal. In: Proceedings of the IEEE International Conference on Computer Vision, pp. 4826–4835 (2017)
13. Galteri, L., et al.: Deep universal generative adversarial compression artifact removal. IEEE Trans. Multimedia (2019)
14. Ghosh, A., Chellappa, R.: Deep feature extraction in the DCT domain. In: 2016 23rd International Conference on Pattern Recognition (ICPR), pp. 3536–3541. IEEE (2016)
15. Independant JPEG Group. libjpeg. http://libjpeg.sourceforge.net

16. Gueguen, L., et al.: Faster neural networks straight from JPEG. In: Advances in Neural Information Processing Systems, pp. 3933–3944 (2018)
17. Guo, J., Chao, H.: Building dual-domain representations for compression artifacts reduction. In: Leibe, B., Matas, J., Sebe, N., Welling, M. (eds.) ECCV 2016. LNCS, vol. 9905, pp. 628–644. Springer, Cham (2016). https://doi.org/10.1007/978-3-319-46448-0_38
18. He, K., et al.: Deep residual learning for image recognition. In: Proceedings of the IEEE Conference on Computer Vision and Pattern Recognition, pp. 770–778 (2016)
19. He, K., et al.: Delving deep into rectifiers: surpassing human-level performance on imagenet classification. In: Proceedings of the IEEE International Conference on Computer Vision, pp. 1026–1034 (2015)
20. Jin, Z., et al.: Dual-stream multi-path recursive residual network for JPEG image compression artifacts reduction. IEEE Trans. Circ. Syst. Video Technol. (2020)
21. Jolicoeur-Martineau, A.: The relativistic discriminator: a key element missing from standard GAN. In: arXiv preprint arXiv:1807.00734 (2018)
22. Kang, D., Dhar, D., Chan, A.B.: Crowd counting by adapting convolutional neural networks with side information. In: arXiv preprint arXiv:1611.06748 (2016)
23. Kim, Y., Soh, J.W., Cho, N.I.K.: AGARNet: adaptively gated JPEG compression artifacts removal network for a wide range quality factor. IEEE Access **8**, 20160–20170 (2020)
24. Kim, Y., et al.: A pseudo-blind convolutional neural network for the reduction of compression artifacts. IEEE Trans. Circuits Syst. Video Technol. **30**(4), 1121–1135 (2019)
25. Kingma, D.P. Ba, J.: Adam: A method for stochastic optimization. In: arXiv preprint arXiv:1412.6980 (2014)
26. LeCun, Y., et al.: Handwritten digit recognition with a back-propagation network. In: Advances in Neural Information Processing Systems, pp. 396–404 (1990)
27. Liu, P., et al.: Multi-level wavelet-CNN for image restoration. In: Proceedings of the IEEE Conference on Computer Vision and Pattern Recognition Workshops, pp. 773–782 (2018)
28. Liu, X., et al.: Data-driven sparsity-based restoration of JPEG- compressed images in dual transform-pixel domain. In: Proceedings of the IEEE Conference on Computer Vision and Pattern Recognition, pp. 5171–5178 (2015)
29. Lo, S.-Y., Hang, H.-M.: Exploring semantic segmentation on the DCT representation. In: Proceedings of the ACM Multimedia Asia on ZZZ, pp. 1–6 (2019)
30. Loshchilov, I., Hutter, F.: SGDR: Stochastic gradient descent with warm restarts. In: arXiv preprint arXiv:1608.03983 (2016)
31. Maas, A.L., Hannun, A.Y., Ng, A.Y.: Rectifier nonlinearities improve neural network acoustic models. In: Proceedings of ICML, vol. 30, no. 1, p. 3 (2013)
32. Paszke, A., et al.: PyTorch: an imperative style, high-performance deep learning library. In: Wallach, H., et al. (eds.) Advances in Neural Information Processing Systems 32, pp. 8024–8035. Curran Associates Inc (2019). http://papers.neurips.cc/paper/9015-pytorch-an-imperative-style-high-performance-deep-learning-library.pdf
33. Radford, A., Metz, L., Chintala, S.: Unsupervised representation learning with deep convolutional generative adversarial networks. In: arXiv preprint arXiv:1511.06434 (2015)
34. Rawzor. Image Compression Benchmark. http://imagecompression.info/

35. Ronneberger, O., Fischer, P., Brox, T.: U-Net: convolutional networks for biomedical image segmentation. In: Navab, N., Hornegger, J., Wells, W.M., Frangi, A.F. (eds.) MICCAI 2015. LNCS, vol. 9351, pp. 234–241. Springer, Cham (2015). https://doi.org/10.1007/978-3-319-24574-4_28
36. Sheikh, H.R., Sabir, M.F., Bovik, A.C.: A statistical evaluation of recent full reference image quality assessment algorithms. IEEE Trans. Image Processing **15**(11), 3440–3451 (2006)
37. Sheikh, H.R., et al.: LIVE image quality assessment database. http://live.ece.utexas.edu/research/quality
38. Smith, B.: Fast software processing of motion JPEG video. In: Proceedings of the Second ACM International Conference on Multimedia, pp. 77–88. ACM. (1994)
39. Sutskever, I., Hinton, G.E., Krizhevsky, A.: Imagenet classification with deep convolutional neural networks. In: Advances in Neural Information Processing Systems, pp. 1097–1105 (2012)
40. Svoboda, P., et al.: Compression artifacts removal using convolutional neural networks. In: arXiv preprint arXiv:1605.00366 (2016)
41. Tadala, T., Narayana, S.E.V.: A Novel PSNR-B Approach for Evaluating the Quality of De-blocked Images (2012)
42. Tai, Y., et al.: Memnet: a persistent memory network for image restoration. In: Proceedings of the IEEE International Conference on Computer Vision, pp. 4539–4547 (2017)
43. Wallace, G.K.: The JPEG still picture compression standard. IEEE Trans. Consumer Electron. **38**(1), xviii–xxxiv (1992)
44. Wang, X., Yu, K., Wu, S., Gu, J., Liu, Y., Dong, C., Qiao, Yu., Loy, C.C.: ESRGAN: enhanced super-resolution generative adversarial networks. In: Leal-Taixé, L., Roth, S. (eds.) ECCV 2018. LNCS, vol. 11133, pp. 63–79. Springer, Cham (2019). https://doi.org/10.1007/978-3-030-11021-5_5
45. Wang, Z., et al.: D3: deep dual-domain based fast restoration of JPEG-compressed images. In: Proceedings of the IEEE Conference on Computer Vision and Pattern Recognition, pp. 2764–2772 (2016)
46. Wang, Z., et al.: Image quality assessment: from error visibility to structural similarity. IEEE Trans. Image Process. **13**(4), 600–612 (2004)
47. Yang, S., et al.: Blocking artifact free inverse discrete cosine transform. In: Proceedings 2000 International Conference on Image Processing (Cat. No. 00CH37101), vol. 3, pp. 869–872. IEEE (2000)
48. Zhang, K., et al.: Beyond a Gaussian denoiser: residual learning of deep CNN for image denoising. IEEE Trans. Image Process. **26**(7), 3142–3155 (2017)
49. Zhang, X., et al.: DMCNN: dual-domain multi-scale convolutional neural network for compression artifacts removal. In: 2018 25th IEEE International Conference on Image Processing (ICIP), pp. 390–394. IEEE (2018)
50. Zhang, Y., et al.: Residual dense network for image restoration. IEEE Trans. Pattern Anal. Mach. Intell. (2020)
51. Zheng, B., et al.: Implicit dual-domain convolutional network for robust color image compression artifact reduction. IEEE Trans. Circ. Syst. Video Technol. (2019)
52. Zheng, B., et al.: S-Net: a scalable convolutional neural network for JPEG compression artifact reduction. J. Electron. Imaging **27**(4), 043037 (2018)
53. Zini, S., Bianco, S., Schettini, R.: Deep Residual Autoencoder for quality independent JPEG restoration. In: arXiv preprint arXiv:1903.06117 (2019)

3PointTM: Faster Measurement of High-Dimensional Transmission Matrices

Yujun Chen[1(✉)], Manoj Kumar Sharma[1], Ashutosh Sabharwal[1], Ashok Veeraraghavan[1], and Aswin C. Sankaranarayanan[2]

[1] Rice University, Houston, TX, USA
{yc67,manoj.sharma,ashu,vashok}@rice.edu
[2] Carnegie Mellon University, Pittsburgh, PA, USA
saswin@andrew.cmu.edu

Abstract. A transmission matrix (TM) describes the linear relationship between input and output phasor fields when a coherent wave passes through a scattering medium. Measurement of the TM enables numerous applications, but is challenging and time-intensive for an arbitrary medium. State-of-the-art methods, including phase-shifting holography and double phase retrieval, require significant amounts of measurements, and post-capture reconstruction that is often computationally intensive. In this paper, we propose 3PointTM, an approach for sensing TMs that uses a minimal number of measurements per pixel—reducing the measurement budget by a factor of two as compared to state of the art in phase-shifting holography for measuring TMs—and has a low computational complexity as compared to phase retrieval. We validate our approach on real and simulated data, and show successful focusing of light and image reconstruction on dense scattering media.

Keywords: Transmission matrix · Inverse scattering · Imaging · Focusing · Optimization · Phase modulation

1 Introduction

When coherent light passes through a highly scattering medium, the photons encounter multiple scattering events, and the resultant interference gives rise to a random speckle pattern. This random input-to-output mapping makes imaging and focusing light through multiple-scattering media a challenging task.

The most general method for imaging through an arbitrarily thick scattering medium is by measuring the transmission matrix (TM) [1,6,30,33,34], which characterizes the relationship between the input and output complex wavefronts. Specifically, the mesoscopic TM of an optical system at a specific wavelength can be represented as a matrix T with complex-valued elements t_{kn} connecting the field in the $k-$th camera pixel (also, commonly referred to as an "output

A. Vedaldi et al. (Eds.): ECCV 2020, LNCS 12353, pp. 310–326, 2020.
https://doi.org/10.1007/978-3-030-58598-3_19

mode") to the field in the $n-$th spatial light modulator (SLM) pixel (or, an "input mode"):

$$u_k^{out} = \sum_n t_{kn} u_n^{in}, \tag{1}$$

where u_k^{out} and u_n^{in} are the complex phasor at the $k-$th camera pixel and $n-$th SLM pixel, respectively. However, while we can control the phase and intensity of the input phasor u_n^{in}, we can only measure the intensity of the output phasor $|u_k^{out}|^2$ and hence, we do not have a direct measurement of its phase. This makes the measurement of TMs a hard and nonlinear inverse problem.

There are two well known methods to measure the TM of an arbitrary medium: phase-shifting holography [34] and phase retrieval [12,27,35,37]. Phase-shifting holography measures the TM by acquiring four images per SLM pixel; it relies on using a portion of the SLM or its surrounding region as a static reference wave and learning the TM against this reference. Phase retrieval methods, on the other hand, acquire multiple intensity measurements by varying the phase pattern on the SLM; the TM is estimated by solving a phase retrieval problem. While phase retrieval methods require a minimum of four measurements per SLM pixel [38], in practice, robustness to measurement noise requires eight to twelve measurements per pixel. Finally, the algorithm to reconstruct the final TM is computationally intensive; for example, [27] reports requiring tens of CPU hours for estimating a TM of size of $128^2 \times 60^2$.

Given a TM of size $M \times N$, corresponding to an SLM with N pixels and an image sensor with M pixels, prior approaches require at least $4N$ intensity images. In contrast, the number of degrees of freedom in a TM is only $M(2N-1)$ since its elements are complex numbers; further, since we can only measure intensities, each row of the TM has a constant phase ambiguity that we cannot resolve. This leads us to pose and subsequently answer the following question: *how do we measure a TM with a minimal number of intensity measurements?*

Contributions. This paper proposes an efficient approach for measuring TMs associated with arbitrary scattering media. We make the following contributions.

- Our main contribution is a near-optimal measurement strategy that measures an $M \times N$ TM from $2N+1$ intensity measurements.
- Our approach provides an analytical expression for estimating the TM from the acquired measurements and is computationally efficient.
- We present detailed empirical analysis of the stability of our method under different operating conditions, including the number of SLM pixels and the signal-to-noise ratio of the measurements, as well as comparisons to prior art.
- We validate our approach using a lab prototype, using which we scan several optical media and demonstrate applications in the form of inverse scattering and focusing through the media.

Limitations. There are three principal limitations of our work, some of which we share with existing approaches. First, since our technique acquires fewer measurements than prior work, our performance is often worse than them in low SNR settings. We characterize this gap in performance with simulations. Second, similar to prior work, our approach requires access to both sides of the scattering medium, something that is unrealistic when the imaging target is biological tissue in-vivo. Third, even with the reduction in the number of measurements, the acquisition of the TM with our approach can take several minutes for high-resolution TMs.

2 Prior Work

The study of light transport in a scene has a long history in the optics and vision community. For incoherent illumination, the propagation of light from the illuminant(s) to the camera is modelled via the light transport matrix, which has found immense applications for image-based relighting [10,36]. In contrast, for coherent light, the propagation also needs to consider the interference between illuminants, which is better modelled via the complex-valued TM. We are particularly interested in propagation of coherent light through or off random scattering media, following the work of Freund [13].

Imaging Based on the Strong Memory Effect. A scatterer is said to possess a strong memory effect if the translation of an input point light source results in a translation of the speckle pattern at the output. This property has been used to enable single shot imaging through a thin multiple-scattering media [19]. However, this method critically relies on the assumption that the medium is thin enough to exhibit the memory effect.

Exploring Time-of-Flight Information. Time-of-flight imaging systems use the photon travel time to differentiate between the scattered and ballistic photons [18,29,31,32,40]. Since these systems only measure ballistic photons, they require high-intensity pulsed lasers as well as time-resolved cameras, with time-resolutions in nano/pico-seconds, which is expensive.

Multi-slice Light Propagation. Multi-slice light-propagation models a scattering media as a sequence of 2D slices [25], thereby modeling the transmission matrix as a composition of linear transformations. However, this method has not been evaluated for thick scatterers that have no memory effect.

2.1 Measurement of Transmission Matrices

Holographic Interferometry. One of the most common methods to measure a TM is holographic interferometry [34]. Here, part of the SLM is held constant and used as a reference and the remainder is modulated, in a Hadamard basis, one input channel at a time using four-step holography. The use of Hadamard basis allows the use of a phase-only SLM. In contrast, we show that it is possible to measure the TM without resorting to probing in Hadamard basis while using a phase-only SLM; crucially, this choice allows to reduce the number of measurements from $4N$ per camera pixel to $2N+1$. Finally, an interesting difference in our approach is that we do not have any area on the SLM plane allocated as a reference.

Quadrature Phase-Shifting Holography. A recent result recovers the phase of a wavefront from two measurements [3,22]. Quadrature phase-shifting holography takes samples at phase 0 and $\frac{\pi}{2}$ for each SLM pixel, and estimates the complex measurement at the camera [22]. However, the method has very stringent requirements on the capture setup; specifically, it requires uniform illumination on the camera from the reference signal, such that a separate reference arm is necessary and cannot be estimated as done in four-step holography [34]. It also assumes that reference intensity is larger than the object wave intensity, which is a stringent condition to fulfill at bright speckle positions [23]. Comparisons between two-step, three-step, and four-step holography methods have shown that the two-step method is less accurate compared with the four-step method because of the above mentioned reasons [23]. It should be noted that quadrature phase-shifting holography has not been applied to TM estimation and its performance in this task is not yet understood.

2.2 Focusing Light Through a Scattering Media

Focusing light through a scattering media [8,9,17,20,24,39,42,43] is one of the intriguing applications for measurement of TMs. We provide a brief overview of the methods most relevant to the ideas of this paper.

Assume that the beam incident on the SLM is collimated with spatially constant intensity. Then, when the $n-$th SLM pixel has the phase shift of ϕ_n, the output on the $k-$th camera pixel can be written as

$$y_k = \left| \sum_{n=1}^{N} t_{kn} e^{i\phi_n} \right|^2 + \epsilon, \tag{2}$$

where ϵ is the measurement noise. Suppose that we seek to focus light on this $k-$th camera pixel. From (2), we observe that the intensity at y_k is largest when all the terms in the summation are in-phase, which is achieved by choosing $\phi_n^* = c - \angle t_{kn}$ where c is a constant; this choice of SLM phase aligns the phasors in the $k-$th row of the TM to form constructive interference at the target pixel.

We can define the effect of focusing using the ratio between the focused intensity and the average intensity prior to it. Assuming the individual TM components t_{mn} are statistically independent and follows the circular Gaussian distribution, the expected maximal enhancement, γ, is defined as

$$\gamma = \frac{\pi}{4}(N-1) + 1, \tag{3}$$

where N is the number of SLM pixels [4,14,15,39].

Vellekoop and Mosk [39] present a method that uses a feedback-based phase modulation technique to focus light through scattering media. This approach works by individually manipulating the phase at each SLM pixel from 0 to 2π in small steps, while fixing the phase on the rest of the pixels. The phase value that produces maximum intensity y_k is chosen as the estimate of the optimal value $\angle\phi_n^*$. This procedure is repeated for all SLM pixels, updating them one at a time. After multiple such iterations over the whole SLM, we can expect to achieve a near-optimal focusing at the desired camera pixel(s).

The work of Vellekoop and Mosk [39] is closely related to our proposed ideas. Specifically, we advance this technique to make only a single pass over the SLM pixel and further, restrict the number of phase patterns to three values at each pixel, of which one is shared across all pixels. In addition to focusing, we also show how to estimate the TM of the system from the measurements.

Genetic Algorithms (GAs). Genetic algorithm interprets the intensity at a chosen point on the camera as the objective function, and optimizes the SLM pixels by randomly selecting patterns and breeding between them [7]. Empirically, we observe that GAs provide steady state solution after $10N$ measurements, and its performance is robust to noise. However, this algorithm requires both amplitude and phase modulation to estimate TMs. Further, the number of measurements required to recover TMs is $10MN$, such that GA is even more computationally intensive than phase retrieval algorithms.

3 Minimal Measurement of Transmission Matrices

We now present an approach for measurement of TMs, that we call *3PointTM*. 3PointTM requires only $2N+1$ images to measure the TM, given an SLM with N pixels, and hence, is nearly minimal.

The core idea underlying 3PointTM relies on an observation that we make about the focusing approach of Vellekoop and Mosk [39]. As before let's consider the intensity at the $k-$th pixel on the sensor, y_k, while optimizing the phase value ϕ_n on the $n-$th SLM pixel. As a small, but important deviation, we set the phase value at the other SLM pixels at zero. Given these, we can write the intensity y_k as a function of ϕ_n as follows:

$$y_k(\phi_n) = \left| t_{kn}e^{i\phi_n} + \sum_{\ell \neq n} t_{k\ell} \right|^2 + \epsilon. \tag{4}$$

Denoting $u_{kn} = \sum_{\ell \neq n} t_{k\ell}$, and dropping the noise term, we get

$$y_k(\phi_n) = |t_{kn}|^2 + |u_{kn}|^2 + 2|t_{kn}||u_{kn}| \cos\left(\phi_n + \angle t_{kn} - \angle u_{kn}\right). \tag{5}$$

This suggests that, as we vary the phase ϕ_n, the intensity at the target pixel varies on a sinusoidal curve with unknown offset $|t_{kn}|^2 + |u_{kn}|^2$, amplitude $2|t_{kn}||u_{kn}|$, and phase offset $\angle t_{kn} - \angle u_{kn}$.

Instead of sweeping a dense set of phase values, as in [39], we propose to sample this sinusoid at three distinct phase values to recover its parameters.

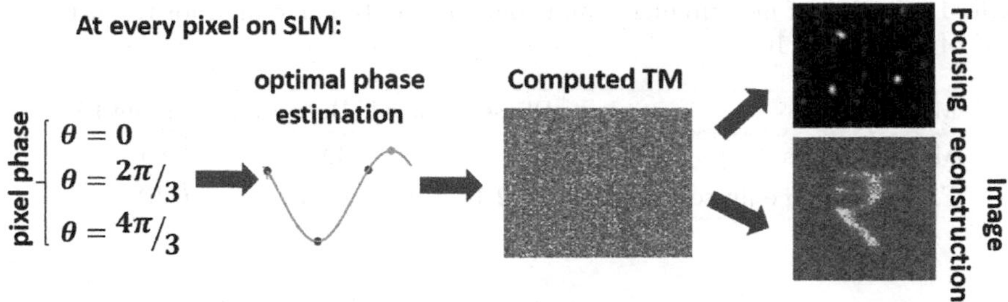

Fig. 1. *Graphical Illustration of 3PointTM.* We vary the phase at each SLM pixel to three values, of which one is shared across all pixels, and measure the resulting intensity image on the camera. The TM matrix is estimated from these intensity images and enables us to perform focusing and imaging operations.

As we will show subsequently, recovering the sinusoidal parameters allows us to estimate the TM and subsequently focus light at any desired sensor pixel (Fig. 1).

Step 1—Sinusoidal Parameter Fitting. We acquire three sensor measurements for each SLM pixel. Specifically, for the n−th SLM pixel, we take three measurements with $\phi_n \in \{0, \frac{2\pi}{3}, \frac{4\pi}{3}\}$, with the phase values at other pixels set to zero. For each of the three phase values, we make full sensor measurements and computationally recover the column of the TM corresponding to the n−th SLM pixel, i.e., the elements $\{t_{kn}, \forall k\}$.

Our technique recovers the TM elements for each sensor pixel in isolation, and so we focus on an arbitrary sensor pixel, say the k−th pixel on the sensor. Given the three intensities measured at this pixel, y_{k_1}, y_{k_2}, and y_{k_3}, corresponding to the ϕ_n set to 0, $\frac{2\pi}{3}$, and $\frac{4\pi}{3}$, respectively, we compute the parameters of the sinusoid. The three measured intensities can be described as

$$y_{k_1} = c_0 + A\cos(\theta), \quad y_{k_2} = c_0 + A\cos(\theta + \frac{2\pi}{3}), \quad y_{k_3} = c_0 + A\cos(\theta + \frac{4\pi}{3}), \tag{6}$$

where

$$c_0 = |t_{kn}|^2 + |u_{kn}|^2, \quad A = 2|t_{kn}||u_{kn}|, \quad \theta = \angle t_{kn} - \angle u_{kn}. \tag{7}$$

We use the technique described in [41] to estimate the three parameters c_0, A and θ from the three measured intensities. It is also worth noting that sinusoidal fitting of this form has found numerous applications in imaging and vision, including polarimetry, and correlation-based time-of-flight sensing.

Step 2—TM Estimation. We now describe the procedure for estimating the TM elements from the sinusoidal parameters, c_0, A and θ. Given c_0 and A defined as in (7), the first step is to solve for $|t_{kn}|$ and $|u_{kn}|$. However, we can immediately observe that the symmetry in the occurrence of $|t_{kn}|$ and $|u_{kn}|$ in both c_0 and A implies that we cannot uniquely recover them. Specifically, while we can estimate the set $\{|t_{kn}|, |u_{kn}|\}$, we cannot associate them without any additional information. One approach is to look beyond a single SLM pixel and

Table 1. Number of measurements and computation time required for reconstructing a TM of size $256^2 \times 16^2$.

Methods	Phase retrieval	Holography	3PointTM
Measurements	4N	4N	2N+1
Processing time [in seconds]	2182.4	1.23	0.16

jointly estimate $\{t_{k\ell}, \forall \ell\}$ since u_{kn} is dependent on TM elements corresponding to other SLM pixels. This can however be quite cumbersome. In practice, we observe that the magnitude of u_{kn} is almost always greater than that of t_{kn}. This can be attributed to the fact that while t_{kn} is a single element of the TM matrix, u_{kn} is the magnitude of sum of $N-1$ random phasors and hence its magnitude is expected to be larger. We also validated this assumption on three previously measured TMs [27]. When no noise is added to the measurements in simulation, 0.0341% of all the TM elements do not satisfy this assumption. When Poisson noise is added at an SNR of 10 dB into the intensity measurements, 0.007% of TM elements do not satisfy this assumption, presumably because the term $|u_{kn}|$ is amplified more as noise is added.

Once we have estimates of $|t_{kn}|$ and $|u_{kn}|$, we define a new variable z_{kn} as

$$\begin{aligned} z_{kn} &= |u_{kn}|e^{-i\theta} + |t_{kn}| = |u_{kn}|e^{i\angle u_{kn} - i\angle t_{kn}} + |t_{kn}| \\ &= \left(|u_{kn}|e^{i\angle u_{kn}} + |t_{kn}|e^{i\angle t_{kn}}\right) e^{-i\angle t_{kn}} = e^{-i\angle t_{kn}} \sum_{\ell} t_{k\ell}. \end{aligned} \tag{8}$$

Hence, if we take the angle of the phasor z_{kn}, we get

$$\angle z_{kn} = -\angle t_{kn} + \angle \sum_{\ell} t_{k\ell} = -\angle t_{kn} + \psi_k, \tag{9}$$

where ψ_k is a constant that is dependent only the sensor pixel but constant for all the SLM pixels. We can now provide an estimate for the TM element as

$$\widehat{t_{kn}} = |t_{kn}|e^{-i\angle z_{kn}} = t_{kn}e^{-\psi_k}.$$

The unknown phase shift does not affect any subsequent processing tasks like focusing as it is independent of the SLM pixels.

Advantages of 3PointTM. Our core contribution is reformulating the problem of TM estimation to this sinusoidal fitting approach, even in the absence of a reference wave. The proposed method has many advantages over its competitors because of the sinusoidal fitting step.

First, given N SLM pixels, we only need to display $2N+1$ SLM patterns (and capture an equal number of images) since we can share the all-zero phase measurements across pixels. Without the assumption of sparsity, this is not just minimal in the sense of the number of unknowns, but also a $2\times$ improvement over holography-based competitors. We highlight these advantages in Table 1, and in the next section provide simulation results that support these claims. Another approach for reducing measurements is the use of compressive sensing where we take advantage of sparsity [28]; however, this imposes priors on the nature of the TM. In contrast, we make no assumption on the unknown medium.

Second, we provide closed-form analytical expressions for estimates of the TM elements. Further, we can do so for each TM element in parallel which can be exploited by parallel processing architectures. This is in sharp contrast to phase retrieval-based methods that require complex optimization.

Third, we do not make any assumptions about the image formation except perhaps for the validity of the TM model. Hence, our method is widely applicable to many different kinds of scattering media—a key improvement over memory-effect based approaches.

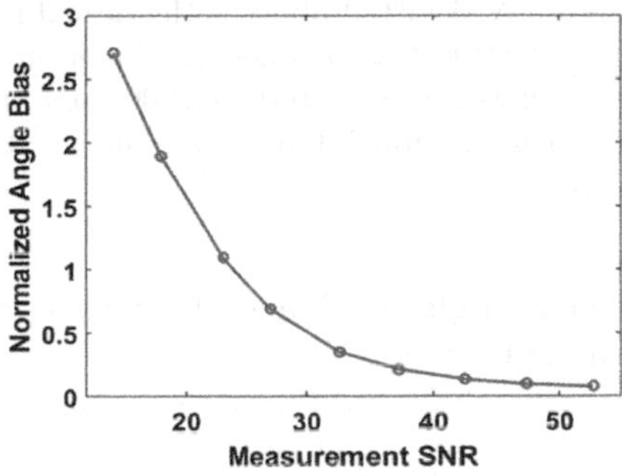

Fig. 2. *Normalized angle bias of the ground truth and TM estimated by 3PointTM.* By comparing the computed TM with the ground truth, it is evident that the bias is low at high SNRs, and the performance of 3PointTM degrades quadratically pass 30 dB.

4 Simulation Results

We perform simulations by generating random transmission matrices from i.i.d. circularly symmetric complex Gaussian distributions of size $256^2 \times 16^2$. For the noise simulations, we mimicked a sensor with a full well-capacity of $10k$ electrons and a read noise standard deviation of 4.7 electrons—these numbers are indicative of the sensor used in our real experiments. Given a noise-free measurement y_0, the simulated noisy measurement is given as

$$\text{poisson}(y_0) + 4.7\mathcal{N}(0,1), \tag{10}$$

which accounts for both photon and read noise. To operate at different SNRs, we adjust the global scene light level by scaling y_0 across all pixels.

Evaluation of Estimated TMs. To characterize how 3PointTM performs against noise, we compute the TM under different levels of noise. Figure 2 plots

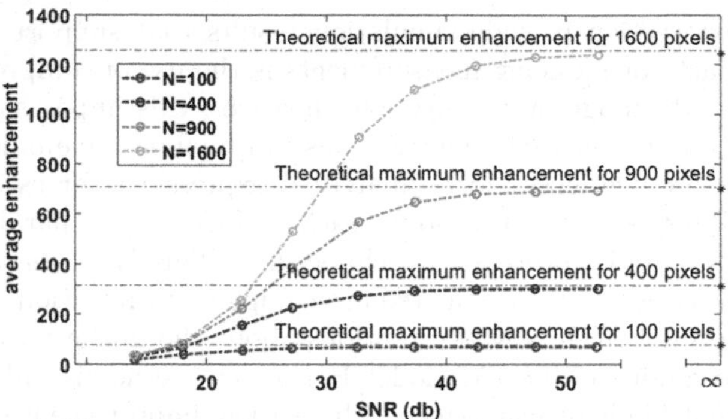

Fig. 3. *Performance characterization of 3PointTM.* The theoretical maximum intensity enhancement that can be obtained using illumination modulation is directly dependent on the number of SLM pixels, N. Plotted above is the actual performance in intensity enhancement obtained by 3PointTM as a function of measurement SNR, each computed over 50 independent trials for $N = 100, 400, 900$, and 1600. 3PointTM asymptotically achieves theoretical maximum intensity enhancement at high SNR regimes, and degrades at lower SNR.

the relationship between average SNR and the measurement error, which is defined as normalized angle bias,

$$\text{error} = \sqrt{\frac{\sum_n (e^{i\text{diag}(\angle \overline{T * \widehat{T}})} T - \widehat{T})^2}{\sum_n T^2}}. \tag{11}$$

The estimation error, or angle bias measured in percentage, is small at low noise region, and increases quadratically as the SNR decreases from 32 dB to 12 dB.

Focusing Enhancement with Respect to Number of Input Channels. To understand the relationship between enhancement and number of input channels, TMs of different sizes are generated and their estimations are used to focus energy at an arbitrarily chosen position, and the ensuing intensity enhancement is computed. We compare this value to the theoretical maximum, described in Eq. (3). From Fig. 3, we observe two distinct trends. First, as expected, a larger number of SLM pixels does lead to a higher intensity enhancement. Second, once the SNR of the measurements are greater than 35dB, the achieved enhancement starts to approach the theoretical maxima indicating that the phase of the TM elements have been accurately estimated. For SNRs smaller than 30dB, errors in estimates lead to a wide gap between the two. The enhancement decays follow a similar pattern across different numbers of SLM pixels.

Comparisons with prior art. To quantitatively compare 3PointTM with the existing methods to recover TMs, we implemented prVAMP from [27] with $12N$ measurements, the GA from [7] with $10NM$ measurements, and the four-step holography from [34] with $4N$ measurements, and compared their respective focusing enhancement abilities.

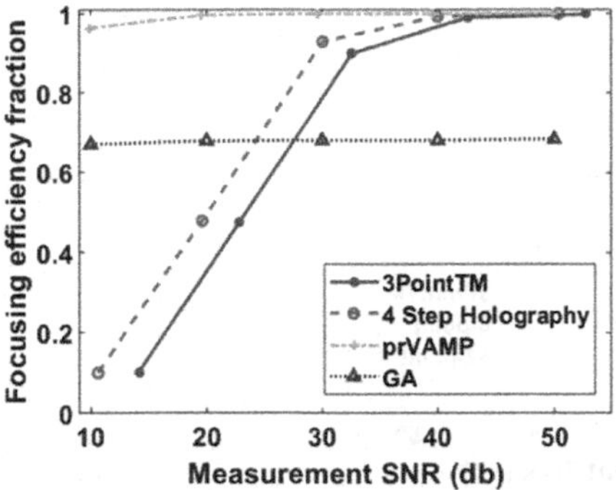

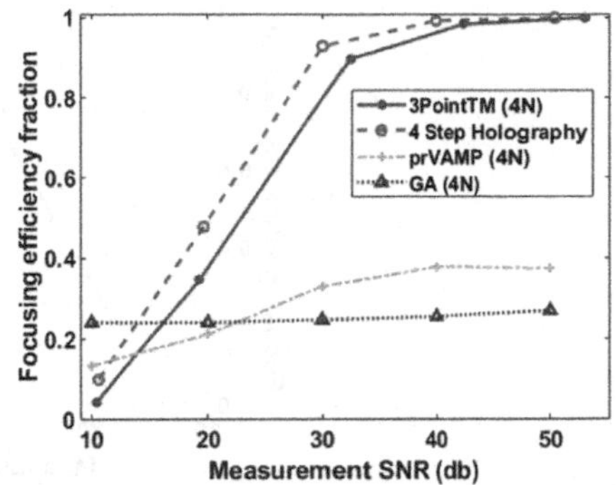

Fig. 4. *Performance comparisons.* (left) We compare the focusing efficiency of 3PointTM, four-step holography, a genetic algorithm (GA) and prVAMP on randomly generated TMs of size $256^2 \times 16^2$. Both GA and prVAMP are robust against noise, in part due to the larger number of measurements, while the focusing ability of both four-step holography and 3PointTM degrades as noise increases. (right) We compare the focusing efficiencies of the aforementioned four methods under the same settings, but at $4N$ image measurements for all of them.

As shown in Fig. 4 (left), phase retrieval is robust against noise, as its focusing efficiency remains high at different measurement SNRs. 3PointTM is comparative to four-step holography in all cases. It performs worse comparing with phase retrieval in high noise scenarios, but the advantage of phase retrieval diminishes when noise is low. On the Grasshopper camera we simulated, a 10 dB operating point would correspond to 33 photoelectrons. In a typical operating point, our measurements have speckle intensity peaks at nearly 9k e-, which would correspond to a measurement SNR of 39 dB. Thus our experiments operate in the regime that 3PointTM could generate reliable estimations. Since 3PointTM requires 6× fewer measurements and is 13640× faster computationally, it could be used with slowly-decorrelating mediums, for which there is not enough time to perform phase retrieval.

Besides running each method until it reaches the optimal solution, we also compared the performances of the methods performed with the same number of measurements. Since four-step holography only works 4 N measurements, all four methods are simulated 4 N measurements. As shown in Fig. 4(right), the performances of prVAMP and GA significantly compromise when no sufficient samples are taken.

Comparisons with 6-point and 52-point Sinusoidal Estimates. In 3Point TM, the accuracy of TM estimation is dependent on the ability to correctly fit sinusoid parameters from the three measurements made at each SLM pixel. To check the benefits to be derived by taking additional measurements, we compare against methods where we obtain more measurements at each SLM and, in particular, 6 and 52 measurements, by uniformly sampling between 0 and 2π and solving the best fitting cosines with the Levenberg algorithm [21,26]. The

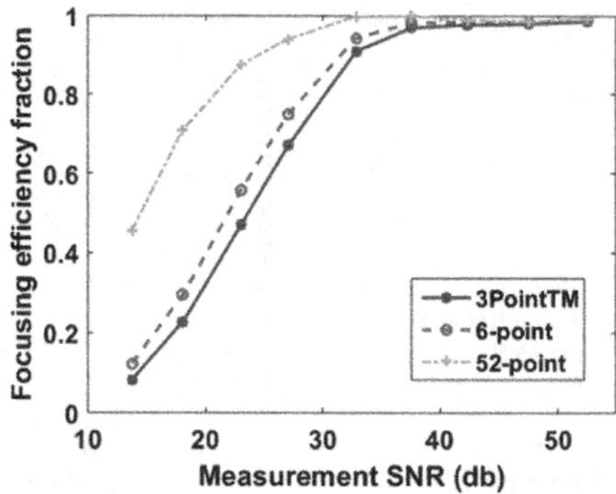

Fig. 5. *Performance comparisons of different measurements.* We show the performance of our approach when we acquire more measurements and fit sinusoids on them.

performances of the methods on focusing is shown in Fig. 5, and we observe the expected improvement in performance with increased measurements. Although the proposed method 52 N measurements takes substantially more measurements compared to prVAMP, prVAMP still outperforms because it uses a richer measurement matrix, where all the phase SLM pixels are changed randomly at each measurement.

Fig. 6. *Experimental setup of 3PointTM. SF:* Spatial filter, *L1:* collimator, *L2:* Focusing lens, *SLM:* Spatial light modulator, *P1* and *P2:* Polarizers, *A1* and *A2:* Apertures, *CCD:* Sensor. The laser beam passes through *SF* and is collimated by *L1*, and hits the reflection mode SLM, on which an input mode pattern is displayed. The modulated beam is then directed onto the diffuser *D*. Finally, the modulated wave is focused onto the CCD sensor.

5 Experimental Results

Setup and Data Collection. The optical system that we use is shown in Fig. 6. It consists of a spatially filtered and collimated laser beam, generated using a laser diode (ZM18GF024) with wavelength $\lambda = 540$ nm. The SLM (Holoeye Leto) used is a phase only modulator with 1920×1080 pixels and $6.4\mu m$ pixel pitch. The modulated light is focused on a strongly scattering sample by the lens $L2$. We have used holographic diffusers from Edmund optics and PDMS tissue phantoms developed in our lab for the experiments; the phantoms are made with polystyrene beads (0.6mL) and isopropyl alcohol (0.6mL). The scattered light is then collected by a microscope objective (Newport 10X, 0.25 NA), and imaged using a CCD camera (Point Grey Grasshopper3 GS3-U3-14S5M).

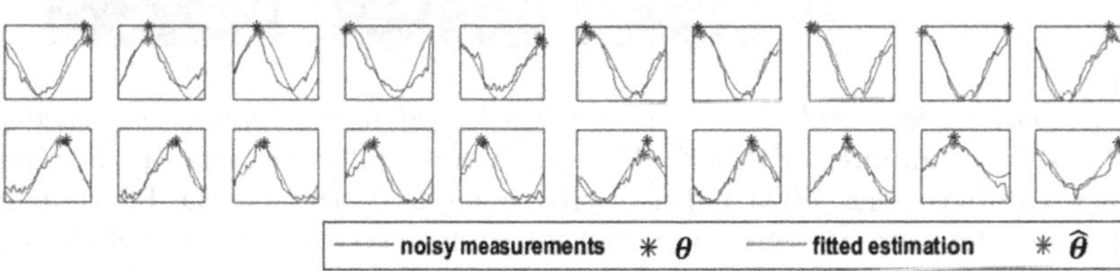

Fig. 7. *Verification of 3PointTM principle: sinusoidal curve fitting on experimental data.* The phase of each SLM pixel is sampled from 0 to 2π in 52 steps, which forms the ground truth measurements (blue curves). The optimal θ is extracted from the 52 samples. 3PointTM is then run on 3 samples to get the estimation (red curves), and thus produces the estimated optimal phase $\hat{\theta}$. The plots show that the estimation could accurately predict the ground truth measurements.

Sinusoidal Fitting on Real Data. We evaluate the accuracy of sinusoidal parameters estimated from intensity measurements made at three phase values $(0, \frac{2\pi}{3}, \frac{4\pi}{3})$ to find the sinusoidal parameters. For comparison, we also measure the actual intensity profile at 52 phase values from 0 to 2π, and visualize this dense measurements along with the sinusoidal fitted using 3PointTM in Fig. 7. The estimated sinusoidal parameters match the dense measurements accurately. Since the TM is computed directly from the estimated cosine, the accuracy of curve fitting in turn suggests that our computed TM have high accuracy.

Image Reconstruction with the Computed TM. With the computed TMs, it is possible to invert the scattering effects and reconstruct images from the captured noisy speckles. To do this, we capture images by displaying an object on the center part of the SLM, and the remaining outer region of the SLM is used as a reference. We compute the complex wavefront on the camera using 3-step phase shifting holography [2], i.e., the phase of the reference is set to 0, $\frac{\pi}{2}$, and π, and the complex wavefront y at the sensor is recovered as:

$$y = \frac{1}{4}[I_0 - I_\pi + i(2I_{\pi/2} - I_0 - I_\pi)], \tag{12}$$

where I_θ is the intensity image at the sensor when the reference phase is θ.

To construct amplitude objects from the phase SLM, two complex measurements $y_{obj}^{(1)}$ and $y_{obj}^{(2)}$ are made, following the steps in [34]. $y_{obj}^{(1)}$ is obtained with

Fig. 8. *Reconstruction results for imaging through a 20 degree diffuser and tissue phantom.* The TM is computed with 60^2 SLM pixels and 625^2 camera pixels. For imaging through the diffuser, the SSIM index of the smiley is 0.678, and the PSNR is 20.20. For the tissue phantom, the SSIM index of the smiley is 0.683, and the PSNR is 20.60.

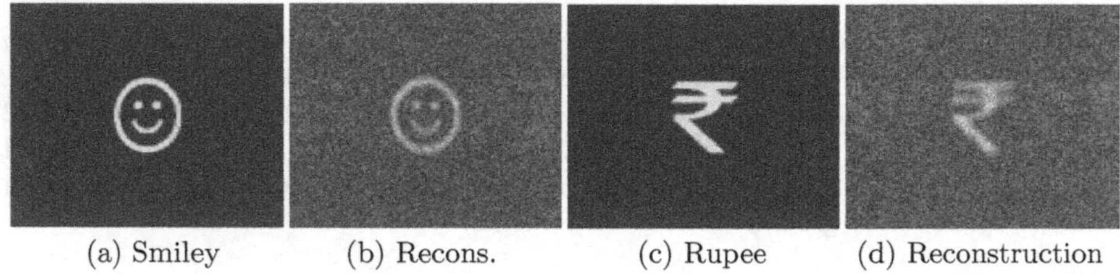

(a) Smiley (b) Recons. (c) Rupee (d) Reconstruction

Fig. 9. *Reconstruction results for imaging through a 20 degree diffuser at higher resolution.* The setup is the same to that in Fig. 8, but the TM has 120^2 SLM pixels, which increases the degrees of freedom by 4 folds. The SSIM index of the reconstructed smiley is 0.447, and that of the rupee is 0.308. The PSNR of the smiley reconstruction is 13.68 dB, and it is 12.89 dB for the rupee reconstruction.

zero phase at the object, and $y_{obj}^{(2)}$ is obtained by flipping the phase of the object from 0 to π. The reconstruction is computed with the least squares solution:

$$\widehat{x} = (T^\top T)^{-1} T^\top (y_{obj}^{(2)} - y_{obj}^{(1)}). \tag{13}$$

To characterize the reconstruction quality, the structural similarity (SSIM) indices and the peak signal-to-noise ratios (PSNR) are computed for each pair of reconstruction $\widehat{x}$ and ground truth x [16,44]. We have computed the TM for a 60^2 input size for a 20° diffuser and for a $270\mu m$ thick tissue phantom (mean free path about 90 µm), and the reconstruction results are shown in Fig. 8. Since we are able to compute a good quality transmission matrix with only $2N+1$ measurements, it is possible for us to compute a large size TM. We could compute the TM for a 120^2 SLM for a 20° diffuser, and the image reconstructions are shown in Fig. 9.

Focusing Light Through Scattering Media. To focus at a single spot p on the camera, its corresponding row t_p in the computed TM is selected, and its conjugate phase $e^{-i\angle t_p}$ is displayed onto the SLM. To focus at multiple spots, the optimal SLM phase pattern is determined by maximizing the sum of intensities observed at those multiple locations. Specifically, the sinusoidal curves for the spots are first computed individually, and then added up. The optimal SLM pattern is then determined to be the phasor which produces maximum intensity at the sum of the curves. As shown in Fig. 10, focus spots can be created at any point on the camera, and at multiple positions. At the single focusing spot through the 20° diffuser, the enhancement is 2059, which is around 75% of the theoretical maximum, which is consistent with our simulation findings.

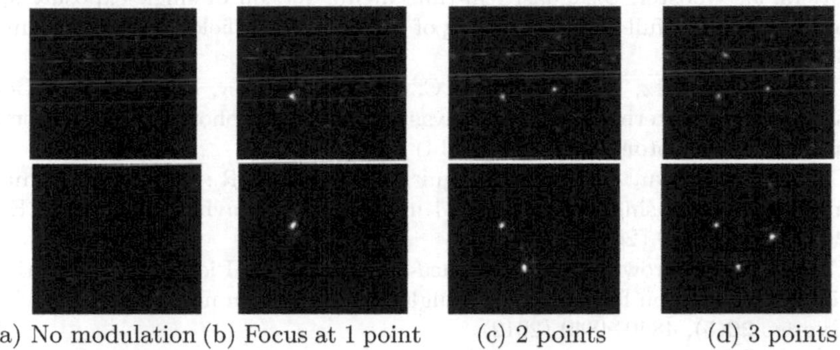

(a) No modulation (b) Focus at 1 point (c) 2 points (d) 3 points

Fig. 10. *Focusing results through scattering materials.* Laser light is focused through a 20° diffuser (top row) and a tissue phantom (bottom row) by displaying different combinations of rows of the computed TM onto the SLM. Without input modulation, only speckle patterns are recorded on the camera.

6 Conclusions

Our proposed system provides a significant speed up to the measurement of TM, and so it can be applied to many scenarios. In our current experiment, the speed of the data capture is limited by the SLM, which operates at 16ms. If the currently used SLM could be substituted by a faster device, such as a nano-mechanical phase modulator, which operates at $1\mu s$ [11], then the data capture process could be faster by 1.6×10^4 times. The system would then be able to successfully compute the TM with a 70×70 SLM for a perfused tissue, which decorrelates on the scale of 10 ms [5]. With such a large TM, focusing at a single point through the tissue can be enhanced by over 3,800 times, and a high resolution image could be reconstructed through the medium.

Acknowledgements. The authors acknowledge support via the NSF Expeditions in Computing Grant for the project "See below the skin" (#1730574, #1730147). ACS also was supported in part by the NSF CAREER award CCF-1652569.

References

1. Andreoli, D., Volpe, G., Popoff, S., Katz, O., Grésillon, S., Gigan, S.: Deterministic control of broadband light through a multiply scattering medium via the multispectral transmission matrix. Sci. Rep. **5**, 10347 (2015)
2. Awatsuji, Y., Fujii, A., Kubota, T., Matoba, O.: Parallel three-step phase-shifting digital holography. Appl. Opt. **45**, 2995–3002 (2006)
3. Awatsuji, Y., et al.: Parallel two-step phase-shifting digital holography. Appl. Opt. **47**, D183–D189 (2008)
4. Beenakker, C.W.J.: Random-matrix theory of quantum transport. Rev. Mod. Phys. **69**, 731–808 (1997)
5. Briers, J., Webster, S.: Quasi real-time digital version of single-exposure speckle photography for full-field monitoring of velocity or flow fields. Opt. Commun. **116**, 36–42 (1995)
6. Chaigne, T., Katz, O., Boccara, A.C., Fink, M., Bossy, E., Gigan, S.: Controlling light in scattering media non-invasively using the photoacoustic transmission matrix. Nat. Photonics **8**(1), 58 (2014)
7. Conkey, D., Brown, A., Caravaca-Aguirre, A., Piestun, R.: Genetic algorithm optimization for focusing through turbid media in noisy environments. Opt. Express **20**(5), 4840–4849 (2012)
8. Conkey, D.B., Brown, A.N., Caravaca-Aguirre, A.M., Piestun, R.: Genetic algorithm optimization for focusing through turbid media in noisy environments. Opt. Express **20**(5), 4840–4849 (2012)
9. Conkey, D.B., Caravaca-Aguirre, A.M., Piestun, R.: High-speed scattering medium characterization with application to focusing light through turbid media. Opt. Express **20**(2), 1733–1740 (2012)
10. Debevec, P., Hawkins, T., Tchou, C., Duiker, H.P., Sarokin, W., Sagar, M.: Acquiring the reflectance field of a human face. In: SIGGRAPH (2000)
11. Dennis, B., Haftel, M., Czaplewski, D.: Compact nanomechanical plasmonic phase modulators. Nat. Photonics **9**, 267–273 (2015)
12. Drémeau, A., et al.: Reference-less measurement of the transmission matrix of a highly scattering material using a DMD and phase retrieval techniques. Opt. Express **23**(9), 11898–11911 (2015). https://doi.org/10.1364/OE.23.011898, http://www.opticsexpress.org/abstract.cfm?URI=oe-23-9-11898
13. Freund, I.: Looking through walls and around corners. Phys. A **168**(1), 49–65 (1990)
14. Garcia, N., Genack, A.Z.: Crossover to strong intensity correlation for microwave radiation in random media. Phys. Rev. Lett. **63**, 1678–1681 (1989)
15. Goodman, J.W.: Statistical Optics. Wiley, New York (2000)
16. Hore, A., Ziou, D.: Image quality metrics: PSNR vs. SSIM. In: 2010 20th International Conference on Pattern Recognition, pp. 2366–2369 (2010)
17. Horstmeyer, R., Ruan, H., Yang, C.: Guidestar-assisted wavefront-shaping methods for focusing light into biological tissue. Nat. Photonics **9**(9), 563 (2015)
18. Indebetouw, G., Klysubun, P.: Imaging through scattering media with depth resolution by use of low-coherence gating in spatiotemporal digital holography. Opt. Lett. **25**(4), 212–214 (2000)
19. Katz, O., Heidmann, P., Fink, M., Gigan, S.: Non-invasive single-shot imaging through scattering layers and around corners via speckle correlations. Nat. Photonics **8**(10), 784–790 (2014)

20. Katz, O., Small, E., Bromberg, Y., Silberberg, Y.: Focusing and compression of ultrashort pulses through scattering media. Nat. Photonics **5**(6), 372 (2011)
21. Levenberg, K.: A method for the solution of certain problems in least-squares. Q. Appl. Math. **2**, 164–168 (1944)
22. Liu, J.P., Poon, T.C.: Two-step-only quadrature phase-shifting digital holography. Opt. Lett. **34**, 250–252 (2009)
23. Liu, J.P., Poon, T.C., Jhou, G.S., Chen, P.J.: Comparison of two-, three-, and four-exposure quadrature phase-shifting holography. Appl. Opt. **50**, 2443–2450 (2011)
24. Ma, C., Xu, X., Liu, Y., Wang, L.V.: Time-reversed adapted-perturbation (trap) optical focusing onto dynamic objects inside scattering media. Nat. Photonics **8**(12), 931 (2014)
25. Ma, X., Xiao, W., Pan, F.: Optical tomographic reconstruction based on multi-slice wave propagation method. Opt. Express **25**(19), 22595–22607 (2017)
26. Marquardt, D.: An algorithm for least-squares estimation of nonlinear parameters. SIAM J. Appl. Math. **11**, 431–441 (1963)
27. Metzler, C.A., Sharma, M.K., Nagesh, S., Baraniuk, R.G., Cossairt, O., Veeraraghavan, A.: Coherent inverse scattering via transmission matrices: efficient phase retrieval algorithms and a public dataset. In: 2017 IEEE International Conference on Computational Photography (ICCP), pp. 1–16. IEEE (2017)
28. Moravec, M.L., Romberg, J.K., Baraniuk, R.G.: Compressive phase retrieval. In: Optical Engineering+ Applications, pp. 670120–670120. International Society for Optics and Photonics (2007)
29. Mosk, A.P., Lagendijk, A., Lerosey, G., Fink, M.: Controlling waves in space and time for imaging and focusing in complex media. Nat. Photonics **6**(5), 283 (2012)
30. Mounaix, M., et al.: Spatiotemporal coherent control of light through a multiple scattering medium with the multispectral transmission matrix. Phys. Rev. Lett. **116**(25), 253901 (2016)
31. Naik, N., Zhao, S., Velten, A., Raskar, R., Bala, K.: Single view reflectance capture using multiplexed scattering and time-of-flight imaging. In: ACM Transactions on Graphics (TOG), vol. 30, p. 171. ACM (2011)
32. Paciaroni, M., Linne, M.: Single-shot, two-dimensional ballistic imaging through scattering media. Appl. Opt. **43**(26), 5100–5109 (2004)
33. Popoff, S., Lerosey, G., Fink, M., Boccara, A.C., Gigan, S.: Image transmission through an opaque material. Nat. Commun. **1**, 81 (2010)
34. Popoff, S., Lerosey, G., Carminati, R., Fink, M., Boccara, A., Gigan, S.: Measuring the transmission matrix in optics: an approach to the study and control of light propagation in disordered media. Phys. Rev. Lett. **104**(10), 100601 (2010)
35. Rajaei, B., Tramel, E.W., Gigan, S., Krzakala, F., Daudet, L.: Intensity-only optical compressive imaging using a multiply scattering material and a double phase retrieval approach. In: 2016 IEEE International Conference on Acoustics, Speech and Signal Processing (ICASSP), pp. 4054–4058, March 2016. https://doi.org/10.1109/ICASSP.2016.7472439
36. Schechner, Y.Y., Nayar, S.K., Belhumeur, P.N.: Multiplexing for optimal lighting. IEEE Trans. Pattern Anal. Mach. Intell. **29**(8), 1339–1354 (2007)
37. Sharma, M., Metzler, C.A., Nagesh, S., Cossairt, O., Baraniuk, R.G., Veeraraghavan, A.: Inverse scattering via transmission matrices: broadband illumination and fast phase retrieval algorithms. IEEE Trans. Comput. Imaging (2019)
38. Shechtman, Y., Eldar, Y.C., Cohen, O., Chapman, H.N., Miao, J., Segev, M.: Phase retrieval with application to optical imaging: a contemporary overview. IEEE Signal Process. Mag. **32**(3), 87–109 (2015)

39. Vellekoop, I.M., Mosk, A.: Focusing coherent light through opaque strongly scattering media. Opt. Lett. **32**(16), 2309–2311 (2007)
40. Velten, A., Willwacher, T., Gupta, O., Veeraraghavan, A., Bawendi, M.G., Raskar, R.: Recovering three-dimensional shape around a corner using ultrafast time-of-flight imaging. Nat. Commun. **3**, 745 (2012)
41. Wu, S.T., Hong, J.L.: Five-point amplitude estimation of sinusoidal signals: With application to LVDT signal conditioning. IEEE Trans. Instrum. Meas. **13**(3), 623–630 (2010)
42. Xu, X., Liu, H., Wang, L.V.: Time-reversed ultrasonically encoded optical focusing into scattering media. Nat. Photonics **5**(3), 154 (2011)
43. Zhou, E.H., Ruan, H., Yang, C., Judkewitz, B.: Focusing on moving targets through scattering samples. Optica **1**(4), 227–232 (2014)
44. Wang, Z., Bovik, A.C., Sheikh, H.R., Simoncelli, E.P.: Image quality assessment: from error visibility to structural similarity. IEEE Trans. Image Process. **13**(4), 600–612 (2004)

Joint Bilateral Learning for Real-Time Universal Photorealistic Style Transfer

Xide Xia[1(✉)], Meng Zhang[2], Tianfan Xue[3], Zheng Sun[3], Hui Fang[3], Brian Kulis[1], and Jiawen Chen[3]

[1] Boston University, Boston, USA
{xidexia,bkulis}@bu.edu
[2] PixelShift.AI, Mountain View, USA
meng@pixelshift.ai
[3] Google Research, New York, USA
{tianfan,zhengs,hfang,jiawen}@google.com

Abstract. Photorealistic style transfer is the task of transferring the artistic style of an image onto a content target, producing a result that is plausibly taken with a camera. Recent approaches, based on deep neural networks, produce impressive results but are either too slow to run at practical resolutions, or still contain objectionable artifacts. We propose a new end-to-end model for photorealistic style transfer that is both fast and inherently generates photorealistic results. The core of our approach is a feed-forward neural network that learns local edge-aware affine transforms that automatically obey the photorealism constraint. When trained on a diverse set of images and a variety of styles, our model can robustly apply style transfer to an arbitrary pair of input images. Compared to the state of the art, our method produces visually superior results and is three orders of magnitude faster, enabling real-time performance at 4K on a mobile phone. We validate our method with ablation and user studies.

Keywords: Style transfer · Bilateral learning · Local affine transform

1 Introduction

Image style transfer has recently received significant attention in the computer vision and machine learning communities [9]. A central problem in this domain is the task of transferring the style of an arbitrary image onto a *photorealistic* target. The seminal work of Gatys et al. [4] formulates this general artistic

X. Xia, M. Zhang and H. Fang—Work done at Google Research.

Electronic supplementary material The online version of this chapter (https://doi.org/10.1007/978-3-030-58598-3_20) contains supplementary material, which is available to authorized users.

A. Vedaldi et al. (Eds.): ECCV 2020, LNCS 12353, pp. 327–342, 2020.
https://doi.org/10.1007/978-3-030-58598-3_20

style transfer problem as an optimization that minimizes both style and content losses, but results often contain spatial distortion artifacts. Luan et al. [20] seek to reduce these artifacts by adding a *photorealism constraint*, which encourages the transformation between input and output to be *locally affine*. However, because the method formulates the problem as a large optimization whereby the loss over a deep network must be minimized for every new image pair, performance is limited. The recent work of Yoo et al. [32] proposes a wavelet-based method which provides stable stylization but is not fast enough to run at practical resolutions. Another line of recent work seeks to pretrain a feed-forward deep model [3,8,10,14,16,18,29,30] that once trained, can produce a stylized result with a single forward pass at test time. While these "universal" [9] techniques are significantly faster than those based on optimization, they may not generalize well to unseen images, may produce non-photorealistic results, and are still to slow to run in real-time on a mobile device (Fig. 1).

Fig. 1. Our method takes an input image (a) and renders it in the style of an arbitrary reference photo (insets) while preserving scene content. Notice that although the stylized outputs (b) and (c) differ dramatically in appearance from the input, one property they share is that nearby pixels of similar color are transformed similarly. We visualize this in the grayscale intensity domain, where the transformation is approximately a curve (zoomed insets). Taking advantage of this property, we propose a feed-forward neural network that directly learns these local curves.

In this work, we introduce a fast end-to-end method for photorealistic style transfer. Our model is a single feed-forward deep neural network that once trained on a suitable dataset, runs in real-time on a mobile phone at full camera resolution (i.e., 12 megapixels or "4K")—significantly faster than the state of the art. Our key observation is that we can guarantee photorealistic results by strictly enforcing Luan et. al's photorealism constraint [20]—locally, regions of similar color in the input must map to a similarly colored region in the output while respecting edges. Therefore, we design a deep learning algorithm in *bilateral space*, where these local affine transforms can be compactly represented. We contribute:

1. A photorealistic style transfer network that learns local affine transforms. The model is robust and degrades gracefully when confronted with unseen or adversarial inputs.
2. An inference implementation that runs in real-time at 4K on a mobile phone.
3. A bilateral-space Laplacian regularizer eliminates spatial grid artifacts.

Fig. 2. Inspiration. Artistic style transfer methods such as AdaIN generalize well to diverse content/style inputs but exhibit distortions on photographic content. HDRnet, designed to reproduce arbitrary imaging operators, learns the desired transform representation but fails to capture universal style transfer. Our work combines ideas from both approaches.

1.1 Related Work

Early work in image style transfer can be classified as either global methods that transfer statistics over the entire image [22,24] or local methods [13,25,26,28, 31] that typically find dense correspondences between content and style. While global methods are efficient, the results are not always faithful to the target style. Local methods can generate high-quality results, but they are computationally expensive and often fail to capture semantics. We highlight that these techniques do produce photorealistic results, albeit not always faithful to the style or well exposed.

Recently, Gatys et al. [4] showed that style can be effectively captured by the statistics of layer activations within deep neural networks trained for discriminative image classification. However, due to its generality, the technique and its successors often contain non-photorealistic painterly spatial distortions. To remove such distortions, He et al. [7] propose to achieve a more accurate color transfer by leveraging semantically-meaningful dense correspondence between images. Luan et al. [20] ameliorate this problem by imposing additional constraints on the loss function. The result is pushed towards photorealism by constraining the transformation to be locally affine in color space.

PhotoWCT [18] imposes a similar constraint as a postprocessing step, while LST [14] appends a spatial propagation network [19] after the main style transfer network to preserves the desired affinity. Similarily, Puy et al. [23] propose a flexible network to perform artistic style transfer, and applies postprocessing after each learned update for photorealistic content. Compared to these ad hoc approaches where the photorealism constraint is a soft penalty, our model directly predicts local affine transforms, guaranteeing that the constraint is satisfied.

Another line of recent work shows that matching the statistics of autoencoders is an effective way to parameterize style transfer [8,14,17,18,32]. Moreover, they show that distortions can be reduced by preserving high frequencies using unpooling [18] or wavelet transform residuals [32].

Our work unifies these two lines of research. Our network architecture builds upon HDRnet [5], which was first employed in the context of learning image enhancement and tone manipulation. Given a large dataset of input/output pairs, it learns local affine transforms that best reproduces the operator. The network is small and the learned transforms that are intentionally constrained to be incapable of introducing artifacts such as noise or false edges. These are exactly the properties we want and indeed, Gharbi et al. demonstrated style transfer in their original paper. However, when we applied HDRnet to our more diverse dataset, we found a number of artifacts (Fig. 2). This is because HDRnet does not explicitly model style transfer and instead learns by memorizing what it sees during training and projecting the function onto local affine transforms. Therefore, it will require a lot of training data and generalize poorly. Since HDRnet learns local affine transforms from low-level image features, our strategy is to start with statistical feature matching using Adaptive Instance Normalization [8] to build a joint distribution. By explicitly modeling the style transformation as a distribution matching process, our network is capable of generalizing to unseen or adversarial inputs (Fig. 2).

2 Method

Our method is based on a single feed-forward deep neural network. It takes as input two images, a *content photo* I_c, and an arbitrary *style image* I_s, producing a photorealistic output O with the former's content but the latter's style. Similar to previous work [14,18,32], our network is "universal"—after training on a diverse dataset of content/style pairs, it can generalize to novel input combinations. Its architecture is centered around the core idea of learning local affine transformations, which inherently enforce the photorealism constraint.

2.1 Background

For completeness, we first summarize the key ideas of recent work.

Content and Style. The Neural Style Transfer [4] algorithm is based on an optimization that minimizes a loss balancing the output image's fidelity to the input images' content and style:

$$\mathcal{L}_g = \alpha\mathcal{L}_c + \beta\mathcal{L}_s \qquad \text{with} \tag{1}$$

$$\mathcal{L}_c = \sum_{i=1}^{N_c} \|F_i[O] - F_i[I_c]\|_2^2 \quad \text{and} \quad \mathcal{L}_s = \sum_{i=1}^{N_s} \|G_i[O] - G_i[I_s]\|_F^2, \tag{2}$$

where N_c and N_s denote the number of intermediate layers selected from a pre-trained VGG-19 network [27] to represent image content and style, respectively. Scene content is captured by the feature maps F_i of intermediate layers of the VGG network, and style is captured by their Gram matrices $G_i[\cdot] = F_i[\cdot]F_i[\cdot]^T$. $||\cdot||_F$ denotes the Frobenius norm.

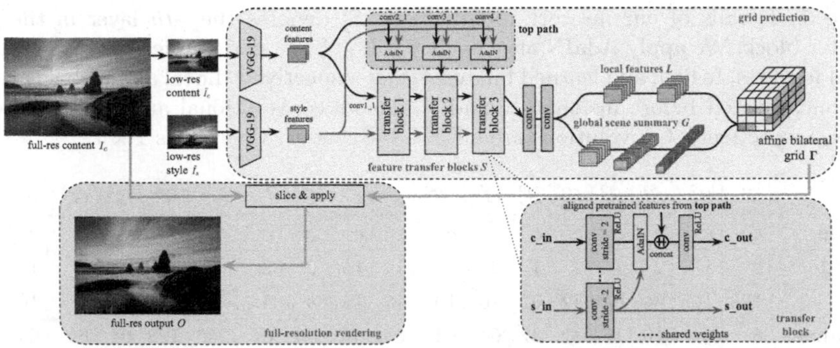

Fig. 3. Model architecture. Our model starts with a low-resolution grid prediction stream that uses three feature transfer blocks $\boldsymbol{S}$ to build a joint distribution between the low-level features of the input content/style pair. This distribution is fed to bilateral learning blocks $\boldsymbol{L}$ and $\boldsymbol{G}$ to predict an affine bilateral grid Γ. Rendering, which runs at full-resolution, performs the minimal per-pixel work of sampling from Γ a 3×4 matrix and then multiplying.

Statistical Feature Matching. Instead of directly minimizing the loss in Eq. 1, followup work shows that it is more effective to match the statistics of feature maps at the bottleneck of an auto-encoder. Variants of the whitening and coloring transform [17,18,32] normalize the singular values of each channel, while Adaptive Instance Normalization (AdaIN) [8] proposes a simple scheme using the mean $\mu(\cdot)$ and the standard deviation $\sigma(\cdot)$ of each channel:

$$\mathrm{AdaIN}(x, y) = \sigma(y) \left(\frac{x - \mu(x)}{\sigma(x)} \right) + \mu(y), \tag{3}$$

where x and y are content and style feature channels, respectively. Due to its simplicity and reduced cost, we also adopt AdaIN layers in our network architecture as well as its induced style loss [8,15]:

$$\mathcal{L}_{sa} = \sum_{i=1}^{N_S} \|\mu(F_i[O]) - \mu(F_i[I_s])\|_2^2 + \sum_{i=1}^{N_S} \|\sigma(F_i[O]) - \sigma(F_i[I_s])\|_2^2. \tag{4}$$

Bilateral Space. Bilateral space was first introduced by Paris and Durand [21] in the context of fast edge-aware image filtering. A 2D grayscale image $I(x, y)$ can be "lifted" into bilateral space as a sparse collection of 3D points $\{x_j, y_j, I_j\}$ in the augmented space. In this space, linear operations are inherently edge-aware because Euclidean distances preserve edges. They prove that bilateral filtering is equivalent to *splatting* the input onto a regular 3D *bilateral grid*, blurring, and *slicing* out the result using trilinear interpolation at the input coordinates $\{x_j, y_j, I_j\}$. Since blurring and slicing are low-frequency operations, the grid can be low-resolution, dramatically accelerating the filter.

Table 1. Details of our network architecture. S_j^i denotes the *i-th* layer in the *j-th* transfer block. We apply AdaIN after each S_j^1. L^i, G^i, F, and Γ refer to local features, global features, fusion, and learned bilateral grid, respectively. Local and global features are concatenated before fusion F. c and f_c denote convolutional and fully-connected layers, respectively. Convolutions are all 3×3 except F, where it is 1×1.

	S_1^1	S_1^2	S_2^1	S_2^2	S_3^1	S_3^2	C^7	C^8	L^1	L^2	G^1	G^2	G^3	G^4	G^5	G^6	F	Γ
Type	c	c	c	c	c	c	c	c	c	c	c	c	f_c	f_c	f_c	f_c	c	c
Stride	2	1	2	1	2	1	2	1	1	1	2	2	–	–	–	–	1	1
Size	128	128	64	64	32	32	16	16	16	16	8	4	–	–	–	–	16	16
Channels	8	8	16	16	32	32	64	64	64	64	64	64	256	128	64	64	64	96

Bilateral Guided Upsampling (BGU) [2] extends the bilateral grid to represent transformations between images. By storing at each cell an affine transformation, an *affine bilateral grid* can encode any image-to-image transformation given sufficient resolution. The pipeline is similar: *splat* both input and output images onto a bilateral grid, blur, and perform a per-pixel least squares fit. To apply the transform, *slice* out a per-pixel affine matrix and multiply by the input color. BGU shows that this representation can accelerate a variety of imaging operators and that the approximation degrades gracefully with resolution when suitably regularized. Affine bilateral grids are constrained to produce an output that is a smoothly varying, edge-ware, and locally affine transformation of the input. Therefore, it fundamentally cannot produce false edges, amplify noise, and inherently obeys the photorealism constraint.

Gharbi et al. [5] showed that slicing and applying an affine bilateral grid are sub-differentiable and therefore can be incorporated as a layer in a deep neural network and learned using gradient descent. They demonstrated that their HDRnet architecture can effectively learn to reproduce many photographic tone mapping and detail manipulation tasks, regardless of whether they are algorithmic or artist-driven.

2.2 Network Architecture

Our end-to-end differentiable network consists of two streams. The *grid prediction* stream takes as input reduced-resolution content $\widetilde{I}_c$ and style $\widetilde{I}_s$ images, learns the joint distribution between their low-level features, and predicts an affine bilateral grid Γ. The *rendering* stream, unmodified from HDRnet, operates at full-resolution. At each pixel (x, y, r, g, b), it uses a learned lookup table to compute a "luma" value $z = g(r, g, b)$, slices out $A = \Gamma(x/w, y/h, z/d)$ (using trilinear interpolation), and outputs $O = A * (r, g, b, 1)^T$. By decoupling grid prediction resolution from that of rendering, our architecture offers a tradeoff between stylization quality and performance. Figure 3 summarizes the entire network and we describe each block below.

Grid Prediction. We aim to first learn a multi-scale model of the joint distribution between content and style features, and from this distribution, predict an affine bilateral grid. Rather than using strided convolutional layers to directly learn from pixel data, we follow recent work [8,10,18] and use a pretrained VGG-19 network to extract low-level features from both images at four scales (*conv*1_1, *conv*2_1, *conv*3_1, and *conv*4_1). We process these multi-resolution feature maps with a sequence of *feature transfer blocks* inspired by the StyleGAN architecture [11] (Fig. 3). Starting from the finest level, each transfer block applies a stride-2 *weight-sharing* convolutional layer to both content and style features, halving spatial resolution while doubling the number of channels (see Table 1). The shared-weight constraint crucially allows the following AdaIN layer to learn the joint content/style distribution without channel-wise correspondence supervision. Once the content feature map is normalized, we concatenate it to the similarly AdaIN-aligned feature maps from the pretrained VGG-19 layer of the same resolution. Since the content feature map now contains more channels, we use a stride-1 convolutional layer to select the relevant channels between learned-and-normalized vs. pretrained-and-normalized features.

We use three transfer blocks in our architecture, corresponding to the three finest-resolution layers of the selected VGG features. While using additional transfer blocks is possible, they are too coarse and replacing them with standard stride-2 convolutions makes little difference in our experiments. Since this component of the network effectively learns the relevant bilateral-space content features based on its corresponding style, it can be thought of as *learned feature transfer*.

Joint Bilateral Learning. With aligned-to-style content features in bilateral space, we seek to learn an affine bilateral grid that encodes a transformation that locally captures style and is aware of scene semantics. Like HDRnet, we split the network into two asymmetric paths: a fully-convolutional *local path* that learns local color transforms and thereby sets the grid resolution, and a *global path*, consisting of both convolutional and fully-connected layers, that learns a summary of the scene and helps spatially regularize the output transforms. The local path consists of two stride 1 convolutional layers, keeping the spatial resolution and number of features constant. This provides enough depth to learn local affine transforms without letting its receptive field grow too large (and thereby discarding any notion of spatial position).

As we aim to perform universal style transfer without any explicit notion of semantics (e.g., per-pixel masks provided by an external pretrained network), we use a small network to learn a global notion of scene category. Our global path consists of two stride 2 convolutional layers to further reduce resolution, followed by four fully-connected layers to produce a 64−element vector "summary". We concatenate the summary at each x, y spatial location output from the local path and use a 1×1 convolutional layer to reduce the final output to 96 channels. These 96 channels can be reshaped into a 8 "luma bins" that separate edges, each storing a 3×4 affine transform. We use the ReLU activation after all but the final 1×1 fusion layer and zero-padding for all convolutional layers.

2.3 Losses

Since our architecture is fully differentiable, we can simply define our loss function on the generated output. We augment the content and style fidelity losses of Huang et al. [8] with a novel *bilateral-space Laplacian regularizer*, similar to the one in [6]:

$$\mathcal{L} = \lambda_c \mathcal{L}_c + \lambda_{sa} \mathcal{L}_{sa} + \lambda_r \mathcal{L}_r, \tag{5}$$

where $\mathcal{L}_c$ and $\mathcal{L}_{sa}$ are the content and style losses defined in Eqs. 2 and 4, and

$$\mathcal{L}_r(\Gamma) = \sum_s \sum_{t \in N(s)} ||\Gamma[s] - \Gamma[t]||_F^2, \tag{6}$$

where $\Gamma[s]$ is one cell of the predicted bilateral grid, and $\Gamma[t]$ one of its neighbors.

The Laplacian regularizer penalizes differences between adjacent cells of the bilateral grid (indexed by s and finite differences computed over its six-connected neighbors $N(s)$) and encourages the learned affine transforms to be smooth in both space and intensity [2,6]. As we show in our ablation study (Sect. 3.1), the Laplacian regularizer is necessary to prevent visible grid artifacts.

We set $\lambda_c = 0.5$, $\lambda_{sa} = 1$, and $\lambda_r = 0.15$ in all experiments.

2.4 Training

We trained our model on high-quality landscape photos using Tensorflow [1], without any explicit notion of semantics. We use the Adam optimizer [12] with hyperparameters $\alpha = 10^{-4}, \beta_1 = 0.9, \beta_2 = 0.999, \epsilon = 10^{-8}$, and a batch size of 12 content/style pairs. For each epoch, we randomly split the data into 50000 content/style pairs. The training resolution is 256×256 and we train for a fixed 25 epochs, taking two days on a single NVIDIA Tesla V100 GPU with 16 GB RAM. Even trained at a low-resolution, our algorithms still performs well with 12 megapixel inputs, as shown in Fig. 8. We attribute this to the fact that our losses are derived from pretrained VGG features, which are relatively invariant with respect to resolution.

3 Results

For evaluation, we collect a test set of 400 high-quality photos of a variety of subjects. We compare our algorithm to the state of the art in photorealistic style transfer, and conduct a user study. Furthermore, we perform a set of ablation studies to better understand the contribution of various components. Detailed comparisons with high-resolution images are included in the supplement.

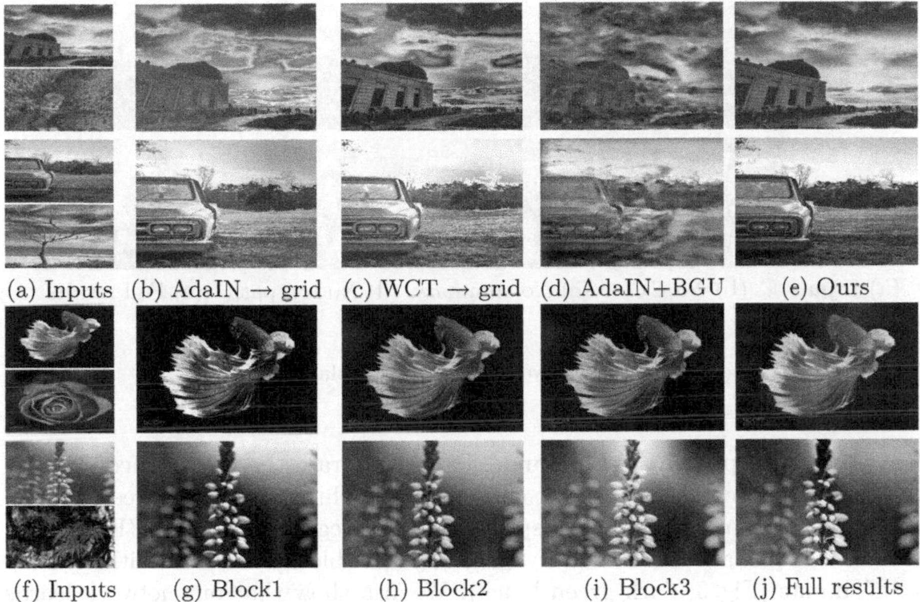

Fig. 4. Ablation studies on feature transfer blocks. (a)–(e): We demonstrate the importance of our feature transfer architecture by replacing it with baseline networks. (f)–(j): Visualization of the contribution of each transfer block by disabling statistical feature matching on the others.

3.1 Ablation Studies

Feature Transfer. We conduct multiple ablations to show the importance of our feature transfer blocks $\boldsymbol{S}$.

First, we consider replacing $\boldsymbol{S}$ with two baseline networks: AdaIN [8] or WCT [17]. Starting with the same features extracted from VGG-19, we perform feature matching using AdaIN or WCT. The rest of the network is unchanged: that is, we attempt to learn local and global features directly from the baseline encoders and predict affine bilateral grids. Figure 4 (b) is the result of using the pretrained AdaIN feature from the "top path" only. The results in Fig. 4 (b) and (c) show that while content is preserved, there is an overall color cast as well as inconsistent blotches when using either AdaIN or WCT. The low-resolution features simply lack the information density to learn even global color correction. Adding our multi-scale feature transfer blocks resolves this issue.

Second, to illustrate the contribution of each transfer block, we visualize our network's output when all but one block is disabled (including the top path inputs). As shown in Fig. 4(f–j), earlier, finer resolution blocks learn texture and local contrast, while later blocks capture more global information such as the style input's dominant color tone, which is consistent with our intuition. By combining all transfer blocks at three different resolutions, our model merges these features at multiple scales into a joint distribution.

(a) Inputs (b) No $\mathcal{L}_r$ (c) No summary (d) No top path (e) Full results

Fig. 5. Network component ablations.

Network Component Ablations. To demonstrate the importance of other network components, in Fig. 5, we compare our architecture with three variants: one trained without the bilateral-space Laplacian regularization loss (Eq. 4), one without the global scene summary (Fig. 3, yellow block), and one without "top path" inputs (Fig. 3, dark green block). We also show that our network learns stylization parameterized as local affine transforms.

Figure 5 (b) shows distinctive dark halos when bilateral-space Laplacian regularization is absent. This is due to the fact that the network can learn to set regions of the bilateral grid to zero where it does not encounter image data (because images occupy a sparse 2D manifold in the grid's 3D domain). When sliced, the result is a smooth transition between zero and the proper transform.

Figure 5 (c) shows that the global summary helps with spatial consistency. For example, in the *mountain* photo, the left part of sky is saturated while the right side of the mountain is slightly washed out. In contrast, the output of our full network in Fig. 5 (e) has more spatially consistent color. This is consistent with the observation in Gharbi et al. [5].

Figure 5 (d) demonstrates the necessity and importance of pretrained-and-normalized features (the output of the "top path" in Fig. 3). The results show distinctive patches of incorrect color characteristic of the network locally overfitting to the style input. Adaptively selecting between learned and pretrained features at multiple resolutions eliminates this inconsistency.

Finally, we show that our network is not just an edge-aware interpolator—it learns stylization parameterized as local affine transforms. We set an edge-aware interpolation baseline by first running the full AdaIN network [8] on our 256×256 content and style images to produce a low-resolution stylized result. We then use BGU [2] to fit a $16 \times 16 \times 8$ affine bilateral grid (the same resolution as our network), and slice it with the full-resolution input to produce a full-resolution output. Figure 4 (d) shows that this strategy works quite poorly: AdaIN's output exhibits spatial distortions even at 256×256, which BGU cannot fix.

Fig. 6. Output using grids with different spatial (top) or luma (bottom) resolutions (w × h × luma bins).

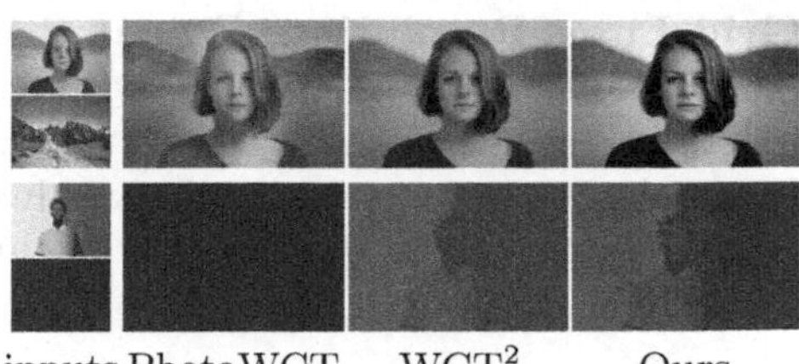

Fig. 7. Our method is robust to adversarial inputs such as when the content image is a portrait (an unseen category) or even a monochromatic "style".

(a) Output at 12 megapixels.

Image Size	PhotoWCT	LST	WCT^2	Ours
512 × 512	0.68s	0.25s	3.85s	**< 5 ms**
1024 × 1024	1.51s	0.84s	6.13s	**< 5 ms**
1000 × 2000	2.75s	OOM	10.94s	**< 5 ms**
2000 × 2000	OOM	OOM	OOM	**< 5 ms**
3000 × 4000	OOM	OOM	OOM	**< 5 ms**

(b) NVIDIA Tesla V100 runtime.

Mean Score	PhotoWCT	LST	WCT^2	Ours
Photorealism	2.02	2.89	**4.21**	4.14
Stylization	3.10	3.19	3.24	**3.49**
Overall quality	2.23	2.84	3.60	**3.79**

(c) User study results (higher is better).

Fig. 8. (a) The output of our method running at 12 megapixels, a typical smartphone camera resolution. Despite being trained at a fixed low resolution, our method produces sharp results while faithfully transferring the style from a significantly different scene. See the supplement for full-resolution images. (b) Performance benchmarks on a NVIDIA Tesla V100 GPU with 16 GB of RAM. OOM indicates out of memory. Note that photorealistic postprocessing adds significant overhead to LST performance. Due to GPU startup and memory I/O time, we were unable to get a precise measurement below 5ms. (c) Mean user study scores from 1200 responses. Raters scored the three output images in each sextet on a scale of 1–5 (higher is better).

Grid Spatial Resolution. Figure 6 (top) shows how the spatial resolution of the grid affects stylization quality. By fixing the number of luma bins at 8, the 1 × 1 case is a single global curve, where the network learns an incorrectly colored compromise. Going up to 2 × 2, the network attempts to spatially vary the transformation, with slightly different colors applied to each quadrant, but the result is still an unsatisfying tradeoff. At 8 × 8, there is sufficient spatial resolution to produce a faithful stylization result.

Grid Luma Resolution. Figure 6 (bottom) also shows how the "luma" resolution affects stylization quality, with a fixed spatial resolution 16×16. With 1 luma bin, the network is restricted to predicting a single affine transform per tile. Interpolating between 2 luma bins reduces to a quadratic spline per tile, which is still insufficient for this image. In our experiments, 8 luma bins is sufficient for most images in our test set.

3.2 Qualitative Results

Visual Comparison. We compare our technique against three state-of-the-art photorealistic style transfer algorithms: PhotoWCT [18], LST [14], and WCT^2 [32], using default settings. Note that for PhotoWCT, we use the most recent version of NVIDIA's FastPhotoStyle code. Comparisons with other algorithms are included in the supplementary material.

Figure 10 features a small sampling of the test set with some challenging examples. Owing to its reliance on unpooling and postprocessing, PhotoWCT results contain noticeable artifacts on nearly all scenes. LST mainly focuses on artistic style transfer, and to generate photorealistic results, uses a compute-intensive spatial propagation network as a postprocessing step to reduce distortion artifacts. Figure 10 shows that there are still noticeable distortions in several instances, even after postprocessing. WCT^2 performs quite well when content and style are semantically similar, but when the scene content is significantly different from the landscapes on which it was trained, the results appear "hazy". Our method performs well even on these challenging cases. Thanks to its restricted output space, our method always produce sharp images which degrades gracefully towards the input (e.g., face, leaves) when given inputs outside the training set. Our primary artifact is a noticeable reduction in contrast along strong edges and is a known limitation of the local affine transform model [2].

Robustness. Thanks to its restricted transform model, our method is significantly more robust than the baselines when confronted with adversarial inputs, as shown in Figure 7. Although our model was trained exclusively on landscapes, the restricted transform model allows it to degrade gracefully on portraits which it has never encountered and even a monochromatic "style".

3.3 Quantitative Results

Runtime and Resolution. As shown in Fig. 8 (b), our runtime on a workstation GPU significantly outperforms the baselines and is essentially invariant to resolution at practical resolutions. This is due to the fact that grid prediction, the "deep" part of the network, runs at a constant low resolution of 256×256. In contrast, our full-resolution stream does minimal work and has hardware acceleration for trilinear interpolation. On a modern smartphone, inference runs comfortably 30 Hz at full 12 megapixel camera resolution when quantized to 16-bit floating point and work distributed between ML acceleration hardware and the GPU. Figure 8 shows one such example. More images and a detailed performance benchmark are included in the supplement.

User Study. The question of whether an image is a faithful rendition of the style of another is inherently a matter of subjective taste. As such, we conducted a user study to judge whether our method delivers subjectively better results compared to the baselines. We recruited 20 users unconnected with the project. Each user was shown 20 sextets of images consisting of the input content, reference style, and four randomly shuffled outputs (PhotoWCT [18], WCT^2 [32], LST [14], and ours). For each output, they were asked to rate the following questions on a scale of 1–5:

- How noticeable are artifacts (i.e., less photorealistic) in the image?
- How similar is the output in style to the reference?
- How would you rate the overall quality of the generated image?

In total, we collected 1200 responses (400 images × 3 questions) and results are shown in Fig. 8 (c). In both stylization and overall quality, our technique outperforms the baselines: PhotoWCT, LST, and WCT^2. In terms of photorealism, our algorithm is on-par with WCT^2, and is significantly better than PhotoWCT and LST.

Video Stylization. Although our network is trained exclusively on images, it generalizes well to video content. Figure 9 shows an example where we transfer the style of a single photo to a video sequence that varies dramatically in appearance. The resulting video has a consistent style and is temporally coherent without any additional regularization or data augmentation.

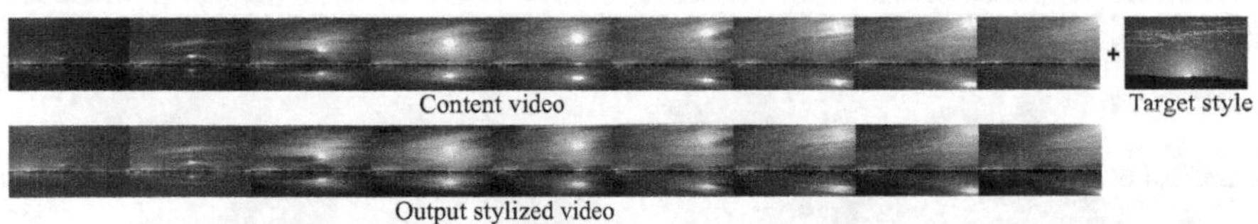

Fig. 9. Transferring the style of a still photo to a video sequence. Although the content frames undergo substantial changes in appearance, our method produces a temporally coherent result consistent with the reference style. Please refer to the supplementary material for the full videos.

Fig. 10. Qualitative comparison of our method against three state of the art baselines on some challenging examples.

4 Conclusion

We presented a feed-forward neural network for universal photorealistic style transfer. The key to our approach is using deep learning to predict affine bilateral grids, which are compact image-to-image transformations that implicitly enforce the photorealism constraint. We showed that our technique is significantly faster than state of the art, runs in real-time on a smartphone at high resolution, and degrades gracefully even in extreme cases. We believe its robustness and fast runtime will lead to practical applications in mobile photography. As future work, we hope to further improve performance by reducing network size, and investigate how to relax the photorealism constraint to generate a continuum between photorealistic images and abstract art.

References

1. Abadi, M., et al.: TensorFlow: a system for large-scale machine learning. In: 12th USENIX Symposium on Operating Systems Design and Implementation (OSDI) (2016)
2. Chen, J., Adams, A., Wadhwa, N., Hasinoff, S.W.: Bilateral guided upsampling. ACM TOG **35**, 1–8 (2016)
3. Dumoulin, V., Shlens, J., Kudlur, M.: A learned representation for artistic style. ICLR (2017)
4. Gatys, L.A., Ecker, A.S., Bethge, M.: Image style transfer using convolutional neural networks. In: CVPR (2016)
5. Gharbi, M., Chen, J., Barron, J.T., Hasinoff, S.W., Durand, F.: Deep bilateral learning for real-time image enhancement. ACM TOG **36**, 1–12 (2017)
6. Gupta, M., et al.: Monotonic calibrated interpolated look-up tables. J. Mach. Learn. Res. **17**, 3790–3836 (2016)
7. He, M., Liao, J., Chen, D., Yuan, L., Sander, P.V.: Progressive color transfer with dense semantic correspondences. ACM TOG **38**, 1–18 (2019)
8. Huang, X., Belongie, S.: Arbitrary style transfer in real-time with adaptive instance normalization. In: ICCV (2017)
9. Jing, Y., Yang, Y., Feng, Z., Ye, J., Yu, Y., Song, M.: Neural style transfer: a review. TVCG (2019)
10. Johnson, J., Alahi, A., Fei-Fei, L.: Perceptual losses for real-time style transfer and super-resolution. In: Leibe, B., Matas, J., Sebe, N., Welling, M. (eds.) ECCV 2016. LNCS, vol. 9906, pp. 694–711. Springer, Cham (2016). https://doi.org/10.1007/978-3-319-46475-6_43
11. Karras, T., Laine, S., Aila, T.: A style-based generator architecture for generative adversarial networks. In: CVPR (2019)
12. Kingma, D.P., Ba, J.: Adam: a method for stochastic optimization. In: ICLR (2015)
13. Laffont, P.Y., Ren, Z., Tao, X., Qian, C., Hays, J.: Transient attributes for high-level understanding and editing of outdoor scenes. ACM TOG **33**, 1–11 (2014)
14. Li, X., Liu, S., Kautz, J., Yang, M.H.: Learning linear transformations for fast image and video style transfer. In: CVPR (2019)
15. Li, Y., Wang, N., Liu, J., Hou, X.: Demystifying neural style transfer. In: IJCAI (2017)
16. Li, Y., Fang, C., Yang, J., Wang, Z., Lu, X., Yang, M.H.: Diversified texture synthesis with feed-forward networks. In: CVPR (2017)

17. Li, Y., Fang, C., Yang, J., Wang, Z., Lu, X., Yang, M.H.: Universal style transfer via feature transforms. In: NeurIPS (2017)
18. Li, Y., Liu, M.-Y., Li, X., Yang, M.-H., Kautz, J.: A closed-form solution to photorealistic image stylization. In: Ferrari, V., Hebert, M., Sminchisescu, C., Weiss, Y. (eds.) ECCV 2018. LNCS, vol. 11207, pp. 468–483. Springer, Cham (2018). https://doi.org/10.1007/978-3-030-01219-9_28
19. Liu, S., De Mello, S., Gu, J., Zhong, G., Yang, M.H., Kautz, J.: Learning affinity via spatial propagation networks. In: NeurIPS (2017)
20. Luan, F., Paris, S., Shechtman, E., Bala, K.: Deep photo style transfer. In: CVPR (2017)
21. Paris, S., Durand, F.: A fast approximation of the bilateral filter using a signal processing approach. In: Leonardis, A., Bischof, H., Pinz, A. (eds.) ECCV 2006. LNCS, vol. 3954, pp. 568–580. Springer, Heidelberg (2006). https://doi.org/10.1007/11744085_44
22. Pitié, F., Kokaram, A.C., Dahyot, R.: N-dimensional probability density function transfer and its application to color transfer. In: ICCV (2005)
23. Puy, G., Pérez, P.: A flexible convolutional solver for fast style transfers. In: CVPR (2019)
24. Reinhard, E., Adhikhmin, M., Gooch, B., Shirley, P.: Color transfer between images. IEEE Comput. Graph. Appl. **21**, 34–41 (2001)
25. Shih, Y., Paris, S., Barnes, C., Freeman, W.T., Durand, F.: Style transfer for headshot portraits. ACM TOG (2014)
26. Shih, Y., Paris, S., Durand, F., Freeman, W.T.: Data-driven hallucination of different times of day from a single outdoor photo. ACM TOG **32**, 1–11 (2013)
27. Simonyan, K., Zisserman, A.: Very deep convolutional networks for large-scale image recognitio. In: ICLR (2015)
28. Tsai, Y.H., Shen, X., Lin, Z., Sunkavalli, K., Yang, M.H.: Sky is not the limit: semantic-aware sky replacement. ACM TOG (2016)
29. Ulyanov, D., Lebedev, V., Vedaldi, A., Lempitsky, V.S.: Texture networks: feed-forward synthesis of textures and stylized images. In: ICML (2016)
30. Ulyanov, D., Vedaldi, A., Lempitsky, V.: Improved texture networks: maximizing quality and diversity in feed-forward stylization and texture synthesis. In: CVPR (2017)
31. Wu, F., Dong, W., Kong, Y., Mei, X., Paul, J.C., Zhang, X.: Content-based colour transfer. In: Computer Graphics Forum (2013)
32. Yoo, J., Uh, Y., Chun, S., Kang, B., Ha, J.W.: Photorealistic style transfer via wavelet transforms. In: ICCV (2019)

Beyond 3DMM Space: Towards Fine-Grained 3D Face Reconstruction

Xiangyu Zhu[1,2], Fan Yang[3], Di Huang[4], Chang Yu[1,2], Hao Wang[1], Jianzhu Guo[1,2], Zhen Lei[1,2(✉)], and Stan Z. Li[4,5]

[1] CBSR & NLPR, Institute of Automation, Chinese Academy of Sciences, Beijing, China
{xiangyu.zhu,chang.yu,jianzhu.guo,zlei}@nlpr.ia.ac.cn, haowang7308@gmail.com

[2] School of Artificial Intelligence, University of Chinese Academy of Sciences, Beijing, China

[3] College of Software, Beihang University, Beijing, China
fanyang@buaa.edu.cn

[4] Beijing Advanced Innovation Center for BDBC, Beihang University, Beijing, China
dhuang@buaa.edu.cn, szli@nlpr.ia.ac.cn

[5] School of Engineering, Westlake University, Hangzhou, China

Abstract. Recently, deep learning based 3D face reconstruction methods have shown promising results in both quality and efficiency. However, most of their training data is constructed by 3D Morphable Model, whose space spanned is only a small part of the shape space. As a result, the reconstruction results lose the fine-grained geometry and look different from real faces. To alleviate this issue, we first propose a solution to construct large-scale fine-grained 3D data from RGB-D images, which are expected to be massively collected as the proceeding of hand-held depth camera. A new dataset Fine-Grained 3D face (FG3D) with 200k samples is constructed to provide sufficient data for neural network training. Secondly, we propose a Fine-Grained reconstruction Network (FGNet) that can concentrate on shape modification by warping the network input and output to the UV space. Through FG3D and FGNet, we successfully generate reconstruction results with fine-grained geometry. The experiments on several benchmarks validate the effectiveness of our method compared to several baselines and other state-of-the-art methods. The proposed method and code will be available at https://github.com/XiangyuZhu-open/Beyond3DMM.

Keywords: 3D face reconstruction · Fine-grained · Deep learning

1 Introduction

With the advent of deep learning and the development of large annotated datasets, recent works have shown results of unprecedented accuracy even on

Electronic supplementary material The online version of this chapter (https://doi.org/10.1007/978-3-030-58598-3_21) contains supplementary material, which is available to authorized users.

A. Vedaldi et al. (Eds.): ECCV 2020, LNCS 12353, pp. 343–358, 2020.
https://doi.org/10.1007/978-3-030-58598-3_21

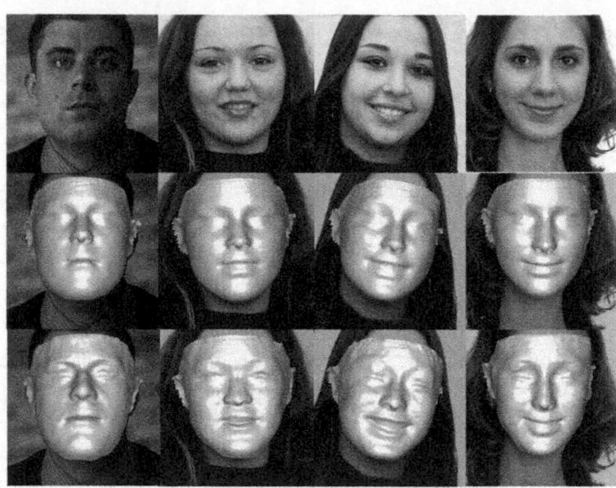

Fig. 1. The first row shows the images, the second row shows the results of the state-of-the-art method PRNet [9] and the third row shows our results.

the most challenging computer vision tasks. In this work, we focus on 3D face reconstruction which recovers the 3D facial geometry from a single 2D image. Despite many years of research, it is still an open problem in vision and graphics research. Since the seminal work of Blanz and Vetter [5], 3D Morphable Model (3DMM) has been widely used to reconstruct 3D face shape. However, most of the popular models like BFM [21] are built from scans of only 200 subjects with a similar ethnicity/age group. They are also captured in well controlled conditions with only neutral expressions. As a result, these models are fragile to large variances in face identity. In more than a decade, almost all the models cover no more than 300 training scans. Such a small training set is far from adequate to describe the full variability of human faces. Recently, there is a surge of interest in 3D face reconstruction using deep Convolution Neural Networks (CNN) rather than the optimization based traditional methods [5,25,26,42]. However, training deep models requires large data with dense 3D annotations, which are expensive and even infeasible in some cases. In most 3D face datasets, the ground truth is constructed by fitting a 3DMM to less than 100 labelled landmarks, which loses the fine-grained geometry, especially on the cheek region. A model trained on such a dataset cannot deal well with the variations that are not present in the 3DMM space. Although recent works bypass 3DMM parameters and use the image-to-volume [11] or image-to-uvmap [9] strategy, the ground truth still comes from the space of 3DMM and the fitting results are still model-like.

In this paper, we aim to overcome the intrinsic limitation of 3D face reconstruction by improving both the training data and the reconstruction method. Firstly, we explore to construct large-scale fine-grained 3D data from RGB-D images. Although complete and high-precision face scans are expensive to acquire, the RGB-D images can be considered as a good alternative, which are much easier to collect and have been popular in face analysis [6,17,20,29,36,36,38]. As the proceeding of hand-held depth camera, we believe medium-precision RGB-D images can be massively collected in the near future. In this paper, we first employ the 3DMM texture and illumination model as a strong constraint to robustly register RGB-D images and perform high-fidelity out-of-plane augmentation, generating a large 3D dataset **F**ine-**G**rained **3D** face (FG3D) from public RGB-D images. Secondly, to reconstruct fine-grained geometry through CNN, we propose a **F**ine-**G**rained reconstruction **Net**work (FGNet) and discuss two possible structures to capture fine-grained shapes: a camera-view structure (FGNet-CV) which directly estimates the shape update from the original image and a model-view structure (FGNet-MV) which normalizes pose variations by UV-space warping to concentrate on shape modification. The two structures are compared experimentally and the better one is adopted for reconstruction.

In summary, our main contributions are: (1) In order to overcome the scarcity of 3D fine-grained training data, we develop a complete solution to generate a large number of "image to 3D face" pairs from RGB-D images. (2) We provide a new fine-grained 3D face dataset FG3D with about 200k samples for neural network training. (3) A novel network structure FGNet is proposed for fine-grained geometry reconstruction. (4) Based on FG3D and FGNet, we finally generate face-like 3D reconstruction results. Extensive experiments show that our method significantly reduces the reconstruction error and achieves the best result. Figure 1 briefly shows some results.

2 Related Works

With the development of deep learning, 3D face reconstruction has witnessed great progress by Convolution Neural Network (CNN). In early years, some methods use CNN to estimate the 3D Morphable Model parameters [14,18,23,24] or its variants [3,4,8,13,31,35], which provide both dense face alignment and 3D face reconstruction results. However, the performance of these methods is restricted due to the limitation of the 3D space defined by the face model basis or the templates [8,10,15,19,28,30]. The required face transformations including perspective projection and 3D thin plate spline transformation are also difficult to estimate. Recently, two end-to-end works [9,11], which bypass the limitation of the PCA model, achieve state-of-the-art performance on their respective tasks. VRN [11] develops a volumetric representation of 3D face and uses a network to regress it from a 2D image. However, this representation discards the semantic meaning of points and the network needs to regress the redundant whole volume in order to restore the face shape. PRNet [9] designs a UV position map, which is a 2D image recording the 3D coordinates of a complete facial point cloud, while at the same time keeps the semantic meaning at each UV place. PRNet uses an encoder-decoder network to regress the UV position map from a single 2D facial image. Although these methods have broken through the limitations of 3DMM, their training sets are still restricted by 3DMM and the reconstruction results are still model-like. Tran et al. [33] achieve a certain breakthrough by utilizing two CNN decoders, instead of two PCA spaces, to learn a nonlinear model from unlabelled images in an weakly-supervised manner. However, the model still needs to be pre-trained on 3DMM data and the learned bilinear model does not go far beyond 3DMM space, making the results still lack fine-grained geometry information. Different from the above methods, our solution can directly obtain fine-grained 3D faces and keep the semantics of vertices.

2.1 3D Morphable Model

The seminal work of Blanz et al. [5] proposes the 3D Morphable Model (3DMM) to describe the 3D face space with PCA:

$$\mathbf{S} = \overline{\mathbf{S}} + \mathbf{A}_{id}\boldsymbol{\alpha}_{id} + \mathbf{A}_{exp}\boldsymbol{\alpha}_{exp}, \tag{1}$$

where $\overline{\mathbf{S}}$ is the mean shape, $\mathbf{A}_{id}$ is the principle axes trained on the 3D face scans with neutral expression and $\boldsymbol{\alpha}_{id}$ is the shape parameter, $\mathbf{A}_{exp}$ is the principle axes trained on the offsets between expression scans and neutral scans and $\boldsymbol{\alpha}_{exp}$ is the expression parameter. The 3D face can be rigidly transformed by:

$$V(\mathbf{p}_{3d}) = f * \mathbf{R} * (\overline{\mathbf{S}} + \mathbf{A}_{id}\boldsymbol{\alpha}_{id} + \mathbf{A}_{exp}\boldsymbol{\alpha}_{exp}) + \mathbf{t}_{3d}, \tag{2}$$

where $V(\mathbf{p}_{3d})$ is the model construction and rigid transformation function, f is the scale factor, $\mathbf{R}$ is the rotation matrix constructed from Euler angles *pitch*, *yaw*, *roll* and $\mathbf{t}_{3d}$ is the translation vector. The collection of 3D geometry parameters is $\mathbf{p}_{3d} = [f, \mathbf{R}, \mathbf{t}_{3d}, \boldsymbol{\alpha}_{id}, \boldsymbol{\alpha}_{exp}]$.

3 Fine-Grained 3D Data Construction

One of the main challenges of fine-grained 3D face reconstruction is the scarcity of training data. However, it is very tedious to acquire complete and high-precision 3D faces. The raw scans must be captured in well controlled conditions and registered to a face template through laborious hand labeling. Differently, RGB-D images are much easier to capture and also contain rich 3D information. In this work, we explore to construct a large 3D face dataset from public RGB-D images.

3.1 Texture Constrained Non-rigid ICP

The first task is registering all the depth images to a template face to get the topology-uniformed shape. Previous methods adopt the Iterative Closest Point (ICP) method [1] for registration. However, most of depth images are collected in semi-controlled environment, suffering from holes, spikes, occlusions and large missing regions due to self-occlusion. Hand labelling such as dense 3D landmarks is needed for robust registration. To improve the robustness of ICP on human faces, we propose to utilize the face texture, from both the RGB-D image and the face model, as a strong constraint in closest point matching. Figure 2 shows the overview of our method.

Firstly, we fit a 3DMM with the detected 240 landmarks [40] to get the initial 3D face $\mathbf{V} = \{\mathbf{v}_i | i = 1, 2, \cdots, N\}$ (Fig. 2(c)), which is defined in Eq. 2. Secondly, we construct the face texture as a template during ICP registration. Specifically, we utilize the PCA raw texture model from BFM [21]:

$$\mathbf{T} = \overline{\mathbf{T}} + \mathbf{B}\boldsymbol{\beta}, \tag{3}$$

where $\overline{\mathbf{T}}$ is the mean texture, $\mathbf{B}$ is the principle axes of the raw texture and $\boldsymbol{\beta}$ is the raw texture parameter. Given 3D vertices $\mathbf{V}$ and its raw texture $\mathbf{T}$, the Phong illumination model is used to produce the final face texture [5]:

$$C_i(\mathbf{p}_{tex}) = \mathbf{Amb} * \mathbf{T}_i + \mathbf{Dir} * \mathbf{T}_i * \langle \mathbf{n}_i, \mathbf{l} \rangle + k_s \cdot \mathbf{Dir} \langle \mathbf{r}_i, \mathbf{ve} \rangle^{\nu}, \tag{4}$$

where C_i is the RGB color of the ith vertex, the diagonal matrix $\mathbf{Amb}$ is the ambient light, the diagonal matrix $\mathbf{Dir}$ is the parallel light from direction $\mathbf{l}$, $\mathbf{n}_i$ is the normal direction of the ith vertex, k_s is the specular reflectance, $\mathbf{ve}$ is the viewing direction, ν controls the angular distribution of the specular reflection and $\mathbf{r}_i = 2 \cdot \langle \mathbf{n}_i, \mathbf{l} \rangle \mathbf{n}_i - \mathbf{l}$ is the direction of maximum specular reflection. The collection of texture parameters is $\mathbf{p}_{tex} = [\boldsymbol{\beta}, \mathbf{Amb}, \mathbf{Dir}, \mathbf{l}, k_s, \nu]$. We fit the illumination model by optimizing Eq. 5 through the Levernberg-Marquardt method:

$$\arg \min_{\mathbf{p}_{tex}} \|\mathbf{Img}(\mathbf{V}) - C(\mathbf{p}_{tex})\|, \tag{5}$$

where $\mathbf{Img}(\mathbf{V})$ is the image pixels at vertex positions. The optimized result $\mathbf{C} = \{\mathbf{c}_i | i = 1, 2, \cdots, N\} = C(\mathbf{p}_{tex})$ is the face texture, shown in Fig. 2(d).

Thirdly, to register the initial shape to the depth image, we propose a **T**exture constrained Nonrigid-**ICP** (**T-ICP**) method to find the vertex correspondence based on both geometry and texture. Suppose the target RGB-D image has the vertices $\mathbf{V}^* = \{\mathbf{v}^*_k | k = 1, 2, \cdots, K\}$ and their corresponding pixels $\mathbf{C}^* = \{\mathbf{c}^*_k | k = 1, 2, \cdots, K\}$. For each vertex $\mathbf{v}_i$ on the initial shape, its closest point $\mathbf{v}^*_{k_{corr}}$ is searched by:

$$\begin{aligned} k_{corr} = \arg\min_k (\|\mathbf{v}_i - \mathbf{v}^*_k\| + \lambda_{tex} \|\mathbf{c}_i - \mathbf{c}^*_k\|) \\ \text{if} \quad \|\mathbf{v}_i - \mathbf{v}^*_k\| < \tau_v \quad \text{and} \quad \|\mathbf{c}_i - \mathbf{c}^*_k\| < \tau_c \end{aligned} \tag{6}$$

where τ_v and τ_c are the distance thresholds in 3D space and color space, respectively. As shown in Fig. 2(e), by incorporating the texture constraint, we improve the robustness not only on the geometry-smooth but texture-rich surfaces like eye-brows, but also on the occluded regions where the matching is filtered out by τ_c due to large texture error. With the correspondence $(\mathbf{v}_i, \mathbf{v}^*_{k_{corr}})$, we perform Optimal Non-rigid ICP [1] to finish the registration, shown in Fig. 2(f). Different from the texture constraint used in scan-to-scan registration [27] where both scans have the texture of the same object, our task is a more challenging model-to-scan registration, where the facial template only has a texture model rather than the real texture. During registration, we must iteratively update the texture parameters and get a more reliable texture constraint.

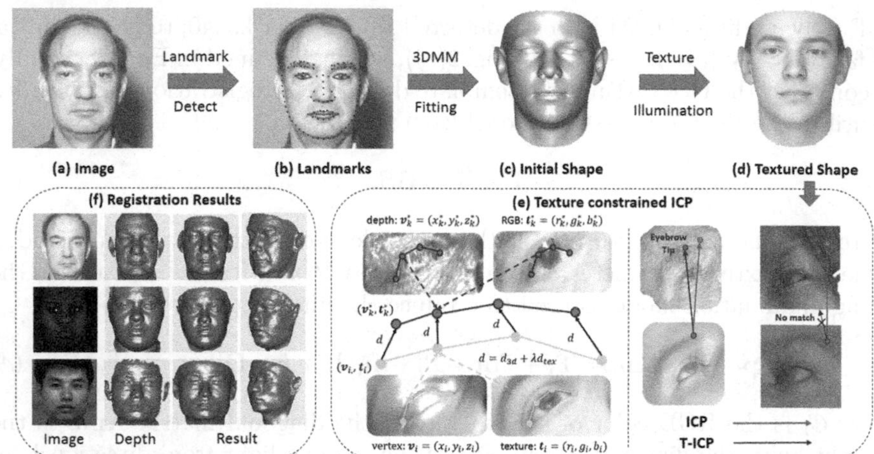

Fig. 2. The overview of 3D face registration. (a) The input image. (b) The detected 240 landmarks. (c) The fitted 3D shape by landmarks. (d) The reconstructed face texture by the texture and illumination model. (e) Left: the T-ICP searches the closest point in both 3D space (x, y, z) and color space (r, g, b). Right: the incorporation of texture improves the robustness on eye-brows and occluded regions. (f) The final registration results.

Finally, we disentangle rigid and non-rigid transformations, getting the ground-truth shape by optimizing the following equation:

$$\mathbf{S}^*_{morph} = \arg \min_{\mathbf{S}_{morph}, \mathbf{R}, f, \mathbf{t3d}} \|\mathbf{V}_{regist} - f * \mathbf{R} * (\overline{\mathbf{S}} + \mathbf{S}_{morph}) + \mathbf{t}_{3d}\| \tag{7}$$

where $\mathbf{V}_{regist}$ is the registered 3D face, $(f, \mathbf{R}, \mathbf{t3d})$ are the rigid transformation parameters, $\overline{\mathbf{S}}$ is the mean shape and $\mathbf{S}^*_{morph}$ is the difference between the target shape and the mean shape, which will be the target of the neural network learning.

3.2 Out-of-Plane Pose Augmentation

Large scale data is crucial for training neural networks. However, there are less than ten thousand public RGB-D samples [16,22,37] and most of them are frontal faces, leading to poor generalization across poses. To address this challenge, we improve the face profiling method [41] for RGB-D data and synthesize hundreds of thousands high-fidelity 3D data for network training. Firstly, we complete the depth channel for the whole image space, where the depth on the face region directly comes from the registered 3D face and the depth on the background is coarsely estimated by some anchors (x_i, y_i), shown in Fig. 3(b). These anchors are triangulated to a background mesh and their depth values d_i are estimated by depth constraints and smoothness constraints, as in Eq. 8:

$$\sum_i Mask(x_i, y_i)\|d_i - Depth(x_i, y_i)\| + \sum_i \sum_j Connect(i, j)\|d_i - d_j\|, \tag{8}$$

where $Depth(x, y)$ is the depth channel of the RGB-D image, $Mask(x, y)$ indicates whether (x, y) is hollow and $Connect(i, j)$ is whether two anchors are connected by the background mesh. By turning the whole image to a 3D mesh (Fig. 3(c)), we can out-of-rotate it (Fig. 3(d)) and render it (Fig. 3(e)) in any views.

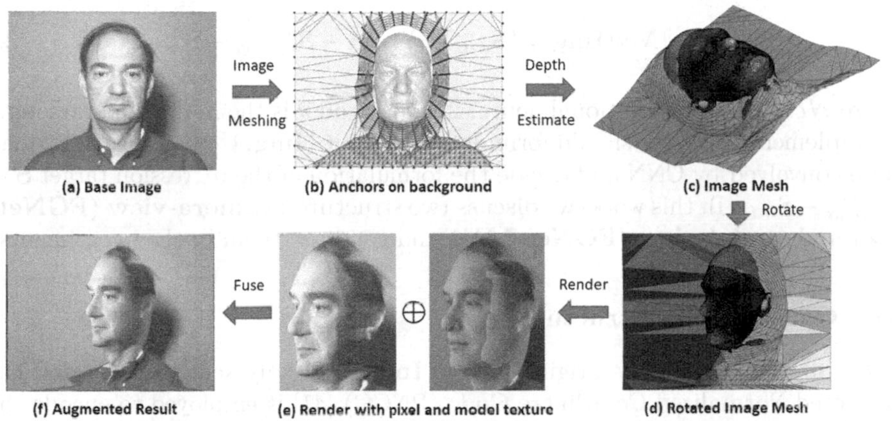

Fig. 3. The overview of out-of-plane pose augmentation. (a) The base image. (b) The original depth image and the anchors on the background. Note that the red anchors locate on the scan and the blue ones locate on the hollow, they have different constraints in Eq. 8. (c) The complete depth of the base image. (d) The rotated 3D mesh of the image. (e) Rendering with the image pixels and the model texture. (f) The augmentation result.

Different from the original face profiling method which aims to generate large poses from medium poses, most of our base images are frontal faces. While a main drawback of face profiling is that when rotating from frontal faces, there are serious artifacts on the side face due to the lack of texture in the original image, shown in Fig. 3(e). In this work, the texture and illumination model is also used as a strong prior to refine the artifacts. With the model texture used in T-ICP (shown in Fig. 2(d)), we render the 3D image mesh with both the image pixels and the model texture, shown in Fig. 3(e). Then we detect the invisible region with the normal directions, and inpaint it with the model texture through Poisson editing, shown in Fig. 3(f). Since the side face is not texture-rich, the model texture is realistic enough to inpaint it and we finally get high-fidelity synthetic samples. In this work, we augment the images by enlarging the *yaw* angle at the step of 15° until 90°, and randomly enlarging the *pitch* angle within ±25°, generating about 200k training samples.

4 Fine-Grained Reconstruction Network

With large scale training data, we are prepared to train a **F**ine-**G**rained reconstruction **Net**work (**FGNet**) to reconstruct the fine-grained geometry. In order to concentrate on face shape modification, we employ a state-of-the-art 3DMM fitting method [12] to get the rigid transformation and an initial 3D shape. Our task can be formulated as follows, given the input image **Img**, the rigid transformation $V(\cdot)$ and the initial shape $\mathbf{S}_{init}$, we aim to estimate the shape update $\Delta\mathbf{S}$ so that the final shape is closer to the ground truth $\overline{\mathbf{S}} + \mathbf{S}_{morph}$ (defined in Eq. 7) after updating:

$$\arg\min_{\theta} \|Net(\mathbf{Img}, V(\mathbf{S}_{init}); \theta) - (\overline{\mathbf{S}} + \mathbf{S}^*_{morph} - \mathbf{S}_{init})\| \tag{9}$$

where $Net(\cdot)$ is a convolutional neural network and θ is the network parameters. To implement Eq. 9, we should formulate the input $(\mathbf{Img}, V(\mathbf{S}_{init}))$ as a 2D map to be convolved by CNN and decide the formulation of the regression target $\overline{\mathbf{S}} + \mathbf{S}^*_{morph} - \mathbf{S}_{init}$. In this work, we discuss two structures: **camera-view (FGNet-CV)** and **model-view (FGNet-MV)** and compare them in the experiments.

4.1 Camera-View Structure

In the first structure, the original image **Img** is directly sent to CNN and the Projected Normalized Coordinate Code (PNCC) [41] is employed to encode the initial fitting result $V(\mathbf{S}_{init})$ as a 2D map for CNN. It is called camera-view since the network observes the face in the same view of the camera. Besides, the shape update $\Delta\mathbf{S} = \overline{\mathbf{S}} + \mathbf{S}^*_{morph} - \mathbf{S}_{init}$ is represented as a UV map [9] and is regressed by a fully convolutional encoder-decoder network. The overview of the structure is in Fig. 4.

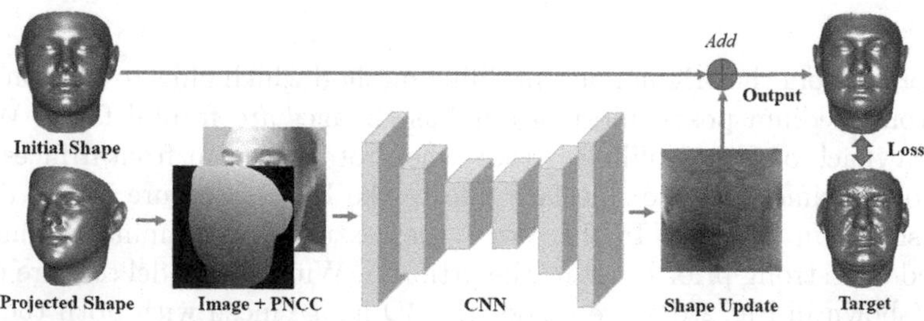

Fig. 4. The Camera-view Fine-grained Reconstruction Network.

The advantage of the camera-view structure is that it does not miss any information provided by the image. However, the structure requires the network to identify small shape variations in any poses, which is hard to learn. Besides, the input and output have different coordinate systems (image space to UV space). For each coordinate on the output, its receptive field may not cover the most related region for reconstruction.

4.2 Model-View Structure

Different from 3DMM fitting where pose estimation is the most important [41], the purpose of fine-grained reconstruction is to modify the shape. To this end, we design a model-view structure to concentrate on shape information. The overview is shown in Fig. 5.

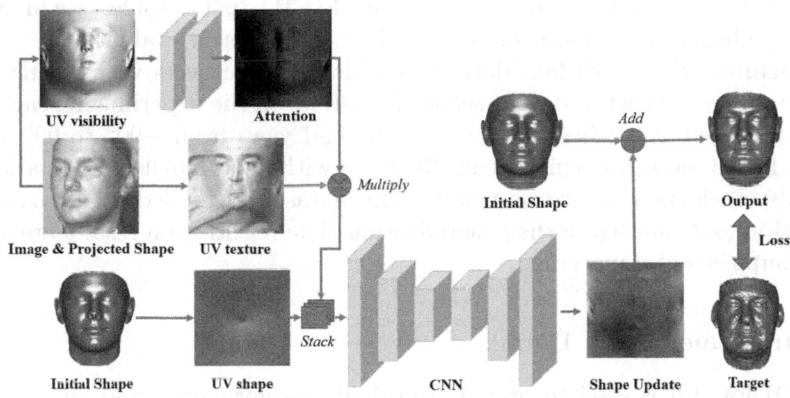

Fig. 5. The Model-view Fine-grained Reconstruction Network.

The input has three parts: we extract the **UV-texture** map according to the projected initial shape $V(\mathbf{S}_{init})$. Given that the UV-texture of a non-frontal face has invalid regions due to self-occlusion, we also construct the **UV-visibility** map which stores the z values of the vertex normals. Besides, a **UV-shape** map which stores the vertex positions of the initial shape $\mathbf{S}_{init}$ is also provided. During inference, the visibility map is first convolved by several layers to an attention map. The UV-texture is then multiplied by the attention, concatenated by the UV-shape, and sent to the backbone to regress the shape update.

The structure is called model-view since the input and the output are both in UV space and the network observes the face through the model vertices. The advantage of the structure is that each 2D position across the network has the same semantic meaning, so that the receptive field of each output coordinate always covers the most related region. Besides, by warping the image pixels to a UV map, the pose variations are implicitly normalized, making the CNN concentrate on shape updating.

5 Experiments

5.1 Datasets

To perform fine-grained reconstruction, multiple datasets listed below are used for training and evaluation in our experiments.

FG3D are constructed from three datasets. The FRGC [22] includes 4,950 samples and each sample has a face image and a 3D scan with pixels in full correspondence. BP4D [37] contains 328 2D+3D videos from 41 subjects, where 3,376 frames are randomly selected. CASIA-3D [16] consists of 4,624 scans of 123 persons and we filter out the non-frontal faces. We register and out-of-plane augment the three datasets, generating a large 3D dataset **FG3D** with 212,579 samples. Among FG3D, 90% subjects are used as the training set **FG3D-train** and the rest 10% subjects are the testing set **FG3D-test**. Besides, we manually delete the bad registration results in FG3D-test for better evaluation.

Florence [2] is a 3D face dataset containing 53 subjects with its 3D mesh acquired from a structured-light scanning system. In the experiments, each subject is rendered at pitches of $-20°$, $0°$, $20°$ and yaws from $-90°$ to $90°$ at the step of $15°$. Besides, we register each 3D mesh with hand-labelled landmarks and carefully check the registration results. This dataset set is used for cross dataset evaluation to demonstrate the generalization. The registration results are shown in the supplemental materials.

5.2 Implementation Details

The 3DDFA [12] is used to provide the rigid transformation and an initial 3D shape for FGNet. The architecture of FGNet is a fully convolutional encoder-decoder network the same as [9]. The models are trained by the SGD optimizer and L1-Loss with a start learning rate of 0.1, which is decayed by 0.1 at epoch 20, 30 and 40, and the model is trained for 50 epochs. The training images are cropped by the bounding boxes of the initial 3DMM fitting results [12] and resized to 256×256 without any perturbation. All of the UV maps in the network are also 256×256. As for the T-ICP, the λ_{tex}, τ_c and τ_t in Eq. 6 are set to 0.013, 0.17 and $2 * 10^{-4}$ times of the interocular distance of the 3D face, respectively.

To evaluate the reconstruction accuracy, we **rigid-align** the result to the ground truth and employ the Normalized Mean Error (NME):

$$NME = \frac{1}{N} \sum_{k=1}^{N} \frac{||\mathbf{v}_k - \mathbf{v}_k^*||}{d}, \tag{10}$$

where $k = 1, 2, \cdots, N$ are the vertices on the face region without neck and ears (the face region is given by [9]), $\mathbf{v}_k$ and $\mathbf{v}_k^*$ are vertices of the reconstructed face and the ground truth and d is the outer interocular distance of 3D coordinates. Note that this NME mainly evaluates the shape error since the pose error is normalized by the rigid alignment.

5.3 Ablation Studies

Camera View vs. Model View. We first discuss the proposed two structures for fine-grained reconstruction: the Camera View (FGNet-CV) and the Model View (FGNet-MV). The FGNet-CV sends the original image to the CNN without losing any image information. While the FGNet-MV warps the image to the UV space and normalizes the pose variations implicitly. Their performance is compared in Table 1.

Table 1. The NME(%) results on FG3D-test with different structures, evaluated by different yaw ranges. The "Initial" indicates the initial 3DMM fitting result by [12].

Method	[30, 60]	[30, 60]	[60, 90]	All
Initial	5.05	5.03	5.14	5.07
FGNet-CV	3.44	3.11	3.12	3.23
FGNet-MV	3.30	3.06	2.95	3.10

Compared with the initial fitting results, both FG3D-CV and FG3D-MV greatly improve the reconstruction accuracy. Among them, FG3D-MV achieves better results by concentrating on shape modification.

Ablation Studies on Input. To perform fine-grained reconstruction, we formulate several inputs to provide the face appearance and the initial fitting result for CNN. FG3D-CV has the original image and the PNCC. FG3D-MV has the UV-texture, UV-visibility and UV-shape. In this part, we analyze the effectiveness of each input, shown in Table 2.

Table 2. The NME(%) results on FG3D-test with different inputs, evaluated by different yaw ranges. UV-tex and UV-vis are short for UV-texture and UV-visibility, respectively.

Input	[30, 60]	[30, 60]	[60, 90]	All
FGNet-CV(img)	3.54	3.30	3.26	3.37
FGNet-CV(img + PNCC)	3.44	3.11	3.12	3.23
FGNet-MV(UV-tex)	3.47	3.18	3.14	3.27
FGNet-MV(UV-tex + UV-vis)	3.39	3.14	3.04	3.19
FGNet-MV(UV-tex + UV-vis + UV-shape)	3.30	3.06	2.95	3.10

In FG3D-CV, PNCC effectively improves the performance by providing the initial fitting result for the CNN and simplifying the reconstruction task. In FG3D-MV, the attention map from the UV-visibility shrinks the self-occluded region of the UV-texture and reduces the shape error. The incorporation of UV-shape further provides the initial shape and gets better results. The combination of UV-texture, UV-visibility and UV-shape achieves the best result, which is used to represent FGNet in comparison experiments. Besides, we provide more visualizations about the attention map from the UV-visibility in the supplemental materials, illustrating the learned knowledge from this map.

Error Reduction Parallel and Orthogonal to Viewing Direction. Intuitively, the shape information orthogonal to the viewing direction is easy to observe, but it is not the case for the shape parallel to the viewing direction. For example, given a frontal face, we can easily know its width and height but have to guess its thickness (such as the height of the nose bridge). We are interested in that the error in which direction is reduced by our method. Given the estimated face rotation matrix $\mathbf{R}$, the viewing direction can be set as $\mathbf{ve} = \mathbf{R}*[0,0,1]^T$, then the errors parallel and orthogonal to the viewing direction are:

$$E_{3d} = \| \mathbf{v} - \mathbf{v}^* \|, \quad E_{pal} = \| (\mathbf{v} - \mathbf{v}^*) \cdot \mathbf{ve} \|, \quad E_{orh} = \sqrt{E_{3d}^2 - E_{pal}^2}, \tag{11}$$

where $\mathbf{v}$ and $\mathbf{v}^*$ are the vertices of the predicted and the ground-truth shapes (shape is always a frontal face in a normalized space), respectively, E_{3d} is the original error measured by Euclidean distance, E_{pal} and E_{orh} are the errors parallel and orthogonal to the viewing direction, respectively. Based on the three types of error, we evaluate the NMEs in Table 3 and find that the error is mainly reduced parallel to the viewing direction, demonstrating that the depth is better recovered. The reason may be that the training set FG3D is constructed from RGB-D images and provides more accurate depth information than the landmark based datasets like 300W-LP [39].

Table 3. The NME(%) parallel and orthogonal to the viewing direction, evaluated on all the samples of FG3D-test.

Input	Orthogonal	Parallel	Euclidean
Initial	3.44	3.11	5.07
FGNet-CV	2.47(28.20% ↓)	1.67(46.30% ↓)	3.23(36.29% ↓)
FGNet-MV	2.38(30.81% ↓)	1.59(48.87% ↓)	3.10(38.86% ↓)

5.4 Comparison Experiments

Qualitative Comparison. We present some visual comparisons to illustrate the identifiability of the reconstructed shapes. Baseline methods include the common used 3DDFA [41] and PRNet [9] which are trained on the landmark based datasets 300W-LP [39], Extreme3D [32] which reconstructs facial details by shape-from-shading, and the released Deng's method [7] as a typical weakly-supervised method [7,33,34] which adaptively learns a nonlinear 3DMM and its fitting strategy from unlabelled images. As shown in Fig. 6, compared with other baselines, our results look more like real scans than blend models due to its better reconstructed shapes. In Extreme3D, even though plausible details are added by shape-from-shading, the face shapes are not modified. Deng's method [7] accurately reconstructs the facial features, but the cheek geometry is not well captured such as the cheekbone and the face silhouette.

Quantitative Comparison. In this part, we firstly compare our method with the state-of-the-art methods including 3DDFA [41], PRNet [9], Extreme3D [32] and Deng's method [7] quantitatively on the FG3D-test. All their inputs are cropped by the ground-truth bounding boxes and only the face region is used for calculating NME. Since these methods share the topology of BFM [21], their results are comparable. As shown in Table 4, our method achieves the best result and outperforms the best of the state-of-the-art methods by 38.86%.

Table 4. The NME(%) on FG3D-test, evaluated by different yaw ranges. The FGNet employs the FGNet-MV structure.

Input	[30, 60]	[30, 60]	[60, 90]	All
3DDFA [41]	5.05	5.03	5.14	5.07
PRNet [9]	5.49	5.89	5.70	5.68
Extreme3D [32]	7.07	7.42	8.03	7.52
Deng et al. [7]	5.26	5.16	5.30	5.24
FGNet	3.30	3.06	2.95	3.10

Considering that FG3D-test shares the same environment with FG3D-train, we also perform cross dataset evaluation on the Florence dataset for fair comparison, shown in Table 5. First, Deng's method [7] performs better than the 300W-LP trained 3DDFA and PRNet, demonstrating that the weakly learned non-linear 3DMM [7,33,34] covers more shape variations than BFM. Second, our method achieves the best result, validating the feasibility of reconstructing fine-grained geometry in a supervised manner.

6 Conclusion

This paper proposes an solution to reconstruct 3D fine-grained face shape, from data construction to neural network training. Firstly, to prepare sufficient training data, we propose a texture constrained non-rigid ICP method to register RGB-D images robustly. Besides, an out-of-plane pose augmentation method

Table 5. The NME(%) on Florence, evaluated by different yaw ranges. The FGNet employs the FGNet-MV structure.

Input	[30, 60]	[30, 60]	[60, 90]	All
3DDFA [41]	6.92	6.89	6.82	6.87
PRNet [9]	6.71	6.98	8.04	7.41
Extreme3D [32]	8.03	8.38	8.53	8.37
Deng et al. [7]	6.05	6.31	6.02	6.12
FGNet	5.62	5.52	5.56	5.56

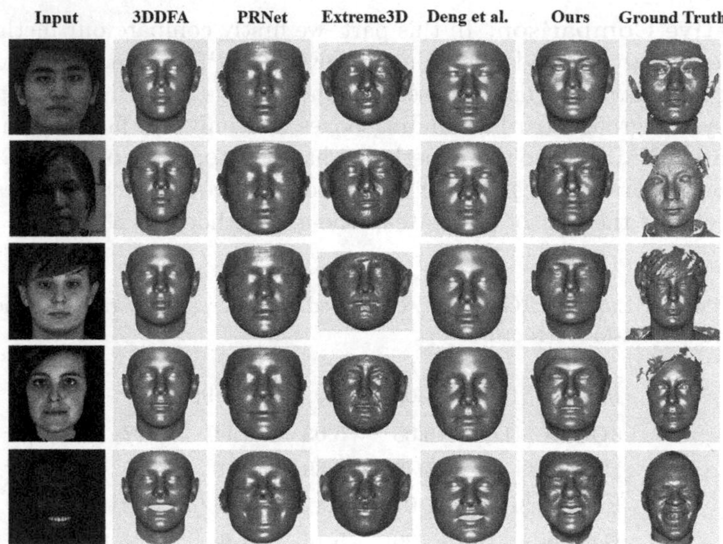

Fig. 6. Qualitative comparison. Baseline methods from left to right: 3DDFA [41] (used as our initial shape), PRNet [9], Extreme3D [32], Deng et al. [7], our FGNet and the ground-truth shape.

specifically designed for RGB-D data is proposed to enrich pose variations and enlarge the scale of data. Secondly we propose a novel network structure FGNet that can concentrate on shape modification by learning the image to shape mapping in UV space. Finally, our method successfully reconstructs fine-grained shape geometry and outperforms other state-of-the-art methods.

Acknowledgment. This work was supported in part by the National Key Research & Development Program (No. 2020YFC2003901), Chinese National Natural Science Foundation Projects #61806196, #61876178, #61872367, #61976229, #61673033.

References

1. Amberg, B., Romdhani, S., Vetter, T.: Optimal step nonrigid ICP algorithms for surface registration. In: 2007 IEEE Conference on Computer Vision and Pattern Recognition, pp. 1–8. IEEE (2007)
2. Bagdanov, A.D., Del Bimbo, A., Masi, I.: The florence 2D/3D hybrid face dataset. In: Proceedings of the 2011 Joint ACM Workshop on Human Gesture and Behavior Understanding, pp. 79–80. ACM (2011)
3. Bas, A., Huber, P., Smith, W.A., Awais, M., Kittler, J.: 3D morphable models as spatial transformer networks. In: Proceedings of the IEEE International Conference on Computer Vision, pp. 904–912 (2017)
4. Bhagavatula, C., Zhu, C., Luu, K., Savvides, M.: Faster than real-time facial alignment: a 3D spatial transformer network approach in unconstrained poses (2017)
5. Blanz, V., Vetter, T.: Face recognition based on fitting a 3D morphable model. IEEE Trans. Pattern Anal. Mach. Intell. **25**(9), 1063–1074 (2003)

6. Cai, Y., Lei, Y., Yang, M., You, Z., Shan, S.: A fast and robust 3D face recognition approach based on deeply learned face representation. Neurocomputing **363**, 375–397 (2019)
7. Deng, Y., Yang, J., Xu, S., Chen, D., Jia, Y., Tong, X.: Accurate 3D face reconstruction with weakly-supervised learning: from single image to image set. In: IEEE Computer Vision and Pattern Recognition Workshops (2019)
8. Dou, P., Shah, S.K., Kakadiaris, I.A.: End-to-end 3D face reconstruction with deep neural networks. In: Proceedings of the IEEE Conference on Computer Vision and Pattern Recognition, pp. 5908–5917 (2017)
9. Feng, Y., Wu, F., Shao, X., Wang, Y., Zhou, X.: Joint 3D face reconstruction and dense alignment with position map regression network. In: Proceedings of the European Conference on Computer Vision (ECCV), pp. 534–551 (2018)
10. Hassner, T.: Viewing real-world faces in 3D. In: Proceedings of the IEEE International Conference on Computer Vision, pp. 3607–3614 (2013)
11. Jackson, A.S., et al.: Large pose 3D face reconstruction from a single image via direct volumetric CNN regression (2017)
12. Guo, J., Zhu, X., Lei, Z.: 3DDFA (2018). https://github.com/cleardusk/3DDFA
13. Jourabloo, A., Liu, X.: Pose-invariant 3d face alignment. In: Proceedings of the IEEE International Conference on Computer Vision, pp. 3694–3702 (2015)
14. Jourabloo, A., Liu, X.: Large-pose face alignment via CNN-based dense 3D model fitting. In: Proceedings of the IEEE Conference on Computer Vision and Pattern Recognition, pp. 4188–4196 (2016)
15. Kemelmacher-Shlizerman, I., Basri, R.: 3D face reconstruction from a single image using a single reference face shape. IEEE Trans. Pattern Anal. Mach. Intell. **33**(2), 394–405 (2011)
16. Li, S.: Casia 3D face database - center for biometrics and security research (2004)
17. Liu, F., Tran, L., Liu, X.: 3D face modeling from diverse raw scan data. In: Proceedings of the IEEE International Conference on Computer Vision, pp. 9408–9418 (2019)
18. Liu, F., Zeng, D., Zhao, Q., Liu, X.: Joint face alignment and 3D face reconstruction. In: Leibe, B., Matas, J., Sebe, N., Welling, M. (eds.) ECCV 2016. LNCS, vol. 9909, pp. 545–560. Springer, Cham (2016). https://doi.org/10.1007/978-3-319-46454-1_33
19. Liu, Y., Jourabloo, A., Ren, W., Liu, X.: Dense face alignment. In: Proceedings of the IEEE International Conference on Computer Vision, pp. 1619–1628 (2017)
20. Mu, G., Huang, D., Hu, G., Sun, J., Wang, Y.: Led3D: a lightweight and efficient deep approach to recognizing low-quality 3D faces. In: Proceedings of the IEEE Conference on Computer Vision and Pattern Recognition, pp. 5773–5782 (2019)
21. Paysan, P., Knothe, R., Amberg, B., Romdhani, S., Vetter, T.: A 3D face model for pose and illumination invariant face recognition. In: Sixth IEEE International Conference on Advanced Video and Signal Based Surveillance, AVSS 2009, pp. 296–301. IEEE (2009)
22. Phillips, P.J., et al.: Overview of the face recognition grand challenge. In: 2005 IEEE Computer Society Conference on Computer Vision and Pattern Recognition (CVPR 2005), vol. 1, pp. 947–954. IEEE (2005)
23. Richardson, E., Sela, M., Kimmel, R.: 3D face reconstruction by learning from synthetic data. In: 2016 Fourth International Conference on 3D Vision (3DV), pp. 460–469. IEEE (2016)
24. Richardson, E., Sela, M., Or-El, R., Kimmel, R.: Learning detailed face reconstruction from a single image. In: Proceedings of the IEEE Conference on Computer Vision and Pattern Recognition, pp. 1259–1268 (2017)

25. Romdhani, S., Vetter, T.: Efficient, robust and accurate fitting of a 3D morphable model. In: Proceedings of the Ninth IEEE International Conference on Computer Vision, pp. 59–66. IEEE (2003)
26. Romdhani, S., Vetter, T.: Estimating 3D shape and texture using pixel intensity, edges, specular highlights, texture constraints and a prior. In: 2005 IEEE Conference on Computer Vision and Pattern Recognition (CVPR), vol. 2, pp. 986–993. IEEE (2005)
27. Saval-Calvo, M., Azorin-Lopez, J., Fuster-Guillo, A., Villena-Martinez, V., Fisher, R.B.: 3D non-rigid registration using color: color coherent point drift. Comput. Vis. Image Underst. **169**, 119–135 (2018)
28. Sela, M., Richardson, E., Kimmel, R.: Unrestricted facial geometry reconstruction using image-to-image translation. In: Proceedings of the IEEE International Conference on Computer Vision, pp. 1576–1585 (2017)
29. Shen, T., Huang, Y., Tong, Z.: FaceBagNet: bag-of-local-features model for multi-modal face anti-spoofing. In: Proceedings of the IEEE Conference on Computer Vision and Pattern Recognition Workshops (2019)
30. Snta, Z., Kato, Z.: 3D Face Alignment Without Correspondences (2016)
31. Tewari, A., et al.: MoFA: model-based deep convolutional face autoencoder for unsupervised monocular reconstruction. In: Proceedings of the IEEE International Conference on Computer Vision, pp. 1274–1283 (2017)
32. Tran, A.T., Hassner, T., Masi, I., Paz, E., Nirkin, Y., Medioni, G.G.: Extreme 3D face reconstruction: seeing through occlusions. In: CVPR, pp. 3935–3944 (2018)
33. Tran, L., Liu, X.: Nonlinear 3D face morphable model. In: The IEEE Conference on Computer Vision and Pattern Recognition (CVPR), June 2018
34. Tran, L., Liu, X.: On learning 3D face morphable model from in-the-wild images. IEEE Trans. Pattern Anal. Mach. Intell. (2019)
35. Tuan Tran, A., Hassner, T., Masi, I., Medioni, G.: Regressing robust and discriminative 3D morphable models with a very deep neural network. In: Proceedings of the IEEE Conference on Computer Vision and Pattern Recognition, pp. 5163–5172 (2017)
36. Zhang, S., et al.: A dataset and benchmark for large-scale multi-modal face anti-spoofing. In: Proceedings of the IEEE Conference on Computer Vision and Pattern Recognition, pp. 919–928 (2019)
37. Zhang, X., et al.: Bp4d-spontaneous: a high-resolution spontaneous 3d dynamic facial expression database. Image Vis. Comput. **32**(10), 692–706 (2014)
38. Zhu, K., Du, Z., Li, W., Huang, D., Wang, Y., Chen, L.: Discriminative attention-based convolutional neural network for 3d facial expression recognition. In: 2019 14th IEEE International Conference on Automatic Face & Gesture Recognition (FG 2019), pp. 1–8. IEEE (2019)
39. Zhu, X., Lei, Z., Liu, X., Shi, H., Li, S.Z.: Face alignment across large poses: a 3D solution. In: Proceedings of the IEEE Conference on Computer Vision and Pattern Recognition, pp. 146–155 (2016)
40. Zhu, X., Lei, Z., Yan, J., Yi, D., Li, S.Z.: High-fidelity pose and expression normalization for face recognition in the wild. In: Proceedings of the IEEE Conference on Computer Vision and Pattern Recognition, pp. 787–796 (2015)
41. Zhu, X., Liu, X., Lei, Z., Li, S.Z.: Face alignment in full pose range: a 3D total solution. IEEE Trans. Pattern Anal. Mach. Intell. **41**(1), 78–92 (2019)
42. Zhu, X., Yi, D., Lei, Z., Li, S.Z.: Robust 3D morphable model fitting by sparse sift flow. In: 2014 22nd International Conference on Pattern Recognition, pp. 4044–4049. IEEE (2014)

World-Consistent Video-to-Video Synthesis

Arun Mallya(✉), Ting-Chun Wang, Karan Sapra, and Ming-Yu Liu

NVIDIA, Santa Clara, USA
{amallya,tingchunw,ksapra,mingyul}@nvidia.com
https://nvlabs.github.io/wc-vid2vid

Abstract. Video-to-video synthesis (vid2vid) aims for converting high-level semantic inputs to photorealistic videos. While existing vid2vid methods can achieve short-term temporal consistency, they fail to ensure the long-term one. This is because they lack knowledge of the 3D world being rendered and generate each frame only based on the past few frames. To address the limitation, we introduce a novel vid2vid framework that efficiently and effectively utilizes all past generated frames during rendering. This is achieved by *condensing* the 3D world rendered so far into a physically-grounded estimate of the current frame, which we call the *guidance image*. We further propose a novel neural network architecture to take advantage of the information stored in the guidance images. Extensive experimental results on several challenging datasets verify the effectiveness of our approach in achieving *world consistency*—the output video is consistent within the entire rendered 3D world.

Keywords: Neural rendering · Video synthesis · GAN

1 Introduction

Video-to-video synthesis [77] concerns generating a sequence of photorealistic images given a sequence of semantic representations extracted from a source 3D world. For example, the representations can be the semantic segmentation masks rendered by a graphics engine while driving a car in a virtual city [77]. The representations can also be the pose maps extracted from a source video of a person dancing, and the application is to create a video of a different person performing the same dance [8]. From the creation of a new class of digital artworks to applications in computer graphics, the video-to-video synthesis task has many exciting practical use-cases. A key requirement of any such video-to-video synthesis model is the ability to generate images that are not only individually

A. Mallya and T.-C. Wang—Equal contribution.

Electronic supplementary material The online version of this chapter (https://doi.org/10.1007/978-3-030-58598-3_22) contains supplementary material, which is available to authorized users.

A. Vedaldi et al. (Eds.): ECCV 2020, LNCS 12353, pp. 359–378, 2020.
https://doi.org/10.1007/978-3-030-58598-3_22

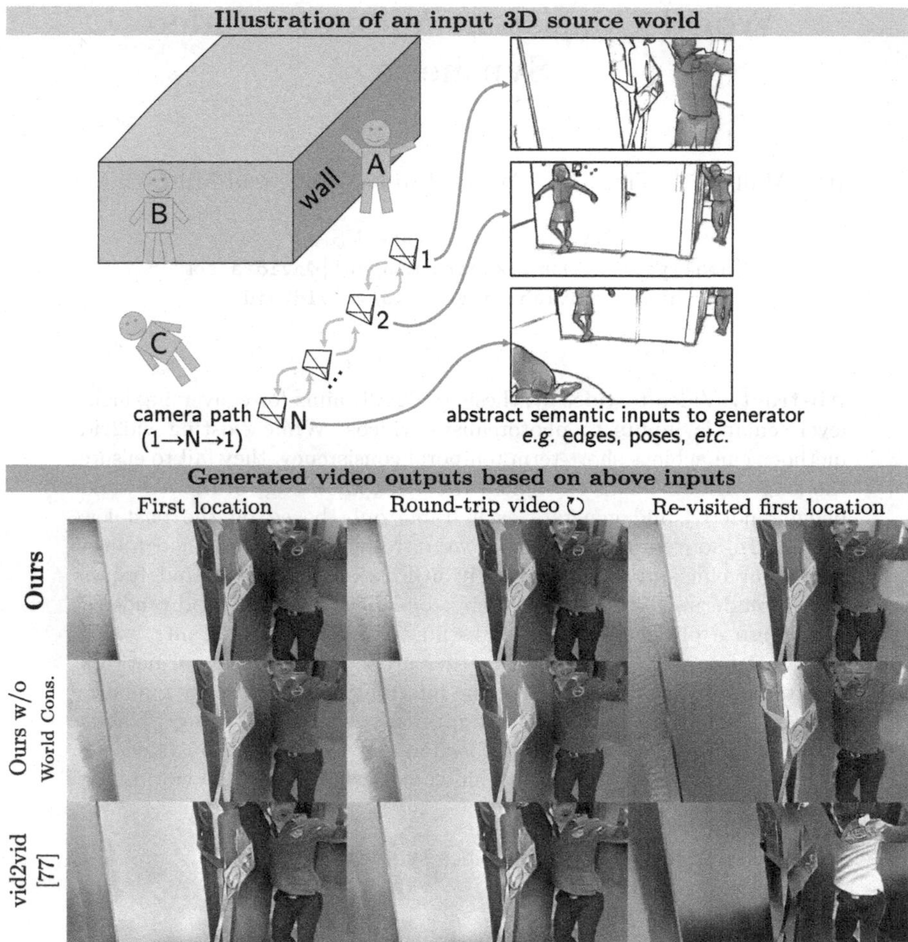

Fig. 1. Imagine moving around in a world such as the one abstracted at the top. As you move from locations 1→N→1, you would expect the appearance of previously seen walls and people to remain unchanged. However, current video-to-video synthesis methods such as vid2vid [77] or our improved architecture combining vid2vid with SPADE [59] cannot produce such world-consistent videos (third and second rows). Only our method is able to produce videos consistent over viewpoints by adding a mechanism for world consistency (first row) (see Supplementary material). *Please view with Acrobat Reader. Click any middle column image to play video.*

photorealistic, but also temporally smooth. Moreover, the generated images have to follow the geometric and semantic structure of the source 3D world.

While we have observed steady improvement in photorealism and short-term temporal stability in the generation results, we argue that one crucial aspect of the problem has been largely overlooked, which is the *long-term temporal consistency* problem. As a specific example, when visiting the same location in the virtual city, an existing vid2vid method [76,77] could generate an image that is very different from the one it generated when the car first visited the location, despite using the same semantic inputs. Existing vid2vid methods rely on optical flow warping and generate an image conditioned on the past few generated images. While such operations can ensure short-term temporal stability, they cannot guarantee long-term temporal consistency. Existing vid2vid models have no knowledge of what they have rendered in the past. Even for a short round-trip in a virtual room, these methods fail to preserve the appearances of the wall and the person in the generated video, as illustrated in Fig. 1.

In this paper, we attempt to address the long-term temporal consistency problem, by bolstering vid2vid models with memories of the past frames. By combining ideas from scene flow [71] and conditional image synthesis models [59], we propose a novel architecture that explicitly enforces consistency in the entire generated sequence. We perform extensive experiments on several benchmark datasets, with comparisons to the state-of-the-art methods. Both quantitative and visual results verify that our approach achieves significantly better image quality and long-term temporal stability. On the application side, we also show that our approach can be used to generate videos consistent across multiple viewpoints, enabling simultaneous multi-agent world creation and exploration.

2 Related Work

Semantic Image Synthesis [11,49,59,60,78] refers to the problem of converting a single input semantic representation to an output photorealistic image. Built on top of the generative adversarial networks (GAN) [24] framework, existing methods [49,59,78] propose various novel network architectures to advance state-of-the-art. Our work is built on the SPADE architecture proposed by Park *et al.* [59] but focuses on the temporal stability issue in video synthesis.

Conditional GANs synthesize data conditioned on user input. This stands in contrast to unconditional GANs that synthesize data solely based on random variable inputs [24,26,37,38]. Based on the input type, there exist label-conditional GANs [6,55,57,87], text-conditional GANs [61,84,88], image-conditional GANs [3,5,14,31,32,41,47,48,63,67,93], scene-graph conditional GANs [33], and layout-conditional GANs [89]. Our method is a video-conditional GAN, where we generate a video conditioned on an input video. We address the long-term temporal stability issue that the state-of-the-art overlooks [8,76,77].

Video synthesis exists in many forms, including 1) unconditional video synthesis [62,70,73], which converts random variable inputs to video clips, 2) future video prediction [17,19,27,30,35,40,42,45,51,52,58,66,72,74,75,85], which generates future video frames based on the observed ones, and 3) video-to-video synthesis [8,12,22,76,77,92], which converts an input semantic video to a real video. Our work belongs to the last category. Our method treats the input video as one from a self-consistent world so that when the agent returns to a spot that it has previously visited, the newly generated frames should be consistent with the past generated frames. While a few works have focused on improving the temporal consistency of an input video [4,39,86], our method does not treat consistency as a post-processing step, but rather as a core part of the video generation process.

Novel-view synthesis aims to synthesize images at unseen viewpoints given some viewpoints of the scene. Most of the existing works require images at multiple reference viewpoints as input [13,20,21,28,34,54,91]. While some works can synthesize novel views based on a single image [65,79,82], the synthesized views are usually close to the reference views. Our work differs from these works in the sense that our input is different – instead of using a set of RGB images, our network takes in a sequence of semantic maps. If we directly treat all past synthesized frames as reference views, it makes the memory requirement grow linearly with respect to the video length. If we only use the latest frames, the system cannot handle long-term consistency as shown in Fig. 1. Instead, we propose a novel framework to keep track of the synthesis history in this work.

The closest related works are those on neural rendering [2,53,64,68], which can re-render a scene from arbitrary viewpoints after training on a set of given viewpoints. However, note that these methods still require RGB images from different viewpoints as input, making it unsuitable for applications such as those to game engines. On the other hand, our method can directly generate RGB images using semantic inputs, so rendering a virtual world becomes more effortless. Moreover, they need to train a separate model (or part of the model) for each scene, while we only need one model per dataset, or domain.

3 World-Consistent Video-to-video Synthesis

Background. Recent image-to-image translation methods perform extremely well when turning semantic images to realistic outputs. To produce videos instead of images, simply doing it frame-by-frame will usually result in severe flickering artifacts [77]. To resolve this, vid2vid [77] proposes to take both the semantic inputs and L previously generated frames as input to the network (*e.g.* $L = 3$). The network then generates three outputs – a hallucinated frame, a flow map, and a (soft) mask. The flow map is used to warp the previous frame and linearly combined with the hallucinated frame using the soft mask. Ideally, the network should reuse the content in the warped frame as much as possible, and only use the disoccluded parts from the hallucinated frame.

While the above framework reduces flickering between neighboring frames, it still struggles to ensure long-term consistency. This is because it only keeps track of the past L frames, and cannot memorize everything in the past. Consider the scenario in Fig. 1, where an object moves out of and back in the field-of-view. In this case, we would want to make sure its appearance is similar during the revisit, but that cannot be handled by existing frameworks like vid2vid [77].

In light of this, we propose a new framework to handle *world-consistency.* It is a superset of *temporal consistency*, which only ensures consistency between frames in a video. A world-consistent video should not only be temporally stable,

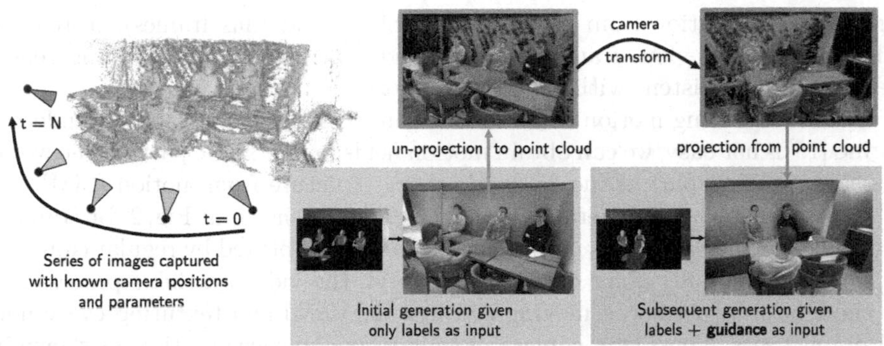

Fig. 2. Overview of *guidance image* generation for training. Consider a scene in which a camera(s) with known parameters and positions travels over time $t = 0, \cdots, N$. At $t = 0$, the scene is textureless and an output image is generated for this viewpoint. The output image is then back-projected to the scene and a *guidance image* for a subsequent camera position is generated by projecting the partially textured point cloud. Using this guidance image, the generative method can produce an output that is consistent across views and smooth over time. Note that the guidance image can be noisy, misaligned, and have holes, and the generation method should be robust to such inputs.

but also be consistent across the entire 3D world the user is viewing. This not only makes the output look more realistic, but also enables applications such as the multi-player scenario where different players can view the same scene from different viewpoints. We achieve this by using a novel *guidance image* conditional scheme, which is detailed below.

Guidance Images and Their Generation. The lack of knowledge about the world structure being generated limits the ability of vid2vid to generate view-consistent outputs. As shown in Fig. 5 and Sect. 4, the color and structure of the objects generated by vid2vid [77] tend to drift over time. We believe that in order to produce realistic outputs that are consistent over time and viewpoint change, an ideal method must be aware of the 3D structure of the world.

To achieve this, we introduce the concept of "*guidance images*", which are physically-grounded estimates of what the next output frame should look like, based on how the world has been generated so far. As alluded to in their name, the role of these "*guidance images*" is to guide the generative model to produce colors and textures that respect previous outputs. Prior works including vid2vid [77] rely on optical flows to warp the previous frame for producing an estimate of the next frame. Our guidance image differs from this warped frame in two aspects. First, instead of using optical flow, the guidance image should be generated by using the motion field, or scene flow, which describes the true motion of each 3D point in the world[1]. Second, the guidance image should

[1] As an example, consider a textureless sphere rotating under constant illumination. In this case, the optical flow would be zero, but the motion field would be nonzero.

aggregate information from *all* past viewpoints (and thus frames), instead of only the direct previous frames as in vid2vid. This makes sure that the generated frame is consistent with the entire history.

While estimating motion fields without an RGB-D sensor [23] or a rendering engine [18] is not easy, we can obtain motion fields for the static parts of the world by reconstructing part of the 3D world using structure from motion (SfM) [50, 69]. This enables us to generate guidance images as shown in Fig. 2 for training our video-to-video synthesis method using datasets captured by regular cameras. Once we have the 3D point cloud of the world, the video synthesis process can be thought of as a camera moving through the world and texturing every new 3D point it sees. Consider a camera moving through space and time as shown in the left part of Fig. 2. Suppose we generate an output image at $t = 0$. This image can be back-projected to the 3D point cloud and colors can be assigned to the points, so as to create a persistent representation of the world. At a later time step, $t = N$, we can obtain the projection of the 3D point cloud to the camera and create a guidance image leveraging estimated motion fields. Our method can then generate an output frame based on the guidance image.

Although we generate guidance images using the projection of 3D point clouds, it can also be generated by any other method that gives a reasonable estimate. This makes the concept powerful, as we can use different sources to generate guidance images at training and test time. For example, at test time we can generate guidance images using a graphics engine, which can provide ground truth 3D correspondences. This enables just-in-time colorization of a virtual 3D world with real-world colors and textures, as we move through the world.

Note that our guidance image also differs from the projected image used in prior works like Meshry *et al.* [53] in several aspects. First, in their case, the 3D point cloud is fixed once constructed, while in our case it is constantly being "colorized" as we synthesize more and more frames. As a result, our guidance image is blank at the beginning, and can become denser depending on the viewpoint. Second, the way we use these guidance images to generate outputs is also different. The guidance images can have misalignments and holes due to limitations of SfM, for example in the background and in the person's head in Fig. 2. As a result, our method also differs from DeepFovea [36], which inpaints sparsely but accurately rendered video frames. In the following subsection, we describe a method that is robust to noises in guidance images, so it can produce outputs consistent over time and viewpoints.

Framework for Generating Videos Using Guidance Images. Once the guidance images are generated, we are able to utilize them to synthesize the next frame. Our generator network is based on the SPADE architecture proposed by Park *et al.* [59], which accepts a random vector encoding the image style as input and uses a series of SPADE blocks and upsampling layers to generate an output image. Each SPADE block takes a semantic map as input and learns to modulate the incoming feature maps through an affine transform $y = x \cdot \gamma_{\text{seg}} + \beta_{\text{seg}}$, where x is the incoming feature map, and γ_{seg} and β_{seg} are predicted from the input segmentation map.

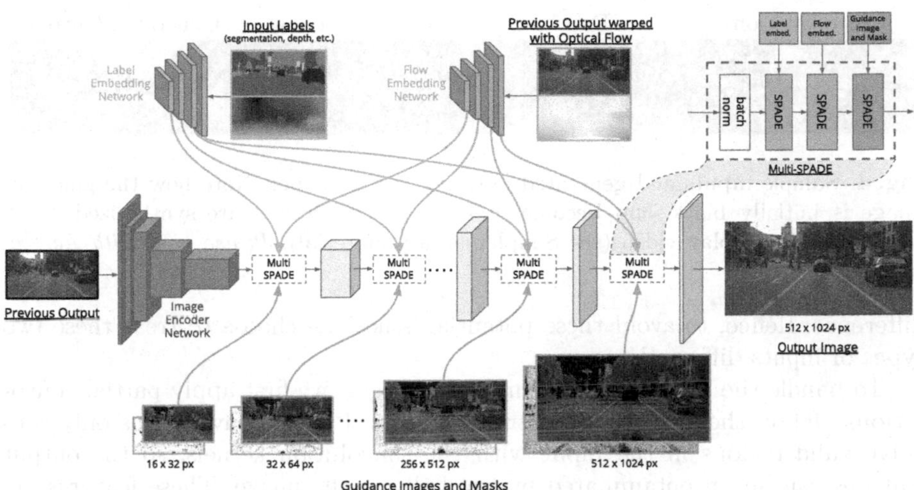

Fig. 3. Overview of our world consistent video-to-video synthesis architecture. Our Multi-SPADE module takes input labels, warped previous frames, and guidance images to modulate the features in each layer of our generator. (Color figure online)

An overview of our method is shown in Fig. 3. At a high-level, our method consists of four sub-networks: 1) an input label embedding network (orange), 2) an image encoder (red), 3) a flow embedding network (green), and 4) an image generator (gray). In our method, we make two modifications to the original SPADE network. First, we feed in the concatenated labels (semantic segmentation, edge maps, *etc.*) to a label embedding network (orange), and extract features in corresponding output layers as input to each SPADE block in the generator. Second, to keep the image style consistent over time, we encode the previously synthesized frame using the image encoder (red), and provide this embedding to our generator (gray) in place of the random vector[2].

Utilizing Guidance Images. Although using this modified SPADE architecture produces output images with better visual quality than vid2vid [77], the outputs are not temporally stable, as shown in Sect. 4. To ensure world-consistency of the output, we would want to incorporate information from the introduced guidance images. Simply linearly combining it with the hallucinated frame from the SPADE generator is problematic, since the hallucinated frame may contain something very different from the guidance images. Another way is to directly concatenate it with the input labels. However, the semantic inputs and guidance images have different physical meanings. Besides, unlike semantic inputs, which are labeled densely (per pixel), the guidance images are labeled sparsely. Directly concatenating them would require the network to compensate for the

[2] When generating the first frame where no previous frame exists, we use an encoder which accepts the semantic map as input.

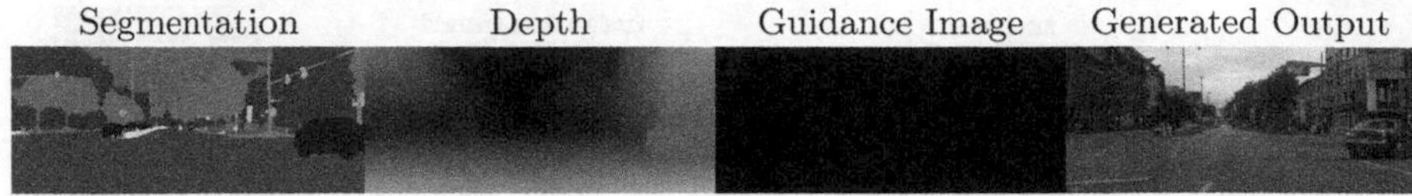

Fig. 4. Sample inputs and generated outputs on Cityscapes. Note how the guidance image is initially black, and becomes denser as more frames are synthesized. Click on any image to play video (see Supplementary material). *Please view with Acrobat Reader.*

difference. Hence, to avoid these potential issues, we choose to treat these two types of inputs differently.

To handle the sparsity of the guidance images, we first apply partial convolutions [46] on these images to extract features. Partial convolutions only convolve valid regions in the input with the convolution kernels, so the output features can be uncontaminated by the holes in the image. These features are then used to generate affine transformation parameters γ_{guidance} and β_{guidance}, which are *inserted* into existing SPADE blocks while keeping the rest of the blocks untouched. This results in a *Multi-SPADE* module, which allows us to use multiple conditioned inputs in sequence, so we can not only condition on the current input labels, but also on our guidance images,

$$y = (x \cdot \gamma_{\text{label}} + \beta_{\text{label}}) \cdot \gamma_{\text{guidance}} + \beta_{\text{guidance}}. \tag{1}$$

Using this module yields several benefits. First, conditioning on these maps generates more temporally smooth and higher quality frames than simple linear blending techniques. Separating the two types of input (semantic labels and guidance images) also allows us to adopt different types of convolutions (i.e. normal vs. partial). Second, since most of the network architecture remains unchanged, we can initialize the weights of the generator with one trained for single image generation. It is easy to collect large training datasets for single image generation by crawling the internet, while video datasets can be harder to collect and annotate. After the single image generator is trained, we can train a video generator by just training the newly added layers (*i.e.* layers generating γ_{guidance} and β_{guidance}) and only finetune the other parts of the network.

Handling Dynamic Objects. The guidance image allows us to generate world-consistent outputs over time. However, since the guidance is generated based on SfM for real-world scenes, it has the inherent limitation that SfM cannot handle dynamic objects. To resolve this issue, we revert to using optical flow-warped frames to serve as additional maps in addition to the guidance images we have from SfM. The complete Multi-SPADE module then becomes

$$y = \big((x \cdot \gamma_{\text{label}} + \beta_{\text{label}}) \cdot \gamma_{\text{flow}} + \beta_{\text{flow}}\big) \cdot \gamma_{\text{guidance}} + \beta_{\text{guidance}}, \tag{2}$$

where γ_{flow} and β_{flow} are generated using a flow-embedding network (green) applied on the optical flow-warped previous frame. This provides additional constraints that the generated frame should be consistent even in the dynamic regions. Note that this is needed only due to the limitation of SfM, and can potentially be removed when ground truth/high quality 3D registrations are available, for example in the case of game engines, or RGB-D data capture.

Figure 4 shows a sample set of inputs and outputs generated by our method on the Cityscapes dataset. More are provided in the supplementary material.

4 Experiments

Implementation Details. We train our network in two stages. In the first stage, we only train our network to generate single images. This means that only the first SPADE layer of our Multi-SPADE block (visualized in Fig. 3) is trained. Following this, we have a network that can generate high-quality single frame outputs. In the second stage, we train on video clips, progressively doubling the generated video length every epoch, starting from 8 frames and stopping at 32 frames. In this stage, all 3 SPADE layers of each Multi-SPADE block are trained. We found that this two-stage pipeline makes the training faster and more stable. We observed that the ordering of the flow and guidance SPADEs did not make a significant difference in the output quality. We train the network for 20 epochs in each stage, and this takes about 10 days on an NVIDIA DGX-1 (8 V-100 GPUs) for an output resolution of 1024×512.

We train our generator with the multi-scale image discriminator using perceptual and GAN feature matching losses as in SPADE [59]. Following vid2vid [77], we add a temporal video discriminator at two temporal scales and a warping loss that encourages the output frame to be similar to the optical flow-warped previous frame. We also add a loss term to encourage the output frame to correspond to the guidance image, and this is necessary to ensure view consistency. Additional details about architecture and loss terms can be found in the supplementary material. Code and trained models will be released upon publication.

Datasets. We train and evaluate our method on three datasets, Cityscapes [15], MannequinChallenge [43], and ScanNet [16], as they have mostly static scenes where existing SfM methods perform well.

- **Cityscapes** [15]. This dataset consists of driving videos of 2048×1024 resolution captured in several German cities, using a pair of stereo cameras. We split this dataset into a training set of 3500 videos with 30 frames each, and a test set of 3 long sequences with 600–1200 frames each, similar to vid2vid [77]. As not all the images are labeled with segmentation masks, we annotate the images using the network from Zhu *et al.* [94], which is based on a DeepLabv3-Plus [10]-like architecture with a WideResNet38 [81] backbone.
- **MannequinChallenge** [43]. This dataset contains video clips captured using hand-held cameras, of people pretending frozen in a large variety of poses, imitating mannequins. We resize all frames to 1024×512 and randomly split this dataset into 3040 train sequences and 292 test sequences, with sequence lengths ranging from 5–140 frames. We generate human body segmentation and part-specific UV coordinate maps using DensePose [25,80] and body poses using OpenPose [7].

- **ScanNet** [16]. This dataset contains multiple video clips captured in a total of 706 indoor rooms. We set aside 50 rooms for testing, and the rest for training. From each video sequence, we extracted 3 sub-sequences of length at most 100, resulting in 4000 train sequences and 289 test sequences, with images of size 512×512. We used the provided segmentation maps based on the NYUDv2 [56] 40 labels.

For all datasets, we also use MegaDepth [44] to generate depth maps and add the visualized inverted depth images as input. As the MannequinChallenge and ScanNet datasets contain a large variety of objects and classes which are not fully annotated, we use edge maps produced by HED [83] in order to better represent the input content. In order to generate guidance images, we performed SfM on all the video sequences using OpenSfM [1], which provided 3D point clouds and estimated cameras poses and parameters as output.

Baselines. We compare our method against the following strong baselines.

- vid2vid [77]. This is the prior state-of-the-art method for video-to-video synthesis. For comparison on Cityscapes, we use the publicly available pretrained model. For the other two datasets, we train vid2vid from scratch using the public code, while providing the same input labels (semantic segmentation, depth, edge maps, *etc.*) as to our method.
- Inpainting [46]. We train a state-of-the-art partial convolution-based inpainting method to fill in the pixels missing from our guidance images. We train the models from scratch for each dataset, using masks obtained from the corresponding guidance images.
- Ours w/o W.C. (World Consistency). As an ablation, we also compare against our model that does not use guidance images. In this case, only the first two SPADE layers in each Multi-SPADE block are trained (label and flow-warped previous output SPADEs). Other details are the same as our full model.

Evaluation Metrics. We use both objective and subjective metrics for evaluating our model against the baselines.

- *Segmentation accuracy and Fréchet Inception Distance (FID).* We adopt metrics widely used in prior work on image synthesis [11,59,78] to measure the quality of generated video frames. We evaluate the output frames based on how well they can be segmented by a trained segmentation network. We report both the mean Intersection-Over-Union (mIOU) and Pixel Accuracy (P.A.) using the PSPNet [90] (Cityscapes) and DeepLabv2 [9] (Mannequin-Challenge & ScanNet). We also use the Fréchet Inception Distance (FID) [29] to measure the distance between the distributions of the generated and real images, using the standard Inception-v3 network.
- *Human preference score.* Using Amazon Mechanical Turk (AMT), we perform a subjective visual test to gauge the relative quality of videos. We evaluate videos on two criteria: 1) *photorealism* and 2) *temporal stability.* The first aims to find which generated video looks more like a real video, while the second

Table 1. Comparison scores. ↓ means lower is better, while ↑ means the opposite.

Method	Cityscapes			MannequinChallenge			ScanNet		
	FID↓	mIOU↑	P.A.↑	FID↓	mIOU↑	P.A.↑	FID↓	mIOU↑	P.A.↑
Image synthesis models									
SPADE [59]	48.25	0.63	0.95	29.99	0.13	0.63	31.46	0.08	0.54
Video synthesis models									
vid2vid [77]	69.07	0.55	0.94	72.25	0.05	0.45	60.03	0.04	0.35
Ours w/o W.C.	51.51	**0.62**	0.95	27.23	0.17	0.67	**20.93**	0.12	0.62
Ours	**49.89**	0.61	**0.95**	**22.69**	**0.19**	**0.69**	21.07	**0.13**	**0.63**

Table 2. Human preference scores. Higher is better.

Compared methods	Cityscapes	MannequinChallenge	ScanNet
Image Realism			
Ours/vid2vid [77]	**0.73**/0.27	**0.83**/0.17	**0.77**/0.23
Temporal Stability			
Ours/vid2vid [77]	**0.75**/0.25	**0.63**/0.37	**0.82**/0.18

Table 3. Forward-backward consistency. △ means difference.

Method	Cityscapes		MannequinChallenge		ScanNet	
	△RGB↓	△LAB↓	△RGB↓	△LAB↓	△RGB↓	△LAB↓
vid2vid [77]	14.90	3.46	37.56	0.42	40.30	12.16
Ours	**8.73**	**2.04**	**12.61**	**3.61**	**11.85**	**3.41**

aims to find which one is more temporally smooth and has lesser flickering. For each question, an AMT participant is shown two videos synthesized by two different methods, and asked to choose the better one according to the current criterion. We generate several hundred questions for each dataset, each of them is answered by 3 different workers. We evaluate an algorithm by the ratio that its outputs are preferred.

- *Forward-Backward consistency.* A major contribution of our work is generating outputs that are consistent over a longer duration of time with the world that was previously generated. All our datasets have videos that explore new parts of the world over time, rarely revisiting previously explored parts. However, a simple way to revisit a location is to play the video in forward and then in reverse, *i.e.* arrange frames from time $t = 0, 1, \cdots, N-1, N, N-1, \cdots, 1, 0$. We can then compare the first produced and last produced frames and measure their difference. We measure the difference per-pixel in both RGB and LAB space, and a lower value would indicate better long-term consistency.

Fig. 5. Comparison of different video generation methods on the Cityscapes dataset. Note that for our results, the textures of the cars, roads, and signboards are stable over time, while they change gradually in vid2vid and other methods. Click on an image to play the video (see Supplementary material). *Please view with Acrobat Reader.*

Main Results. In Table 1, we compare our proposed approach against vid2vid [77], as well as SPADE [59], which is the single image generator that our method builds upon. We also compare against a version of our method that does not use guidance images and is thus not world-consistent (Ours w/o W.C.). Inpainting [46] could not provide meaningful output images without large artifacts, as shown in Fig. 5. We can observe that our method consistently beats vid2vid on all three metrics on all three datasets, indicating superior image quality. Interestingly, our method also improves upon SPADE in FID, probably as a result of reducing temporal variance across an output video sequence. We also see improvements over Ours w/o W.C. on almost all metrics.

In Table 2, we show human evaluation results on metrics of image realism and temporal stability. We observe that the majority of workers rank our method better on both metrics.

Fig. 6. Forward-backward consistency. Click on each image to see the change in output when the viewpoint is revisited. Note how drastically vid2vid results change, while ours remain almost the same (see Supplementary material). *Please view with Acrobat Reader.*

Fig. 7. Qualitative results on the MannequinChallenge and ScanNet datasets. Click on an image to play video. Note the results are consistent over time and viewpoints (see Supplementary material). *Please view with Acrobat Reader.*

In Fig. 5, we visualize some sequences generated by the various methods (please zoom in and play the videos in Adobe Acrobat). We can observe that in the first row, vid2vid [77] produces temporal artifacts in the cars parked to the side and patterns on the road. SPADE [59], which produces one frame at a time, produces very unstable videos, as shown in the second row. The third row shows outputs from the partial convolution-based inpainting [46] method. It clearly has a hard time producing visually and semantically meaningful outputs. The fourth row shows Ours w/o W.C., an intermediate version of our method that uses labels and optical flow-warped previous output as input. While this clearly improves upon vid2vid in image quality and SPADE in temporal stability, it causes flickering in trees, cars, and signboards. The last row shows our method. Note how the textures of the cars, roads, and signboards, which are areas we have guidance images, are stable over time. We also provide high resolution, uncompressed videos for all three datasets in our supplementary material.

In Table 3, we compare the forward-backward consistency of different methods, and it shows that our method beats vid2vid [77] by a large margin, especially on the MannequinChallenge and ScanNet datasets (by more than a factor of 3). Figure 6 visualizes some frames at the start and end of generation. As can

Fig. 8. Stereo results on Cityscapes. Click on an image to see the outputs produced by a pair of stereo cameras. Note how our method produces images consistent across the two views, while they differ in the highlighted regions without using the world consistency (see Supplementary material). *Please view with Acrobat Reader.*

be seen, the outputs of vid2vid change dramatically, while ours are consistent. We show additional qualitative examples in Fig. 7. We also provide additional quantitative results on short-term consistency in the supplementary material.

Generating Consistent Stereo Outputs. Here, we show a novel application enabled by our method through the use of guidance images. We show videos rendered simultaneously for multiple viewpoints, specifically for a pair of stereo viewpoints on the Cityscapes dataset in Fig. 8. For the strongest baseline, Ours w/o W.C., the left-right videos can only be generated independently, and they clearly are not consistent across multiple viewpoints, as highlighted by the boxes. On the other hand, our method can generate left-right videos in sync by sharing the underlying 3D point cloud and guidance maps. Note how the textures on roads, including shadows, move in sync and remain consistent over time and camera locations.

5 Conclusions and Discussion

We presented a video-to-video synthesis framework that can achieve world consistency. By using a novel guidance image extracted from the generated 3D world, we are able to synthesize the current frame conditioned on all the past frames. The conditioning was implemented using a novel Multi-SPADE module, which not only led to better visual quality, but also made transplanting a single image generator to a video generator possible. Comparisons on several challenging datasets showed that our method improves upon prior state-of-the-art methods.

While advancing the state-of-the-art, our framework still has several limitations. For example, the guidance image generation is based on SfM. When SfM fails to register the 3D content, our method will also fail to ensure consistency. Also, we do not consider a possible change in time of the day or lighting in the current framework. In the future, our framework can benefit from improved guidance images enabled by better 3D registration algorithms. Furthermore, the albedo and shading of the 3D world may be disentangled to better model the time effects. We leave these to future work.

Acknowledgements. We would like to thank Jan Kautz, Guilin Liu, Andrew Tao, and Bryan Catanzaro for their feedback, and Sabu Nadarajan, Nithya Natesan, and Sivakumar Arayandi Thottakara for helping us with the compute, without which this work would not have been possible.

References

1. OpenSfM. https://github.com/mapillary/OpenSfM
2. Aliev, K.A., Ulyanov, D., Lempitsky, V.: Neural point-based graphics. arXiv preprint arXiv:1906.08240 (2019)
3. Benaim, S., Wolf, L.: One-shot unsupervised cross domain translation. In: Conference on Neural Information Processing Systems (NeurIPS) (2018)
4. Bonneel, N., Tompkin, J., Sunkavalli, K., Sun, D., Paris, S., Pfister, H.: Blind video temporal consistency. ACM Trans. Graph. (TOG) (2015)
5. Bousmalis, K., Silberman, N., Dohan, D., Erhan, D., Krishnan, D.: Unsupervised pixel-level domain adaptation with generative adversarial networks. In: IEEE Conference on Computer Vision and Pattern Recognition (CVPR) (2017)
6. Brock, A., Donahue, J., Simonyan, K.: Large scale GAN training for high fidelity natural image synthesis. In: International Conference on Learning Representations (ICLR) (2019)
7. Cao, Z., Hidalgo, G., Simon, T., Wei, S.E., Sheikh, Y.: OpenPose: realtime multi-person 2D pose estimation using Part Affinity Fields. arXiv preprint arXiv:1812.08008 (2018)
8. Chan, C., Ginosar, S., Zhou, T., Efros, A.A.: Everybody dance now. In: IEEE International Conference on Computer Vision (ICCV) (2019)
9. Chen, L.C., Papandreou, G., Kokkinos, I., Murphy, K., Yuille, A.L.: DeepLab: semantic image segmentation with deep convolutional nets, atrous convolution, and fully connected CRFs. IEEE Trans. Pattern Anal. Mach. Intell. (TPAMI) **40**, 834–848 (2017)
10. Chen, L.-C., Zhu, Y., Papandreou, G., Schroff, F., Adam, H.: Encoder-decoder with atrous separable convolution for semantic image segmentation. In: Ferrari, V., Hebert, M., Sminchisescu, C., Weiss, Y. (eds.) ECCV 2018. LNCS, vol. 11211, pp. 833–851. Springer, Cham (2018). https://doi.org/10.1007/978-3-030-01234-2_49
11. Chen, Q., Koltun, V.: Photographic image synthesis with cascaded refinement networks. In: IEEE International Conference on Computer Vision (ICCV) (2017)
12. Chen, Y., Pan, Y., Yao, T., Tian, X., Mei, T.: Mocycle-GAN: unpaired video-to-video translation. In: ACM International Conference on Multimedia (MM) (2019)
13. Choi, I., Gallo, O., Troccoli, A., Kim, M.H., Kautz, J.: Extreme view synthesis. In: IEEE International Conference on Computer Vision (ICCV) (2019)
14. Choi, Y., Choi, M., Kim, M., Ha, J.W., Kim, S., Choo, J.: StarGAN: unified generative adversarial networks for multi-domain image-to-image translation. In: IEEE Conference on Computer Vision and Pattern Recognition (CVPR) (2018)

15. Cordts, M., et al.: The Cityscapes dataset for semantic urban scene understanding. In: IEEE Conference on Computer Vision and Pattern Recognition (CVPR) (2016)
16. Dai, A., Chang, A.X., Savva, M., Halber, M., Funkhouser, T., Nießner, M.: ScanNet: Richly-annotated 3D reconstructions of indoor scenes. In: IEEE Conference on Computer Vision and Pattern Recognition (CVPR) (2017)
17. Denton, E.L., Birodkar, V.: Unsupervised learning of disentangled representations from video. In: Conference on Neural Information Processing Systems (NeurIPS) (2017)
18. Dosovitskiy, A., Ros, G., Codevilla, F., Lopez, A., Koltun, V.: CARLA: an open urban driving simulator. In: Conference on Robot Learning (CoRL) (2017)
19. Finn, C., Goodfellow, I., Levine, S.: Unsupervised learning for physical interaction through video prediction. In: Conference on Neural Information Processing Systems (NeurIPS) (2016)
20. Flynn, J., et al.: Deepview: view synthesis with learned gradient descent. In: IEEE Conference on Computer Vision and Pattern Recognition (CVPR) (2019)
21. Flynn, J., Neulander, I., Philbin, J., Snavely, N.: DeepStereo: learning to predict new views from the world's imagery. In: IEEE Conference on Computer Vision and Pattern Recognition (CVPR) (2016)
22. Gafni, O., Wolf, L., Taigman, Y.: Vid2Game: controllable characters extracted from real-world videos. In: International Conference on Learning Representations (ICLR) (2020)
23. Golyanik, V., Kim, K., Maier, R., Nießner, M., Stricker, D., Kautz, J.: Multiframe scene flow with piecewise rigid motion. In: International Conference on 3D Vision (3DV) (2017)
24. Goodfellow, I., et al.: Generative adversarial networks. In: Conference on Neural Information Processing Systems (NeurIPS) (2014)
25. Güler, R.A., Neverova, N., Kokkinos, I.: DensePose: dense human pose estimation in the wild. In: IEEE Conference on Computer Vision and Pattern Recognition (CVPR) (2018)
26. Gulrajani, I., Ahmed, F., Arjovsky, M., Dumoulin, V., Courville, A.C.: Improved training of Wasserstein GANs. In: Conference on Neural Information Processing Systems (NeurIPS) (2017)
27. Hao, Z., Huang, X., Belongie, S.: Controllable video generation with sparse trajectories. In: IEEE Conference on Computer Vision and Pattern Recognition (CVPR) (2018)
28. Hedman, P., Philip, J., Price, T., Frahm, J.M., Drettakis, G., Brostow, G.: Deep blending for free-viewpoint image-based rendering. ACM Trans. Graph. (TOG) **37**, 1–15 (2018)
29. Heusel, M., Ramsauer, H., Unterthiner, T., Nessler, B., Hochreiter, S.: GANs trained by a two time-scale update rule converge to a local Nash equilibrium. In: Conference on Neural Information Processing Systems (NeurIPS) (2017)
30. Hu, Q., Waelchli, A., Portenier, T., Zwicker, M., Favaro, P.: Video synthesis from a single image and motion stroke. arXiv preprint arXiv:1812.01874 (2018)
31. Huang, X., Liu, M.-Y., Belongie, S., Kautz, J.: Multimodal unsupervised image-to-image translation. In: Ferrari, V., Hebert, M., Sminchisescu, C., Weiss, Y. (eds.) ECCV 2018. LNCS, vol. 11207, pp. 179–196. Springer, Cham (2018). https://doi.org/10.1007/978-3-030-01219-9_11
32. Isola, P., Zhu, J.Y., Zhou, T., Efros, A.A.: Image-to-image translation with conditional adversarial networks. In: IEEE Conference on Computer Vision and Pattern Recognition (CVPR) (2017)

33. Johnson, J., Gupta, A., Fei-Fei, L.: Image generation from scene graphs. In: IEEE Conference on Computer Vision and Pattern Recognition (CVPR) (2018)
34. Kalantari, N.K., Wang, T.C., Ramamoorthi, R.: Learning-based view synthesis for light field cameras. ACM Trans. Graph. (TOG) **35**, 1–10 (2016)
35. Kalchbrenner, N., et al.: Video pixel networks. In: International Conference on Machine Learning (ICML) (2017)
36. Kaplanyan, A.S., Sochenov, A., Leimkühler, T., Okunev, M., Goodall, T., Rufo, G.: DeepFovea: neural reconstruction for foveated rendering and video compression using learned statistics of natural videos. ACM Trans. Graph. (TOG) **38**, 1–13 (2019)
37. Karras, T., Aila, T., Laine, S., Lehtinen, J.: Progressive growing of GANs for improved quality, stability, and variation. In: International Conference on Learning Representations (ICLR) (2018)
38. Karras, T., Laine, S., Aila, T.: A style-based generator architecture for generative adversarial networks. In: IEEE Conference on Computer Vision and Pattern Recognition (CVPR) (2019)
39. Lai, W.-S., Huang, J.-B., Wang, O., Shechtman, E., Yumer, E., Yang, M.-H.: Learning blind video temporal consistency. In: Ferrari, V., Hebert, M., Sminchisescu, C., Weiss, Y. (eds.) ECCV 2018. LNCS, vol. 11219, pp. 179–195. Springer, Cham (2018). https://doi.org/10.1007/978-3-030-01267-0_11
40. Lee, A.X., Zhang, R., Ebert, F., Abbeel, P., Finn, C., Levine, S.: Stochastic adversarial video prediction. arXiv preprint arXiv:1804.01523 (2018)
41. Lee, H.-Y., Tseng, H.-Y., Huang, J.-B., Singh, M., Yang, M.-H.: Diverse image-to-image translation via disentangled representations. In: Ferrari, V., Hebert, M., Sminchisescu, C., Weiss, Y. (eds.) ECCV 2018. LNCS, vol. 11205, pp. 36–52. Springer, Cham (2018). https://doi.org/10.1007/978-3-030-01246-5_3
42. Li, Y., Fang, C., Yang, J., Wang, Z., Lu, X., Yang, M.-H.: Flow-grounded spatial-temporal video prediction from still images. In: Ferrari, V., Hebert, M., Sminchisescu, C., Weiss, Y. (eds.) ECCV 2018. LNCS, vol. 11213, pp. 609–625. Springer, Cham (2018). https://doi.org/10.1007/978-3-030-01240-3_37
43. Li, Z., et al.: Learning the depths of moving people by watching frozen people. In: IEEE Conference on Computer Vision and Pattern Recognition (CVPR) (2019)
44. Li, Z., Snavely, N.: MegaDepth: learning single-view depth prediction from internet photos. In: IEEE Conference on Computer Vision and Pattern Recognition (CVPR) (2018)
45. Liang, X., Lee, L., Dai, W., Xing, E.P.: Dual motion GAN for future-flow embedded video prediction. In: Conference on Neural Information Processing Systems (NeurIPS) (2017)
46. Liu, G., Reda, F.A., Shih, K.J., Wang, T.-C., Tao, A., Catanzaro, B.: Image inpainting for irregular holes using partial convolutions. In: Ferrari, V., Hebert, M., Sminchisescu, C., Weiss, Y. (eds.) ECCV 2018. LNCS, vol. 11215, pp. 89–105. Springer, Cham (2018). https://doi.org/10.1007/978-3-030-01252-6_6
47. Liu, M.Y., Breuel, T., Kautz, J.: Unsupervised image-to-image translation networks. In: Conference on Neural Information Processing Systems (NeurIPS) (2017)
48. Liu, M.Y., et al.: Few-shot unsupervised image-to-image translation. In: IEEE International Conference on Computer Vision (ICCV) (2019)
49. Liu, X., Yin, G., Shao, J., Wang, X., et al.: Learning to predict layout-to-image conditional convolutions for semantic image synthesis. In: Conference on Neural Information Processing Systems (NeurIPS) (2019)
50. Longuet-Higgins, H.C.: A computer algorithm for reconstructing a scene from two projections. Nature (1981)

51. Lotter, W., Kreiman, G., Cox, D.: Deep predictive coding networks for video prediction and unsupervised learning. In: International Conference on Learning Representations (ICLR) (2017)
52. Mathieu, M., Couprie, C., LeCun, Y.: Deep multi-scale video prediction beyond mean square error. In: International Conference on Learning Representations (ICLR) (2016)
53. Meshry, M., et al.: Neural rerendering in the wild. In: IEEE Conference on Computer Vision and Pattern Recognition (CVPR) (2019)
54. Mildenhall, B., et al.: Local light field fusion: practical view synthesis with prescriptive sampling guidelines. ACM Trans. Graph. (TOG) **38**, 1–14 (2019)
55. Miyato, T., Koyama, M.: cGANs with projection discriminator. In: International Conference on Learning Representations (ICLR) (2018)
56. Silberman, N., Hoiem, D., Kohli, P., Fergus, R.: Indoor segmentation and support inference from RGBD images. In: Fitzgibbon, A., Lazebnik, S., Perona, P., Sato, Y., Schmid, C. (eds.) ECCV 2012. LNCS, vol. 7576, pp. 746–760. Springer, Heidelberg (2012). https://doi.org/10.1007/978-3-642-33715-4_54
57. Odena, A., Olah, C., Shlens, J.: Conditional image synthesis with auxiliary classifier GANs. In: International Conference on Machine Learning (ICML) (2017)
58. Pan, J., et al.: Video generation from single semantic label map. In: IEEE Conference on Computer Vision and Pattern Recognition (CVPR) (2019)
59. Park, T., Liu, M.Y., Wang, T.C., Zhu, J.Y.: Semantic image synthesis with spatially-adaptive normalization. In: IEEE Conference on Computer Vision and Pattern Recognition (CVPR) (2019)
60. Qi, X., Chen, Q., Jia, J., Koltun, V.: Semi-parametric image synthesis. In: IEEE Conference on Computer Vision and Pattern Recognition (CVPR) (2018)
61. Reed, S., Akata, Z., Yan, X., Logeswaran, L., Schiele, B., Lee, H.: Generative adversarial text to image synthesis. In: International Conference on Machine Learning (ICML) (2016)
62. Saito, M., Matsumoto, E., Saito, S.: Temporal generative adversarial nets with singular value clipping. In: IEEE International Conference on Computer Vision (ICCV) (2017)
63. Shrivastava, A., Pfister, T., Tuzel, O., Susskind, J., Wang, W., Webb, R.: Learning from simulated and unsupervised images through adversarial training. In: IEEE Conference on Computer Vision and Pattern Recognition (CVPR) (2017)
64. Sitzmann, V., Thies, J., Heide, F., Nießner, M., Wetzstein, G., Zollhofer, M.: Deepvoxels: Learning persistent 3d feature embeddings. In: IEEE Conference on Computer Vision and Pattern Recognition (CVPR) (2019)
65. Srinivasan, P.P., Wang, T., Sreelal, A., Ramamoorthi, R., Ng, R.: Learning to synthesize a 4D RGBD light field from a single image. In: IEEE International Conference on Computer Vision (ICCV) (2017)
66. Srivastava, N., Mansimov, E., Salakhudinov, R.: Unsupervised learning of video representations using LSTMs. In: International Conference on Machine Learning (ICML) (2015)
67. Taigman, Y., Polyak, A., Wolf, L.: Unsupervised cross-domain image generation. In: International Conference on Learning Representations (ICLR) (2017)
68. Thies, J., Zollhöfer, M., Nießner, M.: Deferred neural rendering: image synthesis using neural textures. ACM Trans. Graph. (TOG) **38**, 1–12 (2019)
69. Tomasi, C., Kanade, T.: Shape and motion from image streams under orthography: a factorization method. Int. J. Comput. Vis. (IJCV) **9**, 137–154 (1992)

70. Tulyakov, S., Liu, M.Y., Yang, X., Kautz, J.: MoCoGAN: decomposing motion and content for video generation. In: IEEE Conference on Computer Vision and Pattern Recognition (CVPR) (2018)
71. Vedula, S., Baker, S., Rander, P., Collins, R., Kanade, T.: Three-dimensional scene flow. In: IEEE Conference on Computer Vision and Pattern Recognition (CVPR) (1999)
72. Villegas, R., Yang, J., Hong, S., Lin, X., Lee, H.: Decomposing motion and content for natural video sequence prediction. In: International Conference on Learning Representations (ICLR) (2017)
73. Vondrick, C., Pirsiavash, H., Torralba, A.: Generating videos with scene dynamics. In: Conference on Neural Information Processing Systems (NeurIPS) (2016)
74. Walker, J., Doersch, C., Gupta, A., Hebert, M.: An uncertain future: forecasting from static images using variational autoencoders. In: Leibe, B., Matas, J., Sebe, N., Welling, M. (eds.) ECCV 2016. LNCS, vol. 9911, pp. 835–851. Springer, Cham (2016). https://doi.org/10.1007/978-3-319-46478-7_51
75. Walker, J., Marino, K., Gupta, A., Hebert, M.: The pose knows: video forecasting by generating pose futures. In: IEEE International Conference on Computer Vision (ICCV) (2017)
76. Wang, T.C., Liu, M.Y., Tao, A., Liu, G., Kautz, J., Catanzaro, B.: Few-shot video-to-video synthesis. In: Conference on Neural Information Processing Systems (NeurIPS) (2019)
77. Wang, T.C., et al.: Video-to-video synthesis. In: Conference on Neural Information Processing Systems (NeurIPS) (2018)
78. Wang, T.C., Liu, M.Y., Zhu, J.Y., Tao, A., Kautz, J., Catanzaro, B.: High-resolution image synthesis and semantic manipulation with conditional GANs. In: IEEE Conference on Computer Vision and Pattern Recognition (CVPR) (2018)
79. Wiles, O., Gkioxari, G., Szeliski, R., Johnson, J.: SynSin: end-to-end view synthesis from a single image. In: IEEE Conference on Computer Vision and Pattern Recognition (CVPR) (2020)
80. Wu, Y., Kirillov, A., Massa, F., Lo, W.Y., Girshick, R.: Detectron2 (2019). https://github.com/facebookresearch/detectron2
81. Wu, Z., Shen, C., Van Den Hengel, A.: Wider or deeper: revisiting the resnet model for visual recognition. Pattern Recogn. **90**, 119–133 (2019)
82. Xie, J., Girshick, R., Farhadi, A.: Deep3D: fully automatic 2D-to-3D video conversion with deep convolutional neural networks. In: Leibe, B., Matas, J., Sebe, N., Welling, M. (eds.) ECCV 2016. LNCS, vol. 9908, pp. 842–857. Springer, Cham (2016). https://doi.org/10.1007/978-3-319-46493-0_51
83. Xie, S., Tu, Z.: Holistically-nested edge detection. In: IEEE International Conference on Computer Vision (ICCV) (2015)
84. Xu, T., et al.: AttnGAN: fine-grained text to image generation with attentional generative adversarial networks. In: IEEE Conference on Computer Vision and Pattern Recognition (CVPR) (2018)
85. Xue, T., Wu, J., Bouman, K., Freeman, B.: Visual dynamics: probabilistic future frame synthesis via cross convolutional networks. In: Conference on Neural Information Processing Systems (NeurIPS) (2016)
86. Yao, C.H., Chang, C.Y., Chien, S.Y.: Occlusion-aware video temporal consistency. In: ACM International Conference on Multimedia (MM) (2017)
87. Zhang, H., Goodfellow, I., Metaxas, D., Odena, A.: Self-attention generative adversarial networks. In: International Conference on Machine Learning (ICML) (2019)

88. Zhang, H., et al.: StackGAN: text to photo-realistic image synthesis with stacked generative adversarial networks. In: IEEE International Conference on Computer Vision (ICCV) (2017)
89. Zhao, B., Meng, L., Yin, W., Sigal, L.: Image generation from layout. In: IEEE Conference on Computer Vision and Pattern Recognition (CVPR) (2019)
90. Zhao, H., Shi, J., Qi, X., Wang, X., Jia, J.: Pyramid scene parsing network. In: IEEE Conference on Computer Vision and Pattern Recognition (CVPR) (2017)
91. Zhou, T., Tucker, R., Flynn, J., Fyffe, G., Snavely, N.: Stereo magnification: learning view synthesis using multiplane images. In: ACM SIGGRAPH (2018)
92. Zhou, Y., Wang, Z., Fang, C., Bui, T., Berg, T.L.: Dance dance generation: motion transfer for internet videos. arXiv preprint arXiv:1904.00129 (2019)
93. Zhu, J.Y., Park, T., Isola, P., Efros, A.A.: Unpaired image-to-image translation using cycle-consistent adversarial networks. In: IEEE International Conference on Computer Vision (ICCV) (2017)
94. Zhu, Y., et al.: Improving semantic segmentation via video propagation and label relaxation. In: IEEE Conference on Computer Vision and Pattern Recognition (CVPR) (2019)

Commonality-Parsing Network Across Shape and Appearance for Partially Supervised Instance Segmentation

Qi Fan[1], Lei Ke[1], Wenjie Pei[2(✉)], Chi-Keung Tang[1], and Yu-Wing Tai[1,3]

[1] Hong Kong University of Science and Technology, Clear Water Bay, Hong Kong
{qfanaa,lkeab,cktang,yuwing}@cse.ust.hk
[2] Harbin Institute of Technology, Shenzhen, Shenzhen, China
[3] Kwai Inc., Beijing, China
wenjiecoder@gmail.com

Abstract. Partially supervised instance segmentation aims to perform learning on limited mask-annotated categories of data thus eliminating expensive and exhaustive mask annotation. The learned models are expected to be generalizable to novel categories. Existing methods either learn a transfer function from detection to segmentation, or cluster shape priors for segmenting novel categories. We propose to learn the underlying class-agnostic commonalities that can be generalized from mask-annotated categories to novel categories. Specifically, we parse two types of commonalities: 1) shape commonalities which are learned by performing supervised learning on instance boundary prediction; and 2) appearance commonalities which are captured by modeling pairwise affinities among pixels of feature maps to optimize the separability between instance and the background. Incorporating both the shape and appearance commonalities, our model significantly outperforms the state-of-the-art methods on both partially supervised setting and few-shot setting for instance segmentation on COCO dataset. The code is available at https://github.com/fanq15/FewX.

Keywords: Partially supervised · Few-shot · Instance segmentation

1 Introduction

Instance segmentation is a fundamental research topic in computer vision due to its extensive applications ranging from object selection [32], image editing [45,47] to scene understanding [31]. Typical methods [8,21,24,33,37] for instance segmentation have achieved remarkable progress, relying on the fully supervised learning on the precise mask-annotated data. However, this kind of pixel-level mask annotation is extremely labor-consuming and thus expensive to be performed on large amount of data which is typically required for deep learning

Q. Fan and L. Ke—Equal contribution.

A. Vedaldi et al. (Eds.): ECCV 2020, LNCS 12353, pp. 379–396, 2020.
https://doi.org/10.1007/978-3-030-58598-3_23

methods. On the other hand, it is less expensive and more feasible to perform annotation of bounding box for instances, which motivates the newly proposed task: *partially supervised instance segmentation* [23,29]. It aims to learn instance segmentation models on limited mask-annotated categories of data, which can be generalized to new (novel) categories with only bounding-box annotations available. The partially supervised instance segmentation is much more challenging than the typical instance segmentation in full supervision. The major difficulty lies in how to learn the class-agnostic features for instance segmentation that can be generalized from the mask-annotated categories to novel categories.

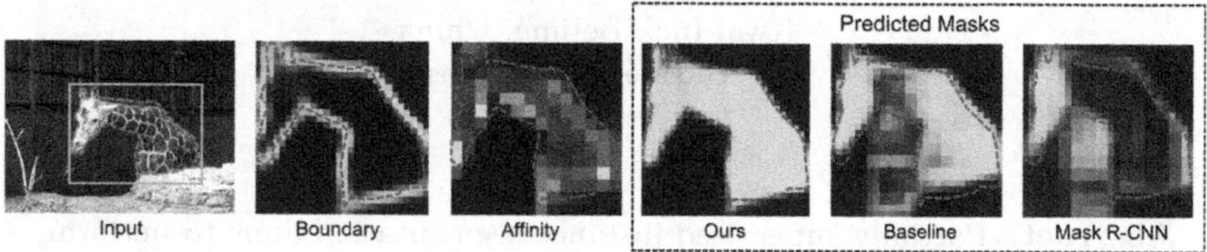

Fig. 1. Given an input image, our model captures shape commonalities by predicting instance boundaries and learns the appearance commonalities by modeling pairwise affinities among all pixels. The learned class-agnostic commonalities in both shape and appearance enable our model to segment more accurate mask than other models.

A straightforward way for partially supervised instance segmentation is to directly extend existing fully supervised algorithms to segmentation of novel categories by class-agnostic training [40,41], which treats all mask-annotated categories of instances involved in training as one foreground category and forces the model to learn to distinguish between foreground and background regions for segmentation. This brute-force way of class-agnostic training expects the model to learn all the generalized features between annotated and novel categories by itself, which is hardly achieved. As the initiator of the partially supervised instance segmentation, Mask^X R-CNN [23] transfers the visual information from the modeling of bounding box to the mask head through a parameterized transfer function. Subsequently, ShapeMask [29] seeks to extract the generic class-agnostic shape features across different categories by summarizing a collection of shape priors as reference for segmenting new categories.

Whilst both Mask^X R-CNN and ShapeMask have distinctly advanced the performance of partially supervised instance segmentation, there are two important features have not been fully exploited. First, the generalized **appearance** features that shared across different categories, e.g., similar hairy body surface between dogs and cats or similar textures on the furniture surface, are not explicitly explored. These class-agnostic appearance features can be potentially generalized from mask-annotated categories of data to novel categories for segmentation. Second, the common **shape** features that can be generalized across different categories are not explicitly learned in a supervised way, though ShapeMask refines the shape priors by simply clustering the annotated masks and adapts them to a given novel object. In this work we intend to tackle the partially supervised instance segmentation by fully exploiting these two features.

We propose to capture the underlying commonalities which can be generalized across different categories by supervised learning for partially supervised instance segmentation. In particular, we aim to learn two types of generalized commonalities: 1) the shape commonalities that can be generalized between different categories like similar instance contour or similar instance boundary features; 2) the appearance commonalities that shared among categories of instances owning similar appearance features such as similar texture or similar color distribution. The resulting model, Commonality-Parsing Network (denoted as CPMask), can be trained in an end-to-end manner. Consider the example in Fig. 1, to segment the giraffe in the red bounding box, our model extracts its shape information by predicting the boundaries of giraffe and captures the appearance information by modeling the pairwise affinities among pixels. Taking into account both the shape and appearance information, our model is able to predict more accurate segmentation mask than other models. It is worth noting that although giraffe is a novel category whose mask-annotation is not provided in the training data, our model is able to accurately predict its boundary and affinity due to the learned class-agnostic commonalities w.r.t. both shape and appearance information.

We evaluate our model on two settings on COCO dataset: 1) partially-supervised instance segmentation, in which partial categories are provided with the ground-truth for both bounding boxes and segmentation masks while the other (novel) categories are only provided with the annotated bounding boxes during training; 2) few-shot instance segmentation, in which each of the novel categories only contain a small number of training samples (with both annotated bounding boxes and masks). Our model outperforms the state-of-the-art models significantly on both settings. We further qualitatively demonstrate the generalization ability of our model by directly applying our trained model on COCO dataset to other 9 datasets with various scenes. It is worth mentioning that our model is more effective given fewer mask-annotated categories of training data compared to methods for fully supervised (routine) instance segmentation. To conclude, our contributions includes:

- We design a supervised learning mechanism for predicting instance boundaries to learn the class-agnostic shape commonalities that can be generalized from mask-annotated categories to novel categories.
- We propose to model the affinities among pixels of feature maps in a supervised way to optimize the separability between the instance region and the background and learn the class-agnostic appearance commonalities that can be generalized to novel objects.
- Incorporating both learned shape and appearance commonalities, our model substantially outperforms state-of-the-art methods on COCO dataset for instance segmentation in both partially supervised and few-shot setting.

2 Related Works

Conventional Instance Segmentation is fully supervised by numerous high-quality pixel-level annotations [9,10,17–20,40,41]. Lots of methods have made great progress on this task by embracing the classical "detect then segment" paradigm, which first generates detection results using the powerful two-stage detector and then segments each object in the bounding box. Mask R-CNN [21] attaches one simple mask predictor on Faster R-CNN [44] to segment each object in the box. PANet [37] merges multi-level features to enhance the performance. FCIS [33] and MaskLab [8] use position-sensitive score maps to encode the segmentation information. Kong and Fowlkes [28] propose to use pairwise pixel affinity for instance segmentation. Mask Scoring R-CNN [24] introduces a mask IoU branch to predict the mask quality and then selects good mask results accordingly. HTC [6] fully leverages the relationship between detection and segmentation to build a successful instance segmentation cascade network. Most recently, some works attempt to build instance segmentation network on the one-stage detector [35,46] for its simplicity and efficiency. In YOLACT [5], a set of prototype masks and coefficients are used to assemble masks for each instance. CenterMask [30] builds an attention-based mask branch on FCOS [46] for fast mask prediction. Compared to these previous works, our model mainly targets for novel objects segmentation, although it also achieves superior performance in the fully supervised task.

Instance Segmentation for Novel Objects. Generalizing instance segmentation model to novel categories with limited annotations is meaningful and challenging, which mainly has three different settings: **Weakly supervised** instance segmentation methods are developed to use weak labels to segment novel categories where the training samples are only annotated with bounding boxes [27,43] or image-level labels [1,59] without pixel-level annotations. **Few-shot supervised** instance segmentation [54] is proposed to solve this problem by imitating the human visual systems to learn new visual concepts with only a few well-annotated samples. **Partially supervised** instance segmentation is formulated in a mixture of strongly and weakly annotated scenario where only a small subset of base categories are well-annotated with both box and mask annotations while the novel categories only have box annotations. In Mask^X R-CNN [23], a parameterized weight transfer function is designed to transfer the visual information from detection to segmentation while ShapeMask [29] learns the intermediate concept of object shape as the prior knowledge. Different from the above two works, which solve the partially supervised segmentation task either from transfer learning perspective or utilizing additional shape priors, our model focuses on learning class-agnostic features with great generalization ability by parsing the shape and appearance commonalities and clearly outperforms the existing methods by a large margin.

3 Commonality-Parsing Network

The crux of performing novel instance segmentation is to learn the underlying commonalities that can be generalized from the mask-annotated categories to novel categories. To surmount this crux, our Commonality-Parsing Network performs class-agnostic learning for partially supervised instance segmentation by two proposed modules: 1) Boundary-Parsing Module for learning shape commonalities and 2) Non-local Affinity-Parsing Module for learning appearance commonalities. We will first present the overall framework of the proposed Commonality-Parsing Network, then we will elaborate on the aforementioned two modules specifically designed for class-agnostic learning.

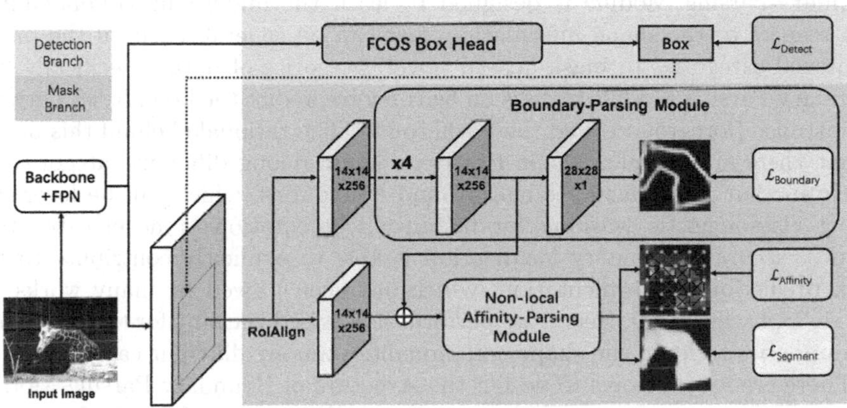

Fig. 2. Architecture of our Commonality-Parsing Network.

3.1 Class-Agnostic Learning Framework

Figure 2 presents the architecture of our Commonality-Parsing Network. Following typical models [8,21,37] for instance segmentation, our model contains two branches: 1) the object detection branch in charge of predicting bounding boxes as instance proposals, and 2) the mask branch for predicting segmented masks for the instance proposals obtained from the object detection branch.

We adopt FCOS [46], which is an excellent one-stage detection model, as our object detection backbone. As illustrated in Fig. 2, a backbone network equipped with FPN [34] is first employed to extract intermediate convolutional features for downstream processing. The object detection branch is then utilized to predict bounding boxes with positions as well as categories for potential instances. In the training phrase, supervision on both the position prediction and the category classification is performed to guide the optimization of the backbone network and FPN as in [46]:

$$\mathcal{L}_{\text{Detect}} = \mathcal{L}_{\text{regression}} + \mathcal{L}_{\text{centerness}} + \mathcal{L}_{\text{classification}}. \tag{1}$$

The mask branch is responsible for segmenting each of target instances predicted by the object detection branch. It is composed of two core modules designed specifically for class-agnostic learning by parsing the commonalities across both the shape and appearance features: Boundary-Parsing Module and Non-local Affinity-Parsing Module. These two modules are trained on a small set of mask-annotated categories of data (termed as base categories) and the learned inter-category commonality of both shape and appearance information enables our model to perform instance segmentation on novel categories of image data.

3.2 Boundary-Parsing Module for Learning Shape Commonality

Boundary-Parsing Module is designed to learn the underlying commonalities with respect to the shape information that can be generalized from the mask-annotated categories to mask-unseen novel categories of data. Specifically, the Boundary-Parsing Module focuses on learning to predict the boundaries between the instance (foreground) and the background. The rationale behind this design is that there are common shape features shared among different categories on discrimination of the instance-background boundaries, which can be leveraged during class-agnostic learning for instance segmentation of novel categories. Besides, accurate boundary localization is able to explicitly contribute to the mask prediction for segmentation, which has been proved by many works [2–4,7,11,38,42,48,56,58]. Hence, we perform supervised learning for the prediction of boundaries to learn the shape commonalities among different categories.

There are several ways to design the structure of Boundary-Parsing Module and we just investigate a straightforward yet effective way: four 3×3 convolutional layers with ReLU as the activation functions, followed by one upsampling layer and one 1×1 convolutional layer to output one channel of feature map as boundary predictions. The Boundary-Parsing Module is trained with the boundary loss:

$$\mathcal{L}_{\text{boundary}} = \mathcal{L}_{\text{BCE}}(\mathcal{F}_B(\mathbf{X}), \mathcal{GT}_B), \tag{2}$$

where $\mathcal{L}_{\text{BCE}}$ denotes the binary cross-entropy loss, $\mathcal{F}_B$ denotes the nonlinear transformation functions by Boundary-Parsing Module, $\mathbf{X}$ is the RoI feature cropped by the *RoIAlign* operation corresponding to a target instance predicted by the object detection branch and $\mathcal{GT}_B$ is the off-the-shelf boundary ground-truth that can be readily obtained from mask annotations.

3.3 Non-local Affinity-Parsing Module for Learning Appearance Commonality

Similar categories tend to share similar appearance commonality, e.g., similar hairy body surface between dogs and cats, or similar texture on the furniture surface. This kind of appearance commonalities can be leveraged for class-agnostic learning to generalize the instance segmentation to novel categories. Therefore, we propose Non-local Affinity-Parsing Module to learn the appearance commonalities across different categories by parsing the affinities among pixels of feature maps in a non-local way. The pixels belonging to an instance (in the foreground region) are expected to have much closer affinities than the affinities between foreground and background pixels.

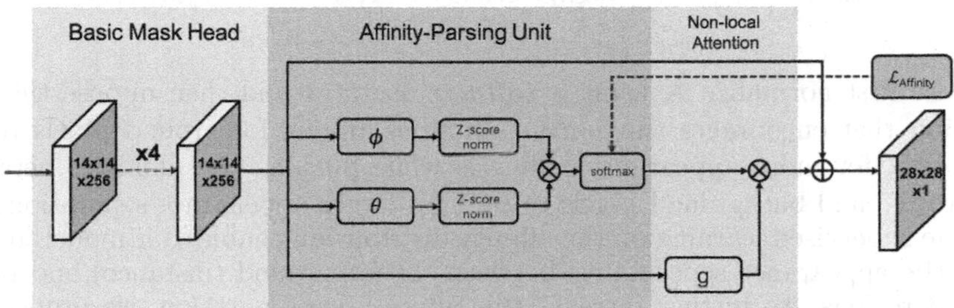

Fig. 3. Architecture of our non-local affinity-parsing module.

Formally, given the RoI feature $\mathbf{X}$ after *RoIAlign* operation for an instance proposal, we first fuse it with the output feature maps $\mathcal{F}_B(\mathbf{X})$ from Boundary-Parsing Module by a simple attention module which incorporates the shape commonality information by weighted element-wise additions. Then the non-linear transformation $\mathcal{G}$ by four convolutional layers is performed on the fused features as a basic mask head of operations:

$$\mathbf{C} = \mathcal{G}(\mathbf{X} \oplus \mathcal{F}_B(\mathbf{X}))). \tag{3}$$

The obtained feature maps $\mathbf{C} \in \mathrm{R}^{c \times h \times w}$, with c feature maps of size $h \times w$, is then fed into the non-local affinity-parsing unit for modeling affinity. Specifically, we model the affinity between the pixel at (i, j) and the pixel at (m, n) in a latent embedding space by:

$$\mathbf{A}_{(<i,j>,<m,n>)} = f\Big[\frac{(\theta(\mathbf{C}_{i,j}) - \mu_{i,j})}{\sigma_{i,j}}, \frac{(\phi(\mathbf{C}_{m,n}) - \mu_{m,n})}{\sigma_{m,n}}\Big], \tag{4}$$

where $\mathbf{C}_{i,j} \in \mathrm{R}^c$ corresponds to the vectorial representation (in channel dimension) for the pixel at (i, j) and the same goes for $\mathbf{C}_{m,n}$. Herein, θ, ϕ are embedding functions and f is a kernel function for encoding affinity. In practice, we opt for the dot-product operator for f, which is a typical way of modeling similarity. μ and σ are the mean value and the standard deviation respectively. Note that here we apply the *z-score* normalization for both $\theta(\mathbf{C}_{i,j})$ and $\phi(\mathbf{C}_{m,n})$ to ease the convergence during optimization.

Larger affinity value indicates closer relationship while smaller affinity value implies larger difference. We expect that the affinities between pixels belonging to an instance (foreground) region are much higher than that between foreground and background pixels. To this end, we introduce a supervision signal to guide the optimization to achieve the desired affinity distribution. In particular, we impose an affinity constraint to maximize the affinities among pixels in the foreground region Fg and minimize the affinities between foreground Fg and background Bg pixels:

$$\begin{aligned} &\mathbf{A} = \text{softmax}(\mathbf{A}), \\ &\mathcal{L}_{\text{Affinity}} = \mathcal{L}_1(1, \sum_{\substack{<i,j>\in Fg \\ <m,n>\in Fg}} \mathbf{A}_{<i,j>,<m,n>}) + \mathcal{L}_1(0, \sum_{\substack{<i,j>\in Fg \\ <m,n>\in Bg}} \mathbf{A}_{<i,j>,<m,n>}). \end{aligned} \tag{5}$$

Here we first normalize $\mathbf{A}$ using a *softmax* operator and then impose the loss function that encourages the sum of affinities among foreground pixels to be close to 1 for more appearance affinities while pushing the affinities between foreground and background pixels to be 0 for larger appearance separation.

The supervised learning on the affinity distribution enables our model to perceive the appearance separability between the foreground (instance) and background regions. To further increase this appearance separation, we propose to coordinate feature maps by explicitly incorporating the learned affinities in a non-local attention manner [49,57]:

$$\widetilde{\mathbf{C}}_{i,j} = \sum_{\forall <m,n>} \mathbf{A}_{<i,j>,<m,n>} \cdot g(\mathbf{C}_{m,n}), \tag{6}$$

where g is a embedding function. Here we coordinate the vectorial representation for the pixel at (i, j) in the feature maps by attending each pixel with the corresponding affinity. Such coordination on feature maps enables our model to perceive the context of whole image region with affinity-based attention, thus resulting in more separation of appearance between foreground and background and closer affinities among pixels in foreground (instance) region, which is beneficial for learning appearance commonalities and instance segmentation.

Together with original feature maps $\mathbf{C}$, the output coordinated feature maps $\tilde{\mathbf{C}}$ from the Non-local Affinity-Parsing Module is subsequently fed into one upsampling layer and one 1×1 convolutional layer for the final prediction of segmented mask:

$$\mathcal{L}_{\text{Segment}} = \mathcal{L}_{\text{BCE}}(\mathcal{F}_{1\times 1conv}(\tilde{\mathbf{C}} \oplus \mathbf{C}), \mathcal{GT}_S), \tag{7}$$

where $\mathcal{F}_{1\times 1conv}$ denotes the nonlinear transformation functions by 1×1 convolutional layer and $\mathcal{GT}_S$ is the ground-truth mask annotations.

3.4 End-to-End Parameter Learning

The whole model of our Commonality-Parsing Network can be trained in an end-to-end manner on two different types of training data:

- For the mask-annotated training data in base categories, the model is optimized by integrating all the aforementioned loss functions:

$$\mathcal{L} = \lambda_1\mathcal{L}_{\text{Detect}} + \lambda_2\mathcal{L}_{\text{Boundary}} + \lambda_3\mathcal{L}_{\text{Affinity}} + \lambda_4\mathcal{L}_{\text{Segment}}, \tag{8}$$

where λ_1, λ_2, λ_3, and λ_4 are hyper-parameter weights to balance the loss functions. In our implementation, they are tuned to be $\{1, 0.5, 0.5, 1\}$ respectively on a validation set.

- For the training data without mask-annotation in novel categories, we train the model with only detection loss, i.e., only the parameters in backbone network, FPN and detection branch are optimized:

$$\mathcal{L} = \mathcal{L}_{\text{Detect}}. \tag{9}$$

4 Experiments

We conduct experiments on MS COCO dataset [36] to evaluate our model. We first perform ablation study to investigate the effect of Boundary-Parsing Module and Non-local Affinity-Parsing Module, then we compare our model with state-of-the-art methods in three different settings for instance segmentation: 1) partially supervised setting, 2) few-shot setting and 3) fully supervised setting.

4.1 Experimental Setup

Evaluation Protocol. We follow the typical data split on COCO in our experiment: *train2017* for training and *val2017* for test. In both of our experiments on partially supervised setting and few-shot setting, we split the 80 COCO categories into "*voc*" and "*non-voc*" category subsets where the *voc* categories are those in PASCAL VOC [12] dataset while the remaining categories are included in the *non-voc* categories. Each time we select classes in one subset as base categories with annotations of both bounding boxes and masks, and those in the other subset as novel categories. Note that the training samples of novel categories have only bounding box annotation (no mask annotation) for partially supervised setting. For few-shot setting [13,26,54], each novel category in the training data only contains a small amount of samples with annotations of both bounding boxes and masks.

Implementation Details. SGD with Momentum is employed for training our model, starting 1 K constant warm-up iterations. The batch size is set to 16 and initial learning rate is set to 0.01. For efficiency, ResNet-50 [22] is used as backbone network for ablation study and the input images are resized in such a way that the short side and long side are no more than 600 and 1000 pixels respectively (denoted as (600, 1000)). For other experiments on comparison with other methods, ResNet-101 [22] backbone with multi-scale training is employed.

4.2 Ablation Study

We investigate the effectiveness of our Boundary-Parsing Module and Non-local Affinity-Parsing Module by carrying out ablation experiments for partially supervised instance segmentation in this section. The *voc* classes is used as base categories and the *non-voc* as novel categories. We refer to the variant of our model

Table 1. Experimental results of ablation studies on the COCO *val* set. The models are trained on the *voc* base categories and evaluated on the *non-voc* novel categories. The "BM" denotes the Boundary-Parsing Module, the "AM" denotes the Non-local Affinity-Parsing Module, the "FF" denotes fusing boundary feature to the mask head and the "AL" denotes the affinity loss.

Model	voc → non-voc					
	AP	AP_{50}	AP_{75}	AP_S	AP_M	AP_L
Baseline	20.7	37.9	20.4	10.6	24.7	27.3
Baseline + BM w/o FF	21.6	38.8	21.1	11.6	26.5	28.8
Baseline + BM	27.4	45.1	28.7	12.4	32.3	39.5
Baseline + AM w/o AL	26.9	45.0	27.8	11.6	31.3	39.2
Baseline + AM	27.2	45.2	28.3	11.7	31.5	40.3
Baseline + BM + AM	28.8	46.1	30.6	12.4	33.1	43.4

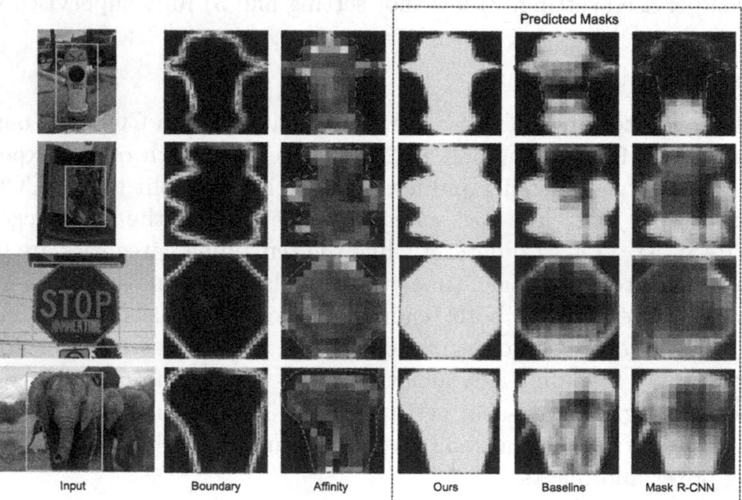

Fig. 4. Visualization of boundary heatmaps and affinity heatmaps learned by our model for four novel categories of cases. The red dash lines indicate the ground-truth mask. (Color figure online)

without Boundary-Parsing Module and Non-local Affinity-Parsing Module as *Baseline* model. The class-agnostic version of Mask R-CNN [23] is compared for reference in this section.

Quantitative Evaluation. Table 1 presents the experimental results. The baseline model obtains 20.7 AP on the novel categories. Boundary-Parsing Module improves the performance by 6.7 AP and explicitly adopting the boundary feature to guide the mask prediction is crucial for the overall performance. Non-local Affinity-Parsing Module promotes the performance by 6.2 AP and the better pixel relationship introduced by the affinity loss further boosts the performance to 27.2 AP. Both the shape and appearance commonalities learned by these two modules from the base categories generalize well to the novel categories. After integrating both modules, our model achieves 28.8 AP which is distinctly better than the performance by each individual module. It implies that the learned shape and appearance commonalities contribute in their own way for instance segmentation.

Qualitative Evaluation. To further reveal the mechanism of these two modules, we visualize boundary and affinity heatmaps on novel categories in Fig. 4. The affinity heatmap is obtained by calculating the mean of the affinity maps for each instance pixel in the Non-local Affinity-Parsing Module. We observe that our model is able to accurately estimate instance boundaries to capture the shape commonalities. Meanwhile, the affinities between instance pixels are evidently higher (closer) than affinities between instance and background pixels, which indicates that appearance commonalities are well learned via affinity modeling for these novel categories. Both of the shape and appearance commonalities can help our model to segment novel instances from background, because the commonalities learned from these modules are successfully generalized from base categories to novel categories. By contrast, the baseline model without these two modules and the Mask R-CNN for fully supervised instance segmentation performs quite poorly on these cases.

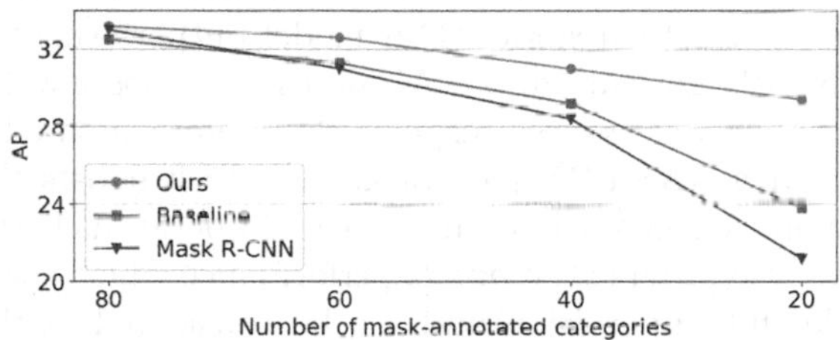

Fig. 5. The segmentation performance of different models on a fixed set of novel categories as a function of number of mask-annotated (base) categories. The novel categories are randomly selected from COCO dataset.

Evaluation of Generalization. To further evaluate the ability of generalization from base (mask-annotated) categories to novel categories for our model, we conduct experiments to investigate the effect of varying the number of mask-annotated categories in Fig. 5. The performances of the both baseline model and Mask R-CNN decay much faster than our model as the number of base categories for training decreases, which indicates that our method is particularly more effective given fewer annotated categories of training data compared to fully supervised methods and benefits from the class-agnostic learning of our model by Boundary-Parsing Module and Non-local Affinity-Parsing Module.

Fig. 6. Qualitative results on novel COCO categories. We use *voc* classes as the base (mask-annotated) categories for training.

4.3 Partially Supervised Instance Segmentation

In this section we compare our model to other state-of-the-art methods for partially supervised instance segmentation.

Table 2 presents the quantitative results on COCO dataset with two sets of experiments: use *voc* or *non-voc* classes as the base categories and treat the remaining classes as novel categories. Our model outperforms the state-of-the-art ShapeMask by a large margin: 3.8 AP on the *non-voc* novel categories and 3.5 AP on the *voc* novel categories respectively. Even compared to its stronger version equipped with NAS-FPN [16] backbone which boosts the performance of both detection and segmentation, our model still performs better than ShapeMask. Besides, we also provide the *oracle* performance which corresponds to the performance under full supervision and can be considered as the performance upper bound for partially supervised learning. We observe that the performance gap between our model and its oracle version is narrowed to 3.6/6.1 AP compared to 4.8/7.6 (4.4/7.4) AP by ShapeMask (ShapeMask with NAS-FPN) and 10.6/9.6 AP by Mask^X R-CNN, indicating the advantages of agnostic learning by our specifically designed modules. Figure 6 shows qualitative results on multiple samples that randomly selected from COCO dataset including various scenes.

Application on Other Datasets. We further qualitatively demonstrate our model on other 9 datasets across various styles and domains [50]: Clipart [25], Comic [25], Watercolor [25], DeepLesions [53], DOTA [51], KITTI [14], LISA [39], Kitchen [15], and WiderFace [55]. It is worth noticing that this is a much harder task due to the cross-dataset generalization. Specifically, we train our model on COCO dataset and feed it ground-truth boxes to obtain the segmentation results on these datasets. As shown in Fig. 7, our model successfully segments novel objects from various domains.

Table 2. Experimental results of partially supervised instance segmentation on the COCO *val* set. The "voc → non-voc" means that we use the *voc* classes as base categories and the *non-voc* as novel categories, and vice versa.

Method	voc → non-voc						non-voc → voc					
	AP	AP_{50}	AP_{75}	AP_S	AP_M	AP_L	AP	AP_{50}	AP_{75}	AP_S	AP_M	AP_L
Mask R-CNN [21]	18.5	34.8	18.1	11.3	23.4	21.7	24.7	43.5	24.9	11.4	25.7	35.1
Mask GrabCut [23]	19.7	39.7	17.0	6.4	21.2	35.8	19.6	46.1	14.3	5.1	16.0	32.4
Mask^X R-CNN [23]	23.8	42.9	23.5	12.7	28.1	33.5	29.5	52.4	29.7	13.4	30.2	41.0
ShapeMask [29]	30.2	49.3	31.5	16.1	38.2	38.4	33.3	56.9	34.3	17.1	38.1	45.4
ShapeMask (NAS-FPN) [29]	33.2	53.1	35.0	18.3	**40.2**	43.3	35.7	60.3	36.6	**18.3**	**40.5**	47.3
CPMask (Ours)	**34.0**	**53.7**	**36.5**	**18.5**	38.9	**47.4**	**36.8**	**60.5**	**38.6**	17.6	37.1	**51.5**
Oracle Mask^X R-CNN [23]	34.4	55.2	36.3	15.5	39.0	52.6	39.1	64.5	41.4	16.3	38.1	55.1
Oracle ShapeMask [29]	35.0	53.9	37.5	17.3	41.0	49.0	40.9	65.1	43.4	18.5	41.9	56.6
Oracle ShapeMask (NAS-FPN) [29]	37.6	57.7	40.2	20.1	44.4	51.1	43.1	67.9	45.8	20.1	44.3	57.8
Oracle CPMask (Ours)	37.6	58.2	40.2	19.9	42.6	54.2	42.9	67.6	46.6	21.6	42.1	58.9

4.4 Few-Shot Instance Segmentation

In this section, we directly apply our model to the challenging few-shot instance segmentation without any network adaption. Few-shot instance segmentation is another challenging task for novel categories. In this task, the model is first trained on base categories with numerous training samples and then generalizes to novel categories with only a few (10 or 20 shots) training samples by direct fine-tuning. Following Meta R-CNN [54], the *non-voc* classes is used as base categories with full samples per category and the *voc* as the novel categories with only 10/20 training samples per category. For fair comparison, we follow Meta R-CNN [54] and use ResNet-50 as backbone and input image size is resized to (600, 1000). Note that the annotations of both bounding box and mask are provided for training samples in novel categories in the few-shot setting.

As shown in Table 3, our model outperforms Meta R-CNN (the state-of-the-art method) by 2.7/3.9 AP in the 10/20-shot settings. Even equipped with the Faster R-CNN detector like Meta R-CNN, our model still performs much better. Although not specifically designed for few-shot learning, our model still obtains the state-of-the-art performance, demonstrating that our proposed model is not limited to the partially supervised learning, and is general for other novel instance segmentation tasks.

Fig. 7. Qualitative results of generalization by our model to 9 different datasets. The model is only trained on COCO and directly applied on these datasets.

Table 3. Experimental results of few-shot instance segmentation on COCO *val* set. The models are trained on the *voc* base categories and fine-tuned on the *non-voc* novel categories with 10/20 instances per category. The evaluation is performed on the held-out *non-voc* novel categories. * denotes using the Faster R-CNN detector.

Method	10-shot						20-shot					
	AP	AP_{50}	AP_{75}	AP_S	AP_M	AP_L	AP	AP_{50}	AP_{75}	AP_S	AP_M	AP_L
Mask R-CNN-ft [54]	1.9	4.7	1.3	0.2	1.4	3.2	3.7	8.5	2.9	0.3	2.5	5.8
Meta R-CNN [54]	4.4	10.6	3.3	**0.5**	3.6	7.2	6.4	14.8	4.4	**0.7**	4.9	9.3
CPMask* (Ours)	6.5	11.6	6.3	0.3	4.1	11.9	9.3	16.0	9.4	0.3	5.8	17.2
CPMask (Ours)	**7.1**	**12.0**	**7.2**	0.3	**5.5**	**12.2**	**10.3**	**16.6**	**10.7**	**0.7**	**8.0**	**17.5**

4.5 Fully Supervised Instance Segmentation

In this section we investigate the performance of our model for fully supervised instance segmentation, namely the routine task for instance segmentation. Table 4 compares our model with other methods on COCO using COCO *train2017* as train set and *test-dev2017* as test set. The experimental results indicate that our model achieves best performance among one-stage methods, although our method focuses on segmenting novel categories. Particularly, our model outperforms the best one-stage model CenterMask [30] by 0.9 AP which is also built on FCOS detection backbone like ours. These encouraging results proves the effectiveness of model on fully supervised instance segmentation.

Table 4. Experimental results of fully supervised instance segmentation on COCO *test-dev* set. The mask AP is reported and all entries are single-model results.

	Method	Backbone	AP	AP_{50}	AP_{75}	AP_S	AP_M	AP_L
Two-stage	Mask R-CNN [21]	ResNet-101	35.7	58.0	37.8	15.5	38.1	52.4
	MaskLab [8]	ResNet-101	37.3	59.8	39.6	19.1	40.5	50.6
	HTC [6]	ResNet-101	39.7	61.8	43.1	21.0	42.2	53.5
	PANet [37]	ResNeXt-101	**42.0**	**65.1**	**45.7**	**22.4**	**44.7**	**58.1**
One-stage	YOLACT [5]	ResNet-101	31.2	50.6	32.8	12.1	33.3	47.1
	PolarMask [52]	ResNet-101	32.1	53.7	33.1	14.7	33.8	45.3
	ShapeMask [29]	ResNet-101	37.4	58.1	40.0	16.1	40.1	53.8
	CenterMask [30]	ResNet-101	38.3	-	-	17.7	40.8	**54.5**
	CPMask (Ours)	ResNet-101	**39.2**	**60.8**	**42.2**	**22.2**	**41.8**	50.1

5 Conclusion

In this paper we present a novel "Commonality-Parsing Network" for partially supervised instance segmentation. Our model learns the class-agnostic commonality knowledge that can be generalized from mask-annotated categories to novel categories without mask annotations. Specifically, we design Boundary-Parsing Module to capture shape commonalities by performing supervised learning on boundary estimation. Further, we propose Non-local Affinity-Parsing Module to model pairwise affinities among pixels in intermediate feature maps to learn appearance commonalities across different categories. Benefiting from these two modules, our model outperforms state-of-the-art methods significantly for instance segmentation in both partially-supervised setting and few-shot setting.

Acknowledgements. This research is supported in part by the Research Grant Council of the Hong Kong SAR under grant no. 1620818.

References

1. Ahn, J., Cho, S., Kwak, S.: Weakly supervised learning of instance segmentation with inter-pixel relations. In: CVPR (2019)
2. Arbelaez, P., Maire, M., Fowlkes, C., Malik, J.: From contours to regions: An empirical evaluation. In: CVPR (2009)
3. Arbelaez, P., Maire, M., Fowlkes, C., Malik, J.: Contour detection and hierarchical image segmentation. IEEE Trans. Pattern Anal. Mach. Intell. **33**(5), 898–916 (2010)
4. Bertasius, G., Shi, J., Torresani, L.: High-for-low and low-for-high: efficient boundary detection from deep object features and its applications to high-level vision. In: ICCV (2015)

5. Bolya, D., Zhou, C., Xiao, F., Lee, Y.J.: Yolact: real-time instance segmentation. In: ICCV (2019)
6. Chen, K., et al.: Hybrid task cascade for instance segmentation. In: CVPR (2019)
7. Chen, L.C., Barron, J.T., Papandreou, G., Murphy, K., Yuille, A.L.: Semantic image segmentation with task-specific edge detection using CNNs and a discriminatively trained domain transform. In: CVPR (2016)
8. Chen, L.C., Hermans, A., Papandreou, G., Schroff, F., Wang, P., Adam, H.: Masklab: instance segmentation by refining object detection with semantic and direction features. In: CVPR (2018)
9. Dai, J., He, K., Sun, J.: Convolutional feature masking for joint object and stuff segmentation. In: CVPR (2015)
10. Dai, J., He, K., Sun, J.: Instance-aware semantic segmentation via multi-task network cascades. In: CVPR (2016)
11. Ding, H., Jiang, X., Liu, A.Q., Thalmann, N.M., Wang, G.: Boundary-aware feature propagation for scene segmentation. In: ICCV (2019)
12. Everingham, M., Van Gool, L., Williams, C.K., Winn, J., Zisserman, A.: The pascal visual object classes (VOC) challenge. Int. J. Comput. Vis. **88**(2), 303–338 (2010)
13. Fan, Q., Zhuo, W., Tang, C.K., Tai, Y.W.: Few-shot object detection with attention-RPN and multi-relation detector. In: CVPR (2020)
14. Geiger, A., Lenz, P., Urtasun, R.: Are we ready for autonomous driving? The kitti vision benchmark suite. In: CVPR (2012)
15. Georgakis, G., Reza, M.A., Mousavian, A., Le, P.H., Košecká, J.: Multiview RGB-D dataset for object instance detection. In: International Conference on 3D Vision (2016)
16. Ghiasi, G., Lin, T.Y., Le, Q.V.: NAS-FPN: learning scalable feature pyramid architecture for object detection. In: CVPR (2019)
17. Girshick, R., Donahue, J., Darrell, T., Malik, J.: Rich feature hierarchies for accurate object detection and semantic segmentation. In: CVPR (2014)
18. Hariharan, B., Arbeláez, P., Girshick, R., Malik, J.: Simultaneous detection and segmentation. In: Fleet, D., Pajdla, T., Schiele, B., Tuytelaars, T. (eds.) ECCV 2014. LNCS, vol. 8695, pp. 297–312. Springer, Cham (2014). https://doi.org/10.1007/978-3-319-10584-0_20
19. Hariharan, B., Arbeláez, P., Girshick, R., Malik, J.: Hypercolumns for object segmentation and fine-grained localization. In: CVPR (2015)
20. Hayder, Z., He, X., Salzmann, M.: Boundary-aware instance segmentation. In: Proceedings of the IEEE Conference on Computer Vision and Pattern Recognition (2017)
21. He, K., Gkioxari, G., Dollár, P., Girshick, R.: Mask R-CNN. In: ICCV (2017)
22. He, K., Zhang, X., Ren, S., Sun, J.: Deep residual learning for image recognition. In: CVPR (2016)
23. Hu, R., Dollár, P., He, K., Darrell, T., Girshick, R.: Learning to segment every thing. In: CVPR (2018)
24. Huang, Z., Huang, L., Gong, Y., Huang, C., Wang, X.: Mask scoring R-CNN. In: CVPR (2019)
25. Inoue, N., Furuta, R., Yamasaki, T., Aizawa, K.: Cross-domain weakly-supervised object detection through progressive domain adaptation. In: CVPR (2018)
26. Kang, B., Liu, Z., Wang, X., Yu, F., Feng, J., Darrell, T.: Few-shot object detection via feature reweighting. In: ICCV (2019)
27. Khoreva, A., Benenson, R., Hosang, J., Hein, M., Schiele, B.: Simple does it: weakly supervised instance and semantic segmentation. In: CVPR (2017)

28. Kong, S., Fowlkes, C.C.: Recurrent pixel embedding for instance grouping. In: CVPR (2018)
29. Kuo, W., Angelova, A., Malik, J., Lin, T.Y.: Shapemask: learning to segment novel objects by refining shape priors. In: ICCV (2019)
30. Lee, Y., Park, J.: Centermask: real-time anchor-free instance segmentation. In: CVPR (2020)
31. Li, L., Huang, W., Gu, I.Y., Tian, Q.: Foreground object detection from videos containing complex background. In: ACM Multimedia (2003)
32. Li, L., Huang, W., Gu, I.Y.H., Tian, Q.: Statistical modeling of complex backgrounds for foreground object detection. IEEE Trans. Image Process. **13**(11), 1459–1472 (2004)
33. Li, Y., Qi, H., Dai, J., Ji, X., Wei, Y.: Fully convolutional instance-aware semantic segmentation. In: CVPR (2017)
34. Lin, T.Y., Dollár, P., Girshick, R., He, K., Hariharan, B., Belongie, S.: Feature pyramid networks for object detection. In: CVPR (2017)
35. Lin, T.Y., Goyal, P., Girshick, R., He, K., Dollár, P.: Focal loss for dense object detection. In: ICCV (2017)
36. Lin, T.-Y., et al.: Microsoft COCO: common objects in context. In: Fleet, D., Pajdla, T., Schiele, B., Tuytelaars, T. (eds.) ECCV 2014. LNCS, vol. 8693, pp. 740–755. Springer, Cham (2014). https://doi.org/10.1007/978-3-319-10602-1_48
37. Liu, S., Qi, L., Qin, H., Shi, J., Jia, J.: Path aggregation network for instance segmentation. In: CVPR (2018)
38. Luo, Z., Mishra, A., Achkar, A., Eichel, J., Li, S., Jodoin, P.M.: Non-local deep features for salient object detection. In: CVPR (2017)
39. Mogelmose, A., Trivedi, M.M., Moeslund, T.B.: Vision-based traffic sign detection and analysis for intelligent driver assistance systems: perspectives and survey. IEEE Trans. Intell. Transp. Syst. **13**(4), 1484–1497 (2012)
40. Pinheiro, P.O., Collobert, R., Dollár, P.: Learning to segment object candidates. In: NeurIPS (2015)
41. Pinheiro, P.O., Lin, T.-Y., Collobert, R., Dollár, P.: Learning to refine object segments. In: Leibe, B., Matas, J., Sebe, N., Welling, M. (eds.) ECCV 2016. LNCS, vol. 9905, pp. 75–91. Springer, Cham (2016). https://doi.org/10.1007/978-3-319-46448-0_5
42. Qin, X., Zhang, Z., Huang, C., Gao, C., Dehghan, M., Jagersand, M.: BASNet: boundary-aware salient object detection. In: CVPR (2019)
43. Remez, T., Huang, J., Brown, M.: Learning to segment via cut-and-paste. In: Ferrari, V., Hebert, M., Sminchisescu, C., Weiss, Y. (eds.) ECCV 2018. LNCS, vol. 11211, pp. 39–54. Springer, Cham (2018). https://doi.org/10.1007/978-3-030-01234-2_3
44. Ren, S., He, K., Girshick, R., Sun, J.: Faster R-CNN: towards real-time object detection with region proposal networks. In: NeurIPS (2015)
45. Rother, C., Kolmogorov, V., Blake, A.: Grabcut: interactive foreground extraction using iterated graph cuts. ACM Trans. Graph. (TOG) **23**(3), 309–314 (2004)
46. Tian, Z., Shen, C., Chen, H., He, T.: FCOS: Fully convolutional one-stage object detection. In: ICCV (2019)
47. Vezhnevets, V., Konouchine, V.: GrowCut: interactive multi-label nd image segmentation by cellular automata. In: Proceedings of Graphicon, vol. 1, pp. 150–156 (2005)
48. Wang, W., Zhao, S., Shen, J., Hoi, S.C., Borji, A.: Salient object detection with pyramid attention and salient edges. In: CVPR (2019)

49. Wang, X., Girshick, R., Gupta, A., He, K.: Non-local neural networks. In: CVPR (2018)
50. Wang, X., Cai, Z., Gao, D., Vasconcelos, N.: Towards universal object detection by domain attention. In: CVPR (2019)
51. Xia, G.S., et al.: DOTA: a large-scale dataset for object detection in aerial images. In: CVPR (2018)
52. Xie, E., et al.: Polarmask: single shot instance segmentation with polar representation. In: CVPR (2020)
53. Yan, K., et al.: Deep lesion graphs in the wild: relationship learning and organization of significant radiology image findings in a diverse large-scale lesion database. In: CVPR (2018)
54. Yan, X., Chen, Z., Xu, A., Wang, X., Liang, X., Lin, L.: Meta R-CNN : towards general solver for instance-level low-shot learning. In: ICCV (2019)
55. Yang, S., Luo, P., Loy, C.C., Tang, X.: Wider face: a face detection benchmark. In: CVPR (2016)
56. Yu, C., Wang, J., Peng, C., Gao, C., Yu, G., Sang, N.: Learning a discriminative feature network for semantic segmentation. In: CVPR (2018)
57. Zhang, S., Yan, S., He, X.: LatentGNN: learning efficient non-local relations for visual recognition. In: ICML (2019)
58. Zhao, J.X., Liu, J.J., Fan, D.P., Cao, Y., Yang, J., Cheng, M.M.: EGNet: edge guidance network for salient object detection. In: ICCV (2019)
59. Zhou, Y., Zhu, Y., Ye, Q., Qiu, Q., Jiao, J.: Weakly supervised instance segmentation using class peak response. In: CVPR (2018)

GMNet: Graph Matching Network for Large Scale Part Semantic Segmentation in the Wild

Umberto Michieli[(✉)], Edoardo Borsato, Luca Rossi, and Pietro Zanuttigh

Department of Information Engineering, University of Padova, Padova, Italy
{michieli,borsatoedo,rossiluc,zanuttigh}@dei.unipd.it

Abstract. The semantic segmentation of parts of objects in the wild is a challenging task in which multiple instances of objects and multiple parts within those objects must be detected in the scene. This problem remains nowadays very marginally explored, despite its fundamental importance towards detailed object understanding. In this work, we propose a novel framework combining higher object-level context conditioning and part-level spatial relationships to address the task. To tackle object-level ambiguity, a class-conditioning module is introduced to retain class-level semantics when learning parts-level semantics. In this way, mid-level features carry also this information prior to the decoding stage. To tackle part-level ambiguity and localization we propose a novel adjacency graph based module that aims at matching the relative spatial relationships between ground truth and predicted parts. The experimental evaluation on the Pascal-Part dataset shows that we achieve state-of-the-art results on this task.

Keywords: Part parsing · Semantic segmentation · Graph matching · Deep learning

1 Introduction

Semantic segmentation is a wide research field and a huge number of approaches have been proposed for this task [5,30,49]. The segmentation and labeling of parts of objects can be regarded as a special case of semantic segmentation that focuses on parts decomposition. The information about parts provides a richer representation for many fine-grained tasks, such as pose estimation [12,47], category detection [2,8,48], fine-grained action detection [42] and image classification [22,39]. However, current approaches for semantic segmentation are not optimized to distinguish between different semantic parts since corresponding parts in different semantic classes often share similar appearance. Additionally,

Electronic supplementary material The online version of this chapter (https://doi.org/10.1007/978-3-030-58598-3_24) contains supplementary material, which is available to authorized users.

A. Vedaldi et al. (Eds.): ECCV 2020, LNCS 12353, pp. 397–414, 2020.
https://doi.org/10.1007/978-3-030-58598-3_24

they only capture limited local context while the precise localization of semantic part layouts and their interactions requires a wider perspective of the image. Thus, it is not sufficient to take standard semantic segmentation methods and treat each part as an independent class. In the literature, object-level semantic segmentation has been extensively studied. Part parsing, instead, has only been marginally explored in the context of a few specific single-class objects, such as humans [14,26,46,52], cars [31,38] and animals [19,40,41]. Multi-class part-based semantic segmentation has only been considered in a recent work [51], due to the challenging scenario of part-level as well as object-level ambiguities. Here, we introduce an approach dealing with the semantic segmentation of an even larger set of parts and we demonstrate that the proposed methodology is able to deal with a large amount of parts contained in the scenes.

Nowadays, one of the most active research directions is the transfer of previous knowledge, acquired on a different but related task, to a new situation. Different interpretations may exist to this regard. In the class-incremental task, the learned model is updated to perform a new task whilst preserving previous capabilities: many methods have been proposed for image classification [11,23,36], object detection [37] and semantic segmentation [33,34]. Another aspect regards the coarse-to-fine refinement at the semantic level, in which previous knowledge acquired on a coarser task is exploited to perform a finer task [20,32,44]. In this paper, instead, we investigate the coarse-to-fine refinement at the spatial level, in which object-level classes are split into their respective parts [41,43,51].

More precisely, we investigate the multi-object and multi-part parsing in the wild, which simultaneously handles all semantic objects and parts within each object in the scene. Even strong recent baselines for semantic segmentation, such as FCN [30], SegNet [3], PSPNet [49] or Deeplab [5,6], face additional challenges when dealing with this task, as shown in [51]. In particular, the simultaneous appearance of multiple objects and the inter-class ambiguity may cause inaccurate boundary localization and severe classification errors. For instance, animals often have homogeneous appearance due to furs on the whole body. Additionally, the appearance of some parts over multiple object classes may be very similar, such as cow legs and sheep legs. Current algorithms heavily suffer from these aspects. To address object-level ambiguity, we propose an object-level conditioning to serve as guidance for part parsing within the object. An auxiliary reconstruction module from parts to objects further penalize predictions of parts in regions occupied by an object which does not contain the predicted parts within it. At the same time, to tackle part-level ambiguity, we introduce a graph-matching module to preserve the relative spatial relationships between ground truth and predicted parts.

When people look at scenes, they tend to locate first the objects and then to refine them via semantic part parsing [43]. This is the same rationale for our class-conditioning approach, which consists of an approach to refine parts localization exploiting previous knowledge. In particular, the object-level predictions of the model serve as a conditioning term for the decoding stage on the part-level. The predictions are processed via an object-level semantic embedding Convolutional Neural Network (CNN) and its features are concatenated with the ones produced by the encoder of the part-level segmentation network. The extracted features are enriched with this type of information prior, guiding the output of the part-level decoding stage. We further propose to address part-level ambiguity via a novel graph-matching technique applied to the segmentation maps. A couple of adjacency graphs are built from neighboring parts both from the ground-truth and from the predicted segmentation maps. Such graphs are weighted with the normalized number of adjacent pixels to represent the strength of connection between the parts. Then, a novel loss function is designed to enforce their similarity. These provisions allow the architecture to discover the differences in appearance between different parts within a single object, and at the same time to avoid the ambiguity across similar object categories.

The main contributions of this paper can be summarized as follows:

- We tackle the challenging multi-class part parsing via an object-level semantic embedding network conditioning the part-level decoding stage.
- We introduce a novel graph-matching module to guide the learning process toward accurate relative localization of semantic parts.
- Our approach (GMNet) achieves new state-of-the-art performance on multi-object part parsing on the Pascal-Part dataset [8]. Moreover, it scales well to large sets of parts.

2 Related Work

Semantic Segmentation is one of the key tasks for automatic scene understanding. Current techniques are based on the Fully Convolutional Network (FCN) framework [30], which firstly enabled accurate and end-to-end semantic segmentation. Recent works based on FCN, such as SegNet [3], PSPNet [49] and Deeplab [5,6], are typically regarded as the state-of-the-art architectures for semantic segmentation. Some recent reviews on the topic are [18,28].

Single-Object Part Parsing has been actively investigated in the recent literature. However, most previous work assumes images containing only the considered object, well-localized beforehand and with no occlusions. Single-object parts parsing has been applied to animals [40], cars [14,31,38] and humans parsing [14,26,46,52]. Traditional deep neural network architectures may also be applied to part parsing regarding each semantic part as a separate class label. However, such strategies suffer from the high similar appearance between parts and from large scale variations of objects and parts. Some coarse-to-fine strategies have been proposed to tackle this issue. Hariharan et al. [20] propose to sequentially perform object detection, object segmentation and part segmentation with different architectures. However, there are some limitations, in particular the complexity of the training and the error propagation throughout the pipeline. An upgraded version of the framework has been presented in [43], where the same structure is employed for the three networks and an automatic adaptation to the size of the object is introduced. In [41] a two-channels FCN is employed to jointly infer object and part segmentation for animals. However, it uses only a single-scale network not capturing small parts and a fully connected CRF is used as post-processing technique to explore the relationship between parts and body to perform the final prediction. In [7] an attention mechanism that learns to softly weight the multi-scale features at each pixel location is proposed.

Some approaches resort to structure-based methodologies, e.g. compositional, to model part relations [16,24,25,27,40,41]. Wang et al. [40] propose a model to learn a mixture of compositional models under various poses and viewpoints for certain animal classes. In [24] a self-supervised structure-sensitive learning approach is proposed to constrain human pose structures into parsing results. In [25,27] graph LSTMs are employed to refine the parsing results of initial over-segmented superpixel maps. Pose estimation is also useful for part parsing task [16,24,35,44,50]. In [44], the authors refine the segmentation maps by supervised pose estimation. In [35] a mutual learning model is built for pose estimation and part segmentation. In [16], the authors exploit anatomical similarity among humans to transfer the parsing results of a person to another person with similar pose. In [50] multi-scale features aggregation at each pixel is combined with a self-supervised joint loss to further improve the feature discriminative capacity. Other approaches utilize tree-based approach to hierarchically partition the parts [31, 45]. Lu et al. [31] propose a method based on tree-structured graphical models from different viewpoints combined with segment appearance consistency for part parsing. Xia et al. [45] firstly generate part segment proposals and then infer the best ensemble of parts-segment through and-or graphs.

Even though single-object part parsing has been extensively studied so far, **Multi-Object and Multi-Part Parsing** has been considered only recently [51]. In this setup, most previous techniques fail struggling with objects that were not previously well-localized, isolated and with no occlusions. Zhao et al. in [51] tackle this task via a joint parsing framework with boundary and semantic awareness for enhanced part localization and object-level guidance. Part boundaries are detected at early stages of feature extraction and then used in an attention mechanism to emphasize the features along the boundaries at the decoding stage. An additional attention module is employed to perform channel selection and is supervised by a supplementary branch predicting the semantic object classes.

3 Method

When we look at images, we often firstly locate the objects and then the more detailed task of semantic part parsing is addressed using mainly two priors: (1) object-level information and (2) relative spatial relationships among parts. Following this rationale, the semantic parts parsing is supported by the information coming from an initial prediction of the coarse object-level set of classes and by a graph-matching strategy working at the parts-level.

An overview of our framework is shown in Fig. 1. We employ two semantic segmentation networks $\mathcal{A}_o$ and $\mathcal{A}_p$ trained for the objects-level and parts-level task respectively, together with a semantic embedding network $\mathcal{S}$ transferring and processing the information of the first network to the second to address the object-level prior. This novel coarse-to-fine strategy to gain insights into

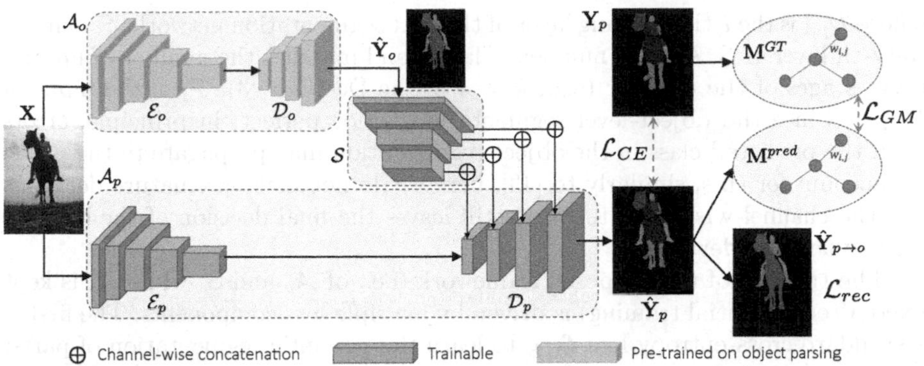

Fig. 1. Architecture of the proposed Graph Matching Network (GMNet) approach. A semantic embedding network takes as input the object-level segmentation map and acts as high level conditioning when learning the semantic segmentation of parts. On the right, a reconstruction loss function rearranges parts into objects and the graph matching module aligns the relative spatial relationships between ground truth and predicted parts.

parts detection will be the subject of this section. Furthermore, we account for the second prior exploiting an adjacency graph structure to mimic the spatial relationship between neighboring parts to allow for a general overview of the semantic parts as described in Sect. 4.

The semantic segmentation networks have an autoencoder structure and can be written as the composition of an encoder and a decoder as $\mathcal{A}_o = \{\mathcal{E}_o, \mathcal{D}_o\}$ and $\mathcal{A}_p = \{\mathcal{E}_p, \mathcal{D}_p\}$ for the object-level and part-level networks, respectively. We employ the Deeplab-v3 [6] segmentation network with Resnet-101 [21] as encoder. The network $\mathcal{A}_o$ is trained using the object-level ground truth labels and then kept fixed. It extracts object-level segmentation maps which serve as a guidance for the decoder of the parts-level network $\mathcal{D}_p$, in order to avoid the ambiguity across similar object categories. We achieve this behavior by feeding the output maps of $\mathcal{A}_o$ to an object-level semantic embedding network. In this work, we used a CNN (denoted with $\mathcal{S}$) formed by a cascade of 4 convolutional layers with stride of 2, square kernel sizes of 7, 5, 3, 3, and channel sizes of 128, 256, 512, 1024.

The parts-level semantic segmentation network $\mathcal{A}_p$ has the same encoder architecture of $\mathcal{A}_o$. Its decoder $\mathcal{D}_p$, instead, merges the features computed on the RGB image and the ones computed on the object-level predicted map via multiple channel-wise concatenations. More in detail, each layer of the decoder considers a different resolution and its feature maps are concatenated with the layer at corresponding resolution of $\mathcal{S}$. In this way, the combination is performed at multiple resolutions in the feature space to achieve higher scale invariance as shown in Fig. 1.

Formally, given an input RGB image $\mathbf{X} \in \mathbb{R}^{W \times H}$, the concatenation between part and object-level aware features is formulated as:

$$\mathcal{F}_i(\mathbf{X}) = \mathcal{D}_{p,i}(\mathbf{X}) \oplus \mathcal{S}_{k+1-i}(\mathcal{A}_o(\mathbf{X})) \qquad i = 1, ..., k \tag{1}$$

where $\mathcal{D}_{p,i}$ is the i-th decoding layer of the part segmentation network, $\mathcal{S}_i$ denotes the i-th layer of $\mathcal{S}$, k is the number of layers and matches the number of upsampling stages of the decoder (e.g., $k = 4$ in the Deeplab-v3), $\mathcal{F}_i$ is the input of $\mathcal{D}_{p,i+1}$. Since the object-level segmentation is not perfect, in principle, errors from the predicted class in the object segmentation may propagate to the parts. To account for this, similarly to [43], here we do not make premature decisions but the channel-wise concatenation still leaves the final decision of the labeling task to the decoder.

The training of the proposed framework (i.e., of $\mathcal{A}_p$ and $\mathcal{S}$, while $\mathcal{A}_o$ is kept fixed after the initial training) is driven by multiple loss components. The first is a standard cross-entropy loss $\mathcal{L}_{CE}$ to learn the semantic segmentation of parts:

$$\mathcal{L}_{CE} = \sum_{c_p=1}^{N_p} \mathbf{Y}_p[c_p] \cdot \log\left(\hat{\mathbf{Y}}_p[c_p]\right) \tag{2}$$

where $\mathbf{Y}_p$ is the one-hot encoded ground truth map, $\hat{\mathbf{Y}}_p$ is the predicted map, c_p is the part-class index and N_p is the number of parts.

The object-level semantic embedding network is further guided by a reconstruction module that rearranges parts into objects. This is done applying a cross-entropy loss between object-level one-hot encoded ground truth maps $\mathbf{Y}_o$ and the summed probability $\hat{\mathbf{Y}}_{p\to o}$ derived from the part-level prediction. More formally, defining l as the parts-to-objects mapping such that object j contains parts from index $l[j-1]+1$ to $l[j]$, we can write the summed probability as:

$$\hat{\mathbf{Y}}_{p\to o}[j] = \sum_{i=l[j-1]+1,\ldots,l[j]} \hat{\mathbf{Y}}_p[i] \qquad j = 1, \ldots, N_o \tag{3}$$

where N_o is the number of object-level classes and $l[0] = 0$. Then, we define the reconstruction loss as:

$$\mathcal{L}_{rec} = \sum_{c_o=1}^{N_o} \mathbf{Y}_o[c_o] \cdot \log\left(\hat{\mathbf{Y}}_{p\to o}[c_o]\right) \tag{4}$$

The auxiliary reconstruction function $\mathcal{L}_{rec}$ acts differently from the usual cross-entropy loss on the parts $\mathcal{L}_{CE}$. While $\mathcal{L}_{CE}$ penalizes wrong predictions of parts in all the portions of the image, $\mathcal{L}_{rec}$ only penalizes for part-level predictions located outside the respective object-level class. In other words, the event of predicting parts outside the respective object-level class is penalized by both the losses. Instead, parts predicted within the object class are penalized only by $\mathcal{L}_{CE}$, i.e., they are considered as a less severe type of error since, in this case, parts only need to be properly localized inside the object-level class.

4 Graph-Matching for Semantic Parts Localization

Providing global context information and disentangling relationships is useful to distinguish fine-grained parts. For instance, upper and lower arms share highly

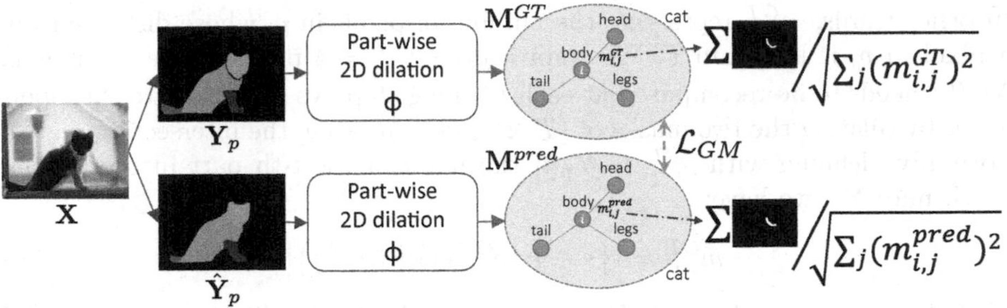

Fig. 2. Overview of the graph matching module. In this case, cat's head and body would be considered as detached without the proper morphological dilation over the parts.

similar appearance. To differentiate between them, global and reciprocal information, like the relationship with neighboring parts, provides an effective context prior. Hence, to further enhance the accuracy of part parsing, we tackle part-level ambiguity and localization by proposing a novel module based on an adjacency graph that matches the parts spatial relationships between ground truth and predicted parts. More in detail, the graphs capture the adjacency relationships between each couple of parts and then we enforce the matching between the ground truth and predicted graph through an additional loss term. Although graph matching is a very well studied problem [13,29], it has never been applied to this context before, i.e. as a loss function to drive deep learning architectures for semantic segmentation. The only other attempt to design a graph matching loss is [9], which however deals with a completely different task, i.e., domain adaptation in classification, and has a different interpretation of the graph, that measures the matching between the source and target domains.

An overview of this module is presented in Fig. 2. Formally, we represent the graphs using two (square) weighted adjacency matrices of size N_p:

$$\tilde{\mathbf{M}}^{GT} = \left\{\tilde{m}_{i,j}^{GT}\right\}_{\substack{i=1,...,N_p\\ j=1,...,N_p}} \qquad \tilde{\mathbf{M}}^{pred} = \left\{\tilde{m}_{i,j}^{pred}\right\}_{\substack{i=1,...,N_p\\ j=1,...,N_p}} \tag{5}$$

The first matrix ($\tilde{\mathbf{M}}^{GT}$) contains the adjacency information computed on ground truth data, while the second ($\tilde{\mathbf{M}}^{pred}$) has the same information computed on the predicted segmentation maps. Each element of the matrices provide a measure of how close the two parts p_i and p_j are in the ground truth and in the predicted segmentation maps, respectively. We do not consider self-connections, hence $\tilde{m}_{i,i}^{GT} = \tilde{m}_{i,i}^{pred} = 0$ for $i = 1, ..., N_p$. To measure the closeness between couples of parts, that is a hint of the strength of connection between them, we consider weighted matrices where each entry $\tilde{m}_{i,j}$ depends on the length of the contour in common between them. Actually, to cope for some inaccuracies inside the dataset where some adjacent parts are separated by thin background regions, the entries of the matrices are the counts of pixels belonging to one part with a distance less or equal than T from a sample belonging to the other part.

In other words, $\tilde{m}_{i,j}^{GT}$ represents the number of pixels in p_i whose distance from a pixel in p_j is less than T. We empirically set $T = 4$ pixels. Since the matrix $\tilde{\mathbf{M}}^{pred}$ needs to be recomputed at each training step, we approximate this operation by dilating the two masks of $\lceil T/2 \rceil$ and computing the intersecting region. Formally, defining with $p_i^{GT} = \mathbf{Y}_p[i]$ the mask of the i-th part in the ground truth map $\mathbf{Y}_p$, we have:

$$\tilde{m}_{i,j}^{GT} = \left|\left\{s \in \Phi\left(p_i^{GT}\right) \cap \Phi\left(p_j^{GT}\right)\right\}\right| \tag{6}$$

Where s is a generic pixel, $\Phi(\cdot)$ is the morphological 2D dilation operation and $|\cdot|$ is the cardinality of the given set. We apply a row-wise L2 normalization and we obtain a matrix of *proximity ratios* $\mathbf{M}_{[i,:]}^{GT} = \tilde{\mathbf{M}}_{[i,:]}^{GT} / ||\tilde{\mathbf{M}}_{[i,:]}^{GT}||_2$ that measures the flow from the considered part to all the others.

With this definition, non-adjacent parts have 0 as entry. The same approach is used for the adjacency matrix computed on the predicted segmentation map $\mathbf{M}^{pred}$ by substituting p_i^{GT} with $p_i^{pred} = \hat{\mathbf{Y}}_p[i]$.

Then, we simply define the Graph-Matching loss as the Frobenius norm between the two adjacency matrices:

$$\mathcal{L}_{GM} = ||\mathbf{M}^{GT} - \mathbf{M}^{pred}||_F \tag{7}$$

The aim of this loss function is to faithfully maintain the reciprocal relationships between parts. On one hand, disjoint parts are enforced to be predicted as disjoint; on the other hand, neighboring parts are enforced to be predicted as neighboring and to match the proximity ratios.

Summarizing, the overall training objective of our framework is:

$$\mathcal{L} = \mathcal{L}_{CE} + \lambda_1 \mathcal{L}_{rec} + \lambda_2 \mathcal{L}_{GM} \tag{8}$$

where the hyper-parameters λ_1 and λ_2 are used to control the relative contribution of the three losses to the overall objective function.

5 Training of the Deep Learning Architecture

5.1 Multi-part Dataset

For the experimental evaluation of the proposed multi-class part parsing framework we employed the Pascal-Part [8] dataset, which is currently the largest dataset for this purpose. It contains a total of 10103 variable-sized images with pixel-level parts annotation on the 20 Pascal VOC2010 [15] semantic object classes (plus the *background* class). We employ the original split from [8] with 4998 images in the *trainset* for training and 5105 images in the *valset* for testing. We consider two different sets of labels for this dataset. Firstly, following [51], which is the only work dealing with the multi-class part parsing problem, we grouped the original semantic classes into 58 part classes in total. Additionally, to further test our method on a even more challenging scenario, we consider the grouping rules proposed by [17] for part detection that, instead, leads to a larger set of 108 parts.

5.2 Training Details

The modules introduced in this work are agnostic to the underlying network architecture and can be extended to other scenarios. For fair comparison with [51] we employ a Deeplab-v3 [6] architecture with ResNet101 [21] as the backbone. We follow the same training schemes of [5,6,51] and we started from the official TensorFlow [1] implementation of the Deeplab-v3 [4,6]. The ResNet101 was pre-trained on ImageNet [10] and its weights are available at [4]. During training, images are randomly left-right flipped and scaled of a factor from 0.5 to 2 times the original resolution with bilinear interpolation. The results in the testing stage are reported at the original image resolution. The model is trained for $50K$ steps with the base learning rate set to $5 \cdot 10^{-3}$ and decreased with a polynomial decay rule with power 0.9. We employ weight decay regularization of 10^{-4}. The atrous rate in the Atrous Spatial Pyramid Poooling (ASPP) is set to $(6, 12, 18)$ as in [5,51]. We use a batch size of 10 images and we set $\lambda_1 = 10^{-3}$ and $\lambda_2 = 10^{-1}$ to balance part segmentation. For the evaluation metric, we employ the mean Intersection over Union (mIoU) since pixel accuracy is dominated by large regions and little sensitive to the segmentation on many small parts, that are instead the main target of this work. The code and the part labels are publicly available at https://lttm.dei.unipd.it/paper_data/GMNet.

6 Experimental Results

In this section we show the experimental results on the multi-class part parsing task in two different scenarios with 58 and 108 parts respectively. We also present some ablation studies to verify the effectiveness of the proposed methodologies.

6.1 Pascal-Part-58

To evaluate our framework we start from the scenario with 58 parts, i.e., the same experimental setting used in [51]. In Table 1 we compare the proposed model with existing semantic segmentation schemes. As evaluation criteria we

Table 1. IoU results on the Pascal-Part-58 benchmark. mIoU: mean per-part-class IoU. Avg: average per-object-class mIoU.

Method	bgr	aero	bike	bird	boat	bottle	bus	car	cat	chair	cow	d. table	dog	horse	mbike	person	plant	sheep	sofa	train	tv	mIoU	Avg.
SegNet[3]	85.4	13.7	40.7	11.3	21.7	10.7	36.7	26.3	28.5	16.6	8.9	16.6	24.2	18.8	44.7	35.4	16.1	17.3	15.7	41.3	26.1	24.4	26.5
FCN[30]	87.0	33.9	51.5	37.7	47.0	45.3	50.8	39.1	45.2	29.4	31.2	32.5	42.4	42.2	58.2	40.3	38.3	43.4	35.7	66.7	44.2	42.3	44.9
DeepLab[5]	89.8	40.7	58.1	43.8	53.9	44.5	62.1	45.1	52.3	36.6	41.9	38.7	49.5	53.9	66.1	49.0	45.3	45.3	40.5	76.8	56.5	49.9	51.9
BSANet[51]	91.6	**50.0**	65.7	**54.8**	60.2	49.2	70.1	**53.5**	**63.8**	36.5	52.8	43.7	58.3*	**66.0**	71.6*	**58.4**	55.0	49.6	43.1	82.2	61.4	58.2	58.9*
Baseline[6]	91.1	45.7	63.2	49.0	54.4	49.8	67.6	49.2	59.8	35.4	47.6	43.0	54.4	62.0	68.0	55.0	48.9	45.9	43.2	79.6	57.7	54.4	55.7
GMNet	**92.7**	46.7	**66.4**	52.0	**70.0**	**55.7**	**71.1**	52.2	63.2	**51.4**	**54.8**	**51.3**	**59.6**	64.4	**73.9**	56.2	**56.2**	**53.6**	**56.1**	**85.0**	**65.6**	**59.0**	**61.8**

*: values different from [51] since they were wrongly reported in the paper.

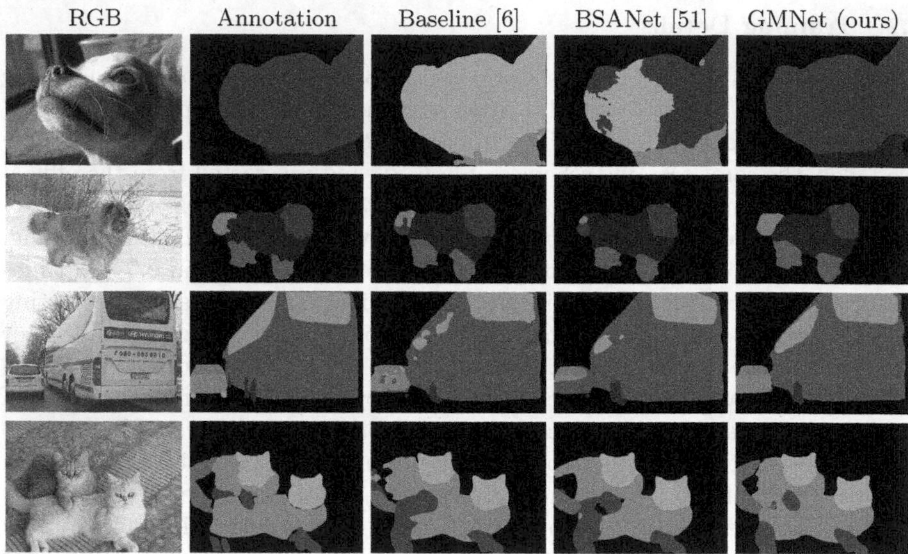

Fig. 3. Segmentation results from the Pascal-Part-58 dataset (*best viewed in colors*). (Color figure online)

employ the mean IoU of all the parts (i.e., mIoU), the average IoU for all the parts belonging to each single object, and the mean of these values (denoted as Avg., i.e., in this case each object has the same weight independently of the number of parts). Part-level metrics are reported in the supplementary material. As expected, traditional semantic segmentation architectures such as FCN [30], SegNet [3] and DeepLab [5] are not able to perform a fully satisfactory part-parsing. We adopt as our baseline network the DeepLab-v3 architecture [6], that is the best performing among the compared standard approaches achieving 54.4% of mIoU. The proposed GMNet approach combining both the object-level semantic embedding and the graph matching module achieves a higher accuracy of 59.0% of mIoU, significantly outperforming all the other methods and in particular the baseline on every class with a gap of 4.6% of mIoU. The only other method specifically addressing part-based semantic segmentation is BSANet [51], which achieves a lower mIoU of 58.2%. Our method achieves higher results over most of the objects, both with many parts (like *cow*, *dog* and *sheep*) and with no or few parts (like *boat*, *bottle*, *chair*, *dining table* and *sofa*).

Some qualitative results are shown in Fig. 3 while additional ones are in the supplementary material. The figure allows to appreciate the effects of the two main contributions, the semantic embedding and the graph matching modules.

From one side, the object-level semantic embedding network brings useful additional information prior to the part-level decoding stage, thus enriching the extracted features to be object discriminative. We can appreciate this aspect from the first and the third row. In the first row, the baseline completely misleads a dog with a cat (light green corresponds to *cat_head* and green to *cat_torso*).

BSANet is able to partially recover the *dog_head* (amaranth corresponds to the proper labeling). Our method, instead, is able to accurately detect and segment the dog parts (*dog_head* in amaranth and *dog_torso* in blue) thanks to object-level priors coming from the semantic embedding module. A similar discussion can be done also on the third image, where the baseline confuses car parts (green corresponds to *window*, aquamarine to *body* and light green to *wheel*) with bus parts (pink is the *window*, brown the *body* and dark green the *wheel*) and BSANet is not able to correct this error. GMNet, instead, can identify the correct object-level class and the respective parts, excluding the very small and challenging *car_wheels*, and at the same time can better segment the *bus_window*.

From the other side, the graph matching module helps in the mutual localization of parts within the same object-level class. The effect of the graph matching module is more evident in the second and fourth row. In the second image, we can verify how both the baseline and BSANet are not able to correctly place the *dog_tail* (in yellow) misleading it with the *dog_head* (in red). Thanks to the graph matching module, GMNet can disambiguate between such parts and correctly exploit their spatial relationship with respect to the *dog_body*. In the fourth image, both the baseline and BSANet tend to overestimate the presence of the *cat_legs* (in dark green) and they miss one *cat_tail*. The constraints on the relative position among the various parts enforced by the graph matching module allow GMNet to properly segment and label the *cat_tail* and to partially correct the estimate of the *cat_legs*.

6.2 Pascal-Part-108

To further verify the robustness and the scalability of the proposed methodology we perform a second set of experiments using an even larger number of parts. The results on the Pascal-Part-108 benchmark are reported in Table 2. Even though we can immediately verify a drop in the overall performance, that is predictable being the task more complex with respect to the previous scenario with an almost double number of parts, we can appreciate that our framework is able to largely surpass both the baseline and [51]. It achieves a mIoU of 45.8%, outperforming the baseline by 4.5% and the other compared standard segmentation networks by an even larger margin. The gain with respect to the main competitor [51]

Table 2. IoU results on the Pascal-Part-108 benchmark. mIoU: mean per-part-class IoU. Avg: average per-object-class mIoU. †: re-trained on the Pascal-Part-108 dataset.

Method	bgr	aero	bike	bird	boat	bottle	bus	car	cat	chair	cow	d. table	dog	horse	mbike	person	plant	sheep	sofa	train	tv	mIoU	Avg.
SegNet[3]	85.3	11.2	32.4	6.3	21.4	10.3	27.9	22.6	22.8	17.0	6.3	12.5	21.1	14.9	12.2	32.2	13.8	12.6	15.2	11.3	27.5	18.6	20.8
FCN[30]	86.8	30.3	35.6	23.6	47.5	44.5	21.3	34.5	35.8	26.6	20.3	24.4	37.7	29.8	14.2	35.6	34.4	28.9	34.0	18.1	45.6	31.6	33.8
Deeplab[5]	90.2	38.3	35.4	29.4	57.0	41.5	27.0	40.1	45.5	36.6	33.3	35.2	41.1	48.8	19.5	40.6	46.0	23.7	40.8	17.5	70.0	35.7	40.8
BSANet† [51]	91.6	45.3	40.9	**41.0**	61.4	48.9	32.2	43.3	50.7	34.1	39.4	45.9	**52.1**	50.0	23.1	52.4	50.6	37.8	44.5	20.7	66.3	42.9	46.3
Baseline [6]	90.9	41.9	44.5	35.3	53.7	47.0	34.1	42.3	49.2	35.4	39.8	33.0	48.2	48.8	23.2	50.4	43.6	35.4	39.2	20.7	60.8	41.3	43.7
GMNet	**92.7**	**48.0**	**46.2**	39.3	**69.2**	**56.0**	**37.0**	**45.3**	**52.6**	**49.1**	**50.6**	**50.6**	52.0	**51.5**	**24.8**	**52.6**	**56.0**	**40.1**	**53.9**	**21.6**	**70.7**	**45.8**	**50.5**

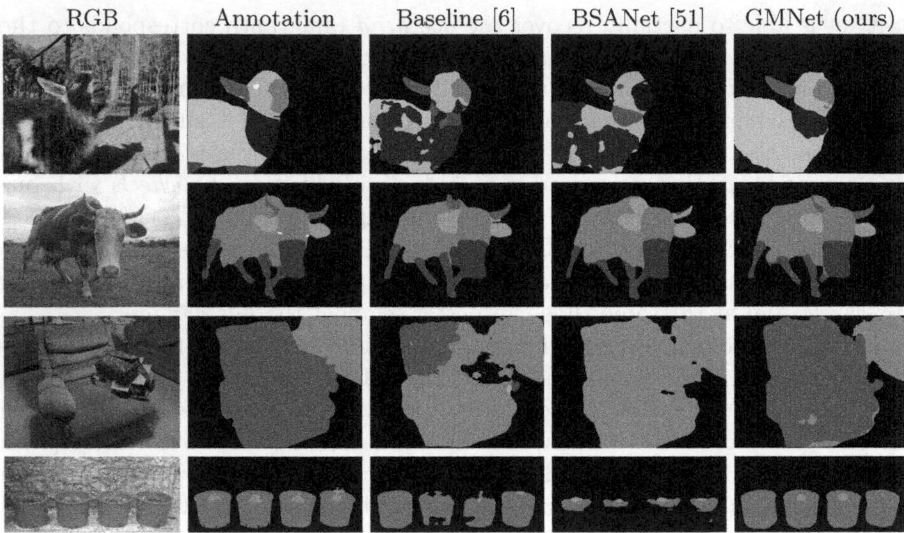

Fig. 4. Segmentation results from the Pascal-Part-108 dataset (*best viewed in colors*). (Color figure online)

is remarkable with a gap of 2.9% of mIoU. In this scenario, indeed, most of the previous considerations holds and are even more evident from the results. The gain in accuracy is stable across the various classes and parts: the proposed framework significantly wins by large margins on almost every per-object-class mIoU. Also for this setup, further results regarding per-part metrics are reported in the supplementary material.

Thanks to the object-level semantic embedding network our model is able to obtain accurate segmentation of all the objects with few or no parts inside, such as *boat, bottle, chair, plant* and *sofa*. On these classes, the gain with respect to [51] ranges from 5.4% for the *plant* class to an impressive 15% on the *chair* class. On the other hand, thanks to the graph matching module, our framework is also able to correctly understand the spatial relationships between small parts, as for example the ones contained in *cat, cow, horse* and *sheep*. Although objects are composed by tiny and difficult parts, the gain with respect to [51] is still significant and ranges between 1.5% on *horse* parts to 11.2% on *cow* ones.

The visual results for some sample scenes presented in Fig. 4 confirm the numerical evaluation (additional samples are shown in the supplementary material). We can appreciate that the proposed method is able to compute accurate segmentation maps both when a few elements or many parts coexist in the scene. More in detail, in the first row we can verify the effectiveness of the object-level semantic embedding in conditioning part parsing. The baseline is not able to localize and segment the body and the neck of the sheep. The BSANet approach [51] achieves even worse segmentation and labeling performance. Such methods mislead the sheep with a dog (in the figure light blue denotes *dog_head*, light purple *dog_neck*, brown *dog_muzzle* and yellow *dog_torso*) or with a cat (purple denotes *cat_torso*). Thanks to the object-level priors, GMNet is able to associate the correct label to each of the parts correctly identifying the sheep as the macro class. In the second row, the effect of the graph matching procedure is more evident. The baseline approach tends to overestimate and badly localize the *cow_horns* (in brown) and BSANet confuses the *cow_horns* with the *cow_ears* (in pink). GMNet, instead, achieves superior results thanks to the graph module which accounts for proper localization and contour shaping of the various parts. In the third row, a scenario with two object-level classes having no sub-parts is reported. Again, we can check how GMNet is able to discriminate between *chair* (in pink) and *sofa* (in light brown). Finally, in the last row we can appreciate how the two parts of the *potted plant* are correctly segmented by GMNet thanks to the semantic embedding module for what concerns object identification and to the graph matching strategy for what concerns small parts localization.

6.3 Ablation Studies

In this section we conduct an accurate investigation of the effectiveness of the various modules of the proposed work on the Pascal-Part-58 dataset.

We start by evaluating the individual impact of the modules and the performance analysis is shown in Table 3. Let us recall that the baseline architecture (i.e., the Deeplab-v3 network trained directly on the 58 parts with only the standard cross-entropy loss enabled) achieves a mIoU of 54.4%. The reconstruction loss on the object-level segmentation maps helps in preserving the object-level shapes rearranging parts into object-level classes and allows to improve the mIoU to 55.2%. The semantic embedding network $\mathcal{S}$ acts as a powerful class-conditioning module to retain object-level semantics when learning parts and allows to obtain a large performance gain: its combination with the reconstruction loss leads to a mIoU of 58.4%. The addition of the graph matching procedure further boost the final accuracy to 59.0% of mIoU. To better understand the contribution of this module we also tried a simpler unweighted graph model whose entries are just binary values representing whether two parts are adjacent or not (column $\mathcal{L}_{GM}^{u}$ in the table). This simplified graph leads to a mIoU of 58.7%, lacking some information about the closeness of adjacent parts.

Then, we present a more accurate analysis of the impact of the semantic embedding module and the results are summarized in Table 4. First of all, the exploitation of the multiple concatenation between features computed by $\mathcal{S}$ and features of $\mathcal{D}_p$ at different resolutions allows object-level embedding at different scales and enhances the scale invariance. Concatenating only the output of $\mathcal{S}$ with the output of $\mathcal{E}_p$ (we refer to this approach with "single concatenation"), the final mIoU slightly decreases to 58.7%. In order to evaluate the usefulness of exploiting features extracted from a CNN, we compared the proposed framework with a variation directly concatenating the output of $\mathcal{E}_p$ with the object-level predicted segmentation maps $\hat{\mathbf{Y}}_o$ after a proper rescaling ("without $\mathcal{S}$"). This approach leads to a quite low mIoU of 55.7%, thus outlining that the embedding network $\mathcal{S}$ is very effective and that a simple stacking of architectures is not the

Table 3. mIoU ablation results on Pascal-Part-58. $\mathcal{L}^u_{GM}$: graph matching with unweighted graph.

$\mathcal{L}_{CE}$	$\mathcal{L}_{rec}$	$\mathcal{S}$	$\mathcal{L}^u_{GM}$	$\mathcal{L}_{GM}$	mIoU
✓					54.4
✓	✓				55.2
✓	✓	✓			58.4
✓	✓	✓	✓		58.7
✓	✓	✓		✓	**59.0**

Table 4. mIoU on Pascal-Part-58 with different configurations for the object-level semantic embedding.

Method	mIoU
Single concatenation	58.7
Without $\mathcal{S}$	55.7
$\mathcal{E}_o$ conditioning	55.7
GMNet	**59.0**
With objects GT	65.6

best option for our task. Additionally, we considered also the option of directly feeding object-level features to the part parsing decoder, i.e., we tried to concatenate the output of $\mathcal{E}_o$ with the output of $\mathcal{E}_p$ and feed these features to $\mathcal{D}_p$ ("$\mathcal{E}_o$ conditioning"). Conditioning the part parsing with this approach does not bring in sufficient object-level indication and it leads to a mIoU of 55.7%, which is significantly lower than the complete proposed framework (59.0%). Finally, to estimate an upper limit of the performance gain coming from the semantic embedding module we fed the object-level semantic embedding network $\mathcal{S}$ with object-level ground truth annotations $\mathbf{Y}_o$ ("with objects GT"), instead of the predictions $\hat{\mathbf{Y}}_o$ (notice that the network $\mathcal{A}_o$ has good performance but introduces some errors, as it has 71.5% of mIoU at object-level). In this case, a mIoU of 65.6% is achieved, showing that there is still room for improvement.

We conclude remarking that GMNet achieves almost always higher accuracy than the starting baseline, even if small and unstructured parts remain the most challenging to be detected. Furthermore, the gain depends also on the amount of spatial relationships that can be exploited.

7 Conclusion

In this paper, we tackled the emerging task of multi-class semantic part segmentation. We propose a novel coarse-to-fine strategy where the features extracted from a semantic segmentation network are enriched with object-level semantics when learning part-level segmentation. Additionally, we designed a novel adjacency graph-based module that aims at matching the relative spatial relationships between ground truth and predicted parts which has shown large improvements particularly on small parts. Combining the proposed methodologies we were able to achieve state-of-the-art results in the challenging task of multi-object part parsing both at a moderate scale and at a larger one.

Further research will investigate the extension of the proposed modules to other scenarios. We will also consider the explicit embedding into the proposed framework of the edge information coming from part-level and object-level segmentation maps. Novel graph representations better capturing part relationships and different matching functions will be investigated.

References

1. Abadi, M., et al.: Tensorflow: a system for large-scale machine learning. In: 12th USENIX Symposium on Operating Systems Design and Implementation (OSDI), pp. 265–283 (2016)
2. Azizpour, H., Laptev, I.: Object detection using strongly-supervised deformable part models. In: Fitzgibbon, A., Lazebnik, S., Perona, P., Sato, Y., Schmid, C. (eds.) ECCV 2012. LNCS, vol. 7572, pp. 836–849. Springer, Heidelberg (2012). https://doi.org/10.1007/978-3-642-33718-5_60
3. Badrinarayanan, V., Kendall, A., Cipolla, R.: Segnet: a deep convolutional encoder-decoder architecture for image segmentation. IEEE Trans. Pattern Anal. Mach. Intell. (PAMI) **39**(12), 2481–2495 (2017)
4. Chen, L.C.: DeepLab official TensorFlow implementation. https://github.com/tensorflow/models/tree/master/research/deeplab. Accessed 01 Mar 2020
5. Chen, L.C., Papandreou, G., Kokkinos, I., Murphy, K., Yuille, A.L.: Deeplab: semantic image segmentation with deep convolutional nets, atrous convolution, and fully connected CRFs. IEEE Trans. Pattern Anal. Mach. Intell. (PAMI) **40**(4), 834–848 (2018)
6. Chen, L.C., Papandreou, G., Schroff, F., Adam, H.: Rethinking atrous convolution for semantic image segmentation. arXiv preprint arXiv:1706.05587 (2017)
7. Chen, L.C., Yang, Y., Wang, J., Xu, W., Yuille, A.L.: Attention to scale: scale-aware semantic image segmentation. In: Proceedings of IEEE Conference on Computer Vision and Pattern Recognition (CVPR), pp. 3640–3649 (2016)
8. Chen, X., Mottaghi, R., Liu, X., Fidler, S., Urtasun, R., Yuille, A.: Detect what you can: detecting and representing objects using holistic models and body parts. In: Proceedings of IEEE Conference on Computer Vision and Pattern Recognition (CVPR), pp. 1971–1978 (2014)
9. Das, D., Lee, C.G.: Unsupervised domain adaptation using regularized hyper-graph matching. In: Proceedings of IEEE International Conference on Image Processing (ICIP), pp. 3758–3762. IEEE (2018)
10. Deng, J., Dong, W., Socher, R., Li, L.J., Li, K., Fei-Fei, L.: Imagenet: a large-scale hierarchical image database. In: Proceedings of IEEE Conference on Computer Vision and Pattern Recognition (CVPR), pp. 248–255. IEEE (2009)
11. Dhar, P., Singh, R.V., Peng, K.C., Wu, Z., Chellappa, R.: Learning without memorizing. In: Proceedings of IEEE Conference on Computer Vision and Pattern Recognition (CVPR), pp. 5138–5146 (2019)
12. Dong, J., Chen, Q., Shen, X., Yang, J., Yan, S.: Towards unified human parsing and pose estimation. In: Proceedings of IEEE Conference on Computer Vision and Pattern Recognition (CVPR), pp. 843–850 (2014)
13. Emmert-Streib, F., Dehmer, M., Shi, Y.: Fifty years of graph matching, network alignment and network comparison. Inf. Sci. **346**, 180–197 (2016)
14. Eslami, S., Williams, C.: A generative model for parts-based object segmentation. In: Neural Information Processing Systems (NeurIPS), pp. 100–107 (2012)
15. Everingham, M., Van Gool, L., Williams, C.K., Winn, J., Zisserman, A.: The pascal visual object classes (VOC) challenge. Int. J. Comput. Vis. (IJCV) **88**(2), 303–338 (2010)

16. Fang, H.S., Lu, G., Fang, X., Xie, J., Tai, Y.W., Lu, C.: Weakly and semi supervised human body part parsing via pose-guided knowledge transfer. In: Proceedings of IEEE Conference on Computer Vision and Pattern Recognition (CVPR) (2018)
17. Gonzalez-Garcia, A., Modolo, D., Ferrari, V.: Do semantic parts emerge in convolutional neural networks? Int. J. Comput. Vis. (IJCV) **126**(5), 476–494 (2018)
18. Guo, Y., Liu, Y., Georgiou, T., Lew, M.S.: A review of semantic segmentation using deep neural networks. Int. J. Multimedia Inf. Retrieval **7**(2), 87–93 (2018)
19. Haggag, H., Abobakr, A., Hossny, M., Nahavandi, S.: Semantic body parts segmentation for quadrupedal animals. In: 2016 IEEE International Conference on Systems, Man, and Cybernetics (SMC), pp. 000855–000860 (2016)
20. Hariharan, B., Arbeláez, P., Girshick, R., Malik, J.: Hypercolumns for object segmentation and fine-grained localization. In: Proceedings of IEEE Conference on Computer Vision and Pattern Recognition (CVPR), pp. 447–456 (2015)
21. He, K., Zhang, X., Ren, S., Sun, J.: Deep residual learning for image recognition. In: Proceedings of IEEE Conference on Computer Vision and Pattern Recognition (CVPR), pp. 770–778 (2016)
22. Krause, J., Jin, H., Yang, J., Fei-Fei, L.: Fine-grained recognition without part annotations. In: Proceedings of IEEE Conference on Computer Vision and Pattern Recognition (CVPR), pp. 5546–5555 (2015)
23. Li, Z., Hoiem, D.: Learning without forgetting. IEEE Trans. Pattern Anal. Mach. Intell. (PAMI) **40**(12), 2935–2947 (2018)
24. Liang, X., Gong, K., Shen, X., Lin, L.: Look into person: joint body parsing & pose estimation network and a new benchmark. IEEE Trans. Pattern Anal. Mach. Intell. (PAMI) **41**(4), 871–885 (2018)
25. Liang, X., Lin, L., Shen, X., Feng, J., Yan, S., Xing, E.P.: Interpretable structure-evolving LSTM. In: Proceedings of IEEE Conference on Computer Vision and Pattern Recognition (CVPR), pp. 1010–1019 (2017)
26. Liang, X., et al.: Deep human parsing with active template regression. IEEE Trans. Pattern Anal. Mach. Intell. (PAMI) **37**(12), 2402–2414 (2015)
27. Liang, X., Shen, X., Feng, J., Lin, L., Yan, S.: Semantic object parsing with graph LSTM. In: Leibe, B., Matas, J., Sebe, N., Welling, M. (eds.) ECCV. pp. 125–143. Springer, Heidelberg (2016). https://doi.org/10.1007/978-3-319-46448-0_8
28. Liu, X., Deng, Z., Yang, Y.: Recent progress in semantic image segmentation. Artif. Intell. Rev. **52**(2), 1089–1106 (2019)
29. Livi, L., Rizzi, A.: The graph matching problem. Pattern Anal. Appl. **16**(3), 253–283 (2013)
30. Long, J., Shelhamer, E., Darrell, T.: Fully convolutional networks for semantic segmentation. In: Proceedings of IEEE Conference on Computer Vision and Pattern Recognition (CVPR), pp. 3431–3440 (2015)
31. Lu, W., Lian, X., Yuille, A.: Parsing semantic parts of cars using graphical models and segment appearance consistency. arXiv preprint arXiv:1406.2375 (2014)
32. Mel, M., Michieli, U., Zanuttigh, P.: Incremental and multi-task learning strategies for coarse-to-fine semantic segmentation. Technologies **8**(1), 1 (2020)
33. Michieli, U., Zanuttigh, P.: Incremental learning techniques for semantic segmentation. In: Proceedings of IEEE Conference on Computer Vision and Pattern Recognition Workshops (CVPRW) (2019)
34. Michieli, U., Zanuttigh, P.: Knowledge distillation for incremental learning in semantic segmentation. arXiv preprint arXiv:1911.03462 (2020)

35. Nie, X., Feng, J., Yan, S.: Mutual learning to adapt for joint human parsing and pose estimation. In: Ferrari, V., Hebert, M., Sminchisescu, C., Weiss, Y. (eds.) ECCV 2018. LNCS, vol. 11209, pp. 519–534. Springer, Cham (2018). https://doi.org/10.1007/978-3-030-01228-1_31
36. Rebuffi, S.A., Kolesnikov, A., Sperl, G., Lampert, C.H.: iCaRL: incremental classifier and representation learning. In: Proceedings of IEEE Conference on Computer Vision and Pattern Recognition (CVPR), pp. 2001–2010 (2017)
37. Shmelkov, K., Schmid, C., Alahari, K.: Incremental learning of object detectors without catastrophic forgetting. In: Proceedings of International Conference on Computer Vision (ICCV), pp. 3400–3409 (2017)
38. Song, Y., Chen, X., Li, J., Zhao, Q.: Embedding 3D geometric features for rigid object part segmentation. In: Proceedings of International Conference on Computer Vision (ICCV), pp. 580–588 (2017)
39. Sun, J., Ponce, J.: Learning discriminative part detectors for image classification and cosegmentation. In: Proceedings of International Conference on Computer Vision (ICCV), pp. 3400–3407 (2013)
40. Wang, J., Yuille, A.L.: Semantic part segmentation using compositional model combining shape and appearance. In: Proceedings of IEEE Conference on Computer Vision and Pattern Recognition (CVPR), pp. 1788–1797 (2015)
41. Wang, P., Shen, X., Lin, Z., Cohen, S., Price, B., Yuille, A.L.: Joint object and part segmentation using deep learned potentials. In: Proceedings of International Conference on Computer Vision (ICCV), pp. 1573–1581 (2015)
42. Wang, Y., Tran, D., Liao, Z., Forsyth, D.: Discriminative hierarchical part-based models for human parsing and action recognition. J. Mach. Learn. Res. **13**(Oct), 3075–3102 (2012)
43. Xia, F., Wang, P., Chen, L.-C., Yuille, A.L.: Zoom better to see clearer: human and object parsing with hierarchical auto-zoom net. In: Leibe, B., Matas, J., Sebe, N., Welling, M. (eds.) ECCV 2016. LNCS, vol. 9909, pp. 648–663. Springer, Cham (2016). https://doi.org/10.1007/978-3-319-46454-1_39
44. Xia, F., Wang, P., Chen, X., Yuille, A.L.: Joint multi-person pose estimation and semantic part segmentation. In: Proceedings of IEEE Conference on Computer Vision and Pattern Recognition (CVPR), pp. 6769–6778 (2017)
45. Xia, F., Zhu, J., Wang, P., Yuille, A.: Pose-guided human parsing with deep learned features. arXiv preprint arXiv:1508.03881 (2015)
46. Yamaguchi, K., Kiapour, M.H., Ortiz, L.E., Berg, T.L.: Parsing clothing in fashion photographs. In: Proceedings of IEEE Conference on Computer Vision and Pattern Recognition (CVPR), pp. 3570–3577. IEEE (2012)
47. Yang, Y., Ramanan, D.: Articulated pose estimation with flexible mixtures-of-parts. In: Proceedings of IEEE Conference on Computer Vision and Pattern Recognition (CVPR), pp. 1385–1392 (2011)
48. Zhang, N., Donahue, J., Girshick, R., Darrell, T.: Part-based R-CNNs for fine-grained category detection. In: Fleet, D., Pajdla, T., Schiele, B., Tuytelaars, T. (eds.) ECCV 2014. LNCS, vol. 8689, pp. 834–849. Springer, Cham (2014). https://doi.org/10.1007/978-3-319-10590-1_54
49. Zhao, H., Shi, J., Qi, X., Wang, X., Jia, J.: Pyramid scene parsing network. In: Proceedings of IEEE Conference on Computer Vision and Pattern Recognition (CVPR), pp. 2881–2890 (2017)
50. Zhao, J., et al.: Self-supervised neural aggregation networks for human parsing. In: Proceedings of IEEE Conference on Computer Vision and Pattern Recognition Workshops (CVPRW), pp. 7–15 (2017)

51. Zhao, Y., Li, J., Zhang, Y., Tian, Y.: Multi-class part parsing with joint boundary-semantic awareness. In: Proceedings of International Conference on Computer Vision (ICCV), pp. 9177–9186 (2019)
52. Zhu, L.L., Chen, Y., Lin, C., Yuille, A.: Max margin learning of hierarchical configural deformable templates (HCDTs) for efficient object parsing and pose estimation. Int. J. Comput. Vis. (IJCV) **93**(1), 1–21 (2011)

Event-Based Asynchronous Sparse Convolutional Networks

Nico Messikommer[1,2(✉)], Daniel Gehrig[1,2], Antonio Loquercio[1,2], and Davide Scaramuzza[1,2]

[1] Department of Informatics, University of Zurich, Zürich, Switzerland
nmessi@ifi.uzh.ch

[2] Department of Neuroinformatics, University of Zurich and ETH Zurich, Zürich, Switzerland

Abstract. Event cameras are bio-inspired sensors that respond to per-pixel brightness changes in the form of asynchronous and sparse "events". Recently, pattern recognition algorithms, such as learning-based methods, have made significant progress with event cameras by converting events into *synchronous* dense, image-like representations and applying traditional machine learning methods developed for standard cameras. However, these approaches discard the spatial and temporal sparsity inherent in event data at the cost of higher computational complexity and latency. In this work, we present a general framework for converting models trained on synchronous image-like event representations into *asynchronous* models with identical output, thus directly leveraging the intrinsic asynchronous and sparse nature of the event data. We show both theoretically and experimentally that this drastically reduces the computational complexity and latency of high-capacity, synchronous neural networks without sacrificing accuracy. In addition, our framework has several desirable characteristics: (i) it exploits spatio-temporal sparsity of events explicitly, (ii) it is agnostic to the event representation, network architecture, and task, and (iii) it does not require any train-time change, since it is compatible with the standard neural networks' training process. We thoroughly validate the proposed framework on two computer vision tasks: object detection and object recognition. In these tasks, we reduce the computational complexity up to 20 times with respect to high-latency neural networks. At the same time, we outperform state-of-the-art *asynchronous* approaches up to 24% in prediction accuracy.

Keywords: Deep Learning: Applications, Methodology, and Theory, Low-level Vision

N. Messikommer and D. Gehrig—Equal contribution.

Electronic supplementary material The online version of this chapter (https://doi.org/10.1007/978-3-030-58598-3_25) contains supplementary material, which is available to authorized users.

A. Vedaldi et al. (Eds.): ECCV 2020, LNCS 12353, pp. 415–431, 2020.
https://doi.org/10.1007/978-3-030-58598-3_25

Multimedia Material

The code of this project is available at https://github.com/uzh-rpg/rpg_asynet. Additional qualitative results can be viewed in this video: https://youtu.be/g_I5k_QFQJA.

1 Introduction

Event cameras are *asynchronous* sensors that operate radically differently from traditional cameras. Instead of capturing dense brightness images at a fixed rate, event cameras measure brightness *changes* (called *events*) for each pixel independently. Therefore, they sample light based on the scene dynamics, rather than on a clock with no relation to the viewed scene. By only measuring brightness changes, event cameras generate an asynchronous signal both sparse in space and time, usually encoding moving image edges[1] [1]. Consequently, they automatically discard redundant visual information and greatly reduce bandwidth. In addition, event cameras possess appealing properties, such as a very high dynamic range, high temporal resolution (in the order of microseconds), and low power consumption.

Due to the sparse and asynchronous nature of events, traditional computer vision algorithms cannot be applied, prompting the development of novel approaches. What remains a core challenge in developing these approaches is how to efficiently extract information from a stream of events. An ideal algorithm should maximize this information while exploiting the signal's spatio-temporal sparsity to allow for processing with minimal latency.

Existing works for processing event data have traded-off latency for prediction accuracy. One class of approaches leverage filtering-based techniques to process events in sequence and thus provide predictions with high temporal resolution and low latency [2–5]. However, these techniques require significant engineering: event features and measurement update functions need to be handcrafted. For this reason, they have difficulties in generalizing to many different tasks, especially high level ones as object recognition and detection. Similarly, other works aim at reducing latency by making inference through a dynamical system, *e.g.* a spiking neural network (SNN)[2]. Despite having low latency, both filtering methods and SNNs achieve limited accuracy in high levels tasks, mainly due to their sensitivity to tuning and their difficult training procedure, respectively. Recently, progress has been made by processing events in batches that are converted into intermediate input representations. Such representations have several advantages. Indeed, they have a regular, *synchronous* tensor-like structure that makes them compatible with conventional machine learning techniques

[1] https://youtu.be/LauQ6LWTkxM?t=4.

[2] Here we use the term SNN as in the neuromorphic literature [6], where it describes continuous-time neural networks. Other networks which are sometimes called SNNs are low precision networks, such as binary networks [7]. However, these are not well suited for asynchronous inputs [6,8,9].

for image-based data (e.g. CNN). This has accelerated the development of new algorithms [3,10–13]. In addition, it has been shown that many of these representations have statistics that overlap with those of natural images, enabling transfer learning with networks pretrained on image data [11,12,14,15]. This last class of approaches achieves remarkable results on several vision benchmarks but at the cost of discarding the asynchronous and sparse nature of event data. By doing so they perform redundant computation at the cost of large inference times, thus losing the inherent low latency property of the event signal.

Contributions. We introduce a general event-based processing framework that combines the advantages of low latency methods and high accuracy batch-wise approaches. Specifically, we allow a neural network to exploit the asynchronous and sparse nature of the input stream and associated representation, thus drastically reducing computation. We mathematically show that the resulting asynchronous network generates identical results to its synchronous variant, while performing strictly less computation. This gives our framework several desirable characteristics: (i) it is agnostic to the event representation, neural network architecture, and task; (ii) it does not require any change in the optimization or training process; (iii) it explicitly models the spatial and temporal sparsity in the events. In addition, we demonstrate both theoretically and experimentally that our approach fully exploits the spatio-temporal sparsity of the data. In order to do so, we relate our framework's computational complexity to the intrinsic dimensionality of the event signal, *i.e.* the events' stream fractal dimension [16]. To show the generality of our framework, we perform experiments on two challenging computer vision tasks: object recognition and object detection. In these tasks, we match the performance of high capacity neural networks but with up 20 times less computation. However, our framework is not limited to these problems and can be applied without any change to a wide range of tasks.

2 Related Work

The recent success of data-driven models in frame-based computer vision [17–19] has motivated the event-based vision community to adopt similar pattern recognition models. Indeed, traditional techniques based on handcrafted filtering methods [2–5] have been gradually replaced with data-driven approaches using deep neural networks [10–13,15]. However, due to their sparse and asynchronous nature, traditional deep models cannot be readily applied to event data, and this has sparked the development of several approaches to event-based learning. In one class of approaches, novel network architecture models directly tailored to the sparse and asynchronous nature of event-based data have been proposed [2,3,6,8,9,20,21]. These include spiking neural networks (SNNs) [2,6,8,9,20] which perform inference through a dynamical system by processing events as asynchronous spike trains. However, due to their novel design and sensitivity to tuning, SNNs are difficult to train and currently achieve limited accuracy on high level tasks. To circumvent this challenge, a second class of approaches has aimed at converting groups of events into image-like representations, which can be either hand-crafted [3,10,11,13] or learned with the task [12]. This makes the sparse and asynchronous event data compatible with conventional machine learning techniques for image data, *e.g.* CNNs, which can be trained efficiently using back-propagation techniques. Due to the higher signal-to-noise ratio of such representations with respect to raw event data, and the high capacity of deep neural networks, these methods achieve state-of-the-art results on several low and high level vision tasks [11–13,22,23]. However, the high performance of these approaches comes at the cost of discarding the sparse and asynchronous property of event data and redunant computation, leading to higher latency and bandwidth. Recently, a solution was proposed that avoids this redundant computation by exploiting sparsity in input data [24]. Graham et al. proposed a technique to process spatially-sparse data efficiently, and used them to develop spatially-sparse convolutional networks. Such an approach brings significant computational advantages to sparse data, in particular when implemented on specific neural network accelerators [25]. Sekikawa et al. [26] showed similar computation gains when generalizing sparse operations to 3D convolutional networks. However, while these methods can address the spatial sparsity in event data, they operate on synchronous data and can therefore not exploit the temporal sparsity of events. This means that they must perform separate predictions for each new event, thereby processing the full representation at each time step. For this reason previous work has focused on finding efficient processing schemes for operations in neural networks to leverage the temporal sparsity of event data. Scheerlinck et al. [27] designed a method to tailor the application of a single convolutional kernel, an essential building block of CNNs, to asynchronous event data. Other work has focused on converting trained models into asynchronous networks by formulating efficient, recursive update rules for newly incoming events [28,29] or converting traditional neural networks into SNNs [30]. However, some of these conversion techniques are limited in the types of representations that can be processed [29,30] or lead to decreases in performance [30]. Other techniques rely on models that do not learn hierarchical features [28] limiting their performance on more complex tasks.

3 Method

In this section we show how to exploit the spatio-temporal sparsity of event data in classical convolutional architectures. In Sect. 3.1 we introduce the working principle of an event camera. Then, in Sect. 3.2 we show how sparse convolutional techniques, such as Submanifold Sparse Convolutional (SSC) Networks [24], can leverage this spatial sparsity. We then propose a novel technique for converting standard *synchronous networks*, into *asynchronous networks* which process events asynchronously and with low computation.

3.1 Event Data

Event cameras have independent pixels that respond to changes in the logarithmic brightness signal $L(\mathbf{u}_k, t_k) \doteq \log I(\mathbf{u}_k, t_k)$. An event is triggered at pixel

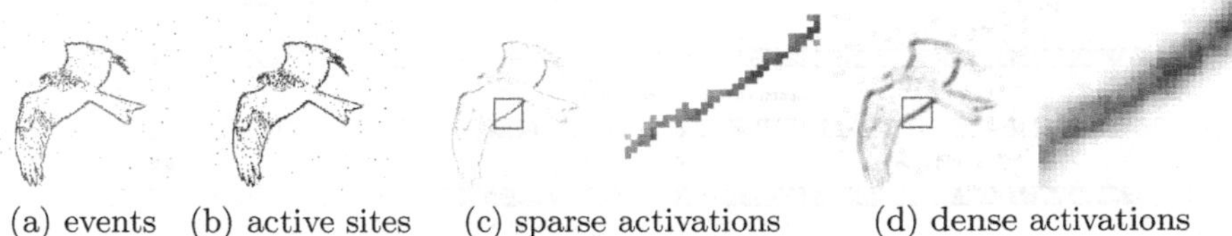

(a) events (b) active sites (c) sparse activations (d) dense activations

Fig. 1. Illustration of Submanifold Sparse Convolutions (SSC) [24]. A sparse event representation (a) is the input the network. SSCs work by only computing the convolution operation at active sites (b), *i.e.* sites that are non-zero, leading to sparse activation maps in the subsequent layers (c). Regular convolutions on the other hand generate blurry activation maps and therefore reduce sparsity (d).

$\mathbf{u}_k = (x_k, y_k)^T$ and at time t_k as soon as the brightness increment since the last event at the pixel reaches a threshold $\pm C$ (with $C > 0$):

$$L(\mathbf{u}_k, t_k) - L(\mathbf{u}_k, t_k - \Delta t_k) \geq p_k C \tag{1}$$

where $p_k \in \{-1, 1\}$ is the sign of the brightness change and Δt_k is the time since the last event at $\mathbf{u}$. Equation (1) is the event generation model for an ideal sensor [4,31]. During a time interval $\Delta\tau$ an event camera produces a sequence of events, $\mathcal{E}(t_N) = \{e_k\}_{k=1}^N = \{(x_k, y_k, t_k, p_k)\}_{k=1}^N$ with microsecond resolution. Inspired by previous approaches [11–13,32] we generate image-like representations $H_{t_N}(\mathbf{u}, c)$ (c denotes the channel) from these sequences, that can be processed by standard CNNs. These representations retain the spatial sparsity of the events, since event cameras respond primarily to image edges, but discard their temporal sparsity. Therefore, previous works only processed them synchronously, reprocessing them from scratch every time a new event is received. This leads of course to redundant computation at the cost of latency. In our framework, we seek to recover the temporal sparsity of the event stream by focusing on the change in $H_{t_N}(\mathbf{u}, c)$ when a new event arrives:

$$H_{t_{N+1}}(\mathbf{u}, c) = H_{t_N}(\mathbf{u}, c) + \sum_i \Delta_i(c)\delta(\mathbf{u} - \mathbf{u}'_i). \tag{2}$$

This recursion can be formulated for arbitrary event representations, making our method compatible with general input representations. However, to maximize efficiency, in this work we focus on a specific class of representations which we term *sparse recursive representations* (SRR). SRRs have the property that they can be *sparsely updated* with each new event, leading to increments $\Delta_i(c)$ at only few positions $\mathbf{u}'_i$ in $H_{t_N}(\mathbf{u}, c)$. There are a number of representations which satisfy this criterion. In fact for the event histogram [11], event queue [32], and time image [33] only single pixels need to be updated for each new event.

3.2 Exploiting the Sparsity of the Event Signal

Event-cameras respond primarily to edges in the scene, which means that event representations are extremely sparse. Submanifold Sparse Convolutions (SSC)

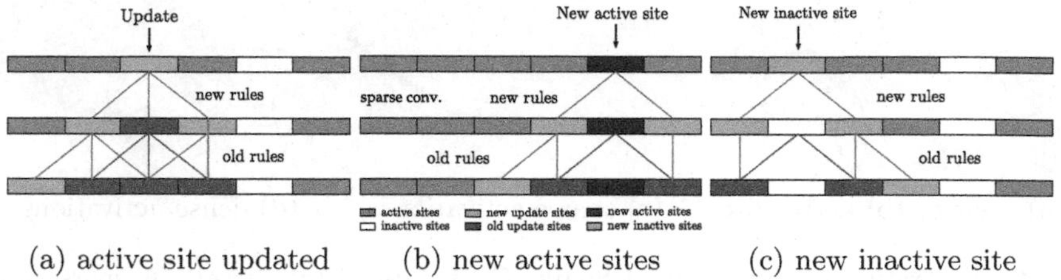

(a) active site updated (b) new active sites (c) new inactive site

Fig. 2. Propagation of the rulebook $\mathcal{R}_{\mathbf{k},n}$ a 1D example. The input is composed of active (gray) and inactive (white) sites. (a) If the value of an active site changes (magenta), the update rules are incrementally added to the rulebook (lines) according to Eq. (5). (b) At newly active sites (blue) the sparse convolution is directly computed using Eq. (3) and repeated at each layer (here 1 to 3). (c) Similarly, new inactive sites (orange) are set to zero at each layer. Thus, new active sites (blue) and new inactive sites (orange) do not contribute to the rulebook propagation. Best viewed in color. (Color figure online)

[24], illustrated in Fig. 1, leverage spatial sparsity in the data to drastically reduce computation. Hence, they are not equivalent to regular convolutions. Compared to regular convolutions, SSCs only compute the convolution at sites $\mathbf{u}$ with a non-zero feature vector, and ignore inputs in the receptive field of the convolution which are 0. These sites with non-zero feature vector are termed *active sites* $\mathcal{A}_t$ (Fig. 1 (b)). Figure 1 illustrates the result of applying an SSC to sparse event data (a). The resulting activation map (c) has the same active sites as its input and therefore, by induction, all SSC layers with the same spatial resolution share the same active sites and level of sparsity. The sparse convolution operation can be written as

$$\tilde{y}_{n+1}^t(\mathbf{u}, c) = b_n(c) + \sum_{c'} \sum_{\mathbf{k} \in \mathcal{K}_n} \sum_{(\mathbf{i},\mathbf{u}) \in R_{t,\mathbf{k}}} W_n(\mathbf{k}, c', c) y_n^t(\mathbf{i}, c'), \quad \text{for } \mathbf{u} \in \mathcal{A}_t \tag{3}$$

$$y_{n+1}^t(\mathbf{u}, c) = \sigma(\tilde{y}_{n+1}^t(\mathbf{u}, c)). \tag{4}$$

Here $y_n^t(\mathbf{u}, c)$ is the activation of layer n at time t and is non-zero only for pixels $\mathbf{u}$, b_n denotes the bias term, W_n and $\mathbf{k} \in \mathcal{K}_n$ the parameters and indices of the convolution kernel, and σ a non-linearity. For the first layer $y_0^t(\mathbf{u}, c) = H_{t_N}(\mathbf{u}, c)$. We also make use of the rulebook $R_{t,\mathbf{k}}$ [24], a data structure which stores a list of correspondences $(\mathbf{i}, \mathbf{j})$ of input and output sites. In particular, a rule $(\mathbf{i}, \mathbf{j})$ is in $R_{t,\mathbf{k}}$ if both $\mathbf{i}, \mathbf{j} \in \mathcal{A}_t$ and $\mathbf{i} - \mathbf{j} = \mathbf{k}$, meaning that the output $\mathbf{j}$ is in the receptive field of the input $\mathbf{i}$ (Fig. 2 (a) lines). The activation at site $\mathbf{i}$ is multiplied with the weight at index $\mathbf{k}$ and added to the activation at output site $\mathbf{j}$. In (3), this summation is performed over rules which have the same output site $\mathbf{j} = \mathbf{u}$. When pooling operations such as max pooling or a strided convolution are encountered, the feature maps' spatial resolution changes and thus the rulebook needs to be recomputed. In this work we only consider max pooling. For sparse input it is the same as regular max pooling but over sites that are active. Importantly, after

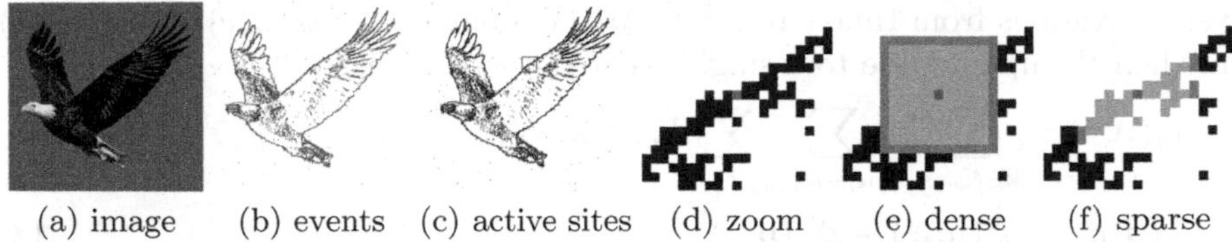

Fig. 3. The difference between asynchronous sparse and dense updates of events (b) is illustrated with an example image (a). The active sites, *i.e.* input sites that are non-zero, are visualized as black pixels (c). (d)-(f) show an asynchronous update (red pixel) processed with traditional convolutions (e) and our proposed asynchronous sparse convolutions (f). The receptive field of traditional convolution (e) grows quadratically with network depth leading to redundant computation. By contrast, our method (f) only updates active sites, which reduces the computational complexity. In both cases the growth frontier is indicated in magenta.

pooling, output sites become active when they have at least one active site in their receptive field. This operation increases the number of active sites.

Asynchronous Processing. While SSC networks leverage the spatial sparsity of the events, they still process event representations synchronously, performing redundant computations. Indeed, for each new event, all layer activations need to be recomputed. Since events are triggered at individual pixels, activations should also be affected locally. We propose to take advantage of this fact by retaining the previous activations $\{y_n^t(\mathbf{u}, c)\}_{n=0}^N$ of the network and formulating novel, efficient update rules $y_n^t(\mathbf{u}, c) \rightarrow y_n^{t+1}(\mathbf{u}, c)$ for each new event. By employing SRRs each new event leads to sparse updates $\Delta_i(c)$ at locations $\mathbf{u}_i'$ in the input layer (Eq. (2)). We propagate these changes to deeper layers by incrementally building a *rulebook* $\mathcal{R}_{\mathbf{k},n}$ and *receptive field* $\mathcal{F}_n$ for each layer, visualized in Fig. 2. The rulebook (lines) are lists of correspondences $(\mathbf{i}, \mathbf{j})$ where $\mathbf{i}$ at the input is used to update the value at $\mathbf{j}$ in the output. The receptive field (colored sited) keeps track of the sites that have been updated by the change at the input. For sites that become newly active or inactive (Fig. 2 (b) and (c)) the active sites $\mathcal{A}_t$ are updated accordingly. At initialization (input layer) the rulebook is empty and the receptive field only comprises the updated pixel locations, caused by new events, *i.e.* $\mathcal{R}_{\mathbf{k},0} = \emptyset$ and $\mathcal{F}_0 = \{\mathbf{u}_i'\}_i$. Then, at each new layer $\mathcal{R}_{\mathbf{k},n}$ and $\mathcal{F}_n$ are expanded:

$$\mathcal{F}_n = \{\mathbf{i} - \mathbf{k} | \mathbf{i} \in \mathcal{F}_{n-1} \text{ and } \mathbf{k} \in \mathcal{K}_{n-1} \text{ if } \mathbf{i} - \mathbf{k} \in \mathcal{A}_t\} \quad (5)$$

$$\mathcal{R}_{\mathbf{k},n} = \{(\mathbf{i}, \mathbf{i} - \mathbf{k}) | \mathbf{i} \in \mathcal{F}_{n-1} \text{ if } \mathbf{i} - \mathbf{k} \in \mathcal{A}_t\}. \quad (6)$$

Rules that have a newly active or inactive site as output (Fig. 2 (b) and (c), blue or orange sites) are ignored.[3] We use Eq. (5) to formulate the update rules to

[3] In the supplement we present an efficient recursive method for computing $\mathcal{R}_{\mathbf{k},n}$ and $\mathcal{F}_n$ by reusing the rules and receptive field from the previous layers.

layer activations from time t to $t+1$. At the input we set $y_0^t(\mathbf{u},c) = H_{t_N}(\mathbf{u},c)$ and then the update due to a single event can written as:

$$\Delta_n(\mathbf{u},c) = \sum_{\mathbf{k}\in\mathcal{K}_{n-1}} \sum_{(\mathbf{i},\mathbf{u})\in\mathcal{R}_{\mathbf{k},n}} \sum_{c'} W_{n-1}(\mathbf{k},c',c)(y_{n-1}^t(\mathbf{i},c') - y_{n-1}^{t-1}(\mathbf{i},c')) \quad (7)$$

$$\tilde{y}_n^t(\mathbf{u},c') = \tilde{y}_n^t(\mathbf{u},c') + \Delta_n(\mathbf{u},c) \quad (8)$$

$$y_n^t(\mathbf{u},c') = \sigma(\tilde{y}_n^t(\mathbf{u},c')). \quad (9)$$

Note, that these equations only consider sites which have not become active (due to a new event) or inactive. For newly active sites, we compute $y_n^t(\mathbf{u},c)$ according to Eq. (3). Finally, sites that are deactivated are set to 0, *i.e.* $y_n^t(\mathbf{u},c) = 0$. In both cases we update the active sites $\mathcal{A}_t$ before the update has been propagated. By iterating over Eqs. (7) and (5), all subsequent layers can be updated recursively. Figure 3 illustrates the update rules above, applied to a single event update (red position) after six layers of both standard convolutions (e) and our approach (f). Note that (e) is a special case of (f) with all pixels being active sites. By using our local update rules we see that computation is confined to a small patch of the image (c). Moreover, it is visible that standard convolutions (e) process noisy or empty regions (green and magenta positions), while our approach (f) focuses computation on sites with events, leading to higher efficiency. Interestingly, we also observe that for traditional convolutions the size of the receptive field grows quadratically in depth while for our approach it grows more slowly, according on the *fractal dimension* of the underlying event data. This point will be explored further in Sect. 3.2.

Equivalence of Asynchronous and Synchronous Operation. By alternating between Eqs. (5) and (7) asynchronous event-by-event updates can be propagated from the input layer to arbitrary network's depth. In the supplement we prove that processing N events by this method is equivalent to processing all events at once and present pseudocode for our method. It follows that a synchronous network, trained efficiently using back-propagation, can be deployed as an asynchronous network. Therefore, our framework does not require any change to the optimization and learning procedure. Indeed, any network architecture, after being trained, can be transformed in its asynchronous version, where it can leverage the high temporal resolution of the events at limited computational complexity. In the next section we explore this reduction in complexity in more detail.

Computational Complexity. In this section we analyze the computational complexity of our approach in terms of floating point operations (FLOPs), and compare it against conventional convolution operations. In general, the number of FLOPs necessary to perform L consecutive convolutions (disregarding non-linearities for the sake of simplicity) is:

$$C_{\text{dense}} = \sum_{l=1}^{L} N(2k^2 c_{l-1} - 1)c_l \quad C_{\text{sparse}} = \sum_{l=1}^{L} N_r^l(2c_l + 1)c_{l-1} \quad (10)$$

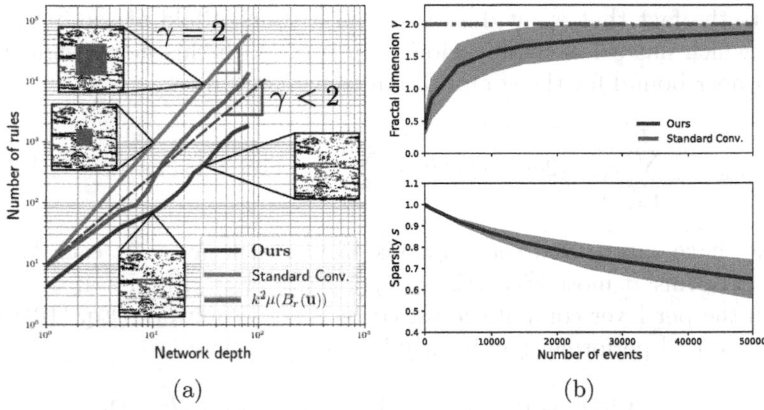

Fig. 4. To bound the complexity of our method, we use the empirical sparsity s and fractal dimension γ of event data (b). The fractal dimension characterizes the rate at which the number of update sites grows from one layer to the next. While for standard convolutions this number grows quadratically (as $(1 + (k-1)n)^2$, k being the kernel size) with layer depth, with our method it grows more slowly, with an exponent $\gamma < 2$.

These formulae are explained in more detail in the supplement. For the sparse case, N_r^l counts the number of rules, *i.e.* input output pairs between layer $l-1$ and l, which corresponds exactly with the size of the rulebook in Eq. (3).

Our method minimizes the number of rules it uses at each layer by only using a subset of the rules used by SSCs and incrementally expanding it from layer to layer (Fig. 2). To characterize the computation from our approach we consider an update at a single pixel. At each layer the computation is proportional to the size of the rulebook in Eq. (5) for which we can find an upper bound. Let n_l be the number of active pixels within a patch of size $1+(k-1)l$ which is an upper bound for the number of updated sites in layer l. If we assume that each pixel can have at most k^2 rules, the number of rules at layer l is at most $n_l k^2$.

For a dense update, this number grows quadratically with the patch size $p = 1+(k-1)l$, however, for sparse updates, this number grows more slowly. To formalize this notion we define a measure $\mu(B(\mathbf{u}, r)))$ which counts the number of active sites within a patch of radius $r = \frac{p}{2}$. This measure can be used to define the *fractal dimension* of event data at pixel $\mathbf{u}$ according to [16]:

$$\gamma(\mathbf{u}) = \lim_{r \to 0} \frac{\log(\mu(B(\mathbf{u}, r)))}{\log 2r} \tag{11}$$

The fractal dimension describes an intrinsic property of the event data, related to its dimensionality and has not been characterized for event data prior to this work. It measures the growth-rate of the number of active sites as the patch size is varied. In particular, it implies that this number grows approximately as $n_l \approx (1+(k-1)l)^\gamma$. To estimate the fractal dimension we consider the slope of $\mu(B(\mathbf{u}, r)))$ over $r = \frac{1+(k-1)l}{2}$ in the log-log domain which we visualize in Fig. 4. Crucially, a slope $\gamma < 2$ indicates that the growth is slower than quadratic. This highlights the fact that event data exists in a submanifold of the image plane ($\gamma = 2$) which has a lower dimension than two. With this new insight we can find an upper bound for the computation using Eq. (10):

$$C_{\text{async. sparse}} \leq \sum_{l=1}^{L} c_{l-1}\,(2c_l+1)\,n_l k^2 \approx \sum_{l=1}^{L} c_{l-1}\,(2c_l+1)\,(1+(k-1)l)^{\gamma} k^2 \quad (12)$$

Where we have substituted the rulebook size at each layer. At each layer, our method performs at most $k^2 c_{l-1}(2c_l+1)\,(1+(k-1)l)^{\gamma}$ FLOPs. If we compare this with the per layer computation used by a dense network (Eq. (10)) we see that our method performs significantly less computation:

$$\frac{C^l_{\text{sparse. anync}}}{C^l_{\text{dense}}} \leq \frac{k^2(2c_l+1)c_{l-1}}{(2k^2c_{l-1}-1)c_l}\frac{(1+(k-1)l)^{\gamma}}{N} \approx \frac{(1+(k-1)l)^{\gamma}}{N} << 1 \quad (13)$$

Where we assume that $2c_l >> 1$ and $k^2c_{l-1} >> 1$ which is the case in typical neural networks[4]. Moreover, as the fractal dimension decreases our method becomes exponentially more efficient. Through our novel asynchronous processing framework, the fractal dimension of event data can be exploited efficiently. It does so with sparse convolution, that can specifically process low-dimensional input data embedded in the image plane, such as points and lines.

4 Experiments

We validate our framework on two computer vision applications: object recognition (Sect. 4.1) and object detection (Sect. 4.2). On these tasks, we show that our framework achieves state-of-the-art results with a fraction of the computation of top-performing models. In addition, we demonstrate that our approach can be applied to different event-based representations. We select the event histogram [11] and the event queue [32] since they can be updated sparsely and asynchronously for each incoming event (see Sect. 3.2).

4.1 Object Recognition

We evaluate our method on two standard event camera datasets for object recognition: Neuromorphic-Caltech101 (N-Caltech101) [34], and N-Cars [10]. The N-Caltech101 dataset poses the task of event-based classification of 300 ms sequences of events. In total, N-Caltech101 contains 8,246 event samples, which are labelled into 101 classes. N-Cars [10] is a benchmark for car recognition. It contains 24,029 event sequences of 100 ms which contain a car or a random scene patch. To evaluate the computational efficiency and task performance we consider two metrics: prediction accuracy and number of floating point operations (FLOPs). While the first indicates the prediction quality, the

[4] In fact, for typical channel sizes $c_l \geq 16$ we incur a $< 3\%$ approximation error.

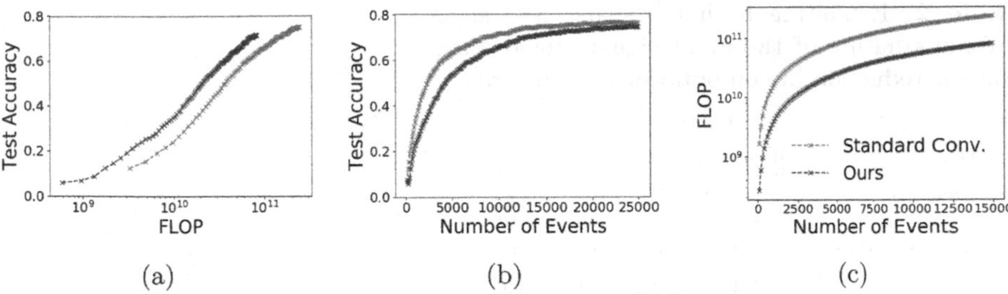

Fig. 5. Our approach requires less cumulative FLOPs w.r.t. the dense method (Standard Conv.) to produce similarly accurate predictions (a). Although dense processing requires fewer events to generate predictions of equal accuracy (b), it needs significantly more computation per event, thus having higher cumulative FLOPs (c).

second one measures the computational complexity of processing an input. The number of FLOPs is commonly used as a complexity metric [17,24,30], since it is independent of both the hardware and the implementation. Details of the FLOP computation are reported in the supplement. Analogously to previous work on sparse data processing [24], we use a VGG13 [35] architecture with 5 convolutional blocks and one final fully connected layer. Each block contains two convolution layers, followed by batch-norm [36], and a max pooling layer. We train the networks with the cross-entropy loss and the ADAM optimizer [37]. The initial learning rate of 10^{-4} is divided by 10 after 500 epochs.

Results. In our first experiment, we compare our sparse and asynchronous processing scheme to the dense and synchronous one. For comparability, both methods share the same VGG13 architecture and the same input representation, which was generated with 25,000 events. This number was empirically found to yield a good trade-off between accuracy and computational efficiency (see supplement). We measure the approaches' computational complexity in terms of required FLOPs per single event update. Classification results shown in Table 1 demonstrate that our processing scheme has similar accuracy to the dense one but requires up to 19.5 times less computations per event. The low-latency of the event signal allows us to make fast predictions. For this reason we compare

Table 1. Our approach matches the performance of the traditional dense and synchronous processing scheme at one order of magnitude less computations.

	Representation	N-Caltech101		N-Cars	
		Accuracy ↑	MFLOP/ev ↓	Accuracy ↑	MFLOP/ev ↓
Standard Conv.	Event Histogram	**0.761**	1621	**0.945**	321
Ours		0.745	**202**	0.944	**21.5**
Standard Conv.	Event Queue	0.755	2014	0.936	419
Ours		0.741	**202**	0.936	**21.5**

Table 2. Exploiting both the sparse and asynchronous nature of the event signal provides significant reduction in computational complexity.

Method	1 event		100 events	
[MFLOP]	Histogram	Queue	Histogram	Queue
Standard Conv.	1621	2014	1621	2014
Sparse Conv. [38]	892	967	892	967
Async. Conv. (Ours)	320	320	958	958
Async. Sparse Conv. (Ours)	**202**	**202**	**690**	**690**

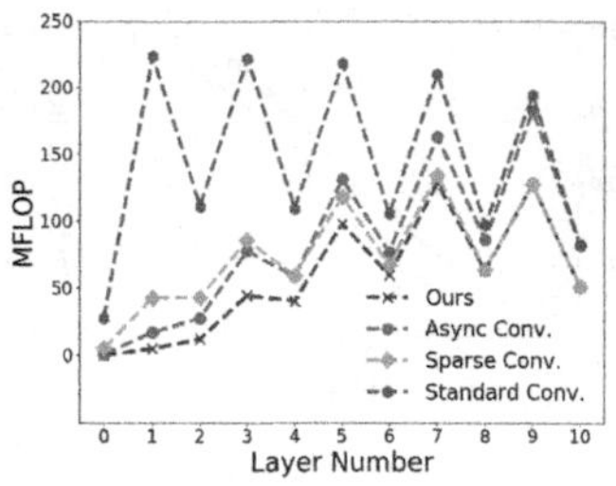

Fig. 6. Layer-wise FLOPs

the maximal prediction accuracy that can be achieved given a fixed computation budget for our method and a standard CNN. This experiment imitates the scenario where an object suddenly appears in the field of view, *e.g.* a pedestrian crossing the street. To do this we use samples from N-Caltech101 [34] and report the multi-class prediction accuracy (Fig. 5 (b)) and total number of FLOPs used (Fig. 5 (c)) as a function of number of events observed for our method and a standard CNN with the same architecture. For each new event we compute the FLOPs needed to generate a new prediction and its accuracy, taking into account all previously seen events. It can be seen that both standard CNN and our method have a higher prediction accuracy as the number of events increases (Fig. 5 (b)). However, compared to standard networks our method performs far less computation (Fig. 5 (c)). This is because for standard networks all activations need to be recomputed for each new event, while our method retains the state and therefore only needs to perform sparse updates. As a result, our method achieves up to 14.8% better accuracy at the same FLOP budget (Fig. 5 (a)), thus improving the prediction latency. Moreover, to show the flexibility of our approach to different update rates, we initialize a representation with 25,000 events and update it either for each new event or for a batch of 100 new events. For comparability, all methods share the same VGG13 architecture and input representation. The results of this experiment are reported in Table 2. For both single and batch event update, exploiting the asynchronous and sparse nature of the signal offers significant computational benefits. Our approach (*Async. Sparse Conv.*) performs on average 8.9 times less computation than standard convolution for one event update, and of 2.60 times for 100 events update. We additionally evaluate the computation per layer of each of previous models in the case of one event update. Figure 6, which presents the results of this evaluation, shows that sparse and asynchronous processing saves the majority of computations in the initial layers of the network. Indeed, the input representations of the event stream are spatially very sparse (see Fig. 4(b)), and are only locally updated for each new event.

Comparison with State-of-the-Art Methods. We finally compare our approach with state-of-the-art methods for event-based object recognition. We consider models that, like ours, perform efficient per-event updates due to a light-

Table 3. Comparison with asynchronous and dense methods for object recognition.

Methods	Async.	N-Caltech101		N-Cars	
		Accuracy ↑	MFLOP/ev ↓	Accuracy ↑	MFLOP/ev ↓
H-First [2]	✓	0.054	-	0.561	-
HOTS [3]	✓	0.210	54.0	0.624	14.0
HATS [10]	✓	0.642	**4.3**	0.902	**0.03**
DART [38]	✓	0.664	-	-	-
YOLE [29]	✓	0.702	3659	0.927	328.16
EST [12]	✗	**0.817**	4150	0.925	1050
SSC [24]	✗	0.761	1621	**0.945**	321
Ours	✓	**0.745**	202	**0.944**	21.5

weight computational model (HATS [10], HOTS [3], DART [38]), a spiking neural network (H-First [2]), or asynchronous processing (YOLE [29]). The results for this evaluation are presented in Table 3. Our method outperforms the state-of-the-art (YOLE) by 4.3% in accuracy on N-Caltech101 and 1.7% on N-Cars at only 6% (on average over datasets) of its computational cost. Finally, we compare against the synchronous state-of-the-art method [12]. Our method achieves a slightly higher accuracy on N-Cars using only 21.5 MFLOPs (47 times reduction). Similarly, our method using the asynchronous processing requires 20 times fewer FLOPs on N-Caltech101 but at the cost of lower accuracy. In addition to the performance evaluation, we timed our experiments conducted on N-Caltech101 by measuring the processing time for a single event on a i7-6900K CPU. In our framework implemented in C++ and Python, our method requires 80.4 ms, while the standard dense CNN needs 202 ms. Therefore, our approach becomes roughly 2.75 times faster by leveraging the sparsity. However, in the highly-optimized framework PyTorch [39], a dense inference takes only 23.4 ms. Given its lower number of FLOPs, we expect that our method will experience a significant run-time reduction with further optimizations and specific hardware.

4.2 Object Detection

Object detection is the task of regressing a bounding box and class probabilities for each object in the image. We evaluate our method on two standard benchmarks for event-based object detection: N-Caltech101 [34] and Gen1 Automotive [40]. While the former contains the N-Caltech101 samples each with a single bounding box, the latter contains 228,123 bounding boxes for cars and 27,658 for pedestrians collected in an automotive scenario. For this task we combine the first convolutional blocks of the object recognition task with the YOLO output layer [41]. The resulting feature maps are processed by the YOLO output layer to generate bounding-boxes and class predictions. As is standard, we report the performance using the mean average precision (mAP) metric [42] using the

Table 4. Accuracies for object detection.

	Representation	N-Caltech101		Gen1 Automotive	
		mAP ↑	MFLOP ↓	mAP ↑	MFLOP ↓
YOLE [29]	Leaky Surface	0.398	3682	-	-
Standard Conv.	Event Queue	0.619	1977	**0.149**	2614
Ours		0.615	**200**	0.119	**205**
Standard Conv.	Event Histogram	0.623	1584	0.145	2098
Ours		**0.643**	**200**	0.129	**205**

implementation of [43]. As for the recognition task, we measure computational complexity with FLOPs.

Results and Comparison with State-of-the-Art. Table 4 shows quantitative results on object detection while Fig. 1 in the supplement illustrates qualitative results. Our approach achieves comparable or superior performance with respect to standard networks at a fraction of the computational cost. Specifically, for the histogram representation our method outperforms dense methods by 2.0% on N-Caltech101. On the Gen1 Automotive dataset, we experience a slight performance drop of 1.8%. The slight performance improvement in N-Caltech101 is probably thanks to the sparse convolution, which give less weight to noisy events. In terms of computation, our method reduces the number of FLOPs per event by a factor of 10.6 with respect to the dense approach, averaged over all datasets and representations. Table 4 also compares our approach to the state-of-the-art method for event-based detection, YOLE [29]. Compared to this baseline, we achieve 24.5% higher accuracy at 5% of the computational costs.

5 Discussion

In the quest of high prediction accuracy, event-based vision algorithms have relied heavily on processing events in synchronous batches using deep neural networks. However, this trend has disregarded the sparse and asynchronous nature of event-data. Our work brings the genuine spatio-temporal sparsity of events back into high-performance CNNs, by significantly decreasing their computational complexity (up to 20 times). By doing this, we achieve up to 15% better accuracy than state-of-the-art synchronous models at the same computational (FLOP) budget. In addition, we outperform existing asynchronous approaches by up to 24.5% in accuracy. Our work highlights the importance for researchers to take into account the intrinsic properties of event data in the pursuit of low-latency and high-accuracy event vision algorithms. Such considerations will open the door to unlock the advantages of event cameras on all tasks that rely on low-latency visual information processing.

Acknowledgements. This work was supported by the Swiss National Center of Competence Research Robotics (NCCR), through the Swiss National Science Foundation, and the SNSF-ERC starting grant.

References

1. Gallego, G., et al.: Event-based vision: A survey. arXiv e-prints, vol. abs/1904.08405 (2019). http://arxiv.org/abs/1904.08405
2. Orchard, G., Meyer, C., Etienne-Cummings, R., Posch, C., Thakor, N., Benosman, R.: HFirst: a temporal approach to object recognition. IEEE Trans. Pattern Anal. Mach. Intell. **37**(10), 2028–2040 (2015)
3. Lagorce, X., Orchard, G., Gallupi, F., Shi, B.E., Benosman, R.: HOTS: a hierarchy of event-based time-surfaces for pattern recognition. IEEE Trans. Pattern Anal. Mach. Intell. **39**(7), 1346–1359 (2017)
4. Gallego, G., Lund, J.E.A., Mueggler, E., Rebecq, H., Delbruck, T., Scaramuzza, D.: Event-based, 6-DOF camera tracking from photometric depth maps. IEEE Trans. Pattern Anal. Mach. Intell. **40**(10), 2402–2412 (2018)
5. Kim, H., Leutenegger, S., Davison, A.J.: Real-time 3D reconstruction and 6-DoF tracking with an event camera. In: Leibe, B., Matas, J., Sebe, N., Welling, M. (eds.) ECCV 2016. LNCS, vol. 9910, pp. 349–364. Springer, Cham (2016). https://doi.org/10.1007/978-3-319-46466-4_21
6. Lee, J.H., Delbruck, T., Pfeiffer, M.: Training deep spiking neural networks using backpropagation. Front. Neurosci. **10**, 508 (2016)
7. Rastegari, M., Ordonez, V., Redmon, J., Farhadi, A.: XNOR-Net: ImageNet classification using binary convolutional neural networks. In: Leibe, B., Matas, J., Sebe, N., Welling, M. (eds.) ECCV 2016. LNCS, vol. 9908, pp. 525–542. Springer, Cham (2016). https://doi.org/10.1007/978-3-319-46493-0_32
8. Perez-Carrasco, J.A., et al.: Mapping from frame-driven to frame-free event-driven vision systems by low-rate rate coding and coincidence processing application to feedforward ConvNets. IEEE Trans. Pattern Anal. Mach. Intell. **35**(11), 2706–2719 (2013)
9. Amir, A., et al.: A low power, fully event-based gesture recognition system. In: IEEE Conference Computer Vision Pattern Recognition (CVPR), pp. 7388–7397 (2017)
10. Sironi, A., Brambilla, M., Bourdis, N., Lagorce, X., Benosman, R.: HATS: histograms of averaged time surfaces for robust event-based object classification. In: IEEE Conference Computer Vision Pattern Recognition (CVPR), pp. 1731–1740 (2018)
11. Maqueda, A.I., Loquercio, A., Gallego, G., García, N., Scaramuzza, D.: Event-based vision meets deep learning on steering prediction for self-driving cars. In: IEEE Conference Computer Vision Pattern Recognition (CVPR), pp. 5419–5427 (2018)
12. Gehrig, D., Loquercio, A., Derpanis, K.G., Scaramuzza, D.: End-to-end learning of representations for asynchronous event-based data. In: International Conference Computer Vision (ICCV) (2019)
13. Zhu, A.Z., Yuan, L., Chaney, K., Daniilidis, K.: EV-FlowNet: self-supervised optical flow estimation for event-based cameras. In: Robotics: Science and Systems (RSS) (2018)

14. Rebecq, H., Horstschäfer, T., Gallego, G., Scaramuzza, D.: EVO: a geometric approach to event-based 6-DOF parallel tracking and mapping in real-time. IEEE Robot. Autom. Lett. **2**(2), 593–600 (2017)
15. Rebecq, H., Ranftl, R., Koltun, V., Scaramuzza, D.: Events-to-video: bringing modern computer vision to event cameras. In: IEEE Conference Computer Vision Pattern Recognition (CVPR) (2019)
16. Xu, Y., Ji, H., Fermüller, C.: Viewpoint invariant texture description using fractal analysis. Int. J. Comput. Vis. **83**(1), 85–100 (2009)
17. He, K., Zhang, X., Ren, S., Sun, J.: Deep residual learning for image recognition. In: IEEE Conference Computer Vision Pattern Recognition (CVPR), pp. 770–778 (2016)
18. Redmon, J., Divvala, S., Girshick, R., Farhadi, A.: You only look once: unified, real-time object detection. In: IEEE Conference Computer Vision Pattern Recognition (CVPR) (2016)
19. Arandjelović, R., Gronat, P., Torii, A., Pajdla, T., Sivic, J.: NetVLAD: CNN architecture for weakly supervised place recognition. In: IEEE Conference Computer Vision Pattern Recognition (CVPR), pp. 5297–5307 (2016)
20. Orchard, G., Benosman, R., Etienne-Cummings, R., Thakor, N.V.: A spiking neural network architecture for visual motion estimation. In: IEEE Biomedical Circuits Systems Conference (BioCAS), pp. 298–301 (2013)
21. Neil, D., Pfeiffer, M., Liu, S.-C.: Phased LSTM: accelerating recurrent network training for long or event-based sequences. In: Conference Neural Information Processing System (NIPS) (2016)
22. Zhu, A.Z., Yuan, L., Chaney, K., Daniilidis, K.: Unsupervised event-based learning of optical flow, depth, and egomotion. In: IEEE Conference Computer Vision Pattern Recognition (CVPR) (2019)
23. Rebecq, H., Ranftl, R., Koltun, V., Scaramuzza, D.: High speed and high dynamic range video with an event camera. IEEE Trans. Pattern Anal. Mach. Intell. (2019)
24. Graham, B., Engelcke, M., van der Maaten, L.: 3D semantic segmentation with submanifold sparse convolutional networks. In: IEEE Conference Computer Vision Pattern Recognition (CVPR), pp. 9224–9232 (2018)
25. Aimar, A., et al.: Nullhop: a flexible convolutional neural network accelerator based on sparse representations of feature maps. IEEE Trans. Neural Netw. Learn. Syst. **30**(3), 644–656 (2019)
26. Sekikawa, Y., et al.: Constant velocity 3D convolution. In: International Conference on 3D Vision (3DV), pp. 343–351 (2018)
27. Scheerlinck, C., Barnes, N., Mahony, R.: Asynchronous spatial image convolutions for event cameras. IEEE Robot. Autom. Lett. **4**(2), 816–822 (2019)
28. Sekikawa, Y., Hara, K., Saito, H.: EventNet: asynchronous recursive event processing. In: IEEE Conference Computer Vision Pattern Recognition (CVPR) (2019)
29. Cannici, M., Ciccone, M., Romanoni, A., Matteucci, M.: Asynchronous convolutional networks for object detection in neuromorphic cameras. In: IEEE Conference Computer Vision Pattern Recognition (CVPR), Workshops (CVPRW) (2019)
30. Rueckauer, B., Lungu, I.-A., Hu, Y., Pfeiffer, M., Liu, S.-C.: Conversion of continuous-valued deep networks to efficient event-driven networks for image classification. Front. Neurosci. **11**, 682 (2017)
31. Gallego, G., Forster, C., Mueggler, E., Scaramuzza, D.: Event-based camera pose tracking using a generative event model (2015) arXiv:1510.01972
32. Tulyakov, S., Fleuret, F., Kiefel, M., Gehler, P., Hirsch, M.: Learning an event sequence embedding for dense event-based deep stereo. In: International Conference Computer Vision (ICCV), October 2019

33. Mitrokhin, A., Fermuller, C., Parameshwara, C., Aloimonos, Y.: Event-based moving object detection and tracking. In: IEEE/RSJ International Conference Intelligent Robots Systems (IROS) (2018)
34. Orchard, G., Jayawant, A., Cohen, G.K., Thakor, N.: Converting static image datasets to spiking neuromorphic datasets using saccades. Front. Neurosci. **9**, 437 (2015)
35. Simonyan, K., Zisserman, A.: Very deep convolutional networks for large-scale image recognition. In: International Conference Learning, Representations (ICLR) (2015)
36. Ioffe, S., Szegedy, C.: Batch normalization: accelerating deep network training by reducing internal covariate shift. In: Proceedings International Conference Machine, Learning (ICML) (2015)
37. Kingma, D.P., Ba, J.L.: Adam: a method for stochastic optimization. In: International Conference Learning, Representations (ICLR) (2015)
38. Ramesh, B., Yang, H., Orchard, G.M., Le Thi, N.A., Xiang, C.: DART: distribution aware retinal transform for event-based cameras, arXiv e-prints, October 2017. http://arxiv.org/abs/1710.10800
39. Paszke, A., et al.: Automatic differentiation in PyTorch. In: NIPS Workshops (2017)
40. de Tournemire, P., Nitti, D., Perot, E., Migliore, D., Sironi, A.: A large scale event-based detection dataset for automotive, ArXiv, vol. abs/2001.08499 (2020)
41. Redmon, J., Divvala, S.K., Girshick, R.B., Farhadi, A.: You only look once: unified, real-time object detection. In: 2016 IEEE Conference on Computer Vision and Pattern Recognition (CVPR), pp. 779–788 (2016)
42. Everingham, M., Van Gool, L., Williams, C.K., Winn, J., Zisserman, A.: The pascal visual object classes (VOC) challenge. Int. J. Comput. Vis. **88**, 303–338 (2010)
43. Padilla, R., Netto, S.L., da Silva, E.A.B.: Survey on performance metrics for object-detection algorithms. In: International Conference on Systems, Signals and Image Processing (IWSSIP) (2020)

AtlantaNet: Inferring the 3D Indoor Layout from a Single 360° Image Beyond the Manhattan World Assumption

Giovanni Pintore[1](✉), Marco Agus[1,2], and Enrico Gobbetti[1]

[1] Visual Computing, CRS4, Cagliari, Italy
{giovanni.pintore,enrico.gobbetti}@crs4.it
[2] College of Science and Engineering, HBKU, Doha, Qatar
magus@hbku.edu.qa

Abstract. We introduce a novel end-to-end approach to predict a 3D room layout from a single panoramic image. Compared to recent state-of-the-art works, our method is not limited to *Manhattan World* environments, and can reconstruct rooms bounded by vertical walls that do not form right angles or are curved – i.e., *Atlanta World* models. In our approach, we project the original gravity-aligned panoramic image on two horizontal planes, one above and one below the camera. This representation encodes all the information needed to recover the *Atlanta World* 3D bounding surfaces of the room in the form of a 2D room footprint on the floor plan and a room height. To predict the 3D layout, we propose an encoder-decoder neural network architecture, leveraging Recurrent Neural Networks (RNNs) to capture long-range geometric patterns, and exploiting a customized training strategy based on domain-specific knowledge. The experimental results demonstrate that our method outperforms state-of-the-art solutions in prediction accuracy, in particular in cases of complex wall layouts or curved wall footprints.

Keywords: 3D floor plan recovery · Panoramic images · 360 images · Data-driven reconstruction · Structured indoor reconstruction · Indoor panorama · Room layout estimation · Holistic scene structure

1 Introduction

Automatic 3D reconstruction of a room's bounding surfaces from a single image is a very active research topic [20].

In this context, 360° capture is very appealing, since it provides the quickest and most complete single-image coverage and is supported by a wide variety of professional and consumer capture devices that make acquisition fast and

Electronic supplementary material The online version of this chapter (https://doi.org/10.1007/978-3-030-58598-3_26) contains supplementary material, which is available to authorized users.

A. Vedaldi et al. (Eds.): ECCV 2020, LNCS 12353, pp. 432–448, 2020.
https://doi.org/10.1007/978-3-030-58598-3_26

Fig. 1. Examples of automatically recovered 3D layouts. Our method returns a 3D room model from a single panorama even in cases not supported by current state-of-the-art methods, such as, for example, vertical walls meeting at non-right angles or with a curved 2D footprints.

cost-effective [31]. Since rooms are full of clutter, single images produce anyway only partial coverage and imperfect sampling, thus reconstruction problem is difficult and ambiguous without prior assumptions. In particular, current approaches, see Sect. 2, are either tuned to simple structures with a limited number of corners [6] or bound by the *Indoor World* assumption [16] (i.e., the environment has a single horizontal floor and ceiling, and vertical walls which all meet at right angles). In this context, recent data-driven approaches [26,30,33] have produced excellent results in recovering the room layout from a single panoramic image [34]. However, state-of-the-art data-driven methods usually follow a costly and constraining framework: a heavy pre-processing to generate Manhattan-aligned panoramas (e.g., edge-based alignment and warping of generated perspective views [16]), a deep neural network that predicts the layout elements on a rectified equirectangular image, and a post-processing that fits the (Manhattan) 3D layout to the predicted elements.

In this work, we present *AtlantaNet*, a novel data-driven solution to estimate a 3D room layout from a single RGB panorama. As its name suggests, we exploit the less restrictive *Atlanta World* model [23], in which the environment is expected to have horizontal floor and ceiling and vertical walls, but without the restriction of walls meeting at right angles or having a limited number of corners (supporting, e.g., curved walls). In our approach, the original equirectangular image, assumed roughly aligned with the gravity vector, is projected on two arbitrary horizontal planes, one above and one below the camera (see Fig. 2a). Exploiting the *Atlanta World* assumption, this representation encodes all the information needed to recover 3D bounding surfaces of the room, i.e., the 2D floor plan and the room height (see Sect. 4). To predict the 3D layout from this representation, we propose an encoder-decoder architecture, leveraging Recurrent Neural Networks (RNNs) to capture the long-range geometric pattern of room layouts. The network maps a projected image, represented as a tensor, to a binary segmentation mask separating the interior and exterior space, respectively for the ceiling and for the floor projection. The walls footprint is found by extracting a polygonal approximation of the contour of the mask generated from the above-camera image (ceiling mask), and the room height is determined by the scale that maximizes the correlation between the lower and upper contour (see Sect. 4). A customized training strategy based on domain-specific knowledge makes it possible to perform data augmentation and reuse the same network for both projected images. For training, we exploit previously released annotated datasets [6,26,30,33]. Our experimental results (see Sect. 5) demonstrate that our method outperforms state-of-the-art methods [26,30,33] in prediction accuracy, especially on rooms with multiple corners or non-Manhattan layouts. Figure 1 shows some 3D layouts predicted by our method.

Our contributions are summarized as follows:

- We introduce a *data encoding based on the* Atlanta World *indoor model*, that allows layout prediction on planar projections free from spherical image deformations, unlike previous approaches that are predominantly based on features extracted from the equirectangular view [6,26,30,33]. As supported by results, working on such a transformed domain simplifies structure detection [19,21,30]. In addition, representative tensors can be treated as conventional 2D images, simplifying, for example, data augmentation and the use of powerful network architectures such as RNNs [1,2].
- We reconstruct the 3D layout, in terms of 2D footprint and room height, by *inferring the 2D layout from the contour of a solid segmentation masks and the room height from the geometric analysis of the correlation between two contours.* Our approach is more stable and well suited to modeling complex structures, such as curved walls, than previous approaches that infer layout from sparse corner positions [6,26,33]. Moreover, we do not need an additional dense network [30] or a post-processing voting scheme [26] to infer the layout height, which can directly determined from a geometric analysis of the masks.
- We propose an *end-to-end network* that, differently from current state-of-the-art approaches [26,30,33], does not require heavy pre-processing, such as detection of main Manhattan-world directions from vanishing lines analysis [16,32,34] and related image warping, nor complex layout post-processing, such as Manhattan-world regularization of detected features [26,30,33]. Our only requirement is that input images are roughly aligned with the gravity vector, a constraint which is easily met by hardware or software means [9], and is verified in all current benchmark databases. As a result, our method, in addition to being faster, does nor require complex per-image deformations that make multi-view analysis difficult (see discussion in Sect. 6).
- We propose a *training strategy* based on feeding both ceiling and floor view on the same network instance, improving inference performance compared to a dual joined branches architecture or on separate training for ceiling and floor (see results and ablation study at Sect. 5.3).

We tested our approach on both conventional benchmarks (see Zou et al. [34]) and more challenging non-Manhattan scenes annotated by us (see Sect. 5.1). Results demonstrate how our method outperforms previous works on both testing sets (Sect. 5). Code and data are made available at https://github.com/crs4/AtlantaNet.

2 Related Work

3D reconstruction and modeling of indoor scenes has attracted a lot of research in recent years. Here, we analyze only the approaches closer to ours, referring the reader to a very recent survey for a general coverage of the subject [20].

A noticeable series of works concentrate on parsing the room layout from a single RGB image. Since man-made interiors often follow very strict rules, several successful approaches have been proposed by imposing specific priors.

Delage et al. [4] presented one the first monocular approaches to automatically recover a 3D reconstruction from a single indoor image. They adopt a dynamic Bayesian network trained to recognize the *floor-wall* boundary in each column of the image, assuming the indoor scene consists only of a flat floor and straight vertical walls. However, in its original formulation, such a reconstruction is limited to partial views (e.g., a room corner).

Full-view *geometric context* (GC) estimation from appearance priors, i.e., the establishment of a correspondence between image pixels and geometric surface labels, was proposed as a method to analyze outdoor scenes by Hoiem et al. [13]. In combination with Orientation Maps (OM) [16], which are map of local belief of region orientations computed from line segments through heuristic rules, GC is the basis for almost all methods based on geometric reasoning on a single image. Hedau et al. [12], in particular, successfully analyzed the labeling of pixels under the cuboid prior, while Lee et al. [16] considered the less constraining *Indoor World Model* (IWM), i.e., a *Manhattan World* with single-floor and single-ceiling, by noting that projections of building interiors under the Indoor World can be fully represented by corners, so a valid structure can be obtained by imposing geometric constraints on corners. Such a geometric reasoning on IWM supports several efficient reconstruction methods. A notable example is the work of Flint et al. [7,8], who, exploiting the *homography* between floor and ceiling, reduce the structure classification problem to the estimation of the y-coordinate of the ceiling-wall boundary in each image column.

One of the main limitations of single-image methods lies, in fact, on the restricted field of view (FOV) of conventional perspective images, which inevitably results in a limited geometric context [32]. With the emergence of consumer-level 360° cameras, a wide indoor context can now be captured with one or at least few shots. As a result, most of the research on reconstruction from sparse imagery is now focused in this direction. Zhang et al. [32] propose a whole-room 3D context model that maps a full-view panorama to a 3D bounding box of the room, also detecting all major objects inside (e.g, *PanoContext*). By combining OM for the top part and GC for the bottom part, they demonstrate that by using panoramas, their algorithm significantly outperforms results on regular-FOV images. More recently, Xu et al. [27] extended this approach of by assuming IWM instead of a box-shaped room, thus obtaining a more accurate shape of the room, and Yang et al. [28] proposed an algorithm that, starting from a single full-view panorama, automatically infers a 3D shape from a collection of partially oriented super-pixel facets and line segments, exploiting the *Manhattan World* constraint. Pintore et al. [19] tackle the problem of recovering room boundaries in a *top-down 2D* domain, in a manner conceptually similar to that of dense approaches. To recover the shape of the room from the single images they combine the ceiling-floor homography [8] to a spatial transform (E2P - i.e., *equirectangular to perspective*) [19], based on the *Unified projection model* for spherical images [10]. Such E2P transform highlights the shape of the room projected on a 2D floorplan, generating two projections, respectively for the floor and for the ceiling edges. Applying the ceiling-floor homography, they recover the height of the walls and enforce the 2D shape estimation from the projected contours. As for all feature-based methods, the effectiveness of these approaches depend on the quality of extracted features (e.g., edges or flat uniform patches). To overcome these problems, more and more solutions are turning towards data-driven approaches [34].

The peculiarity of indoor reconstruction makes generic segmentation solutions (e.g., U-Net [22] or DeepLab [3]) not appropriate. In particular, defining a graphical model at the pixel-level makes it hard to incorporate global shape priors. Recent data-driven approaches have demonstrated impressive performance in recovering the 3D boundary of a single room meeting the Manhattan World constraint. Zou et al. [33] predict the corner probability map and boundary map of directly from a panorama (e..g, *LayoutNet*). They also extend Stanford 2D-3D dataset [25] with annotated layouts for training and evaluation. Yang et al. [30] propose a deep learning framework, called *DuLa-Net*, which exploits features fusion between the original panoramic view and the ceiling E2P transform [19], to output a floor plan probability map. A Manhattan regularization step is then performed to recover the 2D floor plan shape, through a grid aligned to the main Manhattan axes. Similarly to *LayoutNet* approach [33], a number of recent works [6,26] focus on inferring the room layout from the sparse corners position in the panoramic image. Sun et al. [26] represent room layout as three 1D vectors that encode, at each image column, the boundary positions of floor-wall and ceiling-wall, and the existence of wall-wall boundary. The 2D layout is then obtained by fitting Manhattan World segments on the estimated corner positions.

Recently, Zou et al. [34] have presented an extensive evaluation of the latest high-performance methods. In their classification, such methods basically share the same pipeline: a Manhattan World pre-processing step (e.g., based on Zhang et al. [32]), the prediction of layout elements and a post-processing for fitting the 3D model to the predicted elements after a series of regularization. Differently to almost all recent methods [26,30,33], we do not need complex pre-processing steps, such as computation of Manhattan vanishing lines [16] and warping the panoramic image according to them, but only perform projection along the gravity vector. While our method, like many recent ones, shares with HorizonNet [26] and Dula-Net [30] the encoder-decoder concept, we introduce important novelties in the network architecture. In particular, HorizonNet fully works in a 1D domain derived from the equirectangular projection, while we work entirely in a 2D domain derived from projections on horizontal planes. Moreover, in contrast to Dula-Net, we use a single branch working in the transformed domain (both for floor and ceiling), while Dula-Net uses two parallel branches for the ceiling-view probability and for ceiling-floor probability in the equirectangular domain, plus an additional linear branch for deriving the height. Our results show the advantages of our solution. Furthermore, in contrast to many other works, we

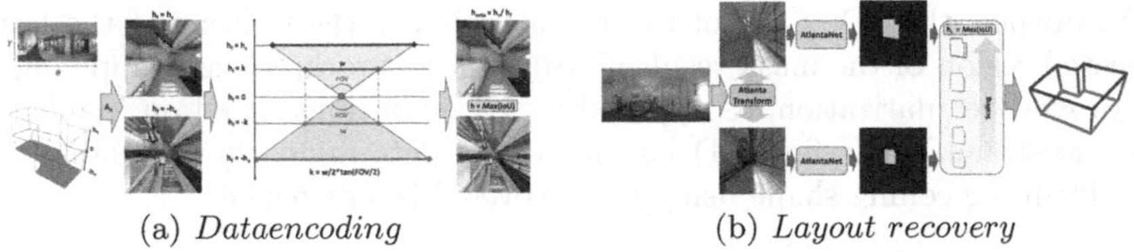

(a) *Dataencoding* (b) *Layout recovery*

Fig. 2. Data encoding and layout prediction. (a): the *Atlanta Transform* A_h maps all the points of the equirectangular image in 3D space as if their height was h_f (focal height), where h_f can assume only two possible values: $-h_e$ (eye height) and h_c (ceiling height). Since at least h_c is an unknown value, we apply the transform by imposing a single, fixed, h_f, which depends by a fixed FOV. (b): We infer through our network the ceiling or floor shapes. The height is directly proportional to the ratio h_r (height ratio) between these shapes, and the 2D footprint of the room is recovered from the ceiling shape.

predict the room layout from dense 2D segmentation maps by simply extracting the largest connected component, rather than from a sparse number of inferred corner positions [6,26]. Such an approach is more robust, particularly in cases of non-Manhattan shapes.

3 Overview

Our method takes as input a single panoramic image, that we assume aligned to the gravity vector. This is easily obtained on all modern mobile devices that have an IMU on board, or can be achieved prior to the application of our pipeline through standard image processing means [9]. Starting from the oriented image, our approach, depicted in Fig. 2 determines the room structure.

The first module generates, from the input equirectangular image (e.g., panorama original size), an *Atlanta Transform* (e.g., $3 \times 1024 \times 1024$) on two horizontal planes placed above and below the camera. For training, the ground truth annotations, conventionally provided on a panoramic image, are transformed in the same way. To simplify discussion, we call the projection on the upper plane the *ceiling projection*, and the projection on the lower plane the *floor projection*. Note, however, that the selected planes do not need to be exactly corresponding to the ceiling or for the floor plane, since the room dimensions are determined automatically by our method and are not known in advance.

During training, the network (see Fig. 3) is fed by alternating ceiling or floor images, according to a probability function (see Sect. 4.3 and Sect. 5.3). In prediction mode, the same trained network is used to infer ceiling or floor shapes.

The height of the layout is directly proportional to the ratio h_r between the ceiling shape and the floor shape (i.e., a scaling factor). Since in real cases, the floor shape is partially occluded by the clutter, we assume as inferred h_r the value that maximizes the intersection-over-union between the contours of the ceiling and the floor shapes (see Fig. 2(b)).

On output, the 2D shape of the room is simply the contour of the largest connected region of the mask resulting from the network, without applying any post-process regularization, as opposed to, e.g., solutions based on Manhattan-world constraints. The final 3D layout is then determined by extruding a 2D shape from the ceiling shape using the recovered layout height.

4 Approach

4.1 Data Encoding

Assuming the *Atlanta World* model [23], we project the panoramic image on two horizontal planes, building, respectively, one representative tensor (i.e., $3 \times 1024 \times 1024$) for the ceiling and one for the floor horizontal plane (see Fig. 2(b)). To transform the equirectangular map we adopt the following relation:

$$A_h(\theta, \gamma, h_f) = \begin{cases} x = h_f / \tan\gamma * \cos\theta \\ y = h_f / \tan\gamma * \sin\theta \\ z = h_f \end{cases} \tag{1}$$

The function A_h, called *Atlanta Transform*, maps all the points of the equirectangular image in 3D space as if their height was h_f [19]. Compared to a classic pin-hole model, h_f can be seen as the focal length for a 180° field-of-view. In the specific case of the Atlanta World model, h_f can assume only two possible values: $-h_e$, that is the floor plane below camera center, and h_c, that is the distance between the camera center and the ceiling plane (see Fig. 2(a)).

Considering h_e a known constant or at most fixed as a scale factor, the 3D layout of an Atlanta model is fully defined by a two-dimensional shape - i.e. the 2D footprint of the layout on the floorplan, and by the ceiling distance h_c. Ideally in order to directly apply Eq. 1 we should also know the value of h_c. Since, in our case, h_c is unknown before reconstruction and must be inferred by the network, we apply a modified version of the transform [30] by imposing a single, fixed, h_f, which depends by a fixed field-of-view (FOV), i.e., $h_f = w/2 * \tan(FOV/2)$, where $w \times w$ is the extent in pixels of each transform (that we assume square). As a consequence, the height of the room is determined by the ratio between h_c and h_e, and is directly proportional to the ratio h_r between the ceiling shape and the floor shape. Ideally, h_r should be the value that makes the floor shape match with the ceiling shape. Since in real cases, the floor shape is heavily occluded by clutter, we assume as inferred h_r the value that maximizes the intersection-over-union between the contours of the ceiling and the floor shapes (see Fig. 2(b)).

4.2 Network Architecture

Figure 3 shows an overview of *AtlantaNet*. The network takes as an input a transform of size $3 \times w \times w$ (see Sect. 4.1) and produces a segmentation mask of size $1 \times w \times w$. We tested different sizes for the input transform, and we found

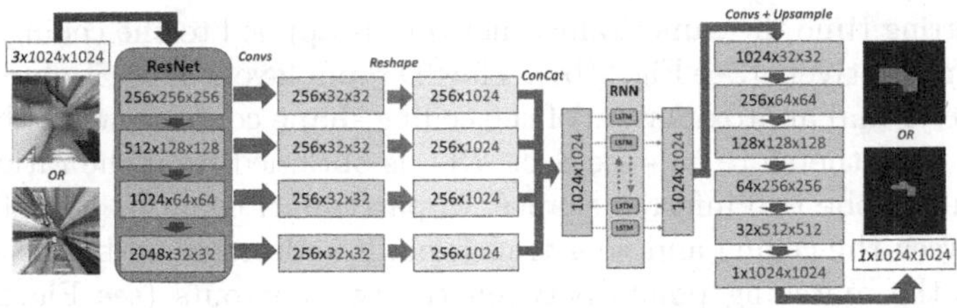

Fig. 3. Network architecture. The network takes as input a transforms of size $3 \times w \times w$ (see Sect. 4.1) and passes it to a ResNet encoder. To capture both low-level and high-level features, we keep the last four feature maps of the encoder. Each feature map is then reduced to the same size, $256 \times 32 \times 32$, through a sequence of convolutional layers and reshaped to 256×1024. The 4 features maps are concatenated to a *sequential feature map* of 1024×1024. We feed such a sequence to a RNN, obtaining, after a reshaping, a $1024 \times 32 \times 32$ map. We upsample such map to recover a $1 \times 1024 \times 1024$ binary segmentation mask describing the ceiling or floor shape.

that 1024×1024 is the best size in terms of performance, so as to guarantee sufficient detail for the most complex forms and not to require large memory resources (see Sect. 5). The size of the output is $1 \times 1024 \times 1024$, that is a binary segmentation mask describing the ceiling or floor shape. We adopt *ResNet* [11] as feature extractor, which has proven to be one of the most effective encoder for both panoramic and perspective images [34]. The output of each *ResNet* block has half spatial resolution compared to that of the previous block. To capture both low-level and high-level features, we keep the last four feature maps of the encoder [26]. Each feature map is then reduced to the same size, $256 \times 32 \times 32$, through a sequence of convolutional layers (*Convs* in Fig. 3), where each layer contains: a 2D convolution having stride 2 (e.g., except for the last block, having stride 1), a batch normalization module and a rectified linear unit function (ReLU). Finally, we reshape the 4 features maps to 256×1024, and we concatenate them layers to obtain a single *sequential feature map* of 1024×1024 (i.e., 1024 layers for a sequence having length 1024).

We feed such a sequence to a RNN, that is exploited to capture the shape of the object and thus make coherent predictions even in ambiguous cases such as occlusions and cluttered scenes. In particular, we employ convolutional LSTM [24] modules in our model as the decoder core. Specifically we adopt a bi-directional LSTM with 512 features in the hidden state and 2 hidden internal layers. The output of the RNN decoder is a 1024×1024 feature map, which collect all the time steps of the RNN layers.

We reshape the RNN output to $1024 \times 32 \times 32$, and, after a a drop-off, we up-sample it through a sequence of 6 convolutional layers (same of *Convs* but with stride 1) each one followed by an interpolation (e.g., factor 2 for each layer). In the final layer of the decoder the ReLU is replaced by Sigmoid. As a result we obtain a prediction mask $1 \times 1024 \times 1024$ of the targeted shape (see Fig. 3).

At inferring time the same trained network is applied to the ceiling and floor transform respectively (See Fig. 2(b)). The 2D room layout $F2D$ is obtained with a simple polygonal approximation of the ceiling shape contour, while the ratio of heights h_r (and therefore h_c - see Sect. 4.1), is obtained from the ratio between the contours of the two inferred shapes. In particular, being h_r actually a scale factor between the ceiling and floor transform, it is determined by the scale that maximizes the matching points between the two contours (see Fig. 2(b)). We build the final 3D model just extruding $F2D$, using h_r to determine h_c and h_e (see Sect. 4.1).

4.3 Training

To train our network, we adopt a specific loss function based on the binary cross entropy error of the predicted pixel probability in the mask M and in its gradient M', compared to ground truth:

$$-\frac{1}{n}\sum_{p=M}(\hat{p}\log p+(1-\hat{p})\log(1-p))-\frac{1}{n}\sum_{q=M'}(\hat{q}\log q+(1-\hat{q})\log(1-q)) \quad (2)$$

where p is the probability of one pixel in M, $\hat{p}$ is the ground truth of p in M, q is the pixel probability in M', $\hat{q}$ is the ground truth, and n is the number of pixels in M and M' which is the transform resolution. The gradient of binary masks is obtained by a Sobel filter of kernel size 3. Even though the gradient component provides a value only near edges, its presence improves the sharpening of the contour in cases of small boundary surface details. This is very important in our case, since in our approach we extract the contour of the largest detected component without performing any post-processing. It also improves noise filtering in highly textured images (see ablation study in Sect. 5.3).

Working completely in a plane-projected domain clearly simplifies data augmentation, compared to panorama augmentation [26]. In practice, for each training iteration, we augment the input panorama set with random rotations and mirrorings, performing all operations in 2D space.

We could separately perform training of an instance for floor prediction and a second instance for ceiling mask prediction, or create an architecture that performs parallel training with a common loss function, or use a single instance capable to handle both ceiling and floors.

In the first case, we experienced, for the ceiling branch training, a tendency to over-fit and a rapid decay of the learning rate after a small number of iterations. At the same time, training the floor branch with only floor images results in rough shapes. This is a predictable behavior, taking into account that the ceiling part usually has cleaner areas but with less features, while in the floor part the architectural structure is more occluded and therefore more difficult to match, alone, with the ground-truth shape of the room [30].

In the second case, we tested two parallel branches by jointly training two instances of *AtlantaNet*, where the loss function is the sum of the ceiling and floor loss respectively. It should be noted that in this case a direct feature fusion is not possible, since this would imply knowledge of the scale factor between the two transformed tensors, which is itself an unknown value. In this case, we obtained an appreciable improvement of the performance compared to single training. However, the resulting shape is not accurate enough, especially in cases of multiple corners or more complex shapes (see results in Sect. 5.3).

We thus adopted a strategy that uses a single *Atlanta Net* instance, but trained to predict indifferently the ceiling or floor shape. To do this, we feed the same network with examples of ceiling and floor transforms, coupled with their respective ground truth. As showed by comparative results (see Sect. 5.3), such a strategy boosts the performances, as it guides the network to find commonalities between clean structures, mostly present in the ceiling transforms, and highly cluttered structures, mostly present in floor transforms.

5 Results

We implemented our method with PyTorch [18], adopting *ResNet50* as feature encoder. The presented results are obtained using the Adam optimizer [14] with $\beta_1 = 0.9$, $\beta_2 = 0.999$ and learning rate 0.0001. We trained the network on 4 NVIDIA RTX 2080Ti GPUs for 300 epochs (best valid around 200 epoch, varying with dataset), with a batch size of 8 ($3 \times 1024 \times 1024$ input size). As an example, training with the *MatterportLayout* [34] dataset takes about 2 min per epoch. The final layout extraction is obtained by applying a simple polygonal approximation [5] to the larger connected region contour (see Sect. 4.2), thus eliminating excess vertices and saving the resulting model as a json file (we adopt the same convention as *MatterportLayout* [34] and *PanoAnnotator* [29]).

5.1 Datasets

We trained *AtlantaNet* using publicly available datasets: *PanoContext* [32], *Stanford 2D-3D* [25] and *Matterport3D* [17]. To simplify comparison, we arrange testing by following the split (*cuboid layout*, or general Manhattan World), adopted by other works [6,26,30,33]. In addiction, we introduce a specific testing set of a hundred images to benchmark more complex Atlanta World cases (*AtlantaLayout*). The testing set was created by annotating a selection of images from *Matterport3D* [17] and *Structured3D* [15]. For cuboid and simple Manhattan layout, we follow the same training/validation/test splitting proposed by LayoutNet [33] and HorizonNet [26], while for general Manhattan World we follow the data split and annotation provided by Zou et al. [34] (e.g., *MatterportLayout*).

To test Atlanta World layouts, we extend existing testing set with annotated 3D layouts having less restrictive assumptions, as, for example, rooms with curved walls or non-right corner angles. In this case, to ensure a fair evaluation we have prepared the test set by combining the new annotations with a subset of test images taken from the *MatterportLayout* testing set.

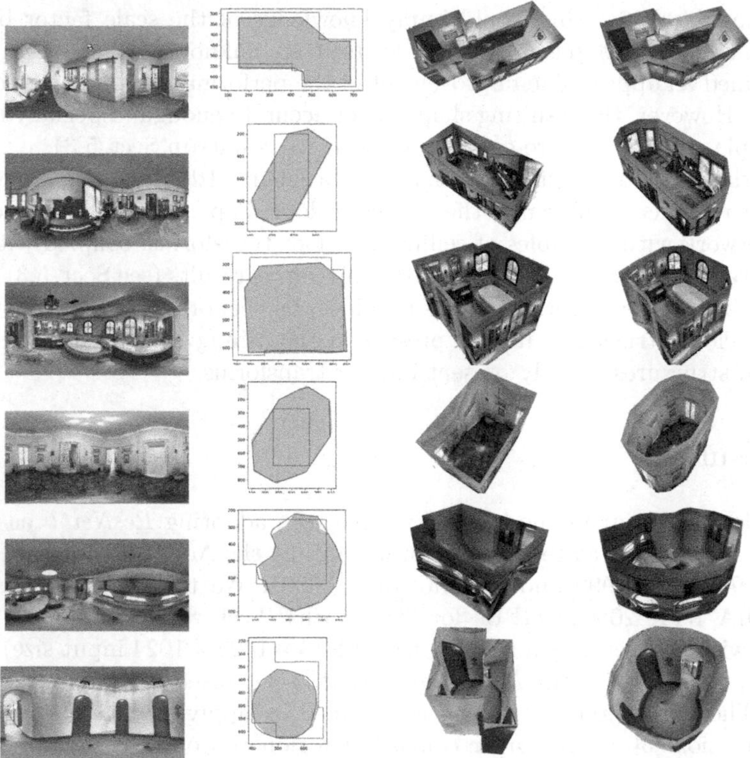

Fig. 4. Qualitative results and comparison. For each row, we show: the original panoramic image annotated with our reconstruction; an intersection-over-union visual comparison, between our approach (green line), HorizonNet [26] (red line) and ground truth (azure mask); the 3D layout obtained with the compared approach [26] (third column) and with ours (fourth column). (Color figure online)

5.2 Performance

We evaluate the performance of our approach by following the standard evaluation metrics proposed by Zou et al. [34] and adopted by others [26,30,33]. Specifically, we considered the following metrics: $3DIoU$ (volumetric intersection-over-union), $2DIoU$ (pixel-wise intersection-over-union), $cornererror$ (L2 distance normalized to bounding box diagonal), $pixelerror$ (floor, ceiling, wall labeling accuracy of the original image) and δ_i (percentage of pixels where the ratio between the prediction label and the ground truth label is within a threshold of 1.25). Following Zou et al. [34], we adopt $3DIoU$, $cornererror$ and $pixelerror$ for cuboid layouts, and $3DIoU$, $2DIoU$, δ_i for other layouts.

We present a comparison with recent state-of-the-art methods [6,26,30,33] for which comparable results are published or for which source code and data are available. For comparison purposes, we adhere to the methodology reported in the mentioned papers, and we split results into *Cuboid layouts*, *General Man-*

Table 1. Cuboid layout performance. All the methods have been tested with the same *S-2D-3D* testing set [25] and trained with the enlisted training sets. Our method, even without Manhattan World pre-processing and regularization, is aligned with the performance of the best state-of-art methods that exploit Manhattan-world constraints.

Training dataset	PanoContext			S-2D-3D			PC+Stanford		
Metrics [%]	3D IoU	Corner error	Pixel error	3D IoU	Corner error	Pixel error	3D IoU	Corner error	Pixel error
CFL [6]	65.13	1.44	4.75	–	–	–	–	–	–
LayoutNet [33]	–	–	–	76.33	1.04	2.70	82.66	0.83	2.59
Dula-Net [30]	–	–	–	79.36	–	–	**86.60**	**0.67**	2.48
HorizonNet [26]	**75.57**	**0.94**	3.18	79.79	0.71	2.39	82.66	0.69	2.27
Ours	75.56	0.96	**3.05**	**82.43**	**0.70**	**2.25**	83.94	0.71	**2.18**

Table 2. General layout performance. All methods are trained with the same *MatterportLayout* dataset [34] and tested on the *MatterportLayout* test set and on a specific set of complex Manhattan and non-Manhattan scenes (e.g., AtlantaLayout). For Dula-Net [30] performance we refer to the latest available results using *MatterportLayout* training [34]. *>10 - corners-odd* row refer to complex layouts, including curved walls.

	Dula-Net [30]			HorizonNet [26]			Ours		
	3D IoU	2D IoU	δ_i	3D IoU	2D IoU	δ_i	3D IoU	2D IoU	δ_i
Manhattan 4 corners	77.02	81.12	0.818	81.88	84.67	0.945	**82.64**	**85.12**	**0.950**
Manhattan 6 corners	78.79	82.69	0.859	**82.26**	**84.82**	**0.938**	80.10	82.00	0.815
Manhattan 8 corners	71.03	74.00	0.823	71.78	73.91	0.903	**71.79**	**74.15**	**0.911**
Manhattan >10 corners	63.27	66.12	0.741	68.32	70.58	0.861	**73.89**	**76.93**	**0.915**
Manhattan Overall	75.05	78.82	0.818	79.11	81.71	0.929	**81.59**	**84.00**	**0.945**
Atlanta 6 corners	–	–	–	74.45	77.13	0.862	**84.26**	**88.78**	**0.972**
Atlanta 8 corners	–	–	–	65.00	66.93	0.820	**78.37**	**80.50**	**0.907**
Atlanta >10 corners-odd	–	–	–	64.40	67.72	0.812	**75.34**	**77.75**	**0.870**
Atlanta Overall	–	–	–	67.08	70.57	0.845	**72.50**	**76.49**	**0.879**
Atlanta FT Overall	–	–	–	73.53	76.38	0.851	**80.01**	**84.33**	**0.924**

hattan World and *Atlanta World*, preserving the same metrics and setup of the original papers. All results are collected with the same $ResNet50$ feature encoder. Missing fields in tables indicate cases not reported in original papers.

Table 1 reports on performance obtained on *Cuboid layouts*, a worst-case comparison for our method, since, in contrast to competitors, we do not assume that walls must meet at right angles. Following the same convention presented by Sun et al. [26] and Zou et al. [34], the networks have been trained with three different datasets (i.e., PanoContext, Stanford 2D-3D-S, both of them), and tested with same testing set - e.g. Stanford 2D-3D-S [25]. Results demonstrate how our approach, on these constrained indoors, has a performance similar to state-of-the-art approaches tuned for Manhattan-world environments, although it does not employ any specific post-processing and cuboid regularization.

In Table 2, we report on performance obtained on *General Manhattan World* and *Atlanta World* layouts (see Sect. 5.1).

We compare our method results with results for methods having best performance in general Manhattan cases [26,30]. All the tested approaches are trained with the same *MatterportLayout* dataset [34] and evaluated both on the *MatterportLayout* testing set (labeled Manhattan in Table 2) and on a specific testing set (labeled Atlanta in Table 2), containing more complex shapes, such as non-right angles and curved walls. For Dula-Net [30] performance, we refer to the latest available results obtained by training with the *MatterportLayout* dataset by Zou et al. [34]. The *Atlanta FT Overall* line presents, in addition, results that have been obtained by augmenting the *MatterportLayout* training dataset with selected Atlanta scenes for fine-tuning. The results demonstrate the accuracy of our approach with both testing sets, and how it outperforms other approaches as the layout complexity grows. It should be noted that a portion of the error depends, for all the approaches, by the approximated ground truth annotation, which clearly affects both training and performance evaluation.

In Fig. 4, we show a selection of scenes for a qualitative evaluation of our method compared to ground truth and *HorizonNet* [26]. At the first column we show the original panoramic image annotated with our results. It should be noted how, in these complex cases, even the manual labeling of an equirectangular image is not trivial, as well as the visual understanding of the room structure. In order to provide a more intuitive comparison, we show, besides, the intersection-over-union of the recovered layout (green) with the ground truth floorplan (azure mask) and the same layout reconstructed by HorizonNet [26] (red). In the third and fourth column, we show the 3D layout obtained, respectively, with our approach and with the *HorizonNet* approach [26]. Visual results confirm numerical performances in terms of footprint and height recovery.

5.3 Ablation Study

Table 3. Ablation. The ablation study demonstrates how our proposed designs improve the accuracy of prediction. Results are sorted by increasing performance, showing only those cases that actually increase it.

Backbone	Setup	Gradient loss	3D IoU	2D IoU	δ_i	Train. params
ResNet50	Two instances trained separately		75.48	78.26	0.856	200M
ResNet50	Two instances trained jointly		76.04	79.92	0.815	200M
ResNet50	One instance and mixed feeding		79.26	83.35	0.854	100M
ResNet50	One instance and mixed feeding	V	80.79	84.12	0.902	100M
Resnet101	One instance and mixed feeding	V	83.22	86.96	0.940	119M

Our ablation experiments are presented in Table 3. We report the results averaged across general Manhattan and Atlanta World testing instances (Table 2).

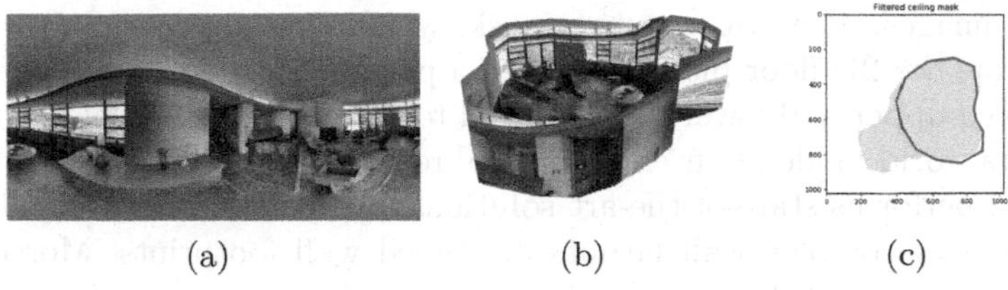

(a) (b) (c)

Fig. 5. Failure case. (a) shows a circular room where the ceiling level is not correctly identified, resulting in the wrong layout of (b) and (c) (ground truth as green line). (Color figure online)

First, we tested, with the same *ResNet50* backbone and without gradient loss function (Sect. 4.3), different configurations: two instances trained separately, two instances trained jointly with a common (overall) loss function and the adopted mixed approach. While the difference between separate and joined training of two instances is quite small, results confirm instead that the mixed feeding approach (see Sect. 4.3) provides a consistent performance boost. For the winning set-up (One instance and mixed feeding), we also evaluate the contribution of the gradient loss component. Including the gradient leads to an accuracy improvement, mainly due to increased performance with more complex shapes.

At last, we show how our method changes its performance by adopting a deeper backbone - i.e., *ResNet101*. While the *ResNet50* encoder (also adopted by compared works) provides consistent results for the given datasets (see Sect. 5.1), increasing the backbone depth appears to be a better option for more complex layouts.

5.4 Limitations and Failure Cases

Our method is trained to return a single connected region for each projection (ceiling and floor), containing the information needed to recover the room layout (see Sect. 4.2). Figure 5 shows an example where the layout of a semi-circular room (Fig. 5(a)) is wrongly predicted. Although geometrically self-consistent (see recovered 3D at Fig. 5(b)), the recovered shape (yellow ceiling mask in Fig. 5(c)) does not describe the real room layout (green annotation). From the topological point-of-view, this happens where the horizontal planes are not clearly identifiable, so, in our example, when the horizontal ceiling is partially occluded by other horizontal structures.

6 Conclusions

We have introduced a novel end-to-end approach to predict the 3D room layout from a single panoramic image. We project the original panoramic image on two horizontal planes, one above and one below the camera, and use a suitably trained deep neural network to recover the inside-outside segmentation mask of these two images. The upper image mask, which contains less clutter, is used to determine the 2D floor plan in form of a polygonal layout, while the correlation between upper and lower mask is used to determine the room height under the Atlanta world model. Our experimental results clearly demonstrate that our method outperforms state-of-the-art solutions in prediction accuracy, in particular in cases of complex wall layouts or curved wall footprints. Moreover, the method requires much less pre- and post-processing than competing solutions based on the more constraining Manhattan world model.

Our current work is concentrating in several directions. In particular, we are planning to exploit multiple images to perform a multi-view recovery of rooms with large amount of clutter or complex convex shapes. Moreover, we are also working on the integration of this approach in a multi-room structured reconstruction environment, in order to automatically reconstruct complete building floors.

Acknowledgments. This work has received funding from Sardinian Regional Authorities under projects VIGECLAB, AMAC, and TDM (POR FESR 2014-2020). We also acknowledge the contribution of the European Union's H2020 research and innovation programme under grant agreements 813170 (EVOCATION).

References

1. Acuna, D., Ling, H., Kar, A., Fidler, S.: Efficient interactive annotation of segmentation datasets with Polygon-RNN++. In: Proceedings of CVPR (2018)
2. Castrejon, L., Kundu, K., Urtasun, R., Fidler, S.: Annotating object instances with a Polygon-RNN. In: Proceedings of CVPR (2017)
3. Chen, L.C., Papandreou, G., Kokkinos, I., Murphy, K., Yuille, A.L.: Deeplab: semantic image segmentation with deep convolutional nets, Atrous convolution, and fully connected CRFs. IEEE TPAMI **40**(4), 834–848 (2017)
4. Delage, E., Honglak Lee, Ng, A.Y.: A dynamic Bayesian network model for autonomous 3D reconstruction from a single indoor image. In: Proceedings of CVPR, vol. 2, pp. 2418–2428 (2006)
5. Douglas, D.H., Peucker, T.K.: Algorithms for the reduction of the number of points required to represent a digitized line or its caricature. Cartographica: Int. J. Geogr. Inf. Geovisual. **10**(2), 112–122 (1973)
6. Fernandez-Labrador, C., Fácil, J.M., Perez-Yus, A., Demonceaux, C., Civera, J., Guerrero, J.J.: Corners for layout: End-to-end layout recovery from 360 images (2019). arXiv:1903.08094
7. Flint, A., Murray, D., Reid, I.: Manhattan scene understanding using monocular, stereo, and 3D features. In: Proceedings of ICCV, pp. 2228–2235 (2011)
8. Flint, A., Mei, C., Murray, D., Reid, I.: A dynamic programming approach to reconstructing building interiors. In: Daniilidis, K., Maragos, P., Paragios, N. (eds.) ECCV 2010. LNCS, vol. 6315, pp. 394–407. Springer, Heidelberg (2010). https://doi.org/10.1007/978-3-642-15555-0_29
9. Gallagher, A.C.: Using vanishing points to correct camera rotation in images. In: Proceedings of CVR, pp. 460–467 (2005)
10. Geyer, C., Daniilidis, K.: A unifying theory for central panoramic systems and practical implications. In: Vernon, D. (ed.) ECCV 2000. LNCS, vol. 1843, pp. 445–461. Springer, Heidelberg (2000). https://doi.org/10.1007/3-540-45053-X_29

11. He, K., Zhang, X., Ren, S., Sun, J.: Deep residual learning for image recognition. In: Proceedings of CVPR, pp. 770–778 (2016)
12. Hedau, V., Hoiem, D., Forsyth, D.: Recovering the spatial layout of cluttered rooms. In: Proceedings of ICCV, pp. 1849–1856 (2009)
13. Hoiem, D., Efros, A.A., Hebert, M.: Recovering surface layout from an image. Int. J. Comput. Vision **75**(1), 151–172 (2007). https://doi.org/10.1007/s11263-006-0031-y
14. Kingma, D.P., Ba, J.: Adam: a method for stochastic optimization (2014)
15. Kujiale.com: Structured3D Data (2019). https://structured3d-dataset.org/. Accessed 25 Sept 2019
16. Lee, D.C., Hebert, M., Kanade, T.: Geometric reasoning for single image structure recovery. In: Proceedings of CVPR, pp. 2136–2143 (2009)
17. Matterport: Matterport3D (2017). https://github.com/niessner/Matterport. Accessed 25 Sept 2019
18. Paszke, A., et al.: Automatic differentiation in pytorch. In: Proceedings of NIPS (2017)
19. Pintore, G., Garro, V., Ganovelli, F., Agus, M., Gobbetti, E.: Omnidirectional image capture on mobile devices for fast automatic generation of 2.5D indoor maps. In: Proceedings of IEEE WACV, pp. 1–9 (2016)
20. Pintore, G., Mura, C., Ganovelli, F., Fuentes-Perez, L., Pajarola, R., Gobbetti, E.: State-of-the-art in automatic 3D reconstruction of structured indoor environments. Comput. Graph. Forum **39**(2), 667–699 (2020)
21. Pintore, G., Pintus, R., Ganovelli, F., Scopigno, R., Gobbetti, E.: Recovering 3D existing-conditions of indoor structures from spherical images. Comput. Graph. **77**, 16–29 (2018)
22. Ronneberger, O., Fischer, P., Brox, T.: U-Net: convolutional networks for biomedical image segmentation. In: Navab, N., Hornegger, J., Wells, W.M., Frangi, A.F. (eds.) MICCAI 2015. LNCS, vol. 9351, pp. 234–241. Springer, Cham (2015). https://doi.org/10.1007/978-3-319-24574-4_28
23. Schindler, G., Dellaert, F.: Atlanta world: an expectation maximization framework for simultaneous low-level edge grouping and camera calibration in complex man-made environments. In: Proceedings of CVPR, vol. 1, p. 1 (2004)
24. Shi, X., Chen, Z., Wang, H., Yeung, D.Y., Wong, W., Woo, W.: Convolutional LSTM network: a machine learning approach for precipitation nowcasting. In: Proceedings of NIPS, pp. 802–810 (2015)
25. Stanford University: BuildingParser Dataset (2017). http://buildingparser.stanford.edu/dataset.html. Accessed 25 Sept 2019
26. Sun, C., Hsiao, C.W., Sun, M., Chen, H.T.: HorizonNet: learning room layout with 1D representation and pano stretch data augmentation. In: Proceedings of CVPR (2019)
27. Xu, J., Stenger, B., Kerola, T., Tung, T.: Pano2CAD: room layout from a single panorama image. In: Proceedings of WACV, pp. 354–362 (2017)
28. Yang, H., Zhang, H.: Efficient 3D room shape recovery from a single panorama. In: Proceedings of CVPR, pp. 5422–5430 (2016)
29. Yang, S.T., Peng, C.H., Wonka, P., Chu, H.K.: PanoAnnotator: a semi-automatic tool for indoor panorama layout annotation. In: Proceedings of SIGGRAPH Asia 2018 Posters, pp. 34:1–34:2 (2018)
30. Yang, S.T., Wang, F.E., Peng, C.H., Wonka, P., Sun, M., Chu, H.K.: DuLa-Net: a dual-projection network for estimating room layouts from a single RGB panorama. In: Proceedings of CVPR (2019)
31. Yang, Y., Jin, S., Liu, R., Yu, J.: Automatic 3D indoor scene modeling from single panorama. In: Proceedings of CVPR, pp. 3926–3934 (2018)

32. Zhang, Y., Song, S., Tan, P., Xiao, J.: PanoContext: a whole-room 3D context model for panoramic scene understanding. In: Fleet, D., Pajdla, T., Schiele, B., Tuytelaars, T. (eds.) ECCV 2014. LNCS, vol. 8694, pp. 668–686. Springer, Cham (2014). https://doi.org/10.1007/978-3-319-10599-4_43
33. Zou, C., Colburn, A., Shan, Q., Hoiem, D.: LayoutNet: reconstructing the 3D room layout from a single RGB image. In: Proceedings of CVPR, pp. 2051–2059 (2018)
34. Zou, C., et al.: 3D Manhattan room layout reconstruction from a single 360 image (2019)

AttentionNAS: Spatiotemporal Attention Cell Search for Video Classification

Xiaofang Wang[2](✉), Xuehan Xiong[1], Maxim Neumann[1], AJ Piergiovanni[1], Michael S. Ryoo[1], Anelia Angelova[1], Kris M. Kitani[2], and Wei Hua[1]

[1] Google, Mountain View, USA
[2] Carnegie Mellon University, Pittsburgh, USA
xiaofan2@cs.cmu.edu

Abstract. Convolutional operations have two limitations: (1) do not explicitly model where to focus as the same filter is applied to all the positions, and (2) are unsuitable for modeling long-range dependencies as they only operate on a small neighborhood. While both limitations can be alleviated by attention operations, many design choices remain to be determined to use attention, especially when applying attention to videos. Towards a principled way of applying attention to videos, we address the task of spatiotemporal attention cell search. We propose a novel search space for spatiotemporal attention cells, which allows the search algorithm to flexibly explore various design choices in the cell. The discovered attention cells can be seamlessly inserted into existing backbone networks, e.g., I3D or S3D, and improve video classification accuracy by more than 2% on both Kinetics-600 and MiT datasets. The discovered attention cells outperform non-local blocks on both datasets, and demonstrate strong generalization across different modalities, backbones, and datasets. Inserting our attention cells into I3D-R50 yields state-of-the-art performance on both datasets.

Keywords: Attention · Video classification · Neural architecture search

1 Introduction

One major contributing factor to the success of neural networks in computer vision is the novel design of network architectures. In early work, most network architectures [10,12,28] were manually designed by human experts based on their knowledge and intuition of specific tasks. Recent work on neural architecture search (NAS) [15,16,21,41,42] proposes to directly learn the architecture

X. Wang—Work done while an intern at Google.

Electronic supplementary material The online version of this chapter (https://doi.org/10.1007/978-3-030-58598-3_27) contains supplementary material, which is available to authorized users.

A. Vedaldi et al. (Eds.): ECCV 2020, LNCS 12353, pp. 449–465, 2020.
https://doi.org/10.1007/978-3-030-58598-3_27

for a specific task from data and discovered architectures have been shown to outperform human-designed ones.

Convolutional Neural Networks (CNNs) have been the *de facto* architecture choice. Most work in computer vision uses convolutional operations as the primary building block to construct the network. However, convolutional operations still have their limitations. It has been shown that attention is complementary to convolutional operations, and they can be combined to further improve performance on vision tasks [2,32,33].

While being complementary to convolution, many design choices remain to be determined to use attention. The design becomes more complex when applying attention to videos, where the following questions arise: *What is the right dimension to apply an attention operation to videos? Should an operation be applied to the temporal, spatial, or spatiotemporal dimension? How to compose multiple attention operations applied to different dimensions?*

Towards a principled way of applying attention to videos, we address the task of spatiotemporal attention cell search, i.e., the automatic discovery of cells that use attention operations as the primary building block. The discovered attention cells can be seamlessly inserted into a wide range of backbone networks, e.g., I3D [5] or S3D [36], to improve the performance on video understanding tasks.

Specifically, we propose a search space for spatiotemporal attention cells, which allows the search algorithm to flexibly explore all of the aforementioned design choices in the cell. The attention cell is constructed by composing several primitive attention operations. Importantly, we consider two types of primitive attention operations: (1) map-based attention [19,33] and (2) dot-product attention (a.k.a., self-attention) [2,30,32]. Map-based attention explicitly models where to focus in videos, compensating for the fact that convolutional operations apply the same filter to all the positions in videos. Dot-product attention enables the explicit modeling of long-range dependencies between distant positions in videos, accommodating the fact that convolutional operations only operate on a small and local neighborhood.

We aim to find an attention cell from the proposed search space such that the video classification accuracy is maximized when adding that attention cell into the backbone network. But the search process can be extremely costly. One significant bottleneck of the search is the need to constantly evaluate different attention cells. Evaluating the performance of an attention cell typically requires training the selected attention cell as well as the backbone network from scratch, which can take days on large-scale video datasets, e.g., Kinetics-600 [4].

To alleviate this bottleneck, we consider two search algorithms: (1) Gaussian Process Bandit (GPB) [25,26], which judiciously selects the next attention cell for evaluation based on the attention cells having been evaluated so far, allowing us to find high-performing attention cells within a limited number of trials; (2) differentiable architecture search [16], where we develop a differentiable formulation of the proposed search space, making it possible to jointly learn the attention cell design and network weights through back-propagation, without explicitly sampling and evaluating different cells. The entire differentiable

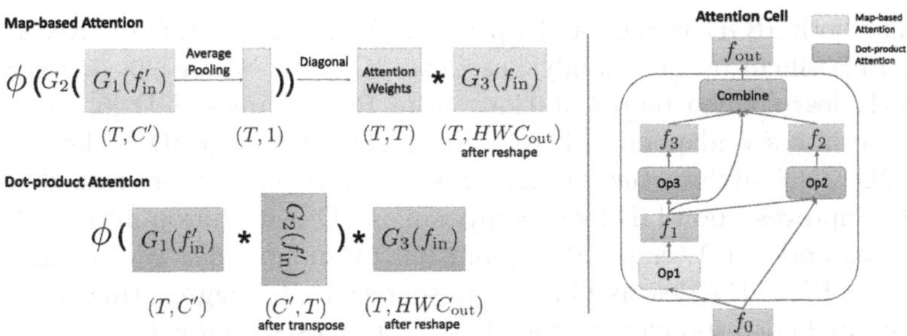

Fig. 1. Illustration of the operation-level search space (left) and cell-level search space (right). The example attention operations use temporal as the attention dimension and the tuple under each feature map denotes its shape.

search process only consumes a computational cost similar to fully training one network on the training videos. This formulation also allows us to learn position-specific attention cell designs with zero extra computational cost (see Sect. 4.2 for details).

We conduct extensive experiments on two benchmark datasets: Kinetics-600 [4] and Moments in Time (MiT) [18]. Our discovered attention cells can improve the performance of two backbone networks I3D [5] and S3D [36] by more than 2% on both datasets, and also outperforms non-local blocks – the state-of-the-art manually designed attention cells for videos. Inserting our attention cells into I3D-R50 [32] yields state-of-the-art performance on both datasets. Notably, our discovered attention cells can also generalize well across modalities (RGB to optical flow), backbones (e.g., I3D to S3D or I3D to I3D-R50), and datasets (MiT to Kinetics-600 or Kinetics-600 to MiT).

Contributions: (1) This is the first attempt to extend NAS beyond discovering convolutional cells to attention cells. (2) We propose a novel search space for spatiotemporal attention cells that use attention operations as the primary building block, which can be seamlessly inserted into existing backbone networks to improve their performance on video classification. (3) We develop a differentiable formulation of the proposed search space, making it possible to learn the attention cell design with back-propagation and learn position-specific attention cell designs with zero extra cost. (4) Our discovered attention cells outperform non-local blocks, on both the Kinetics-600 and MiT dataset. We achieve state-of-the-art performance on both datasets by inserting our discovered attention cells into I3D-R50. Our attention cells also demonstrate strong generalization capability when being applied to different modalities, backbones, or datasets.

2 Related Work

Video Classification. Early work on video classification extends image classification CNNs with recurrent networks [6,38] or two-stream architectures [8,24] that take both RGB frames and optical flow frames as inputs. Recent work on video classification are mainly based on 3D convolution [29] or its variants to directly learn video representations from RGB frames. I3D [5] proposes to inflate the filters and pooling kernels of a 2D CNN into 3D to leverage successful 2D CNN architecture designs and their ImageNet pretrained weights. S3D [36] improves upon I3D by decomposing a 3D convolution into a 2D spatial convolution and a 1D temporal convolution. A similar idea is also explored in P3D [20]. CPNet [17] learns video representations by aggregating information from potential correspondences. SlowFast [7] proposes an architecture operating at two different frame rates, where spatial semantics are learned on low frame rates, and temporal dynamics are learned on high frame rates. Different from them, we do not focus on proposing novel CNN architecture designs for video classification. Instead, we focus on discovering attention cells using attention operations as the primary building block, which are complementary to CNNs.

Attention in Vision. Both map-based attention and dot-product attention are useful for computer vision tasks. Map-based attention [19,33] has been used to improve the performance of CNNs on image recognition, where spatial attention maps are learned to scale the features given by convolutional layers. Dot-product attention [30] is successfully used in sequence modeling and transduction tasks, e.g., machine translation, and is recently used to augment CNNs and enhances their performance on image recognition [2]. Non-local blocks [32] are proposed to capture long-range dependencies in videos and can significantly improve the video classification accuracy of CNNs. Non-local blocks can be viewed as applying one single dot-product attention operation to the spatiotemporal dimension. In contrast, our attention cells can contain multiple attention operations applied to different dimensions of videos. Non-local blocks are a particular case in our proposed search space, and our attention cells are discovered automatically in a data-driven way instead of being manually designed.

NAS - Search Space. Search space is crucial for NAS. Randwire [35] shows that one random architecture from a carefully designed search space can achieve competitive performance on image recognition. NASNet [42] proposes to search for convolutional cells that can be stacked multiple times to form the entire architecture. Auto-DeepLab [14] proposes a two-level hierarchical architecture search space for semantic image segmentation. AssembleNet [23] proposes to search for the connectivity between multi-stream convolutional blocks for video classification. They all focus on finding convolutional cells or networks for the end task. Different from them, our proposed search space uses attention as the primary building component instead of convolution.

NAS - Search Algorithm. Various search algorithms have been explored in NAS, such as random search [13,37], reinforcement learning [1,39,41,42], evolutionary algorithms [21,22,34], Bayesian optimization (BO) [3,11], and differentiable methods [16]. We have tried using GPB (belonging to the category of BO) to search for desired attention cells. We also develop a differentiable formulation of our proposed search space. This makes it possible to conduct the search using differentiable methods and greatly improves the search speed.

3 Attention Cell Search Space

We aim to search for spatiotemporal attention cells, which can be seamlessly inserted into a wide range of backbone networks, e.g., I3D [5] or S3D [36], to improve the performance on video understanding tasks.

Formally, an attention cell takes a 4D feature map of shape (T, H, W, C) as input and outputs a feature map of the same shape. T, H, and W are the temporal dimension, height, and width of the feature map, respectively. C denotes the number of channels. The output of an attention cell is enforced to have the same shape as its input by design, so that the discovered attention cells can be easily inserted after any layers in any existing backbone networks.

An attention cell is composed of K primitive attention operations. The proposed attention cell search space consists of an operation level search space and a cell level search space (see Fig. 1). The operation level search space contains different choices to instantiate an individual attention operation. The cell level search space consists of different choices to compose the K operations to form a cell, i.e., the connectivity between the K operations within a cell. We first introduce the operation level search space and then the cell level search space.

3.1 Operation Level Search Space

An attention operation takes a feature map of shape (T, H, W, C_{in}) as input and outputs an attended featured map of shape $(T, H, W, C_{\text{out}})$. For an attention operation, C_{in} and C_{out} can be different. To construct an attention operation, we need to make two fundamental choices: the dimension to compute the attention weights and the type of the attention operation.

Attention Dimension. For brevity, we term the dimension to compute the attention weights as *attention dimension*. In CNNs for video classification, previous work [7,20,36] has studied when to use temporal convolution (e.g., $3 \times 1 \times 1$), spatial convolution (e.g., $1 \times 3 \times 3$), and spatiotemporal convolution (e.g., $3 \times 3 \times 3$). It is also a valid question to ask for attention what is the right dimension to apply an attention operation to videos: temporal, spatial or spatiotemporal (temporal and spatial together). The choice of the attention dimension is important as computing attention weights for different dimensions represents focusing on different aspects of the video.

Attention Operation Type. We consider two types of attention operations, each of which helps address a specific limitation of convolutional operations, as mentioned in the introduction:

- **Map-based attention** [19,33]: Map-based attention learns a weighting factor for each position in the attention dimension and scales the feature map with the learned attention weights. Map-based attention explicitly models what positions in the attention dimension to attend to in videos.
- **Dot-product attention** [2,30,32]: A dot-product attention operation computes the feature response at a position as a weighted sum of features of all the positions in the attention dimension, where the weights are determined by a similarity function between features of all the positions [2,32]. Dot-product attention explicitly models the long-range interactions among distant positions in the attention dimension.

We now describe the details of the two types of attention operations. Let f_{in} denote the input feature map to an attention operation and denote its shape as (T, H, W, C_{in}). Applying an attention operation consists of three steps, including reshaping the input feature map f_{in}, computing the attention weights, and applying the attention weights.

Reshape f_{in}. We reshape f_{in} into a 2D feature map f'_{in} before computing the attention weights. The first dimension of f'_{in} is the attention dimension and the second dimension contains the remaining dimensions. For example, f'_{in} has the shape of (T, HWC_{in}) when temporal is the attention dimension and has the shape of (THW, C_{in}) when spatiotemporal is the attention dimension. We denote this procedure as a function `ReshapeTo2D`, i.e., $f'_{\text{in}} = \texttt{ReshapeTo2D}(f_{\text{in}})$.

Spatial attention requires extra handling. As video content changes over time, when applying attention to the spatial dimension, each frame f^t_{in} should have its own spatial attention weights, where f^t_{in} is the tth frame in f_{in} and has the shape of (H, W, C_{in}). Therefore, when spatial is the attention dimension, instead of reshaping the entire 4D feature map f_{in}, we reshape f^t_{in} into a 2D feature map f'^t_{in} of shape (HW, C_{in}) for every t, i.e., $f'^t_{\text{in}} = \texttt{ReshapeTo2D}(f^t_{\text{in}})(1 \leq t \leq T)$.

Map-Based Attention. Assuming temporal is the attention dimension, map-based attention generates T attention weights to scale the feature map of each temporal frame. The attention weights are computed as follows:

$$W_{\text{map}} = \texttt{Diag}(\phi(G_2(\texttt{AvgPool}(G_1(f'_{\text{in}}))))). \quad (1)$$

G_1 is a 1D convolutional layer with kernel size as 1, which reduces the dimension of the feature response of each temporal frame from HWC_{in} to C' and gives a feature map of shape (T, C'). `AvgPool` denotes an average pooling operation applied to each temporal dimension and outputs a T-dim vector. The multilayer perceptron G_2 and the activation function ϕ (e.g., the sigmoid function) further transform the T-dim vector to T attention weights. More details about the activation function are discussed later. `Diag` rearranges the T attention weights into a $T \times T$ matrix, where the T attention weights are placed on the diagonal of the matrix. The obtained attention weight matrix W is a diagonal matrix.

Similarly, when spatiotemporal is the attention dimension, map-based attention gives a $THW \times THW$ diagonal matrix containing the attention weights.

When spatial is the attention dimension, we generate one $HW \times HW$ diagonal matrix for every f'^{t}_{in} ($1 \leq t \leq T$) separately, using the above described procedure. Note that while different frames have separate spatial attention weights, G_1 and G_2 are shared among different frames when computing attention weights.

Dot-Product Attention. When applying dot-product attention to the temporal dimension, a $T \times T$ attention weight matrix is generated as follows:

$$W_{\text{dot-prod}} = \phi(G_1(f'_{\text{in}})G_2(f'_{\text{in}})^T). \tag{2}$$

Here, G_1 and G_2 are both a 1D convolutional layer with kernel size as 1 and they both output a feature map of shape (T, C'). Let $Q = G_1(f'_{\text{in}})$ and $K = G_2(f'_{\text{in}})$. QK^T computes an similarity matrix between the features of all the temporal frames. We then use ϕ, an activation function of our choice, e.g., the softmax function, to convert the similarity matrix into attention weights. Note that different from W_{map}, $W_{\text{dot-prod}}$ is a full matrix instead of a diagonal matrix.

When being applied to the spatiotemporal dimension, dot-product attention generates a $THW \times THW$ attention weight matrix. When applying dot-product attention to the spatial dimension, each frame has its own attention weights (a $HW \times HW$ matrix), where G_1 and G_2 are shared among different frames.

Apply the Attention Weights. We apply the attention weight matrix to the input feature map through matrix multiplication to obtain the attended feature map:

$$f_{\text{out}} = \texttt{ReshapeTo2D}^{-1}(W\texttt{ReshapeTo2D}(G_3(f_{\text{in}}))). \tag{3}$$

W is the weight matrix generated by map-based attention (W_{map}) or dot-product attention ($W_{\text{dot-prod}}$). G_3 is a $1 \times 1 \times 1$ convolutional layer to reduce the number of channels of f_{in} from C_{in} to C_{out}. If temporal is the attention dimension, W has the shape of (T, T) and $\texttt{ReshapeTo2D}(G_3(f_{\text{in}}))$ has the shape (T, HWC_{out}). $\texttt{ReshapeTo2D}^{-1}$ is the inverse function of `ReshapeTo2D`, reshaping the attended feature map back to the shape of $(T, H, W, C_{\text{out}})$.

For spatial attention, the attention weights are applied to each frame independently, i.e., $f^t_{\text{out}} = \texttt{ReshapeTo2D}^{-1}(W^t\texttt{ReshapeTo2D}(G_3(f^t_{\text{in}})))$, where W^t is the spatial attention weights for frame t and f^t_{out} has the shape of (H, W, C_{out}). We stack $\{f^t_{\text{out}} \mid 1 \leq t \leq T\}$ along the temporal dimension to form the attended feature map f_{out} of shape $(T, H, W, C_{\text{out}})$. Similar to G_1 and G_2 used for computing attention weights, G_3 is also shared among different frames.

Note that by design G_3 only changes number of channels, i.e., transforms the features at each spatiotemporal position. The spatiotemporal structure of the input f_{in} is preserved. This ensures that after the application of attention weights, f_{out} still follows the original spatiotemporal structure of the input f_{in}.

Activation Function. We empirically find that the activation function ϕ (see Eq. 1 and Eq. 2) used in the attention operation can influence the performance. So, we also include the choice of the activation function in the operation level search space and rely on the search algorithm to choose the right one for each attention operation. We consider the following four choices for the activation function: (1) no activation function, (2) ReLU, (3) sigmoid, and (4) softmax.

3.2 Cell Level Search Space

We define an attention cell as a cell composed of K attention operations. Let f_0 denote the input feature map to the entire attention cell and (T, H, W, C) be the shape of f_0. f_0 is usually the output of a stack of convolutional layers. An attention cell takes f_0 as input and outputs a feature map of the same shape.

The connectivity between convolutional layers is essential to the performance of CNNs, no matter if the network is manually designed, e.g., ResNet [10] and Inception [28], or automatically discovered [35,41,42]. Similarly, to build an attention cell, another critical design choice is how the K attention operations are connected inside the cell, apart from the design of these attention operations.

As shown in Fig. 1, in an attention cell, the first attention operation always takes f_0 as input and outputs feature map f_1. The $k^{th}(2 \leq k \leq K)$ attention operation chooses its input from $\{f_0, f_1, ..., f_{k-1}\}$ and gives feature map f_k based on the selected input. We allow the kth operation to choose multiple feature maps from $\{f_0, f_1, ..., f_{k-1}\}$ and compute a weighted sum of selected feature maps as its input, where the weights are learnable parameters. This process is repeated for all k and allows us to explore all possible connectivities between the K attention operations in the cell.

We combine $\{f_1, f_2, ..., f_K\}$ to obtain the output feature map of the entire attention cell. For all attention operations inside the cell, we set their output shape to be (T, H, W, C_{op}), i.e., f_k has the shape of (T, H, W, C_{op}) for all $k(1 \leq k \leq K)$. C_{op} is usually smaller than C to limit the computation in an attention cell with multiple attention operations. We concatenate $\{f_1, f_2, ..., f_K\}$ along the channel dimension and then employ a $1 \times 1 \times 1$ convolution to transform the concatenated feature map back to the same shape as the input f_0. We denote the feature map after transformation as f_{comb}. Similar to non-local blocks [32], we add a residual connection between the input and output of the attention cell. So the final output of the attention cell is the sum of f_0 and f_{comb}. The combination procedure is the same for all attention cells.

4 Search Algorithm

4.1 Gaussian Process Bandit (GPB)

Given K, i.e., the number of attention operations inside the attention cell, the attention cell design can be parameterized by a fixed number of hyper-parameters, including the attention dimension, the type and the activation function of each attention operation, and the input to each attention operation.

We employ GPB [25,26], a popular hyper-parameter optimization algorithm, to optimize all the hyper-parameters for the attention cell design jointly. Intuitively, GPB can predict the performance of an attention cell at a modest computational cost without actually training the entire network, based on those already evaluated attention cells. Such prediction helps GPB to select promising attention cells to evaluate in the following step and makes it possible to discover high-performing attention cells within a limited number of search steps.

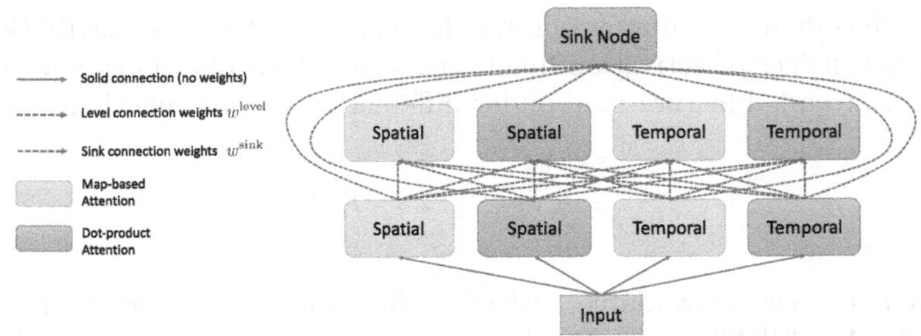

Fig. 2. Illustration of the supergraph used by the differentiable method.

Concretely, in GPB, the performance of an attention cell is modeled as a sample from a Gaussian process. At each search step, GPB selects the attention cell for evaluation by optimizing the Gaussian process upper confidence conditioned on those already evaluated attention cells.

4.2 Differentiable Architecture Search

Inspired by recent progress on differentiable architecture search [16], we develop a differentiable formulation of our proposed search space. The formulation makes it possible to jointly learn the attention cell design and network weights with back-propagation, without explicitly sampling and evaluating different cells.

Differentiable Formulation of Search Space. We propose to represent the attention cell search space as a supergraph, where all the possible attention cells are different subgraphs of this supergraph. The supergraph representation allows us to parameterize the design of an attention cell with a set of continuous and differentiable connection weights between the nodes in the supergraph.

To be more specific, we define the supergraph to have m levels, where each level has n nodes. Each node is an attention operation of a pre-defined type (map-based or dot-product attention) and a pre-defined attention dimension. Figure 2 shows an example supergraph with 2 levels, where each level has 4 nodes. The input feature map to the entire attention cell is passed to all the nodes at the first level. Starting from the second level, the input feature map to a node is a weighted sum of the output feature maps of all the nodes at its previous level:

$$f_{i,j}^{\text{in}} = \sum_{k=1}^{n} w_{i,j,k}^{\text{level}} \cdot f_{i-1,k}^{\text{out}}, \tag{4}$$

where $2 \leq i \leq m$, $1 \leq j \leq n$, $f_{i,j}^{\text{in}}$ is the input to the jth node at i^{th} level, $f_{i-1,k}^{\text{out}}$ is the output of the kth node at $(i-1)^{th}$ level, and $w_{i,j}^{\text{level}}$ are the connection weights between the jth node at i^{th} level and all the nodes at $(i-1)^{th}$ level. In practice, $w_{i,j}^{\text{level}}$ is a probability distribution obtained by softmax.

For each node in the supergraph, we also learn a probability distribution over the possible choices of activation functions. The output of a node is a weighted sum of the attended feature map under different activation functions:

$$f_{i,j}^{\text{out}} = \sum_{k=1}^{|\mathcal{A}|} w_{i,j,k}^{\text{activation}} \cdot f_{i,j}^{\text{out},\phi_k}, \tag{5}$$

where $\mathcal{A}$ is the set of available activation functions, ϕ_k is the kth activation function in $\mathcal{A}$, $w_{i,j,k}^{\text{activation}}$ is the weighting factor to be learned for ϕ_k, and $f_{i,j}^{\text{out},\phi_k}$ is the attended feature map under the activation function ϕ_k. The only difference among these attended feature maps $\{f_{i,j}^{\text{out},\phi_k}\}$ is the activation function ϕ used in Eq. 1 or Eq. 2. The layers G_1, G_2 and G_3 are shared by different activation functions within one node.

The supergraph has a sink node, receiving the output feature maps of all the nodes. The sink node is defined as follows:

$$f_{\text{sink}}^{\text{out}} = \sum_{1 \le i \le m, 1 \le j \le n} w_{i,j}^{\text{sink}} \cdot G_{i,j}(f_{i,j}^{\text{out}}), \tag{6}$$

where $f_{\text{sink}}^{\text{out}}$ is the output of the sink node, $f_{i,j}^{\text{out}}$ is the output of the jth node at i^{th} level, $G_{i,j}$ is a $1 \times 1 \times 1$ convolutional layer changing the number of channels in $f_{i,j}^{\text{out}}$ to C, and $w_{i,j}^{\text{sink}}$ is the weighting factor to be learned. We enforce $f_{\text{sink}}^{\text{out}}$ to have the same shape as the input to the supergraph, so that the supergraph can be inserted into any position of the backbone network. Same as attention cells, a residual connection is added between the input and output of the supergraph.

Attention Cell Design Learning. Both the network weights, e.g., weights of convolutional layers in the network, and the connection weights in the supergraph ($\{w^{\text{level}}, w^{\text{sink}}, w^{\text{activation}}\}$) are differentiable. During the search, we insert supergraphs into the backbone network and jointly optimize the network weights and connection weights by minimizing the training loss using gradient descent. The entire search process only consumes a computational cost similar to fully training one network on the training videos. Once the training is completed, we can derive the attention cell design from the learned connection weights.

Note that we insert the supergraphs at positions where the final attention cells will be inserted. In practice, usually multiple supergraphs or attention cells (e.g., 5) are inserted into the backbone network. If we enforce the inserted supergraphs to share the same set of connection weights, we will obtain one single attention cell design, dubbed as the *position-agnostic* attention cell.

One significant advantage of the differentiable method is that we can also learn separate connection weights for supergraphs inserted at different positions, which will give *position-specific* attention cells (see Table 2). Searching for separate attention cells for different positions results in an exponentially larger search space than searching for one single attention cell. But thanks to the differentiable method, we can learn position-specific attention cells with zero extra cost compared to learning one position-agnostic attention cell.

Attention Cell Design Derivation. We derive the attention cell design from the learned continuous connection weights. We first choose the top α nodes with the highest weights in w^{sink} and add them to the set S. Then for each node in S, we add its top β predecessors in its previous level to S, based on the corresponding connection weights in w^{level}. This process is conducted recursively for every node in S until we reach the first level. α and β are two hyper-parameters.

Recall that each node is an attention operation of a pre-defined type and attention dimension. So, S contains a set of selected attention operations. The construction process of S also determines how these attention operations are connected. For all the selected attention operations, we decide its activation function based on the corresponding weighting factors in $w^{\text{activation}}$.

5 Experiments

5.1 Experimental Setup

Datasets. We conduct experiments on two benchmark datasets: Kinetics-600 [4] and Moments in Time (MiT) [18]. Top-1 and top-5 classification accuracy are used as the evaluation metric for both datasets.

Backbones. We conduct the attention cell search on two backbones: I3D [5] and S3D [36]. Both I3D and S3D are constructed based on the Inception [28] network. When examining the generalization of the found cells, we also consider the backbone I3D-R50 [32], which is constructed based on ResNet-50 [10].

Baselines. Non-local blocks [32] are the state-of-the-art manually designed attention cell for video classification and are the most direct competitor of our automatically searched attention cells. We mainly focus on the relative improvement brought by our attention cells after being inserted into backbones. Besides non-local blocks, we also compare with other state-of-the-art methods for video classification, such as TSN [31], TRN [40], and SlowFast [7].

Table 1. Search results on Kinetics-600 and MiT using GPB. Our attention cells improve the classification accuracy for both backbones and on both datasets.

	Model	Kinetics			MiT		
		Top-1	Top-5	ΔTop-1	Top-1	Top-5	ΔTop-1
I3D	Backbone [5]	75.58	92.93	–	27.38	54.29	–
	Non-local [32]	76.87	93.44	1.29	**28.54**	55.35	**1.16**
	Ours - GPB	**77.39**	**93.63**	**1.81**	28.41	**55.49**	1.03
S3D	Backbone [36]	76.15	93.22	–	27.69	54.68	–
	Non-local [32]	77.56	93.68	1.41	**29.52**	**56.91**	**1.83**
	Ours - GPB	**78.28**	**94.04**	**2.13**	29.23	56.22	1.54

Table 2. Search results on Kinetics-600 and MiT using the differentiable method. Our attention cells consistently outperform non-local blocks on all the combinations of backbones and datasets. Position-specific attention cells ('Pos-Specific') consistently outperform position-agnostic attention cells ('Pos-Agnostic').

	Model	Kinetics			MiT		
		Top-1	Top-5	ΔTop-1	Top-1	Top-5	ΔTop-1
I3D	Backbone [5]	75.58	92.93	–	27.38	54.29	–
	Non-local [32]	76.87	93.44	1.29	28.54	55.35	1.16
	Ours - Pos-Agnostic	77.56	93.63	1.98	28.18	55.01	0.80
	Ours - Pos-Specific	**77.86**	**93.75**	**2.28**	**29.58**	**56.62**	**2.20**
S3D	Backbone [36]	76.15	93.22	–	27.69	54.68	–
	Non-local [32]	77.56	93.68	1.41	29.52	56.91	1.83
	Ours - Pos-Agnostic	77.82	93.72	1.67	29.19	55.96	1.50
	Ours - Pos-Specific	**78.51**	**93.88**	**2.36**	**29.82**	**57.02**	**2.13**

5.2 Search Results

Table 1 shows the search results of GPB and Table 2 summarizes the search results using the differentiable method. Notably, attention cells found by the differentiable method can improve the accuracy of both backbones by more than 2% on both datasets, and consistently outperform non-local blocks on all the combinations of backbones and datasets.

In Table 2, 'Pos-Agnostic' refers to that one attention design is learned for all the positions where the cells are inserted. 'Pos-Specific' means that we learn a separate attention cell design for each position where a cell is inserted, i.e., the cells inserted at different positions can be different. We observe that position-specific attention cells consistently outperform position-agnostic attention cells.

5.3 Generalization of Discovered Cells

We examine how well the discovered attention cells can generalize to new settings. We do not perform any search in the following experiments, but directly apply attention cells searched for one setting to a different setting and see if the attention cells can improve the classification performance. Concretely, we evaluate whether our discovered attentions can generalize across different modalities, different backbones, and different datasets.

Modality. We insert the attention cells discovered on RGB frames into the backbone and train the network on optical flow only. The results are summarized in Table 3. 'GPB' refers to cells discovered by GPB and 'Differentiable' refers to cells discovered by the differentiable method. Our attention cells significantly improve the classification accuracy when being applied on optical flow and consistently outperform non-local blocks for both backbones and on both datasets. For example, our attention cells improve the accuracy of I3D by 5.67%

Table 3. Generalization across different modalities (RGB to Optical flow).

	Model	Kinetics			MiT		
		Top-1	Top-5	ΔTop-1	Top-1	Top-5	ΔTop-1
I3D	Backbone [5]	61.14	82.77	–	20.01	42.42	–
	Non-local [32]	64.88	85.77	3.74	21.86	**46.59**	1.85
	Ours - GPB	65.81	87.04	4.67	21.83	45.45	1.82
	Ours - Differentiable	**66.81**	**87.85**	**5.67**	**21.94**	45.57	**1.93**
S3D	Backbone [36]	62.46	84.59	–	20.50	42.86	–
	Non-local [32]	65.79	86.85	3.33	22.13	**46.48**	1.63
	Ours - GPB	**67.02**	**87.72**	**4.56**	22.29	46.16	1.79
	Ours - Differentiable	66.29	86.97	3.83	**22.52**	46.30	**2.02**

Table 4. Generalization across different backbones.

	Model	Kinetics			MiT		
		Top-1	Top-5	ΔTop-1	Top-1	Top-5	ΔTop-1
I3D	Backbone [5]	75.58	92.93	–	27.38	54.29	–
	S3D - GPB	77.47	93.67	1.89	28.92	56.09	1.54
	S3D - Differentiable	**77.81**	**93.74**	**2.23**	**29.26**	**56.61**	**1.88**
S3D	Backbone [36]	76.15	93.22	–	27.69	54.68	–
	I3D - GPB	78.23	94.07	2.08	29.45	56.50	1.76
	I3D - Differentiable	**78.46**	**94.05**	**2.31**	**29.67**	**57.05**	**1.98**
I3D-R50	Backbone [32]	78.10	93.79	–	30.63	58.15	–
	I3D - Differentiable	**79.83**	**94.37**	**1.73**	**32.48**	**60.31**	**1.85**
	S3D - Differentiable	79.71	94.28	1.61	31.91	59.87	1.28

on Kinetics-600. Note that the cells are discovered by maximizing its performance on RGB frames and no optical flow is involved during search. This demonstrates that our cells discovered on RGB frames can generalize well to optical flow.

Backbone. Table 4 summarizes the results of inserting cells discovered for one backbone to another backbone. The second row shows that cells discovered for S3D can still improve the classification accuracy of I3D by about 2% on both datasets, even though these cells are never optimized to improve the performance of I3D. We observe similar improvement when inserting cells found for I3D to S3D (third row), or cells found for I3D/S3D to I3D-R50 (last row). Notably, our attention cells can still outperform non-local blocks even after being inserted into a different backbone. For example, cells found for S3D achieve 77.81% accuracy on Kinetics-600 after being inserted to I3D, which outperforms non-local blocks (76.87%) and performs similar to cells specifically discovered for I3D (77.86%).

Dataset. We insert attention cells discovered on MiT to the corresponding backbone, fully train the network on Kinetics-600 and report its accuracy on Kinetics-600 in the middle column ('MiT to Kinetics') of Table 5. We observe that cells discovered on MiT can improve the accuracy on Kinetics-600 by more than 2%, although they are never optimized to improve the Kinetics-600 performance during the search. Similarly, the right column ('Kinetics to MiT') demonstrates that the cells searched on Kinetics-600 can also generalize gracefully to MiT. We conclude that our attention cells generalize well across datasets.

5.4 Comparison with State-of-the-Art

We insert our attention cells found on I3D into I3D-R50 ('I3D-R50+Cell') and compare with the state-of-the-art methods in Table 6. On Kinetics-600, we obtain similar performance with SlowFast-R50 [7] with fewer inference FLOPs. On MiT, we achieve 32.48% top-1 accuracy and 60.31% top-5 accuracy only using the

Table 5. Generalization across different datasets.

	Model	MiT to Kinetics			Kinetics to MiT		
		Top-1	Top-5	ΔTop-1	Top-1	Top-5	ΔTop-1
I3D	Backbone [5]	75.58	92.93	–	27.38	54.29	–
	GPB	77.34	93.47	1.76	27.62	56.70	0.24
	Differentiable	**77.85**	**93.89**	**2.27**	**29.45**	**56.83**	**2.07**
S3D	Backbone [36]	76.15	93.22	–	27.69	54.68	–
	GPB	77.54	93.62	1.39	28.80	56.16	1.11
	Differentiable	**78.19**	**93.98**	**2.04**	**29.33**	**56.33**	**1.64**

Table 6. Comparison with the state-of-the-art methods. Our method ('I3D-R50+Cell') obtains similar or higher performance with the state-of-the-art methods on both Kinetics-600 and MiT.

Kinetics-600.

Model	Top-1	Top-5	GFLOPs
I3D [5]	75.58	92.93	1136
S3D [36]	76.15	93.22	656
I3D-R50 [32]	78.10	93.79	938
D3D [27]	77.90	–	–
I3D+NL [32]	76.87	93.44	1305
S3D+NL [32]	77.56	93.68	825
TSN-IRv2 [31]	76.22	–	411
StNet-IRv2 [9]	78.99	–	440
SlowFast-R50 [7]	**79.9**	**94.5**	1971
I3D-R50+Cell	79.83	94.37	1034

MiT.

Model	Top-1	Top-5	Modality
I3D [5]	27.38	54.29	RGB
S3D [36]	27.69	54.68	RGB
I3D+NL [32]	28.54	55.35	RGB
S3D+NL [32]	29.52	56.91	RGB
R50-ImageNet [18]	27.16	51.68	RGB
TSN-Spatial [31]	24.11	49.10	RGB
I3D-R50 [32]	30.63	58.15	RGB
I3D-R50+Cell	**32.48**	**60.31**	RGB
TSN-2stream [31]	25.32	50.10	R+F
TRN-Multiscale [40]	28.27	53.87	R+F
AssembleNet-50 [23]	31.41	58.33	R+F

RGB frames. This significantly outperforms the previous state-of-the-art method AssembleNet-50 [23], which uses both RGB frames and optical flow.

6 Conclusions

We propose a novel search space for spatiotemporal attention cells for the application of video classification. We also propose a differentiable formulation of the search space, allowing us to learn position-specific attention cell designs with zero extra cost compared to learning a single position-agnostic attention cell. We show the significance of our discovered attention cells on two large-scale video classifications benchmarks. The discovered attention cells also outperform non-local blocks and demonstrate strong generalization performance when being applied to different modalities, backbones, or datasets.

Acknowledgement. We thank Guanhang Wu and Yinxiao Li for insightful discussions and the larger Google Cloud Video AI team for the support.

References

1. Baker, B., Gupta, O., Naik, N., Raskar, R.: Designing neural network architectures using reinforcement learning. In: ICLR (2017)
2. Bello, I., Zoph, B., Vaswani, A., Shlens, J., Le, Q.V.: Attention augmented convolutional networks. In: ICCV (2019)
3. Cao, S., Wang, X., Kitani, K.M.: Learnable embedding space for efficient neural architecture compression. In: ICLR (2019)
4. Carreira, J., Noland, E., Banki-Horvath, A., Hillier, C., Zisserman, A.: A short note about kinetics-600. arXiv preprint arXiv:1808.01340 (2018)
5. Carreira, J., Zisserman, A.: Quo vadis, action recognition? A new model and the kinetics dataset. In: CVPR (2017)
6. Donahue, J., et al.: Long-term recurrent convolutional networks for visual recognition and description. In: CVPR (2015)
7. Feichtenhofer, C., Fan, H., Malik, J., He, K.: Slowfast networks for video recognition. In: ICCV (2019)
8. Feichtenhofer, C., Pinz, A., Zisserman, A.: Convolutional two-stream network fusion for video action recognition. In: CVPR (2016)
9. He, D., et al.: StNET: local and global spatial-temporal modeling for action recognition. In: AAAI (2019)
10. He, K., Zhang, X., Ren, S., Sun, J.: Deep residual learning for image recognition. In: CVPR (2016)
11. Kandasamy, K., Neiswanger, W., Schneider, J., Poczos, B., Xing, E.P.: Neural architecture search with Bayesian optimisation and optimal transport. In: NeurIPS (2018)
12. Krizhevsky, A., Sutskever, I., Hinton, G.E.: Imagenet classification with deep convolutional neural networks. In: NeurIPS (2012)
13. Li, L., Talwalkar, A.: Random search and reproducibility for neural architecture search. In: UAI (2019)
14. Liu, C., et al.: Auto-deeplab: hierarchical neural architecture search for semantic image segmentation. In: CVPR (2019)

15. Liu, C.: Progressive neural architecture search. In: Ferrari, V., Hebert, M., Sminchisescu, C., Weiss, Y. (eds.) ECCV 2018. LNCS, vol. 11205, pp. 19–35. Springer, Cham (2018). https://doi.org/10.1007/978-3-030-01246-5_2
16. Liu, H., Simonyan, K., Yang, Y.: DARTS: differentiable architecture search. In: ICLR (2019)
17. Liu, X., Lee, J.Y., Jin, H.: Learning video representations from correspondence proposals. In: CVPR (2019)
18. Monfort, M., et al.: Moments in time dataset: one million videos for event understanding. TPAMI **42**, 502–508 (2019)
19. Park, J., Woo, S., Lee, J.Y., Kweon, I.S.: Bam: bottleneck attention module. In: BMVC (2018)
20. Qiu, Z., Yao, T., Mei, T.: Learning spatio-temporal representation with pseudo-3D residual networks. In: ICCV (2017)
21. Real, E., Aggarwal, A., Huang, Y., Le, Q.V.: Regularized evolution for image classifier architecture search. In: AAAI (2019)
22. Real, E., et al.: Large-scale evolution of image classifiers. In: ICML (2017)
23. Ryoo, M.S., Piergiovanni, A., Tan, M., Angelova, A.: Assemblenet: searching for multi-stream neural connectivity in video architectures. In: ICLR (2020)
24. Simonyan, K., Zisserman, A.: Two-stream convolutional networks for action recognition in videos. In: NeurIPS (2014)
25. Snoek, J., Larochelle, H., Adams, R.P.: Practical Bayesian optimization of machine learning algorithms. In: NeurIPS (2012)
26. Srinivas, N., Krause, A., Kakade, S.M., Seeger, M.W.: Gaussian process optimization in the bandit setting: no regret and experimental design. In: ICML (2009)
27. Stroud, J., Ross, D., Sun, C., Deng, J., Sukthankar, R.: D3d: Distilled 3D networks for video action recognition. In: WACV (2020)
28. Szegedy, C., et al.: Going deeper with convolutions. In: CVPR (2015)
29. Tran, D., Bourdev, L., Fergus, R., Torresani, L., Paluri, M.: Learning spatiotemporal features with 3D convolutional networks. In: ICCV (2015)
30. Vaswani, A., et al.: Attention is all you need. In: NeurIPS (2017)
31. Wang, L., et al.: Temporal segment networks: towards good practices for deep action recognition. In: Leibe, B., Matas, J., Sebe, N., Welling, M. (eds.) ECCV 2016. LNCS, vol. 9912, pp. 20–36. Springer, Cham (2016). https://doi.org/10.1007/978-3-319-46484-8_2
32. Wang, X., Girshick, R., Gupta, A., He, K.: Non-local neural networks. In: CVPR (2018)
33. Woo, S., Park, J., Lee, J.-Y., Kweon, I.S.: CBAM: convolutional block attention module. In: Ferrari, V., Hebert, M., Sminchisescu, C., Weiss, Y. (eds.) ECCV 2018. LNCS, vol. 11211, pp. 3–19. Springer, Cham (2018). https://doi.org/10.1007/978-3-030-01234-2_1
34. Xie, L., Yuille, A.: Genetic CNN. In: ICCV (2017)
35. Xie, S., Kirillov, A., Girshick, R., He, K.: Exploring randomly wired neural networks for image recognition. In: ICCV (2019)
36. Xie, S., Sun, C., Huang, J., Tu, Z., Murphy, K.: Rethinking spatiotemporal feature learning: speed-accuracy trade-offs in video classification. In: Ferrari, V., Hebert, M., Sminchisescu, C., Weiss, Y. (eds.) ECCV 2018. LNCS, vol. 11219, pp. 318–335. Springer, Cham (2018). https://doi.org/10.1007/978-3-030-01267-0_19
37. Yu, K., Sciuto, C., Jaggi, M., Musat, C., Salzmann, M.: Evaluating the search phase of neural architecture search. In: ICLR (2020)

38. Yue-Hei Ng, J., Hausknecht, M., Vijayanarasimhan, S., Vinyals, O., Monga, R., Toderici, G.: Beyond short snippets: Deep networks for video classification. In: CVPR (2015)
39. Zhong, Z., Yan, J., Wu, W., Shao, J., Liu, C.L.: Practical block-wise neural network architecture generation. In: CVPR (2018)
40. Zhou, B., Andonian, A., Oliva, A., Torralba, A.: Temporal relational reasoning in videos. In: Ferrari, V., Hebert, M., Sminchisescu, C., Weiss, Y. (eds.) ECCV 2018. LNCS, vol. 11205, pp. 831–846. Springer, Cham (2018). https://doi.org/10.1007/978-3-030-01246-5_49
41. Zoph, B., Le, Q.V.: Neural architecture search with reinforcement learning. In: ICLR (2017)
42. Zoph, B., Vasudevan, V., Shlens, J., Le, Q.V.: Learning transferable architectures for scalable image recognition. In: CVPR (2018)

REMIND Your Neural Network to Prevent Catastrophic Forgetting

Tyler L. Hayes[1(✉)], Kushal Kafle[2], Robik Shrestha[1], Manoj Acharya[1], and Christopher Kanan[1,3,4]

[1] Rochester Institute of Technology, Rochester, NY 14623, USA
{tlh6792,rss9369,ma7583,kanan}@rit.edu
[2] Adobe Research, San Jose, CA 95110, USA
kkafle@adobe.com
[3] Paige, New York, NY 10036, USA
[4] Cornell Tech, New York, NY 10044, USA

Abstract. People learn throughout life. However, incrementally updating conventional neural networks leads to catastrophic forgetting. A common remedy is replay, which is inspired by how the brain consolidates memory. Replay involves fine-tuning a network on a mixture of new and old instances. While there is neuroscientific evidence that the brain replays compressed memories, existing methods for convolutional networks replay raw images. Here, we propose REMIND, a brain-inspired approach that enables efficient replay with compressed representations. REMIND is trained in an online manner, meaning it learns one example at a time, which is closer to how humans learn. Under the same constraints, REMIND outperforms other methods for incremental class learning on the ImageNet ILSVRC-2012 dataset. We probe REMIND's robustness to data ordering schemes known to induce catastrophic forgetting. We demonstrate REMIND's generality by pioneering online learning for Visual Question Answering (VQA) (https://github.com/tyler-hayes/REMIND).

Keywords: Online learning · Brain-inspired · Deep learning

1 Introduction

The mammalian brain engages in continuous online learning of new skills, objects, threats, and environments. The world provides the brain a temporally structured stream of inputs, which is not independent and identically distributed (iid). Enabling online learning in artificial neural networks from non-iid data is known as lifelong learning. While conventional networks suffer from

T. L. Hayes, K. Kafle, R. Shrestha—Equal Contribution.

Electronic supplementary material The online version of this chapter (https://doi.org/10.1007/978-3-030-58598-3_28) contains supplementary material, which is available to authorized users.

A. Vedaldi et al. (Eds.): ECCV 2020, LNCS 12353, pp. 466–483, 2020.
https://doi.org/10.1007/978-3-030-58598-3_28

catastrophic forgetting [1,57], with new learning overwriting existing representations, a wide variety of methods have recently been explored for overcoming this problem [13,15,27,47,53,58,64,77]. Some of the most successful methods for mitigating catastrophic forgetting use variants of replay [13,22,27,45,64,77], which involves mixing new instances with old ones and fine-tuning the network with this mixture. Replay is motivated by how the brain works: new experiences are encoded in the hippocampus and then these compressed memories are re-activated along with other memories so that the neocortex can learn them [51,60,72]. Without the hippocampus, people lose the ability to learn new semantic categories [48]. Replay occurs both during sleep [31] and when awake [41,74].

For lifelong learning in convolutional neural networks (CNNs), there are two major gaps between existing methods and how animals learn. The first is that replay is implemented by storing and replaying raw pixels, which is not biologically plausible. Based on hippocampal indexing theory [75], the hippocampus stores *compressed* representations of neocortical activity patterns while awake. To consolidate memories, these patterns are replayed and then the corresponding neocortical neurons are re-activated via reciprocal connectivity [51,60,72]. The representations stored in the hippocampus for replay are not veridical (e.g., raw pixels) [31,56], and its visual inputs are high in the visual processing hierarchy [29] rather than from primary visual cortex or retina.

The second major gap with existing approaches is that animals engage in *streaming learning* [20,21], or resource constrained online learning from non-iid (temporally correlated) experiences throughout life. In contrast, the most common paradigm for incremental training of CNNs is to break the training dataset into M distinct batches, where for ImageNet each batch typically has about 100000 instances from 100 classes that are not seen in later batches, and then the algorithm sequentially loops over each batch many times. This paradigm is not biologically plausible. There are many applications requiring online learning of non-iid data streams, where batched learning will not suffice, such as immediate on-device learning. Batched systems also take longer to train, further limiting their utility on resource constrained devices, such as smart appliances, robots, and toys. For example, BiC, a state-of-the-art incremental batch method, requires 65 hours to train in that paradigm whereas our proposed streaming model trains in under 12 hours. The incremental batch setting can be transformed into the streaming learning scenario by using very small batches and performing only a single pass through the dataset; however, this results in a large decrease in performance. As shown in

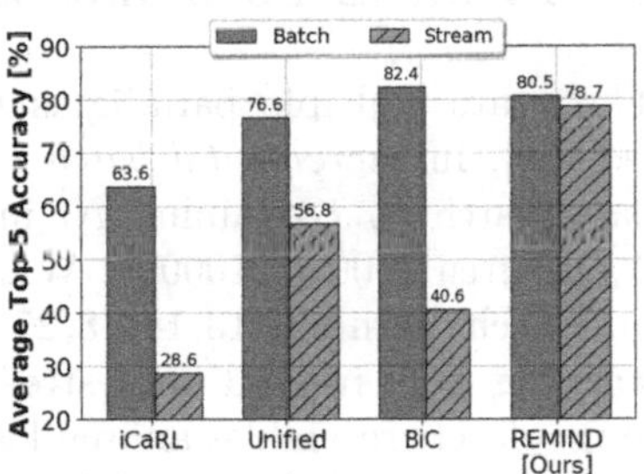

Fig. 1. Average top-5 accuracy results for streaming and incremental batch versions of state-of-the-art models on ImageNet.

Fig. 1, state-of-the-art methods perform poorly on ImageNet in the streaming setting, with the best method suffering an over 19% drop in performance. In contrast, our model outperforms the best streaming model by 21.9% and is only 1.9% below the best batch model.

Here, we propose REMIND, or **re**play using **m**emory **ind**exing, a novel method that is heavily influenced by biological replay and hippocampal indexing theory. **Our main contributions are:**

1. We introduce REMIND, a streaming learning model that implements hippocampal indexing theory using tensor quantization to efficiently store hidden representations (e.g., CNN feature maps) for later replay. REMIND implements this compression using Product Quantization (PQ) [30]. We are the first to test if forgetting in CNNs can be mitigated by replaying hidden representations rather than raw pixels.
2. REMIND outperforms existing models on the ImageNet ILSVRC-2012 [68] and CORe50 [52] datasets, while using the same amount of memory.
3. We demonstrate REMIND's robustness by pioneering streaming Visual Question Answering (VQA), in which an agent must answer questions about images and cannot be readily done with existing models. We establish new experimental paradigms, baselines, and metrics and subsequently achieve strong results on the CLEVR [33] and TDIUC [35] datasets.

2 Problem Formulation

There are multiple paradigms in which incremental learning has been studied [61]. In *incremental batch learning*, at each time step t an agent learns a data batch B_t containing N_t instances and their corresponding labels, where N_t is often 1000 to 100000. While much recent work has focused on incremental batch learning [13,14,18,27,44,45,64,77,81], *streaming learning*, or online learning from non-iid data streams with memory and/or compute constraints, more closely resembles animal learning and has many applications [20,21,49]. In streaming learning, a model learns online in a single pass, i.e., $N_t = 1$ for all t. It cannot loop over any portion of the (possibly infinite) dataset, and it can be evaluated at any point rather than only between large batches. Streaming learning can be approximated by having a system queue up small, temporally contiguous, mini-batches for learning, but as shown in Fig. 1, batch methods cannot easily adapt to this setting.

3 Related Work

Parisi et al. [61] identify three main mechanisms for mitigating forgetting in neural networks, namely 1) replay of previous knowledge, 2) regularization mechanisms to constrain parameter updates, and 3) expanding the network as more data becomes available. Replay has been shown to be one of the most effective methods for mitigating catastrophic forgetting [4,5,13,22,27,44,45,50,59,64,

77]. For ImageNet, all recent state-of-the-art methods for incremental class learning use replay of raw pixels with distillation loss. The earliest was iCaRL [64], which stored 20 images per class for replay. iCaRL used a nearest class prototype classifier to mitigate forgetting. The End-to-End incremental learning model [13] extended iCaRL to use the outputs of the CNN directly for classification, instead of a nearest class mean classifier. Additionally, End-to-End used more data augmentation and a balanced fine-tuning stage during training to improve performance. The Unified classifier [27] extended End-to-End by using a cosine normalization layer, a new loss constraint, and a margin ranking loss. The Bias Correction (BiC) [77] method extended End-to-End by training two additional parameters to correct bias in the output layer due to class imbalance. iCaRL, End-to-End, the Unified classifier, and BiC all: 1) store the same number of raw replay images per class, 2) use the same herding procedure for prototype selection, and 3) use distillation loss to prevent forgetting. REMIND, however, is the first model to demonstrate that storing and replaying quantized mid-level CNN features is an effective strategy to mitigate forgetting.

Regularization methods vary a weight's plasticity based on how important it is to previous tasks. These methods include Elastic Weight Consolidation (EWC) [47], Memory Aware Synapses (MAS) [3], Synaptic Intelligence (SI) [81], Riemannian Walk (RWALK) [14], Online Laplace Approximator [66], Hard Attention to the Task [70], and Learning without Memorizing [16]. The Averaged Gradient Episodic Memory (A-GEM) [15] model extends Gradient Episodic Memory [53], which uses replay with regularization. Variational Continual Learning [58] combines Bayesian inference with replay, while the Meta-Experience Replay model [65] combines replay with meta-learning. All of these regularization methods are typically used for incremental *task* learning, where batches of data are labeled as different tasks and the model must be told which task (batch) a sample came from during inference. When task labels are not available at test time, which is often true for agents operating in real-time, many methods cannot be used or they will fail [14,17,45]. While our main experiments focus on comparisons against state-of-the-art ImageNet models, we compare REMIND against several regularization models in Sect. 7, both with and without task labels. Some regularization methods also utilize cached data, e.g., GEM and A-GEM.

Another approach to mitigating forgetting is to expand the network as new tasks are observed, e.g., Progressive Neural Networks [69], Dynamically Expandable Networks [79], Adaptation by Distillation [26], and Dynamic Generative Memory [59]. However, these approaches also use task labels at test time, have growing memory requirements, and may not scale to thousand-category datasets.

4 REMIND: Replay Using Memory Indexing

REMIND is a novel brain-inspired method for training the parameters of a CNN in the streaming setting using replay. Learning involves two steps: 1) compressing the current input and 2) reconstructing a subset of previously compressed representations, mixing them with the current input, and updating the *plastic*

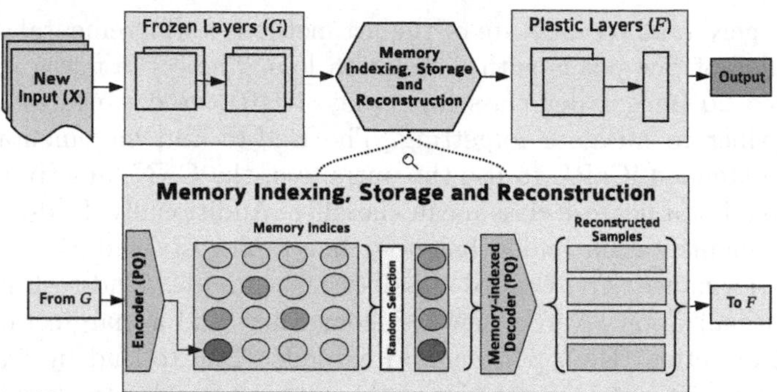

Fig. 2. REMIND takes in an input image and passes it through frozen layers of the network (G) to obtain tensor representations (feature maps). It then quantizes the tensors via product quantization and stores the indices in memory for future replay. The decoder reconstructs tensors from the stored indices to train the plastic layers (F) of the network before a final prediction is made.

weights of the network with this mixture (see Fig. 2). While earlier work for incremental batch learning with CNNs stored raw images for replay [13,27,64,77], by storing compressed mid-level CNN features, REMIND is able to store far more instances with a smaller memory budget. For example, iCaRL [64] uses a default memory budget of 20 K examples for ImageNet, but REMIND can store over 1M compressed instances using the same budget. This more closely resembles how replay occurs in the brain, with high-level visual representations being sent to the hippocampus for storage and re-activation, rather than early visual representations [29]. REMIND does not have an explicit sleep phase, with replay more closely resembling that during waking hours [41,74].

Formally, our CNN $y_i = F(G(\mathbf{X}_i))$ is trained in a streaming paradigm, where $\mathbf{X}_i$ is the input image and y_i is the predicted output category. The network is composed of two nested functions: $G(\cdot)$, parameterized by θ_G, consists of the first J layers of the CNN and $F(\cdot)$, parameterized by θ_F, consists of the last L layers. REMIND keeps θ_G fixed since early layers of CNNs have been shown to be highly transferable [80]. The later layers, $F(\cdot)$, are trained in the streaming paradigm using REMIND. We discuss how $G(\cdot)$ is initialized in Sect. 4.2.

The output of $G(\mathbf{X}_i)$ is a tensor $\mathbf{Z}_i \in \mathbb{R}^{m\times m\times d}$, where m is the dimension of the feature map and d is the number of channels. Using the outputs of $G(\cdot)$, we train a vector quantization model for the $\mathbf{Z}_i$ tensors. As training examples are observed, the quantization model is used to store the $\mathbf{Z}_i$ features and their labels in a replay buffer as an $m \times m \times s$ array of integers using as few bits as necessary, where s is the number of indices that will be stored. For replay, we uniformly select r instances from the replay buffer, which was shown to work well in [14], and reconstruct them. Each of the reconstructed instances, $\hat{\mathbf{Z}}_i$, are mixed with the current input, and then θ_F is updated using backpropagation on this set of $r + 1$ instances. Other selection strategies are discussed in Sect. 8.

During inference, we pass an image through $G(\cdot)$, and then the output, $\mathbf{Z}_i$, is quantized and reconstructed before being passed to $F(\cdot)$.

Our main version of REMIND uses PQ [30] to compress and store $\mathbf{Z}_i$. For high-dimensional data, PQ tends to have much lower reconstruction error than models that use only k-means. The tensor $\mathbf{Z}_i$ consists of $m \times m$ d-dimensional tensor elements, and PQ partitions each d-dimensional tensor element into s sub-vectors, each of size d/s. PQ then creates a separate codebook for each partition by using k-means, where the codes within each codebook correspond to the centroids learned for that partition. Since the quantization is done independently for each partition, each sub-vector of the d-dimensional tensor element is assigned a separate integer, so the element is represented with s integers. If s is equal to one, then this approach is identical to using k-means for vector quantization, which we compare against. For our experiments, we set $s = 32$ and $c = 256$, so that each integer can be stored with 1 byte. We explore alternative values of s and c in supplemental materials (Fig. S4) and use the Faiss PQ implementation [32].

Since lifelong learning systems must be capable of learning from infinitely long data streams, we subject REMIND's replay buffer to a maximum memory restriction. That is, REMIND stores quantization indices in its buffer until this maximum capacity has been reached. Once the buffer is full and a new example comes in, we insert the new sample and randomly remove an example from the class with the most examples, which was shown to work well in [14,77]. We discuss other strategies for maintaining the replay buffer in Sect. 8.

4.1 Augmentation During Replay

To augment data during replay, REMIND uses random resized crops and a variant of manifold mixup [76] on the quantized tensors directly. For random crop augmentation, the tensors are randomly resized, then cropped and bilinearly interpolated to match the original tensor dimensions. To produce more robust representations, REMIND mixes features from multiple classes using manifold mixup. That is, REMIND uses its replay buffer to reconstruct two randomly chosen sets, $\mathcal{A}$ and $\mathcal{B}$, of r instances each ($|\mathcal{A}| = |\mathcal{B}| = r$), which are linearly combined to obtain a set $\mathcal{C}$ of r mixed instances ($|\mathcal{C}| = r$), i.e., a newly mixed instance, $(\mathbf{Z}_{\text{mix}}, y_{\text{mix}}) \in \mathcal{C}$, is formed as:

$$(\mathbf{Z}_{\text{mix}}, y_{\text{mix}}) = (\lambda \mathbf{Z}_a + (1-\lambda)\,\mathbf{Z}_b, \lambda y_a + (1-\lambda)\, y_b)\,, \tag{1}$$

where $(\mathbf{Z}_a, y_a)$ and $(\mathbf{Z}_b, y_b)$ denote instances from $\mathcal{A}$ and $\mathcal{B}$ respectively and $\lambda \sim \beta(\alpha, \alpha)$ is the mixing coefficient drawn from a β-distribution parameterized by hyperparameter α. We use $\alpha = 0.1$, which we found to work best in preliminary experiments. The current input is then combined with the set $\mathcal{C}$ of r mixed samples, and θ_F is updated using this new set of $r+1$ instances.

4.2 Initializing REMIND

During learning, REMIND only updates $F(\cdot)$, i.e., the top of the CNN. It assumes that $G(\cdot)$, the lower level features of the CNN, are fixed. This implies that the low-level visual representations must be highly transferable across image datasets, which is supported empirically [80]. There are multiple methods for training $G(\cdot)$, including supervised pre-training on a portion of the dataset, supervised pre-training on a different dataset, or unsupervised self-taught learning using a convolutional auto-encoder. Here, we follow the common practice of doing a 'base initialization' of the CNN [13,27,64,77]. This is done by training both θ_F and θ_G jointly on an initial subset of data offline, e.g., for class incremental learning on ImageNet we use the first 100 classes. After base initialization, θ_G is no longer plastic. All of the examples $\mathbf{X}_i$ in the base initialization are pushed through the model to obtain $\mathbf{Z}_i = G(\mathbf{X}_i)$, and all of these $\mathbf{Z}_i$ instances are used to learn the quantization model for $G(\mathbf{X}_i)$, which is kept fixed once acquired.

Following [13,27,64,77], we use ResNet-18 [25] for image classification, where we set $G(\cdot)$ to be the first 15 convolutional and 3 downsampling layers, which have 6,455,872 parameters, and $F(\cdot)$ to be the remaining 3 layers (2 convolutional and 1 fully connected), which have 5,233,640 parameters. These layers were chosen for memory efficiency in the quantization model with ResNet-18, and we show the memory efficiency trade-off in supplemental materials (Fig. S1).

5 Experiments: Image Classification

5.1 Comparison Models

While REMIND learns on a per sample basis, most methods for incremental learning in CNNs do multiple loops through a batch. For fair comparison, we train these methods in the streaming setting to fairly compare against REMIND. Results for the incremental batch setting for these models are included in Fig. 1 and supplemental materials (Table S2 and Fig. S2-S3). We evaluate the following:

- **REMIND** – Our main REMIND version uses PQ and replay augmentation. We also explore a version that omits data augmentation and a version that uses k-means rather than PQ.
- **Fine-Tuning (No Buffer)** – Fine-Tuning is a baseline that fine-tunes θ_F of a CNN one sample at a time with a single epoch through the dataset. This approach does not use a buffer and suffers from catastrophic forgetting [45].
- **ExStream** – Like REMIND, ExStream is a streaming learning method, however, it can only train fully connected layers of the network [22]. ExStream uses rehearsal by maintaining buffers of prototypes. It stores the input vector and combines the two nearest vectors in the buffer. After the buffer gets updated, all samples from its buffer are used to train the fully connected layers of a network. We use ExStream to train the final layer of the network, which is the only fully connected layer in ResNet-18.
- **SLDA** – Streaming Linear Discriminant Analysis (SLDA) is a well-known streaming method that was shown to work well on deep CNN features [23]. It maintains running means for each class and a running tied covariance matrix. Given a new input, it assigns the label of the closest Gaussian in feature space. It can be used to compute the output layer of a CNN.

- **iCaRL** – iCaRL is an incremental batch learning algorithm for CNNs [64]. iCaRL stores images from earlier classes for replay, uses a distillation loss to preserve weights, and uses a nearest class mean classifier in feature space.
- **Unified** – The Unified Classifier builds on iCaRL by using the outputs from the network for classification and introducing a cosine normalization layer, a constraint to preserve class geometry, and a margin ranking loss to maximize inter-class separation [27]. Unified also uses replay and distillation.
- **BiC** – The Bias Correction (BiC) method builds on iCaRL by using the output layer of the network for classification and correcting the bias from class imbalance during training, i.e., more new samples than replay samples [77]. The method trains two additional bias correction parameters on the output layer, resulting in improved performance over distillation and replay alone.
- **Offline** – The offline model is trained in a traditional, non-streaming setting and serves as an upper-bound on performance. We train two variants: one with only θ_F plastic and one with both θ_F and θ_G plastic.

Our main experiments focus on comparing state-of-the-art methods on ImageNet and we provide additional comparisons in Sect. 7. Although iCaRL, Unified, and BiC are traditionally trained in the incremental batch paradigm, we conduct experiments with these models in the streaming paradigm for fair comparison against REMIND. To train these streaming variants, we set the number of epochs to 1 and the batch size to $r+1$ instances to match REMIND.

5.2 Model Configurations

In our setup, all models are trained instance-by-instance and have no batch requirements, unless otherwise noted. Because methods can be sensitive to the order in which new data are encountered, all models receive examples in the same order. The same base CNN initialization procedure is used by all models. For ExStream and SLDA, after base initialization, the streaming learning phase is re-started from the beginning of the data stream. All of the parameters except the output layer are kept frozen for ExStream and SLDA, whereas only $G(\cdot)$ is kept frozen for REMIND. All other comparison models do not freeze any layers and incremental training commences with the first new data sample. All models, except SLDA, are trained using cross-entropy loss with stochastic gradient descent and momentum. More parameter settings are in supplemental materials.

5.3 Datasets, Data Orderings, and Metrics

We conduct experiments with ImageNet and CORe50 by dividing both datasets into batches. The first batch is used for base initialization. Subsequently, all models use the same batch orderings, but they are sequentially fed individual samples and they cannot revisit any instances in a batch, unless otherwise noted. For ImageNet, the models are evaluated after each batch on all trained classes. For CORe50, models are evaluated on all test data after each batch.

ImageNet ILSVRC-2012 [68] has 1000 categories each with 732–1300 training samples and 50 validation samples, which we use for testing. During the base initialization phase, the model is trained offline on a set of 100 randomly selected classes. Following [13,27,64,77], each incremental batch then contains 100 random classes, which are not contained within any other batch. We study class incremental (class iid) learning with ImageNet.

CORe50 [52] contains sequences of video frames, with one object in each frame. It has 10 classes, and each sequence is acquired with varied environmental conditions. CORe50 is ideal for evaluating streaming learners since it is naturally non-iid and requires agents to learn from temporally correlated video streams. For CORe50, we follow [22] and sample at 1 frame per second, obtaining 600 training images and 225 test images per class. We use the bounding box crops and splits from [52]. Following [22], we use four training orderings to test the robustness of each algorithm under different conditions: 1) iid, where each batch has a random subset of training images, 2) class iid, where each batch has all of the images from two classes, which are randomly shuffled, 3) instance, where each batch has temporally ordered images from 80 unique object instances, and 4) class instance, where each batch has all of the temporally ordered instances from two classes. All batches have 1200 images across all orderings. Since CORe50 is small, CNNs are first initialized with pre-trained ImageNet weights and then fine-tuned on a subset of 1200 samples for base initialization.

We use the Ω_{all} metric [22,24,45] for evaluation, which normalizes incremental learning performance by offline performance: $\Omega_{\text{all}} = \frac{1}{T}\sum_{t=1}^{T} \frac{\alpha_t}{\alpha_{\text{offline},t}}$, where T is the total number of testing events, α_t is the accuracy of the model for test t, and $\alpha_{\text{offline},t}$ is the accuracy of the optimized offline learner for test t. If $\Omega_{\text{all}} = 1$, then the incremental learner's performance matched the offline model. We use top-5 and top-1 accuracies for ImageNet and CORe50, respectively. Average accuracy results are in supplemental materials (Table S2-S3).

5.4 Results: ImageNet

For ImageNet, we use the pre-trained PyTorch offline model with 89.08% top-5 accuracy to normalize Ω_{all}. We allow the iCaRL, Unified, and BiC models to store 10,000 (224×224 uint8) raw pixel image prototypes in a replay buffer, which is equivalent to 1.51 GB in memory. This allows REMIND to store indices for 959665 examples in its replay buffer. We set $r = 50$ samples. We study additional buffer sizes in Sect. 7. Results for incremental class learning on ImageNet are shown in Table 1 and a learning curve for all models is shown in Fig. 3. REMIND outperforms all other comparison models, with SLDA achieving the sec-

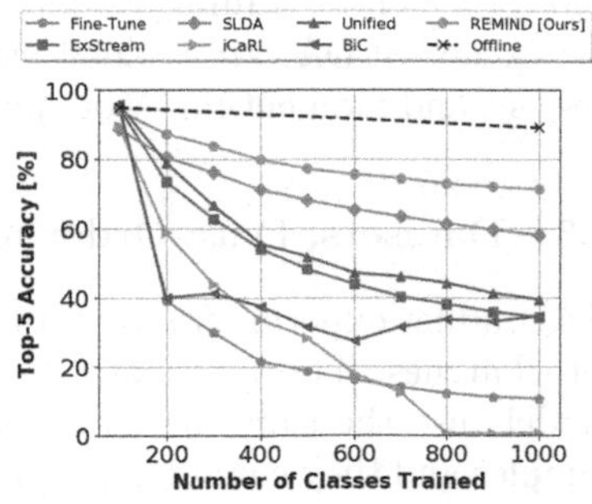

Fig. 3. Performance of streaming ImageNet models.

ond best performance. This is remarkable since REMIND only updates θ_F, whereas iCaRL, Unified, and BiC all update θ_F and θ_G.

REMIND is intended to be used for online streaming learning; however, we also created a variant suitable for incremental batch learning which is described in supplemental materials. Incremental batch results for REMIND and recent methods are given in Fig. 1 and supplemental materials (Table S2 and Fig. S2). While incremental batch methods train much more slowly, REMIND achieves comparable performance to the best methods.

Table 1. ResNet-18 streaming classification results on ImageNet and CORe50 using Ω_{all}. For CORe50, we explore performance across four ordering schemes and report the average of 10 permutations. Upper bounds are at the bottom.

Model	ImageNet	CORe50			
	CLS IID	IID	CLS IID	INST	CLS INST
Fine-Tune (θ_F)	0.288	0.961	0.334	0.851	0.334
ExStream	0.569	0.953	0.873	0.933	0.854
SLDA	0.752	0.976	0.958	0.963	0.959
iCaRL	0.306	–	0.690	–	0.644
Unified	0.614	–	0.510	–	0.527
BiC	0.440	–	0.410	–	0.415
REMIND	**0.855**	**0.985**	**0.978**	**0.980**	**0.979**
Offline (θ_F)	0.929	0.989	0.984	0.985	0.985
Offline	1.000	1.000	1.000	1.000	1.000

5.5 Results: CORe50

We use the CoRe50 dataset to study models under more realistic data orderings. Existing methods including iCaRL, Unified, and BiC assume that classes from one batch do not appear in other batches, making it difficult for them to learn the iid and instance orderings without modifications. To compute Ω_{all}, we use an offline model that obtains 93.11% top-1 accuracy. The iCaRL, Unified, and BiC models use replay budgets of 50 images, which is equivalent to 7.3 MB. This allows REMIND to store replay indices for 4465 examples. Results for other buffer sizes are in supplemental materials (Fig. S3). REMIND replays $r = 20$ samples. Ω_{all} results for CORe50 are provided in Table 1. For CORe50, REMIND outperforms all models for all orderings. In fact, REMIND is only 2.2% below the full offline model in the worst case, in terms of Ω_{all}. Methods that only trained the output layer performed well on CORe50 and poorly on ImageNet. This is likely because the CNNs used for CORe50 experiments are initialized with ImageNet weights, resulting in more robust representations. REMIND's remarkable performance on these various orderings demonstrate its versatility.

6 Experiments: Incremental VQA

In VQA, a system must produce an answer to a natural language question about an image [8,36,54], which requires capabilities such as object detection, scene understanding, and logical reasoning. Here, we use REMIND to pioneer streaming VQA. During training, a streaming VQA model receives a sequence of temporally ordered triplets $\mathcal{D} = \{(X_t, Q_t, A_t)\}_{t=1}^{T}$, where X_t is an image, Q_t is the question (string), and A_t is the answer. If an answer is not provided at time t, then the agent must use knowledge from time 1 to $t-1$ to predict A_t. To use REMIND for streaming VQA, we store each quantized feature along with a question string and answer, which can later be used for replay. REMIND can be used with almost any existing VQA system (e.g., attention-based [6,46,78], compositional [7,28], bi-modal fusion [10,19,71]) and it can be applied to similar tasks like image captioning [12] and referring expression recognition [43,63,67].

6.1 Experimental Setup

For our experiments, we use the TDIUC [35] and CLEVR [33] VQA datasets. TDIUC is composed of natural images and has over 1.7 million QA pairs organized into 12 question types including simple object recognition, complex counting, positional reasoning, and attribute classification. TDIUC tests for generalization across different underlying tasks required for VQA. CLEVR consists of over 700000 QA pairs for 70000 synthetically generated images and is organized into 5 question types. CLEVR specifically tests for multi-step compositional reasoning that is very rarely encountered in natural image VQA datasets. We combine REMIND with two popular VQA algorithms, using a modified version of the stacked attention network (SAN) [42,78] for TDIUC, and a simplified version of the Memory Attention and Control (MAC) [28,55] network for CLEVR. A ResNet-101 model pre-trained on ImageNet is used to extract features for both TDIUC and CLEVR. REMIND's PQ model is trained with 32 codebooks each of size 256. The final offline mean per-type accuracy with SAN on TDIUC is 67.59% and the final offline accuracy with MAC on CLEVR is 94.00%. Our main results with REMIND use a buffer consisting of 50% of the dataset and $r = 50$. Results for other buffer sizes are in supplemental materials (Table S4).

For both datasets, we explore two orderings of the training data: iid and question type (q-type). For iid, the dataset is randomly shuffled and the model is evaluated on all test data when multiples of 10% of the total training set are seen. The q-type ordering reflects a more interesting scenario where QA pairs for different VQA 'skills' are grouped together. Models are evaluated on all test data at the end of each q-type. We perform base initialization by training on the first 10% of the data for the iid ordering and on QA pairs belonging to the first q-type for the q-type ordering. Then, the remaining data is streamed into the model one sample at a time. The buffer is then incrementally updated with PQ encoded features and raw question strings. We use simple accuracy for CLEVR and mean-per-type accuracy for TDIUC.

We compare REMIND to ExStream [22], SLDA [23], an offline baseline, and a simple baseline where models are fine-tuned without a buffer, which causes catastrophic forgetting. To adapt ExStream and SLDA for VQA, we use a variant of the linear VQA model in [34], which concatenates ResNet-101 image features to question features extracted from a universal sentence encoder [73] and then trains a linear classifier. Parameter settings are in supplemental materials.

6.2 Results: VQA

Streaming VQA results for REMIND with a 50% buffer size are given in Table 2. Variants of REMIND with other buffer sizes are in supplemental materials (Table S4). REMIND outperforms the streaming baselines for both datasets, with strong performance on both TDIUC using the SAN model and CLEVR using the MAC model. Interestingly, for CLEVR the results are much greater for q-type than for iid. We hypothesize that the q-type ordering may be acting as a natural curriculum [11], allowing our streaming model to train more efficiently. Our results demonstrate that it is possible to train complex, multi-modal agents capable of attention and compositional reasoning in a streaming manner. Learning curves and qualitative examples are in supplemental materials (Fig. S5-S6).

Table 2. Ω_{all} results for streaming VQA.

Ordering	TDIUC		CLEVR	
	IID	Q-TYPE	IID	Q-TYPE
Fine-Tune	0.716	0.273	0.494	0.260
ExStream	0.676	0.701	0.477	0.375
SLDA	0.624	0.644	0.518	0.496
REMIND	**0.917**	**0.919**	**0.720**	**0.985**
Offline	1.000	1.000	1.000	1.000

7 Additional Classification Experiments

In this section, we study several of REMIND's components. In supplemental materials, we study other factors that influence REMIND's performance (Fig. S4), e.g., where to quantize, number of codebooks, codebook size, and replay samples (r). In supplemental materials, we also explore the performance of iCaRL, Unified, and BiC when only θ_F is updated (Sec. S3.2).

REMIND Components. REMIND is impacted by the size of its overall buffer, using augmentation, and the features used to train $F(\cdot)$. We study these on ImageNet and results are given in Table 3. REMIND (Main) denotes the variant of REMIND from our main experiments that uses augmentation with a buffer size of 959665 and 32 codebooks of size 256. PQ is critical to performance, with PQ (32 codebooks) outperforming k-means (1 codebook) by 7.7% in terms of Ω_{all}. Augmentation

Table 3. REMIND variations on ImageNet with their memory (GB).

Variant	Ω_{all}	Memory
REMIND (Main)	0.855	1.51
100% Buffer	0.856	2.01
No Augmentation	0.818	1.51
k-Means	0.778	0.12
Real Features	0.868	24.08

is the next most helpful component and improves performance by 3.7%. Storing the entire dataset (100% Buffer) does not yield significant improvements. Using real features yields marginal improvements (1.3%) while requiring nearly 16 times more memory.

Replay Buffer Size. Since REMIND and several other models rely on a replay buffer to mitigate forgetting, we studied performance on ImageNet as a function of buffer size. We compared the performance of iCaRL, Unified, and BiC on ImageNet at three different buffer sizes (5K exemplars = 0.75 GB, 10K exemplars = 1.51 GB, and 20 K exemplars = 3.01 GB). To make the experiment fair, we compared REMIND to these models at equivalent buffer sizes, i.e., 479665 compressed samples = 0.75 GB, 959665 compressed samples = 1.51 GB, and 1281167 compressed samples (full dataset) = 2.01 GB. In Fig. 4, we see that more memory generally results in better performance. Overall, REMIND has the best performance and is nearly unaffected by buffer size. A plot with incremental batch models is in supplemental materials (Fig. S2), and follows the same trend: larger buffers yield better performance.

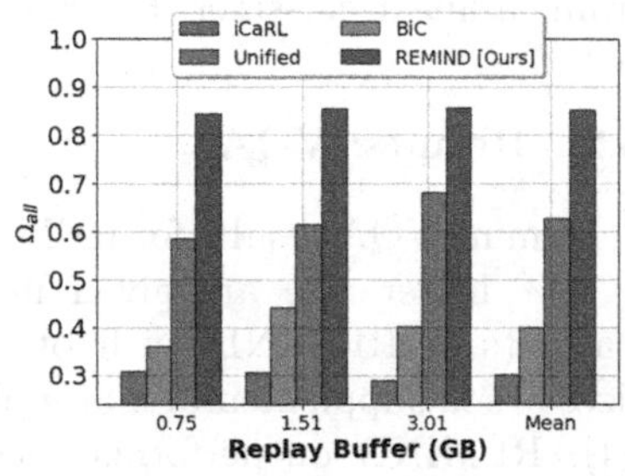

Fig. 4. Ω_{all} as a function of buffer size for streaming ImageNet models.

Regularization Comparisons. In Table 4, we show the results of REMIND and regularization methods for combating catastrophic forgetting on CORe50 class orderings. These regularization methods constrain weight updates to remain close to their previous values and are trained on batches of data, where each batch resembles a *task*. At test time, these models are provided with task labels, denoting which task an unseen sample came from. In our experiments, a *task* consists of several classes, and providing task labels makes classification easier. We analyze performance when task labels are provided and when they are withheld. To evaluate REMIND and Offline with task labels, we mask off probabilities during test time for classes not included in the specific task. Consistent with [14,17,45], we find that regularization methods perform poorly when no task labels are provided. Regardless, REMIND outperforms all comparisons, both with and without task labels.

Table 4. Ω_{all} for regularization models averaged over 10 runs on CORe50 with and without Task Labels (TL).

Model	CLS IID		CLS INST	
	TL	No TL	TL	No TL
SI	0.895	0.417	0.905	0.416
EWC	0.893	0.413	0.903	0.413
MAS	0.897	0.415	0.905	0.421
RWALK	0.903	0.410	0.912	0.417
A-GEM	0.925	0.417	0.916	0.421
REMIND	**0.995**	**0.978**	**0.995**	**0.979**
Offline	1.000	1.000	1.000	1.000

8 Discussion and Conclusion

We proposed REMIND, a brain-inspired replay-based approach to online learning in a streaming setting. REMIND achieved state-of-the-art results for object classification. Unlike iCaRL, Unified, and BiC, REMIND can be applied to iid and instance ordered data streams without modification. Moreover, we showed that REMIND is general enough for tasks like VQA with almost no changes.

REMIND replays compressed (lossy) representations that it stores, rather than veridical (raw pixel) experience, which is more consistent with memory consolidation in the brain. REMIND's replay is more consistent with how replay occurs in the brain during waking hours. Replay also occurs in the brain during slow wave sleep [9,31], and it would be interesting to explore how to effectively create a variant that utilizes sleep/wake cycles for replay. This could be especially beneficial for a deployed agent that is primarily engaged in online learning during certain hours, and is engaged in offline consolidation in other hours.

Several algorithmic improvements could be made to REMIND. We initialized REMIND's quantization model during the base initialization phase. For deployed, on-device learning this could instead be done by pre-training the codebook on a large dataset, or it could be initialized with large amounts of unlabeled data, potentially leading to improved representations. Another potential improvement is using selective replay. REMIND randomly chooses replay instances with uniform probability. In early experiments, we also tried choosing replay samples based on distance from current example, number of times a sample has been replayed, and the time since it was last replayed. While none performed better than uniform selection, we believe that selective replay still holds the potential to lead to better generalization with less computation. Because several comparison models used ResNet-18, we also used ResNet-18 for image classification so that we could compare against these models directly. The ResNet-18 layer used for quantization was chosen to ensure REMIND's memory efficiency, but co-designing the CNN architecture with REMIND could lead to considerably better results. Using less memory, REMIND stores far more compressed representations than competitors. For updating the replay buffer, we used random replacement, which worked well in [14,77]. We tried a queue and a distance-based strategy, but both performed nearly equivalent to random selection with higher computational costs. Furthermore, future variants of REMIND could incorporate mechanisms similar to [62] to explicitly account for the temporal nature of incoming data. To demonstrate REMIND's versatility, we pioneered streaming VQA and established strong baselines. It would be interesting to extend this to streaming chart question answering [37,38,40], object detection, visual query detection [2], and other problems in vision and language [39].

Acknowledgements. This work was supported in part by the DARPA/MTO Lifelong Learning Machines program [W911NF-18-2-0263], AFOSR grant [FA9550-18-1-0121], NSF award #1909696, and a gift from Adobe Research. We thank NVIDIA for the GPU donation. The views and conclusions contained herein are those of the authors and should not be interpreted as representing the official policies or endorsements of any sponsor. We thank Michael Mozer, Ryne Roady, and Zhongchao Qian for feedback on early drafts of this paper.

References

1. Abraham, W.C., Robins, A.: Memory retention-the synaptic stability versus plasticity dilemma. Trends Neurosci. **28**, 73–78 (2005)
2. Acharya, M., Jariwala, K., Kanan, C.: VQD: visual query detection in natural scenes. In: NAACL (2019)
3. Aljundi, R., Babiloni, F., Elhoseiny, M., Rohrbach, M., Tuytelaars, T.: Memory aware synapses: learning what (not) to forget. In: Ferrari, V., Hebert, M., Sminchisescu, C., Weiss, Y. (eds.) ECCV 2018. LNCS, vol. 11207, pp. 144–161. Springer, Cham (2018). https://doi.org/10.1007/978-3-030-01219-9_9
4. Aljundi, R., et al.: Online continual learning with maximal interfered retrieval. In: NeurIPS, pp. 11849–11860 (2019)
5. Aljundi, R., Lin, M., Goujaud, B., Bengio, Y.: Gradient based sample selection for online continual learning. In: NeurIPS, pp. 11816–11825 (2019)
6. Anderson, P., et al.: Bottom-up and top-down attention for image captioning and visual question answering. In: CVPR (2018)
7. Andreas, J., Rohrbach, M., Darrell, T., Klein, D.: Neural module networks. In: CVPR, pp. 39–48 (2016)
8. Antol, S., et al.: VQA: visual question answering. In: ICCV (2015)
9. Barnes, D.C., Wilson, D.A.: Slow-wave sleep-imposed replay modulates both strength and precision of memory. J. Neurosci. **34**(15), 5134–5142 (2014)
10. Ben-Younes, H., Cadene, R., Cord, M., Thome, N.: Mutan: multimodal tucker fusion for visual question answering. In: ICCV (2017)
11. Bengio, Y., Louradour, J., Collobert, R., Weston, J.: Curriculum learning. In: ICML, pp. 41–48 (2009)
12. Bernardi, R., et al.: Automatic description generation from images: a survey of models, datasets, and evaluation measures. J. Artif. Intell. Res. **55**, 409–442 (2016)
13. Castro, F.M., Marín-Jiménez, M.J., Guil, N., Schmid, C., Alahari, K.: End-to-End incremental learning. In: Ferrari, V., Hebert, M., Sminchisescu, C., Weiss, Y. (eds.) ECCV 2018. LNCS, vol. 11216, pp. 241–257. Springer, Cham (2018). https://doi.org/10.1007/978-3-030-01258-8_15
14. Chaudhry, A., Dokania, P.K., Ajanthan, T., Torr, P.H.S.: Riemannian walk for incremental learning: understanding forgetting and intransigence. In: Ferrari, V., Hebert, M., Sminchisescu, C., Weiss, Y. (eds.) ECCV 2018. LNCS, vol. 11215, pp. 556–572. Springer, Cham (2018). https://doi.org/10.1007/978-3-030-01252-6_33
15. Chaudhry, A., Ranzato, M., Rohrbach, M., Elhoseiny, M.: Efficient lifelong learning with A-GEM. In: ICLR (2019)
16. Dhar, P., Singh, R.V., Peng, K.C., Wu, Z., Chellappa, R.: Learning without memorizing. In: CVPR, pp. 5138–5146 (2019)
17. Farquhar, S., Gal, Y.: Towards robust evaluations of continual learning. arXiv:1805.09733 (2018)
18. Fernando, C., et al.: Pathnet: evolution channels gradient descent in super neural networks. arXiv:1701.08734 (2017)
19. Fukui, A., Park, D.H., Yang, D., Rohrbach, A., Darrell, T., Rohrbach, M.: Multimodal compact bilinear pooling for visual question answering and visual grounding. In: EMNLP (2016)

20. Gama, J.: Knowledge Discovery from Data Streams. Chapman and Hall/CRC, Boca Raton (2010)
21. Gama, J., Sebastião, R., Rodrigues, P.P.: On evaluating stream learning algorithms. Mach. Learn. **90**(3), 317–346 (2013). https://doi.org/10.1007/s10994-012-5320-9
22. Hayes, T.L., Cahill, N.D., Kanan, C.: Memory efficient experience replay for streaming learning. In: ICRA (2019)
23. Hayes, T.L., Kanan, C.: Lifelong machine learning with deep streaming linear discriminant analysis. In: CVPRW (2020)
24. Hayes, T.L., Kemker, R., Cahill, N.D., Kanan, C.: New metrics and experimental paradigms for continual learning. In: CVPRW, pp. 2031–2034 (2018)
25. He, K., Zhang, X., Ren, S., Sun, J.: Deep residual learning for image recognition. In: CVPR (2016)
26. Hou, S., Pan, X., Loy, C.C., Wang, Z., Lin, D.: Lifelong learning via progressive distillation and retrospection. In: Ferrari, V., Hebert, M., Sminchisescu, C., Weiss, Y. (eds.) ECCV 2018. LNCS, vol. 11207, pp. 452–467. Springer, Cham (2018). https://doi.org/10.1007/978-3-030-01219-9_27
27. Hou, S., Pan, X., Wang, Z., Change Loy, C., Lin, D.: Learning a unified classifier incrementally via rebalancing. In: CVPR (2019)
28. Hudson, D.A., Manning, C.D.: Compositional attention networks for machine reasoning. In: ICLR (2018)
29. Insausti, R., et al.: The nonhuman primate hippocampus: neuroanatomy and patterns of cortical connectivity. In: Hannula, D.E., Duff, M.C. (eds.) The Hippocampus from Cells to Systems, pp. 3–36. Springer, Cham (2017). https://doi.org/10.1007/978-3-319-50406-3_1
30. Jegou, H., Douze, M., Schmid, C.: Product quantization for nearest neighbor search. TPAMI **33**(1), 117–128 (2010)
31. Ji, D., Wilson, M.A.: Coordinated memory replay in the visual cortex and hippocampus during sleep. Nat. Neurosci. **10**(1), 100–107 (2007)
32. Johnson, J., Douze, M., Jégou, H.: Billion-scale similarity search with GPUs. IEEE Trans. Big Data (2019)
33. Johnson, J., Hariharan, B., van der Maaten, L., Fei-Fei, L., Zitnick, C.L., Girshick, R.: Clevr: a diagnostic dataset for compositional language and elementary visual reasoning. In: CVPR (2017)
34. Kafle, K., Kanan, C.: Answer-type prediction for visual question answering. In: CVPR, pp. 4976–4984 (2016)
35. Kafle, K., Kanan, C.: An analysis of visual question answering algorithms. In: ICCV, pp. 1983–1991 (2017)
36. Kafle, K., Kanan, C.: Visual question answering: datasets, algorithms, and future challenges. Comput. Vis. Image Underst. **163**, 3–20 (2017)
37. Kafle, K., Price, B., Cohen, S., Kanan, C.: DVQA: understanding data visualizations via question answering. In: CVPR, pp. 5648–5656 (2018)
38. Kafle, K., Shrestha, R., Cohen, S., Price, B., Kanan, C.: Answering questions about data visualizations using efficient bimodal fusion. In: WACV, pp. 1498–1507 (2020)
39. Kafle, K., Shrestha, R., Kanan, C.: Challenges and prospects in vision and language research. Front. Artif. Intell. **2**, 28 (2019)
40. Kahou, S.E., Michalski, V., Atkinson, A., Kádár, Á., Trischler, A., Bengio, Y.: Figureqa: An annotated figure dataset for visual reasoning. arXiv preprint arXiv:1710.07300 (2017)
41. Karlsson, M.P., Frank, L.M.: Awake replay of remote experiences in the hippocampus. Nat. Neurosci. **12**(7), 913 (2009)

42. Kazemi, V., Elqursh, A.: Show, ask, attend, and answer: a strong baseline for visual question answering. arXiv:1704.03162 (2017)
43. Kazemzadeh, S., Ordonez, V., Matten, M., Berg, T.: Referitgame: Referring to objects in photographs of natural scenes. In: EMNLP, pp. 787–798 (2014)
44. Kemker, R., Kanan, C.: FearNet: brain-inspired model for incremental learning. In: ICLR (2018)
45. Kemker, R., McClure, M., Abitino, A., Hayes, T.L., Kanan, C.: Measuring catastrophic forgetting in neural networks. In: AAAI (2018)
46. Kim, J.H., Jun, J., Zhang, B.T.: Bilinear attention networks. In: NeurIPS, pp. 1564–1574 (2018)
47. Kirkpatrick, J., et al.: Overcoming catastrophic forgetting in neural networks. In: PNAS (2017)
48. Konkel, A., Warren, D.E., Duff, M.C., Tranel, D., Cohen, N.J.: Hippocampal amnesia impairs all manner of relational memory. Front. Hum. Neurosci. **2**, 15 (2008)
49. Le, T., Stahl, F., Gaber, M.M., Gomes, J.B., Di Fatta, G.: On expressiveness and uncertainty awareness in rule-based classification for data streams. Neurocomputing **265**, 127–141 (2017)
50. Lee, K., Lee, K., Shin, J., Lee, H.: Overcoming catastrophic forgetting with unlabeled data in the wild. In: ICCV, pp. 312–321 (2019)
51. Lewis, P.A., Durrant, S.J.: Overlapping memory replay during sleep builds cognitive schemata. Trends Cogn. Sci. **15**(8), 343–351 (2011)
52. Lomonaco, V., Maltoni, D.: Core50: a new dataset and benchmark for continuous object recognition. In: CoRL, pp. 17–26 (2017)
53. Lopez-Paz, D., Ranzato, M.: Gradient episodic memory for continual learning. In: NeurIPS, pp. 6467–6476 (2017)
54. Malinowski, M., Fritz, M.: A multi-world approach to question answering about real-world scenes based on uncertain input. In: NeurIPS (2014)
55. Marois, V., Jayram, T., Albouy, V., Kornuta, T., Bouhadjar, Y., Ozcan, A.S.: On transfer learning using a mac model variant. arXiv:1811.06529 (2018)
56. McClelland, J.L., Goddard, N.H.: Considerations arising from a complementary learning systems perspective on hippocampus and neocortex. Hippocampus **6**(6), 654–665 (1996)
57. McCloskey, M., Cohen, N.J.: Catastrophic interference in connectionist networks: the sequential learning problem. Psychol. Learn. Motiv. **24**, 109–165 (1989)
58. Nguyen, C.V., Li, Y., Bui, T.D., Turner, R.E.: Variational continual learning. In: ICLR (2018)
59. Ostapenko, O., Puscas, M., Klein, T., Jähnichen, P., Nabi, M.: Learning to remember: a synaptic plasticity driven framework for continual learning. In: CVPR (2019)
60. O'Neill, J., Pleydell-Bouverie, B., Dupret, D., Csicsvari, J.: Play it again: reactivation of waking experience and memory. Trends Neurosci. **33**(5), 220–229 (2010)
61. Parisi, G.I., Kemker, R., Part, J.L., Kanan, C., Wermter, S.: Continual lifelong learning with neural networks: a review. Neural Netw. **113**, 54–71 (2019)
62. Parisi, G.I., Tani, J., Weber, C., Wermter, S.: Lifelong learning of spatiotemporal representations with dual-memory recurrent self-organization. Front. Neurorobot. **12**, 78 (2018)
63. Plummer, B.A., Wang, L., Cervantes, C.M., Caicedo, J.C., Hockenmaier, J., Lazebnik, S.: Flickr30k entities: collecting region-to-phrase correspondences for richer image-to-sentence models. In: ICCV, pp. 2641–2649 (2015)
64. Rebuffi, S.A., Kolesnikov, A., Sperl, G., Lampert, C.H.: icarl: incremental classifier and representation learning. In: CVPR (2017)

65. Riemer, M., et al.: Learning to learn without forgetting by maximizing transfer and minimizing interference. In: ICLR (2019)
66. Ritter, H., Botev, A., Barber, D.: Online structured Laplace approximations for overcoming catastrophic forgetting. In: NeurIPS, pp. 3738–3748 (2018)
67. Rohrbach, A., Rohrbach, M., Hu, R., Darrell, T., Schiele, B.: Grounding of textual phrases in images by reconstruction. In: Leibe, B., Matas, J., Sebe, N., Welling, M. (eds.) ECCV 2016. LNCS, vol. 9905, pp. 817–834. Springer, Cham (2016). https://doi.org/10.1007/978-3-319-46448-0_49
68. Russakovsky, O., et al.: ImageNet large scale visual recognition challenge. IJCV **115**(3), 211–252 (2015). https://doi.org/10.1007/s11263-015-0816-y
69. Rusu, A.A., et al.: Progressive neural networks. arXiv:1606.04671 (2016)
70. Serra, J., Suris, D., Miron, M., Karatzoglou, A.: Overcoming catastrophic forgetting with hard attention to the task. In: ICML, pp. 4555–4564 (2018)
71. Shrestha, R., Kafle, K., Kanan, C.: Answer them all! toward universal visual question answering models. In: CVPR (2019)
72. Stickgold, R., Hobson, J.A., Fosse, R., Fosse, M.: Sleep, learning, and dreams: off-line memory reprocessing. Science **294**(5544), 1052–1057 (2001)
73. Subramanian, S., Trischler, A., Bengio, Y., Pal, C.J.: Learning general purpose distributed sentence representations via large scale multi-task learning. In: ICLR (2018)
74. Takahashi, S.: Episodic-like memory trace in awake replay of hippocampal place cell activity sequences. Elife **4**, e08105 (2015)
75. Teyler, T.J., Rudy, J.W.: The hippocampal indexing theory and episodic memory: updating the index. Hippocampus **17**(12), 1158–1169 (2007)
76. Verma, V., et al.: Manifold mixup: better representations by interpolating hidden states. In: ICML (2019)
77. Wu, Y., et al.: Large scale incremental learning. In: CVPR, pp. 374–382 (2019)
78. Yang, Z., He, X., Gao, J., Deng, L., Smola, A.J.: Stacked attention networks for image question answering. In: CVPR (2016)
79. Yoon, J., Yang, E., Lee, J., Hwang, S.J.: Lifelong learning with dynamically expandable networks. In: ICLR (2018)
80. Yosinski, J., Clune, J., Bengio, Y., Lipson, H.: How transferable are features in deep neural networks? In: NeurIPS, pp. 3320–3328 (2014)
81. Zenke, F., Poole, B., Ganguli, S.: Continual learning through synaptic intelligence. In: ICML, pp. 3987–3995 (2017)

Image Classification in the Dark Using Quanta Image Sensors

Abhiram Gnanasambandam(✉) and Stanley H. Chan

Purdue University, West Lafayette, IN 47907, USA
{agnanasa,stanchan}@purdue.edu

Abstract. State-of-the-art image classifiers are trained and tested using well-illuminated images. These images are typically captured by CMOS image sensors with at least tens of photons per pixel. However, in dark environments when the photon flux is low, image classification becomes difficult because the measured signal is suppressed by noise. In this paper, we present a new low-light image classification solution using Quanta Image Sensors (QIS). QIS are a new type of image sensors that possess photon-counting ability without compromising on pixel size and spatial resolution. Numerous studies over the past decade have demonstrated the feasibility of QIS for low-light imaging, but their usage for image classification has not been studied. This paper fills the gap by presenting a student-teacher learning scheme which allows us to classify the noisy QIS raw data. We show that with student-teacher learning, we can achieve image classification at a photon level of one photon per pixel or lower. Experimental results verify the effectiveness of the proposed method compared to existing solutions.

Keywords: Quanta image sensors · Low light · Classification

1 Introduction

Quanta Image Sensors (QIS) are a type of single-photon image sensors originally proposed by E. Fossum as a candidate solution for the shrinking full-well capacity problem of the CMOS image sensors (CIS) [18,19]. Compared to the CIS which accumulate photons to generate signals, QIS have a different design principle which partitions a pixel into many tiny cells called the jots with each jot being a single-photon detector. By oversampling the space and time, and by using a carefully designed image reconstruction algorithm, QIS can capture very low-light images with signal-to-noise ratio much higher than existing CMOS image sensors of the same pixel pitch [3]. Over the past few years, prototype QIS have been built by researchers at Dartmouth and Gigajot Technology Inc. [48,49], with a number of theoretical and algorithmic contributions by researchers at

Electronic supplementary material The online version of this chapter (https://doi.org/10.1007/978-3-030-58598-3_29) contains supplementary material, which is available to authorized users.

A. Vedaldi et al. (Eds.): ECCV 2020, LNCS 12353, pp. 484–501, 2020.
https://doi.org/10.1007/978-3-030-58598-3_29

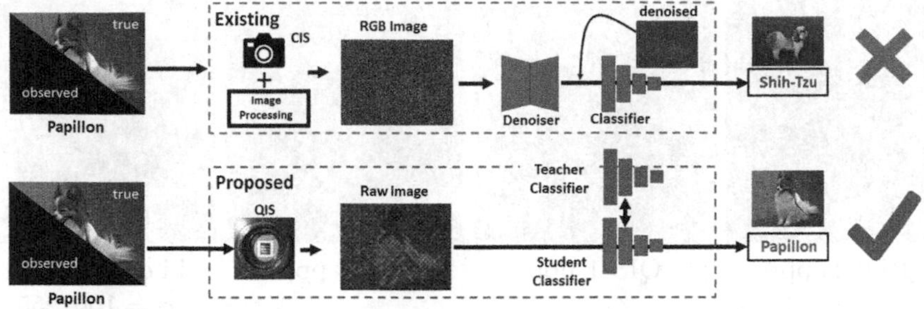

Fig. 1. [Top] Traditional image classification methods are based on CMOS image sensors (CIS), followed by a denoiser-classifier pipeline. [Bottom] The proposed classification method comprises a novel image sensor QIS and a novel student-teacher learning protocol. QIS generates significantly stronger signals, and student-teacher learning improves the robustness against noise.

EPFL [6,64], Harvard [4], and Purdue [11,16,17,27,28]. Today, the latest QIS prototype can perform color imaging with a read noise of $0.25e^-$/pix (compared to at least several electrons in CIS [22]) and dark current of $0.068e^-$/pix/s at room temperature (compared to $>1e^-$/pix/s in CIS) [27,49].

While prior works have demonstrated the effectiveness of using QIS for low-light image formation, there is no systematic study of how QIS can be utilized to perform better image classification in the dark. The goal of this paper is to fill the gap by proposing the first QIS image classification solution. Our proposed method is summarized in Fig. 1. Compared to the traditional CIS-based low-light image classification framework, our solution leverages the unique single-photon sensing capability of QIS to acquire very low-light photon count images. We do not use any image processing, and directly feed the raw Bayer QIS data into our classifier. Our classifier is trained using a novel student-teacher learning protocol, which allows us to transfer knowledge from a teacher classifier to a student classifier. We show that the student-teacher protocol can effectively alleviate the need for a deep image denoiser as in the traditional frameworks. Our experiments demonstrate that the proposed method performs better than the existing solutions. The overall system – QIS combined with student-teacher learning – can achieve image classification on real data at 1 photon per pixel or lower. To summarize, the two contributions of this paper are:

(i) The introduction of student-teaching learning for low-light image classification problems. The experiments show that the proposed method outperforms existing approaches.
(ii) The first demonstration of image classification at a photon level of 1 photon per pixel or lower, on real images. This is a very low photon level compared to other results reported in the image classification literature.

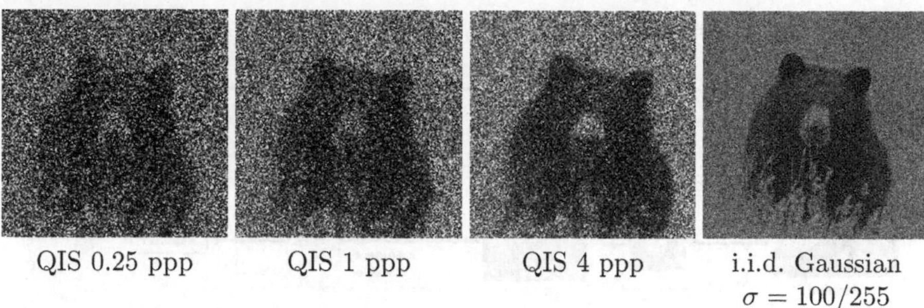

Fig. 2. How dark is one photon per pixel? The first three sub-images in this figure are the real captures by a prototype QIS at various photon levels. The last sub-image is a simulation using additive i.i.d. Gaussian noise of a level of $\sigma = 100/255$, which is often considered as heavy degradation in the denoising literature. Additional examples can be found in Fig. 10.

2 Background

2.1 Quanta Image Sensor

Quanta Image Sensors are considered as one of the candidates for the third generation image sensors after CCD and CMOS. Fabricated using the commercial 3D stacking technology, the current sensor has a pixel pitch of 1.1 μm, with even smaller sensors being developed. The advantage of QIS over the conventional CMOS image sensors is that at 1.1 μm, the read noise of QIS is as low as $0.25e^-$ whereas a typical 1 μm CMOS image sensor is at least several electrons. This low read noise (and also the low dark current) is made possible by the unique non-avalanche design [49] so that pixels can be packed together without causing strong stray capacitance. The non-avalanche design also differentiates QIS from single photon avalanche diodes (SPAD). SPADs are typically bigger >5 μm, have lower fill factor <70%, have lower quantum efficiency <50%, and have significantly higher dark count $>10e^-$. See [27] for a detailed comparison between CIS, SPAD, and QIS. In general, SPADs are useful for applications such as time-of-flight imaging because of their speed [2,24,42,55], although new results in HDR imaging has been reported [31,50]. QIS have better resolution and works well for passive imaging.

2.2 How Dark Is One Photon Per Pixel?

When we say low-light imaging, it is important to clarify the photon level. The photon level is usually measured in terms of lux. However, a more precise definition is the unit of photons per pixel (ppp). "Photons per pixel" is the average number of photons a pixel sees during the exposure period. We use photons per pixel as the metric because the amount of photons detected by a sensor depends on the exposure time and sensor size—A large sensor inherently detects more photons, so does long exposure. For example, under the same low-light condition, images formed by the Keck telescope (aperture diameter = 10 m) certainly has better signal-to-noise than an iPhone camera (aperture diameter = 4.5 mm). A high-end 3.5 μm camera today has a read noise greater than $2e^-$ [59]. Thus, our benchmark choice of 1 ppp is approximately half of the read noise of a high-end sensor today. To give readers an idea of the amount of noise we should expect to see at 1 ppp, we show a set of real QIS images in Fig. 2. Signals at 1 ppp is significantly worse than the so-called "heavy noise" images we observe in the denoising literature and the low-light classification literature.

2.3 Prior Work

Quanta Image Sensors. QIS were proposed in 2005, and since then significant progress has been made over the past 15 years. Readers interested in the sensor development can consult recent keynote reports, e.g., [21]. On the algorithmic side, several theoretical signal processing results and reconstruction algorithms have been proposed [3,5,27], including some very recent methods based on deep learning [6,11]. However, since the sensor is relatively new, computer vision applications of the sensor are not yet common. To the best of our knowledge, the only available method for tracking applications is [32].

Low-Light Classification. The majority of the existing work in classification is based on well-illuminated CMOS images. The first systematic study of the feasibility of low-light classification was presented by Chen and Perona [7], who observed that low-light classification is achievable by using a few photons. In the same year, Diamond et al. [14] proposed the "Dirty Pixels" method by training a denoiser and a classifier simultaneously. They observed that less aggressive denoisers are better for classification because the features are preserved. Other methods adopt similar strategies, e.g., using discrete cosine transform [35], training a classifier to help denoising [61] or using an ensemble method [15], or training a denoiser that are better suited for pre-trained classifiers [44,45].

Low-Light Reconstruction. A closely related area of low-light classification is low-light reconstruction, e.g., denoising. Classical low-light reconstruction usually follows the line of Poisson-based inverse problems [52] and contrast enhancement [23,30,37,53]. Deep neural network methods have recently become the main driving force [47,56,57,62,67,68], including the recent series on "seeing in the dark" by Chen et al. [8,9]. Burst photography [12,33,41,54] (with some older work in [39,43,46]) is related but not directly applicable to us since the methods are developed for multi-frame problems.

3 Method

The proposed method comprises QIS and a novel student-teacher learning scheme. In this section, we first discuss how images are formed by QIS. We will then present the proposed student-teacher learning scheme which allows us to overcome the noise in QIS measurements.

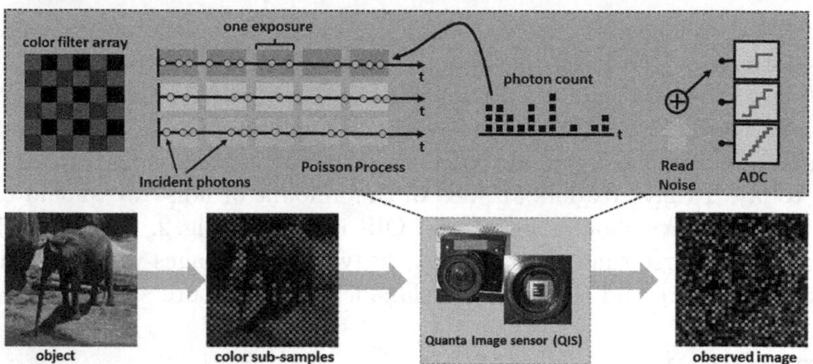

Fig. 3. QIS image formation model. The basic image formation of QIS consists of a color filter array, a Poisson process, read noise, and an analog-to-digital converter (ADC). Additional factors are summarized in (1).

3.1 QIS Image Formation Model

The image formation model is shown in Fig. 3. Given an object in the scene ($\boldsymbol{x}_{\text{rgb}}$), we use a color filter array (CFA) to bandpass the light to subsample the color. Depending on the exposure time and the size of the jots, a sensor gain α is applied to scale the sub-sampled color pixels. The photon arrival is simulated using a Poisson model. Gaussian noise is added to simulate the read noise arising from the circuit. Finally, an analog-to-digital converter (ADC) is used to truncate the real numbers to integers depending on the number of bits allocated by the sensor. For example, a single-bit QIS will output two levels, whereas multi-bit QIS will output several levels. In either case, the signal is clipped to take value in $\{0, 1, \ldots L\}$, where L represents the maximum signal level. The image formation process can be summarized using the following equation

$$\underbrace{\boldsymbol{x}_{\text{QIS}}}_{\mathbb{R}^{M\times N}} = \text{ADC}_{[0,L]}\Big\{ \underbrace{\text{Poisson}}_{\text{photon arrival}} \Big(\underbrace{\alpha}_{\text{sensor gain}} \cdot \text{CFA}(\underbrace{\boldsymbol{x}_{\text{rgb}}}_{\mathbb{R}^{M\times N\times 3}}) \Big) + \underbrace{\eta}_{\text{read noise}} \Big\}, \quad (1)$$

In addition to the basic image formation model described in (1), two other components are included in the simulations. First, we include the dark current which is an additive noise term to $\alpha \cdot \text{CFA}(\boldsymbol{x}_{\text{rgb}})$. The typical dark current of the QIS is $0.068e^{-}$/pix/s. Second, we model the pixel response non-uniformity (PRNU). PRNU is a pixel-wise multiplication applied to $\boldsymbol{x}_{\text{rgb}}$, and is unique for every sensor. Readers interested in details on the image formation model and statistics can consult previous works such as [3,16,20,65].

3.2 Student-Teacher Learning

Inspecting (1), we notice that even if the read noise η is zero, the random Poisson process will still create a fundamental limit due to the shot noise in

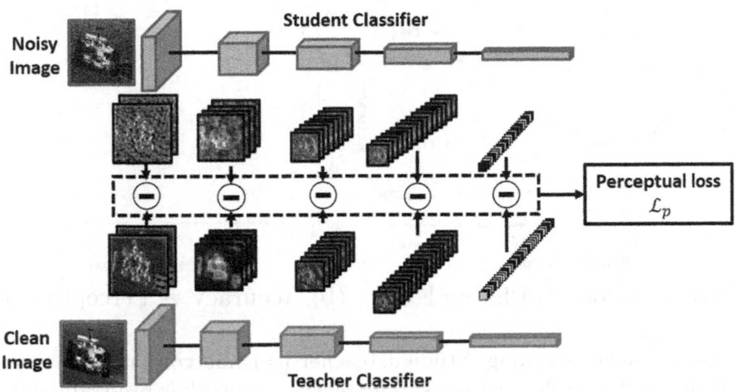

Fig. 4. Proposed method. The proposed method trains a classification network with two training losses: (1) cross-entropy loss to measure the prediction quality, and (2) perceptual loss to transfer knowledge from teacher to student. During testing, only the student is used. We introduce a 2-layer entrance (colored in orange) for the student network so that the classifier can handle the Bayer image. (Color figure online)

$\boldsymbol{x}_{\text{QIS}}$. Therefore, when applying a classification method to the raw QIS data, some capability of removing the shot noise becomes necessary. The traditional solution to this problem (in the context of CIS) is to denoise the images as shown in the top of Fig. 1. The objective of this section is to introduce an alternative approach using the concept of student-teacher learning.

The idea of student-teacher learning can be understood from Fig. 5. There are two networks in this figure: A teacher network and a student network. The teacher network is trained using *clean* samples, and is pre-trained, i.e., its network parameters are fixed during training of the student network. The student network is trained using *noisy* samples with the assistance from the teacher. Because the teacher is trained using clean samples, the features extracted are in principle "good", in contrast to the features of the student which are likely to be "corrupted". Therefore, in order to transfer knowledge from the teacher to the student, we propose minimizing a *perceptual loss* as defined below. We define the j-th layer's feature of the student network as $\phi^j(\boldsymbol{x}_{\text{QIS}})$, where $\phi^j(\cdot)$ maps $\boldsymbol{x}_{\text{QIS}}$ to a feature vector, and we define $\widehat{\phi}^j(\boldsymbol{x}_{\text{rgb}})$ as the feature vector extracted by the teacher network. The perceptual loss is

$$\mathcal{L}_{\text{p}}(\boldsymbol{x}_{\text{QIS}}, \boldsymbol{x}_{\text{rgb}}) = \sum_{j=1}^{J} \underbrace{\frac{1}{N_j} \left\| \widehat{\phi}^j(\boldsymbol{x}_{\text{rgb}}) - \phi^j(\boldsymbol{x}_{\text{QIS}}) \right\|^2}_{j\text{-th layer's perceptual loss}}, \tag{2}$$

where N_j is the dimension of the j-th feature vector. Since the perceptual loss measures the distance between the student and the teacher, minimizing the perceptual loss forces them to be close. This, in turn, forces the network to "denoise" the shot noise and read noise in $\boldsymbol{x}_{\text{QIS}}$ before predicting the label.

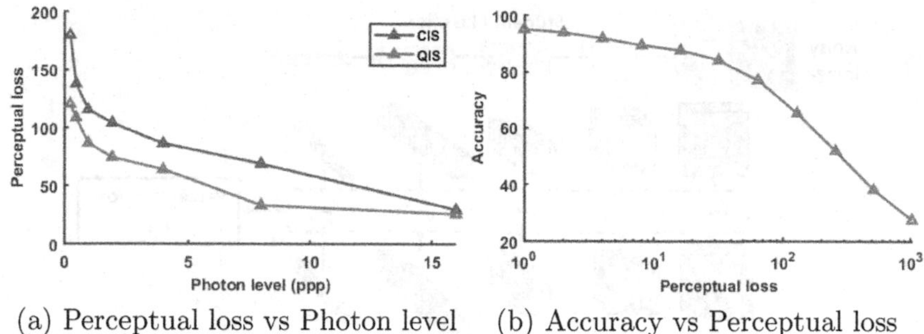

(a) Perceptual loss vs Photon level (b) Accuracy vs Perceptual loss

Fig. 5. Student-teacher learning. Student-teacher learning comprises two networks: a teacher network and a student network. The teacher network is pre-trained using clean samples whereas the student is trained using noisy samples. To transfer knowledge from the teacher to the student, we compare the features extracted by the teacher and the student at different stages of the network. The difference between the features is measured as the perceptual loss.

We conduct a simple experiment to demonstrate the impact of input noise on perceptual loss and classification accuracy. We first consider a pre-trained teacher network by sending QIS data at different photon levels. As photon level drops, the quality of the features also drops, and hence the perceptual loss increases. This is illustrated in Fig. 6(a). Then in Fig. 6(b), we evaluate the classification accuracy by using the synthetic testing data outlined in the Experiment Section. As the perceptual loss increases, the classification accuracy drops. This result suggests that if we minimize the perceptual loss then the classification accuracy can be improved.

Our proposed student-teacher learning is inspired by the knowledge distillation work of Hinton et al. [34] which proposed an effective way to compress networks. Several follow up ideas have been proposed, e.g., [1,29,63,69], including the MobileNet [36]. The concept of perceptual loss has been used in various computer vision applications such as the texture-synthesis and style-transfer by Johnson et al. [38] and Gatys et al. [25,26], among many others [10,44,48,51,58,60,66]. The method we propose here is different because we are not compressing the network. We are not asking the student to mimic the teacher because the teacher and the student are performing two different tasks: The teacher classifies clean data, whereas the student classifies noisy data. In the context of low-light classification, student-teacher learning has not been applied.

3.3 Overall Method

The overall loss function comprises the perceptual loss and the conventional prediction loss using cross-entropy. The cross-entropy loss $\mathcal{L}_c$, measures the difference between true label y and the predicted label $f_\Theta(\boldsymbol{x}_{\text{QIS}})$ generated by the student network, where f_Θ is the student network. The overall loss is mathematically described as

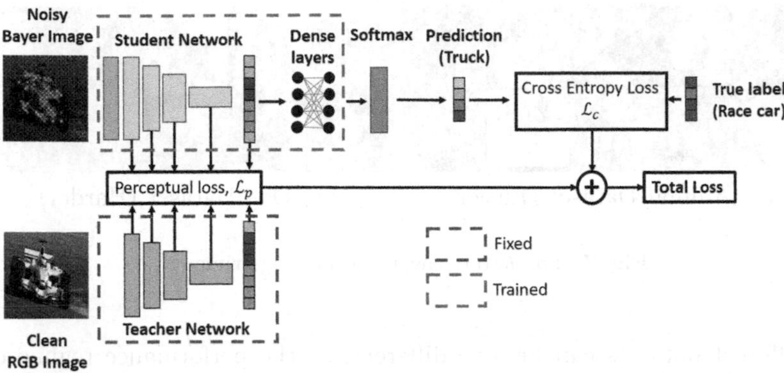

Fig. 6. Effectiveness of student-teacher learning. (a) Perceptual loss as a function of photon level. (b) Classification accuracy as a function of the perceptual loss $\mathcal{L}_p(\boldsymbol{x}_{\text{QIS}}, \boldsymbol{x}_{\text{rgb}})$. The accuracy is measured by repeating the synthetic experiment described in the experiment section. The negative correlation suggests that perceptual loss is indeed an influential factor.

$$\mathcal{L}(\boldsymbol{\Theta}) = \sum_{n=1}^{N} \left\{ \mathcal{L}_{\text{c}}(y^n, f_{\boldsymbol{\Theta}}(\boldsymbol{x}_{\text{QIS}}^n)) + \lambda \mathcal{L}_{\text{p}}(\boldsymbol{x}_{\text{rgb}}^n, \boldsymbol{x}_{\text{QIS}}^n) \right\}, \tag{3}$$

where $\boldsymbol{x}^n$ denotes the n-th training sample with the ground truth label y^n. During the training, we optimize the weights of the student network by solving

$$\widehat{\boldsymbol{\Theta}} = \underset{\boldsymbol{\Theta}}{\text{argmin}}\ \mathcal{L}(\boldsymbol{\Theta}). \tag{4}$$

During testing, we feed a testing sample $\boldsymbol{x}_{\text{QIS}}$ to the student network and evaluate the output:

$$\widehat{y} = f_{\widehat{\boldsymbol{\Theta}}}(\boldsymbol{x}_{\text{QIS}}). \tag{5}$$

Figure 4 illustrates the overall network architecture. In this figure, we emphasize that training is done on the student only. The teacher is fixed and is not trainable. In this particular example, we introduce a very shallow network consisting of 2 convolution layers with 32 and 3 filters respectively. This shallow network is used to perform the necessary demosaicking by converting the raw Bayer pattern to the full RGB before feeding into a standard classification network.

4 Experiments

4.1 Dataset

Dataset. We consider two datasets. The first dataset (Animal) contains visually distinctive images where the class labels are far apart. The second dataset (Dog) contains visually similar images where the class labels are fine-grained. The

(a) Animal Dataset (Easier) (b) Dog Dataset (Harder)

Fig. 7. The two datasets for our experiments.

two different datasets can help to differentiate the performance regime of the proposed method and its benefits over other state-of-the-art networks.

The construction of the two datasets is as follows. For the Animal dataset, we randomly select 10 classes of animals from ImageNet [13], as shown in Fig. 7(a). Each class contains 1300 images, giving a total 13K images. Among these, 9K are used for training, 1K for validation, 3K for testing. For the Dog dataset, we randomly select 10 classes of dogs from the Stanford Dog dataset [40], as shown in Fig. 7(b). Each class has approximately 150 images, giving a total of 1919 images. We use 1148 for training, 292 for validation, and 479 for testing.

4.2 Competing Methods and Our Network

We compare our method with three existing low-light classification methods as shown in Fig. 8. The three competing methods are (a) Vanilla denoiser + classifier, an "off-the-shelf" solution using pre-trained models. The denoiser is pre-trained on the QIS data and the classifier is pre-trained on clean images. (b) Dirty Pixels [14], same as Vanilla denoiser + classifier, but trained end-to-end

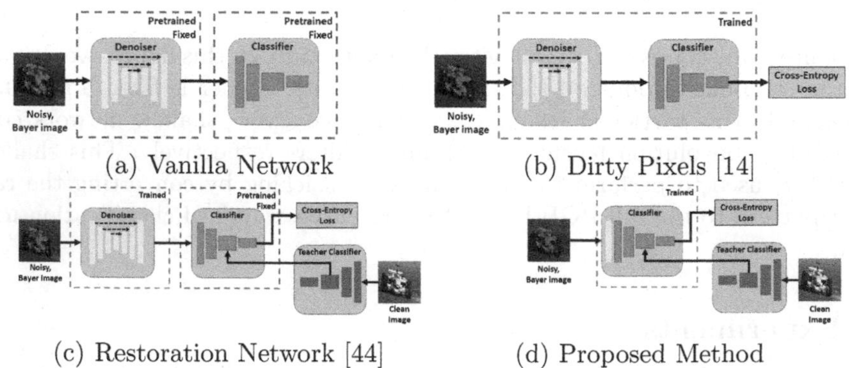

(a) Vanilla Network (b) Dirty Pixels [14]

(c) Restoration Network [44] (d) Proposed Method

Fig. 8. Competing methods. The major difference between the networks are the trainable modules and the loss functions. For Dirty Pixels and our proposed method, we further split it into two versions: using a deep denoiser or using a shallow entrance network.

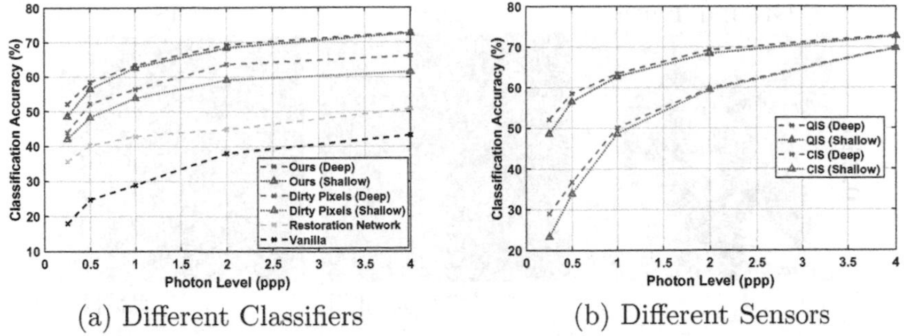

(a) Different Classifiers (b) Different Sensors

Fig. 9. Synthetic data on dog dataset. (a) Comparing different classification methods with QIS images. (b) Comparing QIS and CIS using proposed classifier.

using the QIS data. (c) Restoration Network [44,45], which trains a denoiser but uses a classifier pre-trained on clean images. This can be viewed as a middle-ground solution between Vanilla and Dirty Pixels.

To ensure that the comparison is fair w.r.t. the training protocol and not the architecture, all classifiers in this experiment (including ours) use the same VGG-16 architecture. For methods that use a denoiser, the denoiser is fixed as a UNet. This particular combination of denoiser and classifier will certainly affect the final performance, but the effectiveness of the training protocol can still be observed. Combinations beyond the ones we report here can be found in the ablation study. For Dirty Pixels and our proposed method, we further split them into two versions: (i) Using a deep denoiser as the entrance, i.e., a 20-layer UNet, and (ii) using a shallow two-layer network as the entrance to handle the Bayer pattern, as we described in the proposed method section. We will analyze the influence of this component in the ablation study.

4.3 Synthetic Experiment

The first experiment is based on synthetic data. The training data are created by the QIS model. To simulate the QIS data, we follow Eq. (1) by using the Poisson-Gaussian process. The read noise is $\sigma = 0.25e^-$ according to [49]. The analog-to-digital converter is set to 5 bits so that the number of photons seen by the sensors is between 0 and 31. We use a similar simulation procedure for CIS with the difference being the read noise, which we set to $\sigma = 2.0e^-$ [59].

The experiments are conducted for 5 different photon levels corresponding to 0.25, 0.5, 1, 2, and 4 photons per pixel (ppp). The photon level is controlled by adjusting the value of the multiplier α in Eq. (1). The loss function weights λ in Eq. (3) is tuned for optimal performance.

The results of the synthetic data experiment are shown in Fig. 9. In Fig. 9(a), we observe that our proposed classification is consistently better than competing methods the photon levels we tested. Moreover, since all methods reported in Fig. 9(a) are using QIS as the sensor, the curves in Fig. 9(a) reveal the effec-

Fig. 10. Real image results. This figure shows raw Bayer data obtained from a prototype QIS and a commercially available CIS, and how they are classified using our proposed classifier. The inset images show the denoised images (by [9]) for visualization. Notice the heavy noise at 0.25 and 0.5 ppp, only QIS plus our proposed classification method can produce the correct prediction.

tiveness of just the classification method. In Fig. 9(b), we compare the difference between using QIS and CIS. As we expect, CIS has worse performance compared to QIS.

4.4 Real Experiment

We conduct an experiment using real QIS and CIS data. The real QIS data are collected by a prototype QIS camera Gigajot PathFinder [27], whereas the real CIS data are collected by using a commercially available camera. To set up the experiment, we display the images on a Dell P2314H LED screen (60 Hz). The cameras are 1 m from the display so that the field of view covers 256×256 pixels

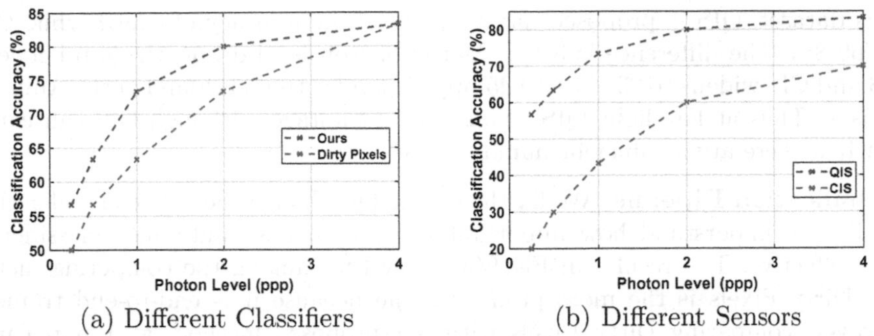

(a) Different Classifiers (b) Different Sensors

Fig. 11. Real data on animal dataset. (a) Comparing different classification methods using QIS as the sensor. (b) Comparing QIS and CIS using our proposed classifier.

of the image. The integration time of the CIS is set to 250 µs and that of QIS is 75 µs. Since the CIS and QIS have different lenses, we control their aperture sizes and the brightness of the screen such that the average number of photons per pixel is equal for both sensors.

The training of the network in this real experiment is still done using the synthetic dataset, with the image formation model parameters matched with the actual sensor parameters. However, since the real image sensors have pixel non-uniformity, during the training we multiply a random PRNU mask to each of the generated images to mimic the process of PRNU. For testing, we collect 30 real images at each photon level, across 5 different photon levels. This corresponds to a total of 150 real testing images.

In Fig. 11 we make two pairs of comparisons: Proposed (shallow) versus Dirty Pixels (shallow), and QIS versus CIS. In Fig. 11(a), where we observe that the proposed method has a consistent improvement over Dirty Pixels. The comparison between QIS and CIS is shown in Fig. 11(b). It is evident that QIS has better performance compared to CIS. Figure 10 shows the visualizations. The ground truth images were displayed on the screen, and the background images in QIS and CIS column are actual measurements from the corresponding cameras, cropped to 256 × 256. The thumbnail images in the front are the denoised images for reference. They are not used during the actual classification. The color bars at the bottom report the confidence level of the predicted class. Note the significant visual difference between QIS and CIS, and the classification results.

4.5 Ablation Study

In this section, we report several ablation study results and highlight the most influencing factors to the design.

Sensor. Our first ablation study is to fix the classifier but change the sensor from QIS to CIS. This experiment will underline the impact of QIS in the overall pipeline. The result of this ablation study can be seen in Fig. 9(b). At 4 ppp of the

Dogs dataset, QIS + proposed has a classification accuracy of 72.9% while CIS has 69.8%. The difference is 3.1%. As the photon level drops, the gap between QIS and CIS widens to 23.1% at 0.25 ppp. A similar trend is found in the Animals dataset. Thus at low light QIS has a clear advantage, although CIS can catch up when there are a sufficient number of photons.

Classification Pipeline. We fix the sensor but change the entire classification pipeline to understand how important the classifier is, and which classifier is more effective. The results in Fig. 9(a) show that among the competing methods, Dirty Pixels is the most promising one because it is end-to-end trained. However, comparing Dirty Pixels with our proposed method, at 1 ppp Dirty Pixels (shallow) achieves an accuracy of 53.9% whereas the proposed (shallow) achieves 62.7%. The trend continues as the photon level increases. This ablation analysis shows that a good sensor (QIS) does not automatically translate to better performance.

Student-Teacher Learning. Let us fix the sensor and the network, but change the training protocol. This will reveal the significance of the proposed student-teacher learning. To conduct this ablation study, we recognize that Dirty Pixels network structure (shallow and deep) is exactly the same as Ours (shallow and deep) since both use the same UNet and VGG-16. The only difference is the training protocol, where ours uses student-teacher learning and Dirty Pixels is a simple end-to-end. The result of this study is summarized in Fig. 9(a). It is evident that our training protocol offers advantages over Dirty Pixels.

We can further analyze the situation by plotting the training and validation error. Figure 12 [Left] shows the comparison between the proposed method (shallow) and Dirty Pixels (shallow). It is evident from the plot that without student-teacher learning (Dirty Pixels), the network overfits. The validation loss drops and then rises whereas the training loss keeps dropping. In contrast, the proposed method appears to mitigate the overfitting issue. One possible reason is that the student-teacher learning is providing some kind of regularization in an implicit form so that the validation loss is maintained at a low level.

Choice of Classification Network. All experiments reported in this paper use VGG-16 as the classifier. In this ablation study, we replace the VGG-16 classifier by other popular classifiers, namely ResNet50 and InceptionV3. These networks are fine-tuned using QIS data. Figure 12 [Right] shows the comparisons. Using the baseline training scheme, i.e., simple fine-tuning as in Dirty Pixels, it is observed that there is a minor gap between the different classifiers. However, by using the proposed student-teacher training protocol, we observe a substantial improvement for all the classifiers. This ablation study confirms that student-teacher learning is not limited to a particular network architecture.

Using a Pre-trained Classifier. This ablation study analyzes the effect of using a pre-trained classifier (trained on clean images). If we do this, then the overall system is exactly the same as the Restoration network [44] in Fig. 8(c). Restoration network has three training losses: (i) MSE to measure the image quality, (ii) Perceptual loss to measure feature quality, and (iii) the cross-entropy

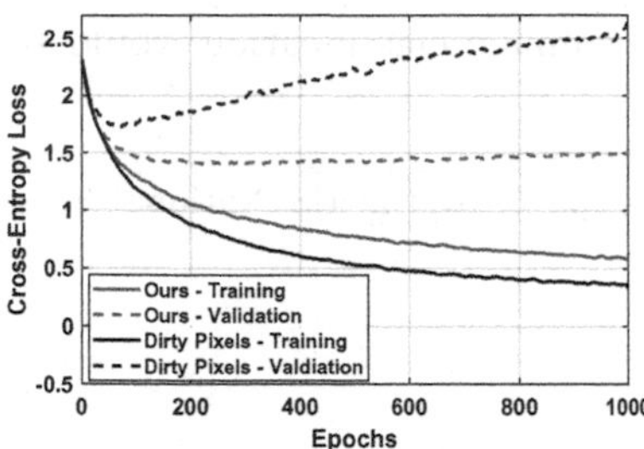

	VGG	ResNet	Inception
Baseline (fine-tuning)	42.1%	43.3%	44.3%
Proposed (student-teacher)	48.6%	49.1%	50.0%
Gap	+6.5%	+5.8%	+5.7%

Fig. 12. [Left] Training and validation loss of our method and Dirty Pixels. Notice that while our training loss is higher, the validation loss is significantly lower than Dirty Pixels. [Right] Ablation study of different classifiers and different training schemes. Reported numbers are based on QIS synthetic experiments at 0.25 ppp for the Dog Dataset.

loss. These three losses are used to just train the denoiser and not the classifier. Since the classifier is fixed, it becomes necessary for the denoiser to produce high-quality images or otherwise the classifier will not work. The results in Fig. 9(a) suggest that when the photon level is low, the denoiser fails to produce high-quality images and so the classification fails. For example, at 0.25 ppp Restoration Network achieves 35.6% but our proposed method achieves 52.1%. Thus it is imperative that we re-train the classifier for low-light images.

Deep or Shallow Denoisers? This ablation study analyzes the impact of using a deep denoiser compared to a shallow entrance layer. The result of this study can be found by comparing Ours (deep) and Ours (shallow) in Fig. 9(a), as well as Dirty (deep) and Dirty (shallow). The deep versions use a 20-layer UNet, whereas the shallow versions use a 2-layer network. The result in Fig. 9(a) suggests that while the deep denoiser has a significant impact on Dirty Pixels, its influence is quite small to the proposed method with the QIS images. Since we are using student-teacher learning, the features are already properly handled. The benefit from a deep denoiser for QIS is therefore marginal. However, for CIS data at low light, the deep denoiser helps in getting better classification performance, especially when the signal level is much below the read noise.

5 Conclusion

We proposed a new low-light image classification method by integrating Quanta Image Sensors (QIS) and a novel student-teacher training protocol. Experimental results confirmed that such combination is effective for low-light image classification, and the student-teacher protocol is a better alternative than the traditional denoise-then-classify framework. This paper also made the first demonstration of low-light image classification at a photon level of 1 photon per pixel or lower. The student-teacher training protocol is transferable to conventional CIS data, however, to achieve the desired performance at low light, QIS must be a part of the overall pipeline. Using multiple frames for image classification would be a fruitful direction for future work.

Acknowledgement. This work is supported, in part, by the US National Science Foundation under grant CCF-1718007.

References

1. Ba, J., Caruana, R.: Do deep nets really need to be deep? In: NeurIPS (2014)
2. Callenberg, C., Lyons, A., den Brok, D., Henderson, R., Hullin, M.B., Faccio, D.: EMCCD-SPAD camera data fusion for high spatial resolution time-of-flight imaging. In: Computational Optical Sensing and Imaging (2019)
3. Chan, S.H., Elgendy, O.A., Wang, X.: Images from bits: non-iterative image reconstruction for quanta image sensors. Sensors **16**(11), 1961 (2016)
4. Chan, S.H., Lu, Y.M.: Efficient image reconstruction for gigapixel quantum image sensors. In: IEEE Global Conference on Signal and Information Processing (2014)
5. Chan, S.H., Wang, X., Elgendy, O.A.: Plug-and-play ADMM for image restoration: fixed-point convergence and applications. IEEE Trans. Comput. Imaging **3**(1), 84–98 (2016)
6. Chandramouli, P., Burri, S., Bruschini, C., Charbon, E., Kolb, A.: A bit too much? High speed imaging from sparse photon counts. In: ICCP (2019)
7. Chen, B., Perona, P.: Seeing into darkness: scotopic visual recognition. In: CVPR (2017)
8. Chen, C., Chen, Q., Do, M.N., Koltun, V.: Seeing motion in the dark. In: ICCV (2019)
9. Chen, C., Chen, Q., Xu, J., Koltun, V.: Learning to see in the dark. In: CVPR (2018)
10. Chen, G., Li, Y., Srihari, S.N.: Joint visual denoising and classification using deep learning. In: ICIP (2016)
11. Choi, J.H., Elgendy, O.A., Chan, S.H.: Image reconstruction for quanta image sensors using deep neural networks. In: ICASSP (2018)
12. Davy, A., Ehret, T., Morel, J.M., Arias, P., Facciolo, G.: A non-local CNN for video denoising. In: ICIP (2019)
13. Deng, J., Dong, W., Socher, R., Li, L.J., Li, K., Fei-Fei, L.: ImageNet: a large-scale hierarchical image database. In: CVPR (2009)
14. Diamond, S., Sitzmann, V., Boyd, S., Wetzstein, G., Heide, F.: Dirty pixels: optimizing image classification architectures for raw sensor data. arXiv preprint arXiv:1701.06487 (2017)
15. Dodge, S., Karam, L.: Quality resilient deep neural networks. arXiv preprint arXiv:1703.08119 (2017)
16. Elgendy, O.A., Chan, S.H.: Optimal threshold design for quanta image sensor. IEEE Trans. Comput. Imaging **4**(1), 99–111 (2017)
17. Elgendy, O.A., Chan, S.H.: Color filter arrays for quanta image sensors. arXiv preprint arXiv:1903.09823 (2019)
18. Fossum, E.R.: Gigapixel digital film sensor (DFS) proposal. In: Nanospace Manipulation of Photons and Electrons for Nanovision Systems (2005)
19. Fossum, E.R.: Some thoughts on future digital still cameras. In: Image Sensors and Signal Processing for Digital Still Cameras (2006)

20. Fossum, E.R.: Modeling the performance of single-bit and multi-bit quanta image sensors. IEEE J. Electron Dev. Soc. **1**(9), 166–174 (2013)
21. Fossum, E.R., Ma, J., Masoodian, S., Anzagira, L., Zizza, R.: The quanta image sensor: every photon counts. Sensors **16**(8), 1260 (2016)
22. Fowler, B., McGrath, D., Bartkovjak, P.: Read noise distribution modeling for CMOS image sensors. In: International Image Sensor Workshop (2013)
23. Fu, Q., Jung, C., Xu, K.: Retinex-based perceptual contrast enhancement in images using luminance adaptation. IEEE Access **6**, 61277–61286 (2018)
24. Gariepy, G., et al.: Single-photon sensitive light-in-fight imaging. Nat. Commun. **6**(1), 1–7 (2015)
25. Gatys, L., Ecker, A., Bethge, M.: A neural algorithm of artistic style. Nat. Commun. (2015)
26. Gatys, L.A., Ecker, A.S., Bethge, M.: Texture synthesis using convolutional neural networks. In: NeurIPS (2015)
27. Gnanasambandam, A., Elgendy, O., Ma, J., Chan, S.H.: Megapixel photon-counting color imaging using quanta image sensor. Opt. Express **27**(12), 17298–17310 (2019)
28. Gnanasambandam, A., Ma, J., Chan, S.H.: High dynamic range imaging using quanta image sensors. In: International Image Sensors Workshop (2019)
29. Guo, T., Xu, C., He, S., Shi, B., Xu, C., Tao, D.: Robust student network learning. IEEE Trans. Neural Netw. Learn. Syst. **31**(7), 2455–2468 (2020)
30. Guo, X., Li, Y., Ling, H.: LIME: low-light image enhancement via illumination map estimation. IEEE Trans. Image Process. **26**(2), 982–993 (2016)
31. Gupta, A., Ingle, A., Gupta, M.: Asynchronous single-photon 3D imaging. In: ICCV, October 2019
32. Gyongy, I., Dutton, N., Henderson, R.: Single-photon tracking for high-speed vision. Sensors **18**(2), 323 (2018)
33. Hasinoff, S.W., et al.: Burst photography for high dynamic range and low-light imaging on mobile cameras. ACM Trans. Graph. **35**(6), 192 (2016)
34. Hinton, G., Vinyals, O., Dean, J.: Distilling the knowledge in a neural network. In: NeurIPS Deep Learning and Representation Learning Workshop (2015)
35. Hossain, M.T., Teng, S.W., Zhang, D., Lim, S., Lu, G.: Distortion robust image classification using deep convolutional neural network with discrete cosine transform. In: ICIP (2019)
36. Howard, A.G., et al.: MobileNets: efficient convolutional neural networks for mobile vision applications. arXiv preprint arXiv:1704.04861 (2017)
37. Hu, Z., Cho, S., Wang, J., Yang, M.H.: Deblurring low-light images with light streaks. In: CVPR (2014)
38. Johnson, J., Alahi, A., Fei-Fei, L.: Perceptual losses for real-time style transfer and super-resolution. In: Leibe, B., Matas, J., Sebe, N., Welling, M. (eds.) ECCV 2016. LNCS, vol. 9906, pp. 694–711. Springer, Cham (2016). https://doi.org/10.1007/978-3-319-46475-6_43
39. Joshi, N., Cohen, M.: Seeing Mt. Rainier: lucky imaging for multi-image denoising, sharpening, and haze removal. In: ICCP (2010)
40. Khosla, A., Jayadevaprakash, N., Yao, B., Li, F.F.: Novel dataset for fine-grained image categorization: stanford dogs. In: CVPR Workshop on Fine-Grained Visual Categorization (2011)
41. Kokkinos, F., Lefkimmiatis, S.: Iterative residual CNNs for burst photography applications. In: CVPR (2019)
42. Lindell, D.B., O'Toole, M., Wetzstein, G.: Single-photon 3D imaging with deep sensor fusion. ACM Trans. Graph. **37**(4), 1–12 (2018)

43. Liu, C., Freeman, W.T.: A high-quality video denoising algorithm based on reliable motion estimation. In: Daniilidis, K., Maragos, P., Paragios, N. (eds.) ECCV 2010. LNCS, vol. 6313, pp. 706–719. Springer, Heidelberg (2010). https://doi.org/10.1007/978-3-642-15558-1_51
44. Liu, D., Wen, B., Jiao, J., Liu, X., Wang, Z., Huang, T.S.: Connecting image denoising and high-level vision tasks via deep learning. IEEE Trans. Image Process. **29**, 3695–3706 (2020)
45. Liu, Z., Zhou, T., Shen, Z., Kang, B., Darrell, T.: Transferable recognition-aware image processing. arXiv preprint arXiv:1910.09185 (2019)
46. Liu, Z., Yuan, L., Tang, X., Uyttendaele, M., Sun, J.: Fast burst images denoising. ACM Trans. Graph. **33**(6), 232 (2014)
47. Lore, K.G., Akintayo, A., Sarkar, S.: LLNet: a deep autoencoder approach to natural low-light image enhancement. Pattern Recogn. **61**, 650–662 (2017)
48. Ma, J., Fossum, E.: A pump-gate jot device with high conversion gain for a quanta image sensor. IEEE J. Electron Dev. Soc. **3**(2), 73–77 (2015)
49. Ma, J., Masoodian, S., Starkey, D., Fossum, E.R.: Photon-number-resolving megapixel image sensor at room temperature without avalanche gain. Optica **4**(12), 1474–1481 (2017)
50. Ma, S., Gupta, S., Ulku, A.C., Brushini, C., Charbon, E., Gupta, M.: Quanta burst photography. ACM Trans. Graph. (TOG) **39**(4) (2020)
51. Mahendran, A., Vedaldi, A.: Understanding deep image representations by inverting them. In: CVPR (2015)
52. Makitalo, M., Foi, A.: Optimal inversion of the Anscombe transformation in low-count Poisson image denoising. IEEE Trans. Image Process. **20**(1), 99–109 (2010)
53. Malm, H., Oskarsson, M., Warrant, E., Clarberg, P., Hasselgren, J., Lejdfors, C.: Adaptive enhancement and noise reduction in very low light-level video. In: ICCV (2007)
54. Mildenhall, B., Barron, J.T., Chen, J., Sharlet, D., Ng, R., Carroll, R.: Burst denoising with kernel prediction networks. In: CVPR (2018)
55. O'Toole, M., Heide, F., Lindell, D.B., Zang, K., Diamond, S., Wetzstein, G.: Reconstructing transient images from single-photon sensors. In: CVPR (2017)
56. Plotz, T., Roth, S.: Benchmarking denoising algorithms with real photographs. In: CVPR (2017)
57. Remez, T., Litany, O., Giryes, R., Bronstein, A.: Deep convolutional denoising of low-light images. arXiv preprint arXiv:1701.01687 (2017)
58. Simonyan, K., Vedaldi, A., Zisserman, A.: Deep inside convolutional networks: visualising image classification models and saliency maps. ICLR (2014)
59. FLIR Sensor Review: Mono Camera. https://www.flir.com/globalassets/iis/guidebooks/2019-machine-vision-emva1288-sensor-review.pdf
60. Talebi, H., Milanfar, P.: Learned perceptual image enhancement. In: ICCP (2018)
61. Wu, J., Timofte, R., Huang, Z., Van Gool, L.: On the relation between color image denoising and classification. arXiv preprint arXiv:1704.01372 (2017)
62. Xu, J., Li, H., Liang, Z., Zhang, D., Zhang, L.: Real-world noisy image denoising: a new benchmark. arXiv preprint arXiv:1804.02603 (2018)
63. Yang, C., Xie, L., Qiao, S., Yuille, A.: Training deep neural networks in generations: a more tolerant teacher educates better students. In: AAAI Conference on Artificial Intelligence, July 2019
64. Yang, F., Lu, Y.M., Sbaiz, L., Vetterli, M.: An optimal algorithm for reconstructing images from binary measurements. In: Proceedings of SPIE, vol. 7533, pp. 158–169 (2010)

65. Yang, F., Lu, Y.M., Sbaiz, L., Vetterli, M.: Bits from photons: Oversampled image acquisition using binary Poisson statistics. IEEE Trans. Image Process. **21**(4), 1421–1436 (2011)
66. Yosinski, J., Clune, J., Nguyen, A., Fuchs, T., Lipson, H.: Understanding neural networks through deep visualization. In: ICML Deep Learning Workshop (2015)
67. Zhang, K., Zuo, W., Chen, Y., Meng, D., Zhang, L.: Beyond a Gaussian denoiser: residual learning of deep CNN for image denoising. IEEE Trans. Image Process. **26**(7), 3142–3155 (2017)
68. Zhang, K., Zuo, W., Zhang, L.: FFDNet: toward a fast and flexible solution for CNN-based image denoising. IEEE Trans. Image Process. **27**(9), 4608–4622 (2018)
69. Zhang, Y., Xiang, T., Hospedales, T.M., Lu, H.: Deep mutual learning. In: CVPR (2018)

n-Reference Transfer Learning for Saliency Prediction

Yan Luo[1(✉)], Yongkang Wong[2], Mohan S. Kankanhalli[2], and Qi Zhao[1]

[1] Department of Computer Science and Engineering, University of Minnesota, Minneapolis, USA
luoxx648@umn.edu, qzhao@cs.umn.edu

[2] School of Computing, National University of Singapore, Singapore, Singapore
{wongyk,mohan}@comp.nus.edu.sg

Abstract. Benefiting from deep learning research and large-scale datasets, saliency prediction has achieved significant success in the past decade. However, it still remains challenging to predict saliency maps on images in new domains that lack sufficient data for data-hungry models. To solve this problem, we propose a few-shot transfer learning paradigm for saliency prediction, which enables efficient transfer of knowledge learned from the existing large-scale saliency datasets to a target domain with limited labeled samples. Specifically, few target domain samples are used as the *reference* to train a model with a source domain dataset such that the training process can converge to a local minimum in favor of the target domain. Then, the learned model is further fine-tuned with the *reference*. The proposed framework is gradient-based and model-agnostic. We conduct comprehensive experiments and ablation study on various source domain and target domain pairs. The results show that the proposed framework achieves a significant performance improvement. The code is publicly available at https://github.com/luoyan407/n-reference.

Keywords: Deep learning · Saliency prediction · n-shot transfer learning

1 Introduction

Saliency prediction is the task that aims to model human attention to predict where people look in the given image. Thanks to the power of deep neural networks [15,24,48] (DNNs), state-of-the-art saliency models [7,50] perform very well in predicting human attention on naturalistic images. Behind the success of this task, a considerable amount of real-world images and corresponding human fixations fuels the process of training the data-hungry DNNs.

However, it is still difficult to predict saliency maps on images in novel domains, which has insufficient or few data to train saliency models with desired

Electronic supplementary material The online version of this chapter (https://doi.org/10.1007/978-3-030-58598-3_30) contains supplementary material, which is available to authorized users.

A. Vedaldi et al. (Eds.): ECCV 2020, LNCS 12353, pp. 502–519, 2020.
https://doi.org/10.1007/978-3-030-58598-3_30

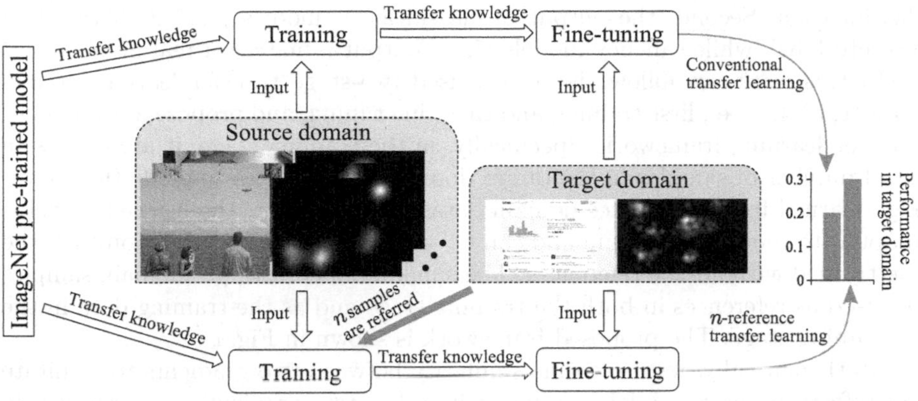

Fig. 1. The proposed n-reference transfer learning framework for saliency prediction. This framework aims to generate a better initialization with n reference samples from the target domain when training on the source domain, followed by fine-tuning to maximize knowledge transfer. It is based on the widely-used two-stage transfer learning framework (i.e., first training and then fine-tuning) and can easily adapt to other fine-tuning strategies

performance. As the time/money cost of collecting human fixations is prohibitive [3,20], a feasible solution is to reuse the existing large-scale saliency datasets along with a few target domain samples to solve this problem. Along this line, we study how to transfer the knowledge learned from the existing large-scale saliency datasets to the target domain in a few-shot transfer learning setting.

The necessity of few-shot transfer learning for saliency prediction lies in the nature of the task. Based on findings drawn from the behavioral experiments, the way that humans attend to regions is significantly affected by the scene context [35,45,47]. The scene context is correlated to the image domain [43]. In other words, each image from a specific domain could be representative of the others from the same domain to some degree, e.g., webpage images generally have a similar layout and design [41]. In visual saliency study, existing datasets [3,42] in non-natural images domain are much smaller than the natural image ones [20, 22]. Moreover, there are numerous images used in the subfields of medicine, biology, etc., which may not have any human fixation data yet. In this work, we assume that it is feasible and viable to collect human fixations on a small number of images to enable few-shot learning.

Compared to n-reference transfer learning for classification task [1], we focus on how to use very few target domain samples as references to learn a better initial model for fine-tuning. Moreover, there exists no such works for saliency prediction task. Models designed for classification may not work for saliency prediction. First, visual samples in existing classification tasks often contain limited visual concepts (i.e., pre-defined object classes), while objects of any class may appear in the images used for saliency prediction. In this sense, saliency prediction often handles images with higher diversity than the ones used for classification. Second, the output of classification models [1,15,24,32,44] is a discrete label, while saliency models [7,50] output a matrix of real numbers.

In this work, we follow the widely-used two-stage transfer learning framework [1,13,41], i.e., first training and then fine-tuning, and propose a n-reference transfer learning framework. Specifically, in the training stage, it aims to use a small number of samples in the target domain as references to guide the knowledge learned from the source domain dataset. In this way, the learned model is adapted to the target domain and can be seen as a better initialization than the one trained without the references. The small number of target domain samples are used as references in both the training stage and as the training data in the fine-tuning stage. The proposed framework is shown in Fig. 1.

Mathematically, we use cosine similarity between two gradients to facilitate the reference aware model training, where the two gradients are respectively computed by samples in the source and target domain. If the angle between the two gradients is greater than 90 degrees, which implies that the directions of the model update are significantly different from each other, we optimize the gradient for the update to have smaller differences with the target-domain referenced gradient in cosine similarity. The intuition behind is to mimic the process of human learning with the reference sample, i.e., we adaptively learn from new information so that the newly absorbed knowledge will not contradict the observation of the reference samples [31,33]. The proposed framework is gradient-based and it is model-agnostic.

To comprehensively evaluate the proposed framework, we employ SALICON [20] and MIT1003 [22] as the source domain datasets (i.e., the knowledge sets), and WebSal [42] and the art subset in CAT2000 [3] as the target domain data. We randomly select 1, 5, or 10 samples from the target domain data as references. The contributions of this work can be summarized as follows:

- To study how humans perceive scenes from a partially explored domain, we propose a model-agnostic few-shot transfer learning paradigm to transfer knowledge from the source domain to the target domain. This is the first work that studies few-shot transfer learning for saliency prediction.
- We propose a n-reference transfer learning framework to adaptively guide the training process. It guarantees that the knowledge learned with the source domain data would not contradict the references in the target domain, and produce a good initialization for further fine-tuning. The proposed framework is model-agnostic and can generally work with existing saliency models.
- Comprehensive experiments show the proposed framework works on various combinations of source domain and target domain pairs. The experiment with various baseline models show that the proposed approach can efficiently transfer the knowledge from the source domain to the target domain.

2 Related Works

2.1 Saliency Prediction

Saliency prediction aims to mimic human vision system to perceive interesting regions in a cluttered visual world. Itti et al. [19] develop the first bottom-up stimulus-driven saliency model. Since then, many works emerge to interpret visual saliency from various perspectives [14,16,22,52]. With the advent of DNNs [15,24,48], saliency prediction benefitted from data-driven discriminative features instead of relying on hand-crafted features [6,25,26,37]. Recently, Cornia et al. [7] introduce a network that integrates ResNet-50 [15] and convolutional LSTMs to better attend to salient regions by iteratively refining the predictions. Yang et al. [50] propose a dilated inception network (DINet) that stacks dilated convolutions with different dilation rates upon ResNet-50 to capture wider spatial information. It achieves state-of-the-art performance on various benchmarks. A widely-used practice to transfer the knowledge learned from image classification to saliency prediction is by using the weights pre-trained on ImageNet as model initialization [6,25,26,37]. In contrast, this work studies the few-shot cross-domain transfer learning problem, which takes place between two domains. Without loss of generality, we follow [33] to adopt both ResNet-50 and DINet as the baseline models in this work.

2.2 Few-Shot Learning

Few-shot learning [11,27,32,44] aims to study how to learn classifiers for unseen visual concepts with only a few samples per class. Lake et al. [28] introduce a Bayesian program learning framework that can learn from one example for predicting character strokes. Matching networks [46] use an attention mechanism that is analogous to a kernel density estimator so that it can learn from a few examples rapidly. Sung et al. [44] propose a relation network to learn a transferable deep metric to compare the relation between the small number samples. In [29], Lee et al. study how to learn feature embeddings with a few samples that can minimize generalization error across a distribution of tasks. As the process of collecting human fixations is prohibitive [20], learning with very few samples is promising for saliency prediction to overcome the need for big data.

2.3 Transfer Learning

Transfer learning, a.k.a. domain adaptation or domain transfer, is a paradigm to utilize training data in the source domain to solve the problem in the target domain [8,9,30,38,41]. In general, it can be seen as a two-stage learning framework, i.e., first training a model with source domain data and then fine-tuning the pre-trained model with target domain data. There are many DNN-based works [1,2,12,13,49] that use this learning framework for classification tasks. Specifically, Guo et al. [13] study and design a variant of the standard fine-tuning method for better transferability. However, it requires many training samples to determine whether it should fine-tune or freeze the parameters in a particular layer. Recently, Bäuml and Tulbure [1] introduce a learning framework that transfers the knowledge learned from the source domain to the target domain with a few samples for tactile material classification. As saliency prediction is by nature class-agnostic, learning to predict human fixations with very few samples (e.g., ≤10) in the target domain is more challenging than the same paradigm for classification and has not been explored yet. Different from the aforementioned methods, we propose the first model-agnostic few-shot transfer learning framework for saliency prediction and conduct comprehensive study on multiple combinations of source domain datasets and target domain datasets.

3 Methodology

In this section, we first formulate the problem and discuss its theoretical generalization bound. Then, we delve into the details of the proposed framework.

3.1 Problem Statement

In this work, we denote the images as $I^S, I^T \in \mathbb{R}^m$ and the human fixation maps as $y^S, y^T \in \mathcal{Y}$ ($\mathcal{Y} \equiv [0,1]^m \subseteq \mathbb{R}^m$), where m is the dimensions of the image and S (T) indicates the source (target) domain. In general, given an image I, the prediction function $f : \mathbb{R}^m \xrightarrow{\theta} \mathcal{Y}$ with parameters θ will predict z and then the loss function $\ell : \mathcal{Y} \times \mathcal{Y} \rightarrow \mathbb{R}_+$ will evaluate the discrepancy between z and y. Transfer learning for saliency prediction task can be considered as a two-stage learning problem. First, the model's parameters are learned with the source domain data through the training process, i.e.,

$$\theta_{\mathsf{TR}} = \arg\min_{\theta} \frac{1}{|D^S|} \sum_{(I_i,y_i)\in D^S} \ell(f(I_i;\theta), y_i)|_{\theta_0} \tag{1}$$

where D^S is the source domain dataset, $|D^S|$ is the number of the samples, and TR stands for training. θ_0 are the initialized parameters and the model is usually pre-trained on ImageNet [10]. Then, θ_{TR} is taken as the initialization for further fine-tuning on the target domain data, i.e.,

$$\theta^*_{\mathsf{FT}} = \arg\min_{\theta} \frac{1}{|D^T|} \sum_{(I_i,y_i)\in D^T} \ell(f(I_i;\theta), y_i)|_{\theta_0=\theta_{\mathsf{TR}}} \tag{2}$$

In this work, we aim to learn a better initialization by the first stage objective (1), which is in favor of the target domain data. Such initialized parameters (i.e., θ_{TR}) are expected to further achieve better performance by fine-tuning on D^T. To this end, we introduce a referencing mechanism that allows the training process fed with D^S to reference the model update w.r.t. the referenced samples $(I^R, y^R) \in D^T(|D^S| \gg |D^T|)$. Mathematically, this can be formulated as

$$\theta_{\mathsf{TR-Ref}} = \arg\min_{\theta} \frac{1}{|D^S|} \sum_{\substack{(I_i, y_i) \in D^S \\ (I_j^R, y_j^R) \in D^T}} \ell(f_{\mathsf{Ref}}(I_i; \theta, (I_j^R, y_j^R)), y_i)|_{\theta_0} \tag{3}$$

where TR−Ref indicates the training process references target domain samples when updating the model. f_{Ref} is a variant of f which has the same forward propagation as f but has more complicated backward propagation. $\theta_{\mathsf{TR-Ref}}$ is taken as the initialization in the second stage objective (2) for further fine-tuning. We denote the resulting parameters as $\theta_{\mathsf{FT|Ref}}$.

3.2 Generalization Bound of Saliency Prediction

Here, we discuss the theoretical guarantee of saliency prediction. Following the setting used in [34], given training data $(I_1, y_1), (I_2, y_2), \ldots \in \mathcal{X} \times \mathcal{Y}$, where $\mathcal{Y} \in [0,1]^m \subseteq \mathbb{R}^m$, we use the L^p loss, i.e., $\ell^p : \mathcal{Y} \times \mathcal{Y} \to \mathbb{R}_+, p \geq 1$. The prediction function $f(\cdot;\theta)$ is denoted as $f(\cdot)$ for simplicity. I is drawn i.i.d. according to the unknown distribution $\mathcal{D}$ and $y = f^*(I)$ where f^* is the target labeling function. Saliency prediction can be considered as a mathematical problem that finds hypothesis $f : \mathbb{R}^m \to [0,1]^m$ in a set H with small generalization error w.r.t. f^*,

$$R_{\mathcal{D}}(f) = E_{I \sim \mathcal{D}}[\ell(f(I), f^*(I))].$$

In practice, as $\mathcal{D}$ is unknown, we use empirical error for approximation, i.e.,

$$\hat{R}_{\mathcal{D}}(f) = \frac{1}{|D|} \sum_{i=1}^{|D|} \ell(f(I_i), y_i),$$

where $|D|$ is the sample number in dataset D for training.

We introduce the generalization bound of saliency prediction as follows. The proof is provided in the supplementary document.

Theorem 1 (Saliency generalization bound). *Denote H as a finite hypothesis set. Given ℓ^p and $y \in [0,1]^m$, for any $\delta > 0$, with probability at least $1 - \delta$, the following inequality holds for all $f \in H$:*

$$|R_{\mathcal{D}}(f) - \hat{R}_{\mathcal{D}}(f)| \leq m^{\frac{1}{p}} \sqrt{\frac{\log|H| + \log\frac{2}{\delta}}{2|D|}}$$

Remark 1. Theorem 1 shows how the training set scale influences the generalization bound. When $|D|$ tends towards infinity, $R_{\mathcal{D}}(f) \equiv \hat{R}_{\mathcal{D}}(f)$. This conforms to the general intuition that it can train a more general model with more data. Contrarily, when $|D| = 1$, it leads to the largest bound for $|R_{\mathcal{D}}(f) - \hat{R}_{\mathcal{D}}(f)|$. Moreover, it demonstrates the task is challenging with small number of samples.

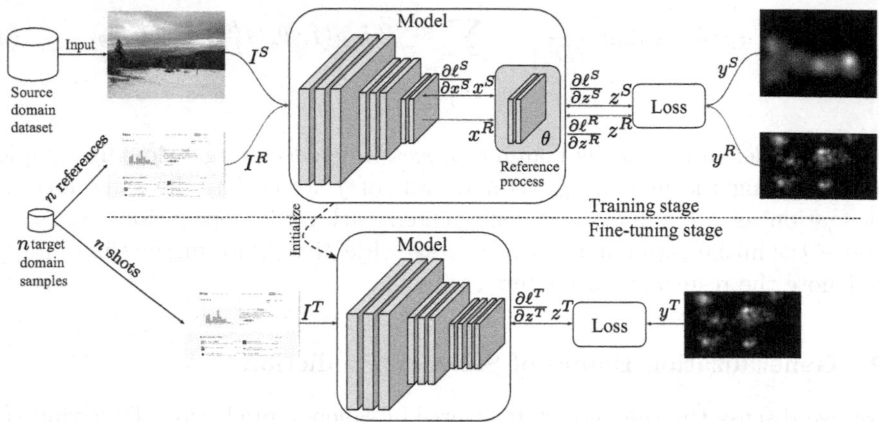

Fig. 2. Proposed n-reference transfer learning framework. Note that we assume that only very few samples from the target domain are available, i.e., $n \leq 10$

3.3 Overall Framework

In this subsection, we introduce the few-shot transfer learning framework that solves the objective function (2) and (3). The overall workflow of the proposed n-reference transfer learning framework is shown in Fig. 2.

Similar to classification model [15,17,48], state-of-the-art saliency models tend to be large. For example, DINet [50] and SAM-ResNet-50 [7] consist of 26M and 70M parameters, respectively. Therefore, instead of inefficiently applying the proposed framework to the whole saliency model, we only apply it to a few downstream layers which are close to the output. The downstream layers produce discriminative features used for prediction with a small number of parameters, and it makes the transfer learning process more cost-effective. Consequently, we split the model into two parts, i.e., the model body $\theta_{\textsf{body}}$ and the model head $\theta_{\textsf{head}}$. This split would be only effective in the training stage and the two parts will be integrated again as they always are in the inference stage. Note that the split is flexible. The effective scope of the proposed framework could cover the whole model and the model body would correspondingly turn to be an empty set. As we only focus on $\theta_{\textsf{head}}$, we simplify it as θ in the following text.

In the forward propagation, as the training image $I^S \in D^S$ and the reference image $I^R \in D_T$ are fed to the model body, the discriminative feature x^S and x^R are generated, respectively. Then, the model head would take x^S and x^R as input to produce prediction z^S and z^R, respectively. Specifically, $z^S = f(x^S; \theta)$. A similar process applies to z^R. The loss function is used to compute the distance between z^S and y^S (and between y^R and y^R as well). In the backward propagation, two gradients are computed by the chain rule

$$\frac{\partial \ell^S}{\partial \theta} = \frac{\partial \ell(f(x^S; \theta), y^S)}{\partial z^S} \frac{\partial z^S}{\partial \theta}, \qquad \frac{\partial \ell^R}{\partial \theta} = \frac{\partial \ell(f(x^R; \theta), y^R)}{\partial z^R} \frac{\partial z^R}{\partial \theta}.$$

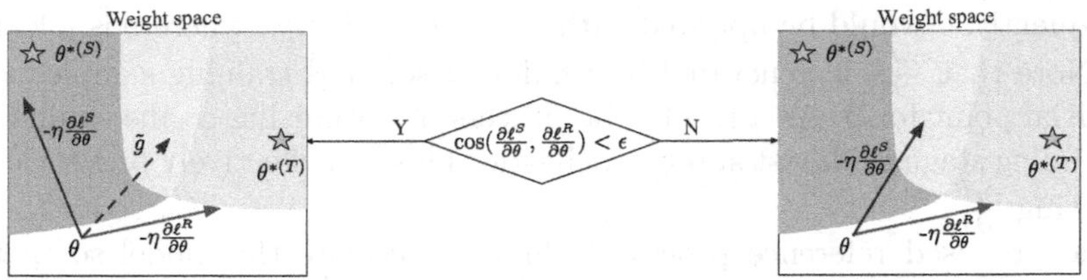

Fig. 3. The reference process computing the gradient that better adapts to the target domain data. $\theta^{*(S)}$ is a local minimum trained by sufficient source domain samples, while $\theta^{*(T)}$ is a local minimum trained by sufficient target domain samples. Given a pre-defined threshold ϵ, if the cosine similarity between the gradient ($\frac{\partial \ell^S}{\partial \theta}$) generated by the source sample and the gradient ($\frac{\partial \ell^R}{\partial \theta}$) generated by the reference sample is smaller than ϵ, it will compute a corrected gradient by optimizing the cosine similarity. It retains $\frac{\partial \ell^S}{\partial \theta}$ otherwise

Specifically, $\frac{\partial \ell^S}{\partial \theta}$ indicates the model update towards a local minimum $\theta^{*(S)}$ which is learned from the samples from D^S, while $\frac{\partial \ell^R}{\partial \theta}$ indicates the model update towards a local minimum $\theta^{*(T)}$ which is learned from the samples from D^T.

As shown in Fig. 2, θ_{head} are updated by the proposed reference process and θ_{body} are updated with the standard gradients in the training stage. During fine-tuning, θ_{head} and θ_{body} are updated with the standard gradients.

3.4 Reference Process

Here, we delve into the formulation of the proposed reference process (Fig. 3). The cosine similarity between $\frac{\partial \ell^S}{\partial \theta}$ and $\frac{\partial \ell^R}{\partial \theta}$ can evaluate the difference of the two gradients. Accordingly, we pre-define a threshold ϵ to determine if the difference is considered as minor and the update with $\frac{\partial \ell^S}{\partial \theta}$ will be close to both $\theta^{*(S)}$ and $\theta^{*(T)}$. If the difference is significant, the proposed reference process will adjust $\frac{\partial \ell^S}{\partial \theta}$ so that it will move more towards $\theta^{*(T)}$. This process is defined as follows

$$\tilde{g} = \begin{cases} \arg\max_g \cos(g, \frac{\partial \ell^R}{\partial \theta}) - \lambda \|g\|_2^2 \big|_{g_0 = \frac{\partial \ell^S}{\partial \theta}} & \text{if} \cos(\frac{\partial \ell^S}{\partial \theta}, \frac{\partial \ell^R}{\partial \theta}) < \epsilon, \\ \frac{\partial \ell^S}{\partial \theta} & \text{otherwise}, \end{cases} \tag{4}$$

where λ is the regularization parameter and $\cos(\cdot, \cdot)$ is the cosine similarity, i.e., $\cos(a, b) = a^\top b/|a||b|$ (a and b are the input vectors), and $\tilde{g}$ is the output gradient. The embedded optimization problem in Eq. (4) aims to find a $\tilde{g}$, which is with an initial point $g_0 = \frac{\partial \ell^S}{\partial \theta}$, to be consistent with the reference gradient $\frac{\partial \ell^R}{\partial \theta}$ in terms of cosine similarity. In other words, the reference gradient $\frac{\partial \ell^R}{\partial \theta}$ provides a reference so that $\tilde{g}$ is able to be aware of a rough direction towards the underlying $\theta^{*(T)}$. In this way, the knowledge learned from D^S is transferred to the target domain. We solve the embedded optimization problem with the gradient ascent method because our goal is to maximize the cosine similarity between $\tilde{g}$ and $\frac{\partial \ell^R}{\partial \theta}$.

Subsequently, θ would be updated with $\tilde{g}$, i.e., $\theta \leftarrow \theta - \eta\tilde{g}$, where η is a learning rate. Note that $\frac{\partial \ell^S}{\partial \theta}$ is generated by randomly selected training samples and is the initial point for $\tilde{g}$. As a result, the process of optimizing cosine similarity in the training stage is almost surely stochastic. This can effectively prevent $\tilde{g}$ from overfitting $\frac{\partial \ell^R}{\partial \theta}$.

The proposed reference process yields $\tilde{g}$ to update the model so that the parameters are close to the underlying $\theta^{*(T)}$. As θ is learned with the references from the target domain, by the chain rule, $\frac{\partial \ell^S}{\partial \theta_{body}} = \frac{\partial \ell^S}{\partial x^S}\frac{\partial x^S}{\partial \theta_{body}}$ and $\frac{\partial \ell^S}{\partial x^S}$ can be considered as a function of θ. So θ_{b} will be affected by the references as well.

As the number of references is expected to be far smaller than the training data, we follow a similar idea of the stochastic process to randomly draw a reference from the reference pool at each iteration.

4 Experiments

In this section, we introduce the experimental protocol, present the experimental results, and then have a discussion about the results.

4.1 Experimental Setup

Datasets. We adopt the large-scale saliency prediction dataset SALICON [20] (the 2017 version) and the MIT1003 [22] as the source domain datasets. Accordingly, we adopt WebSal [42] and the art subset in CAT2000 [3] as the target domain datasets. Specifically, SALICON consists of 10000 real-world images, MIT1003 consists of 1003 natural scene images, and WebSal consists of 149 webpage screenshots. CAT2000 includes 20 categories and each category has 100 images. Art is one of the most common categories, whose images are the pictures of human-made works, like the paintings, handcrafts, and etc.

Baseline Models. To study how well the proposed method would generalize to different models, we use two baseline models, i.e., DINet [51] and ResNet-50 [15].

Settings. There are three dimensions to the experiments in this work, i.e., source domain samples, baseline model, and target domain samples. Specifically, the baseline model is trained with the source domain samples. The learned model is further fine-tuned with the target domain samples. This setting is similar in the case of the proposed method. For convenience, we denote the setting as a combination of the initials of the datasets or the models, e.g., $\langle \mathsf{S}, \mathsf{D}, \mathsf{W} \rangle$ indicates that we use SALICON as the source domain dataset, DINet as the baseline model, and WebSal as the target domain dataset. Similarly, we use initials M, R, and A to represent MIT1003, ResNet-50, and Art, respectively.

To understand how the number of references affects the performance, we evaluate the proposed method with $n = 1, 5, 10$. Moreover, to provide a benchmark of the performance w.r.t. more references, a paradigm that is similar to 3-fold cross validation is applied with more references. For instance, given WebSal as the target domain datasets, we divide it into three subsets, which contain 50, 50, and 49 images, respectively. Then, we alternately use any two subsets as the reference samples and the rest as the validation set. The process is repeated 3 times. We denote the results of this process as an empirical upper bound.

Evaluation Metrics. We adopt the common metrics used in [5] and [20], i.e., normalized scanpath saliency (NSS) [18,40], area under curve (AUC) [4,21], and correlation coefficient (CC) [36]. Higher scores indicate better performance. We use the public implementation[1] provided by [20]. Each experiment is repeated 10 times and the mean metric scores are reported. Due to the space limits, we report the corresponding standard deviation in the supplementary document.

4.2 Training Scheme

We follow the widely-used two-stage transfer learning framework [1,13,41,49], i.e., first train a model with the source domain data and fine-tune with the target domain data. We denote the trained model as TR and the fine-tuned model as FT. In the proposed framework, the n-reference training stage first trains a model with the source domain data and n target domain references (denoted as TR−Ref), and then further fine-tune with the references (denoted as FT|Ref).

Regarding the experimental details, we follow DINet [50] to use Adam optimizer [23] with learning rate $\eta = 5\text{e}{-5}$ and weight decay 1e−4. We use batch size 10 for all the experiments. The number of epochs is 10 and we decrease the learning rate for every 3 epochs by multiplying with 0.2. In TR−Ref, we randomly sample 10 training data without replacement as the training sample at each iteration. Meanwhile, we randomly sample n_r references with replacement as the reference. In this way, the difference between the number of training samples and references will not cause a problem. n_r are 1, 3 and 5 in the experiments with $n = 1, 5, 10$, respectively. This process is the same for the one of FT. We select the model with the best performance over epochs for further fine-tuning. The normalized l_1 loss [50] is used and the threshold ϵ is set to 0 for all the experiments. We implement the proposed framework with PyTorch [39].

4.3 Performance

The experimental results with the following settings, i.e., ⟨S, D, W⟩, ⟨S, R, W⟩, ⟨M, D, W⟩, and ⟨S, D, A⟩, are shown in Table 1. Within setting ⟨S, D, W⟩, the proposed framework (i.e., FT|Ref) achieves better performance than FT over all metrics. Particularly, as the number of references increases, the consequently trained models provide better initializations for fine-tuning. In other words, FT|Ref yields better performance when the dependent trained model uses more reference samples. Using a different baseline model, we experiment it with setting ⟨S, R, W⟩

[1] https://github.com/NUS-VIP/salicon-evaluation.

Table 1. Performance with various settings of ⟨source, model, target⟩. Here, S is SALICON, M is MIT1003, W is WebSal, A is Art subset, D is DINet, and R is ResNet. ↑ implies that a higher score is better. The score in bold font indicates the best result under the respective metric. We report the mean score from 10 runs for conventional training (i.e., $n = 0$) and the proposed method. The empirical upper bound (EUB) is generated by 3-fold cross validation on the target domain. The experimental details are provided in Sect. 4.1 and 4.2

		⟨S, D, W⟩			⟨S, R, W⟩		
		NSS↑	AUC↑	CC↑	NSS↑	AUC↑	CC↑
FT w/o TR	$n = 10$	0.8252	0.7430	0.3635	0.8846	0.7455	0.3852
TR	$n = 0$	1.3330	0.7796	0.5515	1.2950	0.7749	0.5358
TR−Ref	$n = 1$	1.3621	0.7848	0.5628	1.3569	0.7864	0.5611
FT	$n = 1$	1.4731	0.8005	0.5976	1.3722	0.7923	0.5627
FT|Ref	$n = 1$	**1.5077**	**0.8051**	**0.6121**	**1.4272**	**0.7983**	**0.5817**
TR−Ref	$n = 5$	1.3683	0.7874	0.5659	1.3535	0.7837	0.5593
FT	$n = 5$	1.5803	0.8161	0.6355	1.5043	0.8131	0.6139
FT|Ref	$n = 5$	**1.6085**	**0.8200**	**0.6468**	**1.5491**	**0.8149**	**0.6281**
TR−Ref	$n = 10$	1.3647	0.7839	0.5633	1.3583	0.7857	0.5612
FT	$n = 10$	1.6290	0.8247	0.6531	1.5164	0.8103	0.6200
FT|Ref	$n = 10$	**1.6439**	**0.8276**	**0.6605**	**1.5829**	**0.8143**	**0.6414**
TR−Ref	EUB	1.3822	0.7910	0.5708	1.3626	0.7864	0.5645
FT	EUB	1.8695	0.8488	0.7389	1.8325	0.8462	0.7275
FT|Ref	EUB	**1.8831**	**0.8494**	**0.7442**	**1.8500**	**0.8480**	**0.7321**
		⟨M, D, W⟩			⟨S, D, A⟩		
		NSS↑	AUC↑	CC↑	NSS↑	AUC↑	CC↑
FT w/o TR	$n = 10$	0.8252	0.7430	0.3635	1.2183	0.8339	0.5161
TR	$n = 0$	1.3905	0.7991	0.5700	1.5172	0.8225	0.6003
TR−Ref	$n = 1$	1.4405	0.8085	0.5902	1.5651	0.8287	0.6211
FT	$n = 1$	1.4410	0.8023	0.5784	1.6255	0.8324	0.6449
FT|Ref	$n = 1$	**1.4575**	**0.8070**	**0.5838**	**1.6523**	**0.8380**	**0.6564**
TR−Ref	$n = 5$	1.4452	0.8064	0.5908	1.5870	0.8304	0.6274
FT	$n = 5$	1.5795	0.8217	0.6395	1.8049	0.8480	0.7185
FT|Ref	$n = 5$	**1.6136**	**0.8269**	**0.6515**	**1.8314**	**0.8503**	**0.7274**
TR−Ref	$n = 10$	1.4330	0.8060	0.5872	1.5704	0.8288	0.6204
FT	$n = 10$	1.6462	0.8261	0.6660	1.8325	0.8474	0.7288
FT|Ref	$n = 10$	**1.6691**	**0.8283**	**0.6730**	**1.8584**	**0.8503**	**0.7366**
TR−Ref	EUB	1.4402	0.8087	0.5905	1.5980	0.8340	0.6331
FT	EUB	1.8450	0.8466	0.7330	2.1595	0.8636	0.8464
FT|Ref	EUB	**1.8507**	**0.8478**	**0.7344**	**2.1874**	**0.8649**	**0.8519**

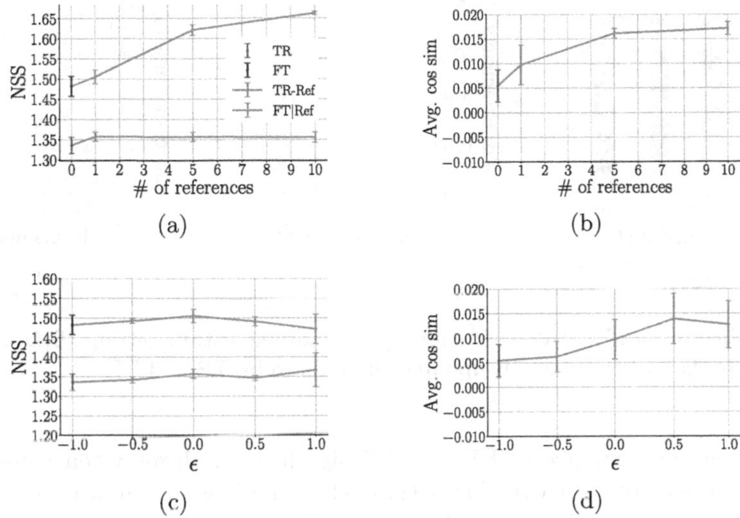

Fig. 4. The effect of the number of references (a, b) and threshold ϵ (c, d) on NSS metric and average cosine similarity within setting $\langle \mathsf{S}, \mathsf{D}, \mathsf{W} \rangle$. $n = 0$ indicates that no reference sample is used. Hence, TR−Ref and FT|Ref turn to be TR and FT. The results of c and d are generated with $n = 1$. ϵ determines whether the gradient needs to be corrected or not (see Fig. 3). Comparing to TR, FT, and FT|Ref, only TR−Ref is able to evaluate the cosine similarity between the samples from the source domain and target domain (see Fig. 2)

which FT|Ref achieves consistent improvement. Moreover, using DINet as the baseline model leads to better performance than using ResNet-50.

We study how well the proposed framework generalizes to different target domain data using setting $\langle \mathsf{S}, \mathsf{D}, \mathsf{A} \rangle$. As seen in Table 1, similar performance improvement can be found, which implies the proposed framework can generalize to a different target domain. Furthermore, the study with MIT1003 as the source domain dataset, i.e., setting $\langle \mathsf{M}, \mathsf{D}, \mathsf{W} \rangle$, shows consistent improvement. The overall performance within setting $\langle \mathsf{M}, \mathsf{D}, \mathsf{W} \rangle$ is slightly lower than the one within setting $\langle \mathsf{S}, \mathsf{D}, \mathsf{W} \rangle$. This implies that SALICON is more efficient than MIT1003 to transfer the knowledge to WebSal. On the other hand, models trained with one sample in target domain have noticeable gaps w.r.t. EUB, and is improved with more training samples. This is consistent with the implication of Theorem 1.

We perform paired t-test and permutation test over images within setting $\langle \mathsf{S}, \mathsf{D}, \mathsf{W} \rangle$ to evaluate the difference between TR−Ref and FT|Ref. Both corresponding p are less than 0.001. This implies that TR−Ref significantly provides a good initialization to FT|Ref to yield high performance. To validate the effect of knowledge transfer in saliency prediction, we conduct the experiment where models are learned using only the target domain samples, i.e., FT w/o TR in Table 1. We set $n = 10$ as $n = 1, 5$ will yield much worse performance. In all

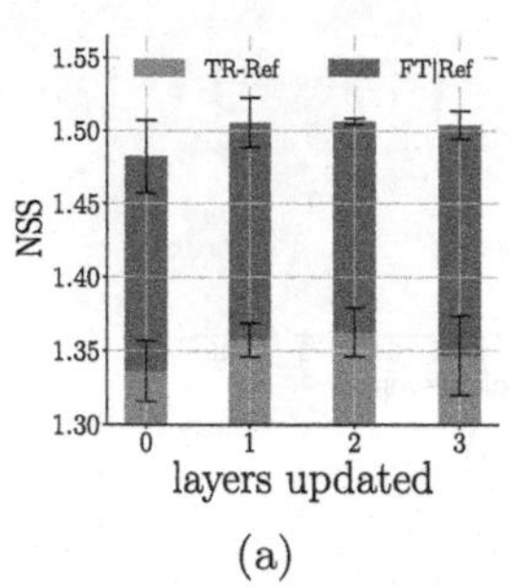

(a)

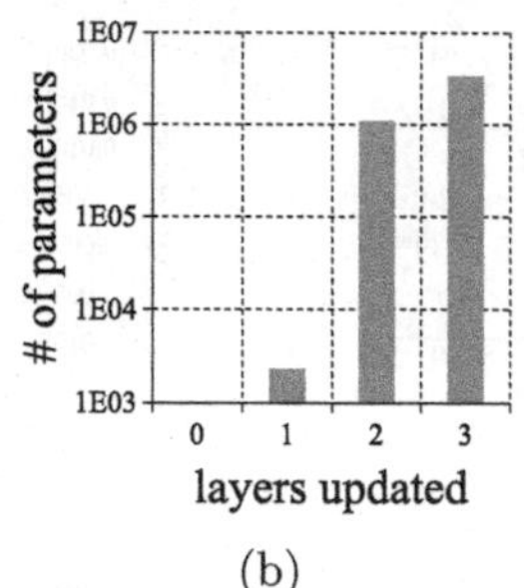

(b)

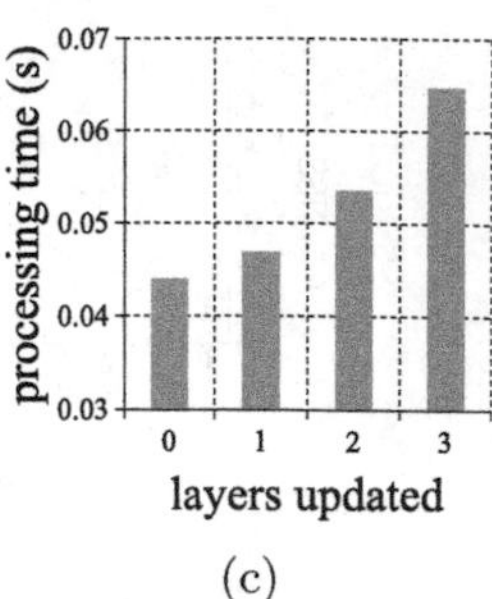

(c)

Fig. 5. Ablation study of downstream layers updated within setting $\langle \mathsf{S}, \mathsf{D}, \mathsf{W} \rangle$ with $n = 1$. Note that when 0 layer is updated, it turns to be TR and FT

settings, the performance of FT w/o TR significantly drops when compare to FT|Ref. These results are even lower than TR and FT, which indicate the importance of efficient initialization with a source domain dataset.

5 Analysis

We study the influences of the number of references, the threshold ϵ, and the layers updated by the proposed framework. All analysis are within setting $\langle \mathsf{S}, \mathsf{D}, \mathsf{W} \rangle$, where the mean score and standard deviation from 3 runs are reported.

5.1 Ablation Study

Effect of Number of References. As shown in Fig. 4, as the number of references increases, the performance of TR−Ref keeps flat or even slightly drops, but the performance of FT|Ref is significantly improved. This implies that the proposed reference process with more reference samples can yield better initialization for fine-tuning. Moreover, the average cosine similarity is increased with more references. This implies that the number of references is helpful to adapt the training process with source domain data to the target domain data.

Effect of Threshold ϵ. We experiment with the proposed framework with $n = 1$, which is more representative and challenging than cases with more references, with various thresholds. An interesting observation in Fig. 4c is that although $\epsilon = 1$ achieves best performance on TR−Ref, it deteriorates the performance of FT|Ref. This shows that when $\epsilon = 1$, all the gradients at each iteration need to be corrected because the cosine similarity between any two gradients is equal or less than 1. As a result, the reference process enforces the training process to overfit the reference samples. This can be verified in Fig. 4d where the average cosine similarity is roughly increased as ϵ is increasing.

Effect of Updated Layers. To understand the effect of layers updated by the proposed 1-reference transfer learning, we experiment with various downstream layers. Consequently, the performance is shown in Fig. 5a, while the number of parameters and the computational cost are reported in Fig. 5b and Fig. 5c, respectively. The layers are downstream layers, which are close to the output. When 0 layer is updated, TR−Ref and FT|Ref are equivalent to TR and FT, respectively. The baseline model in this experiment is DINet.

Figure 5a shows that using the last 2 layers achieves slightly better performance in NSS than using the other numbers of the last layers. However, it takes 69 ms longer in the training process than using the last layers. In light of the trade-off, we use the last layer of the baseline model in Sect. 4.

5.2 Qualitative Comparison

Figure 6 shows the comparison between the predicted saliency maps generated by TR, TR−Ref, FT, and FT|Ref. It can be observed that with the reference process, the proposed framework efficiently leverages the knowledge learned from the source domain, which are based on natural scene images, to subtly identify salience in the new domain. Taking the example in the first row, FT|Ref predicts that the people is salient, which takes the learned knowledge into account, whereas FT predicts that the people is less salient than the text. Figure 7 shows more references lead to better prediction.

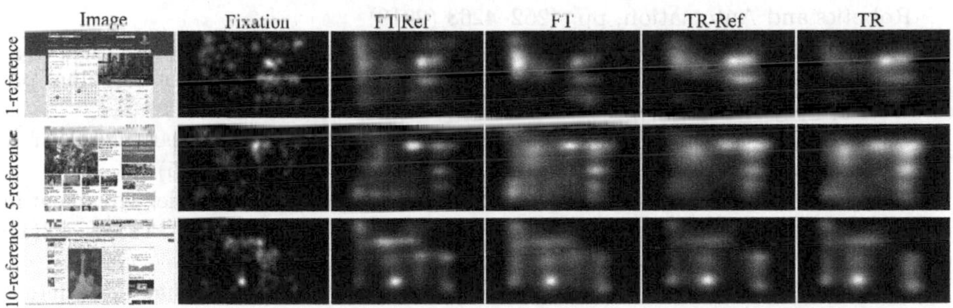

Fig. 6. Qualitative results with human fixations and maps generated by the models trained by the four procedures

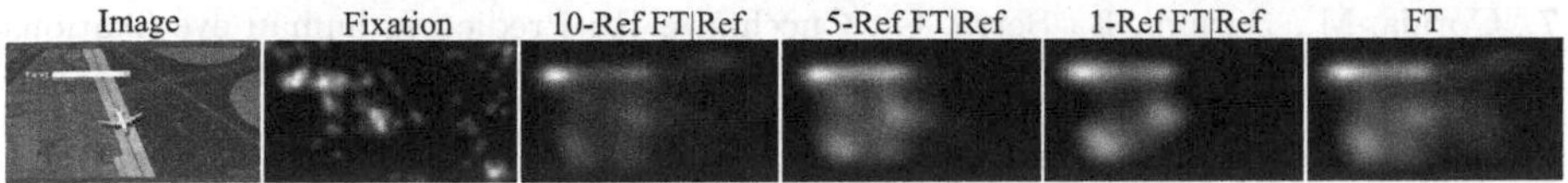

Fig. 7. Qualitative results w.r.t. different n

6 Conclusion

This work studies how to leverage the knowledge learned from a source domain that has adequate images and corresponding human fixations and very few samples (i.e., references) from a new domain (i.e., target domain) to predict saliency maps in the target domain. We propose an n-reference transfer learning framework to guide the training process to converge to a local minimum in favor of the target domain. The proposed framework is gradient-based and model-agnostic. Comprehensive experiments and ablation studies to evaluate the proposed framework are reported. Results show the effectiveness of the framework with a significant performance improvement.

Acknowledgments. This research was funded in part by the NSF under Grants 1908711, 1849107, in part by the University of Minnesota Department of Computer Science and Engineering Start-up Fund (QZ), and in part supported by the National Research Foundation, Singapore under its Strategic Capability Research Centres Funding Initiative. Any opinions, findings and conclusions or recommendations expressed in this material are those of the author(s) and do not reflect the views of National Research Foundation, Singapore.

References

1. Bäuml, B., Tulbure, A.: Deep n-shot transfer learning for tactile material classification with a flexible pressure-sensitive skin. In: International Conference on Robotics and Automation, pp. 4262–4268 (2019)
2. Bengio, Y.: Deep learning of representations for unsupervised and transfer learning. In: Proceedings of ICML Workshop on Unsupervised and Transfer Learning, pp. 17–36 (2012)
3. Borji, A., Itti, L.: CAT2000: a large scale fixation dataset for boosting saliency research. In: CVPR 2015 Workshop on "Future of Datasets" (2015)
4. Borji, A., Sihite, D.N., Itti, L.: Quantitative analysis of human-model agreement in visual saliency modeling: a comparative study. IEEE Trans. Image Process. **22**(1), 55–69 (2013)
5. Bylinskii, Z., Judd, T., Oliva, A., Torralba, A., Durand, F.: What do different evaluation metrics tell us about saliency models? IEEE Trans. Pattern Anal. Mach. Intell. **41**(3), 740–757 (2018)
6. Cornia, M., Baraldi, L., Serra, G., Cucchiara, R.: A deep multi-level network for saliency prediction. In: International Conference on Pattern Recognition, pp. 3488–3493 (2016)
7. Cornia, M., Baraldi, L., Serra, G., Cucchiara, R.: Predicting human eye fixations via an LSTM-based saliency attentive model. IEEE Trans. Image Process. **27**(10), 5142–5154 (2018)
8. Csurka, G.: A comprehensive survey on domain adaptation for visual applications. In: Csurka, G. (ed.) Domain Adaptation in Computer Vision Applications. ACVPR, pp. 1–35. Springer, Cham (2017). https://doi.org/10.1007/978-3-319-58347-1_1
9. Daume III, H., Marcu, D.: Domain adaptation for statistical classifiers. J. Artif. Intell. Res. **26**, 101–126 (2006)

10. Deng, J., Dong, W., Socher, R., Li, L.J., Li, K., Fei-Fei, L.: ImageNet: a large-scale hierarchical image database. In: IEEE Conference on Computer Vision and Pattern Recognition, pp. 248–255 (2009)
11. Fei-Fei, L., Fergus, R., Perona, P.: One-shot learning of object categories. IEEE Trans. Pattern Anal. Mach. Intell. **28**(4), 594–611 (2006)
12. Ge, W., Yu, Y.: Borrowing treasures from the wealthy: deep transfer learning through selective joint fine-tuning. In: IEEE Conference on Computer Vision and Pattern Recognition, pp. 1086–1095 (2017)
13. Guo, Y., Shi, H., Kumar, A., Grauman, K., Rosing, T., Feris, R.: SpotTune: transfer learning through adaptive fine-tuning. In: IEEE Conference on Computer Vision and Pattern Recognition, pp. 4805–4814 (2019)
14. Harel, J., Koch, C., Perona, P.: Graph-based visual saliency. In: Advances in Neural Information Processing Systems, pp. 545–552 (2007)
15. He, K., Zhang, X., Ren, S., Sun, J.: Deep residual learning for image recognition. In: IEEE Conference on Computer Vision and Pattern Recognition, pp. 770–778 (2016)
16. Hou, X., Harel, J., Koch, C.: Image signature: highlighting sparse salient regions. IEEE Trans. Pattern Anal. Mach. Intell. **34**(1), 194–201 (2011)
17. Huang, G., Liu, Z., van der Maaten, L., Weinberger, K.Q.: Densely connected convolutional networks. In: IEEE Conference on Computer Vision and Pattern Recognition, pp. 4700–4708 (2017)
18. Itti, L., Dhavale, N., Pighin, F.: Realistic avatar eye and head animation using a neurobiological model of visual attention. In: Proceedings of SPIE 48th Annual International Symposium on Optical Science and Technology, August 2003
19. Itti, L., Koch, C., Niebur, E.: A model of saliency-based visual attention for rapid scene analysis. IEEE Trans. Pattern Anal. Mach. Intell. **20**(11), 1254–1259 (1998)
20. Jiang, M., Huang, S., Duan, J., Zhao, Q.: SALICON: saliency in context. In: IEEE Conference on Computer Vision and Pattern Recognition, pp. 1072–1080 (2015)
21. Judd, T., Durand, F., Torralba, A.: A benchmark of computational models of saliency to predict human fixations. MIT Technical report (2012)
22. Judd, T., Ehinger, K., Durand, F., Torralba, A.: Learning to predict where humans look. In: IEEE International Conference on Computer Vision, pp. 2106–2113 (2009)
23. Kingma, D.P., Ba, J.: Adam: a method for stochastic optimization. In: International Conference on Learning Representations (2015)
24. Krizhevsky, A., Sutskever, I., Hinton, G.E.: ImageNet classification with deep convolutional neural networks. In: Advances in Neural Information Processing Systems, pp. 1097–1105 (2012)
25. Kruthiventi, S.S., Ayush, K., Babu, R.V.: DeepFix: a fully convolutional neural network for predicting human eye fixations. IEEE Trans. Image Process. **26**(9), 4446–4456 (2017)
26. Kümmerer, M., Theis, L., Bethge, M.: Deep Gaze I: boosting saliency prediction with feature maps trained on imageNet. In: International Conference on Learning Representations (ICLR 2015), pp. 1–12 (2014)
27. Lake, B., Salakhutdinov, R., Gross, J., Tenenbaum, J.: One shot learning of simple visual concepts. In: Proceedings of the Annual Meeting of the Cognitive Science Society (2011)
28. Lake, B.M., Salakhutdinov, R., Tenenbaum, J.B.: Human-level concept learning through probabilistic program induction. Science **350**(6266), 1332–1338 (2015)
29. Lee, K., Maji, S., Ravichandran, A., Soatto, S.: Meta-learning with differentiable convex optimization. In: IEEE Conference on Computer Vision and Pattern Recognition, pp. 10657–10665 (2019)

30. Li, J., Wong, Y., Zhao, Q., Kankanhalli, M.S.: Attention transfer from web images for video recognition. In: ACM Multimedia, pp. 1–9 (2017)
31. Li, J., Xu, Z., Wong, Y., Zhao, Q., Kankanhalli, M.S.: GradMix: multi-source transfer across domains and tasks. In: IEEE Winter Conference on Applications of Computer Vision (2020)
32. Li, W., Wang, L., Xu, J., Huo, J., Gao, Y., Luo, J.: Revisiting local descriptor based image-to-class measure for few-shot learning. In: IEEE Conference on Computer Vision and Pattern Recognition, pp. 7260–7268 (2019)
33. Luo, Y., Wong, Y., Kankanhalli, M., Zhao, Q.: Direction concentration learning: enhancing congruency in machine learning. IEEE Trans. Pattern Anal. Mach. Intell. (2019)
34. Mohri, M., Rostamizadeh, A., Talwalkar, A.: Foundations of Machine Learning. The MIT Press, Cambridge (2012)
35. Neider, M.B., Zelinsky, G.J.: Scene context guides eye movements during visual search. Vis. Res. **46**(5), 614–621 (2006)
36. Ouerhani, N., Von Wartburg, R., Hugli, H., Müri, R.: Empirical validation of the saliency-based model of visual attention. Electron. Lett. Comput. Vis. Image Anal. **3**(1), 13–24 (2004)
37. Pan, J., et al.: SalGAN: visual saliency prediction with generative adversarial networks. arXiv preprint arXiv:1701.01081 (2017)
38. Pan, S.J., Tsang, I.W., Kwok, J.T., Yang, Q.: Domain adaptation via transfer component analysis. IEEE Trans. Neural Netw. **22**(2), 199–210 (2010)
39. Paszke, A., et al.: Automatic differentiation in PyTorch. In: NIPS Autodiff Workshop (2017)
40. Rothenstein, A.L., Tsotsos, J.K.: Attention links sensing to recognition. Image Vis. Comput. **26**(1), 114–126 (2008)
41. Shan, W., Sun, G., Zhou, X., Liu, Z.: Two-stage transfer learning of end-to-end convolutional neural networks for webpage saliency prediction. In: Sun, Y., Lu, H., Zhang, L., Yang, J., Huang, H. (eds.) IScIDE 2017. LNCS, vol. 10559, pp. 316–324. Springer, Cham (2017). https://doi.org/10.1007/978-3-319-67777-4_27
42. Shen, C., Zhao, Q.: Webpage saliency. In: Fleet, D., Pajdla, T., Schiele, B., Tuytelaars, T. (eds.) ECCV 2014. LNCS, vol. 8695, pp. 33–46. Springer, Cham (2014). https://doi.org/10.1007/978-3-319-10584-0_3
43. Shrivastava, A., Malisiewicz, T., Gupta, A., Efros, A.A.: Data-driven visual similarity for cross-domain image matching. ACM Trans. Graph. **30**(6), 154 (2011)
44. Sung, F., Yang, Y., Zhang, L., Xiang, T., Torr, P.H., Hospedales, T.M.: Learning to compare: relation network for few-shot learning. In: IEEE Conference on Computer Vision and Pattern Recognition, pp. 1199–1208 (2018)
45. Torralba, A., Oliva, A., Castelhano, M.S., Henderson, J.M.: Contextual guidance of eye movements and attention in real-world scenes: the role of global features in object search. Psychol. Rev. **113**(4), 766 (2006)
46. Vinyals, O., Blundell, C., Lillicrap, T., Kavukcuoglu, K., Wierstra, D.: Matching networks for one shot learning. In: Advances in Neural Information Processing Systems, pp. 3630–3638 (2016)
47. Wolfe, J.M., Horowitz, T.S.: Five factors that guide attention in visual search. Nat. Hum. Behav. **1**(3), 0058 (2017)
48. Xie, S., Girshick, R., Dollár, P., Tu, Z., He, K.: Aggregated residual transformations for deep neural networks. In: IEEE Conference on Computer Vision and Pattern Recognition, pp. 5987–5995 (2017)

49. Li, X., Grandvalet, Y., Davoine, F.: Explicit inductive bias for transfer learning with convolutional networks. In: International Conference on Machine Learning, pp. 2830–2839 (2018)
50. Yang, S., Lin, G., Jiang, Q., Lin, W.: A dilated inception network for visual saliency prediction. IEEE Trans. Multimedia **22**(8), 2163–2176 (2020)
51. Yang, Y., Ma, Z., Hauptmann, A.G., Sebe, N.: Feature selection for multimedia analysis by sharing information among multiple tasks. IEEE Trans. Multimedia **15**(3), 661–669 (2012)
52. Zhang, J., Sclaroff, S.: Saliency detection: a Boolean map approach. In: IEEE International Conference on Computer Vision, pp. 153–160 (2013)

Progressively Guided Alternate Refinement Network for RGB-D Salient Object Detection

Shuhan Chen[1](✉) and Yun Fu[2]

[1] School of Information Engineering, Yangzhou University, Yangzhou, China
shchen@yzu.edu.cn

[2] Department of ECE and Khoury College of Computer Science, Northeastern University, Boston, USA
yunfu@ece.neu.edu

Abstract. In this paper, we aim to develop an efficient and compact deep network for RGB-D salient object detection, where the depth image provides complementary information to boost performance in complex scenarios. Starting from a coarse initial prediction by a multi-scale residual block, we propose a progressively guided alternate refinement network to refine it. Instead of using ImageNet pre-trained backbone network, we first construct a lightweight depth stream by learning from scratch, which can extract complementary features more efficiently with less redundancy. Then, different from the existing fusion based methods, RGB and depth features are fed into proposed guided residual (GR) blocks alternately to reduce their mutual degradation. By assigning progressive guidance in the stacked GR blocks within each side-output, the false detection and missing parts can be well remedied. Extensive experiments on seven benchmark datasets demonstrate that our model outperforms existing state-of-the-art approaches by a large margin, and also shows superiority in efficiency (**71 FPS**) and model size (**64.9 MB**).

Keywords: RGB-D salient object detection · Lightweight depth stream · Alternate refinement · Progressive guidance

1 Introduction

The goal of salient object detection (SOD) is to detect and segment the objects or regions in an image or video [15] that visually attract human attention most. It is usually serves as a pre-processing step to benefit a lot of vision tasks, such as image-sentence matching [23], weakly-supervised semantic segmentation [51], few-shot learning [54], to name a few. Benefiting from the rapid development of deep convolutional neural networks (CNNs), it has achieved profound progresses

Electronic supplementary material The online version of this chapter (https://doi.org/10.1007/978-3-030-58598-3_31) contains supplementary material, which is available to authorized users.

A. Vedaldi et al. (Eds.): ECCV 2020, LNCS 12353, pp. 520–538, 2020.
https://doi.org/10.1007/978-3-030-58598-3_31

recently. Nevertheless, it is still very challenging in some complex scenes, such as low contrast, objects sharing similar appearance with its surroundings [11].

RGB-D cameras are now easily available with low-price and high-performance, such as RealSense, Kinect, which can provide depth image that contains necessary geometric information. Utilizing depth additional to the RGB image could potentially improve the performance in the above challenging cases, which is also proven to be an effective way in the applications of object detection [27], semantic segmentation [32], and crowd counting [31].

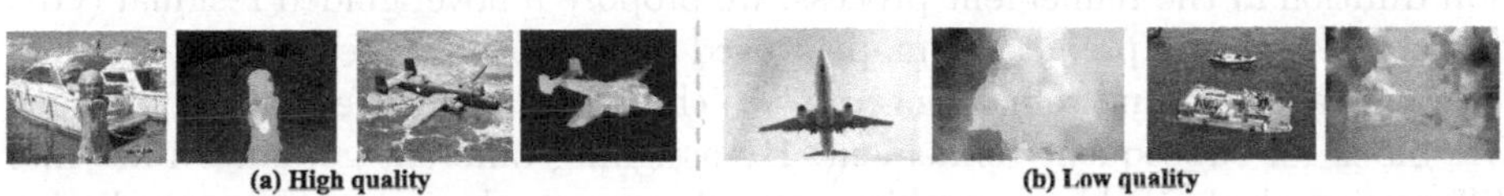

Fig. 1. Example RGB images with corresponding depth images. (a) High quality depth images successfully pop out salient objects, thus can be seen as mid-level or high-level feature maps. (b) Low quality depth images are cluttered thus may be harmful for the prediction.

Although several novel CNN-based SOD approaches [42,62] have been proposed for RGB-D data recently, the optimal way to fuse RGB and depth information remains an open issue, which lies in two aspects: model incompatibility and redundancy, low-quality depth map. Most of the existing fusion strategies can be classified into early fusion [45,47], late fusion [14,20], and middle fusion [1–3,42]. Recent researches mainly focus on the middle fusion where a separate backbone network pre-trained from ImageNet [9] is usually utilized to extract depth features, which may causes incompatible problem due to the inherent modality difference between RGB and depth image [62]. Besides that, such two-stream framework doubles the number of model parameters and computation cost, which is not efficient and also contains much redundancy. Furthermore, depth maps may vary in qualities due to the limitations of the depth sensors, high-quality depth map can well pop-out salient objects with well-defined closed boundaries, while low-quality ones are cluttered and may be noisy to the prediction, as shown in Fig. 1.

To address the above issues, we first construct a depth stream to extract depth features by learning from scratch without using pre-trained backbone network. As we know, RGB and depth have very different properties. Depth only captures the spatial structure and 3D layout of objects, without any texture details, thus contains much less information than RGB image. Since lacking low-level information, it is redundant to use pre-trained backbone network to extract features, especially these shallow layers. As seen in Fig. 1, the high quality depth image can be seen as a mid-level or high-level feature map. Based on this observation, we design a lightweight depth stream to capture complementary high-level features only, specifically, four convolutional layers are sufficient to achieve it. Therefore, it is not only much more compact and efficient than existing two-backbone based models, but also without the incompatible problem.

Instead of directly fusing RGB features and depth features (*e.g.*, concatenation or summation), which may degrade the confident RGB features especially when the input depth is noisy, we propose a alternate refinement strategy to incorporate them separately. The whole architecture of the proposed network follows a coarse-to-fine framework which starts from a initial prediction generated by a proposed Multi-Scale Residual (MSR) block. Then, it is refined progressively from deep side-outputs to shallow ones. To alleviate the information dilution in the refinement process, we propose a novel guided residual (GR) block where input prediction map is used to guide input feature to generate refined prediction and refined feature, which will be further fed into a following GR block for subsequent refinement. By assigning different guidance roles into different stacked GR blocks within each side-output, the input coarse prediction can be refined progressively into more complete and accurate.

Experimental results over 7 benchmark datasets demonstrate that our model significantly outperforms state-of-the-art approaches, and also with advantages in efficiency (**71 FPS**) and compactness (**64.9 MB**). In summary, our main contributions can be concluded as follows:

- We construct a lightweight depth stream to extract complementary depth features, which is much more compact and efficient than using pre-trained backbone network while without the incompatible problem.
- We propose an alternate refinement strategy by feeding RGB feature and depth feature alternately into GR blocks, in this way to avoid breaking the good property of the confident RGB feature when the depth is low quality.
- We further design a guided residual block to address the information dilution issue, where an input prediction is used as a guidance to generate both refined feature and refined prediction. By stacking them with progressive guidance, the missing parts and false predictions can be well remedied.

2 Related Work

2.1 RGB Salient Object Detection

Coarse-to-Fine. Before deep learning, salient regions are usually discovered from less ambiguous regions to difficult regions [7,19,57]. Following this idea, coarse-to-fine frameworks are widely explored in recent CNNs based works. Liu [35] first proposed a hierarchical recurrent convolutional neural network for refinement in a global to local and coarse to fine pipeline. Using a eye fixation map as initial prediction, Wang *et al.* [50] proposed a hierarchy of convolutional LSTMs to progressively optimize it in a top-down manner. Chen *et al.* [4,5] applied side-output residual learning for refinement, which is guided by a novel reverse attention block. While in [17], attentive feedback module was designed for better guidance. Besides the above top-down guidance, Wang *et al.* [49] further integrated bottom-up inference in an iterative and cooperative manner for recurrent refinement. In this paper, we also follow the coarse-to-fine pipeline, the difference is that we learn multiple residuals in each side-output with progressive guidance, which can better remedy the missing object parts and false detection.

Top-Down Guidance. The deep layer contains high-level semantic information, which can be used as a guidance to help shallow layers filter out noisy distraction [6]. Such a top-down guidance manner was also widely applied in existing methods. Deep prediction maps are concatenated with shallow prediction by short connections for guidance in [22], and used to erase its corresponding regions in shallow feature to guide residual learning in [4]. While in [52], it was utilized to weight shallow feature by the proposed holistic attention module. In [36], it was applied into each group of the divided shallow convolutional feature to further promote its guidance role. Liu *et al.* [34] built a global guidance module to transmit the location information into different shallow layers. Zhang *et al.* [58] made a further step by leveraging captioning information to boost SOD. There are also several recent works following this effective strategy, such as [39,56,63]. While in our model, the deep prediction is used as guidance both for feature refinement and prediction refinement. Furthermore, a progressive guidance strategy is proposed to address the feature dilution issue.

2.2 RGB-D Salient Object Detection

Hand-Crafted Based. Early works are all focusing on various hand-crafted features, including multi-contextual contrast [41], anisotropic center-surround difference [24], local background enclosure [16], and so on.

CNNs Based. To increase feature representation ability, CNNs is widely applied and has dominated this area in recent years. As mentioned above, these approaches can be roughly divided into early fusion, middle fusion, and late fusion. Early fusion regards the depth map as a additional channel to concatenate with RGB as initial input, *e.g.*, [47]. Late fusion applies two separate backbone network for RGB and depth to generate individual features or predictions which are fused together for final prediction, such as [14,20]. Most of recent works focused on the middle fusion scheme, which incorporate multi-scale RGB features and depth features in different manners. A complementarity-aware fusion module was proposed in [1]. Piao *et al.* [42] designed a depth refinement block to fuse multi-level paired depth and RGB cues. Instead of cross-modal feature fusion, Zhao *et al.* [62] addressed it in a different manner by integrating enhanced depth cues to weight RGB features for better representation. In [43], an asymmetrical two-stream architecture based on knowledge distillation was proposed for light field SOD. Different from them, we construct a lightweight depth stream by learning from scratch, whose depth features are fed for refinement separately.

3 The Proposed Network

We first present the overall architecture of the proposed alternate refinement network, and then introduce its main components in detail, including MSR block to generate coarse initial prediction, and GR block with progressive guidance. Finally, we also discuss the differences with related networks.

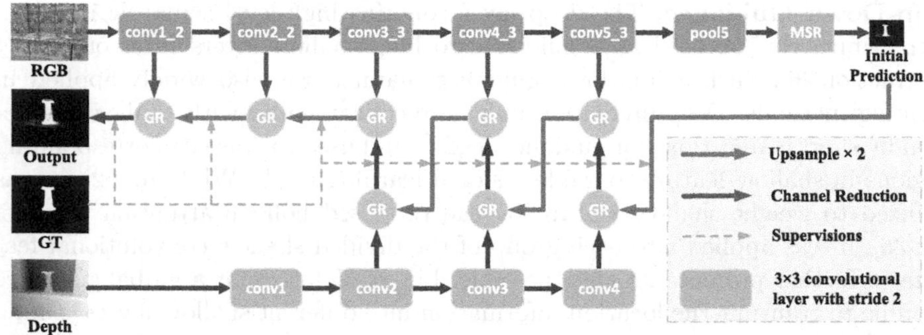

Fig. 2. The overall architecture of the proposed network, where RGB feature and depth feature are fed into GR blocks **alternately** for refinement. Here, we only show single GR block in each side-output for clarity. Detailed structures of MSR and GR are illustrated in Fig. 3 and Fig. 4 respectively.

3.1 The Overall Architecture

Our network follows the existing coarse-to-fine refinement framework as seen in Fig. 2. Given a coarse initial prediction generated by our MSR block, we apply our GR block to refine it progressively by combing multi-level convolutional features from RGB and depth streams alternately. Considering the modal gap between RGB and depth, furthermore, the quality of the depth varies tremendously across different scenarios due to the limitation of depth sensors, we don't directly fuse the RGB and depth features, instead, they are fed into our network alternately to reduce their mutual degradation for better refinement. Finally, we apply deep supervisions on each side-output and train the whole network in an end-to-end manner, only with standard binary cross entropy loss.

RGB Stream. We utilize VGG16 [46] as backbone network to extract multi-level RGB features, where {`conv1_2`, `conv2_2`, `conv3_3`, `conv4_3`, `conv5_3`} are chosen as side-output features, which have {1, 1/2, 1/4, 1/8, 1/16} of the input image resolution respectively. We first apply 1×1 convolutional layers to reduce their dimensions into {16, 32, 64, 64, 64} for efficiency. Then, these side-output features (denoted as F_1, F_2, F_3, F_5, F_7) are used for subsequent refinement.

Depth Stream. Instead of using pre-trained backbone network as most of the existing works did, we construct a light-weight depth stream to extract complementary features, which only consists of four cascaded 3×3 convolutional layers with 64 channels and stride 2. The last three layers are selected as high-level side-output features for refinement, which are denoted as F_4, F_6, F_8, with {1/4, 1/8, 1/16} of the input image resolution respectively.

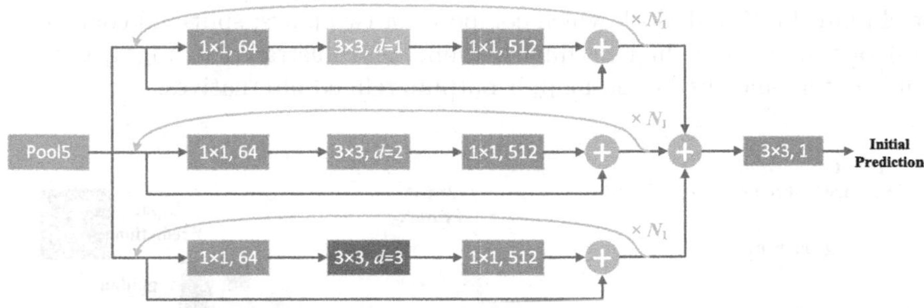

Fig. 3. The proposed multi-scale residual block. "*d*" denotes dilation rate.

3.2 Multi-scale Residual Block

Since the scale of salient objects vary from large to small, which implies that the model needs to capture information at different contexts in order to detect objects reliably. Although the backbone network has a large enough theoretical receptive field to cover most of the large objects, the effective receptive field is smaller than the theoretical receptive field as demonstrated in [37]. Inspired from [30], we design a multi-scale residual block to address the scale issue for SOD. The proposed MSR block is embedded after "pool5" and consists of three parallel branches in which each shares the same residual structure except the dilation rate. Each residual block consists of three convolutions with kernel size 1×1, 3×3, and 1×1. The dilation rates for the 3×3 convolutional layers are 1, 2, and 3 respectively, as shown in Fig. 3. Instead of stacking such residual block to increase receptive field, we implement it in a recurrent manner to reduce the number of parameters, which has been widely applied in previous works [10,29, 60]. The total recurrent iteration is N_1 for each branch. Finally, all the branches are added together then fed into a 3×3 convolutional layer to produce a single channel initial prediction.

Although shares similar structure, the proposed MSR differs from [30] in the following two aspects. Firstly, since the scale-aware ground truth is not easy to obtain in SOD, we don't share weights among different branches but different stacked blocks in each branch. Secondly, these branches are fused together for the initial prediction.

3.3 Guided Residual Block

As we know, different layers of deep CNNs learn different scale features, shallow layers capture low-level structure cues while deep layers capture high-level semantic information. Based on this observation, various fusion strategies were proposed to combine their complementary cues, such as short connection [22], skip connection [33], residual connection [10,21,61]. However, the high-level information of deep layer may be gradually diluted in the fusion process, especially when combing with the noisy shallow features. To address it, we design a novel guided residual block which composes of two parts: split-and-concatenate (SC) operation, and dual residual learning. As illustrated in Fig. 4, given an input feature and prediction map, it outputs refined alternatives.

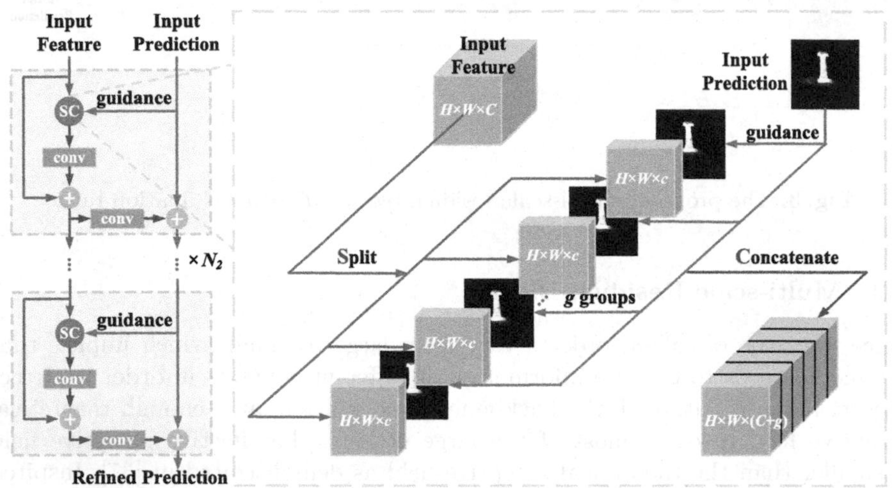

Fig. 4. The proposed guided residual blocks (dashed bounding boxes on the left), which are stacked with progressive guidance. SC denotes split-and-concatenate operation (dashed bounding box on the right).

Split-and-Concatenate. Given the convolutional feature F with C channels and prediction map S as inputs, we first split F into g groups, each of which has c channels. Then S is utilized as a guidance feature map to be concatenated with each split feature maps. After concatenation, we obtain a $C + g$ channel feature. Based on it, the SC operation can be formulated as:

$$F^1, ..., F^j, ..., F^g = \text{Split}(F), j \in 1, 2, ..., g, \tag{1}$$

$$F_{cat} = \text{Cat}(F^1, S, ..., F^j, S, ..., F^g, S), \tag{2}$$

where Cat denotes concatenate operation and F_{cat} is the concatenated feature.

Dual Residual Learning. After SC operation, we feed the concatenated feature F_{cat} into a 3×3 convolutional layer for guided learning and reducing channel number into C, then are added with the input feature as refined output feature. Thus, the first residual learning can be formulated as:

$$\hat{F} = F + \text{Conv}(F_{cat}; \theta_1), \tag{3}$$

where $\text{Conv}(*; \theta)$ is the convolution operation with parameter θ. Another 3×3 convolutional layer is further applied to produce a single channel residual prediction. Based on it, we can obtain the refined prediction map $\hat{S}$ by:

$$\hat{S} = S + \mathrm{Conv}(\hat{F}; \theta_2). \tag{4}$$

$\hat{F}$ and $\hat{S}$ will be further fed into subsequent GR block for guided residual learning.

3.4 Progressive Guidance

Based on the above GR block, we propose a progressive residual refinement framework. Specifically, in each side-output, we stack N_2 GR blocks, which is set to 3 in this paper, and the first GR block takes F_i^0 and S_i^0 as input feature and prediction map:

Table 1. The detailed setting of different guidance styles, including uniform guidance and progressive guidance. {*, *, *} represents the channel number c in each split group from side-output 1 to side-output 3. The rest are the same with side-output 3. GR^r denotes the rth GR block in each side-output.

Guidance style	No.	GR^1	GR^2	GR^3
Uniform guidance	1	c={16, 32, 64}		
	2	c={8, 8, 8}		
	3	c={4, 4, 4}		
	4	c={1, 1, 1}		
Progressive guidance	5	c={16, 32, 64}	c={8, 8, 8}	c={4, 4, 4}
	6	c={16, 32, 64}	c={8, 8, 8}	c={1, 1, 1}
	7	c={16, 32, 64}	c={4,4,4}	c={1, 1, 1}
	8	c={8, 8, 8}	c={4, 4, 4}	c={1, 1, 1}

$$\begin{cases} S_i^0 = \mathrm{Up}(S_{i+1}, F_i), \text{ if } i = 1, 2, 4, 6, 8; & (5) \\ S_i^0 = S_{i+1}, \text{ if } i = 3, 5, 7; & (6) \end{cases}$$

in which $\mathrm{Up}(x, y)$ represents bilinear interpolation operation that upsamples x to the same size as y, and i denotes side-output stage. Then the output of the first GR block can be denoted as $\hat{F}_i^1$ and $\hat{S}_i^1$, which will be fed into the following GR block. The last GR block only outputs the refined prediction $S_i = \hat{S}_i^{N_2}$ as shown in Fig. 4. S_1 is fed into a sigmoid layer as final output.

The channel number c in each split group or group number g is essential for guidance. We can define different guidance styles by varying c or g. If $g = 1$, as [10] did, the guidance role is very weak due to the imbalanced channels (C versus 1). In [36], c is set to 4 in all the side-outputs, which can be seen as a medium guidance. Extremely, the guidance role is very strong when we set c to 1. Different from them, we first define uniform guidance by sharing the same guidance role in all the stacked blocks, as listed in Table 1. Since the prediction map will becomes more and more accurate in the refinement process, we further define progressive guidance by gradually increasing the guidance role. We will conduct ablation experiments to investigate the best setting in Sect. 4.2.

Such a progressive residual refinement inherits the following good properties. The stacked residual units establish multiple shortcut connections between each side-output prediction and the ground truth, which enables it easier to remedy the missing object parts and detection errors. Extremely, with the strong supervision on each side-output, the error is approximately equal to zero if there is no useful information in the input feature, *e.g.*, when the depth image is low quality. In this way, we can greatly reduce its noisy distraction, thus leads to more accurate detection. Furthermore, such residual units also enhance the input feature gradually for better refinement.

3.5 Difference to Other Networks

Although shares the same split-and-concatenate operation, the proposed GR block differs from the group guidance module (GGM) [36] in two aspects. (1) GGM apply group convolution on the output of SC, which only focuses on each split group for guidance. Different from it, the convolution is performed on all the concatenated feature maps in our GR, which benefits the information passing among different groups. (2) The channel number in each split group is fixed to 4 in GGM. While our GR blocks are stacked with progressive guidance by varying different c. Our network also differs other residual learning based architectures, *e.g.*, RAS [4,5], R^3Net [10]. Firstly, we learn dual residuals progressively in each side-output which can better remedy the missing object parts and false detection in the initial prediction. Secondly, the prediction maps are progressively applied for guidance during residual refinement. The effectiveness and superiority will be verified in the following section.

4 Experimental Results

4.1 Experimental Setup

Datasets. We adopt 7 widely used RGB-D benchmark datasets for evaluation, including NJUD [24], NLPR [41], DES [8], STERE [38], LFSD [28], DUT [42], and SIP [14], which contain 1985, 1000, 135, 1000, 100, 1200, 929 well annotated images, respectively. Among them, SIP is a recent collected human activities oriented dataset with high image resolution (744×992). To make a fair comparison, we follow the same training settings as existing works [1,2,14,20], which consists of 1485 samples from NJUD and 700 samples from NLPR. To reduce over-fitting risk, we augment the input image by random horizontal flipping and rotating (0°, 90°, 180°, 270°), which increases the training images by four times.

Evaluation Metrics. We adopt five widely applied metrics for comprehensive evaluation, *i.e.*, precision-recall (PR) curve, F-measure (F_β), S-measure [12]

Table 2. Quantitative comparison of different settings in Table 1.

No.	1	2	3	4	5	6	7	8
$E_\xi \uparrow$	0.903	0.903	0.903	0.904	0.903	**0.908**	0.907	0.905
$S_\alpha \uparrow$	0.866	0.869	0.869	0.870	0.867	**0.875**	0.871	0.871
$F_\beta \uparrow$	0.845	0.845	0.845	0.846	0.845	**0.848**	0.847	0.842
$M \downarrow$	0.062	0.062	0.062	0.060	0.061	0.059	**0.058**	0.059

Table 3. Quantitative comparison with different ablation settings. R and D denote RGB stream and depth stream respectively. St: stacking 7 MSR blocks; Re: proposed recurrent strategy. MS: model size (MB).

	St	Re	Cat	AR	R	R+D	VGG16	Ours
$E_\xi \uparrow$	**0.907**	**0.907**	0.899	**0.908**	0.886	**0.908**	0.896	**0.908**
$S_\alpha \uparrow$	0.871	**0.872**	0.863	**0.875**	0.846	**0.875**	0.857	**0.875**
$F_\beta \uparrow$	0.850	**0.851**	0.840	**0.848**	0.814	**0.848**	0.837	**0.848**
$M \downarrow$	**0.059**	**0.059**	0.064	**0.059**	0.072	**0.059**	0.065	**0.059**
MS $\downarrow$	72.3	**64.9**	**63.0**	64.9	**62.5**	64.9	123.6	**64.9**

(S_α), E-measure [13] (E_ξ), and mean absolute error (M). Specifically, PR curve is plotted via pairs of precision and recall values which are calculated by comparing the binary saliency map with its ground truth. F_β is an overall metric and only its maximum value is reported here, where β^2 is set to 0.3 to emphasize the precision over recall. S_α and E_ξ are two recent proposed metrics which evaluate the spatial structure similarities, local pixel matching and image-level statistics information, respectively. Higher scores of E_ξ, S_α, and F_β indicate better performance, while lower for M.

Implementation Details. We implemented our method in PyTorch [40] and on a PC with single NVIDIA TITAN Xp GPU. All the images are resized to 352×352 both for training and inferring. The depth image needs to be normalized into [0, 1]. The proposed model is trained by Adam optimizer [25] with the following hyper-parameters: batch size (10), epochs (30), initial learning rate (1e−4), which is decreased by 10 after 25 epochs. The parameters of the backbone network in the RGB stream is initialized by VGG16 [46], while the others are using the default setting of the Pytorch. We will release the source code for research purpose on http://shuhanchen.net.

4.2 Ablation Analyses

We first investigate different design options and the effectiveness of different components in the proposed network on a recent challenging dataset SIP [14].

Recurrent Strategy. To verify the effectiveness of the recurrent strategy in MSR, we first made an experiment by comparing with stacking 7 blocks with

Table 4. Quantitative comparison with different side-output depth features.

conv4	conv3	conv2	conv1	$E_\xi \uparrow$	$S_\alpha \uparrow$	$F_\beta \uparrow$	$M \downarrow$
✓				0.892	0.855	0.828	0.067
✓	✓			0.897	0.860	0.834	0.066
✓	✓	✓		**0.908**	**0.875**	**0.848**	**0.059**
✓	✓	✓	✓	0.907	0.873	**0.848**	**0.059**

7 iterations. Here we adopt the No. 1 setting in Table 1 as guidance style. As can be seen in Table 3, their performance are almost the same, but the recurrent strategy achieved more compact model size. We further conduct ablation study to explore how many recurrent iterations are needed in MSR by varying it N_1 from 1 to 7. The results in Fig. 5(f) show that when N_1 grows beyond 5, the performance becomes stable. Therefore, for efficiency, we adopt the recurrent strategy and set N_1 to 5 in the following experiments.

Guidance Style. To investigate the best setting of the guidance style, we compare the performance of all the listed settings in Table 1. From the results in Table 2, we can observe that progressive guidance shows better performance than uniform guidance, which supports our claim that the guidance role should be strengthened with the progressively refined prediction map. The No. 6 setting that achieved best performance was adopt as our final guidance strategy.

Alternate Refinement. We further conduct experiment to verify the proposed alternate refinement (AR) strategy by comparing with directly concatenating (Cat) the RGB and depth features. As shown in Table 3, our proposed AR strategy performs better than Cat, which demonstrates our analysis that Cat may break the good property of the RGB features.

Depth Stream. We first evaluate the effectiveness of the constructed depth stream by removing it for comparison. As seen in Table 3, the performance can be greatly improved when combining the depth stream, which indicates the good ability to capture complementary information by our constructed depth stream. It is also worth to note that the performance is still comparable with state-of-the-art model when only using RGB stream, which further confirms the effectiveness of our progressive residual refinement framework.

Since most of the previous works using two pre-trained backbone networks to extract RGB and depth features respectively, we also made another experiment by replacing the proposed depth stream with VGG16 [46]. As a result, the model size and training time (4 h) are dramatically increased with the decrease of the quantitative performance as shown in Table 3. Therefore, our proposed depth stream is a better choice in extracting complementary depth features.

Finally, to investigate how many depth features are sufficient for the proposed network, we separately evaluate the performance by combing different side-output depth features. We can clearly observe from Table 4 that the performance is gradually improved with the incorporation of more side-output depth features

Table 5. Quantitative comparison including E_ξ, S_α, F_β, and M, over seven widely evaluated datasets. ↑ & ↓ represent higher and lower is better, respectively. $*$ denotes the models are trained on NJUD [24]+NLPR [41]+DUT [42], the rest are trained on NJUD [24]+NLPR [41]. The best three scores are highlighted in red, **blue**, and green respectively.

	Metric	LHM [41]	ACSD [24]	LBE [16]	DF [45]	CTMF [20]	MMCI [3]	TAN [2]	PCAN [1]	CPFP [62]	Ours	DMRA [42]$*$	Ours $*$
NJUD [24]	E_ξ ↑	.711	.790	.796	.839	.864	.878	.893	.896	.895	**.914**	.908	.916
	S_α ↑	.522	.703	.700	.768	.849	.859	.878	.877	.878	**.906**	.886	.909
	F_β ↑	.636	.695	.734	.783	.788	.813	.844	.844	.837	**.883**	.872	.893
	M ↓	.199	.198	.149	.136	.085	.079	.060	.059	.053	**.045**	.051	.042
NLPR [41]	E_ξ ↑	.819	.752	.868	.884	.869	.872	.916	.916	.924	**.948**	.941	.955
	S_α ↑	.631	.684	.777	.806	.860	.856	.886	.874	.888	**.918**	.899	.930
	F_β ↑	.665	.548	.747	.759	.723	.730	.796	.795	.822	**.871**	.854	.885
	M ↓	.103	.171	.073	.079	.056	.059	.041	.044	.036	**.028**	.031	.024
DES [8]	E_ξ ↑	.761	.855	.911	.877	.911	.904	.919	.912	.927	.935	.944	**.939**
	S_α ↑	.578	.728	.703	.752	.863	.848	.858	.842	.872	.894	**.900**	.913
	F_β ↑	.631	.717	.796	.753	.778	.762	.795	.782	.829	**.870**	.866	.880
	M ↓	.114	.169	.208	.093	.055	.065	.046	.049	.038	.032	**.030**	.026
STERE [38]	E_ξ ↑	.770	.793	.749	.838	.864	.901	.906	.897	.903	.917	.920	**.919**
	S_α ↑	.562	.692	.660	.757	.848	.873	.871	.875	.879	**.903**	.886	.907
	F_β ↑	.703	.661	.595	.742	.771	.829	.835	.826	.830	**.872**	.867	.880
	M ↓	.172	.200	.250	.141	.086	.068	.060	.064	.051	**.044**	.047	.041
SIP [14]	E_ξ ↑	.719	.827	.841	.794	.824	.886	.893	**.899**	**.899**	.908	.863	.908
	S_α ↑	.511	.732	.727	.653	.716	.833	.835	.842	.850	**.875**	.806	.876
	F_β ↑	.592	.727	.733	.673	.684	.795	.809	.825	.819	**.848**	.819	.854
	M ↓	.184	.172	.200	.185	.139	.086	.075	.071	.064	**.059**	.085	.055
DUT [42]	E_ξ ↑	.756	.814	.785	.848	.884	.855	.866	.858	.815	.888	**.927**	.944
	S_α ↑	.551	.706	.679	.733	.834	.791	.808	.801	.749	.849	**.889**	.920
	F_β ↑	.683	.699	.668	.764	.792	.753	.779	.760	.736	.829	**.884**	.914
	M ↓	.179	.181	.236	.144	.097	.113	.093	.100	.100	.069	**.048**	.035
LFSD [28]	E_ξ ↑	.736	.801	.770	.844	.851	.840	.845	.842	.867	.869	.899	**.889**
	S_α ↑	.557	.734	.736	.791	.796	.787	.801	.794	.828	.833	**.847**	.853
	F_β ↑	.718	.755	.708	.806	.782	.779	.794	.792	.813	.830	**.840**	.852
	M ↓	.211	.188	.208	.138	.119	.132	.111	.112	.088	.093	**.075**	.074

until "conv2". Further incorporating "conv1" doesn't bring performance gain, which supports our claim that depth image can be seen as a mid-level or high-level feature map, therefore, there is no need to explore low-level features from it with additional convolutional layers. Therefore, three side-output depth features are sufficient to capture the complementary cues.

4.3 Comparison with State-of-the-arts

We compare our model with 10 state-of-the-arts, consisting of 3 traditional methods: LHM [41], ACSD [24], LBE [16]; and 7 CNNs-based methods: DF [45], CTMF [20], MMCI [3], TAN [2], PCAN [1], CPFP [62], DMRA [42]. Note that all the results of the compared approaches are reproduced by running source codes or pre-computed by the authors. In addition, we also trained a model (marked with $*$) using the same trainset with DMRA [42] for fair comparison.

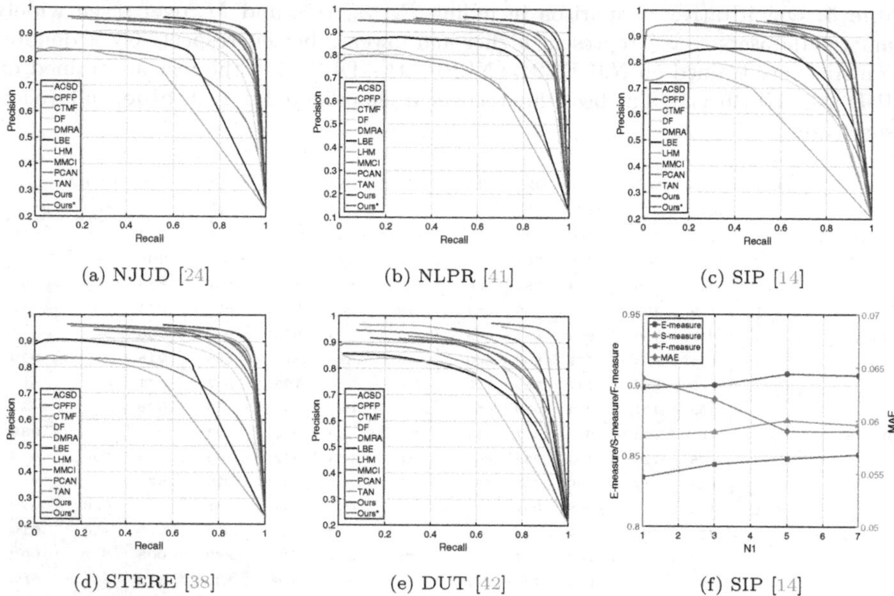

(a) NJUD [24] (b) NLPR [41] (c) SIP [14]

(d) STERE [38] (e) DUT [42] (f) SIP [14]

Fig. 5. (a)–(e): Precision-recall curves comparison. (f) Quantitative comparison of different recurrent iterations in the proposed MSR block.

Table 6. Running speed and model size comparisons with recent models.

Method	Platform	Image Size	FPS ↑	MS (MB) ↓
CPFP [62]	Caffe	400 × 300	10	291.9
DMRA [42]	Pytorch	256 × 256	16	238.8
Ours	Pytorch	352 × 352	**71**	**64.9**

Quantitative Evaluation. The quantitative comparison results in terms of 4 evaluation metrics on 7 datasets are reported in Table 5. As can be clearly observed that the proposed network significantly outperforms the competing methods across all the datasets in all the metrics except E_ξ. Comparing with the recent state-of-the-art model DMRA [42], our approach increases its S_α and F_β scores by an average of **2.8%** and **2.1%**, decreases the M by an average of **1.0%**, which clearly indicates the good consistence with the ground truth. We also perform much better when comparing with CPFP [62] which doesn't use pre-trained backbone network to extract depth features too. We analyze that their proposed contrast prior may also break the good property of the RGB features when the depth image is low quality, while such issue can be well alleviated by our method. It is also worth to note that our model trained on NJUD+NLPR still performs better than DMRA on some datasets, which further demonstrates the superiority and effectiveness of the proposed approach. We also plot the PR curves for comparison on five large datasets. As illustrated in Fig. 5, we consistently achieve the best performance especially at a high level of recall.

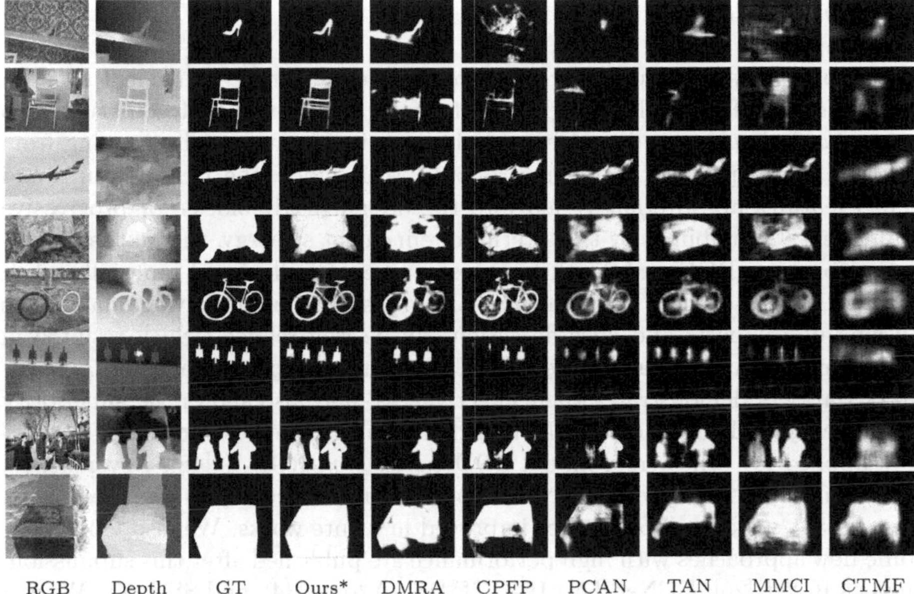

Fig. 6. Visual comparisons with state-of-the-art approaches in different challenging scenarios: low contrast in RGB or depth, complex scene, low quality depth map, multiple (small) objects, and large object. As can be seen, all the salient objects can be completely highlighted while with less false detection.

Qualitative Evaluation. We further illustrate visual examples of several representative images in different challenging scenarios to show the advantage of our method, *i.e.*, low contrast in RGB or depth image (1st–2nd rows), low quality depth map (3rd–4th rows), complex scene (5th row), multiple (small) objects (6th–7th rows), and large object (8th row). As can be seen clearly in Fig. 6 that all these cases are very challenging to the existing methods. Nevertheless, thanks to the proposed alternate refinement strategy, our model can well capture the complementary cues from the depth image, therefore, we can successfully highlight salient objects in these images, and also will not be distracted by the low quality depth maps. Furthermore, contributed by the proposed progressive guidance, the missing object parts and false detection can be well remedied, thus leads to more complete and accurate detection.

Timing and Model Size. The proposed network is also very efficient and compact. When trained on the NJUD+NLPR datasets with 2185 × 4 images, our network only takes about 2.5 h to train for 30 epochs. During the inference stage, contributed by the constructed lightweight depth stream, we can run at **71 FPS** only with **64.9 MB** model size, which is much faster and compact than the existing models as shown in Table 6.

5 Conclusions

In this paper, we developed a progressively guided alternate refinement network for efficient RGB-D SOD. A lightweight depth stream was first constructed to extract complementary depth features by learning from scratch. Starting from a coarse initial prediction by the proposed MSR block, the RGB features and depth features are alternately fed into the designed GR blocks for progressive refinement. Contributed by the alternate refinement strategy, the mutual degradation between RGB and depth features can be well alleviated especially when the depth is low quality. With the help of the proposed progressive guidance, the missing object parts and false detection can be well refined, which resulted in more complete and accurate detection. State-of-the-art performance on 7 benchmark datasets demonstrates their effectiveness, and also shows the superiority in efficiency and compactness. In addition, the proposed network can be flexibly applied for other cross-modal SOD tasks, *e.g.*, RGB-T [48]. Nevertheless, the boundary details are still not accurate enough especially in high resolution images [53], which will be further improved in future works. We also found that some new approaches with high performance are published after this submission, such as ICNet [26], UCNet [55], JL-DCF [18], A2dele [44], and SSF [59]. We will make a more comprehensive comparison in our extended work.

Acknowledgments. This research was supported by the National Nature Science Foundation of China (No. 61802336) and China Scholarship Council (CSC) Program. This work was mainly done when Shuhan Chen was visiting Northeastern University as a visiting scholar.

References

1. Chen, H., Li, Y.: Progressively complementarity-aware fusion network for RGB-D salient object detection. In: Proceedings of the IEEE Conference on Computer Vision and Pattern Recognition, pp. 3051–3060 (2018)
2. Chen, H., Li, Y.: Three-stream attention-aware network for RGB-D salient object detection. IEEE Trans. Image Process. **28**(6), 2825–2835 (2019)
3. Chen, H., Li, Y., Su, D.: Multi-modal fusion network with multi-scale multi-path and cross-modal interactions for RGB-D salient object detection. Pattern Recogn. **86**, 376–385 (2019)
4. Chen, S., Tan, X., Wang, B., Hu, X.: Reverse attention for salient object detection. In: Proceedings of the European Conference on Computer Vision, pp. 234–250 (2018)
5. Chen, S., Tan, X., Wang, B., Lu, H., Hu, X., Fu, Y.: Reverse attention-based residual network for salient object detection. IEEE Trans. Image Process. **29**, 3763–3776 (2020)
6. Chen, S., Wang, B., Tan, X., Hu, X.: Embedding attention and residual network for accurate salient object detection. IEEE Trans. Cybern. **50**(5), 2050–2062 (2020)
7. Chen, S., Zheng, L., Hu, X., Zhou, P.: Discriminative saliency propagation with sink points. Pattern Recogn. **60**, 2–12 (2016)

8. Cheng, Y., Fu, H., Wei, X., Xiao, J., Cao, X.: Depth enhanced saliency detection method. In: Proceedings of International Conference on Internet Multimedia Computing and Service, pp. 23–27 (2014)
9. Deng, J., Dong, W., Socher, R., Li, L.J., Li, K., Fei-Fei, L.: ImageNet: a large-scale hierarchical image database. In: Proceedings of the IEEE Conference on Computer Vision and Pattern Recognition, pp. 248–255 (2009)
10. Deng, Z., et al.: R^3Net: recurrent residual refinement network for saliency detection. In: Proceedings of the International Joint Conference on Artificial Intelligence, pp. 684–690 (2018)
11. Fan, D.P., Cheng, M.M., Liu, J.J., Gao, S.H., Hou, Q., Borji, A.: Salient objects in clutter: bringing salient object detection to the foreground. In: Proceedings of the European Conference on Computer Vision, pp. 186–202 (2018)
12. Fan, D.P., Cheng, M.M., Liu, Y., Li, T., Borji, A.: Structure-measure: a new way to evaluate foreground maps. In: Proceedings of the IEEE International Conference on Computer Vision, pp. 4548–4557 (2017)
13. Fan, D.P., Gong, C., Cao, Y., Ren, B., Cheng, M.M., Borji, A.: Enhanced-alignment measure for binary foreground map evaluation. In: Proceedings of the International Joint Conference on Artificial Intelligence, pp. 698–704 (2018)
14. Fan, D.P., Lin, Z., Zhang, Z., Zhu, M., Cheng, M.M.: Rethinking RGB-D salient object detection: models, datasets, and large-scale benchmarks. IEEE Trans. Neural Netw. Learn. Syst. (2020)
15. Fan, D.P., Wang, W., Cheng, M.M., Shen, J.: Shifting more attention to video salient object detection. In: Proceedings of the IEEE Conference on Computer Vision and Pattern Recognition, pp. 8554–8564 (2019)
16. Feng, D., Barnes, N., You, S., McCarthy, C.: Local background enclosure for RGB-D salient object detection. In: Proceedings of the IEEE Conference on Computer Vision and Pattern Recognition, pp. 2343–2350 (2016)
17. Feng, M., Lu, H., Ding, E.: Attentive feedback network for boundary-aware salient object detection. In: Proceedings of the IEEE Conference on Computer Vision and Pattern Recognition, pp. 1623–1632 (2019)
18. Fu, K., Fan, D.P., Ji, G.P., Zhao, Q.: JL-DCF: joint learning and densely-cooperative fusion framework for RGB-D salient object detection. In: Proceedings of the IEEE Conference on Computer Vision and Pattern Recognition, pp. 3052–3062 (2020)
19. Gong, C., et al.: Saliency propagation from simple to difficult. In: Proceedings of the IEEE Conference on Computer Vision and Pattern Recognition, pp. 2531–2539 (2015)
20. Han, J., Chen, H., Liu, N., Yan, C., Li, X.: CNNs-based RGB-D saliency detection via cross-view transfer and multiview fusion. IEEE Trans. Cybern. **48**(11), 3171–3183 (2017)
21. He, K., Zhang, X., Ren, S., Sun, J.: Deep residual learning for image recognition. In: Proceedings of the IEEE Conference on Computer Vision and Pattern Recognition, pp. 770–778 (2016)
22. Hou, Q., Cheng, M., Hu, X., Borji, A., Tu, Z., Torr, P.: Deeply supervised salient object detection with short connections. IEEE Trans. Pattern Anal. Mach. Intell. **41**(4), 515–828 (2019)
23. Ji, Z., Wang, H., Han, J., Pang, Y.: Saliency-guided attention network for image-sentence matching. In: Proceedings of the IEEE International Conference on Computer Vision, pp. 5754–5763 (2019)

24. Ju, R., Ge, L., Geng, W., Ren, T., Wu, G.: Depth saliency based on anisotropic center-surround difference. In: Proceedings of the IEEE International Conference on Image Processing, pp. 1115–1119 (2014)
25. Kingma, D.P., Ba, J.: Adam: a method for stochastic optimization. In: Proceedings of the International Conference on Learning Representations (2015)
26. Li, G., Liu, Z., Ling, H.: ICNet: information conversion network for RGB-D based salient object detection. IEEE Trans. Image Process. **29**, 4873–4884 (2020)
27. Li, G., Gan, Y., Wu, H., Xiao, N., Lin, L.: Cross-modal attentional context learning for RGB-D object detection. IEEE Trans. Image Process. **28**(4), 1591–1601 (2018)
28. Li, N., Ye, J., Ji, Y., Ling, H., Yu, J.: Saliency detection on light field. In: Proceedings of the IEEE Conference on Computer Vision and Pattern Recognition, pp. 2806–2813 (2014)
29. Li, X., Wu, J., Lin, Z., Liu, H., Zha, H.: Recurrent squeeze-and-excitation context aggregation net for single image deraining. In: Proceedings of the European Conference on Computer Vision, pp. 254–269 (2018)
30. Li, Y., Chen, Y., Wang, N., Zhang, Z.: Scale-aware trident networks for object detection. In: Proceedings of the IEEE International Conference on Computer Vision, pp. 6054–6063 (2019)
31. Lian, D., Li, J., Zheng, J., Luo, W., Gao, S.: Density map regression guided detection network for RGB-D crowd counting and localization. In: Proceedings of the IEEE Conference on Computer Vision and Pattern Recognition, pp. 1821–1830 (2019)
32. Lin, D., Zhang, R., Ji, Y., Li, P., Huang, H.: SCN: switchable context network for semantic segmentation of RGB-D images. IEEE Trans. Cybern. **50**(3), 1120–1131 (2018)
33. Lin, T.Y., Dollár, P., Girshick, R., He, K., Hariharan, B., Belongie, S.: Feature pyramid networks for object detection. In: Proceedings of the IEEE Conference on Computer Vision and Pattern Recognition, pp. 2117–2125 (2017)
34. Liu, J.J., Hou, Q., Cheng, M.M., Feng, J., Jiang, J.: A simple pooling-based design for real-time salient object detection. In: Proceedings of the IEEE Conference on Computer Vision and Pattern Recognition, pp. 3917–3926 (2019)
35. Liu, N., Han, J.: DHSNet: deep hierarchical saliency network for salient object detection. In: Proceedings of the IEEE Conference on Computer Vision and Pattern Recognition, pp. 678–686 (2016)
36. Liu, Y., Han, J., Zhang, Q., Shan, C.: Deep salient object detection with contextual information guidance. IEEE Trans. Image Process. **29**, 360–374 (2019)
37. Luo, W., Li, Y., Urtasun, R., Zemel, R.: Understanding the effective receptive field in deep convolutional neural networks. In: Proceedings of Advances in Neural Information Processing Systems, pp. 4898–4906 (2016)
38. Niu, Y., Geng, Y., Li, X., Liu, F.: Leveraging stereopsis for saliency analysis. In: Proceedings of the IEEE Conference on Computer Vision and Pattern Recognition, pp. 454–461 (2012)
39. Pang, Y., Zhao, X., Zhang, L., Lu, H.: Multi-scale interactive network for salient object detection. In: Proceedings of the IEEE Conference on Computer Vision and Pattern Recognition, pp. 9413–9422 (2020)
40. Paszke, A., et al.: PyTorch: an imperative style, high-performance deep learning library. In: Advances in Neural Information Processing Systems, pp. 8024–8035 (2019)

41. Peng, H., Li, B., Xiong, W., Hu, W., Ji, R.: RGBD salient object detection: a benchmark and algorithms. In: Fleet, D., Pajdla, T., Schiele, B., Tuytelaars, T. (eds.) ECCV 2014. LNCS, vol. 8691, pp. 92–109. Springer, Cham (2014). https://doi.org/10.1007/978-3-319-10578-9_7
42. Piao, Y., Ji, W., Li, J., Zhang, M., Lu, H.: Depth-induced multi-scale recurrent attention network for saliency detection. In: Proceedings of the IEEE International Conference on Computer Vision, pp. 7254–7263 (2019)
43. Piao, Y., Rong, Z., Zhang, M., Lu, H.: Exploit and replace: an asymmetrical two-stream architecture for versatile light field saliency detection. In: Proceedings of the AAAI Conference on Artificial Intelligence, pp. 11865–11873 (2020)
44. Piao, Y., Rong, Z., Zhang, M., Ren, W., Lu, H.: A2dele: adaptive and attentive depth distiller for efficient RGB-D salient object detection. In: Proceedings of the IEEE Conference on Computer Vision and Pattern Recognition, pp. 9060–9069 (2020)
45. Qu, L., He, S., Zhang, J., Tian, J., Tang, Y., Yang, Q.: RGBD salient object detection via deep fusion. IEEE Trans. Image Process. **26**(5), 2274–2285 (2017)
46. Simonyan, K., Zisserman, A.: Very deep convolutional networks for large-scale image recognition. In: Proceedings of the International Conference on Learning Representations (2015)
47. Song, H., Liu, Z., Du, H., Sun, G., Le Meur, O., Ren, T.: Depth-aware salient object detection and segmentation via multiscale discriminative saliency fusion and bootstrap learning. IEEE Trans. Image Process. **26**(9), 4204–4216 (2017)
48. Tang, J., Fan, D., Wang, X., Tu, Z., Li, C.: RGBT salient object detection: benchmark and a novel cooperative ranking approach. IEEE Trans. Circ. Syst. Video Technol. (2019)
49. Wang, W., Shen, J., Cheng, M.M., Shao, L.: An iterative and cooperative top-down and bottom-up inference network for salient object detection. In: Proceedings of the IEEE Conference on Computer Vision and Pattern Recognition, pp. 5968–5977 (2019)
50. Wang, W., Shen, J., Dong, X., Borji, A.: Salient object detection driven by fixation prediction. In: Proceedings of the IEEE Conference on Computer Vision and Pattern Recognition, pp. 1711–1720 (2018)
51. Wei, Y., et al.: STC: a simple to complex framework for weakly-supervised semantic segmentation. IEEE Trans. Pattern Anal. Mach. Intell. **39**(11), 2314–2320 (2016)
52. Wu, Z., Su, L., Huang, Q.: Cascaded partial decoder for fast and accurate salient object detection. In: Proceedings of the IEEE Conference on Computer Vision and Pattern Recognition, pp. 3907–3916 (2019)
53. Zeng, Y., Zhang, P., Zhang, J., Lin, Z., Lu, H.: Towards high-resolution salient object detection. In: Proceedings of the IEEE International Conference on Computer Vision, pp. 7234–7243 (2019)
54. Zhang, H., Zhang, J., Koniusz, P.: Few-shot learning via saliency-guided hallucination of samples. In: Proceedings of the IEEE Conference on Computer Vision and Pattern Recognition, pp. 2770–2779 (2019)
55. Zhang, J., et al.: UC-Net: uncertainty inspired RGB-D saliency detection via conditional variational autoencoders. In: Proceedings of the IEEE Conference on Computer Vision and Pattern Recognition, pp. 8582–8591 (2020)
56. Zhang, J., Yu, X., Li, A., Song, P., Liu, B., Dai, Y.: Weakly-supervised salient object detection via scribble annotations. In: Proceedings of the IEEE Conference on Computer Vision and Pattern Recognition, pp. 12546–12555 (2020)
57. Zhang, L., Yang, C., Lu, H., Ruan, X., Yang, M.H.: Ranking saliency. IEEE Trans. Pattern Anal. Mach. Intell. **39**(9), 1892–1904 (2016)

58. Zhang, L., Zhang, J., Lin, Z., Lu, H., He, Y.: CapSal: leveraging captioning to boost semantics for salient object detection. In: Proceedings of the IEEE Conference on Computer Vision and Pattern Recognition, pp. 6024–6033 (2019)
59. Zhang, M., Ren, W., Piao, Y., Rong, Z., Lu, H.: Select, supplement and focus for RGB-D saliency detection. In: Proceedings of the IEEE Conference on Computer Vision and Pattern Recognition, pp. 3472–3481 (2020)
60. Zhang, X., Wang, T., Qi, J., Lu, H., Wang, G.: Progressive attention guided recurrent network for salient object detection. In: Proceedings of the IEEE Conference on Computer Vision and Pattern Recognition, pp. 714–722 (2018)
61. Zhang, Y., Tian, Y., Kong, Y., Zhong, B., Fu, Y.: Residual dense network for image restoration. IEEE Trans. Pattern Anal. Mach. Intell. (2020)
62. Zhao, J.X., Cao, Y., Fan, D.P., Cheng, M.M., Li, X.Y., Zhang, L.: Contrast prior and fluid pyramid integration for RGBD salient object detection. In: Proceedings of the IEEE Conference on Computer Vision and Pattern Recognition, pp. 3927–3936 (2019)
63. Zhao, J.X., Liu, J.J., Fan, D.P., Cao, Y., Yang, J., Cheng, M.M.: EGNet: edge guidance network for salient object detection. In: Proceedings of the IEEE International Conference on Computer Vision, pp. 8779–8788 (2019)

Bottom-Up Temporal Action Localization with Mutual Regularization

Peisen Zhao[1], Lingxi Xie[2], Chen Ju[1], Ya Zhang[1(✉)], Yanfeng Wang[1], and Qi Tian[2]

[1] Cooperative Medianet Innovation Center, Shanghai Jiao Tong University, Shanghai, China
{pszhao,ju_chen,ya_zhang,wangyanfeng}@sjtu.edu.cn
[2] Huawei Inc., Shenzhen, China
198808xc@gmail.com, tian.qi1@huawei.com

Abstract. Recently, temporal action localization (TAL), *i.e.*, finding specific action segments in untrimmed videos, has attracted increasing attentions of the computer vision community. State-of-the-art solutions for TAL involves evaluating the frame-level probabilities of three action-indicating phases, *i.e.* starting, continuing, and ending; and then post-processing these predictions for the final localization. This paper delves deep into this mechanism, and argues that existing methods, by modeling these phases as individual classification tasks, ignored the potential temporal constraints between them. This can lead to incorrect and/or inconsistent predictions when some frames of the video input lack sufficient discriminative information. To alleviate this problem, we introduce two regularization terms to mutually regularize the learning procedure: the Intra-phase Consistency (IntraC) regularization is proposed to make the predictions verified inside each phase; and the Inter-phase Consistency (InterC) regularization is proposed to keep consistency between these phases. Jointly optimizing these two terms, the entire framework is aware of these potential constraints during an end-to-end optimization process. Experiments are performed on two popular TAL datasets, THUMOS14 and ActivityNet1.3. Our approach clearly outperforms the baseline both quantitatively and qualitatively. The proposed regularization also generalizes to other TAL methods (*e.g.*, TSA-Net and PGCN). Code: https://github.com/PeisenZhao/Bottom-Up-TAL-with-MR.

Keywords: Action localization · Action proposals · Mutual regularization

1 Introduction

Temporal Action Localization (TAL), aiming to locate action instances from untrimmed videos, is a fundamental task in video content analysis. TAL can be

Electronic supplementary material The online version of this chapter (https://doi.org/10.1007/978-3-030-58598-3_32) contains supplementary material, which is available to authorized users.

A. Vedaldi et al. (Eds.): ECCV 2020, LNCS 12353, pp. 539–555, 2020.
https://doi.org/10.1007/978-3-030-58598-3_32

divided into two parts, temporal action proposal and action classification. The latter is relatively well studied with cogent performance achieved by recent action classifiers [6,28,34,35,37]. To improve the performance in standard benchmarks [5,16], how to generate precise action proposals remains a challenge.

Early approaches for generating action proposals mostly adopt a **top-down** approach, i.e., first generate regularly distributed proposals (*e.g.*, multi-scale sliding windows), and then evaluate their confidence. However, the top-down methods [3,4,9,13,32] often suffer from over-generating candidate proposals and rigid proposal boundaries. To solve the above problem, **bottom-up** approaches have been proposed [15,22,23,25,38]. A typical bottom-up method first densely evaluates the frame-level probabilities of three action-indicating phases, i.e. starting, continuing, and ending; then groups action proposals based on the located candidate starting and ending points. This design paradigm enables flexible action proposal generation and achieves a high recall with fewer proposals [38], which has become a more preferred practice in temporal action proposals.

Predicting the frame-level probability of the starting, continuing, and ending phases of actions is crucial for the success of bottom-up approaches. Existing methods model it as three binary classification tasks and use frame-level positive and negative labels converted from action temporal location as supervision, which can suffer the difficulty of learning from limited and/or ambiguous training data. In particular, it is often difficult to determine the accurate time that an action starts or ends, and even when the action continues, there is no guarantee that every frame contains sufficient information of being correctly classified. In other words, one may need to refer to complementary information to judge the status of an action, *e.g.*, if there is no clear sign that an action has ended, the probability that it is continuing is high. Ignoring such temporal relationship may lead to erroneous and inconsistent predictions. Thus, independent classification tasks have the following two drawbacks. **First**, each temporal location is considered as an isolated instance and their probabilities are calculated independently, without considering the temporal relationship among them. In fact, for any of the three phases, the probability is expected to have relatively smooth predictions among contiguous temporal locations. Ignoring the temporal relationship may leads to inconsistent predictions. **Second**, the modeling of the probability for starting, continuing, and ending phases are independent of each other. In fact, for any action, the starting, continuing, and ending phases always come as an ordered triplet. Ignoring the ordering relationship of the three phases could lead to contradictory predictions.

In this paper, we address this issue explicitly by exploring two regularization terms. To enforces the temporal relationship among predictions, **Intra-phase Consistency** (IntraC) regularization is proposed, which targets to minimize the discrepancy inside *positive* or *negative* regions of each phases, and maximize the discrepancy between *positive* and *negative* regions. To meet the ordering constraint of the three phases, we introduce **Inter-phase Consistency** (InterC) regularization, which enforces consistency among the probability of the three phases, by operating between continuing-starting and continuing-ending. When introducing the above two regularization terms to the original loss of bottom-up temporal action localization network, the optimization of IntraC and InterC may be considered as a form of mutual regularization among the three classifiers, since the predictions of the three phases are now coupled via consistency check on classifier outputs. With the above mutual regularization, the entire framework remains end-to-end trainable while enforcing the above constraints.

To validate the effectiveness of the proposed method, we perform experiments on two popular benchmark datasets, THUMOS14 and ActivityNet1.3. Our experimental results have demonstrated that our approach clearly outperforms the state-of-the-arts both quantitatively and qualitatively. Especially on THUMOS14 dataset, we improve absolute 6.8% mAP at a strict IoU of 0.7 settings from the previous best. Moreover, we show that the proposed mutual regularization is independent of the temporal action localization framework. When we introduce IntraC and InterC to other network (TSA-Net [15]) or framework (PGCN [41]), better performance is also achieved.

2 Related Work

Action Recognition. Same as image recognition in image analysis, action recognition is a fundamental task in video domain. Extensive models [6,27,33–35,37] on action recognition have been widely studied. Deeper models [6,21,28], more massive datasets [1,17,18,26], and smarter supervision [10,11] have promoted the development of this direction. These action recognition approaches are based on trimmed videos, which are not suitable for untrimmed videos due to the considerable duration of the background. However, the pre-trained models on action recognition task can provide effective feature representation for temporal action localization task. In this paper, we use the I3D model [6], pre-trained on Kinetics [18], to extract video features.

Temporal Action Localization. Temporal action localization is a mirror problem of image object detection [29,30] in the temporal domain. The TAL task can be decomposed into proposal generation and classification stage, same as the two-stage approach of object detection. Recent methods for proposal generation are divided into two branches, top-down and bottom-up fashions. Top-down approaches [3,4,7–9,13,32,39] generated proposals with pre-defined regularly distributed segments then evaluated the confidence of each proposal. The boundary of top-down proposals are not flexible, and these generation strategies often cause extensive false positive proposals, which will introduce burdens in the classification stage. However, the other bottom-up approaches alleviated this problem and achieved the new state-of-the-art. TAG [38] was an early study of bottom-up fashion, which used frame-level action probabilities to group action proposals. Lin *et al.* proposed the multi-stage BSN [23] and end-to-end BMN [22] models via locating temporal boundaries to generate action proposals. *Gong al.* [15] also predicted action probabilities to generate action proposals from the perspective of multi scales. Zeng *et al.* proposed the PGCN [41] to model the proposal-proposal relations based on bottom-up proposals. Combined top-down

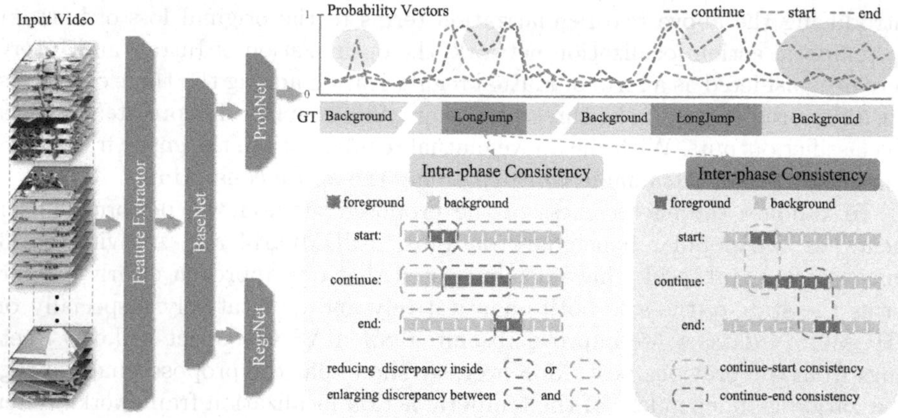

Fig. 1. Schematic of our approach. Three probability phases are predicted by the ProbNet. Intra-phase Consistency loss is built inside each phase by first separating *positive* and *negative* regions, then reduce the discrepancy inside *positive* or *negative*, and enlarge the discrepancy between *positive* and *negative*. Inter-phase Consistency loss is built between the continue-start phase and the continue-end phase. (Color figure online)

and bottom-up fashions, Liu *et al.* proposed a MGG [25] model, which takes advantage of frame-level action probability as well. [40] is relevant to out study that enforced the temporal structure by maximizing the top-K summation of the confidence scores of the starting, continuing, and ending.

3 Method

3.1 Problem and Baseline

Notations. Given an Untrimmed video, we denote $\{\mathbf{f}_t\}_{t=1}^{T}$ as a feature sequence to represent a video, where T is the length of the video and $\mathbf{f}_t$ is the t-th feature vector extracted from continuous RGB frames and optical flows. Annotations are $\varphi = \{(t_{s,n}, t_{e,n}, a_n)\}_{n=1}^{N}$, where $t_{s,n}$, $t_{e,n}$, and a_n are start time, end time, and class label of the action instance n. N is the number of action annotations. Following previous studies [15,22,23,25], we predict continuing, starting, and ending probability vectors $\mathbf{p}^{\mathrm{C}} \in [0,1]^T$, $\mathbf{p}^{\mathrm{S}} \in [0,1]^T$, and $\mathbf{p}^{\mathrm{E}} \in [0,1]^T$ to generate action proposals. Correspondingly, the ground-truth labels are generated via φ, which are notated by $\mathbf{g}^{\mathrm{C}} \in \{0,1\}^T$, $\mathbf{g}^{\mathrm{S}} \in \{0,1\}^T$, and $\mathbf{g}^{\mathrm{E}} \in \{0,1\}^T$, respectively. Continuing ground-truth $\mathbf{g}^{\mathrm{C}}$ has value "1" inside the action instances $[t_{s,n}, t_{e,n}]$, while starting and ending points are expanded to a region $[t_{s,n} - \delta_n, t_{s,n} + \delta_n]$ and $[t_{e,n} - \delta_n, t_{e,n} + \delta_n]$ to assign the ground-truth label $\mathbf{g}^{\mathrm{S}}$ and $\mathbf{g}^{\mathrm{E}}$. δ_n is set to be 0.1 duration of the action instance n, same as [15,22,23].

Baseline. This paper takes the typical bottom-up TAL framework as our baseline, such as BSN [23]. As illustrated in Fig. 1, the baseline network is trained without Intra-phase Consistency and Inter-phase Consistency. We first use 3D convolutional network to extract video features $\{\mathbf{f}_t\}_{t=1}^T$, then feed the feature sequence to several 1D convolutional networks to (i) predict three probability vectors ($\mathbf{p}^{\mathrm{C}}$, $\mathbf{p}^{\mathrm{S}}$, and $\mathbf{p}^{\mathrm{E}}$) by ProbNet, (ii) predict the starting and ending boundary offsets ($\hat{\mathbf{o}}_{\mathrm{S}}$ and $\hat{\mathbf{o}}_{\mathrm{E}}$) by RegrNet. Finally, we generate proposals by combining start-end pairs with high probabilities and classify these candidate proposals.

3.2 Motivation: Avoiding Ambiguity with Temporal Consistency

The first and fundamental procedure in bottom-up TAL is to predict frame-level probabilities of three action-indicating phases, *i.e.* starting, continuing, and ending. Existing approaches use frame-level labels, $\mathbf{g}^{\mathrm{C}} \in \{0,1\}^T$, $\mathbf{g}^{\mathrm{S}} \in \{0,1\}^T$, and $\mathbf{g}^{\mathrm{E}} \in \{0,1\}^T$ to train three binary classification tasks. Since the meaning of "starting", "continuing", and "ending" have certain ambiguity, it is hard to determine the accurate time that an action starts, ends, and continues. Moreover, we find that even in training set, the **False Alarm** of these binary classification tasks reaches 68%, 64%, and 28% for starting, ending, and continuing, respectively. As shown in Fig. 1, we can also observe that the continuing phase in green are not stable inside an action instance "LongJump" or background (*yellow* circles); and different action phases are not support each other (*red* circle). Thus, only supervised by classification labels is hard to optimize these problem, because there is no guarantee that every frame contains sufficient information of being correctly classified.

Therefore, to better regularize the learning process of avoiding ambiguity, we propose two consistency regularization terms during an end-to-end optimization, that consider the relations between different temporal locations inside each probability phase, named **Intra-phase Consistency** (IntraC) and the relations among different probability phases, named **Inter-phase Consistency** (InterC).

3.3 Adding Mutual Regularization

As illustrated in Fig. 1, we add two consistency losses, IntraC and InterC, to regularize the learning process. IntraC is built inside each phase by first separating *positive* and *negative* regions, then reduce the discrepancy inside *positive* or *negative*, and enlarge the discrepancy between *positive* and *negative*. InterC performs consistency among three phases, which operates between continuing-starting and continuing-ending, **(i)** if there were an abrupt rise in the continuing phase, the starting phase should give a high probability, and vise versa; **(ii)** if there were an abrupt drop in the continuing phase, the ending phase should give a high probability, and vise versa.

Intra-phase Consistency. We build our Intra-phase Consistency loss inside each per-frame probability phase of start, end, and continuing. Firstly, we show the detailed operations for continuing phase $\mathbf{p}^{\mathrm{C}}$. The yellow block in Fig. 1 shows an example of the IntraC on continuing phase $\mathbf{p}^{\mathrm{C}}$. To make the per-frame predictions supervised by their context predictions, we first define the *positive* and *negative* regions. The *positive* regions are defined as the locations where action continues by $g_t^{\mathrm{C}} = 1$, and the *negative* regions are the rest of the time where $g_t^{\mathrm{C}} = 0$. In terms of the division of the *positive* and *negative* region, the predicted continuing probabilities $\{p_t^{\mathrm{C}}\}_{t=1}^{T}$ are divided into a positive set $\mathcal{U}^{\mathrm{C}} = \{p_t^{\mathrm{C}} \mid g_t^{\mathrm{C}} = 1\}$ and a negative set $\mathcal{V}^{\mathrm{C}} = \{p_t^{\mathrm{C}} \mid g_t^{\mathrm{C}} = 0\}$. To make each prediction is not only supervised by its own label but other context labels, we optimize this problem by **(i)** $\min f(p_i^{\mathrm{C}}, p_j^{\mathrm{C}}), \forall p_i^{\mathrm{C}} \in \mathcal{U}^{\mathrm{C}}.\forall p_j^{\mathrm{C}} \in \mathcal{U}^{\mathrm{C}}$ **(ii)** $\max f(p_i^{\mathrm{C}}, p_j^{\mathrm{C}}), \forall p_i^{\mathrm{C}} \in \mathcal{U}^{\mathrm{C}}.\forall p_j^{\mathrm{C}} \in \mathcal{V}^{\mathrm{C}}$, where f is a distance function (l_1 distance in our experiments) to measure the difference between p_i^{C} and p_j^{C}. Therefore, the IntraC on continuing probability phase $\mathbf{p}^{\mathrm{C}}$ is formulated in Eq. (1):

$$\mathcal{L}_{\mathrm{Intra}^{\mathrm{C}}} = \frac{1}{N_{\mathrm{U}}} \sum_{i,j} (\mathbf{A} \odot \mathbf{M}_{\mathrm{U}})_{i,j} + \frac{1}{N_{\mathrm{V}}} \sum_{i,j} (\mathbf{A} \odot \mathbf{M}_{\mathrm{V}})_{i,j} + (1 - \frac{1}{N_{\mathrm{UV}}} \sum_{i,j} (\mathbf{A} \odot \mathbf{M}_{\mathrm{UV}})_{i,j}), \tag{1}$$

where $\mathbf{A} \in [0,1]^{T \times T}$ is an adjacency matrix to establish the relationship between predicted probabilities by measuring the distance between them. The elements in $\mathbf{A}$ are formulated as $a_{i,j} = f(p_i^{\mathrm{C}}, p_j^{\mathrm{C}})$. $\mathbf{M}_{\mathrm{U}}$, $\mathbf{M}_{\mathrm{V}}$, and $\mathbf{M}_{\mathrm{UV}} \in \{0,1\}^{T \times T}$ are three masks to select the corresponding pairs $a_{i,j}$ in adjacency matrix $\mathbf{A}$ from $\mathcal{U}^{\mathrm{C}}$ set, $\mathcal{V}^{\mathrm{C}}$ set, and between $\mathcal{U}^{\mathrm{C}}$ and $\mathcal{V}^{\mathrm{C}}$ sets, respectively. The constants N_{U}, N_{V}, and N_{UV} represent the number of "1" in each mask matrix. $\odot$ stand for the element-wise product.

Following this intra consistency between different frame-predictions, we reduce the discrepancy inside *positive* or *negative*, and enlarge the discrepancy between them. Replicating IntraC loss on continuing phase, we can also obtain the $\mathcal{L}_{\mathrm{IC}^{\mathrm{S}}}$ and $\mathcal{L}_{\mathrm{IC}^{\mathrm{E}}}$. Hence, the whole IntraC loss is formulated in Eq. (2):

$$\mathcal{L}_{\mathrm{Intra}} = \mathcal{L}_{\mathrm{Intra}^{\mathrm{C}}} + \mathcal{L}_{\mathrm{Intra}^{\mathrm{S}}} + \mathcal{L}_{\mathrm{Intra}^{\mathrm{E}}}. \tag{2}$$

Inter-phase Consistency. We build our Inter-phase Consistency loss between three probability phases, continuing phase $\mathbf{p}^{\mathrm{C}}$, starting phase $\mathbf{p}^{\mathrm{S}}$, and ending phase $\mathbf{p}^{\mathrm{E}}$. To make the consistency between these probability phases, we propose two hypotheses, (i) if there were an abrupt rise in the continuing phase, the starting phase should give a high probability, and vise versa; (ii) if there were an abrupt drop in the continuing phase, the ending phase should give a high probability, and vise versa. Following these hypotheses, we use the first difference term of $\mathbf{p}^{\mathrm{C}}$ to capture the abrupt rise and drop of the continuing probability phase: $\Delta\mathbf{p}^{\mathrm{C}} = p_{t+1}^{\mathrm{C}} - p_t^{\mathrm{C}}$.

As illustrated in red block of Figure 1, we build two kinds of constraints for InterC, the continue-start constraint in yellow circle and the continue-end constraint in blue circle. We use the positive values in $\Delta\mathbf{p}^{\mathrm{C}}$ to represent continuing probability rise rate, notated as $p_t^{+} = \max\{0, \Delta p_t^{\mathrm{C}}\}$, and use negative values in $\Delta\mathbf{p}^{\mathrm{C}}$ to represent continuing probability drop rate, notated as $p_t^{-} = -\min\{0, \Delta p_t^{\mathrm{C}}\}$. Thus, to make predictions of continuing, starting, and ending support each other, we optimize this problem by **(i)** $\min f(p_t^{+}, p_t^{\mathrm{S}})$ and **(ii)** $\min f(p_t^{-}, p_t^{\mathrm{E}})$, where f is a distance function (l_1 distance in our experiments) to measure the distance. Then the InterC is formulated in Eq. (3):

$$\mathcal{L}_{\text{Inter}} = \frac{1}{T}\sum_{t=1}^{T} \mid p_t^{+} - p_t^{\text{S}} \mid + \mid p_t^{-} - p_t^{\text{E}} \mid . \tag{3}$$

Loss Function. Predicting continuing, starting, and ending probabilities are trained with the cross-entropy loss. We separate the calculation by the *positive* and *negative* regions; then mix them with a ratio of 1:1 to balance the proportion of the *positive* and the *negative*. The loss of predicting the continuing probability is formulated in Eq. (4):

$$\mathcal{L}_{\text{C}} = \frac{1}{T_{\text{C}}^{+}}\sum_{t\in\mathcal{U}^{\text{C}}} \ln(p_t^{\text{C}}) + \frac{1}{T_{\text{C}}^{-}}\sum_{t\in\mathcal{V}^{\text{C}}} \ln(1-p_t^{\text{C}}), \tag{4}$$

where $\mathcal{U}^{\text{C}}$ and $\mathcal{V}^{\text{C}}$ denote the *positive* and *negative* set in $\mathbf{p}^{\text{C}}$, while T_{C}^{+} and T_{C}^{-} are the number of them, respectively. Replacing the script "C" with "S" or "E" in Eq. (4), we can obtain the $\mathcal{L}_{\text{S}}$ and $\mathcal{L}_{\text{E}}$, respectively. Hence, the whole classification loss is formulated as: $\mathcal{L}_{\text{cls}} = \mathcal{L}_{\text{C}} + \mathcal{L}_{\text{S}} + \mathcal{L}_{\text{E}}$.

To make the action boundaries more precise, we also introduce a regression task to predict the starting and ending boundary offsets. Inspired by some object detection studies [20,30], we apply SmoothL1 Loss [14] (SL_1) to our regression task, which is formulated in Eq. (5):

$$\mathcal{L}_{\text{reg}} = \frac{1}{T_{\text{S}}^{+}}\sum_{t\in\mathcal{U}^{\text{S}}} \text{SL}_1(o_t^{\text{S}}, \hat{o_t}^{\text{S}}) + \frac{1}{T_{\text{E}}^{+}}\sum_{t\in\mathcal{U}^{\text{E}}} \text{SL}_1(o_t^{\text{E}}, \hat{o_t}^{\text{E}}), \tag{5}$$

where $\mathcal{U}^{\text{S}}$ and $\mathcal{U}^{\text{E}}$ are the *positive* regions in $\mathbf{p}^{\text{S}}$ and $\mathbf{p}^{\text{E}}$. T_{S}^{+} and T_{E}^{+} are the number of them. $\hat{o_t}^{\text{S}}$ and $\hat{o_t}^{\text{E}}$ are the predicted starting and ending offsets with their ground-truth (o_t^{S} and o_t^{E}). Adding our proposed consistency constrains IntraC and InterC, the overall objective loss function is formulated in Eq. (6):

$$\mathcal{L} = \mathcal{L}_{\text{cls}} + \mathcal{L}_{\text{reg}} + \mathcal{L}_{\text{Intra}} + \mathcal{L}_{\text{Inter}}. \tag{6}$$

3.4 Inference: Proposal Generation and Classification

Following the same rules in BSN [23] and ScaleMatters [15], we select the starting and ending points in terms of $\mathbf{p}^{\text{S}}$ and $\mathbf{p}^{\text{E}}$; then combine them to generate action proposals; finally rank these proposals and classify them with action labels. Operations are conducted sequentially:

Proposal Generation. To generate action proposals, we first select the candidate starting and ending points with predicted $\mathbf{p}^{\text{S}}$ and $\mathbf{p}^{\text{E}}$ by two rules [23]: (i) start points t where $p_t^{\text{S}} > 0.5 \times (\max_{t=1}^{T}\{p_t^{\text{S}}\} + \min_{t=1}^{T}\{p_t^{\text{S}}\})$; (ii) start points t where $p_{t-1}^{\text{S}} < p_t^{\text{S}} < p_{t+1}^{\text{S}}$. The ending points are selected by the same rules. Following these two rules, we obtain starting and ending candidates which have high probability or stay at a peak position. Combining these points under a maximum action duration in training set, we obtain the candidate proposals.

Proposal Ranking. To rank action proposals with a confidence score, we provide two methods: (i) directly use the product of the starting and ending probabilities, $p^{S}_{t_s} \times p^{E}_{t_e}$. (ii) train an additional evaluation network to score candidate proposals [15], which is noted as $\phi(t_s, t_e)$. The detailed information can be found in [15]. Thus, the final confidence score for candidate proposals is $p^{S}_{t_s} \times p^{E}_{t_e} \times \phi(t_s, t_e)$.

Redundant Proposal Suppression. After generating candidate proposals with the confidence score, we need to remove redundant proposals with high overlaps. Standard method such as soft non-maximum suppression (Soft-NMS) [2] is used in our experiments. Soft-NMS decays the confidence score of proposals which are highly overlapped. Finally, we suppress the redundant proposals to achieve a higher recall.

Proposal Classification. The last step of temporal action localization is to classify the candidate proposals. For fair comparison with other temporal localization methods, we use the same classifiers to report our action localization results. Following BSN [23], we use video-level classifier in UntrimmedNet [36] for THUMOS14 dataset. As for ActivityNet1.3 dataset, we use the video-level classification results generated by [42].

3.5 Implementation Details

Network Design. We build our IntraC and InterC on a succinct baseline model with all 1D Convolution layers and the detailed network architecture is shown in Table 1. The input of BaseNet is extracted feature sequence $\{\mathbf{f}_t\}_{t=1}^{T}$ of untrimmed videos. Since untrimmed videos have various video length, we truncate or pad zeros to obtain a fixed length features of window l_{w}. Through BaseNet, the output features are shared by three 2-layer ProbNets to predict probability phases ($\mathbf{p}^{C}$, $\mathbf{p}^{S}$, and $\mathbf{p}^{E}$) and two RegrNets to predict starting and ending boundary offsets ($\hat{\mathbf{o}}_{\mathrm{S}}$ and $\hat{\mathbf{o}}_{\mathrm{E}}$).

Table 1. The detailed network architecture. The output of BaseNet is shared by ProbNet and RegrNet. Three ProbNets (× 3) are used to predict continuing, starting, and ending probability phases. Two RegrNets (× 2) are used to predict starting and ending offsets.

Name	Layer	Kernel	Channels	Activation
BaseNet	Conv1D	9	512	ReLU
	Conv1D	9	512	ReLU
ProbNet (× 3)	Conv1D	5	256	ReLU
	Conv1D	5	1	Sigmoid
RegrNet (× 2)	Conv1D	5	256	ReLU
	Conv1D	5	1	Identity

Network Training. Our BaseNet, ProbNet, and RegrNet are jointly trained from scratch by multiple losses which are the classification loss ($\mathcal{L}_{\text{cls}}$), regression loss ($\mathcal{L}_{\text{reg}}$) and consistency losses ($\mathcal{L}_{\text{Intra}}$ and $\mathcal{L}_{\text{Inter}}$). We find setting the ratio of each loss component equal get relatively proper numerical values and the loss curve can converge well. As mentioned previous, to contain most action instances in a fixed observed window, the input feature length of window l_{w} is set to be 750 for THUMOS14 and scaled to be 100 for ActivityNet1.3. The training process lasts for 20 epochs with a learning rate of 10^{-3} in former 10 epochs and 10^{-4}

in latter 10 epochs. The batch size is set to be 3 for THUMOS14 and 16 for the ActivityNet1.3. We use a SGD optimization method with a momentum of 0.9 to train both datasets. In Sect. 3.4, the additional evaluation network for proposal ranking follows the same settings in [15].

4 Experiments

4.1 Datasets and Evaluation Metrics

Datasets and Features. We validate our proposed IntraC and InterC on two standard datasets: **THUMOS14** includes 413 untrimmed videos with 20 action classes. According to the public split, 200 of them are used for training, and 213 are used for testing. There are more than 15 action annotations in each video; **ActivityNet1.3** is a more considerable action localization dataset with 200 classes annotated. The entire 19,994 untrimmed videos are divided into training, validation, and testing sets by ratio 2:1:1. Each video has around 1.5 action instances. To make a fair comparison with the previous work, we use the same two-stream features of these datasets. The two-stream features, which are provided by [24], are extracted by I3D network [6] pre-trained on Kinetics.

Metric for Temporal Action Proposals. To evaluate the quality of action proposals, we use conventional metrics Average Recall (AR) with different Average Number (AN) of proposals AR@AN for action proposals. On THUMOS14 dataset, the AR is calculated under multiple IoU threshold set from 0.5 to 1.0 with a stride of 0.05. As for ActivityNet1.3 dataset the multiple IoU threshold are from 0.5 to 0.95 with a stride of 0.05. Besides, we also use the area under the AR-AN curve (AUC) to evaluate the performance.

Metric for Temporal Action Localization. To evaluate the performance of action localization, we use mean Average Precision (mAP) metric. On THUMOS14 dataset, we report the mAP with multiple IoUs in set $\{0.3, 0.4, 0.5, 0.6, 0.7\}$. As for ActivityNet1.3 dataset, the IoU set is $\{0.5, 0.7, 0.95\}$. Moreover, we also report the averaged mAP where the IoU is from 0.5 to 0.95 with a stride of 0.05.

4.2 Comparison to the State-of-the-arts

Temporal Action Proposals. We compare the temporal action proposals generated by our IntraC and InterC equipped model on THUMOS14 and ActivityNet1.3 dataset. As illustrated in Table 2, comparing with previous works, we can achieve the best performance especially on AR@50 metric. Our consistency losses help to generate more precise candidate starting and ending points, so we can achieve a high recall with fewer proposals. In Table 4, we also achieve comparable results on ActivityNet1.3, since it is a well studied dataset.

Temporal Action Localization. Classifying the proposed proposals, we obtain the final localization results. As illustrated in Table 3 and Table 5, our method

Table 2. Comparisons in terms of AR@AN (%) on THUMOS14.

Method	@50	@100	@200
TAG [38]	18.55	29.00	39.41
CTAP [12]	32.49	42.61	51.97
BSN [23]	37.46	46.06	53.21
BMN [22]	39.36	47.72	54.70
MGG [25]	39.93	47.75	54.65
TSA-Net [15]	42.83	49.61	54.52
Ours	**44.23**	**50.67**	**55.74**

Table 3. Comparisons in terms of mAP (%) on THUMOS14.

Method	0.3	0.4	0.5	0.6	0.7
SST [3]	41.2	31.5	20.0	0.9	4.7
TURN [13]	46.3	35.3	24.5	14.1	6.3
BSN [23]	53.5	45.0	36.9	28.4	20.0
MGG [25]	53.9	46.8	37.4	29.5	21.3
BMN [22]	**56.0**	47.4	38.8	29.7	20.5
TSA-Net [15]	53.2	48.1	41.5	31.5	21.7
Ours	53.9	**50.7**	**45.4**	**38.0**	**28.5**

Table 4. Comparisons in terms of AUC and AR@100 (%) on ActivityNet1.3.

Method	AUC	AR@100
TCN [8]	59.58	-
CTAP [12]	65.72	73.17
BSN [23]	66.17	74.16
MGG [25]	66.43	74.54
Ours	**66.51**	**75.27**

Table 5. Comparisons in terms of mAP (%) on ActivityNet1.3 (val). "Average" is caculated at the IoU of $\{0.5 : 0.05 : 0.95\}$.

Method	0.5	0.7	0.95	Average
CDC [31]	43.83	25.88	0.21	22.77
SSN [43]	39.12	23.48	5.49	23.98
BSN [23]	**46.45**	29.96	8.02	29.17
Ours	43.47	**33.91**	**9.21**	**30.12**

outperforms the previous studies. Especially at **high IoU settings**, we achieve significant improvements since our consistency loss can make the boundaries more precise. On THUMOS14 dataset, the mAP at IoU of 0.6 is improved from 31.5% to 38.0% and the mAP at IoU of 0.7 is improved from 21.7% to 28.5%. On ActivityNet1.3 dataset, we can achieve the mAP to 9.21% at IoU of 0.95.

Generalizing IntraC&InterC to Other Algorithms. Our proposed two consistency losses, *i.e.*, IntraC and InterC, are effective in generating the probability phases of continuing, starting, and ending. To prove these consistency losses are valid for other network architecture and framework in TAL, we introduce them to TSA-Net [15] and PGCN [41], respectively. **TSA-Net** [15] designed a multi-scale architecture to predict probability phases of continuing, starting and ending. We introduce our IntraC and InterC to their multi-scale networks, TSA-Net-small, TSA-Net-medium, and TSA-Net-large, respectively. As illustrated in Table 6, our IntraC and InterC significantly outperforms the baseline models on all three network architectures. **PGCN** [41] explore the proposal-proposal relations using Graph Convolutional Networks [19] (GCN) to localize action instances. This framework builds upon the prepared proposals from BSN [23] method. We introduce our two consistency losses to generated candidate proposals for PGCN framework. As illustrated in Table 7, introducing IntraC and InterC to PGCN also improves the localization performance.

Table 6. Generalizing IntraC&InterC to multi-scale TSA-Net [15] in terms of AR@AN (%) on THUMOS14. * indicates the results that are implemented by ours.

TSA-Net	AR@50	AR@100	AR@200
Small (Small*)	37.72 (38.32)	45.85 (46.15)	52.03 (52.39)
Small* + IntraC&InterC	**39.73**	**47.69**	**53.48**
Medium (Medium*)	37.77 (39.20)	45.01 (47.17)	50.38 (53.46)
Medium* + IntraC&InterC	**40.05**	**47.53**	**53.88**
Large (Large*)	36.07 (37.91)	44.28 (45.89)	50.80 (52.36)
Large* + IntraC& InterC	**39.68**	**47.47**	**53.50**

Table 7. Generalizing IntraC&InterC to PGCN [41] in terms of mAP (%) on THUMOS14. * indicates the results that are implemented by ours.

Method	0.1	0.2	0.3	0.4	0.5
PGCN	69.50	67.80	63.60	57.80	49.10
PGCN*	69.26	67.76	63.73	58.82	48.88
PGCN* + IntraC& InterC	**71.83**	**70.31**	**66.29**	**60.99**	**50.10**

4.3 Ablation Studies

As mentioned in dataset description, THUMOS has 10 times action instances per video than ActivityNet (only has 1.5 action instances per video) and THUMOS video also contains a larger portion of background. More instances and more background are challenge for detection task. Thus we conduct following detailed ablation studies on THUMOS14 dataset to explore how these constrains, IntraC and InterC, improve the quality of temporal action proposals.

Effectiveness of IntraC. As illustrated in Table 8 "Intra Consistency", we compare the components of IntraC in terms of the AR@AN. The IntraC is introduced to continuing probability phase ($\mathcal{L}_{\text{Intra}^{\text{C}}}$), starting probability phase ($\mathcal{L}_{\text{Intra}^{\text{S}}}$), and ending probability phase ($\mathcal{L}_{\text{Intra}^{\text{E}}}$). Compared with the baseline result without any consistency losses, introducing continuing $\mathcal{L}_{\text{Intra}^{\text{C}}}$ or starting $\mathcal{L}_{\text{Intra}^{\text{S}}}$ and ending $\mathcal{L}_{\text{Intra}^{\text{E}}}$ can both achieve better results. Combined all three IntraC losses, the AR@50 is improved from 39.02% to 41.91%.

Effectiveness of InterC. As illustrated in Table 8 "Inter Consistency", we compare the components of InterC losses in terms of the AR@AN. The InterC is introduced between continue-start (C&S) and continue-end (C&E). InterC on C&S (C&E) makes the consistency between the starting phase (ending phase) and the derivative of continuing phase, which can suppress the false positives only observed from a single probability phase. Only introducing InterC to C&S or C&E obtains around 1% absolute improvement on AR@50. When combined C&S and C&E, it can improve 2.21% on AR@50.

Table 8. Ablation studies on Intra-phase Consistency and Inter-phase Consistency in terms of AR@AN (%) on THUMOS14. The baseline model is define in Table 1.

			AR@50	AR@100	AR@200
Baseline			39.02	46.26	53.09
Continue	Start	End	Intra Consistency		
✓			40.46	47.85	53.87
	✓	✓	40.86	48.26	54.16
✓	✓	✓	41.91	49.06	54.82
C&S	C&E		Inter Consistency		
✓			40.21	47.30	53.38
	✓		40.64	47.85	54.01
✓	✓		41.23	48.81	54.47
IntraC	InterC		Intra&Inter Consistency		
✓			41.91	49.06	54.82
	✓		41.23	48.81	54.47
✓	✓		**42.63**	**49.85**	**55.32**

Table 9. Ablation studies on model structures in terms of AR@AN (%) on THUMOS14. All numbers are the averaged value in the last 10 epochs.

Layers	Kernel Size	AR@50	AR@100	AR@200
2	5	40.98	48.51	54.64
3	5	41.56	49.02	54.88
4	5	41.68	48.93	54.91
5	5	40.98	48.14	54.29
2	3	39.54	47.61	53.84
2	5	40.98	48.51	54.64
2	7	41.49	49.16	55.17
2	9	**42.63**	**49.85**	**55.32**
2	11	42.48	49.32	54.97
2	13	42.17	49.41	55.21

Combining IntraC&InterC. As illustrated in Table 8 "All Consistency", we compare the IntraC and InterC losses in terms of the AR@AN. Both the IntraC and InterC independently achieve more than 2% absolute improvement on AR@50. When combined IntraC and InterC, the AR@50 is improved from 39.02% to 42.63%. Consistency inside each probability phase and between them are coupled, which leads to a positive feedback. It means when we get the better probability phase that fits the IntraC settings, the potential constraint of InterC is more appropriate between three probability phases, and vise versa.

Table 10. Ablation studies on proposal scoring in terms of AR@AN (%) on THUMOS14. Experiments are based on 2 "Layers" and 9 "Kernel Size" model in Table 9. All numbers are the averaged value in the last 10 epochs.

Proposal Scoring	AR@50	AR@100	AR@200
$p^{\mathrm{S}}_{t_s} \times p^{\mathrm{E}}_{t_e}$	42.63	49.85	55.32
$p^{\mathrm{S}}_{t_s} \times p^{\mathrm{E}}_{t_e} \times \phi(t_s, t_e)$	**44.23**	**50.67**	**55.74**

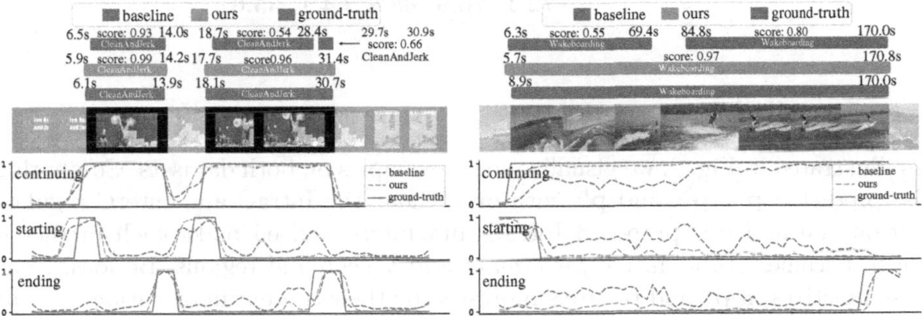

Fig. 2. Qualitative results on THUMOS14 (left) and ActivityNet1.3 (right) datasets. "green" lines are ground-truth, "blue" lines are predicted phases by baseline model and "orange" lines are optimized with IntraC and InterC regularization terms. (Color figure online)

Effectiveness of Kernel Size And Layers. The scale of the receptive field is crucial in temporal action localization tasks. So we explore different scales of receptive field by adjusting the number of layers and the kernel size of the BaseNet. As illustrated in Table 9, we compare results between different kernel sizes and layers in terms of the AR@AN. Deeper layers and larger kernel sizes often lead to a better performance, but using too many layers and/or an over-large kernel size often incurs over-fitting. We also conduct the experiments using different layers with a kernel size of 9 and find that a 2-layer network performs best, so we use this option in the main experiments. This implies that probably increasing the depth is not the best choice here.

Effectiveness of Proposal Scoring. As mentioned in Sect. 3.4, we compare two methods for scoring proposals. Once we get proposals of an untrimmed video, a proper ranking method with convincing scores can achieve the high recall with fewer proposals. As illustrated in Table 10, we compare two scoring functions, $p^{\mathrm{S}}_{t_s} \times p^{\mathrm{E}}_{t_e}$ and $p^{\mathrm{S}}_{t_s} \times p^{\mathrm{E}}_{t_e} \times \phi(t_s, t_e)$. Directly using starting and ending probability at boundaries is simple and effective, however, training a new evaluation network [15,23] to evaluate the confidence of proposals can further improve the performance by a significant margin.

Table 11. Introducing oracle information to TAL in terms of mAP (%) on THUMOS14. O_{rank} is ground-truth rank information and O_{cls} uses ground-truth class label.

O_{rank}	O_{cls}	0.3	0.4	0.5	0.6	0.7
		53.9	50.7	45.4	38.0	28.5
	✓	57.1	53.2	47.3	39.3	29.5
✓		66.4	65.4	63.8	59.9	52.7
✓	✓	**72.1**	**70.9**	**68.8**	**64.1**	**55.6**

4.4 Visualization

As illustrated in Fig. 2, we visualize some examples on both datasets. Comparing the predicted $\mathbf{p}^{\mathrm{C}}$, $\mathbf{p}^{\mathrm{S}}$, and $\mathbf{p}^{\mathrm{E}}$ with or without the IntraC and InterC regularization, we find our proposed IntraC and InterC indeed make each predicted phase becomes stable inside *foreground* and *background* regions. Besides, some false positives in $\mathbf{p}^{\mathrm{S}}$ and $\mathbf{p}^{\mathrm{E}}$ are suppressed by their context information, so that we can remove many candidate proposals of poor quality via these wrong starting and ending points. *e.g.*, the second action "CleanAndJerk" and the action "Wakeboarding" are separate by false positive starting point in baseline model. The visualization results show that only introducing binary classification labels is hard to optimize these probability phases, since it discards the potential constraints between the different temporal locations and action phases. We also perform regularization using the smoothness assumption, *i.e.*, using a Gaussian kernel to penalize local inconsistenies within $\mathbf{p}^{\mathrm{C}}$, $\mathbf{p}^{\mathrm{S}}$, and $\mathbf{p}^{\mathrm{E}}$. In experiments, this kinds of regularization does not necessarily push the positive scores to 1 and negative scores to 0, and we believe smoothness might be useful in the unsupervised or weakly-supervised TAL scenarios.

4.5 Discussion: The Upper Bounds of TAL

Most temporal action localization method can be divided into the following procedures, (i) generating proposals, (ii) ranking proposals, and (iii) classifying proposals. Which one is most awaiting to improve for the intending researchful keystone? We introduce two types of oracle information to reveal the performance gap between the different upper bounds. As illustrated in Table 11, O_{rank} means that each candidate proposal is ranked by the max IoU score with all ground-truth action instances. O_{cls} means that the ground-truth action labels are assigned to candidate proposals. When introducing O_{rank} or/and O_{cls} to our action localization baseline, it is worth to notice that proposal classification has been well solved since there is a small gap when introducing O_{cls}. However, when introducing the oracle ranking information O_{rank}, the upper bound can improve a lot from 53.9% to 66.4% in terms of mAP at IoU of 0.3. That means there is a significant untapped opportunity in how to rank the action proposals.

5 Conclusions

In this paper, we investigate the problem that frame-level probability phases of starting, continuing, and ending are not self-consistent in the bottom-up TAL approach. Our research reveals that state-of-the-art video analysis algorithms, though supervised with classification labels, mostly have a limited understanding in the temporal dimension, which can lead to undesired properties, *e.g.*, inconsistency or discontinuity. To alleviate this problem, we propose two consistency losses (IntraC and InterC) which can mutually regularize the learning process. Experiments reveal that our approach improves the performance of temporal action localization both quantitatively and qualitatively.

Our work reveals that introducing priors for self-regularization is important for learning from high-dimensional data (*e.g.*, videos). We will continue along this direction in the future, and explore the possibility of learning such priors from self-supervised data, *e.g.*, unlabeled videos.

Acknowledgments. This work is supported by the National Key Research and Development Program of China (No. 2019YFB1804304), SHEITC (No. 2018-RGZN-02046), 111 plan (No. BP0719010), and STCSM (No. 18DZ2270700), and State Key Laboratory of UHD Video and Audio Production and Presentation.

References

1. Abu-El-Haija, S., et al.: YouTube-8M: a large-scale video classification benchmark. In: arXiv preprint arXiv:1609.08675 (2016)
2. Bodla, N., Singh, B., Chellappa, R., Davis, L.S.: Soft-NMS-improving object detection with one line of code. In: Proceedings of the International Conference on Computer Vision (ICCV), pp. 5561–5569 (2017)
3. Buch, S., Escorcia, V., Shen, C., Ghanem, B., Carlos Niebles, J.: SST: single-stream temporal action proposals. In: Proceedings of the Conference on Computer Vision and Pattern Recognition (CVPR), pp. 2911–2920 (2017)
4. Caba Heilbron, F., Carlos Niebles, J., Ghanem, B.: Fast temporal activity proposals for efficient detection of human actions in untrimmed videos. In: Proceedings of the Conference on Computer Vision and Pattern Recognition (CVPR), pp. 1914–1923 (2016)
5. Caba Heilbron, F., Escorcia, V., Ghanem, B., Carlos Niebles, J.: ActivityNet: a large-scale video benchmark for human activity understanding. In: Proceedings of the IEEE Conference on Computer Vision and Pattern Recognition (CVPR), pp. 961–970 (2015)
6. Carreira, J., Zisserman, A.: Quo vadis, action recognition? A new model and the kinetics dataset. In: Proceedings of the Conference on Computer Vision and Pattern Recognition (CVPR), pp. 6299–6308 (2017)
7. Chao, Y.W., Vijayanarasimhan, S., Seybold, B., Ross, D.A., Deng, J., Sukthankar, R.: Rethinking the faster R-CNN architecture for temporal action localization. In: Proceedings of the IEEE Conference on Computer Vision and Pattern Recognition (CVPR), pp. 1130–1139 (2018)
8. Dai, X., Singh, B., Zhang, G., Davis, L.S., Qiu Chen, Y.: Temporal context network for activity localization in videos. In: Proceedings of the IEEE International Conference on Computer Vision (ICCV), pp. 5793–5802 (2017)

9. Escorcia, V., Caba Heilbron, F., Niebles, J.C., Ghanem, B.: DAPs: deep action proposals for action understanding. In: Leibe, B., Matas, J., Sebe, N., Welling, M. (eds.) ECCV 2016. LNCS, vol. 9907, pp. 768–784. Springer, Cham (2016). https://doi.org/10.1007/978-3-319-46487-9_47
10. Gan, C., Sun, C., Duan, L., Gong, B.: Webly-supervised video recognition by mutually voting for relevant web images and web video frames. In: Leibe, B., Matas, J., Sebe, N., Welling, M. (eds.) ECCV 2016. LNCS, vol. 9907, pp. 849–866. Springer, Cham (2016). https://doi.org/10.1007/978-3-319-46487-9_52
11. Gan, C., Yao, T., Yang, K., Yang, Y., Mei, T.: You lead, we exceed: labor-free video concept learning by jointly exploiting web videos and images. In: Proceedings of the IEEE Conference on Computer Vision and Pattern Recognition (2016)
12. Gao, J., Chen, K., Nevatia, R.: CTAP: complementary temporal action proposal generation. In: Ferrari, V., Hebert, M., Sminchisescu, C., Weiss, Y. (eds.) ECCV 2018. LNCS, vol. 11206, pp. 70–85. Springer, Cham (2018). https://doi.org/10.1007/978-3-030-01216-8_5
13. Gao, J., Yang, Z., Chen, K., Sun, C., Nevatia, R.: Turn tap: temporal unit regression network for temporal action proposals. In: Proceedings of the International Conference on Computer Vision (ICCV), pp. 3628–3636 (2017)
14. Girshick, R.: Fast R-CNN. In: Proceedings of the International Conference on Computer Vision (ICCV), pp. 1440–1448 (2015)
15. Gong, G., Zheng, L., Bai, K., Mu, Y.: Scale matters: temporal scale aggregation network for precise action localization in untrimmed videos. In: International Conference on Multimedia and Expo (ICME), pp. 1–6 (2020)
16. Jiang, Y.G., et al.: THUMOS challenge: action recognition with a large number of classes (2014)
17. Karpathy, A., Toderici, G., Shetty, S., Leung, T., Sukthankar, R., Fei-Fei, L.: Large-scale video classification with convolutional neural networks. In: Proceedings of the Conference on Computer Vision and Pattern Recognition (CVPR), pp. 1725–1732 (2014)
18. Kay, W., et al.: The kinetics human action video dataset. arXiv preprint arXiv:1705.06950 (2017)
19. Kipf, T.N., Welling, M.: Semi-supervised classification with graph convolutional networks. In: International Conference on Learning Representations (ICLR), pp. 1–14 (2017)
20. Law, H., Deng, J.: CornerNet: detecting objects as paired keypoints. In: Proceedings of the European Conference on Computer Vision (ECCV), pp. 734–750 (2018)
21. Lin, J., Gan, C., Han, S.: TSM: temporal shift module for efficient video understanding. In: ICCV (2019)
22. Lin, T., Liu, X., Li, X., Ding, E., Wen, S.: BMN: boundary-matching network for temporal action proposal generation. In: Proceedings of the International Conference on Computer Vision (ICCV) (2019)
23. Lin, T., Zhao, X., Su, H., Wang, C., Yang, M.: BSN: boundary sensitive network for temporal action proposal generation. In: Ferrari, V., Hebert, M., Sminchisescu, C., Weiss, Y. (eds.) ECCV 2018. LNCS, vol. 11208, pp. 3–19. Springer, Cham (2018). https://doi.org/10.1007/978-3-030-01225-0_1
24. Liu, D., Jiang, T., Wang, Y.: Completeness modeling and context separation for weakly supervised temporal action localization. In: Proceedings of the Conference on Computer Vision and Pattern Recognition (CVPR), pp. 1298–1307 (2019)
25. Liu, Y., Ma, L., Zhang, Y., Liu, W., Chang, S.F.: Multi-granularity generator for temporal action proposal. In: Proceedings of the Conference on Computer Vision and Pattern Recognition (CVPR), pp. 3604–3613 (2019)

26. Monfort, M., et al.: Moments in time dataset: one million videos for event understanding. IEEE Trans. Pattern Anal. Mach. Intell. (T-PAMI) **42**, 502–508 (2019)
27. Peisen, Z., Lingxi, X., Ya, Z., Qi, T.: Universal-to-specific framework for complex action recognition. arXiv preprint arXiv:2007.06149 (2020)
28. Qiu, Z., Yao, T., Mei, T.: Learning spatio-temporal representation with pseudo-3D residual networks. In: Proceedings of the IEEE International Conference on Computer Vision (ICCV), pp. 5534–5542. IEEE (2017)
29. Redmon, J., Divvala, S., Girshick, R., Farhadi, A.: You only look once: unified, real-time object detection. In: Proceedings of the IEEE Conference on Computer Vision and Pattern Recognition (CVPR), pp. 779–788 (2016)
30. Ren, S., He, K., Girshick, R., Sun, J.: Faster R-CNN: towards real-time object detection with region proposal networks. In: Advances in Neural Information Processing Systems (NeurIPS), pp. 91–99 (2015)
31. Shou, Z., Chan, J., Zareian, A., Miyazawa, K., Chang, S.F.: CDC: convolutional-de-convolutional networks for precise temporal action localization in untrimmed videos. In: Proceedings of the IEEE Conference on Computer Vision and Pattern Recognition (CVPR), pp. 5734–5743 (2017)
32. Shou, Z., Wang, D., Chang, S.F.: Temporal action localization in untrimmed videos via multi-stage CNNs. In: Proceedings of the Conference on Computer Vision and Pattern Recognition (CVPR), pp. 1049–1058 (2016)
33. Tran, D., Bourdev, L., Fergus, R., Torresani, L., Paluri, M.: Learning spatiotemporal features with 3D convolutional networks. In: Proceedings of the IEEE International Conference on Computer Vision (ICCV), pp. 4489–4497. IEEE (2015)
34. Tran, D., Wang, H., Torresani, L., Ray, J., LeCun, Y., Paluri, M.: A closer look at spatiotemporal convolutions for action recognition. In: Proceedings of the IEEE Conference on Computer Vision and Pattern Recognition (CVPR), pp. 6450–6459 (2018)
35. Wang, L., Li, W., Li, W., Van Gool, L.: Appearance-and-relation networks for video classification. In: Proceedings of the IEEE Conference on Computer Vision and Pattern Recognition (CVPR), pp. 1430–1439 (2018)
36. Wang, L., Xiong, Y., Lin, D., Van Gool, L.: Untrimmednets for weakly supervised action recognition and detection. In: Proceedings of the Conference on Computer Vision and Pattern Recognition (CVPR). pp. 4325–4334 (2017)
37. Xie, S., Sun, C., Huang, J., Tu, Z., Murphy, K.: Rethinking spatiotemporal feature learning: speed-accuracy trade-offs in video classification. In: Proceedings of the European Conference on Computer Vision (ECCV), pp. 305–321 (2018)
38. Xiong, Y., Zhao, Y., Wang, L., Lin, D., Tang, X.: A pursuit of temporal accuracy in general activity detection. arXiv preprint arXiv:1703.02716 (2017)
39. Xu, H., Das, A., Saenko, K.: R-C3D: region convolutional 3D network for temporal activity detection. In: Proceedings of the IEEE International Conference on Computer Vision (ICCV), pp. 5783–5792 (2017)
40. Yuan, Z., Stroud, J.C., Lu, T., Deng, J.: Temporal action localization by structured maximal sums. In: Proceedings of the IEEE Conference on Computer Vision and Pattern Recognition (CVPR), pp. 3684–3692 (2017)
41. Zeng, R., et al.: Graph convolutional networks for temporal action localization. In: Proceedings of the International Conference on Computer Vision (ICCV) (2019)
42. Zhao, Y., et al.: Cuhk & ethz & siat submission to activitynet challenge 2017. arXiv preprint arXiv:1710.08011 (2017)
43. Zhao, Y., Xiong, Y., Wang, L., Wu, Z., Tang, X., Lin, D.: Temporal action detection with structured segment networks. In: Proceedings of the IEEE International Conference on Computer Vision (ICCV), pp. 2914–2923 (2017)

On Modulating the Gradient for Meta-learning

Christian Simon[1,4(✉)], Piotr Koniusz[1,4], Richard Nock[1,3,4], and Mehrtash Harandi[2,4]

[1] The Australian National University, Canberra, Australia
{christian.simon,piotr.koniusz,richard.nock}@anu.edu.au
[2] Monash University, Melbourne, Australia
mehrtash.harandi@monash.edu
[3] The University of Sydney, Sydney, Australia
[4] Data61-CSIRO, Sydney, Australia
{christian.simon,piotr.koniusz,richard.nock,
mehrtash.harandi}@data61.csiro.au

Abstract. Inspired by optimization techniques, we propose a novel meta-learning algorithm with gradient modulation to encourage fast-adaptation of neural networks in the absence of abundant data. Our method, termed ModGrad, is designed to circumvent the noisy nature of the gradients which is prevalent in low-data regimes. Furthermore and having the scalability concern in mind, we formulate ModGrad via low-rank approximations, which in turn enables us to employ ModGrad to adapt hefty neural networks. We thoroughly assess and contrast ModGrad against a large family of meta-learning techniques and observe that the proposed algorithm outperforms baselines comfortably while enjoying faster convergence.

Keywords: Meta learning · Few shot learning · Adaptive gradients

1 Introduction

Gradient-based algorithms [3,9,31–33,45] can be successfully employed to address a broad set of problems in meta-learning including image classification [4,37], regression [9,45], and reinforcement learning [1] to name a few. Despite the success, employing gradient-based methods in the absence of abundant data (*e.g.*, few-shot learning) is daunting as gradients become *noisy* [44]. To outline the setup, consider training a neural network with parameters $\boldsymbol{\theta} \in \Theta$. Let

$$\mathcal{L}_N(\boldsymbol{\theta}) = \frac{1}{N}\sum_{i=1}^{N} \ell(\boldsymbol{x}_i, y_i, \boldsymbol{\theta}),$$

Electronic supplementary material The online version of this chapter (https://doi.org/10.1007/978-3-030-58598-3_33) contains supplementary material, which is available to authorized users.

A. Vedaldi et al. (Eds.): ECCV 2020, LNCS 12353, pp. 556–572, 2020.
https://doi.org/10.1007/978-3-030-58598-3_33

be the total loss of the network where $\ell : \mathcal{X} \times \mathcal{Y} \times \Theta \rightarrow \mathbb{R}^+$ denotes the loss for the input $\boldsymbol{x}_i \in \mathcal{X}$ and desired output $y_i \in \mathcal{Y}$. In majority of cases, training proceeds by computing the gradient $\boldsymbol{g} = \nabla_{\boldsymbol{\theta}} \mathcal{L}_N(\boldsymbol{\theta}) = \frac{\partial}{\partial \boldsymbol{\theta}} \mathcal{L}_N(\boldsymbol{\theta})$ followed by updates in the form

$$\boldsymbol{\theta}^{(k)} = \boldsymbol{\theta}^{(k-1)} - \alpha \nabla_{\boldsymbol{\theta}^{k-1}} \mathcal{L}_N\big(\boldsymbol{\theta}^{(k-1)}\big). \tag{1}$$

Upon availability of ample data (a big enough N), updates using the gradient will hopefully converge to a good minimum. The dependency of the gradient on N (number of samples) immediately suggests the noisy nature of $\boldsymbol{g}$ in the low-data regime. That is, if we can only see $n \ll N$ samples,

$$\hat{\boldsymbol{g}} = \nabla_{\boldsymbol{\theta}} \mathcal{L}_n(\boldsymbol{\theta}) = \frac{\partial}{\partial \boldsymbol{\theta}} \frac{1}{n} \sum_{i=1}^{n} \ell(\boldsymbol{x}_i, y_i, \boldsymbol{\theta}) \approx \boldsymbol{g}.$$

The issue intensifies if the learning algorithm benefits from the Hessian information or if the distribution of data changes. In meta-learning, the use of former might be tempting to achieve faster convergence while the latter occurs frequently due to the nature of the problem (*e.g.*, adaptation to new tasks).

To combat this issue, one would ideally like to use smaller learning rates for the noisy elements of the gradient, hence reducing their effects. This is indeed the underlying idea of some modern meta-learning algorithms, one way or another.

For example, the Meta-SGD [22] algorithm explicitly formulates the learning problem as finding meta-parameters of the network along their optimal learning rate. In LEO [33], updates are performed in a low-dimensional space and the result is consequently projected to the parameter space using a non-linear mapping.

A tool from optimization, called preconditioning, is a principled way for accelerating the convergence rate of the first-order methods. The idea is that instead of updates in the form of Eq. (1), one would use

$$\boldsymbol{\theta}^{(k)} = \boldsymbol{\theta}^{(k-1)} - \alpha \boldsymbol{P} \boldsymbol{g}\big(\boldsymbol{\theta}^{(k-1)}\big). \tag{2}$$

Obviously, if the preconditioner is chosen to be the inverse of the Hessian matrix (*ie.*, $\boldsymbol{P} = \boldsymbol{H}^{-1}$ for $\boldsymbol{H} = \frac{\partial^2}{\partial \boldsymbol{\theta} \partial \boldsymbol{\theta}^\top}$), or approximates $\boldsymbol{H}^{-1}$ well enough for that matter, preconditioning reduces to the Newton method which enjoys a quadratic convergence rate. Inspecting Eq. 2 raises a question if it *is possible to use the idea of preconditioning to address the noisy gradient problem in meta-learning?* To answer this question, we hypothesize that in contrast to a full preconditioning matrix, using just the diagonal preconditioner may effectively alter the learning rate in an element-wise fashion and provide acceleration of the convergence of the learner. In this paper, we study this particular idea in detail. In particular and inspired by the concept of preconditioning, we make the following contributions.

Contributions

1. We propose a meta-learning algorithm to learn to modulate the gradient in the absence of abundant data. Similarly to preconditioning and as the name

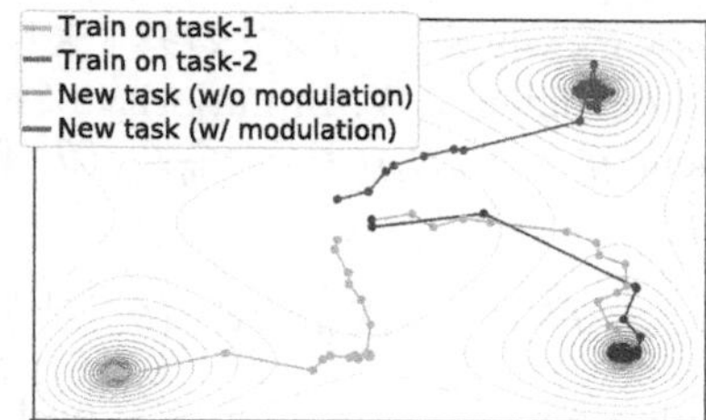

Fig. 1. An illustration of our modulation applied for a new task after learning from two previous tasks (task-1 and task-2).

implies, the gradient modulation is a multiplicative corrective factor albeit task-dependent. Being vigilant about the scalability, we formulate the modulation via low-rank approximation, which in turn let us apply modulation on large models. Such an idea is significantly different from previous works which suffer from poor scalability and require adaptation of parts of the network (usually the classifier).
2. We extensively compare and contrast our algorithm against state-of-the-art methods on several tasks, ranging from image classification to image completion to reinforcement learning. Empirically, our method outperforms various state-of-the-art algorithms and exhibits faster convergence (see Fig. 1 for an illustration).
3. We further study the robustness of the meta-learning algorithms in the presence of corrupted gradients. We observe that our idea of modulating the gradient is able to recover gracefully while other methods severely underperform.

2 Background

The objective of meta-learning is to **1.** achieve rapid convergence for new tasks (task-level) and **2.** generalize beyond previously seen tasks (meta-level). A common approach to meta-learning is to design models that learn from limited data using the concept of episodic training [34,39]. Therein, a model is presented with a set of tasks (*e.g.*, image classification), where for each task, only limited data is available. To put the discussion into context, we first provide a brief overview of the Model Agnostic Meta Learning (MAML) algorithm [9]. Let $\mathcal{D}_\tau^{trn}$ and $\mathcal{D}_\tau^{val}$ be the training and the validation sets of a given task $\tau \sim p(\mathcal{T})$, respectively. We assume that $\mathcal{D} := \{\boldsymbol{x}_i, y_i\}_{i=1}^{|\mathcal{D}|}, \boldsymbol{x}_i \in \mathcal{X},\ y_i \in \mathcal{Y}$ for some small $|\mathcal{D}|$. Furthermore, let $h : \mathcal{X} \times \mathbb{R}^n \rightarrow \mathcal{Y}$ be the predictor function of the model parameterized by $\boldsymbol{\theta} \in \mathbb{R}^n$. The MAML algorithm seeks a universal initialization $\boldsymbol{\theta}^*$ by minimizing:

$$\min_{\boldsymbol{\theta}^*} \sum_{\tau \sim P(\mathcal{T})} \mathcal{L}\Big(D_\tau^{\mathrm{val}}, \boldsymbol{\theta}^* - \alpha \sum_{k=0}^{(K-1)} \nabla \mathcal{L}\big(D_\tau^{\mathrm{trn}}, \boldsymbol{\theta}_\tau^{(k)}\big)\Big). \tag{3}$$

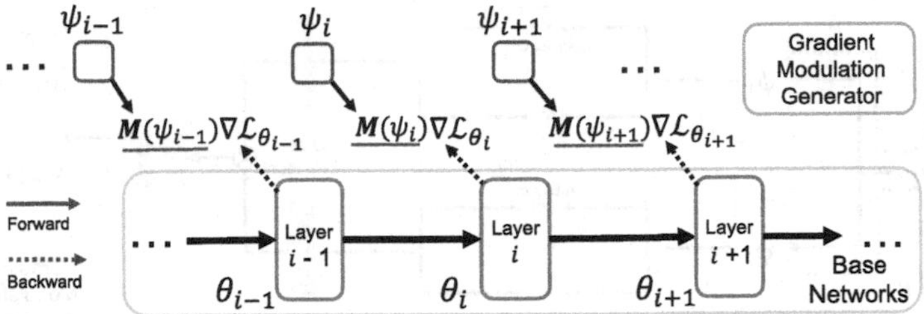

Fig. 2. Every layer is equipped with a gradient correction generator to modulate the incoming gradients.

Here, $\boldsymbol{\theta}_\tau^{(k)} = \boldsymbol{\theta}_\tau^{(k-1)} - \alpha\nabla\mathcal{L}(D_\tau^{\text{trn}}, \boldsymbol{\theta}_\tau^{(k-1)})$ with $\boldsymbol{\theta}_\tau^0 = \boldsymbol{\theta}^*$. The loss terms are:

$$\begin{aligned} \mathcal{L}(D_\tau^{\text{trn}}, \boldsymbol{\theta}) &:= \mathbb{E}_{\boldsymbol{x},y\sim\mathcal{D}_\tau^{\text{trn}}}\big[\ell(h(\boldsymbol{x},\boldsymbol{\theta}), y)\big], \\ \mathcal{L}(D_\tau^{\text{val}}, \boldsymbol{\theta}) &:= \mathbb{E}_{\boldsymbol{x},y\sim\mathcal{D}_\tau^{\text{val}}}\big[\ell(h(\boldsymbol{x},\boldsymbol{\theta}), y)\big]. \end{aligned} \tag{4}$$

Intuitively, given a task τ, the MAML starts from $\boldsymbol{\theta}^*$ and performs K gradient updates on D_τ^{trn} to obtain the adapted parameters $\boldsymbol{\theta}_\tau^{(K)}$ (this is called the inner-loop updates). Then it uses D_τ^{val} and $\boldsymbol{\theta}_\tau^{(K)}$ (which is dependent on $\boldsymbol{\theta}^*$) to improve the universal initialization point $\boldsymbol{\theta}^*$ (this is called the outer-loop update). The extensions of MAML [9] include Meta-SGD [22] and Meta-Curvature [27] with adaptive learning rates, CAVIA [45] with feature modulation, and LEO [33] with low dimensional updates.

3 Proposed Method

In this section, we introduce our meta-learner to accelerate learning process for few- and multi-shot classification, regression, and reinforcement learning. Our proposed method essentially learns to **Mod**ulate the **Grad**ient via so-called meta-learning **ModGrad** such that each task has a specific gradient modulator depending on the context.

3.1 Inner-Loop with Gradient Modulation

We define $\boldsymbol{M} : \mathbb{R}^d \rightarrow \mathbb{R}^n$, a function with parameter $\boldsymbol{\Psi}$ that performs gradient modulation. The purpose of the gradient modulation is two-fold: **1.** to suppress the noise and **2.** accelerate the convergence by amplifying certain elements of the gradient. Inspired by the use of diagonal for preconditioning, we formulate the meta-learning algorithm as follows:

$$\min_{\boldsymbol{\theta}^*,\boldsymbol{\Psi}} \sum_{\tau\sim P(\mathcal{T})} \mathcal{L}\Big(D_\tau^{\text{val}}, \boldsymbol{\theta}^* - \alpha\sum_{k=0}^{K-1} \boldsymbol{M}_\tau^{(k)}(\boldsymbol{\Psi}) \odot \nabla\mathcal{L}(D_\tau^{\text{trn}}, \boldsymbol{\theta}_\tau^{(k)})\Big). \tag{5}$$

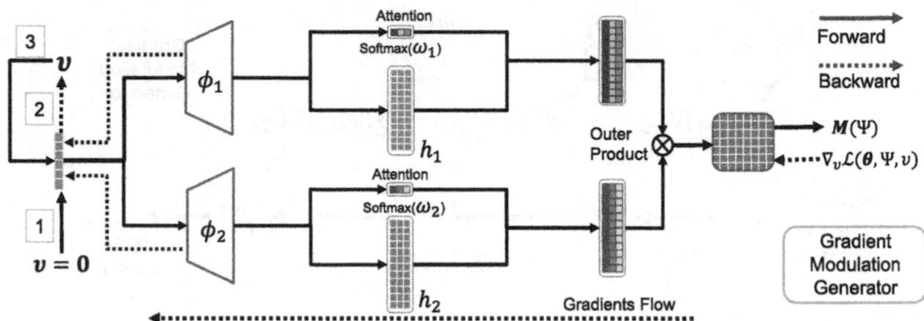

Fig. 3. A gradient modulation generator. Two sister networks (ϕ_1, ϕ_2) produce two tall matrices to generate the correction.

The inner-loop update in Eq. 5 performs an element-wise modulation of the gradient vector by operator '$\odot$'. One can also view this updating scheme as a generalization of meta-learners that adaptively alter the learning rate of the SGD (*e.g.*, [3,22,33]). In what follows next, we build a generative modulator through another neural network.

3.2 Task-Dependent Gradient Modulation Generator

By minimizing Eq. 5, we jointly learn the universal initialization vector $\boldsymbol{\theta}^*$ and the generator $\boldsymbol{M}_\tau^{(k)}(\boldsymbol{\Psi})$ to enrich adaptability of the meta-learner. Figures 2 and 3 illustrate our design to generate the modulation. Without loss of generality, we denote $n = n_1 \times n_2$ and we elaborate on how the function $\boldsymbol{M}_\tau^{(k)}(\boldsymbol{\Psi}) \in \mathbb{R}^n$ is obtained via the generator for a given task τ.

For reasons that become clear shortly, the generator makes use of a context vector $\boldsymbol{\nu} \in \mathbb{R}^d$ to generate the gradient modulation $\boldsymbol{M}_\tau^{(k)}(\boldsymbol{\Psi})$. In doing so, the context vector $\boldsymbol{\nu}$ is first processed by two sister modules ϕ_1 and ϕ_2. This generates,

$$\begin{aligned}(\boldsymbol{\omega}_1, \boldsymbol{h}_1) &= \phi_1(\boldsymbol{\nu}_\tau^{(k)}), \quad \boldsymbol{\omega}_1 \in \mathbb{R}^u, \boldsymbol{h}_1 \in \mathbb{R}^{n_1 \times u},\\ (\boldsymbol{\omega}_2, \boldsymbol{h}_2) &= \phi_2(\boldsymbol{\nu}_\tau^{(k)}), \quad \boldsymbol{\omega}_2 \in \mathbb{R}^u, \boldsymbol{h}_2 \in \mathbb{R}^{n_2 \times u}.\end{aligned} \tag{6}$$

Attention mechanism is employed to re-weigh the matrix produced by two sister modules:

$$\begin{aligned}\boldsymbol{a}_1 &= \mathrm{softmax}(\boldsymbol{\omega}_1),\\ \boldsymbol{a}_2 &= \mathrm{softmax}(\boldsymbol{\omega}_2).\end{aligned} \tag{7}$$

The softmax function is chosen because we observe empirically that it always performs the best compared to other activation functions. The output of the gradient modulator is then computed by the outer product operation:

$$\boldsymbol{M}_\tau^{(k)}(\boldsymbol{\Psi}) = \mathrm{Vec}\big((\boldsymbol{h}_1\boldsymbol{a}_1) \otimes (\boldsymbol{h}_2\boldsymbol{a}_2)\big). \tag{8}$$

Vec$(\cdot)$ operator vectorizes an input matrix. Equation 8 uses a low-rank approximation to generate gradient corrections. This lets us scale up ModGrad to very large networks and simultaneously regularizes gradients. Finally, the produced matrix is passed via ReLU.

Finally, the only remaining detail of the ModGrad algorithm is the context vector generation. To this end, we firstly reset $\boldsymbol{\nu}_{\tau}^{(0)} = \mathbf{0}$ and then we generate an initial modulation $\boldsymbol{M}_{\tau}^{(k)}(\boldsymbol{\Psi})$. We then update $\boldsymbol{\nu}$ as:

$$\boldsymbol{\nu}_{\tau}^{(k)} = -\nabla_{\boldsymbol{\nu}_{\tau}^{(k-1)}} \mathcal{L}\big(\boldsymbol{\theta}^{(k-1)} - \alpha \boldsymbol{M}_{\tau}^{(k-1)}(\boldsymbol{\Psi}) \odot \nabla_{\boldsymbol{\theta}^{(k-1)}} \mathcal{L}(\boldsymbol{\theta}^{(k-1)})\big). \tag{9}$$

The context vector $\boldsymbol{\nu}$ is then fed into two sister networks (Eq. 6) to compute the modulation. Algorithm 1, outlined for classification and regression tasks, provides details of how the parameters of the generator $\boldsymbol{\Psi}$ and the base network $\boldsymbol{\theta}$ are updated. For the reinforcement learning, another variant of ModGrad is provided in the supplementary material.

Algorithm 1. Train ModGrad

Require: $\boldsymbol{\theta}$, $\boldsymbol{\Psi}$, α, $p(\mathcal{T})$
$\boldsymbol{\theta}, \boldsymbol{\Psi} \leftarrow$ Random initialization
while not done **do**
  Sample $\tau_1 \dots \tau_B$ from $p(\mathcal{T})$ ▷ Sample episodes
  for b in $\{1, ..., B\}$ **do**
    $\boldsymbol{\theta}^0 \leftarrow \boldsymbol{\theta}$
    $\mathcal{D}_{\tau}^{trn}, \mathcal{D}_{\tau}^{val}$ from τ_b ▷ Sample training and testing sets
    for k in $\{1, ..., K\}$ **do**
      Reset $\boldsymbol{\nu}$ ▷ Reset context vector to 0
      Compute $\nabla_{\boldsymbol{\theta}^{(k-1)}} \mathcal{L}(\boldsymbol{\theta}^{(k-1)})$
      Compute $\boldsymbol{\nu}$ using Eq. 9 ▷ Update context vector
      Generate the gradient modulation using Eq. 8
      $\boldsymbol{\theta}^{(k)} \leftarrow \boldsymbol{\theta}^{(k-1)} - \alpha \boldsymbol{M}_{\tau}^{(k-1)}(\boldsymbol{\Psi}) \odot \nabla_{\boldsymbol{\theta}^{(k-1)}} \mathcal{L}(\boldsymbol{\theta}^{(k-1)})$
    end for
  end for
  $\boldsymbol{\theta} \leftarrow$ OptimizerStep($\mathcal{D}_{\tau}^{val}, \boldsymbol{\theta}^{(K)}$) ▷ Meta update base networks
  $\boldsymbol{\Psi} \leftarrow$ OptimizerStep($\mathcal{D}_{\tau}^{val}, \boldsymbol{\Psi}$) ▷ Meta update ModGrad
end while

Remark 1. ModGrad uses a lower-dimensional context vector $\boldsymbol{\nu}$ to generate the gradient modulation. This enables us to lower the computational complexity even further. The underlying assumption, based on the smoothness of the gradient updates, is that the gradient field, especially for modern models which are often deep and over-parameterized, should comply with the low-dimensional latent data representation. As such, enforcing the context vector to be low-dimensional implicitly contributes to capturing the geometry of the gradient field.

Remark 2. In contrast to T-Net [21] and natural neural networks [8] that alter the neural networks by inserting additional layers for gradient projection, ModGrad directly produces the modulation vector for the gradients, thus it does not require altering the architecture of the base network to achieve adaptation.

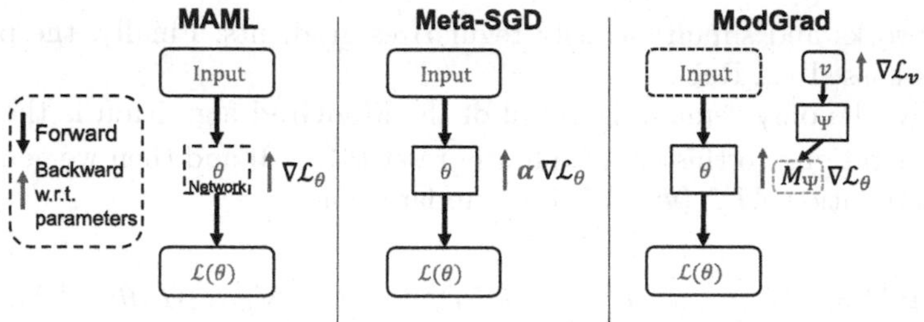

Fig. 4. Comparison of our method to MAML [9] and Meta-SGD [22].

Table 1. Comparison of inner-loop updates of various meta-learners. The red font shows additional parameters updated in the outer-loop.

Method	Inner-Loop Update
MAML [9]	$\boldsymbol{\theta}^{(k+1)} \leftarrow \boldsymbol{\theta}^{(k)} - \alpha \nabla_{\boldsymbol{\theta}^{(k)}} \mathcal{L}(\boldsymbol{\theta}^{(k)})$
Meta-SGD [22]	$\boldsymbol{\theta}^{(k+1)} \leftarrow \boldsymbol{\theta}^{(k)} - \alpha \nabla_{\boldsymbol{\theta}^{(k)}} \mathcal{L}(\boldsymbol{\theta}^{(k)})$
CAVIA [45]	$\boldsymbol{\nu}^{(k+1)} \leftarrow \boldsymbol{\nu}^{(k)} - \alpha \nabla_{\boldsymbol{\nu}^{(k)}} \mathcal{L}(\pi, \boldsymbol{\nu}^{(k)}, \boldsymbol{\theta})$
LEO [33]	$\boldsymbol{z}^{(k+1)} \leftarrow \boldsymbol{z}^{(k)} - \alpha \nabla_{\boldsymbol{z}^{(k)}} \mathcal{L}(\boldsymbol{z}^{(k)}, \boldsymbol{\theta}, \phi_e, \phi_d, \phi_r)$

Remark 3. Meta-SGD benefits from the adaptive learning rate to accelerate its convergence and improve the performance of standard MAML. In Meta-SGD, the modulator is a global parameter for all tasks. Our method differs from Meta-SGD in that we use a task-dependent gradient modulation generated from neural networks.

Remark 4. In practice and to lower the computational complexity, one can make use of a set of distinct ModGrad cells $\boldsymbol{\Psi} = \{\psi_1 \dots \psi_m\}$, each acting on and optimizing a layer of the network (see Fig. 2 for a conceptual diagram). A ModGrad cell may be combined with any part of neural network including fully-connected and convolutional layers. This design requires no changes to the base network and brings flexibility, letting modulate the gradient for selected layers. Implementation details for fully-connected and convolutional layers are left in the supplementary material.

4 Related Work

Enjoying some theoretical guarantees, meta-learning by optimization (*e.g.*, [5, 14]) encompasses methods that learn the parameters of the model, at the meta-level, through gradient updates. The celebrated "model agnostic meta-learning" by Finn *et al.* [9] and "Learning to Learn" by Andrychowicz *et al.* [2] are prime examples of meta-learning by optimization. Note that, our meta-learning approach is an adaptive module acting on the model parameters which differs from relation and metric learning approaches *e.g.*, [17,23,35,36,42,43].

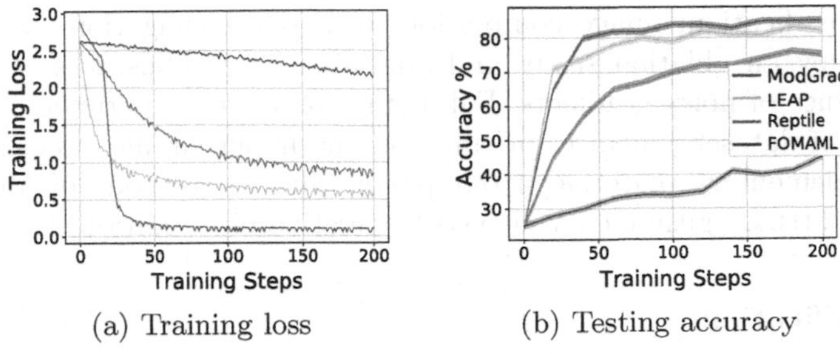

(a) Training loss (b) Testing accuracy

Fig. 5. Training loss and testing accuracy on Omniglot in the multi-shot setting.

Extensions of MAML include Reptile [26] that directly combines the updated parameters in the inner loop with that of the meta-learner, leading to a highly scalable meta-learning algorithm. MAML++ [3] uses a weighted loss in the inner-loop to boost the accuracy. We conclude this section by providing a brief overview on optimization-based meta-learning algorithms (see Table 1 for more details).

Preconditioned Stochastic Gradient Descents (SGD). The general framework of preconditioning for fast adaptation of neural network is introduced in [11]. We discuss Meta-SGD and Meta-Curvature as the specific forms of preconditioned SGD. In Meta-SGD [22], the gradient update is equipped with a universal meta-parameter (α). In Meta-Curvature [27], the meta-parameter has a form of preconditioning matrices which transform the gradient in the inner-loop. During training, the meta-parameters of both methods [22,27] change in each meta-update, however, they remain unchanged when the algorithm is applied to unseen tasks (see Fig. 4 for a conceptual diagram). Furthermore, additional meta-parameters are required if one requires to use more than one update in the inner-loop while ModGrad does not require extra parameters.

Network Modulation. The modulation can be performed in the feature space [40,45]. Borrowing the idea from FiLM [28], a function $f_\pi(.)$ is designed to create the modulation for conditioning layers. In the inner-loop, CAVIA updates only the context vector ($\boldsymbol{\nu}$) which is the input of f_π. The context vector is used as additional inputs to layers of the neural network to facilitate adaptation. In the same spirit, meta-transfer learning [37] learns the functions which scale and shift network parameters for each different task.

Network Augmentation. Other approaches achieve meta-learning through augmenting the fast-adaptive network parameters in addition to the base networks. For instance, LEO [33], (M)T-Nets [21], and Meta-Networks [25] use some parts of the neural networks or generate them for fast-adaptation.

5 Experiments

In this section, we compare and contrast the proposed ModGrad method against baselines and state-of-the-art algorithms on various meta-learning tasks such as few-shot classification, image completion, and robot navigation. We conclude the section by an ablation study and analysis of the robustness of ModGrad in the presence of noisy gradients. Full details of all the experiments, including description of datasets, nuances of training, and hyperparameters can be found in the supplementary material of our paper. The source code of ModGrad is available at https://github.com/chrysts/generative_preconditioner.

5.1 Classification

Multi-shot Classification. For multi-shot classification, we used the Omniglot dataset [19] containing 1623 characters from 50 different alphabets. The task is to classify images given a random number of sampled images (class-imbalance). Following the multi-shot setting in [10], we assess the rate of convergence and the performance of ModGrad in comparison to the LEAP [10], Reptile [26], and FOMAML [9] for a 20-way problem with 25 tasks in total (see Fig. 5 for details). We stress that in this experiment, the number of samples per class varies randomly between 1 and 25. Figure 5 suggests that ModGrad performs adaptation much faster and to a much lower training loss while achieving the highest recognition accuracy of 85.1% compared to 83.3% of LEAP and 76.4% of Reptile. To the best of our knowledge, the performance of ModGrad for the multi-shot classification is the state of the art on this protocol and it tops previous studies by a notable margin.

Few-Shot Classification. As our next experiment, we benchmark the ModGrad method for few-shot classification on the *mini*-ImageNet dataset [32]. In this comparison, we compare ModGrad to the related gradient-based meta-learning algorithms. The *mini*-ImageNet is a subset of the ImageNet [18] with 64, 16, and 20 classes for training, validation, and testing, respectively. We follow the widely-used protocol in the form of *episodes* for 5-way 5-shot and 1-shot with 600 tasks for testing. In all experiments, we performed episodic training using two network architectures, namely 4-convolutional blocks (Conv-4) without augmentation following [9,36] and WideResNet 28-10 (WRN-28-10) [41] with augmentation following [29,33].

As paper [30] suggests that the representations in last layers undergo significant changes during adaptation, we apply updates to the last two convolutional layers of the base network. The model parameters of the base networks and ModGrad are optimized with the Adam optimizer [16]. The learning rate is set to 10^{-3} and then cut by half for 10K episodes. The size of $\boldsymbol{\nu}$ and value of α are set to 300 and 0.1 for all experiments on *mini*-ImageNet.

On this benchmark, ModGrad outperforms existing few-shot methods for various backbones *e.g.*, Conv-4 and WRN-28-10 as presented in Table 2. For a fair comparison, the result is compared to the meta-learning algorithms with the same backbones and experimental setup. Using Conv-4, ModGrad only needs 1-step and 64 filters per layer to outperform CAVIA, which employs 5-steps and 512 filters, by around 1.4% and 3.3% for 5-way 1-shot and 5-shot protocols. Furthermore, ModGrad with WRN-28-10 also performs better than the works

Table 2. The performance of existing gradient-based meta-learning methods for few-shot classification. Reported results are evaluated for 5-way 1- and 5-shot protocols on *mini*-ImageNet. MAML $^{\#}$ is our reimplementation.

Model	Backbone	1-shot	5-shot
ML LSTM [32]	Conv-4	43.44 ± 0.77	60.60 ± 0.71
MAML (64)$^{\#}$ [9]	Conv-4	47.89 ± 1.20	64.59 ± 0.88
Reptile [26]	Conv-4	49.97 ± 0.32	65.99 ± 0.58
Meta-SGD [22]	Conv-4	50.50 ± 1.90	64.00 ± 0.90
R2-D2 [6]	Conv-4	48.70 ± 0.60	65.50 ± 0.60
(M)T-Net [21]	Conv-4	51.70 ± 1.84	–
CAVIA (512) [45]	Conv-4	51.82 ± 0.65	65.85 ± 0.55
ModGrad (1-step)	Conv-4	$\mathbf{53.20 \pm 0.86}$	$\mathbf{69.17 \pm 0.69}$
Qiao et al. [29]	WRN-28-10	59.60 ± 0.41	73.74 ± 0.19
MTL [37]	ResNet-12	61.20 ± 1.80	75.50 ± 0.80
LEO [33]	WRN-28-10	61.76 ± 0.08	77.59 ± 0.12
SCA + MAML++ [4]	DenseNet	62.86 ± 0.79	77.64 ± 0.40
wDAE-MLP [13]	WRN-28-10	62.67 ± 0.15	78.70 ± 0.10
wDAE-GNN [13]	WRN-28-10	62.96 ± 0.15	78.85 ± 0.10
Meta-Curvature [27]	WRN-28-10	64.40 ± 0.10	80.21 ± 0.10
ModGrad (1-step)	WRN-28-10	$\mathbf{65.72 \pm 0.21}$	$\mathbf{81.17 \pm 0.20}$

by [20] and [33]. ModGrad needs only 1-step in the inner-loop compared to other meta-learning methods that need more than 1-step to achieve good results.

Number of Steps. Below, we investigate meta-learning on deeper networks using ResNet [15] and a higher number of shots to capture the relationship between the number of step and these two factors. The *mini*-ImageNet dataset is used for experiments. To this end, we reimplement MAML and use the first-order method as the memory load for the second-order method is enormous given very deep networks. Note that the reported number of step is applied for both training and testing stages for the 5-way classification. To investigate the relationship between the number of shot and the number of step, the number of shot is set to 5, 10, 15, and 20 samples using Conv-4 backbone. On ResNet-34, data augmentation and image size of 224×224 are used *without fine-tuning* the learning rate, following settings in [7]. Training using Conv-4 and ResNet-34 is performed 50K and 100K episodes, respectively. Figure 6 shows that MAML [9] needs more steps to achieve better results for higher shot number and deeper networks.

On the ResNet-34, we observe that the performance gap for 5-steps and 30-steps on Conv-4 (64 filters) is 2.5% but the performance gap reaches 4% on ResNet-34. Using ResNet-34, ModGrad outperforms MAML by ~6.5% and ~2.5% for 5-way 1-shot and 5-shot protocols. Using higher shot numbers, MAML

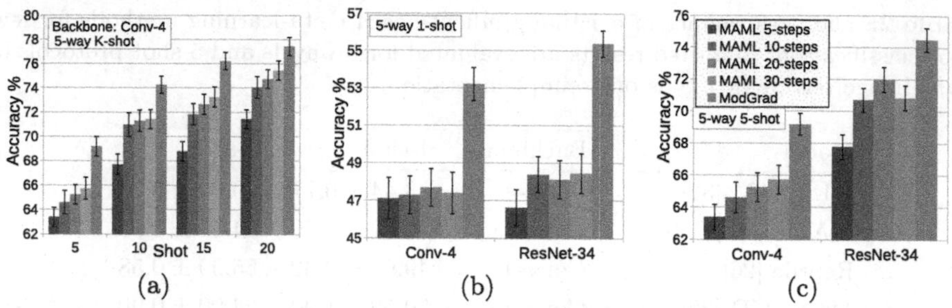

(a) (b) (c)

Fig. 6. The performance of (a) ModGrad with various shot numbers, (b) deeper networks, and (c) MAML with 5, 10, 20, and 30 steps in the inner-loop. In 1-step, ModGrad achieves superior performance given various shots and backbones. MAML cannot perform well with a deeper network and larger shot numbers.

with more additional steps also shows the improvement. The performance gap for 5-shot is about 2.5% between 5-steps and 30-steps but the performance gap increases up to 4% for 5-way 20-shot setting. Furthermore, in 1-step, ModGrad outperforms 30-steps by MAML by 2% in 20-shot classification. We conjecture that ModGrad achieves a good performance in 1-step for both cases because the modulation adaptively scales the gradients to reach a good minimum rapidly.

5.2 Regression

The task of image regression (completion) on the CelebA dataset [24] is adopted from [12]. The goal is to in-paint missing image pixels given only some pixels of images (random and ordered). For inputs pixel locations, models have to perform regression to approximate pixel intensities given 10 and 100 training pixels. The results in Table 3 show that ModGrad has the lowest Mean Square Error (MSE) on the image regression tasks compared to Conditional Neural Process (CNP) [12], CAVIA [45], and MAML [9]. We use the same setup as stated in [45] with five 128 hidden layers and a 128-dimensional input vector ($\boldsymbol{\nu}$). The learning rate is set 0.1. For this regression task, ModGrad is applied only to a single fully-connected layer (preceding the last layer) as we observed no tangible difference if applying ModGrad to several layers. Note that our results use only 1-step while CAVIA [45] and MAML [9] use 5 gradient steps to train from the training pixels. The qualitative results of image completion are shown in Fig. 7.

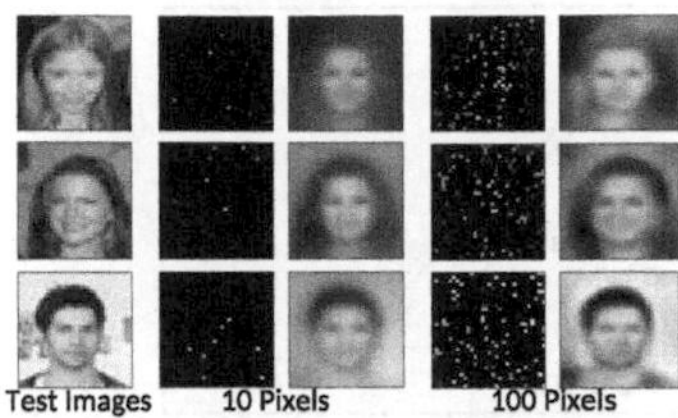

Fig. 7. Qualitative results of ModGrad on the CelebA dataset for image completion with 10 and 100 pixels provided randomly.

Table 3. The MSE for image completion tasks on the CelebA dataset for 10 and 100 pixels provided randomly and in the ordered fashion.

Model	Random Pixels		Ordered Pixels	
	10	100	10	100
CNP [12]	0.039	0.016	0.057	0.047
MAML [9]	0.040	0.017	0.055	0.047
CAVIA [45]	0.037	0.014	0.053	0.047
ModGrad	**0.034**	**0.012**	**0.048**	**0.043**

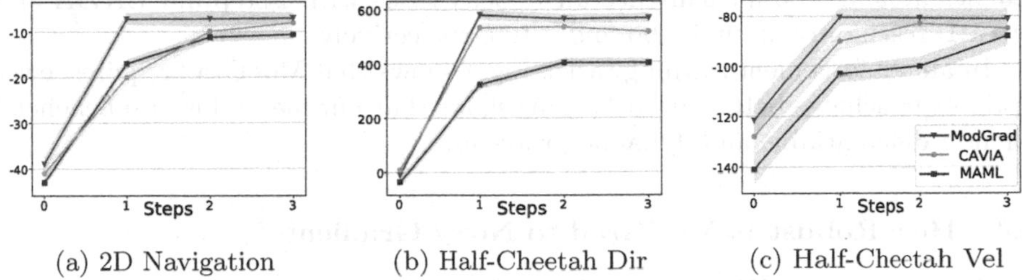

(a) 2D Navigation (b) Half-Cheetah Dir (c) Half-Cheetah Vel

Fig. 8. Results for reinforcement learning on 2D navigation, half-cheetah direction, and velocity.

5.3 Reinforcement Learning

Our experiments on reinforcement learning are adopted from [9,45]. All network architectures, range of parameters, and protocols follow the same setup.

2D Navigation. In this experiment, we evaluate ModGrad on 2D-Navigation tasks from [9]. Every task contains a randomly chosen goal position where an agent has to move towards this position. The goal of this task is to adapt the policy of an agent quickly such that it can maximize the (negative) rewards. The goals of this navigation are within the range $[-0.5, 0.5]$ and the actions are clipped within $[-0.1, 0.1]$. In total, 20 trajectories are used for one gradient update. As in [45], we use the same network with two-layers, 100 hidden units, and a ReLU activation function. In 1-step, ModGrad achieves rewards around -8 while CAVIA and MAML are far below with -15.

Locomotion. We evaluate our method on the half-cheetah locomotion tasks from the MuJoCo simulator [38]. The tasks consist of predicting the direction and the velocity. The velocity ranges between 0.0 and 2.0. Each rollout length is 200, and 20 rollouts are used per gradient step during training. ModGrad reaches rewards around 590 with only 1-step but CAVIA and MAML obtain rewards below 550 only for half-cheetah direction tasks. Furthermore, ModGrad

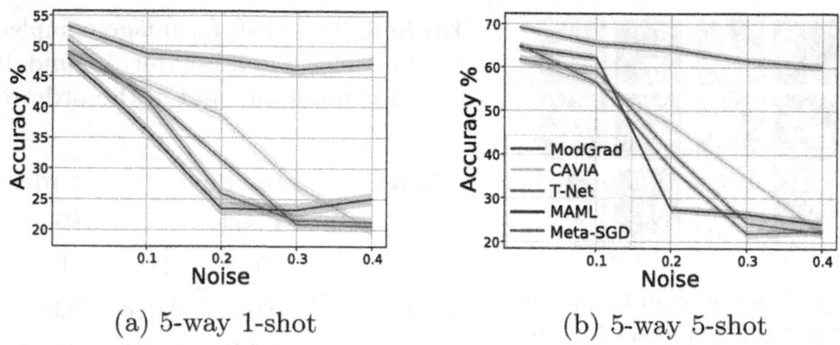

(a) 5-way 1-shot (b) 5-way 5-shot

Fig. 9. The performance comparison on *mini*-ImageNet with various noise level for 5-way 1-shot and 5-way 5-shot protocols.

reaches around −80 for half-cheetah velocity tasks with 1-step but CAVIA and MAML reach only around −90 and −100, respectively.

In all reinforcement learning tasks, Fig. 8 shows that ModGrad requires fewer updates to achieve better rewards. This shows that our method is also beneficial for non-differentiable and dynamic problems.

5.4 How Robust Is ModGrad to Noisy Gradients?

Methods such as CAVIA [45] and T-Net [21] modulate parameters or use an additional layer to transform gradients. These methods receive the gradients directly from the base network. Below, we empirically show that these approaches are fragile in the presence of corrupted gradients (which is a fundamental issue in the few-shot regime). To evaluate the robustness, we corrupted the gradients $\nabla\mathcal{L}(\boldsymbol{\theta})$ by adding noise following $\mathcal{N}(\mathbf{0}_n, \eta\mathbf{1}_n)$ to the gradients in the inner-loop, and we measured the accuracy of ModGrad, CAVIA, T-Net for various η (see Fig. 9 for results) using Conv-4 on *mini*-ImageNet. Compared to other methods, ModGrad degrades gracefully. For example, while our method only degrades about 10%, T-Net, CAVIA, Meta-SGD, and MAML plummet by 30% and 40% for 5-way 1-shot and 5-way 5-shot protocols, respectively.

5.5 Ablation Study

Below, we provide an ablation study on our proposed method using Conv-4 backbone. The experiments show how the number of steps and the column dimension of the tall matrix (u) used by the outer product affect the performance.

Impact of Steps. We run 5-steps in inner-loop to check the performance of 5-way 5-shot and 1-shot protocols on *mini*-ImageNet suing Conv-4. Table 4 shows that ModGrad achieves a good performance in 1-step while running over more steps may vary the performance ±1% on 1-shot and 5-shot protocols. Thus, ModGrad is robust to a varying number of steps in few-shot learning.

Table 4. ModGrad given various numbers of steps on *mini*-ImageNet.

Step	1	2	3	4	5
5-way 5-shot	69.17	68.80	68.42	68.34	68.24
5-way 1-shot	53.20	52.66	53.54	52.87	52.76

The Number of Columns (u). Equation 6 lets us choose the number of columns in the tall matrices. This experiment shows the impact of the number of columns (u) on results. Table 5 shows that the choice of u does not degrade the performance significantly (~0.5% variation on 5-way 1-shot protocol) but it may degrade results by ~1.5% on the 5-way 5-shot protocol. In both cases, our approach outperforms the standard MAML (Conv-4) algorithm. We observe that the lower the value of u is the more the low-pass filtering nature of our gradient modulation is. With a full-rank matrix, the modulator looses its low-pass filtering nature. Our experiments on classification, regression and reinforcement learning use the value of $u = 5$.

Table 5. The impact of column dimensions (u) on *mini*-ImageNet.

Value of u	1	5	10
5-way 5-shot	68.43	69.17	67.62
5-way 1-shot	53.13	53.20	52.53

5.6 Discussion

Time Complexity. ModGrad requires two forward and two backward passes per step. Thus, its complexity is 2× the computation of MAML but ModGrad converges significantly faster than MAML. For example, ModGrad with just one adaptation step comfortably outperforms MAML with five steps (best setting used by MAML). When comparing wall clocks, adaptation in MAML requires 0.05 s while ModGrad performs this step in 0.03 s per adaptation on the 5-way 5-shot protocol given Conv-4 backbone on mini-ImageNet.

Properties of ModGrad. We have observed that the low-rank modulating matrix paired with the Hadamard product acting on gradients have the property to perform adaptive low-pass filtering. This property depends on the context vector of the sister networks and the number of columns in the tall matrix controlled by the value of u. We believe this deems ModGrad a generative adaptive gradient filtering modulator which explains why ModGrad copes so well in the presence of gradients corrupted by the noise. Our y material provides a more detailed theoretical analysis of the low-pass filtering properties of ModGrad. It also provides plots studying this property on both simulated and the real data.

6 Conclusions

This work presents a meta-learner by modulating the gradient via so-called ModGrad. Our approach shows a general ability to address a wide range of problems including few-shot classification, regression and reinforcement learning. Empirical results show that ModGrad is competitive compared with other existing gradient-based meta-learners. Furthermore, ModGrad is designed to be modular (applicable to every layer) in deep neural networks. Thus, it can be utilized for other interesting applications without any extra structural changes to the base network architecture. In practice, our approach copes with the practical optimization matters such as learning rates, gradient step and adaptation to noise. Our gradient-based meta-learning algorithm remains robust in the presence of corrupted gradients while other existing methods have a low tolerance. Another benefit of ModGrad is the accelerated learning of the base network. As a result, ModGrad works also well with deeper networks, higher number of shots and a lower number of steps compared to MAML.

References

1. Al-Shedivat, M., Bansal, T., Burda, Y., Sutskever, I., Mordatch, I., Abbeel, P.: Continuous adaptation via meta-learning in nonstationary and competitive environments. In: International Conference on Learning Representations (2018)
2. Andrychowicz, M., et al.: Learning to learn by gradient descent by gradient descent. In: Advances in Neural Information Processing Systems (2016)
3. Antoniou, A., Edwards, H., Storkey, A.: How to train your MAML. In: International Conference on Learning Representations (2019)
4. Antoniou, A., Storkey, A.J.: Learning to learn by self-critique. In: Advances in Neural Information Processing Systems, pp. 9936–9946 (2019)
5. Balcan, M.F., Khodak, M., Talwalkar, A.: Provable guarantees for gradient-based meta-learning. In: International Conference on Machine Learning, pp. 424–433 (2019)
6. Bertinetto, L., Henriques, J.F., Torr, P., Vedaldi, A.: Meta-learning with differentiable closed-form solvers. In: International Conference on Learning Representations (2019)
7. Chen, W.Y., Liu, Y.C., Kira, Z., Wang, Y.C.F., Huang, J.B.: A closer look at few-shot classification. In: International Conference on Learning Representations (2019)
8. Desjardins, G., Simonyan, K., Pascanu, R., et al.: Natural neural networks. In: Advances in Neural Information Processing Systems, pp. 2071–2079 (2015)
9. Finn, C., Abbeel, P., Levine, S.: Model-agnostic meta-learning for fast adaptation of deep networks. In: International Conference on Machine Learning (2017)
10. Flennerhag, S., Moreno, P.G., Lawrence, N., Damianou, A.: Transferring knowledge across learning processes. In: International Conference on Learning Representations (2019)
11. Flennerhag, S., Rusu, A.A., Pascanu, R., Visin, F., Yin, H., Hadsell, R.: Meta-learning with warped gradient descent. In: International Conference on Learning Representations (2020)
12. Garnelo, M., et al: Neural processes. arXiv preprint arXiv:1807.01622 (2018)

13. Gidaris, S., Komodakis, N.: Generating classification weights with GNN denoising autoencoders for few-shot learning. In: Proceedings of the IEEE Conference on Computer Vision and Pattern Recognition, pp. 21–30 (2019)
14. Grant, E., Finn, C., Levine, S., Darrell, T., Griffiths, T.: Recasting gradient-based meta-learning as hierarchical Bayes. In: International Conference on Learning Representations (2018)
15. He, K., Zhang, X., Ren, S., Sun, J.: Deep residual learning for image recognition. In: IEEE Conference on Computer Vision and Pattern Recognition (2016)
16. Kingma, D.P., Ba, J.L.: Adam: a method for stochastic optimization. In: International Conference on Learning Representations (2015)
17. Koniusz, P., Zhang, H.: Power normalizations in fine-grained image, few-shot image and graph classification. TPAMI (2020)
18. Krizhevsky, A., Sutskever, I., Hinton, G.E.: ImageNet classification with deep convolutional neural networks. In: Advances in Neural Information Processing Systems (2012)
19. Lake, B.M., Salakhutdinov, R., Tenenbaum, J.B.: Human-level concept learning through probabilistic program induction. Science **350**(6266), 1332–1338 (2015)
20. Lee, K., Maji, S., Ravichandran, A., Soatto, S.: Meta-learning with differentiable convex optimization. In: Proceedings of the IEEE Conference on Computer Vision and Pattern Recognition, pp. 10657–10665 (2019)
21. Lee, Y., Choi, S.: Gradient-based meta-learning with learned layerwise metric and subspace. In: International Conference on Machine Learning, pp. 2933–2942 (2018)
22. Li, Z., Zhou, F., Chen, F., Li, H.: Meta-SGD: learning to learn quickly for few shot learning. arXiv preprint arXiv:1707.09835 (2017)
23. Liu, Y., et al.: Learning to propagate labels: transductive propagation network for few-shot learning. In: International Conference on Learning Representations (2019)
24. Liu, Z., Luo, P., Wang, X., Tang, X.: Deep learning face attributes in the wild. In: Proceedings of International Conference on Computer Vision, December 2015
25. Munkhdalai, T., Yu, H.: Meta networks. In: Proceedings of the 34th International Conference on Machine Learning, vol. 70, pp. 2554–2563. JMLR. org (2017)
26. Nichol, A., Achiam, J., Schulman, J.: On first-order meta-learning algorithms. arXiv preprint arXiv:1803.02999 (2018)
27. Park, E., Oliva, J.B.: Meta-curvature. In: Advances in Neural Information Processing Systems, pp. 3309–3319 (2019)
28. Perez, E., Strub, F., De Vries, H., Dumoulin, V., Courville, A.: Film: visual reasoning with a general conditioning layer. In: Thirty-Second AAAI Conference on Artificial Intelligence (2018)
29. Qiao, S., Liu, C., Shen, W., Yuille, A.L.: Few-shot image recognition by predicting parameters from activations. In: IEEE Conference on Computer Vision and Pattern Recognition (2018)
30. Raghu, A., Raghu, M., Bengio, S., Vinyals, O.: Rapid learning or feature reuse? Towards understanding the effectiveness of MAML. In: International Conference on Learning Representations (2020)
31. Rajeswaran, A., Finn, C., Kakade, S., Levine, S.: Meta-learning with implicit gradients. In: Advances in Neural Information Processing Systems (2019)
32. Ravi, S., Larochelle, H.: Optimization as a model for few-shot learning. In: International Conference on Learning Representations (2017)
33. Rusu, A.A., et al.: Meta-learning with latent embedding optimization. In: International Conference on Learning Representations (2019)

34. Santoro, A., Bartunov, S., Botvinick, M., Wierstra, D., Lillicrap, T.: Meta-learning with memory-augmented neural networks. In: International Conference on Machine Learning, pp. 1842–1850 (2016)
35. Simon, C., Koniusz, P., Nock, R., Harandi, M.: Adaptive subspaces for few-shot learning. In: Proceedings of the IEEE/CVF Conference on Computer Vision and Pattern Recognition, pp. 4136–4145 (2020)
36. Snell, J., Swersky, K., Richard, Z.: Prototypical networks for few-shot learning. In: Advances in Neural Information Processing Systems (2017)
37. Sun, Q., Liu, Y., Chua, T.S., Schiele, B.: Meta-transfer learning for few-shot learning. In: The IEEE Conference on Computer Vision and Pattern Recognition (2019)
38. Todorov, E., Erez, T., Tassa, Y.: MuJoCo: a physics engine for model-based control. In: IEEE/RSJ International Conference on Intelligent Robots and Systems, pp. 5026–5033 (2012)
39. Vinyals, O., Blundell, C., Lillicrap, T., Kavukcuoglu, K., Wierstra, D.: Matching networks for one shot learning. In: Advances in Neural Information Processing Systems (2016)
40. Vuorio, R., Sun, S.H., Hu, H., Lim, J.J.: Multimodal model-agnostic meta-learning via task-aware modulation. In: Advances in Neural Information Processing Systems, pp. 1–12 (2019)
41. Zagoruyko, S., Komodakis, N.: Wide residual networks. In: Proceedings of the British Machine Vision Conference, pp. 87.1–87.12 (2016)
42. Zhang, H., Koniusz, P.: Power normalizing second-order similarity network for few-shot learning. In: Winter Conference on Applications of Computer Vision (2019)
43. Zhang, H., Zhang, L., Qui, X., Li, H., Torr, P.H.S., Koniusz, P.: Few-shot action recognition with permutation-invariant attention. In: ECCV (2020)
44. Zhang, Y., Qu, H., Chen, C., Metaxas, D.: Taming the noisy gradient: train deep neural networks with small batch sizes. In: Proceedings of the Twenty-Eighth International Joint Conference on Artificial Intelligence, IJCAI-19, pp. 4348–4354 (2019)
45. Zintgraf, L., Shiarli, K., Kurin, V., Hofmann, K., Whiteson, S.: Fast context adaptation via meta-learning. In: International Conference on Machine Learning, pp. 7693–7702 (2019)

Domain-Specific Mappings for Generative Adversarial Style Transfer

Hsin-Yu Chang, Zhixiang Wang, and Yung-Yu Chuang(✉)

National Taiwan University, Taipei, Taiwan
cyy@csie.ntu.edu.tw

Abstract. Style transfer generates an image whose content comes from one image and style from the other. Image-to-image translation approaches with disentangled representations have been shown effective for style transfer between two image categories. However, previous methods often assume a shared domain-invariant content space, which could compromise the content representation power. For addressing this issue, this paper leverages domain-specific mappings for remapping latent features in the shared content space to domain-specific content spaces. This way, images can be encoded more properly for style transfer. Experiments show that the proposed method outperforms previous style transfer methods, particularly on challenging scenarios that would require semantic correspondences between images. Code and results are available at https://github.com/acht7111020/DSMAP.

1 Introduction

Style transfer has gained great attention recently as it can create interesting and visually pleasing images. It has wide applications such as art creation and image editing. Given a pair of images, the content image x_A and the style image x_B, style transfer generates an image with x_A's content and x_B's style. It is not easy to define the content and style precisely. However, in general, the content involves more in the layout and spatial arrangement of an image while the style refers more to the colors, tones, textures, and patterns. Figure 1 gives examples of our style transfer results in comparison with an existing method, MUNIT [8].

The seminal work of Gatys et al. [4,5] shows that deep neural networks can extract the correlations of the style and content features, and uses an iterative optimization method for style transfer. Since then, many methods have been proposed to address issues such as generalizing to unseen styles, reducing computation overhead, and improving the matching of the style and content. Image-to-image (I2I) translation aims at learning the mapping between images of two domains (categories) and can be employed for style transfer between two image categories naturally. Recently, some I2I translation approaches have shown great

Electronic supplementary material The online version of this chapter (https://doi.org/10.1007/978-3-030-58598-3_34) contains supplementary material, which is available to authorized users.

A. Vedaldi et al. (Eds.): ECCV 2020, LNCS 12353, pp. 573–589, 2020.
https://doi.org/10.1007/978-3-030-58598-3_34

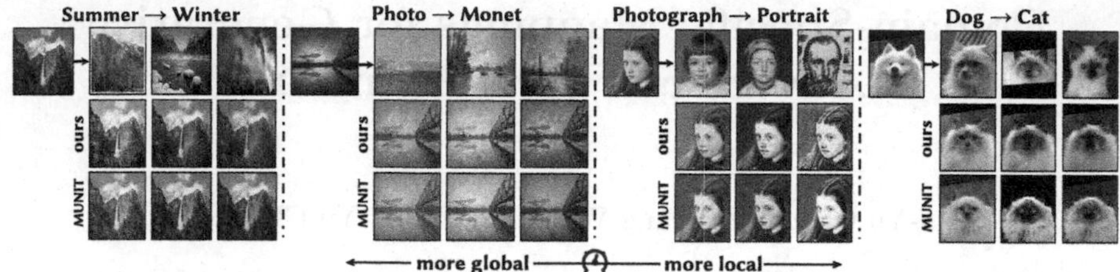

Fig. 1. Our style transfer results in comparison with MUNIT [8]. From left to right, style transfer between two domains becomes "more local" from "more global". For the Summer→Winter task, the style transfer is "more global" as its success mainly counts on adjusting global attributes such as color tones and textures. On the contrary, for the Dog→Cat task, style transfer is "more local" as its success requires more attention on local and structural semantic correspondences, such as "eyes to eyes" and "nose to nose".

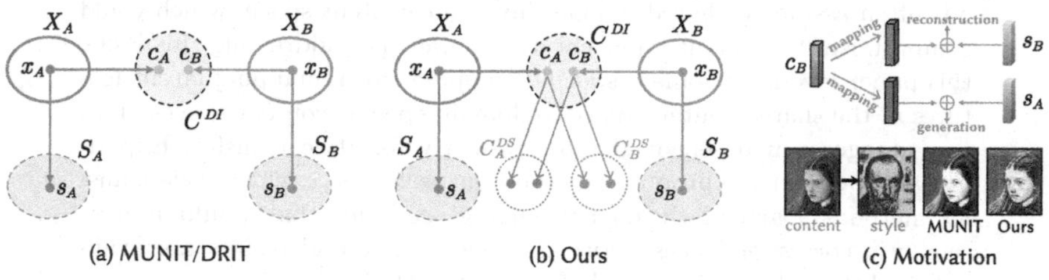

Fig. 2. Main idea and the motivation. (a) MUNIT [8] and DRIT [13] decompose an image into a content feature in the shared domain-invariant content space and a style feature in the domain-specific style space. The shared space could limit representation power. (b) Our method finds the mappings to map a content feature in the shared latent space into each domain's domain-specific content space. This way, the remapped content feature can be better aligned with the characteristics of the target domain. (c) For an image of the source domain, by mapping its content feature to the target domain and combining it with the target domain's style, our method can synthesize better results.

success through disentangled representations, such as MUNIT [8] and DRIT [13]. Although these methods work well for many translation problems, they only give inferior results for some challenging style transfer scenarios, particularly those requiring semantics matches such as eyes to eyes in the Dog-to-Cat transfer.

This paper improves upon the I2I translation approaches with disentangled representations for style transfer. We first give an overview of the I2I approaches using disentangled representations in Fig. 2(a). Inspired by CycleGAN [33] which defines two separated spaces and UNIT [19] which assumes a shared latent space, MUNIT [8] and DRIT [13] encode an image x with two feature vectors, the content feature c and the style feature s. In their setting, although domains X_A and X_B have their own latent domain-specific feature spaces for styles S_A and S_B, they share the same latent domain-invariant space C^{DI} for the content.

Thus, given a content image $x_A \in X_A$ and a style image $x_B \in X_B$, they are encoded as (c_A, s_A) and (c_B, s_B) respectively. The content features c_A and c_B belong to the shared domain-invariant content space C^{DI} while s_A and s_B respectively belong to the style space of its own domain, S_A and S_B. For the cross-domain translation task $A \rightarrow B$, the content feature c_A and the style feature s_B are fed into a generator for synthesizing the result with x_A's content and x_B's style.

Figure 1 shows MUNIT's results for several style transfer tasks. We found that previous I2I methods with disentangled representations often run into problems in "more local" style transfer scenarios. We observe that the shared domain-invariant content space could compromise the ability to represent content since they do not consider the relationship between content and style. We conjecture that the domain-invariant content feature may contain domain-related information, which causes problems in style transfer. To address the issue, we leverage two additional domain-specific mapping functions $\Phi_{C \rightarrow C_A}$ and $\Phi_{C \rightarrow C_B}$ to remap the content features in the shared domain-invariant content space C^{DI} into the features in the domain-specific content spaces C_A^{DS} and C_B^{DS} for different domains (Fig. 2(b)). The domain-specific content space could better encode the domain-related information needed for translation by representing the content better. Also, domain-specific mapping helps the content feature better align with the target domain. Thus, the proposed method improves the quality of translation and handles both local and global style transfer scenarios well.

The paper's main contribution is the observation that both the content and style have domain-specific information and the proposal of the domain-specific content features. Along this way, we design the domain-specific content mapping and propose proper losses for its training. The proposed method is simple yet effective and has the potential to be applied to other I2I translation frameworks.

2 Related Work

Style Transfer. Gatys et al. [5] show that the image representation derived from CNNs can capture the style information and propose an iterative optimization method to calculate the losses between the target and input images. For reducing the substantial computational cost of the optimization problem, some propose to use a feed-forward network [1,10]. Although generating some good results, these methods only have limited ability to transfer to an unseen style.

For general style transfer, Huang et al. [7] propose a novel adaptive instance normalization (AdaIN) layer for better aligning the means and variances of the features between the content and style images. The WCT algorithm [16] embeds a whitening and coloring process to encode the style covariance matrix. Li et al. [15] propose a method that can reduce the matrix computation by learning the transformation matrix with a feed-forward network. Although they can generalize to unseen styles, these networks cannot learn the semantic style translation, such as transferring from a cat to a dog. Cho et al. [2] propose a module

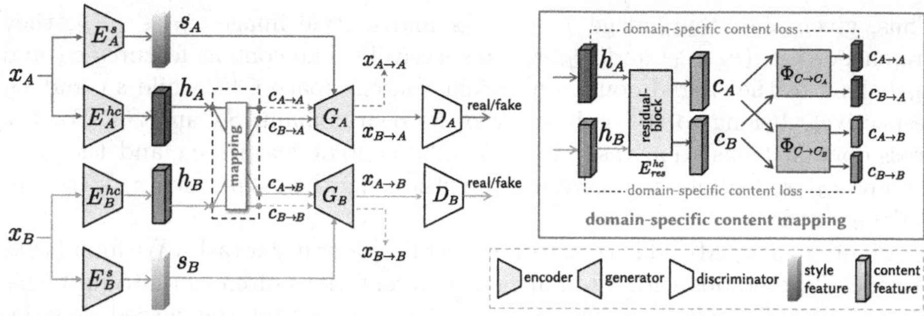

Fig. 3. The framework of the proposed method. For improving cross-domain translation, we use domain-specific mappings for obtaining content features $c_{A\to B}$ and $c_{B\to A}$ instead of domain-invariant content features c_A and c_B for better preserving domain-specific information. The details of the domain-specific content mapping are shown in the box on the top right.

that allows image translation for conveying profound style semantics by using the deep whitening-and-coloring approach.

There are approaches explicitly designed for semantic style transfer. Lu et al. [21] propose a method for generating semantic segmentation maps using VGG, while Luan et al. [22] use the model provided by others for obtaining segmentation masks. The success of these methods heavily depends on the quality of the segmentation masks. For scenarios where there is no good correspondence between masks, these methods can not work well. To reduce computational overhead, Lu et al. [21] decompose the semantic style transfer problem into two sub-problems of feature reconstruction and feature decoding. Liao [17] solve the visual attribute transfer problem when the content and style images have perceptually similar semantic structures. However, it can not work on the images that have less similar semantic structures. Our method learns the semantic correspondences implicitly through content space projection. Thus, it works better even if there is no explicit semantic correspondence between images.

Image-to-Image Translation. Using the cGAN framework, Pix2Pix [9] models the mapping function from the source domain to the target domain. BicycleGAN [34] adds the variational autoencoder (VAE) [12] in cGAN to generate multiple results from a single image, named as multi-modal translation. Both Pix2Pix and BicycleGAN require paired training images. For training with unpaired images, CycleGAN [33] and DualGAN [30] employ a novel constraint, the cycle consistency loss, in the unsupervised GAN framework. The idea is to obtain the same image after transferring to the target domain and then transferring back. Several researchers [3,27,29] use the cycle constraint to generate portraits with different facial attributes. To maintain the background information, some methods [11,25–27] add attention masks in their architectures. In addition to facial images, some deal with images of more classes [11,25,26].

UNIT [19] assumes that two domains map to a shared latent space and learns a unimodal mapping. To deal with multimodal translation problems, DRIT [13] and MUNIT [8] decompose images into a shared content space and different attribute spaces for different domains. To learn individual attribute features, they design a training process for learning the style encoders and the content encoders. However, these methods tend to generate poor style transfer results when there is a large gap between domains. Zheng et al. [32] use a multi-branch discriminator to learn the locations or numbers of objects to be transferred. TransGaGa [28] deals with the translation problem when there are significant geometry variations between domains. It disentangles images into the latent appearance and geometry spaces. Li et al.'s method [14] can learn multiple domains and multimodal translation simultaneously but requires specific labels to generate style-guided results. MSGAN [23] presents a mode seeking regularization term for increasing the diversity of generated results. Recently, some focus on the few-shot and one-shot unsupervised I2I problem [18,20].

3 Proposed Method

For style transfer between two image domains X_A and X_B, our goal is to find two mappings $\Phi_{X_A \to X_B}$ and $\Phi_{X_B \to X_A}$ for the conversion between images in X_A and X_B.

3.1 Model Overview

Figure 3 depicts the basic framework of our method. In addition to basic content encoders $\{E^c_A, E^c_B\}$, style encoders $\{E^s_A, E^s_B\}$, and generators $\{G_A, G_B\}$, our method also learns the domain-specific mapping functions $\{\Phi_{C \to C_A}, \Phi_{C \to C_B}\}$ which map a feature in the shared domain-invariant content space to the domain-specific content spaces of X_A and X_B respectively. For the scenario of $X_A \to X_B$, the content image $x_A \in X_A$ is first encoded into the domain-invariant content feature $c_A = E^c_A(x_A)$, where $c_A \in C^{DI}$. Similarly, the style image $x_B \in X_B$ is encoded into the style feature $s_B = E^s_B(x_B)$, where $s_B \in S_B$. MUNIT and DRIT generate the result simply by $x_{A \to B} = G_B(c_A, s_B)$. We conjecture that the domain-invariant content feature c_A might not be good enough to go with the style feature s_B to generate a good image in the domain X_B. It would be better to align c_A with the content characteristics of the target domain before synthesis. Thus, for improving the results, our method uses the additional domain-specific content mappings to map the content feature into the domain-specific content spaces. In this scenario, the function $\Phi_{C \to C_B}$ is for mapping the content feature c_A to the content space of X_B. At a high level, it obtains the content feature $c_{A \to B}$ by aligning the original content in X_A into the content space of the domain X_B, probably through semantic correspondences between domains. Note that, different from previous approaches that directly use the content feature c_A in X_A, our method aligns the content feature into the content space of X_B, which better matches the style feature and improves results. The generator G_B then takes the content feature $c_{A\rightarrow B}$ and the style feature s_B for synthesizing the output image $x_{A\rightarrow B}$. In sum, for the scenario of $X_A \rightarrow X_B$, our method generates the output $x_{A\rightarrow B}$ by

$$x_{A\rightarrow B} = \Phi_{X_A\rightarrow X_B}(x_A, x_B) = G_B(\Phi_{C\rightarrow C_B}(c_A), s_B).$$

Similarly, for the scenario of $X_B \rightarrow X_A$, we have

$$x_{B\rightarrow A} = \Phi_{X_B\rightarrow X_A}(x_B, x_A) = G_A(\Phi_{C\rightarrow C_A}(c_B), s_A).$$

3.2 Learning Domain-Specific Content Mappings

The key question is how to learn the domain-specific content mappings $\Phi_{C\rightarrow C_A}$ and $\Phi_{C\rightarrow C_B}$ in the latent space. Following the design of MUNIT [8], our content encoder E_A^c is composed of two parts, E_A^{hc} and E_{res}^{hc}, as shown in Fig. 3. The first part E_A^{hc} consists of several strided convolution layers for downsampling. The second part, E_{res}^{hc}, is composed of several residual blocks for further processing. We choose to share E_{res}^{hc} in the encoders of both domains, but using separate residual blocks for different domains also works. Thus, we have

$$c_A = E_A^c(x_A) = E_{res}^{hc}(E_A^{hc}(x_A)), \quad c_B = E_B^c(x_B) = E_{res}^{hc}(E_B^{hc}(x_B)). \tag{1}$$

We opt to use the intermediate content features, $h_A = E_A^{hc}(x_A)$ and $h_B = E_B^{hc}(x_B)$, as the domain-specific content features. In this way, h_A and h_B are domain-specific, and the residual blocks E_{res}^{hc} is responsible for projecting them to the domain-invariant space through minimizing the domain invariant content loss that will be described later in Eq. (4). For finding the domain-specific mapping $\Phi_{C\rightarrow C_A}$, we require that its output resembles the domain-specific content feature h_A so that it can keep domain-specific properties for X_A. Similarly for $\Phi_{C\rightarrow C_B}$. Thus, we have the following domain-specific content (dsc) reconstruction loss. By minimizing the loss, we can obtain the mappings.

$$\begin{aligned} L_1^{dsc_A} &= \mathbb{E}_{x_A}[\|E_A^{hc}(x_A) - \Phi_{C\rightarrow C_A}(E_A^c(x_A))\|_1], \\ L_1^{dsc_B} &= \mathbb{E}_{x_B}[\|E_B^{hc}(x_B) - \Phi_{C\rightarrow C_B}(E_B^c(x_B))\|_1]. \end{aligned} \tag{2}$$

3.3 Losses

In addition to the domain-specific content reconstruction loss introduced in Eq. (2), our formulation consists of several other losses: some for style transfer and the others for image-to-image translation.

3.3.1 Loss for Style Transfer

The style reconstruction loss is employed for training the style encoders.

Style Reconstruction Loss. To ensure the style encoders encode meaningful style features, inspired by the Gaussian priors in DRIT [13], when given a style feature s_A randomly sampled from a Gaussian distribution, we need to reconstruct it back to the original style feature. The loss is defined as the following.

$$L_1^{s_A} = \mathbb{E}_{s_A, c_{B\to A}}[\|E_A^s(G_A(c_{B\to A}, s_A))) - s_A\|_1], \tag{3}$$

where $c_{B\to A} = \Phi_{C\to C_A}(E_B^c(x_B))$.

3.3.2 Losses for Image-to-Image Translation

Similar to other I2I methods [8,13], we adopt adversarial, image reconstruction, domain-invariant content reconstruction, and cycle-consistency losses to facilitate training.

Domain-Invariant Content Loss. Even though we employ domain-specific content features to learn better alignment between domains, we still need the domain-invariant content space for cross-domain translation. We require that x_A and $x_{A\to B}$ in Fig. 3 have the same domain-invariant content feature, *i.e.*,

$$L_1^{dic_A} = \mathbb{E}_{x_A, x_{A\to B}}[\|E_B^c(x_{A\to B})) - E_A^c(x_A)\|_1]. \tag{4}$$

Image Reconstruction Loss. Just like variational autoencoders (VAEs) [12], the image reconstruction loss is used to make sure the generator can reconstruct the original image within a domain.

$$L_1^{x_A} = \mathbb{E}_{x_A}[\|G_A(\Phi_{C\to C_A}(E_A^c(x_A)), E_A^s(x_A)) - x_A\|_1]. \tag{5}$$

Note the original content feature c_A is domain-invariant, we need to map c_A to the domain-specific content feature $c_{A\to A}$. Taking the content feature $c_{A\to A}$ and the style feature s_A, the generator G_A should reconstruct the original image x_A.

Adversarial Loss. Like MUNIT [8], we employ the adversarial loss of LSGAN [24] to minimize the discrepancy between the distributions of the real images and the generated images.

$$\begin{aligned} L_{D_{adv}}^A &= \tfrac{1}{2}\mathbb{E}_{x_A}[((D_A(x_A) - 1))^2] + \tfrac{1}{2}\mathbb{E}_{x_{B\to A}}[(D_A(x_{B\to A}))^2] \\ L_{G_{adv}}^A &= \tfrac{1}{2}\mathbb{E}_{x_{B\to A}}[(D_A(x_{B\to A}) - 1)^2], \end{aligned} \tag{6}$$

where D_A is the discriminator of domain X_A.

Cycle-Consistency Loss. The cycle consistency constraint was proposed in CycleGAN [33] and has been proved useful in unsupervised I2I translation. When a given input x_A passes through the cross-domain translation pipeline $A\to B\to A$, it should be reconstructed back to x_A itself, *i.e.*,

$$L_{cc}^{x_A} = \mathbb{E}_{x_A}[\|x_{A\to B\to A} - x_A\|_1]. \tag{7}$$

3.3.3 Total Loss

Our goal is to perform the cross-domain training, so we need to train on both directions. Losses on the other direction are defined similarly. We combine each pair of dual terms together such as $L_{cc}^{x} = L_{cc}^{x_A} + L_{cc}^{x_B}$, $L_1^{x} = L_1^{x_A} + L_1^{x_B}$ *etc.* Finally, the total loss is defined as

$$\begin{aligned} L_{G_{total}} &= \lambda_{cc} L_{cc}^{x} + \lambda_x L_1^{x} + \lambda_{dsc} L_1^{dsc} + \lambda_{dic} L_1^{dic} + \lambda_s L_1^{s} + \lambda_{adv} L_{G_{adv}} \\ L_D &= \lambda_{adv} L_{D_{adv}} \end{aligned} \tag{8}$$

where λ_{cc}, λ_x, ... are hyper-parameters for striking proper balance among losses.

4 Experiments

4.1 Datasets

We present results on more challenging cross-domain translation problems in the paper. More results can be found in the supplementary document.

Photo ⇄ Monet [33]. This dataset is provided by CycleGAN, which includes real scenic images and Monet-style paintings.

Cat ⇄ Dog [13]. DRIT collects 771 cat images and 1,264 dog images. There is a significant gap between these two domains, and it is necessary to take the semantic information into account for translation.

Photograph ⇄ Portrait [13]. This dataset is also provided by DRIT, which includes portraits and photographs of human faces. The gap is smaller than the Cat ⇄ Dog dataset. However, the difficulty of the translation is to preserve identity while performing style translation.

4.2 Competing Methods

We compare our method with three image-to-image translation methods and three style transfer methods.

Image-to-Image Translation Methods. MUNIT [8] disentangles the images into the domain-invariant content space and domain-specific style spaces. MSGAN [23] extends DRIT [13] by adopting the mode seeking regularization term to improve the diversity of the generated images. Since MSGAN generally generates better results than DRIT, we do not compare with DRIT. We use the pre-trained cat ⇄ dog model provided by MSGAN while training MSGAN on the other two datasets. GDWCT [2] applies WCT [16] to I2I translation and obtains better translation results. For the I2I translation methods: MUNIT, MSGAN, GDWCT and ours, we train them on NVIDIA GTX 1080Ti for two days.

Style Transfer Methods. AdaIN [7] uses the adaptive instance normalization layer to match the mean and variance of content features to those of style features for style transfer. Liao et al. [17] assume that the content and style images have

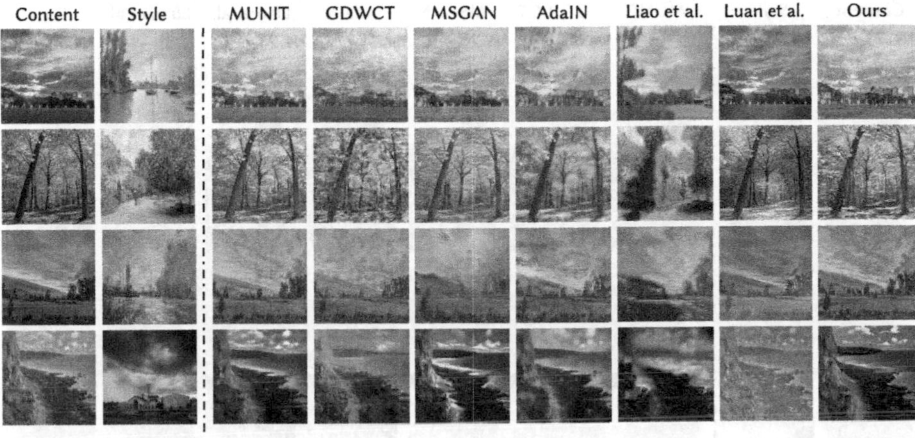

Fig. 4. Comparisons on Photo ⇄ Monet.

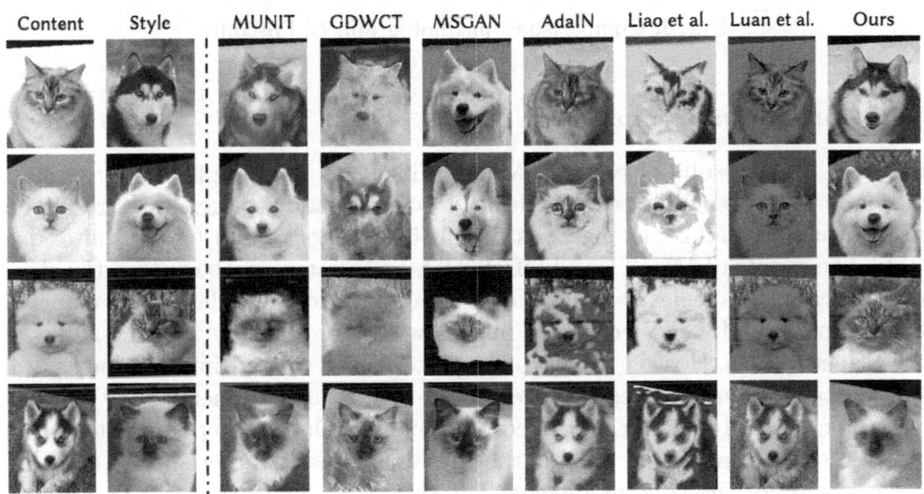

Fig. 5. Comparisons on Cat ⇄ Dog.

perceptually similar semantic structures. Their method first extracts features using DNN, then adopts a coarse-to-fine strategy for computing the nearest-neighbor field and matching features. Luan et al. [22] apply the segmentation masks to segment the semantic of content and style images and employ iterative optimization to calculate the loss functions. Note it is not completely fair to compare I2I methods with the style transfer methods as some of them have different definitions of styles and could use less information than I2I methods.

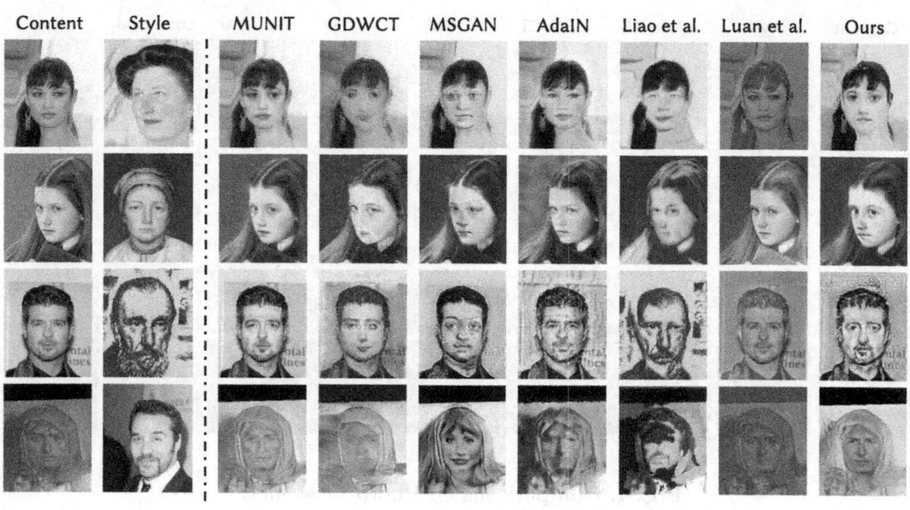

Fig. 6. Comparisons on Photograph ⇄ Portrait.

4.3 Qualitative Comparisons

Figure 4 shows results of photo ⇄ Monet. Since this is a simpler scenario, most methods give reasonable results. However, our method still generates results of higher quality than other methods. Figure 5 presents results for the cat ⇄ dog task. The results of MUNIT and GDWCT have the same problem that the characteristics of the species are not clear. It is not easy to judge the species depicted by the images. MSGAN generates images with more obvious characteristics of the target species. However, it does not preserve the content information as well as our method. In their results, the poses and locations of facial/body features are not necessarily similar to the content images. Our method generates much clearer results that better exhibit the characteristics of target species and preserve layouts of the content images. The style transfer methods [7,17,22] have poor performance due to the different assumption of styles and the use of less information.

Figure 6 shows the comparisons of the photographs ⇄ portraits task. This task is challenging because the identity in the content image must be preserved, thus often requiring better semantic alignment. MUNIT preserves the identity very well but does not transfer the style well. On the contrary, GDWCT transfers styles better, but the identity is often not maintained well. MSGAN transfers the style much better than MUNIT and GDWCT but does not perform well on identity preservation. Our method performs both style transfer and identity preservation well. Style transfer methods again provide less satisfactory results than I2I methods in general. Note that there is subtle expression change in the portraits' expressions in the first two rows of Fig. 6. As other disentangled representations, our method does not impose any high-level constraints or knowledge on the decoupling of the style and content features. Thus, the division between the style and content has to be learned from data alone. There could be ambiguity in the division. Other methods except for MUNIT, exhibit similar or even aggravated expression change. MUNIT is an exception as it tends to preserve content but does little on style transfer.

4.4 Quantitative Comparison

As shown in the previous section, the compared style transfer methods cannot perform cross-domain style transfer well. Thus, we only include image-to-image translation methods in the quantitative comparison.

Table 1. Quantitative comparison. We use the FID score (lower is better) and the LPIPS score (higher is better) to evaluate the quality and diversity of each method on six types of translation tasks. Red texts indicate the best and blue texts indicate the second best method for each task and metric.

	FID↓				LPIPS↑			
	MUNIT	GDWCT	MSGAN	Ours	MUNIT	GDWCT	MSGAN	Ours
Cat → Dog	38.09	91.40	20.80	13.60	0.3501	0.1804	0.5051	0.4149
Dog → Cat	39.71	59.72	28.30	19.69	0.3167	0.1573	0.4334	0.3174
Monet → Photo	85.06	113.16	86.72	81.61	0.4282	0.2478	0.4229	0.5379
Photo → Monet	77.85	71.68	80.37	63.94	0.4128	0.2097	0.4306	0.4340
Portrait → Photograph	93.45	83.69	57.07	62.44	0.1819	0.1563	0.3061	0.3160
Photograph → Portrait	89.97	75.86	57.84	45.81	0.1929	0.1785	0.2917	0.3699
Average	70.69	82.59	55.18	47.85	0.3131	0.1881	0.3978	0.3980

Quality of Images. We use the FID score [6] to measure the similarity between distributions of generated images and real images in the cross-domain translation task. FID is calculated by computing the Fréchet distance through the features extracted from the Inception network. The lower FID score indicates a better quality, and the generated images are closer to the target domain.

We randomly sample 100 test images and generate ten different example-guided results for each image. These results are then used to calculate the FID score for each method. We repeat ten times and report the average scores. As shown in Table 1, our method achieves the best scores except for the task of Portrait→Photograph. Note that it is often more challenging to generate realistic photographs. Thus, the FID scores for the tasks generating photographs are generally worse. For other scenarios, our method often has a significantly lower score than other methods.

Diversity. To measure the diversity among the generated images, we report the LPIPS score [31], which measures feature distances between paired outputs. The higher LPIPS scores indicate better diversity among generated images. We randomly sample 100 content images, and for each of them, we generate 15 paired results. Again, we repeat ten times and report the average scores. As shown in Table 1, even if our mapping function is not designed to increase diversity, our method achieves good diversity and performs very well for the photographs ⇄

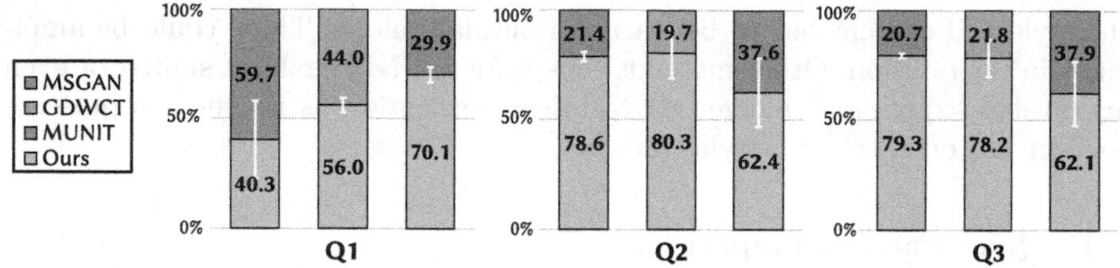

Fig. 7. Result of the user study. The numbers indicate the percentage of users preferring in the pairwise comparison. We conduct the user study on Cat→Dog, Dog→Cat and Photograph→Portrait translation tasks and report their averages. The white error bar indicates the standard deviation. The complete results can be found in the supplementary.

portraits and Monet ⇄ photos tasks. MSGAN [23] specifically adds loss function for promoting the diversity of generated images. Thus, its results also have good diversity, as seen in the cat ⇄ dog tasks.

User Study. For each test set, users are presented with the content image (domain A), the style image (domain B), and two result images, one generated by our method and the other by one of the three compared methods. The result images are presented in random order. The users need to select the better image among the two given results for the following three questions.

- *Q1: Which one preserves content information (identity, shape, semantic) better?*
- *Q2: Which one performs better style translation (in terms of color, pattern)?*
- *Q3: Which one is more likely to be a member of the domain B?*

As shown in Fig. 7, MUNIT preserves visual characteristics of the content very well but does very little on transferring styles. Thus, MUNIT has high scores in *Q1* while having very low scores in *Q2* and *Q3*. GDWCT has a similar performance to MUNIT. On the contrary, MSGAN is less capable of preserving content while performing better in style transfer than MUNIT/GDWCT. For style transfer (*Q2* and *Q3*), MSGAN performs slightly better than our method for the scenario of cats → dogs, but falls significantly behind our method in other scenarios. For content preservation (*Q1*), MSGAN is much worse than ours.

4.5 Discussions

Ablation Studies. Figure 8 gives the results without the individual losses in our method for better understanding their utility. We show the ablation study on Dog→Cat since this task is more challenging. Without the domain-specific content mapping, the spatial layouts of the content images can not be preserved well. With the mapping, poses and sizes of the synthesized cats better resemble those of the dogs in the content images. The proposed loss L_1^{dsc} ensures the remapped feature resembles the domain-specific feature h_A, and it is essential to the learning of $\Phi_{C \to C_A}$. Our model cannot learn the correct mapping without L_1^{dsc}. Without L_1^{dic}, content preservation is less stable. The model can not learn

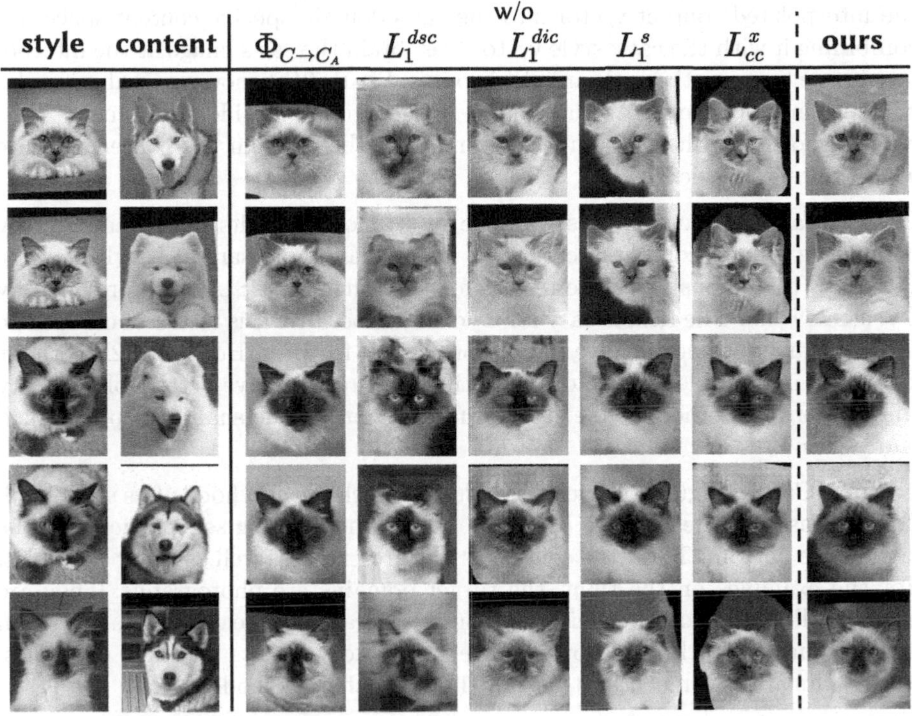

Fig. 8. Visual comparisons of the ablation study. We show the results without (w/o) the domain-specific content mapping $\Phi_{C\to C_A}$, domain-specific content loss L_1^{dsc}, domain-invariant content loss L_1^{dic}, style reconstruction loss L_1^{s}, and cycle-consistency loss L_{cc}^{x} for several examples of Dog→Cat.

the correct style without L_1^{s} because there is no cue to guide proper style encoding. The cycle consistency loss L_{cc}^{x} is essential for unsupervised I2I learning. Without it, content and style cannot be learned in an unsupervised manner.

Interpolation in the Latent Space. Figure 9(a) shows the results of style interpolation for the fixed content at the top and domain-specific content interpolation for the fixed style at the bottom. They demonstrate that our model has nice continuity property in the latent space.

Note that our content space is domain-specific, and thus the content interpolation must be performed in the same domain. It seems that we cannot perform content interpolation between a cat and a dog since their content vectors are in different spaces. However, it is still possible to perform content interpolation between two domains by interpolating in the shared space and then employing the domain-specific mapping. Taking the first row of Fig. 9(b) as an example, the first content image is a cat (domain A) at one pose while the second content is a dog (domain B) at another pose. We first obtain their content vectors in the shared content space and then perform interpolation in the shared space. Next, since the style is a cat (domain A), we employ the mapping $\Phi_{C\to C_A}$ to remap the interpolated content vector into the cat's domain-specific content space. By combining it with the cat's style vector, we obtain the cat's image at the interpolated pose between the cat's and dog's poses. Figure 9(b) shows several examples of linear interpolation between content images from two different domains. The results show that the domain-specific mapping helps align the content feature with the target style.

Other Tasks. Figure 10(a) shows the multi-modal results of our method by combining a fixed content vector with several randomly sampled style vectors. Figure 10(b) demonstrates the results for the Iphone ⇄ DSLR task provided by CycleGAN [33]. The task is easy for CycleGAN because it is a task of one-to-one mapping and does not involve the notion of style features. However, I2I methods could run into problems with color shifting. Our results have better color fidelity than other I2I methods while successfully generating the shallow depth-of-field effects.

Failure Cases. Figure 11 gives examples in which our method is less successful. For the cases on the left, the poses are rare in the training set, particularly the one on the bottom. Thus, the content is not preserved as well as other examples. For the examples on the right, the target domains are photographs. They are more challenging, and our method could generate less realistic images. Even though our approach does not produce satisfactory results for these examples, our results are still much better than those of other methods.

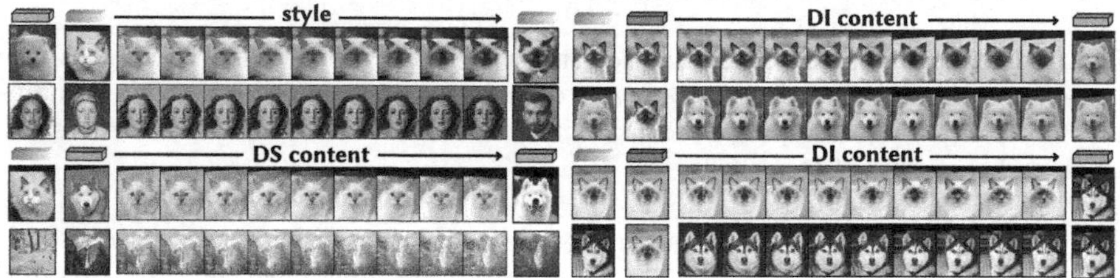

(a) interpolation of style and domain-specific content vectors (b) interpolation of domain-invariant content vectors

Fig. 9. Interpolation in the latent space.

(a) multi-modal results (b) iPhone→DSLR

Fig. 10. Additional results. (a) multi-modal results and (b) iPhone→DSLR.

Fig. 11. Failure examples.

5 Conclusion

This paper proposes to use domain-specific content mappings to improve the quality of the image-to-image translation with the disentangled representation. By aligning the content feature into the domain-specific content space, the disentangled representation becomes more effective. Experiments on style transfer show that the proposed method can better handle more challenging translation problems, which would require more accurate semantic correspondences. In the future, we would like to explore the possibility of applying the domain-specific mapping to other I2I translation frameworks.

Acknowledgments. This work was supported in part by MOST under grant 107-2221-E-002-147-MY3 and MOST Joint Research Center for AI Technology and All Vista Healthcare under grant 109-2634-F-002-032.

References

1. Chen, D., Yuan, L., Liao, J., Yu, N., Hua, G.: StyleBank: an explicit representation for neural image style transfer. In: CVPR (2017)
2. Cho, W., Choi, S., Keetae Park, D., Shin, I., Choo, J.: Image-to-image translation via group-wise deep whitening-and-coloring transformation. In: CVPR (2019)
3. Choi, Y., Choi, M., Kim, M., Ha, J.W., Kim, S., Choo, J.: StarGAN: unified generative adversarial networks for multi-domain image-to-image translation. In: CVPR (2018)
4. Gatys, L., Ecker, A.S., Bethge, M.: Texture synthesis using convolutional neural networks. In: NeurIPS (2015)
5. Gatys, L.A., Ecker, A.S., Bethge, M.: Image style transfer using convolutional neural networks. In: CVPR (2016)
6. Heusel, M., Ramsauer, H., Unterthiner, T., Nessler, B., Hochreiter, S.: GANs trained by a two time-scale update rule converge to a local nash equilibrium. In: NeurIPS (2017)
7. Huang, X., Belongie, S.: Arbitrary style transfer in real-time with adaptive instance normalization. In: ICCV (2017)
8. Huang, X., Liu, M.-Y., Belongie, S., Kautz, J.: Multimodal unsupervised image-to-image translation. In: Ferrari, V., Hebert, M., Sminchisescu, C., Weiss, Y. (eds.) ECCV 2018. LNCS, vol. 11207, pp. 179–196. Springer, Cham (2018). https://doi.org/10.1007/978-3-030-01219-9_11

9. Isola, P., Zhu, J.Y., Zhou, T., Efros, A.A.: Image-to-image translation with conditional adversarial networks. In: CVPR (2017)
10. Johnson, J., Alahi, A., Fei-Fei, L.: Perceptual losses for real-time style transfer and super-resolution. In: Leibe, B., Matas, J., Sebe, N., Welling, M. (eds.) ECCV 2016. LNCS, vol. 9906, pp. 694–711. Springer, Cham (2016). https://doi.org/10.1007/978-3-319-46475-6_43
11. Kim, J., Kim, M., Kang, H., Lee, K.: U-GAT-IT: unsupervised generative attentional networks with adaptive layer-instance normalization for image-to-image translation. arXiv preprint arXiv:1907.10830 (2019)
12. Kingma, D.P., Welling, M.: Auto-encoding variational bayes. arXiv preprint arXiv:1312.6114 (2013)
13. Lee, H.-Y., Tseng, H.-Y., Huang, J.-B., Singh, M., Yang, M.-H.: Diverse image-to-image translation via disentangled representations. In: Ferrari, V., Hebert, M., Sminchisescu, C., Weiss, Y. (eds.) ECCV 2018. LNCS, vol. 11205, pp. 36–52. Springer, Cham (2018). https://doi.org/10.1007/978-3-030-01246-5_3
14. Li, X., et al.: Attribute guided unpaired image-to-image translation with semi-supervised learning. arXiv preprint arXiv:1904.12428 (2019)
15. Li, X., Liu, S., Kautz, J., Yang, M.H.: Learning linear transformations for fast arbitrary style transfer. arXiv preprint arXiv:1808.04537 (2018)
16. Li, Y., Fang, C., Yang, J., Wang, Z., Lu, X., Yang, M.H.: Universal style transfer via feature transforms. In: NeurIPS (2017)
17. Liao, J., Yao, Y., Yuan, L., Hua, G., Kang, S.B.: Visual attribute transfer through deep image analogy. In: ACM SIGGRAPH (2017)
18. Lin, J., Pang, Y., Xia, Y., Chen, Z., Luo, J.: TuiGAN: learning versatile image-to-image translation with two unpaired images. ECCV (2020)
19. Liu, M.Y., Breuel, T., Kautz, J.: Unsupervised image-to-image translation networks. In: NeurIPS (2017)
20. Liu, M.Y., et al.: Few-shot unsupervised image-to-image translation. In: ICCV (2019)
21. Lu, M., Zhao, H., Yao, A., Xu, F., Chen, Y., Zhang, L.: Decoder network over lightweight reconstructed feature for fast semantic style transfer. In: ICCV (2017)
22. Luan, F., Paris, S., Shechtman, E., Bala, K.: Deep photo style transfer. In: CVPR (2017)
23. Mao, Q., Lee, H.Y., Tseng, H.Y., Ma, S., Yang, M.H.: Mode seeking generative adversarial networks for diverse image synthesis. In: CVPR (2019)
24. Mao, X., Li, Q., Xie, H., Lau, R.Y., Wang, Z., Paul Smolley, S.: Least squares generative adversarial networks. In: ICCV (2017)
25. Mejjati, Y.A., Richardt, C., Tompkin, J., Cosker, D., Kim, K.I.: Unsupervised attention-guided image-to-image translation. In: NeurIPS (2018)
26. Mo, S., Cho, M., Shin, J.: InstaGAN: instance-aware image-to-image translation. arXiv preprint arXiv:1812.10889 (2018)
27. Pumarola, A., Agudo, A., Martinez, A.M., Sanfeliu, A., Moreno-Noguer, F.: GANimation: anatomically-aware facial animation from a single image. In: Ferrari, V., Hebert, M., Sminchisescu, C., Weiss, Y. (eds.) ECCV 2018. LNCS, vol. 11214, pp. 835–851. Springer, Cham (2018). https://doi.org/10.1007/978-3-030-01249-6_50
28. Wu, W., Cao, K., Li, C., Qian, C., Loy, C.C.: TransGaGa: geometry-aware unsupervised image-to-image translation. arXiv preprint arXiv:1904.09571 (2019)
29. Xiao, T., Hong, J., Ma, J.: ELEGANT: exchanging latent encodings with GAN for transferring multiple face attributes. In: Ferrari, V., Hebert, M., Sminchisescu, C., Weiss, Y. (eds.) ECCV 2018. LNCS, vol. 11214, pp. 172–187. Springer, Cham (2018). https://doi.org/10.1007/978-3-030-01249-6_11

30. Yi, Z., Zhang, H., Tan, P., Gong, M.: DualGAN: unsupervised dual learning for image-to-image translation. In: ICCV (2017)
31. Zhang, R., Isola, P., Efros, A.A., Shechtman, E., Wang, O.: The unreasonable effectiveness of deep features as a perceptual metric. In: CVPR (2018)
32. Zheng, Z., Yu, Z., Zheng, H., Wu, Y., Zheng, B., Lin, P.: Generative adversarial network with multi-branch discriminator for cross-species image-to-image translation. arXiv preprint arXiv:1901.10895 (2019)
33. Zhu, J.Y., Park, T., Isola, P., Efros, A.A.: Unpaired image-to-image translation using cycle-consistent adversarial networks. In: ICCV (2017)
34. Zhu, J.Y., et al.: Toward multimodal image-to-image translation. In: NeurIPS (2017)

DiVA: Diverse Visual Feature Aggregation for Deep Metric Learning

Timo Milbich[1(✉)], Karsten Roth[1,2], Homanga Bharadhwaj[3,4], Samarth Sinha[2,4], Yoshua Bengio[2,5], Björn Ommer[1], and Joseph Paul Cohen[2]

[1] Heidelberg Collaboratory for Image Processing (HCI), Heidelberg University, Heidelberg, Germany
timo.milbich@iwr.uni-heidelberg.de
[2] Mila, Universite de Montreal, Montreal, Canada
[3] Vector Institute, Toronto Robotics Institute, Toronto, Canada
[4] University of Toronto, Toronto, Canada
[5] CIFAR, Toronto, Canada

Abstract. Visual similarity plays an important role in many computer vision applications. Deep metric learning (DML) is a powerful framework for learning such similarities which not only generalize from training data to identically distributed test distributions, but in particular also translate to *unknown* test classes. However, its prevailing learning paradigm is class-discriminative supervised training, which typically results in representations specialized in separating training classes. For effective generalization, however, such an image representation needs to capture a diverse range of data characteristics. To this end, we propose and study multiple complementary learning tasks, targeting conceptually different data relationships by only resorting to the available training samples and labels of a standard DML setting. Through simultaneous optimization of our tasks we learn a single model to aggregate their training signals, resulting in strong generalization and state-of-the-art performance on multiple established DML benchmark datasets.

Keywords: Deep metric learning · Generalization · Self-supervision

1 Introduction

Many applications in computer vision, such as image retrieval [33,39,64] and face verification [53,54], rely on capturing visual similarity, where approaches are commonly driven by Deep Metric learning (DML) [39,54,64]. These models aim to learn an embedding space which meaningfully reflects similarity between training images and, more importantly, generalizes to test classes which are *unknown* during training. Even though models are evaluated on transfer learning, the prevailing training paradigm in DML utilizes discriminative supervised learning. Consequently, the learned embedding space is specialized to features which help

T. Milbich, K. Roth, B. Ommer and J. P. Cohen—Equal first and last authorship.

A. Vedaldi et al. (Eds.): ECCV 2020, LNCS 12353, pp. 590–607, 2020.
https://doi.org/10.1007/978-3-030-58598-3_35

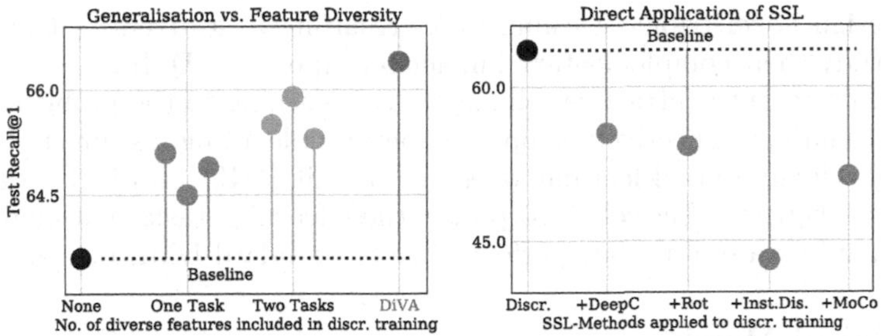

Fig. 1. *DML using diverse learning tasks vs. direct incorporation of self-supervision.* (Left) Generalization performance increases with each task added to training, independent of the exact combination of our proposed tasks (blue: one extra task, orange: two extra tasks, green: all tasks). (Right) Directly combining supervised learning with self-supervised learning techniques such as DeepC(luster) [7], Rot(Net) [18], Inst.Dis(crimination) [65] or Mo(mentum)Co(ntrast) [21] actually hurts DML generalization. (Color figure online)

only in separating among training classes, and may not correctly translate to unseen test classes. Now, if supervised learning does not result in sufficient generalization, how can we exploit the available training data and class labels to provide additional training signals beyond the standard discriminative task?

Recent breakthroughs in self-supervised learning have shown that contrastive image relations inferred from images themselves yield rich feature representations which even surpass the ability of supervised features to generalize to novel downstream task [9,21,43]. However, although DML typically also learns from image relations in the form of pairs [20], triplets [53,64] or more general image tuples [10,42], the complementary benefit of self-supervision in DML is largely unstudied. Moreover, the commonly available class assignments give rise to image relations aside from the standard, supervised learning task of 'pulling' samples with identical class labels together while 'pushing' away samples with different labels. As such ranking-based learning is not limited to discriminative training only, other relations can be exploited to learn beneficial data characteristics which so far have seen little coverage in DML literature.

In this paper, we tackle the issue of generalization in DML by designing diverse learning tasks complementing standard supervised training, leveraging only the commonly provided training samples and labels. Each of these tasks aims at learning features representing different relationships between our training classes and samples: *(i)* features discriminating among classes, *(ii)* features shared across different classes, *(iii)* features capturing variations within classes and *(iv)* features contrasting between individual images. Finally, we present how to effectively incorporate them in a unified learning framework. In our experiments we study mutual benefits of these tasks and show that joint optimization of diverse representations greatly improves generalization performance as shown in Fig. 1 (left), outperforming the state-of-the-art in DML. Our contributions can be summarized as follows:

- We design novel triplet learning tasks resulting in a diverse set of features and study their complementary impact on supervised DML.
- We adopt recent contrastive self-supervised learning to the problem of DML and extend it to effectively support supervised learning, as direct incorporation of self-supervised learning does not benefit DML (cf. Fig. 1) (right).
- We show how to effectively incorporate these learning tasks in a single model, resulting in state-of-the-art performance on standard DML benchmark sets.

2 Related Work

Deep Metric Learning. Deep Metric Learning is one of the primary frameworks for image retrieval [33,39,49,64], zero-shot generalization [2,49,52,53] or face verification [12,24,34]. It is also closely related to recent successful unsupervised representation learning approaches employing contrastive learning [9,21,38]. Commonly, DML approaches are formulated as ranking tasks on data tuples such as image pairs [20], triplets [24,64], quadruplets [10] or higher order relations [42,54,63]. Effective training of these methods is typically promoted by tuple mining strategies alleviating the high sampling complexity, such as distance-based [53,64], hierarchical [17] or learned [50]. Methods like ProxyNCA [39], Softtriple [12], Arcface [12] or Normalized Softmax [67] introduce learnable data proxies which represent entire subsets of the data, thus circumventing the tuple selection process. Orthogonally, DML research has started to pay more emphasis on the training process itself. This involves the generation of artificial training data [33,69] or adversarial objectives [14]. MIC [49] explains away intra-class variance to strengthen the discriminative embedding. [52] propose to separate the input data space to learn subset-specific, yet still only class-discriminative representations similar to other ensemble methods [28,37,44,66]. In contrast, we learn different embeddings on conceptually different tasks to capture diverse image features.

Self-supervised Representation Learning. Commonly, self-supervised representation learning aims to learn transferable feature representations from unlabelled data, and is typically applied as pre-training for downstream tasks [23,36]. Early methods on representation learning are based on sample reconstructions [30,58] which have been further extended by interpolation constraints [4] and generative adversarial networks [3,11,13,15]. Further, introducing manually designed surrogate objectives encourage self-supervised models to learn about data-related properties. Such tasks range from predicting image rotations [18], solving a Jigsaw puzzle [6,40,41] to iteratively refining the initial network bias using clustering algorithms [7]. Recently, self-supervision approaches based on contrastive learning result in strong features performing close to or even stronger than supervised pretraining [9,21,38,55] by leveraging invariance to realistic image augmentations. As the these approaches are essentially defined on pairwise image relations, they share common ground with ranking-based DML. In our work, we extend such a contrastive objective to effectively complement supervised DML training.

Multi-task Learning. Concurrently solving different tasks is also employed by classical multi-task learning which are often based on a divide-and-conquer principle with multiple learner optimizing a given subtask. [5] utilizes additional training data and annotations to capture extra information, while our tasks are defined on standard training data only. [46] learn different classifiers for groups of entire categories, thus following a similar motivation as some DML approaches [44,52]. The latter aims at learning more fine-grained, yet only discriminative features by reducing the data variance for each learner, thus being related to standard hard-negative mining [53]. In contrast, our work formulates various specific learning tasks to target *different* data characteristics of the training data.

3 Method

Let $f_i := f(I_i, \theta) \in \mathbb{R}^N$ be a N-dimensional encoding of an image $I_i \in \mathcal{I}$ represented by a deep neural network with parameters θ. Based on f_i, deep metric learning (DML) aims to learn image embeddings $\phi_i := \phi(f_i, \zeta) : \mathbb{R}^N \mapsto \mathbb{R}^D$ which allow to measure the similarity between images I_i, I_j as $d^\phi_{i,j} := d(\phi_i, \phi_j)$ under a predefined distance metric $d(\cdot,\cdot)$. Typically, $\phi(\cdot, \zeta)$ is a linear layer on the features f representation, parameterized by ζ and normalized to the real hypersphere $\mathbb{S}^{D-1}$ for regularization [64]. $d(\cdot,\cdot)$ is usually chosen to be the Euclidean distance. In standard supervised DML, ϕ is then optimized to reflect semantic similarity between images I_i defined by the corresponding class labels $y_i \in \mathcal{Y}$.

While there are many ways to define training objectives on ϕ, ranking losses, such as variants of the popular triplet loss [42,54,64], are a natural surrogate for the DML problem. Based on image triplets $t = \{I_a, I_p, I_n\}$ with I_a defined as anchor, I_p as a similar, positive and I_n as a negative image, we minimize

$$\mathcal{L}_{\text{tri}}(t) = [d^\phi_{a,p} - d^\phi_{a,n} + \gamma]_+ \ , \tag{1}$$

where $[\cdot]_+$ defines the hinge function which clips any negative value to zero. Hence, we maximize the gap between $d^\phi_{a,p}$ and $d^\phi_{a,n}$ as long as a margin γ is violated.

In supervised DML, $\mathcal{L}_{\text{tri}}(t)$ is typically optimized to discriminate between classes. Thus, f is trained to predominantly capture highly discriminative features while being invariant to image characteristics which do not facilitate training class separation. However, as we are interested in generalizing to unknown test distributions, we should rather aim at maximizing the amount of features captured from the training set $\mathcal{I}$, i.e. additionally formulate learning tasks which yield features beyond mere class-discrimination.

In order to formulate such tasks, we make use of the fact that triplet losses are instance-based. Thus, conceptually they are not restricted to class discrimination, but allow to learn *commonalities between the provided anchor I_a and positive I_p* compared to a negative sample I_n. Following, we will use this observation to also learn those commonalities which are neglected by discriminative training, such as commonalities shared between and within classes.

3.1 Diverse Learning Tasks for DML

We now introduce several tasks for learning a diverse set of features, resorting only to the standard training information provided in a DML problem. Each of these tasks is designed to learn features which are conceptually neglected by the others to be mutually complementary. First, we introduce the intuition behind each feature type, before describing how to learn it based on pairwise or triplet-based image relations.

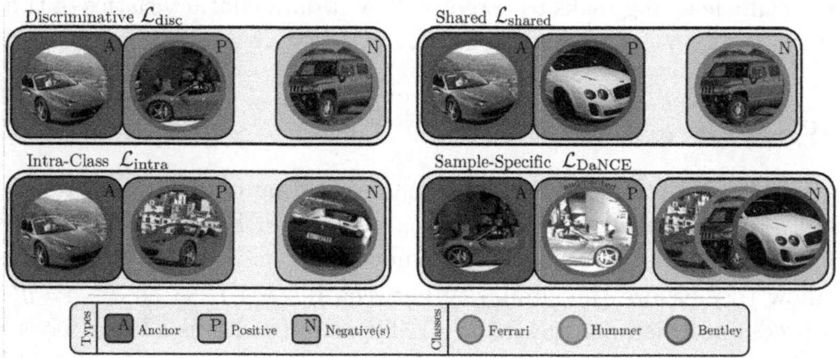

Fig. 2. *Schematic description of each task.* We learn four complementary tasks to capture features focusing on different data characteristics. The standard *class-discriminative* task which learning features separating between samples of different classes, the *shared* task which captures features relating samples across different classes, a *sample-specific* task to enforce image representations invariant to transformations and finally the *intra-class* task modelling data variations within classes.

Class-Discriminative Features. These features are learned by standard class-discriminative optimization of ϕ and focus on data characteristics which allow to accurately separate one class from all others. It is the prevailing training signal of common classification-based [12,61,67], proxy-based [39,47] or ranking-based [44,49] approaches. For the latter, we can formulate the training task using Eq. 1 by means of triplets $\{I_a, I_p, I_n\} \in \mathcal{T}_{\text{disc}}$ with $y_a = y_p$ and $y_a \neq y_n$, as

$$\mathcal{L}_{\text{disc}} = \frac{1}{Z} \sum_{t \sim \mathcal{T}_{\text{disc}}} \mathcal{L}_{\text{tri}}(t) \ , \tag{2}$$

thus minimizing embedding distances between samples of the same class while maximizing it for samples of different classes. Moreover, the discriminative signal is important to learn how to aggregate features into classes, following the intuition of "the whole is more than the sum of its parts" analyzed in Gestalt theory [59].

Class-Shared Features. In contrast to discriminative features which look for characteristics *separating classes*, class-shared features capture commonalities, i.e variations, *shared across classes.* For instance, cars have a certain invariance towards color changes, thus being of little help when separating between them. However, to learn about this characteristic is actually beneficial when generalizing to other colorful object classes like flowers or fishes. Given suitable label information, learning such features would naturally follow the standard discriminative training setup. However, having only class labels available, we must resort to approximations. To this end, we exploit the hypothesis that for most arbitrarily sampled triplets $\{I_a, I_p, I_n\} \in \mathcal{T}_{\text{shared}}$ with each constituent coming from mutually different classes, i.e. $y_a \neq y_p \neq y_n$, the anchor I_a and positive I_p share some common pattern when compared the negative image I_n. Commonalities which are frequently observed between classes y_a, y_p, will occur more often than noisy patterns which are unique to few t_{shared}, which is commonly observed when learning on imbalanced data [8,16,31]. Learning is then performed by optimizing

$$\mathcal{L}_{\text{shared}} = \frac{1}{Z} \sum_{t \sim \mathcal{T}_{\text{shared}}} \mathcal{L}_{\text{tri}}(t). \tag{3}$$

As deep networks learn from frequent backpropagation of similar learning signals resulting in informative gradients, only prominent shared features are captured. Further, since shared features can be learned between any classes, we need to warrant diverse combinations of classes in our triplets $\mathcal{T}_{\text{shared}}$. Thus, enabling triplet constituents to be sampled from the whole embedding space ϕ using distance-based sampling [64] is crucial to avoid any bias towards samples which are mostly far (random sampling) or close (hard-negative sampling) to a given anchor I_a.

Intra-class Features. The tasks defined so far model image relations across classes. In contrast, intra-class features describe variations within a given class. While these variations may also apply to other classes (thus exhibiting a certain overlap with class-shared features), more class-specific details are targeted. Hence, to capture such data characteristics by means of triplet constraints, we can follow a similar intuition as for learning class-shared features. Thus, to learn intra-class features, we define triplets by means of triplets $\{I_a, I_p, I_n\} \in \mathcal{T}_{\text{intra}}$ with $y_a = y_p = y_n$, i.e. each constituent coming from the same class, and minimize

$$\mathcal{L}_{\text{intra}} = \frac{1}{Z} \sum_{t \sim \mathcal{T}_{\text{intra}}} \mathcal{L}_{\text{tri}}(t). \tag{4}$$

Sample-Specific Features. Recent approaches for self-supervised learning [1, 21,43] based on noise contrastive estimation (NCE) [19] show that features exhibiting strong generalization for transfer learning can be learned only from training images themselves. As NCE learns to increase the correlation between embeddings of an anchor sample and a similar positive sample by contrasting against a set of negative samples, it naturally translates to DML. He et al. [21] proposed an efficient self-supervised framework which first applies data augmentation to generate positive surrogates $\tilde{I}_a$ for a given anchor I_a. Next, using NCE we contrast their embeddings $\phi_a := \phi(I_a, \zeta), \tilde{\phi}_a := \phi(\tilde{I}_a, \zeta)$ against randomly sampled negatives $I_n \in \mathcal{N} \subset \mathcal{I}$ by minimizing

$$\mathcal{L}_{\text{NCE}} = \frac{1}{Z} \sum_{I_a \sim \mathcal{I}} - \log \frac{\exp(\phi(I_a, \zeta)^\top \phi(\tilde{I}_a, \zeta)/\tau)}{\sum_{I_n \in \mathcal{N}} \exp(\phi(I_a, \zeta)^\top \phi(I_n, \zeta)/\tau)} \tag{5}$$

where the temperature parameter τ is adjusted during optimization to control the training signal, especially during earlier stages of training. By contrasting each sample against many negatives, i.e. large sets $\mathcal{N}$, this task effectively yields a general, class-agnostic features description of our data. Moreover, as the contrastive objective explicitly increases the similarity of an anchor image with its augmentations, invariance against data transformations and scaling are learned. Figure 2 summarizes and visually explains the different training objectives of each task.

3.2 Improved Generalization by Multi Feature Learning

Following we show how to efficiently incorporate the learning tasks introduced in the previous section into a single DML model. We first extend the objective Eq. 5 using established triplet sampling strategies for improved adjustment to DML, before we jointly train our learning tasks for maximal feature diversity.

Adapting Noise Contrastive Estimation to DML. Efficient strategies for mining informative negatives I_n are a key factor [53] for successful training of ranking-based DML models. Since NCE essentially translates to a ranking between images $I_a, \tilde{I}_a, I_n$, its learning signal is also impaired if $I_n \in \mathcal{N}$ are uninformative, i.e. $d^\phi(I_a, I_n)$ being large. To this end, we control the contribution of each negative $I_n \in \mathcal{N}$ to $\mathcal{L}_{\text{NCE}}$ by a weight factor $w(d) = \min(\lambda, q^{-1}(d))$. Here, $q(d) = d^{D-2}\left[1 - \frac{1}{4}d^2\right]^{\frac{D-3}{2}}$ is the distribution of pairwise distances[1] on the D-dimensional unit hypersphere $\mathbb{S}^{D-1}$ and λ a cut-off parameter. Similar to [64], $w(d)$ helps to equally weigh negatives from the whole range of possible distances in ϕ and, in particular, increases the impact of harder negatives. Thus, our distance-adapted NCE loss becomes

$$\mathcal{L}_{\text{DaNCE}} = \frac{1}{Z} \sum_{I_a \sim \mathcal{I}} - \log \frac{\exp(\phi(I_a)^\top \phi(\tilde{I}_a)/\tau)}{\sum_{I_n \in \mathcal{N}} \exp(w(d^\phi_{a,n}) \cdot \phi^*(I_n)^\top \phi(I_a)/\tau)}. \tag{6}$$

NCE-based objectives learn best using large sets of negatives [21]. However, naively utilizing only negatives from the current mini-batch constrains $\mathcal{N}$ to the available GPU memory. To alleviate this limitation, we follow [21] and realize $\mathcal{N}$ as a large memory queue, which is constantly updated with embeddings $\phi^*(\tilde{I}_a)$ from training iteration t by utilising the running-average network $\phi^{*,t+1} = \mu\phi^{*,t} + (1-\mu)\phi^t$.

[1] To compute d, we use the euclidean distance between samples. Since ϕ is regularized to the unit hypersphere $\mathbb{S}^{D-1}$, the euclidean distance correlates with cosine distance.

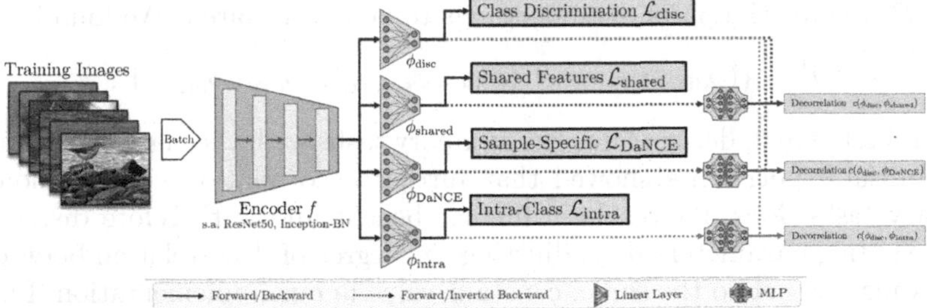

Fig. 3. *Architecture of our propose model.* Each task $\mathcal{L}_\bullet$ optimizes an individual embedding $\phi_\bullet$ implemented as a linear layer with a shared underlying feature encoder f. Pairwise decorrelation $c(\cdot,\cdot)$ of the embeddings utilizing the mapping ψ based on a two-layer MLP encourages each task to further emphasize on its targeted data characteristics. Gradient inversion R is applied during the backward pass to each embedding head.

Joint Optimization for Maximal Feature Diversity. The tasks presented in Sect. 3.1 are formulated to extract mutually complementary information from our training data. In order to capture their learned features in a single model to obtain a rich image representation, we now discuss how to jointly optimize these tasks.

While each task targets a semantically different concept of features, their driving learning signals are based on potentially contradicting ranking constraints on the learned embedding space. Thus, aggregating these signals to optimizing a joint, single embedding function ϕ may entail detrimental interference between them. In order to circumvent this issue, we learn a dedicated embedding space for each task, as often conducted in multi-task optimization [18,49], i.c. $\phi_{\text{disc}}(f), \phi_{\text{shared}}(f), \phi_{\text{intra}}(f)$ and $\phi_{\text{nce}}(f)$ with $\phi_\bullet(f) : \mathbb{R}^N \mapsto \mathbb{R}^D$ (cf. Sect. 3). As all embeddings share the same feature extractor f, each task still benefits from the aggregated learning signals. Additionally, as there may still be redundant overlap in the information captured by each task, we mutually decorrelate these representations, thus maximizing the diversity of the overall training signal. Similar to [44,49] we minimize the mutual information of two embedding functions ϕ^a, ϕ^b by maximizing their correlation c in the embedding space of ϕ^b, followed by a gradient reversal. For that, we learn a mapping $\psi : \mathbb{R}^D \mapsto \mathbb{R}^D$ from ϕ^a_i to ϕ^b_i given an image I_i and compute the correlation $c(\phi^a_i, \phi^b_i) = \|(R(\phi^a_i) \odot \psi(R(\phi^b_i)))\|_2^2$ with $\odot$ being the point-wise product. R denotes a gradient reversal operation, which inverts the resulting gradients during backpropagation. Maximizing c results in ψ aiming to make ϕ^a and ϕ^b comparable. However, through the subsequent gradients reversal, we actually decorrelate the embedding functions. Joint training of all tasks is finally performed by minimizing

$$\mathcal{L} = \mathcal{L}_{\text{disc}} + \alpha_1 \mathcal{L}_{\text{shared}} + \alpha_2 \mathcal{L}_{\text{intra}} + \alpha_3 \mathcal{L}_{\text{DaNCE}} - \sum_{(\phi_a, \phi_b) \in \mathcal{P}} \rho_{a,b} \cdot c(\phi_a, \phi_b) \quad (7)$$

where $\mathcal{P}$ denotes the pairs of embeddings to be decorrelated. We found

$$\mathcal{P} = \{(\phi_{\text{disc}}, \phi_{\text{DaNCE}}), (\phi_{\text{disc}}, \phi_{\text{shared}}), (\phi_{\text{disc}}, \phi_{\text{intra}})\}, \quad (8)$$

to work best, which decorrelates the auxiliary tasks with the class-discriminative task. Initial experiments showed that further decorrelation $c(\bullet, \bullet)$ among the auxiliary tasks does not result in further benefit and is therefore disregarded. The weighting parameters $\rho_{a,b}$ adjusting the degree of decorrelation between the embeddings are set to the same, constant value in our implementation. Figure 3 provides an overview of our model. Finally, we combine our learned embedding to form an ensemble representation to fully make use of all information.

Computational Costs. We train all tasks using the same mini-batch to avoid computational overhead. While optimizing each learner on an individual batch can further alleviate training signal interference [52,66], training time increases significantly. Using a single batch per iteration, we minimize the required extra computations to the extra forward pass through ϕ^* (however without computing gradients) for contrasting against negatives sampled from the memory queue as well as the small mapping networks ψ. Across datasets, we measure an increase in training time by 10–15% per epoch compared to training a standard supervised DML task. This is comparable to or lower than other methods, which perform a full clustering on the dataset [49,52] after each epoch, compute extensive embedding statistics [26] or simultaneously train generative models [33].

4 Experiments

Following we first present our implementation details and the benchmark datasets. Next, we evaluate our proposed model and study how our learning tasks complement each other and improve over baseline performances. Finally, we discuss our results in the context of the current state-of-the-art and conduct analysis and ablation experiments.

Implementation Details. We follow the common training protocol of [49,52, 64] for implementations utilizing a ResNet50-backbone. The shorter image axis is resized to 256, followed by a random crop to 224×224 and a random horizontal flip with $p = 0.5$. During evaluation, only a center crop is taken after resizing. The embedding dimension is set to $D = 128$ for each task embedding. For model variants using the Inception-V1 with Batch-Normalization [25], we follow [26,63] and use $D = 512$. Resizing, cropping and flipping is done in the same way as for ResNet50 versions. The implementation is done using the PyTorch framework [45], and experiments are performed on compute clusters containing NVIDIA Titan X, Tesla V4, P100 and V100, always limited to 12GB VRAM following the standard training protocol [64]. For DiVA, we utilise the triplet-based margin loss [64] with fixed margin $\gamma = 0.2$ and $\beta = 1.2$ and fixed temperature $\tau = 0.1$.

Hyperparameters. For training, we use Adam [29] with learning rate 10^{-5} and a weight decay of $5 \cdot 10^{-4}$. For ablations, we use no learning rate scheduling, while our final model is trained using scheduling values determined by cross-validation. We train for 150 epochs. Our joint training framework can be adjusted by setting the de-correlation weights $\rho_{a,b}$ and weight parameters α_i. In both cases we utilize the same values for all learning task, thus we effectively only adjust two parameters $[\rho, \alpha]$ to each benchmark sets: CUB200-2011 (IBN: $[300, 0.15]$, R50 $[1500, 0.3]$), CARS196 (IBN: $[100, 0.15]$, R50 $[100, 0.1]$), SOP (IBN: $[150, 0.2]$, R50 $[150, 0.2]$). This is comparable to other approaches, e.g. MS [63], SoftTriple [47], D&C [52]. As our auxiliary embeddings generalize better, they are more emphasized for computing $d_{i,j}$ during testing (e.g. double on CUB200 and CARS196).

Table 1. *Comparison of different combinations of learning tasks.* I(nception-V1) B(atch-)N(ormalization), and R(esNet)50 denote the backbone architecture. No learning rate scheduling is used. Our tasks are denoted by D(*iscriminative*), S(*hared*), I(*ntra-Class*) & and Da(*NCE*). The dimensionality per task embedding depends on the number of tasks used, totalling in $D = 512$ (two tasks use 256 each, three 170, four 128.

Dataset →	Dim	CUB200-2011 [60]			CARS196 [32]			SOP [42]		
Approach ↓		R@1	R@2	NMI	R@1	R@2	NMI	R@1	R@10	NMI
Margin [64] (orig, R50)	128	63.6	74.4	69.0	79.6	86.5	69.1	72.7	86.2	90.7
Margin [64] (ours, IBN)	512	63.6	74.7	68.3	79.4	86.6	66.2	76.6	89.2	89.8
DiVA (IBN, D & Da)	512	64.5	76.0	68.8	80.4	87.7	67.2	77.0	89.4	**90.1**
DiVA (IBN, D & S)	512	65.1	76.4	69.0	81.5	88.3	66.8	77.2	89.6	90.0
DiVA (IBN, D & I)	512	64.9	75.8	68.4	80.6	87.9	67.4	76.9	89.4	89.9
DiVA (IBN, D & Da & I)	510	65.3	76.5	68.3	82.2	89.1	67.8	75.8	89.0	89.8
DiVA (IBN, D & S & I)	510	65.5	76.4	68.4	82.1	89.4	67.2	77.0	89.3	89.7
DiVA (IBN, D & Da & S)	510	65.9	76.7	68.9	82.6	89.6	68.0	77.4	89.6	90.1
DiVA (IBN, D & Da & S & I)	512	**66.4**	**77.2**	**69.6**	**83.1**	**90.0**	**68.1**	**77.5**	**90.3**	**90.1**

Datasets. We evaluate the performance on three common benchmark datasets with standard training/test splits (see e.g. [49,52,63,64]): *CARS196* [32], which contains 16,185 images from 196 car classes. The first 98 classes containing 8054 images are used for training, while the remaining 98 classes with 8131 images are used for testing. *CUB200-2011* [60] with 11,788 bird images from 200 classes. Training/test sets contain the first/last 100 classes with 5864/5924 images respectively. *Stanford Online Products (SOP)* [42] provides 120,053 images divided in 22,634 product classes. 11318 classes with 59551 images are used for training, while the remaining 11316 classes with 60502 images are used for testing.

4.1 Performance Study of Multi-feature DML

We now compare our model and the complementary benefit of our proposed feature learning tasks for supervised DML. Table 1 evaluates the performance of our model based on margin loss [64], a triplet based objective with an additionally learnable margin, and distance-weighted triplet sampling [64]. We use

Inception-V1 with Batchnorm and a maximal aggregated embedding dimensionality of 512. Thus, if two tasks are utilized, each embedding has $D = 256$, in case of three tasks 170 and four tasks result in $D = 128$. No learning rate scheduling is used. Evaluation is conducted on CUB200-2011 [60], CARS196 [32] and SOP [42]. Retrieval performance is measured through Recall@k [27] and clustering quality via Normalized Mutual Information (NMI) [35]. While our results vary between possible task combinations, we observe that the generalization of our model consistently increases with each task added to the joint optimization. Our strongest model including all proposed tasks improves the generalization performance by 2.8% on CUB200-2011, 3.7% on CARS196 and 0.9% on SOP. This highlights that *(i)* purely discriminative supervised learning disregards valuable training information and *(ii)* carefully designed learning tasks are able to capture this information for improved generalization to unknown test classes. We further analyze our observations in the ablation experiments.

Table 2. *Comparison to the state-of-the-art methods on CUB200-2011* [60], *CARS196* [32] *and SOP* [42]. DiVA-Arch-Dim describes the backbone used with DiVA (IBN: Inception-V1 with Batchnorm, R50: ResNet50) and the total training and testing embedding dimensionality. For fair comparison, we also ran a standard ResNet50 with embedding dimensionality of 512. Trip-DiVA-Arch-Dim indicates standard triplet loss as base objective.

Dataset →	Dim	CUB200-2011 [60]			CARS196 [32]			SOP [42]		
Approach ↓		R@1	R@2	NMI	R@1	R@2	NMI	R@1	R@2	NMI
HTG [68]	512	59.5	71.8	–	76.5	84.7	–	–	–	–
HDML [69]	512	53.7	65.7	62.6	79.1	87.1	69.7	68.7	83.2	89.3
Margin [64]	128	63.6	74.4	69.0	79.6	86.5	69.1	72.7	86.2	90.8
HTL [17]	512	57.1	68.8	–	81.4	88.0	–	74.8	88.3	–
DVML [33]	512	52.7	65.1	61.4	82.0	88.4	67.6	70.2	85.2	90.8
MultiSim [63]	512	65.7	77.0	–	84.1	90.4	–	78.2	90.5	–
D& C [52]	128	65.9	76.6	69.6	84.6	90.7	70.3	75.9	88.4	90.2
MIC [49]	128	66.1	76.8	69.7	82.6	89.1	68.4	77.2	89.4	90.0
Significant increase in network parameter:										
HORDE [26]+Contr. [20]	512	66.3	76.7	–	83.9	90.3	–	–	–	–
Softtriple [47]	512	65.4	76.4	–	84.5	90.7	70.1	78.3	90.3	**92.0**
Ensemble Methods:										
A-BIER [44]	512	57.5	68.7	–	82.0	89.0	–	74.2	86.9	–
Rank [62]	1536	61.3	72.7	66.1	82.1	89.3	71.8	**79.8**	**91.3**	90.4
DREML [66]	9216	63.9	75.0	67.8	86.0	91.7	**76.4**	–	–	–
ABE [28]	512	60.6	71.5	–	85.2	90.5	–	76.3	88.4	–
Inception-BN										
Ours (Trip-DiVA-IBN-512)	512	66.7	77.1	69.3	83.1	90.3	68.8	76.9	88.9	89.4
Ours (DiVA-IBN-512)	512	66.8	77.7	70.0	84.1	90.7	68.7	78.1	90.6	90.4
ResNet50										
Ours (Margin [64]-R50-512)	512	64.4	75.4	68.4	82.2	89.0	68.1	78.3	90.0	90.1
Ours (Trip-DiVA-R50-512)	512	68.5	78.5	**71.1**	**87.3**	**92.8**	72.1	79.4	90.8	90.3
Ours (DiVA-R50-512)	512	**69.2**	**79.3**	**71.4**	**87.6**	**92.9**	72.2	**79.6**	**91.2**	90.6

4.2 Comparison to State-of-the-art Approaches

Next, we compare our model using fixed learning rate schedules per benchmark to the current state-of-the-art approaches in DML. For fair comparison to the different methods, we report result both using Inception-BN (IBN) and ResNet50 (R50) as backbone architecture. As Inception-BN is typically trained with embedding dimensionality of 512, we restrict each embedding to $D = 128$ for direct comparison with non-ensemble methods. Thus we deliberately impair the potential of our model due to a significantly lower capacity per task, compared to the standard $D = 512$. For comparison with ensemble approaches and maximal performance, we use a ResNet50 [49,52,64] architecture and the corresponding standard dimensionality $D = 128$ per task. Figure 2 summarizes our results using both standard triplet loss [53] and margin loss [64] as our base training objective. In both cases, we significantly improve over methods with comparable backbone architectures and achieve new state-of-the-art results with our ResNet50-ensemble. In particular we outperform the strongest ensemble methods, including DREML [66] which utilize a much higher total embedding dimensionality. The large improvement is explained by the diverse and mutually complementary learning signals contributed by each task in our ensemble. In contrast, previous ensemble methods rely on the same, purely class-discriminative training signal for each learner. Note that some approaches strongly differ from the standard training protocols and architectures, resulting in more parameters and much higher GPU memory consumption, such as Rank [62] (32 GB), ABE [28] (24GB), Softtriple [47] and HORDE [26]. Additionally, Rank [62] employs much larger batch-sizes to increase the number of classes per batch. This is especially crucial on the SOP dataset, which greatly benefits from higher class coverage due to its vast amount of classes [51]. Nevertheless, our model outperforms these methods - in some cases even in its constrained version (IBN-512).

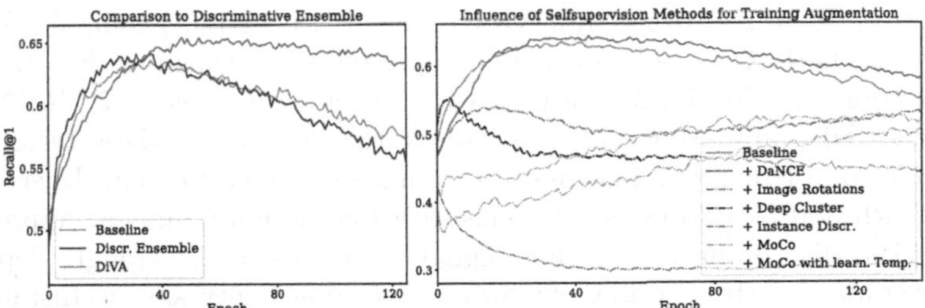

Fig. 4. *Analysis of complementary tasks for supervised learning.* (left): Performance comparison between class-dicsriminative training only (Baseline), ensemble of class-discriminative learners (Discr. Ensemble) and our proposed DiVA, which exhibits a large boost in performance. (right): Evaluation of self-supervised learning approaches combined with standard discriminative DML.

Table 3. *Ablation studies.* We compare standard margin loss as baseline and DiVA performance against ablations of our model: no decorrelation between embeddings (No-Decorrelation.) and training an independent model for each task (Separated models). Total embedding dimensionality is 512.

Methods →	Baseline	DiVA	No De-correlation	Separated models
Recall@1 →	63.6	66.4	65.6	48.7

4.3 Ablation Studies

In this section we conduct ablation experiments for various parts of our model. For every ablation we again use the Inception-BN network. The dimensionality setting follows the performance study in Sect. 4.1. Again, we train each model with a fixed learning rate for fair comparison among ablations.

Influence of Distance-Adaption in DaNCE. To evaluate the benefit of our extension from $\mathcal{L}_{\text{nce}}$ [19,21] to $\mathcal{L}_{\text{DaNCE}}$, we compare both versions in combination with standard supervised DML (i.e. class-discriminative features) in Fig. 4 (right). Our experiment indicates two positive effects: *(i)* The training convergence with our extended objective is much faster and *(ii)* the performance differs greatly between employing $\mathcal{L}_{\text{nce}}$ and $\mathcal{L}_{\text{DaNCE}}$. In fact, using the standard NCE objective is even detrimental to learning, while our extended version improves over the only discriminatively trained baseline. We attribute this to both the slow convergence of $\mathcal{L}_{\text{nce}}$ which is not able to support the faster discriminative learning and to emphasizing harder negatives in $\mathcal{L}_{\text{DaNCE}}$. In particular the latter is an important factor in ranking based DML [53], as during training more and more negatives become uninformative. To tackle this issue, we also experimented with learning the temperature parameter τ. While convergence speed increases slightly, we find no significant benefit in final generalization performance.

Evaluation of Self-supervision Methods. Figure 4 (right) compares DaNCE to other methods from self-supervised representation learning. For that purpose we train the discriminative task with either DeepCluster [7], RotNet [18] or Instance Discrimination [65]. We observe that neither of these tasks is able to provide complementary information to improve generalization. DeepCluster, trained with 300 pseudo classes for classification, actually aims at approximating the class-discriminative learning signal while RotNet is strongly dependent on the variance of the training classes and converges very slowly. Instance discrimination seems to provide a contradictory training signal to the supervised task. These results are in line with previous works [22] which report difficulties to directly combine both supervised and self-supervised learning for improved test performance. In contrast, we explicitly adapt NCE to DML in our proposed objective DaNCE.

Comparison to Purely Class-Discriminative Ensemble. We now compare DiVA to an ensemble of class-discriminative learner (Discr. Ensemble) based on the same model architecture using embedding decorrelation in Fig. 4 (left). While the discriminative ensemble improves over the baseline, the amount of captured data information eventually saturates and, thus, performs significantly worse compared to our multi-feature DiVA ensemble. Further, our ablation reveals that joint optimization of diverse learning tasks regularizes training and reduces overfitting effects which eventually occur during later stages of DML training.

Benefit of Task Decorrelation. The role of decorrelating the embedding representations of each task during learning is analyzed by comparison to a model trained without this constraint. Firstly, Table 3 demonstrates that omitting the decorrelation still outperforms the standard margin loss ('Baseline') by 2.1% while operating on the same total embedding dimensionality. This proves that learning diverse features significantly improves generalization. Adding the decorralation constraint then additionally boosts performance by 1.2%, as now each task is further encouraged to capture distinct data characteristics.

Learning Without Feature Sharing. To highlight the importance of feature sharing among our learning tasks, we train an individual, independent model for the class-discriminative, class-shared, sample-specific and intra-class task. At testing time, we combine their embeddings similar to our proposed model. Table 3 shows a dramatic drop in performance to 48% for the disconnected ensemble ('Separately Trained'), proving that sharing the complementary information captured from different data characteristics is crucial and mutually benefits learning. Without the class-discriminative signal, the other tasks lack the concept of an object class, which hurts the aggregation of embeddings (cf. Sect. 3).

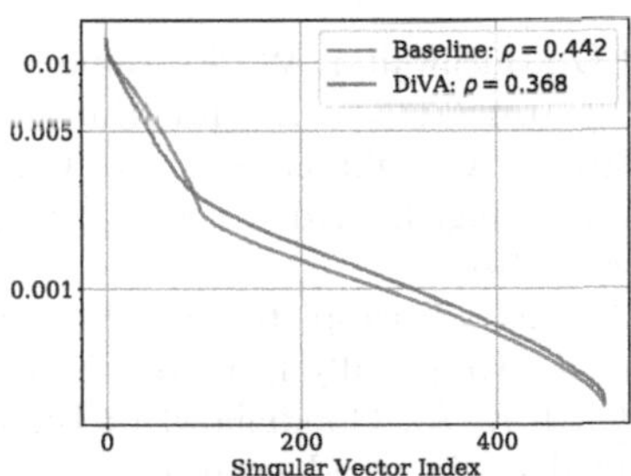

Fig. 5. *Singular Value Spectrum.* We analyze the singular value spectrum of DiVA embeddings and that of a network trained with the standard discriminative task. We find that our gains in generalization performance (Tables 1, 2) are reflected by a reduced spectral decay [51] for our learned embedding space.

Generalization and Embedding Space Compression. Recent work [51] links DML generalization to a decreased compression [56] of the embedding space. Their findings report that the number of directions with significant variance [51, 57] of a representation correlates with the generalization ability in DML. To this end, we analyze our model using their proposed spectral decay ρ (lower is better) which is computed as the KL-divergence between the normalized singular value spectrum and a uniform distribution. Figure 5 compares the spectral decays of our model and a standard supervised baseline model. As expected, due to the diverse information captured, our model learns a more complex representation which results in a significantly lower value of ρ and better generalization.

5 Conclusion

In this paper we propose several learning tasks which complement the class-discriminative training signal of standard, supervised Deep Metric Learning (DML) for improved generalization to unknown test distributions. Each of our tasks is designed to capture different characteristics of the training data: class-discriminative, class-shared, intra-class and sample-specific features. For the latter, we adapt contrastive self-supervised learning to the needs of supervised DML. Jointly optimizing all tasks results in a diverse overall training signal which is further amplified by mutual decorrelation between the individual tasks. Unifying these distinct representations greatly boosts generalization over purely discriminatively trained models.

Acknowledgements. This work has been supported by hardware donations from NVIDIA (DGX-1), resources from Compute Canada, in part by Bayer AG and the German federal ministry BMWi within the project "KI Absicherung".

References

1. Bachman, P., Hjelm, R.D., Buchwalter, W.: Learning representations by maximizing mutual information across views. arXiv preprint arXiv:1906.00910 (2019)
2. Bautista, M.A., Sanakoyeu, A., Tikhoncheva, E., Ommer, B.: Cliquecnn: deep unsupervised exemplar learning. In: Advances in Neural Information Processing Systems, pp. 3846–3854 (2016)
3. Belghazi, M.I., Rajeswar, S., Mastropietro, O., Rostamzadeh, N., Mitrovic, J., Courville, A.: Hierarchical adversarially learned inference (2018)
4. Berthelot, D., Raffel, C., Roy, A., Goodfellow, I.: Understanding and improving interpolation in autoencoders via an adversarial regularizer (2018)
5. Bhattarai, B., Sharma, G., Jurie, F.: Cp-mtml: coupled projection multi-task metric learning for large scale face retrieval. In: 2016 IEEE Conference on Computer Vision and Pattern Recognition (CVPR), pp. 4226–4235 (2016)
6. Büchler, U., Brattoli, B., Ommer, B.: Improving spatiotemporal self-supervision by deep reinforcement learning. In: Proceedings of the European Conference on Computer Vision (ECCV) (2018)
7. Caron, M., Bojanowski, P., Joulin, A., Douze, M.: Deep clustering for unsupervised learning of visual features. In: European Conference on Computer Vision (2018)
8. Chawla, N.V., Japkowicz, N., Kotcz, A.: Editorial: special issue on learning from imbalanced data sets. SIGKDD Explor. Newsl. **6**, 1–6 (2004)
9. Chen, T., Kornblith, S., Norouzi, M., Hinton, G.: A simple framework for contrastive learning of visual representations. arXiv preprint arXiv:2002.05709 (2020)

10. Chen, W., Chen, X., Zhang, J., Huang, K.: Beyond triplet loss: a deep quadruplet network for person re-identification. In: Proceedings of the IEEE Conference on Computer Vision and Pattern Recognition (2017)
11. Chen, X., Duan, Y., Houthooft, R., Schulman, J., Sutskever, I., Abbeel, P.: Infogan: Interpretable representation learning by information maximizing generative adversarial nets (2016)
12. Deng, J., Guo, J., Xue, N., Zafeiriou, S.: Arcface: additive angular margin loss for deep face recognition (2018)
13. Donahue, J., Krähenbühl, P., Darrell, T.: Adversarial feature learning (2016)
14. Duan, Y., Zheng, W., Lin, X., Lu, J., Zhou, J.: Deep adversarial metric learning. In: The IEEE Conference on Computer Vision and Pattern Recognition (CVPR), June 2018
15. Dumoulin, V., Belghazi, I., Poole, B., Mastropietro, O., Lamb, A., Arjovsky, M., Courville, A.: Adversarially learned inference (2016)
16. Gautheron, L., Morvant, E., Habrard, A., Sebban, M.: Metric learning from imbalanced data (2019)
17. Ge, W.: Deep metric learning with hierarchical triplet loss. In: Proceedings of the European Conference on Computer Vision (ECCV), pp. 269–285 (2018)
18. Gidaris, S., Singh, P., Komodakis, N.: Unsupervised representation learning by predicting image rotations. In: International Conference on Learning Representations (2018)
19. Gutmann, M., Hyvärinen, A.: Noise-contrastive estimation: a new estimation principle for unnormalized statistical models. In: Proceedings of the Thirteenth International Conference on Artificial Intelligence and Statistics (2010)
20. Hadsell, R., Chopra, S., LeCun, Y.: Dimensionality reduction by learning an invariant mapping. In: Proceedings of the IEEE Conference on Computer Vision and Pattern Recognition (2006)
21. He, K., Fan, H., Wu, Y., Xie, S., Girshick, R.: Momentum contrast for unsupervised visual representation learning (2019)
22. Hendrycks, D., Mazeika, M., Kadavath, S., Song, D.: Using self-supervised learning can improve model robustness and uncertainty. CoRR abs/1906.12340 (2019). http://arxiv.org/abs/1906.12340
23. Hsu, K., Levine, S., Finn, C.: Unsupervised learning via meta-learning (2018)
24. Hu, J., Lu, J., Tan, Y.: Discriminative deep metric learning for face verification in the wild. In: 2014 IEEE Conference on Computer Vision and Pattern Recognition (2014)
25. Ioffe, S., Szegedy, C.: Batch normalization: accelerating deep network training by reducing internal covariate shift. In: International Conference on Machine Learning (2015)
26. Jacob, P., Picard, D., Histace, A., Klein, E.: Metric learning with horde: high-order regularizer for deep embeddings. In: The IEEE Conference on Computer Vision and Pattern Recognition (CVPR) (2019)
27. Jegou, H., Douze, M., Schmid, C.: Product quantization for nearest neighbor search. IEEE Trans. Pattern Anal. Mach. Intell. **33**(1), 117–128 (2011)
28. Kim, W., Goyal, B., Chawla, K., Lee, J., Kwon, K.: Attention-based ensemble for deep metric learning. In: Proceedings of the European Conference on Computer Vision (ECCV) (2018)
29. Kingma, D.P., Ba, J.: Adam: A method for stochastic optimization (2015)
30. Kingma, D.P., Welling, M.: Auto-encoding variational bayes. In: Proceedings of the International Conference on Learning Representations (ICLR) (2013)

31. Konno, T., Iwazume, M.: Cavity filling: pseudo-feature generation for multi-class imbalanced data problems in deep learning (2018)
32. Krause, J., Stark, M., Deng, J., Fei-Fei, L.: 3D object representations for fine-grained categorization. In: Proceedings of the IEEE International Conference on Computer Vision Workshops, pp. 554–561 (2013)
33. Lin, X., Duan, Y., Dong, Q., Lu, J., Zhou, J.: Deep variational metric learning. In: The European Conference on Computer Vision (ECCV), September 2018
34. Liu, W., Wen, Y., Yu, Z., Li, M., Raj, B., Song, L.: Sphereface: deep hypersphere embedding for face recognition. In: IEEE Conference on Computer Vision and Pattern Recognition (CVPR) (2017)
35. Manning, C., Raghavan, P., Schütze, H.: Introduction to information retrieval. Nat. Lang. Eng. **16**(1), 100–103 (2010)
36. Milbich, T., Ghori, O., Diego, F., Ommer, B.: Unsupervised representation learning by discovering reliable image relations. Pattern Recogn. (PR) **102**, 107107 (2020)
37. Milbich, T., Roth, K., Brattoli, B., Ommer, B.: Sharing matters for generalization in deep metric learning. IEEE Trans. Pattern Anal. Mach. Intell. (2020)
38. Misra, I., van der Maaten, L.: Self-supervised learning of pretext-invariant representations (2019)
39. Movshovitz-Attias, Y., Toshev, A., Leung, T.K., Ioffe, S., Singh, S.: No fuss distance metric learning using proxies. In: Proceedings of the IEEE International Conference on Computer Vision, pp. 360–368 (2017)
40. Noroozi, M., Favaro, P.: Unsupervised learning of visual representations by solving jigsaw puzzles. In: Leibe, B., Matas, J., Sebe, N., Welling, M. (eds.) ECCV 2016. LNCS, vol. 9910, pp. 69–84. Springer, Cham (2016). https://doi.org/10.1007/978-3-319-46466-4_5
41. Noroozi, M., Vinjimoor, A., Favaro, P., Pirsiavash, H.: Boosting self-supervised learning via knowledge transfer. In: Proceedings of the IEEE Conference on Computer Vision and Pattern Recognition, pp. 9359–9367 (2018)
42. Oh Song, H., Xiang, Y., Jegelka, S., Savarese, S.: Deep metric learning via lifted structured feature embedding. In: Proceedings of the IEEE Conference on Computer Vision and Pattern Recognition, pp. 4004–4012 (2016)
43. Oord, A., Li, Y., Vinyals, O.: Representation learning with contrastive predictive coding (2018), arXiv preprint arXiv:1807.03748
44. Opitz, M., Waltner, G., Possegger, H., Bischof, H.: Deep metric learning with bier: boosting independent embeddings robustly. IEEE Trans. Pattern Anal. Mach. Intell. **42**, 276–290 (2018)
45. Paszke, A., et al.: Automatic differentiation in pytorch. In: NIPS-W (2017)
46. Pu, J., Jiang, Y.G., Wang, J., Xue, X.: Which looks like which: exploring inter-class relationships in fine-grained visual categorization. In: Proceedings of the IEEE European Conference on Computer Vision (ECCV), pp. 425–440 (2014)
47. Qian, Q., Shang, L., Sun, B., Hu, J., Li, H., Jin, R.: Softtriple loss: deep metric learning without triplet sampling (2019)
48. Ren, S., He, K., Girshick, R.B., Sun, J.: Faster R-CNN: towards real-time object detection with region proposal networks. In: NeuRips (2015)
49. Roth, K., Brattoli, B., Ommer, B.: Mic: mining interclass characteristics for improved metric learning. In: The IEEE International Conference on Computer Vision (ICCV), October 2019
50. Roth, K., Milbich, T., Ommer, B.: Pads: policy-adapted sampling for visual similarity learning. In: Proceedings of the IEEE/CVF Conference on Computer Vision and Pattern Recognition (CVPR) (2020)

51. Roth, K., Milbich, T., Sinha, S., Gupta, P., Ommer, B., Cohen, J.P.: Revisiting training strategies and generalization performance in deep metric learning. In: Proceedings of the International Conference on Machine Learning (ICML) (2020)
52. Sanakoyeu, A., Tschernezki, V., Buchler, U., Ommer, B.: Divide and conquer the embedding space for metric learning. In: The IEEE Conference on Computer Vision and Pattern Recognition (CVPR) (2019)
53. Schroff, F., Kalenichenko, D., Philbin, J.: Facenet: a unified embedding for face recognition and clustering. In: Proceedings of the IEEE Conference on Computer Vision and Pattern Recognition, pp. 815–823 (2015)
54. Sohn, K.: Improved deep metric learning with multi-class n-pair loss objective. In: Advances in Neural Information Processing Systems, pp. 1857–1865 (2016)
55. Tian, Y., Krishnan, D., Isola, P.: Contrastive multiview coding. arXiv preprint arXiv:1906.05849 (2019)
56. Tishby, N., Zaslavsky, N.: Deep learning and the information bottleneck principle (2015)
57. Verma, V., et al.: Manifold mixup: better representations by interpolating hidden states (2018)
58. Vincent, P., Larochelle, H., Bengio, Y., Manzagol, P.A.: Extracting and composing robust features with denoising autoencoders (2008)
59. Wagemans, J., et al.: A century of gestalt psychology in visual perception: I. perceptual grouping and figure-ground organization. Psychol. Bull. **138**(6), 1172 (2012)
60. Wah, C., Branson, S., Welinder, P., Perona, P., Belongie, S.: The caltech-ucsd birds-200-2011 dataset (2011)
61. Wang, J., Zhou, F., Wen, S., Liu, X., Lin, Y.: Deep metric learning with angular loss. In: Proceedings of the IEEE International Conference on Computer Vision, pp. 2593–2601 (2017)
62. Wang, X., Hua, Y., Kodirov, E., Hu, G., Garnier, R., Robertson, N.M.: Ranked list loss for deep metric learning. In: The IEEE Conference on Computer Vision and Pattern Recognition (CVPR) (2019)
63. Wang, X., Han, X., Huang, W., Dong, D., Scott, M.R.: Multi-similarity loss with general pair weighting for deep metric learning. CoRR abs/1904.06627 (2019), http://arxiv.org/abs/1904.06627
64. Wu, C.Y., Manmatha, R., Smola, A.J., Krahenbuhl, P.: Sampling matters in deep embedding learning. In: Proceedings of the IEEE International Conference on Computer Vision, pp. 2840–2848 (2017)
65. Wu, Z., Xiong, Y., Yu, S., Lin, D.: Unsupervised feature learning via non-parametric instance-level discrimination (2018)
66. Xuan, H., Souvenir, R., Pless, R.: Deep randomized ensembles for metric learning. In: Proceedings of the European Conference on Computer Vision (ECCV), pp. 723–734 (2018)
67. Zhai, A., Wu, H.Y.: Classification is a strong baseline for deep metric learning (2018)
68. Zhao, Y., Jin, Z., Qi, G.J., Lu, H., Hua, X.S.: An adversarial approach to hard triplet generation. In: Proceedings of the European Conference on Computer Vision (ECCV), pp. 501–517 (2018)
69. Zheng, W., Chen, Z., Lu, J., Zhou, J.: Hardness-aware deep metric learning. In: The IEEE Conference on Computer Vision and Pattern Recognition (CVPR) (2019)

DHP: Differentiable Meta Pruning via HyperNetworks

Yawei Li[1(✉)], Shuhang Gu[1,2], Kai Zhang[1], Luc Van Gool[1,3], and Radu Timofte[1]

[1] Computer Vision Lab, ETH Zürich, Zürich, Switzerland
{yawei.li,kai.zhang,vangool,radu.timofte}@vision.ee.ethz.ch, shuhanggu@gmail.com
[2] The University of Sydney, Sydney, Australia
[3] KU Leuven, Leuven, Belgium

Abstract. Network pruning has been the driving force for the acceleration of neural networks and the alleviation of model storage/transmission burden. With the advent of AutoML and neural architecture search (NAS), pruning has become topical with automatic mechanism and searching based architecture optimization. Yet, current automatic designs rely on either reinforcement learning or evolutionary algorithm. Due to the non-differentiability of those algorithms, the pruning algorithm needs a long searching stage before reaching the convergence.

To circumvent this problem, this paper introduces a differentiable pruning method via hypernetworks for automatic network pruning. The specifically designed hypernetworks take latent vectors as input and generate the weight parameters of the backbone network. The latent vectors control the output channels of the convolutional layers in the backbone network and act as a handle for the pruning of the layers. By enforcing ℓ_1 sparsity regularization to the latent vectors and utilizing proximal gradient solver, sparse latent vectors can be obtained. Passing the sparsified latent vectors through the hypernetworks, the corresponding slices of the generated weight parameters can be removed, achieving the effect of network pruning. The latent vectors of all the layers are pruned together, resulting in an automatic layer configuration. Extensive experiments are conducted on various networks for image classification, single image super-resolution, and denoising. And the experimental results validate the proposed method. Code is available at https://github.com/ofsoundof/dhp.

Keywords: Network pruning · Hyperneworks · Meta learning · Differentiable optimization · Proximal gradient

---

Electronic supplementary material The online version of this chapter (https://doi.org/10.1007/978-3-030-58598-3_36) contains supplementary material, which is available to authorized users.

A. Vedaldi et al. (Eds.): ECCV 2020, LNCS 12353, pp. 608–624, 2020.
https://doi.org/10.1007/978-3-030-58598-3_36

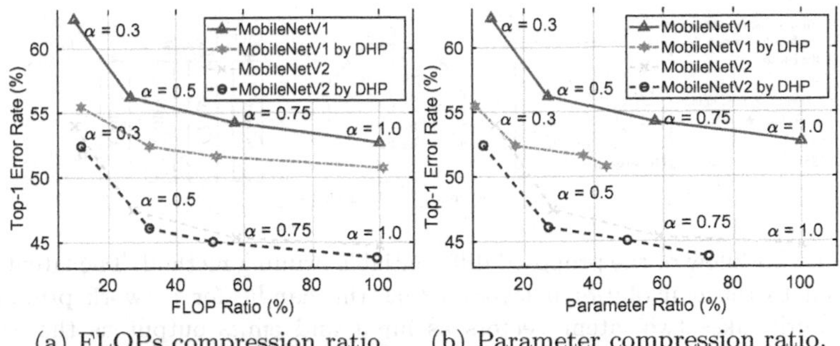

(a) FLOPs compression ratio. (b) Parameter compression ratio.

Fig. 1. Top-1 error *vs.* FLOPs and parameter compression ratio on MobileNets. The original model with different width multipliers α is set as the baseline. The DHP operating points near 100% FLOPs ratio is obtained by pruning the two networks with $\alpha = 2$. The DHP models outperforms the original at all the operating points

1 Introduction

These days, network pruning has become the workhorse for network compression, which aims at lightweight and efficient model for fast inference [12,18,29,38]. This is of particular importance for the deployment of tiny artificial intelligence (Tiny AI) algorithms on smart phones and edge devices [1]. Since the emerging of network pruning a couple of methods have been proposed based on the analysis of gradients, Hessians or filter distribution [9,13,16,25,40]. With the advent of AutoML and neural architecture search (NAS) [6,58], a new trend of network pruning emerges, *i.e.* pruning with automatic algorithms and targeting distinguishing sub-architectures. Among them, reinforcement learning and evolutionary algorithm become the natural choice [17,37]. The core idea is to search a certain fine-grained layer-wise distinguishing configuration among the all of the possible choices (population in the terminology of evolutionary algorithm). After the searching stage, the candidate that optimizes the network prediction accuracy under constrained budgets is chosen.

The advantage of these automatic pruning methods is the final layer-wise distinguishing configuration. Thus, hand-crafted design is no longer necessary. However, the main concern of these algorithms is the convergence property. For example, reinforcement learning is notorious for its difficulty of convergence under large or even middle level number of states [49]. Evolutionary algorithm needs to choose the best candidate from the already converged algorithm. But the dilemma lies in the impossibility of training the whole population till convergence and the difficulty of choosing the best candidate from unconverged population [14,37]. A promising solution to this problem is endowing the searching mechanism with differentiability or resorting to an approximately differentiable algorithm. This is due to the fact that differentiability has the potential to make the searching stage efficient. Actually, differentiability has facilitated a couple of machine learning approaches and the typical one among them is NAS. Early

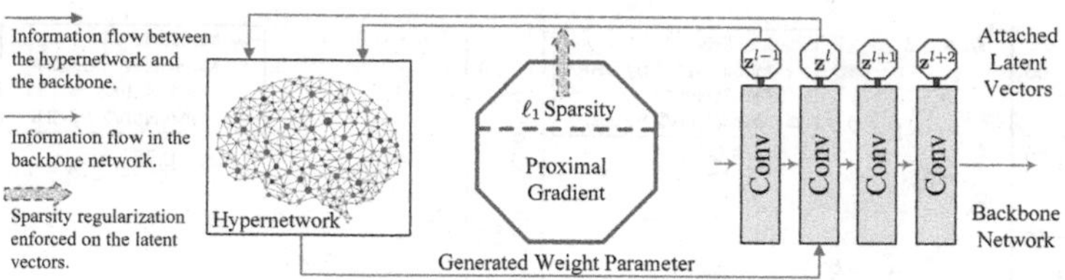

Fig. 2. The workflow of the proposed differentiable pruning method. The latent vectors $\mathbf{z}$ attached to the convolutional layers act as the handle for network pruning. The hypernetwork takes two latent vectors as input and emits output as the weight of the backbone layer. ℓ_1 sparsity regularization is enforced on the latent vectors. The differentiability comes with the hypernetwork tailored to pruning and the proximal gradient exploited to solve problem. After the pruning stage, sparse latent vectors are obtained which result in pruned weights after being passed through the hypernetwork

works on NAS have insatiable demand for computing resources, consuming tens of thousands of GPU hours for a satisfactory convergence [58,59]. Differentiable architecture search (DARTS) reduces the insatiable consumption to tens of GPU hours, which has boosted the development of NAS during the past year [36].

Another noteworthy direction for automatic pruning is brought by MetaPruning [37] which introduces hypernetworks [11] into network compression. The output of the so-called hypernetwork is used as the parameters of the backbone network. During training, the gradients are also back-propagated to the hypernetworks. This method falls in the paradigm of meta learning since the parameters in the hypernetwork act as the meta-data of the parameters in the backbone network. But the problem of this method is that the hypernetworks can only output fixed-size weights, which cannot serve as a layer-wise configuration searching mechanism. Thus, a searching algorithm such as evolutionary algorithm is necessary for the discovery of a good candidate. Although this is quite a natural choice, there is still one interesting question, namely, whether one can design a hypernetwork whose output size depends on the input (termed as latent vector in this paper) so that by only dealing with the latent vector, the backbone network can be automatically pruned.

To solve the aforementioned problem, we propose the differentiable meta pruning approach via hypernetworks (DHP, D – **D**ifferentiable, H – **H**yper, P – **P**runing) shown in Fig. 2. A new design of hypernetwork is proposed to adapt to the requirements of differentiability. Each layer is endowed with a latent vector that controls the output channels of this layer. Since the layers in the network are connected, the latent vector also controls the input channel of the next layer. The hypernetwork takes as input the latent vectors of the current layer and previous layer that controls the output and input channels of the current layer respectively. Passing the latent vectors through the hypernetwork leads to outputs which are used as the parameters of the backbone network. To achieve the effect of automatic pruning, ℓ_1 sparsity regularizer is applied to the latent vectors. A pruned model is discovered by updating the latent vectors with proximal gradient. The searching stage stops when the compression ratio drops to the target level. After the searching stage, the latent vectors are sparsified. Accordingly, the outputs of the hypernetworks that are covariant with the latent vector are also compressed. The advantage of the proposed method is that it is only necessary to deal with the latent vectors, which automates network pruning without the other bells and whistles.

With the fast development of efficient network design and NAS, the usefulness of network pruning is frequently challenged. But by analyzing the performance on MobileNetV1 [19] and MobileNetV2 [47] in Fig. 1, we conclude that automatic network pruning is of vital importance for further exploring the capacity of efficient networks. Efficient network design and NAS can only result in an overall architecture with building blocks endowed with the same sub-architecture. By automatic network pruning, the efficient networks obtained by either human experts or NAS can be further compressed, leading to layer-wise distinguishing configurations, which can be seen as a fine-grained architecture search.

Thus, the contribution of this paper is as follows.

I A new architecture of hypernetwork is designed. Different from the classical hypernetwork composed of linear layers, the new design is tailored to automatic network pruning. By only operating on the input of the hypernetwork, the backbone network can be pruned.

II A differentiable automatic networking pruning method is proposed. The differentiability comes with the designed hypernetwork and the utilized proximal gradient. It accelerates the convergence of the pruning algorithm.

III By the experiments on various vision tasks and modern convolutional neural networks (CNNs) [15,19,20,26,32,46,47,56], the potential of automatic network pruning as fine-grained architecture search is revealed.

2 Related Works

Network Pruning. Aiming at removing the weak filter connections that have the least influence on the accuracy of the network, network pruning attracts increasing attention. Early attempts emphasize more on the storage consumption, various criteria have been explored to remove inconsequential connections in an unstructural manner [12,34]. Despite their success in reducing network parameters, unstructural pruning leads to irregular weight parameters, limited in the actual acceleration of the pruned network. To further address the efficiency issue, structured pruning methods directly zero out structured groups of the convolutional filters. For example, Wen *et al.* [52] and Alvarez *et al.* [2] firstly proposed to resort to group sparsity regularization during training to reduce the number of feature maps in each layer. Since that, the field has witnessed a variety of regularization strategies [3,28,29,50,54]. These elaborately designed regularization methods considerably advance the pruning performance. But they often rely on carefully adjusted hyper-parameters selected for specific network architecture and dataset.

AutoML. Recently, there is an emerging trend of exploiting AutoML for automatic network compression [7,14,17,37]. The rationality lies in the exploration among the total population of network configurations for a final best candidate. He *et al.* exploited reinforcement learning agents to prune the networks where hand-crafted design is not longer necessary [17]. Hayashi *et al.* utilized genetic algorithm to enumerate candidate in the designed hypergraph for tensor network decomposition [14]. Liu *et al.* trained a hypernetwork to generate weights of the backbone network and used evolutionary algorithm to search for the best candidate [37]. The problem of these approachs is that the searching algorithms are not differentiable, which does not result in guaranteed convergence.

NAS. NAS automatizes the manual task of neural network architecture design. Optimally, searched networks achieve smaller test error, require fewer parameters and need less computations than their manually designed counterparts [45,58]. But the main drawback of early strategies is their almost insatiable demand for computational resources. To alleviate the computational burden several methods [35,36,59] are proposed to search for a basic building block, *i.e.* cell, opposed to an entire network. Then, stacking multiple cells with equivalent structure but different weights defines a full network [4,44]. Another recent trend in NAS is differentiable search methods such as DARTS [36]. The differentiability allows the fast convergence of the searching algorithm and thus boosts the fast development of NAS during the past year. In this paper we propose a differentiable counterpart for automatic network pruning.

Meta Learning and Hypernetworks. Meta learning is a broad family of machine learning techniques that deal with the problem of learning to learn. An emerging trend of meta learning uses hypernetworks to predict the weight parameters in the backbone network [11]. Since the introduction of hypernetworks, it has found wide applications in NAS [4], multi-task learning [41], Bayesian neural networks [24], and also network pruning [37]. In this paper, we propose a new design of hypernetwork which is especially suitable for network pruning and makes differentiability possible for automatic network pruning.

3 Methodology

The pipeline of the proposed method is shown in Fig. 2. The two cores of the whole pipeline are the designed hypernetwork and the optimization algorithm. In the forward pass, the designed hypernetwork takes as input the latent vectors and predicts the weight parameters for the backbone network. In the backward pass, the gradients are back-propagated to the hypernetwork. The ℓ_1 sparsity regularizer is enforced on the latent vectors and proximal gradient is used to solve the problem. The dimension of the output of the hypernetwork is covariant with that of the input. Due to this property, the output weights are pruned along with the sparsified latent vectors after the optimization step. The differentiability comes with the covariance property of the hypernetworks, the ℓ_1 sparsity regularization enforced on the latent vectors, and the proximal gradient used to solve the problem. The automation of pruning is due to the fact that all of the latent vectors are non-discriminatively regularized and that proximal gradient discovers the potential less important elements automatically.

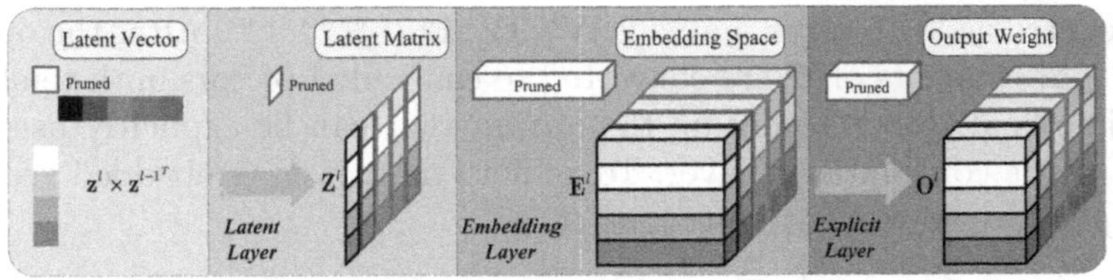

Fig. 3. Illustration of the hypernetwork designed for network pruning. It generates a weight tensor after passing the input latent vector through the latent layer, the embedding layer, and the explicit layer. If one element in $\mathbf{z}^l$ is pruned, the corresponding slice of the output tensor is also pruned. See Subsect. 3.1 for details

3.1 Hypernetwork Design

Notation: Unless otherwise stated, we use the normal (x), minuscule bold ($\mathbf{z}$), and capital bold ($\mathbf{Z}$) letters to denote scalars, vectors, and matrices/high-dimensional tensors. The elements of a matrix/tensor is indexed by the subscript as $\mathbf{Z}_{i,j}$ which could be scalars or vectors depending on the dimension of the indexed subject.

We first introduce the design of the hypernetwork shown in Fig. 3. In summary, the hypernetwork consists of three layers. The latent layer takes as input the latent vectors and computes a latent matrix from them. The embedding layer projects the elements of the latent matrix to an embedding space. The last explicit layer converts the embedded vectors to the final output. This design is inspired by fully connected layers in [11,37] but differs from those designs in that the output dimension is covariant with the input latent vector. This design is applicable to all types of convolutions including the standard convolution, depth-wise convolution, point-wise convolution, and transposed convolution.

Suppose that the given is an L-layer CNN. The dimension of the weight parameter of the l-th convolutional layer is $n \times c \times w \times h$, where n, c, and $w \times h$ denote the output channel, input channel, and kernel size of the layer, respectively. Every layer is endowed with a latent vector $\mathbf{z}^l$. The latent vector has the same size as the output channel of the layer, *i.e.* , $\mathbf{z}^l \in \mathbb{R}^n$. Thus, the previous layer is given a latent vector $\mathbf{z}^{l-1} \in \mathbb{R}^c$. The hypernetwork receives the latent vectors $\mathbf{z}^l$ and $\mathbf{z}^{l-1}$ of the current and the previous layer as input. A latent matrix is first computed from the two latent vectors, namely,

$$\mathbf{Z}^l = \mathbf{z}^l \cdot \mathbf{z}^{l-1^T} + \mathbf{B}_0^l, \tag{1}$$

where $[T]$ and $[\cdot]$ denote matrix transpose and multiplication, $\mathbf{Z}^l, \mathbf{B}_0 \in \mathbb{R}^{n \times c}$. Then every element in the latent matrix is projected to an m dimensional embedding space, namely,

$$\mathbf{E}_{i,j}^l = \mathbf{Z}_{i,j}^l \mathbf{w}_1^l + \mathbf{b}_1^l, i = 1, \cdots, n, j = 1, \cdots, c, \tag{2}$$

where $\mathbf{E}_{i,j}^l, \mathbf{w}_1^l, \mathbf{b}_1^l \in \mathbb{R}^m$. The vectors $\mathbf{w}_1^l$ and $\mathbf{b}_1^l$ are element-wise unique and for the simplicity of notation, the subscript $_{i,j}$ is omitted. $\mathbf{w}_1^l$, $\mathbf{b}_1^l$, and $\mathbf{E}_{i,j}^l$ can be aggregated as 3D tensors, namely $\mathbf{W}_1^l, \mathbf{B}_1^l, \mathbf{E}^l \in \mathbb{R}^{n \times c \times m}$. After the operation in Eq. 2, the elements of $\mathbf{Z}^l$ are converted to embedded vectors in the embedding space. The final step is to obtain the output that can be explicitly used as the weights of the convolutional layer. To achieve that, every embedded vector $\mathbf{E}_{i,j}^l$ is multiplied by an explicit matrix, that is,

$$\mathbf{O}_{i,j}^l = \mathbf{w}_2^l \cdot \mathbf{E}_{i,j}^l + \mathbf{b}_2^l, i = 1, \cdots, n, j = 1, \cdots, c, \tag{3}$$

where $\mathbf{O}_{i,j}^l, \mathbf{b}_2^l \in \mathbb{R}^{wh}$, $\mathbf{w}_2^l \in \mathbb{R}^{wh \times m}$. Again, $\mathbf{w}_2^l$ and $\mathbf{b}_2^l$ are unique for every embedded vector and the subscript $_{i,j}$ is omitted. $\mathbf{w}_2^l$, $\mathbf{b}_2^l$, and $\mathbf{O}_{i,j}^l$ can also be aggregated as high-dimensional tensors, *i.e.* $\mathbf{W}_2^l \in \mathbb{R}^{n \times c \times wh \times m}$ and $\mathbf{B}_2^l, \mathbf{O}^l \in \mathbb{R}^{n \times c \times wh}$. For the sake of simplicity, Eqs. 1, 2 and 3 can be abstracted as

$$\mathbf{O}^l = h(\mathbf{z}^l, \mathbf{z}^{l-1}; \mathbf{W}^l, \mathbf{B}^l), \tag{4}$$

where $h(\cdot)$ denotes the functionality of the hypernetwork. The final output $\mathbf{O}^l$ is used as the weight parameter of the l-th layer. The output $\mathbf{O}^l$ is covariant with the input latent vector because pruning an element in the latent vector removes the corresponding slice of the output $\mathbf{O}^l$ (See Fig. 3).

When designing the hypernetwork, we tried to add batch normalization and non-linear layers after the linear operation in Eq. 2 and Eq. 3. But it did not lead to clearly better results. Thus, we just kept to the simple design. This is also consistent with the previous designs [11,37].

3.2 Sparsity Regularization and Proximal Gradient

The core of differentiability comes with not only the specifically designed hypernetwork but also the mechanism used to search the the potential candidate. To achieve that, we enforce sparsity constraints to the latent vectors. The loss function of the aforementioned L-layer CNN is denoted as

$$\min_{\mathbf{W},\mathbf{B},\mathbf{z}} \mathcal{L}\Big(\mathbf{y}, f\big(\mathbf{x}; h(\mathbf{z}; \mathbf{W}, \mathbf{B})\big)\Big) + \gamma\mathcal{D}(\mathbf{W}) + \gamma\mathcal{D}(\mathbf{B}) + \lambda\mathcal{R}(\mathbf{z}), \tag{5}$$

where $\mathcal{L}(\cdot,\cdot)$, $\mathcal{D}(\cdot)$, and $\mathcal{R}(\cdot)$ are the loss function for a specific vision task, the weight decay term, and the sparsity regularization term, γ and λ are the regularization factors. For the simplicity of notation, the superscript l is omitted. The sparsity regularization takes the form of ℓ_1 norm, namely,

$$\mathcal{R}(\mathbf{z}) = \sum_{l=1}^{L} \|\mathbf{z}^l\|_1. \tag{6}$$

To solve the problem in Eq. 5, the weights $\mathbf{W}$ and and biases $\mathbf{B}$ of the hypernetworks are updated with SGD. The gradients are back-propagated from the backbone network to the hypernetwork. Thus, neither the forward nor the backward pass challenges the information flow between them. The latent vectors are updated with proximal gradient algorithm, *i.e.* ,

$$\mathbf{z}[k+1] = \mathbf{prox}_{\lambda\mu\mathcal{R}}\Big(\mathbf{z}[k] - \lambda\mu\nabla\mathcal{L}\big(\mathbf{z}[k]\big)\Big), \tag{7}$$

where μ is the step size of proximal gradient and is set as the learning rate of SGD updates. As can be seen in the equation, the proximal gradient update contains a gradient descent step and a proximal operation step. When the regularizer has the form of ℓ_1 norm, the proximal operator has closed-form solution, *i.e.*

$$\mathbf{z}[k+1] = \text{sgn}\Big(\mathbf{z}[k+\varDelta]\Big)\Big[\big|\mathbf{z}[k+\varDelta]\big| - \lambda\mu\Big]_+, \tag{8}$$

where $\mathbf{z}[k+\varDelta] = \mathbf{z}[k] - \lambda\mu\nabla\mathcal{L}(\mathbf{z}[k])$ is the intermediate SGD update, the sign operator $\text{sgn}(\cdot)$, the thresholding operator $[\cdot]_+$, and the absolute value operator $|\cdot|$ act element-wise on the vector. Equation 8 is the soft-thresholding function.

The latent vectors first get SGD updates along with the other parameters $\mathbf{W}$ and $\mathbf{B}$. Then the proximal operator is applied. Due to the use of SGD updates and the fact that the proximal operator has closed-form solution, we recognize the whole solution as approximately differentiable (although the ℓ_1 norm is not differentiable at 0), which guarantees the fast convergence of the algorithm compared with reinforcement learning and evolutionary algorithm. The speed-up of proximal gradient lies in that instead of searching the best candidate among the total population it forces the solution towards the best sparse one.

The automation of pruning follows the way the sparsity applied in Eq. 6 and the proximal gradient solution. First of all, all latent vectors are regularized together without distinguishment between them. During the optimization, information and gradients flows fluently between the backbone network and the hypernetwork. The proximal gradient algorithm forces the potential elements of the latent vectors to approach zero quicker than the others without any human effort and interference in this process. The optimization stops immediately when the target compression ratio is reached. In total, there are only two additional hyper-parameters in the algorithm, *i.e.* the sparsity regularization factor and the mask threshold τ in Subsect. 3.3. Thus, running the algorithm is just like turning on the button, which enable the application of the algorithm to all of the CNNs without much interference of domain experts' knowledge.

3.3 Network Pruning

Different from the fully connected layers, the proposed design of hypernetwork can adapt the dimension of the output according to that of the latent vectors. After the searching stage, sparse versions of the latent vectors are derived as $\overline{\mathbf{z}}^{l-1}$ and $\overline{\mathbf{z}}^{l}$. For those vectors, some of their elements are zero or approaching zero. Thus, 1-0 masks can by derived by comparing the sparse latent vectors with a predefined small threshold τ, *i.e.* $\mathbf{m}^l = \mathcal{T}\left(\overline{\mathbf{z}}^l, \tau\right)$, where the function $\mathcal{T}(\cdot)$ element-wise compares the latent vector with the threshold and returns 1 if the element is not smaller than τ and 0 otherwise. Then latent vector $\overline{\mathbf{z}}^l$ is pruned according to the mask $\mathbf{m}^l$. See Subsect. 3.6 for more analysis.

3.4 Latent Vector Sharing

Due to the existence of skip connections in residual networks such as ResNet, MobileNetV2, SRResNet, and EDSR, the residual blocks are interconnected with each other in the way that their input and output dimensions are related. Therefore, the skip connections are notoriously tricky to deal with. But back to the design of the proposed hypernetwork, a quite simple and straightforward solution to this problem is to let the hypernetworks of the correlated layers share the same latent vector. Note that the weight and bias parameters of the hypernetworks are not shared. Thus, sharing latent vectors does not force the correlated layers to be identical. By automatically pruning the single latent vector, all of the relevant layers are pruned together. Actually, we first tried to use different latent vectors for the correlated layers and applied group sparsity to them. But the experimental results showed that this is not a good choice because this strategy shot lower accuracy than the latent vector sharing strategy. (See details in the Supplementary).

3.5 Discussion on the Convergence Property

Compared with reinforcement learning and evolutionary algorithm, proximal gradient may not be the optimal solution for some problems. But as found by previous works [37,38], automatic network pruning serves as an implicit searching method for the channel configuration of a network. In addition, the network is searched and trained from scratch in this paper. The important factor is the number of remaining channels of the convolutional layers in the network. Thus, it is relatively not important which filter is pruned as long as the number of pruned channels are the same. This reduces the number of possible candidates by orders of magnitude. In this case, proximal gradient works quite well.

3.6 Implementation Consideration

Compact Representation of the Hypernetwork. Thanks to the default tensor operations in deep learning toolboxes [42], the operations in the hypernetwork could be represented in a compact form. The embedding operation in Eq. 2 can be written as the following high-dimensional tensor operation

$$\mathbf{E}^l = \mathcal{U}^3\left(\mathbf{Z}^l\right) \circ \mathbf{W}_1^l + \mathbf{B}_1^l, \tag{9}$$

where $\mathcal{U}^3(\mathbf{Z}^l) \in \mathbb{R}^{n \times c \times 1}$, [$\circ$] denotes the broadcastable element-wise tensor multiplication, $\mathcal{U}^3(\cdot)$ inserts a third dimension for $\mathbf{Z}^l$. The operation in Eq. 3 can be easily rewritten as batched matrix multiplication,

$$\mathbf{O}^l = \mathbf{W}_2^l * \mathbf{E}^l + \mathbf{B}_2^l, \tag{10}$$

where [$*$] denotes batched matrix multiplication.

Pruning Analysis. Analyzing the three layers of the hypernetworks together with the masked latent vectors leads to a direct impression on how the backbone layers are automatically pruned. That is,

$$\overline{\mathbf{O}}^l = \mathbf{W}_2^l * \left[\mathcal{U}^3\left((\mathbf{m}^l \circ \overline{\mathbf{z}}^l) \cdot (\mathbf{m}^{l-1} \circ \overline{\mathbf{z}}^{l-1})^T\right) \circ \mathbf{W}_1^l\right] \tag{11}$$

$$= \mathbf{W}_2^l * \left[\mathcal{U}^3(\mathbf{m}^l \cdot \mathbf{m}^{l-1^T}) \circ \mathcal{U}^3(\overline{\mathbf{z}}^l \cdot \overline{\mathbf{z}}^{l-1^T}) \circ \mathbf{W}_1^l\right] \tag{12}$$

$$= \mathcal{U}^3(\mathbf{m}^l \cdot \mathbf{m}^{l-1^T}) \circ \left[\mathbf{W}_2^l * \left(\mathcal{U}^3(\overline{\mathbf{z}}^l \cdot \overline{\mathbf{z}}^{l-1^T}) \circ \mathbf{W}_1^l\right)\right] \tag{13}$$

The equality follows the broadcastability of the the operations [$\circ$] and [$*$]. As shown in the above equations, applying the masks on the latent vectors has the same effect of applying them on the final output. Note that in the above analysis the bias terms $\mathbf{B}_0^l$, $\mathbf{B}_1^l$, and $\mathbf{B}_2^l$ are omitted since they have a really small influence on the output of the hypernetwork. In conclusion, the final output can be pruned according to the same criterion for the latent vectors.

Initialization of the Hypernetwork. All biases are initialized as zero, the latent vector with standard normal distribution, and $\mathbf{W}_1^l$ with Xaiver uniform [10]. The weight of the explicit layer $\mathbf{W}_2^l$ is initialized with Hyperfan-in which guarantees stable backbone network weights and fast convergence [5].

4 Experimental Results

To validate the proposed method, extensive experiments were conducted on various CNN architectures including ResNet [15], DenseNet [20] for CIFAR10 [23] image classification, MobileNetV1 [19], MobileNetV2 [47] for Tiny-ImageNet [8] image classification, SRResNet [26], EDSR [32] for single image super-resolution, and DnCNN [56], UNet [46] for gray image denoising. The proposed DHP algorithm starts from a randomly initialized network with the initialization method detailed in Subsect. 3.6. After pruning, the training of the pruned network continues with the same training protocol used for the original network. All of the experiments are conducted on NVIDIA TITAN Xp GPUs.

Hyperparameters. A target FLOPs compression ratio is set for the pruning algorithm. When the difference between the target compression ratio and the actual compression ratio falls below 2%, the automatic pruning procedure stops. The parameter space of hypernetwork increases in proportion to the dimension m of the embedding space. Thus, m should not be too large. In this paper, m is set to 8. The step size μ of the proximal operator is set as the learning rate of SGD updates. The sparsity regularization factor λ is set by empirical studies. The value is chosen such that the searching epochs constitute arounds 5%–10% of the whole training epochs. This guarantees acceptable convergence during searching while not introducing too much additional computation. Please refer to the supplementary for the detailed training and testing protocol.

4.1 Image Classification

The compression results on image classification networks are shown in Table 1. 'DHP-**' denotes the proposed method with the target FLOPs ratio during

Table 1. Results on image classification networks. The FLOPs ratio and parameter ratio of the pruned networks are reported. DHP outperforms the compared methods under comparable model complexity. On Tiny-ImageNet, DHP-24-2 shoots lower error rates than the original model

Network Top-1 error (%)	Compression method	Top-1 error (%)	FLOPs ratio (%)	Parameter ratio (%)
CIFAR10				
ResNet-20 7.46	[53]	9.10	52.60	62.78
	DHP-50 (Ours)	8.46	51.80	56.13
	FPGM [16]	9.38	46.00	–
ResNet-56 7.05	Variational [57]	7.74	79.70	79.51
	Pruned-B [27]	6.94	72.40	86.30
	GAL-0.6 [33]	6.62	63.40	88.20
	NISP [55]	6.99	56.39	57.40
	DHP-50 (Ours)	6.42	50.96	58.42
	CaP [39]	6.78	50.20	–
	ENC [22]	7.00	50.00	–
	AMC [17]	8.10	50.00	–
	KSE [31]	6.77	48.00	45.27
	FPGM [16]	6.74	47.70	–
	GAL-0.8 [33]	8.42	39.80	34.10
	DHP-38 (Ours)	7.06	39.07	41.10
ResNet-110 5.31	**DHP-62 (Ours)**	5.37	63.66	63.2
	Variational [57]	7.04	63.56	58.73
	Pruned-B [27]	6.70	61.40	67.60
	GAL-0.5 [33]	7.26	51.50	55.20
	DHP-20 (Ours)	6.61	21.63	22.40
ResNet-164 4.97	Hinge [29]	5.40	53.61	70.34
	SSS [21]	5.78	53.53	84.75
	DHP-50 (Ours)	5.22	51.67	50.97
	Variational [57]	6.84	50.92	43.30
	DHP-20 (Ours)	6.30	21.78	20.46
DenseNet-12-40 5.26	Variational [57]	6.84	55.22	40.33
	DHP-38 (Ours)	6.06	39.80	63.76
	DHP-28 (Ours)	6.51	29.52	26.01
	GAL-0.1 [33]	6.77	28.60	25.00
Tiny-ImageNet				
MobileNetV1 52.71	**DHP-24-2 (Ours)**	50.75	101.08	43.58
	MobileNetV1-0.75	54.22	57.42	57.64
	MetaPruning [37]	54.48	56.77	88.14
	DHP-50 (Ours)	51.63	51.91	36.95
MobileNetV2 44.75	**DHP-24-2 (Ours)**	43.82	99.09	72.72
	DHP-10 (Ours)	52.43	11.92	6.50
	MetaPruning [37]	56.72	11.00	90.27
	MobileNetV2-0.3	53.99	10.09	11.64

Table 2. Results on image super-resolution networks. The upscaling factor is ×4. Runtime is averaged for Urban100. Maximum GPU memory consumption is reported for Urban100. FLOPs is reported for a 128 × 128 image patch. DHP achieves significant reduction of runtime

Network	Method	PSNR [dB]					FLOPs [G]	Params [M]	Run-time [ms]	GPU Mem [GB]
		Set5	Set14	B100	Urban100	DIV2K				
SRResNet [26]	Baseline	32.03	28.50	27.52	25.88	28.85	32.83	1.54	34.73	0.6773
	Clustering [48]	31.93	28.44	27.47	25.71	28.75	32.83	0.34	31.07	0.8123
	Factor-SIC3 [51]	31.86	28.38	27.40	25.58	28.65	20.83	0.81	102.51	1.4957
	DHP-60 (Ours)	31.97	28.47	27.48	25.76	28.79	20.29	0.95	27.91	0.5923
	Basis-32-32 [30]	31.90	28.42	27.44	25.65	28.69	19.77	0.74	45.73	0.9331
	Factor-SIC2 [51]	31.68	28.32	27.37	25.47	28.58	18.38	0.66	74.66	1.1201
	Basis-64-14 [30]	31.84	28.38	27.39	25.54	28.63	17.49	0.60	36.75	0.6741
	DHP-40 (Ours)	31.90	28.45	27.47	25.72	28.75	13.71	0.64	22.71	0.4907
	DHP-20 (Ours)	31.77	28.34	27.40	25.55	28.60	7.77	0.36	14.74	0.3795
EDSR [32]	Baseline	32.10	28.55	27.55	26.02	28.93	90.37	3.70	49.73	1.3276
	Clustering [48]	31.93	28.47	27.48	25.77	28.80	90.37	0.82	50.51	1.2838
	Factor-SIC3 [51]	31.96	28.47	27.49	25.81	28.81	65.49	2.19	125.10	1.5007
	Basis-128-40 [30]	32.03	28.45	27.50	25.81	28.82	62.65	2.00	48.19	1.3219
	Factor-SIC2 [51]	31.82	28.40	27.43	25.63	28.70	60.90	1.90	94.94	1.3209
	Basis-128-27 [30]	31.95	28.42	27.46	25.76	28.76	58.28	1.74	45.84	1.3209
	DHP-60 (Ours)	31.99	28.52	27.53	25.92	28.88	55.67	2.28	45.11	0.6950
	DHP-40 (Ours)	32.01	28.49	27.52	25.86	28.85	37.77	1.53	33.50	0.9650
	DHP-20 (Ours)	31.94	28.42	27.47	25.69	28.77	19.40	0.79	22.63	1.1588

pruning stage. As in Fig. 1, the operating point DHP-24-2 is derived by compressing the widened mobile networks with $\alpha = 2$ and the target FLOPs ratio 24%. For ResNet-56, the proposed method is compared with 9 different network compression methods and achieves the best performance, *i.e.* 6.42% Top-1 error rate on the most intensively investigated 50% compression level. The compression of DenseNet-12-40 is reasonable compared with the other methods. The accuracy of the operating points DHP-62 of ResNet-110 and DHP-50 of ResNet-164 is quite close to that of the baseline. More results on ResNet-110 and ResNet-164 are shown in the Supplementary. A comparison between ℓ_1 and ℓ_2 regularization in Eq. 6 is done. The results in the Supplementary shows that ℓ_1 regularization is better than ℓ_2 regularization.

On Tiny-ImageNet, DHP achieves lower Top-1 error rates than MetaPruning [37] under the same FLOPs constraint. DHP results in models that are more accurate than the uniformly scaled networks. The error rate of DHP-10 is 1.56% lower than that of MobileNetV2-0.3 with slightly fewer FLOPs and 5.14% fewer parameters. On MobileNetV1, the accuracy gain of DHP-50 over MobileNetV1-0.75 goes to 2.59% with over 5% fewer FLOPs and 10% fewer parameters. Based on this, we hypothesized that it is possible to derive a model which is more accurate than the original version by pruning the widened mobile networks. And this is confirmed by comparing the accuracy of the operating points DHP-24-2 with the baseline accuracy.

Table 3. Results on image denoising networks. The noise level is 70. Runtime and maximum GPU memory are reported for BSD68. FLOPs is reported for a 128 × 128 image. DHP achieves significant reduction of runtime

Network	Method	PSNR [*dB*]		FLOPs [*G*]	Params [*M*]	Runtime [*ms*]	GPU Mem [*GB*]
		BSD68	DIV2K				
DnCNN [56]	Baseline	24.93	26.73	9.13	0.56	23.38	0.1534
	Clustering [48]	24.90	26.67	9.13	0.12	21.97	0.2973
	DHP-60 (Ours)	24.91	26.69	5.65	0.34	18.90	0.1443
	DHP-40 (Ours)	24.89	26.65	3.83	0.23	14.62	0.1194
	Factor-SIC3 [51]	24.97	26.83	3.54	0.22	125.46	0.5910
	Group [43]	24.88	26.64	3.34	0.20	25.69	0.1807
	Factor-SIC2 [51]	24.93	26.76	2.38	0.15	84.17	0.4149
	DHP-20 (Ours)	24.84	26.58	2.01	0.12	10.72	0.0869
UNet [46]	Baseline	25.17	27.17	3.41	7.76	8.73	0.1684
	Clustering [48]	25.01	26.90	3.41	1.72	10.01	0.6704
	DHP-60 (Ours)	25.14	27.11	2.11	4.76	6.86	0.4992
	Factor-SIC3 [51]	25.04	26.94	1.56	3.42	39.84	0.1889
	Group [43]	25.13	27.08	1.49	2.06	11.20	0.1481
	DHP-40 (Ours)	25.12	27.08	1.43	3.24	4.50	0.4992
	Factor-SIC2 [51]	25.01	26.90	1.22	2.51	30.16	0.1855
	DHP-20 (Ours)	25.04	26.97	0.75	1.61	3.93	0.4992

4.2 Super-Resolution

The results on image super-resolution networks are shown in Table 2. DHP is compared with factorized convolution (Factor) [51], learning filter basis method (Basis) [30], and K-means clustering method (Clustering) [48]. 'SIC*' denotes the number of SIC layers in Factor [51]. The practical FLOPs instead of the theoretical FLOPs is reported for Clustering [48]. To fairly compare the methods and measure the practical compression effectiveness, five metrics are involved including Peak Signal-to-Noise Ratio (PSNR), FLOPs, number of parameters, runtime and GPU memory consumption. Several conclusion can be drawn. **I.** Previous methods mainly focus on the reduction of FLOPs and number of parameter without paying special attention to the actual acceleration. Although Clustering can reduce substantial parameters while maintaining quite good PSNR accuracy, the actual computing resource requirement (GPU memory and runtime) is remained. **II.** Convolution factorization and decomposition methods result in additional CUDA kernel calls, which is not efficient for the actual acceleration. **III.** For the proposed method, the two model complexity metrics, *i.e.* FLOPs and parameters change consistently across different operating points, which leads to consistent reduction of computation resources. **IV.** DHP results in both inference-efficient (DHP-20) and accuracy-preserving (DHP-60) models. The visual results are shown in Fig. 4. As can been seen, the visual quality of the images of DHP is almost indistinguishable from that of the baseline.

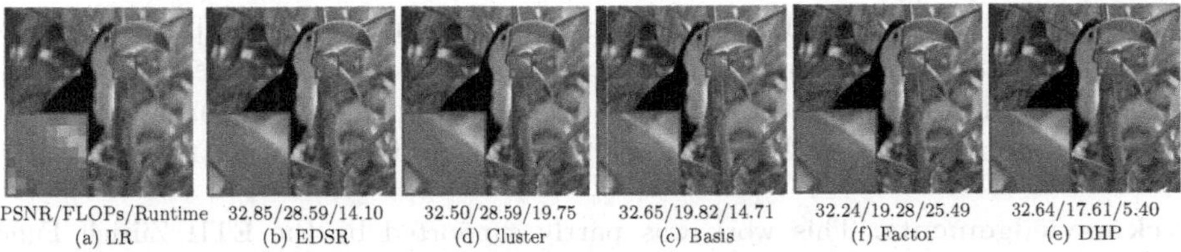

Fig. 4. Single image super-resolution visual results. PSNR and FLOPs measured on the image. Runtime averaged on Set5

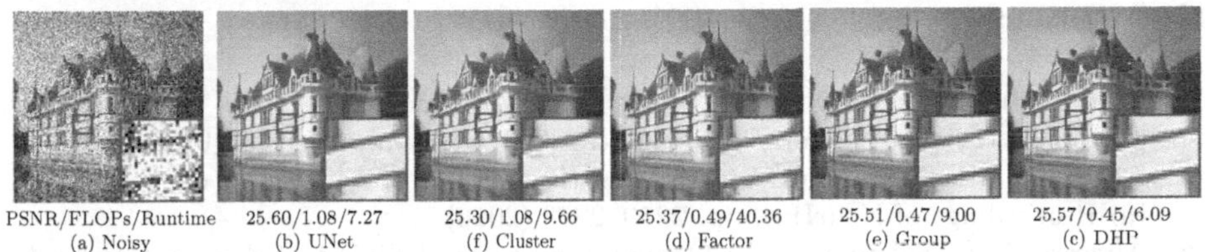

Fig. 5. Image denoising visual results. PSNR and FLOPs measured on the image. Runtime averaged on B100

4.3 Denoising

The results for image denoising networks are shown in Table 3. The same metrics as super-resolution are reported for denoising. An additional method, *i.e.* filter group approximation (Group) [43] is included. In addition to the same conclusion in Subsect. 4.2, another two conclusions are drawn here. **I.** Group [43] fails to reduce the actual computation resources although with quite good accuracy and satisfactory reduction of FLOPs and number of parameters. This might due to the additional 1 × 1 convolution and possibly the inefficient implementation of group convolution in current deep learning toolboxes. **II.** For DnCNN, one interesting phenomenon is that Factor [51] is more accurate than the baseline but has larger appetite for other resources. This is due to two facts. Firstly, Factor [51] has skip connections within the SIC layer. The higher accuracy of Factor [51] just validates the effectiveness of skip connections. Secondly, the SIC layer of Factor [51] introduces more convolutional layers. So Factor-SIC3 has five times more convolutioinal layers than the baseline, which definitely slows down the execution. The visual results are shown in Fig. 5.

5 Conclusion and Future Work

In this paper, we proposed a differentiable automatic meta pruning method via hypernetwork for network compression. The differentiability comes with the specially designed hypernetwork and the proximal gradient used to search the potential candidate network configurations. The automation of pruning lies in the uniformly applied ℓ_1 sparsity on the latent vectors and the proximal gradient that solves the problem. By pruning mobile network with width multiplier

$\alpha = 2$, we obtained models with higher accuracy but lower computation complexity than that with $\alpha = 1$. We hypothesize this is due to the per-layer distinguishing configuration resulting from the automatic pruning. Future work might be investigating whether this phenomenon reoccurs for the other networks.

Acknowledgements. This work was partly supported by the ETH Zürich Fund (OK), a Huawei Technologies Oy (Finland) project, an Amazon AWS grant, and an Nvidia grant.

References

1. MIT technology review: 10 breakthrough technologies 2020. https://www.technologyreview.com/lists/technologies/2020/#tiny-ai. Accessed 01 Mar 2020
2. Alvarez, J.M., Salzmann, M.: Learning the number of neurons in deep networks. In: Proceedings of NeurIPS, pp. 2270–2278 (2016)
3. Alvarez, J.M., Salzmann, M.: Compression-aware training of deep networks. In: Proceedings of NeurIPS, pp. 856–867 (2017)
4. Brock, A., Lim, T., Ritchie, J., Weston, N.: SMASH: one-shot model architecture search through hypernetworks. In: Proceedings of ICLR (2018)
5. Chang, O., Flokas, L., Lipson, H.: Principled weight initialization for hypernetworks. In: Proceedings of ICLR (2020)
6. Chen, C., Tung, F., Vedula, N., Mori, G.: Constraint-aware deep neural network compression. In: Proceedings of ECCV, pp. 400–415 (2018)
7. Chin, T.W., Ding, R., Zhang, C., Marculescu, D.: Towards efficient model compression via learned global ranking. In: Proceedings of CVPR, pp. 1518–1528 (2020)
8. Deng, J., Dong, W., Socher, R., Li, L.J., Li, K., Fei-Fei, L.: ImageNet: a large-scale hierarchical image database. In: Proceedings of CVPR, pp. 248–255. IEEE (2009)
9. Dong, X., Chen, S., Pan, S.: Learning to prune deep neural networks via layer-wise optimal brain surgeon. In: Proceedings of NIPS, pp. 4857–4867 (2017)
10. Glorot, X., Bengio, Y.: Understanding the difficulty of training deep feedforward neural networks. In: Proceedings of AISTATS, pp. 249–256 (2010)
11. Ha, D., Dai, A., Le, Q.V.: HyperNetworks. In: Proceedings of ICLR (2017)
12. Han, S., Mao, H., Dally, W.J.: Deep compression: compressing deep neural networks with pruning, trained quantization and Huffman coding. In: Proceedings of ICLR (2015)
13. Hassibi, B., Stork, D.G.: Second order derivatives for network pruning: optimal brain surgeon. In: Proceedings of NeurIPS, pp. 164–171 (1993)
14. Hayashi, K., Yamaguchi, T., Sugawara, Y., Maeda, S.i.: Einconv: exploring unexplored tensor decompositions for convolutional neural networks. In: Proceedings of NeurIPS, pp. 5553–5563 (2019)
15. He, K., Zhang, X., Ren, S., Sun, J.: Deep residual learning for image recognition. In: Proceedings of CVPR, pp. 770–778 (2016)
16. He, Y., Liu, P., Wang, Z., Hu, Z., Yang, Y.: Filter pruning via geometric median for deep convolutional neural networks acceleration. In: Proceedings of CVPR, pp. 4340–4349 (2019)
17. He, Y., Lin, J., Liu, Z., Wang, H., Li, L.J., Han, S.: AMC: AutoML for model compression and acceleration on mobile devices. In: Proceedings of ECCV, pp. 784–800 (2018)

18. He, Y., Zhang, X., Sun, J.: Channel pruning for accelerating very deep neural networks. In: Proceedings of ICCV, pp. 1389–1397 (2017)
19. Howard, A.G., et al.: MobileNets: Efficient convolutional neural networks for mobile vision applications. arXiv preprint arXiv:1704.04861 (2017)
20. Huang, G., Liu, Z., van der Maaten, L., Weinberger, K.Q.: Densely connected convolutional networks. In: Proceedings of CVPR, pp. 2261–2269 (2017)
21. Huang, Z., Wang, N.: Data-driven sparse structure selection for deep neural networks. In: Proceedings of ECCV, pp. 304–320 (2018)
22. Kim, H., Umar Karim Khan, M., Kyung, C.M.: Efficient neural network compression. In: Proceedings of CVP, June 2019
23. Krizhevsky, A., Hinton, G.: Learning multiple layers of features from tiny images. Technical report, Citeseer (2009)
24. Krueger, D., Huang, C.W., Islam, R., Turner, R., Lacoste, A., Courville, A.: Bayesian hypernetworks. arXiv preprint arXiv:1710.04759 (2017)
25. LeCun, Y., Denker, J.S., Solla, S.A.: Optimal brain damage. In: Proceedings of NeurIPS, pp. 598–605 (1990)
26. Ledig, C., et al.: Photo-realistic single image super-resolution using a generative adversarial network. In: Proceedings of CVPR, pp. 105–114 (2017)
27. Li, H., Kadav, A., Durdanovic, I., Samet, H., Graf, H.P.: Pruning filters for efficient convnets. In: Proceedings of ICLR (2017)
28. Li, J., Qi, Q., Wang, J., Ge, C., Li, Y., Yue, Z., Sun, H.: OICSR: out-in-channel sparsity regularization for compact deep neural networks. In: Proceedings of CVPR, pp. 7046–7055 (2019)
29. Li, Y., Gu, S., Mayer, C., Van Gool, L., Timofte, R.: Group sparsity: the hinge between filter pruning and decomposition for network compression. In: Proceedings of CVPR (2020)
30. Li, Y., Gu, S., Van Gool, L., Timofte, R.: Learning filter basis for convolutional neural network compression. In: Proceedings of ICCV, pp. 5623–5632 (2019)
31. Li, Y., et al.: Exploiting kernel sparsity and entropy for interpretable CNN compression. In: Proceedings of CVPR (2019)
32. Lim, B., Son, S., Kim, H., Nah, S., Lee, K.M.: Enhanced deep residual networks for single image super-resolution. In: Proceedings of CVPRW, pp. 1132–1140 (2017)
33. Lin, S., et al.: Towards optimal structured CNN pruning via generative adversarial learning. In: Proceedings of CVPR, pp. 2790–2799 (2019)
34. Liu, B., Wang, M., Foroosh, H., Tappen, M., Pensky, M.: Sparse convolutional neural networks. In: Proceedings of CVPR, pp. 806–814 (2015)
35. Liu, C., et al.: Progressive neural architecture search. In: Proceedings of ECCV, September 2018
36. Liu, H., Simonyan, K., Yang, Y.: DARTS: differentiable architecture search. In: Proceedings of ICLR (2019)
37. Liu, Z., et a: MetaPruning: meta learning for automatic neural network channel pruning. In: Proceedings of ICCV (2019)
38. Liu, Z., Sun, M., Zhou, T., Huang, G., Darrell, T.: Rethinking the value of network pruning. In: Proceedings of ICLR (2019)
39. Minnehan, B., Savakis, A.: Cascaded projection: end-to-end network compression and acceleration. In: Proceedings of CVPR, June 2019
40. Molchanov, P., Mallya, A., Tyree, S., Frosio, I., Kautz, J.: Importance estimation for neural network pruning. In: Proceedings of CVPR, pp. 11264–11272 (2019)
41. Pan, Z., Liang, Y., Zhang, J., Yi, X., Yu, Y., Zheng, Y.: Hyperst-net: hypernetworks for spatio-temporal forecasting. arXiv preprint arXiv:1809.10889 (2018)

42. Paszke, A., et al.: Automatic differentiation in PyTorch (2017)
43. Peng, B., Tan, W., Li, Z., Zhang, S., Xie, D., Pu, S.: Extreme network compression via filter group approximation. In: Proceedings of ECCV, pp. 300–316 (2018)
44. Pham, H., Guan, M., Zoph, B., Le, Q., Dean, J.: Efficient neural architecture search via parameters sharing. In: Proceedings of ICML, pp. 4095–4104 (2018)
45. Real, E., Aggarwal, A., Huang, Y., Le, Q.V.: Regularized evolution for image classifier architecture search. In: Proceedings of AAAI, vol. 33, pp. 4780–4789 (2019)
46. Ronneberger, O., Fischer, P., Brox, T.: U-Net: convolutional networks for biomedical image segmentation. In: Navab, N., Hornegger, J., Wells, W.M., Frangi, A.F. (eds.) MICCAI 2015. LNCS, vol. 9351, pp. 234–241. Springer, Cham (2015). https://doi.org/10.1007/978-3-319-24574-4_28
47. Sandler, M., Howard, A., Zhu, M., Zhmoginov, A., Chen, L.C.: MobileNetV2: Inverted residuals and linear bottlenecks. In: Proceedings of CVPR, pp. 4510–4520 (2018)
48. Son, S., Nah, S., Mu Lee, K.: Clustering convolutional kernels to compress deep neural networks. In: Proceedings of ECCV, pp. 216–232 (2018)
49. Sutton, R.S., Barto, A.G.: Reinforcement Learning: An Introduction. MIT press, Cambridge (2018)
50. Torfi, A., Shirvani, R.A., Soleymani, S., Nasrabadi, N.M.: GASL: Guided attention for sparsity learning in deep neural networks. arXiv preprint arXiv:1901.01939 (2019)
51. Wang, M., Liu, B., Foroosh, H.: Factorized convolutional neural networks. In: Proceedings of ICCVW, pp. 545–553 (2017)
52. Wen, W., Wu, C., Wang, Y., Chen, Y., Li, H.: Learning structured sparsity in deep neural networks. In: Proceedings of NeurIPS, pp. 2074–2082 (2016)
53. Ye, J., Lu, X., Lin, Z., Wang, J.Z.: Rethinking the smaller-norm-less-informative assumption in channel pruning of convolution layers. In: Proceedings of ICLR (2018)
54. Yoon, J., Hwang, S.J.: Combined group and exclusive sparsity for deep neural networks. In: Proceedings of ICML, pp. 3958–3966. JMLR. org (2017)
55. Yu, R., et al.: NISP: pruning networks using neuron importance score propagation. In: Proceedings of CVPR, pp. 9194–9203 (2018)
56. Zhang, K., Zuo, W., Chen, Y., Meng, D., Zhang, L.: Beyond a Gaussian denoiser: residual learning of deep CNN for image denoising. IEEE TIP **26**(7), 3142–3155 (2017)
57. Zhao, C., Ni, B., Zhang, J., Zhao, Q., Zhang, W., Tian, Q.: Variational convolutional neural network pruning. In: Proceedings of CVPR, pp. 2780–2789 (2019)
58. Zoph, B., Le, Q.V.: Neural architecture search with reinforcement learning. In: Proceedings of ICLR (2017)
59. Zoph, B., Vasudevan, V., Shlens, J., Le, Q.V.: Learning transferable architectures for scalable image recognition. In: Proceedings of CVPR, June 2018

Deep Transferring Quantization

Zheng Xie[1,2], Zhiquan Wen[1], Jing Liu[1], Zhiqiang Liu[1,2], Xixian Wu[3], and Mingkui Tan[1(✉)]

[1] South China University of Technology, Guangzhou, China
{sexiezheng,sewenzhiquan,seliujing,sezhiqiangliu}@mail.scut.edu.cn, mingkuitan@scut.edu.cn

[2] PengCheng Laboratory, Shenzhen, China

[3] HuNan Gmax Intelligent Technology, Changsha, China
wuxixian@gmax-ai.com

Abstract. Network quantization is an effective method for network compression. Existing methods train a low-precision network by fine-tuning from a pre-trained model. However, training a low-precision network often requires large-scale labeled data to achieve superior performance. In many real-world scenarios, only limited labeled data are available due to expensive labeling costs or privacy protection. With limited training data, fine-tuning methods may suffer from the overfitting issue and substantial accuracy loss. To alleviate these issues, we introduce transfer learning into network quantization to obtain an accurate low-precision model. Specifically, we propose a method named deep transferring quantization (DTQ) to effectively exploit the knowledge in a pre-trained full-precision model. To this end, we propose a learnable attentive transfer module to identify the informative channels for alignment. In addition, we introduce the Kullback-Leibler (KL) divergence to further help train a low-precision model. Extensive experiments on both image classification and face recognition demonstrate the effectiveness of DTQ.

Keywords: Quantization · Deep transfer · Knowledge distillation

1 Introduction

Deep convolutional neural networks (CNNs) have been widely applied in various computer vision tasks, such as image classification [17,18,21,48], face recognition [8,12,52], object detection [35,41,42], and semantic segmentation [7,20,44]. However, a deep model often contains millions of parameters and requires billions of floating-point operations (FLOPs) during inference, which restricts its

X. Zheng, Z. Wen, and J. Liu—Authors contributed equally.

Electronic supplementary material The online version of this chapter (https://doi.org/10.1007/978-3-030-58598-3_37) contains supplementary material, which is available to authorized users.

A. Vedaldi et al. (Eds.): ECCV 2020, LNCS 12353, pp. 625–642, 2020.
https://doi.org/10.1007/978-3-030-58598-3_37

applications on resource-limited devices, such as mobile phones. To reduce the computational costs and memory overheads, various studies have been proposed, such as low-rank decomposition [49,63], network pruning [22,37,72] and network quantization [62,68,71]. In this paper, we focus on network quantization, which aims to compress the deep models and reduce the execution latency.

Existing quantization methods can be split into two categories, namely post-training quantization [2,3,67] and training-aware quantization [9,28,68]. Post-training quantization methods quantizes the models by directly converting weights and activations into low-precision ones. However, the performance will degrade severely in regard to low-precision quantization (e.g., 4-bit quantization). To achieve promising performance, training-aware quantization methods fine-tune the low-precision network with a large quantity of data to compensate for the performance loss from quantization. However, in many real-world scenarios, only a small number of labeled data are available due to expensive labeling costs or privacy protection. Since a deep model often contains a large number of parameters, fine-tuning with limited training data may easily suffer from the overfitting problem. Moreover, the training of a low-precision network is very challenging since the training process can easily get trapped in a poor local minimum, resulting in substantial accuracy loss [71]. This issue will be even more severe when the training data are limited.

To alleviate the data burden, we introduce transfer learning [6,32,39], which aims to transfer the knowledge from a source model to a target model. Transfer learning is an important machine learning paradigm that has several general characteristics: 1) the label space of the target task is different from that of the source task; 2) only a small quantity of labeled target data are available. Usually, we have a pre-trained full-precision model trained on the related source large-scale data set (e.g., ImageNet [45]), but the source data are often unavailable. The full-precision model contains rich and useful knowledge, which can be transferred to the low-precision model. Based on this intuition, we study the transferring quantization task, which seeks to obtain a promising low-precision model with a small number of target data while effectively exploiting the knowledge in the pre-trained full-precision model. Since training a low-precision model with limited target data is very challenging, we seek to propose a method to conduct network quantization and transferring simultaneously.

A deep model usually spans the data into a very high-dimensional space. Inspired by [59], the feature representations generated from a pre-trained model can be transferred to the target model. Imposing feature alignment between the two feature maps generated from the intermediate layer of the full-precision and low-precision models is a good way to exploit the knowledge. However, directly using feature alignment has a limitation. Due to the discrepancy between the target and source tasks, some channels of feature maps are irrelevant or even harmful for the discriminative power in the target task. In addition, such a phenomenon also exists between the full-precision and low-precision models. As a result, the low-precision model may have limited performance. Moreover, since the output of the full-precision model contains rich information about how the model discriminates an input image among a large number of classes, we can exploit this knowledge to guide the training of network quantization and improve the performance of the low-precision model under limited training data.

Based on the above intuition, we propose a simple but effective training method named deep transferring quantization (DTQ), which effectively exploits the knowledge in a pre-trained full-precision model. To this end, we devise a learnable attentive transfer module to identify the informative channels and then align them generated from the full-precision and low-precision models for attentive transferring quantization (ATQ). Moreover, we propose probabilistic transferring quantization (PTQ) to force the probability distribution of the low-precision model to mimic that of the full-precision model.

Our main contributions are summarized as follows:

- In this paper, we study the task of transferring quantization, which introduces transfer learning into network quantization to obtain an accurate low-precision model. To our best knowledge, this task has not received enough attention from the community. Nevertheless, we argue and demonstrate that transferring quantization is very necessary when the target data are limited.
- We propose a simple but effective training method named deep transferring quantization (DTQ). This method uses attentive transferring quantization (ATQ) and probabilistic transferring quantization (PTQ) to effectively exploit the knowledge in the full-precision model for transferring quantization under limited training data. Extensive experiments on both image classification and face recognition demonstrate the superior performance of DTQ.

2 Related Work

Network quantization [19,28] obtains a low-precision model that reduces model size and improves inference efficiency. Existing quantization methods can be divided into two categories, namely post-training quantization [2,3,67] and training-aware quantization [9,28,68]. Since post-training quantization does not require fine-tuning, the performance will degrade severely in regard to low-precision quantization (e.g., 4-bit). To compensate for the quantization performance decrease, training-aware quantization methods fine-tune the low-precision models with a large-scale training data set. Existing training-aware methods can be split into two categories: binary quantization [26,40] and fixed-point quantization[9,68]. For binary neural networks (BNNs), the weights and activations are constrained to $\{+1, -1\}$ [26,40]. In this way, BNNs suffer from significant accuracy loss compared with full-precision models. To reduce this accuracy gap, fixed-point methods[9,68] have been proposed to represent weights and activations with higher bitwidths. DoReFa-Net [68] designed quantizers with a constant quantization step and quantizes weights, activations and gradients to arbitrary bitwidths. Based on DoReFa-Net, PACT [9] used an activation clipping parameter that is optimized during the training to find the right quantization scale. Moreover, to achieve efficient integer-arithmetic-only inference, Jacob *et al.* [28] proposed a linear quantization scheme, which can be implemented more efficiently than floating-point inference on hardware. However, existing training-aware quantization methods require a large-scale labeled data set to conduct fine-tuning. In some cases, only limited training data are available, which limits the performance of the low-precision model.

Transfer learning [6,39] seeks to transfer knowledge learned from the source data to the related target tasks. There are several works related to transfer learning, such as domain adaptation [46,64,65] and continual learning [30,34]. To explore the potential factors that affect deep transfer learning performance, Huh *et al.* [27] proposed analyzing features extracted by various networks pre-trained on ImageNet. Recently, several methods have been proposed to improve the transfer performance, such as sparse transfer [36], filter subset selection [10,15] and parameter transfer [66]. Specifically, Li *et al.* proposed the learning without forgetting (LwF) approach [34], which used new target data to retrain models but preserved the knowledge of the source task. Motivated by LwF, Li *et al.* proposed L^2-SP [33] to regularize the parameters between the two models, which forced the target model to approach the source model. However, if the regularization is too weak or too strong, it may hamper the generalization performance of the target model [32]. Recently, DELTA [32] proposed feature alignment at the channel level with channel attention. However, the attention is pre-learned and fixed for all samples. Since the true informative channels may vary for different samples, the shared attention may limit the transfer performance.

Knowledge distillation [23] (KD) is to distill the knowledge of a teacher network down to that of a small student network. The original works [1,23] force the student networks to mimic the teacher networks to generate similar output distribution. Further, some methods [43,72] proposed to align features in intermediate layers of two networks to transfer the knowledge. Recently, several methods [25,58,69] have been proposed to further exploit the knowledge of teacher networks. Specifically, Huang *et al.* formalized distillation as a distribution matching problem to optimize the student models [25]. Furthermore, many studies adopted the KD mechanism to train a compressed model for better performance. For example, Zhuang *et al.* [72] used KD to perform network pruning. Some network quantization methods [31,53,54,71] proposed adopting KD to help train a low-precision model. However, these methods focus on knowledge transfer in the same tasks. In addition, the attention mechanism [24,38,55,56,60] for deep convoutional networks is relevant to this work.

3 Problem Definition

Notation. Throughout this paper, we use the following notations. Specifically, we use bold upper case letters (e.g., $\mathbf{W}$) to denote matrices and bold lower case letters (e.g., $\mathbf{x}$) to denote vectors. Let M_{full} be a pre-trained full-precision model obtained on some related large-scale data sets (e.g., ImageNet) and $\mathbf{W}_{\text{full}}$ be the corresponding parameters. Let $\mathcal{D} = \{(\mathbf{x}_i, y_i)\}_{i=1}^{N}$ be the target training data, where N is the number of samples. M_{low} denotes a low-precision model and

$\mathbf{W}_{\text{low}}$ denotes the corresponding parameters. Let $f(\mathbf{x}, \mathbf{W}_{\text{low}})$ be the prediction of M_{low}. Note that we may need to build a new classifier w.r.t. the target task by introducing a new fully-connected layer with new parameters $\mathbf{W}_{\text{low}}^{\text{FC}}$.

Network Quantization. Given a pre-trained model, network quantization aims to reduce the model size and computational costs by mapping full-precision (i.e., 32-bit) weights and activations to low-precision ones. For each CNN layer, quantization is parameterized by the number of quantization levels and clamping range. Considering k-bit linear quantization [28], the quantization function is:

$$\begin{aligned} \text{clamp}(r; a, b) &= \min(\max(r, a), b)\,, \\ s(a, b, k) &= \frac{b - a}{2^k - 1}\,, \\ q &= \text{round}(\frac{\text{clamp}(r; a, b) - a}{s(a, b, k)})s(a, b, k) + a\,, \end{aligned} \tag{1}$$

where r denotes the full-precision value, q denotes the quantized value, $[a, b]$ is the quantization range, 2^k is the number of quantization levels, and the $\text{round}(\cdot)$ function denotes rounding to the nearest integer.

Transferring Quantization. In many practical scenarios, only limited labeled data are available. In this case, existing quantization methods directly fine-tune a low-precision model, which may easily suffer from the overfitting issue. Moreover, network quantization transforms the continuous values into discrete values, leading to the worse representational ability of the network. Hence, the low-precision training process can easily get trapped in a poor local minimum, resulting in substantial performance degradation. Usually, we have a pre-trained full-precision model M_{full}, which is obtained on some large-scale data sets (e.g., ImageNet) and contains rich knowledge. Hence, effectively exploiting the knowledge of the full-precision model will help the training of quantization. Based on this intuition, we study a task named **transferring quantization**, which aims to obtain a promising low-precision model by effectively exploiting M_{full} and the limited data in $\mathcal{D}$. Note that the source data are often unavailable.

4 Proposed Method

To effectively exploit the knowledge in the full-precision model M_{full}, one feasible method imposes feature alignment on the intermediate feature maps between M_{full} and M_{low}. In this way, the knowledge in M_{full} is expected to be transferred into M_{low}. Since the target task is often different from the source task, some channels of feature maps are irrelevant or even harmful for the discriminative power in the target task. Moreover, such a phenomenon also exists between the full-precision model M_{full} and the low-precision model M_{low}. Thus, directly applying feature alignment may not obtain promising performance. In addition, since the output of the full-precision model contains rich knowledge about discriminating information among a large number of classes [23], the output probability distribution of the full-precision model M_{full}can be also regarded as a

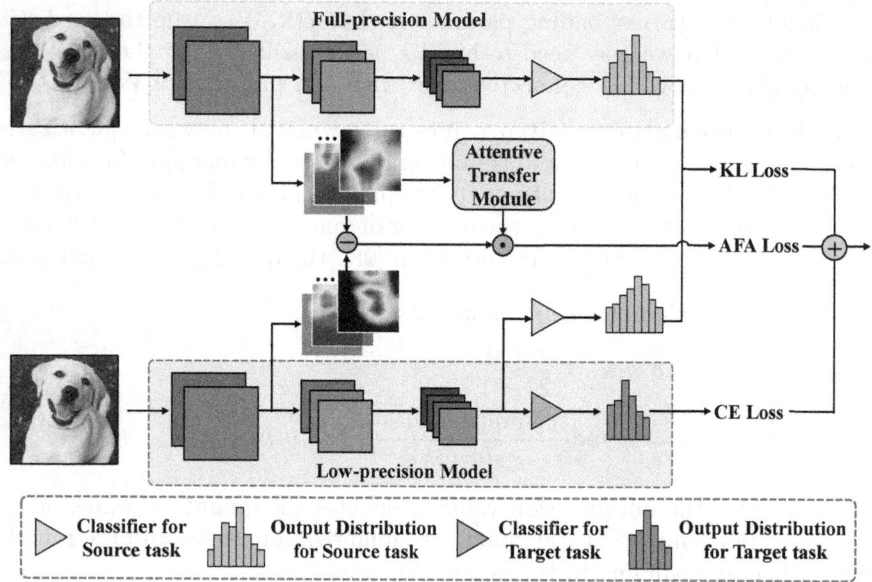

Fig. 1. An overview of DTQ. KL loss is the Kullback-Leibler divergence loss, AFA loss is the attentive feature alignment loss and CE loss is the cross-entropy loss. Note that we evenly take feature maps from four intermediate layers for alignment

guided signal for training the low-precision model M_{low}. Based on this above intuition, in the following, we propose a simple but effective training method named deep transferring quantization (DTQ), which simultaneously performs network quantization and knowledge transfer.

4.1 Deep Transferring Quantization

Motivated by the attention mechanism [55] and knowledge distillation (KD) [43, 60], to effectively exploit the useful knowledge in a pre-trained full-precision model, we first devise a learnable attentive transfer module to identify the informative channels. As mentioned above, it is important to exclude the irrelevant channels and focus on the informative channels for attentive transferring quantization (ATQ). Second, we introduce the Kullback-Leibler (KL) divergence to measure the discrepancy of the probability distribution between the full-precision model and the low-precision model. By minimizing the KL divergence, the useful knowledge learned from the source data can be transferred to the low-precision model for probabilistic transferring quantization (PTQ).

Let P_{full} and P_{low} be the full-precision model and low-precision model predictions for the source task, respectively. With the introduction of ATQ and PTQ, we perform transferring quantization by minimizing this objective w.r.t. $\mathbf{W}_{\text{low}}$ and attention parameters $\mathbf{W}_a$:

Algorithm 1. Deep Transferring Quantization

Input: A pre-trained full-precision model M_{full}, target training data $\mathcal{D} = \{(\mathbf{x}_i, y_i)\}_{i=1}^{N}$, the number of epochs T, the batch size m, and the hyperparameters α and β.
Output: A low-precision model M_{low}.
1: Initialize M_{low} based on M_{full}.
2: **for** $t = 1, \ldots, T$ **do**
3: Randomly sample a mini-batch $(\mathbf{x}, y) \sim \mathcal{D}_m$.
4: Update $\mathbf{W}_{\text{low}}$ and $\mathbf{W}_a$ by minimizing Eq. (2).
5: **end for**

$$\sum_{i=1}^{N}(\mathcal{L}(f(\mathbf{x}_i, \mathbf{W}_{\text{low}}), y_i) + \alpha\Omega(\mathbf{W}_{\text{full}}, \mathbf{W}_{\text{low}}, \mathbf{W}_a, \mathbf{x}_i)) + \beta D_{KL}(\mathrm{P}_{\text{full}}||\mathrm{P}_{\text{low}}), \quad (2)$$

where $\mathcal{L}$ refers to the empirical loss (e.g., cross-entropy loss), Ω denotes the attentive feature alignment (AFA) loss, D_{KL} is the KL divergence loss, and α and β are trade-off hyperparameters. In this way, we can effectively exploit the useful knowledge in the full-precision model M_{full} to obtain a promising low-precision model M_{low}. An overview of DTQ is shown in Fig. 1, and the overall algorithm is shown in Algorithm 1.

4.2 Attentive Transferring Quantization

In this subsection, we introduce attentive transferring quantization (ATQ) in detail. The feature maps derived from the pre-trained full-precision model may contain irrelevant or even detrimental channels to the target low-precision model. To alleviate this issue, we devise an attentive transfer module (ATM) to focus on the discriminative channels of feature maps. As shown in Fig. 2, ATM adopts a two-layer perceptron MLP with a softmax layer to recognize the informative channels. Based on ATM, the attention weight vector $\mathbf{a}^i$ for the i-th sample can be formulated as

$$\mathbf{a}^i = \text{Softmax}(\text{MLP}(\mathbf{W}_a, g(\mathrm{F}(\mathbf{W}_{\text{full}}, \mathbf{x}_i)))), \quad (3)$$

where $\mathrm{F}(\cdot,\cdot)$ denotes a feature map; $g(\cdot)$: $\mathbb{R}^{C\times H\times W} \rightarrow \mathbb{R}^{C\times HW}$ flattens the feature maps in spatial dimension; H, W, C are the height, width and number of channels of the feature maps, respectively; $\mathbf{W}_a$ denotes the parameters of MLP. Based on the attention weight vector, ATQ focuses on the informative channels by minimizing the attentive feature alignment (AFA) loss:

$$\Omega(\mathbf{W}_{\text{full}}, \mathbf{W}_{\text{low}}, \mathbf{W}_a, \mathbf{x}_i) = \sum_{l=1}^{L}\sum_{j=1}^{C} \mathrm{a}_j^i \|\mathrm{F}_{l_j}(\mathbf{W}_{\text{low}}, \mathbf{x}_i) - \mathrm{F}_{l_j}(\mathbf{W}_{\text{full}}, \mathbf{x}_i)\|_{\mathrm{F}}^2, \quad (4)$$

where L is the number of intermediate layers for alignment. Note that DELTA [32] adopted a similar attention mechanism to ATQ, but the attention mechanism of DELTA is pre-learned and then fixed for all samples when training a target model. In our method, each sample has a unique attention weight vector. Last, our method simultaneously updates $\mathbf{W}_{\text{low}}$ and $\mathbf{W}_a$.

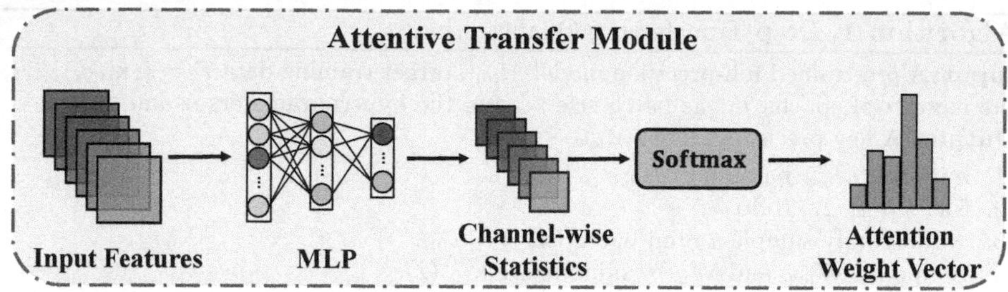

Fig. 2. An overview of our attentive transfer module. For each sample, input features first enter a two-layer perceptron MLP to obtain channel-wise statistics. Then, channel-wise statistics pass the softmax layer to obtain the attention weight vector

4.3 Probabilistic Transferring Quantization

In this subsection, we introduce probabilistic transferring quantization (PTQ) in detail. Except for using attentive feature alignment for ATQ, we also force the probability distribution of the low-precision model to mimic that of the full-precision model by PTQ. However, the number of classes in the target task is often different from that in the source task. It is impossible to directly apply the Kullback-Leibler (KL) divergence to the output of two models. To solve this, we reuse the classifier of the full-precision model on the low-precision model to obtain the probability distribution regarding the source task, as shown in Fig. 1.

Similar to [23, 70], to measure the correlation between the two probability distributions, i.e., P_{full} and P_{low}, we employ the KL divergence:

$$D_{KL}(\mathrm{P}_{\text{full}} \parallel \mathrm{P}_{\text{low}}) = \sum_{i=1}^{N} \mathrm{P}_{\text{full}}(\mathbf{x}_i) \log \frac{\mathrm{P}_{\text{full}}(\mathbf{x}_i)}{\mathrm{P}_{\text{low}}(\mathbf{x}_i)} . \tag{5}$$

By minimizing the KL divergence between the two probability distributions, the knowledge in the output distribution can be transferred from the full-precision model to the low-precision model.

5 Experiments on Image Classification

5.1 Source and Target Data Sets

We choose a large-scale image classification data set as the source data set, namely, ImageNet [11]. The small-scale target data sets are five public data sets with different domains: Stanford Dogs 120 [29], Food-101 [4], CUB-200-2011 [51], Caltech 256-30 and Caltech 256-60 [16]. Similar to [33], we consider Caltech 256-x, where x denotes the number of samples for each class for training (e.g., Caltech 256-10). For validation, we randomly sample 20 images for each class. We show the details of these target data sets in Table 1.

Table 1. Characteristics of the target data sets: name, number of classes, and number of samples of the training set and validation set

Target data sets	# Classes	# Training	# Validation
Stanford Dogs 120	120	12,000	8,580
Caltech 256-30	257	7,710	5,140
Caltech 256-60	257	15,420	5,140
CUB-200-2011	200	5,994	5,794
Food-101	101	75,750	25,250

5.2 Compared Methods

To our best knowledge, the task of transferring quantization has not received enough attention from the community and we fail to find very related baselines for comparison. To evaluate the proposed DTQ, we construct the following methods for comparison: $\boldsymbol{L^2}$**-Q**[1]: directly fine-tune all the parameters of the low-precision model with weight decay on the target data. $\boldsymbol{L^2}$**-*SP*-Q:** based on L^2-SP [33], we regularize the parameters between two models as a part of the loss function to encourage the low-precision model to be similar to the full-precision model. **DELTA-Q**: relying on transferring quantization, we follow DELTA [32] to align feature maps between the full-precision model and the low-precision model with a fixed channel attention mechanism.

5.3 Implementation Details

We adopt MobileNetV2 [47] and ResNet-50 [21] as base models, and use the pre-trained full-precision models from torchvision. We use SGD with a mini-batch size of 64, where the momentum term is set to 0.9. The initial learning rate is set to 0.01. We train low-precision models for 9k iterations, and the learning rate is divided by 10 at the 6k-th iteration. For the hyperparameters α and β in Eq. (2), we fix β to 0.5 and use cross-validation to search for the best α for each experiment. Following in [5], in Eq. (1), we set a and b to the minimum and maximum of the values, respectively. Following [32], we take feature maps from four intermediate layers for attentive feature alignment (i.e., $L = 4$ in Eq. (4)). We repeat each experiment five times, and report the average Top-1 accuracy and the standard deviation on the validation set. The source code and the pre-trained models are available at https://github.com/xiezheng-cs/DTQ.

5.4 Results and Discussions

First, we directly fine-tune the low-precision network on the target data set with different quantization methods, including DoReFa [68], PACT [9] and linear quantization [28]. For convenience, we use L^2-X to denote that directly fine-tune

[1] We follow the naming rule from the transfer learning community to name methods.

Table 2. Comparisons of different methods. We quantize MobileNetV2 and report the Top-1 accuracy (%) on five target data sets. "W" and "A" represent the quantization bitwidth of the weights and activations, respectively

Target data sets	W/A	L^2-DoReFa	L^2-PACT	L^2-Q	L^2-*SP*-Q	DELTA-Q	DTQ
Stanford Dogs 120	5/5	48.4 ± 1.2	48.1 ± 1.0	73.3 ± 0.2	75.2 ± 0.1	79.3 ± 0.2	**80.2 ± 0.2**
	4/4	48.5 ± 1.2	48.2 ± 1.0	69.1 ± 0.3	70.9 ± 0.4	76.0 ± 0.3	**76.1 ± 0.4**
Caltech 256-30	5/5	45.6 ± 1.5	46.0 ± 1.4	74.9 ± 0.3	75.6 ± 0.2	78.3 ± 0.3	**80.1 ± 0.2**
	4/4	44.9 ± 1.0	44.8 ± 0.4	68.2 ± 0.3	69.2 ± 1.3	74.9 ± 0.7	**75.9 ± 1.2**
Caltech 256-60	5/5	58.7 ± 1.3	58.5 ± 1.4	78.3 ± 0.1	79.4 ± 0.2	82.4 ± 0.3	**83.2 ± 0.2**
	4/4	58.1 ± 0.8	58.2 ± 0.7	73.6 ± 0.4	74.1 ± 0.2	79.2 ± 0.6	**79.9 ± 0.3**
CUB-200-2011	5/5	47.3 ± 1.4	46.0 ± 1.5	75.3 ± 0.2	75.1 ± 0.2	**76.2 ± 0.3**	75.4 ± 0.3
	4/4	46.3 ± 1.5	47.3 ± 0.9	69.2 ± 0.9	70.0 ± 1.0	71.9 ± 0.2	**72.1 ± 0.3**
Food-101	5/5	66.3 ± 1.0	66.7 ± 0.8	81.3 ± 0.1	81.1 ± 0.3	**81.7 ± 0.1**	**81.7 ± 0.1**
	4/4	64.6 ± 0.7	64.9 ± 0.7	77.4 ± 0.2	76.9 ± 0.4	77.7 ± 0.5	**78.4 ± 0.4**

Table 3. Comparisons of different methods. We quantize ResNet-50 and report the Top-1 accuracy (%) on five target data sets. "W" and "A" represent the quantization bitwidth of the weights and activations, respectively

Target data sets	W/A	L^2-Q	L^2-*SP*-Q	DELTA-Q	DTQ
Stanford Dogs 120	5/5	78.9 ± 0.3	84.3 ± 0.3	86.5 ± 0.5	**86.5 ± 0.4**
	4/4	75.2 ± 0.2	80.8 ± 0.2	81.7 ± 2.0	**82.3 ± 0.7**
Caltech 256-30	5/5	83.3 ± 0.2	83.3 ± 0.1	85.0 ± 0.2	**85.0 ± 0.1**
	4/4	78.9 ± 0.2	80.6 ± 0.1	82.8 ± 0.5	**83.5 ± 0.6**
Caltech 256-60	5/5	84.4 ± 0.3	86.6 ± 0.2	87.2 ± 0.1	**87.4 ± 0.2**
	4/4	80.8 ± 0.7	84.3 ± 0.3	84.8 ± 1.1	**85.5 ± 0.8**
CUB-200-2011	5/5	80.3 ± 0.1	80.2 ± 0.3	**80.9 ± 0.2**	80.3 ± 0.1
	4/4	76.3 ± 0.2	76.6 ± 0.2	76.4 ± 0.3	**77.8 ± 0.3**
Food-101	5/5	84.1 ± 0.1	84.4 ± 0.2	84.3 ± 0.1	**84.4 ± 0.1**
	4/4	80.5 ± 0.3	80.6 ± 0.2	80.6 ± 0.7	**81.0 ± 1.5**

with X quantization method. From Table 2, compared with L^2-DoReFa and L^2-PACT, L^2-Q achieves much better performance. This indicates that linear quantization is more suitable for transferring quantization under the limited training data. Thus, we adopt linear quantization as our quantization method.

Second, we compare the performance of DTQ with three methods, including L^2-Q, L^2-*SP*-Q and DELTA-Q. We show the results of MobileNetV2 and ResNet-50 in Table 2 and Table 3, respectively. From these results, we make the following observations. 1) For the 5-bit MobileNetV2 and ResNet-50, DTQ outperforms these compared methods in most cases. 2) DTQ achieves significant improvement over the baselines, especially at low-precision (e.g., 4-bit) quantization. Specifically, for 4-bit MobileNetV2, DTQ outperforms DELTA-Q in the Top-1 accuracy by 1.0% on Caltech 256-30; for 4-bit ResNet-50, DTQ outperforms these compared methods in the Top-1 accuracy by at least 0.4% on all

Table 4. Effect of different losses in DTQ. We report the Top-1 accuracy (%) of 5-bit MobileNetV2 on the Caltech 256-30 data set

Model	CE Loss	AFA Loss	KL Loss	Top-1 Accuracy
MobileNetV2 (5-bit)	√			74.9 ± 0.3
	√	√		79.7 ± 0.1
	√	√	√	$\mathbf{80.1 \pm 0.2}$

target data sets. Moreover, DTQ surpasses DELTA-Q in the Top-1 accuracy by 1.4% on CUB-200-2011. In a word, these results show the effectiveness of DTQ.

5.5 Effect of Losses in DTQ

To investigate the effect of the losses in DTQ, we first conduct experiments with different combinations of the losses for the 5-bit MobileNetV2 on Caltech 256-30. Then, we visualize the feature maps of the models with different losses in DTQ. The experimental results are shown in Table 4 and Fig. 3.

Quantitative Comparisons. From the results of 5-bit MobileNetV2 in Table 4, we make the following observations. 1) CE Loss: this baseline just uses the cross-entropy (CE) loss to train a low-precision network. 2) CE Loss + AFA Loss: this experiment increases the Top-1 accuracy by about 5.0% compared with the method just using CE loss. Besides, DTQ achieves 79.7% in Top-1 accuracy, which is significantly better than DELTA-Q (78.3% in Table 2). These results embody the effectiveness of our proposed AFA loss. 3) CE Loss + AFA Loss + KL Loss: compared with the second experiment, it further improves the performance of the low-precision network. These results indicate the effectiveness of our KL loss. In total, both AFA loss and KL loss contribute to better performance of the low-precision models.

Qualitative Comparisons. To further investigate the effect of the losses in DTQ, we visualize the feature maps of the penultimate layer of MobileNetV2 on Caltech 256-30. From Fig. 3, when we only use the CE loss, the 5-bit MobileNetV2 fails to focus on the target object. When we add the AFA loss, the 5-bit MobileNetV2 achieves significantly better performance, whose feature maps activate the information of the target object more accurately. Furthermore, the 5-bit MobileNetV2 equipped with all three losses shows a better concentration on the target object than that equipped with two losses. Due to the page limit, we put more visualization results in the supplementary. This visualization results further demonstrate the effectiveness of the proposed losses in DTQ.

5.6 Further Experiments

Performance on Different Scales of Data Sets. We conduct several experiments to evaluate the proposed DTQ on different scales of data sets. We choose

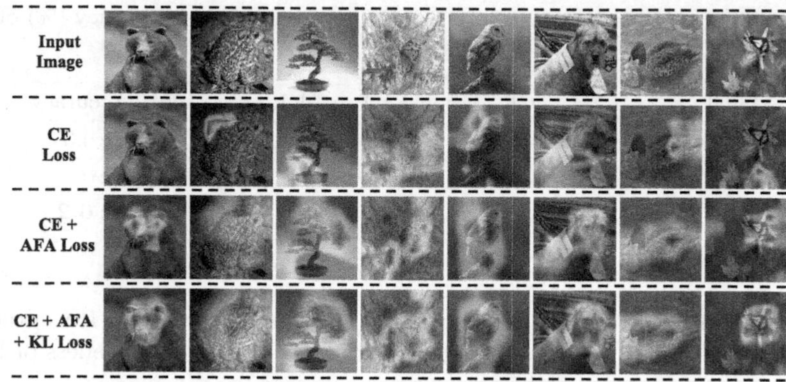

Fig. 3. Visualization of features from models with different losses. Samples are taken from the features of the penultimate layer of 5-bit MobileNetV2 on Caltech 256-30

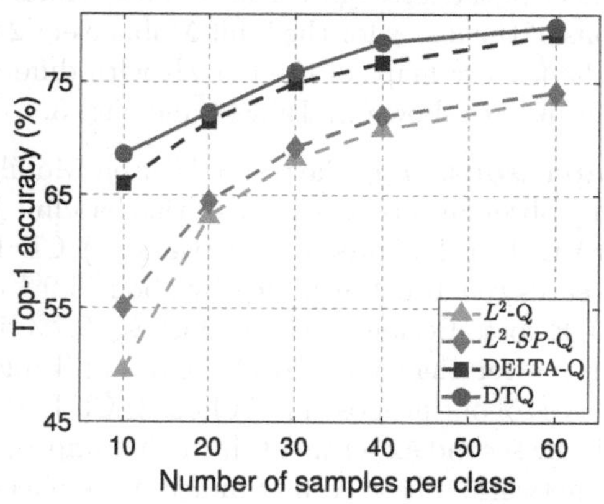

Fig. 4. Performance of different methods in the Top-1 accuracy (%) of 4-bit MobileNetV2 on different scales of the Caltech 256 data set

Caltech 256 with different numbers of training samples for each class, i.e., from 10 to 60. From the results of 4-bit MobileNetV2 in Fig. 4, DTQ outperforms other methods on different scales of Caltech 256, especially on the small scale of data set. For example, on the training data set with 10 training samples for each class, DTQ outperforms DELTA-Q by 2.6% in the Top-1 accuracy. These results indicate the superiority of DTQ under small training data.

Effect of the Attentive Transfer Module. To evaluate the proposed attentive transfer module (ATM), we conduct experiments on the DTQ with and without attention. DTQ without attention means directly using feature alignment without ATM for ATQ. From the results in Table 5, DTQ with attention outperforms the one without attention, especially on Food-101 (1.3% improvement on average Top-1 accuracy), which demonstrates the effectiveness of our ATM.

Table 5. Effect of different training schemes. "One-stage" refers to performing transferring and quantization simultaneously. "Two-stage" denotes that we first perform transferring and then perform quantization. We report the Top-1 accuracy (%) of 4-bit MobileNetV2 on five target data sets. Note that "w/o ATT" means without attention

Target data sets	W/A	One-stage		Two-stage	
		DTQ (w/o ATT)	DTQ	DT → L^2-Q	DT → DTQ
Stanford Dogs 120	4/4	75.2 ± 0.5	76.1 ± 0.4	71.5 ± 0.4	**76.1** ± **0.5**
Caltech 256-30	4/4	75.1 ± 0.4	75.9 ± 1.2	74.2 ± 0.2	**76.7** ± **0.9**
Caltech 256-60	4/4	79.1 ± 0.4	79.9 ± 0.3	77.2 ± 0.1	**80.9** ± **0.3**
CUB-200-2011	4/4	71.6 ± 0.3	72.1 ± 0.3	70.7 ± 0.1	**73.3** ± **0.2**
Food-101	4/4	77.1 ± 0.7	78.4 ± 0.4	77.5 ± 0.4	**79.9** ± **0.4**

Table 6. Performance comparisons of different methods on the PolyU-NIRFD data set. "FAR" denotes the false acceptance rate

LResNet18E-IR (4 bit)	True acceptance rate (%)		
	FAR = 1e−5	FAR = 1e−6	FAR = 1e−7
L^2-Q	84.9	79.9	75.9
L^2-SP-Q	89.1	85.8	83.1
DELTA-Q	91.8	89.9	88.7
DTQ	**93.3**	**91.1**	**90.3**

Effect of Different Training Schemes. To further investigate the effect of training scheme, we extend DTQ to two-stages training scheme. Specifically, we do transferring in the first stage and perform quantization in the second stage. Let "DT" be conducting transferring without quantization by using the DTQ framework. We consider the following methods for comparison. **DT** → L^2**-Q** : we apply DT in the first stage to obtain a full-precision model and then apply L^2-Q to train a low-precision model in the second stage. **DT** → **DTQ** : we apply DT in the first stage and DTQ in the second stage. Note that the pre-trained full-precision model does not change during the training process.

We quantize MobileNetV2 with different methods on five target data sets. From the results in Table 5, compared with two-stage DT → L^2-Q, one-stage DTQ outperforms it by a large margin. For example, one-stage DTQ surpasses two-stage DT → L^2-Q by 4.5% in Top-1 accuracy on Stanford Dogs 120. These results, to some extent, imply the necessity to perform transferring and quantization simultaneously. Furthermore, two-stage DT → DTQ outperforms one-stage DTQ on five target data sets. These results demonstrate that a better initialized point in DTQachieves better performance.

6 Experiments on Face Recognition

In this experiment, we evaluate the proposed DTQ on the face recognition task. We use the visible light (VIS) face data set CASIA-WebFace [57] as the

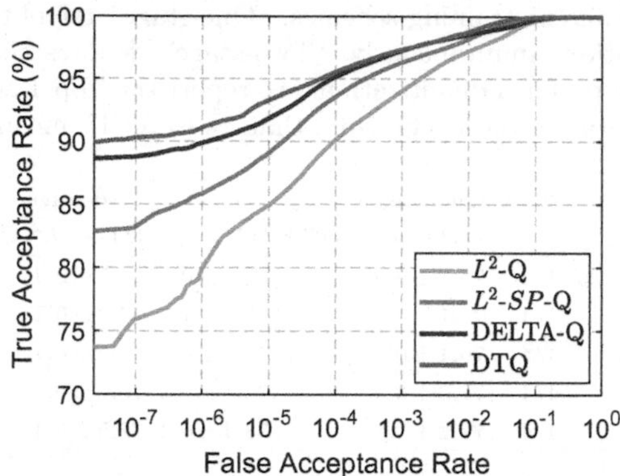

Fig. 5. ROC [14,50] curves of 4-bit LResNet18E-IR on the PolyU-NIRFD data set

source data set, and the PolyU near-infrared ray (NIR) face data set (PolyU-NIRFD) [61] as the target data set. Besides, we adopt LResNet18E-IR [13] as the base model. We report the resutls in Table 6 and Fig. 5. From the results, our DTQ achieves the best performance. Due to the page limit, we put more implementation details and results in the supplementary. These results demonstrate the effectiveness of the proposed DTQ on the face recognition task.

7 Conclusion

In this paper, we have studied the transferring quantization task, which aims to obtain a promising low-precision model by effectively exploiting the pre-trained full-precision model with limited training data. To achieve accurate low-precision models, we have proposed a simple but effective method named deep transferring quantization (DTQ). In our DTQ, we devised an attentive transfer module to identify informative channels, and further proposed attentive transferring quantization (ATQ) to align the informative channels of the low-precision model with that of the full-precision model. In addition, we introduced the Kullback-Leibler (KL) divergence on the probability distribution of two models for probabilistic transferring quantization (PTQ). By minimizing the KL divergence, the useful knowledge learned from the source data can be transferred to the low-precision model. Extensive experimental results on both image classification and face recognition demonstrate the effectiveness of the proposed DTQ.

Acknowledgements. This work was partially supported by the Key-Area Research and Development Program of Guangdong Province 2019B010155002, National Natural Science Foundation of China (NSFC) 61836003 (key project), Program for Guangdong Introducing Innovative and Entrepreneurial Teams 2017ZT07X183, Fundamental Research Funds for the Central Universities D2191240.

References

1. Ba, J., Caruana, R.: Do deep nets really need to be deep? In: Conference on Neural Information Processing Systems, pp. 2654–2662 (2014)
2. Banner, R., Nahshan, Y., Hoffer, E., Soudry, D.: ACIQ: analytical clipping for integer quantization of neural networks. arXiv preprint arXiv:1810.05723 (2018)
3. Banner, R., Nahshan, Y., Soudry, D.: Post training 4-bit quantization of convolutional networks for rapid-deployment. In: Conference on Neural Information Processing Systems, pp. 7948–7956 (2019)
4. Bossard, L., Guillaumin, M., Van Gool, L.: Food-101-mining discriminative components with random forests. In: European Conference on Computer Vision, pp. 446–461 (2014)
5. Cai, Y., Yao, Z., Dong, Z., Gholami, A., Mahoney, M.W., Keutzer, K.: ZeroQ: a novel zero shot quantization framework. arXiv preprint arXiv:2001.00281 (2020)
6. Caruana, R.: Multitask learning. Mach. Learn. **28**(1), 41–75 (1997)
7. Chen, L.C., Papandreou, G., Kokkinos, I., Murphy, K., Yuille, A.L.: DeepLab: semantic image segmentation with deep convolutional nets, atrous convolution, and fully connected crfs. IEEE Trans. Pattern Anal. Mach. Intell. **40**(4), 834–848 (2017)
8. Chen, S., Liu, Y., Gao, X., Han, Z.: MobileFaceNets: efficient CNNs for accurate real-time face verification on mobile devices. In: Chinese Conference on Biometric Recognition, pp. 428–438 (2018)
9. Choi, J., Wang, Z., Venkataramani, S., Chuang, P.I.J., Srinivasan, V., Gopalakrishnan, K.: PACT: parameterized clipping activation for quantized neural networks. arXiv preprint arXiv:1805.06085 (2018)
10. Cui, Y., Song, Y., Sun, C., Howard, A., Belongie, S.: Large scale fine-grained categorization and domain-specific transfer learning. In: IEEE Conference on Computer Vision and Pattern Recognition (2018)
11. Deng, J., Dong, W., Socher, R., Li, L., Li, K., Li, F.: ImageNet: a large-scale hierarchical image database. In: IEEE Conference on Computer Vision and Pattern Recognition, pp. 248–255 (2009)
12. Deng, J., Guo, J., Xue, N., Zafeiriou, S.: ArcFace: additive angular margin loss for deep face recognition. In: IEEE Conference on Computer Vision and Pattern Recognition, pp. 4690–4699 (2019)
13. Deng, J., Guo, J., Zafeiriou, S.: Arcface: additive angular margin loss for deep face recognition. arXiv preprint arXiv:1801.07698v1 (2018)
14. Fawcett, T.: An introduction to ROC analysis. Pattern Recognit. Lett. **27**(8), 861–874 (2006)
15. Ge, W., Yu, Y.: Borrowing treasures from the wealthy: deep transfer learning through selective joint fine-tuning. In: IEEE Conference on Computer Vision and Pattern Recognition, pp. 1086–1095 (2017)
16. Griffin, G., Holub, A., Perona, P.: Caltech-256 object category dataset (2007)
17. Guo, Y., et al.: Breaking the curse of space explosion: towards efficient nas with curriculum search. In: International Conference on Machine Learning (2020)
18. Guo, Y., et al.: NAT: neural architecture transformer for accurate and compact architectures. In: Conference on Neural Information Processing Systems, pp. 735–747 (2019)
19. Han, S., Mao, H., Dally, W.J.: Deep compression: compressing deep neural network with pruning, trained quantization and Huffman coding. In: International Conference on Learning Representations (2016)

20. He, K., Gkioxari, G., Dollár, P., Girshick, R.B.: Mask R-CNN. In: IEEE International Conference on Computer Vision, pp. 2980–2988 (2017)
21. He, K., Zhang, X., Ren, S., Sun, J.: Deep residual learning for image recognition. In: IEEE Conference on Computer Vision and Pattern Recognition, pp. 770–778 (2016)
22. He, Y., Zhang, X., Sun, J.: Channel pruning for accelerating very deep neural networks. In: IEEE International Conference on Computer Vision, pp. 1389–1397 (2017)
23. Hinton, G., Vinyals, O., Dean, J.: Distilling the knowledge in a neural network. arXiv:1503.02531 (2015)
24. Hong, S., Oh, J., Lee, H., Han, B.: Learning transferrable knowledge for semantic segmentation with deep convolutional neural network. In: IEEE Conference on Computer Vision and Pattern Recognition, pp. 3204–3212 (2016)
25. Huang, Z., Wang, N.: Like what you like: knowledge distill via neuron selectivity transfer. arXiv preprint arXiv:1707.01219 (2017)
26. Hubara, I., Courbariaux, M., Soudry, D., El-Yaniv, R., Bengio, Y.: Binarized neural networks. In: Conference on Neural Information Processing Systems, pp. 4107–4115 (2016)
27. Huh, M., Agrawal, P., Efros, A.A.: What makes imagenet good for transfer learning? arXiv:1608.08614 (2016)
28. Jacob, B., et al.: Quantization and training of neural networks for efficient integer-arithmetic-only inference. In: IEEE Conference on Computer Vision and Pattern Recognition, pp. 2704–2713 (2018)
29. Khosla, A., Jayadevaprakash, N., Yao, B., Li, F.F.: Novel dataset for fine-grained image categorization: stanford dogs. In: IEEE Conference on Computer Vision and Pattern Recognition Workshops, vol. 2 (2011)
30. Kirkpatrick, J., Pascanu, R., et al.: Overcoming catastrophic forgetting in neural networks. Proc. Natl. Acad. Sci. **114**(13), 3521–3526 (2017)
31. Leroux, S., Vankeirsbilck, B., Verbelen, T., Simoens, P., Dhoedt, B.: Training binary neural networks with knowledge transfer. Neurocomputing **396**, 534–541 (2020)
32. Li, X., Xiong, H., et al.: DELTA: deep learning transfer using feature map with attention for convolutional networks. In: International Conference on Learning Representations (2019)
33. Li, X., Grandvalet, Y., Davoine, F.: Explicit inductive bias for transfer learning with convolutional networks. In: International Conference on Machine Learning, pp. 2830–2839 (2018)
34. Li, Z., Hoiem, D.: Learning without forgetting. IEEE Trans. Pattern Anal. Mach. Intell. **40**(12), 2935–2947 (2017)
35. Lin, T.Y., Goyal, P., Girshick, R., He, K., Dollár, P.: Focal loss for dense object detection. In: IEEE International Conference on Computer Vision, pp. 2980–2988 (2017)
36. Liu, J., et al.: Sparse deep transfer learning for convolutional neural network. In: AAAI Conference on Artificial Intelligence (2017)
37. Luo, J.H., Wu, J., Lin, W.: ThiNet: a filter level pruning method for deep neural network compression. In: IEEE International Conference on Computer Vision, pp. 5058–5066 (2017)
38. Moon, S., Carbonell, J.G.: Completely heterogeneous transfer learning with attention-what and what not to transfer. In: International Joint Conferences on Artificial Intelligence (2017)

39. Pan, S.J., Yang, Q.: A survey on transfer learning. IEEE Trans. Knowl. Data Eng. **22**(10), 1345–1359 (2009)
40. Rastegari, M., Ordonez, V., Redmon, J., Farhadi, A.: XNOR-Net: imagenet classification using binary convolutional neural networks. In: European Conference on Computer Vision, pp. 525–542 (2016)
41. Redmon, J., Divvala, S.K., Girshick, R.B., Farhadi, A.: You only look once: Unified, real-time object detection. In: IEEE Conference on Computer Vision and Pattern Recognition, pp. 779–788 (2016)
42. Ren, S., He, K., Girshick, R.B., Sun, J.: Faster R-CNN: towards real-time object detection with region proposal networks. In: Conference on Neural Information Processing Systems, pp. 91–99 (2015)
43. Romero, A., Ballas, N., et al.: FitNets: hints for thin deep nets. In: International Conference on Learning Representations (2015)
44. Ronneberger, O., Fischer, P., Brox, T.: U-Net: convolutional networks for biomedical image segmentation. In: International Conference on Medical Image Computing and Computer-Assisted Intervention, pp. 234–241 (2015)
45. Russakovsky, O., et al.: Imagenet large scale visual recognition challenge. Int. J. Comput. Vision **115**(3), 211–252 (2015)
46. Saenko, K., Kulis, B., Fritz, M., Darrell, T.: Adapting visual category models to new domains. In: European Conference on Computer Vision, pp. 213–226 (2010)
47. Sandler, M., Howard, A., Zhu, M., Zhmoginov, A., Chen, L.C.: MobileNetV2: inverted residuals and linear bottlenecks. In: IEEE Conference on Computer Vision and Pattern Recognition (2018)
48. Szegedy, C., Vanhoucke, V., Ioffe, S., Shlens, J., Wojna, Z.: Rethinking the inception architecture for computer vision. In: IEEE Conference on Computer Vision and Pattern Recognition (2016)
49. Tai, C., Xiao, T., Wang, X., Weinan, E.: Convolutional neural networks with low-rank regularization. In: International Conference on Learning Representations (2016)
50. Taigman, Y., Yang, M., Ranzato, M., Wolf, L.: DeepFace: closing the gap to human-level performance in face verification. In: IEEE Conference on Computer Vision and Pattern Recognition, pp. 1701–1708. IEEE Computer Society (2014)
51. Wah, C., Branson, S., Welinder, P., Perona, P., Belongie, S.: The caltech-ucsd birds-200-2011 dataset (2011)
52. Wang, H., et al.: CosFace: large margin cosine loss for deep face recognition. In: IEEE Conference on Computer Vision and Pattern Recognition, pp. 5265–5274 (2018)
53. Wei, Y., Pan, X., Qin, H., Ouyang, W., Yan, J.: Quantization mimic: towards very tiny CNN for object detection. In: European Conference on Computer Vision, pp. 267–283 (2018)
54. Xu, J., Nie, Y., Wang, P., López, A.M.: Training a binary weight object detector by knowledge transfer for autonomous driving. In: International Conference on Robotics and Automation, pp. 2379–2384 (2019)
55. Xu, K., et al.: Show, attend and tell: Neural image caption generation with visual attention. In: International Conference on Machine Learning (2015)
56. Yao, Y., Ren, J., Xie, X., Liu, W., Liu, Y., Wang, J.: Attention-aware multi-stroke style transfer. In: IEEE Conference on Computer Vision and Pattern Recognition, pp. 1467–1475 (2019)
57. Yi, D., Lei, Z., Liao, S., Li, S.Z.: Learning face representation from scratch. arXiv preprint arXiv:1411.7923 (2014)

58. Yim, J., Joo, D., Bae, J., Kim, J.: A gift from knowledge distillation: fast optimization, network minimization and transfer learning. In: IEEE Conference on Computer Vision and Pattern Recognition (2017)
59. Yosinski, J., Clune, J., Bengio, Y., Lipson, H.: How transferable are features in deep neural networks? In: Conference on Neural Information Processing Systems, pp. 3320–3328 (2014)
60. Zagoruyko, S., Komodakis, N.: Paying more attention to attention: Improving the performance of convolutional neural networks via attention transfer. In: International Conference on Learning Representations (2017)
61. Zhang, B., Zhang, L., Zhang, D., Shen, L.: Directional binary code with application to polyu near-infrared face database. Pattern Recognit. Lett. **31**(14), 2337–2344 (2010)
62. Zhang, D., Yang, J., Ye, D., Hua, G.: LQ-Nets: learned quantization for highly accurate and compact deep neural networks. In: European Conference on Computer Vision, pp. 365–382 (2018)
63. Zhang, X., Zou, J., He, K., Sun, J.: Accelerating very deep convolutional networks for classification and detection. IEEE Trans. Pattern Anal. Mach. Intell. **38**(10), 1943–1955 (2015)
64. Zhang, Y., Chen, H., Wei, Y., et al.: From whole slide imaging to microscopy: Deep microscopy adaptation network for histopathology cancer image classification. In: International Conference on Medical Image Computing and Computer-Assisted Intervention, pp. 360–368 (2019)
65. Zhang, Y., et al.: Collaborative unsupervised domain adaptation for medical image diagnosis. IEEE Trans. Image Process. **29**, 7834–7844 (2020)
66. Zhang, Y., Zhang, Y., Yang, Q.: Parameter transfer unit for deep neural networks. In: The Pacific-Asia Conference on Knowledge Discovery and Data Mining, pp. 82–95 (2019)
67. Zhao, R., Hu, Y., Dotzel, J., De Sa, C., Zhang, Z.: Improving neural network quantization without retraining using outlier channel splitting. In: International Conference on Machine Learning, pp. 7543–7552 (2019)
68. Zhou, S., Wu, Y., Ni, Z., Zhou, X., Wen, H., Zou, Y.: DoReFa-Net: training low bitwidth convolutional neural networks with low bitwidth gradients. arXiv preprint arXiv:1606.06160 (2016)
69. Zhu, M., Wang, N., Gao, X., Li, J., Li, Z.: Face photo-sketch synthesis via knowledge transfer. In: International Joint Conferences on Artificial Intelligence, pp. 1048–1054 (2019)
70. Zhuang, B., Liu, J., Tan, M., Liu, L., Reid, I., Shen, C.: Effective training of convolutional neural networks with low-bitwidth weights and activations. arXiv preprint arXiv:1908.04680 (2019)
71. Zhuang, B., Shen, C., Tan, M., Liu, L., Reid, I.: Towards effective low-bitwidth convolutional neural networks. In: IEEE Conference on Computer Vision and Pattern Recognition, pp. 7920–7928 (2018)
72. Zhuang, Z., et al.: Discrimination-aware channel pruning for deep neural networks. In: Conference on Neural Information Processing Systems, pp. 875–886 (2018)

Deep Credible Metric Learning for Unsupervised Domain Adaptation Person Re-identification

Guangyi Chen[1,2,3], Yuhao Lu[1,5], Jiwen Lu[1,2,3(✉)], and Jie Zhou[1,2,3,4]

[1] Department of Automation, Tsinghua University, Beijing, China
chen-gy16@mails.tsinghua.edu.cn, luyuhao998@gmail.com,
{lujiwen,jzhou}@tsinghua.edu.cn

[2] State Key Lab of Intelligent Technologies and Systems, Beijing, China

[3] Beijing National Research Center for Information Science and Technology, Beijing, China

[4] Tsinghua Shenzhen International Graduate School, Tsinghua University, Beijing, China

[5] School of Computer Science, Beijing University of Posts and Telecommunications, Beijing, China

Abstract. The trained person re-identification systems fundamentally need to be deployed on different target environments. Learning the cross-domain model has great potential for the scalability of real-world applications. In this paper, we propose a deep credible metric learning (DCML) method for unsupervised domain adaptation person re-identification. Unlike existing methods that directly finetune the model in the target domain with pseudo labels generated by the source pre-trained model, our DCML method adaptively mines credible samples for training to avoid the misleading from noise labels. Specifically, we design two credibility metrics for sample mining including the k-Nearest Neighbor similarity for density evaluation and the prototype similarity for centrality evaluation. As the increasing of the pseudo label credibility, we progressively adjust the sampling strategy in the training process. In addition, we propose an instance margin spreading loss to further increase instance-wise discrimination. Experimental results demonstrate that our DCML method explores credible and valuable training data and improves the performance of unsupervised domain adaptation.

Keywords: Credible learning · Metric learning · Unsupervised domain adaptation · Person re-identification

1 Introduction

Person re-identification (ReID) aims at identifying a query individual from a large set of candidates under the non-overlapping camera views. As an essential role in various applications of security and surveillance, lots of attempts and dramatic improvements have been witnessed in recent years [22,23,37,48,58].

A. Vedaldi et al. (Eds.): ECCV 2020, LNCS 12353, pp. 643–659, 2020.
https://doi.org/10.1007/978-3-030-58598-3_38

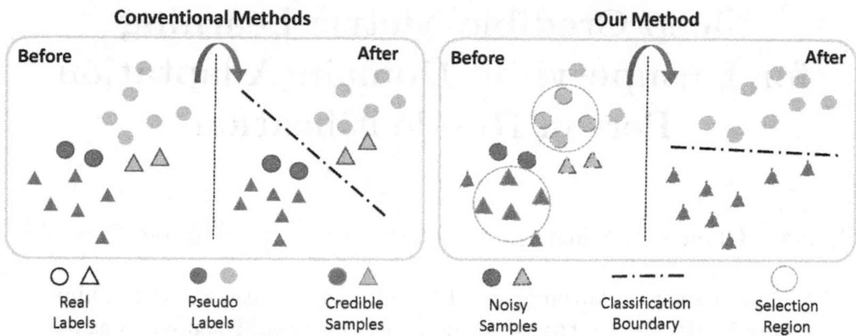

Fig. 1. Difference between our DCML method and conventional methods. The left part shows that conventional metric learning methods treat all samples equally to train the model and thus are easy to be misled by the noise labels. The right part shows that our method adaptively mines credible samples to train the model, which can avoid the damage from these low-quality samples. Best viewed in color.

Despite the satisfactory performance obtained by the supervised deep learning model and some label annotations in the single domain, it is still a challenge to deploy the trained person ReID models on different target environments. It is due to the domain bias between the training and deploying environments, e.g., the model trained on one university dataset need to be applied for airport or underground station. One of the common methods is finetuning the deep model by the image data of the target domain and pseudo labels generated by the source pre-trained model (e.g., clustering [12,34,52], reference comparison [51], or nearest neighborhood [61]). However, the predicted pseudo labels might involve much noise, which misleads the training process in the target domain. As shown in Fig. 1, the noisy labels might generate opposite gradients which undermine the model discrimination.

To address this problem, we propose a deep credible metric learning (DCML) method to avoid the damage from noise pseudo labels by adaptively exploring credible and valuable training samples. Specifically, our DCML method consists of two parts, including adaptively credible anchor sample mining and instance margin spreading. The former is proposed to explore credible samples, which are effective for learning the intra-class compact embeddings. We propose two credibility metrics including the k-Nearest Neighbor similarity and the prototype similarity. We implement two different similarity metrics to demonstrate the generality of the credible anchor sample mining strategy. The k-Nearest Neighbor similarity measures the neighborhood density of the sample by calculating the maximum distance (minimum similarity) between itself and k nearest neighbors. While the prototype similarity calculates the similarity between the sample and class prototype, which denotes the sample's centrality. Using these credibility metrics, we can select samples with higher credibility as anchors. As the training iterations increasing, the credibility of pseudo labels continues to increase too. We therefore, progressively reduce the limitation of anchor sample mining to select more credible training samples. In addition, we propose an instance margin spreading (IMS) loss to increase the instance-wise discrimination, due to the initial embeddings of target samples are always confusing and in-discriminative without supervised training. We regard each sample as an independent individual and learn a spreading embedding apace by pushing the samples away from each other by a large margin. We summarize the contributions of this work as follows:

1) We propose a deep credible metric learning (DCML) method for unsupervised domain adaptation person ReID, which adaptively and progressively mines credible and valuable training samples to avoid the damage from the noise of predicted pseudo labels.
2) We design an instance margin spreading method loss to encourage the instance-wise discrimination by spreading the embeddings of samples with a large margin.
3) We conduct extensive experiments to demonstrate the superiority of our method, and achieve the state-of-the-art performance on several large scale datasets including Market-1501 [57], DukeMTMC-reID [30], and CUHK03 [21].

2 Related Work

Supervised Deep Person ReID: Most existing person ReID methods obtain excellent performance by the supervised deep learning model and a number of label annotations. Some methods are devoted to designing more effective networks by part-based model [3,6,36,37,41] or attention model [1,2,11,22,31,47]. Other methods focus on capture more prior knowledge or supervisory signals, including body structure [18,19,53,54], human pose [29,35], attribute labels [39,55], and other loss functions [4,15,56]. Despite the recent progress in the supervised manner, the deployment of trained models for different target environments is still a challenge due to the large domain bias.

Unsupervised Domain Adaptation Person ReID: To address the above problem, Some works [24,49] study purely unsupervised learning to learn from unlabelled data for Re-ID. However, the performance is limited without any labeled data. Furthermore, many works attempt to learn the unsupervised domain adaptation person ReID model, which leverages the labeled source domain data and unlabeled target domain data. Many existing works [5,7,44] apply the generative model (e.g., GAN) to transform the images of source domain into the target domain as the training data, aiming to reduce the domain bias from data. While other works finetune the deep model with the target domain data and pseudo labels generated by the source pre-trained model. The clustering methods [12,34,52] and reference comparison [51] are widely used to generate the supervisory signal from pre-trained models. Besides, some unsupervised domain adaptation person ReID methods explore other human prior knowledge or auxiliary supervisory signals to improve the adaptation and generalization ability from the source domain to the target domain. EANet [16] employs the human parsing results to assist feature alignment. While TJ-AIDL attempts to learn a joint attribute-identity space which improves the model generalization ability with transferred attribute knowledge. Our work is related to PAST, which randomly selects the positive and negative samples from top k neighbors and k-2k neighbors respectively with all samples as the anchors and employs a cross-entropy loss as the promoting stage. However, PAST applies the fixed sampling strategy for all anchors in the whole training process which ignores the initial low-quality and continuous improvement of pseudo labels. Our DCML method adaptively selects credible anchors by measuring the credibility of each sample and progressively adjusts the sampling strategy for the different stages of the training process.

Deep Metric Learning: Deep metric learning aims to learn the discriminative feature embedding space instead of the final classifier, which generalizes better to the unseen environment [4]. Existing deep metric learning methods mainly focus on design effective loss functions or develop efficient sampling strategies. The loss designing methods focus on utilizing higher order relationships [26,40,42], global information [27,33], or the margin maximum [8,38,50]. While sampling-based methods are devoted to mining the hard negative samples for training efficiency improvement. For instance, TriNet [15] samples the most negative samples in the batch for fast convergence. Harwood et al. [13] found the negative samples from an increasing search space defined by the nearest neighbor distance. However, these mining strategies tend to select the harder samples due to the larger gradient from violating triplet relation defined by the annotations, which is confused with the noise labels, especially for pseudo labels. To address this issue, we adaptively and progressively select the credible anchor samples, which is appropriate for the low-quality predicted pseudo labels.

3 Deep Credible Metric Learning

The goal of our deep credible metric learning method is adaptively and progressively discovering the credible samples to reduce the damage from noise labels. In this section, we will introduce our DCML method from two parts, including adaptively credible sample mining and instance margin spreading.

3.1 Problem Formulation

For the unsupervised domain adaptation person ReID problem, we have a source dataset $\mathcal{S} = \{\mathcal{X}^{\mathcal{S}}, \mathcal{Y}^{\mathcal{S}}\}$, where $\mathcal{X}^{\mathcal{S}}$ denotes the image data and $\mathcal{Y}^{\mathcal{S}}$ is the corresponding labels. Besides, we have another dataset in the deployed environment without any annotations, which is called target dataset $\mathcal{X}^{\mathcal{T}} = \{x_i^t\}_1^N$. The cross-domain person ReID system aims to learn the robust and generalizable representations in the target domain with the supervised source dataset and unsupervised target one. A popular solution for the unsupervised domain adaptation person ReID problem is finetuning the pre-trained model in the target domain with the predicted pseudo labels. Support we have predicted pseudo

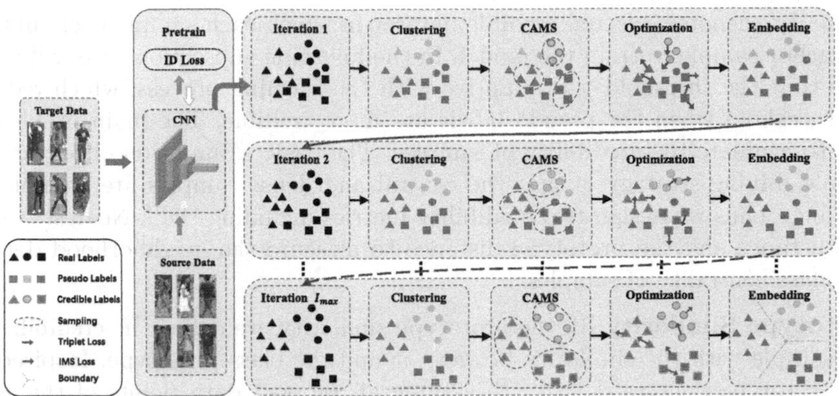

Fig. 2. Illustration of the deep credible metric learning method. The DCML method starts with learning a pre-trained CNN network with the source labeled data. In each iteration, we extract the embeddings of unlabeled target images and generate pseudo labels with the clustering method. To avoid the misleading of noise pseudo labels, we adaptively mine credible samples as the anchor data and optimize the model with these samples. The gradients come from two objective functions including the triplet loss with red arrows and the IMS loss with purple arrows. In addition, we progressively adjust the anchor sample mining strategy to select more anchor samples as iteration increases. Best viewed in color

labels $\hat{\mathcal{Y}}^{\mathcal{T}} = \mathcal{P}(\mathcal{X}^{\mathcal{T}}; \mathcal{X}^{\mathcal{S}}, \mathcal{Y}^{\mathcal{S}})$ generated by the pre-trained model from the source domain, we learn feature embeddings with a convolutional neural network (CNN) $\mathcal{F}_\theta$ as $f_i = \mathcal{F}_\theta(x_i^t)$ with the objective function which is formulated as:

$$\theta = \arg\min_{\theta} \mathcal{L}(\theta; \mathcal{X}^{\mathcal{T}}, \hat{\mathcal{Y}}^{\mathcal{T}}), \tag{1}$$

where the objective is to learn CNN $\mathcal{F}_\theta$ by using pseudo labels as a supervisory signal. However, the performance of this objective function entirely depends on the properties of generated labels without a stable guarantee. The generated labels are always noisy due to the large domain bias between the source and target datasets. These noise labels always mislead the training process by providing wrong gradients. This inevitably leads to the necessity of adaptively credible samples mining for more reliable model learning (Fig. 2).

3.2 Adaptively Credible Sample Mining

The adaptively credible sample mining strategy aims to select the more credible samples to avoid the damage from noise labels. For one target sample and corresponding pseudo label $(x_i^t, \hat{y}_i^t)$, we define a credibility metric $\mathcal{C}(x_i^t, \hat{y}_i^t)$ to evaluate whether a label is credible enough as a supervisory signal. Given a threshold τ, we select the more credible samples as the training data:

$$\mathcal{X}_C^{\mathcal{T}} = \{x_i^t \in \mathcal{X}^{\mathcal{T}} | \mathcal{C}(x_i^t, \hat{y}_i^t) > \tau\}, \tag{2}$$

where $\mathcal{X}_C^{\mathcal{T}}$ denotes selected credible dataset in which each sample is credible as an anchor sample to train the model. In the following subsections, we will introduce that the threshold τ is adaptive with the learning process, which reduces the threshold when the pseudo labels are more credible. The main problem is how to evaluate the credibility of samples. The basic assumption of our anchor sample mining strategy is that the central and dense samples are credible for training. Thus we design two credibility metrics including the k-Nearest Neighbor distance and the prototype distance to measure the neighborhood density and class centrality of samples.

Prototype Similarity: In the prototype similarity, we define the credibility of one sample with the similarity between it and the class prototype. Inspired by the prototypical network [32], we assume all support data points of the same "class" lie in a manifold, and calculate the class prototype as the center of class:

$$\mathcal{P}_k = \frac{1}{|\mathcal{M}_k|} \sum_{x_i^t \in \mathcal{M}_k} \mathcal{F}_\theta(x_i^t), \tag{3}$$

where $\mathcal{M}_k = \{x_i^t \in \mathcal{X}^{\mathcal{T}} | \hat{y}_i^t = k\}$ denotes the set of examples labeled with class k, and $\hat{y}_i^t$ is the pseudo label of x_i^t. Then the intra-class centrality can be calculated with the Euclidean distance as:

$$\mathcal{C}_P(x^t, \hat{y}^t) = -||x^t - \mathcal{P}_{\hat{y}^t}||_2. \tag{4}$$

The larger $\mathcal{C}_P(x^t, \hat{y}^t)$ values correspond to more intra-class consistent samples. When the intra-class centrality $\mathcal{C}_P(x^t)$ is large, the sample x^t is close to the class prototype, which means that its representation as a class is trustworthy. On the contrary, the samples with small credibility values might be mislabeled since these samples are always close to the uncredited classification-plane.

KNN Similarity: Different from prototype similarity measuring the intra-class sample centrality, the KNN similarity calculates the local density by the neighborhood information. For a sample x^t, the neighborhood set $\mathcal{N}(x^t)$ consists of k samples whose distance is nearest with the x^t. The neighborhood set denotes the local neighborhood information of samples, which can be employed to describe the density. We define the KNN distance as

$$\mathcal{C}_N(x^t) = - \max_{x_i^t \in \mathcal{N}(x^t)} d(x^t, x_i^t), \tag{5}$$

where $d(\cdot, \cdot)$ is a distance metric, e.g., the Euclidean distance. We employ the minimal similarity among the k nearest neighborhoods to denote the local density. All the samples in the neighborhood set $\mathcal{N}(x^t)$ are more compact as KNN similarity $\mathcal{C}_N(x^t)$ is large, which denotes that the x^t resides in a high-density region. When the samples are dense in the neighborhood set and far away from other samples, the neighborhood-based pseudo label generation method, e.g., clustering, will give a more reliable result. When the samples are dense and indistinguishable, they are also necessary to pay more attention. Thus, we select the samples with higher KNN similarity as training data.

Progressively Learning: In the whole training stage, we iteratively generate the pseudo labels with the embedding model and train the embedding model with pseudo labels. In each iteration, we first extract the embeddings with current model $\mathcal{F}_\theta$ and cluster on the embedding space to generate the pseudo labels. Then, we apply the pseudo labels as supervisory signal to train and update the embedding model. Though this iterative learning process, the pseudo labels become more and more credible and embeddings become more and more discriminative. In our DCML method, we progressively adjust the anchor sample mining strategy to select more anchor samples by reducing the selection threshold as iteration increases, since the pseudo labels are more credible as the model is finetuned. When the pseudo labels are credible enough, we tend to employ all the data in the target domain to train our model. Specifically, we design a linear threshold adaptation strategy, which progressively reduce the threshold τ with the iterations r. We formulate the threshold adaptation strategy with iterations r as follows:

$$\tau = \arg\min_{\tau} |\mathcal{X}_c^{\mathcal{T}}| \geq (\gamma_0 + r \times \Delta\gamma)|\mathcal{X}^{\mathcal{T}}| \tag{6}$$

where $|\mathcal{X}_c^{\mathcal{T}}|$ and $|\mathcal{X}^{\mathcal{T}}|$ respectively denote the number of samples in the selected and original datasets. γ_0 and $\Delta\gamma$ are the hyperparameters of algorithm which respectively denote the initial sampling rate of anchor samples and the increment in each iteration. The basic goal of this strategy is adapting an appropriate threshold τ to select sufficient credible anchor samples. The number of selected samples progressively increases with the assuming that the credibility of pseudo labels increase as training iterations.

3.3 Instance Margin Spreading

The pre-trained embeddings on the target domain are always confusing and in-discriminative. It is difficult to cluster these in-discriminative samples and generate credible pseudo labels. In order to increase the inter-class discrimination, we propose an instance margin spreading (IMS) loss which spreads the embeddings by pushing the samples a large margin apart from each other for a discriminative embeddings space. Inspirited by the instance discrimination learning [46] which assumes each instance is a independent class, we aim to learn a spreading metric space where the distances between each instance pair are over a large margin. Different from conventional margin-based losses (e.g., triplet loss), our IMS loss doesn't require any labels, which learns the embedding space only by the instance-wise discrimination. The basic formulation of this margin constraint is as follows:

$$\mathcal{L}_{ims}(x_a^t) = \sum_{i \neq a} \max\left(0, m - d_{a,i}\right) \tag{7}$$

where x_a denotes the random selected sample, $d_{a,i}$ denotes the distance between the sample pair $d(x_a^t, x_i^t)$ and $i \neq a$ represents all other samples in the dataset except itself. The m is a margin which denotes the lower bound of distances between each sample pair. As shown in [4] and [33], we can obtain the equivalent

Algorithm 1: DCML

Require: Source dataset $\mathcal{S}$; target dataset $\mathcal{T}$; maximal iterative number R_{max}.
Ensure: The parameters θ of embedding network $\mathcal{F}_\theta$.

```
Obtain the target-style dataset S' by a GAN;
Initialize θ by pre-training on the target-style source dataset S' ;
for r = 1, 2, ..., R_max do
    Extract embedding features of training data by F_θ ;
    Generate pseudo labels Ŷ^T by clustering with extracted features;
    Adjust sampling threshold τ with the number of iterations r as (6)
    Mine credible sample set X_C^T as (2)
    Update F_θ with credible sample set X_C^T and generated pseudo labels Ŷ^T as (9)
end for
return θ
```

loss function by replacing the $\max(0, x)$ with a continuous exponential function and a logarithmic function, which is formulated as:

$$\begin{aligned}\mathcal{L}_{ims}(x_a^t) &= \log\left(1 + \sum_{i \neq a} e^{m - d_{a,i}}\right) \\ &= -\log \frac{e^{-d_{a,a}}}{e^{-d_{a,a}} + \sum_{i \neq a} e^{m - d_{a,i}}} \\ &= -\log \frac{e^{-d_{a,a}}}{\sum_{i=1}^{N} e^{m_a - d_{a,i}}},\end{aligned} \tag{8}$$

where m_a is an adaptive margin. For the same instance, m_a is zero. For others, m_a is large. In this formulation, we assume that the distance between the sample and itself is zero, i.e., $d_{a,a} = 0$. Different from other instance discrimination learning methods (e.g., [46,61]), we learn a spreading metric space with a large margin. This metric space encourages an inter-class discrimination by the margin constraint, which is beneficial for robust clustering and credible sample mining.

3.4 Objective Function

Given the anchor sample set $\mathcal{X}_C^{\mathcal{T}}$ discovered by our adaptively credible sample mining strategy, we train our embedding model $\mathcal{F}_\theta$ with the objective function combining the proposed instance margin spreading loss and conventional metric learning loss:

$$\mathcal{L} = \sum_{x_i^t \in \mathcal{X}_C^{\mathcal{T}}} \mathcal{L}_{tri}(x_i^t) + \lambda \mathcal{L}_{ims}(x_i^t), \tag{9}$$

where $\mathcal{L}_{tri}(x_i^t)$ is the common metric learning loss: Triplet Loss [15], and λ denotes the hyper-parameter that balance the importance of different objectives. The triplet loss aims to learn an embedding space in which an anchor sample is closer to its positive sample than other negative ones by a large margin. We formulated it as follows:

$$\mathcal{L}_{tri}(x_i^t) = [||f_i - f_i^+||_2^2 - ||f_i - f_i^-||_2^2 + m_{tri}]_+, \tag{10}$$

Table 1. The basic statictics of all datasets in experiments.

Datasets	Identities	Images	Cameras	Train IDS	Test IDS	Labeling
Market-1501	1501	32668	6	751	750	Hand/DPM
DukeMTMC-reID	1812	36411	8	702	1110	Hand
CUHK03	1467	14096	2	767	700	DPM

where $[\cdot]_+$ indicates the max function $\max(0, \cdot)$ which denotes that gradients will disappear when the difference between the intra-class and inter-class distances is large enough. f_i, f_i^+, f_i^- respectively denote as features of the anchor, positive and negative sample in a triplet. The positive and negative samples selection strategy follows [15] that only uses the hardest positive and negative points in the mini-batch. m_{tri} is a margin to enhance the discriminative ability, which is similar with m_a in the instance margin spreading loss. For more clear explanation, we provide the Algorithm 1 to introduce the learning process of our DCML method in detail.

3.5 Discussion

Some methods (e.g., PUL [10], UDA [34], PAST [52], and SSG [12]) also apply the clustering algorithm to generate pseudo labels of target domain. However, the pseudo labels might **involve much noise**, which misleads the training process in the target domain. To solve this problem, our DCML method develops a credible sample mining strategy in the metric learning to avoid the noisy labels. PUL [10] have proposed a reliable objective function to regulate the sparsity of samples, and then simultaneously optimized the objective of the discriminative model and the regulation term of the number of samples. However, this regulation term may disturb the original discriminative learning since the valuable samples in the optimization process tend to be removed. Different from PUL, our DCML method proposes a credible sample mining strategy which is inspired by the hard negative mining in the metric learning. The credible data sampling is separated from the metric learning process, **without the disturbance**. As far as we know, DCML is the first metric learning method to adaptively select credible samples, which does not break the discriminative learning.

4 Experiment

In this section, we evaluated our DCML method on three large-scale person ReID datasets: Market-1501 [57], DukeMTMC-reID [30], and CUHK03 [21]. Quantitatively, we compared our DCML method with other state-of-the-art unsupervised domain adaptation person ReID approaches and conducted ablation studies to analyze each component. Besides, we visualized the embedding space to qualitatively analyze our method.

Table 2. Ablation studies show the influences of design choices on mAP and Rank-1,5,10(%), with Market-1501 as the source dataset and DukeMTMC-reID as the target dataset and vice versa. The † denotes that this method is reproduced by ourself with the same backbone and hyperparameters.

Method	M → D				D → M			
	mAP	R1	R5	R10	mAP	R1	R5	R10
UDA†	54.4	72.7	82.1	85.6	56.5	78.4	86.5	89.5
UDA† + GAN	60.4	76.3	85.8	88.4	70.5	85.8	93.2	95.1
UDA† + CAMS+ IMSLoss	60.2	75.9	84.0	86.7	69.2	85.4	92.8	94.8
UDA† + GAN + IMSLoss	62.2	76.9	85.9	88.8	71.3	86.9	92.9	95.1
DCML (KNN)	63.3	79.1	**87.2**	89.4	**72.6**	87.9	**95.0**	**96.7**
DCML (Prototype)	**63.5**	**79.3**	86.7	**89.5**	72.3	**88.2**	94.9	96.4

4.1 Datasets and Experimental Settings

Datasets: Our experiments are conducted on three large-scale datasets including Market-1501 [57], DukeMTMC-reID [30], and CUHK03 [21]. Although all the above datasets are collected from the natural real-world scene of the university environment, there still is a large domain shift among them such as background, illumination, and clothing style. For example, the persons in the Market-1501 and DueMTMC-reID datasets mainly come from Asia and America respectively. For all datasets, we share the same experiment settings with the standard cross-domain person ReID experimental setups in the baseline method UDA [34] and PAST [52]. Specifically, we follow the source/target selection strategy, training/testing ID splitting strategy, and evaluation measuring protocols. For Market-1501 and DukeMTMC-reID datasets, we evaluated our method in the single query mode. While for the CUHK03 dataset, we only use the DPM detected images and choose the new train/test evaluation protocol in [59] for a fair comparison. The detailed information of the datasets are shown in Table 1.

Evaluation Protocol: In our experiments, we employed the standard metrics including cumulative matching characteristic (CMC) curve and the mean average precision (mAP) score to evaluate the performance of the person reID methods. We reported rank-1, rank-5 and rank-10 accuracy and mAP score in our experiments. Note that post-processing methods, e.g., re-ranking [59], are **not** applied for the final evaluation.

4.2 Implementation Details

Source Domain Pre-training: Leveraging the labeled source domain images, we pre-train a CNN model in a supervised manner by following the training strategy described in [2]. Specifically, we use the ImageNet pre-trained ResNet50 [14] without any attention model as the backbone of our model for fairness. The original $stride = 2$ convolution layer in the last block is replaced by a $stride = 1$

one to preserve the image resolution. For image preprocessing, we attempt to use the generative images by the SPGAN [7] and adopt the random horizontal flipping, random cropping, and random erasing data augmentation methods for image diversity. The supervisory signals in the source domain training consist of label smooth cross-entropy loss and triplet loss. Besides, other hyperparameters including image resolution, batch size, learning rate, weight decay factor, learning rate decay strategy, and max epochs are the same as [2].

Pseudo Label Generation: We adopt the DBSCAN clustering method [9] to generate pseudo labels, which is the same as the baseline UDA method [34]. The input of DBSCAN algorithm is the reranked distance matrix of the target domain samples and the output is the clustering result. We give each image cluster containing more than two samples a pseudo-label and then discard the individual images.

DCML: In the process of target domain adaptation, we train our model for 8 iterations and 30 epochs are required in each iteration. For the credible sample mining strategy, we set $\gamma_0 \approx 0.75$ and $\Delta\gamma \approx 0.05$ to update the sample selection threshold. Taking the DukeMTMC-reID datasets as an example, we select 12000 anchor samples in the first iteration and increase 1000 samples each iteration. For objective function, we respectively set the margins $m_a = 0.1$ and $m_{tri} = 0.3$ for instance margin spreading loss and triplet loss. The rate of loss weighting is set as $\lambda = 0.01$. In each mini-batch, we randomly select 224 samples from the credible sample set, in which each individual contains 16 images. We use Adam optimizer with an initial learning rate of 0.0005 and the weight decay of 0.001. The initial learning rate is reduced to 0.1 at 3th and 6th iterations, and in each iteration, it is temporarily reduced in the last 10 epochs. We conducted All our experiments on 4 Nvidia GTX 1080Ti GPUs with PyTorch 1.2.

4.3 Ablation Study

To analyze the effectiveness of individual components in our DCML approach, we conducted comprehensive ablation experiments on the M → D and D → M settings, where M → D denotes that the source dataset is Market-1051 and the target dataset is DukeMTMC-reID. We reproduced the UDA [34] method with the same backbone and hyperparameters of our method as the baseline, and applied the proposed credible anchor mining strategy, instance margin spreading loss, and the GAN based image style transfer on it. Table 2 We exhibited the comparison results in different settings in Table 2 and analyzed different components as follows.

Credible Anchor Mining Strategy: As shown in Table 2, CAMS denotes our credible anchor mining strategy. Compared the performance under the setting of $UDA\dagger + GAN + IMSLoss$ and the full DCML method, we can observe the obvious decline when the CAMS is removed. It illustrates that progressively and adaptively mining credible samples assists the target domain training by discarding samples with noise labels. In addition, we compared the effectiveness

Table 3. Performance comparisons with SOTA unsupervised domain adaptation person Re-ID methods from Market-1501 to DukeMTMC-reID and vice versa.

Method	M → D				D → M			
	mAP	R1	R5	R10	mAP	R1	R5	R10
PTGAN [44]	–	27.4	–	50.7	–	38.6	–	66.1
SPGAN [7]	22.3	41.1	56.6	63.0	22.8	51.5	70.1	76.8
SPGAN+LMP [7]	26.2	46.4	62.3	68.0	26.7	57.7	75.8	82.4
HHL [60]	27.2	46.9	61.0	66.7	31.4	62.2	78.8	84.0
DA2S [17]	30.8	53.5	–	–	27.3	58.5	–	–
CR-GAN [5]	48.6	68.9	80.2	84.7	54.0	77.7	89.7	92.7
TJ-AIDL [43]	23.0	44.3	59.6	65.0	26.5	58.2	74.8	81.1
TAUDL [20]	43.5	61.7	–	–	41.2	63.7	–	–
UCDA [28]	45.6	64.0	–	– 49.6	73.7	–	–	
EANet [16]	48.0	78.0	–	–	51.6	78	–	–
PUL [10]	16.4	30.0	43.4	48.5	20.5	45.5	60.7	66.7
MAR [51]*	48.0	67.1	79.8	–	40.0	67.7	81.9	
CASCL [45]*	37.8	59.3	73.2	77.8	35.5	65.4	80.6	86.2
ENC [61]	40.4	63.3	75.8	80.4	43.0	75.1	87.6	91.6
UDA [34]	49.0	68.4	80.1	83.5	53.7	75.8	89.5	93.2
PAST [52]	54.3	72.4	–	–	54.6	78.4	–	–
SSG++ [12]	60.3	76.0	85.8	89.3	68.7	86.2	94.6	96.5
DCML (KNN)	63.3	79.1	**87.2**	89.4	**72.6**	87.9	**95.0**	**96.7**
DCML (Prototype)	**63.5**	**79.3**	86.7	**89.5**	72.3	**88.2**	94.9	96.4

Table 4. Performance comparisons with other methods from CUHK03 to DukeMTMC-reID and Market-1501.

Methods	C → D		C → M	
	mAP	Rank-1	mAP	Rank-1
PUL [10]	12.0	23.0	18.0	41.9
PTGAN [44]	–	17.6	–	31.5
HHL [60]	23.4	42.7	29.8	56.8
EANet [16]	26.4	45.0	40.6	66.4
PAST [52]	51.8	69.9	57.3	**79.5**
DCML(KNN)	**56.9**	**73.7**	58.0	78.7
DCML(Prototype)	54.6	72.2	**59.5**	78.7

of different credibility similarity methods. The KNN similarity and prototype similarity are comparable to evaluate the credibility, which indicates our sample mining strategy is robust for different credibility evaluation methods.

Instance Margin Spreading Loss: The proposed IMS Loss aims to increase inter-class discrimination by enlarging the margin between the instances. We conducted the ablation studies about IMS Loss on the both "UDA" and "UDA+GAN" baselines, and obtained consistent improvement. Besides, we observed that the improvement on the stronger baseline (GAN+UDA) is lower than the original UDA method. This might be due to the generative images with GAN have a lower domain shift than the original images. The embedding space pre-trained with generative images is more spreading.

Image Style Transfer: In our final system, we employed the domain adaptation generative images with SPGAN [7] to pre-train the model on the source domain. The generator transfers the style of source domain images to the target domain style, which reduces the domain shift between source and target datasets. With the generative images pre-train, the baseline UDA method achieves a large improvement, which demonstrates that the quality of predicted pseudo labels is important for target domain finetuning. It also motivates us to additionally enhance the quality of pseudo labels.

4.4 Comparison with State-of-the-Art Methods

We compared our method with other SOTA unsupervised domain adaptation person ReID methods on the Market-1501, DukeMTMC-ReID and CUHK03 datasets. Specifically, we conducted the experiments following evaluation settings in [52] including M $\rightarrow$ D, D $\rightarrow$ M, C $\rightarrow$ D, and C $\rightarrow$ M tasks, where M, D, C respectively denote Market-1501, DukeMTMC-ReID and CUHK03 datasets. As shown in Table 3 and 4, the bottom groups summarize the performance of methods generating pseudo superiority signal to train the model on the target domain, while the top and middle groups respectively show these methods using GAN or other auxiliary attributes. Our DCML achieved consistent improvement over other comparing methods, which indicates the effectiveness of our credible sample mining strategy and instance margin spreading loss.

M $\rightarrow$ D and D $\rightarrow$ M: As shown in Table 3, we compare our results with 7 methods finetuning meodel by pseudo superiority signal, 5 methods reducing the domain shift with GAN and 4 methods using auxiliary clues. The * in the tables denotes that the method whose source dataset is MSMT17 [44], which is the largest re-ID dataset with large-scale images and multiple cameras. We achieve the state-of-the-art results for both settings.

C $\rightarrow$ D and C $\rightarrow$ M: We also evaluated our DCML method using CUHK03 [21] as the source dataset. The results of our DCML method and other state-of-the-art methods are summarized in Table 4. Our DCML method improved PAST [52] by adaptively and mining credible anchors and progressively adjusting the mining strategy, which avoids the misleading from noise labels. Note that we don't use the complex part model like PCB [37] in our DCML method.

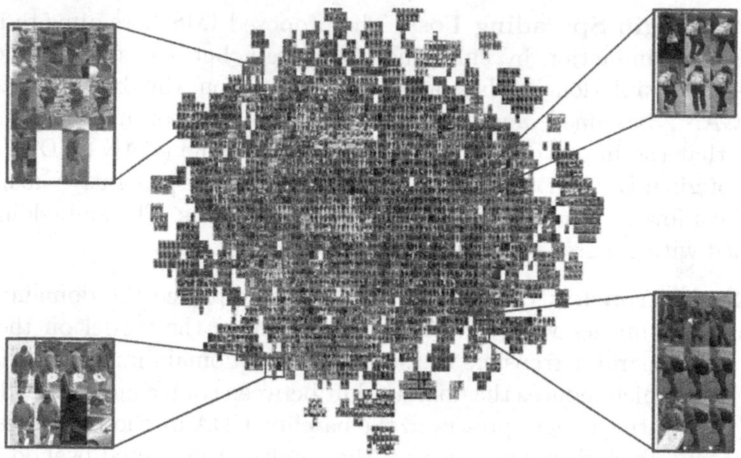

Fig. 3. Barnes-Hut t-SNE visualization [25] of the proposed DCML method on the gallery set of DukeMTMC-ReID, where we zoom in several areas for a clear view.

4.5 Qualitative Analysis

To validate the effectiveness of our DCML method, we qualitatively examined the learned embeddings. As shown in Fig. 3, we visualize the Barnes-Hut t-SNE [25] map of our learned embeddings of the gallery dataset in DukeMTMC-ReID. To observe the details, we magnify several regions in the corners. Despite the large intra-class variations such as illumination, backgrounds, viewpoints and human poses, our DCML method still groups similar individuals on the target domain in an unsupervised manner.

5 Conclusion

In this paper, we have proposed a deep credible metric learning method for unsupervised domain adaptation person re-identification, which adaptively mines credible samples to train the network and progressively adjusts the sample mining strategy with the learning process. It is due to that the generated pseudo labels are always unreliable and the noise will mislead the model training. We present two similarity metrics for the goal of measuring the credibilities of pseudo labels, including the k-Nearest Neighbor distance for density evaluation and the prototype distance for centrality evaluation. With the training process, we progressively reduce the limitation to select more samples. In addition, we propose an instance margin spreading loss to further increase the inter-class discrimination. We have conducted extensive experiments to demonstrate the effectiveness of our DCML method. In the future, we will attempt to design a credible negative mining strategy to further improve the cross-domain metric learning.

Acknowledgement. This work was supported in part by the National Key Research and Development Program of China under Grant 2017YFA0700802, in part by the

National Natural Science Foundation of China under Grant 61822603, Grant U1813218, Grant U1713214, and Grant 61672306, in part by Beijing Natural Science Foundation under Grant No. L172051, in part by Beijing Academy of Artificial Intelligence (BAAI), in part by a grant from the Institute for Guo Qiang, Tsinghua University, in part by the Shenzhen Fundamental Research Fund (Subject Arrangement) under Grant JCYJ20170412170602564, and in part by Tsinghua University Initiative Scientific Research Program.

References

1. Chen, B., Deng, W., Hu, J.: Mixed high-order attention network for person re-identification. In: ICCV, October 2019
2. Chen, G., Lin, C., Ren, L., Lu, J., Jie, Z.: Self-critical attention learning for person re-identification. In: ICCV (2019)
3. Chen, G., Lu, J., Yang, M., Zhou, J.: Spatial-temporal attention-aware learning for video-based person re-identification. TIP **28**(9), 4192–4205 (2019)
4. Chen, G., Zhang, T., Lu, J., Zhou, J.: Deep meta metric learning. In: ICCV, October 2019
5. Chen, Y., Zhu, X., Gong, S.: Instance-guided context rendering for cross-domain person re-identification. In: ICCV, pp. 232–242 (2019)
6. Cheng, D., Gong, Y., Zhou, S., Wang, J., Zheng, N.: Person re-identification by multi-channel parts-based CNN with improved triplet loss function. In: CVPR, pp. 1335–1344 (2016)
7. Deng, W., Zheng, L., Ye, Q., Kang, G., Yang, Y., Jiao, J.: Image-image domain adaptation with preserved self-similarity and domain-dissimilarity for person re-identification. In: CVPR, pp. 994–1003 (2018)
8. Duan, Y., Lu, J., Zhou, J.: Uniformface: learning deep equidistributed representation for face recognition. In: CVPR, pp. 3415–3424 (2019)
9. Ester, M., Kriegel, H.P., Sander, J., Xu, X., et al.: A density-based algorithm for discovering clusters in large spatial databases with noise. In: KDD, vol. 96, pp. 226–231 (1996)
10. Fan, H., Zheng, L., Yan, C., Yang, Y.: Unsupervised person re-identification: clustering and fine-tuning. TOMM **14**(4), 83 (2018)
11. Fang, P., Zhou, J., Roy, S.K., Petersson, L., Harandi, M.: Bilinear attention networks for person retrieval. In: ICCV, October 2019
12. Fu, Y., Wei, Y., Wang, G., Zhou, Y., Shi, H., Huang, T.S.: Self-similarity grouping: a simple unsupervised cross domain adaptation approach for person re-identification. In: ICCV, October 2019
13. Harwood, B., Kumar, B., Carneiro, G., Reid, I., Drummond, T., et al.: Smart mining for deep metric learning. In: ICCV, pp. 2821–2829 (2017)
14. He, K., Zhang, X., Ren, S., Sun, J.: Deep residual learning for image recognition. In: CVPR, pp. 770–778 (2016)
15. Hermans, A., Beyer, L., Leibe, B.: In defense of the triplet loss for person re-identification. arXiv (2017)
16. Huang, H., et al.: EANet: enhancing alignment for cross-domain person re-identification. arXiv preprint arXiv:1812.11369 (2018)
17. Huang, Y., Wu, Q., Xu, J., Zhong, Y.: SBSGAN: suppression of inter-domain background shift for person re-identification. In: ICCV, October 2019
18. Kalayeh, M.M., Basaran, E., Gökmen, M., Kamasak, M.E., Shah, M.: Human semantic parsing for person re-identification. In: CVPR, pp. 1062–1071 (2018)

19. Li, D., Chen, X., Zhang, Z., Huang, K.: Learning deep context-aware features over body and latent parts for person re-identification. In: CVPR (2017)
20. Li, M., Zhu, X., Gong, S.: Unsupervised person re-identification by deep learning tracklet association. In: ECCV, pp. 737–753 (2018)
21. Li, W., Zhao, R., Xiao, T., Wang, X.: DeepReID: deep filter pairing neural network for person re-identification. In: CVPR, pp. 152–159 (2014)
22. Li, W., Zhu, X., Gong, S.: Harmonious attention network for person re-identification. In: CVPR, p. 2 (2018)
23. Liao, S., Hu, Y., Zhu, X., Li, S.Z.: Person re-identification by local maximal occurrence representation and metric learning. In: CVPR, pp. 2197–2206 (2015)
24. Lin, Y., Dong, X., Zheng, L., Yan, Y., Yang, Y.: A bottom-up clustering approach to unsupervised person re-identification. In: AAAI, vol. 33, pp. 8738–8745 (2019)
25. van der Maaten, L., Hinton, G.: Visualizing data using t-SNE. JMLR **9**(Nov), 2579–2605 (2008)
26. Oh Song, H., Jegelka, S., Rathod, V., Murphy, K.: Deep metric learning via facility location. In: CVPR, pp. 5382–5390 (2017)
27. Oh Song, H., Xiang, Y., Jegelka, S., Savarese, S.: Deep metric learning via lifted structured feature embedding. In: CVPR, pp. 4004–4012 (2016)
28. Qi, L., Wang, L., Huo, J., Zhou, L., Shi, Y., Gao, Y.: A novel unsupervised camera-aware domain adaptation framework for person re-identification. In: ICCV, October 2019
29. Qian, X., et al.: Pose-normalized image generation for person re-identification. In: ECCV, pp. 650–667 (2018)
30. Ristani, E., Solera, F., Zou, R., Cucchiara, R., Tomasi, C.: Performance measures and a data set for multi-target, multi-camera tracking. In: Hua, G., Jégou, H. (eds.) ECCV 2016. LNCS, vol. 9914, pp. 17–35. Springer, Cham (2016). https://doi.org/10.1007/978-3-319-48881-3_2
31. Si, J., et al.: Dual attention matching network for context-aware feature sequence based person re-identification. In: CVPR, pp. 5363–5372 (2018)
32. Snell, J., Swersky, K., Zemel, R.: Prototypical networks for few-shot learning. In: NeurIPS, pp. 4077–4087 (2017)
33. Sohn, K.: Improved deep metric learning with multi-class N-pair loss objective. In: NeurIPS, pp. 1857–1865 (2016)
34. Song, L., et al.: Unsupervised domain adaptive re-identification: theory and practice. arXiv preprint arXiv:1807.11334 (2018)
35. Su, C., Li, J., Zhang, S., Xing, J., Gao, W., Tian, Q.: Pose-driven deep convolutional model for person re-identification. In: ICCV (2017)
36. Sun, Y., et al.: Perceive where to focus: learning visibility-aware part-level features for partial person re-identification. In: CVPR, June 2019
37. Sun, Y., Zheng, L., Yang, Y., Tian, Q., Wang, S.: Beyond part models: person retrieval with refined part pooling (and a strong convolutional baseline). In: ECCV, pp. 480–496 (2018)
38. T Ali, M.F., Chaudhuri, S.: Maximum margin metric learning over discriminative nullspace for person re-identification. In: ECCV, pp. 122–138 (2018)
39. Tay, C.P., Roy, S., Yap, K.H.: AANet: attribute attention network for person re-identifications. In: CVPR, pp. 7134–7143 (2019)
40. Ustinova, E., Lempitsky, V.: Learning deep embeddings with histogram loss. In: NeurIPS, pp. 4170–4178 (2016)
41. Wang, G., Yuan, Y., Chen, X., Li, J., Zhou, X.: Learning discriminative features with multiple granularities for person re-identification. In: ACMMM, pp. 274–282 (2018)

42. Wang, J., Zhou, F., Wen, S., Liu, X., Lin, Y.: Deep metric learning with angular loss. In: ICCV, pp. 2593–2601 (2017)
43. Wang, J., Zhu, X., Gong, S., Li, W.: Transferable joint attribute-identity deep learning for unsupervised person re-identification. In: CVPR, pp. 2275–2284 (2018)
44. Wei, L., Zhang, S., Gao, W., Tian, Q.: Person transfer GAN to bridge domain gap for person re-identification. In: CVPR, pp. 79–88 (2018)
45. Wu, A., Zheng, W.S., Lai, J.H.: Unsupervised person re-identification by camera-aware similarity consistency learning. In: ICCV, October 2019
46. Wu, Z., Xiong, Y., Yu, S.X., Lin, D.: Unsupervised feature learning via non-parametric instance discrimination. In: CVPR, pp. 3733–3742 (2018)
47. Xia, B.N., Gong, Y., Zhang, Y., Poellabauer, C.: Second-order non-local attention networks for person re-identification. In: ICCV, October 2019
48. Xiao, T., Li, H., Ouyang, W., Wang, X.: Learning deep feature representations with domain guided dropout for person re-identification. In: CVPR, pp. 1249–1258 (2016)
49. Xiao, T., Li, S., Wang, B., Lin, L., Wang, X.: Joint detection and identification feature learning for person search. In: CVPR, pp. 3415–3424 (2017)
50. Yu, B., Tao, D.: Deep metric learning with tuplet margin loss. In: ICCV, pp. 6490–6499 (2019)
51. Yu, H.X., Zheng, W.S., Wu, A., Guo, X., Gong, S., Lai, J.H.: Unsupervised person re-identification by soft multilabel learning. In: CVPR, pp. 2148–2157 (2019)
52. Zhang, X., Cao, J., Shen, C., You, M.: Self-training with progressive augmentation for unsupervised cross-domain person re-identification. In: ICCV, October 2019
53. Zhang, Z., Lan, C., Zeng, W., Chen, Z.: Densely semantically aligned person re-identification. In: CVPR, pp. 667–676 (2019)
54. Zhao, H., et al.: Spindle net: person re-identification with human body region guided feature decomposition and fusion. In: CVPR (2017)
55. Zhao, Y., Shen, X., Jin, Z., Lu, H., Hua, X.S.: Attribute-driven feature disentangling and temporal aggregation for video person re-identification. In: CVPR, pp. 4913–4922 (2019)
56. Zheng, F., et al.: Pyramidal person re-identification via multi-loss dynamic training. In: CVPR, pp. 8514–8522 (2019)
57. Zheng, L., Shen, L., Tian, L., Wang, S., Wang, J., Tian, Q.: Scalable person re-identification: a benchmark. In: ICCV, pp. 1116–1124 (2015)
58. Zheng, L., et al.: Person re-identification in the wild. In: CVPR, vol. 1, p. 2 (2017)
59. Zhong, Z., Zheng, L., Cao, D., Li, S.: Re-ranking person re-identification with K-reciprocal encoding. In: CVPR (2017)
60. Zhong, Z., Zheng, L., Li, S., Yang, Y.: Generalizing a person retrieval model hetero- and homogeneously. In: ECCV, pp. 172–188 (2018)
61. Zhong, Z., Zheng, L., Luo, Z., Li, S., Yang, Y.: Invariance matters: exemplar memory for domain adaptive person re-identification. In: CVPR, pp. 598–607 (2019)

Temporal Coherence or Temporal Motion: Which Is More Critical for Video-Based Person Re-identification?

Guangyi Chen[1,2,3], Yongming Rao[1,2,3], Jiwen Lu[1,2,3(✉)], and Jie Zhou[1,2,3,4]

[1] Department of Automation, Tsinghua University, Beijing, China
chen-gy16@mails.tsinghua.edu.cn, raoyongming95@gmail.com
{lujiwen,jzhou}@tsinghua.edu.cn
[2] State Key Lab of Intelligent Technologies and Systems, Beijing, China
[3] Beijing National Research Center for Information Science and Technology, Beijing, China
[4] Tsinghua Shenzhen International Graduate School, Tsinghua University, Beijing, China

Abstract. Video-based person re-identification aims to match pedestrians with the consecutive video sequences. While a rich line of work focuses solely on extracting the motion features from pedestrian videos, we show in this paper that the temporal coherence plays a more critical role. To distill the temporal coherence part of video representation from frame representations, we propose a simple yet effective Adversarial Feature Augmentation (AFA) method, which highlights the temporal coherence features by introducing adversarial augmented temporal motion noise. Specifically, we disentangle the video representation into the temporal coherence and motion parts and randomly change the scale of the temporal motion features as the adversarial noise. The proposed AFA method is a general lightweight component that can be readily incorporated into various methods with negligible cost. We conduct extensive experiments on three challenging datasets including MARS, iLIDS-VID, and DukeMTMC-VideoReID, and the experimental results verify our argument and demonstrate the effectiveness of the proposed method.

Keywords: Video-based person re-identification · Temporal coherence · Feature augmentation · Adversarial learning

1 Introduction

Person re-identification (ReID) matches pedestrians in a non-overlapping camera network, which has great potential in surveillance applications [25], such as suspect tracking and missing elderly retrieval. Conventional image-based ReID

G. Chen and Y. Rao—Equal contribution.

A. Vedaldi et al. (Eds.): ECCV 2020, LNCS 12353, pp. 660–676, 2020.
https://doi.org/10.1007/978-3-030-58598-3_39

methods [2,23,48,49] face many challenges, such as pose variations, illumination changes, partial occlusions and clutter background, due to the complicated intra-class variances and the limited clues in the single image. To tackle these challenges, many works [1,3,28,29,54] tend to use videos instead of a single image to identify the persons.

Compared with image data, surveillance videos avoid complex pre-processing and preserve more abundant identity clues from different view angles and poses. To obtain these identity clues from pedestrian videos, many existing works attempt to extract the temporal motion features, such as the gait of person. For example, some works [1,17,18,43,45] extract the shallow motion features like HOG3D [19] or learn the deep motion features from the optical flow [7]. Other methods [26,29,54] further model the motion information with the recurrent model like RNN or LSTM. Besides, Some works [21,24] learn the motion clues with 3D convolution neural network (3D-CNN). Different from them, in this paper, we show that the temporal coherence is more critical than temporal motion, which offers a new perspective on learning better representation for video-based person ReID.

Many video ReID methods capturing the motion clues (e.g. RNN, 3D-CNN, or optical flow based two-stream networks) are inspired by the video recognition tasks. However, different from these video-based recognition tasks (e.g. video classification, action recognition), the video ReID task focuses on the object (person) itself, rather than the pure motion of object. It motivates us to capture the invariant features about the person itself, but not the variant features over time. In Fig. 1, we show a visual comparison between temporal coherence features and optical flows about temporal motion. We can observe that temporal motion features are more instable for different view angles, different actions and other moving occlusions. For example, the optical flows of Tracklet 1 focus on the other occlusion person, while the ones of Tracklet 2 focus on the motion of arm due to the person's action. As a contrast, the temporal coherence features capture the clues that are invariant over time (the cloth of the person), which is more related to the identity and more discriminative. We agree that videos contain the temporal motion clues which are beneficial to person ReID, e.g. the gaits. However, these temporal motion clues also bring intra-class noise like the change of poses, especially for the aggregation stage. As proved in [45], the temporal motion features are harder to be applied to distinguish the person videos, due to the large intra-class variance and small inter-class variance. Thus, we argue that the temporal coherence features are more appropriate for the video-based person ReID task and give a proof-of-concept in this paper.

To distill the temporal coherence feature, we propose a simple yet effective adversarial feature augmentation (AFA) method which generates the adversarial augmented features with the temporal motion noise. In this paper, we use invariant and variant features to represent temporal coherence and temporal motion respectively. Specifically, we disentangle the video representation into the expectation and variance of embeddings of different frames. In the training process, we randomly vary the magnitude of the temporal motion features

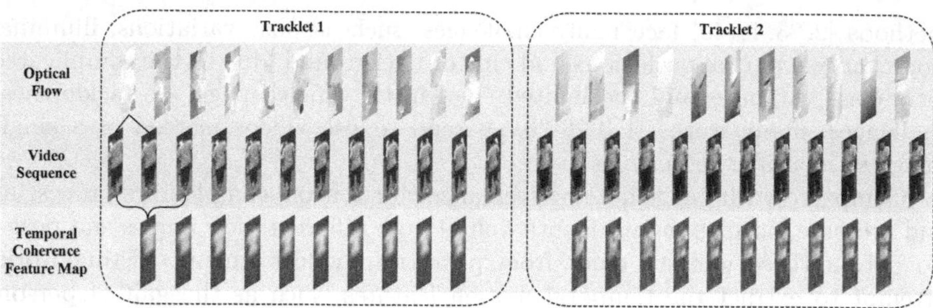

Fig. 1. The visual comparison between temporal coherence features and optical flows of temporal motion. For two tracklets from the same identity under different camera views, the temporal motion optical flows may vary dramatically due to different actions of one pedestrian, while the temporal coherence features are more discriminative since they focus on the invariance of the video.

as an adversarial interference to highlight the temporal coherence feature. On the one hand, the various temporal motion noises break the discriminant of the video representation. On the other hand, the video representation highlights the temporal constance information and adversarially reduce the influence of motion noises. In the testing process, we only use the temporal coherence feature for similarity measuring. AFA is a general module that can be readily incorporated into various video-based person ReID methods. It is lightweight and effective which brings significant performance improvement with negligible computing cost. We conduct experiments to verify our argument that the temporal coherence is more critical than the motion clues for video-based person re-identification. The consistent improvements on three challenging datasets including MARS, iLIDS-VID, and DukeMTMC-VideoReID demonstrate the effectiveness of the proposed AFA method. We summarize the contributions of this work as:

1) We show that the temporal coherence is more critical than the motion clues, which offers a new perspective on learning better representation for video-based person ReID.
2) Based on the observation, we propose a simple yet effective method (AFA) to distill the temporal coherence features in an adversarial manner. The proposed AFA model is a lightweight and efficient component that can be readily incorporated into other methods.
3) We conduct extensive experiments to demonstrate the superiority of our AFA method, and achieve the state-of-the-art performance on several large scale video person ReID benchmarks.

2 Related Work

2.1 Video-Based Person Re-identification

Video sequences provide abundant and diverse person samples, which indicate more real sample distribution. For learning more robust representation from these video sequences, existing video-based person ReID methods mainly take great efforts to: 1) mine the motion clues in the person video; 2) aggregate the video sequence embeddings. To extract discriminative motion clues, many early works [17,18,39,45] directly employ the temporal motion features like HOG3D [19] as the extra features. While some deep learning methods [1,26,29,43] learn the motion features from the optical flow [7]. In addition, many methods [26,29,44,54] model the motion process with the recurrent model like RNN or LSTM. Recently, 3D convolution neural network [21,24] (3D-CNN) has been applied for video person ReID to jointly learn the appearance and motion clues. Different from these methods which are mainly inspired by video (action) recognition methods, we argue that the temporal coherence is more critical than the motion clues for video-based person ReID, since the ReID focuses on the maker of the motion but not the motion itself. Thus, we focus on learning the robust temporal consecutive features of pedestrian videos rather than the temporal motion ones.

The aggregation of embeddings of the video sequence is another popular research field for video-based person ReID, which aims to obtain a discriminative video embedding from a sequence of image embeddings. As the baseline methods, [29,54] apply a temporal pooling layer to average all embeddings of the video. While some attention based methods [1,3,4,22,28,43] select key frames of the video to avoid the misleading from noisy frames. Besides, some works [30,31] sequentially discard confounding frames until the last one, which enlarges the discrimination and reduce the computing cost for video matching. Despite these recent progresses, the aggregation is still difficult due to the large intra-video variance of different frames, especially when the temporal motion features are highlighted. To solve this problem, we distill the temporal coherence feature and reduce the variance of temporal embeddings in an adversarial manner.

2.2 Data Augmentation

Data augmentation is an explicit form of regularization to learn robustness representation and prevent deep models from overfitting by generating extra data. It gains great success in various fields, such as image classification [12,36], object detection [27,32] and video analysis [8,40]. The common data augmentation strategies include flipping, cropping, rotation, color jittering, and adding noises. While Zhang *et al.* [46] proposed to use the convex combinations of pairs as the augmented data. Besides, many works [10,52] apply the GAN to generate the augmented images. Data augmentation have been also applied for person ReID to learn robust representation. For example, Zhong *et al.* [53] selected a rectangle region in an image and erases its pixels with random values. While Huang *et al.* [16] adversarially occluded samples as the data augmentation. Different from these methods which augments data, our proposed AFA method distills the temporal coherence feature from video representation by adversarially augmenting the features. Inspired by these adversarial data augmentation methods, we disentangle the video representation into temporal coherence and temporal motion parts and generate the adversarial augmented features with the variable temporal motion features.

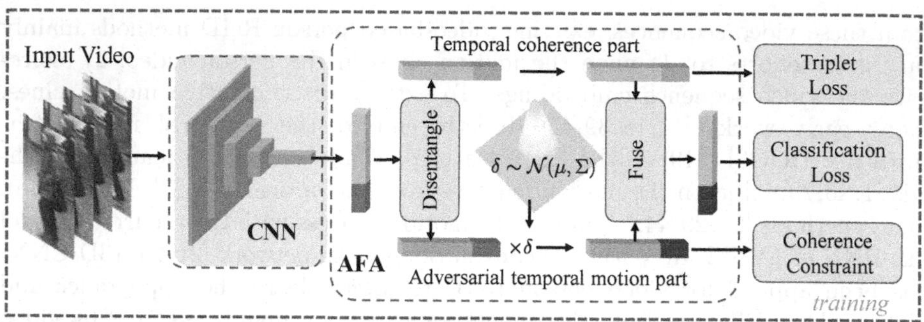

Fig. 2. Illustration of the training process of the adversarial feature augmentation method. The AFA model takes the video representation as input and disentangles it into temporal coherence and temporal motion parts. Then, the AFA model changes the scale of temporal motion features by an adversarial coefficient generated from the Gaussian distribution. The original temporal coherence features and the changed motion features are fused as the new augmented video representation. Finally, the temporal coherence features, temporal motion features and the augmented video representation are feed into the objective functions. The AFA model can be readily incorporated into any video ReID system as a general component. Best viewed in color. (Color figure online)

3 Approach

In this section, we first present our adversarial feature augmentation (AFA) method and then employ it to distill the temporal coherence features for video ReID. Finally, we describe the optimization procedure and implementation details for the proposed AFA method.

3.1 Adversarial Feature Augmentation

Many existing works focus on extracting temporal motion features for video-based person ReID, such as optical flow [1,43], RNN [29,44,54], and 3D-CNN [21,24]. However, in this paper, we argue that the temporal coherence is more critical than the motion clues for the video-based person ReID task, since person ReID focuses on the maker of the motion but not the motion itself. To highlight the temporal coherence of video representation, we propose an adversarial feature augmentation (AFA) method which disentangles the video representation into temporal coherence and motion parts and randomly changes the temporal motion part as an adversarial noise for feature augmentation.

We first describe the feature disentangling process. As shown in Fig. 2, given the video representation which is learned by a CNN model $\mathcal{X} = \mathcal{F}_\theta(\mathcal{V})$, we disentangle the video representation into the temporal coherence and temporal motion parts as:

$$\mathcal{X} = \mathcal{X}_C + \mathcal{X}_D, \tag{1}$$

where $\mathcal{X} = \{X^t \in R^d\}_{t=1:T}$ denotes the video representation, and $\mathcal{X}_C$ and $\mathcal{X}_D$ respectively denote the temporal coherence and motion features. The T is the number of video frames and X^t is the visual embedding of the tth frame, and θ denotes the parameters of the CNN. The temporal coherence feature $\mathcal{X}_C$ represents the invariance in the video sequence. In the pedestrian video, the temporal coherence features mainly focus on the identity information which is invariant for different poses and views in different frames. While the temporal motion features $\mathcal{X}_D$ represent the variety and motion in the video. These temporal motion features not only contain motion clues like gaits but also contain many noises such as pose changing. As quantitatively proved in Fig. 2 of [45], temporal motion features always have more ambiguities than temporal coherence.

Inspired by the prototypical network [33], we assume all embeddings of different frames $\{X^t\}_{t=1:T}$ in the video lie in a manifold, and calculate the video prototype as the center of class:

$$\mathcal{X}_C = \frac{1}{T} \sum_{X^t \in \mathcal{X}} X^t. \tag{2}$$

This prototype $\mathcal{X}_C \in R^d$ denotes the temporal coherence part of the video representation, e.g. the identity information. Then we disentangle the temporal motion features from the video representation as:

$$\mathcal{X}_D = \{X_D^t \in R^d | X_D^t = X^t - \mathcal{X}_C\}. \tag{3}$$

In above definition, we classify all other clues as the temporal motion features except for the identity-related temporal coherence. These temporal motion features may include the motions, the varying backgrounds and other noises. In this paper, we regard temporal motion features as the adversarial noises and utilize them to distill the temporal coherence features.

Given the disentangled temporal coherence and temporal motion features, we design an adversarial coefficient δ to generate the adversarial features as:

$$\mathcal{X}' = \{X'^t | X'^t = \mathcal{X}_C + \delta X_D^t\}, \tag{4}$$

where $\mathcal{X}'$ is the augmented new feature with the various motion noise as shown in Fig. 2. The adversarial coefficient δ randomly varies in the training stage following a Gaussian distribution $\mathcal{N}$:

$$\delta \sim \mathcal{N}(\mu = 1, \Sigma), \tag{5}$$

where $\mu = 1$ indicates the expectation of adversarial coefficient is 1, and the standard deviation Σ is a hyperparameter to control the amplitude of the noise. The larger standard deviation indicates to increase the noises. In the experiments, we set the standard deviation as $\Sigma = 0.025$.

In the training process, we sample δ to generate the adversarial augmented features with the temporal motion noise. These variable temporal motion noises break the discriminant, while the video representation will highlight the temporal constance information and adversarially reduce the influence from motion noise. By the adversarial training for these motion noise, the learned video representation can be robust for the large intra-class variance, including different poses, occlusions, and cluttered background. Note that, compared with the baseline model, our AFA model only introduces a pool layer with negligible computing cost. While in the testing, we fixed $\delta = 0$, which is equal to remove the adversarial feature augmentation and only use the temporal coherence features $\mathcal{X}_C$ for evaluation. It requires **no** extra cost in the inference process.

3.2 Optimization

Given the new video feature $\mathcal{X}'$ augmented by our adversarial feature augmentation method, we optimize it to reduce the influence from motion noise. Instead of calculating the objective function with the single video representation which is aggregated from the embeddings of the video sequence, we separately optimize each augmented feature X'^t to constrain the temporal coherence of the video representation.

The objective function of our method is formulated as follows:

$$\mathcal{L}(\mathcal{X}', \mathcal{X}_C, \mathcal{X}_D) = \mathcal{L}_{cls}(\mathcal{X}') + \mathcal{L}_{tri}(\mathcal{X}_C) + \lambda \mathcal{L}_{coh}(\mathcal{X}_D), \quad (6)$$

which contains three parts: classification loss, triplet loss, and coherence constraint. The λ is a rate to balance different loss functions.

1) Classification Loss: We apply the cross entropy loss function as the classification loss to learn the identify-specific video representation. For each augmented feature $X'^t \in \mathcal{X}'$, we first apply a batch normalization layer before the classifier to normalize the scales, since the classification loss is sensitive to the scale of features. Then we calculate the predicted probabilities of each frame with a linear classifier:

$$p^k(X'^t) = \frac{exp(W_k X'^t)}{\sum_j exp(W_j X'^t)}, \quad (7)$$

where W_k indicate the kth column of the linear classifier, and $p^k(X'^t)$ is the predicted probability of the frame X'^t for kth class. Then we aggregate the classification results of the frames as the video-based classification result:

$$p^k(\mathcal{X}') = \frac{1}{T} \sum_{t=1}^{T} p^k(X'^t). \quad (8)$$

Classification results of all frames are concentrative to the same identity, which constraints the temporal coherence of the video representation. Finally, we apply the cross-entropy loss to supervise the classifier and representation model:

$$\mathcal{L}_{cls}(\mathcal{X}') = \frac{1}{|\Omega|} \sum_{\mathcal{X}' \in \Omega} \sum_{k=1}^{K} y^k(\mathcal{X}') \log(p^k(\mathcal{X}')), \quad (9)$$

where $y^k(\mathcal{X}') = 1$ denotes the ground truth identity of the video clip $\mathcal{X}'$ is k, and 0 denotes not. We use Ω to denote the whole training set.

2) Triplet Loss: We employ the triplet loss function [13] to preserve the rank relationship among a triplet of samples with a large margin, which increases the

Algorithm 1: Adversarial Feature Augmentation

Input: Training video sequences: $\{\mathcal{V}\}$, maximal iterative number I, The standard deviation coefficient Σ.
Output: The parameters of video representation network θ.

Initialize θ;
for $i = 1, 2, \ldots, I$ **do**
  Randomly select a batch of video sequences from $\{\mathcal{V}\}$;
  Obtain the original adversarial coefficient $\mathcal{X}$;
  Disentangle the video representation into temporal coherence and temporal motion parts as (1),(2),(3);
  Generate an adversarial coefficient λ from a Gaussian distribution as (5);
  Obtain the augmentation video representation $\mathcal{X}'$ as (4);
  Update $\theta \leftarrow -\frac{\partial}{\partial\theta}\mathcal{L}(\mathcal{X}', \mathcal{X}_C, \mathcal{X}_D)$ as (6);
end for
return θ

inter-class distance and reduces the intra-class one. The triplet loss is directly applied on the temporal coherence features to increase the discriminative ability:

$$\mathcal{L}_{tri}(\mathcal{X}_C) = \sum_{\mathcal{X}_C \in \Omega} \left[||\mathcal{X}_C - \mathcal{X}_C^{+}||_2^2 - ||\mathcal{X}_C - \mathcal{X}_C^{-}||_2^2 + m\right]_{+}, \tag{10}$$

where $[\cdot]_{+}$ indicates the max function $\max(0, \cdot)$, and $\mathcal{X}_C, \mathcal{X}_C^{+}, \mathcal{X}_C^{-}$ respectively denote as the temporal coherence features of the anchor, positive and negative sample in a triplet. m is a margin to enhance the discriminative ability of learned features. In the experiments, we apply the adaptive soft margin and hard negative mining strategies as [13] and measure the distance in the Euclidean space.

3) Coherence Constraint: To further distill the temporal coherence, we develop a coherence constraint loss, which reduces the influence of temporal motion parts. It is formulated as:

$$\mathcal{L}_{coh}(\mathcal{X}_D) = \sum_{\mathcal{X}_D \in \Omega} ||\mathcal{X}_D||_2, \tag{11}$$

where $||\cdot||_2$ denotes the L2 norm. By this loss function, we aim that the scales of the temporal motion parts are limited. In other view, it is equal to apply a Mean Squared Error (MSE) loss to reduce the intra-class variance of video sequence. To explain the optimization more clearly, we provide Algorithm 1 to detail the learning process of our AFA method.

3.3 Implementation Details

We employed the ResNet-50 [37] as the basic backbone network for our AFA method in the experiments, and initialized it with the ImageNet pre-trained parameters. In order to preserve the resolution of the image, we applied a convolution layer with $stride = 1$, instead of original $stride = 2$ convolution layer in

Table 1. The basic statistics of all datasets in the experiments.

Datasets	Identities	Sequences	Frames	Cameras	Splits	Repetitions
iLIDS-VID [38]	300	600	73	2	150/150	10 times
MARS [51]	1,261	20,715	58	6	625/636	1 time
DukeV [42]	1,404	4,832	168	8	702/702	1 time

the last block of ResNet-50. During training, we apply two data augmentation methods including the horizontal flipping and the designed video-based random erasing. This video random erasing data augmentation method erases the same region for all frames in the same clip, to overcame the partial occlusions. In each mini-batch, we randomly selected 8 individuals and sampled 4 video clips for each individual. Each video clips consists of 7 images for MARS and iLIDS-VID datasets, and 9 images for DukeMTMC-VideoReID dataset, since the video sequences in the DukeMTMC-VideoReID dataset are longer than others. Besides, we only use the optical flows for iLIDS-VID since the better manual alignment. Each input image is resized as 256×128. The standard deviation coefficient Σ in adversarial coefficient distribution and the balance rate λ of loss functions are respectively set as 0.025 and 0.1 in the experiments. We trained our model for 200 epochs in total by the Adam optimizer. The initial learning rate was 0.0001 and was divided by 10 every 50 epochs. The weight decay factor for L2 regularization was set to 0.00001. During evaluation, we removed the feature augmentation part and used the temporal coherence features for evaluation. We employed the Euclidean distance as the metric to measure the similarity of two features. All experiments were implemented with PyTorch 1.3.1 on 2 Nvidia GTX 1080Ti GPUs. Taking MARS dataset as the example, the whole training process took about 2.4 h with data-parallel acceleration.

4 Experiments

In the experiments, we evaluated our method on three public video-based person ReID benchmarks. We compared the proposed method with other state-of-the-art approaches and conducted ablation studies and parameter analysis to analyze our AFA model. In addition, we conducted the transfer testing on the cross-dataset to investigate the generalization ability.

4.1 Datasets and Settings

We conducted experiments on three challenging datasets including iLIDS-VID [38], MARS [51], and DukeMTMC-VideoReID [42]. The detailed statistics and evaluation protocols of all datasets are summarized in Table 1. The iLIDS-VID dataset contains 600 sequences of 300 pedestrians under two camera views. MARS is one of the largest public video ReID dataset, including 1261 persons and around 20000 video sequences captured by 6 cameras. Different from other

Table 2. Comparison with the state-of-the-art video-based person ReID methods on the iLIDS-VID and MARS datasets.

Method	Source	iLIDS-VID			MARS		
		R1	R5	R20	R1	R5	mAP
CNN+XQDA [51]	ECCV 2016	54.1	80.7	95.4	65.3	82.0	47.6
QAN [28]	CVPR 2017	68.0	86.6	97.4	73.7	84.9	51.7
ASTPN [43]	ICCV 2017	62.0	86.0	98.0	44	70	–
RQEN [34]	AAAI 2018	76.1	92.9	99.3	73.7	84.9	51.7
DRSTA [22]	CVPR 2018	80.2	–	–	82.3	–	65.9
CSSA+CASE [1]	CVPR 2018	85.4	96.7	99.5	86.3	94.7	76.1
SDM [47]	CVPR 2018	60.2	84.7	95.2	71.2	85.7	–
STAL [3]	TIP 2019	82.8	95.3	98.8	80.3	90.9	64.5
STA [11]	AAAI 2019	–	–	–	86.3	95.7	80.8
ADFDTA [50]	CVPR 2019	86.3	97.4	**99.7**	87.0	95.4	78.2
DVR [39]	TPAMI 2016	41.3	63.5	83.1	–	–	–
CNN+RNN [29]	CVPR 2016	58.0	84.0	96.0	56	69	–
AMOC+ EpicFlow [26]	TCSVT 2017	68.7	94.3	99.3	68.3	81.4	52.9
TAM+SRM [54]	CVPR 2017	55.2	86.5	97.0	70.6	90.0	50.7
DSAN [41]	TMM 2018	61.2	80.7	97.3	69.7	83.4	–
TRL [5]	TIP 2018	57.7	81.7	94.1	80.5	91.8	69.1
VRSTC [15]	CVPR 2019	83.4	95.5	99.5	88.5	96.5	82.3
COSAM [35]	ICCV 2019	70.6	95.3		84.9	95.5	79.9
GLTR [20]	ICCV 2019	86.0	**98.0**	–	87.0	95.8	78.5
AFA	ours	**88.5**	96.8	**99.7**	**90.2**	**96.6**	**82.9**

datasets, the video sequences of the MARS dataset are detected with DPM detector [9], and tracked by the GMMCP tracker [6], instead of hand-drawn bounding boxes. These bounding boxes are always misaligned which causes the large intra-class variances. DukeMTMC-VideoReID [42] is another large-scale video-based benchmark, which comprises around 4,832 videos from 1,404 identities. In the following description, we use the abbreviation "DukeV" to represent the DukeMTMC-VideoReID dataset for convenience. The video sequences in the DukeV dataset are longer than videos in other datasets, which contain 168 frames on average.

In the experiments, we adopt the protocol of [38] for iLIDS-VID datasets, which repeated experiments 10 times and calculated the average accuracy. In each repeat, the dataset was randomly split into equal-sized training and testing sets, where the videos from the first camera view are regarded as the query set and the other as the gallery set. For a fair comparison, we selected the identical 10 splits as [38], instead of random splits, to avoid the experimental bias from dataset splitting. For MARS and DukeV datasets, we followed the settings

Table 3. Comparison with the state-of-the-art video-based person ReID methods on the DukeMTMC-VideoReID dataset.

Method	Source	DukeMTMC-VideoReID				
		R1	R5	R10	R20	mAP
STA [11]	AAAI 2019	96.2	99.3	99.6	–	94.9
VRSTC [15]	CVPR 2019	95.0	99.1	99.4	–	93.5
COSAM [35]	ICCV 2019	95.4	99.3	–	99.8	94.1
GLTR [20]	ICCV 2019	96.3	99.3	–	99.7	93.7
AFA	ours	**97.2**	**99.4**	**99.7**	**99.9**	**95.4**

as [15,20,35]. Note that, all the experiments are **NOT** applied the re-ranking tricks in the evaluation. We resort to both cumulative matching characteristic (CMC) curves and mean Average Precision (mAP) as evaluation metrics.

4.2 Comparison with the State-of-the-Art Methods

As shown in the Table 2 and Table 3, we respectively compared our method with other SOTA methods on the iLIDS-VID, MARS, and DukeV datasets. We can observe that the proposed AFA method achieves superior performance over other comparing methods by a large margin on all three benchmarks, which confirms the importance of the temporal coherence in the video-based person ReID task.

For iLIDS-VID and MARS datasets, we compared our AFA methods against 10 aggregation-based methods and other 9 methods with temporal feature learning. As shown in Table 2, we summarized the aggregation based methods in the top group and temporal feature learning methods in the bottom group. For both iLIDS-VID and MARS datasets, we achieved consistent improvement on Rank-1 and mAP performance.

DUKEV is a recently proposed large scale video ReID dataset, where only a limited number of works have been evaluated and reported. Table 3 shows the performance of our AFA method and other SOTA video ReID works including STA [11], VRSTC [15], COSAM [35], and GLTR [20]. Our AFA method outperformed all other methods by a large margin, which indicates that our AFA model is also appropriate for the long term videos.

4.3 Assumption Evaluation

In this paper, we argue the temporal coherence is more critical than the temporal motion for the video-based person ReID, and propose an AFA method to distill the main feature. To quantitatively evaluate which is better between temporal coherence and motion, we respectively supposed temporal coherence feature or motion feature is more important and applied AFA method to highlight them. As shown in the part (a) of Fig. 3, we compare the performance under these two assumptions on the MARS dataset. The red and blue curves respectively denote that we apply the AFA method to distill the temporal coherence and motion features. We can observe that using AFA method to distill the temporal coherence features obtains the dramatic improvement than the temporal motion based one. Furthermore, the performance steadily declines when we increase the standard deviation coefficient Σ, (larger Σ indicates larger augmentation). It is because the motion features may contain many noises from the occlusions, pose changing and cluttered background. Compared with the intra-class noise, the effect from beneficial clues of the temporal motion (like gaits) is limited.

4.4 Ablation Studies

In this subsection, we evaluated the generality of our AFA method for different baseline models, and investigated the contributions of different components. We summarized the comparison results on the MARS dataset in different settings in Table 4 and separately analyzed each component as follows:

The Generality for Different Baselines: We compared our AFA methods with two baselines, including original ResNet-50 [12] and QAN [28]. We implemented these two baselines with the same parameters and then added our AFA component. In the QAN* + AFA setting, we apply the quality attention to obtain the temporal coherence features. As shown in the top part of Table 4, our AFA module can obviously improve the baseline network by distilling the temporal coherence features.

Loss Functions: The loss functions of our method including three parts: triplet loss, cross-entropy loss, and MES-based coherence constraint loss. We compared and analyzed the effectiveness of different loss functions. We employed the triplet loss as the basic objective functions in our method and use the original ResNet-50 as the baseline model. As shown the bottom part of Table 4, we achieved a superior performance when we additionally employed the cross-entropy loss to supervise the classification results of all frames are concentrative to the same identity. While the MES-based coherence constraint loss further promotes the performance. Note that both the ResNet-50 + AFA and $L_{tri} + L_{cls} + L_{coh}$ settings denote the full AFA method. We display it twice in the both top and bottom parts of Table 4 for more clear comparison.

4.5 Parameters Analysis

We conducted parameters analysis about the standard deviation coefficient Σ in adversarial coefficient distribution and the balance rate λ of loss functions.

Standard Deviation Coefficient Σ: In the AFA method, we randomly sample the adversarial coefficients δ from a Gaussian distribution $\mathcal{N}$ to generate the adversarial augmented features. The standard deviation coefficient Σ in the Gaussian distribution indicates the scale of the noisy temporal motion feature. In the part (a) of Fig. 3, the abscissa is the reciprocal of Σ and the ordinate

Table 4. Ablation studies on the MARS and DUKEV datasets, including the evaluations of AFA model, different baselines, and loss functions. The * indicates that the method is reproduced by ourself with the same backbone and hyperparameters of AFA.

Method	MARS				DukeV			
	R1	R5	R10	mAP	R1	R5	R10	mAP
ResNet-50	88.1	95.6	96.8	80.1	95.2	99.2	99.7	94.3
QAN*	88.6	95.2	96.9	80.9	95.5	98.8	99.6	94.5
ResNet-50 + AFA	**90.2**	96.6	**97.6**	**82.9**	**97.2**	99.4	**99.7**	95.4
QAN* + AFA	89.7	**96.8**	97.4	82.2	97.2	**99.5**	99.7	**95.5**
L_{tri} only	85.3	93.2	95.4	78.7	94.3	98.9	99.3	93.2
$L_{tri} + L_{cls}$	89.8	96.2	97.2	82.6	96.6	99.3	**99.7**	95.0
$L_{tri} + L_{cls} + L_{coh}$	**90.2**	96.6	**97.6**	**82.9**	**97.2**	99.4	**99.7**	95.4

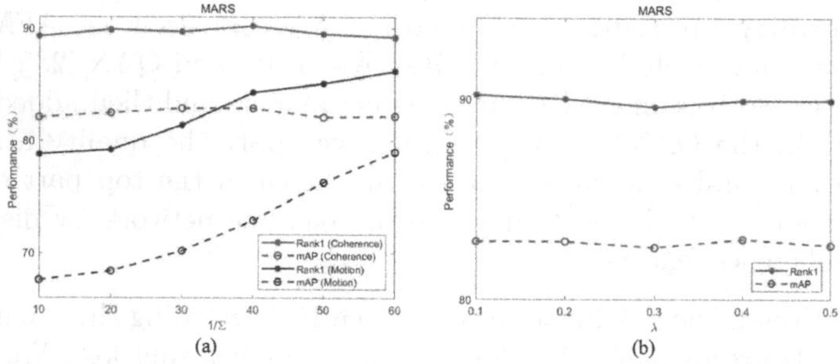

Fig. 3. Parameters analysis on the MARS dataset about (a) the standard deviation coefficient Σ and (b) the balance rate λ.

denotes the performance on the MARS dataset. We observe that the performance is slightly lower when the standard deviation coefficient Σ is too large to interfere the training process.

Balance Rate λ: We apply a trade-off parameter λ to balance different loss functions. Following the [35], we also set and fixed the rate between the triplet loss and cross-entropy loss as 1. In this subsection, we mainly discuss the balance rate on the coherence constraint loss. As shown in the part (b) in Fig. 3, the performance of different balance rates on the MARS dataset is stable, which indicates the robustness of the AFA method for the trade-off parameter of the coherence constraint loss.

4.6 Cross-Dataset Evaluation

In real surveillance systems, it requires intensive human labor to label overwhelming amount of data for training model. Thus, the cross-dataset evaluation is an important evaluation metric for person ReID systems, which measures the

Table 5. Cross dataset evaluations between the MARS and DukeV datasets. The * indicates that the method is reproduced by ourself with the same backbone and hyperparameters of our AFA method.

Method	MARS → DukeV				DukeV → MARS			
	R1	R5	R10	mAP	R1	R5	R10	mAP
Baseline	37.6	56.3	63.9	31.8	43.3	58. 6	64.5	23.8
QAN*	38.9	**59.3**	64.5	33.0	43.3	**58.8**	**64.6**	24.1
AFA	**41.5**	59.1	**67.0**	**34.6**	**44.2**	**58.8**	**64.6**	**24.5**

generalization ability of the ReID model for unseen persons and scenes. Many existing works [3,28,29] have conducted this cross-dataset evaluation, which train the model on the iLIDS-VID [38] dataset and test it on PRID-2011 [14].

However, this experimental setting has two main problems. First, the performance is unstable for different splits of the iLIDS-VID and PRID-2011 datasets. Thus, the comparisons from different works may be unfair. Second, the scales of the iLIDS-VID and PRID-2011 datasets are limited, which are not enough to represent the real surveillance system environment. For above reasons, we propose to use the MARS and DUKEV datasets for cross-dataset evaluation, which both are large-scale benchmarks with the single fixed split.

As shown in Table 5, we trained the model with the data in the MARS dataset and tested it with the samples in the DukeV dataset, and vice versa. We compared the generalization abilities of the ResNet-50 baseline, QAN [28], and our AFA method. All the methods are using the similar backbone network and hyperparameters. For both evaluation settings, our AFA method obtained the superior performance than the baseline method and QAN method.

5 Conclusion

In this work, we have argued that the temporal coherence is more critical than motion clues for the video based person ReID task. To distill these temporal coherence clues, we have proposed an adversarial feature augmentation (AFA) method, which disentangles the video representation into the temporal coherence and temporal motion parts and highlights the temporal coherence features by generating the adversarial augmented features with the variable temporal motion noise. The proposed AFA model can be incorporated into other video ReID methods with negligible cost, as a general lightweight component. Extensive experimental results demonstrate the importance the temporal coherence and validate the effectiveness of our AFA approach.

Acknowledgement. This work was supported in part by the National Key Research and Development Program of China under Grant 2017YFA0700802, in part by the National Natural Science Foundation of China under Grant 61822603, Grant U1813218, Grant U1713214, and Grant 61672306, in part by Beijing Natural Science Foundation under Grant No. L172051, in part by Beijing Academy of Artificial Intelligence

(BAAI), in part by a grant from the Institute for Guo Qiang, Tsinghua University, in part by the Shenzhen Fundamental Research Fund (Subject Arrangement) under Grant JCYJ20170412170602564, and in part by Tsinghua University Initiative Scientific Research Program.

References

1. Chen, D., Li, H., Xiao, T., Yi, S., Wang, X.: Video person re-identification with competitive snippet-similarity aggregation and co-attentive snippet embedding. In: CVPR, pp. 1169–1178 (2018)
2. Chen, G., Lin, C., Ren, L., Lu, J., Zhou, J.: Self-critical attention learning for person re-identification. In: ICCV, pp. 9637–9646 (2019)
3. Chen, G., Lu, J., Yang, M., Zhou, J.: Spatial-temporal attention-aware learning for video-based person re-identification. TIP **28**(9), 4192–4205 (2019)
4. Chen, G., Lu, J., Yang, M., Zhou, J.: Learning recurrent 3D attention for video-based person re-identification. TIP **29**, 6963–6976 (2020)
5. Dai, J., Zhang, P., Wang, D., Lu, H., Wang, H.: Video person re-identification by temporal residual learning. TIP **28**(3), 1366–1377 (2018)
6. Dehghan, A., Modiri Assari, S., Shah, M.: GMMCP tracker: globally optimal generalized maximum multi clique problem for multiple object tracking. In: CVPR, pp. 4091–4099 (2015)
7. Dosovitskiy, A., et al.: Learning optical flow with convolutional networks. In: ICCV, pp. 2758–2766 (2015)
8. Feichtenhofer, C., Fan, H., Malik, J., He, K.: Slowfast networks for video recognition. In: ICCV, pp. 6202–6211 (2019)
9. Felzenszwalb, P.F., Girshick, R.B., McAllester, D., Ramanan, D.: Object detection with discriminatively trained part-based models. TPAMI **32**(9), 1627–1645 (2010)
10. Frid-Adar, M., Klang, E., Amitai, M., Goldberger, J., Greenspan, H.: Synthetic data augmentation using GAN for improved liver lesion classification. In: ISBI, pp. 289–293. IEEE (2018)
11. Fu, Y., Wang, X., Wei, Y., Huang, T.: STA: Spatial-temporal attention for large-scale video-based person re-identification. In: AAAI (2019)
12. He, K., Zhang, X., Ren, S., Sun, J.: Deep residual learning for image recognition. In: CVPR, pp. 770–778 (2016)
13. Hermans, A., Beyer, L., Leibe, B.: In defense of the triplet loss for person re-identification. arXiv (2017)
14. Hirzer, M., Beleznai, C., Roth, P.M., Bischof, H.: Person re-identification by descriptive and discriminative classification. In: SCIA, pp. 91–102 (2011)
15. Hou, R., Ma, B., Chang, H., Gu, X., Shan, S., Chen, X.: VRSTC: occlusion-free video person re-identification. In: CVPR, June 2019
16. Huang, H., Li, D., Zhang, Z., Chen, X., Huang, K.: Adversarially occluded samples for person re-identification. In: CVPR, pp. 5098–5107 (2018)
17. Karanam, S., Li, Y., Radke, R.J.: Person re-identification with discriminatively trained viewpoint invariant dictionaries. In: ICCV, pp. 4516–4524 (2015)
18. Karanam, S., Li, Y., Radke, R.J.: Sparse re-id: block sparsity for person re-identification. In: CVPR Workshops, pp. 33–40 (2015)
19. Klaser, A., Marszałek, M., Schmid, C.: A spatio-temporal descriptor based on 3D-gradients. In: BMVC, pp. 1–10 (2008)
20. Li, J., Wang, J., Tian, Q., Gao, W., Zhang, S.: Global-local temporal representations for video person re-identification. In: ICCV, October 2019

21. Li, J., Zhang, S., Huang, T.: Multi-scale 3D convolution network for video based person re-identification. In: AAAI, vol. 33, pp. 8618–8625 (2019)
22. Li, S., Bak, S., Carr, P., Wang, X.: Diversity regularized spatiotemporal attention for video-based person re-identification. In: CVPR, pp. 369–378 (2018)
23. Li, W., Zhu, X., Gong, S.: Harmonious attention network for person re-identification. In: CVPR, p. 2 (2018)
24. Liao, X., He, L., Yang, Z., Zhang, C.: Video-based person re-identification via 3D convolutional networks and non-local attention. In: ACCV, pp. 620–634. Springer (2018)
25. Lin, J., Ren, L., Lu, J., Feng, J., Zhou, J.: Consistent-aware deep learning for person re-identification in a camera network. In: CVPR (2017)
26. Liu, H., Jie, Z., Jayashree, K., Qi, M., Jiang, J., Yan, S., Feng, J.: Video-based person re-identification with accumulative motion context. TCSVT **28**(10), 2788–2802 (2017)
27. Liu, W., et al.: SSD: single shot multibox detector. In: ECCV, pp. 21–37 (2016)
28. Liu, Y., Yan, J., Ouyang, W.: Quality aware network for set to set recognition. In: CVPR (2017)
29. McLaughlin, N., Martinez del Rincon, J., Miller, P.: Recurrent convolutional network for video-based person re-identification. In: CVPR, pp. 1325–1334, June 2016
30. Ouyang, D., Shao, J., Zhang, Y., Yang, Y., Shen, H.T.: Video-based person re-identification via self-paced learning and deep reinforcement learning framework. In: ACM MM, pp. 1562–1570 (2018)
31. Rao, Y., Lu, J., Zhou, J.: Learning discriminative aggregation network for video-based face recognition and person re-identification. IJCV **127**(6–7), 701–718 (2019)
32. Ren, S., He, K., Girshick, R., Sun, J.: Faster R-CNN: Towards real-time object detection with region proposal networks. In: NIPS, pp. 91–99 (2015)
33. Snell, J., Swersky, K., Zemel, R.: Prototypical networks for few-shot learning. In: NeurIPS, pp. 4077–4087 (2017)
34. Song, G., Leng, B., Liu, Y., Hetang, C., Cai, S.: Region-based quality estimation network for large-scale person re-identification. In: AAAI (2018)
35. Subramaniam, A., Nambiar, A., Mittal, A.: Co-segmentation inspired attention networks for video-based person re-identification. In: ICCV, October 2019
36. Szegedy, C., et al.: Going deeper with convolutions. In: CVPR, pp. 1–9 (2015)
37. Wang, F., et al.: Residual attention network for image classification. In: CVPR (2017)
38. Wang, T., Gong, S., Zhu, X., Wang, S.: Person re-identification by video ranking. In: ECCV, pp. 688–703 (2014)
39. Wang, T., Gong, S., Zhu, X., Wang, S.: Person re-identification by discriminative selection in video ranking. TPAMI **38**(12), 2501–2514 (2016)
40. Wang, X., Girshick, R., Gupta, A., He, K.: Non-local neural networks. In: CVPR, pp. 7794–7803 (2018)
41. Wu, L., Wang, Y., Gao, J., Li, X.: Where-and-when to look: deep siamese attention networks for video-based person re-identification. TMM **21**(6), 1412–1424 (2018)
42. Wu, Y., Lin, Y., Dong, X., Yan, Y., Ouyang, W., Yang, Y.: Exploit the unknown gradually: one-shot video-based person re-identification by stepwise learning. In: CVPR, pp. 5177–5186 (2018)
43. Xu, S., Cheng, Y., Gu, K., Yang, Y., Chang, S., Zhou, P.: Jointly attentive spatial-temporal pooling networks for video-based person re-identification. In: ICCV (2017)
44. Yan, Y., Ni, B., Song, Z., Ma, C., Yan, Y., Yang, X.: Person re-identification via recurrent feature aggregation. In: ECCV, pp. 701–716 (2016)

45. You, J., Wu, A., Li, X., Zheng, W.S.: Top-push video-based person re-identification. In: CVPR, pp. 1345–1353, June 2016
46. Zhang, H., Cisse, M., Dauphin, Y.N., Lopez-Paz, D.: mixup: beyond empirical risk minimization. arXiv preprint arXiv:1710.09412 (2017)
47. Zhang, J., Wang, N., Zhang, L.: Multi-shot pedestrian re-identification via sequential decision making. In: CVPR (2018)
48. Zhang, L., Xiang, T., Gong, S.: Learning a discriminative null space for person re-identification. In: CVPR, pp. 1239–1248 (2016)
49. Zhao, H., et al.: Spindle net: person re-identification with human body region guided feature decomposition and fusion. In: CVPR (2017)
50. Zhao, Y., Shen, X., Jin, Z., Lu, H., Hua, X.S.: Attribute-driven feature disentangling and temporal aggregation for video person re-identification. In: CVPR, June 2019
51. Zheng, L., Bie, Z., Sun, Y., Wang, J., Su, C., Wang, S., Tian, Q.: Mars: a video benchmark for large-scale person re-identification. In: ECCV, pp. 868–884 (2016)
52. Zheng, Z., Zheng, L., Yang, Y.: Unlabeled samples generated by GAN improve the person re-identification baseline in vitro. In: ICCV, pp. 3754–3762 (2017)
53. Zhong, Z., Zheng, L., Kang, G., Li, S., Yang, Y.: Random erasing data augmentation. arXiv preprint arXiv:1708.04896 (2017)
54. Zhou, Z., Huang, Y., Wang, W., Wang, L., Tan, T.: See the forest for the trees: joint spatial and temporal recurrent neural networks for video-based person re-identification. In: CVPR, July 2017

Arbitrary-Oriented Object Detection with Circular Smooth Label

Xue Yang[1,2] and Junchi Yan[1,2(✉)]

[1] Department of Computer Science and Engineering, Shanghai Jiao Tong University, Shanghai, China
{yangxue-2019-sjtu,yanjunchi}@sjtu.edu.cn

[2] MoE Key Lab of Artificial Intelligence, AI Institute, Shanghai Jiao Tong University, Shanghai, China

Abstract. Arbitrary-oriented object detection has recently attracted increasing attention in vision for their importance in aerial imagery, scene text, and face etc. In this paper, we show that existing regression-based rotation detectors suffer the problem of discontinuous boundaries, which is directly caused by angular periodicity or corner ordering. By a careful study, we find the root cause is that the ideal predictions are beyond the defined range. We design a new rotation detection baseline, to address the boundary problem by transforming angular prediction from a regression problem to a classification task with little accuracy loss, whereby high-precision angle classification is devised in contrast to previous works using coarse-granularity in rotation detection. We also propose a circular smooth label (CSL) technique to handle the periodicity of the angle and increase the error tolerance to adjacent angles. We further introduce four window functions in CSL and explore the effect of different window radius sizes on detection performance. Extensive experiments and visual analysis on two large-scale public datasets for aerial images i.e. DOTA, HRSC2016, as well as scene text dataset ICDAR2015 and MLT, show the effectiveness of our approach. The code is public available at https://github.com/Thinklab-SJTU/CSL_RetinaNet_Tensorflow.

Keywords: Oriented object detection · Circular smooth label

1 Introduction

Object detection is one of the fundamental tasks in computer vision. In particular, rotation detection has played a huge role in the field of aerial images

The work is supported by National Key Research and Development Program of China (2018AAA0100704), National Natural Science Foundation of China (61972250, U19B2035).

Electronic supplementary material The online version of this chapter (https://doi.org/10.1007/978-3-030-58598-3_40) contains supplementary material, which is available to authorized users.

A. Vedaldi et al. (Eds.): ECCV 2020, LNCS 12353, pp. 677–694, 2020.
https://doi.org/10.1007/978-3-030-58598-3_40

[2,4,41,42,44], scene text [12,18,19,24,27,49] and face [11,33,34]. The rotation detector can provide accurate orientation and scale information, which will be helpful in applications such as object change detection in aerial images and recognition of sequential characters for multi-oriented scene texts.

Recently, a line of advanced rotation detectors evolved from classic detection algorithms [3,7,20,21,32] have been proposed. Among these methods, detectors based on region regression occupy the mainstream, and the representation of multi-oriented object is achieved by rotated bounding box or quadrangles. Although these rotation detectors have achieved promising results, there are still some fundamental problems. Specifically, we note both the five-parameter regression and the eight-parameter regression methods suffer the problem of discontinuous boundaries, as often caused by angular periodicity or corner ordering. However, the inherent reasons are not limited to the particular representation of the bounding box. In this paper, we argue that the root cause of boundary problems based on regression methods is that the ideal predictions are beyond the defined range. Thus, the model's loss value suddenly increase at the boundary situation so that the model cannot obtain the prediction result in the simplest and most direct way, and additional more complicated treatment is often needed. Therefore, these detectors often have difficulty in boundary conditions. For detection using rotated bounding boxes, the accuracy of angle prediction is critical. A slight angle deviation leads to important Intersection-over-Union (IoU) drop, resulting in inaccurate object detection, especially in case of large aspect ratios.

There have been efforts addressing the boundary problem. IoU-smooth L1 [44] loss introduces the IoU factor, and modular rotation loss [30] increases the boundary constraint to eliminate the sudden increase in boundary loss and reduce the difficulty of model learning. Yet these methods are still regression-based detection methods, and still have not solved the root cause as mentioned above.

In this paper, we are aimed to find a more fundamental rotation detection baseline to solve the boundary problem. Specifically, we consider the prediction of the object angle as a classification problem to better limit the prediction results, and then we design a circular smooth label (CSL) to address the periodicity of the angle and increase the error tolerance between adjacent angles. Although the conversion from continuous regression to discrete classification , the impact of the lost accuracy on the rotation detection task is negligible. We also introduce four window functions in CSL and explore the effect of different window radius sizes on detection performance. After a lot of experiments and visual analysis, we find that CSL-based rotation detection algorithm is indeed a better baseline choice than the angle regression-based method on different detectors and datasets.

In summary, the main contribution of this paper are four-folds:

- We summarize the boundary problems in different regression-based rotation detection methods [2,4,41,42] and show the root cause is that the ideal predictions are beyond the defined range.

- We design a new rotation detection baseline, which transforms angular prediction from a regression problem to a classification problem. Specifically, to our best knowledge, we devise the first high-precision angle (less than 1 degree) classification based pipeline in rotation detection, in contrast to previous coarse classification granularity (around 10-degree) methods [33]. Our method has little accuracy loss compared with regression-based methods and can effectively eliminate the boundary problem.
- We also propose the circular smooth label (CSL) technique, as an independent module which can also be readily reused in existing regression based methods by replacing the regression with classification, to address angular prediction for boundary conditions and objects with large aspect ratio.
- Extensive experimental results on DOTA and HRSC2016 show the state-of-the-art performance of our detector, and the efficacy of our CSL technique as an independent component has been verified across different detectors.

2 Related Work

Horizontal Region Object Detection. Classic object detection aims to detect general objects in images with horizontal bounding boxes, and many high-performance general-purpose object detections have been proposed. R-CNN [8] pioneers a method based on CNN detection. Subsequently, region-based models such as Fast R-CNN [7], Faster R-CNN [32], and R-FCN [3] are proposed, which improve the detection speed while reducing computational storage. FPN [20] focus on the scale variance of objects in images and propose feature pyramid network to handle objects at different scales. SSD [23], YOLO [31] and RetinaNet [21] are representative single-stage methods, and their single-stage structure allows them to have faster detection speeds. Compared to anchor based methods, many anchor-free have become extremely popular in recent years. CornerNet [15], CenterNet [5] and ExtremeNet [48] attempt to predict some keypoints of objects such as corners or extreme points, which are then grouped into bounding boxes. However, horizontal detector does not provide accurate orientation and scale information, which poses problem in real applications such as object change detection in aerial images and recognition of sequential characters for multi-oriented scene texts.

Arbitrary-Oriented Object Detection. Aerial images and scene text are the main application scenarios of the rotation detector. Recent advances in multi-oriented object detection are mainly driven by adaption of classical object detection methods using rotated bounding boxes or quadrangles to represent multi-oriented objects. Due to the complexity of the remote sensing image scene and the large number of small, cluttered and rotated objects, multi-stage rotation detectors are still dominant for their robustness. Among them, ICN [2], ROI-Transformer [4], SCRDet [41], R^{3}Det [41] are state-of-the-art detectors. Gliding Vertex [40] and RSDet [30] achieve more accurate object detection through quadrilateral regression prediction. For scene text detection, RRPN [27] employ

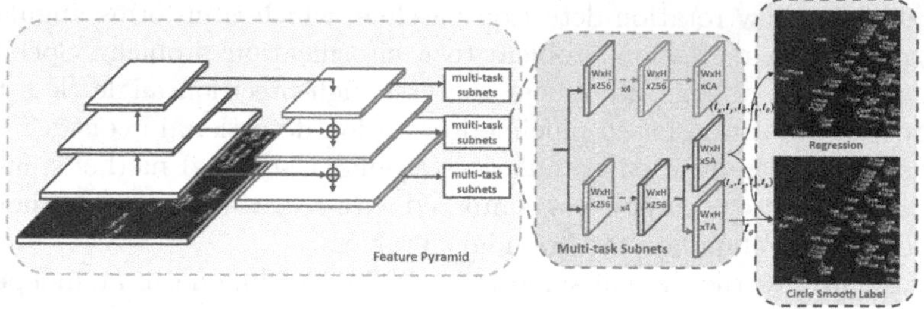

Fig. 1. Architecture of the proposed rotation detector (RetinaNet as an embodiment). 'C' and 'T' represent the number of object and angle categories, respectively.

rotated RPN to generate rotated proposals and further perform rotated bounding box regression. TextBoxes++ [18] adopts vertex regression on SSD. RRD [19] further improves TextBoxes++ by decoupling classification and bounding box regression on rotation-invariant and rotation sensitive features, respectively. Although the regression-based arbitrary-oriented object detection method occupies the mainstream, we have found that most of these methods have some boundary problems due to the situations beyond the defined range. Therefore, we design a new rotation detection baseline, which basically eliminates the boundary problem by transforming angular prediction from a regression problem to a classification problem with little accuracy loss.

Classification for Orientation Information. The method of obtaining orientation information through classification is earlier used for multi-view face detection with arbitrary rotation-in-plane (RIP) angles. Divide-and-Conquer is adopted in [11], which use several small neural networks to deal with a small range of face appearance variations individually. In [33], a router network is firstly used to estimate each face candidate's RIP angle. PCN [34] progressively calibrates the RIP orientation of each face candidate and shrinks the RIP range by half in early stages. Finally, PCN makes the accurate final decision for each face candidate to determine whether it is a face and predict the precise RIP angle. In other research areas, [14] adopts ordinal regression for or effective future motion classification. [43] obtains the orientation information of the ship by classifying the four sides. The above methods all obtain the approximate orientation range through classification, but cannot be directly applied to scenarios that require precise orientation information such as aerial images and scene text.

3 Proposed Method

We give an overview of our method as sketched in Fig. 1. The embodiment is a single-stage rotation detector based on the RetinaNet [21]. The figure shows a multi-tasking pipeline, including regression-based prediction branch and CSL-based prediction branch, to facilitate the comparison of the performance of the two methods. It can be seen from the figure that CSL-based method is more

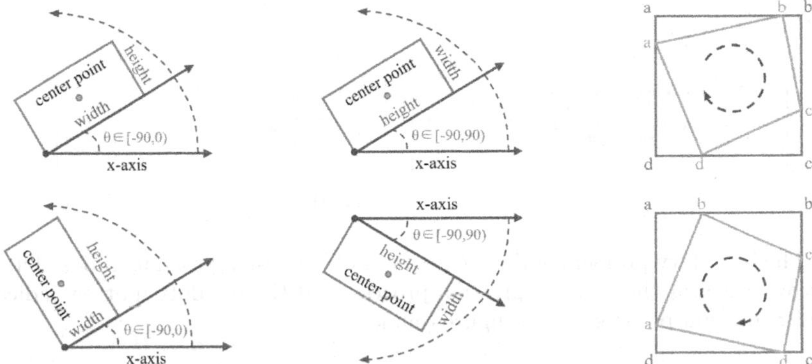

(a) Five-parameter method with 90° angular range. (b) Five-parameter method with 180° angular range. (c) Ordered quadrilateral representation.

Fig. 2. Several definitions of bounding boxes.

accurate for learning the orientation and scale information of the object. It should be noted that the method proposed in this paper is applicable to most regression-based methods, which has been verified in the FPN [20] detector in experiments.

3.1 Regression-Based Rotation Detection Method

Parametric regression is currently a popular method for rotation object detection, mainly including five-parameter regression-based methods [4,12,27,41,42,44] and eight-parameter regression-based methods [18,25,30,40]. The commonly used five-parameter regression-based methods realize arbitrary-oriented bounding box detection by adding an additional angle parameter θ. Figure 2(a) shows one of the rectangular definition (x, y, w, h, θ) with 90° angular range [27,41,42,44], θ denotes the acute angle to the x-axis, and for the other side we refer it as w. It should be distinguished from another definition (x, y, h, w, θ) illustrated in Fig. 2b, with 180° angular range [4,27], whose θ is determined by the long side (h) of the rectangle and x-axis. The eight-parameter regression-based detectors directly regress the four corners $(x_1, y_1, x_2, y_2, x_3, y_3, x_4, y_4)$ of the object, so the prediction is a quadrilateral. The key step to the quadrilateral regression is to sort the four corner points in advance, which can avoid a very large loss even if the prediction is correct, as shown in Fig. 2(c).

3.2 Boundary Problem of Regression Method

Although the parametric regression-based rotation detection method has achieved competitive performance in different vision tasks, and has been a building block for a number of excellent detection methods, these methods essentially suffer the discontinuous boundaries problem [30,44]. Boundary discontinuity problems are often caused by angular periodicity in the five-parameter method

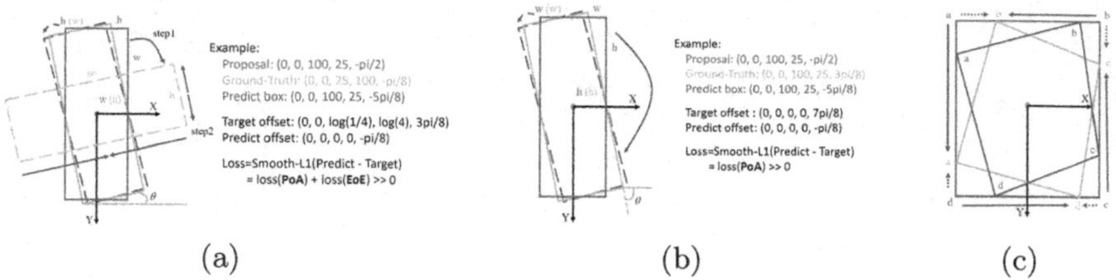

Fig. 3. The boundary problem of three categories of regression based methods. The red solid arrow indicates the actual regression process, and the red dotted arrow indicates the ideal regression process. (Color figure online)

and corner ordering in the eight-parameter method, but there exist more fundamental root cause regardless the representation choices of the bounding box.

The boundary discontinuity problem often makes the model's loss value suddenly increase at the boundary situation. Thus methods have to resort to particular and often complex tricks to mitigate this issue. Therefore, these detection methods are often inaccurate in boundary conditions. We describe the boundary problem in three typical categories of regression-based methods according to their different representation forms (the first two refer to the five-parameter methods):

- 90°**-regression-based method, as sketched in Figure 3(a)** . It shows that an ideal form of regression (the blue box rotates counterclockwise to the red box), but the loss of this situation is very large due to the periodicity of angular (PoA) and exchangeability of edges (EoE), see the example in Fig. 3(a) and Eq. 3, 4 for detail. Therefore, the model has to be regressed in other complex forms (such as the blue box rotating clockwise to the gray box while scaling w and h), increasing the difficulty of regression.
- 180°**-regression-based method, as illustrated in Figure 3(b).** Similarly, this method also has a problem of sharp increase of loss caused by the PoA at the boundary. The model will eventually choose to rotate the proposal a large angle clockwise to get the final predicted bounding box.
- **Point-based method, as shown in Figure 3(c).** Through further analysis, the boundary discontinuity problem still exists in the eight-parameter regression method due to the advance ordering of corner points. Consider the situation of an eight-parameter regression in the boundary case, the ideal regression process should be $\{(a \rightarrow b), (b \rightarrow c), (c \rightarrow d), (d \rightarrow a)\}$, but the actual regression process from the blue reference box to the green ground truth box is $\{(a \rightarrow a), (b \rightarrow b), (c \rightarrow c), (d \rightarrow d)\}$. In fact, this situation also belongs to PoA. By contrast, the actual and ideal regression of the blue to red bounding boxes is consistent.

Some approaches have been proposed to solve these problems based on the above analysis. For example, IoU-smooth L1 [44] loss introduces the IoU factor, and modular rotation loss [30] increases the boundary constraint to eliminate

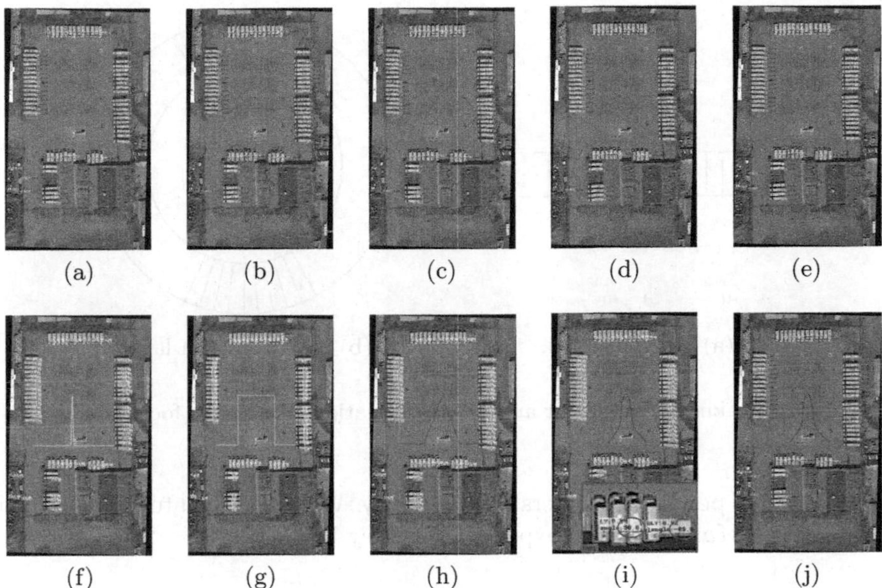

Fig. 4. Comparison of five regression-based rotation detection methods and CSL in the boundary case. (a) RetinaNet-H [41]. (b) RetinaNet-R [41]. (c) FPN-H [20]. (d) R^{3}Det [41]. (e) IoU-Smooth L1 [44]. (f) 180-CSL-Pulse. (g) 180°-CSL-Rectangular. (h) 180°-CSL-Triangle. (i) 180°-CSL-Gaussian. (j) 90°-CSL-Gaussian. 'H' and 'R' represent the horizontal and rotating anchors. Red dotted circles indicate some bad cases. (Color figure online)

the sudden increase in boundary loss and reduce the difficulty of model learning. However, these methods are still regression-based detection methods, and no solution is given from the root cause. In this paper, we will start from a new perspective and replace regression with classification to achieve better and more robust rotation detectors. We reproduce some classic rotation detectors based on regression and compare them visually under boundary conditions, as shown in Fig. 4(a) to Fig. 4(e). In contrast, CLS-based methods have no boundary problem, as shown in Fig. 4(i).

3.3 Circular Smooth Label for Angular Classification

The main cause of boundary problems based on regression methods is that the ideal predictions are beyond the defined range. Therefore, we consider the prediction of the object angle as a classification problem to better limit the prediction results. A simple solution is to use the object angle as its category label, and the number of categories is related to the angle range. Figure 5(a) shows the label setting for a standard classification problem (one-hot label encoding). The conversion from regression to classification can cause certain accuracy loss. Taking the five-parameter method with 180° angle range as an example, ω (default

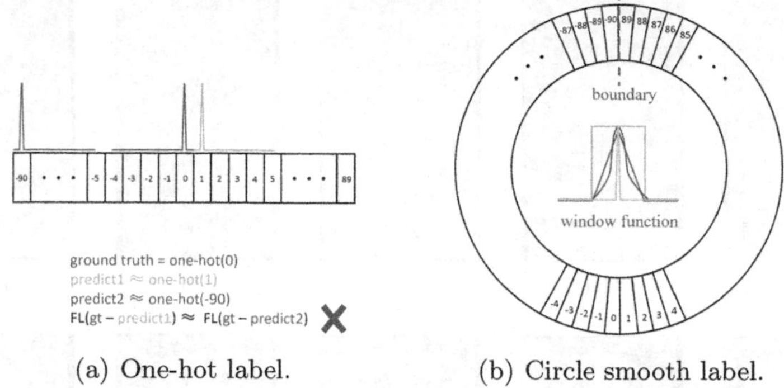

(a) One-hot label. (b) Circle smooth label.

Fig. 5. Two kind of labels for angular classification. FL means focal loss [21].

$\omega = 1°$) degree per interval refers to a category. We can calculate the maximum accuracy loss $Max(loss)$ and expected accuracy loss $E(loss)$:

$$Max(loss) = \frac{\omega}{2}, \quad E(loss) = \int_a^b x * \frac{1}{b-a} dx = \int_0^{\omega/2} x * \frac{1}{\omega/2 - 0} dx = \frac{\omega}{4} \tag{1}$$

Based on the above equations, one can see the loss is slight for a rotation detector. For example, when two rectangles with a 1:9 aspect ratio differ by 0.25° and 0.5° (default expected and maximum accuracy loss), the Intersection over Union (IoU) between them only decreases by 0.02 and 0.05. However, one-hot label has two drawbacks for rotation detection:

- The EoE problem still exists when the bounding box uses the 90°-regression-based method. In addition, 90°-regression-based method has two different border cases (vertical and horizontal), while 180°-regression-based method has only vertical border cases.
- Note vanilla classification loss is agnostic to the angle distance between the predicted label and ground truth label, thus is inappropriate for the nature of the angle prediction problem. As shown in Fig. 5(a), when the ground-truth is 0° and the prediction results of the classifier are 1° and −90° respectively, their prediction losses are the same, but the prediction results close to ground-truth should be allowed from a detection perspective.

Therefore, we design a circular smooth label (CSL) technique to obtain more robust angular prediction through classification without suffering boundary conditions, including EoE and PoA. It can be clearly seen from Fig. 5(b) that CSL involves a circular label encoding with periodicity, and the assigned label value is smooth with a certain tolerance. The expression of CSL is as follows:

$$CSL(x) = \begin{cases} g(x), & \theta - r < x < \theta + r \\ 0, & otherwise \end{cases} \tag{2}$$

where $g(x)$ is a window function. r is the radius of the window function. θ represents the angle of the current bounding box. An ideal window function $g(x)$ is required to hold the following properties:

- **Periodicity**: $g(x) = g(x + kT), k \in N$. $T = 180/\omega$ represents the number of bins into which the angle is divided, and the default value is 180.
- **Symmetry**: $0 \leq g(\theta + \varepsilon) = g(\theta - \varepsilon) \leq 1, |\varepsilon| < r$. θ is the center of symmetry.
- **Maximum**: $g(\theta) = 1$.
- **Monotonic**: $0 \leq g(\theta \pm \varepsilon) \leq g(\theta \pm \varsigma) \leq 1, |\varsigma| < |\varepsilon| < r$. The function presents a monotonous non-increasing trend from the center point to both sides

We give four efficient window functions that meet the above three properties: pulse functions, rectangular functions, triangle functions, and Gaussian functions, as shown in Fig. 5(b). Note that the label value is continuous at the boundary and there is no arbitrary accuracy loss due to the periodicity of CSL. In addition, one-hot label is equivalent to CSL when the window function is a pulse function or the radius of the window function is very small.

3.4 Loss Function

Our multi-tasking pipeline contains regression-based prediction branch and CSL-based prediction branch, to facilitate the performance comparison of the two methods on an equal footing. The regression of the bounding box is:

$$\begin{aligned}
t_x &= (x - x_a)/w_a, t_y = (y - y_a)/h_a \\
t_w &= \log(w/w_a), t_h = \log(h/h_a), \\
t_\theta &= (\theta - \theta_a) \cdot \pi/180 \quad (only\ for\ regression\ branch) \\
t_x^{'} &= (x^{'} - x_a)/w_a, t_y^{'} = (y^{'} - y_a)/h_a \\
t_w^{'} &= \log(w^{'}/w_a), t_h^{'} = \log(h^{'}/h_a), \\
t_\theta^{'} &= (\theta^{'} - \theta_a) \cdot \pi/180 \quad (only\ for\ regression\ branch)
\end{aligned} \tag{3}$$

where x, y, w, h, θ denote the box's center coordinates, width, height and angle, respectively. Variables $x, x_a, x^{'}$ are for the ground-truth box, anchor box, and predicted box, respectively (likewise for y, w, h, θ). The multi-task loss is:

$$\begin{aligned}
L = &\frac{\lambda_1}{N} \sum_{n=1}^{N} obj_n \cdot \sum_{j \in \{x,y,w,h,\theta_{reg}\}} L_{reg}(v_{nj}^{'}, v_{nj}) \\
&+ \frac{\lambda_2}{N} \sum_{n=1}^{N} L_{CSL}(\theta_n^{'}, \theta_n) + \frac{\lambda_3}{N} \sum_{n=1}^{N} L_{cls}(p_n, t_n)
\end{aligned} \tag{4}$$

where N indicates the number of anchors, obj_n is a binary value ($obj_n = 1$ for foreground and $obj_n = 0$ for background, no regression for background). $v_{*j}^{'}$ denotes the predicted offset vectors, v_{*j} is the targets vector of ground-truth. θ_n, $\theta_n^{'}$ denote the label and predict of angle respectively. t_n represents the label of object, p_n is the probability distribution of various classes calculated by Sigmoid function. The hyper-parameter λ_1, λ_2, λ_3 control the trade-off and are set to $\{1, 0.5, 1\}$ by default. The classification loss L_{cls} and L_{CSL} is focal loss [21] or sigmoid cross-entropy loss depend on detector. The regression loss L_{reg} is smooth L1 loss as used in [7].

4 Experiments

We use Tensorflow [1] to implement the proposed methods on a server with GeForce RTX 2080 Ti and 11G memory. The experiments in this article are initialized by ResNet50 [10] by default unless otherwise specified. Weight decay and momentum are set 0.0001 and 0.9, respectively. We employ MomentumOptimizer over 4 GPUs with a total of 4 images per minibatch (1 images per GPU). At each pyramid level we use anchors at seven aspect ratios $\{1, 1/2, 2, 1/4, 4, 1/6, 6\}$, and the remaining anchor settings are the same as the original RetinaNet and FPN.

4.1 Benchmarks and Protocls

DOTA [39] is one of the largest aerial image detection benchmarks. There are two detection tasks for DOTA: horizontal bounding boxes (HBB) and oriented bounding boxes (OBB). DOTA contains 2,806 aerial images from different sensors and platforms and the size of image ranges from around 800×800 to $4,000 \times 4,000$ pixels. The fully annotated DOTA benchmark contains 15 common object categories and 188,282 instances, each of which is labeled by an arbitrary quadrilateral. Half of the original images are randomly selected as the training set, 1/6 as the validation set, and 1/3 as the testing set. We divide the training and validation images into 600×600 subimages with an overlap of 150 pixels and scale it to 800×800. With all these processes, we obtain about 27,000 patches. **ICDAR2015** [13] is the Challenge 4 of ICDAR 2015 Robust Reading Competition, which is commonly used for oriented scene text detection and spotting. This dataset includes 1,000 training images and 500 testing images. In training, we first train our model using 9,000 images from ICDAR 2017 MLT training and validation datasets, then we use 1,000 training images to fine-tune our model. **ICDAR 2017 MLT** [28] is a multi-lingual text dataset, which includes 7,200 training images, 1,800 validation images and 9,000 testing images. The dataset is composed of complete scene images in 9 languages, and text regions in this dataset can be in arbitrary orientations, being more diverse and challenging. **HRSC2016** [26] contains images from two scenarios including ships on sea and ships close inshore. All images are collected from six famous harbors. The training, validation and test set include 436, 181 and 444 images, respectively.

All datasets are trained by 20 epochs (the number of image iterations per epoch is e) in total, and learning rate was reduced tenfold at 12 epochs and 16 epochs, respectively. The initial learning rates for RetinaNet and FPN are 5e−4 and 1e−3 respectively. The value of e for DOTA, ICDAR2015, MLT and HRSC2016 are 27k, 10k, 10k and 5k, and doubled if data augmentation and multi-scale training are used.

Table 1. Comparison of four window functions on the DOTA dataset. 5-mAP refers to the mean average precision of the five categories with large aspect ratio. mAP means mean average precision of all 15 categories. EoE indicates the issue of exchangeability of edges and a tick in table means the method suffers from EoE.

Based method	Angle range	EoE	Label mode	BR	SV	LV	SH	HA	5-mAP	mAP
RetinaNet-H (CSL-Based)	90	✓	Pulse	9.80	28.04	11.42	18.43	23.35	18.21	39.52
	90	✓	Rectangular	37.62	54.28	48.97	62.59	50.26	50.74	58.86
	90	✓	Triangle	37.25	54.45	44.01	60.03	52.20	49.59	60.15
	90	✓	Gaussian	**41.03**	**59.63**	**52.57**	**64.56**	**54.64**	**54.49**	**63.51**
	180		Pulse	13.95	16.79	6.50	16.80	22.48	15.30	42.06
	180		Rectangular	36.14	60.80	50.01	65.75	53.17	53.17	61.98
	180		Triangle	32.69	47.25	44.39	54.11	41.90	44.07	57.94
	180		Gaussian	**41.16**	**63.68**	**55.44**	**65.85**	**55.23**	**56.21**	**64.50**

Table 2. Comparison of detection results under different radius.

Based method	Angle range	Label mode	r = 0	r = 2	r = 4	r = 6	r = 8
RetinaNet-H(CSL-Based)	180	Gaussian	40.78	59.23	62.12	**64.50**	63.99
FPN-H(CSL-Based)	180	Gaussian	48.08	70.18	70.09	**70.92**	69.75

4.2 Ablation Study

Comparison of Four Window Functions. Table 1 shows the performance comparison of the four window functions on the DOTA dataset. It also details the accuracy of the five categories with larger aspect ratio and more border cases in the dataset. We believe that these categories can better reflect the advantages of our method. In general, the Gaussian window function performs best, while the pulse function performs worst because it has not learned any orientation and scale information. Figures 4(f)–4(i) show the visualization of the four window functions. According to Fig. 4(i)–4(j), the 180°-CSL-based method obviously has better boundary prediction due to the EoE problem still exists in the 90°-CSL-based method. The visualization results in Fig. 4 are consistent with the data analysis results in Table 1.

Suitable Window Radius. The Gaussian window form has shown best performance, while here we study the effect of radius of the window function. When the radius is too small, the window function tends to a pulse function. Conversely, the discrimination of all predictable results becomes smaller. Therefore, we choose a suitable radius range from 0 to 8, Table 2 shows the performance of the two detectors in this range. Although both detectors achieve the best performance with a radius of 6, the single-stage detection method is more sensitive to radius. We speculate that the instance-level feature extraction capability (like RoI Pooling [7] and RoI Align [9]) in the two-stage detector is stronger than the image-level in the single-stage detector. Therefore, the two-stage detection method can distinguish the difference between the two approaching angles. Figure 6 compares visualizations at different window raduis. When the radius is

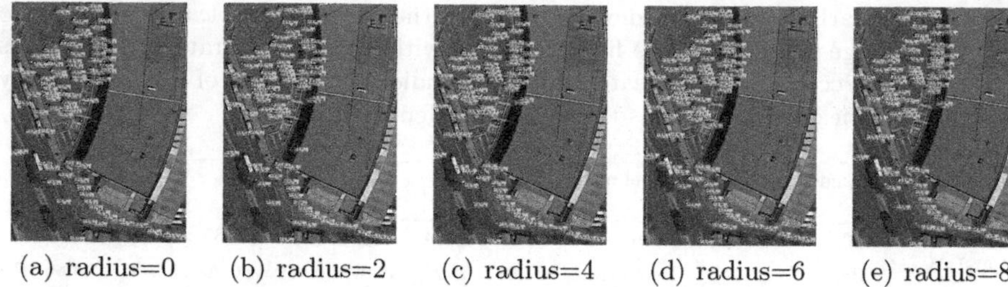

(a) radius=0 (b) radius=2 (c) radius=4 (d) radius=6 (e) radius=8

Fig. 6. Visualization of detection results (RetinaNet-H CSL-Based) under different radius. The red bounding box indicates that no orientation and scale information has been learned, and the green bounding box is the correct detection result. (Color figure online)

Table 3. Comparison between CSL-based and regression-based methods on DOTA. Improvement by CSL-based methods have been made under the same configuration.

Based method	Angle range	Angle pred.	PoA	EoE	Label m ode	BR	SV	LV	SH	HA	5-mAP	mAP
RetinaNet-H	90	regression-based	✓	✓	-	41.15	53.75	48.30	55.92	55.77	50.98	63.18
	90	CSL-based		✓	Gaussian	41.03	59.63	52.57	64.56	54.64	54.49 (**+3.51**)	63.51 (**+0.33**)
	180	regression-based	✓		-	38.47	54.15	47.89	60.87	53.63	51.00	64.10
	180	CSL-based			Gaussian	41.16	63.68	55.44	65.85	55.23	56.21 (**+5.21**)	64.50 (**+0.40**)
RetinaNet-R	90	regression-based	✓	✓	-	32.27	64.64	71.01	68.62	53.52	58.01	62.76
	90	CSL-based		✓	Gaussian	35.14	63.21	73.92	69.49	55.53	59.46 (**+1.45**)	65.45 (**+2.69**)
FPN-H	90	regression-based	✓	✓	-	44.78	70.25	71.13	68.80	54.27	61.85	68.25
	90	CSL-based		✓	Gaussian	45.46	70.22	71.96	76.06	54.84	63.71 (**+1.86**)	69.02 (**+0.77**)
	180	regression-based	✓		-	45.88	69.37	72.06	72.96	62.31	64.52	69.45
	180	CSL-based			Gaussian	47.90	69.66	74.30	77.06	64.59	66.70 (**+2.18**)	70.92 (**+1.47**)

Table 4. Comparison between CSL-based and regression-based methods on the text dataset ICDAR2015, MLT, and another remote sensing dataset HRSC2016. 07 or 12 means use the 2007 or 2012 evaluation metric.

Method	ICDAR2015			MLT			HRSC2016	
	Recall	Precision	Hmean	Recall	Precision	Hmean	mAP (07)	mAP (12)
FPN-regression-based	81.81	83.07	82.44	56.15	80.26	66.08	88.33	94.70
FPN-CSL-based	83.00	84.30	83.65 (**+1.21**)	56.72	80.77	66.64 (**+0.56**)	89.62 (**+1.29**)	96.10 (**+1.40**)

0, the detector cannot learn any orientation and scale information, which is consistent with the performance of the pulse function above. As the radius becomes larger and optimal, the detector can learn the angle in any direction.

Classification Is Better Than Regression. Three rotation detectors in Table 3, including RetinaNet-H, RetinaNet-R and FPN-H, are used to compare the performance differences between CSL-based and regression-based methods. The former two are single-stage detectors, whose anchor format is different. The latter is a classic two-stage detection method. It can be clearly seen that CSL has better detection ability for objects with large aspect ratios and more boundary conditions. It also should be noted that CSL is designed to solve the boundary

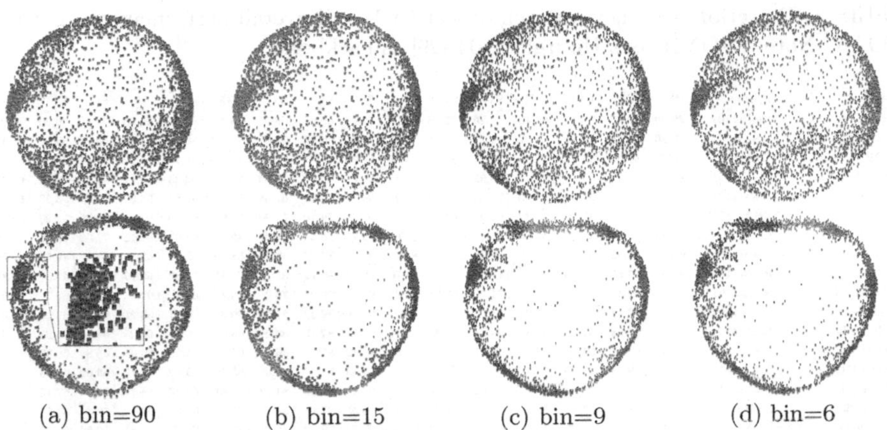

(a) bin=90 (b) bin=15 (c) bin=9 (d) bin=6

Fig. 7. Angular feature visualization of the 90-CSL-FPN detector on the DOTA dataset. First, we divide the entire angular range into several bins, and bins are different between columns. The two rows show two-dimensional feature visualizations of pulse and gaussian function, respectively. Each point represents a RoI of the test set with a index of the bin it belongs to.

problem, whose proportion in the entire dataset is relatively small, so the overall performance (mAP) is not as obvious as the five categories listed (5-mAP). Overall, the CSL-based rotation detection algorithm is indeed a better baseline choice than the angle regression-based method.

CSL Performance on Other Datasets. In order to further verify that CSL-based method is a better baseline model, we have also verified it in other datasets, including the text dataset ICDAR2015, MLT, and another remote sensing dataset HRSC2016. These three datasets are single-class object detection datasets, whose objects have a large aspect ratio. Although boundary conditions still account for a small proportion of these data sets, CSL still shows a stronger performance advantage. As shown in Table 4, the CSL-based method is improved by 1.21%, 0.56%, and 1.29% (1.4%) respectively compared with the regression-based method under the same experimental configuration. These experimental results provide strong support for demonstrating the versatility of the CSL-based method.

Visual Analysis of Angular Features. By zooming in on part of Fig. 6(i), we find that the prediction of the boundary conditions became continuous (for example, two large vehicle in the same direction predicted 90° and −88°, respectively). This phenomenon reflects the purpose of designing the CSL: the labels are periodic (circular) and the prediction of adjacent angles has a certain tolerance. In order to confirm that the angle classifier has indeed learned this property, we visually analyze the angular features of each region of interest (RoI) in the FPN detector by principal component analysis (PCA) [38], as shown in Fig. 7. The detector does not learn the orientation information of well when we use the pulse window function. It can be seen from the first row of Fig. 7 that

Table 5. Detection accuracy on each object (AP) and overall performance (mAP) on DOTA. Note O^2-DNet uses Hourglass104 [29] as backbone.

Method	Backbone	PL	BD	BR	GTF	SV	LV	SH	TC	BC	ST	SBF	RA	HA	SP	HC	mAP
FR-O [39]	ResNet101	79.09	69.12	17.17	63.49	34.20	37.16	36.20	89.19	69.60	58.96	49.4	52.52	46.69	44.80	46.30	52.93
IENet [22]	ResNet101	80.20	64.54	39.82	32.07	49.71	65.01	52.58	81.45	44.66	78.51	46.54	56.73	64.40	64.24	36.75	57.14
R-DFPN [42]	ResNet101	80.92	65.82	33.77	58.94	55.77	50.94	54.78	90.33	66.34	68.66	48.73	51.76	55.10	51.32	35.88	57.94
R^2CNN [12]	ResNet101	80.94	65.67	35.34	67.44	59.92	50.91	55.81	90.67	66.92	72.39	55.06	52.23	55.14	53.35	48.22	60.67
RRPN [27]	ResNet101	88.52	71.20	31.66	59.30	51.85	56.19	57.25	90.81	72.84	67.38	56.69	52.84	53.08	51.94	53.58	61.01
ICN [2]	ResNet101	81.40	74.30	47.70	70.30	64.90	67.80	70.00	90.80	79.10	78.20	53.60	62.90	67.00	64.20	50.20	68.20
RADet [17]	ResNeXt101	79.45	76.99	48.05	65.83	65.46	74.40	68.86	89.70	78.14	74.97	49.92	64.63	66.14	71.58	62.16	69.09
RoI-Transformer [4]	ResNet101	88.64	78.52	43.44	75.92	68.81	73.68	83.59	90.74	77.27	81.46	58.39	53.54	62.83	58.93	47.67	69.56
P-RSDet [47]	ResNet101	89.02	73.65	47.33	72.03	70.58	73.71	72.76	90.82	80.12	81.32	59.45	57.87	60.79	65.21	52.59	69.82
CAD-Net [45]	ResNet101	87.8	82.4	49.4	73.5	71.1	63.5	76.7	90.9	79.2	73.3	48.4	60.9	62.0	67.0	62.2	69.9
O^2-DNet [37]	Hourglass104	89.31	82.14	47.33	61.21	71.32	74.03	78.62	90.76	82.23	81.36	60.93	60.17	58.21	66.98	61.03	71.04
SCRDet [44]	ResNet101	89.98	80.65	52.09	68.36	68.36	60.32	72.41	90.85	**87.94**	**86.86**	65.02	66.68	66.25	68.24	65.21	72.61
SARD [36]	ResNet101	89.93	84.11	54.19	72.04	68.41	61.18	66.00	90.82	87.79	86.59	65.65	64.04	66.68	68.84	68.03	72.95
FADet [16]	ResNet101	90.21	79.58	45.49	76.41	73.18	68.27	79.56	90.83	83.40	84.68	53.40	65.42	74.17	69.69	64.86	73.28
R^3Det [41]	ResNet152	89.49	81.17	50.53	66.10	70.92	78.66	78.21	90.81	85.26	84.23	61.81	63.77	68.16	69.83	67.17	73.74
RSDet [30]	ResNet152	90.1	82.0	53.8	68.5	70.2	**78.7**	73.6	**91.2**	87.1	84.7	64.3	68.2	66.1	69.3	63.7	74.1
Gliding Vertex [40]	ResNet101	89.64	85.00	52.26	**77.34**	73.01	73.14	**86.82**	90.74	79.02	86.81	59.55	**70.91**	72.94	70.86	57.32	75.02
Mask OBB [35]	ResNeXt-101	89.56	**85.95**	54.21	72.90	**76.52**	74.16	85.63	89.85	83.81	86.48	54.89	69.64	73.94	69.06	63.32	75.33
FFA [6]	ResNet101	90.1	82.7	54.2	75.2	71.0	79.9	83.5	90.7	83.9	84.6	61.2	68.0	70.7	**76.0**	63.7	75.7
APE [50]	ResNeXt-101	89.96	83.62	53.42	76.03	74.01	77.16	79.45	90.83	87.15	84.51	67.72	60.33	**74.61**	71.84	65.55	75.75
CSL (FPN based)	ResNet152	**90.25**	85.53	**54.64**	75.31	70.44	73.51	77.62	90.84	86.15	86.69	**69.60**	68.04	73.83	71.10	**68.93**	**76.17**

Table 6. Detection accuracy on HRSC2016 dataset.

Method	R^2CNN [12]	RC1 & RC2 [26]	RRPN [27]	R^2PN [46]	RetinaNet-H [41]	RRD [19]
mAP (07)	73.07	75.7	79.08	79.6	82.89	84.30
Method	RoI-Transformer [4]	RSDet [30]	Gliding Vertex [40]	RetinaNet-R [41]	R^3Det [41]	FPN-CSL-based
mAP (07)	86.20	86.5	88.20	89.18	89.33	**89.62**

the feature distribution of RoI is relatively random, and the prediction results of some angles occupy the vast majority. For the gaussian function, the feature distribution is obvious a ring structures, and the features of adjacent angles are close to each other and have a certain overlap. It is this property that helps CSL-based detectors to eliminate boundary problems and accurately obtain the orientation and scale information of the object.

4.3 Comparison with the State-of-the-Art

Results on DOTA. Although CSL is only a theoretical improvement on the original regression-based rotation detection method, it can still show competitive performance through data augmentation and multi-scale training and test that are widely used. We chose DOTA as the main validation dataset due to the complexity of the remote sensing image scene and the large number of small, cluttered and rotated objects. Our data augmentation methods mainly include random horizontal, vertical flipping, random graying, and random rotation. Training and testing scale set to $[400, 600, 720, 800, 1000, 1100]$. As shown in Table 5, FPN-CSL-based method shows competitive performance, at 76.17%.

Results on HRSC2016. The HRSC2016 contains lots of large aspect ratio ship instances with arbitrary orientation, which poses a huge challenge to the positioning accuracy of the detector. Experimental results show that our model achieves state-of-the-art performances, about 89.62%.

5 Conclusions

We study and summarize the boundary problems on different regression-based rotation detection methods. The main cause of boundary problems based on regression methods is that the ideal predictions are beyond the defined range. Therefore, consider the prediction of the object angle as a classification problem to better limit the prediction results, and then we design a circular smooth label (CSL) to adapt to the periodicity of the angle and increase the tolerance of classification between adjacent angles with little accuracy loss. We also introduce four window functions in CSL and explore the effect of different window radius sizes on detection performance. Extensive experiments and visual analysis on different detectors and datasets show that CSL-based rotation detection algorithm is indeed an effective baseline choice.

Acknowledgment. Corresponding author is Junchi Yan. The work is supported by National Key Research and Development Program of China (2018AAA0100704), National Natural Science Foundation of China (61972250, U19B2035). The author Xue Yang is partly supported by Wu Wen Jun Honorary Doctoral Scholarship, AI Institute, Shanghai Jiao Tong University.

References

1. Abadi, M.: Tensorflow: a system for large-scale machine learning. In: 12th {USENIX} Symposium on Operating Systems Design and Implementation ({OSDI} 16), pp. 265–283 (2016)
2. Azimi, S.M., Vig, E., Bahmanyar, R., Körner, M., Reinartz, P.: Towards multi-class object detection in unconstrained remote sensing imagery. In: Jawahar, C.V., Li, H., Mori, G., Schindler, K. (eds.) ACCV 2018. LNCS, vol. 11363, pp. 150–165. Springer, Cham (2019). https://doi.org/10.1007/978-3-030-20893-6_10
3. Dai, J., Li, Y., He, K., Sun, J.: R-FCN: Object detection via region-based fully convolutional networks. In: Advances in Neural Information Processing Systems, pp. 379–387 (2016)
4. Ding, J., Xue, N., Long, Y., Xia, G.S., Lu, Q.: Learning ROI transformer for oriented object detection in aerial images. In: The IEEE Conference on Computer Vision and Pattern Recognition, pp. 2849–2858 (2019)
5. Duan, K., Bai, S., Xie, L., Qi, H., Huang, Q., Tian, Q.: Centernet: keypoint triplets for object detection. In: Proceedings of the IEEE International Conference on Computer Vision, pp. 6569–6578 (2019)
6. Fu, K., Chang, Z., Zhang, Y., Xu, G., Zhang, K., Sun, X.: Rotation-aware and multi-scale convolutional neural network for object detection in remote sensing images. ISPRS J. Photogram. Remote Sens. **161**, 294–308 (2020)
7. Girshick, R.: Fast R-CNN. In: Proceedings of the IEEE International Conference on Computer Vision, pp. 1440–1448 (2015)

8. Girshick, R., Donahue, J., Darrell, T., Malik, J.: Rich feature hierarchies for accurate object detection and semantic segmentation. In: Proceedings of the IEEE Conference on Computer Vision and Pattern Recognition, pp. 580–587 (2014)
9. He, K., Gkioxari, G., Dollár, P., Girshick, R.: Mask r-cnn. In: Proceedings of the IEEE International Conference on Computer Vision, pp. 2961–2969 (2017)
10. He, K., Zhang, X., Ren, S., Sun, J.: Deep residual learning for image recognition. In: Proceedings of the IEEE Conference on Computer Vision and Pattern Recognition, pp. 770–778 (2016)
11. Huang, C., Ai, H., Li, Y., Lao, S.: High-performance rotation invariant multiview face detection. IEEE Trans. Pattern Anal. Mach. Intell. **29**(4), 671–686 (2007)
12. Jiang, Y., et al.: R2CNN: rotational region CNN for orientation robust scene text detection. arXiv preprint arXiv:1706.09579 (2017)
13. Karatzas, D., et al.: Icdar 2015 competition on robust reading. In: 2015 13th International Conference on Document Analysis and Recognition, pp. 1156–1160. IEEE (2015)
14. Kim, K.R., Choi, W., Koh, Y.J., Jeong, S.G., Kim, C.S.: Instance-level future motion estimation in a single image based on ordinal regression. In: Proceedings of the IEEE International Conference on Computer Vision, pp. 273–282 (2019)
15. Law, H., Deng, J.: Cornernet: detecting objects as paired keypoints. In: Proceedings of the European Conference on Computer Vision, pp. 734–750 (2018)
16. Li, C., Xu, C., Cui, Z., Wang, D., Zhang, T., Yang, J.: Feature-attentioned object detection in remote sensing imagery. In: 2019 IEEE International Conference on Image Processing, pp. 3886–3890. IEEE (2019)
17. Li, Y., Huang, Q., Pei, X., Jiao, L., Shang, R.: Radet: refine feature pyramid network and multi-layer attention network for arbitrary-oriented object detection of remote sensing images. Remote Sens. **12**(3), 389 (2020)
18. Liao, M., Shi, B., Bai, X.: Textboxes++: a single-shot oriented scene text detector. IEEE Trans. Image Process. **27**(8), 3676–3690 (2018)
19. Liao, M., Zhu, Z., Shi, B., Xia, G.S., Bai, X.: Rotation-sensitive regression for oriented scene text detection. In: Proceedings of the IEEE Conference on Computer Vision and Pattern Recognition, pp. 5909–5918 (2018)
20. Lin, T.Y., Dollár, P., Girshick, R., He, K., Hariharan, B., Belongie, S.: Feature pyramid networks for object detection. In: Proceedings of the IEEE Conference on Computer Vision and Pattern Recognition, pp. 2117–2125 (2017)
21. Lin, T.Y., Goyal, P., Girshick, R., He, K., Dollár, P.: Focal loss for dense object detection. In: Proceedings of the IEEE International Conference on Computer Vision, pp. 2980–2988 (2017)
22. Lin, Y., Feng, P., Guan, J.: Ienet: interacting embranchment one stage anchor free detector for orientation aerial object detection. arXiv preprint arXiv:1912.00969 (2019)
23. Liu, W., et al.: SSD: single shot multibox detector. In: Leibe, B., Matas, J., Sebe, N., Welling, M. (eds.) ECCV 2016. LNCS, vol. 9905, pp. 21–37. Springer, Cham (2016). https://doi.org/10.1007/978-3-319-46448-0_2
24. Liu, X., Liang, D., Yan, S., Chen, D., Qiao, Y., Yan, J.: FOTS: fast oriented text spotting with a unified network. In: Proceedings of the IEEE Conference on Computer Vision and Pattern Recognition, pp. 5676–5685 (2018)
25. Liu, Y., Zhang, S., Jin, L., Xie, L., Wu, Y., Wang, Z.: Omnidirectional scene text detection with sequential-free box discretization. arXiv preprint arXiv:1906.02371 (2019)

26. Liu, Z., Yuan, L., Weng, L., Yang, Y.: A high resolution optical satellite image dataset for ship recognition and some new baselines. In: Proceedings of the International Conference on Pattern Recognition Applications and Methods, vol. 2, pp. 324–331 (2017)
27. Ma, J., et al.: Arbitrary-oriented scene text detection via rotation proposals. IEEE Trans. Multimedia **20**(11), 3111–3122 (2018)
28. Nayef, N., et al.: Icdar 2017 robust reading challenge on multi-lingual scene text detection and script identification-RRC-MLT. In: 2017 14th IAPR International Conference on Document Analysis and Recognition, vol. 1, pp. 1454–1459. IEEE (2017)
29. Newell, A., Yang, K., Deng, J.: Stacked hourglass networks for human pose estimation. In: European Conference on Computer Vision, pp. 483–499. Springer (2016)
30. Qian, W., Yang, X., Peng, S., Guo, Y., Yan, J.: Learning modulated loss for rotated object detection. arXiv preprint arXiv:1911.08299 (2019)
31. Redmon, J., Divvala, S., Girshick, R., Farhadi, A.: You only look once: unified, real-time object detection. In: Proceedings of the IEEE Conference on Computer Vision and Pattern Recognition, pp. 779–788 (2016)
32. Ren, S., He, K., Girshick, R., Sun, J.: Faster r-cnn: towards real-time object detection with region proposal networks. In: Advances in Neural Information Processing Systems, pp. 91–99 (2015)
33. Rowley, H.A., Baluja, S., Kanade, T.: Rotation invariant neural network-based face detection. In: Proceedings of 1998 IEEE Computer Society Conference on Computer Vision and Pattern Recognition (Cat. No. 98CB36231), pp. 38–44. IEEE (1998)
34. Shi, X., Shan, S., Kan, M., Wu, S., Chen, X.: Real-time rotation-invariant face detection with progressive calibration networks. In: Proceedings of the IEEE Conference on Computer Vision and Pattern Recognition, pp. 2295–2303 (2018)
35. Wang, J., Ding, J., Guo, H., Cheng, W., Pan, T., Yang, W.: Mask obb: a semantic attention-based mask oriented bounding box representation for multi-category object detection in aerial images. Remote Sens. **11**(24), 2930 (2019)
36. Wang, Y., Zhang, Y., Zhang, Y., Zhao, L., Sun, X., Guo, Z.: Sard: towards scale-aware rotated object detection in aerial imagery. IEEE Access **7**, 173855–173865 (2019)
37. Wei, H., Zhou, L., Zhang, Y., Li, H., Guo, R., Wang, H.: Oriented objects as pairs of middle lines. arXiv preprint arXiv:1912.10694 (2019)
38. Wold, S., Esbensen, K., Geladi, P.: Principal component analysis. Chemom. Intell. Lab. Syst. **2**(1–3), 37–52 (1987)
39. Xia, G.S., et al.: Dota: a large-scale dataset for object detection in aerial images. In: Proceedings of the IEEE Conference on Computer Vision and Pattern Recognition, pp. 3974–3983 (2018)
40. Xu, Y., et al.: Gliding vertex on the horizontal bounding box for multi-oriented object detection. IEEE Trans. Pattern Anal. Mach. Intell. (2020)
41. Yang, X., Liu, Q., Yan, J., Li, A., Zhang, Z., Yu, G.: R3det: refined single-stage detector with feature refinement for rotating object. arXiv preprint arXiv:1908.05612 (2019)
42. Yang, X., et al.: Automatic ship detection in remote sensing images from google earth of complex scenes based on multiscale rotation dense feature pyramid networks. Remote Sens. **10**(1), 132 (2018)
43. Yang, X., Sun, H., Sun, X., Yan, M., Guo, Z., Fu, K.: Position detection and direction prediction for arbitrary-oriented ships via multitask rotation region convolutional neural network. IEEE Access **6**, 50839–50849 (2018)

44. Yang, X., et al.: Scrdet: towards more robust detection for small, cluttered and rotated objects. In: Proceedings of the IEEE International Conference on Computer Vision, pp. 8232–8241 (2019)
45. Zhang, G., Lu, S., Zhang, W.: Cad-net: a context-aware detection network for objects in remote sensing imagery. IEEE Trans. Geosci. Remote Sens. **57**(12), 10015–10024 (2019)
46. Zhang, Z., Guo, W., Zhu, S., Yu, W.: Toward arbitrary-oriented ship detection with rotated region proposal and discrimination networks. IEEE Geosci. Remote Sens. Lett. **15**(11), 1745–1749 (2018)
47. Zhou, L., Wei, H., Li, H., Zhang, Y., Sun, X., Zhao, W.: Objects detection for remote sensing images based on polar coordinates. arXiv preprint arXiv:2001.02988 (2020)
48. Zhou, X., Zhuo, J., Krahenbuhl, P.: Bottom-up object detection by grouping extreme and center points. In: Proceedings of the IEEE Conference on Computer Vision and Pattern Recognition, pp. 850–859 (2019)
49. Zhou, X., et al.: East: an efficient and accurate scene text detector. In: Proceedings of the IEEE Conference on Computer Vision and Pattern Recognition, pp. 5551–5560 (2017)
50. Zhu, Y., Du, J., Wu, X.: Adaptive period embedding for representing oriented objects in aerial images. IEEE Trans. Geosci. Remote Sens. (2020)

Learning Event-Driven Video Deblurring and Interpolation

Songnan Lin[1], Jiawei Zhang[2(✉)], Jinshan Pan[3], Zhe Jiang[2], Dongqing Zou[2], Yongtian Wang[1], Jing Chen[1(✉)], and Jimmy Ren[2]

[1] Beijing Institute of Technology, Beijing, China
chen74jing29@bit.edu.cn
[2] SenseTime Research, Shenzhen, China
zhjw1988@gmail.com
[3] Nanjing University of Science and Technology, Nanjing, China

Abstract. Event-based sensors, which have a response if the change of pixel intensity exceeds a triggering threshold, can capture high-speed motion with microsecond accuracy. Assisted by an event camera, we can generate high frame-rate sharp videos from low frame-rate blurry ones captured by an intensity camera. In this paper, we propose an effective event-driven video deblurring and interpolation algorithm based on deep convolutional neural networks (CNNs). Motivated by the physical model that the residuals between a blurry image and sharp frames are the integrals of events, the proposed network uses events to estimate the residuals for the sharp frame restoration. As the triggering threshold varies spatially, we develop an effective method to estimate dynamic filters to solve this problem. To utilize the temporal information, the sharp frames restored from the previous blurry frame are also considered. The proposed algorithm achieves superior performance against state-of-the-art methods on both synthetic and real datasets.

1 Introduction

Slow-motion analysis of fast-moving objects is crucial for numerous applications but challenging for conventional intensity cameras which only capture low frame-rate blurry videos. To catch the high-speed motion, some recent works, *e.g.* [9,10], attempt to generate a high frame-rate video given a low frame-rate blurry one by deblurring [24,26,28] and interpolation [1,13,17]. Despite their success in certain scenarios, they may fail to deal with severely-blurred videos (see Fig. 1(e)).

S. Lin–This work was done when Songnan Lin was an intern at SenseTime.

Electronic supplementary material The online version of this chapter (https://doi.org/10.1007/978-3-030-58598-3_41) contains supplementary material, which is available to authorized users.

A. Vedaldi et al. (Eds.): ECCV 2020, LNCS 12353, pp. 695–710, 2020.
https://doi.org/10.1007/978-3-030-58598-3_41

Fig. 1. Challenging case for video reconstruction. (a) The input blurry image. (b) The corresponding event data. The color pair (red, blue) represents its polarity $(1, -1)$ throughout this paper. (c) Ground truth. (d) Our reconstruction result. (e) Result of the image-based video construction [9]. (f) Result of the event-based video generation [15]. (g) Result of conventional BHA [18]. (h) Result of deep learning-based LEMD [8]. The proposed method restores high-quality images via an end-to-end network based on the physical event-based video reconstruction model. (Color figure online)

Instead of purely relying on an intensity camera, this work utilizes event-based one with a high temporal resolution to compensate for the lost information in intensity frames. Event cameras [5,12] are biologically-inspired sensors capable of asynchronously encoding the changes of pixel intensity, *i.e.*, events, with microsecond accuracy. Significant efforts [2,3,15,21] have been devoted to directly converting event streams into intensity videos. However, videos reconstructed from these event-dependent solutions tend to lack textures and seem to be non-photorealistic without intensity information (see Fig. 1(f)).

Therefore, it would be desirable to use both advantages of the intensity and event-based sensors for high-speed video generation. Little attention [6,22,23] has been paid to considering both sources of information. However, as they do not take blur into consideration, the generated videos are blurry sometimes. To solve this problem, Pan *et al.* [18] physically model the relationship among a blurry image, events and latent frames and propose an Event-based Double Integral (EDI) model. Therefore, sharp latent images can be obtained given blurry frames and corresponding event streams. After deblurring, other latent video frames are interpolated from the above initial deblurred one by estimating the residuals between them from the events. This method naturally connects intensity images and event data and shows promising results on high frame-rate video generation. However, as the triggering threshold of an event camera varies spatially and temporally with hardware and scene conditions [4,6,20], it is less effective to consider it as constant as in [18], which introduces strong accumulated noises (see Fig. 1(g)). Jiang *et al.* [8] propose to utilize the large capacity of deep convolutional neural networks (CNNs) to refine the estimated frames from [18] and recover finer details. However, as the deblurring and refinement are separately considered, their method fails to make full use of the model capacity of the CNNs, which makes it less effective for high-speed video generation

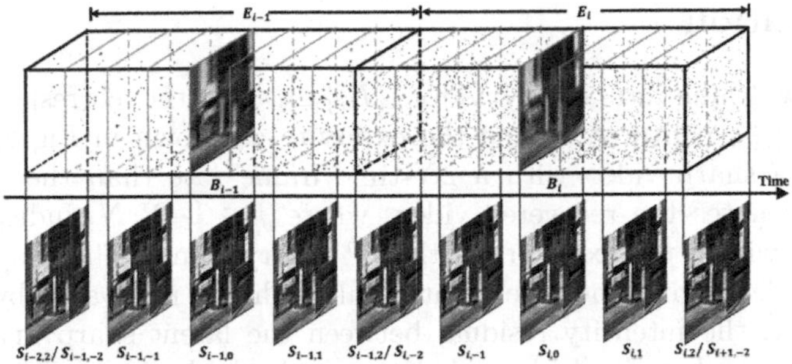

Fig. 2. Sample frames reconstructed from the proposed method. Given a low frame-rate blurry video $\{B_i\}$ and the corresponding event streams $\{E_i\}$ during the exposure time, the proposed method recovers a sharp video $\{S_{i,j}\}_{j\in[-N,N]}$ in a $2N$-time frame rate than the original. $N = 2$ in this example.

(see Fig. 1(h)). Moreover, the algorithms [8,18] above bring one blurry frame alive without exploiting the additional information from previous frames.

In this paper, we propose an effective event-driven video deblurring and interpolation algorithm to generate sharp high frame-rate videos based on deep CNNs and the physical model of event-based video reconstruction. Motivated by [18] which estimates the residual between sharp and blurry images for deblurring as well as that between sharp frames for interpolation, we propose to use a deep CNN to effectively predict them. Moreover, as the triggering threshold is spatially variant, it is inappropriate to use a uniform one as in [18]. Therefore, we propose to use the dynamic filtering layer [7,14,17] to handle this spatially variant threshold. Besides, the proposed network can also help remove the noises from the events when predicting the residuals. To better exploit the additional information across the frames, we further utilize the previously recovered frames together with the previous blurry frame as well as the event stream to estimate current frames, which can enforce the temporal consistency. Our method incorporates the physical properties of event-based video reconstruction compactly and can be trained in an end-to-end manner.

The main contributions of this paper are summarized as follows:

- We propose an end-to-end trainable neural network to generate high-speed videos from the hybrid intensity and event-based sensors. Our algorithm hinges on the physical event-based video reconstruction model with a compact network architecture.
- We propose to use dynamic filtering to handle the events triggered by the spatially variant threshold.
- We quantitatively and qualitatively evaluate our network on both synthetic and real-world videos and show that it performs favorably against state-of-the-art high-speed video generation algorithms.

2 Motivation

Given a low frame-rate blurry video $\{B_i\}_{i\in\mathbb{N}}$ and the corresponding event streams $\{E_i\}_{i\in\mathbb{N}}$ captured during the exposure as shown in Fig. 2, we aim to reconstruct a sharp video with a $2N$-time frame rate than the original. Let $\{S_{i,j}\}_{i,j\in\mathbb{N}}$ denotes the recovered video, where $j \in [-N, N]$ indicates the jth sharp frame within the exposure of the ith blurry frame. The proposed event-based video deblurring and interpolation algorithm is motivated by two observations. First, the intensity residual between the latent sharp images as well as that between sharp and blurry images are both the integral of the events. As a result, we can use the network to estimate accurate integrals from noisy events and then reconstruct high frame-rate sharp videos. Second, even though the intensity residuals can be estimated from the integrals of events, the triggering threshold c_m is spatially and temporally variant. We propose to integrate dynamic filters [7] to handle this spatially variant issue. This section will discuss the above motivations in details.

2.1 Physical Model of Event-based Video Reconstruction

To better motivate our algorithm, we first revisit the physical model of event-based video reconstruction.

Once a log intensity change exceeds a preset threshold c_m, an event e_m[1] is triggered, represented as

$$e_m = (x_m, y_m, t_m, p_m), \tag{1}$$

in which x_m, y_m and t_m denote the spatio-temporal coordinates of the mth event respectively and $p_m \in \{-1, 1\}$ denotes the direction (decrease or increase) of the change. Regardless of quantization, the sum of the events captured in a time interval represents the proportional change in intensity. And thus, given an interval $\Omega_{i,j\to i',j'} = [iT + \frac{j}{2N}T, i'T + \frac{j'}{2N}T]$ of events e_m and a latent sharp frame $S_{i,j}$, we can reconstruct the latent sharp frame $S_{i',j'}$ at pixel (x, y) using:

$$\begin{aligned} S_{i',j'}(x,y) &= S_{i,j}(x,y) \cdot \exp\Big(\sum_{t_m\in\Omega_{i,j\to i',j'}} c_m \cdot p_m \cdot 1(x_m, y_m, x, y)\Big) \\ &= S_{i,j}(x,y) \cdot I_{i,j\to i',j'}(x,y), \end{aligned} \tag{2}$$

in which T denotes the exposure time of blurry frames, $\cdot$ is Hadamard product, the indicator function $1(\cdot)$ equals to 1 if $x_m = x$ & $y_m = y$, and 0 otherwise, and $I_{i,j\to i',j'}$ represents the intensity residual between $S_{i,j}$ and $S_{i',j'}$.

For the blurry image B_i, it can be modeled as the average of discrete latent sharp frames $S_{i,j}$ by:

$$B_i(x,y) = \frac{1}{2N+1} \sum_{j=-N}^{N} S_{i,j}(x,y). \tag{3}$$

[1] When $t_m \in \Omega_{i,-N\to i,N}$, e_m is in the event stream E_i.

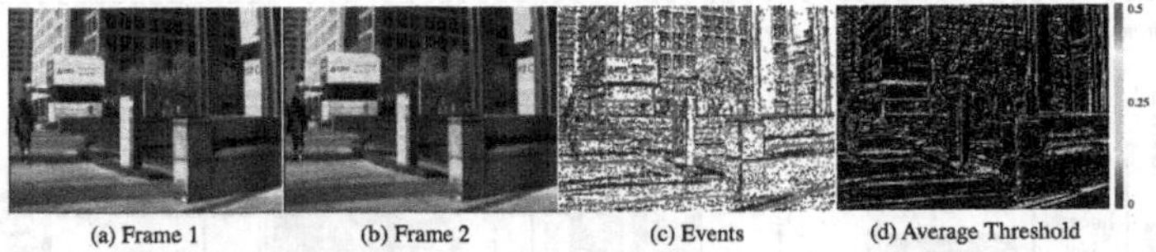

Fig. 3. Demonstration of the spatially variant event triggering threshold. Given two latent sharp frames (a)(b) and their interval of the events (c) captured with an event camera, we estimate the average threshold of each pixel in this interval using Eq. 2. The valid threshold where events occur is spatially variant across the image plane.

Then, we can represent B_i according to Eq. 2 and Eq. 3 as:

$$\begin{aligned} B_i(x,y) &= S_{i,j_0}(x,y) \cdot [\frac{1}{2N+1} \sum_{j=-N}^{N} \exp(\sum_{t_m \in \Omega_{i,j_0 \to i,j}} c_m \cdot p_m \cdot 1(x_m, y_m, x, y))] \\ &= S_{i,j_0}(x,y) \cdot D^{-1}_{i \to i,j_0}(x,y), \end{aligned} \tag{4}$$

where S_{i,j_0} is the key latent sharp frame related to the blurry frame B_i and $D_{i \to i,j_0}$ is the intensity residual between B_i and S_{i,j_0} and it is actually the discrete version of the Event-based Double Integral (EDI) in [18].

Therefore, it is physically possible to first deblur latent keyframe S_{i,j_0} based on Eq. 4, and then interpolate all other video frames $S_{i,j}$ using Eq. 2. Also, Eq. 2 can be used to generate $S_{i,j}$ from the previously estimated latent frames $S_{i-1,j}$. In [18], they estimate the residual I and D directly according to Eq. 2 and Eq. 4 from the events e. However, the estimation is inaccurate since the events contain severe noises. In this paper, we propose to use a deep neural network to predict the residuals by utilizing its strong capacity and flexibility to compensate for the imperfectness of event data.

2.2 Spatially Variant Triggering Threshold

In previous works, *e.g.* [18], they estimate a fixed triggering threshold and apply it to the whole frame sequence. However, this threshold c_m is both spatially and temporally variant according to [4,6,20]. As can be seen in Fig. 3(d), the estimated thresholds given sharp frames and the respective events by Eq. 2 are not uniform. Therefore, it is inappropriate to use a network composed of convolution layers which are spatially invariant to estimate the residual I and D. We propose to integrate dynamic filters [7], which are estimated at every position, to handle this spatially variant issue.

3 Proposed Methods

3.1 Network Architecture

The overall framework of the proposed video deblurring and interpolation algorithm is illustrated in Fig. 4. It consists of four parts:

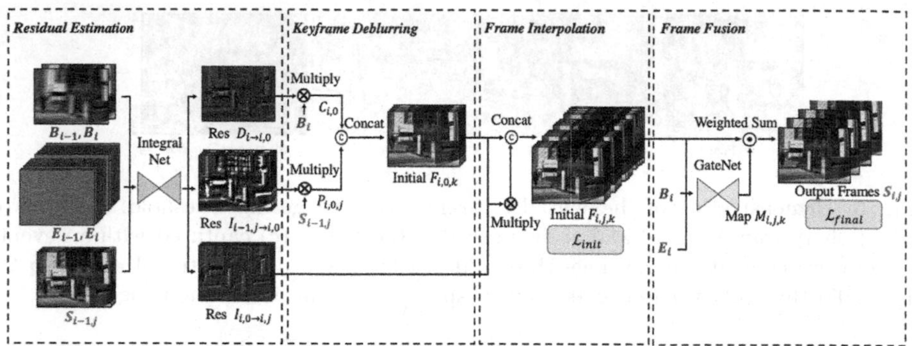

Fig. 4. Overview of our framework. For $2N$-time frame-rate video reconstruction, the previous and current blurry frames B_{i-1}, B_i, their corresponding event streams E_{i-1}, E_i and $2N$ previously recovered sharp frames $S_{i-1,j}$ are fed into an *IntegralNet* to predict the residuals $D_{i\to i,0}$, $I_{i-1,j\to i,0}$ and $I_{i,0\to i,j}$. Given the learned residuals, an initial deblurred keyframe $C_{i,0}$ is estimated from the blurry frame via Eq. 4. Moreover, with $2N$ previously recovered sharp frames, the other initial sharp keyframes $P_{i,0,j}$, where $j \in (-N, N]$, are inferred via Eq. 2. Therefore we obtain $2N+1$ initial keyframes, denoted as $F_{i,0,k}$ by concatenating $C_{i,0}$ and $P_{i,0,j}$. Afterward, the other initial latent sharp frames $F_{i,j,k}$, where $j \in (-N, 0) \cup (0, N]$ and $k \in [0, 2N]$, are interpolated from $F_{i,0,k}$ via Eq. 2. At last, to adaptively select the initial reconstructed frames, *GateNet* is utilized to predict the weights $M_{i,j,k}$ and the final results are obtained by weight summation of the initial results. Please see the manuscript for more details.

- *Residual Estimation:* It aims to estimate the residuals, including $D_{i\to i,0}$ and $I_{i-1,j\to i,0}$ for keyframe deblurring and $I_{i,0\to i,j}$ for video frame interpolation.
- *Keyframe Deblurring:* It utilizes the learned residual $D_{i\to i,0}$ and the blurry frame B_i to estimate a keyframe $C_{i,0}$ via Eq. 4. Then it generates $2N$ keyframes $P_{i,0,j}$ from $I_{i-1,j\to i,0}$ and $2N$ previously recovered sharp frames $S_{i-1,j}$ via Eq. 2, where $j \in (-N, N]$. And there are totally $2N+1$ initial estimated keyframes $F_{i,0,k}$ which are the concatenation of $C_{i,0}$ and $P_{i,0,j}$.
- *Frame Interpolation:* It interpolates the latent sharp frames $F_{i,j,k}$ from every initial deblurred keyframe $F_{i,0,k}$ and $I_{i,0\to i,j}$ according to Eq. 2, where $j \in (-N, 0) \cup (0, N]$.
- *Frame Fusion:* It fuses the $2N+1$ initial sharp frames at (i, j) in an adaptive selection manner and restores the final results $S_{i,j}$ with finer details.

Residual Estimation. We estimate the residuals $D_{i\to i,0}$, $I_{i-1,j\to i,0}$ and $I_{i,0\to i,j}$ in Eq. 4 and Eq. 2 via an *IntegtalNet.* As discussed above, we need to deal with the spatially and temporally variant triggering contrast threshold. However, the convolution is translation invariant across the feature plane, which is less effective to solve this problem. We apply the dynamic filtering [7] whose pixel-wise filters are estimated by the dynamic filter generation module in the proposed network. Moreover, the proposed network can also help remove the noises from the events when predicting the residuals.

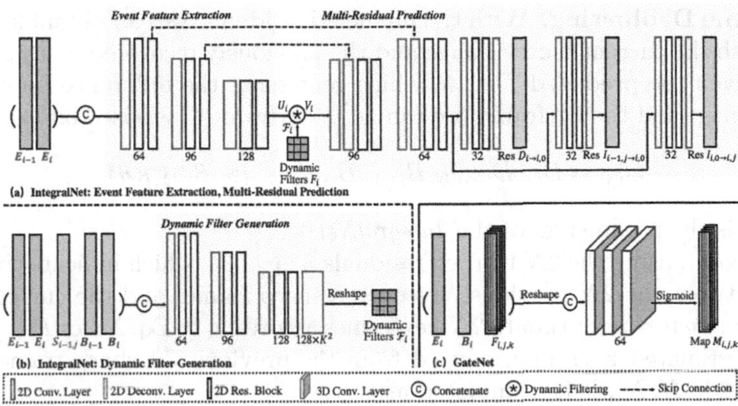

Fig. 5. Structures of the sub-networks. *IntegraltNet* contains the event feature extraction, the dynamic filter generation and the multi-residual prediction. *GateNet* contains three 3D convolution layers. The detailed configurations are provided in the supplemental material.

As shown in Fig. 5(a)(b), the *IntegralNet* is composed of three modules: event feature extraction, dynamic filter generation and multi-residual prediction.

As discussed in Sec. 2.1, the residual D and I are the integral of events. Therefore, the events E_{i-1} and E_i are the only input of the event feature extraction module which extracts features U_i. To feed asynchronous events into the neural network, we divide every event stream E_i into $2N$ equal-time-interval bins. To hold more temporal information, we further divide each bin into M equal-size chunks and stack them as M-channel input images as stated in [25] ($N = 2, M = 2$ in Fig. 2 for example). The stacked event data E_i is passed through three convolution layers followed by two residual blocks. The extracted event features U_i are transformed by dynamic filters in the following process.

We generate different filters for each position in feature maps and perform a spatially variant convolution using the filters. Specifically, for each position (h, w, c) in the extracted feature map $U_i \in \mathbb{R}^{H_U \times W_U \times C_U}$, a specific local filter $\mathcal{F}_i^{(h,w,c)} \in \mathbb{R}^{K \times K \times 1}$ is applied to the region centered around $U_i(h, w, c)$ as

$$V_i(h, w, c) = \mathcal{F}_i^{(h,w,c)} * U_i(h, w, c), \tag{5}$$

where $*$ denotes convolution operation. Filters are dynamically generated given the current and previous blurry frames B_i, B_{i-1}, the corresponding event streams E_i, E_{i-1} and the previously recovered sharp frames $S_{i-1,j}$ by the dynamic filter generation module.

The multi-residual prediction module is used to estimate $D_{i \to i,0}$, $I_{i-1,j \to i,0}$ and $I_{i,0 \to i,j}$ taken the transformed event features V_i. As shown in Fig. 5(a), it first upsamples the features back to the full resolution and then generates the residuals respectively. The skip-connections are also adopted in *IntegralNet*.

Keyframe Deblurring. With the predicted residuals, we can obtain keyframes from both the current blurry image and the previously recovered images. Specifically, given the predicted $D_{i\to i,0}$, which represents the difference between the blurry image and the keyframe, we can get a keyframe $C_{i,0}$ based on Eq. 4 using:

$$C_{i,0} = B_i \cdot D_{i\to i,0}(B_{i-1}, B_i, E_{i-1}, E_i, S_{i-1,j}; \theta), \tag{6}$$

where θ is the parameters of the *IntegralNet.*

Moreover, using the $2N$ learned residuals $I_{i-1,j\to i,0}$, which indicate the differences between the $2N$ previously recovered sharp frames and the current sharp keyframe, we further estimate $2N$ keyframes according to Eq. 2. Let $P_{i,0,j}$ denote the jth estimated keyframe inferred from the previous j^{th} sharp frame $S_{i-1,j}$, where $j \in (-N, N]$, it is formulated as:

$$P_{i,0,j} = S_{i-1,j} \cdot I_{i-1,j\to i,0}(B_{i-1}, B_i, E_{i-1}, E_i, S_{i-1}; \theta). \tag{7}$$

After all, we obtain a keyframe $C_{i,0}$ from the current blurry frame and $2N$ ones $P_{i,0,j}$ from the previously recovered frames. For simplification, the estimated keyframes are concatenated and represented as $F_{i,0,k}$, in which $k \in [0, 2N]$ indicates the index of the keyframe.

Frame Interpolation. Given the $2N+1$ initial deblurred keyframes $F_{i,0,k}$ and the $2N-1$ learned residuals $I_{i,0\to i,j;j\neq 0}$ between the latent keyframe and the interpolated frames, the interpolated frames can be estimated according to Eq. 2:

$$F_{i,j,k;j\neq 0} = F_{i,0,k} \cdot I_{i,0\to i,j;j\neq 0}(B_{i-1}, B_i, E_{i-1}, E_i, S_{i-1}; \theta). \tag{8}$$

Frame Fusion. After frame interpolation, there are $2N+1$ latent images for each frame. To utilize the merits and remove the flaws of all these latent images $F_{i,j,k}$, we conduct the frame fusion module to integrate them by an adaptive selection scheme. We feed them into *GateNet* to generate a soft gate map $M_{i,j,k}$ together with the blurry frame B_i and the corresponding event stream E_i. We first transform the inputs into four dimensions. The initial results $F_{i,j,k}$ are divided into $2N$ chunks by the timestamps j, generating a feature with the size $(2N+1) \times 2N \times H \times W$, in which H and W represent the resolution of the video frame. As for event data E_i, the events in the intervals between two adjacent sharp frames are stacked together as $M \times 2N \times H \times W$. The blurry input is expanded to $2N$ times along a new dimension as $1 \times 2N \times H \times W$. After that, these transformed features are fed into three 3D convolution layers to generate a gate map $M_{i,j,k}$, as shown in Fig. 5(c). Thus, the final reconstructed frames can be estimated by:

$$S_{i,j} = \sum_{k=0}^{2N} F_{i,j,k} \cdot M_{i,j,k}(B_i, E_i, F_{i,j,k}; \mu), \tag{9}$$

where μ represents the parameters of the $GateNet$.

3.2 Loss Function

We consider two loss functions to measure the differences between the reconstructed frames and the ground-truth ones $G_{i,j}$ for both intermediate and final estimations. Specifically, as for the initial recovered frames $F_{i,j,k}$, we constrain $IntegralNet$ using MSE loss:

$$\mathcal{L}_{init}(B_{i-1}, B_i, E_{i-1}, E_i, S_{i-1}, \theta) = \frac{1}{(2N+1)HW} \sum_{k=0}^{2N} \|F_{i,j,k} - G_{i,j}\|^2. \quad (10)$$

The other one is defined between the final results $S_{i,j}$ and the ground-truth ones $G_{i,j}$ to constrain both $IntegralNet$ and $GateNet$:

$$\mathcal{L}_{final}(B_{i-1}, B_i, E_{i-1}, E_i, S_{i-1}, \theta, \mu) = \frac{1}{HW} \|S_{i,j} - G_{i,j}\|^2. \quad (11)$$

The overall loss function is:

$$\mathcal{L} = w_1 \mathcal{L}_{init} + w_2 \mathcal{L}_{final}, \quad (12)$$

where w_1, w_2 are set to 0.01, 1 in our experiment, respectively.

4 Experiment

4.1 Implementation Details

Training Dataset. We train the proposed method on two synthetic datasets: GoPro [16] and the synthetic subset of Blur-DVS [8]. Low frame-rate blurred inputs, high frame-rate sharp videos, and event streams are required during training. GoPro [16], a widely used video deblurring dataset, provides ground-truth sharp videos and we use them to generate blurred frames. We simulate event data by first increasing the video frame rate from 240 fps to 960 fps via a high-quality frame interpolation algorithm [17] and then applying an event simulator ESIM [20] to the videos. To add noise diversity, we set different contrast thresholds for each pixel from a Gaussian distribution $\mathcal{N}(0.18, 0.03)$ similar to [21]. We also use the synthetic subset of Blur-DVS [8] for training, which is captured with slow camera movement in relatively static scenes and thus provides ground-truth sharp videos and event streams. Blurry images are obtained in the same manner as on GoPro. We split the training and testing datasets as suggested.

Experimental Settings. Our network is implemented using Pytorch [19] and trained in an end-to-end manner supervised by Eq. 12 on a GeForce GTX 1080 GPU. For both datasets, we utilize a batch size of 4 training pairs and Adam [11] optimizer with momentum and momentum2 as 0.9 and 0.999. The network is trained for 60 epochs with the learning rate initialized as 0.0001 for the first 10 epochs and then decayed to zero linearly. We set the parameters M and N as 4, 5 on the GoPro dataset, and 3, 3 on the synthetic Blur-DVS. As for initialization, we recover the first blurry frame B_0 of a video sequence by replacing B_{i-1} and $S_{i-1,j}$ in Fig. 4 with B_0. Moreover, we repeatedly input E_0 to substitute E_{i-1}.

Table 1. Video deblurring and reconstruction performance on GoPro [16] dataset, in terms of average PSNR, SSIM and parameter numbers($\times 10^6$) of different networks.

Average results of video deblurring							Average results of video Deblurring and interpolation					
Methods	STFAN [28]	STFAN*	E2V*	BHA [18]	LEMD [8]	Ours	TNTT	TNTT*	E2V*	BHA [18]	LEMD [8]	Ours
PSNR	30.28	38.17	35.38	29.06	31.79	**38.74**	32.47	35.90	34.89	28.49	29.67	**37.99**
SSIM	0.901	0.973	0.959	0.943	0.949	**0.982**	0.936	0.965	0.953	0.920	0.927	**0.981**
Params	5.36	5.38	10.71	-	5.37	**4.80**	10.68	10.88	10.71	-	9.13	**5.00**

* denotes the enhanced version of the corresponding single-sensor algorithm. See text for more details.

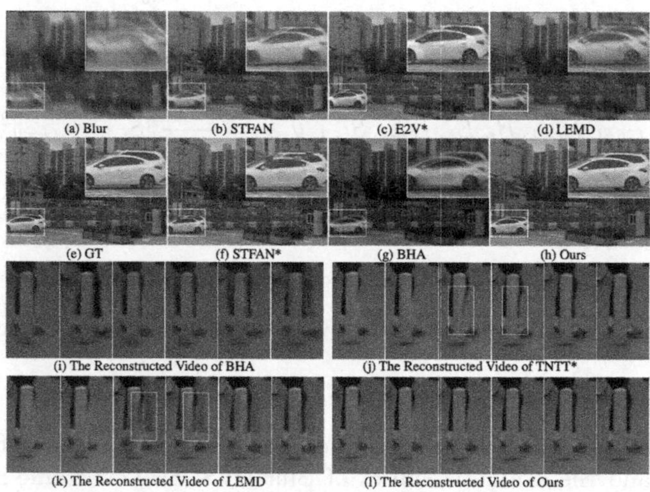

Fig. 6. Visual comparisons on video deblurring (above) and high frame-rate video reconstruction (below) with the state-of-the-art on GoPro [16] datasets. The proposed method generates much clearer frames with fewer noises and artifacts. Zoom in for a better view.

4.2 Experimental Results

We quantitatively and qualitatively evaluate our video deblurring network (*i.e.* recovering videos with the original frame rate) and the simultaneous deblurring and interpolation network on both GoPro and Blur-DVS.

We conduct extensive comparisons with state-of-the-art algorithms including image-based methods on video deblurring [28] and high-speed video generation [9], event-based video generation methods [15,21], conventional video reconstruction methods from hybrid intensity and event-based sensors [18,22] and a deep learning-based method with hybrid sensors [8]. To demonstrate the effectiveness of the proposed framework, we also compare the enhanced versions of the single-sensor algorithms. As for image-based STFAN [28], we feed additional event data into the spatio-temporal filter adaptive network to assist frame alignment and deblurring (denoted as 'STFAN*'). TNTT [9] inputs events and blurry images for both keyframe deblurring network and frame interpolation network (denoted as 'TNTT*'). We also feed events together with intensity

Table 2. Video deblurring and reconstruction performance on the synthetic subset of Blur-DVS [8], in terms of average PSNR, SSIM.

Average results of video deblurring									
Methods	E2V [21]	E2V*	STFAN [28]	STFAN*	MRL [15]	CIE [22]	BHA [18]	LEMD [8]	Ours
PSNR	16.89	24.81	19.03	30.18	10.59	19.02	22.43	26.48	**30.57**
SSIM	0.597	0.790	0.518	0.897	0.195	0.478	0.715	0.839	**0.904**
Average results of video deblurring and interpolation									
Methods	E2V [21]	E2V*	TNTT [9]	TNTT*	MRL [15]	CIE [22]	BHA [18]	LEMD [8]	Ours
PSNR	16.60	24.10	19.05	29.02	10.57	18.94	22.06	25.33	**29.65**
SSIM	0.587	0.777	0.521	0.875	0.194	0.473	0.699	0.827	**0.890**

* denotes the enhanced version of the corresponding single-sensor algorithm. See text for more details.

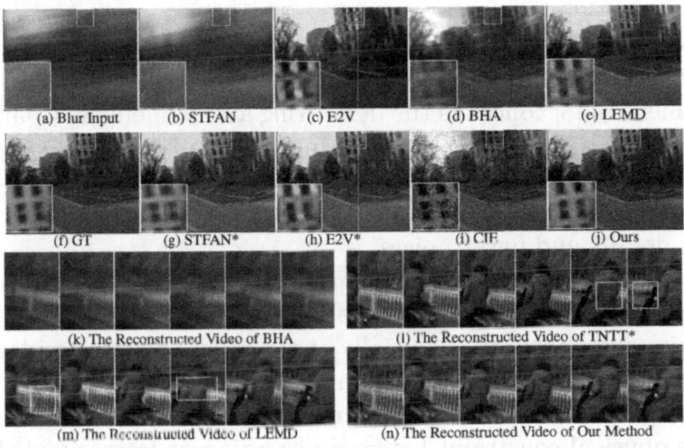

Fig. 7. Visual comparisons on video deblurring (above) and high frame-rate video reconstruction (below) on the synthetic subset of Blur-DVS [8]. The proposed method generates much sharper results with fewer noises and artifacts. More results are provided in our supplementary material. Zoom in for a better view.

frames into the event-based E2V [21] for each of its recurrent reconstruction step (denoted as 'E2V*').

We evaluate PSNR and SSIM on the video deblurring task on two synthetic datasets in Table 1 and Table 2. The proposed network performs favorably against state-of-the-art methods. Figure 6 and Fig. 7 show some examples in the testing sets. The image-based method [28] purely relies on intensity images, thus it is less effective on severely-blurred videos. As event data encodes dense temporal information, it facilitates STFAN* to capture motion information across the frames and makes it more effective on video deblurring. E2V [21], which purely relies on event data, restores images with wrong contrast. However, its enhanced version E2V* keeps the correct contrast with the assistance of intensity frames. These significant improvements demonstrate the inherent advantage of each sensor and the effectiveness of utilizing both advantages for video deblurring. As for existing intensity and event-based algorithms, CIE [22] and BHA [18]

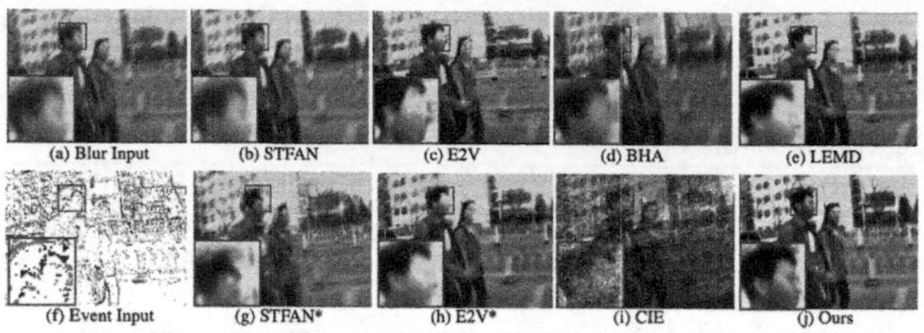

Fig. 8. Visual comparisons with the state-of-the-art on real-word blurry videos. The recovered results of the proposed method have fewer noises and more details. More results are provided in our supplementary material. Zoom in for a better view.

adopt simplified physical models without considering the blur or the non-uniform threshold, which leads to blurry results and introduces accumulated noises. The CNN-based method [8] conducts the deblurring and refinement separately, which makes the approach sensitive to deblurring and leads to limited performance. On the contrary, the proposed method hinges on the physical event-based video reconstruction model via an end-to-end architecture. The restored video frames present finer details and fewer noises.

We also report the results on simultaneous video deblurring and interpolation in Table 1, Table 2, Fig. 6 and Fig. 7. The conventional method BHA [18] is prone to noises during interpolation, especially at the object edges. Besides, its threshold choosing scheme is not robust, which introduces unaddressed blur when estimating a wrong threshold. The deep learning-based LEMD and TNTT* neglect to utilize the physical constraint between two adjacent frames and thus interpolate blurry frames with undesirable artifacts, especially at occlusion. Figure 6(l) and Fig. 7(n) show that the proposed method can restore sharp and artifact-free frames.

To validate the generalization capacity of the proposed method, we qualitatively compare the proposed network with other algorithms on real-world blurry videos in the real subset of Blur-DVS [8]. As shown in Fig. 8, our method restores more visually pleasing frames than the state-of-the-art.

5 Ablation Study

We have shown that the proposed algorithm performs favorably against state-of-the-art methods. In this section, we further analyze the effectiveness of each component in video deblurring and interpolation.

5.1 Effectiveness of Physical-Based Framework

The proposed algorithm is designed based on the physical model of the event-based video reconstruction. We predict the residuals I and D and apply multiply operation to them according to Eq. 2 and Eq. 4. To demonstrate the effectiveness

Table 3. Ablation Study. 'Addition' replaces the multiplication with addition to verify the effectiveness of the physical-based network. 'w/o DF', 'w/o Pre' and 'w/o ASF' represent removing the dynamic filtering, the previous information in keyframe estimation and the adaptively-selected fusion. Our method achieves the highest quantitative results, which demonstrates the effectiveness of each component. See text for details.

Methods	Addition	w/o DF	w/o Pre	w/o ASF	Ours
PSNR	29.24	28.64	29.42	29.34	**29.65**
SSIM	0.882	0.872	0.855	0.885	**0.890**

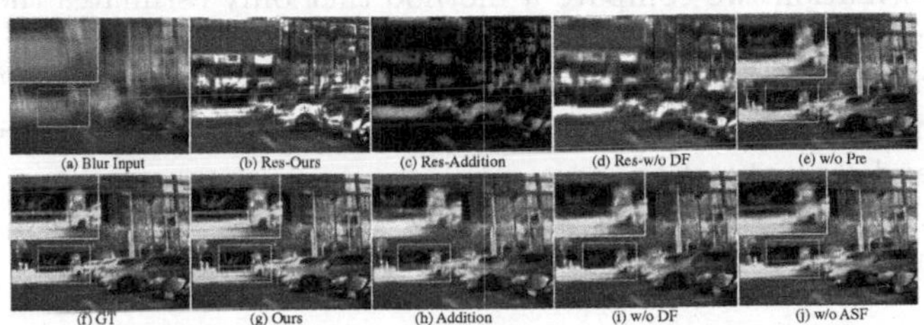

Fig. 9. Ablation Study. 'Res-' in (b)(c)(d) denotes the learned residual between the keyframe and the interpolated frame. 'Addition' replaces the multiplication with addition to verify the effectiveness of the physical-based framework. 'w/o DF', 'w/o Pre' and 'w/o ASF' represent removing the dynamic filtering, the previous information in keyframe estimation step and the adaptively-selected fusion. The proposed method restores clearer images with more details and fewer artifacts. See text for details.

of the physical-based framework, we compare the method that adds the residuals and the intensity images up (denoted as 'Addition'), as already used in pure image-based algorithms [9,27,28]. The results in Table 3 show that using multiplication achieves higher performance than 'Addition'. As shown in Fig. 9(c), 'Addition' predicts a blurry addition residual and thus generates a smooth result but with more artifacts (Fig. 9(h)). However, as the proposed method is based on the physical model, which makes it easy to calculate the multiplication residuals (Fig. 9(b)) from event data, it is robust to severely-blurred frames and restores images with more details and fewer artifacts (Fig. 9(g)).

5.2 Effectiveness of Dynamic Filtering

To handle the events triggered by the spatially variant threshold, we propose to integrate the dynamic filters when estimating residuals. To validate the above discussions, we remove the dynamic filter generation module and feed its inputs $(B_{i-1}, B_i, E_{i-1}, E_i, S_{i-1})$ into the event feature extraction directly for a fair comparison (denoted as 'w/o DF'). Table 3 shows that 'w/o DF' is less effective. Due to the lack of compensation for the spatially variant triggering threshold, it provides an overly-smooth residual (Fig. 9(d)) compared to ours (Fig. 9(b)). And thus, it cannot restore the lost details in the final results (Fig. 9(i)), which demonstrates that using dynamic filtering facilitates to minimize the effects of the non-uniform threshold. Besides, generated filters are illustrated in the supplementary materials for visual interpretation.

5.3 Effectiveness of Previous Information

We note that the existing event-based video deblurring and interpolation algorithms [8,18] bring one blurry frame alive without considering additional information that exists across adjacent frames. To verify the effectiveness of utilizing previous information, we compare a method that only estimates the keyframes $C_{i,0}$ from current blurry inputs without the ones $P_{i,0,j}$ from the previously recovered frames (denoted as 'w/o Pre'). The final results shown in Table 3 and Fig. 9(e) indicate that involving previous information is more effective for video deblurring and reconstruction.

5.4 Effectiveness of Frame Fusion

To integrate the $2N+1$ initial recovered results $F_{i,j,k}$ in an adaptive selection manner, the proposed frame fusion step utilizes the information from the blurry frame, event data and the initial results to generate a gate map and then obtains the final results by weighted summation. To demonstrate the effectiveness of this design, we compare the method that removes the estimation of the gate map but feeds the initial results into three 3D convolution layers to estimate the final results directly (denoted as 'w/o ASF'). The final results in Table 3 and Fig. 9(j) indicate that the proposed frame fusion module can integrate the initial results in an adaptive selection scheme and keep more details, which is more effective for video deblurring and interpolation.

6 Concluding Remarks

In this paper, we propose to learn event-driven video frame deblurring and interpolation to solve high frame-rate video generation. The whole framework hinges on the physical model of the event-based video reconstruction, which estimates the residual between the latent sharp frames as well as that between sharp and blurry frames, and integrates the model into a compact architecture. Benefiting from this design, the proposed method can generate physically-correct results and handle severely-blurred videos. Furthermore, we show that using dynamic filters when predicting residuals can deal with event data triggered by the spatially variant threshold. By training the proposed network in an end-to-end manner, the proposed algorithm is able to reconstruct high-quality and high frame-rate videos. Experiments on the synthetic datasets and real images demonstrate that the proposed method achieves superior performance against the existing image and event-based approaches.

We note that one limitation of the proposed method is that the network need be retrained if we aim to further increase the frame rate. However, we can solve it by applying an additional interpolation network recursively between pairs of restored sharp frames. Further research will be devoted to arbitrary frame-rate video reconstruction.

Acknowledgments. This project was supported by the 863 Program of China (No. 2013AA013802), NSFC (Nos. 61872421, 61922043) and NSF of Jiangsu Province (No. BK20180471).

References

1. Bao, W., Lai, W., Zhang, X., Gao, Z., Yang, M.: Memc-net: motion estimation and motion compensation driven neural network for video interpolation and enhancement. IEEE Trans. Pattern Anal. Mach. Intell. (2019)
2. Bardow, P., Davison, A.J., Leutenegger, S.: Simultaneous optical flow and intensity estimation from an event camera. In: Proceedings of the IEEE Conference on Computer Vision and Pattern Recognition, pp. 884–892 (2016)
3. Barua, S., Yoshitaka, M., Ashok, V.: Direct face detection and video reconstruction from event cameras. In: 2016 IEEE winter conference on applications of computer vision (WACV), pp. 1–9. IEEE (2016)
4. Brandli, C.: Event-Based Machine Vision. Ph.D. thesis, ETH Zurich (2015)
5. Brandli, C., Berner, R., Yang, M., Liu, S.C., Delbruck, T.: A 240$\times$ 180 130 db 3 μs latency global shutter spatiotemporal vision sensor. IEEE J. Solid-State Circuits **49**(10), 2333–2341 (2014)
6. Brandli, C., Muller, L., Delbruck, T.: Real-time, high-speed video decompression using a frame-and event-based davis sensor. In: 2014 IEEE International Symposium on Circuits and Systems (ISCAS), pp. 686–689. IEEE (2014)
7. Jia, X., De Brabandere, B., Tuytelaars, T., Gool, L.V.: Dynamic filter networks. In: Advances in neural information processing systems, pp. 667–675 (2016)
8. Jiang, Z., Zhang, Y., Zou, D., Ren, J., Lv, J., Liu, Y.: Learning event-based motion deblurring. In: Proceedings of the IEEE/CVF Conference on Computer Vision and Pattern Recognition, pp. 3320–3329 (2020)
9. Jin, M., Hu, Z., Favaro, P.: Learning to extract flawless slow motion from blurry videos. In: Proceedings of the IEEE Conference on Computer Vision and Pattern Recognition, pp. 8112–8121 (2019)
10. Jin, M., Meishvili, G., Favaro, P.: Learning to extract a video sequence from a single motion-blurred image. In: Proceedings of the IEEE Conference on Computer Vision and Pattern Recognition, pp. 6334–6342 (2018)
11. Kingma, D.P., Ba, J.: Adam: a method for stochastic optimization. arXiv preprint arXiv:1412.6980 (2014)
12. Lichtsteiner, P., Christoph, P., Tobi, D.: A 128$\times$128 120 db 15 μs latency asynchronous temporal contrast vision sensor. IEEE J. Solid-State Circ. **43**(2), 566–576 (2008)
13. Meyer, S., Djelouah, A., McWilliams, B., Sorkine-Hornung, A., Gross, M., Schroers, C.: Phasenet for video frame interpolation. In: Proceedings of the IEEE Conference on Computer Vision and Pattern Recognition, pp. 498–507 (2018)
14. Mildenhall, B., Barron, J.T., Chen, J., Sharlet, D., Ng, R., Carroll, R.: Burst denoising with kernel prediction networks. In: Proceedings of the IEEE Conference on Computer Vision and Pattern Recognition, pp. 2502–2510 (2018)

15. Munda, G., Reinbacher, C., Pock, T.: Real-time intensity-image reconstruction for event cameras using manifold regularisation. arXiv preprint arXiv:1607.06283 (2018)
16. Nah, S., Hyun Kim, T., Mu Lee, K.: Deep multi-scale convolutional neural network for dynamic scene deblurring. In: Proceedings of the IEEE Conference on Computer Vision and Pattern Recognition, pp. 3883–3891 (2017)
17. Niklaus, S., Mai, L., Liu, F.: Video frame interpolation via adaptive separable convolution. In: Proceedings of the IEEE International Conference on Computer Vision, pp. 261–270 (2017)
18. Pan, L., Scheerlinck, C., Yu, X., Hartley, R., Liu, M., Dai, Y.: Bringing a blurry frame alive at high frame-rate with an event camera. In: Proceedings of the IEEE Conference on Computer Vision and Pattern Recognition, pp. 6820–6829 (2019)
19. Paszke, A., et al.: Automatic differentiation in pytorch (2017)
20. Rebecq, H., Gehrig, D., Scaramuzza, D.: ESIM: an open event camera simulator. In: Conference on Robot Learning, pp. 969–982 (2018)
21. Rebecq, H., Ranftl, R., Koltun, V., Scaramuzza, D.: Events-to-video: bringing modern computer vision to event cameras. In: Proceedings of the IEEE Conference on Computer Vision and Pattern Recognition, pp. 3857–3866 (2019)
22. Scheerlinck, C., Barnes, N., Mahony, R.: Continuous-time intensity estimation using event cameras. In: Jawahar, C.V., Li, H., Mori, G., Schindler, K. (eds.) ACCV 2018. LNCS, vol. 11365, pp. 308–324. Springer, Cham (2019). https://doi.org/10.1007/978-3-030-20873-8_20
23. Shedligeri, P., Mitra, K.: Photorealistic image reconstruction from hybrid intensity and event-based sensor. J. Electron. Imaging **28**(6), 063012 (2019)
24. Su, S., Delbracio, M., Wang, J., Sapiro, G., Heidrich, W., Wang, O.: Deep video deblurring for hand-held cameras. In: Proceedings of the IEEE Conference on Computer Vision and Pattern Recognition, pp. 1279–1288 (2017)
25. Wang, L., et al.: Event-based high dynamic range image and very high frame rate video generation using conditional generative adversarial networks. In: Proceedings of the IEEE Conference on Computer Vision and Pattern Recognition, pp. 10081–10090 (2019)
26. Wang, X., Chan, K.C., Yu, K., Dong, C., Change Loy, C.: Edvr: video restoration with enhanced deformable convolutional networks. In: Proceedings of the IEEE Conference on Computer Vision and Pattern Recognition Workshops (2019)
27. Zhang, H., Dai, Y., Li, H., Koniusz, P.: Deep stacked hierarchical multi-patch network for image deblurring. In: Proceedings of the IEEE Conference on Computer Vision and Pattern Recognition, pp. 5978–5986 (2019)
28. Zhou, S., Zhang, J., Pan, J., Xie, H., Zuo, W., Ren, J.: Spatio-temporal filter adaptive network for video deblurring. In: Proceedings of the IEEE International Conference on Computer Vision, pp. 2482–2491 (2019)

Vectorizing World Buildings: Planar Graph Reconstruction by Primitive Detection and Relationship Inference

Nelson Nauata(✉) and Yasutaka Furukawa(✉)

Simon Fraser University, Burnaby, Canada
{nnauata,furukawa}@sfu.ca

Abstract. This paper tackles a 2D architecture vectorization problem, whose task is to infer an outdoor building architecture as a 2D planar graph from a single RGB image. We provide a new benchmark with ground-truth annotations for 2,001 complex buildings across the cities of Atlanta, Paris, and Las Vegas. We also propose a novel algorithm utilizing 1) convolutional neural networks (CNNs) that detects geometric primitives and infers their relationships and 2) an integer programming (IP) that assembles the information into a 2D planar graph. While being a trivial task for human vision, the inference of a graph structure with an arbitrary topology is still an open problem for computer vision. Qualitative and quantitative evaluations demonstrate that our algorithm makes significant improvements over the current state-of-the-art, towards an intelligent system at the level of human perception. We will share code and data.

Keywords: Vectorization · Remote sensing · Deep learning · Planar graph

1 Introduction

Human vision has a stunning perceptual capability in inferring geometric structure from raster imagery. What is remarkable is the holistic nature of our geometry perception. Imagine a task of inferring a building structure as a 2D graph from a satellite image (See Fig. 1). We learn structural patterns from examples quickly, utilize the learned patterns to augment the reconstruction process from incomplete data.

Computer Vision is still at its infancy in holistic reasoning of geometric structure. For low-level geometric primitives such as corners [20] or junctions [16], Convolutional Neural Networks (CNNs) have been an effective detector. Unfortunately, the task of high-level geometry reasoning, for example, the construction

Electronic supplementary material The online version of this chapter (https://doi.org/10.1007/978-3-030-58598-3_42) contains supplementary material, which is available to authorized users.

A. Vedaldi et al. (Eds.): ECCV 2020, LNCS 12353, pp. 711–726, 2020.
https://doi.org/10.1007/978-3-030-58598-3_42

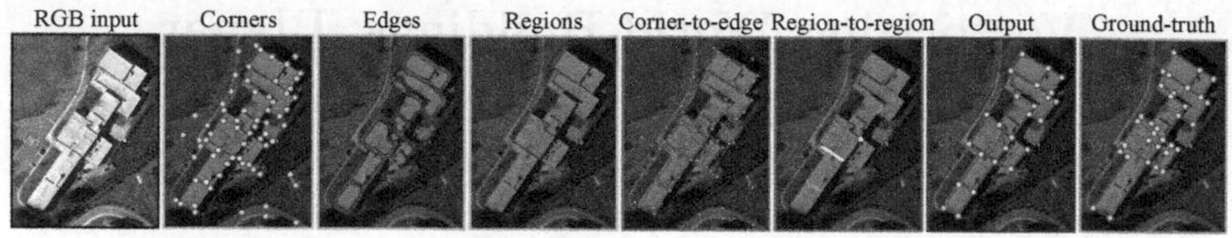

Fig. 1. The paper takes a RGB image, detects three geometric primitives (i.e., corners, edges, and regions), infers their relationships (i.e., corner-to-edge and region-to-region), and fuses the information via Integer Programming to reconstruct a planar graph.

of CAD-quality geometry, is often only possible by the hands of expert modelers. CAD-level building reconstruction at a city-scale would enable richer architectural modeling and analysis across the globe, opening doors for broad applications in digital mapping, architectural study, or urban visualization/planning.

In an effort towards more holistic structured reconstruction techniques, this paper proposes a new 2D outdoor architecture vectorization problem, whose task is to reconstruct a 2D planar graph of outdoor building architecture from a single RGB image. While building segmentation from a satellite image has been a popular problem [1], their task is to extract only the external boundary as a 1D polygonal loop. Our problem seeks to reconstruct a planar graph of an arbitrary topology, including internal building feature lines that separate roof components and yield more high-fidelity building models (See Fig. 1). The inference of graph topology is the challenge in our problem, which is exacerbated by the fact that buildings on satellite images do not follow the Manhattan geometry due to the foreshortening effects through perspective projection.

Our approach combines CNNs and integer programming (IP) to tackle the challenge. CNNs extract low- to mid-level topology information, in particular, detecting three types of geometric primitives (i.e., corners, edges, and regions) and inferring two types of pairwise primitive relationships (i.e., corner-to-edge and region-to-region relationships). IP consolidates all the information and reconstructs a planar graph.

We downloaded high-resolution satellite images from SpaceNet [9] corpus and annotated 2,001 complex buildings across the cities of Atlanta, Paris and Las Vegas as 2D polygonal graphs including internal and external architectural feature lines. Our qualitative and quantitative evaluations demonstrate significant improvements in our approach over the competing methods.

In summary, the contribution of the paper is two-fold: 1) A new outdoor architecture reconstruction problem as a 2D planar graph with a benchmark; 2) A hybrid algorithm combining primitive detectors, their relationship inference, and IP, which makes significant improvements over the existing state-of-the-art. We will share our code and data to promote further research.

2 Related Work

Architectural reconstruction has a long history in Computer Vision. We first review building footprint extraction methods then focus our description on vector-graphics reconstruction techniques.

Building Footprint Extraction: In the SpaceNet Building Footprint Extraction challenge [1], a ground-truth building is represented as a set of pixels, ignoring the underlying vector graphics building structure. The winning method by Hamaguchi *et al.* [13] utilizes a multi-task U-Net for segmenting roads and buildings of different sizes, producing a binary building segmentation mask. Cheng *et al.* [8] utilizes CNNs for defining energy maps and optimizing polygon-based contours for building footprints. Acuna and Ling *et al.* [2] formulates the footprint extraction as the boundary tracing problem, finding a sequence of vertices forming a polygonal loop. Their method is designed for general object segmentation and tends to over-estimate vertices. All these methods extract the building footprint (i.e., external boundary) and ignores internal architectural feature lines.

Low-Level Reconstruction: Harris corner detection [26], Canny edge detection [3], and LSD line segment extractor [27] are popular traditional methods for low-level geometry detection. Recently, deep neural network (DNN) based approaches have been an active area of research [4], being also effective for junction detection [16], by classifying the combination of incident edge directions.

Mid-Level Reconstruction: Room layout estimation infers a graph of architectural feature lines from a single image, where nodes are room corners and edges are wall boundaries. Most approaches assume that the room shape is a 3D box, then solves an optimization problem with hand-engineered cost functions [6,15,17,25]. For a room beyond a box shape, Markov Random Field (MRF) infers detailed architectural structures [12] and Dynamic Programming (DP) searches for an optimal room shape [10,11], again via hand-engineered cost functions.

High-Level Reconstruction (Knowledge): Given a prior knowledge about the overall geometric structure, corner detection alone suffices to reconstruct a complex graph structure. Human pose estimation is one of the most successful examples, where DNN is trained to detect human junctions with body types such as heads, right arms, and left legs [5]. Their connections come from a prior knowledge (e.g. a right hand is connected to a right arm).

High-Level Reconstruction (Optimization): A classical approach for CAD-quality 3D reconstruction is to inject domain knowledge as ad-hoc cost functions or processes in the optimization formulation [18]. The emergence of deep learning enabled robust solutions for low-level primitive detection. However, mid to high level geometric reasoning still relies on hand-crafted optimization [19,20].

Floor-SP [7] is the closest to our work, utilizing CNN-based corner, edge, and region detection with a sophisticated optimization technique to reconstruct floorplans. However, their method suffers from two limitations. First, Floor-SP does not allow any mistake in the region detection.[1] Second, Floor-SP requires principal directions and mostly Manhattan scenes, which is hardly true in our problem due to severe foreshortening effects. Our approach handles non-Manhattan scenes and utilizes region detection as soft-constraints.

[1] Rooms are regions in their problem and can be detected easily. Our regions are roof segments and much less distinguishable.

Fig. 2. Sample input RGB images and their corresponding planar graph annotations.

High-Level Reconstruction (Shape-Grammar): A shape grammar defines rules of procedural shape generation [24]. Procedural reconstruction exploits the shape grammar in constraining the reconstruction process. Rectified building facade parsing is a good example, where heuristics and hand-engineered cost functions control the process [21,22]. More recently, DNNs learn to drive the procedural reconstruction for building facades [23] or top-down residential houses [29]. However, these shape grammars are too restrictive and do not scale to more complex large buildings in our problem.

3 2D Architecture Vectorization Problem

This paper proposes a new building vectorization problem, where a building is to be reconstructed as a 2D planar graph from a single RGB image. We retrieved high-resolution satellite images from the SpaceNet [9] corpus, which are hosted as an Amazon Web Services Public Dataset [1].

We annotated 2D planar graphs for 1010, 670, and 321 buildings from the cities of Atlanta, Paris, and Las Vegas, respectively. The average and the standard deviation of the number of corners, edges, and regions are 12.56/8.23, 14.15/9.53 and 2.8/2.19, respectively. Roughly 60% of the buildings have either 1 or 2 regions. 30% have 3 to 10 regions. The remaining 10% have more than 10 regions. We randomly chose 1601 training and 400 testing samples. We refer to the supplementary document for the complete distribution of samples per the number of corners, edges, or regions. Note a region is a space bounded by the edges, which is well-defined in our planar graphs. When multiple satellite images cover the same city region, we chose the one with the least off-Nadir angle to minimize the foreshortening effects.

For each building instance, we crop a tight axis-aligned bounding-box with 24 pixels margin, and paste to the center of a 256×256 image patch. The white

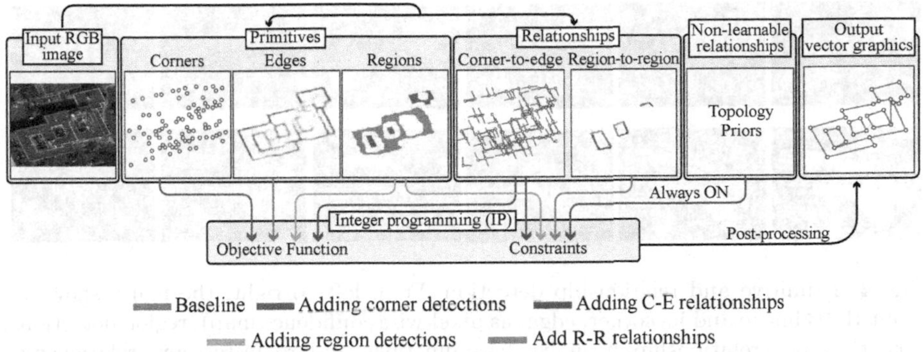

Fig. 3. System overview. Our pipeline detects geometric primitives, infers their relationships, and fuses all the information into integer programming to reconstruct a planar graph.

color is padded at the background. We apply uniform shrinking if the bounding box is larger than 256 × 256. Figure 2 shows sample building annotations. We borrow the metrics introduced for the floorplan reconstruction [7] (except for room++ metric), measuring the precision, recall, and f1-scores for the corner, edge, and region primitives.[2]

4 Algorithm

Our architecture vectorization algorithm consists of three modules: CNN-based primitive detection, CNN-based primitive relationship inference, and IP optimization (See Fig. 3). We now explain these three modules.

4.1 Primitive Detection

We follow Floor-SP [7] and obtain corner candidates, an edge confidence image, and region candidates by standard CNN architecture (See Fig. 4), in particular, Fully Convolutional Network (FCN) for corners [16], Dilated Residual Networks (DRN) [28] for edges, and Mask-RCNN [14] for regions. Corner detections are thresholded at 0.2, where the confidence scores will also be used in the optimization. Edge information is estimated as a pixel-wise confidence score without thresholding. Every pair of corner candidates is considered to be an edge candidate. Region detections are thresholded at 0.5. We refer the full architectural specification to the supplementary document.

[2] In short, a corner is declared to be correct if there exists a ground-truth corner within a certain distance. An edge is declared to be correct if both corners are declared to be correct. A region is declared to be correct if there exists a ground-truth region with more than 0.7 IOU. Our only change is to tighten the distance tolerance on the corner detection from 10 pixels to 8 pixels.

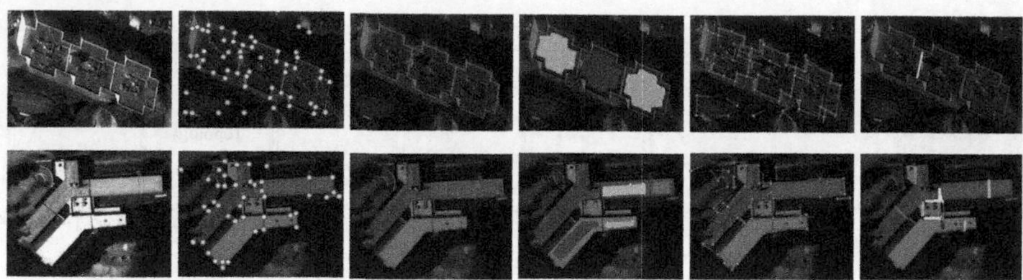

Fig. 4. Primitive and relationship detection. From left to right, the figure shows an input RGB image and its corner, edge (as pixel-wise confidence map), region detections, corner-to-edge relationships visualized as junctions and room-to-room relationships visualized as common boundaries.

4.2 Primitive Relationship Classification

We classify two types of pairwise primitive relationships by CNNs (See Fig. 4).

Corner-to-Edge Relationships: For every pair of a corner and an incident edge, we compute the confidence score by utilizing the junction-type inference technique by Huang *et al.* [16] without any changes. In short, we discretize 360 degrees around each detected corner into 15 angular bins, and add a module at the end of the corner detection head to estimate the presence of an edge in each bin. A corner to edge confidence score is simply set to the edge presence score in the corresponding bin. This score will be used by the objective function and the corner-to-edge relationship constraints in the optimization.

Region-to-Region Relationships: Given a RGB image and a pair of regions, we use Mask R-CNN [14] to find their common boundary as a pixel-wise segmentation mask. More precisely, we represent the input as a 5-channel image, where the two regions are represented as binary masks. The output is a set of common edges of the two regions, each of which is represented as a segmentation instance. When 2 regions do not have a shared boundary, Mask R-CNN should not output any segments. Detected segments are thresholded at 0.5, which are often reliable and the confidence scores will not be used in the next optimization. The common boundaries will be used by the region-to-region relationship constraints in the optimization.

4.3 Geometric Primitive Assembly via IP

Integer Programming (IP) fuses detected primitives and their relationship information into a planar graph, where the inspiration of our formulation comes from the floorplan vectorization works by Liu et al. [19].

Objective Function: Indicator variables are defined for each primitive: (1) I_{cor} for a corner $c \in \mathcal{C}$; (2) I_{edg} for an edge $e \in \mathcal{E}$; and (3) I_{reg} for a region $r \in \mathcal{R}$. After the optimization, we collect the set of primitives whose indicator variables are 1 as a building reconstruction. We also have an indicator variable I_{dir} for a corner to an incident edge direction relationship. The variable becomes 1, if a corner has an incident edge along the direction (with binning).

The objective function consists of the three terms:

$$\max_{\{I_{cor},I_{edg},I_{reg},I_{dir}\}} \sum_{e\in\mathcal{E}} \underbrace{(e_{conf}c'_{conf}c''_{conf} - 0.5^3)I_{edg}(e)}_{\text{corner and edge primitives}} + 0.1\sum_{c\in\mathcal{C}}\sum_{\theta\in\mathcal{D}_c} \underbrace{(\theta_{conf}c_{conf} - 0.5^2)I_{dir}(\theta,c)}_{\text{corner-to-edge relationship}} + \sum_{r\in\mathcal{R}} \underbrace{I_{reg}(r)}_{\text{region primitive}} . \quad (1)$$

c_{conf} and e_{conf} denotes the confidence scores for the corner and the edge detections, respectively. θ_{conf} denotes the corner-to-edge relationship confidence. With abuse of notation, c' and c'' denotes the end-points of an edge e.

The first objective term states that if an edge and its two end-points have high confidence scores (i.e., their product is at least 0.5^3), there is an incentive to select that edge. The second term suggests to select a corner and its incident edge direction if their confidence scores are high. The third objective term suggests to select as many regions as possible.

The maximization of the function is subject to four constraints, which are intuitive but require complex mathematical formulations. We here focus on explaining the ideas, while referring the details to the supplementary material. Note that we describe constraints as hard constraints, but turn them into soft constraints via slack variables before solving the problem for robustness. Lastly, after reconstructing a graph with IP, we apply a simple post-processing and eliminate a corner when it has two incident edges that are colinear with an error tolerance of 5°.

Topology Prior Constraints: There are domain-specific constraints. First, if an edge is active, its two end-points must also be active. Second, no dangling edges are allowed, and every corner must have at least two incident edges. Third, no two edges can intersect, which ensures the planarity of the reconstructed graphs.

Region Primitive Constraints: Suppose a region is selected. All the edges that intersect with the region should be off. Similarly, the region must be surrounded by edges. We take a point at the region boundary and cast a ray outwards the region, collecting all edges that intersect the ray and enforce that at least one edge must be on. We sample points at every 2 pixels around the region boundary.

Region-to-Region Relationship Constraints: This constraint is similar in spirit to the region primitive constraint but is more powerful. Suppose a pair of regions have a common boundary prediction as a segmentation mask. The constraint enforces that at least one edge is selected nearby the mask. Precisely, we fit a line segment to the boundary segment and consider an orthogonal line

Table 1. Quantitative evaluations. P_C, P_E, and P_R denote corner, edge, and region primitive information, respectively. R_{CE} and R_{RR} denote corner-to-edge and region-to-region relationship information, respectively. The cyan, orange, and magenta indicate the top 3 results.

Model	Corner			Edge			Region		
	Prec.	Recall	F1	Prec.	Recall	F1	Prec.	Recall	F1
PolyRNN++ [2]	49.6	43.7	46.4	19.5	15.2	17.1	39.8	13.7	20.4
PPGNet [30]	78.0	69.2	73.3	55.1	50.6	52.8	32.4	30.8	31.6
Hamaguchi *et al.* [13]	58.3	57.8	58.0	25.4	22.3	23.8	51.0	36.7	42.7
L-CNN [31]	66.7	86.2	75.2	51.0	71.2	59.4	25.9	41.5	31.9
Floor-SP [7]	55.0	51.4	53.1	29.0	26.9	27.9	39.0	32.5	35.5
Ours (P_E)	75.0	41.5	53.4	52.4	15.6	24.1	66.7	0.5	1.0
Ours ($P_E + P_C$)	85.3	57.9	69.0	66.8	29.8	41.2	81.6	6.9	12.6
Ours ($P_E + P_C + R_{CE}$)	81.3	66.1	72.9	62.5	38.8	47.9	71.7	15.6	25.6
Ours ($P_E + P_C + R_{CE} + P_R$)	91.7	61.6	73.7	68.0	44.2	53.6	71.8	46.6	56.5
Ours ($P_E + P_C + R_{CE} + P_R + R_{RR}$)	91.1	64.6	75.6	68.1	48.0	56.3	70.9	53.1	60.8

segment (16 pixels in length) at the center. We collect all the edge primitives that intersect with this line segment and enforce that one edge will be chosen.

Corner-to-Edge Relationship Constraints: If the incident indicator is on, the corresponding corner must be on, and one of the edges in the corresponding directional bin must be on. If two edges are incident to the same corner and within 5 degrees in angular distance, both edges cannot be on at the same time. If corner-to-edge compatibility score from the relationship inference is below 0.2, we do not allow any edges in that direction bin to be on.

5 Experimental Results

We have implemented the proposed DNNs in PyTorch and the IP optimization in Python with Gurobi (a quadratic integer programming solver). We have used a workstation with Intel Xeon processors (2.2 GHz) and NVIDIA GTX 1080 GPU with 11 GB of RAM. The training usually takes 2 days for the primitive detectors and relationship classifiers. At test time, more than 80% of the buildings have at most 20 edges as shown in the edge histogram from supplementary document (Fig. 1). For these cases, our method runs quickly (< 5 min). However, our method slows down significantly for large buildings with 50 or 60 edges, where the running time reaches nearly an hour.

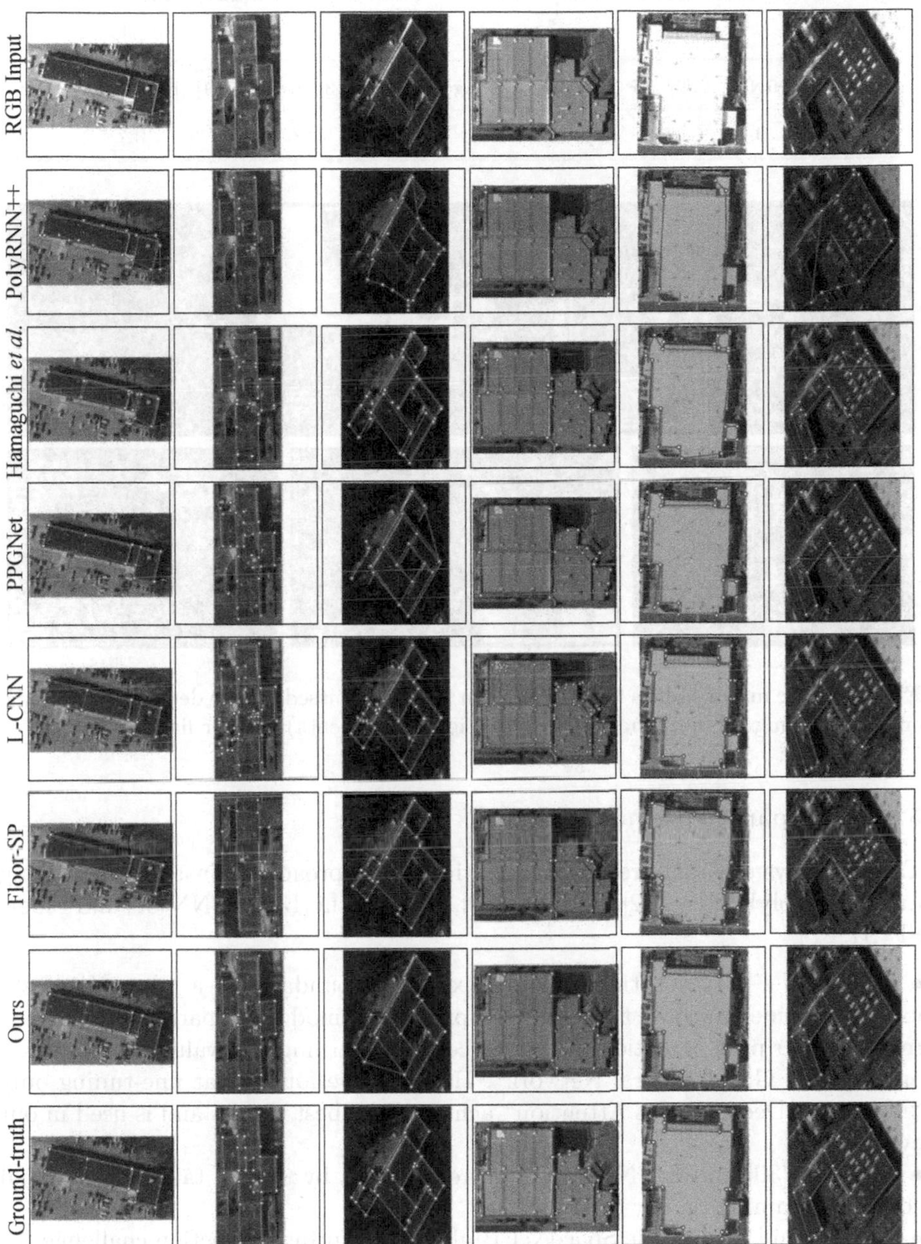

Fig. 5. Comparisons against five competing methods. The top row is the input image. The last row is the ground-truth. The supplementary document contains results for the entire testing set.

Table 2. Planarity evaluation. $\mathcal{G}_P$ indicates the proportion of predicted planar graphs for the entire test set. The cyan, orange, and magenta indicate the top 3 results.

Model	PolyRNN++ [2]	PPGNet [30]	Hamaguchi *et al.* [13]	L-CNN [31]	Floor-SP [7]	Ours
$\mathcal{G}_P$	0.17	0.69	0.96	0.58	0.70	1.00

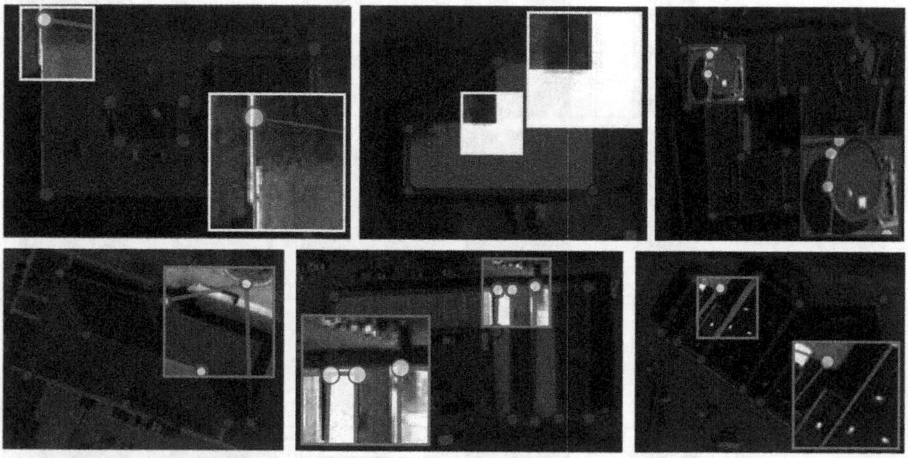

Fig. 6. Three major failure modes from our method: missed corner detections (yellow), curved buildings (blue), and weak image signals (magenta). (Color figure online)

5.1 Comparative Evaluations

Table 1 shows our main results, comparing our approach against five competing methods: PolyRNN++ [2], PPGNet [30], Hamaguchi [13], L-CNN [31] and Floor-SP [7].

• PolyRNN++ traces the building external boundary in a recurrent fashion [2]. We fine-tuned all their released pretrained models, in particular, "Recurrent Decoder plus Attention", "Reinforcement Learning", "Evaluator Network", and "Gated Graph Neural Network". However, we found that fine-tuning only "Recurrent Decoder plus Attention" achieved the best results and is used in our evaluation.
• PPGNet [30] and L-CNN [31] were reproduced by simply taking the official code and training on our data.
• Hamaguchi [13] won the SpaceNet Building Footprint Extraction challenge [1]. The authors graciously trained their model and produced results using our data. Since their method produces pixel-wise binary masks of building footprints [13], which performs poorly in our metrics, we utilized the OpenCV implementation of the Ramer-Douglas-Peucker algorithm with a threshold of 10 to simplify the boundary curve.

• Floor-SP [7] is a state-of-the-art floorplan reconstruction system. Their algorithm is sensitive to the principal direction extraction (PDE), which becomes challenging against severe foreshortening effects in our problem. We tried to improve their PDE implementation without much success and used their default code, which extracts a mixture of 4 Manhattan frames (8 directions).

All the models were trained and tested on the same split. In the table, the last row (our system with all the features) has the best f1-scores for the corner and the region metrics, and the second best f1-score for the edge metric. Overall, our model makes steady improvements over the competing methods when more features are added, especially on the region metric, which is the most challenging and consistent with the visual reconstruction quality.

PPGNet and L-CNN achieve compelling f1-scores for the corners and the edges. L-CNN even outperforms our method for the edge f1-score. However, this metric is not a good indicator as illustrated in Figs. 5 and 7. The figures show that the reconstruction results by PPGNet and L-CNN "look" reasonable at first sight. However, close examinations reveal that their results suffer from thin triangles, self-intersecting edges (i.e., the graph is not actually planar), and colinear edges. Their limitation comes from the fact that they infer edges independently. Their region metrics are far behind ours, which requires more holistic structure reasoning.

Floor-SP, on the other hand, performs poorly on all the metrics. There are two reasons. First, their shortest path algorithm at the core requires accurate principal directions, whose extraction is difficult without the Manhattan constraints in our problem. Second, they assume region detections to be 100% correct and cannot recover from region detection mistakes. As a result, the method often generates too many extraneous corners or completely miss parts of the graphs.

In Table 2, we present the proportion of planar graphs predicted by different methods for the entire test set. The closest method to ours is Hamaguchi *et al.*, however, it tends to output disconnected components for the building regions. L-CNN, PPGNet and Floor-SP are outperformed by a large margin, as these methods are more prone to contain self intersections.

5.2 Ablation Study

The bottom half of Table 1 and Fig. 8 verify the contributions of various components in our system: the three primitive detections and two relationship inference. We use symbols $\mathrm{P}_C, \mathrm{P}_E$, and P_R to denote if the corner, edge, and region primitive detections are used by our system. Similarly, R_{CE} and R_{RR} denote if the corner-to-edge and region-to-region relationships are used by our system.

Edge Detections Only ($\mathbf{P}_E$): Our first baseline utilizes only the edge detection results, that is, seeking to maximize $(\sum_{e\in\mathcal{E}}(c^e - 0.5)I_{edg}(e))$. In short, this baseline accepts all the edges whose score are above 0.5.

Adding Corner Detections ($\mathbf{P}_E, \mathbf{P}_C$): The second baseline ($\mathrm{P}_E$ + P_C) adds the corner detection results, seeking to maximize $\sum_{e\in\mathcal{E}}(e_{conf}c'_{conf}c''_{conf} -$

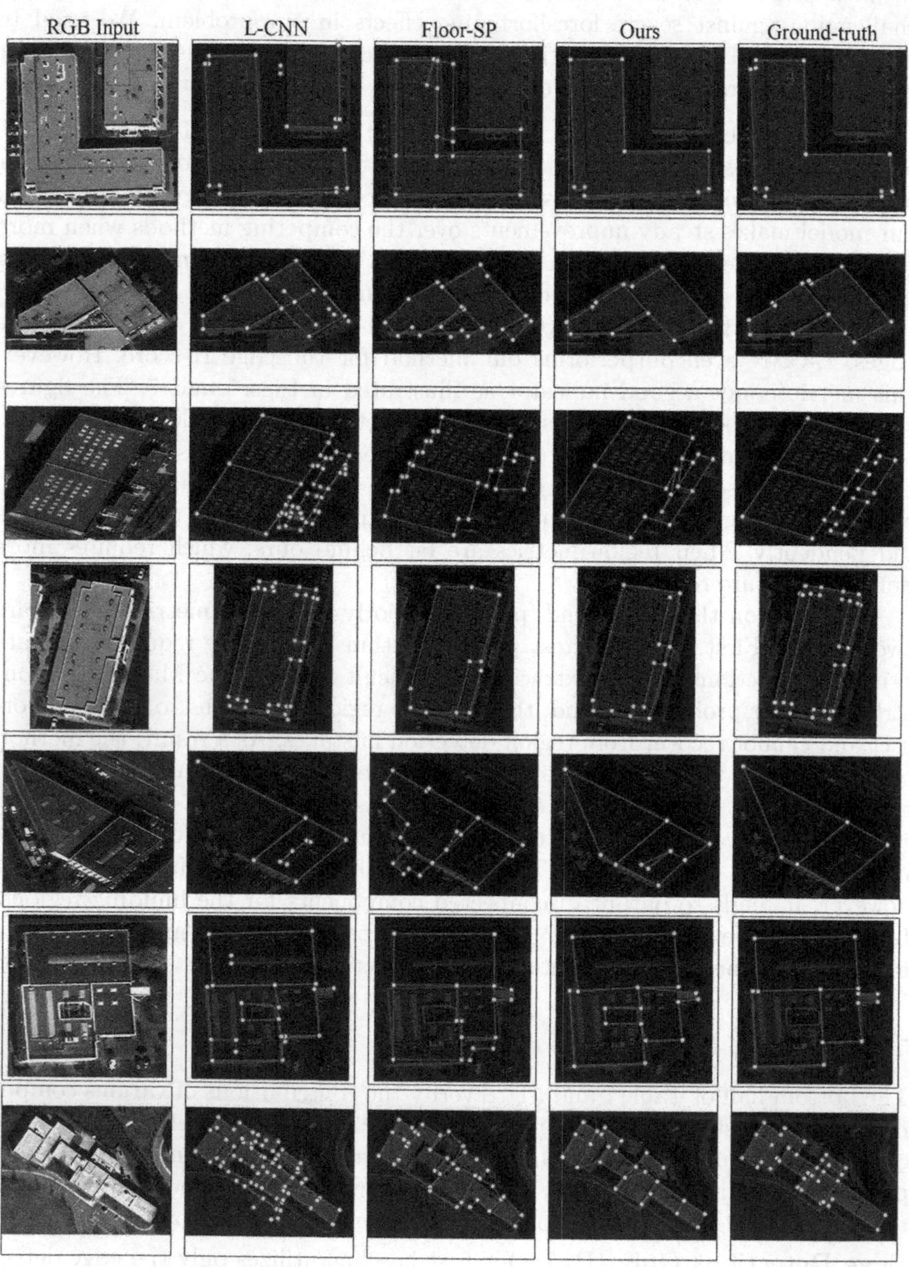

Fig. 7. More detailed comparisons against L-CNN [31] and Floor-SP [7].

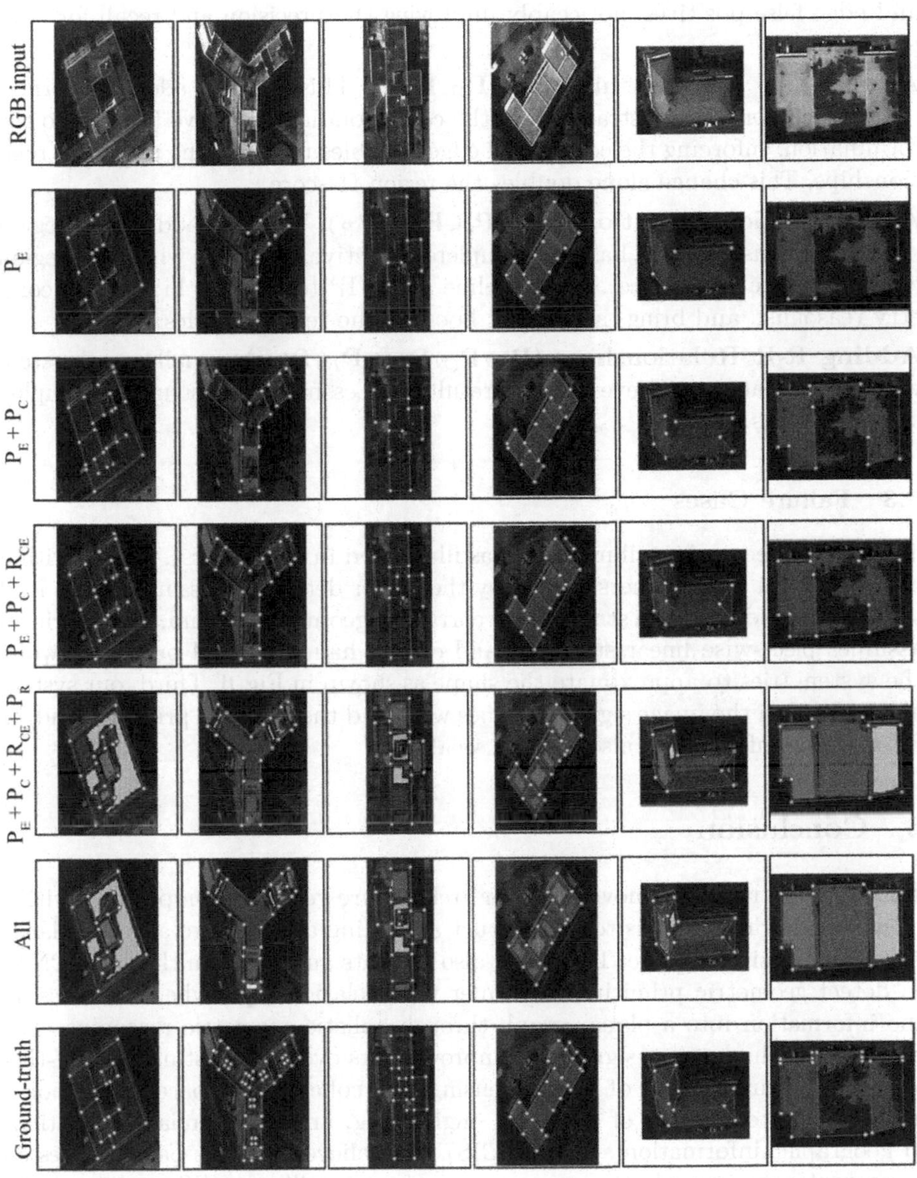

Fig. 8. Qualitative evaluation for ablation study. Figure shows results for input RGB image displayed in the first row for ablation experiments presented in Sect. 5.2 (same order) for multiple target buildings (columns) followed by ground-truth in the last row.

$0.5^3)I_{edg}(e)$. This baseline accepts all the edges, whose scores based on the corner and the edge detection are above 0.5^3. This effectively suppresses the corner and edge false positives, noticeably improving the precision and recall for both primitives.

Adding C-E Relationships ($\mathbf{P}_E, \mathbf{P}_C, \mathbf{R}_{CE}$): This baseline adds the corner-to-edge relationship constraints and the corresponding objective in Eq. 1 to the formulation, enforcing the corner and edge variables to follow the predicted relationships. This change alone doubles the region f1-score.

Adding Region Detections ($\mathbf{P}_E, \mathbf{P}_C, \mathbf{R}_{CE}, \mathbf{P}_R$): With the addition of region detections, this baseline has the complete objective function, while the region constraints are also added. This baseline allows IP to conduct high-level geometry reasoning, and brings significant boost to the region metrics.

Adding R-R Relationships ($\mathbf{P}_E, \mathbf{P}_C, \mathbf{R}_{CE}, \mathbf{P}_R, \mathbf{R}_{RR}$): Finally, our system with all the features achieve the best results, successfully reconstructing complex large buildings.

5.3 Failure Cases

There are three major failure modes as illustrated in Fig. 6. First, our algorithm cannot recover from corners missed by the corner detector. Missing corners lead to missing incident graph structure or corrupted geometry. Second, our algorithm assumes piece-wise linear structure and cannot handle curved buildings, while the system tries to approximate the shape as shown in Fig. 6. Third, our system also fails when the image signal becomes weak and the detected primitive and/or relationship information also become weak.

6 Conclusion

This paper introduces a novel outdoor architecture vectorization problem with a benchmark, whose task is to reconstruct a building architecture as a 2D planar graph from a single image. The paper also presents an algorithm that uses CNNs to detect geometric primitives and infer their relationships, where IP fuses all the information into a planar graph through holistic geometric reasoning. The proposed method makes significant improvements over the existing state-of-the-art. The growing volume of remote sensing data collected by space and airborne assets facilitates myriad of scientific, engineering, and commercial applications in geographic information systems (GIS). We believe that this paper makes an important step towards the construction of an intelligent GIS system at the level of human perception. We will share our code and data to promote further research.

Acknowledgement. This research is partially supported by NSERC Discovery Grants, NSERC Discovery Grants Accelerator Supplements, and DND/NSERC Discovery Grant Supplement. This research is also supported by the Intelligence Advanced

Research Projects Activity (IARPA) via Department of Interior/Interior Business Center (DOI/IBC) contract number D17PC00288. The U.S. Government is authorized to reproduce and distribute reprints for Governmental purposes notwithstanding any copyright annotation thereon. The views and conclusions contained herein are those of the authors and should not be interpreted as necessarily representing the official policies or endorsements, either expressed or implied, of IARPA, DOI/IBC, or the U.S. Government.

References

1. SpaceNet on Amazon Web Services (AWS). "Datasets." The SpaceNet Catalog. Last modified April 30, 2018. https://spacenetchallenge.github.io/datasets/datasetHomePage.html Accessed 19 Oct 2018
2. Acuna, D., Ling, H., Kar, A., Fidler, S.: Efficient interactive annotation of segmentation datasets with polygon-rnn++. In: Proceedings of the IEEE conference on Computer Vision and Pattern Recognition, pp. 859–868 (2018)
3. Canny, J.: A computational approach to edge detection. IEEE Trans. Pattern Anal. Mach. Intell. **6**, 679–698 (1986)
4. Cao, Z., Hidalgo, G., Simon, T., Wei, S.E., Sheikh, Y.: Openpose: realtime multi-person 2d pose estimation using part affinity fields. arXiv preprint arXiv:1812.08008 (2018)
5. Cao, Z., Simon, T., Wei, S.E., Sheikh, Y.: Realtime multi-person 2D pose estimation using part affinity fields. In: Proceedings of the IEEE Conference on Computer Vision and Pattern Recognition, pp. 7291–7299 (2017)
6. Chao, Y.-W., Choi, W., Pantofaru, C., Savarese, S.: Layout estimation of highly cluttered indoor scenes using geometric and semantic cues. In: Petrosino, A. (ed.) ICIAP 2013. LNCS, vol. 8157, pp. 489–499. Springer, Heidelberg (2013). https://doi.org/10.1007/978-3-642-41184-7_50
7. Chen, J., Liu, C., Wu, J., Furukawa, Y.: Floor-sp: inverse cad for floorplans by sequential room-wise shortest path. In: Proceedings of the IEEE International Conference on Computer Vision, pp. 2661–2670 (2019)
8. Cheng, D., Liao, R., Fidler, S., Urtasun, R.: Darnet: deep active ray network for building segmentation. In: Proceedings of the IEEE Conference on Computer Vision and Pattern Recognition, pp. 7431–7439 (2019)
9. Etten, A.V., Lindenbaum, D., Bacastow, T.M.: Spacenet: A remote sensing dataset and challenge series. arXiv preprint arXiv:1807.01232 (2018)
10. Flint, A., Mei, C., Murray, D., Reid, I.: A dynamic programming approach to reconstructing building interiors. In: Daniilidis, K., Maragos, P., Paragios, N. (eds.) ECCV 2010. LNCS, vol. 6315, pp. 394–407. Springer, Heidelberg (2010). https://doi.org/10.1007/978-3-642-15555-0_29
11. Flint, A., Murray, D., Reid, I.: Manhattan scene understanding using monocular, stereo, and 3D features. In: 2011 International Conference on Computer Vision, pp. 2228–2235. IEEE (2011)
12. Furukawa, Y., Curless, B., Seitz, S.M., Szeliski, R.: Manhattan-world stereo. In: 2009 IEEE Conference on Computer Vision and Pattern Recognition, pp. 1422–1429. IEEE (2009)
13. Hamaguchi, R., Hikosaka, S.: Building detection from satellite imagery using ensemble of size-specific detectors. In: 2018 IEEE/CVF Conference on Computer Vision and Pattern Recognition Workshops (CVPRW), pp. 223–2234. IEEE (2018)

14. He, K., Gkioxari, G., Dollár, P., Girshick, R.: Mask r-cnn. In: Computer Vision (ICCV), 2017 IEEE International Conference on, pp. 2980–2988. IEEE (2017)
15. Hedau, V., Hoiem, D., Forsyth, D.: Recovering the spatial layout of cluttered rooms. In: Computer vision, 2009 IEEE 12th international conference on, pp. 1849–1856. IEEE (2009)
16. Huang, K., Wang, Y., Zhou, Z., Ding, T., Gao, S., Ma, Y.: Learning to parse wireframes in images of man-made environments. In: Proceedings of the IEEE Conference on Computer Vision and Pattern Recognition, pp. 626–635 (2018)
17. Lee, C.Y., Badrinarayanan, V., Malisiewicz, T., Rabinovich, A.: Roomnet: end-to-end room layout estimation. arXiv preprint arXiv:1703.06241 (2017)
18. Lin, H., et al.: Semantic decomposition and reconstruction of residential scenes from lidar data. ACM Trans. Graph. (TOG) **32**(4), 66 (2013)
19. Liu, C., Wu, J., Kohli, P., Furukawa, Y.: Raster-to-vector: revisiting floorplan transformation. In: Proceedings of the IEEE International Conference on Computer Vision, pp. 2195–2203 (2017)
20. Liu, C., Wu, J., Furukawa, Y.: Floornet: a unified framework for floorplan reconstruction from 3D scans. In: Proceedings of the European Conference on Computer Vision (ECCV), pp. 201–217 (2018)
21. Liu, H., Zhang, J., Zhu, J., Hoi, S.: Deepfacade: a deep learning approach to facade parsing. pp. 2301–2307 (2017) https://doi.org/10.24963/ijcai.2017/320
22. Martinović, A., Mathias, M., Weissenberg, J., Van Gool, L.: A three-layered approach to facade parsing. In: Fitzgibbon, A., Lazebnik, S., Perona, P., Sato, Y., Schmid, C. (eds.) ECCV 2012. LNCS, vol. 7578, pp. 416–429. Springer, Heidelberg (2012). https://doi.org/10.1007/978-3-642-33786-4_31
23. Nishida, G., Bousseau, A., Aliaga, D.G.: Procedural modeling of a building from a single image. Comput. Graph. Forum **37**, 415–429 (2018)
24. Parish, Y.I., Müller, P.: Procedural modeling of cities. In: Proceedings of the 28th annual conference on Computer graphics and interactive techniques, pp. 301–308. ACM (2001)
25. Schwing, A.G., Hazan, T., Pollefeys, M., Urtasun, R.: Efficient structured prediction for 3D indoor scene understanding. In: Computer Vision and Pattern Recognition (CVPR), 2012 IEEE Conference on, pp. 2815–2822. IEEE (2012)
26. Szeliski, R.: Computer Vision: Algorithms and Applications. Springer Science & Business Media, Springer, London (2010). https://doi.org/10.1007/978-1-84882-935-0
27. Von Gioi, R.G., Jakubowicz, J., Morel, J.M., Randall, G.: Lsd: a line segment detector. Image Process. Line **2**, 35–55 (2012)
28. Yu, F., Koltun, V., Funkhouser, T.A.: Dilated residual networks. In: 2017 IEEE Conference on Computer Vision and Pattern Recognition (CVPR), pp. 636–644 (2017)
29. Zeng, H., Wu, J., Furukawa, Y.: Neural procedural reconstruction for residential buildings. In: The European Conference on Computer Vision (ECCV), pp. 737–753 (2018)
30. Zhang, Z., et al.: Ppgnet: learning point-pair graph for line segment detection. arXiv preprint arXiv:1905.03415 (2019)
31. Zhou, Y., Qi, H., Ma, Y.: End-to-end wireframe parsing. arXiv preprint arXiv:1905.03246 (2019)

Learning to Combine: Knowledge Aggregation for Multi-source Domain Adaptation

Hang Wang, Minghao Xu, Bingbing Ni(✉), and Wenjun Zhang

Shanghai Jiao Tong University, Shanghai 200240, China
{wang--hang,xuminghao118,nibingbing,zhangwenjun}@sjtu.edu.cn

Abstract. Transferring knowledges learned from multiple source domains to target domain is a more practical and challenging task than conventional single-source domain adaptation. Furthermore, the increase of modalities brings more difficulty in aligning feature distributions among multiple domains. To mitigate these problems, we propose a Learning to Combine for Multi-Source Domain Adaptation (LtC-MSDA) framework via exploring interactions among domains. In the nutshell, a knowledge graph is constructed on the prototypes of various domains to realize the information propagation among semantically adjacent representations. On such basis, a graph model is learned to predict query samples under the guidance of correlated prototypes. In addition, we design a Relation Alignment Loss (RAL) to facilitate the consistency of categories' relational interdependency and the compactness of features, which boosts features' intra-class invariance and inter-class separability. Comprehensive results on public benchmark datasets demonstrate that our approach outperforms existing methods with a remarkable margin. Our code is available at https://github.com/ChrisAllenMing/LtC-MSDA.

Keywords: Multi-Source Domain Adaptation · Learning to Combine · Knowledge graph · Relation Alignment Loss

1 Introduction

Deep Neural Network (DNN) is expert at learning discriminative representations under the support of massive labeled data, and it has achieved incredible successes in many computer-vision-related tasks, *e.g.* object classification [11,17], object detection [23,34] and semantic segmentation [3,10]. However,

H. Wang and M. Xu—Equal contribution.

Electronic supplementary material The online version of this chapter (https://doi.org/10.1007/978-3-030-58598-3_43) contains supplementary material, which is available to authorized users.

A. Vedaldi et al. (Eds.): ECCV 2020, LNCS 12353, pp. 727–744, 2020.
https://doi.org/10.1007/978-3-030-58598-3_43

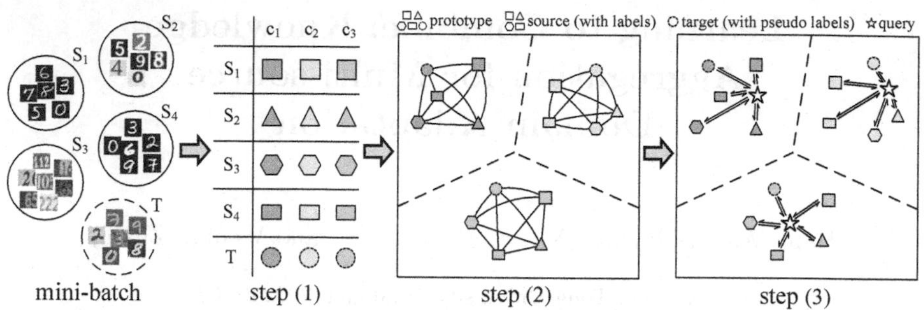

Fig. 1. Given a randomly sampled mini-batch, in step (1), our model first updates each category's global prototype for all domains. In step (2), a knowledge graph is constructed on these prototypes. Finally, in step (3), a bunch of query samples are inserted into the graph and predicted via knowledge aggregation.

when directly deploying the model trained on a specific dataset to the scenarios with distinct backgrounds, weather or illumination, undesirable performance decay commonly occurs, due to the existence of domain shift [49].

Unsupervised Domain Adaptation (UDA) is an extensively explored technique to address such problem, and it focuses on the transferability of knowledge learned from a labeled dataset (source domain) to another unlabeled one (target domain). The basic intuition behind these attempts is that knowledge transfer can be achieved by boosting domain-invariance of feature representations from different domains. In order to realize such goal, various strategies have been proposed, including minimizing explicitly defined domain discrepancy metrics [24,39,46], adversarial-training-based domain confusion [4,25,40] and GAN-based domain alignment [2,6,37].

However, in real-world applications, it is unreasonable to deem that the labeled images are drawn from a single domain. Actually, these samples can be collected under different deployment environments, *i.e.* from multiple domains, which reflect distinct modal information. Integrating such factor into domain alignment, a more practical problem is *Multi-Source Domain Adaptation* (MSDA), which dedicates to transfer the knowledges learned from multiple source domains to an unlabeled target domain.

Inspired by the theoretical analysis [12,29], recent works [33,45,52] predict target samples by combining the predictions of source classifiers. However, the interaction of feature representations learned from different domains has not been explored to tackle MSDA tasks. Compared to combining classifiers' predictions using hand-crafted or model-induced weights, knowledge propagation among multiple domains enables related feature representations to interact with each other before final prediction, which makes the operation of domain combination learnable. In addition, although category-level domain adaptation has been extensively studied in the literature, *e.g.* maximizing dual classifiers discrepancy [19,36] and prototype-based alignment [32,42], the relationships among categories are not constrained in these works. For instance, the source domain's knowledges that truck is more similar to car than person should also be applicable to target domain. Motivated by these limitations, we propose a novel framework and loss function for MSDA as follows.

Learning to Combine. We propose a new framework, *Learning to Combine for MSDA* (LtC-MSDA), which leverages the knowledges learned from multiple source domains to assist model's inference on target domain. In the training phase, three major steps are performed, which are graphically illustrated in Fig. 1. (1) *Global prototype*[1] *maintenance*: Based on a randomly sampled mini-batch containing samples from source and target domains, we estimate the prototype representation of each category for all domains. In order to mitigate the randomness of these estimations, global prototypes are maintained through a moving average scheme. (2) *Knowledge graph construction*: In this step, a knowledge graph is constructed on the global prototypes of different domains, and the connection weight between two global prototypes is determined by their similarity. (3) *Knowledge-aggregation-based prediction*: Given a bunch of query samples from arbitrary domains, we first extend the knowledge graph with these samples. After that, a graph convolutional network (GCN) is employed to propagate feature representations throughout the extended graph and output the classification probability for each node. After training, the knowledge graph is saved, and only step (3) is conducted for model's inference.

Class-Relation-Aware Domain Alignment. During the process of domain adaptation, in order to exploit the relational interdependency among categories, we propose a *Relation Alignment Loss* (RAL), which is composed of a global and a local term. (1) *Global relation constraint*: In this term, based on the adjacency matrix of knowledge graph, we constrain the connection weight between two arbitrary classes to be consistent among all domains, which refines the relative position of different classes' features in the latent space. (2) *Local relation constraint*: This term facilitates the compactness of various categories' features. In specific, we restrain the feature representation of a sample to be as close as possible to its corresponding global prototype, which makes the features belonging to distinct categories easier to be separated.

Our contributions can be summarized as follows:

1. We propose a Learning to Combine for MSDA (LtC-MSDA) framework, in which the knowledges learned from source domains interact with each other and assist model's prediction on target domain.
2. In order to better align the feature distributions of source and target domains, we design a Relation Alignment Loss (RAL) to constrain the global and local relations of feature representations.
3. We evaluate our model on three benchmark datasets with different domain shift and data complexity, and extensive results show that the proposed method outperforms existing approaches with a clear margin.

[1] Prototype is the mean embedding of all samples within the same class.

2 Related Work

Unsupervised Domain Adaptation (UDA). UDA seeks to generalize a model learned from a labeled source domain to a new target domain without labels. Many previous methods achieve such goal via minimizing an explicit domain discrepancy metric [19,24,39,41,46]. Adversarial learning is also employed to align two domains on feature level [4,25,40] or pixel level [2,6,37,44]. Recently, a group of approaches performs category-level domain adaptation through utilizing dual classifier [19,36], or domain prototype [32,42,43]. In this work, we further explore the consistency of category relations on all domains.

Multi-source Domain Adaptation (MSDA). MSDA assumes data are collected from multiple source domains with different distributions, which is a more practical scenario compared to single-source domain adaptation. Early theoretical analysis [1,29] gave strong guarantees for representing target distribution as the weighted combination of source distributions. Based on these works, Hoffman *et al.* [12] derived normalized solutions for MSDA problems. Recently, Zhao *et al.* [51] aligned target domain to source domains globally using adversarial learning. Xu *et al.* [45] deployed multi-way adversarial learning and combined source-specific perplexity scores for target predictions. Peng *et al.* [33] proposed to transfer knowledges by matching the moments of feature representations. In [52], source distilling mechanism is introduced to fine-tune the separately pre-trained feature extractor and classifier.
Improvements Over Existing Methods. In order to derive the predictions of target samples, former works [33,45,52] utilize the ensemble of source classifiers to output weighted classification probabilities, while such combination scheme prohibits the end-to-end learnable model. In this work, we design a *Learning to Combine* framework to predict query samples based on the interaction of knowledges learned from source and target domains, which makes the whole model end-to-end learnable.

Knowledge Graph. A knowledge graph describes entities and their interrelations, organized in a graph. Learning knowledge graphs and using attribute relationships has recently been of interest to the vision community. Several works [8,16] utilize knowledge graphs based on the defined semantic space for natural language understanding. For multi-label image classification [20,30], knowledge graphs are applied to exploit explicit semantic relations. In this paper, we construct a knowledge graph on global prototypes of different domains, which lays foundation for our method.

Graph Convolutional Network (GCN). GCN [15] is designed to compute directly on graph-structured data and model the inner structural relations. Such structures typically come from some prior knowledges about specific problems. Due to its effectiveness, GCNs have been widely used in various tasks, *e.g.* action recognition [47], person Re-ID [22,48] and point cloud learning [21]. For MSDA task, we employ GCN to propagate information on the knowledge graph.

3 Method

In Multi-Source Domain Adaptation (MSDA), there are M source domains $\mathcal{S}_1$, $\mathcal{S}_2$, $\cdots$, $\mathcal{S}_M$. The domain $\mathcal{S}_m = \{(x_i^{\mathcal{S}_m}, y_i^{\mathcal{S}_m})\}_{i=1}^{N_{\mathcal{S}_m}}$ is characterized by $N_{\mathcal{S}_m}$ i.i.d. labeled samples, where $x_i^{\mathcal{S}_m}$ follows one of the source distributions $\mathbb{P}_{\mathcal{S}_m}$ and $y_i^{\mathcal{S}_m} \in \{1, 2, \cdots, K\}$ (K is the number of classes) denotes its corresponding label. Similarly, target domain $\mathcal{T} = \{x_j^{\mathcal{T}}\}_{j=1}^{N_{\mathcal{T}}}$ is represented by $N_{\mathcal{T}}$ i.i.d. unlabeled samples, where $x_j^{\mathcal{T}}$ follows target distribution $\mathbb{P}_{\mathcal{T}}$. In the training phase, a randomly sampled mini-batch $B = \{\widehat{\mathcal{S}}_1, \widehat{\mathcal{S}}_2, \cdots, \widehat{\mathcal{S}}_M, \widehat{\mathcal{T}}\}$ is used to characterize source and target domains, and $|B|$ denotes the batch size.

3.1 Motivation and Overview

For MSDA, the core research topic is how to achieve more precise predictions for target samples through fully utilizing the knowledges among different domains. In order to mitigate the error of single-source prediction, recent works [33,45,52] express the classification probabilities of target samples as the weighted average of source classifiers' predictions. However, such scheme requires prior knowledges about the relevance of different domains to obtain combination weights, which makes the whole model unable to be end-to-end learnable.

In addition, learning to generalize from multiple source domains to target domain has a "double-edged sword" effect on model's performance. From one perspective, samples from multiple domains provide more abundant modal information of different classes, and thus the decision boundaries are refined according to more support points. From the other perspective, the distribution discrepancy among distinct source domains increases the difficulty of learning domain-invariant features. Off-the-shelf UDA techniques might fail in the condition that multi-modal distributions are to be aligned, since the relevance among different modalities, *i.e.* categories of various domains, are not explicitly constrained in these methods. Such constraints [7,38] are proved to be necessary when large amounts of clusters are formed in the latent space.

To address above issues, we propose a *Learning to Combine for MSDA* (LtC-MSDA) framework. In specific, a knowledge graph is constructed on the prototypes of different domains to enable the interaction among semantically adjacent entities, and query samples are added into this graph to obtain their classification probabilities under the guidance of correlated prototypes. In this process, the combination of different domains' knowledges is achieved via information propagation, which can be learned by a graph model. On the basis of this framework, a *Relation Alignment Loss* (RAL) is proposed, which facilitates the consistency of categories' relational interdependency on all domains and boosts the compactness of feature embeddings within the same class.

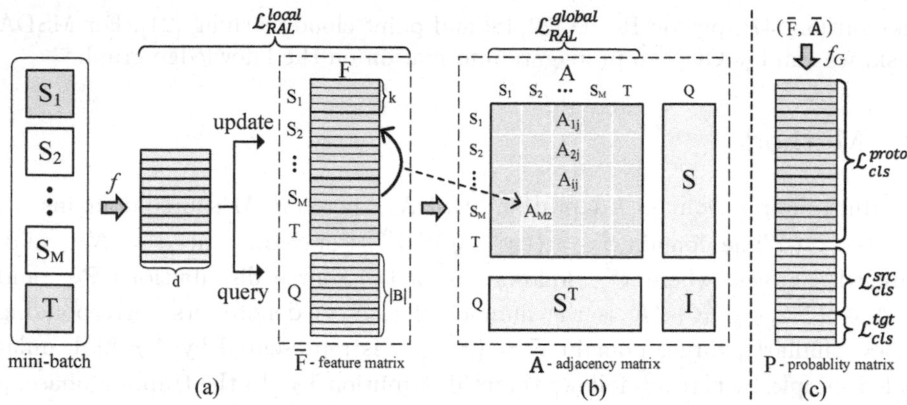

Fig. 2. Framework overview. (a) A randomly sampled mini-batch is utilized to update global prototypes and also serves as query samples, and the local relation loss $\mathcal{L}_{RAL}^{local}$ is constrained to promote feature compactness. (b) A knowledge graph is constructed on prototypes, whose adjacency matrix $\mathbf{A}$ embodies the relevance among different domains' categories. On the basis of block matrices in $\mathbf{A}$, global relation loss $\mathcal{L}_{RAL}^{global}$ is derived. (c) Extended by query samples, feature matrix $\bar{\mathbf{F}}$ and adjacency matrix $\bar{\mathbf{A}}$ are fed into a GCN model f_G to produce final predictions $\mathbf{P}$. On such basis, three kinds of classification losses are defined.

3.2 Learning to Combine for MSDA

In the proposed LtC-MSDA framework, for each training iteration, a mini-batch containing samples from all domains is mapped to latent space, and the produced feature embeddings are utilized to update global prototypes and also served as queries. After that, global prototypes and query samples are structured as a knowledge graph. Finally, a GCN model is employed to perform information propagation and output classification probability for each node of knowledge graph. Figure 2 gives a graphical illustration of the whole framework, and its details are presented in the following parts.

Global Prototype Maintenance. This step updates global prototypes with mini-batch statistics. Based on a mini-batch B, we estimate the prototype of each category for all domains. For source domain $\mathcal{S}_m$, the estimated prototype $\widehat{c}_k^{\mathcal{S}_m}$ is defined as the mean embedding of all samples belonging to class k in $\widehat{\mathcal{S}}_m$:

$$\widehat{c}_k^{\mathcal{S}_m} = \frac{1}{|\widehat{\mathcal{S}}_m^k|} \sum_{(x_i^{\mathcal{S}_m}, y_i^{\mathcal{S}_m}) \in \widehat{\mathcal{S}}_m^k} f(x_i^{\mathcal{S}_m}), \tag{1}$$

where $\widehat{\mathcal{S}}_m^k$ is the set of all samples with class label k in the sampling $\widehat{\mathcal{S}}_m$, and f represents the mapping from image to feature embedding.

For target domain $\mathcal{T}$, since ground truth information is unavailable, we first assign pseudo labels for the samples in $\widehat{\mathcal{T}}$ using the strategy proposed by [50], and the estimated prototype $\widehat{c}_k^{\mathcal{T}}$ of target domain is defined as follows:

$$\widehat{c}_k^{\mathcal{T}} = \frac{1}{|\widehat{\mathcal{T}}_k|} \sum_{(x_i^{\mathcal{T}}, \widehat{y}_i^{\mathcal{T}}) \in \widehat{\mathcal{T}}_k} f(x_i^{\mathcal{T}}), \tag{2}$$

where $\widehat{y}_i^{\mathcal{T}}$ is the pseudo label assigned to $x_i^{\mathcal{T}}$, and $\widehat{\mathcal{T}}_k$ denotes the set of all samples labeled as class k in $\widehat{\mathcal{T}}$. In order to correct estimation bias brought by the randomness of mini-batch samplings, we maintain the global prototypes for source and target domains with an exponential moving average scheme:

$$c_k^{\mathcal{S}_m} := \beta c_k^{\mathcal{S}_m} + (1-\beta)\widehat{c}_k^{\mathcal{S}_m} \quad m = 1, 2, \cdots, M, \tag{3}$$

$$c_k^{\mathcal{T}} := \beta c_k^{\mathcal{T}} + (1-\beta)\widehat{c}_k^{\mathcal{T}}, \tag{4}$$

where β is the exponential decay rate which is fixed as 0.7 in all experiments. Such moving average scheme is broadly used in the literature [9,14,42] to stabilize the training process through smoothing global variables.

Knowledge Graph Construction. In order to further refine category-level representations with knowledges learned from multiple domains, this step structures the global prototypes of various domains as a knowledge graph $\mathcal{G} = (\mathcal{V}, \mathcal{E})$. In this graph, the vertex set $\mathcal{V}$ corresponds to $(M+1)K$ prototypes, and the feature matrix $\mathbf{F} \in \mathbb{R}^{|\mathcal{V}| \times d}$ (d: the dimension of feature embedding) is defined as the concatenation of global prototypes:

$$\mathbf{F} = \Big[\underbrace{c_1^{\mathcal{S}_1}\, c_2^{\mathcal{S}_1} \cdots c_K^{\mathcal{S}_1}}_{\text{prototypes of } \mathcal{S}_1} \cdots \underbrace{c_1^{\mathcal{S}_M}\, c_2^{\mathcal{S}_M} \cdots c_K^{\mathcal{S}_M}}_{\text{prototypes of } \mathcal{S}_M} \underbrace{c_1^{\mathcal{T}}\, c_2^{\mathcal{T}} \cdots c_K^{\mathcal{T}}}_{\text{prototypes of } \mathcal{T}}\Big]^{\mathrm{T}}. \tag{5}$$

The edge set $\mathcal{E} \subseteq \mathcal{V} \times \mathcal{V}$ describes the relations among vertices, and an adjacency matrix $\mathbf{A} \in \mathbb{R}^{|\mathcal{V}| \times |\mathcal{V}|}$ is employed to model such relationships. In specific, we derive the adjacency matrix by applying a Gaussian kernel $\mathcal{K}_G$ over pairs of global prototypes:

$$\mathbf{A}_{i,j} = \mathcal{K}_G(\mathbf{F}_i^{\mathrm{T}}, \mathbf{F}_j^{\mathrm{T}}) = \exp\Big(-\frac{||\mathbf{F}_i^{\mathrm{T}} - \mathbf{F}_j^{\mathrm{T}}||_2^2}{2\sigma^2}\Big), \tag{6}$$

where $\mathbf{F}_i^{\mathrm{T}}$ and $\mathbf{F}_j^{\mathrm{T}}$ denote the i-th and j-th global prototype in feature matrix $\mathbf{F}$, and σ is the standard deviation parameter controlling the sparsity of $\mathbf{A}$.

Knowledge-Aggregation-Based Prediction. In this step, we aim to obtain more accurate predictions for query samples under the guidance of multiple domains' knowledges. We regard the mini-batch B as a bunch of query samples and utilize them to establish an extended knowledge graph $\bar{\mathcal{G}} = (\bar{\mathcal{V}}, \bar{\mathcal{E}})$. In this graph, the vertex set $\bar{\mathcal{V}}$ is composed of the original vertices in $\mathcal{V}$, *i.e.* global prototypes, and query samples' feature embeddings, which yields an extended feature matrix $\bar{\mathbf{F}} \in \mathbb{R}^{|\bar{\mathcal{V}}| \times d}$ as follows:

$$\bar{\mathbf{F}} = \left[\mathbf{F}^{\mathrm{T}}\ f(q_1)\ f(q_2)\ \cdots\ f(q_{|B|})\right]^{\mathrm{T}}, \tag{7}$$

where q_i $(i = 1, 2, \cdots, |B|)$ denotes the i-th query sample.

The edge set $\bar{\mathcal{E}}$ is expanded with the edges of new vertices. Concretely, an extended adjacency matrix $\bar{\mathbf{A}}$ is derived by adding the connections between global prototypes and query samples:

$$\mathbf{S}_{i,j} = \mathcal{K}_G(\mathbf{F}_i^{\mathrm{T}}, f(q_j)) = \exp\Big(-\frac{\|\mathbf{F}_i^{\mathrm{T}} - f(q_j)\|_2^2}{2\sigma^2}\Big), \tag{8}$$

$$\bar{\mathbf{A}} = \begin{bmatrix} \mathbf{A} & \mathbf{S} \\ \mathbf{S}^{\mathrm{T}} & \mathbf{I} \end{bmatrix}, \tag{9}$$

where $\mathbf{S} \in \mathbb{R}^{|\mathcal{V}| \times |B|}$ is the similarity matrix measuring the relevance between original and new vertices. Considering that the semantic information from a single sample is not precise enough, we ignore the interaction among query samples and use an identity matrix $\mathbf{I}$ to depict their relations.

After these preparations, a Graph Convolutional Network (GCN) is employed to propagate feature representations throughout the extended knowledge graph, such that the representations within the same category are encouraged to be consistent across all domains and query samples. In specific, inputted with the feature matrix $\bar{\mathbf{F}}$ and adjacency matrix $\bar{\mathbf{A}}$, the GCN model f_G outputs the classification probability matrix $\mathbf{P} \in \mathbb{R}^{|\bar{\mathcal{V}}| \times K}$ as follows:

$$\mathbf{P} = f_G\,(\bar{\mathbf{F}},\ \bar{\mathbf{A}}). \tag{10}$$

Model Inference. After training, we store the feature extractor f, GCN model f_G, feature matrix $\mathbf{F}$ and adjacency matrix $\mathbf{A}$. For inference, only the *knowledge-aggregation-based prediction* step is conducted. Concretely, based on the feature embeddings extracted by f, the extended feature matrix $\bar{\mathbf{F}}$ and adjacency matrix $\bar{\mathbf{A}}$ are derived by Eq. 7 and Eq. 9 respectively. Using these two matrices, the GCN model f_G produces the classification probabilities for test samples.

3.3 Class-Relation-Aware Domain Alignment

In the training phase, our model is optimized by two kinds of losses which facilitate the domain-invariance and distinguishability of feature representations. The details are stated below.

Relation Alignment Loss (RAL). This loss aims to conduct domain alignment on category level. During the domain adaptation process, except for promoting the invariance of same categories' features, it is necessary to constrain the relative position of different categories' feature embeddings in the latent space, especially when numerous modalities exist in the task, *e.g.* MSDA. Based on this idea, we propose the RAL which consists of a global and a local constraint:

$$\mathcal{L}_{RAL} = \lambda_1 \mathcal{L}_{RAL}^{global} + \lambda_2 \mathcal{L}_{RAL}^{local}, \tag{11}$$

where λ_1 and λ_2 are trade-off parameters.

For the global term, we facilitate the relevance between two arbitrary classes to be consistent on all domains, which is implemented through measuring the similarity of block matrices in $\mathbf{A}$:

$$\mathcal{L}_{RAL}^{global} = \frac{1}{(M+1)^4} \sum_{i,j,m,n=1}^{M+1} ||\mathbf{A}_{i,j} - \mathbf{A}_{m,n}||_F, \tag{12}$$

where the block matrix $\mathbf{A}_{i,j}$ $(1 \leqslant i,j \leqslant M+1)$ evaluates all categories' relevance between the i-th and j-th domain, which is shown in Fig. 2(b), and $||\cdot||_F$ denotes Frobenius norm. In this loss, features' intra-class invariance is boosted by the constraints on block matrices' main diagonal elements, and the consistency of different classes' relational interdependency is promoted by the constraints on other elements of block matrices.

For the local term, we enhance the feature compactness of each category via impelling the feature embeddings of samples in mini-batch B to approach their corresponding global prototypes, which derives the following loss function:

$$\begin{aligned} \mathcal{L}_{RAL}^{local} = \frac{1}{|B|} \sum_{k=1}^{K} \Bigg(\sum_{m=1}^{M} \sum_{(x_i^{\mathcal{S}_m}, y_i^{\mathcal{S}_m}) \in \widehat{\mathcal{S}}_m^k} ||f(x_i^{\mathcal{S}_m}) - c_k^{\mathcal{S}_m}||_2^2 \\ + \sum_{(x_i^{\mathcal{T}}, \widehat{y}_i^{\mathcal{T}}) \in \widehat{\mathcal{T}}_k} ||f(x_i^{\mathcal{T}}) - c_k^{\mathcal{T}}||_2^2 \Bigg). \end{aligned} \tag{13}$$

Classification Losses. This group of losses aims to enhance features' distinguishability. Based on the predictions of all vertices in extended knowledge graph $\bar{\mathcal{G}}$, the classification loss is defined as the composition of three terms for global prototypes, source samples and target samples respectively:

$$\mathcal{L}_{cls} = \mathcal{L}_{cls}^{proto} + \mathcal{L}_{cls}^{src} + \mathcal{L}_{cls}^{tgt}. \tag{14}$$

For the global prototypes and source samples, since their labels are available, two cross-entropy losses are employed for evaluation:

$$\mathcal{L}_{cls}^{proto} = \frac{1}{(M+1)K} \Bigg(\sum_{m=1}^{M} \sum_{k=1}^{K} \mathcal{L}_{ce}\big(p(c_k^{\mathcal{S}_m}), k\big) + \sum_{k=1}^{K} \mathcal{L}_{ce}\big(p(c_k^{\mathcal{T}}), k\big) \Bigg), \tag{15}$$

$$\mathcal{L}_{cls}^{src} = \frac{1}{M} \sum_{m=1}^{M} \Big(\mathbb{E}_{(x_i^{\mathcal{S}_m}, y_i^{\mathcal{S}_m}) \in \widehat{\mathcal{S}}_m} \mathcal{L}_{ce}\big(p(x_i^{\mathcal{S}_m}), y_i^{\mathcal{S}_m}\big) \Big), \tag{16}$$

Table 1. Classification accuracy (mean ± std %) on *Digits-five* dataset.

Standards	Methods	→ **mm**	→ **mt**	→ **up**	→ **sv**	→ **syn**	Avg
Single Best	Source-only	59.2 ± 0.6	97.2 ± 0.6	84.7 ± 0.8	77.7 ± 0.8	85.2 ± 0.6	80.8
	DAN [24]	63.8 ± 0.7	96.3 ± 0.5	94.2 ± 0.9	62.5 ± 0.7	85.4 ± 0.8	80.4
	CORAL [39]	62.5 ± 0.7	97.2 ± 0.8	93.5 ± 0.8	64.4 ± 0.7	82.8 ± 0.7	80.1
	DANN [5]	71.3 ± 0.6	97.6 ± 0.8	92.3 ± 0.9	63.5 ± 0.8	85.4 ± 0.8	82.0
	ADDA [40]	71.6 ± 0.5	97.9 ± 0.8	92.8 ± 0.7	75.5 ± 0.5	86.5 ± 0.6	84.8
Source Combine	Source-only	63.4 ± 0.7	90.5 ± 0.8	88.7 ± 0.9	63.5 ± 0.9	82.4 ± 0.6	77.7
	DAN [24]	67.9 ± 0.8	97.5 ± 0.6	93.5 ± 0.8	67.8 ± 0.6	86.9 ± 0.5	82.7
	DANN [5]	70.8 ± 0.8	97.9 ± 0.7	93.5 ± 0.8	68.5 ± 0.5	87.4 ± 0.9	83.6
	JAN [27]	65.9 ± 0.7	97.2 ± 0.7	95.4 ± 0.8	75.3 ± 0.7	86.6 ± 0.6	84.1
	ADDA [40]	72.3 ± 0.7	97.9 ± 0.6	93.1 ± 0.8	75.0 ± 0.8	86.7 ± 0.6	85.0
	MCD [36]	72.5 ± 0.7	96.2 ± 0.8	95.3 ± 0.7	78.9 ± 0.8	87.5 ± 0.7	86.1
Multi-Source	MDAN [51]	69.5 ± 0.3	98.0 ± 0.9	92.4 ± 0.7	69.2 ± 0.6	87.4 ± 0.5	83.3
	DCTN [45]	70.5 ± 1.2	96.2 ± 0.8	92.8 ± 0.3	77.6 ± 0.4	86.8 ± 0.8	84.8
	M^{3}SDA [33]	72.8 ± 1.1	98.4 ± 0.7	96.1 ± 0.8	81.3 ± 0.9	89.6 ± 0.6	87.7
	MDDA [52]	78.6 ± 0.6	98.8 ± 0.4	93.9 ± 0.5	79.3 ± 0.8	89.7 ± 0.7	88.1
	LtC-MSDA	$\mathbf{85.6} \pm 0.8$	$\mathbf{99.0} \pm 0.4$	$\mathbf{98.3} \pm 0.4$	$\mathbf{83.2} \pm 0.6$	$\mathbf{93.0} \pm 0.5$	**91.8**

where $\mathcal{L}_{ce}$ denotes the cross-entropy loss function, and $p(x)$ represents the classification probability of x.

For the target samples, it is desirable to make their predictions more deterministic, and thus an entropy loss is utilized for measurement:

$$\mathcal{L}_{cls}^{tgt} = -\mathbb{E}_{(x_i^{\mathcal{T}}, \widehat{y}_i^{\mathcal{T}}) \in \widehat{\mathcal{T}}} \sum_{k=1}^{K} p(\widehat{y}_i^{\mathcal{T}} = k | x_i^{\mathcal{T}}) \log p(\widehat{y}_i^{\mathcal{T}} = k | x_i^{\mathcal{T}}), \tag{17}$$

where $p(y = k|x)$ is the probability that x belongs to class k.

Overall Objectives. Combining the classification and domain adaptation losses defined above, the overall objectives for feature extractor f and GCN model f_G are as follows:

$$\min_{f} \mathcal{L}_{cls} + \mathcal{L}_{RAL}, \qquad \min_{f_G} \mathcal{L}_{cls}. \tag{18}$$

4 Experiments

In this section, we first describe the experimental settings and then compare our model with existing methods on three Multi-Source Domain Adaptation datasets to demonstrate its effectiveness.

Table 2. Classification accuracy (%) on *Office-31* dataset.

Standards	Methods	→ D	→ W	→ A	Avg
Single Best	Source-only	99.0	95.3	50.2	81.5
	RevGrad [4]	99.2	96.4	53.4	83.0
	DAN [24]	99.0	96.0	54.0	83.0
	RTN [26]	**99.6**	96.8	51.0	82.5
	ADDA [40]	99.4	95.3	54.6	83.1
Source Combine	Source-only	97.1	92.0	51.6	80.2
	DAN [24]	98.8	96.2	54.9	83.3
	RTN [26]	99.2	95.8	53.4	82.8
	JAN [27]	99.4	95.9	54.6	83.3
	ADDA [40]	99.2	96.0	55.9	83.7
	MCD [36]	99.5	96.2	54.4	83.4
Multi-Source	MDAN [51]	99.2	95.4	55.2	83.3
	DCTN [45]	**99.6**	96.9	54.9	83.8
	M^{3}SDA [33]	99.4	96.2	55.4	83.7
	MDDA [52]	99.2	97.1	56.2	84.2
	LtC-MSDA	**99.6**	**97.2**	**56.9**	**84.6**

4.1 Experimental Setup

Training Details. For all experiments, a GCN model with two graph convolutional layers is employed, in which the dimension of feature representation is $d \rightarrow d \rightarrow K$ (d: the dimension of feature embedding; K: the number of classes). Unless otherwise specified, the trade-off parameters λ_1, λ_2 are set as 20, 0.001 respectively, and the standard deviation σ is set as 0.005. In addition, "$\rightarrow D$" denotes the task of transferring from other domains to domain D.

Performance Comparison. We compare our approach with state-of-the-art methods to verify its effectiveness. For the sake of fair comparison, we introduce three standards. (1) *Single Best*: We report the best performance of single-source domain adaptation algorithm among all the sources. (2) *Source Combine*: All the source domain data are combined into a single source, and domain adaptation is performed in a traditional single-source manner. (3) *Multi-Source*: The knowledges learned from multiple source domains are transferred to target domain. For the first two settings, previous single-source UDA methods, *e.g.* DAN [24], JAN [27], DANN [5], ADDA [40], MCD [36], are introduced for comparison. For the *Multi-Source* setting, we compare our approach with four existing MSDA algorithms, MDAN [51], DCTN [45], M^{3}SDA [33] and MDDA [52].

Table 3. Classification accuracy (mean ± std %) on *DomainNet* dataset.

Standards	Methods	→ clp	→ inf	→ pnt	→ qdr	→ rel	→ skt	Avg
Single Best	Source-only	39.6 ± 0.6	8.2 ± 0.8	33.9 ± 0.6	11.8 ± 0.7	41.6 ± 0.8	23.1 ± 0.7	26.4
	DAN [24]	39.1 ± 0.5	11.4 ± 0.8	33.3 ± 0.6	16.2 ± 0.4	42.1 ± 0.7	29.7 ± 0.9	28.6
	JAN [27]	35.3 ± 0.7	9.1 ± 0.6	32.5 ± 0.7	14.3 ± 0.6	43.1 ± 0.8	25.7 ± 0.6	26.7
	DANN [5]	37.9 ± 0.7	11.4 ± 0.9	33.9 ± 0.6	13.7 ± 0.6	41.5 ± 0.7	28.6 ± 0.6	27.8
	ADDA [40]	39.5 ± 0.8	14.5 ± 0.7	29.1 ± 0.8	14.9 ± 0.5	41.9 ± 0.8	30.7 ± 0.7	28.4
	MCD [36]	42.6 ± 0.3	19.6 ± 0.8	42.6 ± 1.0	3.8 ± 0.6	50.5 ± 0.4	33.8 ± 0.9	32.2
Source Combine	Source-only	47.6 ± 0.5	13.0 ± 0.4	38.1 ± 0.5	13.3 ± 0.4	51.9 ± 0.9	33.7 ± 0.5	32.9
	DAN [24]	45.4 ± 0.5	12.8 ± 0.9	36.2 ± 0.6	15.3 ± 0.4	48.6 ± 0.7	34.0 ± 0.5	32.1
	JAN [27]	40.9 ± 0.4	11.1 ± 0.6	35.4 ± 0.5	12.1 ± 0.7	45.8 ± 0.6	32.3 ± 0.6	29.6
	DANN [5]	45.5 ± 0.6	13.1 ± 0.7	37.0 ± 0.7	13.2 ± 0.8	48.9 ± 0.7	31.8 ± 0.6	32.6
	ADDA [40]	47.5 ± 0.8	11.4 ± 0.7	36.7 ± 0.5	14.7 ± 0.5	49.1 ± 0.8	33.5 ± 0.5	32.2
	MCD [36]	54.3 ± 0.6	22.1 ± 0.7	45.7 ± 0.6	7.6 ± 0.5	58.4 ± 0.7	43.5 ± 0.6	38.5
Multi-Source	MDAN [51]	52.4 ± 0.6	21.3 ± 0.8	46.9 ± 0.4	8.6 ± 0.6	54.9 ± 0.6	46.5 ± 0.7	38.4
	DCTN [45]	48.6 ± 0.7	23.5 ± 0.6	48.8 ± 0.6	7.2 ± 0.5	53.5 ± 0.6	47.3 ± 0.5	38.2
	M^{3}SDA [33]	58.6 ± 0.5	26.0 ± 0.9	52.3 ± 0.6	6.3 ± 0.6	62.7 ± 0.5	49.5 ± 0.8	42.6
	MDDA [52]	59.4 ± 0.6	23.8 ± 0.8	53.2 ± 0.6	12.5 ± 0.6	61.8 ± 0.5	48.6 ± 0.8	43.2
	LtC-MSDA	**63.1** ± 0.5	**28.7** ± 0.7	**56.1** ± 0.5	**16.3** ± 0.5	**66.1** ± 0.6	**53.8** ± 0.6	**47.4**

4.2 Experiments on Digits-Five

Dataset. Digits-five dataset contains five digit image domains, including MNIST (**mt**) [18], MNIST-M (**mm**) [5], SVHN (**sv**) [31], USPS (**up**) [13], and Synthetic Digits (**syn**) [5]. Each domain contains ten classes corresponding to digits ranging from 0 to 9. We follow the setting in DCTN [45] to sample the data.

Results. Table 1 reports the performance of our method compared with other works. Source-only denotes the model trained with only source domain data, which serves as the baseline. From the table, it can be observed that the proposed LtC-MSDA surpasses existing methods on all five tasks. In particular, a performance gain of 7.0% is achieved on the "→ **mm**" task. The results demonstrate the effectiveness of our approach on boosting model's performance through integrating multiple domains' knowledges.

4.3 Experiments on Office-31

Dataset. Office-31 [35] is a classical domain adaptation benchmark with 31 categories and 4652 images. It contains three domains: Amazon (A), Webcam (W) and DSLR (D), and the data are collected from office environment.

Results. In Table 2, we report the performance of our approach and existing methods on three tasks. The LtC-MSDA model outperforms the state-of-the-art method, MDDA [52], with 0.4% in the term of average classification accuracy, and a 0.7% performance improvement is obtained on the hard-to-transfer task,

Table 4. Ablation study for domain adaptation losses on global and local levels.

$\mathcal{L}_{RAL}^{global}$	$\mathcal{L}_{RAL}^{local}$	→ **mm**	→ **mt**	→ **up**	→ **sv**	→ **syn**	Avg
		74.85	98.60	97.95	74.56	88.54	86.90
✓		82.49	98.97	98.06	81.64	91.70	90.57
	✓	79.57	98.64	98.06	78.66	90.16	89.02
✓	✓	85.56	98.98	98.32	83.24	93.04	91.83

Table 5. Ablation study for three kinds of classification losses.

$\mathcal{L}_{cls}^{src}$	$\mathcal{L}_{cls}^{proto}$	$\mathcal{L}_{cls}^{tgt}$	→ **mm**	→ **mt**	→ **up**	→ **sv**	→ **syn**	Avg
✓			73.65	98.47	96.61	78.20	88.93	87.17
✓	✓		78.44	98.64	96.77	79.24	89.05	88.43
✓		✓	81.36	98.76	97.93	81.26	91.70	90.20
✓	✓	✓	85.56	98.98	98.32	83.24	93.04	91.83

"→ A". On this dataset, our approach doesn't have obvious superiority, which probably ascribes to two reasons. (1) First, domain adaptation models exhibit saturation when evaluated on "→ D" and "→ W" tasks, in which Source-only models achieve performance higher than 95%. (2) Second, the Webcam and DSLR domains are highly similar, which restricts the benefit brought by multiple domains' interaction in our framework, especially in "→ A" task.

4.4 Experiments on DomainNet

Dataset. DomainNet [33] is by far the largest and most difficult domain adaptation dataset. It consists of around 0.6 million images and 6 domains: clipart (clp), infograph (inf), painting (pnt), quickdraw (qdr), real (rel) and sketc.h (skt). Each domain contains the same 345 categories of common objects.

Results. The results of various methods on DomainNet are presented in Table 3. Our model exceeds existing works with a notable margin on all six tasks. In particular, a 4.2% performance gain is achieved on mean accuracy. The major challenges of this dataset are two-fold. (1) Large domain shift exists among different domains, *e.g.* from real images to sketc.hes. (2) Numerous categories increase the difficulty of learning discriminative features. Our approach tackles these two problems as follows. For the first issue, the global term of *Relation Alignment Loss* constrains the similarity between two arbitrary categories to be consistent on all domains, which encourages better feature alignment in the latent space. For the second issue, the local term of *Relation Alignment Loss* promotes the compactness of the same categories' features, which eases the burden of feature separation among different classes.

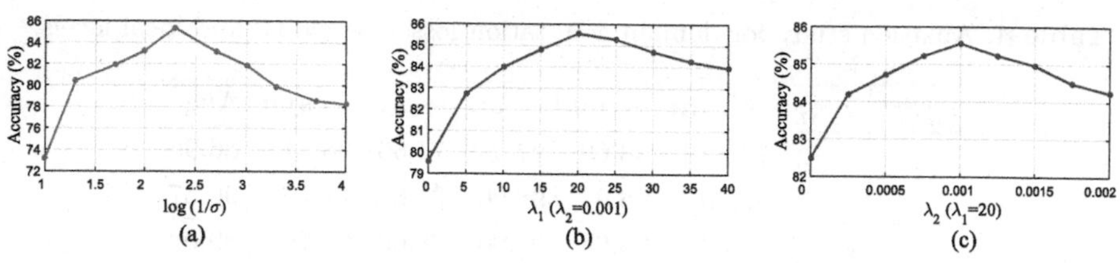

Fig. 3. Sensitivity analysis of standard deviation σ (left) and trade-off parameters λ_1, λ_2 (middle, right). (All results are reported on the "$\rightarrow$ **mm**" task.)

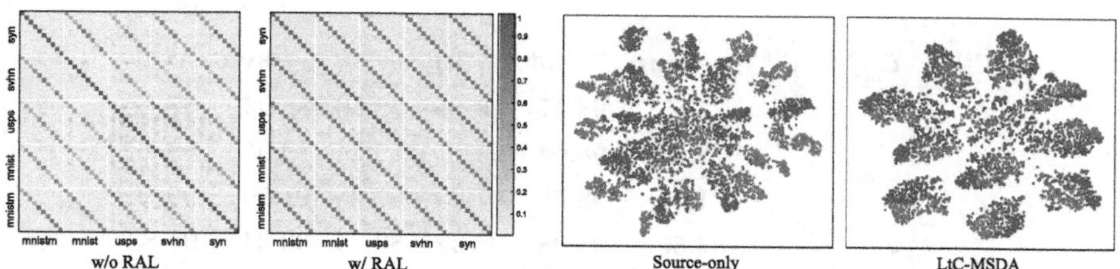

(a) Adjacency matrix **A** with and without RAL constraint.

(b) Feature distributions of source domain ("blue") and target domain ("red").

Fig. 4. Visualization of adjacency matrix and feature embeddings. (All results are evaluated on the "$\rightarrow$ **mm**" task.) (Color figure online)

5 Analysis

In this section, we provide more in-depth analysis of our method to validate the effectiveness of major components, and both quantitative and qualitative experiments are conducted for verification.

5.1 Ablation Study

Effect of Domain Adaptation Losses. In Table 4, we analyze the effect of global and local *Relation Alignment Loss* on Digits-five dataset.

On the basis of baseline setting (1st row), the global consistency loss (2nd rows) can greatly promote model's performance by promoting category-level domain alignment. For the local term, after adding it to the baseline configuration (3rd row), a 2.12% performance gain is achieved, which demonstrates the effectiveness of $\mathcal{L}_{RAL}^{local}$ on enhancing the separability of feature representations. Furthermore, the combination of $\mathcal{L}_{RAL}^{global}$ and $\mathcal{L}_{RAL}^{local}$ (4th row) obtains the best performance, which shows the complementarity of global and local constraints.

Effect of Classification Losses. Table 5 presents the effect of different classification losses on Digits-five dataset. The configuration of using only source samples' classification loss $\mathcal{L}_{cls}^{src}$ (1st row) serves as the baseline. After adding the entropy constraint for target samples (3rd row), the accuracy increases by

3.03%, which illustrates the effectiveness of $\mathcal{L}_{cls}^{tgt}$ on making target samples' features more discriminative. Prototypes' classification loss $\mathcal{L}_{cls}^{proto}$ is able to further boost the performance by constraining prototypes' distinguishability (4th row).

5.2 Sensitivity Analysis

Sensitivity of Standard Deviation σ. In this part, we discuss the selection of parameter σ which controls the sparsity of adjacency matrix. In Fig. 3(a), we plot the performance of models trained with different σ values. The highest accuracy on target domain is achieved when the value of σ is around 0.005. Also, it is worth noticing that obvious performance decay occurs when the adjacency matrix is too dense or sparse, *i.e.* $\sigma > 0.05$ or $\sigma < 0.0005$.

Sensitivity of Trade-Off Parameters λ_1, λ_2. In this experiment, we evaluate our approach's sensitivity to λ_1 and λ_2 which trade off between domain adaptation and classification losses. Figure 3(b) and Fig. 3(c) show model's performance under different λ_1 (λ_2) values when the other parameter λ_2 (λ_1) is fixed. From the line charts, we can observe that model's performance is not sensitive to λ_1 and λ_2 when they are around 20 and 0.001, respectively. In addition, performance decay occurs when these two parameters approach 0, which demonstrates that both global and local constraints are indispensable.

5.3 Visualization

Visualization of Adjacency Matrix. Figure 4(a) shows the adjacency matrix $\mathbf{A}$ before and after applying the *Relation Alignment Loss* (RAL), in which each pixel denotes the relevance between two categories from arbitrary domains. It can be observed that, after adding RAL, the relevance among various categories is apparently more consistent across different domains, which is compatible with the relational structure constrained by the global term of RAL.

Visualization of Feature Embeddings. In Fig. 4(b), we utilize t-SNE [28] to visualize the feature distributions of one of source domains (SVHN) and target domain (MNIST-M). Compared with the Source-only baseline, the proposed LtC-MSDA model makes the features of target domain more discriminative and better aligned with those of source domain.

6 Conclusion

In this paper, we propose a Learning to Combine for Multi-Source Domain Adaptation (LtC-MSDA) framework. In this framework, the knowledges learned from multiple domains are aggregated to assist the prediction for query samples. Furthermore, we conduct class-relation-aware domain alignment via constraining global category relationships and local feature compactness. Extensive experiments and analytical studies demonstrate the prominent performance of our approach under various domain shift settings.

Acknowledgement. This work was supported by National Science Foundation of China (61976137, U1611461, U19B2035) and STCSM(18DZ1112300). Authors would like to appreciate the Student Innovation Center of SJTU for providing GPUs.

References

1. Blitzer, J., Crammer, K., Kulesza, A., Pereira, F., Wortman, J.: Learning bounds for domain adaptation. In: Advances in Neural Information Processing Systems (2007)
2. Bousmalis, K., Silberman, N., Dohan, D., Erhan, D., Krishnan, D.: Unsupervised pixel-level domain adaptation with generative adversarial networks. In: IEEE Conference on Computer Vision and Pattern Recognition (2017)
3. Chen, L., Papandreou, G., Kokkinos, I., Murphy, K., Yuille, A.L.: Semantic image segmentation with deep convolutional nets and fully connected CRFs. In: International Conference on Learning Representations (2015)
4. Ganin, Y., Lempitsky, V.S.: Unsupervised domain adaptation by backpropagation. In: International Conference on Machine Learning (2015)
5. Ganin, Y., et al.: Domain-adversarial training of neural networks. J. Mach. Learn. Res. **17**(1), 2030–2096 (2016)
6. Ghifary, M., Kleijn, W.B., Zhang, M., Balduzzi, D., Li, W.: Deep reconstruction-classification networks for unsupervised domain adaptation. In: Leibe, B., Matas, J., Sebe, N., Welling, M. (eds.) ECCV 2016. LNCS, vol. 9908, pp. 597–613. Springer, Cham (2016). https://doi.org/10.1007/978-3-319-46493-0_36
7. Hadsell, R., Chopra, S., LeCun, Y.: Dimensionality reduction by learning an invariant mapping. In: IEEE Conference on Computer Vision and Pattern Recognition (2006)
8. Hakkani-Tür, D., Heck, L.P., Tür, G.: Using a knowledge graph and query click logs for unsupervised learning of relation detection. In: IEEE International Conference on Acoustics, Speech and Signal Processing (2013)
9. He, K., Fan, H., Wu, Y., Xie, S., Girshick, R.B.: Momentum contrast for unsupervised visual representation learning. In: IEEE Conference on Computer Vision and Pattern Recognition (2020)
10. He, K., Gkioxari, G., Dollár, P., Girshick, R.B.: Mask R-CNN. In: IEEE International Conference on Computer Vision (2017)
11. He, K., Zhang, X., Ren, S., Sun, J.: Deep residual learning for image recognition. In: IEEE Conference on Computer Vision and Pattern Recognition (2016)
12. Hoffman, J., Mohri, M., Zhang, N.: Algorithms and theory for multiple-source adaptation. In: Advances in Neural Information Processing Systems (2018)
13. Hull, J.J.: A database for handwritten text recognition research. IEEE Trans. Pattern Anal. Mach. Intell. **16**(5), 550–554 (1994)
14. Kingma, D.P., Ba, J.: Adam: a method for stochastic optimization. In: International Conference on Learning Representations (2015)
15. Kipf, T.N., Welling, M.: Semi-supervised classification with graph convolutional networks. In: International Conference on Learning Representations (2017)
16. Krishnamurthy, J., Mitchell, T.: Weakly supervised training of semantic parsers. In: Joint Conference on Empirical Methods in Natural Language Processing and Computational Natural Language Learning, pp. 754–765 (2012)
17. Krizhevsky, A., Sutskever, I., Hinton, G.E.: Imagenet classification with deep convolutional neural networks. In: Advances in Neural Information Processing Systems (2012)

18. LeCun, Y., Bottou, L., Bengio, Y., Haffner, P.: Gradient-based learning applied to document recognition. Proc. IEEE **86**(11), 2278–2324 (1998)
19. Lee, C., Batra, T., Baig, M.H., Ulbricht, D.: Sliced Wasserstein discrepancy for unsupervised domain adaptation. In: IEEE Conference on Computer Vision and Pattern Recognition (2019)
20. Lee, C., Fang, W., Yeh, C., Wang, Y.F.: Multi-label zero-shot learning with structured knowledge graphs. In: IEEE Conference on Computer Vision and Pattern Recognition (2018)
21. Liu, J., Ni, B., Li, C., Yang, J., Tian, Q.: Dynamic points agglomeration for hierarchical point sets learning. In: IEEE International Conference on Computer Vision (2019)
22. Liu, J., Ni, B., Yan, Y., Zhou, P., Cheng, S., Hu, J.: Pose transferrable person re-identification. In: IEEE Conference on Computer Vision and Pattern Recognition (2018)
23. Liu, W., et al.: SSD: single shot multibox detector. In: Leibe, B., Matas, J., Sebe, N., Welling, M. (eds.) ECCV 2016. LNCS, vol. 9905, pp. 21–37. Springer, Cham (2016). https://doi.org/10.1007/978-3-319-46448-0_2
24. Long, M., Cao, Y., Wang, J., Jordan, M.I.: Learning transferable features with deep adaptation networks. In: International Conference on Machine Learning (2015)
25. Long, M., Cao, Z., Wang, J., Jordan, M.I.: Conditional adversarial domain adaptation. In: Advances in Neural Information Processing Systems (2018)
26. Long, M., Zhu, H., Wang, J., Jordan, M.I.: Unsupervised domain adaptation with residual transfer networks. In: Advances in Neural Information Processing Systems (2016)
27. Long, M., Zhu, H., Wang, J., Jordan, M.I.: Deep transfer learning with joint adaptation networks. In: International Conference on Machine Learning (2017)
28. Maaten, L.V.D., Hinton, G.: Visualizing data using t-SNE. J. Mach. Learn. Res. **9**(2605), 2579–2605 (2008)
29. Mansour, Y., Mohri, M., Rostamizadeh, A.: Domain adaptation with multiple sources. In: Advances in Neural Information Processing Systems (2008)
30. Marino, K., Salakhutdinov, R., Gupta, A.: The more you know: using knowledge graphs for image classification. In: IEEE Conference on Computer Vision and Pattern Recognition (2017)
31. Netzer, Y., Wang, T., Coates, A., Bissacco, A., Wu, B., Ng, A.Y.: Reading digits in natural images with unsupervised feature learning. In: NIPS Workshops (2011)
32. Pan, Y., Yao, T., Li, Y., Wang, Y., Ngo, C., Mei, T.: Transferrable prototypical networks for unsupervised domain adaptation. In: IEEE Conference on Computer Vision and Pattern Recognition (2019)
33. Peng, X., Bai, Q., Xia, X., Huang, Z., Saenko, K., Wang, B.: Moment matching for multi-source domain adaptation. In: IEEE International Conference on Computer Vision (2019)
34. Ren, S., He, K., Girshick, R.B., Sun, J.: Faster R-CNN: towards real-time object detection with region proposal networks. In: Advances in Neural Information Processing Systems (2015)
35. Saenko, K., Kulis, B., Fritz, M., Darrell, T.: Adapting visual category models to new domains. In: Daniilidis, K., Maragos, P., Paragios, N. (eds.) ECCV 2010. LNCS, vol. 6314, pp. 213–226. Springer, Heidelberg (2010). https://doi.org/10.1007/978-3-642-15561-1_16
36. Saito, K., Watanabe, K., Ushiku, Y., Harada, T.: Maximum classifier discrepancy for unsupervised domain adaptation. In: IEEE Conference on Computer Vision and Pattern Recognition (2018)

37. Sankaranarayanan, S., Balaji, Y., Castillo, C.D., Chellappa, R.: Generate to adapt: aligning domains using generative adversarial networks. In: IEEE Conference on Computer Vision and Pattern Recognition (2018)
38. Schroff, F., Kalenichenko, D., Philbin, J.: FaceNet: a unified embedding for face recognition and clustering. In: IEEE Conference on Computer Vision and Pattern Recognition (2015)
39. Sun, B., Saenko, K.: Deep CORAL: correlation alignment for deep domain adaptation. In: Hua, G., Jégou, H. (eds.) ECCV 2016. LNCS, vol. 9915, pp. 443–450. Springer, Cham (2016). https://doi.org/10.1007/978-3-319-49409-8_35
40. Tzeng, E., Hoffman, J., Saenko, K., Darrell, T.: Adversarial discriminative domain adaptation. In: IEEE Conference on Computer Vision and Pattern Recognition (2017)
41. Tzeng, E., Hoffman, J., Zhang, N., Saenko, K., Darrell, T.: Deep domain confusion: Maximizing for domain invariance. CoRR abs/1412.3474 (2014)
42. Xie, S., Zheng, Z., Chen, L., Chen, C.: Learning semantic representations for unsupervised domain adaptation. In: International Conference on Machine Learning (2018)
43. Xu, M., Wang, H., Ni, B., Tian, Q., Zhang, W.: Cross-domain detection via graph-induced prototype alignment. In: IEEE Conference on Computer Vision and Pattern Recognition (2020)
44. Xu, M., et al.: Adversarial domain adaptation with domain mixup. In: AAAI Conference on Artificial Intelligence (2020)
45. Xu, R., Chen, Z., Zuo, W., Yan, J., Lin, L.: Deep cocktail network: multi-source unsupervised domain adaptation with category shift. In: IEEE Conference on Computer Vision and Pattern Recognition (2018)
46. Yan, H., Ding, Y., Li, P., Wang, Q., Xu, Y., Zuo, W.: Mind the class weight bias: weighted maximum mean discrepancy for unsupervised domain adaptation. In: IEEE Conference on Computer Vision and Pattern Recognition (2017)
47. Yan, S., Xiong, Y., Lin, D.: Spatial temporal graph convolutional networks for skeleton-based action recognition. In: AAAI Conference on Artificial Intelligence (2018)
48. Yan, Y., Zhang, Q., Ni, B., Zhang, W., Xu, M., Yang, X.: Learning context graph for person search. In: IEEE Conference on Computer Vision and Pattern Recognition (2019)
49. Yosinski, J., Clune, J., Bengio, Y., Lipson, H.: How transferable are features in deep neural networks? In: Advances in Neural Information Processing Systems (2014)
50. Zhang, W., Ouyang, W., Li, W., Xu, D.: Collaborative and adversarial network for unsupervised domain adaptation. In: IEEE Conference on Computer Vision and Pattern Recognition (2018)
51. Zhao, H., Zhang, S., Wu, G., Moura, J.M.F., Costeira, J.P., Gordon, G.J.: Adversarial multiple source domain adaptation. In: Advances in Neural Information Processing Systems (2018)
52. Zhao, S., et al.: Multi-source distilling domain adaptation. In: AAAI Conference on Artificial Intelligence (2020)

CSCL: Critical Semantic-Consistent Learning for Unsupervised Domain Adaptation

Jiahua Dong[1,2,3], Yang Cong[1,2](✉), Gan Sun[1,2], Yuyang Liu[1,2,3], and Xiaowei Xu[4]

[1] State Key Laboratory of Robotics, Shenyang Institute of Automation, Chinese Academy of Sciences, Shenyang 110016, China
{dongjiahua,liuyuyang}@sia.cn, congyang81@gmail.com, sungan1412@gmail.com
[2] Institutes for Robotics and Intelligent Manufacturing, Chinese Academy of Sciences, Shenyang 110016, China
[3] University of Chinese Academy of Sciences, Beijing 100049, China
[4] Department of Information Science, University of Arkansas at Little Rock, Little Rock, USA
xwxu@ualr.edu

Abstract. Unsupervised domain adaptation without consuming annotation process for unlabeled target data attracts appealing interests in semantic segmentation. However, 1) existing methods neglect that not all semantic representations across domains are transferable, which cripples domain-wise transfer with untransferable knowledge; 2) they fail to narrow category-wise distribution shift due to category-agnostic feature alignment. To address above challenges, we develop a new Critical Semantic-Consistent Learning (CSCL) model, which mitigates the discrepancy of both domain-wise and category-wise distributions. Specifically, a critical transfer based adversarial framework is designed to highlight transferable domain-wise knowledge while neglecting untransferable knowledge. Transferability-critic guides transferability-quantizer to maximize positive transfer gain under reinforcement learning manner, although negative transfer of untransferable knowledge occurs. Meanwhile, with the help of confidence-guided pseudo labels generator of target samples, a symmetric soft divergence loss is presented to explore inter-class relationships and facilitate category-wise distribution alignment. Experiments on several datasets demonstrate the superiority of our model.

G. Sun—The author contributes equally to this work.

Electronic supplementary material The online version of this chapter (https://doi.org/10.1007/978-3-030-58598-3_44) contains supplementary material, which is available to authorized users.

A. Vedaldi et al. (Eds.): ECCV 2020, LNCS 12353, pp. 745–762, 2020.
https://doi.org/10.1007/978-3-030-58598-3_44

Keywords: Unsupervised domain adaptation · Semantic segmentation · Adversarial learning · Reinforcement learning · Pseudo label

1 Introduction

Convolutional neural networks relying on a large amount of annotations have achieved significant successes in many computer vision tasks, *e.g.*, semantic segmentation [2,31,45]. Unfortunately, the learned models could not generalize well to the unlabeled target domain, especially when there is a large distribution gap between the training and evaluation datasets [5,12,19]. Unsupervised domain adaptation [11,29,33,47] shows appealing segmentation performance for unlabeled target domain by transferring effective domain-invariant knowledge from labeled source domain. To this end, enormous related state-of-the-art models [9,20,24,34,35] are developed to mitigate the distribution discrepancy between different datasets.

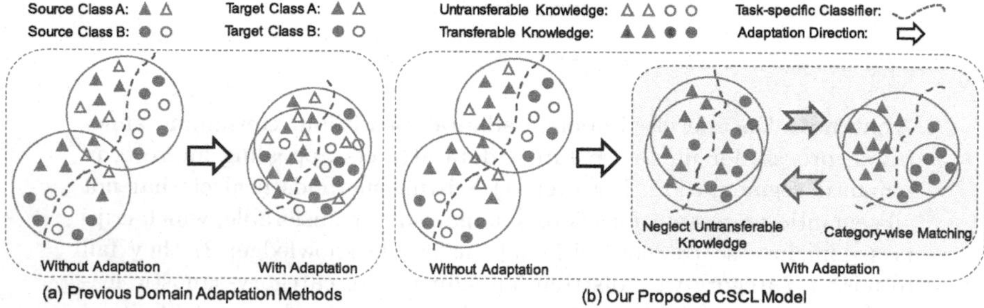

Fig. 1. Illustration of previous domain adaptation methods and our proposed CSCL model. (a): Existing models neglect the negative transfer brought by untransferable knowledge, and result in the category-wise distribution mismatch due to the lack of valid labels of target samples. (b): As for our CSCL model, a critical transfer based adversarial framework is developed to prevent the negative transfer of untransferable knowledge. Moreover, a symmetric soft divergence loss is designed to align category-wise distributions with the assistance of a confidence-guided pseudo labels generator.

However, most existing methods [10,13,20,21,24,34] ignore the fact that not all semantic representations among source and target datasets are transferable, while forcefully taking advantage of untransferable knowledge leads to negative domain-wise transfer, as shown in Fig. 1(a). In other words, semantic representations across domains cannot contribute equally to narrowing domain-wise distribution shift. For example, unbalanced object categories and objects with various appearances in different datasets are not equally essential to facilitate the semantic transfer. Moreover, category-wise distributions cannot be matched well across domains [4,17,29,30,33,38,39] since the lack of valid labels of target samples results in the category-agnostic feature alignment. As depicted in

Fig. 1(a), the semantic features of same class among different datasets are not mapped nearby. Therefore, exploring efficient semantic knowledge to narrow both domain-wise and category-wise distributions discrepancy is a crucial challenge.

To tackle above mentioned challenges, as presented in Fig. 1(b), a new Critical Semantic-Consistent Learning (CSCL) model is developed for unsupervised domain adaptation. **1)** On one hand, we design a critical transfer based adversarial framework to facilitate the exploration of transferable domain-wise representations while preventing the negative transfer of untransferable knowledge. To be specific, the transferability-quantizer highlights domain-wise semantic representations with high transfer scores, and transferability-critic evaluates the quality of corresponding transfer process. With a direct supervision from the transferability-critic, positive transfer feedback guides the transferability-quantizer to maximize the transfer gain, even though negative transfer of untransferable knowledge occurs. Both the transferability-critic and transferability-quantizer are trained under the reinforcement learning manner to narrow the marginal distribution gap. **2)** On the other hand, confident soft pseudo labels for target samples are progressively produced by the confidence-guided pseudo label generator, which efficiently attenuates the misleading effect brought by incorrect or ambiguous pseudo labels. With the guidance of generated soft pseudo labels, a symmetric soft divergence loss is developed to explore inter-class relationships across domains, and further bridge category-wise conditional distribution shift across domains. Therefore, our proposed CSCL model could not only learn discriminative transferable knowledge with high transfer scores for target samples prediction, but also well match both domain-wise and category-wise semantic consistence by minimizing marginal and conditional distributions gaps. Experiments on several datasets illustrate the effectiveness of our model with state-of-the-art performance.

The main contributions of this work are summarized as follows:

- A new Critical Semantic-Consistent Learning (CSCL) model is proposed to facilitate the exploration of both domain-wise and category-wise semantic consistence for unsupervised domain adaptation. To our best knowledge, this is the first attempt to highlight transferable knowledge via a critical transfer mechanism while neglecting untransferable representations.
- A critical transfer based adversarial framework is developed to explore transferable knowledge via a transferability-quantizer, while preventing the negative transfer of irrelevant knowledge via a transferability-critic.
- A symmetric soft divergence loss is designed to explore inter-class relationships and narrow category-wise distribution shift, under the supervision of confident pseudo labels mined by confidence-guided pseudo labels generator.

2 Related Work

Semantic Segmentation: Deep neural network [36,37,42] based semantic segmentation has caught growing research attention recently. [31] develops the first fully convolutional network for pixel-level prediction. To enlarge the receptive field, [1,40] propose the dilated convolution operator. Later, encoder-decoder networks, such as U-net [27], Deeplab [2], etc., are proposed to fuse low-level and high-level information. [22,23,44] aim to capture long-range contextual dependencies. However, they require large-scale labeled samples to attain remarkable performance, which is time-consuming to manually annotate the data.

Unsupervised Domain Adaptation: After generative adversarial network [14] is first employed by Hoffman *et al.* [16] to narrow distribution discrepancy for domain adaptation, enormous adversarial learning based variants [4,10,11, 17,24,29,30,33,34,38] are proposed to learn domain-invariant knowledge with a domain classifier. Another important strategy is curriculum learning [21,43], which predicts target labels according to source distribution properties. [46,47] develops a self-training non-adversarial model to mitigate the discrepancy shift with the assistance of pseudo labels. Li *et al.* [20] design a bidirectional learning model for knowledge transfer, which incorporates appearance translation module and semantic adaptation module. [13] considers intermediate optimal transfer domain for semantic segmentation. Moreover, some novel discriminative losses, such as sliced Wasserstein discrepancy [18], adversarial entropy minimization [35] and adaptive category-level adversarial loss [25], are proposed to minimize distribution discrepancy between different datasets.

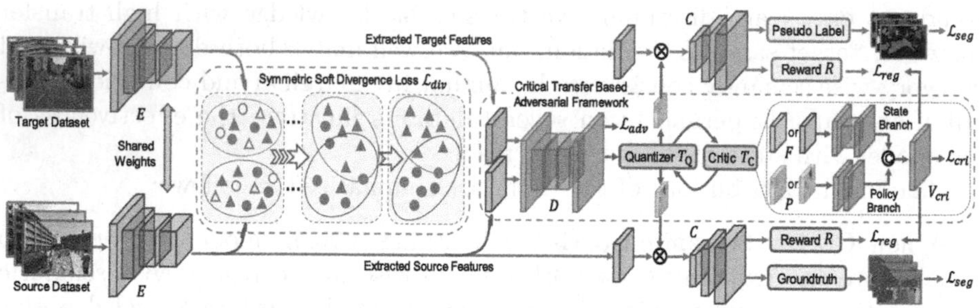

Fig. 2. Overview of our CSCL model. A critical transfer based adversarial framework is developed to highlight transferable knowledge while preventing the negative transfer of untransferable knowledge, where T_C guides T_Q to maximize the transfer gain. A symmetric soft divergence loss $\mathcal{L}_{div}$ aims to explore inter-class relationships and align category-wise distribution with the help of confidence-guided pseudo labels generator.

3 The Proposed CSCL Model

3.1 Overview

Given a labeled source dataset $D_s = \{x_i^s, y_i^s\}_{i=1}^{n_s}$ with n_s samples, where x_i^s and y_i^s denote the image and corresponding pixel label, respectively. Define

$D_t = \{x_j^t\}_{j=1}^{n_t}$ as the unlabeled target domain with n_t images. In unsupervised domain adaptation task, source and target domains share the consistent annotations. Our goal is to explore the transferable knowledge across domains for classifying unlabeled target data, even though there is a large distribution discrepancy.

The overview of our proposed CSCL model is depicted in Fig. 2. The source images x_i^s with annotations y_i^s are first forwarded into the segmentation network to optimize the encoder E and pixel classifier C via the loss $\mathcal{L}_{seg}$. Then the semantic features of both source and target samples extracted by encoder E are employed to update discriminator D via adversarial objective $\mathcal{L}_{adv}$. Unfortunately, such strategy cannot align domain-wise distribution well to some degree, since not all semantic knowledge across domains are transferable. Untransferable knowledge significantly degrades the transfer performance, which is obviously neglected by most existing models [18,21,24,25,34,35]. In light of this issue, a critical transfer based adversarial framework in Sect. 3.2 is developed to explore transferable knowledge while neglecting irrelevant knowledge. Furthermore, category-wise distribution shift is difficult to narrow due to lack of valid pixel labels of target samples. Therefore, a symmetric soft divergence loss $\mathcal{L}_{div}$ in Sect. 3.4 is designed to facilitate category-wise distribution alignment, with the assistance of pseudo labels generator in Sect. 3.3.

3.2 Critical Transfer Based Adversarial Framework

Adversarial learning is utilized to encourage semantic features among different datasets share consistent domain-wise distribution, as presented in Fig. 2. The features extracted from source and target samples are forwarded into discriminator D for distinguishing which one is from the source or target. Meanwhile, encoder E aims to fool D by encouraging the target features to look like the source one. Thus, the generative adversarial objective $\mathcal{L}_{adv}$ is formulated as:

$$\begin{aligned}\min_{\theta_E}\max_{\theta_D}\mathcal{L}_{adv} = \ & \mathbb{E}_{x_i^s \in D_s}\big[1 - \log(D(E(x_i^s;\theta_E);\theta_D))\big] + \\ & \mathbb{E}_{x_j^t \in D_t}\big[\log(D(E(x_j^t;\theta_E);\theta_D))\big],\end{aligned} \tag{1}$$

where θ_D and θ_E are the parameters of D and E. $D(E(x_i^s;\theta_E);\theta_D)$ and $E(x_i^s;\theta_E)$ denote the probability output of D and the features extracted by E for source sample x_i^s, respectively. Furthermore, the definitions of $D(E(x_j^t;\theta_E);\theta_D)$ and $E(x_j^t;\theta_E)$ are similar to $D(E(x_i^s;\theta_E);\theta_D)$ and $E(x_i^s;\theta_E)$ as well.

However, Eq. (1) cannot efficiently align domain-wise distributions between different datasets, since not all semantic representations across domains are transferable. To prevent the negative transfer brought by untransferable knowledge, we develop a critical transfer based adversarial framework, as depicted in Fig. 2. Specifically, transferability-quantizer T_Q highlights transferable knowledge while neglecting untransferable knowledge. Transferability-critic T_C examines the quality of quantified transferability via T_Q, and feedbacks positive supervision to guide T_Q when negative transfer of untransferable knowledge occurs. The details about our critical transfer based adversarial framework are as follows:

Transferability-Quantizer T_Q**:** According to the capacity of discriminator D in identifying whether the input is from source or target datasets, we can easily distinguish which representations across domains are transferable, already transferred or untransferable. In other words, the output probabilities of D can determine whether the corresponding features are transferable or not for domain adaptation. For an intuitive example, the features that are already adapted across different datasets will confuse D to distinguish whether the input is from the source or target. Consequently, T_Q encodes the output probability of D via one convolutional layer to highlight the relevance transferability of transferable knowledge, and then utilizes uncertainty measure function of information theory $\mathcal{U}(f) = -\sum_b f_b \log(f_b)$ to quantify the transferability of corresponding semantic features. Generally, the quantified transferability P_i^s and P_j^t for source image x_i^s and target sample x_j^t are respectively defined as:

$$P_i^s = 1 - \mathcal{U}\big(T_Q(D(E(x_i^s;\theta_E);\theta_D);\theta_{T_Q})\big); P_j^t = 1 - \mathcal{U}\big(T_Q(D(E(x_j^t;\theta_E);\theta_D);\theta_{T_Q})\big), \tag{2}$$

where $T_Q(D(E(x_i^s;\theta_E);\theta_D);\theta_{T_Q})$ and $T_Q(D(E(x_j^t;\theta_E);\theta_D);\theta_{T_Q})$ are the outputs of T_Q with network parameters θ_{T_Q} for source and target samples. Since the false or misleading transferability highlighted by T_Q may degrade transfer performance to some degree, transferability-critic T_C is developed to evaluate the transfer quality of T_Q and maximize the positive transfer gain.

Transferability-Critic T_C**:** Due to the non-differentiability of most evaluation strategies (*e.g.*, the transfer gain), transferability-critic T_C is trained under the reinforcement learning manner. Specifically, the input sample x (x_i^s or x_j^t) is regarded as the state in each training step, and the extracted feature F via encoder E is formulated as $F = E(x;\theta_E)$. Afterwards, based on the input feature F, T_Q predicts the quantified transferability P (P_i^s or P_j^t) for sample x, *i.e.*, $P = 1 - \mathcal{U}\big(T_Q(D(E(x;\theta_E);\theta_D);\theta_{T_Q})\big)$. To conduct the maximum transfer gain and evaluate the quality of transferability, the transferability-critic T_C takes both F and P as the inputs, as depicted in Fig. 2. It contains two branches, where the state branch utilizes three convolutional block to extract state information, and the policy branch applies two convolutional layers to encode relevance transferability of transferable representations. Afterwards, the outputs of both state and policy branches are concatenated, which are forwarded into a convolution layer to evaluate the critic value $V_{cri} = T_C(F, P;\theta_{T_C})$, where θ_{T_C} denotes the network weights of T_C. For transferability-quantizer T_Q, $\mathcal{L}_{cri}$ is proposed to maximize the transfer gain (*i.e.*, critic value V_{cri}), which is defined as follows:

$$\min_{\theta_{T_Q}} \mathcal{L}_{cri} = \mathbb{E}_{x\in(D_s,D_t)}[-V_{cri}] = \mathbb{E}_{x\in(D_s,D_t)}[-T_C(F, P;\theta_{T_C})]. \tag{3}$$

Intuitively, $\mathcal{L}_{cri}$ encourages T_Q to highlight knowledge with high transfer scores by generating higher critic value V_{cri}, which maximizes the positive transfer gain.

To guide T_C feedback positive transfer gain, a new reward mechanism R is developed, which consists of the segmentation reward R^s and the amelioration reward R^a. To be specific, R^s measures whether the transferability quantified via T_Q leads to the correct prediction. The value of R^s at the n-th pixel is:

$$R_n^s = \begin{cases} 1, & \text{if } \operatorname{argmax}\big(C(F;\theta_C)_n\big) = \operatorname{argmax}(y_n), \\ 0, & \text{otherwise}, \end{cases} \tag{4}$$

where θ_C denotes the parameters of classifier C. $C(F;\theta_C)_n$ denotes the probability outputs at the n-th pixel in sample x (x_i^s or x_j^t). y_n is the one-hot groundtruth y_i^s of x_i^s or soft pseudo label $\hat{y}_j^t$ of x_j^t mined in Sect. 3.3 at the n-th pixel. Moreover, the amelioration reward R^a examines whether the transferability quantified via T_Q facilitates positive transfer, where the value of R^a at the n-th pixel is:

$$R_n^a = \begin{cases} 1, & \text{if } C(F;\theta_C)_n^k \; > \; C^b(F;\theta_C)_n^k \text{ and } k = \operatorname{argmax}(y_n) \\ 0, & \text{otherwise}, \end{cases} \tag{5}$$

where $C(F;\theta_C)_n^k$ and $C^b(F;\theta_C)_n^k$ respectively represent the probability of the k-th class predicted by C with quantified transferability P or not. The overall reward is represented as $R = R^s + R^a$. Thus, to conduct T_C feedback accurate supervision, we develop a reward regression loss $\mathcal{L}_{reg}$, which minimizes the gap between the estimated critic value V_{cri} and the defined reward R, *i.e.*,

$$\min_{\theta_{T_C}} \mathcal{L}_{reg} = \mathbb{E}_{x\in(D_s,D_t)}\big[(V_{cri} - R)^2\big] = \mathbb{E}_{x\in(D_s,D_t)}\big[\big(T_C(F,P;\theta_{T_C}) - R\big)^2\big]. \tag{6}$$

3.3 Confidence-Guided Pseudo Labels Generator

To guide the critical transfer based adversarial framework in Sect. 3.2 and facilitate the category-wise alignment in Sect. 3.4, we employ a confidence-guided pseudo labels generator [46,47] to produce soft pseudo labels for unlabeled target data. It efficiently attenuates the enormous deviation brought by unconfident or invalid supervision of pseudo labels. A feasible solution is to progressively mine valid labels relying on their confidence score along the training process. Specifically, after the source sample x_i^s with label y_i^s are utilized to train E and C, the probability output of target sample x_j^t via C is regarded as confidence score. The generation of valid soft pseudo labels has a preference for pixels with high confidence scores. Generally, the joint objective function for training with source data and pseudo labels selection is formulated as follows:

$$\begin{aligned} \min_{\theta_E,\theta_C} \mathcal{L}_{seg} &= \mathbb{E}_{(x_i^s,y_i^s)\in D_s}\Big[\sum_{i=1}^{n_s}\sum_{m=1}^{|x_i^s|}\sum_{k=1}^{K}\big(-(y_i^s)_m^k \log(C(x_i^s;\theta_C)_m^k)\big)\Big] + \\ &\mathbb{E}_{(x_j^t,\hat{y}_j^t)\in D_t}\Big[\sum_{j=1}^{n_t}\sum_{n=1}^{|x_j^t|}\sum_{k=1}^{K}\big(-(\hat{y}_j^t)_n^k \log(\frac{C(x_j^t;\theta_C)_n^k}{\delta_k}) + \gamma(\hat{y}_j^t)_n^k \log((\hat{y}_j^t)_n^k)\big)\Big], \\ &s.t. \; (\hat{y}_j^t)_n = [(\hat{y}_j^t)_n^1, \ldots, (\hat{y}_j^t)_n^K] \in \{\{\mathbb{C}^K | \sum_{k=1}^{K}(\hat{y}_j^t)_n^k = 1, \mathbb{C}^K \in \mathbb{R}^K\} \cup \mathbf{0}\}, \end{aligned} \tag{7}$$

Algorithm 1. The Determination of δ_k in Eq. (7)

Input: The number of classes K, selection amount $\mathcal{S_A}$ of soft pseudo labels, classifier C, target dataset $D_t = \{x_j^t\}_{j=1}^{n_t}$;
Output: δ_k $(k = 1, \ldots, K)$

for $j = 1, \ldots, n_t$ **do**
  $MP_j = C(x_j^t, \theta_C)$; # Maximum probability output of each pixel;
  $Y_j = \text{argmax}(C(x_j^t, \theta_C), \text{axis} = 3)$; # Category prediction of each pixel;
  for $k = 1, \ldots, K$ **do**
    $L_k = [L_k, \text{matrix\_to\_vector}(MP_j(Y_j == k))]$; # Probabilities predicted as k;
  end for
end for
for $k = 1, \ldots, K$ **do**
  $S_k = \text{sorting}(L_k, \text{descending})$; # Sort in a descending order
  $\delta_k = S_k[\mathcal{S_A} \cdot \text{length}(S_k)]$; # Select probability ranked at $\mathcal{S_A} \cdot \text{length}(S_k)$ as δ_k;
end for
return δ_k $(k = 1, \ldots, K)$.

where the first term is supervised training for source data D_s, and the second term is used for pseudo labels selection of target data D_t. $\gamma \geq 0$ is a balanced parameter. θ_C denotes the parameters of classifier C. $\mathbb{C}^K$ represents the K dimensional vector in continuous probability space. K is the number of classes. y_i^s and $\hat{y}_j^t$ denote the one-hot encoding groundtruth of x_i^s and soft pseudo label of x_j^t. Note that the values of $\hat{y}_j^t$ are in continuous space rather than discrete space (*i.e.*, one-hot labels). $(y_i^s)_m^k$ and $(\hat{y}_j^t)_n^k$ denote the values of the k-th category in the m-th pixel of x_i^s and the n-th pixel of x_j^t, respectively. Likewise, $C(x_i^s; \theta_C)_m^k$ and $C(x_j^t; \theta_C)_n^k$ are the probabilities predicted as the k-th class via C for the m-th pixel of x_i^s and the n-th pixel of x_j^t. $\delta_k (k = 1, \ldots, K)$ are the selection thresholds of pseudo labels for each class, which are adaptively determined in Algorithm 1. The last part in the second item of Eq. (7) encourages the continuity of output probabilities, and prevents the misleading supervision of invalid or unconfident pseudo labels.

Intuitively, δ_k determines the selection amount of confident pseudo labels belonging to the k-th class, and the larger value of δ_k facilitates the selection of more confident pseudo labels along the training process. More importantly, as shown in Eq. (7), it forcefully regards unconfident pseudo labels as invalid labels (*i.e.*, $(\hat{y}_j^t)_n = \mathbf{0}$) while preventing the trivial solution from assigning all pseudo pixel labels as $\mathbf{0}$. Note that allocating $(\hat{y}_j^t)_n$ as $\mathbf{0}$ can neglect this unconfident pseudo label of the n-th pixel in the training phase. The determination of δ_k is summarized in Algorithm 1, where $\mathcal{S_A}$ is initialized as 35% and increased by 5% in each epoch until the maximum value 50%.

To progressively mine confident soft pseudo labels, the first step is to optimize the second term of Eq. (7) by employing a Lagrange multiplier ψ. For simplification, the formulation in term of the n-th pixel of x_j^t is rewritten as:

$$\min_{(\hat{y}_j^t)_n^k} \sum_{k=1}^{K} \left(-(\hat{y}_j^t)_n^k \log(\frac{C(x_j^t;\theta_C)_n^k}{\delta_k}) + \gamma(\hat{y}_j^t)_n^k \log((\hat{y}_j^t)_n^k)\right) + \psi(\sum_{k=1}^{K}(\hat{y}_j^t)_n^k - 1), \quad (8)$$

where the first item of Eq. (8) is defined as selection cost $\mathcal{S}_C((\hat{y}_j^t)_n)$. The optimal solution of $(\hat{y}_j^t)_n^k$ can be relaxedly achieved by setting the gradient equal to 0 with respect to the k-th category, *i.e.*,

$$(\hat{y}_j^t)_n^k = e^{-\frac{\gamma+\psi}{\gamma}} (\frac{C(x_j^t;\theta_C)_n^k}{\delta_k})^{\frac{1}{\gamma}}, \ \forall k = 1, \ldots, K, \quad (9)$$

where $e^{-\frac{\gamma+\psi}{\gamma}} = 1/\left[\sum_{k=1}^{K} (\frac{C(x_j^t;\theta_C)_n^k}{\delta_k})^{\frac{1}{\gamma}}\right]$ is determined via the Eq. (9) and the constrain $\sum_{k=1}^{K}(\hat{y}_j^t)_n^k = 1$. The second step involves selecting $(\hat{y}_j^t)_n$ or $\mathbf{0}$ as pseudo label by judging which one leads to a lower selection cost $\mathcal{S}_C$. Thus, the final solution for selecting pseudo labels is:

$$(\hat{y}_j^t)_n^k = \begin{cases} (\frac{C(x_j^t;\theta_C)_n^k}{\delta_k})^{\frac{1}{\gamma}} / \left[\sum_{k=1}^{K} (\frac{C(x_j^t;\theta_C)_n^k}{\delta_k})^{\frac{1}{\gamma}}\right], & \text{if } \mathcal{S}_C((\hat{y}_j^t)_n) < \mathcal{S}_C(\mathbf{0}), \\ 0, & \text{otherwise.} \end{cases} \quad (10)$$

3.4 Symmetric Soft Divergence Loss $\mathcal{L}_{div}$

With the confident pseudo labels mined by Eq. (10) in Sect. 3.3 as reliable guidance, a symmetric soft divergence loss $\mathcal{L}_{div}$ is proposed to align category-wise distribution. In other words, $\mathcal{L}_{div}$ encourages semantic features of same class to be compactly clustered together via category labels regardless of domains, and drives features from different classes across domains to satisfy the inter-category relationships. Different from [7,41] that align the feature centroids directly, $\mathcal{L}_{div}$ aims to mitigate the conditional distributions shift by minimizing the symmetric soft Kullback-Leibler (KL) divergence among source and target features, with respect to each class k. Furthermore, $\mathcal{L}_{div}$ enforces category ambiguities across domains to be more consistent, which is concretely written as:

$$\min_{\theta_E} \mathcal{L}_{div} = \frac{1}{K}\sum_{k=1}^{K}\frac{1}{2}(D_{\mathrm{KL}}(F_s^k||F_t^k) + D_{\mathrm{KL}}(F_t^k||F_s^k)); \quad (11)$$

where $D_{\mathrm{KL}}(F_s^k||F_t^k) = \sum_q (F_s^k)_q \log\frac{(F_s^k)_q}{(F_t^k)_q}$ denotes the KL divergence between source feature centroid F_s^k and target soft feature centroid F_t^k, in terms of the k-th category. Likewise, $D_{\mathrm{KL}}(F_t^k||F_s^k) = \sum_q (F_t^k)_q \log\frac{(F_t^k)_q}{(F_s^k)_q}$ shares the similar definition with $D_{\mathrm{KL}}(F_s^k||F_t^k)$. According to the source label y_i^s, the source feature centroid F_s^k of the k-th class is computed by the following equation, *i.e.*,

$$F_s^k = \mathbb{E}_{(x_i^s,y_i^s)\in D_s}\left[\frac{1}{N_s^k}\sum_{i=1}^{n_s}\sum_{m=1}^{|x_i^s|}\left(E(x_i^s;\theta_E)_m \cdot \mathbf{1}_{\mathrm{argmax}((y_i^s)_m)=k}\right)\right]; \quad (12)$$

Algorithm 2. The Optimization Procedure of Our CSCL Model

Input: The source and target data (D_s and D_t), maximum iteration I, γ, ξ_1, ξ_2, ξ_3;
Output: The network parameters $\theta_E, \theta_C, \theta_D, \theta_{T_Q}$ and θ_{T_C}

Initialize all parameters of network architecture;
for $i = 1, \ldots, I$ **do**
    Randomly select a batch of samples from both D_s and D_t;
    Update θ_E and θ_C via minimizing the first term of $\mathcal{L}_{seg}$;
    Generate confident soft pseudo labels for target samples via Eq. (10);
    Update θ_E and θ_D via optimizing $\min_{\theta_E} \max_{\theta_D} \ \xi_2 \mathcal{L}_{adv} + \xi_3 \mathcal{L}_{div}$;
    Update θ_{T_Q} via minimizing $\mathcal{L}_{seg} + \xi_1 \mathcal{L}_{cri}$;
    Update θ_{T_C} via minimizing $\mathcal{L}_{reg}$;
end for
Return $\theta_E, \theta_C, \theta_D, \theta_{T_Q}$ and θ_{T_C}.

where $N_s^k = \sum_{i=1}^{n_s} \sum_{m=1}^{|x_i^s|} \mathbf{1}_{\text{argmax}((y_i^s)_m)=k}$ is the number of pixels annotated as class k in dataset D_s. $E(x_i^s; \theta_E)_m$ represents the feature extracted by E at the m-th pixel in source sample x_i^s. Similarly, with the assistance of generated pseudo labels $\hat{y}_j^t$, the target soft feature centroid F_t^k of the k-th category is defined as:

$$F_t^k = \mathbb{E}_{(x_j^t, \hat{y}_j^t) \in D_t} \Big[\frac{1}{N_t^k} \sum_{j=1}^{n_t} \sum_{n=1}^{|x_j^t|} (\hat{y}_j^t)_n^k E(x_j^t; \theta_E)_n \cdot \mathbf{1}_{\text{argmax}((\hat{y}_j^t)_n)=k} \Big], \tag{13}$$

where $N_t^k = \sum_{j=1}^{n_t} \sum_{n=1}^{|x_j^t|} (\hat{y}_j^t)_n^k \cdot \mathbf{1}_{\text{argmax}((\hat{y}_j^t)_n)=k}$ is the sum of the probabilities predicted as class k in dataset D_t. $E(x_j^t; \theta_E)_n$ denotes the feature extracted by E at the n-th pixel in target image x_j^t.

3.5 Implementation Details

Network Architecture: The DeepLab-v2 [2] with ResNet-101 [15] and FCN-8s [31] with VGG-16 [32] are employed as our backbone network, *i.e.*, E and C. They are initially pre-trained by the ImageNet [8]. The discriminator D contains 5 convolutional layers with channels as {64, 128, 256, 512, 1}, where each block excluding the last one is activated by the leaky ReLU with parameter as 0.2. As for T_C, the state branch includes 3 layers whose channels are {64, 32, 16}, and the channels of two convolutional blocks in policy branch are both 16. Moreover, there is only one convolution layer with the channel as 1 in T_Q.

Training and Evaluating: In the training phase, the overall optimization objective $\mathcal{L}_{obj}$ for our proposed CSCL model is formulated as:

$$\min_{\theta_E, \theta_C, \theta_{T_Q}, \theta_{T_C}} \max_{\theta_D} \mathcal{L}_{obj} = \mathcal{L}_{seg} + \mathcal{L}_{reg} + \xi_1 \mathcal{L}_{cri} + \xi_2 \mathcal{L}_{adv} + \xi_3 \mathcal{L}_{div}, \tag{14}$$

where ξ_1, ξ_2, $\xi_3 \geq 0$ are balanced weights. The optimization procedure is summarized in **Algorithm** 2. For DeepLab-v2 with ResNet-101, SGD is utilized as optimizer whose the initial learning rate is 2.5×10^{-4} and decreased via poly policy with power as 0.9. For FCN-8s with VGG-16, we employ Adam as optimizer whose initial learning rate is 1.0×10^{-4} and the momentum is set as 0.9 and 0.99. Moreover, Adam is also used to optimize the discriminator D. The initial learning rate is respectively set as 1.5×10^{-4} and 1.5×10^{-6} for ResNet-101 and VGG-16. We empirically set $\gamma = 0.25$ in Eq. (10), and set $\xi_1 = 0.3, \xi_2 = 0.001$ and $\xi_3 = 10$ in Eq. (14) for all experiments. The batch size for training is set as 1. In the evaluating stage, we directly forward x_j^t into E and C for prediction.

4 Experiments

4.1 Datasets and Evaluation Metric

Cityscapes [6] is a real-world dataset about street scenes from 50 different European cities, which is divided into a training subset with 2993 samples, a testing subset with 1531 images and a validation subset with 503 samples. There are total 34 distinct finely-annotated categories in this dataset.

GTA [26] with 24996 images is generated from a fictional computer game called Grand Theft Auto V, whose 19 classes are compatible with Cityscapes [6].

SYNTHIA [28] is a large-scale automatically-labeled synthetic dataset for semantic segmentation task of urban scenes, whose the subset named SYNTHIA-RAND-CITYSCAPES with 9400 images is used in our experiments.

NTHU [4] consists of four real world datasets, which are respectively collected from Rome, Rio, Tokyo and Taipei. Every dataset contains the training and testing subsets, and shares 13 common categories with Cityscapes [6].

Evaluation Metric: Intersection over Union (IoU) is employed as evaluation metric, and mIoU denotes the mean value of IoU.

4.2 Experiments on GTA → Cityscapes Task

When transferring from GTA [26] to Cityscapes [6], GTA and the training subset of Cityscapes are respectively considered as source and target domains. The validation subset of Cityscapes is used for evaluation.

Comparisons Performance: Table 1 reports the adaptation performance of our model compared with state-of-the-art methods on GTA → Cityscapes task. From the Table 1, we have the following observations: 1) Our model outperforms all existing state-of-the-art methods about 1.2%~14.3% mIoU, which efficiently matches both domain-wise and category-wise semantic consistency across

Table 1. Adaptation performance of transferring from GTA [26] to Cityscapes [6].

Method	Net	road	sidewalk	building	wall	fence	pole	light	sign	veg	terrain	sky	person	rider	car	truck	bus	train	mbike	bike	mIoU
Source only [32]	VGG	18.1	6.8	64.1	7.3	8.7	21.0	14.9	16.8	45.9	2.4	64.4	41.6	17.5	55.3	8.4	5.0	6.9	4.3	13.8	22.3
Wild [16]		70.4	32.4	62.1	14.9	5.4	10.9	14.2	2.7	79.2	21.3	64.6	44.1	4.2	70.4	8.0	7.3	0.0	3.5	0.0	27.1
CDA [43]		74.9	22.0	71.7	6.0	11.9	8.4	16.3	11.1	75.7	11.3	66.5	38.0	9.3	55.2	18.8	18.9	0.0	16.8	14.6	28.9
MCD [29]		86.4	8.5	76.1	18.6	9.7	14.9	7.8	0.6	82.8	32.7	71.4	25.2	1.1	76.3	16.1	17.1	1.4	0.2	0.0	28.8
CBST [47]		66.7	26.8	73.7	14.8	9.5	28.3	25.9	10.1	75.5	15.7	51.6	47.2	6.2	71.9	3.7	2.2	5.4	18.9	**32.4**	30.9
CLAN [25]		88.0	30.6	79.2	23.4	20.5	26.1	23.0	14.8	81.6	34.5	72.0	45.8	7.9	80.5	26.6	29.9	0.0	10.7	0.0	36.6
SWD [18]		**91.0**	35.7	78.0	21.6	21.7	**31.8**	**30.2**	**25.2**	80.2	23.9	74.1	53.1	15.8	79.3	22.1	26.5	1.5	17.2	30.4	39.9
ADV [35]		86.9	28.7	78.7	28.5	25.2	17.1	20.3	10.9	80.0	26.4	70.2	47.1	8.4	81.5	26.0	17.2	**18.9**	11.7	1.6	36.1
DPR [34]		87.3	35.7	79.5	**32.0**	14.5	21.5	24.8	13.7	80.4	32.0	70.5	50.5	16.9	81.0	20.8	28.1	4.1	15.5	4.1	37.5
LDF [19]		88.8	36.9	76.9	20.9	15.4	19.6	21.8	7.9	82.9	26.7	76.1	51.7	9.4	76.1	22.4	28.9	1.7	15.2	0.0	35.8
SSF [12]		88.7	32.1	79.5	29.9	22.0	23.8	21.7	10.7	80.8	29.8	72.5	49.5	16.1	**82.1**	23.2	18.1	3.5	24.4	8.1	37.7
PyCDA [21]		86.7	24.8	**80.9**	21.4	27.3	30.2	26.6	21.1	**86.6**	28.9	58.8	**53.2**	17.9	80.4	18.8	22.4	4.1	9.7	6.2	37.2
Ours-w/oTC		88.7	43.5	74.2	29.4	26.1	10.8	17.9	16.0	78.9	38.7	75.2	38.4	17.5	73.3	29.5	37.8	0.4	26.7	29.3	39.6
Ours-w/oCG		88.3	39.1	76.4	28.3	22.4	11.9	17.3	18.7	76.2	36.0	74.5	39.3	18.6	73.0	**30.7**	**36.1**	0.8	27.4	30.1	39.2
Ours-w/oSD		89.1	45.0	77.6	28.8	27.5	12.1	18.6	21.2	77.4	37.7	76.5	40.7	**19.8**	74.4	29.3	33.8	0.6	28.2	29.2	40.4
Ours		89.8	**46.1**	75.2	30.1	**27.9**	15.0	20.4	18.9	82.6	**39.1**	**77.6**	47.8	17.4	76.2	28.5	33.4	0.5	**29.4**	30.8	**41.4**
Source only [15]	ResNet	75.8	16.8	77.2	12.5	21.0	25.5	30.1	20.1	81.3	24.6	70.3	53.8	26.4	49.9	17.2	25.9	6.5	25.3	36.0	36.6
LtA [33]		86.5	36.0	79.9	23.4	23.3	23.9	35.2	14.8	83.4	33.3	75.6	58.5	27.6	73.7	32.5	35.4	3.9	30.1	28.1	42.4
CGAN [17]		89.2	49.0	70.7	13.5	10.9	38.5	29.4	33.7	77.9	37.6	65.8	**75.1**	32.4	77.8	**39.2**	45.2	0.0	25.2	35.4	44.5
CBST [47]		88.0	**56.2**	77.0	27.4	22.4	**40.7**	**47.3**	40.9	82.4	21.6	60.3	50.2	20.4	83.8	35.0	**51.0**	15.2	20.6	37.0	46.2
DLOW [13]		87.1	33.5	80.5	24.5	13.2	29.8	29.5	26.6	82.6	26.7	81.8	55.9	25.3	78.0	33.5	38.7	0.0	22.9	34.5	42.3
CLAN [25]		87.0	27.1	79.6	27.3	23.3	28.3	35.5	24.2	83.6	27.4	74.2	58.6	28.0	76.2	33.1	36.7	6.7	31.9	31.4	43.2
SWD [18]		92.0	46.4	82.4	24.8	24.0	35.1	33.4	34.2	83.6	30.4	80.9	56.9	21.9	82.0	24.4	28.7	6.1	25.0	33.6	44.5
ADV [35]		89.4	33.1	81.0	26.6	26.8	27.2	33.5	24.7	83.9	36.7	78.8	58.7	30.5	84.8	38.5	44.5	1.7	31.6	32.5	45.5
SWLS [10]		**92.7**	48.0	78.8	25.7	27.2	36.0	42.2	**45.3**	80.6	14.6	66.0	62.1	30.4	**86.2**	28.0	45.6	**35.9**	16.8	34.7	47.2
DPR [34]		92.3	51.9	82.1	29.2	25.1	24.5	33.8	33.0	82.4	32.8	82.2	58.6	27.2	84.3	33.4	46.3	2.2	29.5	32.3	46.5
SSF [12]		90.3	38.9	81.7	24.8	22.9	30.5	37.0	21.2	84.8	38.8	76.9	58.8	30.7	85.7	30.6	38.1	5.9	28.3	36.9	45.4
PyCDA [21]		90.5	36.3	84.4	32.4	**28.7**	34.6	36.4	31.5	**86.8**	37.9	78.5	62.3	21.5	85.6	27.9	34.8	18.0	22.9	**49.3**	47.4
Ours-w/oTC		91.6	47.9	83.4	35.0	23.6	31.6	36.5	31.9	82.9	36.6	76.4	58.7	25.6	81.5	37.1	46.6	0.5	26.0	34.0	46.7
Ours-w/oCG		89.6	40.8	**84.6**	30.4	22.7	32.0	37.4	33.7	82.3	39.6	80.7	57.4	28.7	82.8	27.4	48.2	1.0	27.0	29.5	46.1
Ours-w/oSD		89.3	47.8	82.4	31.3	25.1	31.2	37.3	34.9	83.9	37.9	83.0	59.4	31.4	79.0	35.7	42.0	0.2	34.1	34.6	47.4
Ours		89.6	50.4	83.0	**35.6**	26.9	31.1	37.3	35.1	83.5	**40.6**	**84.0**	60.6	**34.3**	80.9	35.1	47.3	0.5	**34.5**	33.7	**48.6**

domains. 2) For the classes with various appearances among different datasets (*e.g.*, rider, motorbike, terrain, fence and sidewalk), our model achieves better performance to mitigate the large distribution shifts by exploring transferable representations.

Ablation Studies: In this subsection, we investigate the importance of different components in our model by conducting variant experiments, as shown in the gray part of Table 1. Training the model without critical transfer based adversarial framework, confidence-guided pseudo labels generator and symmetric soft divergence loss are denoted as Ours-w/oTC, Ours-w/oCG and Ours-w/oSD, respectively. The transfer performance degrades 1.0%$\sim$2.5% when any component of our model is removed. Table 1 validates that all designed components play an importance role in exploring transferable representations while neglecting irrelevant knowledge. Moreover, they can well narrow both marginal and conditional distributions shifts across domains.

Convergence Investigation: The convergence curves of our model with respect to mIoU and domain gap are respectively depicted in Fig. 3(a) and Fig. 3(b). Specifically, our model equipped with VGG-16 or ResNet-101 can achieve efficient convergence when the iterative epoch number is 5. More importantly, domain gap among different datasets is iteratively minimized to a stable value via our model, which efficiently narrows the domain discrepancy.

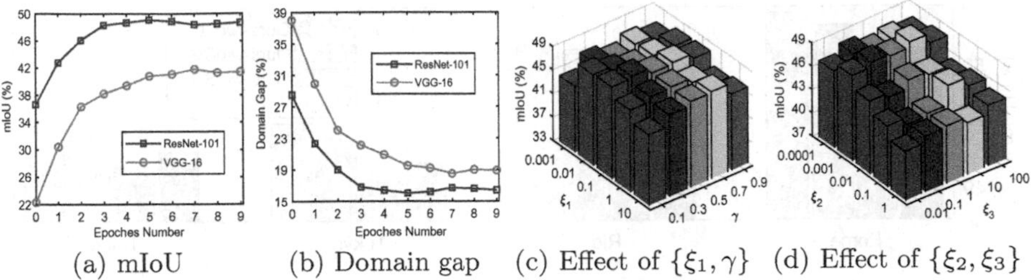

(a) mIoU (b) Domain gap (c) Effect of $\{\xi_1, \gamma\}$ (d) Effect of $\{\xi_2, \xi_3\}$

Fig. 3. Experimental results on GTA → Cityscapes task. (a): Convergence curves about mIoU; (b): Convergence curves about domain gap; (c): The effect of $\{\xi_1, \gamma\}$ when $\xi_2 = 0.001$ and $\xi_3 = 10$; (d): The effect of $\{\xi_2, \xi_3\}$ when $\xi_1 = 0.3$ and $\gamma = 0.25$.

Parameter Sensitivity: Extensive parameter experiments are empirically conducted to investigate the parameter sensitivity, and determine the optimal selection of hyper-parameters. Figure 3(c) and Fig. 3(d) present the effects of hyper-parameters $\{\xi_1, \gamma\}$ and $\{\xi_2, \xi_3\}$ on our model with ResNet-101 as basic network, respectively. We can conclude that our model achieves stable transfer performance even though hyper-parameters have a wide range of selection. Moreover, γ is essential to mine confident pseudo labels for target samples, which provides positive guidance for transferability-critic T_C and $\mathcal{L}_{div}$.

Table 2. Adaptation performance of transferring from SYNTHIA [28] to Cityscapes [6].

Method	Net	road	sidewalk	building	wall	fence	pole	light	sign	veg	sky	person	rider	car	bus	mbike	bike	mIoU
Source only [32]		6.4	17.7	29.7	1.2	0.0	15.1	0.0	7.2	30.3	66.8	51.1	1.5	47.3	3.9	0.1	0.0	17.4
Wild [16]		11.5	19.6	30.8	4.4	0.0	20.3	0.1	11.7	42.3	68.7	51.2	3.8	54.0	3.2	0.2	0.6	20.2
CDA [43]		65.2	26.1	74.9	0.1	0.5	10.7	3.7	3.0	76.1	70.6	47.1	8.2	43.2	20.7	0.7	13.1	29.0
LSD [30]		80.1	29.1	77.5	2.8	0.4	26.8	11.1	18.0	78.1	76.7	48.2	15.2	70.5	17.4	8.7	16.7	36.1
DCAN [38]		79.9	30.4	70.8	1.6	0.6	22.3	6.7	**23.0**	76.9	73.9	41.9	16.7	61.7	11.5	10.3	**38.6**	35.4
CBST [47]	VGG	69.6	28.7	69.5	**12.1**	0.1	25.4	11.9	13.6	82.0	**81.9**	49.1	14.5	66.0	6.6	3.7	32.4	35.4
ADV [35]		67.9	29.4	71.9	6.3	0.3	19.9	0.6	2.6	74.9	74.9	35.4	9.6	67.8	21.4	4.1	15.5	31.4
TGCF [5]		**90.1**	**48.6**	**80.7**	2.2	0.2	27.2	3.2	14.3	**82.1**	78.4	**54.4**	16.4	**82.5**	12.3	1.7	21.8	38.5
DPR [34]		72.6	29.5	77.2	3.5	0.4	21.0	1.4	7.9	73.3	79.0	45.7	14.5	69.4	19.6	7.4	16.5	33.7
PyCDA [21]		80.6	26.6	74.5	2.0	0.1	18.1	**13.7**	14.2	80.8	71.0	48.0	19.0	72.3	22.5	12.1	18.1	35.9
Ours-w/oTC		79.4	36.9	76.1	5.3	0.6	25.1	12.4	12.5	76.7	78.5	44.6	22.4	72.8	28.6	12.5	25.7	38.1
Ours-w/oCG		75.0	32.1	78.2	2.7	**0.8**	22.0	4.2	7.4	74.3	81.1	40.7	23.6	72.3	34.0	15.9	27.6	37.0
Ours-w/oSD		71.2	30.4	75.7	6.3	0.6	26.7	15.6	17.4	77.1	80.6	50.3	21.4	72.0	29.6	16.7	26.3	38.6
Ours		70.9	30.5	77.8	9.0	0.6	**27.3**	8.8	12.9	74.8	81.1	43.0	**25.1**	73.4	**34.5**	**19.5**	38.2	**39.2**
Source only [15]		55.6	23.8	74.6	9.2	0.2	24.4	6.1	12.1	74.8	79.0	55.3	19.1	39.6	23.3	13.7	25.0	33.5
CGAN [17]		85.0	25.8	73.5	3.4	**3.0**	31.5	19.5	21.3	67.4	69.4	**68.5**	25.0	76.5	41.6	17.9	29.5	41.2
DCAN [38]		81.5	33.4	72.4	7.9	0.2	20.0	8.6	10.5	71.0	68.7	51.5	18.7	75.3	22.7	12.8	28.1	36.5
CBST [47]		53.6	23.7	75.0	12.5	0.3	**36.4**	23.5	26.3	**84.8**	74.7	67.2	17.5	84.5	28.4	15.2	**55.8**	42.5
ADV [35]		**85.6**	**42.2**	79.7	8.7	0.4	25.9	5.4	8.1	80.4	84.1	57.9	23.8	73.3	36.4	14.2	33.0	41.2
SWLS [10]		68.4	30.1	74.2	21.5	0.4	29.2	29.3	25.1	80.3	81.5	63.1	16.4	75.6	13.5	26.1	51.9	42.9
MSL [3]	ResNet	82.9	40.7	80.3	10.2	0.8	25.8	12.8	18.2	82.5	82.2	53.1	18.0	79.0	31.4	10.4	35.6	41.4
DPR [34]		82.4	38.0	78.6	8.7	0.6	26.0	3.9	11.1	75.5	84.6	53.5	21.6	71.4	32.6	19.3	31.7	40.0
PyCDA [21]		75.5	30.9	**83.3**	20.8	0.7	32.7	27.3	**33.5**	84.7	**85.0**	64.1	25.4	85.0	**45.2**	21.2	32.0	46.7
Ours-w/oTC		71.8	32.4	80.5	22.6	0.4	28.6	29.4	27.9	83.1	83.7	65.5	19.8	84.5	25.6	24.3	46.1	45.4
Ours-w/oCG		74.5	33.4	78.2	**24.1**	0.7	30.8	**31.7**	25.2	76.5	81.4	58.7	25.6	75.7	24.8	25.3	41.6	44.3
Ours-w/oSD		79.2	38.1	79.8	21.3	0.8	27.6	31.0	24.3	81.5	81.7	62.4	22.1	**86.4**	31.5	23.8	44.2	46.0
Ours		80.2	41.1	78.9	23.6	0.6	31.0	27.1	29.5	82.5	83.2	62.1	**26.8**	81.5	37.2	**27.3**	42.9	**47.2**

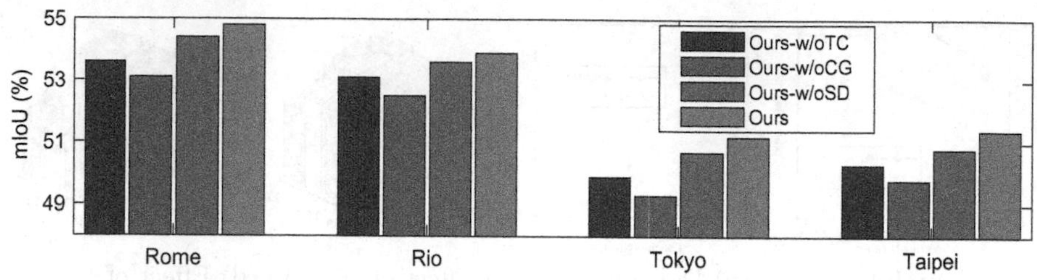

Fig. 4. Ablation studies of our model with ResNet-101 on Cityscapes → NTHU task.

Table 3. Performance of our model with ResNet-101 on Cityscapes [6] → NTHU [4].

City	Method	road	sidewalk	building	light	sign	veg	sky	person	rider	car	bus	mbike	bike	mIoU
Rome	Source only [15]	83.9	34.3	87.7	13.0	41.9	84.6	92.5	37.7	22.4	80.8	38.1	39.1	5.3	50.9
	NMD [4]	79.5	29.3	84.5	0.0	22.2	80.6	82.8	29.5	13.0	71.7	37.5	25.9	1.0	42.9
	CBST [47]	**87.1**	**43.9**	89.7	14.8	**47.7**	85.4	90.3	45.4	26.6	**85.4**	20.5	49.8	**10.3**	53.6
	LtA [33]	83.9	34.2	88.3	18.8	40.2	**86.2**	93.1	47.8	21.7	80.9	47.8	48.3	8.6	53.8
	MSL [3]	82.9	32.6	86.7	20.7	41.6	85.0	93.0	47.2	22.5	82.2	**53.8**	50.5	9.9	54.5
	SSF [12]	84.2	38.4	87.4	**23.4**	43.0	85.6	88.2	**50.2**	23.7	80.6	38.1	51.6	8.6	54.1
	Ours	85.7	36.5	**92.1**	19.4	42.6	84.8	**95.0**	46.9	**28.3**	79.4	40.5	**54.2**	7.5	**54.8**
Rio	Source only [15]	76.6	47.3	82.5	12.6	22.5	77.9	86.5	43.0	19.8	74.5	36.8	29.4	16.7	48.2
	NMD [4]	74.2	43.9	79.0	2.4	7.5	77.8	69.5	39.3	10.3	67.9	**41.2**	27.9	10.9	42.5
	CBST [47]	**84.3**	**55.2**	**85.4**	**19.6**	**30.1**	80.5	77.9	55.2	28.6	**79.7**	33.2	37.6	11.5	52.2
	LtA [33]	76.2	44.7	84.6	9.3	25.5	**81.8**	87.3	55.3	32.7	74.3	28.9	43.0	27.6	51.6
	MSL [3]	76.9	48.8	85.2	13.8	18.9	81.7	**88.1**	54.9	34.0	76.8	39.8	44.1	29.7	53.3
	SSF [12]	74.2	43.7	82.5	10.3	21.7	79.4	86.7	55.9	36.1	74.9	33.7	**52.6**	33.7	52.7
	Ours	79.5	52.7	83.6	12.4	23.0	80.9	79.7	**56.1**	**37.7**	72.4	36.0	51.6	**34.1**	**53.8**
Tokyo	Source only [15]	82.9	31.3	78.7	14.2	24.5	81.6	89.2	48.6	33.3	70.5	7.7	11.5	45.9	47.7
	NMD [4]	83.4	35.4	72.8	12.3	12.7	77.4	64.3	42.7	21.5	64.1	**20.8**	8.9	40.3	42.8
	CBST [47]	**85.2**	33.6	80.4	8.3	**31.1**	**83.9**	78.2	53.2	28.9	**72.7**	4.4	27.0	47.0	48.8
	LtA [33]	81.5	26.0	77.8	17.8	26.8	82.7	**90.9**	55.8	38.0	72.1	4.2	24.5	50.8	49.9
	MSL [3]	81.2	30.1	77.0	12.3	27.3	82.8	89.5	58.2	32.7	71.5	5.5	**37.4**	48.9	50.5
	SSF [12]	82.1	27.4	78.0	**18.4**	26.6	83.0	90.8	57.1	35.8	72.0	4.6	27.3	**52.8**	50.4
	Ours	83.1	**35.5**	**81.2**	15.8	24.9	81.3	86.4	**58.8**	**39.2**	68.1	6.7	30.4	51.2	**51.0**
Taipei	Source only [15]	83.5	33.4	86.6	12.7	16.4	77.0	**92.1**	17.6	13.7	70.7	37.7	44.4	18.5	46.5
	NMD [4]	78.6	28.6	80.0	13.1	7.6	68.2	82.1	16.8	9.4	60.4	34.0	26.5	9.9	39.6
	CBST [47]	**86.1**	35.2	84.2	15.0	**22.2**	75.6	74.9	22.7	**33.1**	**78.0**	37.6	**58.0**	30.9	50.3
	LtA [33]	81.7	29.5	85.2	26.4	15.6	76.7	91.7	31.0	12.5	71.5	41.1	47.3	27.7	49.1
	MSL [3]	80.7	32.5	85.5	**32.7**	15.1	**78.1**	91.3	32.9	7.6	69.5	**44.8**	52.4	**34.9**	50.6
	SSF [12]	84.5	**35.3**	86.4	17.7	16.9	77.7	91.3	31.8	22.3	73.7	41.1	55.9	28.5	51.0
	Ours	83.4	33.7	**87.5**	24.3	17.2	75.8	90.6	**33.2**	24.1	75.3	35.8	56.4	31.2	**51.4**

4.3 Experiments on SYNTHIA → Cityscapes Task

For the experimental configurations, we regard SYNTHIA [28] and the training subset of Cityscape [6] as source and target domains, and utilize the validation subset of Cityscapes to evaluate the transfer performance. From the presented comparison results in Table 2, we can notice that: 1) Compared with other competing methods such as [3,10,21,34,35], our model improves the performance about 0.5%∼19.0% mIoU to bridge both domain-wise and category-wise distributions shifts across domains. 2) Ablation studies verify that each component is indispensable to boost semantic knowledge transfer. 3) A critical transfer based adversarial framework efficiently highlights transferable domain-wise knowledge while neglecting irrelevant knowledge.

4.4 Experiments on Cityscapes → NTHU Task

As for Cityscapes → NTHU task, we respectively regard the training subset of Cityscapes and NTHU as source and target domains. The testing subset of every city in NTHU is utilized for evaluation. Table 3 reports the comparison adaptation results, and the corresponding ablation studies are depicted in Fig. 4. Some conclusions are drawn from Table 3 and Fig. 4: 1) Our model achieves about 0.3%∼11.9% improvement than other methods across all the evaluated cities. 2) Ablation experiments validate the rationality and effectiveness of each module. 3) With the guidance of confident soft pseudo labels, $\mathcal{L}_{div}$ could map the semantic features of same class across domains compactly nearby.

5 Conclusion

In this paper, we propose a new Critical Semantic-Consistent Learning (CSCL) model, which aims to narrow both domain-wise and category-wise distributions shifts. A critical transfer based adversarial framework is developed to highlight transferable domain-wise knowledge while preventing the negative transfer of irrelevant knowledge. Specifically, transferability-critic feedbacks positive guidance to transferability quantizer when negative transfer occurs. Moreover, with confidence-guided pseudo label generator assigning soft pseudo labels for target samples, a symmetric soft divergence loss is designed to minimize category-wise discrepancy. Experiments on public datasets verify the effectiveness of our model.

Acknowledgment. This work is supported by Ministry of Science and Technology of the People's Republic of China (2019YFB1310300), National Nature Science Foundation of China under Grant (61722311, U1613214, 61821005, 61533015) and National Postdoctoral Innovative Talents Support Program (BX20200353).

References

1. Chen, L.C., Papandreou, G., Kokkinos, I., Murphy, K., Yuille, A.: Semantic image segmentation with deep convolutional nets and fully connected crfs. arXiv preprint arXiv:1412.7062, December 2014
2. Chen, L.-C., Zhu, Y., Papandreou, G., Schroff, F., Adam, H.: Encoder-decoder with atrous separable convolution for semantic image segmentation. In: Ferrari, V., Hebert, M., Sminchisescu, C., Weiss, Y. (eds.) ECCV 2018. LNCS, vol. 11211, pp. 833–851. Springer, Cham (2018). https://doi.org/10.1007/978-3-030-01234-2_49
3. Chen, M., Xue, H., Cai, D.: Domain adaptation for semantic segmentation with maximum squares loss. In: The IEEE International Conference on Computer Vision (ICCV), October 2019
4. Chen, Y.H., Chen, W.Y., Chen, Y.T., Tsai, B.C., Frank Wang, Y.C., Sun, M.: No more discrimination: cross city adaptation of road scene segmenters. In: The IEEE International Conference on Computer Vision (ICCV), October 2017

5. Choi, J., Kim, T., Kim, C.: Self-ensembling with GAN-based data augmentation for domain adaptation in semantic segmentation. In: The IEEE International Conference on Computer Vision (ICCV), October 2019
6. Cordts, M., et al.: The cityscapes dataset for semantic urban scene understanding. In: The IEEE Conference on Computer Vision and Pattern Recognition (CVPR), June 2016
7. Courty, N., Flamary, R., Habrard, A., Rakotomamonjy, A.: Joint distribution optimal transportation for domain adaptation. In: Guyon, I., et al. (eds.) Advances in Neural Information Processing Systems, vol. 30, pp. 3730–3739. Curran Associates, Inc. (2017)
8. Deng, J., Dong, W., Socher, R., Li, L., Li, K., Fei-Fei, L.: ImageNet: a large-scale hierarchical image database. In: 2009 IEEE Conference on Computer Vision and Pattern Recognition, pp. 248–255, June 2009
9. Ding, Z., Li, S., Shao, M., Fu, Y.: Graph adaptive knowledge transfer for unsupervised domain adaptation. In: Ferrari, V., Hebert, M., Sminchisescu, C., Weiss, Y. (eds.) ECCV 2018. LNCS, vol. 11206, pp. 36–52. Springer, Cham (2018). https://doi.org/10.1007/978-3-030-01216-8_3
10. Dong, J., Cong, Y., Sun, G., Hou, D.: Semantic-transferable weakly-supervised endoscopic lesions segmentation. In: The IEEE International Conference on Computer Vision (ICCV), October 2019
11. Dong, J., Cong, Y., Sun, G., Zhong, B., Xu, X.: What can be transferred: unsupervised domain adaptation for endoscopic lesions segmentation. In: IEEE/CVF Conference on Computer Vision and Pattern Recognition (CVPR), June 2020
12. Du, L., et al.: SSF-DAN: separated semantic feature based domain adaptation network for semantic segmentation. In: The IEEE International Conference on Computer Vision (ICCV), October 2019
13. Gong, R., Li, W., Chen, Y., Gool, L.V.: DLOW: domain flow for adaptation and generalization. In: The IEEE Conference on Computer Vision and Pattern Recognition (CVPR), June 2019
14. Goodfellow, I.J., et al.: Generative adversarial nets. In: Proceedings of the 27th International Conference on Neural Information Processing Systems - Volume 2, pp. 2672–2680 (2014)
15. He, K., Zhang, X., Ren, S., Sun, J.: Deep residual learning for image recognition. In: The IEEE Conference on Computer Vision and Pattern Recognition (CVPR), June 2016
16. Hoffman, J., Wang, D., Yu, F., Darrell, T.: Fcns in the wild: Pixel-level adversarial and constraint-based adaptation. arXiv preprint arXiv:1612.02649 (2016)
17. Hong, W., Wang, Z., Yang, M., Yuan, J.: Conditional generative adversarial network for structured domain adaptation. In: The IEEE Conference on Computer Vision and Pattern Recognition (CVPR), June 2018
18. Lee, C.Y., Batra, T., Baig, M.H., Ulbricht, D.: Sliced wasserstein discrepancy for unsupervised domain adaptation. In: The IEEE Conference on Computer Vision and Pattern Recognition (CVPR), June 2019
19. Lee, S., Kim, D., Kim, N., Jeong, S.G.: Drop to adapt: learning discriminative features for unsupervised domain adaptation. In: The IEEE International Conference on Computer Vision (ICCV), October 2019
20. Li, Y., Yuan, L., Vasconcelos, N.: Bidirectional learning for domain adaptation of semantic segmentation. In: The IEEE Conference on Computer Vision and Pattern Recognition (CVPR), June 2019

21. Lian, Q., Lv, F., Duan, L., Gong, B.: Constructing self-motivated pyramid curriculums for cross-domain semantic segmentation: a non-adversarial approach. In: The IEEE International Conference on Computer Vision (ICCV), October 2019
22. Liu, S., De Mello, S., Gu, J., Zhong, G., Yang, M.H., Kautz, J.: Learning affinity via spatial propagation networks. In: Guyon, I., et al. (eds.) Advances in Neural Information Processing Systems, vol. 30, pp. 1520–1530. Curran Associates, Inc. (2017)
23. Liu, Z., Li, X., Luo, P., Loy, C.C., Tang, X.: Semantic image segmentation via deep parsing network. In: The IEEE International Conference on Computer Vision (ICCV), December 2015
24. Luo, Y., Liu, P., Guan, T., Yu, J., Yang, Y.: Significance-aware information bottleneck for domain adaptive semantic segmentation. In: The IEEE International Conference on Computer Vision (ICCV), October 2019
25. Luo, Y., Zheng, L., Guan, T., Yu, J., Yang, Y.: Taking a closer look at domain shift: category-level adversaries for semantics consistent domain adaptation. In: The IEEE Conference on Computer Vision and Pattern Recognition (CVPR), June 2019
26. Richter, S.R., Vineet, V., Roth, S., Koltun, V.: Playing for data: ground truth from computer games. In: Leibe, B., Matas, J., Sebe, N., Welling, M. (eds.) ECCV 2016. LNCS, vol. 9906, pp. 102–118. Springer, Cham (2016). https://doi.org/10.1007/978-3-319-46475-6_7
27. Ronneberger, O., Fischer, P., Brox, T.: U-net: Convolutional networks for biomedical image segmentation. arXiv preprint arXiv:1505.04597, August 2015
28. Ros, G., Sellart, L., Materzynska, J., Vazquez, D., Lopez, A.M.: The synthia dataset: a large collection of synthetic images for semantic segmentation of urban scenes. In: The IEEE Conference on Computer Vision and Pattern Recognition (CVPR), June 2016
29. Saito, K., Watanabe, K., Ushiku, Y., Harada, T.: Maximum classifier discrepancy for unsupervised domain adaptation. In: The IEEE Conference on Computer Vision and Pattern Recognition (CVPR), June 2018
30. Sankaranarayanan, S., Balaji, Y., Jain, A., Nam Lim, S., Chellappa, R.: Learning from synthetic data: addressing domain shift for semantic segmentation. In: The IEEE Conference on Computer Vision and Pattern Recognition (CVPR), June 2018
31. Shelhamer, E., Long, J., Darrell, T.: Fully convolutional networks for semantic segmentation. IEEE Trans. Pattern Anal. Mach. Intell. **39**(4), 640–651 (2017)
32. Simonyan, K., Zisserman, A.: Very deep convolutional networks for large-scale image recognition. In: International Conference on Learning Representations (2015)
33. Tsai, Y.H., Hung, W.C., Schulter, S., Sohn, K., Yang, M.H., Chandraker, M.: Learning to adapt structured output space for semantic segmentation. In: The IEEE Conference on Computer Vision and Pattern Recognition (CVPR), June 2018
34. Tsai, Y., Sohn, K., Schulter, S., Chandraker, M.: Domain adaptation for structured output via discriminative patch representations. In: The IEEE International Conference on Computer Vision (ICCV), October 2019
35. Vu, T.H., Jain, H., Bucher, M., Cord, M., Perez, P.: ADVENT: adversarial entropy minimization for domain adaptation in semantic segmentation. In: The IEEE Conference on Computer Vision and Pattern Recognition (CVPR), June 2019

36. Wang, Q., Fan, H., Sun, G., Cong, Y., Tang, Y.: Laplacian pyramid adversarial network for face completion. Pattern Recogn. **88**, 493–505 (2019)
37. Wang, Q., Fan, H., Sun, G., Ren, W., Tang, Y.: Recurrent generative adversarial network for face completion. IEEE Trans. Multimed. (2020)
38. Wu, Z., et al.: DCAN: dual channel-wise alignment networks for unsupervised scene adaptation. In: Ferrari, V., Hebert, M., Sminchisescu, C., Weiss, Y. (eds.) ECCV 2018. LNCS, vol. 11209, pp. 535–552. Springer, Cham (2018). https://doi.org/10.1007/978-3-030-01228-1_32
39. Xia, H., Ding, Z.: Structure preserving generative cross-domain learning. In: IEEE/CVF Conference on Computer Vision and Pattern Recognition (CVPR), June 2020
40. Yu, F., Koltun, V.: Multi-scale context aggregation by dilated convolutions. arXiv preprint arXiv:1511.07122, November 2015
41. Zhang, K., Schölkopf, B., Muandet, K., Wang, Z.: Domain adaptation under target and conditional shift. In: Proceedings of the 30th International Conference on International Conference on Machine Learning - Volume 28 (ICML 2013), pp. III-819–III-827. JMLR.org (2013)
42. Zhang, T., Cong, Y., Sun, G., Wang, Q., Ding, Z.: Visual tactile fusion object clustering. In: AAAI Conference on Artificial Intelligence (2020)
43. Zhang, Y., David, P., Gong, B.: Curriculum domain adaptation for semantic segmentation of urban scenes. In: The IEEE International Conference on Computer Vision (ICCV), October 2017
44. Zhao, H., et al.: PSANet: point-wise spatial attention network for scene parsing. In: Ferrari, V., Hebert, M., Sminchisescu, C., Weiss, Y. (eds.) ECCV 2018. LNCS, vol. 11213, pp. 270–286. Springer, Cham (2018). https://doi.org/10.1007/978-3-030-01240-3_17
45. Zhu, Z., Xu, M., Bai, S., Huang, T., Bai, X.: Asymmetric non-local neural networks for semantic segmentation. In: The IEEE International Conference on Computer Vision (ICCV), October 2019
46. Zou, Y., Yu, Z., Liu, X., Kumar, B.V., Wang, J.: Confidence regularized self-training. In: Proceedings of the IEEE/CVF International Conference on Computer Vision (ICCV), October 2019
47. Zou, Y., Yu, Z., Vijaya Kumar, B.V.K., Wang, J.: Unsupervised domain adaptation for semantic segmentation via class-balanced self-training. In: Ferrari, V., Hebert, M., Sminchisescu, C., Weiss, Y. (eds.) ECCV 2018. LNCS, vol. 11207, pp. 297–313. Springer, Cham (2018). https://doi.org/10.1007/978-3-030-01219-9_18

Prototype Mixture Models for Few-Shot Semantic Segmentation

Boyu Yang, Chang Liu, Bohao Li, Jianbin Jiao, and Qixiang Ye(✉)

University of Chinese Academy of Sciences, Beijing, China
{yangboyu18,liuchang615}@mails.ucas.ac.cn,
libohao1998@gmail.com, {jiaojb,qxye}@ucas.ac.cn

Abstract. Few-shot segmentation is challenging because objects within the support and query images could significantly differ in appearance and pose. Using a single prototype acquired directly from the support image to segment the query image causes semantic ambiguity. In this paper, we propose prototype mixture models (PMMs), which correlate diverse image regions with multiple prototypes to enforce the prototype-based semantic representation. Estimated by an Expectation-Maximization algorithm, PMMs incorporate rich channel-wised and spatial semantics from limited support images. Utilized as representations as well as classifiers, PMMs fully leverage the semantics to activate objects in the query image while depressing background regions in a duplex manner. Extensive experiments on Pascal VOC and MS-COCO datasets show that PMMs significantly improve upon state-of-the-arts. Particularly, PMMs improve 5-shot segmentation performance on MS-COCO by up to 5.82% with only a moderate cost for model size and inference speed (Code is available at github.com/Yang-Bob/PMMs.).

Keywords: Semantic segmentation · Few-shot segmentation · Few-shot learning · Mixture models

1 Introduction

Substantial progress has been made in semantic segmentation [1–8]. This has been broadly attributed to the availability of large datasets with mask annotations and convolutional neural networks (CNNs) capable of absorbing the annotation. However, annotating object masks for large-scale datasets is laborious, expensive, and can be impractical [9–11]. It is also not consistent with cognitive learning, which builds a model upon few-shot supervision [12].

Given a few examples, termed *support* images, and the related segmentation masks [13], few-shot segmentation aims to segment the *query* images based on

Electronic supplementary material The online version of this chapter (https://doi.org/10.1007/978-3-030-58598-3_45) contains supplementary material, which is available to authorized users.

A. Vedaldi et al. (Eds.): ECCV 2020, LNCS 12353, pp. 763–778, 2020.
https://doi.org/10.1007/978-3-030-58598-3_45

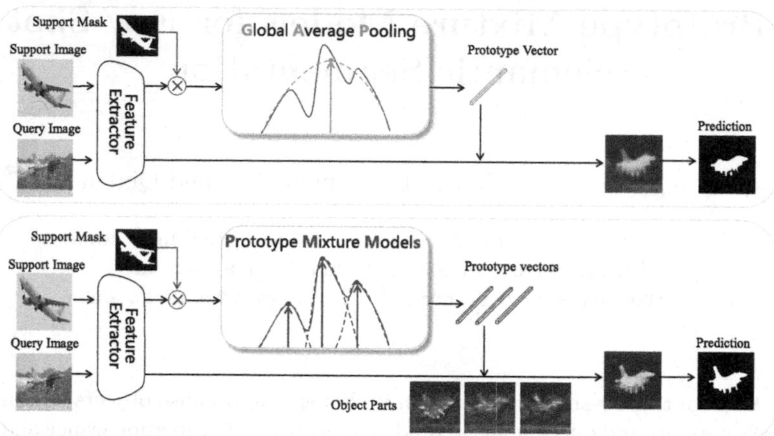

Fig. 1. The single prototype model (upper) based on global average pooling causes semantic ambiguity about object parts. In contrast, prototype mixture models (lower) correlate diverse image regions, *e.g.*, object parts, with multiple prototypes to enhance few-shot segmentation model. (Best viewed in color) (Color figure online)

a feature representation learned on training images. It remains a challenging problem when we consider that target category is not included in the training data while objects within the support and query image significantly differ in appearance and pose.

By introducing the metric learning framework, Shaban *et al.* [14], Zhang *et al.* [15], and Dong *et al.* [16] contributed early few-shot semantic segmentation methods. They also introduced the concept of "prototype" which refers to a weight vector calculated with global average pooling guided by ground-truth masks embedded in feature maps. Such a vector squeezing discriminative information across feature channels is used to guide the feature comparison between support image(s) and query images for semantic segmentation.

Despite clear progress, we argue that the commonly used prototype model is problematic when the spatial layout of objects is completely dropped by global average pooling, Fig. 1(upper). A single prototype causes semantic ambiguity around various object parts and deteriorates the distribution of features [17]. Recent approaches have alleviated this issue by prototype alignment [18], feature boosting [13], and iterative mask refinement [19]. However, the semantic ambiguity problem caused by global average pooling remains unsolved.

In this paper, we propose prototype mixture models (PMMs) and focus on solving the semantic ambiguity problem in a systematic manner. During the training procedure, the prototypes are estimated using an Expectation-Maximization (EM) algorithm, which treats each deep pixel (a feature vector) within the mask region as a positive sample. PMMs are primarily concerned with representing the diverse foreground regions by estimating mixed prototypes for various object parts, Fig. 1(lower). They also enhance the discriminative capacity of features by modeling background regions.

The few-shot segmentation procedure is implemented in a metric learning framework with two network branches (a support branch and a query branch), Fig. 2. In the framework, PMMs are utilized in a duplex manner to segment a query image. On the one hand, they are regarded as spatially squeezed representation, which match (P-Match) with query features to activate feature channels related to the object class. On the other hand, each vector is regarded as a C-dimensional linear classifier, which multiplies (P-Conv) with the query features in an element-wised manner to produce a probability map. In this way, the channel-wised and spatial semantic information of PMMs is fully explored to segment the query image.

The contributions of our work are summarized as follows:

- We propose prototype mixture models (PMMs), with the target to enhance few-shot segmentation by fully leveraging semantics of limited support image(s). PMMs are estimated using an EM algorithm, which is integrated with feature learning by a plug-and-play manner.
- We propose a duplex strategy, which treats PMMs as representations and classifiers, to activate spatial and channel-wised semantics for segmentation.
- We assemble PMMs to RPMMs using a residual structure and significantly improve upon the state-of-the-arts.

2 Related Work

Semantic Segmentation. Semantic segmentation, which performs per-pixel classification of a class of objects, has been extensively investigated. State-of-the-art methods, such as UNet [2], PSPNet [1], DeepLab [3–5], are based on fully convolutional networks (FCNs) [20]. Semantic segmentation has been updated to instance segmentation [8] and panoptic segmentation [21], which shared useful modules, *e.g.*, Atrous Spatial Pyramid Pooling (ASPP) [4] and multi-scale feature aggregation [1], with few-shot segmentation. The clustering method used in SegSort [7], which partitioned objects into parts using a divide-and-conquer strategy, provides an insight for this study.

Few-Shot Learning. Existing methods can be broadly categorized as either: metric learning [17,22,23], meta-learning [24–27], or data argumentation. Metric learning based methods train networks to predict whether two images/regions belong to the same category. Meta-learning based approaches specify optimization or loss functions which force faster adaptation of the parameters to new categories with few examples. The data argumentation methods learn to generate additional examples for unseen categories [28,29].

In the metric learning framework, the effect of prototypes for few-shot learning has been demonstrated. With a simple prototype, *e.g.*, a linear layer learned on top of a frozen CNN [30], state-of-the-art results can be achieved based on a simple baseline. This provides reason for applying prototypes to capture representative and discriminative features.

Few-Shot Segmentation. Existing few-shot segmentation approaches largely followed the metric learning framework, *e.g.*, learning knowledge using a prototype vector, from a set of support images, and then feed learned knowledge to a metric module to segment query images [18].

In OSLSM [14], a two-branch network consisting of a support branch and a query branch was proposed for few-shot segmentation. The support branch is devoted to generating a model from the support set, which is then used to tune the segmentation process of an image in the query branch. In PL [16], the idea of prototypical networks was employed to tackle few-shot segmentation using metric learning. SG-One [15] also used a prototype vector to guide semantic segmentation procedure. To obtain the squeezed representation of the support image, a masked average pooling strategy is designed to produce the prototype vector. A cosine similarity metric is then applied to build the relationship between the guidance features and features of pixels from the query image. PANet [18] further introduced a prototype alignment regularization between support and query branches to fully exploit knowledge from support images for better generalization. CANet [19] introduced a dense comparison module, which effectively exploits multiple levels of feature discriminativeness from CNNs to make dense feature comparison. With this approach comes an iterative optimization module which re?nes segmentation masks. The FWB approach [13] focused on discriminativeness of prototype vectors (support features) by leveraging foreground-background feature differences of support images. It also used an ensemble of prototypes and similarity maps to handle the diversity of object appearances.

As a core of metric learning in few-shot segmentation, the prototype vector was commonly calculated by global average pooling. However, such a strategy typically disregards the spatial extent of objects, which tends to mix semantics from various parts. This unintended mixing seriously deteriorates the diversity of prototype vectors and feature representation capacity. Recent approaches alleviated this problem using iterative mask refinement [19] or model ensemble [13]. However, issues remain when using single prototypes to represent object regions and the semantic ambiguity problem remains unsolved.

Our research is inspired by the prototypical network [31], which learns a metric space where classification is performed using distances to the prototype of each class. The essential differences are twofold: (1) A prototype in prototypical network [31] represents a class of samples while a prototype in our approach represents an object part; (2) The prototypical network does not involve mixing prototypes for a single sample or a class of samples.

3 The Proposed Approach

3.1 Overview

The few-shot segmentation task is to classify pixels in query images into foreground objects or backgrounds by solely referring to few labeled support images containing objects of the same categories. The goal of the training procedure is

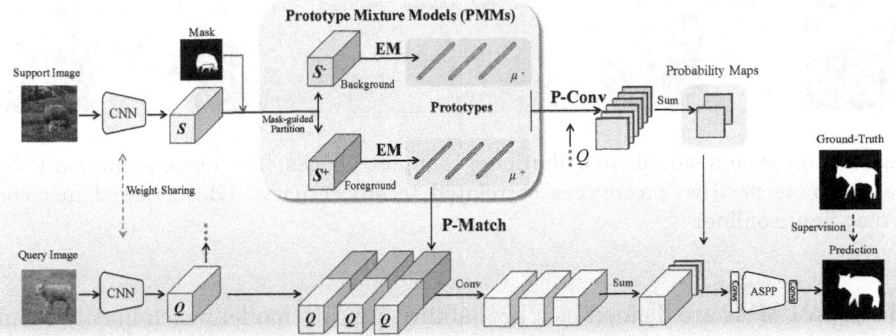

Fig. 2. The proposed approach consists of two branches *i.e.*, the support branch and the query branch. During training, the feature set S of a support image is partitioned into a positive sample set S^+ and a negative sample set S^- guided by the ground-truth mask. S^+ and S^- are respectively used to train μ^+ and μ^-, which are used to activate query features in a duplex way (P-Conv and P-Match) for semantic segmentation. "ASPP" refers to the Atrous Spatial Pyramid Pooling (ASPP) [4].

to learn a segmentation model that is trained by numbers of images different from the task query image categories. The training image set is split into many small subsets and within every subset one image serves as the query and the other(s) as the support image(s) with known ground-truth(s). Once the model is trained, the segmentation model is fixed and requires no optimization when tested on a new dataset [19]. The proposed few-shot segmentation model follows a metric learning framework, Fig. 2, which consists of two network branches *i.e.*, the support branch (above) and the query branch (below). Over the support branch, PMMs are estimated for the support image(s). In the support and query branches, two CNNs with shared weights are used as the backbone to extract features. Let $S \in \mathcal{R}^{W \times H \times C}$ denote the features of the support image where $W \times H$ denotes the resolution of feature maps and C the number of feature channels. The features for a query image are denoted as $Q \in \mathcal{R}^{W \times H \times C}$.

Without loss of generality, the network architecture and models are illustrated for 1-shot setting, which can be extended to 5-shot setting by feeding five support images to the PMMs to estimate prototypes.

3.2 Prototype Mixture Models

During training, features $S \in \mathcal{R}^{W \times H \times C}$ for the support image are considered as a sample set with $W \times H$ C-dimensional samples. S is spatially partitioned into foreground samples S^+ and background samples S^-, where S^+ corresponds to feature vectors within the mask of the support image and S^- those outside the mask. S^+ is used to learn foreground PMMs$^+$ corresponding to object parts, Fig. 3, and S^- to learn background PMMs$^-$. Without loss generality, the models and learning procedure are defined for PMMs, which represent either PMMs$^+$ or PMMs$^-$.

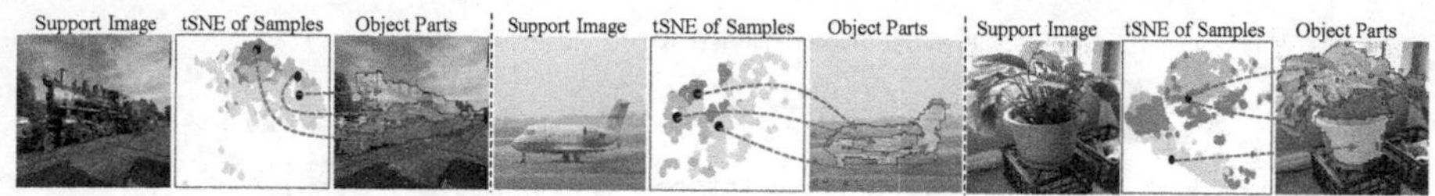

Fig. 3. Foreground sample distribution of support images. The black points on tSNE maps denote positive prototypes correlated to object parts. (Best viewed in color) (Color figure online)

Models. PMMs are defined as a probability mixture model which linearly combine probabilities from base distributions, as

$$p(s_i|\theta) = \sum_{k=1}^{K} w_k p_k(s_i|\theta) \tag{1}$$

where w_k denotes the mixing weights satisfying $0 \leq w_k \leq 1$ and $\sum_{k=1}^{K} w_k = 1$. θ denotes the model parameters which are learned when estimating PMMs. $s_i \in S$ denotes the ith feature sample and $p_k(s_i|\theta)$ denotes the kth base model, which is a probability model based on a Kernel distance function, as

$$p_k(s_i|\theta) = \beta(\theta) e^{Kernel(s_i, \mu_k)}, \tag{2}$$

where $\beta(\theta)$ is the normalization constant. $\mu_k \in \theta$ is one of the parameter. For the Gaussian mixture models (GMMs) with fixed co-variance, the Kernel function is a radial basis function (RBF), $Kernel(s_i, \mu_k) = -||(s_i - \mu_k)||_2^2$. For the von Missies-Fisher (VMF) model [32], the kernel function is defined as a cosine distance function, as $Kernel(s_i, \mu_k) = \frac{\mu_k^T s_i}{||\mu_k||_2 ||s_i||_2}$, where μ_k is the mean vector of the kth model. Considering the metric learning framework used, the vector distance function is more appropriate in our approach, as is validated in experiments. Based on the vector distance, PMMs are defined as

$$p_k(s_i|\theta) = \beta_c(\kappa) e^{\kappa \mu_k^T s_i}, \tag{3}$$

where $\theta = \{\mu, \kappa\}$. $\beta_c(\kappa) = \frac{\kappa^{c/2-1}}{(2\pi)^{c/2} I_{c/2-1}(\kappa)}$ is the normalization coefficient, and $I_\nu(\cdot)$ denotes the Bessel function. κ denotes the concentration parameter, which is empirically set as $\kappa = 20$ in experiments.

Model Learning. PMMs are estimated using the EM algorithm which includes iterative E-steps and M-steps. In each E-step, given model parameters and sample features extracted, we calculate the expectation of the sample s_i as

$$E_{ik} = \frac{p_k(s_i|\theta)}{\sum_{k=1}^{K} p_k(s_i \theta)} = \frac{e^{\kappa \mu_k^T s_i}}{\sum_{k=1}^{K} e^{\kappa \mu_k^T s_i}}. \tag{4}$$

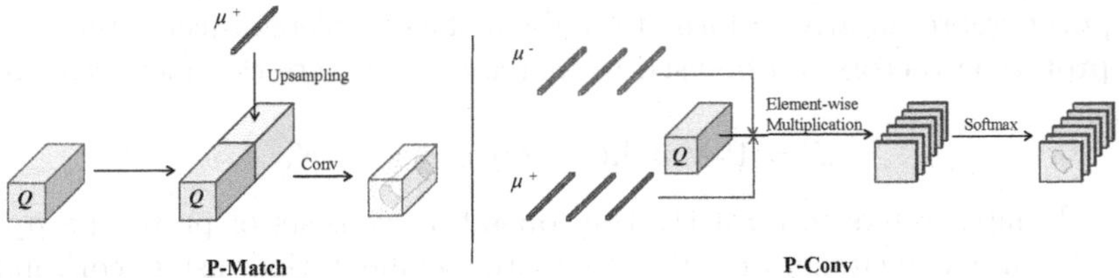

Fig. 4. Illustration of P-Match and P-Conv opeartions for query feature (Q) activation.

Algorithm 1. Learning PMMs for few-shot segmentation

Input:
Support images, mask M_s for each support image, query image;
Output:
Network parameters α, prototypes μ^+ and μ^- for each support image;
for (each support image) **do**
  Calculate support and query features S and Q by forward propagation with θ;
  Partition S into S^+ and S^- according to M_s;
  Learn PMMs;
    Estimate μ^+ upon S^+ by iterative EM steps defined by Eqs. 4 and 5;
    Estimate μ^- upon S^- by iterative EM steps defined by Eqs. 4 and 5;
  Activate Q using P-Match and P-Conv defined on μ^+ and μ^-, Fig. 4;
  Predict a segmentation mask and calculate the segmentation loss;
  Update α to minimize the cross-entropy loss at the query branch, Fig. 2.
end for

In each M-step, the expectation is used to update PMMs' mean vectors, as

$$\mu_k = \frac{\sum_{i=1}^{N} E_{ik} s_i}{\sum_{i=1}^{N} E_{ik}}, \tag{5}$$

where $N = W \times H$ denotes the number of samples.

After model learning, the mean vectors $\mu^+ = \{\mu_k^+, k = 1, ..., K\}$ and $\mu^- = \{\mu_k^-, k = 1, ..., K\}$ are used as prototype vectors to extract convolutions features for the query image. The mixture coefficient w_k is ignored so that each prototype vectors have same importance for semantic segmentation. Obviously, each prototype vector is the mean of a cluster of samples. Such a prototype vector can represent a region around an object part in the original image for the reception field effect, Fig. 3.

3.3 Few-Shot Segmentation

During inference, the learned prototype vectors $\mu^+ = \{\mu_k^+, k = 1, ..., K\}$ and $\mu^- = \{\mu_k^-, k = 1, ..., K\}$ are duplexed to activate query features for semantic segmentation, Fig. 2.

PMMs as Representation (P-Match). Each positive prototype vector squeezes representative information about an object part and all prototypes incorporate representative information about the complete object extent. Therefore, prototype vectors can be used to match and activate the query features Q, as

$$Q' = \text{P-Match}(\mu_k^+, Q), k = 1, ..., K, \tag{6}$$

where P-Match refers to an activation operation consists of prototype upsampling, feature concatenation, and semantic segmentation using convolution, Fig. 4. The convolution operation on concatenated features implements a channel-wise comparison, which activates feature channels related to foreground while suppressing those associated with backgrounds. With P-Match, semantic information about the extent of the complete object is incorporated into the query features for semantic segmentation.

PMMs as Classifiers (P-Conv). On the other hand, each prototype vector incorporating discriminative information across feature channels can be seen as a classifier, which produces probability maps $M_k = \{M_k^+, M_k^-\}$ using the P-Conv operation, as

$$M_k = \text{P-Conv}(\mu_k^+, \mu_k^-, Q), k = 1, ..., K. \tag{7}$$

As shown in Fig. 4, P-Conv first multiplies each prototype vector with the query feature Q in an element-wise manner. The output maps are then converted to probability maps M_k by applying Softmax across channels.

After P-Conv, the produced probability maps $M_k^+, k = 1, ..., K$ and $M_k^-, k = 1, ..., K$ are respectively summarized to two probability maps, as

$$\begin{aligned} M_p^+ &= \sum_k M_k^+, \\ M_p^- &= \sum_k M_k^-, \end{aligned} \tag{8}$$

which are further concatenated with the query features to activate objects of interest, as

$$Q'' = M_p^+ \oplus M_p^- \oplus Q', \tag{9}$$

where $\oplus$ denotes the concatenation operation.

After the P-Match and P-Conv operations, the semantic information across channels and discriminative information related to object parts are collected from the support feature S to activate the query feature Q. in a dense comparison manner [19]. The activated query features Q'' are further enhanced with Atrous Spatial Pyramid Pooling (ASPP) and fed to a convolutional module to predict the segmentation mask, Fig. 2.

Segmentation Model Learning. The segmentation model is implemented as an end-to-end network, Fig. 2. The learning procedure for the segmentation model is described in Algorithm 1. In the feed forward procedure, the support features are partitioned into backgrounds and foreground sample sets S^+ and S^- according to the ground-truth mask. PMMs are learned on S^+ and S^- and the learned prototype vectors μ^+ and μ^- are leveraged to activate query features to

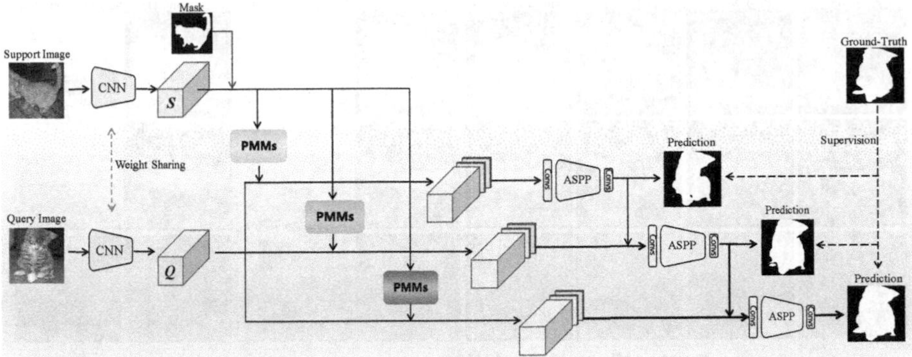

Fig. 5. Network architecture of residual prototype mixture models (RPMMS).

predict segmentation mask of the query image. In the back-propagation procedure, the network parameters θ are updated to optimize the segmentation loss at the query branch. With multiple training iterations, rich feature representation about diverse object parts is absorbed into the backbone network. During the inference procedure, the learned feature representation together with PMMs of the support image(s) is used to segment the query image.

3.4 Residual Prototype Mixture Models

To further enhance the model capacity, we implement model ensemble by stacking multiple PMMs, Fig. 5. Stacked PMMs, termed residual PMMs (RPMMs), leverage the residual from the previous query branch to supervise the next query branch for fine-grained segmentation. RPMMs not only further improve the performance but also defines a new model ensemble strategy. This incorporates the advantages of model residual learning, which is inspired by the idea of side-output residual [33] but has the essential difference to handle models instead of features. This is also different from the ensemble of experts [13], which generates an ensemble of the support features guided by the gradient of loss.

4 Experiments

4.1 Experimental Settings

Implementation Details. Our approach utilizes CANet [19] without iterative optimization as the baseline, which uses VGG16 or ResNet50 as backbone CNN for feature extraction. During training, four data augmentation strategies including normalization, horizontal flipping, random cropping and random resizing are used [19]. Our approach is implemented upon the PyTorch 1.0 and run on Nvidia 2080Ti GPUs. The EM algorithm iterates 10 rounds to calculate PMMs for each image. The network with a cross-entropy loss is optimized by SGD with the initial learning of 0.0035 and the momentum of 0.9 for 200,000 iterations with

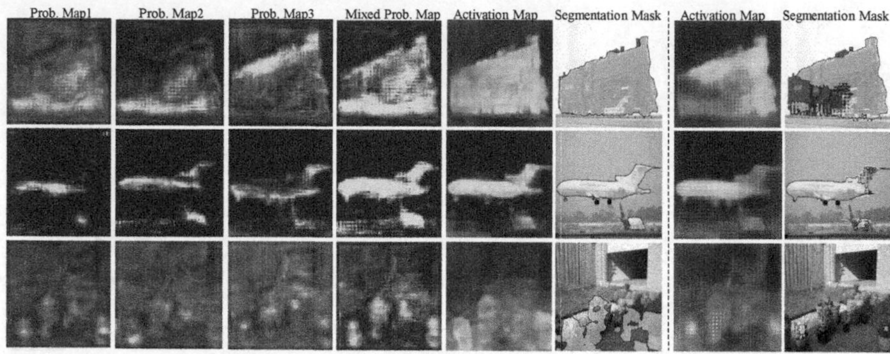

Fig. 6. Activation maps by PMMs and CANet [19]. PMMs produce multiple probability maps and fuse them to a mixed map, which facilities activating and segmenting complete object extent (first two rows) or multiple objects (last row). CANet that uses a single prototype to segment object tends to miss object parts. (Best viewed in color) (Color figure online)

8 pairs of support-query images per batch. The learning rate reduces following the "poly" policy defined in DeepLab [4]. For each training step, the categories in the train split are randomly selected and then the support-query pairs are randomly sampled in the selected categories.

Datasets. We evaluate our model on Pascal-5^i and COCO-20^i. Pascal-5^i is a dataset specified for few-shot semantic segmentation in OSLSM [14], which consists of the Pascal VOC 2012 dataset with extra annotations from extended SDS [34]. 20 object categories are partitioned into four splits with three for training and one for testing. At test time, 1000 support-query pairs were randomly sampled in the test split [19]. Following FWB [13], we create COCO-20^i from MSCOCO 2017 dataset. The 80 classes are divided into 4 splits and each contains 20 classes and the *val* dataset is used for performance evaluation. The other setting is the same as that in Pascal-5^i.

Evaluation Metric. Mean intersection over-union (mIoU) which is defined as the mean IoUs of all image categories was employed as the metric for performance evaluation. For each category, the IoU is calculated by $\text{IoU}=\frac{\text{TP}}{\text{TP+FP+FN}}$, where TP, FP and FN respectively denote the number of true positive, false positive and false negative pixels of the predicted segmentation masks.

4.2 Model Analysis

In Fig. 6, we visualize probability maps produced by positive prototypes of PMMs. We also visualize and compare the activation maps and segmentation masks produced by PMMs and CANet. PMMs produce multiple probability maps and fuse them to a mixed probability map, which facilities activating complete object extent (first two rows). The advantage in terms of representation

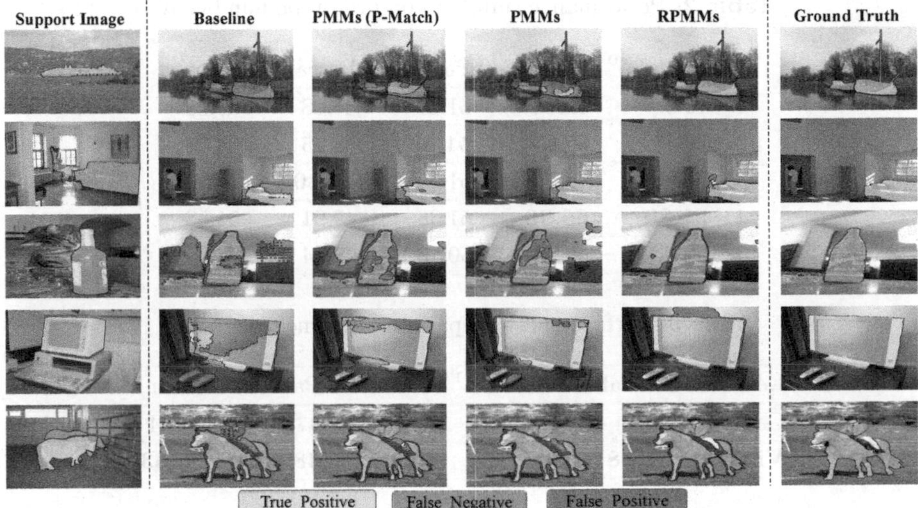

Fig. 7. Semantic segmentation results. 'Baseline' refers to the CANet method [19] without iterative optimization. (Best viewed in color) (Color figure online)

Table 1. Ablation study. 'Mean' denotes mean mIoU on Pascal-5^i with PMMs using three prototypes. The first row is the baseline method without using the PMMs or the RPMMs method.

PMMs^+(P-Match)	PMMs(P-Match & P-Conv)	RPMMs	Mean
			51.93
✓			54.63
✓	✓		55.27
✓	✓	✓	56.34

capacity is that PMMs perform better than CANet when segmenting multiple objects within the same image (last row). By comparison, CANet using a single prototype to activate object tends to miss object parts or whole objects. The probability maps produced by PMMs validate our idea, *i.e.*, prototypes correlated to multiple regions and alleviate semantic ambiguity.

In Fig. 7, we compare the segmentation results by the baseline method and the proposed modules. The segmentation results show that PMMs^+(P-Match) can improve the recall rate by segmenting more target pixels. By introducing background prototypes, PMMs reduce the false positive pixels, which validates that the background mixture models can improve the discriminative capability of the model. RPMMs further improve the segmentation results by refining object boundaries about hard pixels.

Table 2. Performance (mIoU%) on prototype number K.

K	Pascal-5^0	Pascal-5^1	Pascal-5^2	Pascal-5^3	Mean
1	49.38	66.42	51.29	47.68	53.69
2	50.85	66.65	**51.89**	48.25	54.41
3	51.88	66.72	51.14	**48.80**	**54.63**
4	**51.89**	**66.96**	51.36	47.91	54.53
5	50.76	66.89	50.76	47.94	54.09

Table 3. Performance comparison of Kernel functions.

Kernal	Pascal-5^0	Pascal-5^1	Pascal-5^2	Pascal-5^3	Mean
Gaussian	50.94	66.70	50.59	47.91	54.04
VMF	**51.88**	**66.72**	**51.14**	**48.80**	**54.63**

4.3 Ablation Study

PMMs. In Table 1, with P-Match modules, PMMs improve segmentation performance by 2.70% (54.63% vs. 51.93%), which validates that the prototypes generated by the PMMs perform better than the prototype generated by global average pooling. By introducing the duplex strategy, PMMs further improve the performance by 0.64% (55.27% vs. 54.63%), which validates that the probability map generated by the combination of foreground and background prototypes can suppress backgrounds and reduce false segmentation. In total, PMMs improve the performance by 3.34% (55.27% vs. 51.93%), which is a significant margin in semantic segmentation. This clearly demonstrates the superiority of the proposed PMMs over previous prototype methods.

RPMMs. RPMMs further improve the performance by 1.07% (56.34% vs. 55.27%), which validates the effectiveness of the residual ensemble strategy. Residual from the query prediction output of the previous branch of PMMs can be used to supervise the next branch of PMMs, enforcing the stacked PMMs to reduce errors, step by step.

Number of Prototypes. In Table 2, ablation study is carried out to determine the number of prototypes using PMMs$^+$ with P-Match. $K = 2$ significantly outperforms $K = 1$. The best Pascal-5^i performance occurs at $K = 2, 3, 4$. When $K = 3$ the best mean performance is obtained. When $K = 4, 5$, the performance slightly decreases. One reason lies in that the PMMs are estimated on a single support image, which includes limited numbers of samples. The increase of K substantially decreases the samples of each prototype and increases the risk of over-fitting.

Kernel Functions. In Table 3, we compare the Gaussian and VMF kernels for sample distance calculation when estimating PMMs. The better results from VMF kernel show that the cosine similarity defined by VMF kernel is preferable.

Table 4. Performance of 1-way 1-shot semantic segmentation on Pascal-5^i. CANet reports multi-scale test performance.

Backbone	Method	Pascal-5^0	Pascal-5^1	Pascal-5^2	Pascal-5^3	Mean
VGG16	OSLSM [14]	33.60	55.30	40.90	33.50	40.80
	co-FCN [35]	36.70	50.60	44.90	32.40	41.10
	SG-One [15]	40.20	58.40	48.40	38.40	46.30
	PANet [18]	42.30	58.00	51.10	41.20	48.10
	FWB [13]	47.04	59.64	**52.51**	48.27	51.90
	RPMMs (ours)	**47.14**	**65.82**	50.57	**48.54**	**53.02**
Resnet50	CANet [19]	49.56	64.97	49.83	**51.49**	53.96
	PMMs (ours)	51.98	**67.54**	51.54	49.81	55.22
	RPMMs (ours)	**55.15**	66.91	**52.61**	50.68	**56.34**

Table 5. Performance of 1-way 5-shot semantic segmentation on Pascal-5^i.

Backbone	Method	Pascal-5^0	Pascal-5^1	Pascal-5^2	Pascal-5^3	Mean
VGG16	OSLSM [14]	35.90	58.10	42.70	39.10	43.95
	SG-One [15]	41.90	58.60	48.60	39.40	47.10
	FWB [13]	50.87	62.86	56.48	**50.09**	55.08
	PANet [18]	**51.80**	64.60	**59.80**	46.05	**55.70**
	RPMMs (ours)	50.00	**66.46**	51.94	47.64	54.01
Resnet50	CANet [19]	–	–	–	–	55.80
	PMMs (ours)	55.03	**68.22**	52.89	**51.11**	56.81
	RPMMs (ours)	**56.28**	67.34	**54.52**	51.00	**57.30**

Table 6. Performance of 1-shot and 5-shot semantic segmentation on MS COCO. FWB uses the ResNet101 backbone while other approaches use the ResNet50 backbone.

Settings	Method	COCO-20^0	COCO-20^1	COCO-20^2	COCO-20^3	Mean
1-shot	PANet [18]	–	–	–	–	20.90
	FWB [13]	16.98	17.98	20.96	28.85	21.19
	Baseline	25.08	30.25	24.45	24.67	26.11
	PMMs (ours)	29.28	34.81	27.08	27.27	29.61
	RPMMs (ours)	**29.53**	**36.82**	**28.94**	**27.02**	**30.58**
5-shot	FWB [13]	19.13	21.46	23.93	30.08	23.65
	PANet [18]	–	–	–	–	29.70
	Baseline	25.95	32.38	26.11	26.98	27.86
	PMMs (ours)	33.00	40.55	30.29	33.27	34.28
	RPMMs (ours)	**33.82**	**41.96**	**32.99**	**33.33**	**35.52**

Inference Speed. The size of PMMs model is 19.5M, which is slightly larger than that of the baseline CANet [19] (19M) but much smaller than that of OSLSM [14] (272.6M). Because the prototypes are $1 \times 1 \times C$ dimensional vectors, they do not significantly increase the model size or computational cost. In one shot setting, with $K = 3$, our inference speed on single 2080Ti GPU is 26 FPS, which is slightly lower than that of CANet (29 FPS). With RPMMs the speed decreases to 20 FPS while the model size (19.6M) does not significantly increase.

4.4 Performance

PASCAL-5^i. In Table 4 and Table 5, PMMs and RPMMs are compared with the state-of-the-art methods. They outperform state-of-the-art methods in both 1-shot and 5-shot settings. With the 1-shot setting and a Resnet50 backbone, RPMMs achieve 2.38% (56.34% vs. 53.96%) performance improvement over the state-of-the-art, which is a significant margin.

With the 5-shot setting and a Resnet50 backbone, RPMMs achieve 1.50% (57.30% vs. 55.80%) performance improvement over the state-of-the-art, which is also significant. With the VGG16 backbone, our approach is comparable with the state-of-the-arts. Note that the PANet and FWB used additional k-shot fusion strategy while we do not use any post-processing strategy to fuse the predicted results from five shots.

MS COCO. Table 6 displays the evaluation results on MS COCO dataset following the evaluation metric on COCO-20^i [13]. Baseline is achieved by running CANet without iterative optimization. PMMs and RPMMs again outperform state-of-the-art methods in both 1-shot and 5-shot settings. For the 1-shot setting, RPMMs improves the baseline by 4.47%, respectively outperforms the PANet and FWB methods by 9.68% and 9.39%.

For the 5-shot setting, it improves the baseline by 7.66%, and respectively outperforms the PANet and FWB by 5.82% and 11.87%, which are large margins for the challenging few-shot segmentation problem. Compared to PASCAL VOC, MS COCO has more categories and images for training, which facilities learning richer representation related to various object parts and backgrounds. Thereby, the improvement on MS COCO is larger than that on Pascal VOC.

5 Conclusion

We proposed prototype mixture models (PMMs), which correlate diverse image regions with multiple prototypes to solve the semantic ambiguity problem. During training, PMMs incorporate rich channel-wised and spatial semantics from limited support images. During inference, PMMs are matched with query features in a duplex manner to perform accurate semantic segmentation. On the large-scale MS COCO dataset, PMMs improved the performance of few-shot segmentation, in striking contrast with state-of-the-art approaches. As a general method to capture the diverse semantics of object parts given few support examples, PMMs provide a fresh insight for the few-shot learning problem.

Acknowledgement. This work was supported in part by the National Natural Science Foundation of China (NSFC) under Grant 61836012, 61671427, and 61771447.

References

1. Zhao, H., Shi, J., Qi, X., Wang, X., Jia, J.: Pyramid scene parsing network. In: IEEE CVPR, pp. 6230–6239 (2017)
2. Ronneberger, O., Fischer, P., Brox, T.: U-Net: convolutional networks for biomedical image segmentation. In: Navab, N., Hornegger, J., Wells, W.M., Frangi, A.F. (eds.) MICCAI 2015. LNCS, vol. 9351, pp. 234–241. Springer, Cham (2015). https://doi.org/10.1007/978-3-319-24574-4_28
3. Chen, L., Papandreou, G., Kokkinos, I., Murphy, K., Yuille, A.L.: Semantic image segmentation with deep convolutional nets and fully connected CRFs. In: ICLR, pp. 6230–6239 (2015)
4. Chen, L., Papandreou, G., Kokkinos, I., Murphy, K., Yuille, A.L.: Deeplab: semantic image segmentation with deep convolutional nets, atrous convolution, and fully connected CRFs. IEEE Trans. Pattern Anal. Mach. Intell. **40**(4), 834–848 (2018)
5. Chen, L., Papandreou, G., Schroff, F., Adam, H.: Rethinking atrous convolution for semantic image segmentation. CoRR abs/1706.05587 (2017)
6. Chen, L., Zhu, Y., Papandreou, G., Schroff, F., Adam, H.: Encoder-decoder with atrous separable convolution for semantic image segmentation. In: ECCV, pp. 833–851 (2018)
7. Hwang, J., Yu, S.X., Shi, J., Collins, M.D., Yang, T., Zhang, X., Chen, L.: SegSort: segmentation by discriminative sorting of segments. In: IEEE ICCV, pp. 7334–7344 (2019)
8. He, K., Gkioxari, G., Dollár, P., Girshick, R.B.: Mask R-CNN. IEEE Trans. Pattern Anal. Mach. Intell. **42**(2), 386–397 (2020)
9. Wan, F., Liu, C., Ke, W., Ji, X., Jiao, J., Ye, Q.: C-MIL: continuation multiple instance learning for weakly supervised object detection. In: IEEE CVPR, pp. 2199–2208 (2019)
10. Wan, F., Wei, P., Han, Z., Jiao, J., Ye, Q.: Min-entropy latent model for weakly supervised object detection. IEEE Trans. Pattern Anal. Machine Intell. **41**(10), 2395–2409 (2019)
11. Zhang, X., Wan, F., Liu, C., Ji, R., Ye, Q.: Freeanchor: learning to match anchors for visual object detection. In: NeurIPS, pp. 147–155 (2019)
12. Tokmakov, P., Wang, Y., Hebert, M.: Learning compositional representations for few-shot recognition. In: IEEE ICCV, pp. 6372–6381 (2019)
13. Nguyen, K., Todorovic, S.: Feature weighting and boosting for few-shot segmentation. In: IEEE ICCV, pp. 622–631 (2019)
14. Shaban, A., Bansal, S., Liu, Z., Essa, I., Boots, B.: One-shot learning for semantic segmentation. In: BMVC (2017)
15. Zhang, X., Wei, Y., Yang, Y., Huang, T.: SG-One: similarity guidance network for one-shot semantic segmentation. CoRR abs/1810.09091 (2018)
16. Dong, N., Xing, E.P.: Few-shot semantic segmentation with prototype learning. In: BMVC, p. 79 (2018)
17. Hao, F., He, F., Cheng, J., Wang, L., Cao, J., Tao, D.: Collect and select: semantic alignment metric learning for few-shot learning. In: IEEE ICCV, pp. 8460–8469 (2019)
18. Wang, K., Liew, J., Zou, Y., Zhou, D., Feng, J.: PANet: few-shot image semantic segmentation with prototype alignment, pp. 622–631 (2019)

19. Zhang, C., Lin, G., Liu, F., Yao, R., Shen, C.: CANet: class-agnostic segmentation networks with iterative refinement and attentive few-shot learning. In: IEEE CVPR, pp. 5217–5226 (2019)
20. Shelhamer, E., Long, J., Darrell, T.: Fully convolutional networks for semantic segmentation. IEEE Trans. Pattern Anal. Mach. Intell. **39**(4), 640–651 (2017)
21. Kirillov, A., Girshick, R., He, K., Dollar, P.: Panoptic feature pyramid networks. In: IEEE CVPR, pp. 6399–6408 (2019)
22. Vinyals, O., Blundell, C., Lillicrap, T., Kavukcuoglu, K., Wierstra, D.: Matching networks for one shot learning. In: NeurIPS, pp. 3630–3638 (2016)
23. Sung, F., Yang, Y., Zhang, L., Xiang, T., Torr, P.H.S., Hospedales, T.M.: Learning to compare: Relation network for few-shot learning. In: IEEE CVPR, pp. 1199–1208 (2018)
24. Wang, Y.-X., Hebert, M.: Learning to learn: model regression networks for easy small sample learning. In: Leibe, B., Matas, J., Sebe, N., Welling, M. (eds.) ECCV 2016. LNCS, vol. 9910, pp. 616–634. Springer, Cham (2016). https://doi.org/10.1007/978-3-319-46466-4_37
25. Ravi, S., Larochelle, H.: Optimization as a model for few-shot learning. In: ICLR (2017)
26. Finn, C., Abbeel, P., Levine, S.: Model-agnostic meta-learning for fast adaptation of deep networks. In: ICML, pp. 1126–1135 (2017)
27. Jamal, M.A., Qi, G.J.: Task agnostic meta-learning for few-shot learning. In: IEEE ICCV, pp. 111719–111727 (2019)
28. Hariharan, B., Girshick, R.B.: Low-shot visual recognition by shrinking and hallucinating features. In: IEEE ICCV, pp. 3037–3046 (2017)
29. Wang, Y., Girshick, R.B., Hebert, M., Hariharan, B.: Low-shot learning from imaginary data. In: IEEE CVPR, pp. 7278–7286 (2018)
30. Chen, W.Y., Liu, Y.C., Kira, Z., Wang, Y.C.: A closer look at few-shot classification. In: IEEE ICLR (2019)
31. Snell, J., Swersky, K., Zemel, R.S.: Prototypical networks for few-shot learning. In: NeurIPS, pp. 4077–4087 (2017)
32. Banerjee, A., Dhillon, I.S., Ghosh, J., Sra, S.: Clustering on the unit hypersphere using von Mises-Fisher distributions. J. Mach. Learn. Res. **6**, 1345–1382 (2005)
33. Ke, W., Chen, J., Jiao, J., Zhao, G., Ye, Q.: SRN: side-output residual network for object symmetry detection in the wild. In: IEEE CVPR, pp. 302–310 (2017)
34. Hariharan, B., Arbelaez, P., Bourdev, L.D., Maji, S., Malik, J.: Semantic contours from inverse detectors. In: IEEE ICCV, pp. 991–998 (2011)
35. Rakelly, K., Shelhamer, E., Darrell, T., Efros, A.A., Levine, S.: Conditional networks for few-shot semantic segmentation. In: ICLR Workshop (2018)

Webly Supervised Image Classification with Self-contained Confidence

Jingkang Yang[1,2(✉)], Litong Feng[1], Weirong Chen[1,3], Xiaopeng Yan[1], Huabin Zheng[1], Ping Luo[4], and Wayne Zhang[1]

[1] SenseTime Research, Hong Kong SAR, China
{yangjingkang,fenglitong,chenweirong,yanxiaopeng, zhenghuabin,wayne.zhang}@sensetime.com

[2] Rice University, Houston, TX, USA

[3] The Chinese University of Hong Kong, Hong Kong SAR, China

[4] The University of Hong Kong, Hong Kong SAR, China

Abstract. This paper focuses on webly supervised learning (WSL), where datasets are built by crawling samples from the Internet and directly using search queries as web labels. Although WSL benefits from fast and low-cost data collection, noises in web labels hinder better performance of the image classification model. To alleviate this problem, in recent works, self-label supervised loss $\mathcal{L}_s$ is utilized together with webly supervised loss $\mathcal{L}_w$. $\mathcal{L}_s$ relies on pseudo labels predicted by the model itself. Since the correctness of the web label or pseudo label is usually on a case-by-case basis for each web sample, it is desirable to adjust the balance between $\mathcal{L}_s$ and $\mathcal{L}_w$ on sample level. Inspired by the ability of Deep Neural Networks (DNNs) in confidence prediction, we introduce Self-Contained Confidence (SCC) by adapting model uncertainty for WSL setting, and use it to sample-wisely balance $\mathcal{L}_s$ and $\mathcal{L}_w$. Therefore, a simple yet effective WSL framework is proposed. A series of SCC-friendly regularization approaches are investigated, among which the proposed graph-enhanced mixup is the most effective method to provide high-quality confidence to enhance our framework. The proposed WSL framework has achieved the state-of-the-art results on two large-scale WSL datasets, WebVision-1000 and Food101-N. Code is available at https://github.com/bigvideoresearch/SCC.

Keywords: Webly supervised learning · Noisy labels · Model uncertainty

J. Yang and W. Chen—Work done during an internship at SenseTime EIG Research.

Electronic supplementary material The online version of this chapter (https://doi.org/10.1007/978-3-030-58598-3_46) contains supplementary material, which is available to authorized users.

A. Vedaldi et al. (Eds.): ECCV 2020, LNCS 12353, pp. 779–795, 2020.
https://doi.org/10.1007/978-3-030-58598-3_46

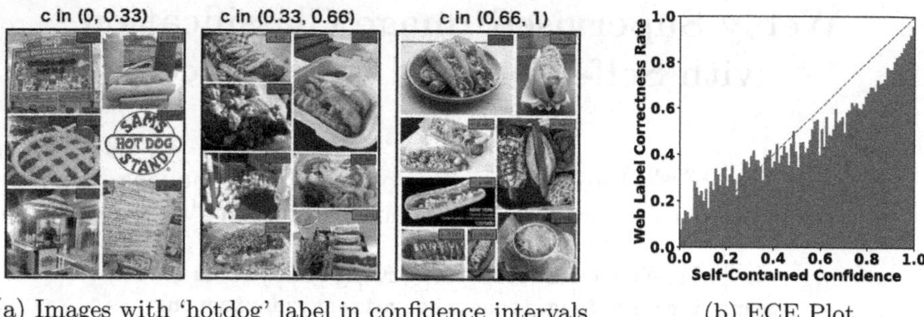

(a) Images with 'hotdog' label in confidence intervals (b) ECE Plot

Fig. 1. Exemplary images and ECE plot showing self-contained confidence (SCC) generally reflects web label correctness. A standard ResNet-50 is pretrained on the Food-101N training set for SCC extraction. (a) shows image samples grouped by low/medium/high confidences. The upper right tag on each image shows SCC value and the tag color indicates web label correctness (red: wrong, green: correct). (b) is the ECE plot using Food-101N human-verification set ($M = 100$) (Color figure online)

1 Introduction

Large-scale human-labeled data plays a vital role in deep learning-based applications such as image classification [3], scene recognition [41], face recognition [30], etc. However, high-quality human annotations require significant cost in labor and time. Webly supervised learning (WSL), therefore, has attracted more attention recently as a cost-effective approach for developing learning systems from abundant web data. Generally, search queries fed into image crawlers are directly used as web labels for crawled images, which also introduce label noise due to semantic ambiguity and search engine bias. How to deal with these unreliable and noisy web labels becomes a key task in WSL.

A straight-forward approach of WSL is to treat web labels as ground truth and all web samples are directly used to train DNNs [20,29]. Some previous methods [14,17] require additional clean subsets to learn a guidance model to judge the correctness of web labels and adopt a sample reweighting strategy for robust training of DNNs. CurriculumNet [8] avoids extra clean set by leveraging density assumption that samples from the high-density region are more reliable, and trains the model in a curriculum learning manner. As all the above works only use webly supervised loss $\mathcal{L}_w$, recent works attempt to combine self-label supervised loss $\mathcal{L}_s$ with $\mathcal{L}_w$ [9,32]. $\mathcal{L}_s$ comes from the predictions of the model itself in a fashion of self-distillation [13] or prototype-based rectification [28].

Although it is promising to utilize $\mathcal{L}_s$ together with $\mathcal{L}_w$, we argue that the ratio balancing $\mathcal{L}_w$ and $\mathcal{L}_s$ should not be a constant across the entire dataset as in previous works [9,32]. The correctness of web labels varies on a case-by-case basis, due to various causes of real-world label noise. Motivated by this observation, we design a framework that adaptively balances $\mathcal{L}_w$ and $\mathcal{L}_s$ on sample level.

Inspired by the uncertainty prediction ability of DNNs [5], we use DNN's prediction confidence, termed as self-contained confidence (SCC), to achieve a sample-wise balance between $\mathcal{L}_w$ and $\mathcal{L}_s$. Model uncertainty shows how unsure the model considers its correctness on its own prediction, which is revealed by DNN's soft label output. When the model is trained with binary cross entropy (BCE) loss, the model uncertainty can be estimated independently across all categories. Here, we regard model uncertainty corresponding to the category of the sample's web label as SCC, reflecting the likelihood of web label correctness from the model's scope [5]. Figure 1a vividly shows a strong positive correlation between SCC and the correctness of web labels. This association is further confirmed by Expected Calibrated Error (ECE) plot [7], who groups samples with SCC scores within an interval and calculates their average web label correctness rate using a human-annotated verification set from Food-101N [17]. According to Fig. 1b, samples who lie in higher SCC intervals generally have larger probabilities of correct web labels.

With SCC as an effective indicator of web label correctness, a generic SCC-based WSL framework is proposed. Intuitively, with the help of SCC, our framework enforces a webly supervised loss $\mathcal{L}_w$ if a web label is considered reliable, and a self-label supervised loss $\mathcal{L}_s$ otherwise. The self-label supervised loss utilizes the soft label predicted by a model pretrained on the WSL dataset as a self-supervised target. SCC, which is also extracted from the pretrained model, balances the ratio between $\mathcal{L}_w$ and $\mathcal{L}_s$ for each web sample. Our SCC is emphasized as 'self-contained', as no extra guidance model or labeled clean dataset is needed. Following the uncertainty calibration approaches [7,33], we also investigate the relationship between statistical metrics (e.g. ECE metric) and image classification accuracy.

Our contributions are summarized as follows:

- A generic noise-robust WSL framework that does not require a human-verified clean dataset is proposed, novelly featured by sample-level confidence from the perspective of model uncertainty.
- Based on our framework, we further design a graph-enhanced mixup method that stands out among a series of SCC-friendly regularization methods to achieve better classification performance.
- We empirically conclude that under our framework, the statistical metrics of SCC are positively correlated with final classification accuracy, and self-label supervision is superior to consistency regularization for WSL tasks.
- The proposed framework achieves state-of-the-art results on two large-scale realistic WSL datasets, WebVision-1000 and Food-101N.

2 Related Work

2.1 Webly Supervised Learning

Learning with noisy labels can be divided into two categories of problems according to sources of label noise, i.e., synthetic or realistic. For synthetic label noise,

Table 1. Highlighting the principal differences between other WSL methods and ours

Method	Clean set?	Prior knowledge	How to suppress label noise?
MentorNet [14]	✓	Clean Set	Low weight on $\mathcal{L}_w$ for noisy samples
CleanNet [17]	✓	Clean Set	Low weight on $\mathcal{L}_w$ for noisy samples
CurriculumnNet [8]	✗	Density	Schedule noisy samples to later stages
Joint Optim. [32]	✗	Self-training	Replace $\mathcal{L}_w$ with $\mathcal{L}_s$
Self-Learning [9]	✗	Density	Combine $\mathcal{L}_w$ and $\mathcal{L}_s$ with constant-ratio
Ours	✗	Uncertainty	Balance $\mathcal{L}_w$ and $\mathcal{L}_s$ sample-wisely

some works estimate a noisy channel (e.g., a transition matrix) to model the label noise [23,35,38]. However, the designed or estimated channels might not stay effective in the real-world scenario. WSL lies in the realistic noisy label problem. Seminal WSL works attempted to leverage a subset of human-verified samples, referred as 'clean set'. MentorNet [14] learns a dynamic curriculum from the clean set for the sample-reweighting scheme, making the StudentNet only focus on probably correct samples. CleanNet [17] transfers knowledge of label noise learned from a clean set with partial categories towards all categories, and adjust sample weights accordingly to alleviate the impact of noisy labels. In contrast to 'clean set' prior, CurriculumNet [8] assumes that samples with correct labels usually locate at high-density regions in visual feature space and designs a three-stage training strategy to train the model with data stratified by cleanness-levels.

Self labeling is another solution to purify noisy labels by replacing unreliable web labels with predictions by a model. Joint Optimization [32] uses DNN's predictions as self labels, and Self-Learning [9] generates self labels by prototype voting and combines web labels and pseudo-labels using constant ratio. Compared to them, we balance self labels and web labels using sample-wise confidence, which relies on our observation that DNNs are capable of perceiving noisy labels with self-contained confidences. Self labels and confidences are unified in a single pretrained model in our approach. Table 1 clarifies the differences between other WSL solutions and ours.

2.2 Semi-supervised Learning

Semi-supervised learning (SSL) utilizes a small fraction of labeled data and a large unlabeled data set altogether [42]. Solutions to SSL are basically within two main categories. One uses consistency regularization to ensure the model robustness by forcing networks producing identical predictions upon inputs with different augmentations, which is used in MixMatch [1] and UDA [36]. Another uses pseudo-labeling in the representative methods of Billion Scale [37] and data distillation [25], which firstly trains models on the clean labeled set and then provides pseudo-labels for unlabeled data.

The differences between WSL and SSL settings lead to key differences between our method and SSL methods. First, the self-label supervision in our method has a close connection with pseudo-labeling. However, our method utilizes all samples with both web labels and self labels, and SSL methods utilize a subset of unlabeled data with pseudo-labels only. The model for self-labeling in our method is learned from the entire noisy dataset, and the model for pseudo-labeling in SSL methods is trained on a small 'clean' labeled set. Second, our self-label supervised loss has a similar form to consistency regularization. However, consistency regularization may be less powerful to correct the bias caused by label noise than cleaning the labels with self-labeling explicitly. Details are discussed in Sect. 3.2.

2.3 Model Uncertainty

Model uncertainty refers to the level of distrust that the model considers its own prediction, which is vital for real-world applications. For classification tasks, the calculation is as simple as leveraging the highest score of the softmax output. To quantify the quality of model uncertainty, expected calibration error (ECE) is one widely used metric that claims an accurate uncertainty should align model predictions with classification accuracy [7,22]. For instance, if a network predicts a group of samples with a probability of 0.6, we expect exactly 60% samples of this group are classified correctly.

Following this path, several methods were proposed to improve the quality of uncertainty. Post-hoc calibration such as temperature scaling is one family of methods, which optimizes the mapping of produced uncertainty on the verification set [7]. However, such data-dependent rescaling methods cannot improve confidence quality fundamentally. Some other works explored within-training strategies that can provide high-quality model uncertainty, such as label smoothing [21], dropout [6], mixup [33], Bayesian models [16], etc. AugMix [12] is directly designed to improve uncertainty estimates through a data augmentation approach. However, few research works utilized the model confidence to architect model training.

In our work, model uncertainty is adapted for web label confidence estimation. Instead of using the maximum of the model's output probabilities, we pick the value on the exact web label from the probability distribution, which estimates the correctness of the sample's web label.[1] Metrics such as ECE can also be adapted, i.e., web label confidence is considered well-calibrated if a model predicts all samples in a group with web label confidences of 0.6, 60% samples in this group have correct web labels. Being aware that the extraction of web label confidence requires the probability of each class to be calculated independently, binary cross-entropy (BCE) rather than softmax cross-entropy loss is used for training the network.

[1] Web label confidence and self-contained confidence are used interchangeably throughout the paper.

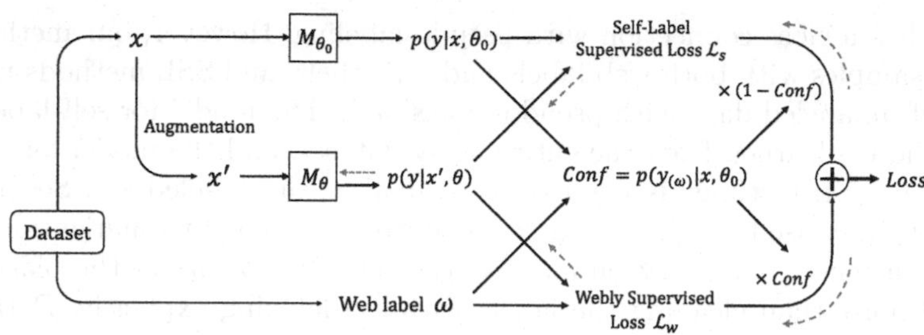

Fig. 2. Diagram of the proposed framework. Backward gradients pass through dashed arrows to update the only trainable parameter θ. Pretrained model M_{θ_0} learns from entire WSL dataset to provide self label and SCC. M_θ is initialized by M_{θ_0}

3 Proposed Method

In this section, after a formal description of WSL task, we introduce two loss functions and the proposed framework with highlighted SCC. Our framework is compatible with various regularization methods. Especially, we propose graph-based aggregation (GBA) to enhance SCC for network training. The diagram of our framework is shown in Fig. 2.

3.1 Webly Supervised Learning: Problem Statement and Notations

Webly Supervised Learning (WSL) aims at training an optimal deep neural network $\mathcal{M}_\theta$ from a dataset $\mathcal{D} = \{(x_1, y_1^*), \ldots, (x_N, y_N^*)\}$ collected from the Internet. x_i denotes the i-th sample in the dataset, and the one-hot web label y_i^* is the one-hot encoding of the web label ω_i (referring to ω_i-th category). The web label ω_i is obtained from the search query of crawling the image x_i. Consider the massive noise in retrieved images from a search engine, ω_i or y_i^* might not reflect the correct category that x_i belongs to. Therefore, suppressing the noise in unreliable web labels becomes the main challenge in WSL.

For convenience, we use symbols x, y^*, ω directly to represent an arbitrary sample, its one-hot web label and its web label, respectively. For the multi-label problem, $y_{(j)}$ denotes sample's label on j-th class. $p(y|x,\theta)$ denotes the label prediction of sample x by the model $\mathcal{M}_\theta$.

3.2 Webly Supervised Loss and Self-label Supervised Loss

Webly supervised loss and self-label supervised loss are two widely adopted loss functions in WSL [8,9,26,32]. Webly supervised loss utilizes web labels as supervision information, and self-label supervised loss [9,32] utilizes predictions of a pretrained model instead. Formally, we define them as follows.

For webly supervised loss, given x' augmented from x with web label ω, the loss function can be expressed as

$$\mathcal{L}_w = -\left[\log\left(p(y_{(\omega)}|x',\theta)\right) + \sum\nolimits_{j\in\mathcal{S}\setminus\omega} \log\left(1 - p(y_{(j)}|x',\theta)\right)\right]. \tag{1}$$

Notice that webly supervised loss is in the form of binary cross-entropy (BCE) loss, because a webly-crawled image probably has multi-label semantics.

For self-label supervised loss, we use the prediction of the pretrained model $\mathcal{M}_{\theta_0}$, which is trained directly on the original web label dataset. As the predictions on samples will be used for finetuning $\mathcal{M}_{\theta_0}$ itself, We call them self labels. Therefore, with self label $p(y|x, \theta_0)$ from model $\mathcal{M}_{\theta_0}$, the self-label supervised loss is

$$\mathcal{L}_s = -\sum_{j \in \mathcal{S}} \Big[p(y_{(j)}|x, \theta_0) \log \left(p(y_{(j)}|x', \theta) \right) + \left(1 - p(y_{(j)}|x, \theta_0)\right) \log \left(1 - p(y_{(j)}|x', \theta)\right) \Big], \tag{2}$$

where $y_{(j)}$ represents the prediction for j-th class in label set $\mathcal{S}$ for multi-class classification problem.

A similar loss to self-label supervised loss is consistency loss [1,36], which provides an auxiliary regularization by enforcing a model to output similar predictions on different augmented counterparts of the same image. Consistency loss is proven to be effective on a large number of unlabeled images for semi-supervised learning. In WSL, however, as the quality of self labels can be guaranteed by feeding a pretrained model with weak augmented images, we found that the high-quality self-supervised loss is more effective than auxiliary consistency loss. An experimental comparison will be shown in Sect. 4.4.

3.3 Self-contained Confidence

It is desirable to adaptively balance webly supervised loss and self-label supervised loss on sample level. Intuitively, we should trust webly supervised loss more on samples with reliable web labels, while self-label supervised loss would dominate the total loss confronting incorrect web labels.

In our method, model $\mathcal{M}_{\theta_0}$ provides only self labels, but also the reliability of web labels. Notice that with BCE loss, model $\mathcal{M}_{\theta_0}$ predicts the probability that x belongs to class i as $p(y_{(i)}|x, \theta_0)$. Specially, we focus on the model prediction on the one-hot web label y^* whose category index is ω, denoted as $p(y_{(\omega)}|x, \theta_0)$. Therefore, the only trainable parameter θ would be updated by minimizing the final loss

$$\mathcal{L} = c \times \mathcal{L}_w + (1 - c) \times \mathcal{L}_s, \quad \text{where } c = p(y_{(\omega)}|x, \theta_0). \tag{3}$$

The confidence c is named as *self-contained confidence* (SCC), as it is self contained in the pretrained model and requires no extra data or knowledge.

3.4 Graph-Based Aggregation

A key component in the proposed method is the pretrained model $\mathcal{M}_{\theta_0}$ for estimating both SCC and self labels. As the model is trained on noisy web labels, we employ mixup [39], which is known as an effective regularization to make DNNs less prone to over-confident predictions and predicted scores of DNNs better calibrated to the actual confidence of a correct prediction [33].

In addition, we propose a graph-based aggregation (GBA) method to further boost the confidence quality and classification performance. GBA does a smoothing operation on a visual similarity graph spanned by image features. By viewing every image as a node, a k-nearest-neighbor (k-NN) graph is firstly constructed based on features located before *fc* layer of pretrained model $\mathcal{M}_{\theta_0}$. Cosine similarity of features is computed across every pair in the neighborhood as edge weight. Hereby, an undirected k-NN graph with weighted adjacent matrix $\mathbf{A}$ is obtained. Let $\mathbf{P}$ denote a matrix of self labels, and the corrected self labels after GBA are denoted as

$$\hat{\mathbf{P}} = \mathbf{D}^{-\frac{1}{2}} (\lambda \mathbf{I} + \mathbf{A}) \mathbf{D}^{-\frac{1}{2}} \mathbf{P}, \tag{4}$$

where $D(i,i) = \lambda + \sum_{j=1}^{N} A(i,j)$. λ controls the portion of original self labels in the post-GBA self labels. SCC will also be extracted from $\hat{\mathbf{P}}$. GBA is a post-processing step with graph filtering [15] and complementary to other methods such as mixup. We evaluate several potential methods and conclude mixup + GBA leads to the optimal performance in Sect. 4.2.

4 Experiments

In this section, we firstly introduce three public WSL datasets. Then, we investigate several SCC-friendly methods, among which GBA-enhanced mixup stands out as the best one in both statistical metrics and classification accuracy. More ablation studies demonstrate the effectiveness of both sample-wise adaptive loss and self-label supervision. Finally, we show that the proposed method reaches the state-of-the-art on the public WSL datasets. We leave the exploration of robustness of our framework and formal algorithm in the Appendix.

4.1 Datasets and Configurations

WebVision-1000 [19] contains $2.4M$ noisy-labeled training images crawled from Flickr and Google, with keywords from 1000 class-labels in ILSVRC-2012 [3]. The estimated web label accuracy is 48% [8]. The ILSVRC-2012 validation set is also utilized along with WebVision-1000's own validation set.

WebVision-500 is a quarter-sized version of WebVision-1000 for evaluation and ablation study in low cost without losing generalization. We randomly sample one-half categories with one-half samples in the training set, and keep the full validation set of the selected 500 categories. This dataset is used for our ablation study in Sects. 4.2, 4.3, 4.4.

Food-101N [17] is another web dataset with $310k$ images classified into 101 food categories. Images are crawled from Google, Yelp, etc. We evaluate our model on the test set of Food-101 [2], Food-101N's clean dataset counterpart. $60k$ human verification labels are provided, indicating the correctness of web labels. The estimated label accuracy is around 80%.

Configuration details. ResNet50 is selected as our CNN model in all experiments [10]. For more efficient training on WebVision, a minor-revised ResNet50-D is utilized [11]. Food101N uses standard ResNet50 for a fair comparison. We use the following settings that completely refer to [11]. Batch size is set as 256 and mini-batch size as 32. We use the standard SGD with the momentum of 0.9 and weight decay of 10^{-4}. A warm-start linearly reaches the initial learning rate (LR) in the first 10 epochs. The remained epochs are ruled by a cosine learning rate scheduler. A simple class reweighting is performed to deal with class imbalance. The initial LR is 0.1 with total L epochs for pretrained models. The main model has initial LR of 0.05 with identical epoch numbers. $L = 120$ for WebVision-500 and Food101N, $L = 150$ for WebVision-1000.

4.2 Exploring Optimal Regularization Method

In this section, we experiment with seven different confidence-friendly regularization methods for $\mathcal{M}_\theta$ under our framework. We conclude that GBA-enhanced mixup (mixup+GBA) is the most efficient one for the best performance. However, as the main contribution of our work is the simple yet effective noise-robust pipeline with SCC, regularization is not a necessary part of our model.

Besides the standard setting with BCE loss, which is denoted as '**Vanilla**', we introduce the following regularization methods for model $\mathcal{M}_\theta$.

Label Smoothing prevents over-confidence problems by adding a small value of ϵ on the zero-values in one-hot encoding labels [31]. We use $\epsilon = 0.1$.

Entropy Regularizer discourages over-confident model prediction by adding a penalizing term to standard loss functions [24]. Regularizer weight is set as 0.1.

MC Dropout is selected as the representation of Bayesian methods. It approximates Bayesian inference by randomness in dropout operation [6]. Dropout rate p is set 0.5. When testing, we infer 50 times and average the predictions.

Mixup is a simple but effective pre-processing method that convexly combines every pair of two sampled images and labels [39]. [33] proves its strong uncertainty calibration capability beyond its label smoothing effects.

AugMix is another data augmentation method with consistency loss, which produces well-calibrated model uncertainty [12].

Ensemble utilizes several models with identical tasks to boost the ultimate performance [4]. With E vanilla models with different random initializations, we average their predictions on every sample.

Graph-based Aggregation (GBA) is introduced in Sect. 3.4. We use $k = 10$ and $\lambda = 0.5$ as hyper-parameters.

Result Analysis. Table 2 reports the results. S1 is short for the pretraining stage for $\mathcal{M}_{\theta_0}$, S2 for the finetuning stage using our framework. Generally, good performance in S1 favors S2. Mixup and Ensemble are the two most effective regularizers. As mentioned in Sect. 3.4, Mixup smooths discriminative spaces and ensemble averages models' biases. The advantages of these two methods are

Table 2. Performance of the pretrained model (S1) and finetuned model (S2)

Method	S1-WebVision		S1-ImageNet		S2-WebVision		S2-ImageNet	
	Top-1	Top-5	Top-1	Top-5	Top-1	Top-5	Top-1	Top-5
Vanilla	75.42	88.65	68.84	84.62	76.46	89.63	69.78	85.32
Label Smoothing	75.81	89.32	69.11	85.54	77.02	90.33	70.86	86.71
Entropy Regularizer	74.77	89.44	68.80	85.81	73.78	88.76	67.99	85.36
MC Dropout ($p = 0.5$)	75.16	88.90	68.73	84.60	76.00	89.50	69.78	86.01
Mixup ($\alpha = 0.2$)	76.35	90.31	71.15	**87.36**	77.47	91.02	72.25	88.47
AugMix	76.61	89.58	69.06	84.30	76.96	90.10	69.61	85.32
Ensemble ($E = 5$)	**78.98**	**91.27**	**72.45**	87.26	**79.12**	**91.73**	**72.73**	87.96
Vanilla + GBA	75.42	88.65	68.84	84.62	77.12	90.73	71.56	87.78
Mixup + GBA	76.35	90.31	71.15	87.36	77.76	91.43	72.59	**88.65**

combined in GBA design, as a graph smoothing operator for neighbor predictions, which is proven effective empirically. Improvement from GBA is weaker on mixup compared to vanilla since mixup offers the same effect of smoothing space with GBA. However, mixup+GBA still reaches the optimal result besides the costly ensemble method.

4.3 Understanding Self-contained Confidence

As SCC plays a critical role in our framework, we explore an interesting question: how great the SCC quality affects the final accuracy reported in the previous section? We also show the relationship between three statistical metrics adapted from uncertainty theories and our accuracy-based metric.

For statistical metrics, we manually create a verification set $\mathcal{V} = \{v_1, \ldots, v_n\}$ for WebVision-500 by annotating whether the web label is correct on $n = 12500$ samples, with 50 randomly sampled cases from 250 random classes.

To evaluate the quality of SCC, The following metrics are utilized.

Second-stage Accuracy on Vanilla (SAV). To empirically evaluate different SCCs, we use an identical vanilla pretrained model for self-labeling and finetuning under our framework. Therefore, the accuracy of second-stage finetuned model is only determined by the quality of SCC. Note that the models for producing SCC are different and with different regularization methods.

Mean Square Error (MSE). Verification set $\mathcal{V}$ can be considered as a set of ground-truth confidence since it values 1 with the correct web label and values 0 when incorrect. Thus, MSE estimates the squared difference between the given confidence and the ground-truth, which is defined as

$$MSE = \frac{1}{n}\sum_{i=1}^{n}(v_i - c_i)^2. \tag{5}$$

Expected Calibration Error (ECE). Calibration error is originally used to evaluate the model interpretability on their predictions [7], while we slightly

Table 3. Evaluations of SCC provided by different methods. Column 1-3 reports statistical metrics MSE, ECE and OCE. Column 4-7 reports model-based metric SAV

Confidence Provider	MSE	ECE	OCE	SAV-WebVision		SAV-ImageNet	
				Top-1	Top-5	Top-1	Top-5
Vanilla	0.2795	0.2371	0.1518	76.46	89.63	69.78	85.32
Label Smoothing	0.2786	0.2280	0.1200	76.82	89.86	70.06	85.76
Entropy Regularizer	0.4137	0.4138	0.0370	76.40	90.17	70.31	86.36
MC Dropout ($p = 0.5$)	0.2807	0.2431	0.1193	76.68	89.89	70.34	85.97
Mixup ($\alpha = 0.2$)	**0.2510**	**0.1828**	0.0135	77.14	90.18	**71.00**	86.48
Augmix	0.2869	0.2366	0.1757	76.67	89.65	69.89	85.63
Ensemble ($E = 5$)	0.2687	0.2233	0.1537	76.49	89.68	70.06	85.86
Vanilla + GBA	0.2612	0.2494	**0.0002**	**77.17**	**90.55**	70.89	**86.84**

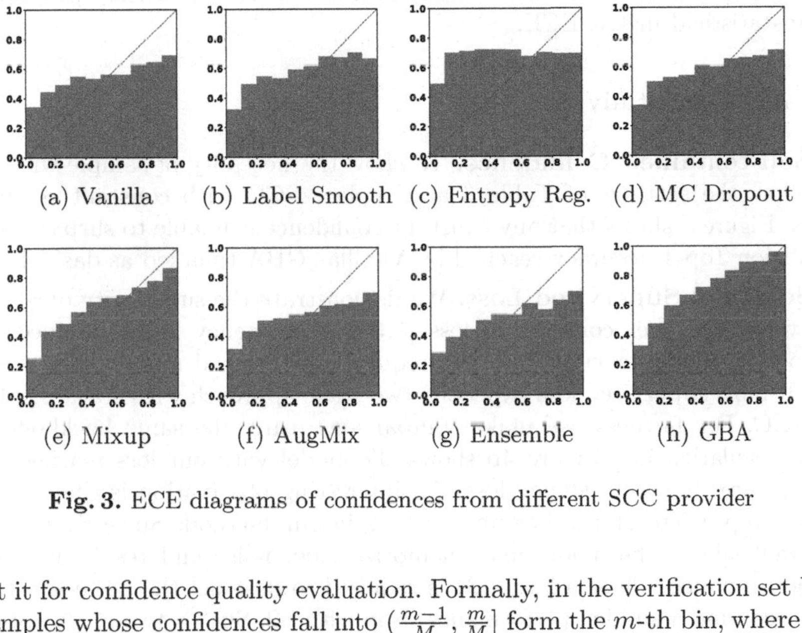

(a) Vanilla (b) Label Smooth (c) Entropy Reg. (d) MC Dropout

(e) Mixup (f) AugMix (g) Ensemble (h) GBA

Fig. 3. ECE diagrams of confidences from different SCC provider

adapt it for confidence quality evaluation. Formally, in the verification set $\mathcal{V}$, for all samples whose confidences fall into $(\frac{m-1}{M}, \frac{m}{M}]$ form the m-th bin, where average confidence $conf(B_m) = \frac{1}{|B_m|}\sum_{i \in B_m} c_i$ and the average web-label reliability $rel(B_m) = \frac{1}{|B_m|}\sum_{i \in B_m} v_i$ are calculated. Thus, ECE is defined as

$$ECE = \sum_{m=1}^{M} \frac{|B_m|}{n} \left| rel(B_m) - conf(B_m) \right|. \tag{6}$$

Over-Confidence Error (OCE). Samples with high SCC but incorrect web labels are especially harmful to our framework, since introducing the wrong web label is much worse than using self labels. OCE evaluates the level of over-confidence by punishing more on higher-confident bins with low reliability, defined as

$$OCE = \sum_{m=1}^{M} \frac{|B_m|}{n} \left[conf(B_m) \times \max\left\{ conf(B_m) - rel(B_m), 0 \right\} \right]. \quad (7)$$

In this work, we calculate ECE and OCE with $M = 100$. For visualization in Fig. 3, we use $M = 10$. Figure 1b uses $M = 100$.

Result Analysis. Best metric performance is reached by either mixup or Vanilla+GBA, while the ensemble also produces a good result. Figure 3 visualizes ECE diagrams, where GBA and mixup look more calibrated than any other model. A similar result is shown in Tabel 3 Column 2–4. Column 5–8 presents the metric of SAV which shows GBA provides good quality confidence that favors our proposed framework. According to our exploration of SCC, we conclude the following insights: (1) SCC can reflect the reliability of the web label according to Fig. 3; (2) SCC plays a key role in our pipeline through adaptively balancing two losses on the sample level since empirical metric SAV is generally proportional to the statistical metric ECE.

4.4 Ablation Study

On Self-contained Confidence. To show the necessity of sample-wise SCC, we follow the settings of Table 3 and replace SCC with constant confidence values. Figure 4 shows that any constant confidence is unable to surpass 77.17% WebVision Top-1 accuracy reached by Vanilla+GBA (marked as dashed line).

On Self-label Supervised Loss. We demonstrate the superiority of self-label supervised loss over consistency loss [1,36]. Consistency loss is trained in an end-to-end fashion since it does not require a pretrained model $\mathcal{M}_{\theta_0}$, whereas our self-label supervised loss expects a two-stage approach with static self labels and SCC. For fairness, we make comparisons using the same backbone with mixup regularization. Figure 4b shows the model with our loss reaches better performance than consistency loss. An interesting observation is that a performance drop exists at the beginning of S2 in our method. Since S1 is trained with web labels, the model may memorize label noise and result in suboptimal performance. Thus, a large LR is required to destruct the noise-affected S1 model, causing a sudden performance drop with S2. Such a two-stage approach

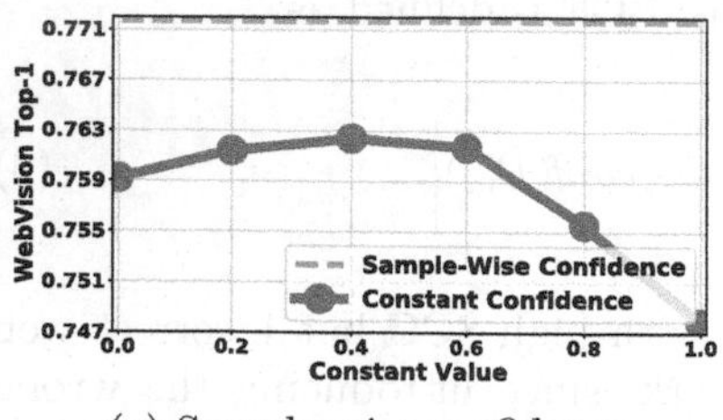

(a) Sample-wise confidence

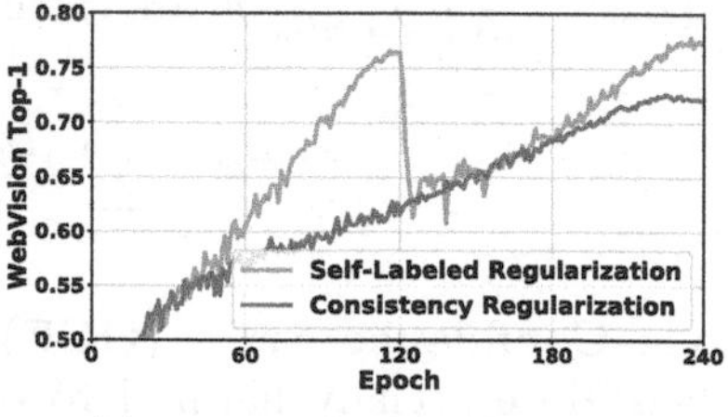

(b) Self-label supervised loss

Fig. 4. Ablation studies of sample-wise confidence and self-label supervised loss

is adopted, because we find the end-to-end approach unsuitable for our method: in the early stage, inaccurate pseudo labels and SCC mislead the model, and in the late stage, the model finally obtains reliable SCC, however, small LR cannot correct the accumulated errors.

4.5 Real-World Experiments

WebVision-1000. Table 4 reports experiments on WebVision-1000 using both vanilla and mixup models. With the vanilla model, using our pipeline, a 0.3% improvement is achieved for WebVision top-1 accuracy, and 0.7% increase on ImageNet top-1/5. GBA can further improve the performance of every metric. When enabling mixup operation ($\alpha = 0.2$), although on top of a high-accuracy pretrained model, our method can still improve both ImageNet top-1 and top-5 accuracy by 1.9%. The WebVision top-5 accuracy is improved by 0.7%. The WebVision top-1 accuracy is improved a little. More improvements on ImageNet prove a good generalization ability of the proposed method. The larger improvement than vanilla may attribute to the higher SCC quality achieved by mixup. GBA advances an average 0.3% extra improvement on every metric.

We also show the superiority of our method over state-of-the-art methods. Note that both MentorNet [14] and CleanNet [17] use extra human-verified datasets to train a guidance network first, and MentorNet [14] chooses a backbone of InceptionResNetV2 stronger than ResNet50-D. Multimodal image classification [27] uses ImageNet data for training visual embedding and a query-image pairs dataset for training phrase generation. Stronger InceptionV3 is also selected as the backbone. Although with these disadvantages, our ResNet50-D still works the best among all.

Table 4. The state-of-the-art results on WebVision-1000

Method	Backbone network	WebVision		ImageNet	
		Top-1	Top-5	Top-1	Top-5
MentorNet [14]	InceptionResNetV2	72.60	88.90	64.20	84.80
CleanNet [17]	ResNet50	70.31	87.77	63.42	84.59
CurriculumNet [8]	InceptionV2	72.10	89.20	64.80	84.90
Multimodal [27]	InceptionV3	73.15	89.73	–	–
Initial Vanilla Model	ResNet50-D	75.08	89.22	67.23	84.09
Ours (Vanilla)	ResNet50-D	75.36	89.38	67.93	84.77
Ours (Vanilla+GBA)	ResNet50-D	75.69	89.42	68.35	85.24
Initial Mixup Model	ResNet50-D	75.54	90.36	68.77	86.59
Ours (Mixup)	ResNet50-D	75.74	90.78	70.38	88.25
Ours (Mixup+GBA)	ResNet50-D	**75.78**	**91.07**	**70.66**	**88.46**

Table 5. The state-of-the-art results on Food-101N

Method	Top-1
CleanNet [17]	83.95
Guidance Learning [18]	84.20
MetaCleaner [40]	85.05
Deep Self-Learning [9]	85.11
SOMNet [34]	87.50
Initial Vanilla Model	84.08
Ours (Vanilla)	84.87
Ours (Vanilla+GBA)	85.76
Initial Mixup Model	86.00
Ours (Mixup)	87.43
Ours (Mixup+GBA)	**87.55**

Food-101N. According to Table 5, we significantly advance the state-of-the-art model without any usage of human annotations. For vanilla model, the second stage of our method pushes 0.8% higher accuracy than the first stage, and the usage of GBA even double the improvement. For the mixup model ($\alpha = 0.5$), the second stage increases a higher 1.4% accuracy as mixup provides better SCC and self labels than vanilla, but the advance of GBA is deducted due to the overlapping effects of mixup and GBA. Rather than our normally used ResNet50-D, we use standard ResNet50 here for fair comparisons with others. While all the other methods (except [40]) train Food-101N from ImageNet pretrained model, we train our model from scratch and still reach optimal performance.

5 Conclusion

We propose a generic noise-robust framework featured by sample-level confidence balancing webly supervised loss and self-label supervised loss. Our framework is compatible with model regularization methods, among which our proposed mixup+GBA is the most effective.

Here we recall two main takeaway messages from our extensive experiments: (1) Reliability of the web label can be reflected by SCC (ref. Fig. 3), and empirical metric SAV is generally proportional to the statistical metrics like ECE (ref. Table 3). (2) Our framework is in favor of high-quality confidence provided by the pretrained model, and mixup and ensemble are the two most effective regularizers (ref. Fig. 2 and 3). Considering that mixup smooths discriminative spaces and ensemble averages models' biases, both advantages are combined in GBA design, as a graph smoothing operator for neighbor predictions.

We also leave a valuable discussion in the appendix for readers of interests, which basically shows: although the performance is largely dependent on the quality of SCC, the framework still works on Food101N even with a weak DNN backbone.

Acknowledgement. The work described in this paper was partially supported by Innovation and Technology Commission of the Hong Kong Special Administrative Region, China (Enterprise Support Scheme under the Innovation and Technology Fund B/E030/18).

References

1. Berthelot, D., Carlini, N., Goodfellow, I., Papernot, N., Oliver, A., Raffel, C.: Mixmatch: a holistic approach to semi-supervised learning. In: NeurIPS (2019)
2. Bossard, L., Guillaumin, M., Van Gool, L.: Food-101 – mining discriminative components with random forests. In: Fleet, D., Pajdla, T., Schiele, B., Tuytelaars, T. (eds.) ECCV 2014. LNCS, vol. 8694, pp. 446–461. Springer, Cham (2014). https://doi.org/10.1007/978-3-319-10599-4_29
3. Deng, J., Dong, W., Socher, R., Li, L.J., Li, K., Fei-Fei, L.: Imagenet: a large-scale hierarchical image database. In: CVPR, pp. 248–255 (2009)
4. Dietterich, T.G.: Ensemble methods in machine learning. In: Kittler, J., Roli, F. (eds.) MCS 2000. LNCS, vol. 1857, pp. 1–15. Springer, Heidelberg (2000). https://doi.org/10.1007/3-540-45014-9_1
5. Gal, Y.: Uncertainty in deep learning. Univ. Camb. **1**(3) (2016)
6. Gal, Y., Ghahramani, Z.: Dropout as a bayesian approximation: representing model uncertainty in deep learning. In: ICML, pp. 1050–1059 (2016)
7. Guo, C., Pleiss, G., Sun, Y., Weinberger, K.Q.: On calibration of modern neural networks. In: ICML, pp. 1321–1330 (2017)
8. Guo, S., et al.: Curriculumnet: weakly supervised learning from large scale web images. In: ECCV, pp. 135–150. Springer (2018)
9. Han, J., Luo, P., Wang, X.: Deep self-learning from noisy labels. In: ICCV, pp. 5138–5147 (2019)
10. He, K., Zhang, X., Ren, S., Sun, J.: Deep residual learning for image recognition. In: CVPR, pp. 770–778 (2016)
11. He, T., Zhang, Z., Zhang, H., Zhang, Z., Xie, J., Li, M.: Bag of tricks for image classification with convolutional neural networks. In: CVPR, pp. 558–567 (2019)
12. Hendrycks, D., Mu, N., Cubuk, E.D., Zoph, B., Gilmer, J., Lakshminarayanan, B.: AugMix: a simple data processing method to improve robustness and uncertainty. In: ICLR (2020)
13. Hinton, G., Vinyals, O., Dean, J.: Distilling the knowledge in a neural network. In: NIPS Deep Learning and Representation Learning Workshop (2015)
14. Jiang, L., Zhou, Z., Leung, T., Li, L.J., Fei-Fei, L.: Mentornet: learning data-driven curriculum for very deep neural networks on corrupted labels. In: ICML, pp. 2304–2313 (2018)
15. Kipf, T.N., Welling, M.: Semi-supervised classification with graph convolutional networks. In: ICLR (2017)
16. Lakshminarayanan, B., Pritzel, A., Blundell, C.: Simple and scalable predictive uncertainty estimation using deep ensembles. In: NIPS, pp. 6402–6413 (2017)

17. Lee, K.H., He, X., Zhang, L., Yang, L.: Cleannet: transfer learning for scalable image classifier training with label noise. In: CVPR, pp. 5447–5456 (2018)
18. Li, Q., Peng, X., Cao, L., Du, W., Xing, H., Qiao, Y.: Product image recognition with guidance learning and noisy supervision. arXiv preprint arXiv:1907.11384 (2019)
19. Li, W., Wang, L., Li, W., Agustsson, E., Van Gool, L.: Webvision database: visual learning and understanding from web data. arXiv preprint arXiv:1708.02862 (2017)
20. Mahajan, D., et al.: Exploring the limits of weakly supervised pretraining. In: Ferrari, V., Hebert, M., Sminchisescu, C., Weiss, Y. (eds.) ECCV 2018. LNCS, vol. 11206, pp. 185–201. Springer, Cham (2018). https://doi.org/10.1007/978-3-030-01216-8_12
21. Müller, R., Kornblith, S., Hinton, G.E.: When does label smoothing help? In: NeurIPS, pp. 4694–4703 (2019)
22. Ovadia, Y., et al.: Can you trust your model's uncertainty? evaluating predictive uncertainty under dataset shift. In: NeurIPS, pp. 13991–14002 (2019)
23. Patrini, G., Rozza, A., Krishna Menon, A., Nock, R., Qu, L.: Making deep neural networks robust to label noise: a loss correction approach. In: CVPR, pp. 1944–1952 (2017)
24. Pereyra, G., Tucker, G., Chorowski, J., Kaiser, Ł., Hinton, G.: Regularizing neural networks by penalizing confident output distributions. arXiv preprint arXiv:1701.06548 (2017)
25. Radosavovic, I., Dollár, P., Girshick, R., Gkioxari, G., He, K.: Data distillation: towards omni-supervised learning. In: CVPR, pp. 4119–4128 (2018)
26. Reed, S., Lee, H., Anguelov, D., Szegedy, C., Erhan, D., Rabinovich, A.: Training deep neural networks on noisy labels with bootstrapping. In: ICLR (2015)
27. Shah, M., et al.: Inferring context from pixels for multimodal image classification. In: CIKM, pp. 189–198. ACM (2019)
28. Snell, J., Swersky, K., Zemel, R.: Prototypical networks for few-shot learning. In: NIPS, pp. 4077–4087 (2017)
29. Sun, C., Shrivastava, A., Singh, S., Gupta, A.: Revisiting unreasonable effectiveness of data in deep learning era. In: ICCV, pp. 843–852 (2017)
30. Sun, Y., Wang, X., Tang, X.: Deep learning face representation from predicting 10,000 classes. In: CVPR, pp. 1891–1898 (2014)
31. Szegedy, C., Vanhoucke, V., Ioffe, S., Shlens, J., Wojna, Z.: Rethinking the inception architecture for computer vision. In: CVPR, pp. 2818–2826 (2016)
32. Tanaka, D., Ikami, D., Yamasaki, T., Aizawa, K.: Joint optimization framework for learning with noisy labels. In: CVPR, pp. 5552–5560 (2018)
33. Thulasidasan, S., Chennupati, G., Bilmes, J.A., Bhattacharya, T., Michalak, S.: On mixup training: improved calibration and predictive uncertainty for deep neural networks. In: NeurIPS, pp. 13888–13899 (2019)
34. Tu, Y., Niu, L., Chen, J., Cheng, D., Zhang, L.: Learning from web data with self-organizing memory module. In: CVPR, pp. 12846–12855 (2020)
35. Xia, X., et al.: Are anchor points really indispensable in label-noise learning? In: NeurIPS, pp. 6838–6849 (2019)
36. Xie, Q., Dai, Z., Hovy, E., Luong, M.T., Le, Q.V.: Unsupervised data augmentation for consistency training. arXiv preprint arXiv:1904.12848 (2019)
37. Yalniz, I.Z., Jégou, H., Chen, K., Paluri, M., Mahajan, D.: Billion-scale semi-supervised learning for image classification. arXiv preprint arXiv:1905.00546 (2019)
38. Yu, X., Liu, T., Gong, M., Batmanghelich, K., Tao, D.: An efficient and provable approach for mixture proportion estimation using linear independence assumption. In: CVPR, pp. 4480–4489 (2018)

39. Zhang, H., Cisse, M., Dauphin, Y.N., Lopez-Paz, D.: mixup: Beyond empirical risk minimization. In: ICLR (2018)
40. Zhang, W., Wang, Y., Qiao, Y.: Metacleaner: Learning to hallucinate clean representations for noisy-labeled visual recognition. In: CVPR, pp. 7373–7382 (2019)
41. Zhou, B., Lapedriza, A., Khosla, A., Oliva, A., Torralba, A.: Places: a 10 million image database for scene recognition. In: TPAMI (2017)
42. Zhu, X.J.: Semi-supervised learning literature survey. University of Wisconsin-Madison Department of Computer Sciences, Technical report (2005)

Author Index

Printed in the United States
By Bookmasters